International Congress of Mathematicians (ICM 2022)

Sections 5-8
Volume 4 of 7

Online
6-14 July 2022

Part A

Editors:

Dmitry Beliaev
Stanislav Smirnov

ISBN: 978-1-7138-9871-9

ICM INTERNATIONAL CONGRESS OF MATHEMATICIANS 2022 JULY 6–14

SECTIONS 5–8

EDITED BY D. BELIAEV AND S. SMIRNOV

INTERNATIONAL MATHEMATICAL UNION IMU

EMS PRESS

Editors

Dmitry Beliaev
Mathematical Institute
University of Oxford
Andrew Wiles Building
Radcliffe Observatory Quarter
Woodstock Road
Oxford OX2 6GG, UK

Email: belyaev@maths.ox.ac.uk

Stanislav Smirnov
Section de mathématiques
Université de Genève
rue du Conseil-Général 7–9
1205 Genève, Switzerland

Email: stanislav.smirnov@unige.ch

2020 Mathematics Subject Classification: 00B25

ISBN 978-3-98547-058-7, eISBN 978-3-98547-558-2, DOI 10.4171/ICM2022

Volume 1. Prize Lectures
ISBN 978-3-98547-059-4, eISBN 978-3-98547-559-9, DOI 10.4171/ICM2022-1
Volume 2. Plenary Lectures
ISBN 978-3-98547-060-0, eISBN 978-3-98547-560-5, DOI 10.4171/ICM2022-2
Volume 3. Sections 1–4
ISBN 978-3-98547-061-7, eISBN 978-3-98547-561-2, DOI 10.4171/ICM2022-3
⟶ Volume 4. Sections 5–8
ISBN 978-3-98547-062-4, eISBN 978-3-98547-562-9, DOI 10.4171/ICM2022-4
Volume 5. Sections 9–11
ISBN 978-3-98547-063-1, eISBN 978-3-98547-563-6, DOI 10.4171/ICM2022-5
Volume 6. Sections 12–14
ISBN 978-3-98547-064-8, eISBN 978-3-98547-564-3, DOI 10.4171/ICM2022-6
Volume 7. Sections 15–20
ISBN 978-3-98547-065-5, eISBN 978-3-98547-565-0, DOI 10.4171/ICM2022-7

This edition printed by Curran Associates, Inc.
ISBN 978-1-7138-9871-9

e-Book available: https://ems.press/books/standalone/276

No changes have been made to the content of these proceedings. There may be changes to pagination and minor adjustments for aesthetics.

Bibliographic information published by the Deutsche Nationalbibliothek
The Deutsche Nationalbibliothek lists this publication in the Deutsche Nationalbibliografie; detailed bibliographic data are available on the Internet at http://dnb.dnb.de.

Published by EMS Press, an imprint of the

European Mathematical Society – EMS – Publishing House GmbH
Institut für Mathematik
Technische Universität Berlin
Straße des 17. Juni 136
10623 Berlin, Germany

https://ems.press

CONTENTS

VOLUME 1

PRIZE LECTURES

POPULAR SCIENTIFIC EXPOSITIONS

SUMMARIES OF PRIZE WINNERS' WORK

VOLUME 2

SPECIAL PLENARY LECTURES

PLENARY LECTURES

VOLUME 3

1. LOGIC

2. ALGEBRA

3. NUMBER THEORY - SPECIAL LECTURE

3. NUMBER THEORY

4. ALGEBRAIC AND COMPLEX GEOMETRY - SPECIAL LECTURE

4. ALGEBRAIC AND COMPLEX GEOMETRY

VOLUME 4

5. GEOMETRY – SPECIAL LECTURES

5. GEOMETRY

6. TOPOLOGY

7. LIE THEORY AND GENERALIZATIONS

8. ANALYSIS – SPECIAL LECTURE

8. ANALYSIS

VOLUME 5

9. DYNAMICS

10. PARTIAL DIFFERENTIAL EQUATIONS

11. MATHEMATICAL PHYSICS – SPECIAL LECTURE

11. MATHEMATICAL PHYSICS

VOLUME 6

12. PROBABILITY – SPECIAL LECTURE

12. PROBABILITY

13. COMBINATORICS – SPECIAL LECTURE

13. COMBINATORICS

14. MATHEMATICS OF COMPUTER SCIENCE – SPECIAL LECTURES

14. MATHEMATICS OF COMPUTER SCIENCE

VOLUME 7

15. NUMERICAL ANALYSIS AND SCIENTIFIC COMPUTING

16. CONTROL THEORY AND OPTIMIZATION – SPECIAL LECTURE

16. CONTROL THEORY AND OPTIMIZATION

17. STATISTICS AND DATA ANALYSIS

5. GEOMETRY

SPECIAL LECTURES

DEVELOPMENTS IN 3D RICCI FLOW SINCE PERELMAN

BRUCE KLEINER

ABSTRACT

This is a report on progress in 3D Ricci flow since Perelman's work 20 years ago.

MATHEMATICS SUBJECT CLASSIFICATION 2020

53C44

KEYWORDS

Ricci flow, singularities

INTERNATIONAL CONGRESS OF MATHEMATICIANS 2022 JULY 6–14

Proc. Int. Cong. Math. 2022, Vol. 4, pp. 2376–2390
DOI 10.4171/ICM2022/150

1. INTRODUCTION

A smooth family of Riemannian metrics $(g(t))_{t\in[0,T)}$ is a *Ricci flow* if it satisfies the equation

$$\partial_t g(t) = -2\operatorname{Ric}\big(g(t)\big)$$

for every $t \in [0, T)$, where $\operatorname{Ric}(g(t))$ denotes the Ricci tensor of $g(t)$ [34]. The Ricci flow equation is a fundamental partial differential equation in mathematics—it is the natural analog of the heat equation for Riemannian metrics, just as mean curvature flow and harmonic map heat flow are the heat equation analogs for submanifolds and mappings, and the (elliptic) Einstein equation, minimal surface equation, and harmonic map equation are the respective analogs of Laplace's equation. Ricci flow may be used to canonically smooth a metric, and, in favorable situations, deform it into an optimal shape. For this reason, it has had a profound impact on geometry and topology as a powerful tool for solving many problems, including many longstanding conjectures which had resisted all other techniques. Singularity formation has been a great challenge central to the topic, one which has required a wide range of ingredients from PDE, differential geometry, metric geometry, and topology. While Ricci flow is fascinating from many points of view, it is especially interesting from the PDE viewpoint because it has some features in common with other geometric evolution equations (e.g., mean curvature flow and harmonic map heat flow); in particular, for the last 40 years the treatment of singularities has been a common theme, and has led to important cross-fertilization.

In his preprints from 2002–2003, Perelman made a series of landmark contributions to Ricci flow, some specific to flow on 3-manifolds, and some applicable in any dimension. He introduced a number of new ingredients which opened the way to subsequent progress in many directions, including (in particular) flow on 3-manifolds, Kähler–Ricci flow, and Ricci flow under certain curvature assumptions. The aim of this article is to present the advances in 3D Ricci flow from a bird's-eye view, for a general mathematical audience. The technical nature of the subject forces some compromises in the exposition, both in the precision of statements and in the coverage of accompanying history and conceptual background. Also, in writing for a broad audience it was unavoidable to make some choices of material and emphasis which may be unsatisfactory to the experts; I hope that any such readers will be understanding.

By convention, all 3-manifolds will be orientable.

2. PERELMAN'S WORK ON 3D RICCI FLOW

In this section, we briefly review what was known about 3D Ricci flow up through 2003, when Perelman posted his preprints. We refer the interested reader to the introductions in [21, 28, 39, 50, 51] for more detailed overviews.

Hamilton showed that if h is a smooth Riemannian metric on a compact n-manifold M, then there exists a unique solution $(g(t))_{t\in[0,T)}$ to the Ricci flow equation

$$\partial_t g = -2\operatorname{Ric}\big(g(t)\big)$$

with initial condition $g(0) = h$, and which is defined on a maximal time interval $[0, T)$; moreover, if the time T is finite, then the norm of the curvature tensor Rm becomes unbounded as t approaches T [34]. When M is a 2-sphere, the behavior of the Ricci flow $(g(t))_{[0,T)}$ is very simple: it blows up in finite time, and as $t \to T$ the volume-renormalized metric $\hat{g}(t) := (\mathrm{vol}(g(t)))^{-\frac{2}{n}} g(t)$ converges smoothly as $t \to T$ to a metric of constant curvature 4π [29, 35]. Ricci flow in 3D is more complicated. Consider, for instance, the case when $(M, g(0))$ is obtained from two compact Riemannian 3-manifolds (M, h_M), (N, h_N) by performing a geometric connected sum, i.e., by choosing $r > 0$ small, removing r-balls, and attaching a round cylindrical "neck" (interpolating appropriately near the gluing locus). Heuristically, one expects that when r is small, the Ricci flow $(g(t))_{t\in[0,T)}$ will blow up after a short time due to the large positive Ricci curvature in the neck region, and that away from the neck region $g(t)$ will have a smooth limit as $t \to T$. The occurrence of such localized "neck pinch" singularities leads to the idea of prolonging the evolution by performing "surgery", i.e., by removing an open set U diffeomorphic to $(0, 1) \times S^2$ which contains the set where the metric goes singular as $t \to T$, gluing approximate hemispherical caps onto the two 2-sphere boundary components, and restarting the Ricci flow from the resulting compact smooth Riemannian manifold. By iterating this procedure, one might hope to obtain a Ricci flow with surgery defined for all time, allowing for the possibility that the manifold may be empty from some time onward. Noting that the surgery has the potential to simplify topology by undoing connected sums, around the time of Hamilton's original paper Yau suggested that Ricci flow with surgery might be used to address fundamental questions in three-dimensional topology—the Poincaré Conjecture and, more generally, Thurston's Geometrization Conjecture. Pursuing this idea, Hamilton developed many tools for analyzing singularities, and implemented a version of Ricci flow with surgery in an analogous 4-dimensional setting.

In 2003, Perelman completed the program in a breakthrough result:

Theorem 2.1 (Informal statement). *For any compact smooth Riemannian* 3*-manifold* (M, h)*, there exists a Ricci flow with surgery with initial condition* (M, h) *which is defined for all time.*

In addition to many fresh insights, the proof involved numerous ingredients, including most of Hamilton's prior results on Ricci flow, as well as a variety of other tools from geometric analysis. We will only touch on a few key points here, treating some aspects differently from Perelman.

Perelman's Ricci flow with surgery consists of a sequence of Ricci flows

$$\big(M_1, (g_1(t))_{t\in[T_0,T_1)}\big), \big(M_2, (g_2(t)))_{t\in[T_1,T_2)}\big), \big(M_3, (g_3(t))_{t\in[T_2,T_3)}\big), \ldots$$

where the time intervals $[T_{i-1}, T_i)$ are consecutive and $\bigcup_i [T_{i-1}, T_i) = [0, \infty)$. For every $0 < T_i < \infty$, the Ricci flow $(g_i(t))_{t\in[T_{i-1},T_i)}$ goes singular as $t \to T_i$ and has a smooth limit $\bar{g}_i$ on an open (possibly empty) subset $\Omega_i \subset M_i$. The initial condition $(M_{i+1}, g_{i+1}(T_i))$ for the next flow is obtained from $(\Omega_i, \bar{g}_i)$ by a geometric surgery procedure—cutting along 2-spheres, capping off boundary components, and throwing away some connected components—which generalizes the simple neck removal described above. The cumulative effect of the surgery process on the topology is easy to describe: for every $i > 1$, the original manifold M_1 is diffeomorphic to a connected sum

$$M \overset{\text{diff}}{\simeq} M_i\#(\#_j N_j) \tag{2.2}$$

where for every j the summand N_j is either a copy of $S^2 \times S^1$ or a spherical space form; recall that a spherical space form is a manifold of the form S^3/Γ where $\Gamma \subset O(4)$ is a finite subgroup acting freely on S^3.

A central issue in Perelman's argument is controlling the structure of the flow near singularities. This control is implemented as a set of conditions collectively referred to as the *Canonical Neighborhood Assumption*. Informally speaking, the Canonical Neighborhood Assumption asserts that near points with large curvature the flow has a restricted form, i.e., it is well approximated by a flow belonging to a family of model Ricci flows Perelman called *κ-solutions*; examples include:

(A) A shrinking round metric on S^3 or a spherical space form;

(B) A shrinking round cylindrical metric on $S^2 \times \mathbb{R}$ or the quotient $(S^2 \times \mathbb{R})/\mathbb{Z}_2$;

(C) A special Ricci flow solution $(g_{\mathrm{Bry}}(t))_{t\in\mathbb{R}}$ on $\mathbb{R}^3$ called the *Bryant soliton*.

In the schematic diagram shown in Figure 1, these provide models near the points A, B, and C, respectively.

After proving the existence of a Ricci flow with surgery, Perelman analyzed the behavior as $t \to \infty$, and used the geometry of the flow to deduce a topological conclusion:

Theorem 2.3 ([54]). *For every t sufficiently large, if $t \in [T_{i-1}, T_i)$, there is a finite disjoint collection $\{N_j\}$ of embedded incompressible tori in M_i such that each connected component of $M_i \setminus \bigcup_j N_j$ is diffeomorphic to either a complete, finite-volume hyperbolic manifold or a graph manifold.*

A *hyperbolic manifold* is a Riemannian manifold with universal cover isometric to hyperbolic 3-space $\mathbb{H}^3$. A connected embedded surface N in a 3-manifold X is *incompressible* if the inclusion map $N \to X$ induces an injective homomorphism of fundamental groups. A 3-manifold X is a *graph manifold* if there is a finite disjoint collection $\{N_k\}$ of embedded tori such that every connected component of $X \setminus \bigcup_k N_k$ is diffeomorphic to (the total space of) a circle bundle over a surface.

A few years prior to the appearance of Perelman's preprints, Hamilton proved an assertion roughly similar to Theorem 2.3, assuming an additional bound on the curvature tensor [36]. The proof of Theorem 2.3 uses several key contributions from [36] in identifying the hyperbolic piece, as well as several fundamental new ideas.

The results on Ricci flow with surgery have many applications to problems in geometry and topology. Combining Theorem 2.3 with well-known results from 3-manifold topology, Perelman proved Thurston's Geometrization Conjecture:

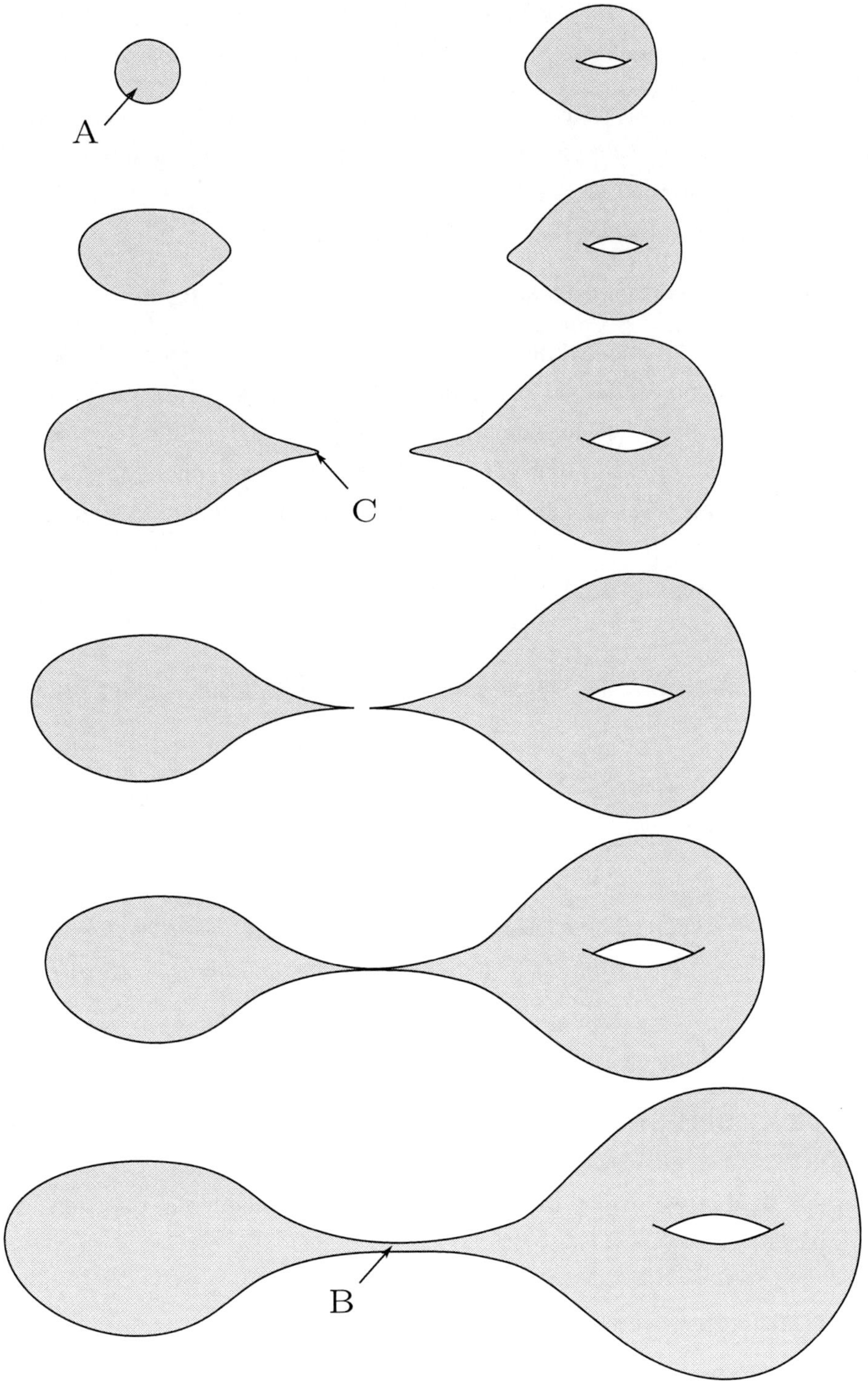

FIGURE 1
A Ricci flow with surgery with a neckpinch.

Theorem 2.4. *Every closed 3-manifold is a connected sum of manifolds that can be cut along embedded, incompressible copies of T^2 into geometrizable pieces.*

A connected 3-manifold is *geometrizable* if it admits a finite-volume Riemannian metric with universal cover isometric to one of the eight Thurston geometries $S^3, H^3, \mathbb{R}^3, \mathbb{H}^2 \times \mathbb{R}, S^2 \times \mathbb{R}$, Nil, Solv, $\widetilde{\mathrm{SL}(2, \mathbb{R})}$ [61].

The Poincaré Conjecture is an immediate corollary:

Theorem 2.5. *Any closed, simply connected 3-manifold is diffeomorphic to S^3.*

Perelman and Colding–Minicozzi showed that if the initial manifold M_1 of a Ricci flow with surgery has no aspherical summands in its prime decomposition, then the flow eventually becomes extinct, i.e., for some i, we have $[T_{i-1}, T_i) = [T_{i-1}, \infty)$ and $M_i = \emptyset$ [30, 53]; it then follows from (2.2) that M_1 is a connected sum of spherical space forms and copies of $S^2 \times S^1$. This gives an alternative approach to the Poincaré Conjecture avoiding Theorem 2.3.

In addition to settling central conjectures in topology, Perelman solved longstanding problems in geometry:

Theorem 2.6. *Let M be a closed 3-manifold, and $\sigma(M)$ denote the Yamabe invariant of M* [43, 55].

- *Manifold M admits a Riemannian metric with positive scalar curvature if and only if it is a connected sum of spherical space forms and copies of $S^2 \times S^1$* [33, 56].
- *If M is irreducible and $\sigma(M) \leq 0$, then $(-\frac{1}{6}\sigma(M))^{\frac{3}{2}}$ is total volume of the hyperbolic pieces appearing in the geometric decomposition of M, as in Theorem 2.4* [2, 39].

3. DEVELOPMENTS BASED ON QUESTIONS RAISED BY PERELMAN'S WORK

In this section we review the progress on some fundamental questions arising in Perelman's papers [52, 54].

3.1. Large-time behavior

Let $\{(M_i, ((g_i(t))_{t \in [T_{i-1}, T_i)}\}$ be a Ricci flow with surgery as constructed by Perelman.

One basic question concerns the set of surgery times $\{T_i \mid 0 < T_i < \infty\}$; in the statement of Theorem 2.1, this set could potentially be infinite. Although Perelman discussed finiteness of surgeries in his preprints, he did not settle the issue or give any indication how it might be addressed, because he was able to find an approach to Theorem 2.3 (and the Geometrization Conjecture) which circumvented the matter altogether. The problem of finiteness of surgeries was also noted earlier: Hamilton had expressed the hope that it would be possible to define a Ricci flow with surgery such that only finitely many surgeries would be necessary and that the curvature would then remain bounded for large t, after appropriate normalization [36]. In a tour de force, Bamler was able to confirm this:

Theorem 3.1 ([8–12]). *In any Ricci flow with surgery as constructed by Perelman, there are only finitely many surgery times, and there exist $C, T \in (0, \infty)$ such that the curvature tensor satisfies the bound $|\mathrm{Rm}_{g(t)}| < Ct^{-1}$ for all $t > T$.*

In the theorem and in what follows, we let $g(t) := g_i(t)$ for $t \in [T_{i-1}, T_i)$.

It is natural to ask: beyond the assertions in Theorem 3.1, how much more can be said about the asymptotic behavior of the Ricci flow at $t \to \infty$? First, the results of [30,53] imply that each connected component of M_i is prime and aspherical when $T_i > T$. On general principles, one might expect that Ricci flow improves the geometry, and therefore as $t \to \infty$ the asymptotic behavior should be very simple. For instance, when M is geometrizable, then one might expect that as $t \to \infty$ the Ricci flow would converge (in some appropriate sense) to the geometric structure, and when M is not geometrizable, then the Ricci flow would construct the JSJ decomposition—a system of embedded incompressible tori which are canonical up to isotopy—as well the geometric structure on the pieces. This speculation has been confirmed only when M admits a hyperbolic metric, in which case Perelman's proof of Theorem 2.3 implies that $\frac{1}{4}t^{-1}g(t)$ converges to a hyperbolic metric as $t \to \infty$. In other cases there has been progress in this direction. For instance, Lott has shown:

Theorem 3.2 ([48]). *Let N be a connected component of M_i, where $[T_{i-1}, T_i) = [T_{i-1}, \infty)$. If the quantity $t^{-\frac{1}{2}} \operatorname{diam}(N, g(t))$ remains bounded as $t \to \infty$, then the pullback of the rescaled metric $t^{-1}g(t)$ to the universal cover $\tilde{N}$ converges to a homogeneous expanding soliton.*

Bamler has a number of results covering both geometrizable and nongeometrizable cases. The simplest case is the torus:

Theorem 3.3 ([12]). *If M is diffeomorphic to T^3, then either $g(t)$ converges to a flat metric as $t \to \infty$, or the quantity $t^{-\frac{1}{2}} \operatorname{diam}(g(t))$ is unbounded and for large t the metric $g(t)$ is well approximated by another metric $g'(t)$ with T^2-symmetry and T^2-orbits of diameter $\ll t^{\frac{1}{2}}$.*

A similar alternative holds for 3-manifolds modeled on Thurston's Nil or Solv geometries. We refer the reader to [12] for these and other results, as well as a discussion of open questions.

3.2. Classification of singularity models

As described in Section 2, Perelman's treatment of Ricci flow with surgery involved a family of Ricci flows called κ-solutions, which model the formation of singularities. Building on Hamilton's work on singularity formation, in his first preprint Perelman established many properties of κ-solutions, including:

(a) (Topological classification) Every κ-solution is diffeomorphic to a spherical space form S^3/Γ, the cylinder $S^2 \times \mathbb{R}$, the $\mathbb{Z}_2$-quotient $(S^2 \times \mathbb{R})/\mathbb{Z}_2$, or $\mathbb{R}^3$.

(b) Any κ-solution not diffeomorphic to $\mathbb{R}^3$, S^3, or RP^3 is isometrically covered by a shrinking round sphere or shrinking round cylinder.

(c) In a quantitative sense, κ-solutions are "mostly necklike." For instance, any κ-solution diffeomorphic to $\mathbb{R}^3$ is asymptotically cylindrical near infinity.

In the exceptional cases in (b), Perelman's work provided both qualitative and quantitative information, but not a complete classification. In the $\mathbb{R}^3$ case, he made the following conjecture:

Conjecture 3.4 ([52]). *Any κ-solution diffeomorphic to $\mathbb{R}^3$ is isometric to a Bryant soliton, up to rescaling.*

He also constructed a κ-solution on S^3 which is $O(3)$-symmetric, and becomes more and more elongated as $t \to -\infty$; this descends to a κ-solution on RP^3.

Recently, in the culmination of a long development in the theory of ancient solutions, Brendle, Daskalopoulos, and Sesum have completed the classification of κ-solutions:

Theorem 3.5 ([25]). *Conjecture 3.4 holds.*

Theorem 3.6 ([26]). *Any compact κ-solution is isometrically covered by a shrinking round metric or a rescaling of the nonround κ-solution constructed by Perelman.*

3.3. Ricci flow through singularities

Although the construction of Ricci flow with surgery had a spectacular impact on mathematics, in both of his preprints Perelman indicated that he had a further objective in mind:

> *"It is likely that by passing to the limit in this construction [of Ricci flow with surgery] one would get a canonically defined Ricci flow through singularities, but at the moment I don't have a proof of that."* [52, P. 37]

> *"Our approach ...is aimed at eventually constructing a canonical Ricci flow, defined on a largest possible subset of space-time,—a goal, that has not been achieved yet in the present work."* [54, P. 1]

From the PDE perspective, one may interpret Perelman's notion of a "Ricci flow through singularities" as a kind of generalized solution to the Ricci flow equation; his stated goal then fits into a long-established theme in PDE—the existence and uniqueness of weak or generalized solutions. A further motivation for pursuing such a program comes from applications in geometry and topology involving families of Ricci flows depending continuously on a parameter, which necessitate well-behaved unique solutions.

In recent years Perelman's goal was attained in the papers **[16,41]**. The first step was a definition Ricci flow through singularities, which was given in **[41]**. This is uses the following spacetime version of Ricci flow:

Definition 3.7 (**[41]**). A *Ricci flow spacetime* is a tuple $(\mathcal{M}, \mathfrak{t}, \partial_{\mathfrak{t}}, g)$, where:

- $\mathcal{M}$ is a smooth 4-manifold with boundary.
- $\mathfrak{t} : \mathcal{M} \to [0, \infty)$ is a smooth function called the *time function*; its level sets $\mathcal{M}_t := \mathfrak{t}^{-1}(t)$ are called *time-slices*.
- $\partial_{\mathfrak{t}}$ is a smooth vector field satisfying $\partial_t \mathfrak{t} \equiv 1$; it is called the *time vector field*, and its trajectories are called *worldlines*.
- g is a Riemannian metric on the subbundle of the tangent bundle $T\mathcal{M}$ defined by $\ker d\mathfrak{t}$, and hence induces a Riemannian metric g_t on the time slice $\mathcal{M}_t$.
- g satisfies the Ricci flow equation
$$\mathcal{L}_{\partial_t} g = -2\,\mathrm{Ric}(g).$$
- The time slice $\mathcal{M}_0$ is the boundary of $\mathcal{M}$.

For brevity, we typically denote the entire spacetime by $\mathcal{M}$.

An ordinary Ricci flow $(g(t))_{t\in[0,T)}$ on a manifold M gives rise to a Ricci flow spacetime $(\mathcal{M}, \mathfrak{t}, \partial_t, g)$ where $\mathcal{M} = M \times [0, T)$, the time function $\mathfrak{t}$ is projection onto the second factor, the time vector field $\partial_{\mathfrak{t}}$ projects to the unit vector field ∂_x on the second factor, and g induces the metric $g(t)$ on the time slice $\mathcal{M}_t = M \times \{t\}$ corresponding to $g(t)$. Up to diffeomorphism, a general Ricci flow spacetime looks locally like such a product Ricci flow spacetime.

A Ricci flow spacetime by itself is too general to be useful; one obtains a good notion of Ricci flow through singularities by imposing some extra conditions on a Ricci flow spacetime:

Definition 3.8 (**[41]**). A *singular Ricci flow* is a Ricci flow spacetime $(\mathcal{M}, \mathfrak{t}, \partial_{\mathfrak{t}}, g)$ where:

(1) The initial time slice $\mathcal{M}_0$ is compact.

(2) $\mathcal{M}$ satisfies the Canonical Neighborhood Assumption.

(3) $\mathcal{M}$ is 0-complete.

Here condition (2) is similar to the Canonical Neighborhood Assumption in Perelman's Ricci flow with surgery, and asserts that around a point $x \in \mathcal{M}_t$ with large curvature, the time slice $\mathcal{M}_t$ is well approximated by a κ-solution. The 0-completeness requirement in condition (3) is a replacement for the conventional notion of completeness. A generic neck pinch gives rise to a Ricci flow spacetime exhibiting both spatial and temporal incompleteness: if T is the time at which the pinch occurs, then the time slice $\mathcal{M}_T$ will be an incomplete Riemannian manifold, and the trajectories of the time vector field ∂_t which go into the singularity are incomplete.

Definition 3.9. A Ricci flow spacetime $\mathcal{M}$ is *0-complete* if the following holds. Suppose $\gamma : [0, s_0) \to \mathcal{M}$ is either an integral curve of $\pm\partial_t$, or a unit speed curve in some time slice of $\mathcal{M}$. If $\sup |\operatorname{Rm}|(\gamma(s)) < \infty$, then $\lim_{s \to s_0} \gamma(s)$ exists.

Ricci flow in dimension 3 is globally well posed in the setting of singular Ricci flows [16, 41]:

Theorem 3.10. (1) *If (N, h) is a compact Riemannian 3-manifold, then there exists a singular Ricci flow $\mathcal{M}$ with initial time slice $\mathcal{M}_0$ isometric to (N, h).*

(2) *A singular Ricci flow is determined uniquely by its initial condition: if $\mathcal{M}$, $\mathcal{M}'$ are singular Ricci flows then any isometry $\mathcal{M}_0 \to \mathcal{M}'_0$ extends to an isometry of $\mathcal{M} \to \mathcal{M}'$ of Ricci flow spacetimes (i.e., a diffeomorphism respecting the tuples).*

The methods of [16] also imply that singular Ricci flows depend continuously on their initial condition. Perelman's assertion about convergence of Ricci flow with surgery also holds:

Theorem 3.11 ([16]). *Let (N, h) be a compact Riemannian 3-manifold, which by Theorem 3.10(1) we may identify with the time 0 slice $\mathcal{M}_0$ of some singular Ricci flow $\mathcal{M}$. Then the family of Ricci flows with surgery with initial condition $\mathcal{M}_0$ converges to $\mathcal{M}$ as the surgery parameter δ tends to zero.*

Here δ is a parameter appearing in Perelman's construction of Ricci flow with surgery; when δ is small then in particular the surgery process involves cutting along necks with small cross-section.

The results above show that there is a well-behaved notion of Ricci flow through singularities in dimension three, for arbitrary smooth initial conditions. It is natural to ask:

Question 3.12. Is there a good notion of Ricci flow through singularities in higher dimensions, for arbitrary initial conditions?

This currently seems to be a significant challenge already in dimension 4; see, however, [4–6] and the references therein for recent progress in this direction. Note that the answer to Question 3.12 is "yes" if one imposes restrictions the initial condition (see, for instance, [37, 60]); also, starting in dimension 5 there are examples of Angenent–Knopf showing that one should not expect uniqueness [3]. We remark that the problem of constructing a well-behaved generalized solutions to a closely related PDE—the mean curvature flow equation—has been a major topic of research in geometric analysis for more than 40 years [24].

We now state a few results concerning the structure of singular Ricci flows.

Theorem 3.13 ([15, 41, 42]). *Let $\mathcal{M}$ be a singular Ricci flow.*

- *If C is a connected component of some time slice, then C is diffeomorphic to a compact manifold punctured at finitely many points.*

- *Let $\hat{M}_t$ be the manifold obtained from some time slice $\mathcal{M}_t$ by filling in punctures and throwing away components diffeomorphic to S^3. Then $\hat{M}_t$ is compact and its prime decomposition is part of the prime decomposition of $\mathcal{M}_0$.*

- *The set of times $t \in [0, \infty)$ such that the time slice $\mathcal{M}_t$ is noncompact has Minkowski dimension $\leq \frac{1}{2}$.*

At present it is unknown if time slices could have infinitely many connected components, or if there could be uncountably many noncompact time slices. In this direction we have the following conjecture:

Conjecture 3.14. *If $\mathcal{M}$ is a singular Ricci flow, then the set of times t for which $\mathcal{M}_t$ is noncompact is finite. Moreover, if $\mathcal{M}_t$ is noncompact, then as $\bar{t} \nearrow t$ each connected component of $\mathcal{M}_{\bar{t}}$ either goes extinct, or experiences finitely many (possibly degenerate) neckpinch singularities.*

4. FURTHER RESULTS

We conclude by listing a number of other directions which have seen progress involving 3D Ricci flow.

- Ricci flow with surgery and/or singular Ricci flow can be extended, or partially extended, to noncompact manifolds [19, 20, 22, 47].

- There is a large literature on various types of special Ricci flow solutions, including shrinking, expanding, and steady solitons, ancient solutions, and eternal solutions. Many of these solutions arise as potential singularity models for finite time singularities or as blow-up limits of type I, II, or III [36]. There does not seem to be a good single source covering these developments, so we recommend searching the internet for "Ricci soliton."

- Perelman's results on Ricci flow with surgery (Theorems 2.1 and 2.3) extend to orbifolds, giving a Ricci flow proof of the Orbifold Theorem, see [23, 40].

- Singular Ricci flow may be used to understand the topology of the space $\mathrm{Diff}(M)$ of diffeomorphisms $M \to M$ with the smooth topology, in particular, settling the Generalized Smale Conjecture, and completing the determination of the topology of $\mathrm{Diff}(M)$ when M is a prime 3-manifold. See [13, 15, 17, 18], and also [38] for history and background.

- Ricci flow with surgery and singular Ricci flow may be used to study the topology of the space $\mathrm{Met}_{\mathrm{PSC}}(M)$ of Riemannian metrics of positive scalar curvature on a 3-manifold M, and the moduli space $\mathrm{Met}_{\mathrm{PSC}}(M)/\,\mathrm{Diff}(M)$. It was shown in [49] that the moduli space $\mathrm{Met}_{\mathrm{PSC}}(M)/\mathrm{Diff}(M)$ is empty or path-connected, and [17] extended this, proving that $\mathrm{Met}_{\mathrm{PSC}}(M)$ is empty or contractible. See [17, 49] for more background and references.

- In [1] Ricci flow with surgery was used to give sharp volume estimates for hyperbolic 3-manifolds with minimal surface boundary.

- Although the topics are not specific to dimension 3, we mention that there are a number of papers studying Ricci flow starting from rough initial conditions [14, 31, 44–46, 57–59], and papers using Ricci flow to study scalar curvature lower bounds in a C^0-setting [7, 27, 32].

FUNDING

Supported by NSF grant DMS-2005553, and a Simons Collaboration grant.

REFERENCES

[1] I. Agol, P. A. Storm, and W. P. Thurston, Lower bounds on volumes of hyperbolic Haken 3-manifolds. With an appendix by Nathan Dunfield. *J. Amer. Math. Soc.* **20** (2007), no. 4, 1053–1077.

[2] M. T. Anderson, Scalar curvature and geometrization conjectures for 3-manifolds. In *Comparison geometry (Berkeley, CA, 1993–94)*, pp. 49–82, Math. Sci. Res. Inst. Publ. 30, Cambridge Univ. Press, Cambridge, 1997.

[3] S. Angenent and D. Knopf, Ricci solitons, conical singularities, and nonuniqueness. 2021, arXiv:1909.08087.

[4] R. Bamler, Compactness theory of the space of super Ricci flows. 2020, arXiv:2008.09298.

[5] R. Bamler, Entropy and heat kernel bounds on a Ricci flow background. 2020, arXiv:2008.07093.

[6] R. Bamler, Structure theory of non-collapsed limits of Ricci flows. 2020, arXiv:2009.03243.

[7] R. H. Bamler, A Ricci flow proof of a result by Gromov on lower bounds for scalar curvature. *Math. Res. Lett.* **23** (2016), no. 2, 325–337.

[8] R. H. Bamler, Long-time behavior of 3-dimensional Ricci flow: A: generalizations of Perelman's long-time estimates. *Geom. Topol.* **22** (2018), no. 2, 775–844.

[9] R. H. Bamler, Long-time behavior of 3-dimensional Ricci flow: B: evolution of the minimal area of simplicial complexes under Ricci flow. *Geom. Topol.* **22** (2018), no. 2, 845–892.

[10] R. H. Bamler, Long-time behavior of 3-dimensional Ricci flow: C: 3-manifold topology and combinatorics of simplicial complexes in 3-manifolds. *Geom. Topol.* **22** (2018), no. 2, 893–948.

[11] R. H. Bamler, Long-time behavior of 3-dimensional Ricci flow: D: proof of the main results. *Geom. Topol.* **22** (2018), no. 2, 949–1068.

[12] R. H. Bamler, Long-time behavior of 3-dimensional Ricci flow: introduction. *Geom. Topol.* **22** (2018), no. 2, 757–774.

[13] R. H. Bamler, Some recent developments in Ricci flow. In *ICM 2022 Proceedings, Vol. 4*, pp. 2432–2455, EMS Press, 2022.

[14] R. H. Bamler, E. Cabezas-Rivas, and B. Wilking, The Ricci flow under almost non-negative curvature conditions. *Invent. Math.* **217** (2019), no. 1, 95–126.

[15] R. Bamler and B. Kleiner, Ricci flow and diffeomorphism groups of 3-manifolds. 2017, arXiv:1712.06197.

[16] R. Bamler and B. Kleiner, Uniqueness and stability of Ricci flow through singularities. 2018, arXiv:1709.04122.

[17] R. Bamler and B. Kleiner, Ricci flow and contractibility of spaces of metrics. 2019, arXiv:1909.08710.

[18] R. Bamler and B. Kleiner, Diffeomorphism groups of prime 3-manifolds. 2021, arXiv:2108.03302.

[19] L. Bessières, G. Besson, and S. Maillot, Ricci flow on open 3-manifolds and positive scalar curvature. *Geom. Topol.* **15** (2011), no. 2, 927–975.

[20] L. Bessières, G. Besson, and S. Maillot, Long time behaviour of Ricci flow on open 3-manifolds. *Comment. Math. Helv.* **90** (2015), no. 2, 377–405.

[21] L. Bessières, G. Besson, S. Maillot, M. Boileau, and J. Porti, *Geometrisation of 3-manifolds*. EMS Tracts Math. 13, European Mathematical Society (EMS), Zürich, 2010.

[22] L. Bessières, G. Besson, S. Maillot, and F. C. Marques, Deforming 3-manifolds of bounded geometry and uniformly positive scalar curvature. *J. Eur. Math. Soc. (JEMS)* **23** (2021), no. 1, 153–184.

[23] M. Boileau, B. Leeb, and J. Porti, Geometrization of 3-dimensional orbifolds. *Ann. of Math. (2)* **162** (2005), no. 1, 195–290.

[24] K. A. Brakke, *The motion of a surface by its mean curvature*. Math. Notes 20, Princeton University Press, Princeton, NJ, 1978.

[25] S. Brendle, Ancient solutions to the Ricci flow in dimension 3. *Acta Math.* **225** (2020), no. 1, 1–102.

[26] S. Brendle, P. Daskalopoulos, and N. Sesum, Uniqueness of compact ancient solutions to three-dimensional Ricci flow. 2020, arXiv:2002.12240.

[27] P. Burkhardt-Guim, Pointwise lower scalar curvature bounds for C^0 metrics via regularizing Ricci flow. *Geom. Funct. Anal.* **29** (2019), no. 6, 1703–1772.

[28] H.-D. Cao and X.-P. Zhu, A complete proof of the Poincaré and geometrization conjectures—application of the Hamilton–Perelman theory of the Ricci flow. *Asian J. Math.* **10** (2006), no. 2, 165–492.

[29] B. Chow, The Ricci flow on the 2-sphere. *J. Differential Geom.* **33** (1991), no. 2, 325–334.

[30] T. H. Colding and W. P. Minicozzi II, Estimates for the extinction time for the Ricci flow on certain 3-manifolds and a question of Perelman. *J. Amer. Math. Soc.* **18** (2005), no. 3, 561–569.

[31] P. Gianniotis and F. Schulze, Ricci flow from spaces with isolated conical singularities. *Geom. Topol.* **22** (2018), no. 7, 3925–3977.

[32] M. Gromov, Dirac and Plateau billiards in domains with corners. *Cent. Eur. J. Math.* **12** (2014), no. 8, 1109–1156.

[33] M. Gromov and H. Blaine Lawson Jr., Spin and scalar curvature in the presence of a fundamental group. I. *Ann. of Math. (2)* **111** (1980), no. 2, 209–230.

[34] R. S. Hamilton, Three-manifolds with positive Ricci curvature. *J. Differential Geom.* **17** (1982), no. 2, 255–306.

[35] R. S. Hamilton, The Ricci flow on surfaces. In *Mathematics and general relativity (Santa Cruz, CA, 1986)*, pp. 237–262, Contemp. Math. 71, Amer. Math. Soc., Providence, RI, 1988.

[36] R. S. Hamilton, Non-singular solutions of the Ricci flow on three-manifolds. *Comm. Anal. Geom.* **7** (1999), no. 4, 695–729.

[37] R. Haslhofer, Uniqueness and stability of singular Ricci flows in higher dimensions. 2021, arXiv:2110.03412.

[38] S. Hong, J. Kalliongis, D. McCullough, and J. H. Rubinstein, *Diffeomorphisms of elliptic 3-manifolds*. Lecture Notes in Math. 2055, Springer, Heidelberg, 2012.

[39] B. Kleiner and J. Lott, Notes on Perelman's papers. *Geom. Topol.* **12** (2008), no. 5, 2587–2855.

[40] B. Kleiner and J. Lott, Geometrization of three-dimensional orbifolds via Ricci flow. *Astérisque* **365** (2014), 101–177.

[41] B. Kleiner and J. Lott, Singular Ricci flows I. *Acta Math.* **219** (2017), no. 1, 65–134.

[42] B. Kleiner and J. Lott, Singular Ricci flows II. 2018, arXiv:1804.03265.

[43] O. Kobayashi, Scalar curvature of a metric with unit volume. *Math. Ann.* **279** (1987), no. 2, 253–265.

[44] H. Koch and T. Lamm, Geometric flows with rough initial data. *Asian J. Math.* **16** (2012), no. 2, 209–235.

[45] H. Koch and T. Lamm, Parabolic equations with rough data. *Math. Bohem.* **140** (2015), no. 4, 457–477.

[46] Y. Lai, Ricci flow under local almost non-negative curvature conditions. *Adv. Math.* **343** (2019), 353–392.

[47] Y. Lai, Producing 3d Ricci flows with non-negative Ricci curvature via singular Ricci flows. 2020, arXiv:2004.05291.

[48] J. Lott, Dimensional reduction and the long-time behavior of Ricci flow. *Comment. Math. Helv.* **85** (2010), no. 3, 485–534.

[49] F. Marques, Deforming three-manifolds with positive scalar curvature. *Ann. of Math. (2)* **176** (2012), no. 2, 815–863.

[50] J. Morgan and G. Tian, *Ricci flow and the Poincaré conjecture*. Clay Math. Monogr. 3, American Mathematical Society, Providence, RI; Clay Mathematics Institute, Cambridge, MA, 2007.

[51] J. Morgan and G. Tian, *The geometrization conjecture*. Clay Math. Monogr. 5, American Mathematical Society, Providence, RI; Clay Mathematics Institute, Cambridge, MA, 2014.

[52] G. Perelman, The entropy formula for the Ricci flow and its geometric applications. 2002, arXiv:math/0211159.

[53] G. Perelman, Finite extinction time for the solutions to the Ricci flow on certain three-manifolds. 2003, arXiv:math/0307245.

[54] G. Perelman, Ricci flow with surgery on three-manifolds. 2003, arXiv:math/0303109v1.

[55] R. M. Schoen, Variational theory for the total scalar curvature functional for Riemannian metrics and related topics. In *Topics in calculus of variations (Montecatini Terme, 1987)*, pp. 120–154, Lecture Notes in Math. 1365, Springer, Berlin, 1989.

[56] R. Schoen and S. T. Yau, On the structure of manifolds with positive scalar curvature. *Manuscripta Math.* **28** (1979), no. 1–3, 159–183.

[57] M. Simon, Ricci flow of almost non-negatively curved three manifolds. *J. Reine Angew. Math.* **630** (2009), 177–217.

[58] M. Simon, Ricci flow of non-collapsed three manifolds whose Ricci curvature is bounded from below. *J. Reine Angew. Math.* **662** (2012), 59–94.

[59] M. Simon and P. M. Topping, Local mollification of Riemannian metrics using Ricci flow, and Ricci limit spaces. *Geom. Topol.* **25** (2021), no. 2, 913–948.

[60] J. Song and G. Tian, The Kähler–Ricci flow through singularities. *Invent. Math.* **207** (2017), no. 2, 519–595.

[61] W. P. Thurston, *Three-dimensional geometry and topology. Vol. 1*. Princeton Math. Ser. 35, Princeton University Press, Princeton, NJ, 1997.

BRUCE KLEINER

Courant Institute of Mathematical Sciences, New York University, 251 Mercer St., New York, NY 10012, USA, bkleiner@cims.nyu.edu

This Page Intentionally Left Blank

SURVEY LECTURE ON BILLIARDS

RICHARD EVAN SCHWARTZ

ABSTRACT

In this survey of billiards, I will discuss a variety of topics: rational polygonal billiards, irrational polygonal billiards, polygonal outer billiards, billiards in smooth ovals, and a bit about billiards in tables with scatterers.

MATHEMATICS SUBJECT CLASSIFICATION 2020

Primary 38C83; Secondary 32G15

KEYWORDS

Billiards, outer billiards, polygons, Riemann surfaces, geometry, dynamics

ICM INTERNATIONAL CONGRESS OF MATHEMATICIANS 2022 JULY 6–14

Proc. Int. Cong. Math. 2022, Vol. 4, pp. 2392–2429
DOI 10.4171/ICM2022/66

1. INTRODUCTION

In one popular version of billiards, called *eightball*, the game is initialized with a triangular array of 10 polyester balls placed towards the back of a rectangular table that is about the length of a horse. The first player impacts the white *cue ball* with a tapered *pool stick*. The cue ball slides then rolls across green cloth, striking the other 10 balls and scattering them. This fateful start, which sets the game of eightball in motion, is called *the break*.

There are some who play billiards and there are some who think about billiards. Those who play care about the quality of the cloth, the weight of the balls, the feel of the stick. They grind the cushion that covers the front end of their pool stick into a tub of powdered chalk, trying to get just the right conditions of contact with the cue ball. Those who think about billiards usually free their minds from these physical properties and contemplate games of a more abstract nature.

This survey concerns the abstract games of mathematical billiards, henceforth simply called *billiards*. In billiards, the table might be a regular pentagon, or an ellipse, or the planar region bounded by a closed loop that is once but not twice differentiable. The tables might have obstacles in them, smaller shapes that the balls can bounce off as they move through the big table. Often there is just one ball in the game, a single point that slides without friction, but not always. There might be many balls, or charged particles influenced by magnetic fields. The game might be played on a sphere or in hyperbolic space.

Billiards is a huge, sprawling subject with deep connections to topics such as mathematical physics, ergodic theory, surface dynamics, Teichmüller theory, and algebraic geometry. There are many other surveys on billiards, most more comprehensive than this article. I will point some of these out later on. There would be no way for me to give a comprehensive survey of the whole business, even if I actually knew more than a little bit of it.

Rather, I will take the point of view that I am the proprietor of a Platonic pool hall. I built the establishment based on excellent advice but had to work quickly and on a limited budget. There are some beautiful rooms but also some of the plumbing and wiring is not quite up to code. Some doors lead nowhere at all. You have come to my establishment and I will show you around. I may entice you to stick around and play some games, maybe spend some money... The topics are organized mostly according to table type: square (Section 2), regular polygons (Section 3), rational polygons (Section 4), irrational polygons (Section 5), polygonal outer billiards (Section 6), convex ovals (Section 7), and tables with obstacles (Section 8).

This subdivision does not really cover it all, so sometimes I may drift off topic or omit important things. Some of the tables here, like rational polygons, are extremely crowded. You can barely hear yourself think above the shouting and the excitement. Other tables, like irrational triangles or convex ovals of intermediate differentiability, are much quieter. There are just a few patrons wandering around them and scratching their heads. I will also march you past the tables I have played on and (like it or not) regale you with tales of my own exploits from half-remembered "glory days." My own work is concentrated in

Sections 5 and 6. If I had more space, I would also discuss magnetic billiards, Minkoswski billiards, symplectic billiards, polyhedral billiards, and billiards in hyperbolic space.

2. THE SQUARE

In the most common kind of billiards, a point moves along the table at unit speed and bounces off the sides according to the usual "angle in equals angle out" rule. The ball is not allowed to go into the corners. The square is the classic billiards table. I like the square $P = [0, 1/2]^2$ because it is nicely covered by the square torus $Y = \boldsymbol{R}^2/\boldsymbol{Z}^2$. There is a piece-wise isometric map $f : Y \to P$, as indicated in Figure 1, which gives a bijection between geodesics (which miss the corners of the square tiling of Y) and billiard paths.

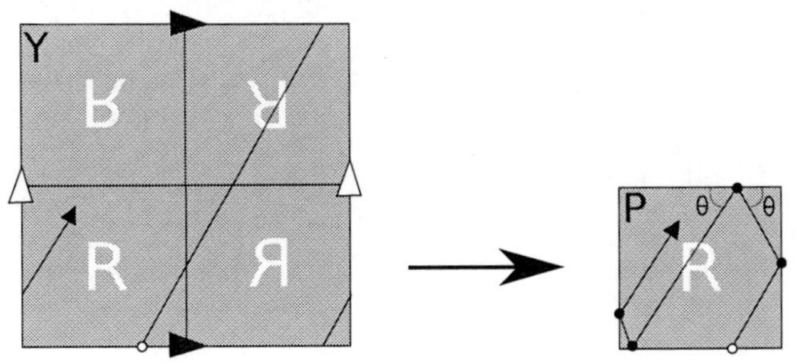

FIGURE 1
The torus covering the square.

2.1. Periodic billiard paths

A *periodic billiard path* is one that retraces itself. Each periodic billiard path on P corresponds to a closed geodesic on Y, which in turn corresponds to a line segment in the plane connecting $(0, 0)$ to some integer lattice point (m, n). There are infinitely many periodic billiard paths, but they come in a discrete set of maximal parallel families and there is one lattice point per family. The number $N'(L)$ of maximal parallel families consisting of periodic billiard paths of length at most L satisfies a beautiful asymptotic formula. Number $N'(L)$ counts the nonzero lattice points in the disk of radius L centered at $(0, 0)$, so

$$\lim_{L\to\infty} \frac{N'(L)}{L^2} = \pi. \tag{2.1}$$

How large is the error $E'(L) = N'(L) - \pi L^2$? A really crisp answer would be:

$$\lim_{L\to\infty} \frac{|E'(L)|}{L^{1/2}} = \infty, \quad \lim_{L\to\infty} \frac{E'(L)}{L^{(1/2)+\varepsilon}} = 0, \quad \forall \varepsilon > 0. \tag{2.2}$$

The first equation is a theorem proved independently by Hardy and Landau. The second equation is a famous open problem called *the Gauss Circle Problem*. See [56] for a survey.

The periodic billiard path is *primitive* if it does not trace several times over a smaller periodic billiard path. The lattice points (m, n) corresponding to primitive periodic billiard paths are coprime: m and n have no common divisors. Let $N(L)$ be the same count as above, but only for the primitive periodic billiard paths. We also ignore the orientation, which cuts the count in half. Estimating $N(L)$ recalls a happy exercise in number theory. The chances that a prime p does not divide (m,n) is $1-p^{-2}$. So, the proportion of coprime lattice points is asymptotically

$$\prod_p 1-p^{-2} = \left(\sum_{n=1}^{\infty} \frac{1}{n^2}\right)^{-1} = \frac{1}{\zeta(2)} = \frac{6}{\pi^2}.$$

This gives us

$$\lim_{L\to\infty} \frac{N(L)}{L^2} = \frac{\pi}{2\zeta(2)} = \frac{c_4}{\text{area}(P)}, \quad c_4 = \frac{3}{4\pi}. \tag{2.3}$$

2.2. Equidistribution

The aperiodic billiard paths on P correspond to geodesics γ on Y having irrational slope. Let γ_n denote the initial portion of γ having length n. We say that γ is *equidistributed* if, for all open $U \subset Y$,

$$\lim_{n\to\infty} \frac{\text{length}(\gamma_n \cap U)}{n} = \text{area}(U). \tag{2.4}$$

The following result establishes a dichotomy for square billiards. A geodesic is either closed or equidistributed.

Theorem 2.1. *Each irrational geodesic is equidistributed and hence dense.*

Proof. (Sketch.) The is equivalent to the statement that the orbits of an irrational rotation T of $\boldsymbol{R}/\boldsymbol{Z}$ are equidistributed in $\boldsymbol{R}/\boldsymbol{Z}$. That is, the fraction A_n/n converges to $|I|$, the length of I. Here A_n is the number of the first n orbit points contained in the interval I. Let I be an interval of length p/q. We can find powers $n_1,\dots,n_q$ such that the union

$$T^{n_1}(I) \cup \cdots \cup T^{n_q}(I)$$

covers $\boldsymbol{R}/\boldsymbol{Z}$ a total of p times, up to tiny overlaps and gaps that we can make as small as we like. Relatively speaking, very few orbit points fall into the tiny overlaps and gaps. So, by symmetry, $A_n/n \to p/q$. The case when $|I|$ is irrational follows from the case when $|I|$ is rational by a similar kind of limiting argument. ■

2.3. Connection to hyperbolic geometry

The group $\mathrm{SL}_2(\boldsymbol{Z})$ of integer 2×2 matrices of determinant 1 acts on $\boldsymbol{R}^2$ in such a way as to preserve $\boldsymbol{Z}^2$. Hence $\mathrm{SL}_2(\boldsymbol{Z})$ acts as affine automorphisms of Y. These maps permute the closed geodesics. One can study this action in terms of hyperbolic geometry. Let $\boldsymbol{H}^2$ denote the hyperbolic plane, given as the upper half-plane in $\boldsymbol{C}$. The *ideal boundary* of $\boldsymbol{H}^2$ is the extended real line $\boldsymbol{R} \cup \infty$. The group $\mathrm{SL}_2(\boldsymbol{Z})$ acts on $\boldsymbol{H}^2$ by the *linear fractional action*,

$$\begin{bmatrix} a & b \\ c & d \end{bmatrix}(z) = \frac{az+b}{cz+d}. \tag{2.5}$$

When $z = \infty$, the right-hand side is set to a/c. The modular group, $\mathrm{SL}_2(\boldsymbol{Z})$, is an example of a *lattice*: $\boldsymbol{H}^2/\,\mathrm{SL}_2(\boldsymbol{Z})$ has finite hyperbolic area.

A *parabolic* element of the larger group $\mathrm{SL}_2(\boldsymbol{R})$ of real determinant-one matrices is one which is conjugate in $\mathrm{SL}_2(\boldsymbol{R})$ to an upper-triangular matrix. Such elements act on $\boldsymbol{H}^2$ without fixed points and they fix one point on the ideal boundary. The *cusps* of a subgroup of $\mathrm{SL}_2(\boldsymbol{R})$ are the fixed points of the parabolics. For $\mathrm{SL}_2(\boldsymbol{Z})$, the set of cusps is $\boldsymbol{Q} \cup \infty$. Thus, the periodic billiard directions are bijective with the cusps of $\mathrm{SL}_2(\boldsymbol{Z})$. The periodic direction of slope s corresponds to the cusp $1/s$.

To see more geometry of the correspondence, let us consider the closed geodesics on Y whose slopes lie in $[1, \infty]$. There is a familiar pattern of horodisks in $\boldsymbol{H}^2$ associated to the modular group. The horodisk associated to ∞ is the half-plane $\{z \mid \mathrm{Im}(z) \geq 1\}$. The horodisk tangent to the ideal boundary at $a/c \in [0, 1]$ is the round disk whose diameter is c^{-2}. This horodisk corresponds to the primitive closed geodesic whose slope is $c/a \in [1, \infty]$ and whose length is $\sqrt{a^2 + c^2} \in [c, 2c]$. Thus, the primitive closed geodesics of length about L and slope in $[1, \infty]$ correspond to those horodisks in $[0, 1]$ of diameter about L^{-2}.

2.4. Symbolic dynamics

Given an aperiodic billiard path in P, we can associate a biinfinite periodic binary sequence β. We record a 0 every time the path hits the horizontal side and a 1 every time the path hits the vertical side. Which sequences occur? I will explain the approach taken in **[98]** and also discussed in **[27, 29, 100]**.

Call a biinfinite binary sequence *derivable* if either 00 never occurs or 11 never occurs in the sequence. Our sequence β is derivable. When the slope of the billiard path is less than (respectively greater than) 1, we never see 00 (respectively 11). If 00 does not occur, we let β' be the sequence obtained from β by removing a single 1 from every consecutive run of (1)s. The new *derived sequence* β' corresponds to the image of β under the element of $\mathrm{SL}_2(\boldsymbol{Z})$ which is the lower triangular matrix consisting of all 1s. In particular, β' is also derivable. We play the same game in the 11 case, with the roles of the digits swapped.

This analysis shows that the derivation process $\beta \to \beta' \to \beta'' \to \cdots$ can be continued forever, producing an infinite list of derivable sequences. Such sequences are called *Sturmian*. With just a few easily classifiable exceptions, every Sturmian sequences arises as the symbolic sequence for an irrational geodesic on Y. See **[98]**.

3. REGULAR POLYGONS

Let P_n be the regular n-gon. To avoid exceptional cases, I will take $n \neq 3, 4, 6$. W. Veech **[108, 109]** noticed that many features of billiards on P_n resemble square billiards. At the end of the section, I will also explain a subtle dynamical difference.

3.1. The covering surface

As with the square, there is a surface X_n and an isometric map $f : X_n \to P_n$ which gives a bijection between geodesics on X_n and billiard paths on P_n.

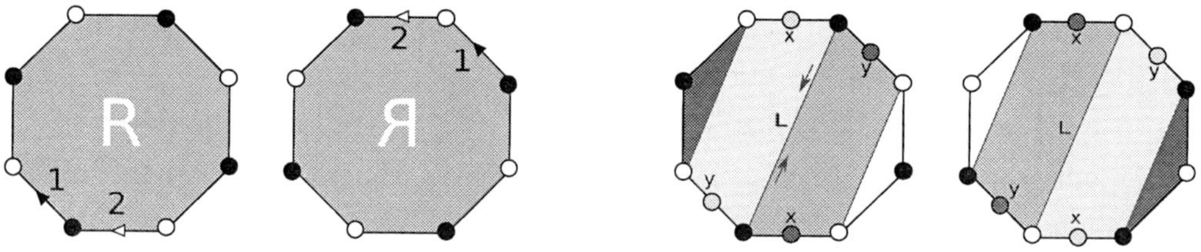

FIGURE 2
The octagon surface X_8.

The left-hand side of Figure 2 indicates the genus 3 surface X_8 made from the union of two octagons by gluing their sides together in the pattern indicated. The pattern is meant to continue to all 8 pairs of sides, and it is more natural if you think of the two octagons as stacked on top of each other in space. The white points are all identified and the black points are all identified. Away from these two special points of X_8, the surface is locally isometric to the plane. At each special *cone point*, X_8 is locally isometric to the space made by gluing together 8 sectors, each having angle $3\pi/4$. Each cone point has *cone angle* 6π. Our surface has a well-defined notion of direction, because the parallel rays pointing in any given direction in the plane induce a corresponding parallel vector field on X_8. The R-letters indicate the map $f : X_8 \to P_8$.

3.2. Connection to hyperbolic geometry

Let $A(X_n)$ denote the group of affine automorphisms of X_n. Such maps preserve the cone points and are locally affine away from them. There is a homomorphism

$$\phi : A(X_n) \to \mathrm{Isom}(\boldsymbol{H}^2). \tag{3.1}$$

Here $\phi(f)$ is defined to be the action of df, when df is interpreted as acting on $\boldsymbol{H}^2$. Because there is a well-defined notion of direction on X_n, there is a canonical identification of all the tangent spaces of X_n (away from the cone points). So, we can interpret df as acting on the same copy of $\boldsymbol{R}^2$ and then we get the hyperbolic action as above. Let

$$\Gamma(X_n) = \phi\big(A(X_n)\big). \tag{3.2}$$

This group is often called the *Veech group*, though sometimes one restricts to the orientation-preserving subgroup. Nontrivial affine maps of X_n which preserve the cusps cannot be too close to the identity, so $\Gamma(X_n)$ is a discrete group.

Theorem 3.1. *The group $\Gamma(X_n)$ is generated by reflections in the sides of a hyperbolic triangle with angles $0, 0, 2\pi/n$ when n is even and with angles $0, \pi/2, \pi/n$ when n is odd.*

Proof. (Sketch.) Consider the case $n = 8$. Simultaneous reflection in the vertical bisectors of our two octagons induces an affine automorphism corresponding to the hyperbolic reflection I_1 in the hyperbolic geodesic γ_1 connecting 0 to ∞. Likewise, simultaneous reflection in the bisector marked L on the right side of Figure 2 corresponds to an affine automorphism ι_2 and $I_2 = \phi(\iota_2)$ reflects in a geodesic γ_2 which makes an angle of $2\pi/8$ with γ_1.

The right-hand side of Figure 2 shows a decomposition of X_8 into 4 cylinders, all of the same modulus. On each of the big cylinders, there is an affine transformation, called a *Dehn twist*, which is the identity on the boundary and maps the vertex marked x to the same colored vertex marked y. The grey arrows show the motion of points near L. Thanks to the same-modulus condition, these maps extend to all of X_8, giving an affine automorphism of X_8 which restricts to Dehn twists on each cylinder. Call this map β. Let $\iota_3 = \iota_2 \circ \beta$ and $I_3 = \phi(\iota_3)$. Note that ι_3 has order 2 and preserves the directions parallel to the vertical and also parallel to L. So, I_3 is the reflection in the geodesic γ_3 connecting the appropriate endpoints of γ_1 and γ_2. The three geodesics bound the desired triangle.

The group Γ' generated by I_1, I_2, I_3 is the triangle group, and it is a subgroup of $\Gamma(X_n)$. The only discrete subgroup of $\mathrm{Isom}(\boldsymbol{H}^2)$ properly containing Γ' is the one generated by Γ' and the reflection in the bisector of our hyperbolic triangle. In the even case, this is not in $\Gamma(X_8)$, so $\Gamma' = \Gamma(X_8)$. In the odd case this extra element would be in the Veech group. ■

3.3. The Veech dichotomy

The Veech dichotomy [109] establishes the same kind of periodic/equistributed dichotomy for regular polygons that we saw for the square in Section 2.2.

Theorem 3.2. *(1) In each direction of X_n corresponding to a cusp of $\Gamma(X_n)$, there is a partition of X_n into metric cylinders foliated by parallel closed geodesics. (2) Conversely, any direction containing a closed geodesic corresponds to a cusp of $\Gamma(X_n)$. (3) Every geodesic in a noncusp direction is equidistributed.*

Proof. (First statement.) Let $X = X_n$. Let h be the parabolic affine automorphism of X corresponding to the cusp. Replacing h by h^2 if necessary we can assume that h fixes both cone points. Consider a geodesic ray γ emanating from one of the cone points in the direction fixed by h. Then h is the identity on γ. If γ does not return to a cone point then some small Euclidean disk $D \subset X$ intersect γ in at least 2 parallel strands. Inside D the map h acts as a shear and therefore shifts one strand of $\gamma \cap D$ with respect to the other in a nontrivial way. But h is the identity on both strands. This is a contradiction. Hence all geodesic rays emanating from the cone points return to cone points. The union of these *saddle connections* divides X into open cylinders foliated by parallel closed geodesics. ■

In the next section I will prove the second statement and a weak version of the third.

3.4. Periodic billiard paths

The Veech dichotomy relates the count of the periodic billiard paths to the enumeration of cusps in the Veech group. Veech [108, 109] makes this enumeration and proves that equation (2.3) holds for all regular polygons P_n with some constant c_n in place of c_4. The *Siegel–Veech constant* c_n in general is complicated, but here is the nice formula when respectively p is an odd prime and n is a power of 2:

$$c_p = \frac{p(p-1)(p^2+1)}{48(p-2)\pi}, \quad c_n = \frac{n^2(n-1)}{16(n-2)\pi}. \tag{3.3}$$

Veech's argument is a subtle mix of number theory and dynamics, but the horodisk picture discussed in Section 2.3 gives some geometric intuition. Rotate so that the horizontal direction is periodic. As with the modular group, consider a $\Gamma(X)$-invariant pattern of disjoint horodisks in $\boldsymbol{H}^2$, one per cusp. The primitive periodic directions of slope in $[1, \infty]$ and having length at most L essentially correspond to horodisks of diameter larger than about L^{-2} that are based at points in $[0, 1]$. There are certainly at most $O(L^2)$ of these horodisks, and it is at least plausible (and not too hard to prove) that there are at least $O(L^2)$ of them.

3.5. Symbolic dynamics and cusps

In [100], J. Smillie and C. Ulcigrai use the Veech group to show that the symbolic sequences for billiards on P_8 follow a derivation rule much like that for Sturmian sequences discussed above. Subsequently D. Davis [27] worked this out for all P_n.

In [73], A. Leutbecher proves the following result:

Theorem 3.3. *A point of the ideal boundary of the hyperbolic plane is a cusp of $\Gamma(X_5)$ if and only if it lies in $\boldsymbol{Q}(\cos(2\pi/5)) \cup \infty$.*

Similar results cover $n = 3, 4, 5, 6, 8$. Compare Theorem 1.5 in [80]. Is it an open problem to characterize the cusps of $\Gamma(X_n)$ for the cases other than these. The case $n = 7$ is the first unknown case. See [80] for a discussion of all this.

In [28], D. Davis and S. Lelièvre obtain many additional results about coding the billiards in P_5 and the cusps of $\Gamma(X_5)$.

3.6. Mixing

Here is one way billiards in P_n different from billiards in the square. On the square, the billiards map (which keeps track of the billiard paths at the bounce points) typically equidistributes the points, but it does not *mix them up*. A transformation T of a measure space (X, μ) is called respectively *mixing* and *weak mixing* if

$$\lim_{n\to\infty} \left|\mu\big(U \cap T^n(V)\big) - \mu(U)\mu(V)\right| = 0, \quad \lim_{n\to\infty} \frac{1}{n} \sum_{j=0}^{n-1} \left|\mu\big(U \cap T^j(V)\big) - \mu(U)\mu(V)\right| = 0. \tag{3.4}$$

A weak mixing transformation is "mixing at most times." On the square, the billiards map (see Section 7.1) is not weak mixing, but A. Avila and V. Delecroix [4] show that with respect to a typical aperiodic direction on P_n, the billiards map is weak mixing (but never mixing).

4. RATIONAL POLYGONS

A *rational polygon* is a polygon whose angles are all rational multiples of π. It is difficult to overstate (and to adequately survey) the spectacular development of rational billiards, the study of billiards in rational polygons and more generally the straight line flow in translation surfaces. For additional sources, see [46,77,115,118,119].

4.1. Translation surfaces

A *translation surface* is any oriented surface made by gluing together a finite collection of disjoint polygons in such a way that the side identifications are given by translations. As above, such a surface is locally Euclidean and has a well-defined sense of direction away from a finite number of cone points whose angle is an integer multiple of 2π. The geodesics on a translation surface are locally straight lines and they avoid the cone points. Here are 3 basic definitions.

Strata. The set of all translation surfaces with a fixed topological type and a fixed list of cone angles is called a *stratum*.

Veech group. The affine automorphism group $A(X)$ and the *Veech group* $\Gamma(X) = \phi(A(X))$ are defined for a general translation surface X just as in Section 3.2. We let $A_+(X)$ and $\Gamma_+(X)$ denote the respective orientation preserving subgroups. We sometimes think of $\Gamma_+(X)$ as a subgroup of $\mathrm{SL}_2(\boldsymbol{R})$ rather than as a subgroup of $\mathrm{Isom}(\boldsymbol{H}^2)$.

Lattice property. Group $\Gamma(X)$ need not be a lattice in $\mathrm{Isom}(\boldsymbol{H}^2)$, but when it is a lattice we say that X *has the lattice property*.

Lemma 4.1 (Katok–Zemylakov construction). *Let P be rational polygon. Then there is a translation surface X_P and a piecewise isometric map $f : X_P \to P$ which carries geodesics on X_P to billiard paths on P.*

Proof. For each edge e of P there is a reflection R_e in the line through the origin parallel to e. The group G generated by these reflections is finite, thanks to the rationality assumption. For each $g \in G$, define $P_g = g(P) + V_g$. Here V_g is a vector included so that all the polygons $\{P_g \mid g \in G\}$ are disjoint. The final answer is independent of these auxiliary vectors.

We form an identification space on the union of these polygons by gluing together every pair of edges of the form

$$e_1 = g(e) + V_g, \quad e_2 = gr(e) + V_{gr}, \quad r = R_e \tag{4.1}$$

by a translation. By construction, X_P is a translation surface. The map $X_P \to P$ is defined to be the inverse of the map $g + V_g$ on the piece P_g. ■

This all-purpose construction is not necessarily the most efficient one in special situations. For instance, when applied to the regular n-gon it produces a cyclic cover of the surface X_n considered in the previous chapter.

4.2. Affine action

Each stratum has an action of $\mathrm{GL}_2(\boldsymbol{R})$ on it. If we have a translation surface Y made by gluing together a finite list $P_1, \ldots, P_k$ of polygons and an element $g \in \mathrm{GL}_2(\boldsymbol{R})$, we get a new translation surface $g(Y)$ by gluing together $g(P_1), \ldots, g(P_k)$ in the same pattern.

When Y is the square torus, and $g \in \mathrm{SL}_2(\boldsymbol{Z})$, the surface $g(Y)$ is obtained by gluing the opposite sides of an area 1 parallelogram whose vertices have integer coordinates. This is just Y again, presented differently. More generally, if $g \in A_+(Y)$ then $dg(Y)$ is the same surface as Y. Conversely, $g \in \mathrm{SL}_2(\boldsymbol{R})$ and $g(Y) = Y$ then $g \in \Gamma_+(Y)$. This lets us identity the orbit $\mathrm{SL}_2(\boldsymbol{R}) \cdot Y$ with $\mathrm{SL}_2(\boldsymbol{R})/\Gamma_+(Y)$. We may identify this latter space with the (orbifold) unit tangent bundle of $\boldsymbol{H}^2/\Gamma_+(X)$. Equipped with this point of view, let us prove the second statement of the Veech dichotomy theorem.

Lemma 4.2. *Suppose that X is a translation surface and $\Gamma_+(X)$ is a lattice in $\mathrm{SL}_2(\boldsymbol{R})$. If X has a closed geodesic with direction δ, then δ corresponds to a cusp of $\Gamma(X)$.*

Proof. For ease of exposition, assume that elements of $\Gamma_+(X)$ act on $\boldsymbol{H}^2$ without fixed points, so that the quotient $\Sigma = \boldsymbol{H}^2/\Gamma_+(X)$ is a finite-area hyperbolic surface. In the general case, we would pass to a finite-index subgroup. A geodesic ray in Σ either goes to a cusp or else recurs infinitely often to a compact subset. We rotate X so that δ is horizontal. Let

$$g_t = \begin{bmatrix} e^{-t} & 0 \\ 0 & e^t \end{bmatrix}. \tag{4.2}$$

Since $g_t(X)$ has a closed loop whose length tends to 0 as $t \to \infty$, the surface $g_t(X)$ exits every compact subset of $\mathrm{SL}_2(\boldsymbol{R})/\Gamma_+(X)$ as $t \to \infty$. But the set $\{g_t(X) \mid t \geq 0\}$ projects to a geodesic ray on Σ. If ∞ (the point on the ideal boundary corresponding to the horizontal direction) is not a cusp of Σ then this ray recurs infinitely often to a compact subset of Σ. But this is a contradiction. ∎

Here is a weaker version of the third statement. The reason that the aperiodic directions are dense is that the boundary of the complement of a nondense geodesic would have a closed loop, and then Lemma 4.2 would say that the direction corresponds to a cusp. The equidistribution statement follows from Theorem 1.1 in H. Masur's paper [76] and the fact that a geodesic which does not exit the cusp of $\boldsymbol{H}^2/\Gamma(C)$ is recurrent.

4.3. Connection to Teichmüller space

The space $\boldsymbol{M}_g$ is the space of Riemann surfaces of genus g. The universal orbifold cover of $\boldsymbol{M}_g$ is called *Teichmüller space* and denoted $\boldsymbol{T}_g$. To define $\boldsymbol{T}_g$, we fix a background genus g surface Σ_0. A point in $\boldsymbol{T}_g$ is then an equivalence class of pairs (Σ, ψ), where Σ is a genus g Riemann surface and $\psi : \Sigma_0 \to \Sigma$ is a homeomorphism. Two pairs (Σ_1, ψ_1) and (Σ_2, ψ_2) are equivalent if there is a biholomorphic map $f : \Sigma_1 \to \Sigma_2$ such that $f \circ \psi_1$ and ψ_2 are homotopic maps. The map ψ is often called the *marking* of Σ.

One can think of a genus g translation surface as a Riemann surface equipped with a holomorphic 1-form, an expression which looks like $f(z)dz$ in local coordinates. These objects are also called *abelian differentials*. Thus a translation surface Y naturally gives rise to a point in the "vector bundle" of abelian differentials over moduli space. I put "vector bundle" in quotes just because $\boldsymbol{M}_g$ is an orbifold rather than a manifold. More simply, each marked translation surface corresponds to a point in the vector bundle of abelian differentials over Teichmüller space.

We can interpret the $SL_2(\boldsymbol{R})$ action as giving a group action on the abelian differential bundle over Teichmüller space. For any translation surface Y, the orbit $SL_2(\boldsymbol{R}) \cdot Y$ projects to the hyperbolic plane in $\boldsymbol{T}_g$. When Y has the lattice property, this hyperbolic plane further projects to an isometric copy of $\boldsymbol{H}^2/\Gamma_+(Y)$ in $\boldsymbol{M}_g$ called a *Teichmüller curve*.

4.4. Structure of strata

The following lemma gives the basic structure of the strata.

Lemma 4.3. *A stratum having genus g and v cone points is a complex orbifold of (complex) dimension $n = 2g + v - 1$. The manifold cover of the orbifold has an atlas of coordinate charts with transition functions in* $GL_n(\boldsymbol{C})$.

Proof. (Sketch.) Let Σ be the stratum and let $X \in \Sigma$ be a translation surface. We construct local coordinates called *period coordinates* to describe a neighborhood of X in Σ. Triangulate X so that the v vertices of the triangulation are the cone points. This realizes X as the quotient of a union of f triangles with e pairs of edges glued together by translations. Orient each pair of edges and pick one edge from each pair and record the complex number that describes its direction and magnitude.

A nearby assignment of e complex numbers gives a recipe for a new collection of triangles provided that, around each triangle, the corresponding sum of the complex numbers (perhaps with signs in front) is 0. This gives f relations, but one relation is redundant because the closing conditions on all but one triangle determine the last closing condition. So, the space of valid choices has dimension $n = d - f + 1 = 2g - v + 1$.

These coordinates might not give a local homeomorphism into $\boldsymbol{C}^n$ because (thanks to symmetries) different assignments can give rise to the same translation surface. By considering the same marking trick as with the definition of Teichmüller space, you can construct a cover of the stratum which is a manifold and for which the above coordinates are a coordinate chart in the usual sense. If X is triangulated in a different way then the transition functions between the coordinate charts are complex linear. ∎

Lemma 4.3 has a kinship with W. Thurston's paper [105], in which he considers the moduli space of flat cone metrics on the sphere with $n + 2 \geq 4$ prescribed cone angles. He constructs "period coordinates" in which these spaces are orbifolds whose transitions functions lie in $PU(1, n-1) \subset GL_n(\boldsymbol{C})$. Under certain arithmetic conditions on the cone points, Thurston shows that the subset of unit area structures is open dense in a complex hyperbolic orbifold coming from a Deligne–Mostow [31] lattice.

4.5. Periodic billiard paths

M. Boshernitzan, G. Galperin, T. Krüger, and S. Troubetzkoy [14] prove that the periodic billiard positions and directions are dense in a rational polygon.

Theorem 4.4. *For any rational polygon P, the set of periodic billiard paths lifts to a dense subset of the unit tangent bundle of P.*

The closed geodesics on a translation surface Y come in parallel families and sweep out maximal metric cylinders. Let $N(L, Y)$ denote the number of these maximal cylinders of length less than L. Here is a result of H. Masur's [74, 75]:

$$0 < A_Y \leq \liminf_{L \to \infty} \frac{N(L, Y)}{L^2} \leq \limsup_{L \to \infty} \frac{N(L, Y)}{L^2} \leq B_Y < \infty. \tag{4.3}$$

This result implies a similar-looking result for periodic billiard paths on rational polygons.

Here is one of the main open problems in the field.

Conjecture 4.5. $A_Y = B_Y$ *for all translation surfaces* Y.

A. Eskin and Masur [42] prove Conjecture 4.5 for almost all surfaces within each stratum, and they show that the common value, called the *Siegel–Veech constant*, just depends on the stratum. One application of the main result in [43, 44] (discussed below) is that Conjecture 4.5 is true in an average sense,

$$\lim_{L \to \infty} \frac{1}{L} \int_0^L N(Y, e^t) e^{-2t}\, dt = C_Y, \tag{4.4}$$

for all surfaces Y. The constant depends on the surface.

These strong asymptotic counting results rely on powerful dynamical results about the action of $SL_2(\boldsymbol{R})$ and its subgroups on the bundle of abelian differentials on Teichmuller space. This survey does not really touch on these ideas. See [115] for details.

4.6. Classification problem

Which polygons and translation surfaces have the lattice property? Which Teichmüller curves arise? How does the Teichmüller curve depend on the translation surface (with the lattice property)? Here is some progress on these questions.

Genus 2. In the genus 2 stratum with 2 cone points, only the regular decagon with opposite sides identified has the lattice property. See [79]. In the genus 2 stratum with 1 cone point, there is an infinite family, all coming from L-shaped polygons in which a rectangle is cut out the corner of the square. The corresponding Veech groups are classified by an invariant (± 1), called the *spin*, and square-free integer D congruent to 0 or 1 mod 4 called the *discriminant*. The corresponding quotient $\Sigma_D = \boldsymbol{H}^2 / \Gamma(X_D)$ embeds in a Hilbert modular surface $(\boldsymbol{H}^2 \times \boldsymbol{H}^2) / PSL_2(O_D)$. Here O_D is the ring of algebraic integers in $\boldsymbol{Q}(\sqrt{D})$. See [78] and [20]. The topological features of Σ_D, namely its Euler characteristic (M. Bainbridge [6]) and orbifold points (R. Mukamel [85]), are known.

Genus 3 and 4. C. McMullen, R. Mukamel, and A. Wright [81] recently discovered new infinite families of primitive translation surfaces, i.e., not covers of other translation surfaces in lower genus, in genus 3 and 4 which have the lattice property. These surfaces correspond to certain dart-shaped quadrilaterals. One family corresponds to darts of the form $(1, 1, 1, 9)$, and the other corresponds to darts of the form $(1, 1, 2, 8)$. These numbers describe the relative proportions of the interior angles of the darts.

Triangle groups. The Veech groups all have cusps. With Theorem 3.1 in mind, one can ask whether every triangle group with a cusp arises as a Veech group. Subject to certain congruence conditions and an index 2 ambiguity, this turns out to be true. I. Bouw and M. Möller [15] discovered translation surfaces with this property and defined them in terms of algebraic geometry. Later, W. Hooper gave concrete combinatorial descriptions for them [61]. Compare also [29] and [116].

Computation. J. Athreya, D. Aulicino, and W. Hooper [3] explicitly compute the quotient $\boldsymbol{H}^2/\Gamma(X)$ where X is the translation surface associated to the regular dodecahedron. This canonical object has genus 131, 19 cone singularities, and 362 cusps. Relatedly, the authors show that modulo the action of $A(X)$ there are exactly 31 closed geodesics on X which contain just a single cone point. The point of these results is not just the surprising complexity, but also their vital reliance on software. In this case, the results use Sage-based software written by Hooper and V. Delacroix [30]. See [16] and [41] for computational Veech group algorithms.

4.7. Orbit closures

Let g_t be as in equation (4.2). When Y has the lattice property, the flow $t \to g_t(Y)$ traces out the lift of a geodesic to the unit tangent bundle U of $\boldsymbol{H}^2/\Gamma(Y)$. The closure of this set could be the lift to U_1 of a geodesic lamination, a set which looks locally like the product of a geodesic and a Cantor set. When Y does not have the lattice property, the orbit closure could be even wilder. A huge breakthrough in rational billiards rules out this kind of wildness for the closure of $\mathrm{GL}_2(\boldsymbol{R}) \cdot Y$. The following result is a combination of results in A. Eskin and M. Mirzakhani [43] and Eskin–Mirzakhani–A. Mohammadi [44].

Theorem 4.6. *For any translation surface Y, the closure of $\mathrm{GL}_2(\boldsymbol{R}) \cdot Y$ is a locally affine orbifold (possibly with self-intersections). The support of any ergodic $\mathrm{SL}_2(\boldsymbol{R})$-invariant measure on the orbit closure has the same structure, and the measure is affine.*

Theorem 4.6, which A. Zorich [119] calls the *Magic Wand Theorem*, has spurred a huge amount of activity. I have already mentioned equation (4.4) as an application. Here are three more developments.

Algebraic structure. The bundle of abelian differentials over moduli space has an algebraic structure, but the Magic Wand Theorem only says that the orbit closures are real analytic in this structure. (The periodic coordinates are transcendental.) In [45], S. Filip shows that nonetheless the orbit closures are algebraic.

The illumination problem. Given a billiard table P and two points $x, y \in P$, one says that *x illuminates y* if there is a billiard path starting at x and containing y. In [72], S. Lelièvre, T. Monteil, and B. Weiss use the Magic Wand Theorem to prove that for any rational polygon P and for any point $x \in P$, there are at most finitely many $y \in P$ such that x does not illuminate y. Compare also [1]. In a recent refinement [114], A. Wolecki shows that there are at most finitely many *pairs* in any rational polygon P such that x does not illuminate y.

Wild horocyclic closures. Let P be the parabolic subgroup consisting of the upper-triangular matrices. J. Chaika, J. Smillie, and B. Weiss [22] prove that the closure of $P \cdot Y$ can be wild. For instance, it can have fractional Hausdorff dimension. The surprise here is that when Y has the lattice property, the closure $P \cdot Y$ is either a single curve or all of U.

5. IRRATIONAL POLYGONS

Now we leave the excitement of rational billiards and consider the case of billiards on irrational polygons, those whose angles are not all rational multiples of π. Given the incredible depth and precision of the results about rational billiards, it is remarkable that we do not even know if every obtuse triangle has a periodic billiard path! We also do not know if every acute triangle has a periodic billiard path that is not of the kind shown in Figure 3 below. Are there polygons without periodic billiard paths? Nobody knows. Problems such as the illumination problem discussed above are wide open.

5.1. Easy examples

Figure 3 shows periodic billiard paths which exist on all acute triangles and all right triangles, respectively. The periodic path with 6 bounces shown on the left-hand side of Figure 3 is part of an infinite parallel family of such paths. This family degenerates to the periodic billiard path having 3 bounces. This special periodic billiard path, called the *Fagnano path*, is the inscribed triangle of minimum length. The billiard path on the right starts and ends perpendicular to the side. Call such periodic billiard paths *orthogonal*.

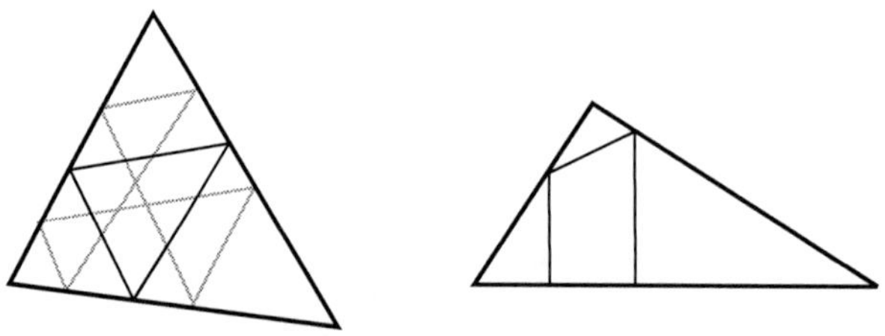

FIGURE 3
Periodic billiard paths on acute and right triangles.

5.2. Right triangles

When applied to irrational polygons, Lemma 4.1 produces an infinite translation surface. When P is a right triangle, the resulting surface X_P is really neat. Let Q denote the rhombus that is tiled by 4 copies of P. Now glue a countable collection of copies of Q around a single vertex, in a kind of spiral pattern, as indicated in Figure 4.

Finally, glue together the remaining sides of the infinite union in the pattern indicated. This surface is constructed in the abstract so that the different rhombi do not really lie in the plane and overlap. Surface X_P has 4 infinite cone points, all of which have the same

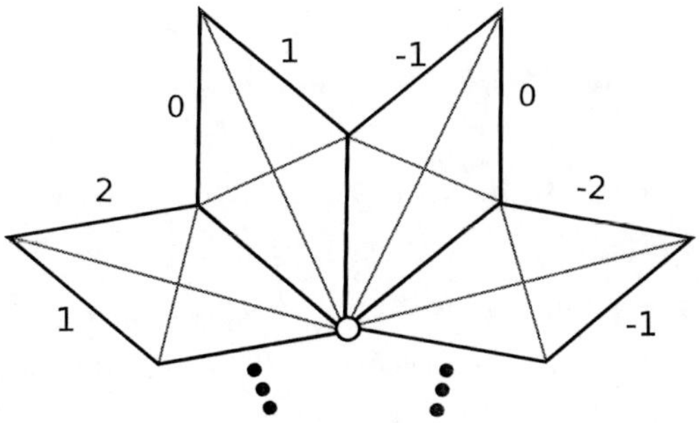

FIGURE 4
The translation surface associated to an irrational right triangle.

structure. The construction above favors one of the cone points, and so you have to stare at the picture for a while to see the other three.

In [107], S. Troubetzkoy analyzes these surfaces and proves the following result:

Theorem 5.1. *Any irrational right triangle has a side such that all but countably many points in that side are the start points of orthogonal periodic billiard paths.*

Compare the work of B. Cipra, R. Hanson, and A. Kolan [26]. A quick corollary of Theorem 5.1 is that periodic billiard paths on a right triangle are dense, as in Theorem 4.4.

A periodic billiard path on a triangle is called *stable* if a periodic billiard path of the same combinatorial type exists on all nearby triangles. W. Hooper [58, 60] proved the following result:

Theorem 5.2. *No combinatorial type of periodic billiard path exists on both acute and obtuse triangles. In particular, periodic billiard paths on right triangles are unstable.*

There are still lots of open problems about right triangular billiards. For instance,

Question 5.3. Does the number of maximal families of periodic billiard paths in a right triangle have quadratic growth, as in Masur's theorem?

5.3. Existence results for obtuse triangles

In [49], G. A. Galperin, A. M. Stepin, and Y. B. Vorobets construct some infinite families of periodic billiard paths in irrational polygons. In [57], L. Halbeisen and N. Hungerbuhler construct additional infinite families of periodic billiard paths in obtuse triangles. The examples in [49] and [57] are stable.

Here is one of my results, proved in a series of two papers [92, 94].

Theorem 5.4. *An obtuse triangle has a stable periodic billiard path provided all its angles are at most* 100 *degrees.*

Proof. (Rough sketch.) Let $\boldsymbol{P}$ denote the parameter space of similarity classes of obtuse triangles. Then $\boldsymbol{P}$ itself is a triangle. Let $\boldsymbol{P}_{100}$ denote the subset corresponding to triangles having less than 100 degrees.

Suppose that w is some finite word that corresponds to a stable periodic billiard path on some triangle. Let $O(w) \subset \boldsymbol{P}$ be the set corresponding to triangles which support a periodic billiard path with sequence w. We call $O(w)$ the *orbit tile*. To estimate $O(w)$, we successively reflect the initial triangle according to the digits of the word, as in Figure 5. The result is called the *unfolding*. The stability condition guarantees that the first and last sides of the unfolding are parallel, no matter which triangle we use for the construction. Rotate so that the translation carrying the first side to the last side is horizontal.

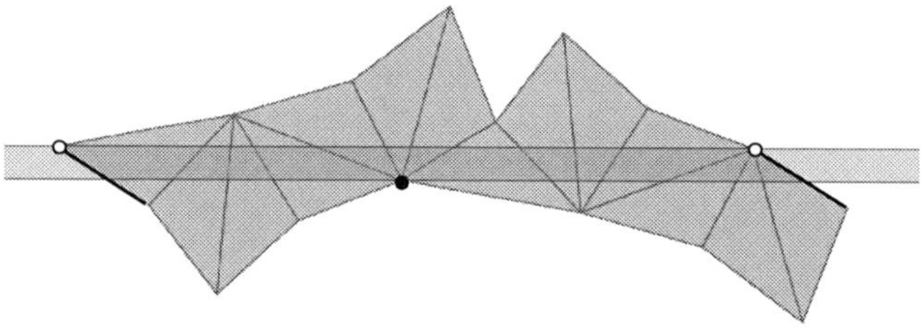

FIGURE 5
An unfolding and a corridor.

There is a maximal strip, with horizontal sides, such that any horizontal line in the strip corresponds to a periodic billiard path. We call this a *corridor*. As the triangle changes shape, the corridor widens or narrows according to how the vertices move. The billiard path disappears when the height of one of the vertices along the top of the unfolding has the same height as one of the vertices along the bottom. This condition is given by the vanishing of a finite trigonometric sum. Using some mixture of analytic and numerical methods, one can approximate $O(w)$ in a rigorous way.

Showing that a given region in $\boldsymbol{P}$ consists entirely of triangles with stable periodic billiard paths amounts to proving that one can cover this region with orbit tiles. First, I found some nice infinite families of orbit tiles which in a systematic way cover certain regions of $\boldsymbol{P}_{100}$, *trouble spots*, which have no finite cover. (See the discussion and also Theorem 5.8 below.) Then I do repeated depth-first searches through the tree of words, up to a suitable depth, in order to cover the remainder $\boldsymbol{P}_{100}$ with about 200 more orbit tiles. ■

W. Hooper and I wrote a computer program, called *McBilliards*, which does these searches and also plots the orbit tiles to a high degree of accuracy. The search looks like it is exponential in the depth—which would be very bad—but in fact it is much faster. Given a triangle and a word, McBilliards performs the unfolding until it appears that the unfolding is so crooked that no continuation of the word would produce a nonempty corridor. This pruning vastly increases the speed.

Recently, in a preprint [106], J. Garber, B. Marinov, K. Moore, and G. Tokarsky improved my 100-degree result to 112.3 degrees, though they do not have the stability conclusion. They discovered that certain special triangles are quite difficult to cover with stable orbit tiles, but nonetheless have unstable periodic billiard paths. As I mentioned above, my dataset involved several infinite families of words and about 200 additional "sporadic" words. The dataset in [106] involved bazillions of sporadic words. I do not know the number, but it took me several hours just to download the dataset from the internet!

One serious difficulty in using the orbit tile approach to prove that every triangle has a periodic billiard path is that regions near the boundary of the parameter space, corresponding to thin triangles, are extremely hard to cover with orbit tiles. In my 100-degree theorem, I got lucky and found an infinite family of periodic billiard paths which cover a neighborhood of the boundary, up to $5\pi/8$. Beyond this, there is no known point on the boundary which has a neighborhood covered by orbit tiles. Note that $5\pi/8$ radians is 112.5 degrees, so the 112.3 result of [106] cannot be much improved without more luck at the boundary.

Here are a few more existence results. In [62], Hooper and I used similar ideas to prove the following result.

Theorem 5.5. *If $\{P_n\}$ is any sequence of triangles converging to an isosceles triangle, then P_n has a periodic billiard path once n is sufficiently large.*

There is only one counting result, due to Hooper [59], which even vaguely resembles equation (2.3).

Theorem 5.6. *There exists an open subset of obtuse triangles such that for each triangle in the set the number $N(L)$ of primitive periodic billiard paths has the property that*

$$\lim_{L\to\infty} \frac{N(L)}{L\log(L)^k} = \infty$$

for any k.

Numerical experiments with McBilliards lead to the following conjecture:

Conjecture 5.7. *Orbit tiles are connected and simply connected.*

It would also be interesting to know how the area of the orbit tile depends on the length of the word. One approach to showing that some triangles do not have any periodic billiard paths would be to show that in general the area decays very rapidly and the number of words does not grow quickly. See the result of D. Scheglov discussed below in Section 5.6.

5.4. Recalcitrance

Call a triangle T *recalcitrant* if for any $\varepsilon > 0$ there are triangles within ε of T (in terms of angle differences) supporting no periodic billiard paths of length less than $1/\varepsilon$. The corresponding point in the parameter space has no neighborhood covered by finitely many orbit tiles. In [92] I proved the following result:

Theorem 5.8. *The $(2, 3, 6)$ right triangle is recalcitrant.*

Theorems 5.8 and 5.4 complement each other. Basically, Theorem 5.8 says that a result like Theorem 5.4 is intrinsically hard. A neighborhood of the (2, 3, 6) triangle, on the obtuse side of the parameter space, is one of the trouble spots I mentioned above in connection with the proof of Theorem 5.4.

Numerical experiments with McBilliards lead to the following conjectures:

Conjecture 5.9. *Every obtuse Veech triangle is recalcitrant.*

Conjecture 5.10. *Once the Fagnano orbit tile is removed, the acute triangle parameter space is not covered by any finite union of orbit tiles.*

Conjecture 5.11. *For any N there is some $\varepsilon > 0$ so that no triangle within ε of the equilateral triangle has an orthogonal periodic billiard path of length less than N.*

5.5. Bounce rigidity

One of the few sweeping geometric results about billiards in any polygon is *bounce rigidity*. Every polygon P gives rise to a collection $B(P)$ of biinfinite words corresponding to biinfinite billiard paths. These billiard paths may or may not be periodic. The set $B(P)$ is called the *bounce spectrum*. In [40], M. Duchin, V. Erlandsson, C. Leininger, and C. Sadanand prove that the bounce spectrum essentially determines the shape of P.

Theorem 5.12. *If two polygons P_1, P_2 are such that $B(P_1) = B(P_2)$, then either P_1 and P_2 are related by a similarity or else P_1 and P_2 have all right angles and are affinely equivalent.*

A very similar result is proved independently by A. Calderon, S. Coles, D. Davis, J. Lanier, and A. Oliveira in [19]. These results are the culmination of many works on this topic. See [40] and [19] for further references.

5.6. Ergodicity and complexity

A. Katok [68] has called the ergodicity and orbit growth for irrational polygonal billiards one of the five most resistent problems in dynamics. Here are two subproblems of Problem 3 on Katok's list.

Question 5.13. Is the billiard flow ergodic with respect to almost every polygon? What about with respect to almost every triangle?

Conjecture 5.14. *With respect to any polygon the number $S(L)$ of saddle connections of length less than L is eventually less than $L^{2+\varepsilon}$ for any $\varepsilon > 0$.*

The work of S. Kerkhoff, H. Masur, and J. Smillie [69] gives a G_δ-set of ergodic tables. Recently, J. Chaika and G. Forni [21] proved a similar result about weak mixing. Compare [5]. Ya. Vorobets [112] gives an explicit (but crazily impractical) criterion for ergodicity:

Theorem 5.15. *If the polygon Q admits approximation by rational polygons at the rate*

$$\phi(N) = \left(2^{2^{2^{2^N}}}\right)^{-1}.$$

Then the billiard flow is ergodic on Q.

See [77] for many other references about ergodicity of the billiard flow.

In [67], Katok proves that $S(L)$ grows subexponentially. D. Scheglov [91] has the best refinement on this result to date:

Theorem 5.16. *With respect to almost every irrational triangle T, the following estimate on $S(L)$ holds:*

$$\lim_{L\to\infty} \frac{S(L)}{\exp(L^{\varepsilon})} = 0 \quad \forall \varepsilon > 0.$$

6. POLYGONAL OUTER BILLIARDS

B. H. Neumann [86] introduced outer billiards in the late 1950*s* and then J. Moser [83, 84] popularized it in the 1970s as a toy model for planetary motion. Outer billiards is a game that is played on the outside of a billiard table. Given a compact convex set $K \subset \boldsymbol{R}^2$ and a point $x_0 \in \boldsymbol{R}^2 - K$, one defines x_1 to be the point such that the segment $\overline{x_0 x_1}$ is tangent to K at its midpoint and K lies to the right of the ray $\overrightarrow{x_0 x_1}$. See Figure 6 for an example.

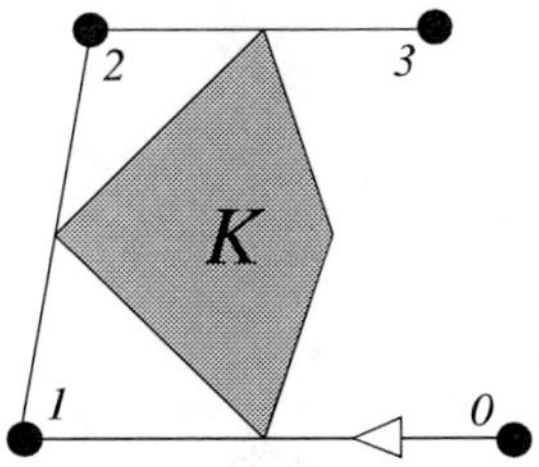

FIGURE 6
Polygonal outer billiards.

The iteration $x_0 \to x_1 \to x_2 \to \cdots$ is called the *forward outer billiards orbit* of x_0. The backward orbit is defined similarly. When K is a polygon, this map is well defined in the complement of finitely many rays which extend the sides of K.

6.1. Periodic orbits

When K is a polygon, the second iterate ψ of the outer billiards map is a piecewise translation. The translation vectors all have the form $2(v_i - v_j)$, where v_i and v_j are vertices. Every finite power of ψ is defined in the complement of finitely many lines. In particular, if ψ has a periodic orbit, then there is a maximal open convex set containing this point, often called a *periodic island*, which consists entirely of periodic points of the same period. This is somewhat akin to the phenomenon that periodic billiard paths on polygons come in infinite parallel families.

Periodic orbits are easier to come by in polygonal outer billiards. C. Culter proved the following pretty easy result. See [104].

Theorem 6.1. *Outer billiards with respect to any convex polygon has infinitely many periodic islands.*

Three teams of authors, namely Vivaldi–Shaidenko [110], Kolodziej [70], and Gutkin–Simanyi [55] proved the following result:

Theorem 6.2. *If K is a convex polygon with rational vertices then all outer billiards orbits on K are bounded. Hence all orbits are periodic.*

Proof. (Sketch of both results.) Let P be a convex polygon. Scale so that P has integer vertices. For simplicity, assume that P has no parallel sides. For each edge e of P, consider the strip Σ_e with the following description. One side of Σ_e contains e, and the unique vertex of P farthest from e lies on the centerline of the strip. This Σ_e is twice as wide as P in some sense. Let $e_1, \ldots, e_n$ be the edges of P ordered according to their slopes. Let $\Sigma_1, \ldots, \Sigma_n$ be the corresponding strips.

There are partitions of each strip Σ_j into parallelograms translation equivalent to $\Sigma_j \cap \Sigma_{j+1}$ such that, for p far away from P, some power of ψ maps each parallelogram in Σ_j isometrically into Σ_{j+1}. Figure 7 shows this.

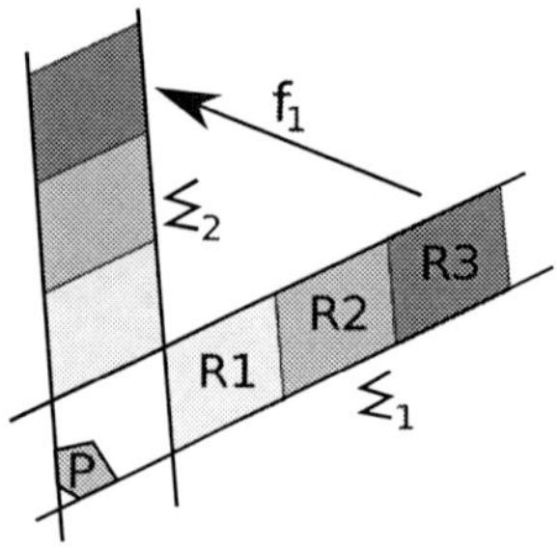

FIGURE 7
The strip map.

Let f_j be the map which maps the nth parallelogram in Σ_j to the nth parallelogram in Σ_{j+1}, as indicated by Figure 7. Let Ψ denote the first return map to the strip Σ_1. Outside a large compact set, Ψ agrees with $f_n \circ \cdots \circ f_1$. In particular, Ψ (and hence ψ) has a periodic orbit provided that, far away from P, there are parallelograms $R_{i_j} \subset \Sigma_j$ such that

$$H(i_1, \ldots, i_n) = R_1 \cap f_1^{-1}(R_2) \cap \cdots \cap f_1^{-1} \circ \cdots \circ f_n^{-1}(R_n) \neq \emptyset. \qquad (6.1)$$

I like to think of these as "resonances."

If P has integer vertices, then for certain lists of integers $i_1, \ldots, i_n$, the set $H(i_1, \ldots, i_n)$ is convex set that completely spans Σ_1, like a tennis ball in a can, so that $\Sigma_1 - H$ is disconnected. The corresponding periodic island separates P from infinity, like a necklace. These special integer lists recur periodically, so there is an infinite sequence of these necklace barriers marching out to ∞. The other orbits are trapped between these barriers. Hence all orbits are bounded. But since ψ is locally a translation by integer vectors, all orbits are periodic by the pidgeonhole principle.

When P is arbitrary, these exact resonances do not occur, but infinitely often they nearly occur. Hence there are infinitely many periodic islands which very nearly span the strip Σ_1. ■

Question 6.3. Does outer billiards have a dense set of periodic islands with respect to almost every polygon?

6.2. Aperiodic orbits

Let us examine the proof sketch just given more carefully. The existence of the resonances producing the necklace orbits just discussed does not really depend on the polygon P having integer vertices. The important thing is that the consecutive parallelograms $\Sigma_j \cap \Sigma_{j+1}$ have commensurable areas. A polygon which has this property is called *quasirational*. Thus, the stronger version of Theorem 6.2 is that every quasirational polygon has all orbits bounded.

The regular polygons certainly are quasirational. Hence all the outer billiards orbits are bounded for regular polygons. In [102], S. Tabachnikov proved the following result:

Theorem 6.4. *Outer billiards orbits on the regular pentagon has some aperiodic orbits.*

Figure 8 shows the pattern of periodic islands for the regular pentagon. The outer 5 periodic islands, not entirely shown, form the first necklace. Theorem 6.4 is proved by establishing that the first return map to a certain triangular region T in the plane is a *renormalizable* polygon exchange map. In this case, this means that the first return map to some smaller triangle $T' \subset T$ is conjugate, via a similarity, to the first return map to T.

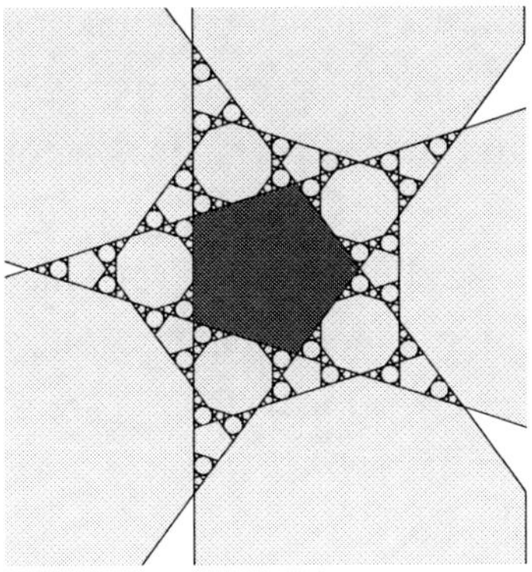

FIGURE 8
Periodic islands for the regular pentagon.

The cases $n = 8, 10, 12$ also have a self-similar structure. Without having a reference, I have the sense that the case $n = 7$ is somewhat understood in the sense that there are some regions of renormalization. I think that the cases $n = 9, 11$ are not understood at all. G. Hughes [63] has made beautiful and detailed pictures of outer billiards on regular polygons. These pictures (and earlier ones) suggest

Conjecture 6.5. *Outer billiards on the regular n-gon has an aperiodic orbit if $n \neq 3, 4, 6$.*

I think that this is not known aside from $n = 5, 8, 10, 12$, and perhaps $n = 7$.

D. Genin [50] made a thorough study of outer billiards on trapezoids, and found examples of open subsets of aperiodic orbits.

6.3. Unbounded orbits

One central problem in the subject is the *Moser–Neumann Problem:* Do there exist any outer billiards systems with unbounded orbits? In [84] and [83], Moser discussed this problem in terms of the stability of a toy problem for planetary motion.

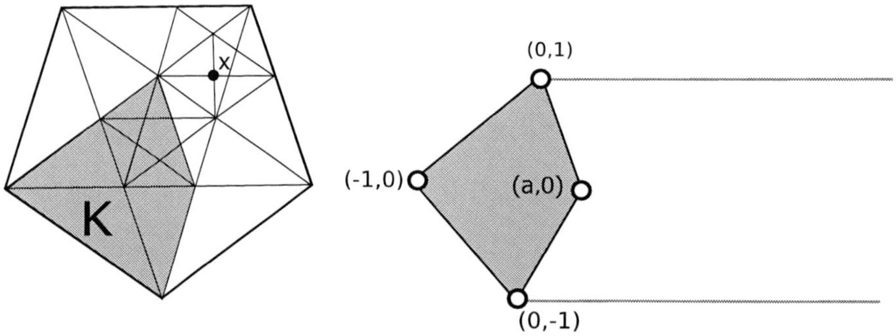

FIGURE 9
The Penrose kite (left) and the kite K_a (right).

In [93] I answered this question by showing that outer billiards with respect to the Penrose kite has an unbounded orbit. The left side of Figure 9 shows the Penrose kite and a point x with an unbounded orbit. The auxiliary lines are just scaffolding to show the construction.

Later, I proved a more general theorem in [95] which I will now describe. A *kite* is a convex quadrilateral with a line of symmetry that is a diagonal. The other diagonal divides K into two triangles, and the kite is *irrational* if these areas are irrationally related. Call an outer billiards orbit with respect to K *erratic* if it exits every compact subset of the plane and enters every open neighborhood of K.

Theorem 6.6. *There exist erratic orbits with respect to any irrational kite.*

Proof. (Very rough sketch.) Outer billiards is affinely natural, so it suffices to consider the kite K_a shown on the right side of Figure 9. Let Λ denote the two rays $[0, \infty) \times \{-1, 1\}$. Let Ψ denote the first return map to Λ. It suffices to prove that Ψ has unbounded orbits. Much like the continued fraction approximation, there is a canonical sequence of odd/odd rationals

$\{p_n/q_n\} \to a$ such that

$$|p_n q_{n+1} - p_{n+1} q_n| = 2. \tag{6.2}$$

Let $K_n = K_{p_n/q_n}$. Let Ψ_n be the corresponding first return map to Λ.

We partition Λ into intervals of length $2/q_n$ having centers $(1/q_n, \pm 1)$, $(3/q_n, \pm 1)$, etc. The map Ψ_n permutes these infinitely many intervals. We encode the combinatorial structure of this permutation as follows. There is a map from $\boldsymbol{Z}^2$ to our intervals defined as follows:

$$\Phi(m,n) = \left(\frac{2mp_n}{q_n} + 2n, (-1)^{m+n+1}\right).$$

I discovered that for each Ψ_n orbit there is an embedded nearest neighbor path on $\boldsymbol{Z}^2$ such that Φ maps consecutive vertices of the path to consecutive points of the orbit. I call this path the *arithmetic graph* of the orbit.

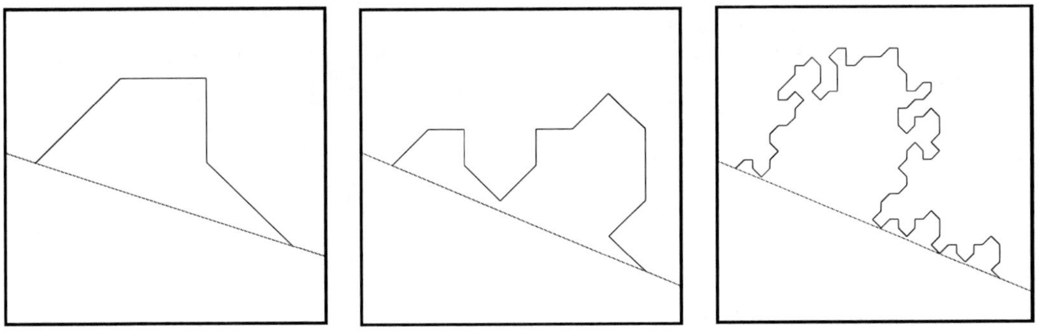

FIGURE 10
The arithmetic graphs $\Gamma(1/3)$ and $\Gamma(3/7)$ and $\Gamma(13/31)$.

Let $\Gamma_n = \Gamma(p_n/q_n)$ be the arithmetic graph of the orbit of $(1/q_n, 1)$. It is useful to think of Γ_n as a biinfinite path. One period of Γ_n connects $(0,0)$ to $(q_n, -p_n)$. (This statement requires p_n, q_n both to be odd.) The distance that Γ_n rises up above the line L_n of slope $-p_n/q_n$ through $(0,0)$ is comparable to the excursion distance of the corresponding Ψ_n orbit. Figure 10 shows one period of three of these graphs. Each pair of rationals satisfies equation (6.2). Notice that each graph copies at least one period of the previous one. Let us call this property *coherence*.

There are two main steps in the proof. The first one is to establish the coherence. The second step involves showing that the graph Γ_n rises up at least $q_n/2$ above the line L_n. Once we have these properties, we can take a limit as $n \to \infty$ and get an unbounded orbit. The fact that these graphs copy each other makes them oscillate away and back to L_n on the order of n times, with some of the oscillations being very large. This is the mechanism for the erratic orbits.

For each parameter $b \in (0,1)$ there is a 3-dimensional polyhedron exchange transformation $(\widehat{\Lambda}_b, \widehat{\Psi}_b)$ and a locally affine map $\Upsilon_b : \Lambda \to \widehat{\Lambda}_b$ such that

$$\widehat{\Psi}_b \circ \Upsilon_b = \Upsilon_b \circ \Psi.$$

In other words, Υ_b is a semiconjugacy between a 1-dimensional noncompact dynamical system and a 3-dimensional compact dynamical system. For almost every b, the image of Υ_b is dense. Thus, the 3-dimensional system is typically a kind of a dynamical compactification of the 1-dimensional system. Each dynamical system $(\widehat{\Lambda}_b, \Psi_b)$ in turn sits as a slice of a 4-dimensional integral affine polytope exchange transformation.

Step 1: For two parameters $b = p_n/q_n$ and $b' = p_{n+1}/q_{n+1}$ satisfying equation (6.2), the corresponding polyhedron exchange maps and semiconjugacies are close in the sense needed for the coherence phenomenon.

Step 2: The polyhedron exchange map $(\widehat{\Lambda}_{p_n/q_n}, \widehat{\Psi}_{p_n/q_n})$ is determined by some partition of $\widehat{\Lambda}_{p_n/q_n}$ into smaller polyhedra, and the walls of this partition give rise to two infinite grids H_n and H_n' in $\boldsymbol{R}^2$. It turns out that Γ_n can only cross lines of H_n near points where H_n and H_n' intersect. In particular, some line from H_n separates the endpoints of (one period of) Γ_n, and the only point of this line on H_n' is at least $q_n/2$ from the line L_n. ∎

It turns out that the grid phenomenon in Step 2 above was just the tip of the iceberg. I eventually found a kind of combinatorial model for the arithmetic graphs discussed in the proof above. See [96].

Question 6.7. Does outer billiards have unbounded orbits with respect to almost every polygon?

6.4. Nonpolygonal domains

Let me say a few words about nonpolygonal outer billiards. D. Dolgopyat and B. Fayad [34] prove the following result:

Theorem 6.8. *Outer billiards has unbounded orbits with respect to the domain obtained by cutting a disk in half.*

There is some stability to the argument. Dolgopyat and Fayad more generally prove their result for domains obtained by nearly cutting a disk in half. The unbounded orbits look much different from my erratic orbits: there is just an open set of them which escapes straight to infinity. Kites and (near) semidisks are (up to affine transformations) the only known examples of shapes with respect to which outer billiards has unbounded orbits.

Question 6.9. Is there a strictly convex domain or a C^1-domain with respect to which outer billiards has unbounded orbits?

Using KAM theory (in an argument outlined by J. Moser), R. Douady [35] proved the following result:

Theorem 6.10. *Outer billiards has all bounded orbits with respect to C^6-oval of positive curvature.*

In a different direction, P. Boyland [17] gives examples of C^1-domains which have orbits that enter every open neighborhood of the domain. These orbits crash into the domain, in an asymptotic sense. The domains are not C^2.

One other kind of domain I got curious about is a closed convex domain in the projective plane that is invariant with respect to a surface group of projective automorphisms. The interiors of such domains are universal covers of the so-called *convex projective surfaces*. Typically, such curves are C^1 with a Hölder-continuous derivative.

7. OVALS

I will start by describing billiards inside an ellipse and related topics. See [36] for a comprehensive reference. Following this, I will move on to more exotic kinds of tables.

7.1. Billiards in an ellipse

Let E_1 be a noncircular ellipse. There are 2 special billiard paths on E_1 having period 2 and then a third special one which goes through the foci of E in-between bounces. Every other orbit is tangent to a confocal conic section E_2, either an ellipse or a hyperbola, called a *caustic*. The *phase portrait* nicely organizes all the billiard paths. The *phase space* Φ is the open cylinder of pairs (p, ℓ) where $p \in E_1$ and ℓ is a line through p which is not tangent to E_1 at p. The *billiards map* carries (p_1, ℓ_1) to (p_2, ℓ_2), where $\{p_1, p_2\}$ are the two points of $l_1 \cap E_2$ and $\{l_1, l_2\}$ are the two lines making the same angle with (the tangent line to) E_1 at p_1.

The left-hand side of Figure 11 shows the phase portrait and indicates both the special and generic orbits by letters. The right-hand side shows what the corresponding billiard paths look like. Note that the billiards map preserves the curves corresponding to orbits with ellipse caustic and the square of the billiards map preserves the curves corresponding to orbits with hyperbola caustic.

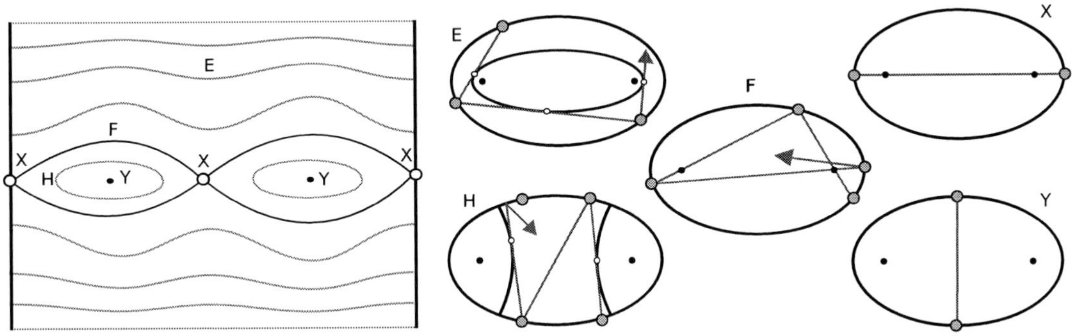

FIGURE 11
The phase portrait for elliptical billiards.

One well-known fact is that periodic billiard paths in an ellipse having the same caustic have the same perimeter. Experimenting with the computer recently, Dan Reznik has discovering an avalanche of related results. For instance, within a family of periodic billiard paths corresponding to the same caustic, the product of the cosines of the interior angles is constant. See [89] for an exposition of some of these results.

7.2. Poncelet and Cayley

The next result is a special case of the Poncelet Porism, which (when suitably phrased to avoid mentioning billiards) works for any pair of conics, confocal or not.

Theorem 7.1. *Let Ω be one of the smooth curves of the billiards phase space. If Ω corresponds to an ellipse caustic then the restriction of f to Ω is a rotation in suitable coordinates.*

Proof. (Sketch.) Let $E_1(\boldsymbol{C})$ and $E_2(\boldsymbol{C})$ denote the spheres which are extensions of E_1 and E_2 to the complex projective plane. Let $\Omega(\boldsymbol{C})$ denote the set of pairs (p, ℓ) where $p \in E_1(\boldsymbol{C})$ and ℓ is a complex line through p tangent to $E_2(\boldsymbol{C})$. The map $(p, \ell) \to p$ is a 2-fold holomorphic branched cover, branched over the 4 intersection points of the two spheres. Like all complex tori, $\Omega(\boldsymbol{C})$ is biholomorphic to a flat torus. The billiards map is an isometric rotation in these coordinates because it is the product of 2 holomorphic involutions, $(p, \ell)) \to (p, \ell')$ and $(p, \ell)) \to (p', \ell)$. Here ℓ' is the other line through p tangent to $E_2(\boldsymbol{C})$ and p' is the other point where ℓ intersects $E_1(\boldsymbol{C})$. See [53] for more details. ■

There is another approach to Theorem 7.1 which works specifically in the case when E_1, E_2 are confocal. Let E_1^* be the region bounded by E_1. There is a uniformizing change of coordinates [36, 38] somewhat akin to the Schwarz–Christoffel transform, which carries the relevant component of $E_1^* - E_2$ to either a flat cylinder or a rectangle. In these coordinates, the billiard paths with caustic E_2 transform to ordinary billiards which move parallel to the directions $(\pm 1, \pm 1)$. Changing the caustic E_2 changes the rectangle/cylinder.

Which caustics give rise to periodic billiard paths? Cayley's amazing answer works for any pair (E_1, E_2) of conics, confocal or not. Say that (E_1, E_2) *supports a Poncelet n-gon* if there exists a closed n-gon whose vertices are in E_1 and whose edges are contained in lines tangent to E_2. In homogeneous coordinates, E_k is the zero-set of an equation $\sum d_{ij} x_i x_j = 0$ encoded by a 3×3 matrix $D_k = \{d_{ij}\}$. Take the Taylor series expansion

$$\sqrt{\det(tD_1 + D_2)} = A_0 + A_1 t + A_2 t^2 + \cdots . \tag{7.1}$$

Theorem 7.2. *Let Δ_n to be the left (respectively right) determinant when $n = 2m + 1$ (respectively $n = 2m$)*

$$\begin{vmatrix} A_2 & \cdots & A_{m+1} \\ \vdots & \ddots & \vdots \\ A_{m+1} & \cdots & A_{2m} \end{vmatrix}, \quad \begin{vmatrix} A_3 & \cdots & A_{m+1} \\ \vdots & \ddots & \vdots \\ A_{m+1} & \cdots & A_{2m} \end{vmatrix}. \tag{7.2}$$

Then (E_1, E_2) supports a Poncelet n-gon if and only if $\Delta_n = 0$.

See [54] for a modern proof of Cayley's Theorem. In [39], V. Dragović and M. Radnović give the complete analogue of Theorem 7.2 for billiards in a higher-dimensional ellipsoid.

7.3. Piecewise elliptical tables

In [37, 38] Dragović and Radnović use the transformation mentioned in connection with the Poncelet Porism to study a more exotic situation in which the table is made from pieces of two confocal ellipses, as in Figure 12. They term this kind of billiards *pseudo-integrable*. (This term also refers to rational billiards in the physics literature.)

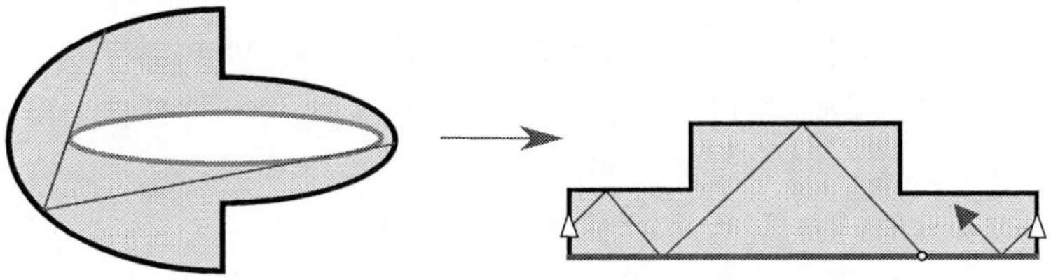

FIGURE 12
A pseudointegrable table.

Since the pieces are confocal, the billiard paths still have caustics. For each choice of caustic, the uniformizing map carries the domain which is either a right-angled polygon or a topological cylinder with a right-angled boundary. The billiard paths on these tables move parallel to $(\pm 1, \pm 1)$ as above. (My picture is just a cartoon; I did not compute the uniformizing map.) These systems exhibit a wider variety of behavior than integrable billiards [39], such as orbits which are dense but not equidistributed. See [48] and [47] for further developments.

7.4. The stadium

Figure 13 shows the *Bunimovich stadium*, another billiard table that has been intensely studied. This domain is convex but not strictly convex, and only C^1. The boundary of the stadium is a union of two semicircles and two line segments. This is really a 1-parameter family of examples. The parameter is the ratio of the line segment length to the semicircle length.

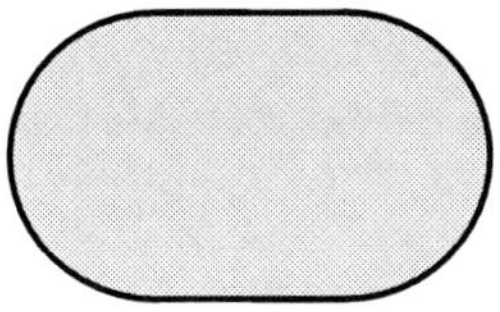

FIGURE 13
A Bunimovich stadium.

Here is a version of the theorem of Bunimovich which is easy to state:

Theorem 7.3. *Billiards in any stadium is ergodic. In particular, almost every billiard path is dense.*

This result is quite surprising because billiards in the disk is completely integrable. Once you introduce even the tiniest line segment, the billiard map changes completely. The full theorem of Bunimovich has more conclusions. See the paper of Misiurewicz and Zhang [82] for recent results about stadium billiards, and many other references.

7.5. Periodic orbits

Now I will talk about billiards in a general oval. From now on, by an *oval* I mean an infinitely differentiable and strictly convex closed curve. Many authors care about the exact level of differentiability. I am going to sweep this under the rug; you should consult the original sources for the precise generality needed for the results.

When C is an oval, we can define the phase space just as in the ellipse case. There is always an invariant area form on the phase space. It is given locally by $\sin(\theta) d\theta ds$, where θ is the angle that the relevant line ℓ in the pair (p, ℓ) makes with C, and ds is arc length. The following theorem of Birkhoff [13] vitally uses the area-preserving nature of the billiard map on phase space.

Theorem 7.4. *If C is an oval then, for every $n > 1$ and every integer rotation number $|r| < n$, there are at least 2 periodic billiard paths in C having period n and rotation number r.*

Area-preserving maps are special cases of symplectic maps. Sometimes one can use symplectic geometry to get results about billiards which seem (to me) impossible to get in a different way. Let me discuss one striking result along these lines, due to Y. Ostrover and S. Artstein-Avidan [2]. Let $\xi(K)$ denote the length of the shortest periodic billiard path in K. Given two sets K_1, K_2, define $K_1 + K_2$ to be the set of sums $v_1 + v_2$ with $v_1 \in K_1$ and $v_2 \in K_2$.

Theorem 7.5. *For any ovals K_1, K_2, we have $\xi(K_1 + K_2) \geq \xi(K_1) + \xi(K_2)$.*

The result in [2] is stated and proved for smooth convex domains in all dimensions.

7.6. Two guiding conjectures

Here I will discuss two geometric conjectures about billiards in convex ovals. Motivated by his theorem about the asymptotics of the eigenvalues of the Laplacian in a convex domain, V. Ya. Ivrii [64] made the following conjecture:

Conjecture 7.6. *Almost every billiard path in an oval is aperiodic.*

Ivrii's conjecture is wide open, but here is some partial progress. Y. Baryshnikov and V. Zharniksky [9] prove that there cannot exist an open set of 3-periodic orbits for an oval. M. Rychlik [90] proves the following result with the assistance of the computer:

Theorem 7.7. *The set of 3-periodic billiard paths in any oval has measure 0.*

L. Stojanov [101] removes the computer dependence, then M. Wojtkowski [113] and Ya. B. Vorobets [111] give different and independent proofs for more general domains.

In [52], A. Glutsyuk and Y. Kudryashov prove the following result:

Theorem 7.8. *No oval has an open set of* 4*-periodic billiard paths.*

In [71], V. F. Lazutkin proves that for any strictly convex and sufficiently smooth oval, there is a positive Lebesgue measure union of caustics for billiard paths in the oval. However, there are generally gaps between the caustics. The Birkhoff–Poritsky Conjecture is a rigidity conjecture which says essentially that if there are no gaps between the caustics then the table is an ellipse. Let Φ denote the phase space for billiards on the oval C.

Conjecture 7.9. *Let C be an oval. Suppose that some neighborhood of $\partial\Phi$ is foliated by invariant curves. Then C is an ellipse.*

The first progress is due to M. Bialy [10]:

Theorem 7.10. *If Φ is completely foliated by invariant curves then C is a circle.*

Say that an invariant curve in Φ is a q-curve if every orbit in the curve has period q. Recently, M. Bialy and A. Mironov [12] proved the following result:

Theorem 7.11. *Suppose C is centrally symmetric and there is a neighborhood N of $\partial\Phi$, foliated by invariant curves, such that ∂N is a union of two* 4*-curves. Then C is an ellipse.*

The smaller the neighborhood of the boundary, the higher the period of the orbit, so the neighborhood needed for this theorem is sort of medium-sized. One really neat fact proved along the way is that, given the hypotheses of the theorem, the billiard paths corresponding to points on the 4-curves are all parallelograms.

In another direction, A. Glutsyuk [51], extending work in Bialy–Mironov [11], has given a solution to the conjecture in a restricted case where the objects are not just smooth but algebraic. See [103] for a related result in the outer billiards case.

In [66], V. Kaloshin and A. Sorrentino have proved a completely general version of the Birkhoff Conjecture for ovals which are perturbations of ellipses. The main result in [66] pays careful attention to the level of differentiability; here is a corollary.

Theorem 7.12. *Suppose that C is sufficiently close to an ellipse in the C^∞-sense and also Φ has a q-curve for each $q = 3, 4, 5, 6, \ldots$ Then C is an ellipse.*

Referring to the discussion about Lazutkin's theorem, this last result allows there to be gaps between the caustics, but it still supposes a very particular kind of structure to certain of them.

7.7. The pentagram rigidity conjecture

I cannot resist bringing up a question I have been curious about for 30 years. The question intertwines the Poncelet Porism and the so-called *deep diagonal pentagram maps*. To me it seems like a discrete variant of the Birkhoff–Poritsky Conjecture. I will state the conjecture for the pair $(3, 8)$ just for simplicity. Figure 14 shows two 8-gons O_1 and O_2 related by a construction involving the 3-diagonals of O_1. The right-hand side indicates how

O_2 might not be convex even if O_1 is convex. This operation is best defined in the projective plane. (Convexity still makes sense there.)

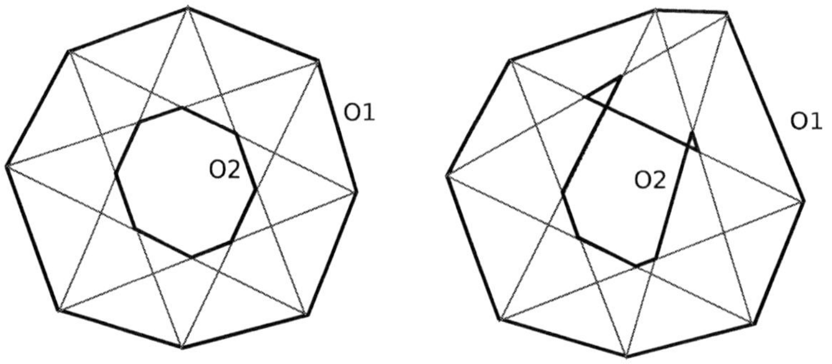

FIGURE 14
The 8-gons O_1 and O_2.

Starting with O_0 we can construct the biinfinite sequence $\{O_n\}$ of 8-gons, in which successive ones are related by the construction. If O_0 is a convex Poncelet polygon, then O_k is a projectively equivalent convex Poncelet polygon for all $k \in \boldsymbol{Z}$.

Conjecture 7.13. *If O_k is convex for all k then O_0 is a Poncelet polygon.*

A. Izosimov [65] has made a bit of general progress related to this conjecture by showing that if n is odd and two convex polygons related by the $(2, n)$ construction are projectively equivalent then they both are Poncelet polygons. I recently [97] established the very special case when the octagons have 4-fold rotational symmetry.

8. TABLES WITH OBSTACLES

The subject of dispersive billiards is an enormous one and this small chapter just gives you a taste. See [24] for a survey. One of the main themes in the subject is understanding the mixing properties of the system. Another main theme is the attempt to give rigorous mathematical foundations for physical processes like Brownian motion and the transfer of mass and heat in a gas.

8.1. Mixing

To build some intuition for dispersive billiards, consider how some given element $T \in \mathrm{SL}_2(\boldsymbol{Z})$, with an eigenvalue $\lambda > 1$, acts on the square torus $Y = \boldsymbol{R}^2/\boldsymbol{Z}^2$. Let us show that T is mixing in the sense given by the left-hand side of equation (3.4).

Lemma 8.1. *T is mixing.*

Proof. (Sketch). I will just consider the case when U is a parallelogram with sides parallel to the eigenvectors of T. Let $|U|$ be the side length of U. Corresponding to λ there is an irrational invariant geodesic foliation of Y. For large n, the set $T^n(U)$ is a long thin parallelogram smeared out along this foliation. The long side of $T^n(U)$ is about $|U|\lambda^n$ in length and the short side is about $|U|\lambda^{-n}$ in length. So, $T^n(U)$ is essentially a really thin strip that closely follows an irrational geodesic for a really long time. We have already seen that the irrational geodesics in Y are equidistributed. Given this property, $T^n(U)$ spends about $\mu(U)$ percent of the time in V. ∎

Given the exponential growth of the length of $T^n(U)$, the quantity on the left-hand side of equation (3.4) decays exponentially in n. That is, at least when U and V are nice sets like rectangles or disks the quantity on the left side of equation (3.4) is of the order of e^{-Cn} for some $C > 0$. This kind of decay, suitably generalized and formalized, is called *exponential mixing*. See [88] for a definition. Mixing is stronger than ergodicity, and exponential mixing is even stronger than that.

8.2. The Lorentz gas

The classic *Lorentz gas*, also known as a *Sinai billiard* [99], is a billiard ball bouncing around on the table you get by removing a round disk D from the center of a square.

Theorem 8.2. *The billiard map on* $[0, 1]^2 \setminus D$ *is mixing.*

Proof. (Very rough sketch.) Ignoring the measure zero set of billiard paths which avoid D, we can define the phase space Φ of the system to be the cylinder of pairs (p, ℓ) where $p \in \partial D$ and ℓ is a line through p not tangent to ∂D. The same measure as for smooth ovals is an invariant one for the system.

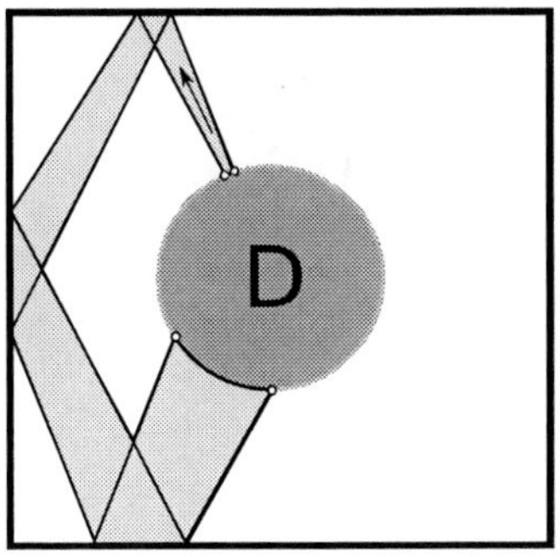

FIGURE 15
Scattering property in action.

Each billiard path that leaves ∂D bounces some number of times on the square and then returns to ∂D. Some word records the intermediate bounces. Partition Φ by the elements that correspond to the same word. Consider an arc of elements of the same partition which leave ∂D at the same angle, as shown in Figure 15. These elements spread out before returning to ∂D. Since the billiards map is area preserving, it stretches each partition piece in one direction and compresses it in the other. The longer the word, the more dramatic the effect. So, the local behavior is like that considered for the map T considered in Lemma 8.1. ■

More generally, you could remove finitely many disjoint strictly convex scatterers from the interior of the unit square or from a flat torus. The table has *finite horizon* if all billiard paths hit the scatterers. The mixing properties of billiards on these tables—or at least when the properties were established—depend on the finite horizon property and also on whether one considers the billiards map or the billiards *flow* on the unit tangent bundle. Here is a rundown of the results:

(1) Y. Young [117] shows that the billiard map is exponentially mixing in the finite horizon case, and then N. Chernov [23] establishes this in the infinite horizon case.

(2) V. Baladi, M. Demers, and C. Liverani [7] establish the exponential mixing for the flow in the finite horizon case and P. Bálint, O. Butterley, and I. Melbourne [8] establish polynomial mixing in the infinite horizon case.

A more complicated situation arises when the scatterers are allowed to touch. J. de Simoi and P. Toth [32] prove that the billiards map is exponentially mixing in the finite horizon case when no scatterers are tangent. In [25], N. Chernov and H.-K. Zhang show that the billiard map is polynomially mixing in the finite horizon case when tangencies are allowed.

Here are some poorly understood situations in this area. One thing you can do is play billiards in the plane, after removing an infinite number of scatterers but not in a periodic pattern. (The periodic case is the universal cover of the kind of the example considered above.) Another thing you can do is replace a single bouncing point with several or many bouncing disks of finite size. D. Dolgopyat and P. Nándori [33] make some recent progress in the case of 2 disks.

8.3. Breakout

Let me close this survey with some whimsical questions. Inspired by the video game *Breakout* [18,87], one could imagine a ball bouncing around an infinite periodic array of disk scatterers but with the twist that a scatterer disappears as soon as it is hit.

Question 8.3. Does a typical billiard path erase all the scatters eventually?

Here is one thing I noticed about the breakout game when it is played on the 1-skeleton of the infinite square tiling. (Again, a reflector disappears as soon as it is hit.)

Conjecture 8.4. *If you start the ball moving with slope $\sqrt{2}$, the billiard path eventually erases all the reflectors.*

These systems remind me a little bit of Langton's ant, and the questions about them seem to verge on the territory of cellular automata.

ACKNOWLEDGMENTS

Many people helped me write this survey. I would like to think Jayadev Athreya, Misha Bialy, Dan Cristofaro-Gardiner, Mark Demers, Vladimir Dragovic, Giovanni Forni, Pat Hooper, Howard Masur, Curtis McMullen, Péter Nándori, Dan Reznik, Sergei Tabachnikov, and Barak Weiss for very helpful discussions. I listened to all these folks but I did not always follow their advice; any errors in the exposition and omissions are my fault.

FUNDING

This work is supported by the U. S. National Science Foundation grant DMS-2102802.

REFERENCES

[1] P. Apisa and A. Wright, Marked points on translation surfaces. 2021, arXiv:1708.03411.

[2] S. Artstein-Avidan and Y. Ostrover, Bounds for Minkowski billiard trajectories in convex bodies. 2012, arXiv:1111.2353.

[3] J. S. Athreya, D. Aulicino, W. Hooper (appendix by A. Randecker), Platonic solids and high genus covers of lattice surfaces. *Exp. Math.* (2018), 1–51.

[4] A. Avila and V. Delacroix, Weak mixing directions in non-arithmetic Veech surfaces. *J. Amer. Math. Soc.* **29** (2016).

[5] A. Avila and G. Forni, Weak mixing for interval exchange transformations and translation flows. *Ann. of Math.* **165** (2007), 637–664.

[6] M. Bainbridge, Euler characteristics of Teichmüller curves in genus two. *Geom. Topol.* **11** (2007), 1887–2073.

[7] V. Baladi, M. Demers, and C. Liverani, Exponential decay of correlations of finite horizon Sinai billiard flows. *Invent. Math.* 2018.

[8] P. Bálint, O. Butterley, and I. Melbourne, Polynomial decay of correlations for flows, including Lorentz gas examples. *Comm. Math. Phys.* **368** (2019), 55–111.

[9] Y. Baryshnikov and V. Zharniksky, Billiards and nonholonomic distributions. *Zap. Nauchn. Sem. S.-Peterburg. Otdel. Mat. Inst. Steklov. (POMI)* **300** (2004), 56–64.

[10] M. Bialy, Convex billiards and a theorem of E. Hopf. *Math. Z.* **214** (1993), 147–154.

[11] M. Bialy and A. Mironov, Angular billiard and algebraic Birkhoff conjecture. *Adv. Math.* **313** (2017), 102–126.

[12] M. Bialy and A. Mironov, The Birkhoff–Poritsky conjecture for centrally symmetric billiard tables. 2020, arXiv:2008.03566.

[13] G. D. Birkhoff, On the periodic motions of dynamical systems. *Acta Math.* (1927).

[14] M. Boshernitzan, G. Galperin, T. Krüger, and S. Troubetzkoy, Periodic billiard orbits are dense in rational polygons. *Trans. Amer. Math. Soc.* **350** (1998), no. 9, 3523–3535.

[15] I. Bouw and M. Möller, Teichmüller curves, triangle groups, and Lyapunov exponents. *Ann. of Math.* (2010).

[16] J. P. Bowman, Teichmüller geodesics, Delaunay triangulations, and Veech groups, Teichmüller theory and moduli problems. *Ramanujan Math. Soc. Lect. Notes Ser.* **10** (2020), 113–129.

[17] P. Boyland, Dual billiards, twist maps, and impact oscillators. *Nonlinearity* **9** (1996), 1411–1438.

[18] X. Bressaud and M.-C. Fournier, Casse-Briques. *Exp. Math.* (2015).

[19] A. Calderon, S. Coles, D. Davis, J. Lanier, and A. Oliveira, How to hear the shape of a billiard table. 2018, arXiv:1806.09644.

[20] C. Calta, Veech surfaces and complete periodicity in genus 2. *J. Amer. Math. Soc.* **17** (2004), 871–908.

[21] J. Chaika and G. Forni, Weakly mixing polygonal billiards. 2020, arXiv:2003.00890.

[22] J. Chaika, J. Smillie, and B. Weiss, Tremors and horocycle dynamics on the moduli space of translation surfaces. 2020, arXiv:2004.04027.

[23] N. Chernov, Decay of correlations in dispersing billiards. *J. Stat. Phys.* **94** (1999), 513–556.

[24] N. Chernov and R. Markarian, Chaotic billiards. *Math. Surveys Monogr.* **127**, 2006.

[25] N. Chernov and H.-K. Zhang, Improved estimates for correlations in billiards. *Comm. Math. Phys.* **277** (2008), 305–321.

[26] B. Cipra, R. Hanson, and A. Kolan, Periodic trajectories in right-triangle billiards. *Phys. Rev. E* **52** (1995), no. 2.

[27] D. Davis, Cutting sequences, regular polygons, and the Veech group. *Geom. Dedicata* **162** (2013), no. 1, 231–261.

[28] D. Davis and S. Lelièvre, Periodic paths on the pentagon, double pentagon, and golden L. 2018, arXiv:1810.11310.

[29] D. Davis, I. Pasquinelli, and C. Ulcigrai, Cutting sequences on Bouw–Möller surfaces: an S-adic characterization. *Ann. Sci. Éc. Norm. Supér.* **52** (2019), no. 4, 927–1023.

[30] V. Delacroix and W. P. Hooper, Flat surfaces in SageMath github, 2020, https://github.com/videlec/sage-flatsurf.

[31] P. Deligne and G. D. Mostow, Commensurabilities among lattices in PU$(1, n)$. *Ann. of Math. Stud.* **132** (1993).

[32] J. De Simoi and I. P. Toth, An expansion estimate for dispersing planar billiards with corner points. *Ann. Henri Poincaré* **15** (2014), 1223–1243.

[33] D. Dolgopyat and P. Nándori, The first encounter of two billiard particles of small radius. 2016, arXiv:1603.07590.

[34] D. Dolyopyat and B. Fayad, Unbounded orbits for semicircular outer billiards. *Ann. Henri Poincaré* **10** (2009), 357–375.

[35] R. Douady, *These de 3-eme cycle*, Université de Paris 7, 1982.

[36] V. Dragović and M. Radnović, *Poncelet porisms and beyond: integrable billiards, hyperelliptic Jacobians, and pencils of quadrics*. Springer, 2011.

[37] V. Dragović and M. Radnović, Pseudo-integrable billiards and arithmetic dynamics. *J. Mod. Dyn.* **8** (2014), no. 1, 109–132.

[38] V. Dragović and M. Radnović, Periods of pseudo-integrable billiards. *Arnold Math. J.* **1** (2015), no. 1, 69–73.

[39] V. Dragović and M. Radnović, Periodic ellipsoidal polynomials and extremal polynomials. *Comm. Math. Phys.* **372** (2019), no. 1, 183–211.

[40] M. Duchin, V. Erlandsson, C. J. Leininger, and C. Sadanand, You can hear the shape of a billiard table: symbolic dynamics and rigidity for flat surfaces. 2019, arXiv:1804.05690.

[41] B. Edwards, S. Sanderson, and T. A. Schmidt, Computing Veech groups. 2021, arXiv:2012.12444.

[42] A. Eskin and H. Masur, Asymptotic formulas on flat surfaces. *Ergodic Theory Dynam. Systems* **21** (2001), no. 02, 443–478.

[43] A. Eskin and M. Mirzakhani, Invariant and stationary measures for the $\mathrm{SL}_2(\boldsymbol{R})$ action on moduli space. *Publ. Math. Inst. Hautes Études Sci.* **127** (2018), 95–324.

[44] A. Eskin, M. Mirzakhani, and A. Mohammadi, Isolation, equidistribution, and orbit closures for the $\mathrm{SL}_2(\boldsymbol{R})$ action on moduli space. *Ann. of Math.* **182** (2015).

[45] S. Filip, Splitting mixed Hodge structures over affine invariant manifolds. *Ann. of Math.* **183** (2016), no. 2, 681–713.

[46] G. Forni and C. Matheus, Introduction to Teichmüller theory and its applications to dynamics of interval exchange transformations, flows of surfaces, and billiards. 2013, arXiv:1311.2758.

[47] K. Fraczek, Recurrence for smooth curves in the moduli space and application to the billiard flow on nibbled ellipses. *Anal. PDE* **14** (2021), no. 4, 793–821.

[48] K. Fraczek, R. Shi, and C. Ulcigrai, Genericity on curves and applications: pseudo-integrable billiards, Eaton lenses and gap distributions. *J. Mod. Dyn.* **12** (2018).

[49] G. A. Galperin, A. M. Stepin, and Y. B. Vorobets, Periodic billiard trajectories in polygons. *Russian Math. Surveys* **47** (1001), 5–80.

[50] D. Genin, *Regular and chaotic dynamics of outer billiards*. Ph.D. thesis, Pennsylvania State University, State College, 2005.

[51] A. Glutsyuk, On polynomially integrable Birkhoff billiards on surfaces of constant curvature. 2019, arXiv:1706.04030.

[52] A. Glutsyuk and Y. Kudryashov, No planar billiard possesses an open set of quadrilateral trajectories. *J. Mod. Dyn.* **6** (2012), no. 3.

[53] P. Griffiths and J. Harris, Poncelet theorem in space. *Comment. Math. Helv.* **52** (1977), no. 2, 145–160.

[54] P. Griffiths and J. Harris, On Cayley's explicit solution to the Poncelet porism L'Enseignement. *Enseign. Math. (2)* **24** (1978), 1–2, 31–40.

[55] E. Gutkin and N. Simanyi, Dual polygonal billiard and necklace dynamics. *Comm. Math. Phys.* **143** (1991), 431–450.

[56] R. K. Guy, In *Unsolved problems in number theory*, pp. 365–366, Springer, 2004.

[57] L. Halbeisen and N. Hungerbuhler, On periodic billiard trajectories in obtuse triangles. *SIAM Rev.* **42** (1986), no. 2, 657–670.

[58] W. P. Hooper, *On the stability of periodic billiard paths in triangles*. Ph.D. thesis, Stony Brook, 2006.

[59] W. P. Hooper, Lower bounds on growth rates of periodic billiard trajectories in some irrational triangles. *J. Mod. Dyn.* **1** (2007), no. 4, 649–663.

[60] W. P. Hooper, Periodic billiard paths in right triangles are unstable. *Geom. Dedicata* **125** (2007), 39–46.

[61] W. P. Hooper, Grid graphs and lattice surfaces. *Int. Math. Res. Not.* **2013** (2013), 2657–2698.

[62] W. Hooper and R. E. Schwartz, Billiards in nearly isosceles triangles. *J. Mod. Dyn.* **3** (2009), no. 2, 159–231.

[63] G. Hughes, Dynamics of polygons. http://www.dynamicsofpolygons.org.

[64] V. Ya. Ivrii, The second term of the spectral asymptotics for a Laplace–Beltrami operator on manifolds with boundary. *Funct. Anal. Appl.* **14** (1980), no. 2, 98–106.

[65] A. Izosimov, The pentagram map, Poncelet polygons, and commuting difference operators. 2021, arXiv:1906.10749.

[66] V. Kaloshin and A. Sorrentino, On local Birkhoff conjecture for convex billiards. *Ann. of Math.* **188** (2018), 1–66.

[67] A. Katok, The growth rate for the number of singular and periodic orbits of a polygonal billiard. *Comm. Math. Phys.* **111** (1987), 151–160.

[68] A. Katok, Five more resistent problems in dynamics. http://www.personal.psu.edu/axk29/pub/5problems-expanded.pdf.

[69] S. Kerkhoff, H. Masur, and J. Smillie, Ergodicity of billiard flows and quadratic differentials. *Ann. of Math.* **124** (1986), 293–311.

[70] Kolodziej, The antibilliard outside a polygon. *Bull. Pol. Acad. Sci. Math.* **37** (1994), 163–168.

[71] V. F. Lazutkin, The existence of caustics for a billiard problem in a convex domain. *Izv. Akad. Nauk SSSR Ser. Mat.* **37** (1973).

[72] S. Lelièvre, T. Monteil, and B. Weiss, Everything is illuminated. *Geom. Topol.* **20** (2016), no. 3, 1737–1762.

[73] A. Leutbecher, Über die Heckeschen Gruen (λ) II. *Math. Ann.* **211** (1974), 63–86.

[74] H. Masur, Lower bounds for the number of saddle connections and closed trajectories of a quadratic differential. In *Holomorphic Functions and Moduli 1*, edited by D. Drasin, pp. 215–228, Springer, 1988.

[75] H. Masur, The Growth rate of trajectories of a quadratic differential. *Ergodic Theory Dynam. Systems* **10** (1990), 151–176.

[76] H. Masur, Hausdorff dimension of the set of nonergodic foliations of a quadratic differential. *Duke Math. J.* **66** (1992), 387–442.

[77] H. Masur and S. Tabachnikov, Rational billiards and flat structures. In *Handbook of Dynamical Systems 1, Part A, Chapter 13*, pp. 1015–1089, Elsevier, 2002.

[78] C. McMullen, Teichmüller curves in genus two: discriminant and spin. *Math. Ann.* **333** (2005), 87–130.

[79] C. McMullen, Teichmüller curves in genus two: torsion divisors and ratios of sines. *Invent. Math.* **165** (2006).

[80] C. McMullen, Billiards, heights, and the arithmetic of non-arithmetic groups. Preprint, 2020.

[81] C. McMullen, R. Mukamel, and A. Wright, Cubic curves and totally geodesic subvarieties of moduli space. *Ann. of Math.* **185** (2017), 957–990.

[82] M. Misiurewicz and H.-K. Zhang, Topological entropy of Bunimovich stadium billiards. 2020, arXiv:2006.03882.

[83] J. Moser, *Stable and random motions in dynamical systems, with special emphasis on celestial mechanics*. Ann. of Math. Stud. 77, Princeton University Press, Princeton, NJ, 1973.

[84] J. Moser, Is the solar system stable? *Math. Intelligencer* **1** (1978), 65–71.

[85] R. E. Mukamel, Orbifold points on Teichmüller curves and Jacobians with complex multiplication. *Geom. Topol.* **18** (2014), no. 2, 779–829.

[86] B. H. Neumann, *Sharing ham and eggs*, Summary of a Manchester Mathematics Colloquium, 25 Jan 1959, published in Iota, the Manchester University Mathematics. Students' Journal.

[87] E. Newkirk, Billiards with bombs. 2015, arXiv:1506.00683.

[88] M. Pollicott, Exponential mixing for the geodesic flow on hyperbolic three-manifolds. *J. Stat. Phys.* **67** (1992), 667–673.

[89] D. Reznik, R. Garcia, and J. Koillar, Can the elliptic billiard still surprise us? *Math. Intelligencer* **42** (2020), no. 1.

[90] M. R. Rychlik, Periodic points of the billiard ball map in a convex domain. *J. Differential Geom.* **30** (1989), 191–205.

[91] D. Scheglov, Complexity growth of a typical triangular billiard is weakly exponential. *J. Anal. Math.* **142** (2020), 105–124.

[92] R. E. Schwartz, Obtuse triangular billiards I: near the $(2, 3, 6)$ triangle. *Exp. Math.* (2006).

[93] R. E. Schwartz, Unbounded orbits for outer billiards. *J. Mod. Dyn.* **3** (2007).

[94] R. E. Schwartz, Obtuse triangular billiards II: 100 degrees of periodic billiard paths. *Exp. Math.* (2008).

[95] R. E. Schwartz, Outer billiards on kites. *Ann. of Math. Stud.* **171** (2009).

[96] R. E. Schwartz, The plaid model. *Ann. of Math. Stud.* **198** (2018).

[97] R. E. Schwartz, A textbook case of pentagram rigidity. 2021, arXiv:2108.07604.

[98] C. Series, The modular surface and continued fractions. *J. Lond. Math. Soc.* **31** (1985), no. 1.

[99] Ya. G. Sinai, Dynamical systems with elastic reflections. Ergodic properties of dispersing billiards. *Russian Math. Surveys* **25** (1970), 137–189.

[100] J. Smillie and C. Ulcigrai, Beyond Sturmian sequences: coding linear trajectories in the regular octagon. *Proc. Lond. Math. Soc.* **101** (2020), no. 1.

[101] L. Stojanov, Note on the periodic points of the billiard. *J. Differential Geom.* **34** (1991).

[102] S. Tabachnikov, *Billiards*, Soc. Math. de Fr., "Panoramas et Syntheses" 1, 1995.

[103] S. Tabachnikov, On algebraically integrable outer billiards. 2007, arXiv:0708.0255.

[104] S. Tabachnikov, A proof of Culter's theorem on the existence of periodic orbits in polygonal outer billiards. 2007, arXiv:0706.1003.

[105] W. P. Thurston, *Shapes of polyhedra and triangulations of the sphere*. Geom. Topol. Monogr., 1998.

[106] G. Tokarsky, J. Garber, B. Marinov, and K. Moore, One hundred and twelve point three degree theorem. 2018, arXiv:1808.06667.

[107] S. Troubetzkoy, Periodic billiard orbits in right triangle II. 2005, arXiv:math/0505637.

[108] W. A. Veech, Teichmüller curves in moduli space, Eisenstein series, and an application to triangular billiards. *Invent. Math.* **97** (1989), 553–583.

[109] W. A. Veech, The billiard in a regular polygon. *Geom. Funct. Anal.* **2** (1992), no. 3.

[110] F. Vivaldi and A. Shaidenko, Global stability of a class of discontinuous dual billiards. *Comm. Math. Phys.* **110** (1987), 625–640.

[111] Ya. Vorobets, On the measure of the set of periodic points of a billiard. *Math. Notes* **55** (1994).

[112] Ya. Vorobets, Ergodicity of billiards in polygons. *Mat. Sb.* **188** (1997), no. 3, 117–171.

[113] M. P. Wojtkowski, Two applications of Jacobi fields to the billiard ball problem. *J. Differential Geom.* **40** (1994), 155–164.

[114] A. Wolecki, Illumination in rational billiards. 2019, arXiv:1905.09358.

[115] A. Wright, From rational billiards to dynamics on moduli spaces. *Bull. Amer. Math. Soc.* **53** (2016), no. 1, 41–56.

[116] A. Wright, Schwarz triangle mappings and Teichmüller curves: the Veech–Ward–Bouw–Möller curves. 2013, arXiv:1203.2685.

[117] Y. Young, Statistical properties of dynamical systems with some hyperbolicity. *Ann. of Math.* **147** (1998), 585–650.

[118] A. Zorich, Flat surfaces. In *Frontiers in number theory, physics, and geometry 1*, edited by P. Cartier, B. Julia, P. Moussa, and P. Vanhove, pp. 439–586, Springer, 2006.

[119] A. Zorich, The magic wand theorem of A. Eskin and M. Mirzakhani. *Gaz. Math.* **142** (2014), 39–54.

RICHARD EVAN SCHWARTZ
Department of Mathematics, Brown University, 151 Thayer Street, Box 1917, Providence, RI 02912, USA, Richard.Evan.Schwartz@gmail.com

5. GEOMETRY

SOME RECENT DEVELOPMENTS IN RICCI FLOW

RICHARD H. BAMLER

ABSTRACT

In this article we survey some of the recent developments in Ricci flow. We present a new theory of weak, 3-dimensional Ricci flows "through singularities," which can be viewed as an improvement of Perelman's Ricci flow with surgery. We point out two topological applications: the resolution of the Generalized Smale Conjecture regarding the diffeomorphism groups of 3-manifolds and the resolution of a conjecture regarding the space of positive scalar curvature metrics on 3-manifolds. We also describe ongoing research on the formation of singularities in higher dimensions, which may yield further interesting applications in the future.

MATHEMATICS SUBJECT CLASSIFICATION 2020

Primary 53E20; Secondary 57K30, 57S05, 53C21, 35K67, 30L99

KEYWORDS

Ricci flow, singular Ricci flow, Ricci flow with surgery, diffeomorphism groups, Generalized Smale Conjecture, scalar curvature, well-posedness, parabolic equation, heat equation, compactness theory, partial regularity theory, solitons, Einstein metrics

INTERNATIONAL CONGRESS OF MATHEMATICIANS 2022 JULY 6–14

Proc. Int. Cong. Math. 2022, Vol. 4, pp. 2432–2455
DOI 10.4171/ICM2022/58

1. INTRODUCTION

The Ricci flow has proven to be a powerful tool, as it was used by Perelman in the early 2000s to resolve two of the most important conjectures in 3-manifold topology: the Poincaré Conjecture and the Geometrization Conjecture [46–48]. These applications were far from coincidental, as they provide a new perspective on 3-manifold topology using the geometric-analytic language of Ricci flow. Since then there have been further advances in the study of Ricci flow, which have led to new topological applications in dimension 3. In addition, more applications in higher dimensions may be forthcoming. The goal of this article is to survey some of these developments, particularly those relating to questions in (geometric) topology.

This article[1] is structured as follows. We will first provide a brief introduction to Ricci flow, review some of the earlier results in dimension 2 and Perelman's work in dimension 3. Next, we will discuss more recent results in dimension 3 regarding singular "Ricci flows through singularities," their uniqueness and continuous dependence on the initial data and describe their topological applications. Lastly, we present a new approach towards the study of Ricci flows in higher dimensions and point out potential future directions and applications.

2. RICCI FLOW

A Ricci flow (introduced by Hamilton [31]) on a manifold M is given by a smooth family $g(t)$, $t \in [0, T)$, of Riemannian metrics satisfying the evolution equation

$$\partial_t g(t) = -2 \operatorname{Ric}_{g(t)}, \tag{2.1}$$

where $\operatorname{Ric}_{g(t)}$ denotes the Ricci curvature of the metric $g(t)$, i.e., the trace of its Riemann curvature tensor $\operatorname{Rm}_{g(t)}$. Equation (2.1) is weakly parabolic and it implies an evolution equation for the curvature tensor $\operatorname{Rm}_{g(t)}$ of the form

$$\partial_t \operatorname{Rm}_{g(t)} = \Delta \operatorname{Rm}_{g(t)} + Q(\operatorname{Rm}_{g(t)}), \tag{2.2}$$

where the last term denotes a quadratic term; its exact form will not be important for this survey. Equation (2.2) suggests that the metric $g(t)$ becomes "smoother" or "more homogeneous" as time moves on, similar to solutions of heat equations. On the other hand, the last term in (2.2) seems to indicate that – possibly at larger scales or in regions of large curvature – this diffusion property may be outweighed by some other nonlinear effects, which could lead to singularities.

If M is compact, then for any initial metric g_0 the Ricci flow equation (2.1) has a unique solution $g(t)$, $t \in [0, T)$, with initial condition $g(t) = g_0$ and for some maximal $T \in (0, \infty]$ [31]. If $T < \infty$, then the flow $g(t)$ must develop a singularity at time T and the curvature must blow up: $\max_M |\operatorname{Rm}_{g(t)}| \xrightarrow[t \nearrow T]{} \infty$.

1 This article has appeared in a modified form in the Notices of the AMS [8].

The most basic examples of Ricci flows are those in which g_0 is Einstein, i.e., $\mathrm{Ric}_{g_0} = \lambda g_0$. In this case the flow evolves by rescaling,

$$g(t) = (1 - 2\lambda t) g_0. \tag{2.3}$$

So, for example, a round sphere ($\lambda > 0$) shrinks under the flow, develops a singularity in finite time, and its diameter goes to 0. On the other hand, if we start with a hyperbolic metric ($\lambda < 0$), then the flow is immortal (i.e., $T = \infty$) and the metric expands linearly. In the following we will consider more general initial metrics g_0 and hope that – at least in some cases – the flow is asymptotic to a solution of the form (2.3). This will then allow us to understand the topology of the underlying manifold in terms of the limiting geometry.

3. DIMENSION 2

In dimension 2, Ricci flows are very well understood [24, 26, 32]:

Theorem 3.1. *Any Ricci flow on a compact 2-dimensional manifold converges, modulo rescaling, to a metric of constant curvature.*

In addition, one can show that the flow in dimension 2 preserves the conformal class, i.e., for all times t we have $g(t) = f(t) g_0$ for some smooth positive function $f(t)$ on M. This observation, combined with Theorem 3.1, can in fact be used to reprove the Uniformization Theorem:[2]

Theorem 3.2. *Each compact surface M admits a metric of constant curvature in each conformal class.*

In order to obtain *new* applications, however, we will need to study the flow in higher dimensions.

4. DIMENSION 3

In dimension 3, the behavior of the flow – and its singularity formation – becomes far more complicated. In the following, we will first review prior work on Ricci flow in dimension 3, which is mostly due to Hamilton and Perelman and which led to the resolution of the Poincaré and Geometrization Conjectures. We will keep this part short and only focus on aspects that will become important later; for a more in-depth discussion see, for example, [1]. Next, we will focus on more recent work by Kleiner, Lott, and the author on singular Ricci flows and their uniqueness and continuous dependence, which led to the resolution of several longstanding topological conjectures.

2 The original proof of Theorem 3.1 relied on the Uniformization Theorem. This dependence was later removed by Chen, Lu, and Tian.

4.1. Singularity formation – an example

To get an idea of the possible singularity formation of 3-dimensional Ricci flows, it is useful to consider the famous dumbbell example [2, 3] (see Figure 1). In this example, the initial manifold (M, g_0) is the result of connecting two round spheres of radii r_1, r_3 by a certain type of rotationally symmetric neck of radius r_2 (see Figure 1). So $M \approx S^3$ and $g_0 = f^2(s)g_{S^2} + ds^2$ is a warped product away from the two endpoints.

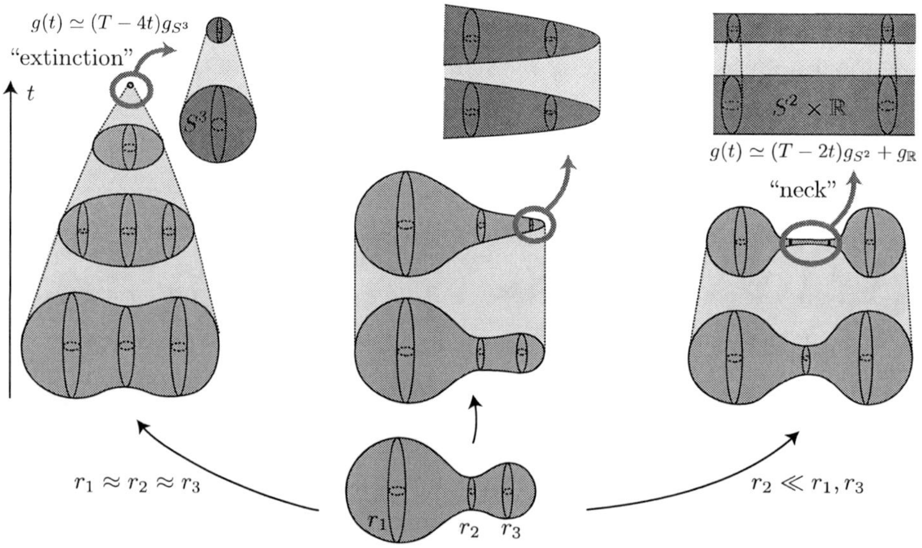

FIGURE 1

Different singularity formations in the rotationally symmetric case, depending on the choice of the radii r_1, r_2, r_3. The flows depicted on the top are the corresponding singularity models. These turn out to be the only singularity models, even in the nonrotationally symmetric case (see Section 4.3).

It can be shown that any flow starting from a metric of this form must develop a singularity in finite time. The singularity *type,* however, depends on the choice of the radii r_1, r_2, r_3. More specifically, if the radii r_1, r_2, r_3 are comparable (Figure 1, left), then the diameter of the manifold converges to zero and, after rescaling, the flow becomes asymptotically round – just as in Theorem 3.1. This case is called *extinction.* On the other hand, if $r_2 \ll r_1, r_3$ (Figure 1, right), then the flow develops a *neck singularity,* which looks like a round cylinder ($S^2 \times \mathbb{R}$) at small scales. Note that in this case the singularity only occurs in a certain region of the manifold, while the metric converges to a smooth limit everywhere else. Lastly, there is also an intermediate case (Figure 1, center), in which the flow develops a singularity that is modeled on the Bryant soliton – a one-ended paraboloid-like model [16].[3]

3 We have omitted a less important nongeneric case, called the *peanut solution.* In this case the diameter converges to zero in finite time. However, after rescaling, the metric looks like a long cylinder with a slight indentation that is capped off on each side by regions whose geometry is close to Bryant soliton. See [2] for more details.

4.2. Blow-up analysis

Perelman's work implied that the previous example is in fact prototypical for the singularity formation of *general* (not necessarily rotationally symmetric) 3-dimensional Ricci flow. In order to make this statement more precise, let us first recall a method called *blow-up analysis,* which is used frequently to study singularities in geometric analysis.

Suppose that $(M, (g(t))_{t\in[0,T)})$ is a Ricci flow that develops a singularity at time $T < \infty$ (see Figure 2). Then we can find a sequence of spacetime points $(x_i, t_i) \in M \times [0, T)$ such that $\lambda_i := |\mathrm{Rm}|(x_i, t_i) \to \infty$ and $t_i \nearrow T$. Our goal will be to understand the local geometry at small scales near (x_i, t_i), for large i. For this purpose, we consider the sequence of pointed, parabolically rescaled flows

$$\big(M, \big(g'_i(t) := \lambda_i g\big(\lambda_i^{-1} t + t_i\big)\big)_{t\in[-\lambda_i t_i, 0]}, (x_i, 0)\big).$$

Geometrically, the flows $(g'_i(t))$ are the result of rescaling distances by $\lambda_i^{1/2}$, times by λ_i and an application of a time-shift such that the points $(x_i, 0)$ in the new flows corresponds to the points (x_i, t_i) in the old flows. The new flows $(g'_i(t))$ still satisfy the Ricci flow equation and are defined on larger and larger backwards time-intervals of size $\lambda_i t_i \to \infty$. Moreover, we have $|\mathrm{Rm}|(x_i, 0) = 1$ on these new flows. Observe also that the geometry of the original flows near (x_i, t_i) at scale $\lambda_i^{-1/2} \ll 1$ is a rescaling of the geometry of $(g'_i(t))$ near $(x_i, 0)$ at scale 1.

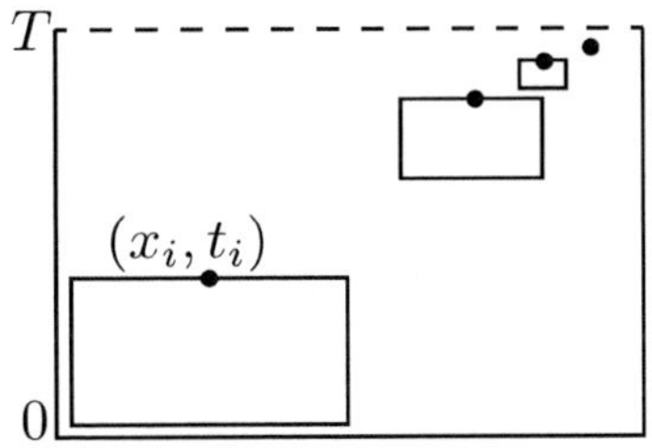

FIGURE 2
A Ricci flow $M \times [0, T)$ that develops a singularity at time T and a sequence of points (x_i, t_i) that "run into a singularity." The geometry in the parabolic neighborhoods around (x_i, t_i) (rectangles) is close to the singularity model modulo rescaling if $i \gg 1$.

Under certain additional assumptions, we may now apply a compactness theorem (à la Arzela–Ascoli) such that, after passing to a subsequence, we have convergence

$$\big(M, \big(g'_i(t)\big)_{t\in[-\lambda_i^2 t_i, 0]}, (x_i, 0)\big) \xrightarrow[i\to\infty]{} \big(M_\infty, \big(g_\infty(t)\big)_{t\le 0}, (x_\infty, 0)\big). \tag{4.1}$$

The limit is called a *blow-up* or *singularity model,* as it gives valuable information on the singularity formation near the points (x_i, t_i). This model is a Ricci flow that is defined for all times $t \le 0$; it is therefore called *ancient.* So in summary, a blow-up analysis reduces the study of singularity *formation* to the classification of ancient singularity *models.*

The notion of the convergence in (4.1) is a generalization of Cheeger–Gromov convergence to Ricci flows; it is due to Hamilton [33]. Instead of demanding global convergence of the metric tensors, as in Theorem 3.1, we only require convergence up to diffeomorphisms here. More specifically, we roughly require that we have convergence

$$\phi_i^* g_i'(t) \xrightarrow[i\to\infty]{C^\infty_{\mathrm{loc}}} g_\infty \tag{4.2}$$

on $M_\infty \times (-\infty, 0]$ of the pullbacks of $g_i'(t)$ via (time-independent) diffeomorphisms $\phi_i : U_i \to V_i \subset M$ that are defined over larger and larger subsets $U_i \subset M_\infty$ and satisfy $\phi_i(x_\infty) = x_i$. We will see later (in Section 5) that this notion of convergence is too strong to capture the more subtle singularity formation of higher dimensional Ricci flows and we will discuss necessary refinements. Luckily, in dimension 3 the current notion is still sufficient for our purposes, though.

4.3. Singularity models and canonical neighborhoods

One of key discoveries of Perelman's work was the classification of singularity models of 3-dimensional Ricci flows and the resulting structural description of the flow near a singularity. The following theorem[4] summarizes this classification.

Theorem 4.1. *Any singularity model $(M_\infty, (g_\infty(t))_{t\le 0})$ obtained as in (4.1) is isometric, modulo rescaling, to one the following:*

(1) *a quotient of the round shrinking sphere $(S^3, (1-4t)g_{S^3})$,*

(2) *the Bryant soliton $(M_{\mathrm{Bry}}, (g_{\mathrm{Bry}}(t)))$,*

(3) *the round shrinking cylinder $(S^2 \times \mathbb{R}, (1-2t)g_{S^2} + g_{\mathbb{R}})$ or its quotient $(S^2 \times \mathbb{R})/\mathbb{Z}_2$.*

Note that these three models correspond to the three cases in the rotationally symmetric dumbbell example from Section 4.1 (see Figure 1). The Bryant soliton in (2) is a rotationally symmetric solution to the Ricci flow on $\mathbb{R}^3$ with the property that all its time-slices are isometric to a metric of the form

$$g_{\mathrm{Bry}} = f^2(r) g_{S^2} + dr^2, \quad f(r) \sim \sqrt{r}.$$

The name *soliton* refers to the fact that all time-slices of the flow are isometric, so the flow merely evolves by pullbacks of a family of diffeomorphisms.

The next theorem describes the structure of the flow near *any* point of the flow that is close to a singularity – not just along a single blow-up sequence. In order to state this result efficiently, we will need to consider the class of *κ-solutions.* This class consists of all solutions listed in Theorem 4.1, plus an additional compact, ellipsoidal solution **[15]** (the details of this solution won't be important here[5]). Then we have:

---

4 Perelman proved a version of Theorem 4.1 that contained a more qualitative characterization in Case (2), which was sufficient for most applications. Later, Brendle **[14]** showed that the only possibility in Case (2) is the Bryant soliton.

5 This solution does not occur as a singularity of a single flow, but can be observed as a transitional model in families of flows that interpolate between two different singularity models.

Theorem 4.2 (Canonical neighborhood theorem). *If $(M, (g(t))_{t\in[0,T)})$, $T < \infty$, is a 3-dimensional Ricci flow and $\varepsilon > 0$, then there is a constant $r_{\mathrm{can}}(g(0), T, \varepsilon) > 0$ such that for any $(x,t) \in M \times [0,T)$ with the property that*

$$r := |\mathrm{Rm}|^{-1/2}(x,t) \le r_{\mathrm{can}}$$

the geometry of the metric $g(t)$ restricted to the ball $B_{g(t)}(x, \varepsilon^{-1}r)$ is ε-close[6] to a time-slice of a κ-solution.

4.4. Ricci flow with surgery

Our understanding of the structure of the flow near a singularity allows us to carry out a so-called *surgery construction.* Under this construction, (almost) singularities of the flow are removed, resulting in a "less singular" geometry, from which the flow can be restarted. This leads to a new type of flow that is defined beyond its singularities and which will provide important information on the underlying manifold.

Let us be more precise. A *(3-dimensional) Ricci flow with surgery* (see Figure 3) consists of a sequence of Ricci flows

$$\big(M_1, (g_1(t))_{t\in[0,T_1]}\big), \quad \big(M_2, (g_2(t))_{t\in[T_1,T_2]}\big), \quad \big(M_3, (g_3(t))_{t\in[T_2,T_3]}\big), \quad \ldots,$$

which live on manifolds $M_1, M_2, \ldots$ of possibly different topology and are parameterized by consecutive time-intervals of the form $[0, T_1], [T_1, T_2], \ldots$ whose union equals $[0, \infty)$.

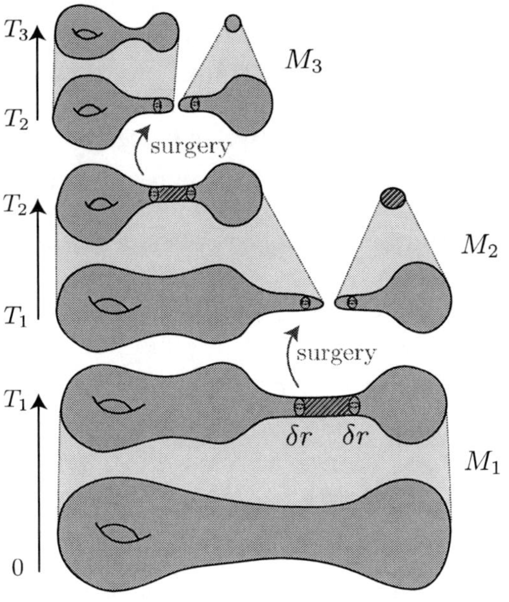

FIGURE 3
A schematic depiction of a Ricci flow with surgery. The almost-singular parts $M_{\text{almost-sing}}$, i.e., the parts that are discarded under each surgery construction, are hatched.

---

6 Similar to the definition of (4.2), this roughly means that there is a diffeomorphism between an ε^{-1}-ball in a κ-solution and this ball such that the pullback of $r^{-2}g(t)$ is ε-close in the $C^{[\varepsilon^{-1}]}$-sense to the metric on the κ-solution.

The time-slices $(M_i, g_i(T_i))$ and $(M_{i+1}, g_{i+1}(T_i))$ are related by a *surgery process,* which can be roughly summarized as follows. Consider the set $M_{\text{almost-sing}} \subset M_i$ of all points of high enough curvature, such that they have a canonical neighborhood as in Theorem 4.2. Cut M_i open along approximate cross-sectional 2-spheres of diameter $r_{\text{surg}}(T_i) \ll 1$ near the cylindrical ends of $M_{\text{almost-sing}}$, discard most of the high-curvature components (including the closed, spherical components of $M_{\text{almost-sing}}$), and glue in cap-shaped 3-disks to the cutting surfaces. In doing so we have constructed a new, "less singular," Riemannian manifold $(M_{i+1}, g_{i+1}(T_i))$, from which we can restart the flow. Stop at some time $T_{i+1} > T_i$, shortly before another singularity occurs and repeat the process.

The precise surgery construction is quite technical and more delicate than presented here. The main difficulty in this construction is to ensure that the surgery times T_i do not accumulate, i.e., that the flow can be extended for all times. It was shown by Perelman that this and other difficulties can, indeed, be overcome:

Theorem 4.3. *Let (M, g) be a closed, 3-dimensional Riemannian manifold. If the surgery scales $r_{\text{surg}}(T_i) > 0$ are chosen sufficiently small (depending on (M, g) and T_i), then a Ricci flow with surgery with initial condition $(M_1, g_1(0)) = (M, g)$ can be constructed.*

Note that the topology of the underlying manifold M_i may change in the course of a surgery, but only in a controlled way. In particular, it is possible to show that for any i the initial manifold M_1 is diffeomorphic to a connected sum of components of M_i and copies of spherical space forms S^3/Γ and $S^2 \times S^1$. So if the flow goes *extinct* in finite time, meaning that $M_i = \emptyset$ for some large i, then

$$M_1 \approx \#_{j=1}^{k}(S^3/\Gamma_j)\#m(S^2 \times S^1). \tag{4.3}$$

Perelman, moreover, showed that if M_1 is simply connected, then the flow *must* go extinct and therefore M_1 must be of the form (4.3). This implies the Poincaré Conjecture:

Theorem 4.4 (Poincaré Conjecture). *Any simply connected, closed 3-manifold is diffeomorphic to S^3.*

On the other hand, Perelman showed that if the Ricci flow with surgery does not go extinct, meaning if it exists for all times, then for large times $t \gg 1$ the flow decomposes the manifold (at time t) into a thick and a thin part:

$$M_{\text{thick}}(t) \cup M_{\text{thin}}(t), \tag{4.4}$$

such that the metric on $M_{\text{thick}}(t)$ is asymptotic to a hyperbolic metric, the metric on $M_{\text{thin}}(t)$ is locally collapsed and the boundary of $M_{\text{thick}}(t)$ consists of incompressible 2-tori. A further topological analysis of this collapse implied the Geometrization Conjecture:

Theorem 4.5 (Geometrization Conjecture). *Every closed 3-manifold is a connected sum of manifolds that can be cut along embedded, incompressible copies of T^2 into pieces which each admit a locally homogeneous geometry.*

4.5. Ricci flows through singularities

Despite their spectacular applications, Ricci flows with surgery have one major drawback: their construction is not canonical. In other words, each surgery step depends on a number of auxiliary parameters, for which there does not seem to be a canonical choice, such as:

- The surgery scales $r_{\mathrm{surg}}(T_i)$, i.e., the diameters of the cross-sectional spheres along which the manifold is cut open. These scales need to be positive and small.
- The precise locations of these surgery spheres.

Different choices of these parameters may influence the future development of the flow significantly (as well as the space of future surgery parameters). Hence a Ricci flow with surgery is not *uniquely* determined by its initial metric.

This disadvantage was already recognized in Perelman's work, where he conjectured that there should be another flow, in which surgeries are effectively carried out automatically at an infinitesimal scale (think "$r_{\mathrm{surg}} = 0$"), or which, in other words, "flows through singularities."

Perelman's conjecture was recently resolved by Kleiner, Lott, and the author (see [39] for the "Existence" and [10] for the "Uniqueness" part; part (2) of Theorem 4.6 follows from a combination of both papers):

Theorem 4.6. *There is a notion of singular Ricci flow (through singularities) such that:*

(1) *For any compact, 3-dimensional Riemannian manifold (M, g), there is a unique singular Ricci flow $\mathcal{M}$ whose initial time-slice $(\mathcal{M}_0, g_0)$ is (M, g).*

(2) *Any Ricci flow with surgery starting from (M, g) can be viewed as an approximation of $\mathcal{M}$. More specifically, if we consider a sequence of Ricci flows with surgery starting from (M, g) with surgery scales $\max_t r_{\mathrm{surg}}(t) \to 0$, then these flows converge to $\mathcal{M}$ in the local C^∞-sense. (More details on this convergence will be given in the end of this subsection.)*

Before continuing, let us compare this result with past work on the mean curvature flow – a close cousin of the Ricci flow. There are two important weak formulations of the mean curvature flow, namely Brakke and level set flows. Existence theories [13,25,28,38] rely heavily on the fact that a mean curvature flow concerns *embedded* submanifold in an ambient space. Uniqueness of these flows, on the other hand, is false in general [50], but true in the mean convex case [51]; so the analogous statement to Part (1) holds in this case. Moreover, under the more restrictive condition of 2-convexity, which guarantees the existence of a surgery procedure, an equivalent of Part (2) holds as well [35,41].

The concept of a singular Ricci flow is less technical than that of a Ricci flow with surgery—in fact, we will be able to state its full definition here. To do this, we will first define the concept of a Ricci flow spacetime. In short, this is a smooth 4-manifold that locally looks like a Ricci flow, but which may have non-trivial *global* topology (see Figure 4).

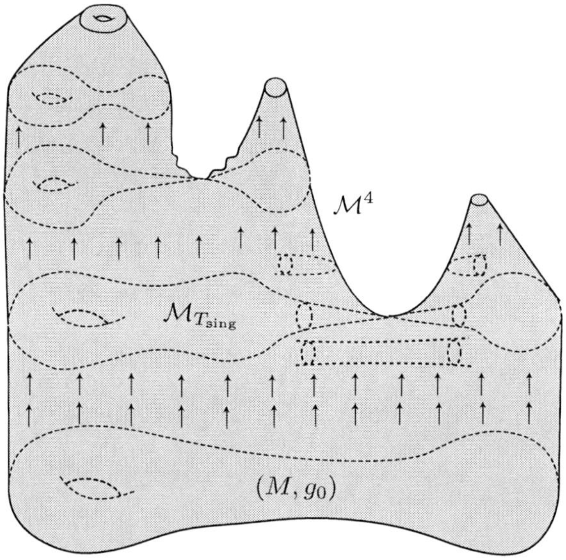

FIGURE 4

Illustration of a singular Ricci flow given by a Ricci flow spacetime. The arrows indicate the time-vector field $\partial_{\mathfrak{t}}$.

Definition 4.7. A *Ricci flow spacetime* consists of:

(1) a smooth 4-dimensional manifold $\mathcal{M}$ with boundary, called *spacetime;*

(2) a *time-function* $\mathfrak{t} : \mathcal{M} \to [0, \infty)$; its level sets $\mathcal{M}_t := \mathfrak{t}^{-1}(t)$ are called *time-slices* and we require that $\mathcal{M}_0 = \partial \mathcal{M}$;

(3) a *time-vector field* $\partial_{\mathfrak{t}}$ on $\mathcal{M}$ with $\partial_{\mathfrak{t}} \cdot \mathfrak{t} \equiv 1$; trajectories of $\partial_{\mathfrak{t}}$ are called *world-lines;*

(4) a family g of inner products on $\ker d\mathfrak{t} \subset T\mathcal{M}$, which induce a Riemannian metric g_t on each time-slice $\mathcal{M}_t$; we require that the Ricci flow equation holds:

$$\mathcal{L}_{\partial_{\mathfrak{t}}} g_t = -2 \operatorname{Ric}_{g_t} .$$

By abuse of notation, we will often write $\mathcal{M}$ instead of $(\mathcal{M}, \mathfrak{t}, \partial_{\mathfrak{t}}, g)$.

A classical, 3-dimensional Ricci flow $(M, (g(t))_{t \in [0,T)})$ can be converted into a Ricci flow spacetime by setting $\mathcal{M} := M \times [0, T)$, letting $\mathfrak{t}, \partial_{\mathfrak{t}}$ be the projection onto the second factor and the pullback of the unit vector field on the second factor, respectively, and letting g_t be the metric corresponding to $g(t)$ on $M \times \{t\} \approx M$. Hence worldlines correspond to curves of the form $t \mapsto (x, t)$.

Likewise, a Ricci flow with surgery, given by flows

$$\big(M_1, (g_1(t))_{t \in [0,T_1]}\big), \quad \big(M_2, (g_2(t))_{t \in [T_1,T_2]}\big), \quad \dots,$$

can be converted into a Ricci flow spacetime as follows. Consider first the Ricci flow spacetimes $M_1 \times [0, T_1]$, $M_2 \times [T_1, T_2], \dots$ arising from each single flow. We can now glue these flows together by identifying the set of points $U_i^- \subset M_i \times \{T_i\}$ and $U_i^+ \subset M_{i+1} \times \{T_i\}$ that survive each surgery step via maps $\phi_i : U_i^- \to U_i^+$. The resulting space has a boundary that consists of the time-0-slice $M_1 \times \{0\}$ and the points

$$\mathcal{S}_i = \big(M_i \times \{T_i\} \setminus U_i^-\big) \cup \big(M_{i+1} \times \{T_i\} \setminus U_i^+\big),$$

which were removed and added during each surgery step. After removing these points, we obtain a Ricci flow spacetime of the form:

$$\mathcal{M} = \big(M_1 \times [0, T_1] \cup_{\phi_1} M_2 \times [T_1, T_2] \cup_{\phi_2} \cdots\big) \setminus (\mathcal{S}_1 \cup \mathcal{S}_2 \cup \cdots). \tag{4.5}$$

Note that, for any regular time $t \in (T_{i-1}, T_i)$, the time-slice $\mathcal{M}_t$ is isometric to $(M_i, g_i(t))$. On the other hand, the time-slices $\mathcal{M}_{T_i}$ corresponding to surgery times are incomplete; they have cylindrical open ends of scale $\approx r_{\text{surg}}(T_i)$.

The following definition captures this incompleteness:

Definition 4.8. A Ricci flow spacetime is *r-complete,* for some $r \geq 0$, if the following holds. Consider a smooth path $\gamma : [0, s_0) \to \mathcal{M}$ with the property that

$$\inf_{s\in[0,s_0)} |\text{Rm}|^{-1/2}\big(\gamma(s)\big) > r$$

and:

(1) $\gamma([0, l)) \subset \mathcal{M}_t$ is contained in a single time-slice and its length measured with respect to the metric g_t is finite, or

(2) γ is a worldline, i.e., a trajectory of $\pm\partial_t$.

Then the limit $\lim_{s\nearrow s_0} \gamma(s)$ exists.

So $\mathcal{M}$ being r-complete roughly means that it has only "holes" of scale $\lesssim r$. For example, the flow from (4.5) is $C \max_t r_{\text{surg}}(t)$-complete for some universal $C < \infty$.

In addition, Theorem 4.2 motivates the following definition:

Definition 4.9. A Ricci flow spacetime is said to satisfy the *ε-canonical neighborhood assumption at scales* (r_1, r_2) if for any point $x \in \mathcal{M}_t$ with $r := |\text{Rm}|^{-1/2}(x) \in (r_1, r_2)$ the metric g_t restricted to the ball $B_{g_t}(x, \varepsilon^{-1} r)$ is ε-close, after rescaling by r^{-2}, to a time-slice of a κ-solution.

We can finally define singular Ricci flows (through singularities), as used in Theorem 4.6:

Definition 4.10. A *singular Ricci flow* is a Ricci flow spacetime $\mathcal{M}$ with the following two properties:

(1) It is 0-complete.

(2) For any $\varepsilon > 0$ and $T < \infty$, there is an $r(\varepsilon, T) > 0$ such that the flow $\mathcal{M}$ restricted to $[0, T)$ satisfies the ε-canonical neighborhood assumption at scales $(0, r)$.

See again Figure 4 for a depiction of a singular Ricci flow. The time-slices $\mathcal{M}_t$ for $t < T_{\text{sing}}$ develop a cylindrical region, which collapses to some sort of topological double cone singularity in $\mathcal{M}_{T_{\text{sing}}}$ at time T_{sing}. This singularity is immediately resolved and the flow is smooth for some $t > T_{\text{sing}}$.

Let us digest the definition of a singular Ricci flow a bit more. It is tempting to think of the time function $\mathfrak{t}$ as a Morse function and compare critical points with infinitesimal surgeries. However, this comparison is flawed: First, by definition $\mathfrak{t}$ cannot have critical points since $\partial_{\mathfrak{t}} \mathfrak{t} = 1$. In fact, a singular Ricci flow is a completely smooth object. The "singular points" of the flow are not part of $\mathcal{M}$, but can be obtained after metrically completing each time-slice by adding a discrete set of points. Second, it is currently unknown whether the set of singular times, i.e., the set of times whose time-slices are incomplete, is discrete. In addition, the approach taken in the definition of singular Ricci flows is different from that of weak solutions to the mean curvature flow. While the Brakke and level set flows characterize the flow equation at singular points via integral or barrier conditions, a singular Ricci flow only characterizes the flow on its regular part. In lieu of a weak formulation of the Ricci flow equation on the singular set, we have to impose the canonical neighborhood assumption, which serves as an asymptotic characterization near the incomplete ends.

Finally, let us briefly explain how singular Ricci flows are constructed and convey the meaning of Part (2) of Theorem 4.6. Fix an initial time-slice (M, g) and consider a sequence of Ricci flow spacetimes $\mathcal{M}^j$ that correspond to Ricci flows with surgery starting from (M, g), with surgery scale $\max_t r_{\text{surg},j}(t) \to 0$. It can be shown that these flows are $C \max_t r_{\text{surg},j}(t)$-complete and satisfy the ε-canonical neighborhood assumption at scales $(C_\varepsilon \max_t r_{\text{surg},j}(t), r_\varepsilon)$, where $C, C_\varepsilon, r_\varepsilon$ do not depend on j. A compactness theorem implies that a subsequence of the spacetimes $\mathcal{M}^j$ converges to a spacetime $\mathcal{M}$, which is a singular Ricci flow. This implies the existence of $\mathcal{M}$; the proof of uniqueness uses other techniques, which are outside the scope of this article.

4.6. Continuous dependence

The proof of the uniqueness property in Theorem 4.6, due to Kleiner and the author, implies an important continuity property, which leads to further topological applications. To state this property, let M be a compact 3-manifold and for every Riemannian metric g on M let $\mathcal{M}^g$ be the singular Ricci flow with initial condition $(\mathcal{M}^g_0, g) = (M, g)$.

Theorem 4.11 ([11]). *The flow $\mathcal{M}^g$ depends continuously on g.*

Together with Theorem 4.6, this implies that the Ricci flow equation in dimension 3 is well-posed within the class of singular Ricci flows.

Note that the topology of the flow $\mathcal{M}^g$ may change as we vary g and the continuity holds for the entire flows – past potential singularities. We therefore have to choose an appropriate sense of continuity in Theorem 4.11 that allows such a topological change. This is roughly done via a topology and lamination structure on the disjoint union $\bigsqcup_g \mathcal{M}^g$, transverse to which the variation of the flow can be studied locally.

Instead of elaborating on these technicalities, let us discuss the example illustrated in Figure 5. In this example $(g_s)_{s\in[0,1]}$ denotes a continuous family of metrics on S^3 such that the corresponding flows $\mathcal{M}^s := \mathcal{M}^{g_s}$ interpolate between a round and a cylindrical sin-

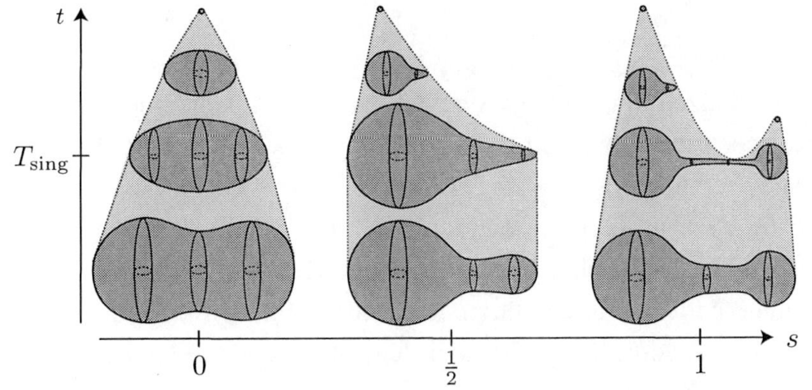

FIGURE 5
A family of singular Ricci flows starting from a continuous family of initial conditions.

gularity. For $s \in [0, \frac{1}{2})$, the flow $\mathcal{M}^s$ can be described in terms of a conventional, nonsingular Ricci flow (g^s_t) on M and the continuity statement in Theorem 4.11 is equivalent to continuous dependence of this flow on s. Likewise, the flows $\mathcal{M}^s$ restricted to $[0, T_{\text{sing}})$ can again be described by a continuous family of conventional Ricci flows. The question is now what happens at the critical parameter $s = \frac{1}{2}$, where the type of singularity changes. The uniqueness property guarantees that the flows $\mathcal{M}^s$ for $s \nearrow \frac{1}{2}$ and $s \searrow \frac{1}{2}$ must limit to the same flow $\mathcal{M}^{\frac{1}{2}}$. The convergence is locally smooth, but the topology of the spacetime manifold $\mathcal{M}^s$ may still change.

4.7. Topological applications

Theorem 4.11 provides us a tool to deduce the first topological applications of Ricci flow since Perelman's work. Before stating these, let us make the following definitions. We denote by $\mathrm{Met}(M)$ the space of all Riemannian metrics on a manifold M, equipped with the C^∞-topology. Let $\mathrm{Met}_{K\equiv k}(M), \mathrm{Met}_{\mathrm{PSC}}(M) \subset \mathrm{Met}(M)$ be the subsets of metrics of constant sectional curvature k and of positive scalar curvature, respectively. Furthermore, we denote by $\mathrm{Diff}(M)$ the group of diffeomorphisms $\phi : M \to M$, again equipped with the C^∞-topology, and for a Riemannian metric $g \in \mathrm{Met}(M)$ we denote by $\mathrm{Isom}(M, g) \subset \mathrm{Diff}(M)$ the isometry group of (M, g).

Theorem 4.12 ([11]). *For any closed 3-manifold M, the space $\mathrm{Met}_{\mathrm{PSC}}(M)$ is either contractible or empty.*

Theorem 4.13 (Generalized Smale Conjecture, [9, 11]). *Suppose that (M^3, g) is a closed manifold of constant curvature $K \equiv \pm 1$. Then the inclusion map $\mathrm{Isom}(M, g) \hookrightarrow \mathrm{Diff}(M)$ is a homotopy equivalence.*

The study of the spaces $\mathrm{Met}_{\mathrm{PSC}}(M)$ was initiated by Hitchin in the 1970s and has led to many interesting results – based on index theory – which show that these spaces have nontrivial topology when M is high dimensional. Theorem 4.12 provides the first examples of manifolds of dimension ≥ 3 for which the homotopy type of $\mathrm{Met}_{\mathrm{PSC}}(M)$ is completely understood; see also prior work by Marques [42].

The Generalized Smale Conjecture has had a long history and many interesting special cases have been established using topological methods, including the case $M = S^3$ by Hatcher [34] and the hyperbolic case by Gabai [30]. However, the full conjecture remained open until recently. For more background, see the first chapter of [37]. The proof of Theorem 4.13 is independent of Hatcher's and Gabai's proof, so it provides an alternative approach to the S^3 and hyperbolic case. In addition, it provides a uniform treatment of all topological cases and the same method can also be used characterize the homotopy type of other prime 3-manifolds (see, for example, [11] for the case $S^2 \times S^1$). For many of these manifolds, this was already accomplished using topological methods; however, the following result is new:

Theorem 4.14 ([12]). *Let g be a compact, orientable, non-Haken 3-manifold modeled on the Thurston geometry* Nil *and let g be a* Nil*-metric on M. Then the inclusion* $\mathrm{Isom}(M, g) \hookrightarrow \mathrm{Diff}(M)$ *is a homotopy equivalence.*

Combining Theorems 4.13, 4.14 with the previously known characterization of $\mathrm{Diff}(M)$ in all other cases, this completes the project of understanding the topology of $\mathrm{Diff}(M)$ when M is a closed 3-manifold.

There are two proofs for Theorem 4.13: a short proof and a long proof. The short proof [9] requires the additional assumption that $M \not\approx \mathbb{R}P^3$ and relies on Hatcher's resolution of the Smale Conjecture. The long proof [11] establishes both Theorems 4.12, 4.13 in their full form.

Both proofs rely on two basic observations:

- The positive scalar curvature condition is preserved by the flow.
- Theorem 4.13 is equivalent to the contractibility of the space $\mathrm{Met}_{K\equiv\pm1}(M)$ of constant curvature metrics. This can be seen via a standard argument involving the long-exact homotopy sequence for the fiber bundle $\mathrm{Isom}(M) \to \mathrm{Diff}(M) \to \mathrm{Met}_{K\equiv\pm1}(M)$.

Let us simplify our setting for a moment and suppose that M was the 2-dimensional sphere. Then by Theorem 3.1, Ricci flow can be seen as a deformation retraction of $\mathrm{Met}(M)$ or $\mathrm{Met}_{\mathrm{PSC}}(M)$ to $\mathrm{Met}_{K\equiv\pm1}(M)$ – modulo rescaling and reparameterization. This shows that the spaces $\mathrm{Met}(M)$, $\mathrm{Met}_{\mathrm{PSC}}(M)$ and $\mathrm{Met}_{K\equiv\pm1}(M)$ are homotopy equivalent, and since the first space is contractible (it is a convex subset of a vector space), we obtain that all spaces are contractible.

Unfortunately, the strategy in the 2-dimensional case does not readily generalize to dimension 3, because singular flows cannot be viewed as trajectories in $\mathrm{Met}(M)$ as they are defined by metrics on different time-slices – possibly of different topology. Therefore, the proofs of Theorems 4.12, 4.13 have to follow a different strategy, which we will outline now. To this end we first observe that, since $\mathrm{Met}(M)$ is contractible, it is enough to show that

$\pi_k(\mathrm{Met}(M), \mathrm{Met}_X(M))$ is trivial, where X may stand for "PSC" or "$K \equiv \pm1$."

Let us now fix a representative $g : D^{k+1} \to \mathrm{Met}(M)$ of this relative homotopy group, i.e., $g(s) \in \mathrm{Met}_X(M)$ for all $s \in \partial D^{k+1}$. Our goal will be to construct a null-homotopy $\hat{g} : D^{k+1} \times [0,1] \to \mathrm{Met}(M)$, where $\hat{g}(\cdot, 0) = g$ and $\hat{g}(s,t) \in \mathrm{Met}_X(M)$ if $s \in \partial D^{k+1}$ or $t = 1$.

If all the Ricci flows starting from each metric $g(s)$ were to converge to a round metric (modulo rescaling), then $\hat{g}(s,\cdot)$ could simply be constructed using these flows (as we did in dimension 2). In general, however, the family g only induces a continuous family of singular Ricci flows $(\mathcal{M}^s := \mathcal{M}^{g(s)})_{s \in D^{k+1}}$. In a second step, this family of flows has to be converted to the desired null-homotopy $\hat{g}$ within $\mathrm{Met}(M)$. In the long proof, this is achieved via a new notion called *partial homotopy*. This notion is a hybrid between a null-homotopy in $\mathrm{Met}(M)$ and a continuous family of Ricci flows, which permits variation of underlying topology. A partial homotopy allows the construction of a null-homotopy via backwards induction in time via certain modification moves that roughly correspond to the singularities of the flows $\mathcal{M}^s$. The short proof, on the other hand, uses the flow of the time-vector field ∂_t on each $\mathcal{M}^s$ to push forward the metrics g^s_t to its initial time-slice $\mathcal{M}^s_0 = M$. This flow is not defined everywhere and thus such a construction only offers a continuous family of metrics $\tilde{g}^s$ defined on open subsets $U^s \subset M$, where $M \setminus U^s$ can be covered by pairwise disjoint 3-disks. These metrics then have to be extended onto all of M via an obstruction theoretic argument, which relies on Hatcher's resultion of the S^3-case.

5. DIMENSIONS $n \geq 4$

For a long time, most of the known results of Ricci flows in higher dimensions concerned special cases, such as Kähler–Ricci flows or flows that satisfy certain preserved curvature conditions. *General* flows, on the other hand, were relatively poorly understood. Recently, however, there has been some movement on this topic – in part, thanks to a different geometric perspective on Ricci flows [5–7]. The goal of this section is to convey some of these new ideas and to provide an outlook on possible geometric and topological applications.

5.1. Gradient shrinking solitons

Gradient shrinking solitons (GSSs) comprise an important class of singularity models in Ricci flow, especially in higher dimensions. The GSS equation concerns Riemannian manifolds (M, g) equipped with a potential function $f \in C^\infty(M)$ and reads

$$\mathrm{Ric} + \nabla^2 f - \frac{1}{2} g = 0.$$

This generalization of the Einstein equation gives rise to an associated self-similar Ricci flow

$$g(t) := |t| \phi_t^* g, \quad t < 0,$$

where $(\phi_t : M \to M)_{t<0}$ is the flow of the vector field $|t| \nabla f$.

A basic class of examples for GSSs are the round cylinders $S^{k \geq 2} \times \mathbb{R}^{n-k}$, where

$$g = 2(k-1) g_{S^k} + g_{\mathbb{R}^{n-k}}, \quad f = \frac{1}{4} \sum_{i=k+1}^{n} x_i^2.$$

In this case, $|t|\nabla f$ generates a family of dilations on the $\mathbb{R}^{n-k}$ factor and

$$g(t) = 2(k-1)|t|g_{S^k} + g_{\mathbb{R}^{n-k}},$$

which is isometric to $|t|g$. A special case of this is the round shrinking sphere ($k = n$). In dimensions $n \leq 3$, all nontrivial[7] GSSs are quotients of round spheres or cylinders. However, more complicated GSSs exist in dimensions $n \geq 4$ (see, for example, [29]).

By construction, GSSs (or their associated flows, to be precise) are invariant under parabolic rescaling. So the blow-up singularity model of the singularity at time 0 (taken along an appropriately chosen sequence of basepoints) is equal to the flow itself. Therefore every GSS does indeed occur as a singularity model, at least of its own flow.

Vice versa, the following conjecture, which will be kept vague for now, predicts that the converse should also be true in a certain sense.

Conjecture 5.1. *For any Ricci flow, "most" singularity models are gradient shrinking solitons.*

This conjecture has been implicit in Hamilton's work from the 1990s, and a similar result is known to be true for mean curvature flow. In the remainder of this section, we will present a resolution of a version of this conjecture.

5.2. Examples of singularity formation

Let us first discuss an example in order to adjust our expectations in regards to Conjecture 5.1. In [4], Appleton constructs a class of 4-dimensional Ricci flows[8] that develop a singularity in finite time, which can be studied via the blow-up technique from Section 4.2 – this time we even allow the rescaling factors to be *any* sequence of numbers $\lambda_i \to \infty$, not just $\lambda_i = |\mathrm{Rm}|^{1/2}(x_i, t_i)$. Appleton obtains the following classification of all nontrivial blow-up singularity models:

(1) the Eguchi–Hanson metric, which is Ricci flat and asymptotic to the flat orbifold $\mathbb{R}^4/\mathbb{Z}_2$,

(2) the flat orbifold $\mathbb{R}^4/\mathbb{Z}_2$,

(3) the quotient $M_{\mathrm{Bry}}/\mathbb{Z}_2$ of the Bryant soliton, which has an isolated orbifold singularity at its tip,

(4) the cylinder $\mathbb{R}P^3 \times \mathbb{R}$.

Here the models (1) and (2) *have* to occur as singularity models, and it is unknown whether the models (3) and (4) actually do show up. The only GSSs in this list are (2) and (4). Note that the flow on $\mathbb{R}^4/\mathbb{Z}_2$ is constant, but each time-slice is a metric cone, and therefore invariant

7 Euclidean space $\mathbb{R}^n$ equipped with $f = \frac{1}{4}r^2$ is called a trivial GSS.

8 The flows are defined on noncompact manifolds, but the geometry at infinity is well controlled.

under rescaling. So we may also view this model as a (degenerate) gradient shrinking soliton (in this case $f = \frac{1}{4}r^2$).

It is conceivable that there are Ricci flow singularities whose only blow-up models are of type (1) or (2). In addition, there are further examples in higher dimensions [49] whose only blow-up models that are GSSs must be singular and possibly degenerate. This motivates the following revision of Conjecture 5.1.

Conjecture 5.2. *For any Ricci flow, "most" singularity models are gradient shrinking solitons. These may be degenerate and may have a singular set of codimension* ≥ 4.

5.3. A compactness and partial regularity theory for Ricci flows

The previous example suggests that in higher dimensions we may need to consider *nonsmooth* blow-up limits. The usual convergence and compactness theory of Ricci flows due to Hamilton (see Section 4.2) is too restrictive for such purposes, as it relies on curvature bounds and only produces smooth limits. Instead, we need a fundamentally new *compactness and partial regularity theory* for Ricci flows, which will enable us to take limits of arbitrary Ricci flows and study their structural properties. This theory was recently found by the author [5–7] and will lie at the heart of a resolution of Conjecture 5.2.

An important related compactness and partial regularity theory is that for Einstein metrics due to Cheeger, Colding, Gromov, Naber, and Tian [17–23, 27]. This theory roughly states that any noncollapsed sequence of pointed Einstein metrics subsequentially converges in the pointed Gromov–Hausdorff sense to a metric space whose singular set has Minkowski dimension $\leq n - 4$. Similar theories also exist for other geometric equations (e.g., minimal surfaces, harmonic maps, mean curvature flow). What these theories have in common is that their proofs all rely on only a few basic ingredients (e.g., a monotonicity formula, an almost cone rigidity theorem, and an ε-regularity theorem), which can be verified in each setting. A similar theory for Ricci flows, however, is more complicated, mainly due to two reasons:

- The basic ingredients mentioned above are – at least *a priori* – not available for Ricci flows. This necessitates a different approach for proving partial regularity.
- Parabolic versions of notions like "metric space", "Gromov–Hausdorff convergence", etc., did not exist until recently. So these – and a theory surrounding them – first had to be developed.

Let us now state the main compactness and partial regularity results for Ricci flows. We will remain somewhat vague on the new terminologies for now and defer a more detailed discussion to Section 5.5. Consider a sequence of pointed, n-dimensional Ricci flows

$$\big(M_i, (g_i(t))_{t\in(-T_i,0]}, (x_i, 0)\big),$$

where we imagine the basepoints $(x_i, 0)$ to live in the final time-slices, and suppose that $T_\infty := \lim_{i\to\infty} T_i > 0$. Then we have:

Theorem 5.3. *After passing to a subsequence, these flows $\mathbb{F}$-converge to a pointed metric flow*

$$\big(M_i, \big(g_i(t)\big), (x_i, 0)\big) \xrightarrow[i\to\infty]{\mathbb{F}} \big(\mathcal{X}, (\nu_{x_\infty;t})\big).$$

Here the terms "metric flow" and "$\mathbb{F}$-convergence" can be thought as a parabolic versions of "metric space" and "Gromov–Hausdorff convergence," respectively.

Next, we impose the following noncollapsing condition:

$$\mathcal{N}_{x_i,0}\big(r_*^2\big) \geq -Y_*. \tag{5.1}$$

Here $\mathcal{N}_{x,t}(r^2)$ is the pointed Nash-entropy, which is a natural quantity in Ricci flow and related to Perelman's $\mathcal{W}$-functional and rediscovered by work of Hein and Naber [36]. It and can be thought of as the parabolic analogue of the normalized volume of a ball.

Theorem 5.4. *Assuming* (5.1)*, we have a regular–singular decomposition*

$$\mathcal{X} = \mathcal{R} \uplus \mathcal{S}$$

such that:

(1) *The flow on $\mathcal{R}$ can be described by a smooth Ricci flow spacetime structure (see Definition 4.7). The entire flow $\mathcal{X}$ is uniquely determined by this structure.*

(2) *We have the following dimensional estimate on the singular set:*

$$\dim_{\mathcal{M}^*} \mathcal{S} \leq (n+2) - 4.$$

(3) *Tangent flows (i.e., blow-ups based at a fixed point of $\mathcal{X}$) are (possibly singular) gradient shrinking solitons.*

(4) *There is a filtration $\mathcal{S}^0 \subset \cdots \subset \mathcal{S}^{n-2} = \mathcal{S}$ such that $\dim_{\mathcal{H}^*} \mathcal{S}^k \leq k$ and such that every $x \in \mathcal{S}^k \setminus \mathcal{S}^{k-1}$ has a tangent cone that either splits off an $\mathbb{R}^k$-factor or it splits off an $\mathbb{R}^{k-2}$-factor and is static.*

Let us make a few remarks. First, note that the fact that $\mathcal{X}$ is uniquely determined by the smooth Ricci flow spacetime structure on $\mathcal{R}$ is comparable to what we have observed in dimension 3 (see Section 4.5), where we did not even *consider* the entire flow $\mathcal{X}$.

Second, property (2) involves a parabolic version of the Minkowski dimension that is natural for Ricci flows; a precise definition would be beyond the scope of this article. Note that the time direction accounts for 2 dimensions, which is natural. In dimension 3, this implies that the set of singular times has dimension $\leq \frac{1}{2}$; this what was previously known in this dimension [40]. In Appleton's 4-dimensional example, the singular set $\mathcal{S}$ may consist of an isolated orbifold point in every time-slice; so its parabolic dimension is $2 = (4+2) - 4$. On the other hand, a flow on $S^2 \times T^2$ develops a singularity at a single time and collapses to the 2-torus T^2, which again has parabolic dimension 2. This shows that the dimensional bounds in Theorem 5.4 are optimal.

Lastly, the "tangent flows" in property (3) can be viewed as parabolic versions of "tangent cones," as both are invariant under rescaling.

5.4. Applications

Theorems 5.3 and 5.4 enable us to study the finite-time singularity formation and long-time behavior of Ricci flows in higher dimensions.

Regarding Conjecture 5.2, we roughly obtain:

Theorem 5.5. *Suppose that $(M, (g(t))_{t\in[0,T)})$ develops a singularity at time $T < \infty$. Then we can extend this flow by a "singular time-T-slice" (M_T, d_T) such that the tangent flows at any $(x, T) \in M_T$ are (possibly singular) gradient shrinking solitons.*

Regarding the long-time asymptotics, we obtain the following picture, which closely resembles that in dimension 3; compare with (4.4) in Section 4.4:

Theorem 5.6. *Suppose that $(M, (g(t))_{t\geq 0})$ is immortal. Then for $t \gg 1$ we have a thick–thin decomposition*

$$M = M_{\text{thick}}(t) \mathbin{\dot\cup} M_{\text{thin}}(t)$$

such that the flow on $M_{\text{thick}}(t)$ converges, after rescaling, to a singular Einstein metric ($\mathrm{Ric}_{g_\infty} = -g_\infty$) *and the flow on $M_{\text{thin}}(t)$ is collapsed in the opposite sense of* (5.1).

In dimension 3, these theorems essentially recover Perelman's results; so they can be seen as generalizations to higher dimensions.

5.5. Metric flows

The definition of a metric flow and associated concepts require a new perspective on the geometry of Ricci flows. In the following we will briefly convey some of the rough ideas behind this perspective.

Let us first imitate the process of passing from a (smooth) Riemannian manifold (M, g) to its metric length space (M, d_g). So our goal will be to turn a Ricci flow $(M, (g(t))_{t\in I})$ into a synthetic object, which we call "metric flow." To do this, we consider the spacetime $\mathcal{X} := X \times I$ and the time-slices $\mathcal{X}_t := X \times \{t\}$ equipped with the length metrics $d_t := d_{g(t)}$. It may be tempting to retain the product structure $X \times I$ on $\mathcal{X}$, i.e., to record the set of worldlines $t \mapsto (x, t)$. However, this turns out to be unnatural. Instead, we will view the time-slices $(\mathcal{X}_t, d_t)$ as separate metric spaces whose points may not even be in 1–1 correspondence to some given space X.

It remains to record some relation between these metric spaces $(\mathcal{X}_t, d_t)$. This will be done via the conjugate heat kernel $K(x, t; y, s)$ – an important object in the study of Ricci flows. For fixed $(x, t) \in M \times I$ and $s < t$, this kernel satisfies the backwards conjugate[9] heat equation on a Ricci flow background,

$$(-\partial_s - \Delta_{g(s)} + R_{g(s)})K(x, t; \cdot, s) = 0, \tag{5.2}$$

9 Equation (5.2) is the L^2-conjugate of the standard (forward) heat equation and $K(\cdot, \cdot; y, t)$ is a heat kernel centered at (y, t).

centered at (x, t). This kernel has the property that, for any (x, t) and $s < t$,

$$\int_M K(x,t;\cdot,s)dg(s) = 1,$$

which motivates the definition of the following probability measures:

$$d\nu_{(x,t);s} := K(x,t;\cdot,s)dg(s), \quad \nu_{(x,t);t} = \delta_x.$$

This is the additional information that we will record. So we define:

Definition 5.7. A *metric flow* is (essentially[10]) given by a pair

$$\big((\mathcal{X}_t, d_t)_{t\in I}, (\nu_{x;s})_{x\in\mathcal{X}_t, s<t, s\in I}\big)$$

consisting of a family of metric spaces $(\mathcal{X}_t, d_t)$ and probability measures $\nu_{x;s}$ on $\mathcal{X}_s$, which satisfy certain (basic) compatibility relations.

So given points $x \in \mathcal{X}_t$, $y \in \mathcal{X}_s$ at two times $s < t$, it is not possible to say whether "y corresponds to x." Instead, we only know that "y belongs to the past of x with a probability density of $d\nu_{x;s}(y)$." This definition is surprisingly fruitful. For example, it is possible to use the measures $\nu_{x;s}$ to define a natural topology on $\mathcal{X}$ and to understand when and in what sense the geometry of time-slices $\mathcal{X}_t$ depends continuously on t.

The concept of metric flows also allows the definition of a natural notion of geometric convergence – $\mathbb{F}$-convergence – which is similar to Gromov–Hausdorff convergence. Even better, this notion can be phrased on terms of a certain $d_{\mathbb{F}}$-distance, which is similar to the Gromov–Hausdorff distance, and the Compactness Theorem 5.3 can be expressed as a statement on the compactness of a certain subset of metric flow (pairs),[11] similar to the definition of Gromov–Hausdorff compactness.

Lastly, we sketch an example that illustrates why it was so important that we have divorced ourselves from the concept of worldlines. Consider the Bryant soliton $(M_{\mathrm{Bry}}, (g_{\mathrm{Bry}}(t))_{t\le 0})$ (see Figure 6). Recall that every time-slice $(M_{\mathrm{Bry}}, g_{\mathrm{Bry}}(t))$ is isometric to the same rotationally symmetric model with center x_{Bry}. By Theorems 5.3 and 5.4, any pointed sequence of blow-*downs* ($\lambda_i \to 0$),

$$\big(M_{\mathrm{Bry}}, (\lambda_i^2 g_{\mathrm{Bry}}(\lambda_i^{-2}t))_{t\le 0}, (x_{\mathrm{Bry}}, 0)\big),$$

$\mathbb{F}$-converges to a pointed metric flow $\mathcal{X}$ that is regular on a large set. What is this $\mathbb{F}$-limit $\mathcal{X}$? For any fixed time $t < 0$, the sequence of pointed Riemannian manifolds $(M_{\mathrm{Bry}}, \lambda_i^2 g_{\mathrm{Bry}}(\lambda_i^{-2}t), x_{\mathrm{Bry}})$ converges to a pointed ray of the form $([0,\infty), 0)$. This seems to contradict Theorem 5.4. However, here we have implicitly used the concept of worldlines, because we have used the point (x_{Bry}, t) corresponding to the "official" basepoint $(x_{\mathrm{Bry}}, 0)$ at time t. Instead, we have to focus on the "past" of $(x_{\mathrm{Bry}}, 0)$, i.e., the region of $(M_{\mathrm{Bry}}, \lambda_i^2 g_{\mathrm{Bry}}(\lambda_i^{-2}t))$ where the conjugate heat kernel $\nu_{(x_{\mathrm{Bry}},0);\lambda_i^{-2}t}$ is concentrated. This region is cylindrical of

10 This is a simplified definition.

11 Strictly speaking, $\mathbb{F}$-convergence and $d_{\mathbb{F}}$-distance concern metric flow *pairs*, $(\mathcal{X}, (\nu_{x;t}))$, where the second entry serves as some kind of substitute of a basedpoint.

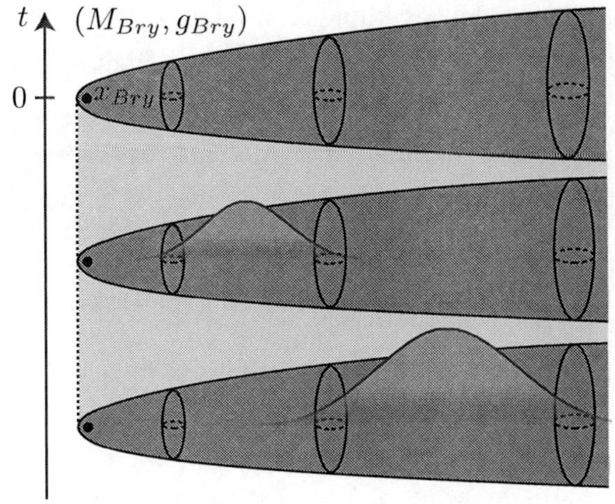

FIGURE 6
The Bryant soliton (green) and a conjugate (backward) heat kernel (orange) starting at the point x_{Bry} at time 0.

scale $\sim \sqrt{|t|}$, because the conjugate heat kernel "drifts away from the tip" at an approximate linear rate. In fact, one can show that the blow-down limit $\mathcal{X}$ is isometric to a round shrinking cylinder that develops a singularity at time 0. While this may seem slightly less intuitive at first, it turns out to be a much more natural way of looking at it.

5.6. Outlook

Our new theory of higher-dimensional Ricci flows demonstrates that, at least on an analytical level, Ricci flows behave similarly in higher dimension as they do in dimension 3. However, while there are only a handful of possible singularity models in dimension 3, gaining a full understanding of all such models in higher dimensions (e.g., classifying gradient shrinking solitons) may be impossible. Some past work in dimension 4 (e.g., by Munteanu and Wang **[43–45]**) has demonstrated that most *noncompact* gradient shrinking solitons have ends that are either cylindrical or conical. This motivates the following conjecture:

Conjecture 5.8. *Given a closed Riemannian 4-manifold (M, g), there is a "Ricci flow through singularities" in which topological change occurs along cylinders or cones and in which time-slices are allowed to have isolated orbifold singularities.*

The term "Ricci flow through singularities" is still left somewhat vague. Most likely, it should denote an object that is similar to a metric flow and that has the same partial regularity properties as described in Theorem 5.4, but with the exception that time-slices may consist of several components (i.e., we allow distances to be infinite). It may also be useful to require some sort of topological monotonicity property, meaning that the topology becomes "simpler" after the resolution of a singularity.

The existence of such a flow may have interesting consequences. For example, it may be used to decompose 4-manifolds with positive scalar curvature into certain building blocks. It may also offer an approach to proving the $\frac{11}{8}$-Conjecture. Note here that this conjecture holds for both important asymptotic models – gradient shrinking solitons and Einstein metrics – due to Lichnerowicz' Theorem and the Hitchin–Thorpe inequality. Lastly, there also seems to be potential applications in Kähler geometry, for example towards the Minimal Model Program and the Abundance Conjecture, assuming a similar flow could be constructed in higher dimensions.

ACKNOWLEDGMENTS

The author thanks Robert Bamler, Paula Burkhardt-Guim, Bennett Chow, Bruce Kleiner, Yi Lai, John Lott, and the anonymous referees for helpful feedback on an early version of the manuscript.

FUNDING

The author was supported by NSF grant DMS-1906500.

REFERENCES

[1] M. T. Anderson, Geometrization of 3-manifolds via the Ricci flow. *Notices Amer. Math. Soc.* **51** (2004), no. 2, 184–193.

[2] S. B. Angenent, J. Isenberg, and D. Knopf, Degenerate neckpinches in Ricci flow. *J. Reine Angew. Math.* **709** (2015), 81–117.

[3] S. Angenent and D. Knopf, An example of neckpinching for Ricci flow on S^{n+1}. *Math. Res. Lett.* **11** (2004), no. 4, 493–518.

[4] A. Appleton, Eguchi–Hanson singularities in $U(2)$-invariant Ricci flow. 2019, arXiv:1903.09936.

[5] R. H. Bamler, Compactness theory of the space of super Ricci flows. 2020, arXiv:2008.09298.

[6] R. H. Bamler, Entropy and heat kernel bounds on a Ricci flow background. 2020, arXiv:2008.07093.

[7] R. H. Bamler, Structure theory of non-collapsed limits of Ricci flows. 2020, arXiv:2009.03243.

[8] R. H. Bamler, Recent developments in Ricci flows. *Notices Amer. Math. Soc.* **68** (2021), no. 9, 1486–1498.

[9] R. H. Bamler and B. Kleiner, Ricci flow and diffeomorphism groups of 3-manifolds. 2017, arXiv:1712.06197.

[10] R. H. Bamler and B. Kleiner, Uniqueness and stability of Ricci flow through singularities. *Acta Math.* (to appear), 2017, arXiv:1709.04122.

[11] R. H. Bamler and B. Kleiner, Ricci flow and contractibility of spaces of metrics. 2019, arXiv:1909.08710.

[12] R. H. Bamler and B. Kleiner, Diffeomorphism groups of prime 3-manifolds. 2021, arXiv:2108.03302.

[13] K. A. Brakke, *The motion of a surface by its mean curvature.* Math. Notes 20, Princeton University Press, Princeton, NJ, 1978.

[14] S. Brendle, Ancient solutions to the Ricci flow in dimension 3. *Acta Math.* **225** (2020), no. 1, 1–102.

[15] S. Brendle, P. Daskalopoulos, and N. Sesum, Uniqueness of compact ancient solutions to three-dimensional Ricci flow. *Invent. Math.* **226** (2020), no. 2, 579–651.

[16] R. Bryant, Ricci flow solitons in dimension three with SO(3)-symmetries. 2005. http://www.math.duke.edu/~bryant/3DRotSymRicciSolitons.pdf

[17] J. Cheeger and T. H. Colding, Lower bounds on Ricci curvature and the almost rigidity of warped products. *Ann. of Math. (2)* **144** (1996), no. 1, 189–237.

[18] J. Cheeger and T. H. Colding, On the structure of spaces with Ricci curvature bounded below. I. *J. Differential Geom.* **46** (1997), no. 3, 406–480.

[19] J. Cheeger and T. H. Colding, On the structure of spaces with Ricci curvature bounded below. II. *J. Differential Geom.* **54** (2000), no. 1, 13–35.

[20] J. Cheeger and T. H. Colding, On the structure of spaces with Ricci curvature bounded below. III. *J. Differential Geom.* **54** (2000), no. 1, 37–74.

[21] J. Cheeger, T. H. Colding, and G. Tian, On the singularities of spaces with bounded Ricci curvature. *Geom. Funct. Anal.* **12** (2002), no. 5, 873–914.

[22] J. Cheeger and A. Naber, Lower bounds on Ricci curvature and quantitative behavior of singular sets. *Invent. Math.* **191** (2013), no. 2, 321–339.

[23] J. Cheeger and A. Naber, Regularity of Einstein manifolds and the codimension 4 conjecture. *Ann. of Math.* (2015), 1093–1165.

[24] X. Chen, P. Lu, and G. Tian, A note on uniformization of Riemann surfaces by Ricci flow. *Proc. Amer. Math. Soc.* **134** (2006), no. 11, 3391–3393.

[25] Y. G. Chen, Y. Giga, and S. Goto, Uniqueness and existence of viscosity solutions of generalized mean curvature flow equations. *J. Differential Geom.* **33** (1991), no. 3, 749–786.

[26] B. Chow, The Ricci flow on the 2-sphere. *J. Differential Geom.* **33** (1991), no. 2, 325–334.

[27] T. H. Colding, Ricci curvature and volume convergence. *Ann. of Math. (2)* **145** (1997), no. 3, 477–501.

[28] L. C. Evans and J. Spruck, Motion of level sets by mean curvature. I. *J. Differential Geom.* **33** (1991), no. 3, 635–681.

[29] M. Feldman, T. Ilmanen, and D. Knopf, Rotationally symmetric shrinking and expanding gradient Kähler–Ricci solitons. *J. Differential Geom.* **65** (2003), no. 2, 169–209.

[30] D. Gabai, The Smale conjecture for hyperbolic 3-manifolds: $\mathrm{Isom}(M^3) \simeq \mathrm{Diff}(M^3)$. *J. Differential Geom.* **58** (2001), no. 1, 113–149.

[31] R. S. Hamilton, Three-manifolds with positive Ricci curvature. *J. Differential Geom.* **17** (1982), no. 2, 255–306.

[32] R. S. Hamilton, The Ricci flow on surfaces. In *Mathematics and general relativity (Santa Cruz, CA, 1986)*, pp. 237–262, Contemp. Math. 71, Amer. Math. Soc., Providence, RI, 1988.

[33] R. S. Hamilton, A compactness property for solutions of the Ricci flow. *Amer. J. Math.* **117** (1995), no. 3, 545–572.

[34] A. E. Hatcher, A proof of the Smale conjecture, $\mathrm{Diff}(S^3) \simeq \mathrm{O}(4)$. *Ann. of Math. (2)* **117** (1983), no. 3, 553–607.

[35] J. Head, On the mean curvature evolution of two-convex hypersurfaces. *J. Differential Geom.* **94** (2013), no. 2, 241–266.

[36] H.-J. Hein and A. Naber, New logarithmic Sobolev inequalities and an ϵ-regularity theorem for the Ricci flow. *Comm. Pure Appl. Math.* **67** (2014), no. 9, 1543–1561.

[37] S. Hong, J. Kalliongis, D. McCullough, and J. H. Rubinstein, *Diffeomorphisms of elliptic 3-manifolds*. Lecture Notes in Math. 2055, Springer, Heidelberg, 2012.

[38] T. Ilmanen, Elliptic regularization and partial regularity for motion by mean curvature. *Mem. Amer. Math. Soc.* **108** (1994), no. 520, x+90.

[39] B. Kleiner and J. Lott, Singular Ricci flows I. *Acta Math.* **219** (2017), no. 1, 65–134.

[40] B. Kleiner and J. Lott, Singular Ricci flows II. In *Geometric analysis*, pp. 137–155, Progr. Math. 333, Birkhäuser/Springer, Cham, 2020.

[41] J. Lauer, Convergence of mean curvature flows with surgery. *Comm. Anal. Geom.* **21** (2013), no. 2, 355–363.

[42] F. C. Marques, Deforming three-manifolds with positive scalar curvature. *Ann. of Math. (2)* **176** (2012), no. 2, 815–863.

[43] O. Munteanu and J. Wang, Geometry of shrinking Ricci solitons. *Compos. Math.* **151** (2015), no. 12, 2273–2300.

[44] O. Munteanu and J. Wang, Conical structure for shrinking Ricci solitons. *J. Eur. Math. Soc. (JEMS)* **19** (2017), no. 11, 3377–3390.

[45] O. Munteanu and J. Wang, Structure at infinity for shrinking Ricci solitons. *Ann. Sci. Éc. Norm. Supér. (4)* **52** (2019), no. 4, 891–925.

[46] G. Perelman, The entropy formula for the Ricci flow and its geometric applications. 2002, arXiv:math/0211159.

[47] G. Perelman, Finite extinction time for the solutions to the Ricci flow on certain three-manifolds. 2003, arXiv:math/0307245.

[48] G. Perelman, Ricci flow with surgery on three-manifolds. 2003, arXiv:math/0303109.

[49] M. Stolarski, Curvature blow-up in doubly-warped product metrics evolving by Ricci flow. 2019, arXiv:1905.00087.

[50] B. White, Evolution of curves and surfaces by mean curvature. In *Proceedings of the International Congress of Mathematicians, Vol. I (Beijing, 2002)*, pp. 525–538, Higher Ed. Press, Beijing, 2002.

[51] B. White, The nature of singularities in mean curvature flow of mean-convex sets. *J. Amer. Math. Soc.* **16** (2003), no. 1, 123–138 (electronic).

RICHARD H. BAMLER

Department of Mathematics, University of California, Berkeley, Berkeley, CA 94720, USA, rbamler@berkeley.edu

EMERGENT COMPLEX GEOMETRY

ROBERT J. BERMAN

ABSTRACT

This is a double exposure of the probabilistic construction of Kähler–Einstein metrics on a complex projective algebraic variety X – where the Kähler–Einstein metric emerges from a canonical random point process on X – and the variational approach to the Yau–Tian–Donaldson conjecture, highlighting their connections. The final section is a report on joint work in progress with Sébastien Boucksom and Mattias Jonsson on how the non-Archimedean geometry of X (with respect to the trivial absolute value) also emerges from the probabilistic framework.

MATHEMATICS SUBJECT CLASSIFICATION 2020

Primary 53C55; Secondary 58E11, 60G55

ICM INTERNATIONAL CONGRESS OF MATHEMATICIANS 2022 JULY 6–14

Proc. Int. Cong. Math. 2022, Vol. 4, pp. 2456–2483
DOI 10.4171/ICM2022/59

1. INTRODUCTION

A recurrent theme in geometry is the quest for canonical metrics on a given manifold X. The prototypical case is when X is a compact orientable two-dimensional surface, which can be endowed with a metric of constant scalar curvature, essentially uniquely determined by a complex structure J on X. On the other hand, from a physical point of view, geometrical shapes – as we know them from everyday experience – are, of course, not fundamental physical entities. They merely arise as macroscopic *emergent* features of ensembles of microscopic point particles in the limit as the number N of particles tends to infinity. In mathematical terms such microscopical ensembles are *random point processes*, i.e., they are represented by a probability measure on the configuration space of N points on X or, equivalently, a symmetric probability measure $\mu^{(N)}$ on the N-fold product X^N. One is thus led to ask whether a given manifold X may be endowed with a canonical random point process – defined without reference to any metric – from which a canonical metric g emerges as $N \to \infty$? Here we shall focus on Kähler metrics with constant Ricci curvature. From the physics perspective, these arise as solutions to Einstein's equations in vacuum (with Euclidean signature). The Kähler condition means that X is compatible with an integrable complex structure J on X (in that parallel translation preserves the complex structure J). Such metrics – known as *Kähler–Einstein metrics* – play a central role in current complex geometry and the study of complex algebraic varieties, in particular in the context of the Yau–Tian–Donaldson conjecture [38] and the Minimal Model Program in birational algebraic geometry [45]. When a projective algebraic variety X admits a Kähler–Einstein metric, it is essentially unique, i.e., canonically attached to X and can thus be leveraged to probe X using differential-geometric techniques (as, for example, in the construction of moduli spaces [61]).

One virtue of the probabilistic approach is that it leads to essentially explicit period type integral formulas for canonical Kähler metrics converging towards the Kähler–Einstein metric as $N \to \infty$ (see formula (2.7)). These formulas are reminiscent of the few explicit formulas for Kähler–Einstein metrics that are available on special complex curves, involving hypergeometric integrals (notably the modular curve, the Klein curve, and Fermat curves; see [6, SECTION 2.1]). The probabilistic approach also generates new connections between Kähler geometry and algebraic geometry in the context of the Yau–Tian–Donaldson conjecture on Fano varieties, through the concept of Gibbs stability and the related stability threshold (δ-invariant) [19, 41]. The present contribution to the 2022 ICM proceedings attempts a double exposure of the probabilistic approach in [2, 4, 5] and the variational approach to the Yau–Tian–Donaldson conjecture in [14], highlighting their connections. For more details and background, the reader is referred to the survey [6]. See also [15] for connections between the present probabilistic approach to Kähler geometry and quantum gravity in the context of the AdS/CFT correspondence, and [7, 39] for connections to polynomial approximation theory and pluripotential theory in $\mathbb{C}^n$.

2. EMERGENT KÄHLER GEOMETRY

Let X be a compact complex manifold, whose dimension over $\mathbb{C}$ will be denoted by n. The existence of a Kähler–Einstein metric ω_{KE} on X, i.e., a Kähler metric with constant Ricci curvature,

$$\operatorname{Ric}\omega = -\beta\omega, \tag{2.1}$$

implies that the *canonical line bundle* K_X of X (the top exterior power of the cotangent bundle of X) has a definite sign, when $\beta \neq 0$,

$$\operatorname{sign}(K_X) = \operatorname{sign}(\beta). \tag{2.2}$$

We will be using the standard terminology of positivity in complex geometry: a line bundle is said to be *positive*, $L > 0$, if L carries some Hermitian metric with strictly positive curvature (or, equivalently, L is ample in the algebro-geometric sense). The standard additive notation for tensor products of line bundles will be adopted. Accordingly, the dual of L is expressed as $-L$, and L is thus said be negative, $L < 0$, if $-L > 0$. In general, when $\beta \neq 0$, the manifold X is automatically a complex projective algebraic manifold, and after a rescaling of the Kähler–Einstein metric we may as well assume that $\beta = \pm 1$. For example, in the case when X is a hypersurface in $\mathbb{P}^{n+1}_{\mathbb{C}}$, cut out by a homogeneous polynomial of degree d, $K_X > 0$ when $d > n+2$, and $-K_X > 0$ when $d < n+2$.

Remark 2.1. In the more general "logarithmic" setup, X is replaced by a *log pair* (X, Δ) consisting of a $\mathbb{Q}$-divisor Δ on a normal variety X and K_X is replaced by $K_X + \Delta$, assumed to be a $\mathbb{Q}$-line bundle. The corresponding log Kähler–Einstein equation (2.1) is obtained by replacing $\operatorname{Ric}\omega$ with $\operatorname{Ric}\omega - [\Delta]$, where $[\Delta]$ denotes the current of integration corresponding to Δ. For simplicity we will stick to the case when X is nonsingular and Δ is trivial (but all the results surveyed in this and the following section generalize to the logarithmic setting, assuming that (X, Δ) is klt (Kawamata log terminal) [5, 8, 13]).

Coming back to the question of emergence of geometry, discussed in the introduction, a Kähler–Einstein metric g_{KE} has the crucial property that it can be readily recovered from its volume form dV_{KE}, in the case $\beta \neq 0$. Indeed, in local terms g_{KE} is proportional to the complex Hessian of the logarithm of the local density of dV_{KE} (see formula (3.4)). Thus in order to probalistically construct the Kähler–Einstein metric, one just needs to construct a random point process on X with N particles such that the empirical measure

$$\delta_N := \frac{1}{N}\sum_{i=1}^{N}\delta_{x_i}, \tag{2.3}$$

viewed as a random discrete probability measure on X, converges in probability to dV_{KE}, as $N \to \infty$.

2.1. The case $K_X > 0$ ($\beta = 1$)

The starting point for the probabilistic approach is the observation that there is a canonical symmetric probability measure $\mu^{(N)}$ on the N-fold product X^N of X. More precisely, the integers N are taken to be of the special form

$$N = N_k := \dim_{\mathbb{C}} H^0(X, kK_X),$$

where $H^0(X, kK_X)$ denotes the complex vector space of all holomorphic section of the kth tensor power of the canonical line bundle $K_X \to X$. Recall that the elements $s^{(k)}$ of $H^0(X, kK_X)$ are called pluricanonical forms and may be represented by local holomorphic functions transforming as $dz^{\otimes k}$, in terms of local holomorphic coordinates $z \in \mathbb{C}^n$ on X. As a consequence, $|s^{(k)}(z)|^{2/k}$ transforms as a local density on X and thus defines a global measure on X. Replacing X with X^{N_k}, the canonical symmetric probability measure $\mu^{(N_k)}$ on X^{N_k} is now defined by

$$\mu^{(N_k)} = \frac{1}{\mathcal{Z}_{N_k}} |\det S^{(k)}|^{2/k}, \quad \mathcal{Z}_{N_k} := \int_{X^{N_k}} |\det S^{(k)}|^{2/k}, \tag{2.4}$$

where $\det S^{(k)}$ is the holomorphic section of the line bundle $(kK_{X^{N_k}}) \to X^{N_k}$, expressed as the Slater determinant

$$(\det S^{(k)})(x_1, x_2, \dots, x_N) := \det(s_i^{(k)}(x_j)), \tag{2.5}$$

in terms of a given basis $s_i^{(k)}$ in $H^0(X, kK_X)$. Under a change of bases, the section $\det S^{(k)}$ only changes by a multiplicative complex constant (the determinant of the change of bases matrix on $H^0(X, kK_X)$). As a consequence, $\mu^{(N_k)}$ is independent of the choice of bases in $H^0(X, kK_X)$ and, since $\det S^{(k)}$ is antisymmetric, this means that the probability measure $\mu^{(N_k)}$ indeed defines a canonical symmetric probability measure on X^{N_k}. Moreover, it is completely encoded by algebro-geometric data in the following sense: realizing X as a projective algebraic subvariety, the section $\det S^{(k)}$ can be identified with a homogeneous polynomial, determined by the coordinate ring of X.

The assumption that $K_X > 0$ ensures that $N_k \to \infty$ as $k \to \infty$. To simplify the notation, we will often drop the subindex k on N_k and consider the large-N limit. The following convergence result was shown in [4]:

Theorem 2.2. *Let X be a compact complex manifold with positive canonical line bundle K_X. Then the empirical measures δ_N of the corresponding canonical random point processes on X (formula (2.3)) converge in probability, as $N \to \infty$, towards the normalized volume form dV_{KE} of the unique Kähler–Einstein metric ω_{KE} on X.*

In fact, the proof shows that the convergence holds at an exponential rate, in the sense of large deviation theory: for any given $\varepsilon > 0$, there exists a positive constant C_ε such that

$$\mathrm{Prob}\left(d\left(\frac{1}{N} \sum_{i=1}^{N} \delta_{x_i}, dV_{\mathrm{KE}} \right) > \varepsilon \right) \le C_\varepsilon e^{-N\varepsilon}, \tag{2.6}$$

where d denotes any metric on the space $\mathcal{P}(X)$ of probability measures on X compatible with the weak topology. The convergence in probability in the previous theorem implies, in particular, that the measures dV_k on X, defined by the expectations $\mathbb{E}(\delta_{N_k})$ of the empirical measure δ_{N_k}, converge to dV_{KE} in the weak topology of measures on X. Concretely, dV_k

is obtained by integrating $\mu^{(N_k)}$ over the fibers of the projection from X^{N_k} onto the first factor X, that is,

$$dV_k := \int_{X^{N_k-1}} \mu^{(N_k)} \to dV_{\mathrm{KE}}, \quad k \to \infty.$$

For k sufficiently large (ensuring that kK_X is very ample), the measures dV_k are, in fact, volume forms on X and induce a sequence of canonical Kähler metrics ω_k on X [5, PROP. 5.3]:

$$\omega_k := \frac{i}{2\pi}\partial\bar{\partial}\log dV_k = \frac{i}{2\pi}\partial\bar{\partial}\log \int_{X^{N_k-1}} \left|\det S^{(k)}\right|^{2/k}. \tag{2.7}$$

The convergence above also implies that the canonical Kähler metrics ω_k converge, as $k \to \infty$, towards the Kähler–Einstein metric ω_{KE} on X, in the weak topology. More generally, as shown in [5], the convergence holds on any variety X of positive Kodaira dimension (i.e., such that $N_k \to \infty$, as $k \to \infty$) if dV_{KE} and ω_{KE} are replaced by the canonical measure and current on X, respectively, introduced by Song-Tian and Tsuji in different geometric contexts [5] (in the case when X is singular it is assumed that X is klt and k is assumed to be sufficently divisible to ensure that kK_X is a bona fide line bundle).

2.2. The Fano case, $K_X < 0$ ($\beta = -1$)

When $-K_X$ is positive, which means that X is a Fano manifold, any Kähler–Einstein metric on X has positive Ricci curvature. However, not all Fano manifolds X carry Kähler–Einstein metrics; according to the Yau–Tian–Donaldson conjecture (discussed in Section 4) a Fano manifold admits a Kähler–Einstein-metric if and only if X is K-polystable. In the probabilistic approach, a new type of stability assumption naturally appears, as is explained next. First note that when $-K_X > 0$ the spaces $\dim H^0(X, kK_X)$ are trivial for all positive integers k. On the other hand, the dimensions tend to infinity as $k \to -\infty$. Thus it is natural to replace k with $-k$ in the previous constructions. In particular, given a positive integer k, we set

$$N_k := \dim H^0(X, -kK_X)$$

and attempt to define a probability measure on X^{N_k} as

$$\mu^{(N_k)} := \frac{|\det S^{(k)}|^{-2/k}}{\mathcal{Z}_{N_k}}, \quad \mathcal{Z}_{N_k} := \int_{X^{N_k}} \left|\det S^{(k)}\right|^{-2/k},$$

where the numerator defines a measure on the complement in X^{N_k} of the zero-locus of $\det S^{(k)}$. However, it may happen that the normalizing constant $\mathcal{Z}_{N_k}$ diverges, since the integrand of $\mathcal{Z}_{N_k}$ blows-up along the zero-locus in X^{N_k} of $\det S^{(k)}$. Accordingly, a Fano manifold X is called *Gibbs stable at level* k if $\mathcal{Z}_{N_k} < \infty$ and *Gibbs stable* if it is Gibbs stable at level k for k sufficiently large. We thus arrive at the following probabilistic analog of the Yau–Tian–Donaldson conjecture posed in [5]:

Conjecture 2.3. *Let X be Fano manifold. Then*

- *X admits a unique Kähler–Einstein metric ω_{KE} if and only if X is Gibbs stable.*
- *If X is Gibbs stable, the empirical measures δ_N of the corresponding canonical point processes converge in probability to the normalized volume form of ω_{KE}.*

It should be stressed that the Gibbs stability of X implies that the group $\mathrm{Aut}(X)$ of automorphisms of X is finite [5, PROP. 6.5]. Accordingly, when comparing Conjecture 2.3 with the Yau–Tian–Donaldson conjecture, one should view Gibbs stability as the analog of K-stability. There is also a natural analog of the stronger notion of uniform K-stability [24, 36]. To see this, first note that Gibbs stability can be given a purely algebro-geometric formulation, saying that the $\mathbb{Q}$-divisor D_{N_k} in X^{N_k} cut out by the (multivalued) holomorphic section $(\det S^{(k)})^{1/k}$ of $-K_{X^{N_k}}$ has mild singularities in the sense of the Minimal Model Program [44]. More precisely, X is Gibbs stable at level k iff D_{N_k} is *klt (Kawamata log terminal)*. This means that the *log canonical threshold (lct)* of D_{N_k} satisfies $\mathrm{lct}(D_{N_k}) > 1$, as follows directly from the standard analytic representation of the log canonical threshold of a $\mathbb{Q}$-divisor as an integrability threshold [44]. Accordingly, X is called *uniformly Gibbs stable* if the there exists $\varepsilon > 0$ such that, for k sufficiently large, $\mathrm{lct}(D_{N_k}) > 1 + \varepsilon$. One is thus led to pose the following purely algebro-geometric conjecture:

Conjecture 2.4. *Let X be a Fano manifold. Then X is (uniformly) K-stable iff X is (uniformly) Gibbs stable.*

One direction of the uniform version of the previous conjecture was established in [40, 41], using techniques from the Minimal Model Program:

Theorem 2.5 ([41]). *Uniform Gibbs stability implies uniform K-stability.*

Let us briefly recall the elegant argument in [41], introducing the invariant $\delta(X)$, which has come to play a key role in recent developments around the Yau–Tian–Donaldson conjecture. First, by [41, THM. 2.5],

$$\mathrm{lct}(D_{N_k}) \leq \delta_k(X) := \inf_{\Delta_k} \mathrm{lct}(\Delta_k), \tag{2.8}$$

where the inf is taken over all anticanonical $\mathbb{Q}$-divisors Δ_k on X of k-basis type, i.e., Δ_k is the normalized sum of the N_k zero-divisors on X defined by the members of a given basis in $H^0(X, -kK_X)$. Finally, by [41, THM. 0.3], if the invariant $\delta(X)$ defined as

$$\delta(X) := \limsup_{k\to\infty} \delta_k(X) \tag{2.9}$$

satisfies $\delta(X) > 1$, then X is uniformly K-stable [40] and thus admits a unique Kähler–Einstein metric by the solution of the (uniform) Yau–Tian–Donaldson conjecture recalled in Section 4.2. In particular, this means that uniform Gibbs stability implies the existence of a Kähler–Einstein metrics (in line with Conjecture 2.3). For a direct analytic proof of this implication see [9]. However, the converse implication, that we shall come back to in Section 5, is still open. Anyhow, even if confirmed, it is a separate analytic problem to prove the convergence in Conjecture 2.3. "Tropicalized" analogs of Conjecture 2.3 are established on toric varieties in [18] and on tori in [43].

In [6] a variational approach to the convergence problem was introduced, further developed in [8], where the convergence was settled on log Fano curves. In the general case the approach yields, in particular, the following conditional convergence result:

Theorem 2.6 ([6,8]). *Let X be a Fano manifold and assume that X admits a Kähler–Einstein metric ω_{KE}. Take the basis $s_i^{(k)}$ in formula (2.5) to be orthonormal with respect to the Hermitian metric on $H^0(X, -kK_X)$ induced by ω_{KE} and assume that*

$$\lim_{N\to\infty} \frac{1}{N} \log \mathcal{Z}_N = 0. \tag{2.10}$$

Then $\mathrm{Aut}(X)$ *is finite and the empirical measures δ_N converge in probability to the normalized volume form dV_{KE} of the unique Kähler–Einstein-metric ω_{KE} on X.*

In [6] two different types of hypotheses were put forth, ensuring that the convergence (2.10) holds, one of which will be recalled in Section 2.3.1. The other assumes, in particular, that the partition function $\mathcal{Z}_N(\beta)$, discussed in the following section, is zero-free in some N-independent neighborhood Ω of $]-1,0]$ in $\mathbb{C}$ (when $\mathcal{Z}_N(\beta)$ is analytically continued to a holomorphic function on Ω). This allows one to "analytically continue" the convergence when $\beta > 0$ to $\beta < 0$. This is discussed in detail in [8], where some intriguing connections between this zero-free hypothesis and the zero-free property of the local L-functions appearing in the Langlands program are also pointed out.

2.3. The statistical mechanical formalism and outlines of the proofs

Theorem 2.2 (or more precisely, the exponential convergence in formula (2.6)) is deduced from a large deviation principle (LDP), which may be symbolically expressed as

$$\mathrm{Prob}\left(\frac{1}{N}\sum_{i=1}^{N}\delta_{x_i} \in B_\varepsilon(\mu)\right) \sim e^{-NR(\mu)}, \quad N\to\infty,\ \varepsilon\to 0, \tag{2.11}$$

where $B_\varepsilon(\mu)$ denotes the ball of radius ε centered at a given μ in the space $\mathcal{P}(X)$ of all probability measures on X, endowed with a metric d compatible with the weak topology. In probabilistic terminology, the functional $R(\mu)$ is called the *rate functional*. By general principles, any rate functional of an LDP is lower-semicontinuous and its infimum vanishes. In the present setup, the volume form dV_{KE} of the Kähler–Einstein metric is the unique minimizer of $R(\mu)$, which yields the exponential convergence in formula (2.6).

As next explained, the proof of the LDP is inspired by statistical mechanics. Fix a Kähler metric on X. It induces a volume form dV on X and a Hermitian metric $\|\cdot\|$ on K_X. The canonical probability measure (2.4) may then be decomposed as

$$\mu^{(N)} = \frac{1}{\mathcal{Z}_{N_k}} \|\det S^{(k)}\|^{2/k} dV^{\otimes N},$$

where the basis $s_i^{(k)}$ in formula (2.5) is taken to be orthonormal with respect to the Hermitian metric on $H^0(X, kK_X)$ induced by dV and $\|\cdot\|$. Introducing the *energy per particle* as

$$E^{(N)}(x_1,\dots,x_N) := -\frac{1}{kN}\log\left\|\det S^{(k)}(x_1,\dots,x_{N_k})\right\|^2, \tag{2.12}$$

we can thus express $\mu^{(N)}$ as the following *Gibbs measure*, at *inverse temperature* $\beta = 1$:

$$\mu_\beta^{(N)} = \frac{e^{-\beta N E^{(N)}}}{\mathcal{Z}_N(\beta)} dV^{\otimes N}, \quad \mathcal{Z}_N(\beta) := \int_{:X^N} e^{-\beta N E^{(N)}} dV^{\otimes N}. \tag{2.13}$$

In statistical mechanical terms, the Gibbs measures represent the microscopic thermal equilibrium state of N interacting identical particles on X. The normalizing constant $\mathcal{Z}_N(\beta)$ is called the *partition function.*

The starting point of the proof of the LDP (2.11) is a classical result of Sanov in probability, going back to Boltzmann, saying that in the "noninteracting case" $\beta = 0$ (where the positions x_i define independent random variables on X) the LDP holds with rate functional given by the entropy $\mathrm{Ent}(\mu)$ of μ relative to dV, i.e., the functional on $\mathcal{P}(X)$ defined by

$$\mathrm{Ent}(\mu) := \int_X \log\left(\frac{\mu}{dV}\right)\mu,$$

if μ is absolutely continuous with respect to dV, and otherwise $\mathrm{Ent}(\mu) := +\infty$.[1] The strategy to handle the "interacting case" $\beta \neq 0$ is to first show that there exists a functional $E(\mu)$ on $\mathcal{P}(X)$ such that the energy per particle, $E^{(N)}(x_1, \ldots, x_N)$, may be approximated as

$$E^{(N)}(x_1, \ldots, x_N) \to E(\mu), \tag{2.14}$$

when $\frac{1}{N}\sum_{i=1}^{N}\delta_{x_i} \to \mu$, in an appropriate sense, as $N \to \infty$. Formally combining this result with Sanov's LDP suggests that, for any $\beta > 0$, the corresponding rate functional is given by

$$R_\beta(\mu) = F_\beta(\mu) - \inf_{\mathcal{P}(X)} F_\beta, \quad F_\beta(\mu) = \beta E(\mu) + \mathrm{Ent}(\mu) \in]0, \infty], \tag{2.15}$$

In thermodynamical terms, the functional $F_\beta(\mu)$ is the *free energy,* at inverse temperature β (strictly speaking, it is $\beta^{-1}F_\beta$ which is the free energy, i.e., the energy that is free to do work once the disordered thermal energy has been subtracted). In the present setting the role of the "macroscopic" energy $E(\mu)$ is played by the *pluricomplex energy* of the measure μ (introduced in **[12]** and discussed in Section 3). Briefly, it is first shown in **[4]** that the convergence (2.14) holds in the sense of *Gamma-convergence*. This means that

$$\frac{1}{N_j}\sum_{i=1}^{N_j}\delta_{x_i} \to \mu \implies \liminf_{N_j \to \infty} E^{(N_j)}(x_1, \ldots, x_{N_j}) \geq E(\mu) \tag{2.16}$$

and, for any μ, there exists some sequence of configurations in X^N saturating the previous inequality. The Gamma-convergence is deduced from the convergence of weighted transfinite diameters established in **[11]** using a duality argument (where $E(\mu)$ arises as a Legendre–Fenchel transform; compare formula (3.12)). The combination with Sanov's theorem is then made rigorous using an effective submean inequality on small balls in the Riemannian orbifold X^N/S_N, established using geometric analysis.

The free energy functional F_β has a unique minimizer μ_β in $\mathcal{P}(X)$ for any $\beta > 0$ (as discussed in Section 3.3). As a consequence, the empirical measures δ_N converge in probability to μ_β, as $N \to \infty$. The LDP proved in **[4]** also implies that for $\beta > 0$,

$$\lim_{N\to\infty} -\frac{1}{N}\log \mathcal{Z}_N(\beta) = \inf_{\mathcal{P}(X)} F_\beta. \tag{2.17}$$

Incidentally, the free energy functional F_β on $\mathcal{P}(X)$ may be identified with the (twisted) Mabuchi functional in Kähler geometry, as explained in Section 3.4.

---

1 In the physics literature, the opposite sign convention for $\mathrm{Ent}(\mu)$ is used.

2.3.1. The case $\beta < 0$

The Gibbs measure $\mu_\beta^{(N)}$ can, alternatively, be viewed as a Gibbs measure at *unit* temperature, if $E^{(N)}$ is replaced with with the rescaled energy $\beta E^{(N)}$ (thus treating β as a coupling constant). For $\beta > 0$, this energy is *repulsive*, since it tends to ∞ as any two particle positions merge (due to the vanishing of the determinant $\det S^{(k)}(x_1, \dots, x_{N_k})$). However, when β changes sign, the rescaled energy $\beta E^{(N)}$ becomes *attractive*; it tends to $-\infty$ as any two points merge, which leads to subtle concentration phenomena and various new technical difficulties. For example, one reason that the proof of the LDP does not generalize to $\beta < 0$ is that the Gamma-convergence in formula (2.14) is not preserved when $E^{(N)}$ is replaced by $-E^{(N)}$. In order to bypass this difficulty, a variational approach was introduced in [6]. The starting point is the classical Gibbs variational principle, which yields

$$-\frac{1}{N}\log \mathcal{Z}_N(\beta) = \inf_{\mathcal{P}(X^N)} F_\beta^{(N)}, \quad F_\beta^{(N)}(\cdot) := \beta\langle E^{(N)}, \cdot\rangle + N^{-1}\mathrm{Ent}(\cdot), \tag{2.18}$$

where the functional $F_\beta^{(N)}$ on $\mathcal{P}(X^N)$ is called the *N-particle mean free energy* and $\mathrm{Ent}(\cdot)$ denotes the entropy relative to $dV^{\otimes N}$. When its infimum is finite, it is uniquely attained at the corresponding Gibbs measure $\mu_\beta^{(N)}$. In [6, 8] this variational formulation is leveraged to show that, if X admits a Kähler–Einstein metric dV_{KE}, then δ_N converge in probability to dV_{KE}, under the assumption that the convergence of the partition functions (2.17) holds at $\beta = -1$. In particular, when the fixed metric on X is taken to be a Kähler–Einstein metric, this proves Theorem 2.6, since $F_{-1}(dV_{\mathrm{KE}}) = 0$. Moreover, the convergence (2.17) of the partition functions at $\beta = -1$ is shown to be implied by the following hypothesis:

$$\lim_{N_j\to\infty} (\delta_{N_j})_*\mu_{-1}^{(N_j)} = \Gamma \in \mathcal{P}(\mathcal{P}(X)) \implies \limsup_{N_j\to\infty}\langle E^{(N_j)}, \mu_{-1}^{(N_j)}\rangle \leq \langle E, \Gamma\rangle, \tag{2.19}$$

where $(\delta_N)_*\mu_{-1}^{(N)}$ is the probability measure on the infinite-dimensional $\mathcal{P}(X)$, defined as the pushforward of the canonical probability measure $\mu_{-1}^{(N)}$ on X^N to $\mathcal{P}(X)$ under the map δ_N (the reversed inequality holds for *any* sequence μ_N in $\mathcal{P}(X^N)$, as follows from the inequality (2.16)). If the hypothesis holds, then it follows that Γ is the Dirac mass at dV_{KE}, which is equivalent to the convergence in Theorem 2.6. In fact, as shown in [8], the previous hypothesis is "almost" equivalent to the convergence in Conjecture 2.3.

Finally, we note that the conjectural extension of formula (2.17) to any $\beta < 0$ also suggests the following conjecture posed in [4] (the definition of the log canonical threshold $\mathrm{lct}(D_N)$ was discussed after Conjecture 2.3):

Conjecture 2.7. *For any Fano manifold X,*

$$\lim_{N\to\infty} \mathrm{lct}(D_N) = \Gamma(X), \quad \Gamma(X) := \sup_{\beta<0}\Big\{-\beta : \inf_{\mathcal{P}(X)} F_\beta > -\infty\Big\}. \tag{2.20}$$

3. THE THERMODYNAMICAL FORMALISM AND PLURIPOTENTIAL THEORY

The pluricomplex energy $E(\mu)$, appearing as the "energy part" of the free energy functional $F_\beta(\mu)$ in formula (2.15), may be defined as the greatest lower semicontinuous extension to the space $\mathcal{P}(X)$ of the functional whose first variation on the subspace of volume forms is given by

$$dE(\mu) = -u_\mu, \tag{3.1}$$

with $u_\mu \in C^\infty(X)$ denoting the solution to the following complex Monge–Ampère equation (known as the *Calabi–Yau equation*)

$$\mathrm{MA}(u) = \mu, \tag{3.2}$$

expressed in terms of the complex Monge–Ampère measure $\mathrm{MA}(u)$, whose definition we next recall.

3.1. Kähler geometry recap

Assume that we are given a line bundle L endowed with a Hermitian metric $\|\cdot\|$ (in the present setup, $L = \pm K_X$ and $\|\cdot\|$ is the metric on L induced by a fixed Kähler metric on X). Then any smooth function u on X induces a metric $\|\cdot\|e^{-u/2}$ on L, whose curvature form, multiplied by $i/2\pi$, will be denoted by ω_u; it is a real closed two-form on X, representing the first Chern class $c_1(L) \in H^2(X, \mathbb{Z})$ of L. Concretely,

$$\omega_u = \omega_0 + \frac{i}{2\pi}\partial\bar{\partial}u_\beta, \quad \partial\bar{\partial}u := \sum_{i,j\leq n} \frac{\partial^2 u}{\partial z_i \partial \bar{z}_i} dz_i \wedge d\bar{z}_j, \tag{3.3}$$

in terms of local holomorphic coordinates, where ω_0 is the normalized curvature form of the fixed metric $\|\cdot\|$ on L. The complex Monge–Ampère measure $\mathrm{MA}(u)$ is the normalized volume form on X defined by

$$\mathrm{MA}(u) := \omega_u^n / V, \quad V := \int_X \omega_u^n = \int_X \omega_0^n.$$

By the Calabi–Yau theorem, there exists a smooth solution u_μ to the Calabi–Yau equation (3.2), uniquely determined up to an additive constant. It has the property that ω_{u_μ} is a *Kähler form*. Recall that a J-invariant closed real form ω on X is said to be Kähler if $\omega > 0$ in the sense that the corresponding symmetric two-tensor

$$g := \omega(\cdot, J\cdot)$$

is positive definite, i.e., defines a Riemannian metric (where J denotes the complex structure on X). In practice, one then identifies the Kähler form ω with the corresponding *Kähler metric* g. Likewise, the Ricci curvature of a Kähler metric ω may be identified with the two-form

$$\mathrm{Ric}\,\omega = -\frac{i}{2\pi}\partial\bar{\partial}\log dV, \tag{3.4}$$

where dV denotes the volume form of ω. In other words, $\mathrm{Ric}\,\omega$ is the curvature of the metric on $-K_X$ induced by ω. If the Kähler form ω is of the form ω_u (as in formula (3.3)), then u is said to be a *Kähler potential* for ω (relative to ω_0). We will denote by $\mathcal{H}(X, \omega_0)$ the space of all Kähler potentials relative to ω_0, and by $\mathcal{H}(X, \omega_0)_0$ the subspace of all sup-normalized u, $\sup_X u = 0$. The map

$$u \mapsto \omega_u, \quad \mathcal{H}(X, \omega_0)_0 \hookrightarrow c_1(L)$$

yields a one-to-one correspondence between $\mathcal{H}(X,\omega_0)_0$ and the space of all Kähler forms in the first Chern class $c_1(L)$ of L. Similarly, the Calabi–Yau theorem yields the "Calabi–Yau correspondence"

$$u \mapsto \mathrm{MA}(u), \quad \mathcal{H}(X,\omega_0)_0 \hookrightarrow \mathcal{P}(X) \tag{3.5}$$

between $\mathcal{H}(X,\omega_0)_0$ and the space of all volume forms in $\mathcal{P}(X)$, where u corresponds to the normalized volume form of the Kähler metric ω_u. The one-form on $\mathcal{H}(X,\omega_0)$ induced by MA is exact, i.e., there exists a functional $\mathcal{E}$ on $\mathcal{H}(X,\omega_0)$ such that

$$d\mathcal{E} = \mathrm{MA}, \quad \text{i.e.,} \quad \left.\frac{d\mathcal{E}(u+t\dot{u})}{dt}\right|_{t=0} = \langle \mathrm{MA}(u), \dot{u} \rangle.$$

(this functional is often denoted by E in the literature **[22]**, but here we shall reserve capital letters for functionals defined on $\mathcal{P}(X)$). The functional $\mathcal{E}(u)$ is uniquely determined up to an additive a constant and may be explicitly defined by

$$\mathcal{E}(u) := \frac{1}{V(n+1)} \sum_{j=0}^{n} \int_X u\omega_u^j \wedge \omega_0^{n-j}. \tag{3.6}$$

3.2. Pluripotential theory recap

The analysis of the minimizers of F_β involves some pluripotential theory that we briefly recall. The space $\mathrm{PSH}(X,\omega)$ of all ω_0-psh functions on X may be defined as the closure of $\mathcal{H}(X,\omega_0)$ in $L^1(X)$ (more precisely, any $u \in \mathrm{PSH}(X,\omega)$ is the decreasing limit of elements $u_k \in \mathcal{H}(X,\omega_0)$). The corresponding sup-normalized subspace $\mathrm{PSH}(X,\omega_0)_0$ is compact in $L^1(X,\omega_0)$. By **[12]**, the "Calabi–Yau correspondence" (3.5) extends to a correspondence between the subspace of probability measures μ with finite energy and a subspace of $\mathrm{PSH}(X,\omega_0)$ denoted by $\mathcal{E}^1(X,\omega_0)$, that is,

$$\mathrm{MA}: \quad \mathcal{E}^1(X,\omega_0)_0 \leftrightarrow \{\mu \in \mathcal{P}(X) : E(\mu) < \infty\}, \tag{3.7}$$

where $\mathrm{MA}(u)$ is defined on $\mathcal{E}^1(X,\omega_0)$ using the notion of nonpluripolar products introduced in **[22]**. The space $\mathcal{E}^1(X,\omega_0)$ was originally introduced in **[42]**, but, as shown in **[12]**, it may also be defined as the space of all $u \in \mathrm{PSH}(X,\omega_0)$ such that $\mathcal{E}(u) > -\infty$, where $\mathcal{E}$ denotes the smallest upper semicontinuous extension of $\mathcal{E}$ to $\mathrm{PSH}(X,\omega_0)$.

3.3. Back to the free energy functional F_β

The free energy functional F_β, defined in formula (2.15), $F_\beta = \beta E + \mathrm{Ent}$, is lsc and convex on $\mathcal{P}(X)$ when $\beta > 0$ (since both terms are). In the case when $\beta < 0$, we define $F_\beta(\mu)$ by the same expression when $E_{\omega_0}(\mu) < \infty$ and otherwise we set $F_\beta(\mu) = \infty$. The definition is made so that we still have $F_\mu(\mu) \in]-\infty,\infty]$ with $F_\mu(\mu) < \infty$ iff both $E(\mu) < \infty$ and $\mathrm{Ent}(\mu) < \infty$.

The following lemma follows readily from the first variation (3.1) and formula (3.4) for Ricci curvature of a Kähler metric.

Lemma 3.1. *A volume form μ on X is a critical point of the functional F_β on $\mathcal{P}(X)$ iff the function*

$$u_\beta := \frac{1}{\beta} \log \frac{\mu}{dV}$$

solves the complex Monge–Ampère equation

$$\mathrm{MA}(u) = e^{\beta u} dV \tag{3.8}$$

iff $\omega_\beta := \omega_{u_\beta}$ *is a Kähler form solving the twisted Kähler–Einstein equation*

$$\mathrm{Ric}\,\omega + \beta\omega = \theta, \quad \theta := (\beta \mp 1)\omega_0. \tag{3.9}$$

In the Fano case, the previous equation coincides with Aubin's continuity equation with "time-parameter" $t := -\beta$. When $\beta > 0$, it follows directly from the lower semicontinuity of F_β on the compact space $\mathcal{P}(X)$ that F_β admits a minimizer.

Theorem 3.2 ([2]). *The following are true:*

(regularity) *Any minimizer* μ_β *of the functional* F_β *on* $\mathcal{P}(X)$ *is a volume form and thus of the form in Lemma* 3.1.

(existence) *If* F_{β_0} *is bounded from below for some* $\beta_0 < 0$, *then for any* $\beta > \beta_0$ *the functional* F_β *on* $\mathcal{P}(X)$ *admits a minimizer. In other words, if* F_β *is* coercive *(with respect to* E*) in the sense that there exists* $\varepsilon > 0$ *and* $C > 0$ *such that*

$$F_\beta \geq \varepsilon E + C, \tag{3.10}$$

then F_β *admits a minimizer.*

Moreover, by the Bando–Mabuchi theorem, if $\beta > -1$, the minimizer is uniquely determined and, if $\beta = -1$, it is uniquely determined iff the automorphism group $\mathrm{Aut}(X)$ of X is finite (see [10] for generalizations). The proof of the previous theorem employs a duality argument, which fits naturally into the thermodynamical formalism, when combined with pluripotential theory and the variational approach to complex Monge–Ampère equation developed in [12]. The strategy is to show that any minimizer satisfies the Monge–Ampère equation (3.8) in the weak sense of pluripotential theory, so that the regularity theory for Monge–Ampère equations (going back to Aubin and Yau) can be invoked. In the case when $\beta > 0$, the proof of Theorem 3.2 follows from the strict convexity of F_β, resulting from the convexity of $E(\mu)$ and the strict convexity of $\mathrm{Ent}(\mu)$ on $\mathcal{P}(X)$, combined with the Aubin–Yau theorem [1, 69] (showing that there exists a unique smooth solution to equation (3.8)). The proof in the case when $\beta < 0$ exploits the *Legendre–Fenchel transform*. Recall that, in general, this transform yields a correspondence between lsc convex functions on a locally convex topological vector space V and its dual V^*. In order to facilitate the comparison to the standard functionals in Kähler geometry (discussed in the following section), it will, however, be convenient to use a slightly nonstandard sign convention where an lsc convex function f on V corresponds to the usc concave function f^* on V^* defined by

$$f^*(w) := \inf_{v \in V} \big(\langle v, w\rangle + f(v)\big). \tag{3.11}$$

Conversely, if Λ is a functional on V^*, we define $\Lambda^*(v)$ as the lsc convex function

$$\Lambda^*(v) = \sup_{w \in V^*} \big(-\langle v, w\rangle + f(w)\big).$$

We take V to be the space of all signed measures μ on X, so that $V^* = C^0(X)$. We can then view E and Ent as convex lsc functions on V, which, by definition, are equal to ∞ on the complement of $\mathcal{P}(X)$ in V. Under the Legendre–Fenchel transform, these correspond to the usc convex functions E^* and Ent^*, respectively, on $C^0(X)$, which turn out to be Gateaux differentiable. Indeed, by a classical result (which follows from Jensen's inequality),

$$\mathrm{Ent}^*(u) = -\log \int e^{-u} dV.$$

Moreover, as shown in [11, 12], the functional E^* on $C^0(X)$ is Gateaux differentiable and

$$E^*(u) = \mathcal{E}(u), \quad dE^*|_u = \mathrm{MA}(u), \quad \text{for } u \in \mathcal{H}(X, \omega_0). \tag{3.12}$$

Now consider, for simplicity, the case $\beta = -1$ (the general case is obtained by a simple scaling). It follows directly from the fact that the Legendre–Fenchel transform is increasing and involutive that

$$\inf_{\mathcal{P}(X)} F_{-1} := \inf_{\mathcal{P}(X)} (-E + \mathrm{Ent}) = \inf_{C^0(X)} (-E^* + \mathrm{Ent}^*). \tag{3.13}$$

Moreover, it readily from the definitions that

$$F_{-1}\big(\mathrm{MA}(u)\big) = (-E + \mathrm{Ent})\big(dE^*|_u\big) \geq (-E^* + \mathrm{Ent}^*)(u).$$

Hence, if μ minimizes F_{-1} and we express $\mu = \mathrm{MA}(u_\mu)$, then u_μ minimizes the functional $-E^* + \mathrm{Ent}^*$ on $C^0(X)$. However, in the present setup u_μ is not, a priori, in $C^0(X)$, but only in $\mathcal{E}^1(X, \omega_0)$. This problem is circumvented using a simple approximation argument to deduce that u_μ minimizes the extension of the functional $(-E^* + \mathrm{Ent}^*)$ to $\mathcal{E}^1(X, \omega_0)$. Finally, by the Gateaux differentiability of the functional $-E^* + \mathrm{Ent}^*$ on $C^0(X)$ (or more precisely, on $\{u\} + C^0(X)$ for any given $u \in \mathcal{E}^1(X, \omega_0)$), it then follows that u_μ is a critical point of the functional $-E^* + \mathrm{Ent}^*$. Thus, after perhaps adding a constant to u_μ, it satisfies the complex Monge–Ampère equation (3.8) in the weak sense of pluripotential theory.

The proof of the first point in Theorem 3.2 can now be concluded by invoking the regularity results for pluripotential solutions to Monge–Ampère equations (which, by [13, APPENDIX B], hold in the general setup of log Fano varieties). As for the second point, it is shown in [2] by proving that any minimizing sequence μ_j in $\mathcal{P}(X)$ (i.e., a sequence μ_j such that $F_\beta(\mu_j)$ converges to the infimum of F_β) converges (after perhaps passing to a subsequence) to a minimizer of F_β. This is shown using a duality argument, as above. Alternatively, as shown in [13] in a more general singular context (including singular log Fano varieties), the existence of a minimizer for $F_\beta(\mu)$ follows from the following result in [13]:

Theorem 3.3 (energy/entropy compactness). *The functional $E(\mu)$ is continuous on any sublevel set $\{\mathrm{Ent} \leq C\} \subset \mathcal{P}(X)$. As a consequence, if F_β is coercive on $\mathcal{P}(X)$, then it is lower semicontinuous and thus admits a minimizer.*

This result has come to play a prominent role in recent developments in Kähler geometry, as discussed in Section 4.1.1.

3.4. The Mabuchi and Ding functionals

Under the "Calabi–Yau correspondence" (3.5), the free energy functional F_β on $\mathcal{P}(X)$ corresponds to a functional $\mathcal{M}_\beta(u)$ on $\mathcal{E}^1(X,\omega_0)$ defined by

$$\mathcal{M}_\beta(u) := F_\beta\big(\mathrm{MA}(u)\big). \tag{3.14}$$

Also, the functional $E(\mu)$ on $\mathcal{P}(X)$ corresponds to the functional $E(\mathrm{MA}(u))$ on $\mathrm{PSH}(X,\omega_0)$ which induces an exhaustion function on $\mathcal{E}^1(X,\omega_0)_0$, comparable to $-\mathcal{E}(u)$, defining a notion of coercivity on $\mathcal{E}^1(X,\omega_0)$ (in terms of the standard functionals I and J in Kähler geometry $E(\mathrm{MA}(u)) = (I-J)(u)$).

As is turns out, when restricted to $\mathcal{H}(X,\omega_0)$ the functional $\mathcal{M}_\beta(u)$ coincides with the *(twisted) Mabuchi functional.* The Mabuchi functional $\mathcal{M}$ associated to a general polarized manifold (X,L) was originally defined (up to normalization) by the property that its first variation is proportional to the scalar curvature of the Kähler metric ω_u minus the average scalar curvature **[53]**. An "energy+entropy" formula for $\mathcal{M}$, similar to formula (3.14), holds for a general polarized manifold, as first discovered in **[29,64]**. Likewise, the functional on $\mathcal{E}^1(X,\omega_0)$ induced by $-E^* + \mathrm{Ent}^*$ coincides with the *Ding functional* $\mathcal{D}(u)$ in Kähler geometry, extended to $\mathcal{E}^1(X,\omega_0)$ in **[12]**. For a general β, the corresponding twisted Ding functional $\mathcal{D}_\beta$ on $\mathcal{E}^1(X,\omega_0)$ is given by

$$\mathcal{D}_\beta(u) := -\mathcal{E}(u) + \frac{1}{\beta}\log\int e^{\beta u}\,dV.$$

An extension of the argument used to prove formula (3.13) (concerning the boundedness statement) now gives

Theorem 3.4 (**[2]**). *The functional $\mathcal{M}_\beta$ is bounded from below (coercive) on $\mathcal{E}^1(X,\omega_0)_0$ iff $\mathcal{D}_\beta$ is bounded from below (coercive) on $\mathcal{E}^1(X,\omega_0)_0$. Moreover, by the regularization result in* **[16]**, *these properties are equivalent to the corresponding boundedness/coercivity properties on the dense subspace $\mathcal{H}(X,\omega_0)_0$ of $\mathcal{E}^1(X,\omega_0)_0$.*

For $\beta=-1$, the first statement was first established in **[46, 57]**. The proof in **[46]** shows that the difference $\mathcal{M}_\beta - \mathcal{D}_\beta$ is bounded along the Kähler–Ricci flow, thanks to Perelman's estimates, while the proof in **[57]** utilizes the Ricci iteration. In the case $\beta=-1$, the coercivity of $\mathcal{M}_\beta$ is, in fact, equivalent to the existence of unique Kähler–Einstein metric, as first shown in **[65]**, using Aubin's method of continuity (discussed above in connection to Lemma 3.1). More recently, this result has been given a new proof using the notion of geodesics in $\mathcal{E}^1(X)$ and extended in various directions, as discussed in Section 4.1.1.

4. THE YAU–TIAN–DONALDSON CONJECTURE

4.1. The Yau–Tian–Donaldson conjecture for polarized manifolds (X,L)

Let (X,L) be a polarized projective algebraic manifold, i.e., L is a holomorphic line bundle over X whose first Chern class $c_1(L)$ contains some Kähler form.

Conjecture 4.1 (Yau–Tian–Donaldson, YTD). *There exists a Kähler metric in $c_1(L)$ with constant scalar curvature iff (X,L) is K-polystable.*

We will briefly recall the notion of K-polystability (see the survey [38] for more background on the Yau–Tian–Donaldson conjecture and its relation to geometric invariant theory (GIT)). The notion of K-polystability can be viewed as a "large-N_k limit" of the classical notion of Chow polystability in GIT with respect to the action of complex reductive group $\mathrm{GL}(N_k, \mathbb{C})$ on the Chow variety, induced from the action of $\mathrm{GL}(N_k, \mathbb{C})$ on the N_k-dimensional complex vector space $H^0(X, kL)$. Recall that in GIT the stability in question is equivalent to the positivity of the GIT-weight of all one-parameter subgroups (by the Mumford–Hilbert criterion). In the definition of K-polystability, the role of a one-parameter subgroup ρ_k of $\mathrm{GL}(N_k, \mathbb{C})$ is played by a *test configuration* ρ for (X, L). In a nutshell, this is a $\mathbb{C}^*$-equivariant embedding

$$\rho : (X \times \mathbb{C}^*, L) \hookrightarrow (\mathcal{X}, \mathcal{L})$$

of the polarized trivial fibration $(X \times \mathbb{C}^*, L)$ over $\mathbb{C}^*$ into a normal variety $\mathcal{X}$ fibered over $\mathbb{C}$ endowed with a relatively ample $\mathbb{Q}$-line bundle $\mathcal{L}$. To any test configuration ρ is attached an invariant, called the *Donaldson–Futaki invariant* $\mathrm{DF}(\rho) \in \mathbb{R}$, and (X, L) is said to be *K-semistable* if $\mathrm{DF}(\rho) \geq 0$ for any test configuration, *K-polystable* if, moreover, equality only holds when $\mathcal{X}$ is biholomorphic to $X \times \mathbb{C}$, and *K-stable* if the equality only holds when $\mathcal{X}$ is equivariantly biholomorphic to $X \times \mathbb{C}$. The Donaldson–Futaki invariant of ρ may be defined as a limit of the GIT-weights of a sequence of one-parameter subgroups ρ_k of $\mathrm{GL}(N_k, \mathbb{C})$ induced by ρ. But it may also be expressed directly as an intersection number [54, 66]:

$$\mathrm{DF}(\rho) = \frac{1}{L^n(n+1)}\left(a\mathcal{L}^{n+1} + (n+1)K_{\overline{\mathcal{X}}/\mathbb{P}^1} \cdot \mathcal{L}^n\right), \quad a := -nK_X \cdot L^{n-1}/L^n,$$

where we have identified a test configuration $(\mathcal{X}, \mathcal{L})$ with its $\mathbb{C}^*$-equivariant compactification over $\mathbb{P}^1$ (obtained by replacing the base $\mathbb{C}$ of $\mathcal{X}$ with $\mathbb{P}^1$) and the intersection numbers are computed on the compactification $\overline{\mathcal{X}}$ of the total space $\mathcal{X}$.

4.1.1. The uniform YTD and geodesic stability

The "only if" direction of the YTD conjecture was established in [60] in the case when the group $\mathrm{Aut}(X, L)$ of all automorphisms of X that lift to L is finite and in [16], in general. However, for the converse implication, there are indications that the notion of K-polystability needs to be strengthened, in general. Here we will, for simplicity, focus on the case when $\mathrm{Aut}(X, L)$ is finite. Then K-polystability is equivalent to K-stability and, moreover, if $c_1(L)$ contains a Kähler metric with constant curvature then it is uniquely determined [10, 37]. Following [24, 36], (X, L) is said to be *uniformly K-stable* (in the L^1-sense) if there exists $\varepsilon > 0$ such that

$$\mathrm{DF}(\rho) \geq \varepsilon \|\rho\|_{L^1}, \tag{4.1}$$

where the L^1-norm $\|\rho\|_{L^1}$ is defined as the normalized limit of the l^1-norms of the weights of the $\mathbb{C}^*$-action on the central fiber of $(\mathcal{X}, \mathcal{L})$. The "only if" direction of the *"uniform YTD conjecture"* – where K-stability is replaced by uniform K-stability (in the L^1-sense) – was established in [24], by leveraging the connection to the "metric space analog" of the uniform

YTD conjecture, to which we next turn. Denote by d_1 the metric on $\mathcal{H}(X,\omega_0)$ induced by the intrinsic L^1-Finsler metric

$$\int_X |\dot{u}|^1 \omega_{u_0}^n, \quad \dot{u} := \left.\frac{du}{dt}\right|_{t=0}, \quad u_0 \in \mathcal{H}.$$

As shown in [32], the metric space completion $(\overline{\mathcal{H}(X,\omega_0)_0, d_1})$ may be identified with the space $\mathcal{E}^1(X,\omega_0)_0$ (discussed in Section 3.2) and $d_1(u,0)$ is comparable to $-\mathcal{E}(u)$, which, equivalently, means that there exists a constant c such that

$$-c + c^{-1} d_1(u,0) \leq E\big(\mathrm{MA}(u)\big) \leq c d_1(u,0) + c. \tag{4.2}$$

The relevant constant speed geodesics u_t in the metric space $(\mathcal{E}^1(X,\omega_0)_0, d_1)$ have the property that

$$U(x,\tau) := u_{-\log|\tau|}(x) \in \mathrm{PSH}(X \times D^*, \omega_0), \tag{4.3}$$

where we are using the same notation ω_0 for the pullback of ω_0 to the product $X \times D^*$ of X with the punctured unit-disc D^* in $\mathbb{C}$. In fact, u_t may be characterized by a maximality property of the corresponding ω_0-psh function U [14]. Any test configuration ρ induces a geodesic ray u_t in $\mathcal{E}^1(X,\omega_0)_0$, emanating from $0 \in \mathcal{H}(X,\omega_0)$ (such that U extends, after removing divisorial singularities, to a bounded function on $\mathcal{X}$) [32, 55]. Moreover,

$$\|\rho\|_{L^1} = \frac{d}{dt} d_1(u_t, 0) = t^{-1} d(u_t, 0)$$

for any $t > 0$. As conjectured in [29], and confirmed in [10], the Mabuchi functional $\mathcal{M}$ (Section 3.4) is convex along geodesic u_t such that $\omega_U \in L^\infty_{\mathrm{loc}}$. More generally, the extension of $\mathcal{M}$ to $\mathcal{E}^1(X,\omega_0)$ is also convex along geodesics u_t [16]. In particular, its (asymptotic) slope

$$\dot{\mathcal{M}}(u_t) := \lim_{t\to\infty} t^{-1} \mathcal{M}(t) \in \,]-\infty, \infty]$$

is well defined. In the case when u_t is the geodesic ray attached to a test configuration ρ the slope $\dot{\mathcal{M}}(u_t)$ is closely related to $\mathrm{DF}(\rho)$ (the two invariants coincide after a base change [49, 59]).

Theorem 4.2 ([17, 30, 33]). *Let (X, L) be a polarized manifold. The following are equivalent:*

(1) *(X, L) admits a unique Kähler metric with constant scalar curvature.*

(2) *(X, L) is geodesically stable, i.e., $\dot{\mathcal{M}}(u_t) > 0$ for any nontrivial geodesic ray u_t in $\mathcal{E}^1(X,\omega_0)_0$.*

(3) *$\mathcal{M}$ is coercive on $\mathcal{E}^1(X,\omega_0)_0$ (or, equivalently, on $\mathcal{H}(X,\omega_0)_0 \subset \mathcal{E}^1(X,\omega_0)_0$).*

The equivalence "2 $\iff$ 3" is implicit in [33] (see [14, THM. 2.16] for a generalization). It can be seen as an analog of the classical fact that a convex function on Euclidean $\mathbb{R}^n$ is comparable to the distance to the origin iff all its slopes are positive. In the proof of "2 $\iff$ 3" a substitute for the compactness of the unit-sphere in $\mathbb{R}^n$ (parametrizing all unit speed geodesics) is provided by the energy–entropy compactness in Theorem 3.3. The implication "1 $\implies$ 3" follows directly from the convexity of $\mathcal{M}$ combined with the weak-strong uniqueness result in [17], showing, in particular, that if (X, L) admits a unique Kähler metric with constant scalar curvature ω, then any minimizer of $\mathcal{M}$ in $\mathcal{E}^1$ coincides with the Kähler potential of ω. The final implication "3 $\implies$ 1" was recently settled in [30], using a new a priori estimate for a generalization of Aubin's continuity method for constant scalar curvature metrics (bounding the C^0-norm of the solutions by the entropy of the corresponding Monge–Ampère measures, which, in turn, is uniformly bounded under the coercivity assumption).

4.2. The variational approach to the uniform YTD conjecture in the "Fano case"

The "Fano case" of the YTD conjecture, i.e., the case when X is Fano and $L = -K_X$, was settled in [31], by establishing Tian's partial C^0-estimate [63] along a singular version of Aubin's continuity method. Here we will focus on the variational proof of the uniform YTD conjecture on Fano manifolds in [14], which, in particular, exploits the notion of Ding stability originating in [3] (as further developed in [14, 24]; see the survey [20] for more background).

Theorem 4.3 ([14]). *Let X be a Fano manifold. The following are equivalent:*

(1) *X admits a unique Kähler–Einstein metric.*

(2) *X is uniformly Ding stable.*

(3) *X is uniformly K-stable.*

The implication "1 $\implies$ 2" follows from the convexity of the Ding functional along geodesics, as in [3] – here we shall focus on the converse implication. By Theorem 4.2, it is enough to show that if X is uniformly Ding stable, then X it geodesically stable. This is achieved in [14], using a valuative (non-Archimedean) language. For simplicity, it may be helpful to briefly first describe the argument with the non-Archimedean language stripped away. The starting point is the observation that the function U on $X \times D^*$ corresponding to a geodesic u_t in $\mathcal{E}^1(X, \omega_0)_0$ (formula (4.3)) extends to a sup-normalized ω_0-psh function U on $X \times D$, which, however, is highly singular on $X \times \{0\}$, unless u_t is trivial. But employing Demailly's approximation procedure [35] (involving the multiplier ideal sheaves $\mathfrak{J}(kU)$, whose definition is recalled in the following section) the function U may be expressed as a decreasing limit of S^1-invariant ω_0-psh functions U_k with analytic (algebraic) singularities, which define $\mathbb{C}^*$-invariant ideals $\mathfrak{J}_k$ supported in $X \times \{0\}$. Accordingly, by the standard resolution of singularities, there exists a $\mathbb{C}^*$-equivariant holomorphic surjection π_k from a nonsingular variety $\mathcal{X}_k$ to $X \times \mathbb{C}$ such that $E_k := \pi_k^* \mathfrak{J}_k$ is a principal ideal, i.e., defines a divisor on $\mathcal{X}_k$. This procedure yields a sequence of test configurations $\rho_k = (\mathcal{X}_k, \mathcal{L}_k)$ where $\mathcal{L}_k$ is the pullback to $\mathcal{X}_k$ of $L \to X$ with an appropriate multiple of $\mathcal{O}(E_k)$ subtracted. To show that "3 $\implies$ 1," it would, essentially, be enough show that the slope $\mathcal{M}(u_t)$ dominates the Donaldson–Futaki invariants $\mathrm{DF}(\rho_k)$. However, this leads to technical problems that are bypassed by exploiting that $\mathcal{M} \geq \mathcal{D}$, where $\mathcal{D}$ is the Ding functional on $\mathcal{H}_0$ (discussed in

Section 3.4) which behaves better under the approximation procedure above, giving

$$\dot{\mathcal{D}}(u_t) \geq \liminf_{k \to \infty} \mathcal{D}(\rho_k), \tag{4.4}$$

where $\mathcal{D}(\rho_k)$ is the "Ding invariant" originating in [3] (that we shall come back to in Section 4.3.2). Assuming that $\mathcal{X}$ is uniformly Ding stable this shows that "2 $\Longrightarrow$ 1" (after a twist of the argument which amounts to replacing $\mathcal{D}$ with $\mathcal{D}_\beta$ for $\beta = -(1+\varepsilon)$).

Finally, the equivalence "2 $\Longleftrightarrow$ 3" is shown in the first preprint version of [14], using techniques from the Minimal Model Program, inspired by [51] (the proof can, loosely speaking, be interpreted as a non-Archimedean analog of the Kähler–Ricci flow argument in [46] mentioned in connection to Theorem 3.4). The equivalence "2 $\Longleftrightarrow$ 3" in the general setup of log Fano varieties is established in [40].

4.2.1. Twisted Kähler–Einstein metrics

The results in [14] apply more generally to Kähler–Einstein metrics twisted by a positive klt current θ, showing that such a metric exists iff $\delta_\theta(X) > 1$, where $\delta_\theta(X)$ is a twisted generalization of the invariant $\delta(X)$ appearing in formula (2.9). This part of the proof does not need any results from the Minimal Model Program (as discussed in the following section). As a corollary, it is also shown that

$$\min\{1, \delta(X)\} = \min\{1, \Gamma(X)\} = R(X), \tag{4.5}$$

where $\Gamma(X)$ is the invariant appearing in Conjecture 2.7 and $R(X)$ denotes the greatest lower bound on the Ricci curvature (independently shown in [28]).

4.3. Non-Archimedean pluripotential theory and the variational formula for $\delta(X)$

The only properties of the geodesic u_t that actually entered into the proof outlined above concerned the multiplier ideal sheaves $\mathfrak{J}(kU)$ of the ω_0-psh function U on $X \times D$, whose stalks consist of all germs of holomorphic functions f such that $|f|^2 e^{-2kU}$ is locally integrable. In turn, the multiplier ideal sheaves $\mathfrak{J}(kU)$ only depend on the Lelong numbers of U on all modifications (blow-ups) of $X \times \mathbb{C}$ (see [23, THM. A] and [14, THM. B.5]). The Lelong numbers in question can be packaged into a function $U(v)$ on the space $[X \times \mathbb{C}]_{\mathrm{div}}$ of all divisorial valuations v on $X \times \mathbb{C}$, as follows. First recall that, by definition, a divisorial valuation v on variety Y is encoded by a positive number c and a prime divisor E_v over Y, i.e., a prime divisor on some blow-up of Y (which may be assumed to be a nonsingular hypersurface). Such a valuation v acts on rational (meromorphic) function $f \in \mathbb{C}(Y)$ by $v(f) := c \operatorname{ord}_{E_v}(f) \in \mathbb{R}$, where $\operatorname{ord}_{E_v}(f)$ denotes the order of vanishing at a generic point of E_v of the pullback of f. Now, if U is, locally, of the form $U = \log|f| + O(1)$ for a holomorphic function, one defines

$$U(v) := -v(f) := -c \operatorname{ord}_{E_v}(f).$$

In the general definition of $U(v)$, one replaces $\operatorname{ord}_{E_v}(f)$ with the Lelong number of U at a generic point p of E_v (i.e., the sup of all $\lambda \in [0, \infty[$ such that $f \leq \lambda \log|z| + O(1)$

with respect to local holomorphic coordinates z centered at p). In this context, Demailly's approximation procedure yields

$$U_k(v) := k^{-1} \max_i\left(-\mathrm{ord}_{E_v}\left(f_i^{(k)}\right)\right) \to U(v), \tag{4.6}$$

where $f_i^{(k)}$ denote local generators of the multiplier ideal sheaf $\mathcal{J}(kU)$. In fact, after passing to a subsequence (replacing k with 2^k), the sequence U_k is decreasing in k (by the subaditivity of multiplier ideals).

4.3.1. Pluripotential theory on the Berkovich space X_{NA}

In the present setup, the valuative procedure above is initially applied to $Y = X \times \mathbb{C}$. However, exploiting that we are only interested in the value $U(w)$ at a divisorial valuation w on $X \times \mathbb{C}$ which is $\mathbb{C}^*$-invariant, we can identify $[U](w)$ with the function on $u(v)$ on X_{div}, defined by

$$u(v) := U(w), \quad v \in X_{\mathrm{div}},\ w \in (X \times \mathbb{C})_{\mathrm{div}},$$

where w is the Gauss extension of v, defining a $\mathbb{C}^*$-equivariant valuation over $X \times \mathbb{C}$ normalized by $w(\tau) = 1$ (where τ denotes the coordinate on the factor $\mathbb{C}$) [24, SECTION 4.1]. Next, by identifying a valuation v on X with the corresponding non-Archimedean absolute value on $\mathbb{C}(X)$, i.e., with $|\cdot|_v := e^{-v(\cdot)}$, the space X_{div} injects as a dense subspace of the Berkovich analytification X_{NA} of the projective variety X over the field $\mathbb{C}$, induced by the trivially valued absolute value on the ground field $\mathbb{C}$ (locally consisting of all multiplicative seminorms extending the trivially valued absolute value, $|\cdot|_v \equiv 1$, on the field $\mathbb{C}$). The notation X_{NA} (with NA a shorthand for non-Archimedean) is used here to distinguish X_{NA} from X which is the Berkovich analytification in the "Archimedean case," i.e., the case of the standard absolute value $|\cdot|$ on the ground field $\mathbb{C}$.

The topological space X_{NA} has the virtue of being both compact and connected. Moreover, the function $u(v)$ on X_{div} extends to a plurisubharmonic (psh) function on X_{NA} in the sense of [25], denoted by u_{NA}. Indeed, in analogy to the Archimedean case, one can first define $\mathcal{H}(X_{\mathrm{NA}})_0$ to be the space of all functions u_{NA} on X_{NA} induced by test configurations ρ as above, and then define $\mathrm{PSH}(X_{\mathrm{NA}})$ as the space of all functions that can be written as decreasing nets of functions in $\mathcal{H}(X_{\mathrm{NA}})_0$ plus constants (functions in $\mathrm{PSH}(X_{\mathrm{NA}})$ are called L-psh in [25] to emphasize their global dependence on L). There is a Monge–Ampère operator MA on $\mathcal{H}(X_{\mathrm{NA}})$ taking values in the space of probability measures on X_{NA} [24, 25] (which, in a very general setup can be defined in terms of the non-Archimedean generalization of exterior products of curvature forms introduced in [27]). Concretely, $\mathrm{MA}(u_{\mathrm{NA}})$ is a discrete probability measure supported on the valuations $v_i \in X_{\mathrm{div}}$ induced by irreducible components of the central fiber of the test configuration corresponding to u_{NA} [24, SECTION 6.7]. Anyhow, in the present setup, one may directly define MA on $\mathcal{H}(X_{\mathrm{NA}})$ as the differential of the functional

$$\mathcal{E}_{\mathrm{NA}}(u_{\mathrm{NA}}) := \frac{\mathcal{L}^{n+1}}{(n+1)L^n},$$

whose definition mimics formula (3.6) (with $\omega_0 = 0$); this analogy becomes more clear when both $\mathcal{E}$ and $\mathcal{E}_{\mathrm{NA}}$ are expressed in terms of Deligne pairings [21]. As in the usual Archimedean setup (Section 3.2), the function $\mathcal{E}_{\mathrm{NA}}$ on $\mathcal{H}(X_{\mathrm{NA}})$ has a unique smallest usc extension to $\mathrm{PSH}(X_{\mathrm{NA}})$; the subspace $\{\mathcal{E}_{\mathrm{NA}} > -\infty\}$ of $\mathrm{PSH}(X_{\mathrm{NA}})$ is denoted by $\mathcal{E}^1(X_{\mathrm{NA}})$ and MA extends to $\mathcal{E}^1(X_{\mathrm{NA}})$, as the differential of the functional $\mathcal{E}^1_{\mathrm{NA}}$.

Remark 4.4. The map $u_t \mapsto u_{\mathrm{NA}}$ from geodesic rays in $\mathcal{E}^1(X,\omega_0)_0$ to the space $\mathcal{E}^1(X_{\mathrm{NA}})_0$, described above, has the property that $\dot{\mathcal{E}}(u_t) \leq \mathcal{E}(u_{\mathrm{NA}})$ and is, in general, not injective. The geodesic rays satisfying $\dot{\mathcal{E}}(u_t) = \mathcal{E}(u_{\mathrm{NA}})$ are precise those called *maximal* in [14, SECTION 6.4] and they are in one-to-one correspondence with $\mathcal{E}^1(X_{\mathrm{NA}})$.

4.3.2. The thermodynamical formalism

The non-Archimedean formalism naturally ties in with the thermodynamical formalism (discussed in Section 3). For example, as shown in [24–26], up to a base change of ρ,[2]

$$\mathrm{DF}(\rho) = \mathcal{M}_{\mathrm{NA}}(U_{\mathrm{NA}}) := F_{\mathrm{NA}}\big(\mathrm{MA}(U_{\mathrm{NA}})\big), \tag{4.7}$$

where F_{NA} is the non-Archimedean analog on $\mathcal{P}(X_{\mathrm{NA}})$ of the free energy functional F on $\mathcal{P}(X)$ defined by

$$F_{\mathrm{NA}}(\mu) = -E_{\mathrm{NA}}(\mu) + \mathrm{Ent}_{\mathrm{NA}}(\mu),$$

where the non-Archimedean energy $E_{\mathrm{NA}}(\mu)$ may be defined as a Legendre–Fenchel transform of the functional $\mathcal{E}_{\mathrm{NA}}$ and the non-Archimedean entropy $\mathrm{Ent}_{\mathrm{NA}}(\mu)$ is defined by

$$\mathrm{Ent}_{\mathrm{NA}}(\mu) := \int_{X_{\mathrm{NA}}} A(v)\mu, \quad A(v) := c\big(1 + \mathrm{ord}_{E_v}(K_{Y_v/X})\big) v \in X_{\mathrm{div}}$$

where $A(v)$ is the *log discrepancy*, defined as the greatest lsc extension to X_{NA} of the function on X_{div} defined above. Thus, in contrast to the usual entropy functional on $\mathcal{P}(X)$, the non-Archimedean entropy is a linear functional. Likewise, the "Ding invariant" appearing in formula (4.4) may be expressed as follows in terms of the Legendre–Fenchel transform

$$\mathcal{D}(\rho) = \mathcal{D}_{\mathrm{NA}}(u_{\mathrm{NA}}) := -E^*_{\mathrm{NA}}(u_{\mathrm{NA}}) + \mathrm{Ent}^*_{\mathrm{NA}}(u_{\mathrm{NA}})$$

in analogy with the usual Archimedean setup in Section 3.4. Inequality (4.4) is then obtained by showing that the slope $\dot{\mathcal{D}}(u_t)$ is bounded from below by $\mathcal{D}(u_{\mathrm{NA}})$, which, in turn, equals the limit of $\mathcal{D}(\rho_k)$ (where ρ_k is the test configuration corresponding to U_k defined by formula (4.6)).

As shown in [26] (and [14] in the general twisted setting) the thermodynamical formalism can be leveraged to prove the following theorem ("1 $\Longleftrightarrow$ 3" is shown in [40] using the Minimal Model Program):

---

2 The base change is needed as the righ-hand side in formula (4.7) is one-homogeneous under the natural action of $\mathbb{R}_{>0}$ on X_{NA}, corresponding to a base change of ρ.

Theorem 4.5 ([26]). *Let X be a Fano manifold. The following are equivalent:*

(1) $\delta(X) > 1$.

(2) *X is uniformly K-stable on $\mathcal{E}^1(X_{\mathrm{NA}})$ (i.e., inequality (4.1) extends from $\mathcal{H}(X_{\mathrm{NA}})$ to $\mathcal{E}^1(X_{\mathrm{NA}})$).*

(3) *X is uniformly Ding stable.*

The starting point of the proof of "$1 \iff 2$" is the following variational formula for $\delta(X)$ established in [19, 26], realizing $\delta(X)$ as a "stability threshold" (where δ_v denotes the Dirac measure at a point v in X_{NA}):

$$\delta(X) = \inf_{v \in X_{\mathrm{div}}} \frac{\mathrm{Ent}_{\mathrm{NA}}(\delta_v)}{E_{\mathrm{NA}}(\delta_v)} = \inf_{v \in X_{\mathrm{NA}}} \frac{\mathrm{Ent}_{\mathrm{NA}}(\delta_v)}{E_{\mathrm{NA}}(\delta_v)} = \inf_{\mu \in \mathcal{P}(X_{\mathrm{NA}})} \frac{\mathrm{Ent}_{\mathrm{NA}}(\mu)}{E_{\mathrm{NA}}(\mu)} \tag{4.8}$$

using, in the second equality, that X_{div} is dense in X_{NA} (together with a semicontinuity argument) and in the last equality (shown in [26]) that $\mathrm{Ent}_{\mathrm{NA}}(\mu)$ and $E_{\mathrm{NA}}(\mu)$ are linear and convex, respectively, on $\mathcal{P}(X_{\mathrm{NA}})$. The function $v \mapsto E_{\mathrm{NA}}(\delta_v)$ is usually denoted by $S(v)$ and can be shown to coincide with the "expected order of vanishing along v" [19]. In terms of the non-Archimedean version of the free energy functional at inverse temperature β, denoted by $F_{\mathrm{NA},\beta}(\mu)$, formula (4.8) yields

$$\delta(X) \geq 1+\varepsilon \iff \inf_{\mu \in \mathcal{P}(X_{\mathrm{NA}})} F_{\mathrm{NA},-1-\varepsilon}(\mu) \geq 0 \iff \inf_{\mu \in \mathcal{P}(X_{\mathrm{NA}})} \frac{F_{\mathrm{NA}}(\mu)}{E_{\mathrm{NA}}(\mu)} \geq \varepsilon.$$

Finally, expressing $\mu = \mathrm{MA}(U_{\mathrm{NA}})$ for $U_{\mathrm{NA}} \in \mathcal{E}^1(X_{\mathrm{NA}})$, using the non-Archimedean version of the "Calabi–Yau correspondence" (3.5), and invoking the non-Archimedean version of inequalities (4.2) (established in [24]) proves the equivalence "$1 \iff 2$". Next, using the Legendre–Fenchel transform, just as in the proof of Theorem 3.4, one sees that uniform K-stability on $\mathcal{E}^1(X_{\mathrm{NA}})$ is equivalent to uniform Ding stability on $\mathcal{E}^1(X_{\mathrm{NA}})$. Finally, "$2 \iff 3$" follows from the fact that $\mathcal{D}_{\mathrm{NA}}$ is continuous under approximation of $U_{\mathrm{NA}} \in \mathcal{E}^1(X_{\mathrm{NA}})$ by a decreasing sequence in $\mathcal{H}(X_{\mathrm{NA}})$ (e.g., using multiplier ideal sheaves as in formula (4.6)).

In order to deduce the equivalence "$2 \iff 3$" in Theorem 4.3 from the previous theorem, it would be enough to prove the following non-Archimedean analog of the regularization property shown in [16, SECTION 3].

Conjecture 4.6 ([26]). *Given any $u \in \mathcal{E}^1(X_{\mathrm{NA}})$, there exists a sequence of $u_j \in \mathcal{H}(X_{\mathrm{NA}})$ converging weakly to u such that $E_{\mathrm{NA}}(\mathrm{MA}(u_j))$ and $\mathrm{Ent}_{\mathrm{NA}}(\mathrm{MA}(u_j))$ converge to $E_{\mathrm{NA}}(\mathrm{MA}(u))$ and $\mathrm{Ent}_{\mathrm{NA}}(\mathrm{MA}(u))$, respectively.*

Remark 4.7. Combining Theorem 4.3 and Theorem 4.5 reveals that a Fano manifold X is uniformly K-stable iff $\delta(X) > 1$, as first shown in [19, 40, 41]. More precisely, the "if statement" was shown in [41], where the "only if" statement was also conjectured. The conjecture was then settled in [19]. It should also be pointed out that if one defines δ(X) as a stability threshold (see the first equality in formula (4.8)), then the equivalence between the uniform K-stability of X and the criterion $\delta(X) > 1$ is essentially equivalent to the valuative criterion for uniform K-stability established in [40]. A closely related valuative criterion for K-semistability was established in [47].

4.4. Recent developments

Recently there has been an explosion of exciting further developments. In [48, 50], Theorem 4.3 and its variational proof were extended to general singular (log) Fano varieties using, in particular, the singular version of Theorem 3.2 established in [13]. Moreover, very recently it was shown in [52], using techniques from the Minimal Model Program, that the infimum over X_{div} in formula (4.8) is (when $\delta(X) \leq 1$) attained at some $v \in X_{\mathrm{div}}$. Moreover, any such minimizing divisorial valuation v has the property that associated graded ring is finitely generated and defines a special test configuration ρ for $(X, -K_X)$. In particular, the central fiber of ρ is irreducible (the relation between test configurations, filtrations, and finitely graded rings originates in [62, 67]). In non-Archimedean terms, the result in [52] can be formulated as a regularity result for the minimizer in question, saying that $\delta_v = \mathrm{MA}(U_{\mathrm{NA}})$ for some $U_{\mathrm{NA}} \in \mathcal{H}(X_{\mathrm{NA}})$ (in analogy to the regularity result in Theorem 3.2; cf. the appendix in [40]). As a corollary it is shown in [52] that uniform K-stability is equivalent to K-stability. In fact, these results are shown to hold in the general setup of (log) Fano varieties. When combined with the aforementioned results in [48, 50] this settles the YTD conjecture in the general setting of (log) Fano varieties (the "only if" implication was previously shown in [3]). In another direction, a new variational proof of the uniform YTD conjecture in the nonsingular Fano case is given in [70], using the quantized Ding-functional (leveraging the result in [58] saying that the algebro-geometric invariant $\delta_k(X)$ in formula (2.8) coincides with coercivity threshold of the quantized Ding-functional). More generally, the results in [70] imply that the first equality in formula (4.5) holds without taking the minimum with 1 (by combining [70] with Theorem 3.4)

The variational/non-Archimedean approach is extended to polarized manifolds (X, L) in [49] to show that, if X is uniformly K-stable on $\mathcal{E}^1(X_{\mathrm{NA}})$ (as in Theorem 4.5), then X is geodesically stable and thus by Theorem 4.2 (i.e., by [30]) (X, L) admits a Kähler metric with constant scalar curvature. The converse statement is, however, still open. The complete solution of the uniform YTD conjecture for (X, L) is thus reduced to Conjecture 4.6. An important ingredient in [49] is the notion of maximal geodesic rays u_t introduced in [14] (see Remark 4.4). The theory of maximal geodesic rays is further developed in in [34] and related to singularity types of quasi-psh functions and the Legendre transform construction of geodesic rays introduced in [56]. In [68], analytic variants of stability thresholds are introduced, expressed in terms of singularity types of quasi-psh functions.

5. A NON-ARCHIMEDEAN APPROACH TO GIBBS STABILITY

This final section is a report on joint work in progress with Sébastien Boucksom and Mattias Jonsson to prove the converse of Theorem 2.5 or, more generally, to prove that

$$\lim_{N\to\infty} \mathrm{lct}(D_N) = \delta(X) \tag{5.1}$$

(which, when combined with results in [70], would also settle Conjecture 2.7). The strategy is to adapt the variational approach to the convergence in Conjecture 2.3, discussed in Section 2.3.1, to the non-Archimedean setup. The starting point is the standard valuative expression for the log canonical threshold of a divisor that yields (using the notation in Section 4.3)

$$\operatorname{lct}(D_N) = \inf_{v^{(N)}\in[X^N]_{\mathrm{div}}} \frac{A(v^{(N)})}{k^{-1}(v^{(N)}(\det S^{(k)}))} := \frac{N^{-1}A(v^{(N)})}{E_{\mathrm{NA}}^{(N)}(v^{(N)})}, \tag{5.2}$$

where we have introduced the *non-Archimedean energy per particle* as the following function on $[X^N]_{\mathrm{div}}$:

$$E_{\mathrm{NA}}^{(N)}(v^{(N)}) := N^{-1}k^{-1}\big(v^{(N)}(\det S^{(k)})\big) =: -N^{-1}k^{-1}\log\big|\det S^{(k)}\big|_{v^{(N)}}$$

(which is proportional to the negative of the psh function on $[X^N]_{\mathrm{NA}}$ induced by the quasi-psh function $\log\|\det S^{(k)}\|^2$ on X^N). In this notation, formula (5.2) can be viewed as a non-Archimedean analog of Gibbs variational principle (2.18) (since $\operatorname{lct}(D_N)-1$ is equal to the one-homogeneous non-Archimedean "N-particle free energy" $-E_{\mathrm{NA}}^{(N)}+N^{-1}A$, normalized by $E_{\mathrm{NA}}^{(N)}$). There are standard inclusions i_N and surjections π_N,

$$i_N : (X_{\mathrm{NA}})^N \hookrightarrow [X^N]_{\mathrm{NA}}, \quad \pi_N : [X^N]_{\mathrm{NA}} \twoheadrightarrow (X_{\mathrm{NA}})^N.$$

(the map i_N is, however, not surjective). The non-Archimedean version of the empirical measure δ_N mapping $(X_{\mathrm{NA}})^N$ to $\mathcal{P}(X_{\mathrm{NA}})$ (obtained by replacing X with X_{NA} in formula (2.3)) thus induces a map

$$\pi_N^*\delta_N : [X^N]_{\mathrm{div}} \to \mathcal{P}(X_{\mathrm{NA}}), \quad v^{(N)} \mapsto N^{-1}\sum_{i=1}^{N}\delta_{(\pi_N(v^{(N)}))_i}.$$

It follows from the results in [70] (which are non-Archimedean versions of results in [11]) that the restriction of $E_{\mathrm{NA}}^{(N)}$ to $(X_{\mathrm{NA}})^N$ Gamma-converges towards $E_{\mathrm{NA}}(\mu)$ (in analogy with the convergence (2.14)). In particular,

$$\lim_{N\to\infty}\delta_N(v_1,\dots,v_N) = \mu \in \mathcal{P}([X]_{\mathrm{NA}}) \implies \liminf_{N\to\infty} E_{\mathrm{NA}}^{(N)}\big(i_N(v_1,\dots,v_N)\big) \geq E_{\mathrm{NA}}(\mu). \tag{5.3}$$

Moreover, $N^{-1}A(i_N(v_1,\dots,v_N)) = \int_{X_{\mathrm{NA}}} A(v)\delta_N(v_1,\dots,v_N)$, as follows readily from the definitions. Hence, restricting the inf in formula (5.2) to v_N of the form $v_N = i_N(v,\dots,v)$ for $c \in X_{\mathrm{div}}$ reveals that the lim sup of $\operatorname{lct}(D_N)$ is bounded from above by $A(v)/E(\delta_v)$, proving the upper bound in formula (5.1). This proof essentially amounts to a reformulation of the proof of Theorem 2.5 in [41] into a non-Archimedean language. But the main point of the non-Archimedean formulation is that it opens the door for a non-Archimedean approach to the missing lower bound. Indeed, it can be shown that

$$\lim_{N_j\to\infty}\big(\pi_{N_j}^*\delta_{N_j}\big)(v^{(N_j)}) = \mu \in \mathcal{P}([X]_{\mathrm{NA}}) \implies \liminf_{N_j\to\infty} N_j^{-1}A(v_{N_j}) \geq \mathrm{Ent}_{\mathrm{NA}}(\mu).$$

Hence, all that remains is to establish the following hypothesis for any valuation $v_*^{(N)}$ realizing the infimum in formula (5.2) (which is a non-Archimedean analog of hypothesis (2.19)):

$$\lim_{N_j\to\infty}\big(\pi_{N_j}^*\delta_{N_j}\big)(v_*^{(N_j)}) = \mu_* \in \mathcal{P}(X_{\mathrm{NA}}) \implies \limsup_{N_j\to\infty} E_{\mathrm{NA}}^{(N_j)}\big(v_*^{(N_j)}\big) \leq E_{\mathrm{NA}}(\mu_*), \tag{5.4}$$

(by (5.3) the opposite inequality holds). Indeed, if the hypothesis holds then we get

$$\inf_{\mu\in\mathcal{P}([X]_{\mathrm{NA}})}\frac{\mathrm{Ent}_{\mathrm{NA}}(\mu)}{E_{\mathrm{NA}}(\mu)}\leq\liminf_{N\to\infty}\operatorname{lct}(D_N)\leq\limsup_{N\to\infty}\operatorname{lct}(D_N)\leq\inf_{v\in[X]_{\mathrm{div}}}\frac{\mathrm{Ent}_{\mathrm{NA}}(\delta_v)}{E_{\mathrm{NA}}(\delta_v)},\quad(5.5)$$

which, when combined with identity (4.8), yields the desired formula (5.1).

It remains to verify the inequality in the hypothesis above. It would be enough to establish the following *"restriction hypothesis"*: the minimizer $v_*^{(N)}$ can, asymptotically, be taken to be of the form $i_N(v_*,v_*,\dots,v_*)$ for a fixed divisorial valuation v_* on X:

$$\exists v_*\in X_{\mathrm{div}}\quad\text{such that }\liminf_{N\to\infty}\operatorname{lct}(D_N)=\liminf_{N\to\infty}\frac{N^{-1}A(i_N(v_*,v_*,\dots,v_*))}{E_{\mathrm{NA}}^{(N)}(i_N(v_*,v_*,\dots,v_*))}.$$

Indeed, it follows from the convergence of Fekete points on X_{NA} in **[21]** that

$$\lim_{N\to\infty}E_{\mathrm{NA}}^{(N)}\big(i_N(v,v,\dots,v)\big)=E(\delta_v)\quad(5.6)$$

for *any* divisorial valuation v on X (or, more generally, for any nonpluripolar point v in X_{NA}). In particular, it then follows that any v_* satisfying the "restriction hypothesis" above computes $\delta(X)$. For instance, it can be verified that the "restriction hypothesis" does hold for log Fano curves (X,Δ). Anyhow, for any given divisorial valuation v on X, formula (5.6) yields a "microscopic" formula for the non-Archimedean free-energy $F_{\mathrm{NA}}(\delta_v)$ (coinciding with the invariant $\beta(v)$ introduced in **[40]**) of independent interest:

$$F_{\mathrm{NA}}(\delta_v):=-E(\delta_v)+A(\delta_v)=\lim_{N\to\infty}\big(-E_{\mathrm{NA}}^{(N)}\big(i_N(v,v,\dots,v)\big)+N^{-1}A\big(i_N(v,v,\dots,v)\big)\big).$$

In particular, if ρ is a given test configuration, whose central fiber $\mathcal{X}_0$ is irreducible, this gives a new formula for the Donaldson–Futaki invariant $\mathrm{DF}(\rho)$, using that $\mathrm{DF}(\rho)=F_{\mathrm{NA}}(\delta_v)$, where v is the divisorial valuation on X corresponding to $\mathcal{X}_0$. Comparing with the formula for $\mathrm{DF}(\rho)$ in terms of Chow weights thus suggests that the divisorial valuation $i_N(v,v,\dots,v)$ on X^N, attached to v, plays the role of the one-parameter subgroup of $\mathrm{GL}(N,\mathbb{C})$ attached to ρ. Accordingly, the "restriction hypothesis" is an analog of the Hilbert–Mumford criterion for stability in Geometric Invariant Theory.

Finally, coming back to the statistical mechanical point of view discussed in Section 2.3, it may be illuminating to point out that the "restriction hypothesis" essentially amounts to a concentration phenomenon which may be pictured as follows. Let us decrease the inverse temperature β from a given positive value towards the critical negative inverse temperature β_N where $Z_N(\beta)=\infty$. As β changes sign from positive to negative, all the particles start to mutually attract each other and, as $\beta\to\beta_N$, a large number of particles concentrate along the subvariety of X defined by the center of the valuation v_*.

ACKNOWLEDGMENTS

It is a great pleasure to thank Bo Berndtsson, Sébastien Boucksom, Tamas Darvas, Philippe Eyssidieux, Vincent Guedj, Mattias Jonsson, Chinh Lu, David Witt-Nyström, and Ahmed Zeriahi for the stimulating collaborations that paved the way for the work exposed here. Also thanks to Sébastien Boucksom, Jakob Hultgren, and Mingchen Xia for helpful comments on a draft of the present manuscript.

REFERENCES

[1] T. Aubin, Equations du type Monge–Ampère sur les varietes Kahleriennes compactes. *Bull. Sci. Math. (2)* **102** (1978), no. 1, 63–95.

[2] R. J. Berman, A thermodynamical formalism for Monge–Ampère equations, Moser–Trudinger inequalities and Kähler–Einstein metrics. *Adv. Math.* **248** (2013), 1254.

[3] R. J. Berman, K-polystability of Q-Fano varieties admitting Kähler–Einstein metrics. *Invent. Math.* **203** (2016), no. 3, 973–1025.

[4] R. J. Berman, Large deviations for Gibbs measures with singular Hamiltonians and emergence of Kähler–Einstein metrics. *Comm. Math. Phys.* **354** (2017), no. 3, 1133–1172.

[5] R. J. Berman, Kähler–Einstein metrics, canonical random point processes and birational geometry. In *Algebraic Geometry, Salt Lake City 2015 (Part 1)*, pp. 29–74, Proc. Sympos. Pure Math. 97.1, Amer. Math. Soc., Providence, RI, 2018.

[6] R. J. Berman, An invitation to Kähler–Einstein metrics and random point processes. *Surv. Differ. Geom.* **23** (2018), 35–87.

[7] R. J. Berman, Statistical mechanics of interpolation nodes, pluripotential theory and complex geometry. *Ann. Polon. Math.* **123** (2019), 71–153.

[8] R. J. Berman, Kähler–Einstein metrics and Archimedean zeta functions. 2021, arXiv:2112.04791.

[9] R. J Berman, The probabilistic vs the quantization approach to Kähler–Einstein geometry. 2021, arXiv:2109.06575.

[10] R. J. Berman and B. Berndtsson, Convexity of the K-energy on the space of Kähler metrics. *J. Amer. Math. Soc.* **30** (2017), 1165–1196.

[11] R. J. Berman and S. Boucksom, Growth of balls of holomorphic sections and energy at equilibrium. *Invent. Math.* **181** (2010), no. 2, 337.

[12] R. J. Berman, S. Boucksom, V. Guedj, and A. Zeriahi, A variational approach to complex Monge–Ampère equations. *Publ. Math. Inst. Hautes Études Sci.* **117** (2013), 179–245.

[13] R. J. Berman, P. Eyssidieu, S. Boucksom, V. Guedj, and A. Zeriahi, Kähler–Einstein metrics and the Kähler–Ricci flow on log Fano varieties. *J. Reine Angew. Math.* (2016), published online.

[14] R. J. Berman, S. Boucksom, and M. Jonsson, A variational approach to the Yau–Tian–Donaldson conjecture. *J. Amer. Math. Soc.* **34** (2021), 605–652. arXiv:1509.04561.

[15] R. J. Berman, T. Collins, and D. Persson, The AdS/CFT correspondence and emergent Sasaki–Einstein metrics. *Nat. Commun.* (to appear), arXiv:2008.12004.

[16] R. J. Berman, T. Darvas, and C. H. Lu, Convexity of the extended K-energy and the long time behavior of the Calabi flow. *Geom. Topol.* **21** (2017), no. 5, 2945–2988.

[17] R. J. Berman, T. Darvas, and C. H. Lu, Regularity of weak minimizers of the K-energy and applications to properness and K-stability. *Ann. Sci. Éc. Norm. Supér.* **53** (2020), no. 2, 267–289.

[18] R. J. Berman and M. Önnheim, Propagation of chaos, Wasserstein gradient flows and toric Kähler–Einstein metrics. *Ann. PDE* **11** (2018), no. 6, 1343–1380.

[19] H. Blum and M. Jonsson, Thresholds, valuations, and K-stability. *Adv. Math.* **365** (2020). pp. 57

[20] S. Boucksom, Variational and non-Archimedean aspects of the Yau–Tian–Donaldson conjecture. In *Proc. of the International Congress of Mathematicians—Rio de Janeiro 2018. Vol. II. Invited lectures*, pp. 591–617, World Sci. Publ., Hackensack, NJ, 2018.

[21] S. Boucksom and D. Eriksson, Spaces of norms, determinant of cohomology and Fekete points in non-Archimedean geometry. *Adv. Math.* **378** (2021), 12.

[22] S. Boucksom, P. Essidieux, V. Guedj, and A. Zeriahi, Monge–Ampère equations in big cohomology classes. *Acta Math.* **205** (2010), no. 2, 199–262.

[23] S. Boucksom, C. Favre, and M. Jonsson, Valuations and plurisubharmonic singularities. *Publ. Res. Inst. Math. Sci.* **44** (2008), no. 2, 449–494.

[24] S. Boucksom, T. Hisamoto, and M. Jonsson, Uniform K-stability, Duistermaat–Heckman measures and singularities of pairs. *Ann. Inst. Fourier* **67** (2017), 743–841.

[25] S. Boucksom and M. Jonsson, Global pluripotential theory over a trivially valued field. 2018, arXiv:1801.08229.

[26] S. Boucksom and M. Jonsson, A non-Archimedean approach to K-stability. 2018, arXiv:1805.11160v1.

[27] A. Chambert-Loir and A. Ducros, Formes différentielles réelles et courants sur les espaces deBerkovich. 2012, arXiv:1204.6277.

[28] I. A. Cheltsov, Y. A. Rubinstein, and K. Zhang, Basis log canonical thresholds, local intersection estimates, and asymptotically log del Pezzo surfaces. *Selecta Math. (N.S.)* **25** (2019), 34. arXiv:1807.07135v2.

[29] X. X. Chen, On the lower bound of the Mabuchi energy and its application. *Int. Math. Res. Not.* **2000** (2000), no. 12, 607–623.

[30] X. X. Chen and J. Cheng, On the constant scalar curvature Kähler metrics (II)—Existence results. 2018, arXiv:1801.00656.

[31] X. X. Chen, S. Donaldson, and S. Sun, Kähler–Einstein metrics on Fano manifolds, I, II, III. *J. Amer. Math. Soc.* **28** (2015).

[32] T. Darvas, The Mabuchi geometry of finite energy classes. *Adv. Math.* **285** (2015), 182–219.

[33] T. Darvas and Y. Rubinstein, Tian's properness conjectures and Finsler geometry of the space of Kähler metrics. *J. Amer. Math. Soc.* **30** (2017), 347–387.

[34] T. Darvas and M. Xia, The closures of test configurations and algebraic singularity types. 2020, arXiv:2003.04818.

[35] J.-P. Demailly, Regularization of closed positive currents and intersection theory. *J. Algebraic Geom.* **1** (1992), 361–409.

[36] R. Dervan, Uniform stability of twisted constant scalar curvature Kähler metrics. *Int. Math. Res. Not. IMRN* (2016), no. 15, 4728–4783.

[37] S. K. Donaldson, Scalar curvature and projective embeddings. I. *J. Differential Geom.* **59** (2001), no. 3, 479–522.

[38] S. K. Donaldson, Stability of algebraic varieties and Kähler geometry. In *Algebraic geometry: Salt Lake City 2015*, pp. 199–221, Proc. Sympos. Pure Math. 97.1, Amer. Math. Soc., Providence, RI, 2018.

[39] R. Dujardin, Theorie globale de pluripotentiel, equidistributions et processes ponctuels [d'après Berman, Boucksom, Witt Nyström,...]. *Séminaire Bourbaki* 2018–2019, no. 1153. http://www.bourbaki.ens.fr/TEXTES/Exp1153-Dujardin.pdf

[40] K. Fujita, A valuative criterion for uniform K-stability of $\mathbb{Q}$-Fano varieties. *J. Reine Angew. Math.* **751** (2019), 309–338.

[41] K. Fujita and Y. Odaka, On the K-stability of Fano varieties and anticanonical divisors. *Tohoku Math. J. (2)* **70** (2018), no. 4, 511–521.

[42] V. Guedj and A. Zeriahi, The weighted Monge–Ampère energy of quasiplurisubharmonic functions. *J. Funct. Anal.* **250** (2007), 442–482.

[43] J. Hultgren, Permanental point processes on real tori, theta functions and Monge–Ampère equations. *Ann. Fac. Sci. Toulouse Math. (6)* **28** (2019), no. 1, 11–65.

[44] J. Kollár, Singularities of pairs. In *Algebraic geometry—Santa Cruz 1995*, pp. 221–287, Proc. Sympos. Pure Math. 62, Part 1, Amer. Math. Soc., Providence, RI, 1997.

[45] J. Kollár, The structure of algebraic varieties. In *Proceedings of ICM, Seoul, 2014, Vol. I.*, pp. 395–420, Kyung Moon SA, 2014, http://www.icm2014.org/en/vod/proceedings.html.

[46] H. Li, On the lower bound of the K-energy and F-functional. *Osaka J. Math.* **45** (2008), no. 1, 253–264.

[47] C. Li, K-semistability is equivariant volume minimization. *Duke Math. J.* **166** (2017), no. 16, 3147–3218.

[48] C. Li, G-uniform stability and Kähler–Einstein metrics on Fano varieties, 2019. arXiv:1907.09399.

[49] C. Li, Geodesic rays and stability in the cscK problem. *Ann. Sci. Éc. Norm. Supér.* (to appear), arXiv:2001.01366.

[50] C. Li, G. Tian, and F. Wang, The uniform version of Yau–Tian–Donaldson conjecture for singular Fano varieties. 2019, arXiv:1903.01215.

[51] C. Li and C. Xu, Special test configuration and K-stability of Fano varieties. *Ann. of Math.* **180** (2014), no. 1, 197–232.

[52] Y. Liu, C. Xu, and Z. Zhuang, Finite generation for valuations computing stability thresholds and applications to K-stability. 2021, arXiv:2102.09405.

[53] T. Mabuchi, K-energy maps integrating Futaki invariants. *Tohoku Math. J. (2)* **38** (1986), no. 4, 575–593.

[54] Y. Odaka, A generalization of the Ross–Thomas slope theory. *Osaka J. Math.* **50** (2013), no. 1, 171–185.

[55] D. H. Phong and J. Sturm, Test configurations for K-stability and geodesic rays. *J. Symplectic Geom.* **5** (2007), no. 2, 221–247.

[56] J. Ross and D. Witt Nyström, Analytic test configurations and geodesic rays. *J. Symplectic Geom.* **12** (2014), no. 1, 125–169.

[57] Y. A. Rubinstein, Some discretizations of geometric evolution equations and the Ricci iteration on the space of Kähler metrics. *Adv. Math.* **218** (2008), 1526–1565.

[58] Y. A. Rubinstein, G. Tian, and K. Zhang, Basis divisors and balanced metrics. *J. Reine Angew. Math.* **778** (2021), 171–218. 2020, arXiv:2008.08829.

[59] Z. Sjöström Dyrefelt, K-semistability of cscK manifolds with transcendental cohomology class. *J. Geom. Anal.* **28** (2018), 2927–2960.

[60] J. Stoppa, K-stability of constant scalar curvature Kähler manifolds. *Adv. Math.* **221** (2009), no. 4, 1397–1408.

[61] S. Sun, Degenerations and moduli spaces in Kähler geometry. In *Proceedings of the International Congress of Mathematicians (ICM 2018)*, pp. 993–1012, World Sci. Publ., Hackensack, NJ, 2018.

[62] G. Székelyhidi, Filtrations and test configurations. With an appendix by S. Boucksom. *Math. Ann.* **362** (2015), 451–484.

[63] G. Tian, On Calabi's conjecture for complex surfaces with positive first Chern class. *Invent. Math.* **101** (1990), no. 1, 101–172.

[64] G. Tian, Kähler-Einstein metrics on algebraic manifolds. In *Transcendental methods in algebraic geometry (Cetraro, 1994)*, pp. 143–185, Lecture Notes in Math. 1646, Fond. CIME/CIME Found. Subser., Springer, Berlin, 1996.

[65] G. Tian, Kähler–Einstein metrics with positive scalar curvature. *Invent. Math.* **130** (1997), no. 1, 1–37.

[66] X. Wang, Height and GIT weight. *Math. Res. Lett.* **19** (2012), no. 04, 909–926.

[67] D. Witt Nyström, Test configurations and Okounkov bodies. *Compos. Math.* **148** (2012), 1736–1756.

[68] M. Xia, Pluripotential-theoretic stability thresholds. 2020, arXiv:2012.12039.

[69] S.-T. Yau, On the Ricci curvature of a compact Kähler manifold and the complex Monge–Ampère equation. I. *Comm. Pure Appl. Math.* **31** (1978), no. 3, 339–411.

[70] K. Zhang, A quantization proof of the uniform Yau–Tian–Donaldson conjecture. 2021, arXiv:2102.02438.

ROBERT J. BERMAN

Mathematical Sciences, Chalmers University of Technology and the University of Gothenburg, SE-412 96 Göteborg, Sweden, robertb@chalmers.se

SAUSAGES

DANNY CALEGARI

ABSTRACT

The *shift locus* is the space of normalized polynomials in one complex variable for which every critical point is in the attracting basin of infinity. The method of *sausages* gives a (canonical) decomposition of the shift locus in each degree into (countably many) codimension 0 submanifolds, each of which is homeomorphic to a complex algebraic variety. In this paper we explain the method of sausages, and some of its consequences.

MATHEMATICS SUBJECT CLASSIFICATION 2020

Primary 37F10; Secondary 37C15, 20F05, 37E30, 20F36, 20F67

ICM INTERNATIONAL CONGRESS OF MATHEMATICIANS 2022 JULY 6–14

Proc. Int. Cong. Math. 2022, Vol. 4, pp. 2484–2503
DOI 10.4171/ICM2022/4

1. SAUSAGES

For each integer $q \geq 2$, the *shift locus* $\mathcal{S}_q$ is the set of degree q polynomials f in one complex variable of the form

$$f(z) := z^q + a_2 z^{q-2} + a_3 z^{q-3} + \cdots + a_q,$$

for which every critical point of f is in the attracting basin of ∞. One can think of $\mathcal{S}_q$ as a open submanifold of $\mathbb{C}^{q-1}$; understanding its topology is a fundamental problem in complex dynamics. For example, when $q = 2$, the complement of $\mathcal{S}_2$ in $\mathbb{C}$ is the Mandelbrot set. Knowing that $\mathcal{S}_2$ is homeomorphic to a cylinder implies the famous theorem of Douady–Hubbard that the Mandelbrot set is connected.

Although the $\mathcal{S}_q$ are highly transcendental spaces, the method of *sausages* (which we explain in this section) shows that each $\mathcal{S}_q$ has a canonical decomposition into codimension 0 submanifolds whose interiors are homeomorphic to certain explicit algebraic varieties. From this one can deduce a considerable amount about the topology of $\mathcal{S}_q$, especially in low degree.

The construction of sausages has several steps, and goes via an intermediate construction that associates, to each polynomial f in $\mathcal{S}_q$, a certain combinatorial object called a *dynamical elamination.*

1.1. Green's function

Let K be a compact subset of $\mathbb{C}$ with connected complement $\Omega_K := \mathbb{C} - K$. If K has positive logarithmic capacity (for example, if the Hausdorff dimension is positive) then there is a canonical *Green's function* $g : \Omega_K \to \mathbb{R}^+$ satisfying

(1) g is harmonic;

(2) g extends continuously to 0 on K; and

(3) g is asymptotic to $\log|z|$ near infinity (in the sense that $g(z) - \log|z|$ is harmonic near infinity).

There is a unique germ near infinity of a holomorphic function ϕ, tangent to the identity at ∞, for which $g = \log|\phi(z)|$.

1.2. Filled Julia set

Let f be a degree q complex polynomial. After conjugacy by a complex affine transformation $z \to \alpha z + \beta$, we may assume that f is *normalized*; i.e., of the form

$$f(z) := z^q + a_2 z^{q-2} + a_3 z^{q-3} + \cdots + a_q.$$

The *filled Julia set* $K(f)$ is the set of complex numbers z for which the iterates $f^n(z)$ are (uniformly) bounded. It is a fact that $K(f)$ is compact, and its complement $\Omega_f := \mathbb{C} - K(f)$ is connected. The union $\widehat{\Omega}_f := \Omega_f \cup \infty$ is the attracting basin of ∞.

Böttcher's Theorem (see, e.g., [20, THM. 9.1]) says that f is holomorphically conjugate near infinity to the map $z \to z^q$. For normalized f, the germ of the conjugating map ϕ (i.e.,

ϕ so that $\phi(f(z)) = \phi(z)^q$) is uniquely determined by requiring that ϕ is tangent to the identity at infinity. The (real-valued) function $g(z) := \log|\phi(z)|$ is harmonic, and satisfies the functional equation $g(f(z)) = q \cdot g(z)$. We may extend g via this functional equation to all of Ω_f and observe that g so defined is the Green's function of $K(f)$.

1.3. Maximal domain of ϕ^{-1}

Let $\overline{\mathbb{D}} \subset \mathbb{C}$ denote the closed unit disk, and $\mathbb{E} := \mathbb{C} - \overline{\mathbb{D}}$ the exterior. We will use logarithmic coordinates $h = \log(|z|)$ and $\theta = \arg(z)$ on $\mathbb{E}$ and on Riemann surfaces obtained from $\mathbb{E}$ by cut and paste. Note that $g = h\phi$ where g and ϕ are as in Section 1.1.

For any K with Green's function g and associated ϕ, we can analytically continue ϕ^{-1} from infinity along radial lines of $\mathbb{E}$. The image of these radial lines under ϕ^{-1} are the descending gradient flowlines of g (i.e., the integral curves of $-\text{grad}(g)$), and we can analytically continue ϕ^{-1} until the gradient flowlines run into critical points of g. Figure 1 shows some gradient flowlines of g for a Cantor set K.

FIGURE 1
Gradient flowlines of g for a Cantor set K.

Note that some critical points of g might have multiplicity greater than one; however, because g is harmonic, the multiplicity of every critical point is finite, and the critical points of g are isolated and can accumulate (in $\widehat{\mathbb{C}}$) only on K. With this proviso about multiplicity, we want to do a sort of "Morse theory" for the function g.

Let L' be the union of the segments of the gradient flowlines of g descending from all the critical points of g; in Figure 1 these are in red (gray, for black and white reproduction). Then $\Omega_K - L'$ is the image of the maximal (radial) analytic extension of ϕ^{-1}. The domain of this maximal extension ϕ^{-1} may be described as follows. For $w \in \mathbb{E}$, define the *radial segment* $\sigma(w) \subset \mathbb{E}$ to be the set of points z with $\arg(z) = \arg(w)$ and $|z| \leq |w|$. The *height* of σ, denoted $h(\sigma)$, is $\log(|w|)$. The domain of ϕ^{-1} is $\mathbb{E} - L$ where L is the union of a countable proper (in $\mathbb{E}$) collection of radial segments.

If $K = K(f)$ for a polynomial f, the critical points of g are the critical points *and critical preimages* of f, i.e., points z for which $(f^n)'(z) = 0$ for some positive n. Thus L' is nearly f-invariant: the image $f(L')$ is equal to $L' \cup \ell'$ where ℓ' is the (finite!) set of descending flowlines from the critical *values* of f in Ω_f (which are themselves not typically critical).

Likewise, the map $z \to z^q$ on $\mathbb{E}$ takes L to $L \cup \ell$ where ℓ is a finite set of radial segments mapped by ϕ^{-1} to ℓ'.

1.4. Cut and paste

Let c be a critical point of g and let L'_c be the union of the gradient flowlines of g descending from c (and, for simplicity, here and in the sequel let us suppose these flowlines do not run into another critical point). Then L'_c is the union of $n + 1$ proper embedded rays from c to K where n is the multiplicity of c as a critical point (these rays extend continuously to K when the components of K are locally connected; otherwise they may "limit to" a *prime end* of a component of K). There is a corresponding collection L_c of $n + 1$ radial segments $\sigma_j := \sigma(w_j)$ all of the same height, where indices are circularly ordered according to the arguments of the w_j. The map ϕ^{-1} extends continuously along radial lines from infinity all the way to the w_j: the w_j all map to c. But any "extension" of ϕ^{-1} over L_c will be multivalued. We can repair this multivaluedness by *cut and paste*: cut open $\mathbb{E}$ along the segments L_c to create two copies σ_j^+ (resp. σ_j^-) for each σ_j on the "left" (resp. "right") in the circular order. Then glue each segment σ_j^- to σ_{j+1}^+ by a homeomorphism respecting absolute value. Under this operation the collection of segments L_c are reassembled into an "asterisk" which resembles the cone on $n + 1$ points; see Figure 2.

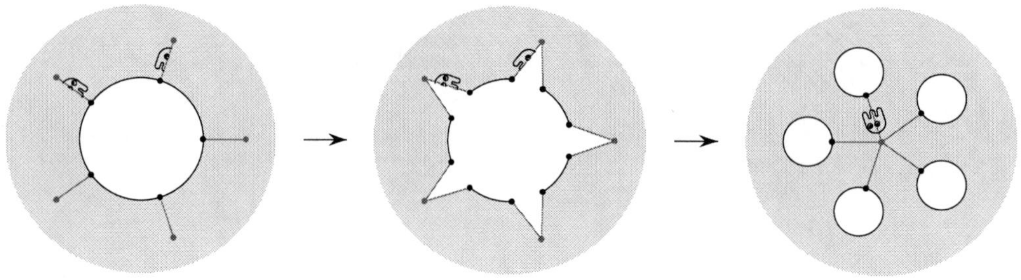

FIGURE 2
Cut and paste over L_c of multiplicity 4.

The result is a new Riemann surface for which the map ϕ^{-1} now extends (analytically and single-valuedly) over the (cut-open and reglued) image of L_c, whose image is exactly L'_c.

If we perform this cut and paste operation simultaneously for all the different L_c making up L, the Riemann surface $\mathbb{E}$ is reassembled into a new Riemann surface Ω so that ϕ^{-1} extends to an *isomorphism* $\phi^{-1} : \Omega \to \Omega_K$.

If $K = K(f)$ for a polynomial f, then the map $z \to z^q$ on $\mathbb{E}$ descends to a well-defined degree q holomorphic self-map $F : \Omega \to \Omega$ and ϕ^{-1} conjugates $F|\Omega$ to $f|\Omega_f$.

1.5. Elaminations

It is useful to keep track of the partition of L' and L into finite collections L'_c and L_c associated to the critical points c of g.

For each critical point c of multiplicity n we span the $n + 1$ segments of L_c by an ideal hyperbolic $(n + 1)$-gon in $\overline{\mathbb{D}}$. The segments of L_c become the *tips* and the ideal polygon becomes the *vein* of a *leaf* of multiplicity n in an object called an *extended lamination*—or *elamination* for short. When every critical point has multiplicity 1, we say the elamination is *simple*. See Figures 3 and 7 for examples of simple elaminations. The key topological property of elaminations is that the veins associated to different leaves *do not cross*. This is equivalent to the fact that the result Ω of cut and paste along L is a *planar* surface (because it is isomorphic to $\Omega_X \subset \mathbb{C}$).

Elaminations are introduced and studied in [11]. The set $\mathcal{EL}$ of elaminations becomes a space with respect to a certain topology (the *collision topology*), and can be given the structure of a disjoint union of (countable dimensional) complex manifolds. For example, the space of elaminations with $n - 1$ leaves (counted with multiplicity) is homeomorphic (but *not* biholomorphic) to the space of degree n normalized polynomials with no multiple roots.

1.6. Dynamical elaminations

Figure 3 depicts the elamination associated to $K(f)$ for a degree 3 polynomial f. The *critical leaves,* i.e., the leaves with tips L_c associated to c a critical point of f, are in red. Every other leaf corresponds to a *precritical* point of f (which are critical points of the Green's function). This elamination is *simple*: every leaf has exactly two tips.

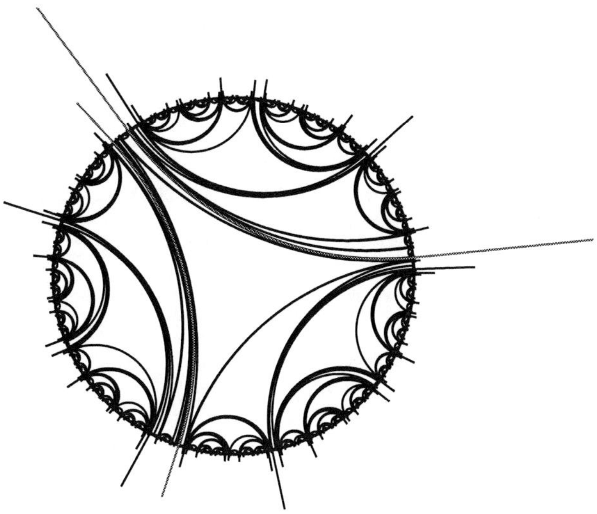

FIGURE 3
Simple dynamical elamination of degree 3; critical leaves are in red.

Let Λ denote the elamination associated to L. Note that Λ depends not just on L as a set of segments, but also on their partition into subsets L_c.

The map $z \to z^q$ on $\mathbb{E}$ acts on segments and therefore also on leaves, with the following exception. If ℓ is a leaf whose tips have arguments that all differ by integer multiples of $2\pi/q$ then these segments will have the same image under $z \to z^q$. Since leaves should have at least two tips (by convention), if ℓ is a leaf *all* of whose tips have arguments that differ by integer multiples of $2\pi/q$ then the image of ℓ under $z \to z^q$ is undefined.

Suppose $K = K(f)$ for a degree q polynomial. Let C denote the critical leaves of L (those associated to critical points of f). The map $z \to z^q$ takes leaves to leaves in the obvious sense, and takes $\Lambda - C$ to Λ.

We say an elamination Λ is a *degree q dynamical elamination* if

(1) it has finitely many leaves C each of whose arguments differ by integer multiples of $2\pi/q$ (the *critical leaves*);

(2) the map $z \to z^q$ takes $\Lambda - C$ to Λ; and

(3) every leaf has exactly q preimages.

A degree q dynamical elamination is *maximal* if there are $q - 1$ critical leaves, counted with multiplicity.

The elamination Λ associated to a degree q polynomial f is a degree q dynamical elamination. It is maximal if and only if all the critical points of f are in Ω_f.

A set of (noncrossing) leaves C, each with arguments that differ by integer multiples of $2\pi/q$ is called a *degree q critical set*. A critical set is *maximal* if there are $q - 1$ leaves counted with multiplicity. It turns out that every maximal degree q critical set C is exactly the set of critical leaves of a *unique* (maximal) degree q dynamical elamination Λ; see [11, Prop. 5.3]. The set of maximal degree q dynamical elaminations is denoted $\mathcal{DL}_q$. As a subset of $\mathcal{EL}$, it has the structure of an open complex manifold of dimension $q - 1$ with local coordinates coming from the (endpoints of) segments of C (at least at a generic Λ).

1.7. The shift locus

For each degree q, the *shift locus* $\mathcal{S}_q$ is the space of degree q normalized polynomials $f(z) := z^q + a_2 z^{q-2} + a_3 z^{q-3} + \cdots + a_q$ for which every critical point is in the basin of infinity Ω_f. The coefficients $a_2, \dots, a_q$ are coordinates on $\mathcal{S}_q$ realizing it as an open subset of $\mathbb{C}^{q-1}$.

A polynomial f is in $\mathcal{S}_q$ if and only if the Julia set of f is a Cantor set on which f is uniformly expanding (for some metric). Thus for such polynomials, $J(f) = K(f)$ and Ω_f is the entire Fatou set (i.e., the maximal domain of normality of f and its iterates; see, e.g., [20]).

Suppose $f \in \mathcal{S}_q$ with associated dynamical elamination Λ. Since all critical points of f are in Ω_f, it follows that Λ is maximal; thus there is a map $\mathcal{S}_q \to \mathcal{DL}_q$ called the *Böttcher map*. Conversely, if Λ is any maximal degree q dynamical elamination, and Ω is obtained from $\mathbb{E}$ by cut and paste along Λ, then $F|\Omega$ extends (topologically) over the space of ends of Ω to define a degree q self-map $\bar{F}$ of a topological sphere $\overline{\Omega} \cong S^2$. It turns out that there is a canonical conformal structure on $\overline{\Omega}$ extending that on Ω so that $\bar{F}$ is holomorphic. After choosing suitable coordinates on $\overline{\Omega}$ near ∞, the map $\bar{F}$ becomes a degree q normalized polynomial, which is contained in $\mathcal{S}_q$. The analytic content of this theorem is essentially due to de Marco–McMullen; see, e.g., [16, THM. 7.1] or [11, THM. 5.4] for a different proof.

Thus the Böttcher map $\mathcal{S}_q \to \mathcal{DL}_q$ is a homeomorphism (and, in fact, an isomorphism of complex manifolds).

1.8. Stretching and spinning

There is a (multiplicative) $\mathbb{R}^+$ action on $\mathcal{EL}$ called *stretching* where $t \in \mathbb{R}^+$ acts on Λ by multiplying the h coordinate of every leaf by t. This action is free and proper. It preserves $\mathcal{DL}_q$ for each q, and shows that $\mathcal{DL}_q$ (and therefore also $\mathcal{S}_q$) is homeomorphic to the product of $\mathbb{R}$ with a manifold of (real) dimension $2q - 3$. It is convenient for what follows to define $\mathcal{DL}'_q$ to be the open subspace of $\mathcal{DL}_q$ for which the highest critical leaf has $\log_q(h) \in (-1/2, 1/2)$. By suitably "compressing" orbits of the $\mathbb{R}^+$ action, we see there is a homeomorphism $\mathcal{DL}_q \to \mathcal{DL}'_q$.

There is also an $\mathbb{R}$ action on $\mathcal{EL}$ called *spinning* where $t \in \mathbb{R}$ simultaneously rotates the arguments of leaves of height h by th. This makes literal sense for the (finitely many) leaves of greatest height. When leaves of lesser height are collided by those of greater height the shorter leaf is "pushed over" the taller one; the precise details are explained in [11, § 3.2]. This $\mathbb{R}$ action also preserves each $\mathcal{DL}_q$. The closure of the $\mathbb{R}$-orbits in each $\mathcal{DL}_q$ are real tori, and the $\mathbb{R}$-orbits sit in these tori as parallel lines of constant slope. A typical $\mathbb{R}$-orbit has closure which is a torus of real dimension $q - 1$, but if some critical leaves have multiplicity > 1 or if distinct critical leaves have rationally related heights, the closure will be a torus of lower dimension.

Stretching and spinning combine to give an action of the (oriented) affine group $\mathbb{R} \rtimes \mathbb{R}^+$ of the line on $\mathcal{EL}$ and on each individual $\mathcal{DL}_q$.

1.9. Sausages

Suppose $K = K(f)$ for a degree q polynomial. The map f is algebraic, but the domain Ω_f is transcendental. When we move to the elamination side, the map $z \to z^q$ and the domain $\mathbb{E}$ are (semi)algebraic, but the combinatorics of L is hard to understand. Sausages are a way to find a substitute for (f, Ω_f) for which both the map and domain are algebraic and more comprehensible.

The idea of sausages is to find a dynamically-invariant way to cut up the domain Ω into a *tree of Riemann spheres*, so that F induces polynomial maps between these spheres. The sausage map is *not* holomorphic, but it induces homeomorphisms between certain codimension 0 submanifolds of $\mathcal{DL}'_q$ and certain explicit algebraic varieties whose topology is in some ways much easier to understand.

Now let us discuss the details of the construction. First, consider the map $z \to z^q$ on $\mathbb{E}$ alone. Let $h := \log(|z|)$ and $\theta = \arg(z)$ be cylindrical coordinates on $\mathbb{E}$, so that $\mathbb{E}$ becomes the half-open cylinder $S^1 \times \mathbb{R}^+$ in (θ, h)-coordinates, and $z \to z^q$ becomes the map which is multiplication by q which we denote $\times q$. For each integer n, let I_n denote the open interval $(q^{n-1/2}, q^{n+1/2})$ and let A_n be the annulus in $\mathbb{E}$ where $h \in I_n$ and let $A = \bigcup_n A_n \subset \mathbb{E}$; the complement of A is the countable set of circles with $\log_q(h) \in 1/2 + \mathbb{Z}$. Then $\times q$ takes A_n to A_{n+1}.

This data is holomorphic but not algebraic. So let us choose (rather arbitrarily) an orientation-preserving diffeomorphism $\nu_0 : I_0 \to \mathbb{R}$ and for each n define $\nu_n : I_n \to \mathbb{R}$ by $\nu_n(h) = q^n \nu_0(q^{-n}h)$ (so that by induction the ν_n satisfy $\nu_{n+1}(qh) = q\nu_n(h)$ for all n and $h \in I_n$), and define $\mu : A \to S^1 \times \mathbb{R}$ to be the map that sends (θ, h) to $(\theta, \nu_n(h))$ if $(\theta, h) \in A_n$. By construction, μ commutes with multiplication by q:

$$\mu(q\theta, qh) = \big(q\theta, \nu_{n+1}(qh)\big) = \big(q\theta, q\nu_n(h)\big) = q\mu(\theta, h).$$

In other words, μ semiconjugates $\times q$ on A to $\times q$ on $S^1 \times \mathbb{R}$, which (by exponentiating) becomes the map $z \to z^q$ on $\mathbb{C}^*$, an algebraic map on an algebraic domain. Actually, it is better to keep a separate copy $\mathbb{C}_n^* := \mu(A_n)$ of $\mathbb{C}^*$ for each n, so that μ *conjugates* $\times q$ on A to the self-map of $\bigcup_n \mathbb{C}_n^*$ which sends each $\mathbb{C}_n^*$ to $\mathbb{C}_{n+1}^*$ by $z \to z^q$.

1.10. Sausages and dynamics

Now suppose we have a dynamical elamination Λ with critical leaves C invariant under $z \to z^q$. For each A_n, the tips of Λ intersect A_n in a finite collection of vertical segments L_n (some of which will pass all the way through A_n) and we can perform cut-and-paste separately on each A_n to produce a (typically disconnected) surface B_n. Furthermore, we can perform cut-and-paste on $\mathbb{C}_n^*$ along the image $\mu(L_n)$ which, by construction, is compatible with the Riemann surface structure. The result is to cut and paste $\mathbb{C}_n^*$ into a finite collection of algebraic Riemann surfaces, each individually isomorphic to $\mathbb{C}$ minus a finite set of points and which can be canonically completed to Riemann spheres in such a way that the map F on Ω descends to a map $\mathfrak{f}$ from this union of Riemann spheres to itself; see Figure 4.

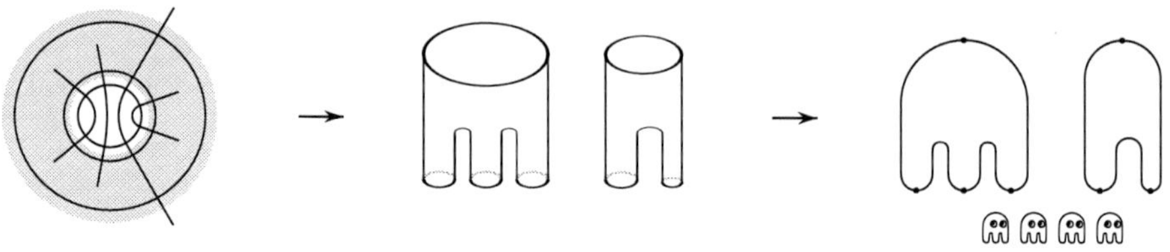

FIGURE 4
A_n is cut and paste into B_n which in turn maps to a disjoint union of Riemann spheres.

Denote the individual Riemann spheres by X_v and, by abuse of notation, write $\mathfrak{f}_v : X_v \to X_{\mathfrak{f}(v)}$ for the restriction of $\mathfrak{f}$ to the component X_v. By the previous discussion, each map $\mathfrak{f}_v$ is *holomorphic*, so that if we choose suitable coordinates on X_v and $X_{\mathfrak{f}(v)}$, the map $\mathfrak{f}_v$ becomes a polynomial. There is almost a canonical choice of coordinates, which we explain in the next two sections.

Each X_v corresponds to a component B_v of some B_n, and gets a canonical finite set of *marked points* Z'_v which correspond to the "boundary circles" of B_v. The unique boundary circle with the greatest h coordinate picks out a point that we can identify with $\infty \in X_v$; we denote by Z_v the set consisting of the rest of the marked points. The collection of individual Riemann spheres X_v can be glued up along their marked points into an infinite genus-zero nodal Riemann surface so that the indices v are parameterized by the vertices v of the tree of gluings T. This tree is oriented, so that an edge v goes to w if X_v is glued along ∞ to one of the (finite) marked points of X_w. We call w the *parent* of v and v one of the *children* of w. If we make the assumption that no boundary component of any B_v contains a critical point (this is the generic case) then each $\zeta \in Z_w \subset X_w$ is attached to a unique X_v for v some child of w. If v is a child of w, and X_v is glued to X_w at the point $\zeta \in Z_w$, then if ζ is a critical point of $\mathfrak{f}_w$ of multiplicity m, the degree of $\mathfrak{f}_v$ is $m+1$. By abuse of notation, we denote the induced (simplicial, orientation-preserving) map on T also by $\mathfrak{f}$.

If Λ is empty, then T is just a line, and each vertex has a unique child. If Λ is nonempty, then since there are only finitely many leaves of greatest height, there is a unique highest vertex v of T with more than one child. Let w be the parent of v. The uppermost boundary components of B_v and B_w are canonically identified with the unit circle $S^1 := \mathbb{R}/2\pi\mathbb{Z}$. By identifying these circles with the unit tangent circles at ∞ in X_v and X_w, we can choose coordinates on these Riemann spheres so that the tangent to the positive real axis corresponds to the angle $0 \in S^1$. In these coordinates X_v and X_w are identified with copies $\widehat{\mathbb{C}}_v$ and $\widehat{\mathbb{C}}_w$ of the Riemann sphere $\widehat{\mathbb{C}}$, and after precomposing with a suitable complex affine translation, $\mathfrak{f}_v$ becomes a normalized degree q polynomial map $\mathfrak{f}_v : z \to z^q + b_2 z^{q-2} + \cdots + b_q$, and the (finite) marked points of X_v become the roots of $\mathfrak{f}_v$ in $\widehat{\mathbb{C}}_v$.

Vertices of T above v and the maps between their respective Riemann surfaces do not carry any information. Let $w_1 := w$ denote the parent of v, and inductively let w_n be the parent of w_{n-1}. Then each X_{w_n} has exactly two marked points, which we can canonically identify with ∞ and 0, and the map $\mathfrak{f}_{w_{n-1}} : \widehat{\mathbb{C}}_{w_{n-1}} \to \widehat{\mathbb{C}}_{w_n}$ is canonically normalized as $z \to z^q$.

Since these vertices carry no information, we discard them. Thus we make the convention that T is the *rooted* tree consisting of v together with its (iterated) children, and we let $\mathfrak{X}$ denote the nodal Riemann surface corresponding to the union of X_w with w in T. We record the data of the polynomial $\mathfrak{f}_v$ associated to the root v, though we do not interpret this any more as a map between Riemann spheres, so that $\mathfrak{f}$ is now a map from $\mathfrak{X} - X_v$ to $\mathfrak{X}$ and $\mathfrak{f}_v$ is a polynomial function on $X_v \cong \widehat{\mathbb{C}}$.

1.11. Tags and sausage polynomials

The choice of a distinguished point on a boundary S^1 component of some B_u is called a *tag*. Tags are the data we need to choose coordinates on $\mathfrak{X}$ so that every $\mathfrak{f}_u$ becomes a polynomial. We may identify this boundary circle with the unit tangent circle at a marked point on X_u, and think of the tag as data on X_u. By induction, we can choose tags on X_u in the preimage of the tags of $X_{\mathfrak{f}(u)}$ under the map $\mathfrak{f}_u : X_u \to X_{\mathfrak{f}(u)}$ and inductively define coordinates $\widehat{\mathbb{C}}_u$ on X_u for which $\mathfrak{f}_u$ is represented by a normalized polynomial map (in general of degree $\leq q$).

Suppose u has parent u', and ∞ in $\widehat{\mathbb{C}}_u$ is attached at some point $\zeta \in Z_{u'} \in \widehat{\mathbb{C}}_{u'}$. Suppose ζ is a critical point of $\mathfrak{f}_{u'}$ with multiplicity m. Then $\mathfrak{f}_u$ has degree $m+1$. There are $m+1$ different choices of tag at ζ that map to the tag at $\mathfrak{f}_{u'}(\zeta)$, and the different choices affect the normalization of $\mathfrak{f}_u$ by precomposing with multiplication by an $(m+1)$st root of unity.

The endpoint of this discussion is that we can recover $\mathfrak{X}, \mathfrak{f}$ from the data of a rooted tree T, and a set of equivalence classes of pair (tag, normalized polynomial $\mathfrak{f}_u$). Call this data a (degree q) *sausage polynomial*.

A dynamical elamination Λ is *generic* if the critical points of F are all contained in A, i.e., if no critical (or by induction, precritical) point has h coordinate with $\log(h) \in 1/2 + \mathbb{Z}$. The *sausage map* is the map that associates a sausage polynomial to a degree q dynamical elamination. A sausage polynomial is *generic* (resp. *maximal*) if it is in the image of a generic (resp. maximal) dynamical elamination.

A polynomial $\mathfrak{f}_w$ associated to a (generic) sausage polynomial has two kinds of critical points. The *genuine* critical points are those in $\widehat{\mathbb{C}}_w - Z'_w$ (recall that Z'_w is $Z_w \cup \infty$). The *fake* critical points are those in Z'_w ($= \infty \cup Z_w$) which correspond to circle components of B_w mapping with degree > 1. For a generic dynamical elamination, the genuine critical points of the associated sausage polynomial are exactly the images of the critical points of the elamination (i.e., the endpoints of the critical leaves) under the sausage map. Thus for a generic maximal sausage polynomial of degree q, there are exactly $q-1$ genuine critical points, counted with multiplicity.

For a generic, maximal sausage polynomial, all but finitely many $\mathfrak{f}_v$ have degree one. A degree-one map uniquely pulls back tags, and has only one possible normalized polynomial representative, namely the identity map $z \to z$. Thus a generic, maximal sausage polynomial is described by a finite amount of combinatorial data, together with a finite collection of normalized polynomials. The reader who wants to see some examples should look ahead to Sections 2.1 and 2.3.

Let $\mathcal{X}_q$ denote the space of generic maximal degree q sausage polynomials. Then $\mathcal{X}_q$ is the disjoint union of countably infinitely many components, indexed by the combinatorics of T and the degrees of the normalized polynomials between the associated Riemann spheres. Each component of $\mathcal{X}_q$ is a *quasiprojective complex variety* of complex dimension $q-1$. In fact, each component is an iterated fiber bundle whose base and fibers are certain *affine* (complex) varieties called *Hurwitz varieties*, which we shall describe in more detail in Section 2.6.

1.12. Sausage space

Recall that $\mathcal{DL}'_q \subset \mathcal{DL}_q$ denotes the set of maximal degree q dynamical elaminations for which the highest critical point has $\log_q(h) \in (-1/2, 1/2)$. Let $\mathcal{DL}''_q \subset \mathcal{DL}'_q$ denote the subspace of *generic* maximal degree q dynamical elaminations. Then the construction of the previous few sections defines a map $\mathcal{DL}''_q \to \mathcal{X}_q$.

In fact, this map is invertible. Given a sausage polynomial $\mathfrak{X}, \mathfrak{f}$ over a tree T with root v, we can inductively construct (singular) vertical (resp. horizontal) foliations on each $\widehat{\mathbb{C}}_w$ as follows. On $\widehat{\mathbb{C}}_v$ we pull back the foliations of $\mathbb{C}^*$ by lines (resp. circles) of constant argument (resp. absolute value) under the polynomial $\mathfrak{f}_v$. Then on every other w, we inductively pull back these foliations under $\mathfrak{f}_w : \widehat{\mathbb{C}}_w \to \widehat{\mathbb{C}}_{\mathfrak{f}(w)}$. These foliations all carry coordinates pulled back from $\mathbb{C}^*$, and $\widehat{\mathbb{C}}_w$ minus infinity and its marked points become isomorphic to a branched Euclidean Riemann surface with ends isomorphic to the ends of (infinite) Euclidean cylinders. We can reparameterize the vertical coordinates on each of these Riemann surfaces by the inverses of the maps ν_n, and then glue together the result by matching up boundary circles using tags. This defines an inverse to the map $\mathcal{DL}_q'' \to \mathcal{X}_q$ and shows that this map is a homeomorphism; see **[11, THM. 9.20]** for details.

1.13. Decomposition of the shift locus

Putting together the various constructions we have discussed so far, we obtain the following summary:

(1) Section 1.7 describes the map that associates to $f \in \mathcal{S}_q$ a maximal degree q dynamical elamination Λ gives an isomorphism of complex manifolds $\mathcal{S}_q \to \mathcal{DL}_q$.

(2) Section 1.8 elaborates on how, by compressing orbits of the free $\mathbb{R}^+$ action on $\mathcal{DL}_q$, we obtain a homeomorphism $\mathcal{DL}_q \to \mathcal{DL}_q'$ to the subspace whose largest critical leaf has log-height $\log_q(h) \in (-1/2, 1/2)$.

(3) Section 1.12 discusses how the open dense subset $\mathcal{DL}_q'' \subset \mathcal{DL}_q'$ of generic dynamical elaminations maps homeomorphically by the sausage map $\mathcal{DL}_q'' \to \mathcal{X}_q$.

(4) Section 1.11 tells that the space $\mathcal{X}_q$ is the disjoint union of countably many quasiprojective complex varieties, each of which has the structure of an iterated bundle of affine (Hurwitz) varieties.

In words, the shift locus $\mathcal{S}_q$ of degree q has a canonical decomposition into codimension 0 submanifolds whose interiors are homeomorphic to certain explicit algebraic varieties. It is a fact that we do not explain here (see **[11, § 8 ESPECIALLY THM. 8.11]**) that the abstract cell complex which combinatorially parameterizes the decomposition of $\mathcal{S}_q$ into these pieces is *contractible*, so that all the interesting topology of $\mathcal{S}_q$ is localized in the components of $\mathcal{X}_q$.

In the remainder of the paper we give examples, and explore some of the consequences of this structure.

2. SAUSAGE MODULI

Each component Y of $\mathcal{X}_q$ parameterizes sausages of a fixed combinatorial type. The combinatorial type determines finitely many vertices u for which the (normalized) polynomial $\mathfrak{f}_u$ has degree > 1. The combinatorics constrains these polynomials by imposing conditions on their critical values, for instance, that the critical values are required to lie outside a certain (finite) set. Thus, each component has the structure of an algebraic variety which is an iterated fiber bundle, and so that the base and each fiber is something called a *Hurwitz variety*.

For this and other reasons, the spaces $\mathcal{S}_q$ and the components Y of which they are built bear a close family resemblance to the kinds of discriminant complements that arise in the study of classical braid groups. The full extent of this resemblance is an open question, partially summarized in Table 2.

2.1. Degree 2

Let $\mathfrak{X}, \mathfrak{f}$ be a generic maximal sausage polynomial of degree 2. The root polynomial $\mathfrak{f}_v$ is quadratic and normalized. It has one critical point, necessarily genuine. Thus $\mathfrak{f}_v(z) := z^2 + c$ for some $c \neq 0$. Every other vertex w has a polynomial $\mathfrak{f}_w$ of degree one; since polynomials are normalized, $\mathfrak{f}_w(z) := z$. Thus all the information is contained in the choice of the (nonzero) constant coefficient c of $\mathfrak{f}_v$, so that $\mathcal{X}_2 = \mathbb{C}^*$. The tree T is an infinite dyadic rooted tree, where every vertex is attached to its parent at the points $\pm\sqrt{-c}$; see Figure 5.

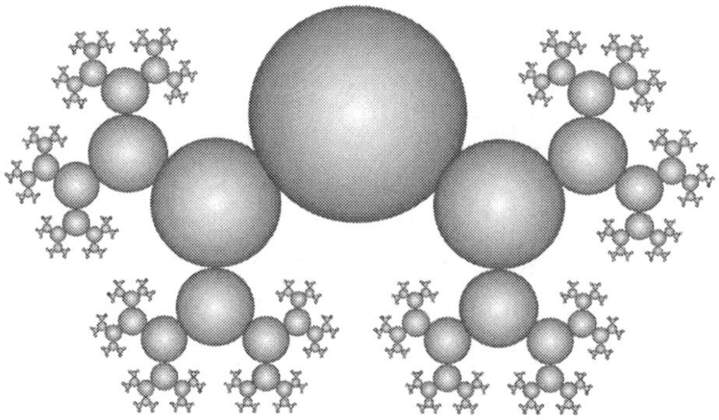

FIGURE 5
A degree 2 sausage; each vertex is attached to its parent at the points $\pm\sqrt{-c}$.

Furthermore, in this case $\mathcal{DL}_2' = \mathcal{DL}_2''$ so that $\mathcal{S}_2$ is homeomorphic (but *not* holomorphically isomorphic) to $\mathbb{C}^*$. As a corollary, one deduces the famous theorem of Douady–Hubbard [17] that the Mandelbrot set $\mathcal{M}$ (i.e., $\mathbb{C} - \mathcal{S}_2$) is connected.

2.2. Discriminant locus

In any degree q, there is a unique component of $\mathcal{X}_q$ for which all the (genuine) critical points are in the root vertex. Thus $\mathfrak{f}_v$ is a degree q normalized polynomial with no fake critical points. Since the marked points Z_v of the root vertex are exactly the *roots* of $\mathfrak{f}_v$, this means that $\mathfrak{f}_v$ is a normalized polynomial with no critical roots. Equivalently, $\mathfrak{f}_v$ has q distinct roots, so that $\mathfrak{f}_v$ is in $Y_q := \mathbb{C}^{q-1} - \Delta_q$ where Δ_q is the *discriminant locus*. As is well known, Y_q is a $K(B_q, 1)$ where B_q denotes the braid group on q strands.

2.3. Degree 3

Let $\mathcal{X}, \mathfrak{f}$ be a generic maximal sausage polynomial of degree 3. If the root polynomial $\mathfrak{f}_v$ has two genuine critical points, we are in the case discussed in Section 2.2 and the corresponding component of $\mathcal{X}_3$ is a $K(B_3, 1)$. Otherwise, since the root polynomial must have at least one genuine critical point, if it does not have two, it must have exactly one and $\mathfrak{f}_v$ is of the form $z \to (z-c)^2(z+2c)$ for some $c \in \mathbb{C}^*$.

The (finite) marked points Z_v of $\widehat{\mathbb{C}}_v$ are c and $-2c$, and the root vertex correspondingly has two children w_1, w_2 where $\widehat{\mathbb{C}}_{w_1}$ is attached at c and $\widehat{\mathbb{C}}_{w_2}$ is attached at $-2c$. Because c is a double root, the polynomial $\mathfrak{f}_{w_1}$ has degree 2; because $-2c$ is a simple root, the polynomial $\mathfrak{f}_{w_2}$ has degree 1.

Write $\mathfrak{f}_{w_1} : z \to z^2 + d$. If $d \neq c, -2c$ then Z_{w_1} has four (noncritical) points (the distinct square roots of $c - d$ and $-2c - d$) and every other $\mathfrak{f}_u$ is degree 1. See Figure 6. Thus c and d are moduli parameterizing a single component of $\mathcal{X}_3$, and topologically this component is a bundle over $\mathbb{C}^*$ whose fiber is homeomorphic to $\mathbb{C} - \{c, -2c\}$.

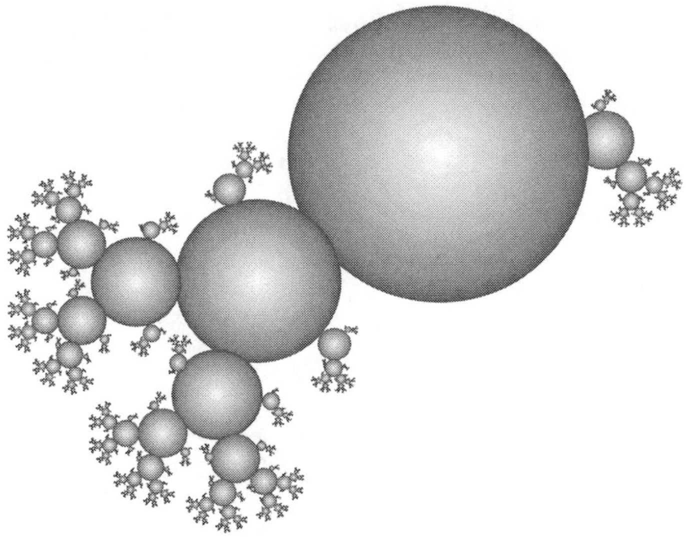

FIGURE 6
A degree 3 sausage; the root v has $Z_v := \{c, -2c\}$. The child w_1 has $Z_{w_1} := \{\pm\sqrt{c-d}, \pm\sqrt{-2c-d}\}$.

If $d = c$ or $d = -2c$ then 0 is a fake critical point for $\mathfrak{f}_{w_1}$, and if u is the child of w_1 for which $\widehat{\mathbb{C}}_u$ is attached at 0 then $\mathfrak{f}_u$ has degree 2. Since $\mathfrak{f}$ is maximal, there is always some vertex u' at finite combinatorial distance from the root for which $\mathfrak{f}_{u'}$ has degree 2 and for which the critical point 0 of $\mathfrak{f}_{u'}$ is genuine. Thus each component of $\mathcal{X}_3$ is a bundle over $\mathbb{C}^*$ with fiber homeomorphic to $\mathbb{C}$ minus finitely many points.

2.4. The tautological elamination

The combinatorics of the components of $\mathcal{X}_3$ is quite complicated. Each component of $\mathcal{X}_3$ (other than the discriminant complement, cf. Section 2.2) is a punctured plane bundle over the curve $\mathbb{C}^*$ with parameter c, and these components glue together in $\mathcal{S}_3$ to form a bundle over $\mathbb{C}^*$ whose fiber Ω_T is homeomorphic to a plane minus a Cantor set.

Actually, there is another description of Ω_T in terms of elaminations. For each degree 3 critical leaf C, there is a certain elamination $\Lambda_T(C)$ called the *tautological elamination* which can be defined as follows. Let us suppose that we have a maximal degree 3 dynamical elamination with two critical leaves C and C', and that C has the greater height. If we fix C, then Ω_T parameterizes the space of configurations of C'.

The elamination $\Lambda_T(C)$ is defined as follows. With C fixed, each choice of (non-crossing) C' determines a dynamical elamination Λ. By hypothesis, $h(C') < h(C)$ and there are only finitely many (perhaps zero) precritical leaves P of C with $h(P) > h(C')$. As we vary C', the laminations Λ also vary (in rather a complicated way), but while $h(P) > h(C')$ the leaves P stay fixed under continuous variations of C'. It might happen that, as we vary the leaf C', it collides with a leaf P with $h(P) > h(C')$; the elamination $\Lambda_T(C)$ consists of the *cubes* P^3 of all such P (there is a similar, though more complicated construction in higher degrees). The fact that $\Lambda_T(C)$ is an elamination is not obvious from this definition.

The result of cut and paste (as in Section 1.4) on the annulus $1 < |z| < |C|$ (thought of as a subset of $\mathbb{E}$) along $\Lambda_T(C)$ is a Riemann surface $\Omega_T(C)$ holomorphically isomorphic to the moduli space of degree 3 maximal dynamical elaminations for which C is the unique critical leaf of greatest height. Figure 7 depicts the elamination $\Lambda_T(C)$ for a particular value of C whose tips have angles $\pm\pi/3$.

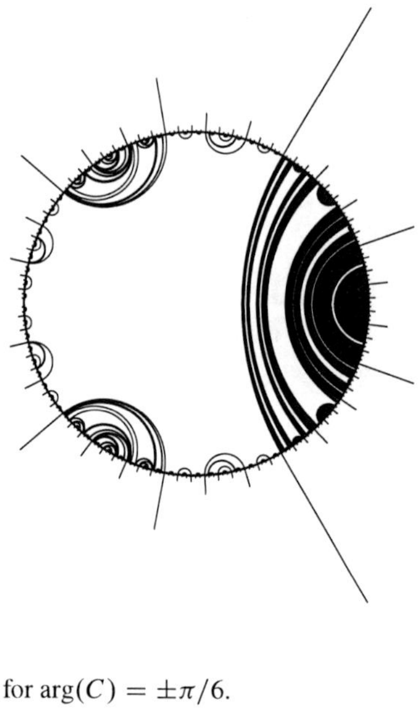

FIGURE 7
The tautological elamination $\Lambda_T(C)$ for $\arg(C) = \pm\pi/6$.

These $\Omega_T(C)$ are the leaves of a (singular) one complex dimensional holomorphic foliation of $\mathcal{S}_3$.

Although it is not a dynamical elamination, the tautological elamination $\Lambda_T(C)$ is in a natural way the increasing union of finite elaminations Λ_n, namely the leaves of the form

P^3 as above where P is a depth n preimage of C. Let $\overline{\mathbb{E}}$ denote the closure of $\mathbb{E}$ in $\mathbb{C}$ so that $\overline{\mathbb{E}} = \mathbb{E} \cup S^1$, the union of $\mathbb{E}$ with the unit circle. The result Ω_n of cut and pasting $\mathbb{E}$ along Λ_n is partially compactified by a finite set of circles, obtained from S^1. By abuse of notation, we denote this finite set of circles by $S^1 \mod \Lambda_n$. It turns out that the the components of $\mathcal{X}_3 \cap \Omega_T$ corresponding to sausage polynomials with fixed $c \in \mathbb{C}^*$ and for which the second genuine critical point is in a vertex at depth $n+1$ are in bijection with the set of components of $S^1 \mod \Lambda_n$. In fact, more is true.

For each combinatorial type $\mathfrak{X}, \mathfrak{f}$, let u be the vertex containing the second genuine critical point (the first, by hypothesis, is contained in the root). We define the *depth n* of $\mathfrak{X}, \mathfrak{f}$ to be the combinatorial distance of u to the root. There is another invariant of $\mathfrak{X}, \mathfrak{f}$: the *$\ell$-value*, defined as follows. Under iteration of $\mathfrak{f}$ (acting on the tree), the vertex u has a length n orbit terminating in the root (note that $\mathfrak{f}(u)$ is *not* typically equal to the parent of u, but it does have the same depth as the parent). The point ∞ in $\widehat{\mathbb{C}}_u$ is mapped to ∞ in $\widehat{\mathbb{C}}_{\mathfrak{f}(u)}$ and so on. The product of the degrees of the polynomials $\mathfrak{f}_{\mathfrak{f}^i(u)}$ up to but not including the root is some power of 2; by definition, ℓ is this number divided by 2. The invariants n and ℓ, taking discrete values, are really invariants of the components of $\mathcal{X}_3$ and ipso facto of the components of $\mathcal{X}_3 \cap \Omega_T$.

Here is the relation to $\Lambda_T(C)$. Components of $\mathcal{X}_3 \cap \Omega_T$ of depth $n+1$ are in bijective correspondence with components of $S^1 \mod \Lambda_n$, and a component of $\mathcal{X}_3 \cap \Omega_T$ with ℓ-value ℓ corresponds to a component of $S^1 \mod \Lambda_n$ of length $2\pi\ell \cdot 3^{-n}$.

2.5. Combinatorics

Let $N_3(n, m)$ denote the number of components of $S^1 \mod \Lambda_n$ with depth $n+1$ and $\ell = 2^m$. We do not know a simple closed form for $N_3(n, m)$ and perhaps none exists—one subtle issue is that there are several combinatorially different ways that a component can have a particular ℓ-value. However, an ℓ-value of 1 is special, since it corresponds to an $\mathfrak{f}$ for which $\mathfrak{f}_{\mathfrak{f}^i(u)}$ has degree 1 for all positive i. Correspondingly, there is an explicit formula for $N_3(n, 0)$ that we now give; see [10, THM. 3.6] for a proof.

First of all, $N_3(n, 0)$ satisfies the recursion $N_3(0, 0) = 1$, $N_3(1, 0) = 1$, and

$$N_3(2n, 0) = 3N_3(2n-1, 0) \quad \text{and} \quad N_3(2n+1, 0) = 3N_3(2n, 0) - 2N_3(n, 0).$$

Knowing this, one can write down an explicit generating function for $N_3(n, 0)$; the generating function is $(\beta(t) - 1)/3t$ where

$$\beta(t) = \left(\sum_{n=0}^{\infty} h(n)t^n\right) \prod_{j=0}^{\infty} \frac{1}{(1 - 3t^{2^j})}$$

and where the numbers $h(n)$ are defined by

$$h(0) = 1 \quad \text{and} \quad h(n) = (-3)^{s(n)}\left(1 - (-2)^{k(n)}\right)$$

with $2^{k(n)}$ being the biggest power of 2 dividing n, and $s(n)$ the sum of the binary digits of n.

Table 1 gives values of $N_3(n, m)$ for $0 \le n, m \le 12$. Note that $N_3(n, m) = 0$ for $n/2 < m < n$; see [10, THM 5.9].

n	ℓ												
	1	2	2^2	2^3	2^4	2^5	2^6	2^7	2^8	2^9	2^{10}	2^{11}	2^{12}
0	1												
1	1	1											
2	3	1	1										
3	7	6	0	1									
4	21	16	3	0	1								
5	57	51	13	0	0	1							
6	171	149	39	5	0	0	1						
7	499	454	117	23	0	0	0	1					
8	1497	1348	360	66	9	0	0	0	1				
9	4449	4083	1061	207	41	0	0	0	0	1			
10	13347	12191	3252	591	126	17	0	0	0	0	1		
11	39927	36658	9738	1799	370	81	0	0	0	0	0	1	
12	119781	109898	29292	5351	1125	240	33	0	0	0	0	0	1

TABLE 1
Number of components of length $\ell/3^n$ at depth n.

2.6. Hurwitz varieties

Let X be a component of $\mathcal{X}_q$ parameterizing sausage polynomials of a fixed combinatorial type. Then X is an iterated bundle whose base and fibers Y are all of the following sort. There are specific vertices u, w with $\mathfrak{f}(u) = w$. The set $Z_w \subset \widehat{\mathbb{C}}_w$ is fixed, as is the degree p of $\mathfrak{f}_u : \widehat{\mathbb{C}}_u \to \widehat{\mathbb{C}}_w$. Furthermore, for each $\zeta \in Z_w$ the *ramification data* of $\mathfrak{f}_u$ at ζ is specified, i.e., the monodromy of $\mathfrak{f}_u^{-1}$ in a small loop around each ζ, thought of as a conjugacy class in the symmetric group on p letters. Then Y is the space of normalized degree p polynomials with the specified ramification data. We call Y a *Hurwitz variety*, and observe that each X is an iterated bundle with total (complex) dimension $q-1$ whose base and fibers are all Hurwitz varieties.

The generic case is that the monodromy of $\mathfrak{f}_u^{-1}$ in a small loop around each $\zeta \in Z_w$ is trivial, i.e., each ζ is a regular value. In that case, Y is a Zariski open subset of $\mathbb{C}^{p-1}$. In fact, we can say something more precise. Let $\Delta_p \subset \mathbb{C}^{p-1}$ be the discriminant variety, i.e., the set of normalized degree p polynomials with a multiple root. For each $\zeta \in Z_w$, let $\Delta_{p,\zeta} := \Delta_p + \zeta$ be the *translate* of Δ_p which parameterizes the set of normalized degree p polynomials f for which ζ is a critical value. Then

$$Y = \mathbb{C}^{p-1} - \bigcup_{\zeta \in Z_w} \Delta_{p,\zeta}.$$

It turns out that the topology of Y depends only on the cardinality of Z_w; see **[11, PROP. 9.14]**. This is not obvious, since the $\Delta_{p,\zeta}$ are singular, and they do not intersect in general position.

2.7. $K(\pi, 1)$s

For a finite set $Z \subset \mathbb{C}$ and degree p, let $Y_p(Z)$ denote the Hurwitz variety of normalized degree p polynomials for which no element of Z is a critical value.

As we remarked already in Section 2.2, when $|Z| = 1$ the space $Y_p(Z)$ is a $K(B_p, 1)$ where B_p denotes the braid group on p strands. Furthermore, when $p = 2$ the space $Y_2(Z)$ may be identified with $\mathbb{C} - Z$ in the obvious way, so that $Y_2(Z)$ is a $K(F_n, 1)$ where F_n is the free group on n elements, and $n = |Z|$.

It turns out (see [11, THM. 9.17]) that $Y_3(Z)$ is a $K(\pi, 1)$ for any finite set Z. This is proved by exhibiting an explicit CAT(0) 2-complex with the homotopy type of each $Y_3(Z)$. One component of $\mathcal{X}_4$ is a $K(B_4, 1)$ and all the others are nontrivial iterated fibrations where the fibers are $Y_2(Z)$ or $Y_3(Z)$s. It follows that every component of $\mathcal{X}_4$ is a $K(\pi, 1)$, and, in fact, so is the shift locus $\mathcal{S}_4$ itself (the same is true for simpler reasons of $\mathcal{S}_3$ and $\mathcal{S}_2$).

One knows few examples of algebraic varieties which are $K(\pi, 1)$s, and fewer methods to construct or certify them (one of the few general methods, which applies to certain complements of hyperplane arrangements, is due to Deligne [15]). Is $Y_p(Z)$ a $K(\pi, 1)$ for all p and all Z?

2.8. Monodromy

For each p and $|Z|$, there is a natural representation (well defined up to conjugacy) $\pi_1(Y_p(Z)) \to B_{p|Z|}$ defined by the braiding of the $p|Z|$ points $f^{-1}(Z)$ in $\mathbb{C}$ as f varies in $Y_p(Z)$. This map is evidently injective when $p = 2$ or when $|Z| = 1$. Is it injective in any other case? I do not know the answer even when $p = 3$ and $|Z| = 2$.

Here is one reason to be interested. There is a monodromy representation of $\pi_1(\mathcal{S}_q)$ into the "Cantor braid group", i.e., the mapping class group of a disk minus a Cantor set, defined by the braiding of the (Cantor) Julia set J_f in $\mathbb{C}$ as f varies in $\mathcal{S}_q$. A priori this representation lands in the mapping class group of the plane minus a Cantor set, but it lifts canonically to the Cantor braid group (which is a central extension) because every $f \in \mathcal{S}_q$ acts in a standard way at infinity. If one forgets the braiding and only considers the permutation action on the Cantor set itself, the image in Aut(Cantor set) is known to be precisely equal to the automorphism group of the full (one-sided) shift on a q element alphabet, by a celebrated theorem of Blanchard–Devaney–Keen [5]. However, this action of $\pi_1(\mathcal{S}_q)$ on the Cantor set alone is very far from faithful.

The automorphism group of the Cantor set is to the Cantor braid group as a finite symmetric group is to a (finite) braid group. It is natural to ask: Is the monodromy representation from $\pi_1(\mathcal{S}_q)$ to the Cantor braid group injective? It turns out that the restriction of the monodromy representation to the image of $\pi_1(Y_p(Z))$ in $\pi_1(\mathcal{S}_q)$ factors through the representation to $B_{p|Z|}$. So a precondition for the monodromy representation to the Cantor braid group to be injective is that each $\pi_1(Y_p(Z)) \to B_{p|Z|}$ should be injective.

When $q = 2$, we have $\pi_1(\mathcal{S}_2) = \mathbb{Z}$ and the monodromy representation is evidently injective, since the Cantor braid group is torsion-free. With Yan Mary He and Juliette Bavard, we have shown that the monodromy representation is injective in degree 3 (work in progress).

2.9. Big mapping class groups

The Cantor braid group and the (closely related) mapping class group of the plane minus a Cantor set are quintessential examples of what are colloquially known as *big mapping class groups*. The study of these groups is an extremely active area of current research; for an excellent recent survey, see Aramayona–Vlamis [1]. There are connections to the theory of finite-type mapping class groups (particularly to stability and uniformity phenomena in such groups); to taut foliations of 3-manifolds; to pruning theory and the de-Carvalho–Hall theory of endomorphisms of planar trees; to Artinizations of Thompson-like groups and universal algebra, etc. (see [1] for references).

One major goal of this theory – largely unrealized as of yet – is to develop new tools for applications to dynamics in 2 real and 1 complex dimension. Cantor sets appear in surfaces as attractors of hyperbolic systems (e.g., in Katok–Pesin theory [18]), and big mapping class groups (and some closely related objects) are relevant to the study of their moduli. The paper [11] and the theory of sausages is an explicit attempt to work out some of these connections in a particular case.

2.10. Rays

Let Γ denote the mapping class group of the plane (which we identify with $\mathbb{C}$) minus a Cantor set K. The Cantor braid group $\widehat{\Gamma}$ is the universal central extension of Γ. Some of the tools discussed in this paper may be used to study $\widehat{\Gamma}$ and its subgroups in some generality; for instance, components of $\mathcal{EL}$ are classifying spaces for subgroups of $\widehat{\Gamma}$.

The group Γ acts in a natural way on the set $\mathcal{R}$ of *isotopy classes of proper simple rays* in $\mathbb{C} - K$ from ∞ to a point in K. Associated to this action are two natural geometric actions of Γ:

(1) there is a natural circular order on $\mathcal{R}$, so that Γ acts faithfully by order-preserving homeomorphisms on a certain completion of $\mathcal{R}$, the *simple circle*; see [3,7,12]; and

(2) the elements of $\mathcal{R}$ are the vertices of a (connected) graph (the *ray graph*) whose edges correspond to pairs of rays that may be realized disjointly; this graph is connected, has infinite diameter, and is Gromov-hyperbolic; see [2,9].

(Landing) Rays are also a critical tool in complex dynamics, and in the picture developed in the previous two sections. For K a Cantor Julia set, nonsingular gradient flowlines of the Green's function extend continuously to K; the set of distinct isotopy classes of nonsingular flowlines associated to single K form a clique in the ray graph. Because the ray graph is Gromov-hyperbolic, there is (up to bounded ambiguity) a canonical path in the ray graph between any two such cliques; one can ask whether such paths are coarsely realized by paths in $\mathcal{S}_q$, and if so what geometric properties such paths have, and how this geometry manifests itself in algebraic properties of $\pi_1(\mathcal{S}_q)$. For example, does $\pi_1(\mathcal{S}_q)$ admit a (bi)automatic structure? (To make sense of this, one should work with a locally finite groupoid presentation for $\pi_1(\mathcal{S}_q)$.) One piece of evidence in favor of this is that $\mathcal{S}_3$ (and, for trivial reasons, $\mathcal{S}_2$)

is homotopy equivalent to a locally CAT(0) complex, and it is plausible that the same holds for all $\mathcal{S}_q$. Although there are known examples of groups which are locally CAT(0) but not biautomatic [19], nevertheless in practice these two properties often go hand in hand.

2.11. Left orderability

A group is left-orderable if it admits a total order that is preserved under left multiplication. The left-orderability of braid groups (see [14]) is key to some of their most important properties (e.g., faithfulness of the Lawrence–Kraamer–Bigelow representations [4]). Left-orderability of 3-manifold groups is also conjecturally ([6]) related to both symplectic topology (via Heegaard Floer homology) and to big mapping class groups via the theory of taut foliations and universal circles; see, e.g., [8, 13]. The Cantor braid group is left-orderable (via the faithful action of Γ on the simple circle) so, to show that $\pi_1(\mathcal{S}_q)$ is left-orderable, it would suffice to prove injectivity of the monodromy representation as in Section 2.8.

2.12. Comparison with finite braids

Define $Y_q := \mathbb{C}^{q-1} - \Delta_q$, the space of normalized degree q polynomials without multiple roots. Our study of $\mathcal{S}_q$ has been guided by a heuristic that one should think of $\mathcal{S}_q$ as a sort of "dynamical cousin" to Y_q, and that they ought to share many key algebraic and geometric properties. Table 2 compares some of what is known about the topology of Y_q and $\mathcal{S}_q$.

	Y_q	$\mathcal{S}_2, \mathcal{S}_3$	$\mathcal{S}_4$	$\mathcal{S}_q, q > 4$
locally CAT(0)	yes for $q \leq 6$	yes	unknown	unknown
$K(\pi, 1)$	yes	yes	yes	unknown
H_* vanishes below middle dimension	yes	yes	yes	yes
π_1 is mapping class group	yes	yes	unknown	unknown
π_1 is left-orderable	yes	yes	unknown	unknown
π_1 is biautomatic	yes	yes	yes	unknown

TABLE 2
Comparison of $\mathcal{S}_q$ with discriminant complements Y_q.

ACKNOWLEDGMENTS

I would like to thank Lvzhou Chen, Toby Hall, Sarah Koch, Curt McMullen, Sandra Tilmon, Alden Walker, and Amie Wilkinson for useful feedback on early drafts of this paper.

REFERENCES

[1] J. Aramayona and N. Vlamis, Big mapping class groups: an overview. arXiv:2003.07950.

[2] J. Bavard, Hyperbolicité du graphe des rayons et quasi-morphismes sur un gros groupe modulaire. *Geom. Topol.* **20** (2016), 491–535.

[3] J. Bavard and A. Walker, The Gromov boundary of the ray graph. *Trans. Amer. Math. Soc.* **370** (2018), no. 11, 7647–7678.

[4] S. Bigelow, Braid groups are linear. *J. Amer. Math. Soc.* **14** (2000), no. 2, 471–486.

[5] P. Blanchard, L. Devaney, and L. Keen, The dynamics of complex polynomials and automorphisms of the shift. *Invent. Math.* **104** (1991), 545–580.

[6] S. Boyer, C. Gordon, and L. Watson, On L-spaces and left-orderable fundamental groups. *Math. Ann.* **356** (2013), no. 4, 1213–1245.

[7] D. Calegari, Circular groups, planar groups and the Euler class. In *Proceedings of the Casson fest*, pp. 431–491, Geom. Topol. Monogr. 7, Geom. Topol. Publ., Coventry, 2004.

[8] D. Calegari, *Foliations and the geometry of 3-manifolds*. Oxford Math. Monogr., Oxford University Press, Oxford, 2007.

[9] D. Calegari, Big mapping class groups and dynamics, blog post, 2009, https://lamington.wordpress.com/2009/06/22/big-mapping-class-groups-and-dynamics/

[10] D. Calegari, Combinatorics of the tautological lamination. arXiv:2106.00578.

[11] D. Calegari, Sausages and Butcher paper. arXiv:2105.11265.

[12] D. Calegari and L. Chen, Big mapping class groups and rigidity of the simple circle. *Ergodic Theory Dynam. Systems* **41** (2021), no. 7, 1961–1987.

[13] D. Calegari and N. Dunfield, Laminations and groups of homeomorphisms of the circle. *Invent. Math.* **152** (2003), no. 1, 149–204

[14] P. Dehornoy, I. Dynnikov, D. Rolfsen and B. Wiest, *Ordering braids*. Math. Surveys Monogr. 148, AMS, Providence, RI, 2008.

[15] P. Deligne, Les immeubles des groupes de tresses généralisés, *Invent. Math.* **17** (1972), 273–302.

[16] L. DeMarco and C. McMullen, Trees and the dynamics of polynomials. *Ann. Sci. Éc. Norm. Supér. (4)* **41** (2008), no. 3, 337–382.

[17] A. Douady and J. Hubbard, Itération des polynômes quadratiques complexes. *C. R. Acad. Sci. Paris Sér. I Math.* **294** (1982), no. 3, 123–126.

[18] A. Katok, Lyapunov exponents, entropy and periodic orbits for diffeomorphisms. *Publ. Math. Inst. Hautes Études Sci.* **51** (1980), 137–173.

[19] I. Leary and A. Minasyan, Commensurating HNN-extensions: non-positive curvature and biautomaticity. arXiv:1907.03515.

[20] J. Milnor, *Dynamics in one complex variable*. Third edition, Ann. of Math. Stud. 160, Princeton University Press, Princeton NJ, 2006.

DANNY CALEGARI

University of Chicago, Department of Mathematics, Eckhart Hall, 5734 S University Ave, Chicago IL, 60637, USA, dannyc@math.uchicago.edu

LAGRANGE MULTIPLIER FUNCTIONALS AND THEIR APPLICATIONS IN SYMPLECTIC GEOMETRY AND STRING TOPOLOGY

KAI CIELIEBAK

ABSTRACT

This note discusses the role of Lagrange multiplier functionals in mathematics and physics. The main focus is on Rabinowitz' action functional and its usage in symplectic geometry, as well as recent applications in string topology and the study of closed geodesics.

MATHEMATICS SUBJECT CLASSIFICATION 2020

Primary 53D40; Secondary 55N45, 55P35, 57R17, 57R19

ICM INTERNATIONAL CONGRESS OF MATHEMATICIANS 2022 JULY 6–14

Proc. Int. Cong. Math. 2022, Vol. 4, pp. 2504–2528
DOI 10.4171/ICM2022/46

Published by EMS Press

1. INTRODUCTION

The purpose of this note is to tell the story of how an old and simple idea—Lagrange multipliers—has led to new insights in symplectic geometry and loop space topology.

The beginning of our story is the observation by Joseph-Louis Lagrange in 1804 [58] that critical points of a functional $f(x)$ subject to a constraint $h(x) = 0$ correspond to unconstrained critical points of the function $F(x, \lambda) = f(x) - \lambda h(x)$ depending on a Lagrange multiplier λ. In modern terms, $f : X \to \mathbb{R}$ and $h : X \to V$ should be sufficiently smooth maps, where X is a Banach manifold and V a Banach space. Denoting by $\langle \cdot, \cdot \rangle$ the canonical pairing between V and its topological dual V^*, we consider the *Lagrange multiplier functional*

$$F : X \times V^* \to \mathbb{R}, \quad F(x, \lambda) = f(x) - \langle \lambda, h(x) \rangle.$$

Then (x, λ) is a critical point of F if and only if

$$df(x) = \langle \lambda, dh(x) \rangle \quad \text{and} \quad h(x) = 0.$$

Assuming that 0 is a regular value of h, so that $Z = h^{-1}(0) \subset X$ is a Banach submanifold, this is equivalent to x being a critical point of the restriction $f|_Z$. The Lagrange multiplier λ at x is uniquely determined by the first equation. Although it was introduced as an auxiliary parameter, the Lagrange multiplier often has mathematical or physical meaning.

Example 1.1 (Eigenvalues). Let X be a complex Hilbert space and $A : X \to X$ a self-adjoint bounded linear operator. Consider the functions

$$f, h : X \to \mathbb{R}, \quad f(x) = \langle x, Ax \rangle, \quad h(x) = \|x\|^2 - 1.$$

Then the critical points of the restriction of f to the unit sphere $S = h^{-1}(0)$ correspond to solutions $(x, \lambda) \in X \times \mathbb{R}$ of the equations $\|x\| = 1$ and $Ax = \lambda x$, so the Lagrange multiplier $\lambda \in \mathbb{R}$ is an eigenvalue of A with eigenvector x. If A is compact (e.g., if X is finite dimensional), then f attains its maximum and minimum on S and it follows that $\|A\|$ or $-\|A\|$ is an eigenvalue.

The Hessian of F at a critical point (x, λ) is given by

$$\text{Hess}\, F(x, \lambda) = \begin{pmatrix} \text{Hess}\, f(x) & dh(x)^* \\ dh(x) & 0 \end{pmatrix}.$$

If X and V are finite dimensional, it follows that the Hessians Hess $F(x, \lambda)$ and Hess $f|_Z(x)$ have the same nullity and signature (the number of positive minus the number of negative eigenvalues). These relations also hold in some infinite-dimensional cases where nullity and signature can be defined; one such case arises for Hamiltonian systems where the role of the signature is played by the Conley–Zehnder index, see Section 2.

We see in particular that the Hessian of F is never positive or negative definite, so its critical points cannot be detected by direct maximization or minimization methods and one needs to resort to indirect variational methods. Of particular relevance for this note will be *Morse homology* (see, e.g., [70, 77]). This is the homology of the chain complex whose generators are critical points of F and whose differential counts gradient trajectories $(x, \lambda) : \mathbb{R} \to X \times V^*$ between critical points. Based on the preceding discussion, we expect that the Morse homology of F equals the Morse homology of $f|_Z$ if both are graded by the signature rather than the Morse index. However, even in finite dimensions it is not obvious that both Morse homologies are defined, and in addition equal, due to the possible escape of gradient trajectories to infinity. This issue will be a recurring theme in this note, which is structured as follows.

Section 2 focuses on a specific Lagrange multiplier functional, the Rabinowitz action functional, and its applications in symplectic geometry. Section 3 presents some recent applications of the ideas from Section 2 in string topology. Section 4 discusses some further occurrences of Lagrange multiplier functionals in mathematics and physics. Besides established results, I will also discuss some work in progress, as well as open questions.

2. RABINOWITZ FLOER HOMOLOGY

In this section we will focus on one particular Lagrange multiplier functional, the Rabinowitz action functional, and discuss properties and applications of the corresponding Floer homology. For more details and background, see the original references or the survey by P. Albers and U. Frauenfelder [9].

2.1. Definition and basic properties

Let (W, λ) be a *Liouville manifold* of dimension $2n$, i.e., a connected manifold with a 1-form such that $\omega = d\lambda$ is symplectic and W is exhausted by compact sets W_k with smooth boundary such that $\lambda|_{\partial W_k}$ is a positive contact form. Examples of Liouville manifolds include $\mathbb{C}^n$, cotangent bundles, and, more generally, Stein and Weinstein manifolds (see [18]).

To a 1-periodic time-dependent Hamilton function $H : S^1 \times W \to \mathbb{R}$, we associate its Hamiltonian vector field X_{H_t} by $dH_t = \omega(\cdot, X_{H_t})$, where $H_t = H(t, \cdot)$. Then 1-periodic solutions $x : S^1 \to W$ of the Hamiltonian system $\dot{x} = X_{H_t}(x)$ are the critical points of the Hamiltonian action

$$\mathcal{A}_H : C^\infty(S^1, W) \to \mathbb{R}, \quad \mathcal{A}_H(x) = \int_x \lambda - \int_0^1 H(t, x)dt.$$

Assume now that $H : W \to \mathbb{R}$ is time-independent. Then we have conservation of energy and it is natural to look for solutions of prescribed energy rather than prescribed period. For this, suppose that 0 is a regular value of H and consider the *Rabinowitz action functional*

$$\mathcal{A}^H : C^\infty(S^1, W) \times \mathbb{R} \to \mathbb{R}, \quad \mathcal{A}^H(x, \eta) = \int_x \lambda - \eta \int_0^1 H(x)dt.$$

Its critical points satisfy the equations

$$\dot{x} = \eta X_H(x), \quad \int_0^1 H(x)dt = 0.$$

By the first equation, $H(x(t))$ is constant and, by the second equation, this constant equals zero, so the critical point equations become

$$\dot{x} = \eta X_H(x), \quad H\big(x(t)\big) \equiv 0.$$

Critical points of $\mathcal{A}^H$ thus correspond to orbits $t \mapsto x(t/\eta)$ of X_H of period η and energy 0 (if $\eta > 0$), such orbits run backwards (if $\eta < 0$), or to constant loops on $H^{-1}(0)$ (if $\eta = 0$). As we will see in Section 3, the appearance of solutions with negative η is responsible for an additional symmetry of the corresponding Floer homology.[1] In 1978, P. Rabinowitz used this functional to prove existence of periodic orbits on star-shaped energy hypersurfaces in $\mathbb{C}^n$ **[66]**.[2]

To define the Floer homology of $\mathcal{A}^H$, we pick an ω-compatible almost complex structure J on W and equip $C^\infty(S^1, W) \times \mathbb{R}$ with the metric

$$m_{(x,\eta)}\big((\hat{x}_1, \hat{\eta}_1), (\hat{x}_2, \hat{\eta}_2)\big) = \int_0^1 \omega(\hat{x}_1, J\hat{x}_2)dt + \hat{\eta}_1\hat{\eta}_2.$$

Then gradient flow lines of $\mathcal{A}^H$ are maps $(u, \eta) : \mathbb{R} \to C^\infty(S^1, W) \times \mathbb{R}$, satisfying

$$\partial_s u + J(u)\big(\partial_t u - \eta X_H(u)\big) = 0, \quad \partial_s \eta + \int_0^1 H(u)dt = 0, \tag{2.1}$$

where (s, t) are the coordinates on $\mathbb{R} \times S^1$. This is a coupled system of an elliptic PDE and a nonlocal ODE. Its solutions exhibit three potential sources of noncompactness: explosion of the gradient of u, which is excluded by exactness of ω; escape of u to infinity, which can be prevented by suitable conditions on J; and escape of the Lagrange multiplier η to $\pm\infty$. To prevent the latter, we need to impose some geometric condition on the hypersurface $\Sigma = H^{-1}(0)$.

A hypersurface $\Sigma \subset W$ is of *restricted contact type* if it admits a contact form α such that $\alpha - \lambda|_\Sigma$ is exact (for $H^1(\Sigma; \mathbb{R}) = 0$ this agrees with A. Weinstein's contact type condition **[78]**). We also assume that Σ is connected and bounds a compact subset. Then it admits a (nonunique) *defining Hamiltonian*, i.e., a smooth function $H : W \to \mathbb{R}$ which is constant outside a compact set such that $H^{-1}(0) = \Sigma$ and $X_H = R$ along Σ, where R is the Reeb vector field of α.

Theorem 2.1 (**[19]**). (a) *Given a hypersurface $\Sigma \subset W$ of restricted contact type and a defining Hamiltonian H, the Floer homology* $\mathrm{FH}_*(\mathcal{A}^H)$ *is well defined; it is independent of the defining Hamiltonian and called the* Rabinowitz Floer homology $\mathrm{RFH}_*(\Sigma)$.

(b) *For a smooth family of hypersurfaces Σ_s, $s \in [0, 1]$, of restricted contact type there is a canonical isomorphism* $\mathrm{RFH}_*(\Sigma_0) \cong \mathrm{RFH}_*(\Sigma_1)$.

(c) *If a hypersurface $\Sigma \subset W$ of restricted contact type is displaceable from itself by a Hamiltonian isotopy, then* $\mathrm{RFH}_*(\Sigma) = 0$.

---

1 Though of course unrelated, this phenomenon is reminiscent of the appearance of negative energy solutions in Dirac's equation.

2 As H. Hofer pointed out, the functional appeared already in a 1976 article by J. Moser **[61]**, where he concluded that the corresponding variational principle "is certainly not suitable for an existence proof."

Remark 2.2 (Grading and coefficients). (i) For simplicity, we will assume throughout this note that *the first Chern class of TW vanishes* and we have made choices so that $\mathrm{RFH}_*(\Sigma)$ and all other Floer homologies below are $\mathbb{Z}$-graded by their Conley–Zehnder indices.

(ii) Coefficients are in a principal ideal domain R, which will sometimes be specialized to $\mathbb{Z}$ or a field or generalized to twisted coefficients.

(iii) The notation $\mathrm{RFH}_*(\Sigma)$ is chosen to emphasize the dependence on Σ, but it a priori depends also on the ambient Liouville manifold (W, λ); we will return to this question below.

If Σ carries no periodic orbits, then the only generators of $\mathrm{RFH}_*(\Sigma)$ are the constant loops on Σ with Lagrange multiplier $\eta = 0$ and it follows that $\mathrm{RFH}_*(\Sigma) \cong H^{n-*}(\Sigma)$. In view of Theorem 2.1(c), such a hypersurface cannot be displaceable. This implies the Weinstein conjecture for hypersurfaces of restricted contact type in Liouville manifolds where all compact sets are displaceable such as $\mathbb{C}^n$, subcritical Stein manifolds, or products of a Liouville manifold with $\mathbb{C}$ (see [76,78]).

2.2. Stability and Mañé's critical values

One may wonder whether Theorem 2.1 can be extended to a larger class of hypersurfaces Σ. The preceding discussion shows that some condition on Σ is needed: for example, it cannot apply to closed hypersurfaces in $\mathbb{C}^n$ without periodic orbits as constructed in [44,46].

In [21], Theorem 2.1 is generalized to the case that (W, ω) is a geometrically bounded symplectic manifold with $\omega|_{\pi_2(W)} = 0$, and the hypersurface $\Sigma \subset W$ and homotopy Σ_s are *tame and stable*. Here *stability* was introduced by Hofer and Zehnder [53] as a condition, generalizing contact type, under which existence results for periodic orbits continue to hold; it appeared again in [14,29] as the hypothesis for compactness in symplectic field theory.

An intriguing class of Hamiltonian systems is given by Hamiltonians $H(q, p) = \frac{1}{2}|p|^2 + U(q)$ on a cotangent bundle $W = T^*M$ with a twisted symplectic form

$$\omega = dp \wedge dq + \tau^*\sigma,$$

where $\tau : T^*M \to M$ is the projection and σ is a closed 2-form on M whose physical significance is that of a *magnetic field*. It has long been known that the dynamics on a level set $\Sigma_k = H^{-1}(k)$ can change drastically with the level k, even in the case $U = 0$ where all level sets are diffeomorphic (see [45] and the references therein). A famous example is that of a hyperbolic surface M with its area form σ and $U = 0$: Here Σ_k is foliated by contractible periodic orbits for $k < 1/2$, all periodic orbits on Σ_k are noncontractible for $k > 1/2$, and $\Sigma_{1/2}$ (the horocycle flow) does not possess any periodic orbits. The value $1/2$ at which the dynamics changes is the *Mañé critical value*, and at this value also the geometric type of the hypersurfaces Σ_k changes: above $1/2$ they are of contact type, below $1/2$ they are stable and tame but not of contact type, and $\Sigma_{1/2}$ is unstable. The Rabinowitz Floer homology of Σ_k is well defined and zero for $k < 1/2$, it is well defined and nonzero for $k > 1/2$, and it is

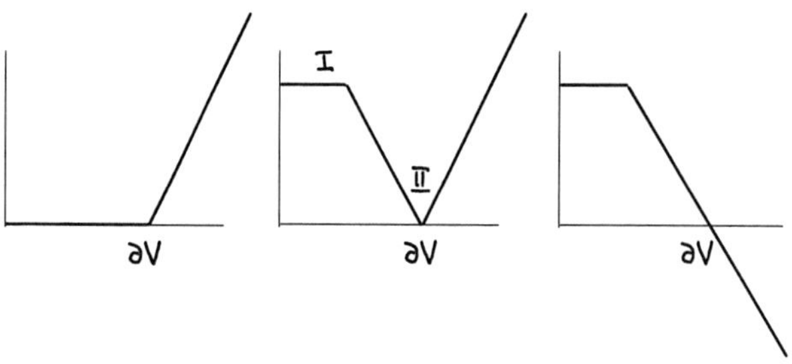

FIGURE 1
Three shapes of Hamiltonans.

undefined for $k = 1/2$. Using the generalization of Theorem 2.1, it is shown in [21] that this picture persists for large classes of magnetic systems in arbitrary dimension.

2.3. Relation to symplectic homology

Let us return to the setup in Section 2.1, so Σ is a hypersurface of restricted contact type in a Liouville manifold (W, λ). Recall that, by assumption, $\Sigma = \partial V$ for a compact subdomain $V \subset W$. After modifying λ near Σ, we may assume that $\lambda|_{\partial V}$ is a contact form, so that (V, λ) is a *Liouville domain.* It is shown in [20] that $\mathrm{RFH}_*(\partial V)$ depends only on the completion $\hat{V} = V \cup [1, \infty) \times \partial V$ of V (which is a Liouville manifold with the 1-form $\widehat{\lambda}$ that equals λ on V and $r\lambda|_{\partial V}$ on $[1, \infty) \times \partial V$, with r the coordinate on $[1, \infty)$). Moreover, $\mathrm{RFH}_*(\partial V)$ is closely related to another invariant of V that we now recall.

Symplectic homology was introduced in 1994 by A. Floer and H. Hofer [43]. We will use the version defined by C. Viterbo [76] as the direct limit

$$\mathrm{SH}_*(V) = \varinjlim \mathrm{FH}_*(H)$$

over Hamiltonians $H : \hat{V} \to \mathbb{R}$ that are zero on V and linearly increasing in r outside a compact set, as shown (up to smoothing) on the left of Figure 1. Dualizing, we obtain symplectic cohomology as the inverse limit

$$\mathrm{SH}^*(V) = \varprojlim \mathrm{FH}^*(H) = \varprojlim \mathrm{FH}_{-*}(-H).$$

These groups have refinements where the action is restricted to some interval (a, b),

$$\mathrm{SH}_*^{(a,b)}(V) = \varinjlim \mathrm{FH}_*^{(a,b)}(H), \quad \mathrm{SH}^*_{(a,b)}(V) = \varprojlim \mathrm{FH}^*_{(a,b)}(H) = \varprojlim \mathrm{FH}_{-*}^{(-b,-a)}(-H).$$

In [20], a new *V-shaped symplectic homology* was introduced as the direct–inverse limit

$$\check{\mathrm{SH}}_*(V) = \varinjlim_b \varprojlim_a \check{\mathrm{SH}}_*^{(a,b)}(V), \quad \check{\mathrm{SH}}_*^{(a,b)}(V) = \varinjlim_H \mathrm{FH}_*^{(a,b)}(H),$$

where the second direct limit is taken over "V-shaped" Hamiltonians $H : \hat{V} \to \mathbb{R}$ as shown (up to smoothing) in the middle of Figure 1. For any given $-\infty < a < b < \infty$ and sufficiently large H, the orbits in group I have action outside (a, b), so $\check{\mathrm{SH}}_*(V)$ is generated by the orbits in group II, which are in one-to-one correspondence with generators of $\mathrm{RFH}_*(\partial V)$. This observation combined with a technical tour de force leads to

Theorem 2.3 ([20]). *For each Liouville domain, we have*

$$\check{\mathrm{SH}}_*(V) = \mathrm{RFH}_*(\partial V).$$

Moreover, this group fits into a commuting diagram with exact row

$$\begin{array}{ccccccc} \dots \mathrm{SH}^{-*}(V) & \xrightarrow{\varepsilon} & \mathrm{SH}_*(V) & \xrightarrow{\iota} & \mathrm{RFH}_*(\partial V) & \xrightarrow{\pi} & \mathrm{SH}^{1-*}(V)\dots \\ \downarrow & & \uparrow & & & & \\ H^{n-*}(V,\partial V) & \xrightarrow{e} & H^{n-*}(V). & & & & \end{array} \tag{2.2}$$

In this diagram, e is the canonical map in the long exact sequence of the pair $(V, \partial V)$ and the vertical arrows correspond to the action zero (constant loop) part. It allows the computation of Rabinowitz Floer homology in terms of symplectic homology and singular cohomology. The following example will play a fundamental role in Section 3.

Example 2.4 (Cotangent bundles). Let T^*M be the cotangent bundle of a closed manifold M with its canonical Liouville form $\lambda = p\,dq$. Its unit disk bundle $D^*M = \{(q,p) \in T^*M \mid |p| \le 1\}$ with respect to some Riemannian metric is a Liouville domain with boundary $S^*M = \{(q,p) \in T^*M \mid |p| = 1\}$. Viterbo's isomorphism (proved by a joint effort of many people, see [2,4,55,68,75])

$$\mathrm{SH}_*(D^*M) \cong H_*(\Lambda) \tag{2.3}$$

expresses its symplectic homology in terms of the singular homology (with suitably twisted coefficients) of the loop space $\Lambda = C^\infty(S^1, M)$. Hence diagram (2.2) becomes

$$\begin{array}{ccccccc} \dots H^{-*}(\Lambda) & \xrightarrow{\varepsilon} & H_*(\Lambda) & \xrightarrow{\iota} & \mathrm{RFH}_*(S^*M) & \xrightarrow{\pi} & H^{1-*}(\Lambda)\dots \\ \downarrow & & \uparrow & & & & \\ H^{-*}(M) & \xrightarrow{e} & H^{n-*}(M), & & & & \end{array} \tag{2.4}$$

where e is the canonical map in the Gysin sequence of the sphere bundle $S^*M \to M$. We see that the map e (and therefore ε) lives only in degree zero and multiplies the class of a basepoint $q_0 \in M$ by the Euler characteristic χ of M. So

$$\mathrm{RFH}_*(S^*M) \cong H_*(\Lambda, \chi q_0) \oplus H^{1-*}(\Lambda, \chi q_0)$$

is the direct sum of "reduced" loop space homology $H_*(\Lambda, \chi q_0) = \operatorname{coker} \varepsilon$ (in degrees ≥ 0) and cohomology $H^{1-*}(\Lambda, \chi q_0) = \ker \varepsilon$ (in degrees ≤ 1).

2.4. Applications in symplectic topology

Over the past ten years, Rabinowitz Floer homology has found numerous applications in symplectic topology and Hamiltonian dynamics. One circle of applications was touched in Section 2.2, and three more are discussed in this subsection. I apologize for the omission, due to space constraints, of many other beautiful applications, such as [7,10], that would also have deserved to be included.

Leafwise intersections. The proof of Theorem 2.1(c) is based on more general action functionals

$$\mathcal{A}_F^{\chi H}(x,\eta) = \int_x \lambda - \eta \int_0^1 \chi(t) H(x) dt - \int_0^1 F(t,x) dt,$$

where $H : W \to \mathbb{R}$ is a defining Hamiltonian for a hypersurface $\Sigma = H^{-1}(0)$ of restricted contact type, $\chi \in C^\infty(S^1, \mathbb{R})$ has support in $(0, 1/2)$ and integral 1, and $F : S^1 \times W \to \mathbb{R}$ has compact support and vanishes for $t \in [0, 1/2]$. Critical points of $\mathcal{A}_F^{\chi H}$ correspond to *leafwise intersections*, i.e., points on Σ whose image under the time-one-map of X_F lands on the same X_H-orbit on Σ. P. Albers and U. Frauenfelder [8] have proved that the Floer homology of any such functional equals $\mathrm{RFH}_*(\Sigma)$. Applied to a Hamiltonian F whose time-one-map displaces Σ from itself, and which therefore has no leafwise intersections, this implies Theorem 2.1(c). Conversely, it proves the existence of leafwise intersections for any F if $\mathrm{RFH}_*(\Sigma) \neq 0$. See [8,9] for further results in this direction.

Exact contact embeddings. We now return to the question of dependence of $\mathrm{RFH}_*(\Sigma)$ on the ambient Liouville manifold W. Using neck-stretching from symplectic field theory, independence of W is proved in [20] if $\pi_1(\Sigma) = 0$ and all periodic orbits on Σ have Conley–Zehnder index $> 3 - n$. For example, this holds if Σ is the unit cotangent bundle S^*M of a closed simply connected manifold with $\dim M > 3$ with its standard contact structure. Since $\mathrm{RFH}_*(S^*M) \neq 0$ by Example 2.4, it follows that the image of an *exact contact embedding* $S^*M \hookrightarrow (W, \lambda)$ (i.e., an embedding such that the pullback of λ defines the standard contact structure) cannot be displaceable. Thus no such embedding exists if all compact sets are displaceable in W (e.g., for $\mathbb{C}^n$, subcritical Stein manifolds, or products of a Liouville manifold with $\mathbb{C}$), and if W is a cotangent bundle the image of such an embedding must intersect each fiber. Since an exact Lagrangian embedding $M \hookrightarrow W$ gives rise to an exact contact embedding $S^*M \hookrightarrow W$, these results generalize Gromov's theorem [49] that there are no closed exact Lagrangian submanifolds $M \subset \mathbb{C}^n$, under the assumptions that M is simply connected of dimension > 3. The nonexistence of exact contact embeddings $S^*M \hookrightarrow \mathbb{C}^n$ without these assumptions on M appears to be unknown.

Periodic Reeb flows. In many examples for which symplectic homology has been explicitly computed, such as Brieskorn manifolds (see, e.g., [42, 56]), it exhibits some kind of periodicity. P. Uebele [74] has found a beautiful explanation of this phenomenon in terms of Rabinowitz Floer homology. It uses the graded commutative associative products on symplectic homology and Rabinowitz Floer homology that will be discussed in the next section.

Theorem 2.5 (P. Uebele [74]). *Let V be a Liouville domain such that ∂V is simply connected and all periodic orbits on ∂V have Conley–Zehnder index $> 3 - n$. Assume that the Reeb flow on ∂V is periodic with minimal common period $T > 0$. Let $s \in \mathrm{RFH}_*(\partial V)$ be the class of a principal orbit (corresponding to the maximum on the Bott manifold of orbits of period T). Its Conley–Zehnder index has the form $n + 2b$ for some $b \in \mathbb{Z}$. If $b \neq 0$, then multiplication* *with s makes $\mathrm{RFH}_{*+n}(\partial V)$ with coefficients in a field $\mathbb{K}$ a free and finitely generated module over the ring of Laurent polynomials $\mathbb{K}[s, s^{-1}]$.*[3]

Besides establishing periodicity, this theorem allows the computation of the ring structure on $\mathrm{RFH}_*(\partial V)$ for many such Liouville domains. It also implies that the symplectic homology of such a Liouville domain with $\mathbb{K}$-coefficients is finitely generated as a $\mathbb{K}$-algebra. This finiteness result does not hold for all Liouville domains, counterexamples arising, e.g., from unit disk cotangent bundles of closed hyperbolic manifolds of dimension ≥ 3. It would be interesting to understand for which Liouville domains the finiteness result holds.

3. POINCARÉ DUALITY FOR LOOP SPACES

Let M be a closed connected manifold of dimension n and $\Lambda = C^\infty(S^1, M)$ its free loop space. Let us assume for simplicity that M is oriented, although everything in this section remains true in the unoriented case with suitable twisted coefficients. In their 1999 article **[17]** and its sequels, M. Chas and D. Sullivan introduced a wealth of operations on the homology of Λ that gave rise to a whole new research area named *string topology*. We will focus on the following two operations:

- the *loop product* $\mu = \bullet$ on $H_*\Lambda$, which is graded commutative, associative, and unital of degree $-n$ **[17]**, and
- the *loop coproduct* λ on the homology $H_*(\Lambda, \Lambda_0)$ relative to the subspace $\Lambda_0 \subset \Lambda$ of constant loops, which is graded cocommutative and coassociative of degree $1 - n$ **[73]**.

The loop coproduct is dual to a product $\circledast$ of degree $n - 1$ on cohomology $H^*(\Lambda, \Lambda_0)$ that was extensively studied in **[48]** and is often referred to as the *Goresky–Hingston product*. Subsequent studies of these products led to a number of puzzles, including the following two (see **[25]** for more details):

(a) Sullivan **[73]** has conjectured the following relation which we will refer to as *Sullivan's relation*:

$$\lambda\mu = (1 \otimes \mu)(\lambda \otimes 1) + (\mu \otimes 1)(1 \otimes \lambda). \tag{3.1}$$

How and on which space is this relation to be interpreted and proved?

(b) Many results concerning $\bullet$ and $\circledast$ arise in dual pairs. For example, the *critical levels* $\mathrm{Cr}(X)$ for $X \in H_*\Lambda$ and $\mathrm{cr}(x)$ for $x \in H^*(\Lambda, \Lambda_0)$ defined in **[48]** satisfy

3 In **[74]**, the result is stated with $\mathbb{Z}_2$-coefficients, but the extension to an arbitrary field $\mathbb{K}$ is straightforward. The restriction to field coefficients is essential because the proof uses the fact that $\mathbb{K}[s, s^{-1}]$ is a principal ideal domain.

the dual inequalities

$$\mathrm{Cr}(X \bullet Y) \le \mathrm{Cr}(X) + \mathrm{Cr}(Y), \quad \mathrm{cr}(x \circledast y) \ge \mathrm{cr}(x) + \mathrm{cr}(y).$$

Can these phenomena be explained by some kind of "Poincaré duality"?

We will see in this section that these puzzles get naturally resolved in terms of the Rabinowitz Floer homology of the unit sphere cotangent bundle S^*M, which we will call the *Rabinowitz loop homology* and denote by

$$\check{H}_*\Lambda = \mathrm{RFH}_*(S^*M) = \check{\mathrm{SH}}_*(D^*M).$$

3.1. Product and coproduct on symplectic homology

We begin by describing the analogues of μ and λ on symplectic homology of a Liouville domain V. They are based on topological quantum field theory (TQFT) operations that were introduced by M. Schwarz [71] on Floer homology over closed symplectic manifolds, and extended by P. Seidel [72] to symplectic homology. Let us recall the construction, following the exposition of A. Ritter [67]. It takes the following inputs: a nonnegative asymptotically linear Hamiltonian $H : \hat{V} \to \mathbb{R}_{\ge 0}$; a Riemann surface (S, j) with q positive ends modeled over $\mathbb{R}_+ \times S^1$ and p negative ends modeled over $\mathbb{R}_- \times S^1$; and a 1-form β on S with $d\beta \le 0$ which equals $A_k dt$ in canonical coordinates $s + it$ on the negative ends and $B_\ell dt$ on the positive ends, for some positive weights A_k, B_ℓ. By Stokes' theorem, such a β exists if and only if the weights satisfy

$$\sum_{k=1}^{p} A_k \ge \sum_{\ell=1}^{q} B_\ell. \tag{3.2}$$

The algebraic count of maps $u : S \to \hat{V}$ satisfying $(du - \beta \otimes X_H)^{0,1} = 0$ gives a map

$$\psi_S : \bigotimes_{\ell=1}^{q} \mathrm{FH}_*(B_\ell H) \to \bigotimes_{k=1}^{p} \mathrm{FH}_*(A_k H).$$

(Here we need to use field coefficients in order to have a Künneth formula). Applied to Hamiltonians as on the left of Figure 1, these maps induce maps on symplectic homology

$$\overline{\psi}_S : \mathrm{SH}_*(V)^{\otimes q} \to \mathrm{SH}_*(V)^{\otimes p}$$

which depend only on the topological type of S and satisfy the usual TQFT composition rules. Note, however, that (3.2) forces $p \ge 1$, so we only get a *noncompact TQFT structure*. As part of this structure, we get on $\mathrm{SH}_*(V)$ the unital, graded commutative and associative *pair-of-pants product* μ of degree $-n$.

The TQFT structure on $\mathrm{SH}_*(V)$ also includes a coproduct of degree $-n$, which is, however, not very interesting. Namely, by deforming the weight at one of the outputs to 0, we can force the corresponding output to land in the action zero (constant loop) part $\mathrm{SH}_*^{=0}(V)$, hence to vanish in the quotient $\mathrm{SH}_*^{>0}(V)$. Since the coproduct vanishes in two different ways, interpolating the weights at the two outputs gives rise to a *secondary pair-of-pants coproduct* λ on $\mathrm{SH}_*^{>0}(V)$ of degree $1 - n$ (this was first pointed out by P. Seidel and further explored by T. Ekholm and A. Oancea [38]).

The following theorem relates the operations on symplectic homology to those in string topology. Here the assertion concerning the products is due to A. Abbondandolo and M. Schwarz [3], and that concerning the coproduct is proved in [24].

Theorem 3.1 (Relation to string topology operations [3, 24]). *Viterbo's isomorphism* (2.3) *intertwines the pair-of-pants product with the loop product (both denoted μ). It descends to an isomorphism* $\mathrm{SH}_*^{>0}(D^*M) \cong H_*(\Lambda, \Lambda_0)$ *which intertwines the secondary pair-of-pants coproduct with the loop coproduct (both denoted λ).*

3.2. Product and coproduct on Rabinowitz Floer homology

It was observed in [30] that applying the arguments of the previous subsection to Hamiltonians as in the middle of Figure 1 also equips Rabinowitz Floer homology $\mathrm{RFH}_*(V)$ with a noncompact TQFT structure. In particular, it carries a unital, graded commutative and associative product of degree $-n$ which we will denote by $\boldsymbol{\mu}$. Moreover, the map ι in diagram (2.2) is a ring homomorphism.

It turns out that Rabinowitz Floer homology also carries a canonical coproduct. To describe the resulting algebraic structure, let us introduce the degree shifted (co)homology groups[4]

$$\mathbb{SH}_*(V) = \mathrm{SH}_{*+n}(V), \quad \mathbb{SH}^*(V) = \mathrm{SH}^{*+n}(V), \quad \mathrm{RF}\mathbb{H}_*(\partial V) = \mathrm{RFH}_{*+n}(\partial V).$$

With respect to the shifted gradings, the products μ and $\boldsymbol{\mu}$ have degree 0. Let us call *involutive infinitesimal bialgebra*[5] the structure consisting of a graded commutative associative product and a graded cocommutative coassociative coproduct satisfying $\mu\lambda = 0$ and

$$\lambda\mu = (1 \otimes \mu)(\lambda \otimes 1) + (\mu \otimes 1)(1 \otimes \lambda) - (\mu \otimes \mu)(1 \otimes \lambda(1) \otimes 1). \tag{3.3}$$

Theorem 3.2 (Involutive infinitesimal bialgebra structure on Rabinowitz Floer homology [25]). *There exists a degree* $1-2n$ *coproduct* $\boldsymbol{\lambda}$ *on* $\mathrm{RF}\mathbb{H}_*(\partial V)$ *making* $(\mathrm{RF}\mathbb{H}_*(\partial V), \boldsymbol{\mu}, \boldsymbol{\lambda})$ *an involutive infinitesimal bialgebra.*

To define $\boldsymbol{\lambda}$, we modify the construction of the secondary pair-of-pants coproduct λ described above by deforming the weights at the outputs to *negative weights* -1 rather than 0. This has the effect of splitting off the chain-level continuation map $\varepsilon : SC^{-*}(V) \to SC_*(V)$ at the corresponding output. Since the continuation map is induced by monotone homotopies from $-H$ to H as on the left of Figure 1, which factor through the zero Hamiltonian which has only constant orbits, this shows again that λ induces an operation on positive symplectic homology $\mathrm{SH}_*^{>0}(V)$. Applying the same reasoning on the Rabinowitz Floer complex we are in for a pleasant surprise: the chain-level continuation map is now induced by monotone homotopies from $-H$ to H with H as in the middle of Figure 1, which factor through the Hamiltonian on the right of Figure 1 which has no 1-periodic orbits at all in a given

4 This degree shift is also common in string topology, see [17].

5 The structure has further properties which will not be discussed here. Similar structures have appeared in the work of Aguiar [6], Joni-Rota [54], Ehrenborg-Readdy [36], and Loday-Ronco [60].

action interval! Thus the secondary pair-of-pants coproduct induces an operation $\boldsymbol{\lambda}$ on all of $\mathrm{RFH}_*(\partial V)$.

Remark 3.3. The operations $\boldsymbol{\mu}, \boldsymbol{\lambda}$ are defined using the interpretation of Rabinowitz Floer homology as V-shaped symplectic homology. It would be interesting to find a definition in terms of the original definition of Rabinowitz Floer homology.

The next result relates the operations on Rabinowitz Floer homology to those on symplectic homology. Here we denote by $\mu^\vee$ the coproduct on $\mathrm{SH}^*(V)$ dual to μ, and by $\lambda^\vee$ the product on $\mathrm{SH}^*_{>0}(V)$ dual to λ.

Theorem 3.4 (Almost splitting [25]). *The long exact sequence* (2.2) *fits into the canonical commuting diagram*

$$\begin{array}{cccccc}
 & & & & (S\mathbb{H}^{1-2n-*}_{>0}(V),\lambda^\vee) & \\
 & & & \overset{i}{\swarrow} & \downarrow j & \\
\xrightarrow{\varepsilon} (S\mathbb{H}_*(V),\mu) & \xrightarrow{\iota} & (\mathrm{RF}\mathbb{H}_*(\partial V),\boldsymbol{\mu},\boldsymbol{\lambda}) & \xrightarrow{\pi} & (S\mathbb{H}^{1-2n-*}(V),\mu^\vee) & \xrightarrow{\varepsilon} \\
q \downarrow & \overset{p}{\swarrow} & & & & \\
(S\mathbb{H}_*^{\geq 0}(V),\lambda) & & & & &
\end{array}$$

in which the maps ι and i intertwine the products μ, $\boldsymbol{\mu}$, and $\lambda^\vee$, and the maps p and π intertwine the coproducts λ, $\boldsymbol{\lambda}$, and $\mu^\vee$.

We can interpret this as saying that the long exact sequence (2.2) "almost splits" in the sense that it splits up to some discrepancy in the action zero part. Note that, while the map ι preserves the products, the corresponding splitting map p preserves the coproducts, and similarly for π and i.

To apply the preceding discussion to string topology, we introduce the corresponding degree shifted (co)homology groups

$$\mathbb{H}_*\Lambda = H_{*+n}\Lambda, \quad \mathbb{H}^*\Lambda = H^{*+n}\Lambda, \quad \check{\mathbb{H}}_*\Lambda = \check{H}_{*+n}\Lambda.$$

Then Theorems 3.2 and 3.4 yield

Corollary 3.5 (Involutive infinitesimal bialgebra structure on Rabinowitz loop homology [25]). *There exists a degree $1-2n$ coproduct $\boldsymbol{\lambda}$ on $\check{\mathbb{H}}_*\Lambda$ making $(\check{\mathbb{H}}_*\Lambda, \boldsymbol{\mu}, \boldsymbol{\lambda})$ an involutive infinitesimal bialgebra. Moreover, the long exact sequence* (2.4) *fits into the canonical commuting diagram*

$$\begin{array}{cccccc}
 & & & & (\mathbb{H}^{1-2n-*}(\Lambda,\Lambda_0),\lambda^\vee = \circledast) & \\
 & & & \overset{i}{\swarrow} & \downarrow j & \\
\xrightarrow{\varepsilon} (\mathbb{H}_*\Lambda,\mu = \bullet) & \xrightarrow{\iota} & (\check{\mathbb{H}}_*\Lambda,\boldsymbol{\mu},\boldsymbol{\lambda}) & \xrightarrow{\pi} & (\mathbb{H}^{1-2n-*}\Lambda,\mu^\vee) & \xrightarrow{\varepsilon} \\
q \downarrow & \overset{p}{\swarrow} & & & & \\
(\mathbb{H}_*(\Lambda,\Lambda_0),\lambda) & & & & &
\end{array}$$

in which the maps ι and i intertwine the products $\mu = \bullet$, $\boldsymbol{\mu}$ and $\lambda^\vee = \circledast$, and the maps p and π intertwine the coproducts λ, $\boldsymbol{\lambda}$ and $\mu^\vee$.

In this case the maps i, j are injective and the maps p, q are surjective. The map ι becomes injective after replacing $H_*\Lambda$ by its "reduced" version $H_*(\Lambda, \chi q_0)$ from Example 2.4, and the map π becomes surjective after replacing $H^*\Lambda$ by $H^*(\Lambda, \chi q_0)$.

This provides the full solution to puzzle (a): By the left triangle, μ and λ extend to operations $\boldsymbol{\mu}$ and $\boldsymbol{\lambda}$ on the common domain $\check{\mathbb{H}}_*\Lambda$ satisfying the generalized form (3.3) of Sullivan's relation. Note that the right triangle gives the same conclusion for the dual operations $\lambda^\vee = \circledast$ and $\mu^\vee$.

On the other hand, the products $\bullet$ and $\circledast$ both appear as components of the product $\boldsymbol{\mu}$ on $\check{H}_*\Lambda$. This provides an unexpected alternative interpretation of Sullivan's relation for μ and λ, as part of associativity for the product $\boldsymbol{\mu}$ on $\check{H}_*\Lambda$! See [27] for further discussion of this topic.

3.3. Poincaré duality for Rabinowitz Floer homology

The main motivation for introducing Rabinowitz Floer homology into string topology was that it satisfies a form of Poincaré duality. This was proved in [30] on the level of vector spaces, and in [25] with the additional algebraic structure. To formulate it, note that the operations $\boldsymbol{\lambda}^\vee, \boldsymbol{\mu}^\vee$ dual to $\boldsymbol{\lambda}, \boldsymbol{\mu}$ define again the structure of an involutive infinitesimal bialgebra on the degree shifted Rabinowitz Floer cohomology $\mathrm{RF}\mathbb{H}^*(\partial V) = \mathrm{RFH}^{*+n}(\partial V)$ (the notion of an involutive infinitesimal bialgebra is "self-dual").

Theorem 3.6 (Poincaré duality for Rabinowitz Floer homology [25]). *With field coefficients, there exists for each Liouville domain V a canonical isomorphism of involutive infinitesimal bialgebras*

$$\mathrm{PD} : (\mathrm{RF}\mathbb{H}_*(\partial V), \boldsymbol{\mu}, \boldsymbol{\lambda}) \xrightarrow{\cong} (\mathrm{RF}\,\mathbb{H}^{1-2n-*}(\partial V), \boldsymbol{\lambda}^\vee, \boldsymbol{\mu}^\vee).$$

In the zero action range, the left-hand side is $H^{-*}(\partial V)$ with $\boldsymbol{\mu}$ the cup product, and the right-hand side is $H_{2n-1+*}(\partial V)$ with $\boldsymbol{\lambda}^\vee$ the intersection product. These are related by Poincaré duality on ∂V, and the other two operations are their algebraic duals. Theorem 3.6 thus extends classical Poincaré duality on ∂V to Rabinowitz Floer homology.

On the level of vector spaces, Poincaré duality is most transparent in the original definition of Rabinowitz Floer homology via a Lagrange multiplier functional: it arises from the simple observation that, under the canonical involution

$$(x, \eta) \mapsto (\bar{x}, \bar{\eta}), \quad \bar{x}(t) = x(-t), \quad \bar{\eta} = -\eta,$$

the Rabinowitz action functional changes sign,

$$\mathcal{A}^H(\bar{x}, \bar{\eta}) = -\mathcal{A}^H(x, \eta).$$

It follows that the involution maps positive gradient lines of $\mathcal{A}^H$ to negative gradient lines of $\mathcal{A}^H$, and thus induces an isomorphism from Floer homology to Floer cohomology which implies Poincaré duality on the level of vector spaces. To show compatibility with the involutive infinitesimal bialgebra structures, one needs to reprove Poincaré duality using the description of Rabinowitz Floer homology as V-shaped symplectic homology (which is less intuitive).

3.4. Applications in Riemannian geometry

Applied to degree shifted Rabinowitz loop homology and cohomology $\check{\mathbb{H}}^*\Lambda = \check{H}^{*+n}\Lambda = \mathrm{RFH}^{*+n}(S^*M)$, Theorem 3.6 becomes

Corollary 3.7 (Poincaré duality for free loop spaces [25]). *With field coefficients, there exists a canonical isomorphism of involutive infinitesimal bialgebras*

$$\mathrm{PD} : (\check{\mathbb{H}}_*\Lambda, \boldsymbol{\mu}, \boldsymbol{\lambda}) \xrightarrow{\cong} \big(\check{\mathbb{H}}^{1-2n-*}\Lambda, \boldsymbol{\lambda}^\vee, \boldsymbol{\mu}^\vee\big).$$

It is shown in [25] that Corollary 3.7 resolves puzzle (b): each classical pair of theorems for the products $\bullet$ and $\circledast$ on loop (co)homology extends to a pair of theorems for the products $\boldsymbol{\mu}$ and $\boldsymbol{\lambda}^\vee$ on Rabinowitz loop (co)homology which are related via Poincaré duality. This leads to unified proofs for each classical pair of theorems. While the original proofs were topological, the unified proofs are always symplectic. More specifically, the following applications are discussed in [25, 26, 28]:

- the behavior of critical levels of the length and action functionals with respect to products;
- the computation of the Rabinowitz loop homology ring of manifolds all of whose geodesics are closed using Uebele's theorem [74], with applications to the question of string point invertibility of constant rank one symmetric spaces, resonances, and a conjecture of Viterbo concerning spectral norms;
- a duality between index and index + nullity for closed geodesics as a consequence of an iteration formula due to Liu and Long;
- the characterization of level-potent (co)homology classes in terms of symplectically degenerate maxima and minima, with dynamical implications for the existence of infinitely many closed geodesics and the Conley conjecture.

The question of homotopy invariance of Rabinowitz loop homology will be addressed in upcoming work. This question has become particularly interesting due to the recent discovery by Naef [63] that, in contrast to the loop product, the loop coproduct is *not* homotopy invariant.

3.5. Topological descriptions of Rabinowitz loop homology

Rabinowitz loop homology $\check{H}_*\Lambda$ was defined above as Rabinowitz Floer homology $\mathrm{RFH}_*(S^*M)$ and the operations $\boldsymbol{\mu}, \boldsymbol{\lambda}$ were constructed Floer theoretically. This subsection outlines four purely topological constructions of $\check{H}_*\Lambda$ and its operations which are the subject of joint work in progress with A. Oancea, N. Hingston, M. Abouzaid, and T. Kragh.

Construction via cones. In [31], Rabinowitz Floer homology $\mathrm{RFH}_*(\partial V)$ of a Liouville domain V is described in terms of the cones of chain-level continuation maps $\varepsilon : \mathrm{FC}_*(-H) \to$

$\mathrm{FC}_*(H)$ for Hamiltonians H as on the left of Figure 1. Moreover, the operations $\boldsymbol{\mu}, \boldsymbol{\lambda}$ are derived from an *A_2^+-structure* on the Floer chain complex $\mathrm{FC}_*(H)$, which consists of a chain-level product and coproduct satisfying suitable compatibility conditions with the continuation map. This description carries over to the Morse chain complex $\mathrm{MC}_*(S)$ of a Lagrangian action functional of the form $S(\gamma) = \int_0^1 (|\dot\gamma|^2 - V(t,\gamma))dt$ on $H^1(S^1, M)$. In view of the discussion in Example 2.4, we define the chain-level continuation map $\varepsilon : \mathrm{MC}^{-*}(S) \to \mathrm{MC}_*(S)$ to live only in degree 0 and send the basepoint q_0 to χq_0. It is proved in [24] that suitable chain-level versions of the loop product and coproduct give rise to an A_2^+-structure on $\mathrm{MC}_*(S)$, and the homology of the cone of ε is isomorphic to $\check{H}_*\Lambda$ as an infinitesimal bialgebra.

Construction via spectra. Since Rabinowitz loop homology generally lives in arbitrarily positive and negative degrees, it cannot be the homology of a topological space. It can, however, be obtained as the homology of a *spectrum* (see, e.g., [5] for background on spectra). The construction uses the Spanier–Whitehead dual and a cone construction on the level of spectra.

Construction via constant speed loops. It was suggested by N. Hingston that the cohomology product $\circledast$ should correspond to a "Chas–Sullivan product on cochains of constant speed loops." This leads to the following conjectural description of $\check{H}_*\Lambda$. Let $\mathcal{CS} \subset H^1(S^1, M)$ be the subspace of *constant speed loops*, i.e., loops parametrized with constant speed. For an action functional $S : \mathcal{CS} \to \mathbb{R}$ as above, we consider the chain complex generated by the stable and unstable manifolds of its critical points, viewed as chains of finite dimension resp. finite codimension, with a differential such that its homology equals $\check{H}_*\Lambda$. Chas–Sullivan-type products and coproducts between these chains should then recover its infinitesimal bialgebra structure. One difficulty in making this approach rigorous is that $\mathcal{CS}$ does not appear to be a Hilbert manifold, but only (away from constant loops) an sc-manifold in the sense of [52] (see [64]).

Construction via a Lagrange multiplier functional. The space of constant speed loops is defined by the constraint $|\dot\gamma(t)| = \text{const}$. Morse homology with such a constraint can be described by a Lagrange multiplier functional with an infinite-dimensional space of Lagrange multipliers. Imitating the construction of Rabinowitz Floer homology, we can replace this by an integrated constraint with a 1-dimensional Lagrange multiplier. More precisely, we fix an $\epsilon > 0$ and define the *Rabinowitz energy functional*

$$\check{E} : \Lambda \times \mathbb{R} \to \mathbb{R}, \quad \check{E}(\gamma, \eta) = \eta \int_0^1 |\dot\gamma|^2 - \frac{\eta^3}{3} + \eta\epsilon^2.$$

Its critical points are pairs $(\gamma, \pm\eta_\gamma)$ with γ a (possibly constant) closed geodesic and $\eta_\gamma = (\int_0^1 |\dot\gamma|^2 + \epsilon^2)^{1/2}$, so they are in one-to-one correspondence with the generators of the complex computing $\check{H}_*\Lambda$. It should not be too hard to prove that the Morse homology of $\check{E}$ equals Rabinowitz loop homology, but it remains unclear how to recover its product and coproduct from this description.

4. OTHER LAGRANGE MULTIPLIER FUNCTIONALS

This section is devoted to some further examples of Lagrange multiplier functionals.

Example 4.1 (Constrained Lagrangian systems). Let M be a Riemannian manifold and $L : TM \to \mathbb{R}$ a smooth Lagrange function. For $a < b$ and $A, B \in M$, consider the path space

$$X = \{x \in C^\infty([a,b], M) \mid x(a) = A, x(b) = B\}$$

and the Lagrangian action

$$S_L : X \to \mathbb{R}, \quad S_L(x) = \int_a^b L(x(t), \dot{x}(t))dt.$$

Given a smooth function $k : M \to W$ to a vector space W with 0 as a regular value, consider the Lagrange multiplier functional

$$\hat{S}_L : X \times C^\infty([a,b], W^*) \to \mathbb{R}, \quad \hat{S}_L(x, \lambda) = S_L(x) - \int_a^b \langle \lambda(t), k(x(t)) \rangle dt.$$

Its critical points are solutions of the equations

$$k(x(t)) \equiv 0, \quad \frac{\partial L}{\partial x} - \nabla_t \frac{\partial L}{\partial \dot{x}} = \langle \lambda, \nabla k(x) \rangle.$$

and correspond to critical points of S_L subject to the pointwise constraint $k(x(t)) \equiv 0$. Here $\frac{\partial L}{\partial x} - \nabla_t \frac{\partial L}{\partial \dot{x}}$ is the familiar term from the Euler–Lagrange equation of L and $-\langle \lambda, \nabla k(x) \rangle$ is the constraint force.

Example 4.2 (Euler's equation for rigid bodies). Let us now specialize Example 4.1 to the case that M is a Lie group G, and $H = k^{-1}(0)$ is a subgroup defined by a function $k : G \to W$ satisfying $k(gh) = k(g)$ for all $g \in G, h \in H$. We also specialize the Lagrange function to $L(g, \dot{g}) = \frac{1}{2}|\dot{g}|_g^2$ for a right-invariant Riemannian metric on G. Then the critical point equation for $(g, \lambda) \in C^\infty([a,b], G) \times C^\infty([a,b], W^*)$ is equivalent to the first-order equations

$$v(t) \in \mathfrak{h}, \quad \dot{v} + B(v, v) + \langle \lambda, \nabla k(e) \rangle = 0 \tag{4.1}$$

on the Lie algebra $\mathfrak{g} = T_e G$ (see [12, 13]). Here $v(t) = \dot{g}(t)g(t)^{-1}$ is the "body angular velocity," $e \in G$ is the unit, $\mathfrak{h}$ is the Lie algebra of H, and $B : \mathfrak{g} \times \mathfrak{g} \to \mathfrak{g}$ is the bilinear form defined by $\langle B(c, a), b \rangle = \langle [a, b], c \rangle$. In the case $G = H = SO(3)$, this becomes Euler's equation for the motion of a free rigid body [41].

Example 4.3 (Euler's equations of hydrodynamics). We also owe L. Euler the equations of motion for the velocity field v and the pressure p of an inviscous incompressible fluid [40], namely

$$\operatorname{div} v = 0, \quad \dot{v} + \nabla_v v + \nabla p = 0. \tag{4.2}$$

The general setup for these equations is a closed Riemannian manifold M equipped with a volume form vol, so that v is a vector field and p a function on M (both time dependent). In 1966, V. I. Arnold [11] derived these equations by formally applying Example 4.2 to the diffeomorphism group $G = \mathrm{Diff}(M)$ and its subgroup $H = \mathrm{Diff}(M, \mathrm{vol})$ of volume preserving diffeomorphisms. Here the right-invariant metric on $\mathrm{Diff}(M)$ is defined for $g \in \mathrm{Diff}(M)$ and vector fields $v, w \in \mathfrak{g} = \mathfrak{X}(M)$ by

$$\langle v \circ g, w \circ g \rangle = \int_M \langle v, w \rangle \,\mathrm{vol},$$

and the subgroup $\mathrm{Diff}(M, \mathrm{vol})$ is the zero set of the function

$$k : \mathrm{Diff}(M) \to \Omega^n(M), \quad k(g) = (g^{-1})^* \,\mathrm{vol} - \mathrm{vol}\,.$$

The Lagrange multiplier λ, now denoted by p, is a function from $[a, b]$ to $\Omega^n(M)^* = C^\infty(M, \mathbb{R})$. Short computations (see [13]) yield $B(v, v) = \nabla_v v$ and $\langle p, \nabla k(e)\rangle = \nabla p$, so that equation (4.1) becomes equation (4.2).

A famous open problem (unfortunately not worth a million dollars) concerns the existence for all times of smooth solutions of (4.2) with smooth initial conditions. This is only known in dimension 2 where it was first proved by O. Ladyzhenskaya [57]. In 1970, D. Ebin and J. Marsden [34] reproved this result using Arnold's geometric interpretation of (4.2) as rigid body motion on the group $\mathrm{Diff}(M, \mathrm{vol})$. This interpretation also allows us to consider Euler equations on other subgroups of $\mathrm{Diff}(M)$. For example, long-time existence has been proved for the Euler equations on groups of symplectomorphisms (Ebin [33]) and contactomorphisms (Ebin and Preston [35]). See also [50] for results on the group of diffeomorphisms preserving a stable Hamiltonian structure. It would be interesting to investigate the interaction of the Euler equations with other geometric structures on these groups such as Hofer's metric on symplectomorphisms, or partial orders on contactomorphisms.

Example 4.4 (Gauge theories). One often encounters the situation that a function $f : X \to \mathbb{R}$ is invariant under the action of a Lie group G on X, and we are interested in its critical G-orbits. Suppose there exists a function $h : X \to \mathfrak{g}$ such that $Z = h^{-1}(0)$ meets all G-orbits and for each $x \in Z$ the map

$$\mathfrak{g} \to \mathfrak{g}, \quad \xi \mapsto dh(x) \cdot X_\xi(x)$$

is an isomorphism, where $\xi \mapsto X_\xi(x) = \frac{d}{dt}|_{t=0} \exp(t\xi)x$ denotes the infinitesimal action of the Lie algebra. Then Z is a slice for the G-action and the critical G-orbits correspond to critical points of the Lagrange multiplier functional $F(x, \lambda) = f(x) - \langle \lambda, h(x)\rangle$ with $\lambda \in \mathfrak{g}^*$. While such a slice Z usually does not exist globally, it often exists at least locally near a given G-orbit.

For example, let G be a compact connected simple Lie group and consider the Chern–Simons action

$$S(A) = \frac{1}{4\pi} \int_M \mathrm{Tr}\Big(A \wedge dA + \frac{2}{3} A \wedge A \wedge A \Big)$$

on connections $\nabla = d + A$, $A \in \Omega^1(M, \mathfrak{g})$ on the trivial principal G-bundle over a closed 3-manifold M. Its critical points are the flat connections, and it is invariant up to integer multiples of 2π under the action of the gauge group $\mathcal{G} = C^\infty(M, G)$ (see [69,79]). Let us fix a flat connection A and consider the complex

$$\Omega^0(M, \mathfrak{g}) \stackrel{d_A}{\to} \Omega^1(M, \mathfrak{g}) \stackrel{d_A}{\to} \Omega^2(M, \mathfrak{g}).$$

Here the first map is the infinitesimal action of the Lie algebra Lie $\mathcal{G} = \Omega^0(M, \mathfrak{g})$ and the second map is the linearization of S at A. Let $h = d_A^* : \Omega^1(M, \mathfrak{g}) \to \Omega^0(M, \mathfrak{g})$ be the adjoint of d_A with respect to some Riemannian metric on M. This function satisfies the condition above and thus defines a slice $Z = h^{-1}(0)$ for the $\mathcal{G}$-action iff the d_A-cohomology $H_A^0(M, \mathfrak{g})$ vanishes, and the restriction $S|_Z$ has a nondegenerate critical point at A iff $H_A^1(M, \mathfrak{g})$ vanishes. Writing a general connection as $A + \beta$, $\beta \in \Omega^1(M, \mathfrak{g})$, the Lagrange multiplier functional

$$F(\beta, \lambda) = S(A + \beta) - \langle \lambda, d_A^* \beta \rangle = \frac{1}{4\pi} \int_M \mathrm{Tr}\Big(\beta \wedge d_A \beta + \frac{2}{3} \beta \wedge \beta \wedge \beta + d_A^* \beta \wedge * \lambda \Big)$$

corresponds to the first three terms of the Fadeev–Popov action, where $\lambda \in \Omega^*(M, \mathfrak{g})$ is the gauge fixing boson. With an additional fermionic term, this becomes the relevant functional for the perturbative expansion of the Chern–Simons partition function at the flat connection A (see [69,79]).

Example 4.5 (Symplectic vortex equations). This example follows [23]; it was the original motivation leading to Rabinowitz Floer homology. Consider a Hamiltonian action of a compact connected Lie group G on a symplectic manifold (M, ω) with an equivariant moment map $\mu : M \to \mathfrak{g}^*$. Let $\Lambda = C^\infty(S^1, M)$, where $S^1 = \mathbb{R}/\mathbb{Z}$, and denote by $\Lambda^{\mathrm{contr}} \subset \Lambda$ the subspace of contractible loops. Consider the action

$$\mathcal{A} : \Lambda^{\mathrm{contr}} \to \mathbb{R}, \quad \mathcal{A}(x) = \int_{\bar{x}} \omega,$$

where $\bar{x} : D \to M$ is an extension of x to the closed unit disk. This is independent of the choice of $\bar{x}$ if ω vanishes on $\pi_2(M)$, which we will assume for simplicity. Suppose that 0 is a regular value of μ and consider the Lagrange multiplier functional

$$\mathcal{A}^\mu : \Lambda^{\mathrm{contr}} \times C^\infty(S^1, \mathfrak{g}) \to \mathbb{R}, \quad \mathcal{A}^\mu(x, \eta) = \mathcal{A}(x) - \int_0^1 \langle \mu(x), \eta \rangle dt$$

with a Lagrange multiplier $\eta \in C^\infty(S^1, \mathfrak{g})$. To describe its gradient flow, we pick a compatible almost complex structure J on (M, ω) and an Ad-invariant inner product on $\mathfrak{g}$, and define a metric m on $\Lambda^{\mathrm{contr}} \times C^\infty(S^1, \mathfrak{g})$ by

$$m_{(x,\eta)}\big((\hat{x}_1, \hat{\eta}_1), (\hat{x}_2, \hat{\eta}_2)\big) = \int_0^1 \big(\omega(\hat{x}_1, J(x)\hat{x}_2) + \langle \hat{\eta}_1, \hat{\eta}_2 \rangle\big) dt.$$

Then gradient flow lines of $\mathcal{A}^\mu$ are maps $(u, \eta) : \mathbb{R} \times S^1 \to M \times \mathfrak{g}$, satisfying

$$\frac{\partial u}{\partial s} + J(u)\Big(\frac{\partial u}{\partial t} + X_\eta(u)\Big) = 0, \quad \frac{\partial \eta}{\partial s} + \mu(u) = 0,$$

where (s, t) are the coordinates on $\mathbb{R} \times S^1$ and $X_\eta(x) = \frac{d}{dt}|_{t=0} \exp(t\eta)x$. To interpret these equations more geometrically, we view the Lagrange multiplier as a connection

$$A = \eta(s, t) dt \in \Omega^1(Z, \mathfrak{g})$$

on the cylinder $Z = \mathbb{R} \times S^1$. Its curvature is $F_A = \frac{\partial \eta}{\partial s} ds \wedge dt$, which can be converted to a function $Z \to \mathbb{R}$ using the Hodge $*$ operator $*(ds \wedge dt) = 1$. Moreover, the connection induces a covariant derivative

$$d_A u = du + X_A(u) = du + X_\eta(u) dt : TZ \to TM$$

and its complex antilinear part

$$\bar{\partial}_{J,A}(u) = \frac{1}{2}\big(d_A u + J(u) \circ d_A u \circ j\big)$$

with respect to the standard complex structure j on $\mathbb{R} \times S^1$ sending ∂_s to ∂_t. Then the above equations for gradient flow lines of $\mathcal{A}^\mu$ become the *symplectic vortex equations*

$$\bar{\partial}_{J,A}(u) = 0, \quad *F_A + \mu(u) = 0.$$

These equations were discovered independently by D. Salamon and I. Mundet i Riera [23,62] for an arbitrary Riemann surface in place of the cylinder Z. They give rise to invariants of Hamiltonian group actions and a quantum Kirwan map, with applications to the quantum cohomology of symplectic quotients [22,32,47,65]. Applied to suitable infinite-dimensional symplectic manifolds, the symplectic vortex equations comprise other well-known equations of mathematical physics such as the anti-self-dual Yang–Mills equations and the Seiberg–Witten equations [23].

The functional $\mathcal{A}^\mu$ is still invariant under the action of the gauge group $\mathcal{G} = C^\infty(S^1, G)$. If this action has a global slice, we can remove the gauge symmetry as in Example 4.4 by introducing another Lagrange multiplier in the dual Lie algebra $C^\infty(S^1, \mathfrak{g}^*)$. One situation where such a slice exists is an $\mathbb{R}$-action generated by a single Hamiltonian $\mu = H$ such that $\Sigma = H^{-1}(0)$ is of contact type, in which case the functional $\mathcal{A}^\mu$ restricts on the slice to the Rabinowitz action functional [9]. On the other hand, considering the action $\mathcal{A}$ on the loop space $C^\infty(S^1, \Sigma)$ and removing the gauge symmetry leads to the equations in Example 4.6 below. It would be interesting to further explore the various Lagrange multiplier functionals and their Floer homologies in this situation.

Example 4.6 (Symplectic field theory). Let (ω, λ) be a stable Hamiltonian structure on a closed $(2n-1)$-manifold M with Reeb vector field R (see [14]).[6] Assume for simplicity that $\omega = d\theta$ is exact, so that we have a well-defined action functional

$$\mathcal{A} : \Lambda = C^\infty(S^1, M) \to \mathbb{R}, \quad x \mapsto \int_x \theta.$$

This functional is invariant under the group $G = \mathrm{Diff}_+(S^1)$ of orientation-preserving diffeomorphisms of the circle. One can break this symmetry by imposing the gauge fixing condition $\lambda(\dot{x}) = \text{const}$. (The constant should not be fixed because we cannot expect closed Reeb orbits of prescribed period). Critical points of $\mathcal{A}$ subject to this constraint are critical points of the Lagrange multiplier functional

$$\hat{\mathcal{A}} : \Lambda \times \mathfrak{g}_0 \to \mathbb{R}, \quad \hat{\mathcal{A}}(x, \eta) = \mathcal{A}(x) + \int_0^1 \eta(t)\lambda\big(\dot{x}(t)\big)dt,$$

with the codimension-one subspace $\mathfrak{g}_0$ of the Lie algebra of G given by

$$\mathfrak{g}_0 = \left\{ \eta \in C^\infty(S^1, \mathbb{R}) \mid \int_0^1 \eta(t)dt = 0 \right\}.$$

6 Even in the contact case $\omega = d\lambda$, separating the roles of ω and λ clarifies the discussion.

The derivative of $\hat{\mathcal{A}}$ is given by

$$d\hat{\mathcal{A}}(x,\eta)(\hat{x},\hat{\eta}) = \int_0^1 \big[\omega(\hat{x},\dot{x}) + \eta\, d\lambda(\hat{x},\dot{x}) - \dot{\eta}\lambda(\hat{x}) + \hat{\eta}\lambda(\dot{x})\big]dt,$$

so its critical points are solutions of the equations

$$\lambda(\dot{x}) = T = \text{const}, \quad i_{\dot{x}}(\omega + \eta\, d\lambda) + \dot{\eta}\lambda = 0, \quad \int_0^1 \eta = 0.$$

Inserting R in the second equation yields $\dot{\eta} = 0$ and thus $\eta \equiv 0$, so critical points are pairs $(x, 0)$ where $\dot{x} = TR(x)$ for some $T \in \mathbb{R}$. Now consider, for an ω-compatible complex structure J on $\xi = \ker\lambda$, the "metric" on $\Lambda^{\text{contr}} \times \mathfrak{g}_0$ defined by

$$m_{(x,\eta)}\big((\hat{x}_1,\hat{\eta}_1),(\hat{x}_2,\hat{\eta}_2)\big) = \int_0^1 \big[(\omega + \eta\, d\lambda)(\pi\hat{x}_1, J\pi\hat{x}_2) + \lambda(\hat{x}_1)\lambda(\hat{x}_2) + \hat{\eta}_1\hat{\eta}_2\big]dt,$$

where $\pi : TM \to \xi$ is the projection along R. Note that in the contact case $\omega = d\lambda$, the bilinear form $m_{(x,\eta)}$ is symmetric, but it is only positive definite as long as $\eta > -1$. Nevertheless, $\hat{\mathcal{A}}$ has a well-defined "gradient" with respect to m given by

$$\nabla_m \hat{\mathcal{A}}(x,\eta) = \left(-J(x)\pi\dot{x} - \dot{\eta}R(x), \lambda(\dot{x}) - \int_x \lambda\right),$$

where the term $-\int_x \lambda$ comes from projecting $\lambda(\dot{x})$ onto $\mathfrak{g}_0$. So a gradient flow line $u = (\eta, f) : \mathbb{R} \times S^1 \to \mathbb{R} \times M$ of $\hat{\mathcal{A}}$ satisfies the equations

$$\begin{cases} \partial_s\eta - \lambda(\partial_t f) + \int_0^1 f(s,\cdot)^*\lambda = 0, \\ \lambda(\partial_s f) + \partial_t\eta = 0, \\ \pi\partial_s f + J(f)\pi\partial_t f = 0. \end{cases} \tag{4.3}$$

Replacing η by $a(s,t) = \eta(s,t) + \alpha(s)$ for a function $\alpha : \mathbb{R} \to \mathbb{R}$ (unique up to a constant) satisfying $\alpha'(s) = \int_0^1 f(s,\cdot)^*\lambda$, we obtain the familiar equations

$$\begin{cases} \partial_s a - \lambda(\partial_t f) = 0, \\ \lambda(\partial_s f) + \partial_t a = 0, \\ \pi\partial_s f + J(f)\pi\partial_t f = 0, \end{cases} \tag{4.4}$$

for $\hat{J}$-holomorphic curves $u = (a, f) : \mathbb{R} \times S^1 \to \mathbb{R} \times M$ with respect to the almost complex structure $\hat{J}$ restricting to J on ξ and mapping the unit vector in $\mathbb{R}$ to R. On the plane $\mathbb{C}$ instead of the cylinder $\mathbb{R} \times S^1$, these equations were introduced by H. Hofer in his 1993 paper **[51]** on the Weinstein conjecture in dimension three. They make sense for a domain being any punctured Riemann surface, giving rise to *symplectic field theory (SFT)* **[39]**, a general theory of punctured holomorphic curves in symplectic cobordisms which has found numerous applications in contact and symplectic topology.

The description via Lagrange multipliers raises some interesting questions concerning symplectic field theory. Let us begin with a brief comparison of equations (4.3) and (4.4). Note first that $\eta \to 0$ as $s \to \pm\infty$, whereas a grows linearly with slope the asymptotic periods as $s \to \pm\infty$. Moreover, shifting a by a constant yields again a solution, which is not the case for η. The Floer energy of (η, f) equals the ω-energy $\int_{\mathbb{R}\times S^1} f^*\omega = \mathcal{A}(x^+) - \mathcal{A}(x^-)$, which in the contact case equals the difference $T_+ - T_-$ of the asymptotic periods. Moreover, in the contact case the action $\mathcal{A}(x^+)$ at $+\infty$ is equivalent to the Hofer energy (see [29]).

It would be interesting to give a direct proof of compactness modulo breaking of solutions of (4.3) (say, in the absence of finite energy planes) without appealing to the SFT compactness theorem [14,29]. Generalizing this proof to λ being replaced by a loop of contact forms λ_t may lead to a description of nonequivariant contact homology (see [15,38]) in terms of such loops, analogous to the definition of Hamiltonian Floer homology in terms of loops of Hamiltonians. In a different direction, this may also shed some light on the variant of (4.4) introduced in [1] where the first two equations are replaced by a harmonic 1-form $(f^*\lambda) \circ j$, for which the compactness question is still wide open.

Another interesting feature of (4.3) is the fact that the asymptotic periods $T_\pm$ can also be negative or zero. This parallels the corresponding feature in Rabinowitz Floer homology (see Section 2) and suggests that suitable counts of solutions of (4.3) (or equivalently (4.4)) compute the equivariant Rabinowitz Floer homology of the symplectization $\mathbb{R} \times M$ whenever the latter is defined. This should lead to an interpretation of algebraic structures on Rabinowitz Floer homology such as its involutive infinitesimal bialgebra structure and Poincaré duality in terms of symplectic field theory. In the Lagrangian setting, N. Legout has constructed an A_∞-structure on the corresponding SFT-type complex ([59], see also [16, 37]), whose relation to Rabinowitz Floer homology is still conjectural.

ACKNOWLEDGMENTS

The ideas and results described in this note have emanated from long-term collaborations with several coauthors, of whom I wish to specifically mention Urs Frauenfelder, Nancy Hingston, Janko Latschev, Alexandru Oancea, and Evgeny Volkov.

FUNDING

This work was partially supported by Deutsche Forschungsgemeinschaft, grants CI 45/5 and CI 45/6.

REFERENCES

[1] C. Abbas, K. Cieliebak, and H. Hofer, The Weinstein conjecture for planar contact structures in dimension three. *Comment. Math. Helv.* **80** (2005), no. 4, 771–793.

[2] A. Abbondandolo and M. Schwarz, On the Floer homology of cotangent bundles. *Comm. Pure Appl. Math.* **59** (2006), no. 2, 254–316.

[3] A. Abbondandolo and M. Schwarz, Floer homology of cotangent bundles and the loop product. *Geom. Topol.* **14** (2010), no. 3, 1569–1722.

[4] M. Abouzaid, Symplectic cohomology and Viterbo's theorem. In *Free loop spaces in geometry and topology*, pp. 271–485, IRMA Lect. Math. Theor. Phys. 24, Eur. Math. Soc., Zürich, 2015.

[5] J. F. Adams, *Stable homotopy and generalised homology*. Chicago Lectures in Math., University of Chicago Press, Chicago, IL–London, 1974.

[6] M. Aguiar, Infinitesimal Hopf algebras. In *New trends in Hopf algebra theory (La Falda, 1999)*, pp. 1–29, Contemp. Math. 267, Amer. Math. Soc., Providence, RI, 2000.

[7] P. Albers and U. Fauenfelder, A variational approach to Givental's nonlinear Maslov index. *Geom. Funct. Anal.* **22** (2012), no. 5, 1033–1050.

[8] P. Albers and U. Frauenfelder, Infinitely many leaf-wise intersections on cotangent bundles. *Expo. Math.* **30** (2012), no. 2, 168–181.

[9] P. Albers and U. Frauenfelder, Rabinowitz Floer homology: a survey. In *Global differential geometry*, pp. 437–461, Springer Proc. Math. 17, Springer, Heidelberg, 2012.

[10] P. Albers and W. J. Merry, Orderability, contact non-squeezing, and Rabinowitz Floer homology. *J. Symplectic Geom.* **16** (2018), no. 6, 1481–1547.

[11] V. I. Arnold, Sur la géométrie différentielle des groupes de Lie de dimension infinie et ses applications à l'hydrodynamique des fluides parfaits. *Ann. Inst. Fourier (Grenoble)* **16** (1966), 319–361.

[12] V. I. Arnold, *Mathematical methods of classical mechanics*. Grad. Texts in Math. 60, Springer, New York, 1989.

[13] V. I. Arnold and B. A. Khesin, *Topological methods in hydrodynamics. Second edn*. Appl. Math. Sci. 125, Springer, Cham, 2021.

[14] F. Bourgeois, Y. Eliashberg, H. Hofer, K. Wysocki, and E. Zehnder, Compactness results in symplectic field theory. *Geom. Topol.* **7** (2003), 799–888.

[15] F. Bourgeois and A. Oancea, An exact sequence for contact- and symplectic homology. *Invent. Math.* **175** (2009), no. 3, 611–680.

[16] B. Chantraine, G. Dimitroglou Rizell, P. Ghiggini, and R. Golovko, Floer theory for Lagrangian cobordisms. *J. Differential Geom.* **114** (2020), no. 3, 393–465.

[17] M. Chas and D. Sullivan, String topology. 1999, arXiv:math/9911159.

[18] K. Cieliebak and Y. Eliashberg, *From Stein to Weinstein and Back*. Amer. Math. Soc. Colloq. Publ. 59, Amer. Math. Soc., Providence, RI, 2012.

[19] K. Cieliebak and U. Frauenfelder, A Floer homology for exact contact embeddings. *Pacific J. Math.* **239** (2009), no. 2, 251–316.

[20] K. Cieliebak, U. Frauenfelder, and A. Oancea, Rabinowitz Floer homology and symplectic homology. *Ann. Sci. Éc. Norm. Supér. (4)* **43** (2010), no. 6, 957–1015.

[21] K. Cieliebak, U. Frauenfelder, and G. P. Paternain, Symplectic topology of Mañé's critical values. *Geom. Topol.* **14** (2010), no. 3, 1765–1870.

[22] K. Cieliebak, A. R. Gaio, I. Mundet i Riera, and D. Salamon, The symplectic vortex equations and invariants of Hamiltonian group actions. *J. Symplectic Geom.* **1** (2002), no. 3, 543–645.

[23] K. Cieliebak, A. R. Gaio, and D. A. Salamon, J-holomorphic curves, moment maps, and invariants of Hamiltonian group actions. *Int. Math. Res. Not.* **16** (2000), 831–882.

[24] K. Cieliebak, N. Hingston, and A. Oancea, Loop coproduct in Morse and Floer homology. 2020, arXiv:2008.13168.

[25] K. Cieliebak, N. Hingston, and A. Oancea, Poincaré duality for loop spaces. 2020, arXiv:2008.13161.

[26] K. Cieliebak, N. Hingston, and A. Oancea, Index growth and level-potency (in preparation).

[27] K. Cieliebak, N. Hingston, and A. Oancea, Reduced symplectic homology (in preparation).

[28] K. Cieliebak, N. Hingston, A. Oancea, and E. Shelukhin, Resonances and string point invertibility for compact rank one symmetric spaces (in preparation).

[29] K. Cieliebak and K. Mohnke, Compactness for punctured holomorphic curves. *J. Symplectic Geom.* **3** (2005), no. 4, 589–654.

[30] K. Cieliebak and A. Oancea, Symplectic homology and the Eilenberg–Steenrod axioms. *Algebr. Geom. Topol.* **18** (2018), no. 4, 1953–2130.

[31] K. Cieliebak and A. Oancea, Multiplicative structures on cones and duality. 2020, arXiv:2008.13165.

[32] K. Cieliebak and D. Salamon, Wall crossing for symplectic vortices and quantum cohomology. *Math. Ann.* **335** (2006), no. 1, 133–192.

[33] D. G. Ebin, Geodesics on the symplectomorphism group. *Geom. Funct. Anal.* **22** (2012), no. 1, 202–212.

[34] D. G. Ebin and J. Marsden, Groups of diffeomorphisms and the motion of an incompressible fluid. *Ann. of Math. (2)* **92** (1970), 102–163.

[35] D. G. Ebin and S. C. Preston, Riemannian geometry of the contactomorphism group. *Arnold Math. J.* **1** (2015), no. 1, 5–36.

[36] R. Ehrenborg and M. Readdy, Coproducts and the cd-index. *J. Algebraic Combin.* **8** (1998), no. 3, 273–299.

[37] T. Ekholm, Rational SFT, linearized Legendrian contact homology, and Lagrangian Floer cohomology. In *Perspectives in analysis, geometry, and topology*, pp. 109–145, Progr. Math. 296, Springer, New York, 2012.

[38] T. Ekholm and A. Oancea, Symplectic and contact differential graded algebras. *Geom. Topol.* **21** (2017), no. 4, 2161–2230.

[39] Y. Eliashberg, A. Givental, and H. Hofer, Introduction to symplectic field theory. *Geom. Funct. Anal.* **Special Volume, Part II** (2000), 560–673.

[40] L. Euler, Principes généraux du mouvement des fluides. *Acad. R. Sci. B.-Lett. Berlin, Mém.* **11** (1757), 274–315.

[41] L. Euler, *Leonhardi Euleri theoria motus corporum solidorum seu rigidorum ex primis nostrae cognitionis principiis stabilita et ad onmes motus qui in huiusmodi corpora cadere possunt accommodata. Vol. prius/posterius.* Leonhardi Euleri Opera Omnia (Series Secunda, Opera Mechanica et Astronomica) III/IV, Orell Füssli, Zürich; B. G. Teubner, Leipzig, 1948/1950.

[42] A. Fauck, Rabinowitz–Floer homology on Brieskorn spheres. *Int. Math. Res. Not. IMRN* **14** (2015), 5874–5906.

[43] A. Floer, H. Hofer, and K. Wysocki, Applications of symplectic homology. I. *Math. Z.* **217** (1994), no. 4, 577–606.

[44] V. L. Ginzburg, An embedding $S^{2n-1} \to \mathbb{R}^{2n}$, $2n - 1 \geq 7$, whose Hamiltonian flow has no periodic trajectories. *Int. Math. Res. Not.* **2** (1995), 83–97.

[45] V. L. Ginzburg, On closed trajectories of a charge in a magnetic field. An application of symplectic geometry. In *Contact and symplectic geometry (Cambridge, 1994)*, pp. 131–148, Publ. Newton Inst. 8, Cambridge Univ. Press, Cambridge, 1996.

[46] V. L. Ginzburg and B. Z. Gürel, A C^2-smooth counterexample to the Hamiltonian Seifert conjecture in $\mathbb{R}^4$. *Ann. of Math. (2)* **158** (2003), no. 3, 953–976.

[47] E. González and C. T. Woodward, Quantum cohomology and toric minimal model programs. *Adv. Math.* **353** (2019), 591–646.

[48] M. Goresky and N. Hingston, Loop products and closed geodesics. *Duke Math. J.* **150** (2009), no. 1, 117–209.

[49] M. Gromov, Pseudo holomorphic curves in symplectic manifolds. *Invent. Math.* **82** (1985), no. 2, 307–347.

[50] K. Helmsauer, *Riemannian geometry of groups of diffeomorphisms preserving a stable hamiltonian structure*. Ph.D. thesis, Universität Augsburg, Germany, 2020.

[51] H. Hofer, Pseudoholomorphic curves in symplectizations with applications to the Weinstein conjecture in dimension three. *Invent. Math.* **114** (1993), 515–563.

[52] H. Hofer, K. Wysocki, and E. Zehnder, A general Fredholm theory. I. A splicing-based differential geometry. *J. Eur. Math. Soc. (JEMS)* **9** (2007), no. 4, 841–876.

[53] H. Hofer and E. Zehnder, *Symplectic invariants and Hamiltonian dynamics*. Birkhäuser Advanced Texts: Basler Lehrbücher, Birkhäuser, Basel, 1994.

[54] S. A. Joni and G.-C. Rota, Coalgebras and bialgebras in combinatorics. *Stud. Appl. Math.* **61** (1979), no. 2, 93–139.

[55] T. Kragh, The Viterbo transfer as a map of spectra. *J. Symplectic Geom.* **16** (2018), no. 1, 85–226.

[56] M. Kwon and O. van Koert, Brieskorn manifolds in contact topology. *Bull. Lond. Math. Soc.* **48** (2016), no. 2, 173–241.

[57] O. A. Ladyzhenskaya, *The mathematical theory of viscous incompressible flow*. Math. Appl. 2, Gordon and Breach, Science Publishers, New York–London–Paris, 1969.

[58] J. L. Lagrange, *Leçons sur le calcul des fonctions*. Courcier, 1804.

[59] N. Legout, A_∞-category of Lagrangian cobordisms in the symplectization of $P \times \mathbb{R}$. 2020, arXiv:2012.08245.

[60] J.-L. Loday and M. Ronco, On the structure of cofree Hopf algebras. *J. Reine Angew. Math.* **592** (2006), 123–155.

[61] J. Moser, Periodic orbits near an equilibrium and a theorem by Alan Weinstein. *Comm. Pure Appl. Math.* **29** (1976), no. 6, 724–747.

[62] I. Mundet i Riera, A Hitchin–Kobayashi correspondence for Kähler fibrations. *J. Reine Angew. Math.* **528** (2000), 41–80.

[63] F. Naef, The string coproduct "knows" Reidemeister/Whitehead torsion. 2021, arXiv:2106.11307.

[64] O. Neumeister, The curve shrinking flow, compactness and its relation to scale manifolds. 2021, arXiv:2104.12906.

[65] K. L. Nguyen, C. Woodward, and F. Ziltener, Morphisms of CohFT algebras and quantization of the Kirwan map. In *Symplectic, Poisson, and noncommutative geometry*, pp. 131–170, Math. Sci. Res. Inst. Publ. 62, Cambridge Univ. Press, New York, 2014.

[66] P. H. Rabinowitz, Periodic solutions of Hamiltonian systems. *Comm. Pure Appl. Math.* **31** (1978), no. 2, 157–184.

[67] A. F. Ritter, Topological quantum field theory structure on symplectic cohomology. *J. Topol.* **6** (2013), no. 2, 391–489.

[68] D. Salamon and J. Weber, Floer homology and the heat flow. *Geom. Funct. Anal.* **16** (2006), no. 5, 1050–1138.

[69] J. Sawon, Perturbative expansion of Chern–Simons theory. In *The interaction of finite-type and Gromov–Witten invariants (BIRS 2003)*, pp. 145–166, Geom. Topol. Monogr. 8, Geom. Topol. Publ., Coventry, 2006.

[70] M. Schwarz, *Morse homology*. Progr. Math. 111, Birkhäuser, Basel, 1993.

[71] M. Schwarz, Cohomology operations from S^1-cobordisms in Floer homology. Ph.D. thesis, ETH, Zürich, 1995.

[72] P. Seidel, A biased view of symplectic cohomology. In *Current developments in mathematics, 2006*, pp. 211–253, Int. Press, Somerville, MA, 2008.

[73] D. Sullivan, Open and closed string field theory interpreted in classical algebraic topology. In *Topology, geometry and quantum field theory*, pp. 344–357, London Math. Soc. Lecture Note Ser. 308, Cambridge Univ. Press, Cambridge, 2004.

[74] P. Uebele, Periodic Reeb flows and products in symplectic homology. *J. Symplectic Geom.* **17** (2019), no. 4, 1201–1250.

[75] C. Viterbo, Functors and computations in Floer homology with applications. II. 1998, arXiv:1805.01316.

[76] C. Viterbo, Functors and computations in Floer homology with applications. I. *Geom. Funct. Anal.* **9** (1999), no. 5, 985–1033.

[77] J. Weber, The Morse–Witten complex via dynamical systems. *Expo. Math.* **24** (2006), no. 2, 127–159.

[78] A. Weinstein, On the hypotheses of Rabinowitz' periodic orbit theorems. *J. Differential Equations* **33** (1979), no. 3, 353–358.

[79] E. Witten, Quantum field theory and the Jones polynomial. *Comm. Math. Phys.* **121** (1989), no. 3, 351–399.

KAI CIELIEBAK
Institut für Mathematik, Universität Augsburg, 86135 Augsburg, Germany, kai.cieliebak@math.uni-augsburg.de

This Page Intentionally Left Blank

REAL GROMOV–WITTEN THEORY

PENKA GEORGIEVA

ABSTRACT

In this note we survey some of the recent developments in real Gromov–Witten theory. In particular, we discuss the main difficulties of the construction and important structural results.

MATHEMATICS SUBJECT CLASSIFICATION 2020

Primary 53D45; Secondary 14N35

KEYWORDS

Real Gromov–Witten invariants, orientations, Klein TQFT

INTERNATIONAL CONGRESS OF MATHEMATICIANS 2022 JULY 6–14

Proc. Int. Cong. Math. 2022, Vol. 4, pp. 2530–2550
DOI 10.4171/ICM2022/189

Published by EMS Press

1. INTRODUCTION

The Gromov–Witten invariants can be viewed as a modern counterpart of the classical enumeration of curves in projective varieties. They arise from integration over the moduli spaces of pseudoholomorphic maps into a symplectic manifold introduced in the seminal work of Gromov [21]. An influential perspective proposed by Witten interprets them as the coefficients of a partition function of a topological string theory. As such they play a central role in striking dualities relating them to mathematical objects of completely different nature. Understanding these relations has and continues to generate substantial amount of high-level research.

The real Gromov–Witten invariants arise in a similar way from integration over moduli spaces of pseudoholomorphic maps. In the real case these maps are required to be equivariant with respect to an antisymplectic involution on the target and one on the domain. The antisymplectic involution corresponds to an intrinsic symmetry of the theory and is preserved under dualities thus providing conjectures relating the real Gromov–Witten invariants with the dual equivariant objects. In particular, relations with SO/Sp gauge theory and the Gaussian orthogonal/symplectic ensembles are expected.

In this note we present an overview of the construction of the real Gromov–Witten invariants, based on a joint work [18] with Aleksey Zinger, and discuss structural results for the local real Gromov–Witten theory, based on a joint work [16] with Eleny Ionel.

2. REAL GROMOV–WITTEN INVARIANTS

The foundations of (complex) Gromov–Witten theory, i.e., of counts of J-holomorphic curves in symplectic manifolds, were established in the 1990s and have been spectacularly applied ever since. On the other hand, the progress in establishing the foundations of real GW theory, i.e., of counts of J-holomorphic curves in symplectic manifolds preserved by antisymplectic involutions, has been much slower. The two main difficulties in developing real GW theory are the potential nonorientability of the moduli space $\mathfrak{M}_{g,l}(X, B; J)^{\phi,\sigma}$, defined in (2.2), and the fact that its virtual boundary strata have real codimension 1. This is in contrast with the complex GW theory, where the moduli spaces have canonical orientations and the "boundary" strata have real codimension of at least 2. These two ingredients are crucial for the construction of a (virtual) fundamental class, integration upon which defines the invariants.

The difficulty arising from the existence of a real codimension 1 boundary strata can be resolved by considering the larger moduli space (2.3) that is a union over all topological types of involutions on the domain. As explained in Section 2.1, inside this space all codimension 1 strata form a hypersurface rather than boundary and the definition of the invariants becomes a question about the orientability of this moduli space. We introduce the notion of real orientation on a symplectic manifold in Section 2.2 – these are topological conditions on the symplectic manifold which ensure the orientability of the real moduli space (2.3). We define the primary and descendant real GW invariants in Section 2.3 and give examples of large collections of real-orientable symplectic manifolds.

Invariant counts of real curves were first constructed by Welschinger [38, 39] following a different approach. They are defined in genus 0, for real symplectic 4- and 6-folds, and under certain topological conditions ruling out maps from type (E) nodal symmetric surfaces (later removed in [13]). In the Gromov–Witten-style approach to these counts developed in [11, 35], the invariance corresponds to the relevant moduli spaces being orientable outside of (virtual) hypersurfaces which are shown not to be crossed by the paths of stable maps induced by paths between two generic almost complex structures and two generic collections of constraints.

Many methods have been developed for the computation of the real invariants, notably by employing methods from tropical geometry [5–8, 23, 24, 29, 34], establishing WDVV-type formulas [10, 17, 36], and through localization techniques [19, 30, 33]. In particular, the result of [33] provides the first instance of a real mirror symmetry phenomenon and that of [30] the first real enumerative bounds in higher genus.

2.1. Moduli spaces of real maps

A real symplectic manifold is a triple (X, ω, ϕ) consisting of a symplectic manifold (X, ω) and an antisymplectic involution ϕ. For such a triple, denote by $\mathcal{J}_\omega^\phi$ the space of ω-compatible almost complex structures J on X such that $\phi^* J = -J$. The fixed locus X^ϕ of ϕ is then a Lagrangian submanifold of (X, ω) which is totally real with respect to any $J \in \mathcal{J}_\omega^\phi$.

Example 2.1. An example of a real Kähler manifold (X, ω, ϕ, J) is the complex projective space $\mathbb{P}^{n-1}$. The maps

$$\tau_n : \mathbb{P}^{n-1} \to \mathbb{P}^{n-1}, \qquad [z_1, \dots, z_n] \to [\bar{z}_1, \dots, \bar{z}_n],$$
$$\eta_{2m} : \mathbb{P}^{2m-1} \to \mathbb{P}^{2m-1}, \qquad [z_1, z_2, \dots, z_{2m-1}, z_{2m}] \to [-\bar{z}_2, \bar{z}_1, \dots, -\bar{z}_{2m}, \bar{z}_{2m-1}],$$

are antisymplectic involutions with respect to the standard Fubini–Study symplectic form ω_n on $\mathbb{P}^{n-1}$. Another important example is a real quintic threefold X_5, i.e., a smooth hypersurface in $\mathbb{P}^4$ cut out by a real equation.

A symmetric surface (Σ, σ) is a connected oriented, possibly nodal, surface Σ with an orientation-reversing involution σ. There are $\lfloor \frac{3g+4}{2} \rfloor$ topological types of smooth symmetric genus g surfaces; the type is determined by the number of fixed components and the orientability of the quotient. A symmetric Riemann surface $(\Sigma, \sigma, \mathfrak{j})$ is a symmetric surface (Σ, σ) with an almost complex structure $\mathfrak{j}$ on Σ such that $\sigma^* \mathfrak{j} = -\mathfrak{j}$. We denote by $\mathcal{J}_\Sigma^\sigma$ the space of such complex structures.

A continuous map

$$u : (\Sigma, \sigma) \to (X, \phi)$$

is called real if $u \circ \sigma = \phi \circ u$; see Figure 1. It is said to be of degree $B \in H_2(X; \mathbb{Z})$ if $u_*[\Sigma] = B$. We denote the space of such maps by $\mathfrak{B}_g(X)^{\phi,\sigma}$, with g denoting the genus of the domain Σ of σ.

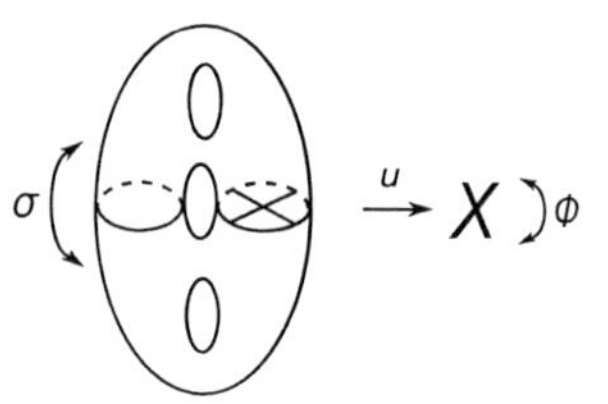

FIGURE 1
Here the domain Σ has 1 fixed circle and 1 cross-cap circle; the quotient Σ/σ is a nonorientable surface with 1 boundary and 1 cross-cap.

For $J \in \mathcal{J}_\omega^\phi$, $\mathfrak{j} \in \mathcal{J}_\Sigma^\sigma$, and $u \in \mathfrak{B}_g(X)^{\phi,\sigma}$, let

$$\bar{\partial}_{J,\mathfrak{j}} u = \frac{1}{2}(\mathrm{d}u + J \circ \mathrm{d}u \circ \mathfrak{j})$$

be the $\bar{\partial}_J$-operator on $\mathfrak{B}_g(X)^{\phi,\sigma}$.

Let $g, l \in \mathbb{Z}^{\geq 0}$, (Σ, σ) be a genus g symmetric surface, $B \in H_2(X; \mathbb{Z}) - 0$, and $J \in \mathcal{J}_\omega^\phi$. Let $\Delta^{2l} \subset \Sigma^{2l}$ be the big diagonal, i.e., the subset of $2l$-tuples with at least two coordinates equal. Denote by

$$\begin{aligned}\mathfrak{M}_{g,l}(X, B; J)^{\phi,\sigma} =& \big\{\big(u, (z_1^+, z_1^-), \ldots, (z_l^+, z_l^-), \mathfrak{j}\big) \in \mathfrak{B}_g(X)^{\phi,\sigma} \times (\Sigma^{2l} - \Delta^{2l}) \times \mathcal{J}_\Sigma^\sigma : \\ & z_i^- = \sigma(z_i^+)\ \forall i = 1, \ldots, l,\ u_*[\Sigma]_{\mathbb{Z}} = B,\ \bar{\partial}_{J,\mathfrak{j}} u = 0\big\}/\sim \end{aligned} \tag{2.1}$$

the (uncompactified) moduli space of equivalence classes of degree B real J-holomorphic maps from (Σ, σ) to (X, ϕ) with l conjugate pairs of marked points. Two marked J-holomorphic (ϕ, σ)-real maps determine the same element of this moduli space if they differ by an orientation-preserving diffeomorphism of Σ commuting with σ. We denote by

$$\overline{\mathfrak{M}}_{g,l}(X, B; J)^{\phi,\sigma} \supset \mathfrak{M}_{g,l}(X, B; J)^{\phi,\sigma} \tag{2.2}$$

Gromov's convergence compactification of $\mathfrak{M}_{g,l}(X, B; J)^{\phi,\sigma}$ obtained by including stable real maps from nodal symmetric surfaces. The (virtually) codimension-one boundary strata of

$$\overline{\mathfrak{M}}_{g,l}(X, B; J)^{\phi,\sigma} - \mathfrak{M}_{g,l}(X, B; J)^{\phi,\sigma} \subset \overline{\mathfrak{M}}_{g,l}(X, B; J)^{\phi,\sigma}$$

consist of real J-holomorphic maps from one-nodal symmetric surfaces to (X, ϕ). Each stratum is either a (virtual) hypersurface in $\overline{\mathfrak{M}}_{g,l}(X, B; J)^{\phi,\sigma}$ or a (virtual) boundary. The existence of boundary is what prevents us from defining invariants for each topological type of involutions σ. However, one-nodal symmetric surfaces can always be smoothed out in (real) one-parameter family to symmetric surfaces. Thus, each boundary stratum appears in the compactification of precisely two of the moduli spaces $\overline{\mathfrak{M}}_{g,l}(X, B; J)^{\phi,\sigma}$ corresponding to two different topological types of orientation-reversing involutions σ on Σ. This means that the union over all topological types of involutions on Σ forms a space without boundary. Let

$$\overline{\mathfrak{M}}_{g,l}(X, B; J)^{\phi} = \bigcup_{\sigma} \overline{\mathfrak{M}}_{g,l}(X, B; J)^{\phi,\sigma} \tag{2.3}$$

denote the union of the compactified real moduli spaces taken over all topological types of orientation-reversing involutions σ on Σ. Furthermore, denote by

$$\mathbb{R}\overline{\mathcal{M}}_{g,l} \equiv \overline{\mathfrak{M}}_{g,l}(\mathrm{pt}, 0)^{\mathrm{id}}$$

the Deligne–Mumford moduli space of marked real curves. If $g + l \geq 2$, there is a natural forgetful morphism

$$\mathfrak{f} : \overline{\mathfrak{M}}_{g,l}(X, B; J)^{\phi} \to \mathbb{R}\overline{\mathcal{M}}_{g,l} \equiv \overline{\mathfrak{M}}_{g,l}(\mathrm{pt}, 0)^{\mathrm{id}}. \quad (2.4)$$

In order to study the orientability of these spaces, it is crucial to understand their codimension one strata which consist of maps from one-nodal domains. As described in **[27, SECTION 3]**, there are four types of nodes a one-nodal symmetric surfaces (Σ, σ) may have:

(E) the node is an isolated point of the fixed locus $\Sigma^{\sigma} \subset \Sigma$;

(H) the node is a nonisolated point of the fixed locus Σ^{σ} and

- (H1) the topological component of Σ^{σ} containing the node is algebraically irreducible (its normalization is connected);
- (H2) the topological component of Σ^{σ} containing the node is algebraically reducible, but Σ is algebraically irreducible;
- (H3) Σ is algebraically reducible.

In the genus 0 case, the degenerations (E) and (H3) are known as the codimension 1 sphere bubbling and disk bubbling, respectively; the degenerations (H1) and (H2) cannot occur in the genus 0 case.

As a one-nodal symmetric surface is smoothed out in one-parameter family of symmetric surfaces, we observe the transition of a smooth symmetric surface through one-nodal degeneration. A transition through a degeneration (H3) does not change the topological type of the involution. Thus, each stratum of morphisms from a one-nodal symmetric surface of type (H3) to (X, ϕ) is a hypersurface inside of $\overline{\mathfrak{M}}_{g,l}(X, B; J)^{\phi,\sigma}$ for some genus g involution σ.

A transition through a degeneration (H2) also does not change the number of fixed components. The transformation of the real locus is the same as in the (H3) case, but an (H2) transition also inserts or removes two cross-caps. This transition may or may not change the topological type of the involution. The former occurs when the fixed locus is separating in which case the transition changes the topological type of the involution and thus each stratum of morphisms from such one-nodal surfaces to (X, ϕ) is a boundary of the spaces $\overline{\mathfrak{M}}_{g,l}(X, B; J)^{\phi,\sigma}$ for precisely two topological types of genus g involutions σ. If the fixed locus is nonseparating, then the transition does not change the topological type of the involution and each stratum of morphisms from such one-nodal surfaces to (X, ϕ) is a hypersurface inside of $\overline{\mathfrak{M}}_{g,l}(X, B; J)^{\phi,\sigma}$ for some genus g involution σ. A degeneration (H2) cannot occur in genus 0 or 1, but does occur in genus 2 and higher.

A transition through a degeneration (E) or (H1) changes the number of fixed components by one. In particular, each stratum of morphisms from a one-nodal symmetric surface of type (E) or (H1) to (X, ϕ) is a boundary of the spaces $\overline{\mathfrak{M}}_{g,l}(X, B; J)^{\phi,\sigma}$ for precisely two topological types of genus g involutions σ. A degeneration (H1) cannot occur in genus 0, but does occur in genus 1 and higher.

2.2. Real orientations

Let (X, ϕ) be a topological space with an involution. A conjugation on a complex vector bundle $V \to X$ lifting an involution ϕ is a vector bundle homomorphism $\varphi\colon V \to V$ covering ϕ (or equivalently a vector bundle homomorphism $\varphi\colon V \to \phi^* V$ covering id_X) such that the restriction of φ to each fiber is anticomplex linear and $\varphi \circ \varphi = \mathrm{id}_V$.

A real bundle $(V, \varphi) \to (X, \phi)$ consists of a complex vector bundle $V \to X$ and a conjugation φ on V lifting ϕ.

Example 2.2. (1) If X is a smooth manifold with a smooth involution ϕ, then $(TX, d\phi)$ is a real bundle over (X, ϕ).

(2) If $L \to X$ is a complex vector bundle then $L \oplus \phi^* \bar{L} \to X$ with the conjugation $\widetilde{\phi} : (x, v, w) \mapsto (\phi(x), w, v)$ is also a real bundle over (X, ϕ).

For any real bundle (V, φ) over (X, ϕ), the fixed locus

$$V^{\varphi} \to X^{\phi}$$

of φ is a real vector bundle over X^{ϕ}. We denote by

$$\Lambda_{\mathbb{C}}^{\mathrm{top}}(V, \varphi) = \big(\Lambda_{\mathbb{C}}^{\mathrm{top}} V, \Lambda_{\mathbb{C}}^{\mathrm{top}} \varphi\big)$$

the top exterior power of V over $\mathbb{C}$ with the induced conjugation. Direct sums, duals, and tensor products over $\mathbb{C}$ of real bundles over (X, ϕ) are again real bundles over (X, ϕ).

Definition 2.3 ([15, 18]). Let (X, ϕ) be a topological space with an involution and (V, φ) be a real bundle over (X, ϕ). A real orientation on (V, φ) consists of

(RO1) a complex line bundle $L \to X$ such that

$$w_2\big(V^{\varphi} \oplus L^*_{|X^{\phi}}\big) = 0 \quad \text{and} \quad \Lambda_{\mathbb{C}}^{\mathrm{top}}(V, \varphi) \approx \Lambda_{\mathbb{C}}^{\mathrm{top}}(L \oplus \phi^* \bar{L}, \widetilde{\phi}), \qquad (2.5)$$

(RO2) a homotopy class of isomorphisms of real bundles in (2.5), and

(RO3) a spin structure on the real vector bundle $V^{\varphi} \oplus L^*$ over X^{ϕ} compatible with the orientation induced by (RO2).

An isomorphism in (2.5) restricts to an isomorphism $\Lambda_{\mathbb{R}}^{\mathrm{top}} V^{\varphi} \approx \Lambda_{\mathbb{R}}^{\mathrm{top}} L$ of real line bundles over X^{ϕ}. Since L is a complex vector bundle it is canonically oriented, and thus (RO2) determines orientations on V^{φ} and $V^{\varphi} \oplus L^*$. By the first assumption in (2.5), the real vector bundle $V^{\varphi} \oplus L^*$ over X^{ϕ} admits a spin structure.

A real orientation on a real symplectic manifold (X, ω, ϕ) is a real orientation on the real bundle $(TX, d\phi)$. We call a real symplectic manifold (X, ω, ϕ) real-orientable if it admits a real orientation. As established in [18] a real orientation on (X,ϕ) determines a canonical orientation of the uncompactified moduli spaces when X is of odd complex dimension. This orientation extends across the codimension 1 boundary strata of types (H2) and (H3) and changes across the codimension 1 boundary strata of types (E) and (H1). The parity of $|\pi_0(\Sigma^\sigma)|$ behaves in the same way. This allows us to readjust this canonical orientation by the parity of the number of fixed components of the domain and thus obtain an orientation on the compactified moduli space.

Theorem 2.4 ([18, THEOREM 1.3]). *Let (X,ω,ϕ) be a real-orientable $2n$-manifold, $g,l \in \mathbb{Z}^{\geq 0}$, $B \in H_2(X;\mathbb{Z})$, and $J \in \mathcal{J}_\omega^\phi$.*

(1) *If $n \notin 2\mathbb{Z}$, a real orientation on (X,ω,ϕ) orients $\overline{\mathfrak{M}}_{g,l}(X,B;J)^\phi$.*

(2) *If $n \in 2\mathbb{Z}$ and $g+l \geq 2$, a real orientation on (X,ω,ϕ) orients the real line bundle*

$$\Lambda_{\mathbb{R}}^{\mathrm{top}}(T\overline{\mathfrak{M}}_{g,l}(X,B;J)^\phi) \otimes \mathfrak{f}^*\Lambda_{\mathbb{R}}^{\mathrm{top}}(T\mathbb{R}\overline{\mathcal{M}}_{g,l}) \to \overline{\mathfrak{M}}_{g,l}(X,B;J)^\phi.$$

Examples of real-orientable manifolds include $\mathbb{P}^{2n-1}$, X_5, many other projective complete intersections, and simply-connected real symplectic Calabi–Yau and real Kähler Calabi–Yau manifolds with spin fixed locus as described by the following propositions.

Proposition 2.5 ([19, PROPOSITION 1.2]). *Let (X,ω,ϕ) be a real symplectic manifold with $w_2(X^\phi)=0$. If*

(1) *$H_1(X;\mathbb{Q})=0$ and $c_1(X)=2(\mu-\phi^*\mu)$ for some $\mu \in H^2(X;\mathbb{Z})$ or*

(2) *X is compact Kähler, ϕ is antiholomorphic, and $\mathcal{K}_X = 2([D]+[\overline{\phi_* D}])$ for some divisor D on X,*

then (X,ω,ϕ) is a real-orientable symplectic manifold.

Corollary 2.6 ([19, COROLLARY 1.3]). *Let $n \in \mathbb{Z}^+$ and $\mathbf{a} \equiv (a_1,\dots,a_{n-4}) \in (\mathbb{Z}^+)^{n-4}$ be such that*

$$a_1+\dots+a_{n-4} \equiv n \mod 4.$$

If $X_{n;\mathbf{a}} \subset \mathbb{P}^{n-1}$ is a complete intersection of multidegree $\mathbf{a}$ preserved by τ_n, then $(X_{n;\mathbf{a}},\omega_{n;\mathbf{a}},\tau_{n;\mathbf{a}})$ is a real-orientable symplectic manifold.

Proposition 2.7 ([19, PROPOSITION 1.4]). *Let $m,n \in \mathbb{Z}^+$, $k \in \mathbb{Z}^{\geq 0}$, and $\mathbf{a} \equiv (a_1,\dots,a_k) \in (\mathbb{Z}^+)^k$.*

(1) *If $X_{n;\mathbf{a}} \subset \mathbb{P}^{n-1}$ is a complete intersection of multidegree $\mathbf{a}$ preserved by τ_n,*

$$a_1+\dots+a_k \equiv n \mod 2, \quad \text{and} \quad a_1^2+\dots+a_k^2 \equiv a_1+\dots+a_k \mod 4, \tag{2.6}$$

then $(X_{n;\mathbf{a}},\omega_{n;\mathbf{a}},\tau_{n;\mathbf{a}})$ is a real-orientable symplectic manifold.

(2) *If* $X_{2m;\mathbf{a}} \subset \mathbb{P}^{2m-1}$ *is a complete intersection of multidegree* $\mathbf{a}$ *preserved by* η_{2m} *and*

$$a_1 + \cdots + a_k \equiv 2m \mod 4,$$

then $(X_{2m;\mathbf{a}}, \omega_{2m;\mathbf{a}}, \eta_{2m;\mathbf{a}})$ *is a real-orientable symplectic manifold.*

2.3. Real Gromov–Witten theory

The moduli space $\overline{\mathfrak{M}}_{g,l}(X, B; J)^{\phi}$ is not smooth in general and its tangent bundle in Theorem 2.4 should be viewed in the usual moduli-theoretic (or virtual) sense. Since the (virtual) boundary of $\overline{\mathfrak{M}}_{g,l}(X, B; J)^{\phi}$ is empty, Theorem 2.4(1) implies that this moduli space carries a virtual fundamental class over $\mathbb{Q}$ (determined by the choice of orientation) and thus gives rise to real GW-invariants in arbitrary genus.

Theorem 2.8 ([18, THEOREM 1.4]). *Let* (X, ω, ϕ) *be a compact real-orientable* $2n$*-manifold with* $n \notin 2\mathbb{Z}$, $g, l \in \mathbb{Z}^{\geq 0}$, $B \in H_2(X; \mathbb{Z})$, *and* $J \in \mathcal{J}_{\omega}^{\phi}$. *Then a real orientation on* (X, ω, ϕ) *endows the moduli space* $\overline{\mathfrak{M}}_{g,l}(X, B; J)^{\phi}$ *with a virtual fundamental class and thus gives rise to genus g real GW-invariants of* (X, ω, ϕ) *that are independent of the choice of* $J \in \mathcal{J}_{\omega}^{\phi}$.

If $n \in 2\mathbb{Z}$ and $g + l \geq 2$, Theorem 2.4 implies that a real orientation on (X, ω, ϕ) induces an orientation on the real line bundle

$$\Lambda_{\mathbb{R}}^{\text{top}}\big(T\overline{\mathfrak{M}}_{g,l}(X, B; J)^{\phi}\big) \otimes \mathfrak{f}^*\big(\Lambda_{\mathbb{R}}^{\text{top}}(T\mathbb{R}\overline{\mathcal{M}}_{g,l})\big) \to \overline{\mathfrak{M}}_{g,l}(X, B; J)^{\phi}, \tag{2.7}$$

where $\mathfrak{f}$ is the forgetful morphism (2.4). This orientation can be used to construct GW invariants of (X, ω, ϕ) with classes twisted by the orientation system of $\mathbb{R}\overline{\mathcal{M}}_{g,l}$.

For each $i = 1, \ldots, l$, let

$$\mathrm{ev}_i : \overline{\mathfrak{M}}_{g,l}(X, B; J)^{\phi} \to X, \quad \big[u, (z_1^+, z_1^-), \ldots, (z_l^+, z_l^-)\big] \to u(z_i^+),$$

be the evaluation at the first point in the ith pair of conjugate points. For $\mu_1, \ldots, \mu_l \in H^*(X)$, the numbers

$$\langle \mu_1, \ldots, \mu_l \rangle_{g,B}^{\phi} \equiv \int_{[\overline{\mathfrak{M}}_{g,l}(X,B;J)^{\phi}]} \mathrm{ev}_1^*\mu_1 \cdots \mathrm{ev}_l^*\mu_l \in \mathbb{Q}$$

are virtual counts of real J-holomorphic curves in X passing through generic cycle representatives for the Poincaré duals of $\mu_1, \ldots, \mu_l$, i.e., real GW invariants of (X, ω, ϕ) with conjugate pairs of insertions. They are independent of the choices of cycles representatives and of J.

Moreover, for each $i = 1, \ldots, l$, let

$$\psi_i \in H^2\big(\overline{\mathfrak{M}}_{g,l}(X, B; J)^{\phi}; \mathbb{Q}\big)$$

be the Chern class of the universal cotangent line bundle for the marked point z_i^+. For $a_1, \ldots, a_l \in \mathbb{Z}^{\geq 0}$ and $\mu_1, \ldots, \mu_l \in H^*(X; \mathbb{Q})$, let

$$\big\langle \tau_{a_1}(\mu_1), \ldots, \tau_{a_l}(\mu_l) \big\rangle_{g,B}^{\phi} = \int_{[\overline{\mathfrak{M}}_{g,l}(X,B;J)^{\phi}]^{\text{vir}}} \psi_1^{a_1}\big(\mathrm{ev}_1^*\mu_1\big) \cdots \psi_l^{a_l}\big(\mathrm{ev}_l^*\mu_l\big) \tag{2.8}$$

be the associated real descendant GW invariant. This number is again independent of the choices of cycle representatives and of $J \in \mathcal{J}_{\omega}^{\phi}$.

Given the existence of a full descendant theory, there are many natural questions that arise and that are not well understood at the moment. In particular, they are related to finding structures governing the invariants. One expects to find a real cohomological field theory behind them and a Givental–Teleman-type classification result would be very valuable for reconstruction results, mirror symmetry, and connections to Dubrovin–Zhang-type integrable hierarchies. Further connections to integrable systems that parallel those established in the classical case for KdV, KP, and Toda [25, 31, 32, 40] are also expected.

3. STRUCTURAL RESULTS

Here we consider the real Gromov–Witten theory of real 3-folds which are the total space of real bundles over curves with an antisymplectic involution. The motivation for considering 3-folds of this type comes from the virtual contribution to the real GW invariants of a real elementary curve in a compact real Calabi–Yau 3-fold, sometimes referred to as multiple-covers contribution, and the real Gopakumar–Vafa conjecture [37] expressing the connected real Gromov–Witten invariants in terms of integer invariants.

The invariants associated with this setup are called local real Gromov–Witten (RGW) invariants and are discussed in Section 3.1. They give rise to a semisimple 2D Klein TQFT defined on an extension of the category of unorientable surfaces. This structure allows us to completely solve the theory by providing a closed formula for the local RGW invariants in terms of representation-theoretic data, extending earlier results of Bryan and Pandharipande [9]. The local version of the real Gopakumar–Vafa formula is obtained as a consequence of the structural results. Furthermore, in the case of the resolved conifold, we find that the partition function of the RGW invariants agrees with that of the SO/Sp Chern–Simons theory [3].

3.1. Local real Gromov–Witten invariants

Let (Σ, c) be a symmetric Riemann surface and $L \to \Sigma$ a holomorphic line bundle. Then the total space of

$$L \oplus c^*\bar{L} \to \Sigma, \quad c_{tw}(z; u, v) = \big(c(z); v, u\big) \tag{3.1}$$

is a real manifold with an antiholomorphic involution c_{tw}. An $U(1)$-action on the line bundle $L \to \Sigma$ induces an action on the 3-fold (3.1) compatible with the real structure. We define local (relative) RGW invariants associated to the real 3-fold (3.1) as pairings between the $U(1)$-equivariant Euler class of the index bundle Ind $\bar{\partial}_L$ (regarded as an element in K-theory) and the virtual fundamental class of the (relative) real moduli space $\overline{\mathcal{M}}_{d,\chi}^{c,\bullet}(\Sigma)$ discussed below.

Definition 3.1. Let (Σ, c) be a marked symmetric surface, with r pairs of conjugate marked points $(x_1^+, c(x_1^+)), \ldots, (x_r^+, c(x_r^+))$, and $\vec{\lambda} = (\lambda^1, \ldots, \lambda^r)$ be a collection of r partitions of d. Denote by

$$\overline{\mathcal{M}}_{d,\chi}^{\bullet,c}(\Sigma)_{\lambda^1,\ldots,\lambda^r} \tag{3.2}$$

the relative real moduli space of degree d stable real maps $f : (C, \sigma) \to (\Sigma, c)$ such that

- f has ramification pattern λ^i over x_i^+ (and thus also over $x_i^- = c(x_i^+)$), for all $i = 1, \dots, r$;
- the domain C is possibly disconnected and has total Euler characteristic χ;
- f is nontrivial on each connected component of C.

The moduli space $\overline{\mathcal{M}}_{d,\chi}^{\bullet,c}(\Sigma)_{\lambda^1,\dots,\lambda^r}$ has virtual dimension b, where

$$b = d\chi(\Sigma) - \chi - 2\delta(\vec{\lambda}) \quad \text{and} \quad \delta(\vec{\lambda}) = \sum_{i=1}^{r} (d - \ell(\lambda^i)). \tag{3.3}$$

Here $\ell(\lambda^i)$ is the length of the partition λ^i, i.e., the cardinality of $f^{-1}(x_i^+)$.

These real moduli spaces are orientable, but a priori the local RGW invariants depend on the choice of real orientation (cf. Definition 2.3) and on the topological type of the real structure c on Σ. We show in [15] that there is a canonical choice of orientation for the local RGW invariants, compatible with the splitting formula (3.21), and, moreover, that they do not depend on the real structure c. We therefore omit these choices from the notation below.

If $L \to \Sigma$ is a holomorphic bundle, the operator $\bar{\partial}_L$ determines a family of complex operators over the moduli spaces of maps to Σ; the fiber at a stable map $f : C \to \Sigma$ is the pullback operator $\bar{\partial}_{f^*L}$. Denote by Ind $\bar{\partial}_L$ the index bundle associated to this family of operators, regarded as an element in K-theory.

Let $\bar{\partial}_{(L\oplus c^*\bar{L}, c_{tw})}$ denote the restriction of $\bar{\partial}_{L\oplus c^*\bar{L}}$ to the invariant part of its domain and target, cf. [18, SECTION 4.3]. Via the projection onto the first factor, the kernel and cokernel of $\bar{\partial}_{(L\oplus c^*\bar{L}, c_{tw})}$ are canonically identified with the kernel and cokernel of $\bar{\partial}_L$ and

$$\operatorname{Ind} \bar{\partial}_{(L\oplus c^*\bar{L}, c_{tw})} \cong \operatorname{Ind} \bar{\partial}_L. \tag{3.4}$$

The right-hand side carries a natural complex structure, which pulls back to one on the left-hand side. An $U(1)$-action on L induces one on $(L \oplus c^*\bar{L}, c_{tw})$, compatible with the real structure. In turn, these induce $U(1)$-actions on Ind $\bar{\partial}_L$ and Ind $\bar{\partial}_{(L\oplus c^*\bar{L}, c_{tw})}$ and the isomorphism (3.4) identifies their equivariant Euler classes.

Definition 3.2. Assume (Σ, c) is a symmetric surface with r pairs of marked points. Let $L \to \Sigma$ be a holomorphic line bundle and $\vec{\lambda} = (\lambda^1, \dots, \lambda^r)$ a collection of r partitions of d. The local real relative GW invariants associated with the real 3-fold $(L \oplus c^*\bar{L}, c_{tw}) \to (\Sigma, c)$ are the equivariant pairings

$$RZ_{d,\chi}^c(\Sigma, L)_{\vec{\lambda}} = \int_{[\overline{\mathcal{M}}_{d,\chi}^{c,\bullet}(\Sigma)_{\vec{\lambda}}]^{\text{vir}}} e_{U(1)}(-\operatorname{Ind} \bar{\partial}_L). \tag{3.5}$$

We further consider the shifted generating function

$$\mathrm{RGW}_d(\Sigma, L)_{\vec{\lambda}} = \sum_{\chi} u^{d(\frac{\chi(\Sigma)}{2} + c_1(L)[\Sigma]) - \frac{\chi}{2} - \delta(\vec{\lambda})} RZ_{d,\chi}^c(\Sigma, L)_{\vec{\lambda}} \in \mathbb{Q}(t)((u)), \tag{3.6}$$

where $\delta(\vec{\lambda})$ is as in (3.3). It takes values in the localized equivariant cohomology ring of $U(1)$ generated by t.

3.2. TQFT and Klein TQFT

Let 2**Cob** be the usual (oriented, closed) 2-dimensional cobordism category. It is the symmetric monoidal category with objects given by compact oriented 1-manifolds (without boundary) and morphisms given by (diffeomorphism classes of) oriented cobordisms. A 2-dimensional topological quantum field theory (2D TQFT) with values in a commutative ring R is a symmetric monoidal functor

$$F : 2\mathbf{Cob} \to R\,\mathrm{mod},$$

where R mod is the category of R-modules. This is equivalent to a commutative Frobenius algebra over R; the product and coproduct correspond to the pair of pants while the unit and counit to the cap and cup, respectively. In [9], Bryan and Pandharipande enlarge the category 2**Cob** to a category $2\mathbf{Cob}^{L_1,L_2}$ with the same objects, but with morphisms decorated by a pair of complex line bundles (L_1, L_2) trivialized over the boundary; the Euler numbers (k_1, k_2) of these bundles determine the level of the theory. Restricting the morphisms to $k_1 = k_2 = 0$ defines an embedding

$$2\mathbf{Cob} \subset 2\mathbf{Cob}^{L_1,L_2}.$$

Bryan and Pandharipande use the local GW invariants to define a symmetric monoidal functor

$$\mathbf{GW} : 2\mathbf{Cob}^{L_1,L_2} \to R\,\mathrm{mod} \tag{3.7}$$

on this larger category. The functor (3.7) extends the classical 2D TQFT that appeared in the work of Dijkgraaf–Witten [12] and Freed–Quinn [14], and whose Frobenius algebra is the center $\mathbb{Q}[S_d]^{S_d}$ of the group algebra of the symmetric group S_d. It is used to completely solve the local Gromov–Witten theory.

A different extension of 2**Cob** is obtained by allowing unoriented and possibly unorientable surfaces as cobordisms; see [2,4]. We refer to this category as 2**KCob**, where **K** stands for Klein (surface). The objects are closed unoriented 1-manifolds and the morphisms are diffeomorphism classes of *unoriented* (and possibly unorientable) cobordisms. An equivalent point of view is to consider the orientation double covers of both the objects and the morphisms: the objects are then closed oriented 1-manifolds with an orientation-reversing involution (deck transformation) exchanging the sheets of the cover and the morphisms are compact oriented 2-dimensional manifolds with a *fixed-point free* orientation-reversing involution extending the one on the boundary. Such 2-dimensional manifolds are called symmetric surfaces, and we denote this category by 2**SymCob**. Moreover,

$$2\mathbf{KCob} \equiv 2\mathbf{SymCob},$$

where the identification is obtained by passing to the orientation double cover in one direction and taking the quotient by the involution in the other. Working from the perspective of 2**SymCob** allows us to construct an extension $2\mathbf{SymCob}^L$ of this category related to that of [9] and completely solve the local real Gromov–Witten theory.

The category **2Cob** can be regarded as a subcategory of **2KCob** with the same objects, but fewer morphisms

$$\mathbf{2Cob} \subset \mathbf{2KCob}.$$

Note that even if a cobordism in **2KCob** is orientable, there may not be way to orient it in a way compatible with the boundary identifications.

The generators of **2Cob** ⊂ **2KCob** are the usual cap, cup, tube, twist, and pair of pants cobordisms, and the corresponding elements of **2SymCob** are their orientation double covers. The category **2KCob** has two extra generators, the cross-cap (a Möbius band) and the involution

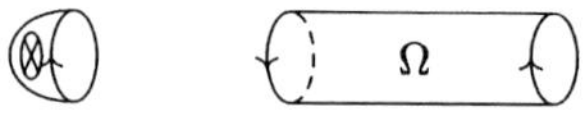

(3.8)

respectively. In **2SymCob** these correspond to their orientation double covers

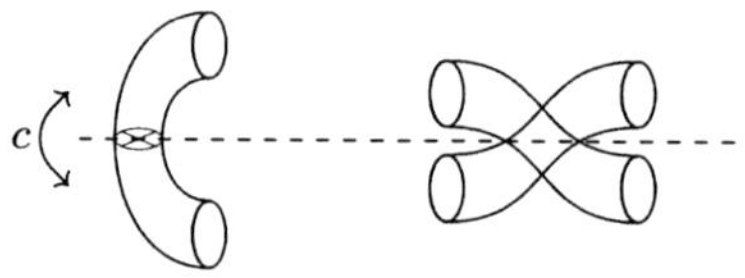

(3.9)

Note that in **2SymCob** the involution swaps the two outgoing circles – this distinguishes it from the tube which acts as the identity.

The extra generators satisfy certain relations in **2KCob** (see [4, PP. 1840–1841]). For example, moving a puncture once around the Möbius band changes the orientation of the puncture, cf. Figure 2; equivalently, the involution acts trivially on the product of the cross-cap with another element, cf. (3.13). Another relation comes from decomposing the product of two cross-caps as in Figure 3, cf. (3.14).

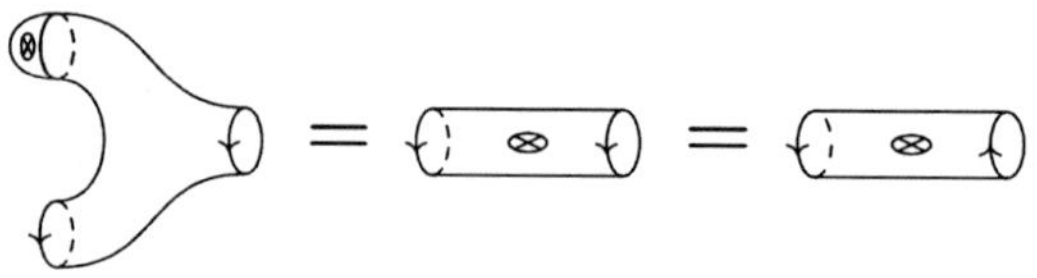

FIGURE 2
Relation in **2KCob**: involution acts trivially on products with a cross-cap.

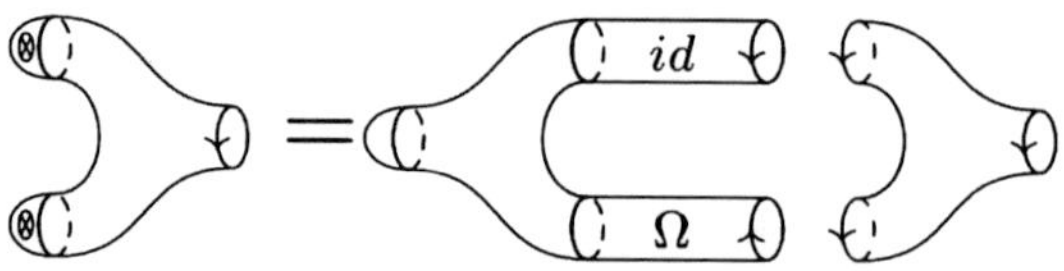

FIGURE 3
Relation in **2KCob**: decomposing the punctured Klein bottle.

3.2.1. Semisimple Klein TQFT

Definition 3.3. A (closed) 2D Klein TQFT is a symmetric monoidal functor

$$F : \mathbf{2KCob} \to R\,\mathrm{mod}\,. \tag{3.10}$$

When (3.10) is regarded as a morphism on $\mathbf{2SymCob} \equiv \mathbf{2KCob}$ via the orientation double cover construction, we denote it by

$$\tilde{F} : \mathbf{2SymCob} \to R\,\mathrm{mod}\,. \tag{3.11}$$

In fact, cf. [4, PROPOSITION 1.11], a (closed) 2D Klein TQFT is equivalent to a commutative Frobenius algebra $H = F(S^1)$ together with two extra structures:

(a) an involutive (anti)automorphism Ω of the Frobenius algebra H, denoted $x \mapsto x^*$. This means

$$(x^*)^* = x, \quad (xy)^* = y^* x^* \quad \text{and} \quad \langle x^*, y^* \rangle = \langle x, y \rangle \quad \text{for all } x, y \in H. \tag{3.12}$$

(b) an element $U \in H$ such that

$$(aU)^* = aU \quad \text{for all } a \in H \quad \text{and} \tag{3.13}$$

$$U^2 = m(id \otimes \Omega)\big(\Delta(1)\big) = \sum \alpha_i \beta_i^*. \tag{3.14}$$

Here the coproduct is $\Delta(1) = \sum \alpha_i \otimes \beta_i$. The involution Ω and the element U correspond to the cobordisms (3.8). The relations (b) correspond to Figures 2 and 3.

Definition 3.4. A semisimple Klein TQFT is a Klein TQFT whose associated Frobenius algebra is semisimple.

A semisimple TQFT is determined by the structure constants $\{\lambda_\rho\}$, i.e., the coefficients of the comultiplication $\Delta(v_\rho) = \lambda_\rho v_\rho \otimes v_\rho$ in the idempotent basis $\{v_\rho\}$. Moreover,

Proposition 3.5 ([15, PROPOSITION 7.4]). *Assume* (3.10) *is a semisimple KTQFT with idempotent basis $\{v_\rho\}$ and structure constants $\{\lambda_\rho\}$, and assume that the ground ring R has no zero divisors. Then*

(i) *Ω defines an involution on the idempotent basis $\Omega(v_\rho) = v_{\rho^*}$.*

(ii) *If $U = \sum_\rho U_\rho v_\rho$ then $U_\rho^2 = \lambda_\rho$ if $\rho = \rho^*$, and $U_\rho = 0$ if $\rho \neq \rho^*$.*

Assume Σ is a closed symmetric surface, considered as a morphism in $\mathbf{2SymCob}$ from the ground ring to the ground ring.

Corollary 3.6 ([15, COROLLARY 7.5]). *With the notation of Proposition* 3.5, *the morphism* (3.11) *is given by:*

$$\tilde{F}(\Sigma) = \sum_{\rho=\rho^*} U_\rho^{g-1}, \quad \textit{when } \Sigma \textit{ is a connected genus } g \textit{ surface, and}$$

$$\tilde{F}(\Sigma \sqcup \overline{\Sigma}) = \sum_{\rho} \lambda_\rho^{g-1}, \quad \textit{when } \Sigma \sqcup \overline{\Sigma} \textit{ is a } g\textit{-doublet.}$$

3.2.2. The category $\mathbf{2SymCob}^L$

Consider the category $2\mathbf{SymCob}^L$ whose objects are disjoint unions of copies of $\mathcal{S} = (S^1 \sqcup \overline{S^1}, \varepsilon)$, where ε swaps the two components, and morphisms correspond to isomorphism classes relative boundary of decorated cobordisms $W = (\Sigma, c, L)$, where Σ is an oriented cobordism with a fixed-point free orientation-reversing involution c, extending ε, and L is a complex line bundle over Σ, trivialized along the boundary of Σ.

The level 0 theory corresponds to a trivial bundle L, and defines embeddings

$$2\mathbf{Cob} \subset 2\mathbf{KCob} \equiv 2\mathbf{SymCob} \subset 2\mathbf{SymCob}^L. \tag{3.15}$$

The doubling procedure defines an embedding

$$2\mathbf{Cob}^{L_1,L_2} \subset 2\mathbf{SymCob}^L, \quad (\Sigma, L_1, L_2) \mapsto (\Sigma \sqcup \overline{\Sigma}, L_1 \sqcup \bar{L}_2). \tag{3.16}$$

The category $2\mathbf{Cob}^{L_1,L_2}$ has 4 extra generators, the level $(\pm 1, 0)$, $(0, \pm 1)$-caps, besides those of $2\mathbf{Cob}$, cf. [9, SECTION 4.3]. Similarly, the generators of the category $2\mathbf{SymCob}^L$ are those of $2\mathbf{SymCob}$ together with the images of the $(\pm 1, 0)$, $(0, \pm 1)$-caps under (3.16).

Proposition 3.7 ([15, PROPOSITION 7.6]). *A symmetric monoidal functor*

$$F : 2\mathbf{SymCob}^L \to R\,\mathrm{mod} \tag{3.17}$$

is uniquely determined by the level 0 *theory and the images* η *and* $\bar{\eta}$ *of the level* $(-1, 0)$ *and* $(0, -1)$*-caps.*

If the restriction of (3.17) to the level 0 theory defines a semisimple KTQFT with idempotent basis $\{v_\rho\}$ let

$$\eta = \sum_\rho \eta_\rho v_\rho \quad \text{and} \quad \bar{\eta} = \sum_\rho \bar{\eta}_\rho v_\rho. \tag{3.18}$$

As in Corollary 3.6, then the value of F on a closed connected genus g symmetric surface Σ at level $k = c_1(L)[\Sigma]$ is equal to

$$F(\Sigma|L) = \sum_{\rho=\rho^*} U_\rho^{g-1} \eta_\rho^{-k}. \tag{3.19}$$

The value of F on a g-doublet $\Sigma \sqcup \overline{\Sigma}$ with a line bundle $L_1 \sqcup L_2$ is similarly equal to

$$F(\Sigma \sqcup \overline{\Sigma}|L_1, L_2) = \sum_\rho \lambda_\rho^{g-1} \eta_\rho^{-k_1} \bar{\eta}_\rho^{-k_2},$$

where $k_1 = c_1(L_1)[\Sigma]$ and $k_2 = c_1(L_2)[\overline{\Sigma}]$.

3.3. Splitting formulas

Let (Σ_0, c_0) be a nodal symmetric surface with a pair of conjugate nodes and r pairs of conjugate marked points. It has a normalization $(\widetilde{\Sigma}, \tilde{c})$ which has $r + 2$ pairs of conjugate marked points. Similarly, (Σ_0, c_0) has a family of smooth deformations $(\mathcal{F}, c_{\mathcal{F}}) = \bigcup_s (\Sigma_s, c_s)$, simultaneously smoothing out the conjugate nodes using complex conjugate gluing parameters. The generic fiber (Σ_s, c_s) of the family is a symmetric surface with r

pairs of conjugate marked points, and a pair of "splitting circles" (disjoint vanishing cycles) swapped by the involution; as the gluing parameters converge to 0, these circles pinch to produce the two complex conjugate nodes of Σ_0; see Figure 4. Any complex line bundle L over Σ_s can be deformed to the nodal surface and then lifted to its normalization to give a line bundle $\tilde{L}$ over $\widetilde{\Sigma}$.

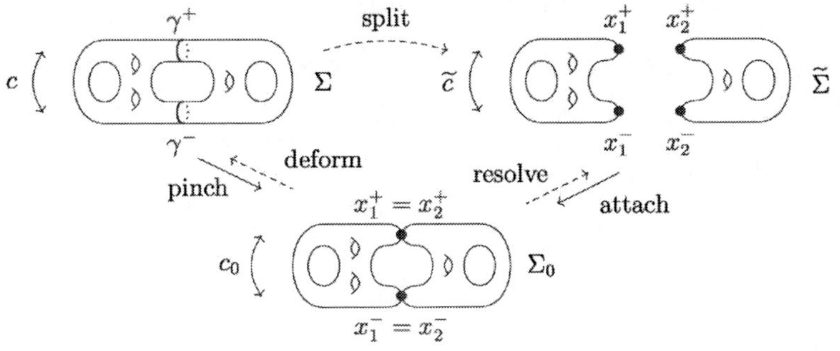

FIGURE 4
Splitting Σ along a pair of conjugate circles (γ^+, γ^-).

In order to state the splitting theorem in a more compact form, we define the raising of the indices by the formula

$$\mathrm{RGW}(\Sigma, L)^{\nu^1 \dots \nu^s}_{\mu^1 \dots \mu^r} = \mathrm{RGW}(\Sigma, L)_{\mu^1 \dots \mu^r, \nu^1 \dots \nu^s} \left(\prod_{i=1}^{s} \zeta(\nu^i) t^{2\ell(\nu^i)} \right), \tag{3.20}$$

where $\zeta(\lambda) = \prod m_k! k^{m_k}$ for a partition $\lambda = (1^{m_1}, 2^{m_2}, \dots)$.

Theorem 3.8 (RGW splitting theorem, [16]). *Assume (Σ, c) is a marked symmetric surface with r pairs of conjugate points and L is a complex line bundle over Σ. Let $(\widetilde{\Sigma}, \tilde{c})$ denote the symmetric surface obtained as described above from (Σ, c) by splitting it along two conjugate circles, and let $\tilde{L}$ be the corresponding line bundle over $\widetilde{\Sigma}$.*

Then for any collection $\vec{\mu} = (\mu^1, \dots, \mu^r)$ of r partitions of d, the RGW invariants (3.6) satisfy

$$\mathrm{RGW}_d(\Sigma, L)_{\vec{\mu}} = \sum_{\lambda \vdash d} \mathrm{RGW}_d(\widetilde{\Sigma}, \tilde{L})^{\lambda}_{\vec{\mu}, \lambda}. \tag{3.21}$$

This result is used to show that the local RGW theory gives rise to (an extension of) a KTQFT; it corresponds to compatibility of cobordism decompositions.

3.4. The RGW Klein TQFT

In this section we use the local RGW invariants (3.6) to define an extension of a Klein TQFT, i.e., a functor **RGW** from the category $2\mathbf{SymCob}^L$ described in Section 3.2.2. This extends the Bryan–Pandharipande TQFT constructed from the GW theory for the anti-diagonal action.

Let $R = \mathbb{C}(t)((u))$ be the ring of Laurent series in u whose coefficients are rational functions of t and d be a positive integer. Denote by $\mathcal{S} = (S^1 \sqcup \overline{S^1}, \varepsilon)$ the disjoint union of two copies of a circle with opposite orientations and with the involution ε swapping them. To the object $\mathcal{S}$ we associate

$$\mathbf{RGW}_d(\mathcal{S}) = H = \bigoplus_{\alpha \vdash d} Re_\alpha, \tag{3.22}$$

the free module with basis $\{e_\alpha\}_{\alpha \vdash d}$ indexed by partitions α of d. Let

$$\mathbf{RGW}_d(\mathcal{S} \sqcup \cdots \sqcup \mathcal{S}) = H \otimes \cdots \otimes H.$$

To each cobordism $W = (\Sigma, c, L)$ in $2\mathbf{SymCob}^L$ from n copies of $\mathcal{S}$ to m copies of $\mathcal{S}$, associate the R-module homomorphism

$$\mathbf{RGW}_d(W) : H^{\otimes n} \to H^{\otimes m} \tag{3.23}$$

defined by

$$e_{\lambda^1} \otimes \cdots \otimes e_{\lambda^n} \mapsto \sum_{\mu^i \vdash d} \mathrm{RGW}_d(\Sigma_W | L_W)^{\mu^1..\mu^m}_{\lambda^1..\lambda^n} \, e_{\mu^1} \otimes \cdots \otimes e_{\mu^m}.$$

Here Σ_W is a closed marked symmetric Riemann surface whose topological type is that of Σ after removing small disks around the pairs of marked points and $L_W \to \Sigma_W$ is a holomorphic line bundle whose first Chern class corresponds to the Euler class of $L \to \Sigma$.

Theorem 3.9 ([15, THEOREM 8.1]). *The assignment* (3.23) *defines a symmetric monoidal functor*

$$\mathbf{RGW}_d : 2\mathbf{SymCob}^L \to R \,\mathrm{mod}\,. \tag{3.24}$$

Its restriction to $2\mathbf{KCob}$ *under* (3.15) *is a Klein TQFT, while its restriction to* $2\mathbf{Cob}^{L_1,L_2}$ *under* (3.16) *is*

$$\mathbf{RGW}_d(\Sigma \sqcup \overline{\Sigma} | L_1 \sqcup \bar{L}_2)(u,t) = (-1)^{dk_2} \mathbf{GW}_d(\Sigma | L_1, L_2)(iu, it). \tag{3.25}$$

Here k_i is the total degree of L_i and $\mathbf{GW}_d$ *is the TQFT* (3.7) *considered by Bryan–Pandharipande (for the antidiagonal action).*

The KTQFT determined by the level 0 local RGW invariants is semisimple, cf [15]. It corresponds in fact to *signed* counts of degree d real Hurwitz covers. The idempotent basis is indexed by irreducible representations of the symmetric group S_d and $\Omega(v_\rho) = v_{\rho'}$ where ρ' is the conjugate representation. In order to calculate the coefficients of U in the idempotent basis, we introduced in [15] the signed Frobenius–Schur indicator (SFS). The SFS takes values $0, \pm 1$ on irreducible real representations, unlike the standard FS indicator which is $+1$ on them. The SFS is 0 if and only if the representation is *not* self-conjugate and the sign of a self-conjugate representation is given as a function of its characters. While these considerations are valid for real representations of any finite group, in the case of the symmetric group we find a simpler expression for the latter function using the Weyl formula for B_n. In particular, for an irreducible self-conjugate representation ρ of S_d,

$$\mathrm{SFS}(\rho) = (-1)^{(d-r(\rho))/2},$$

where $r(\rho)$ is the rank of ρ, i.e., the length of the main diagonal of the Young diagram associated to ρ. This is precisely the sign that appears in the partition function of the SO/Sp Chern–Simons theory **[3, (6.1)]**; in the case of the resolved conifold, Theorems 3.12 and 3.13 below recover the partition function **[3, (6.3)]** and the free energy **[3, (3.2)]**, respectively.

The idempotent basis for the theory is given by

$$v_\rho = \frac{\dim \rho}{d!} \sum_\alpha (-t)^{\ell(\alpha)-d} \chi_\rho(\alpha) e_\alpha, \tag{3.26}$$

indexed by the irreducible representations ρ of S_d. We then have the following results.

Lemma 3.10 (**[15, Lemma 9.2]**). *In the idempotent basis $\{v_\rho\}$, the structure constants $\{\lambda_\rho\}$ and the coefficients $\{\eta_\rho\}$, $\{\bar{\eta}_\rho\}$ of the $(-1,0)$ and $(0,-1)$-caps are given by*

$$\lambda_\rho = t^{2d}\left(\frac{d!}{\dim \rho}\right)^2, \quad \eta_\rho = t^d Q^{c_\rho/2}\left(\frac{\operatorname{dimh}_Q \rho}{\dim \rho}\right), \quad \bar{\eta}_\rho = t^d Q^{-c_\rho/2}\left(\frac{\operatorname{dimh}_Q \rho}{\dim \rho}\right). \tag{3.27}$$

Here $Q = e^u$, c_ρ is the total content of the Young diagram associated to ρ, and

$$\operatorname{dimh}_Q \rho = d! \prod_{\square \in \rho}\left(2\sinh\frac{h(\square)u}{2}\right)^{-1} = d! \prod_{\square \in \rho}\left(Q^{\frac{h(\square)}{2}} - Q^{-\frac{h(\square)}{2}}\right)^{-1}, \tag{3.28}$$

where $h(\square)$ denotes the hooklength of the square $\square$ in the Young diagram associated to ρ.

Proposition 3.11 (**[15, Corollary 9.7]**). *In the idempotent basis, the level 0 cross-cap U is given by*

$$U = \sum_{\substack{\rho \vdash d \\ \rho = \rho'}} (-1)^{(d-r(\rho))/2}\, t^d \frac{d!}{\dim \rho} v_\rho, \tag{3.29}$$

where $r(\rho)$ is the length of the main diagonal of the Young diagram of ρ.

Combining these results with the results of Section 3.2 we obtain a closed expression for the local RGW theory of the 3-folds (3.1) in terms of representation theoretic data. The Calabi–Yau case is given in the following theorem.

Theorem 3.12 (**[15, Lemma 9.14]** (Local CY)). *Let Σ be a connected genus g symmetric surface and $L \to \Sigma$ a holomorphic line bundle with Chern number $g-1$. Then the generating function of the degree d local RGW invariants is equal to*

$$\mathrm{RGW}_d(\Sigma, L) = \sum_{\rho = \rho'} \left((-1)^{\frac{d-r(\rho)}{2}} \prod_{\square \in \rho} 2\sinh\frac{h(\square)u}{2}\right)^{g-1}.$$

Here the sum is over all self-conjugate partitions ρ of d, the product is over all boxes $\square$ in the Young diagram of ρ, $h(\square)$ is the hooklength of $\square$, and $r(\rho)$ is the length of the main diagonal of the Young diagram of ρ.

3.5. Real Gopakumar–Vafa formula

The local RGW invariants correspond to possibly disconnected counts. As usual they can be expressed in terms of more basic invariants. In the real GW theory, these basic counts come in two flavors, $\mathrm{CRGW}_d(\Sigma, L)$ and $\mathrm{DRGW}_d(\Sigma, L)$, corresponding to maps from connected real domains and respectively from doublet domains, i.e., domains consisting of two copies of a connected surface with opposite complex structures and the real structure exchanging the two copies. In fact,

$$1+\sum_{d=1}^{\infty} \mathrm{RGW}_d(\Sigma, L)q^d = \exp\left(\sum_{d=1}^{\infty} \mathrm{CRGW}_d(\Sigma, L)q^d + \sum_{d=1}^{\infty} \mathrm{DRGW}_{2d}(\Sigma, L)q^{2d}\right).$$

Furthermore, the doublet invariants are related to half of the complex GW invariants whenever the target Σ is connected,

$$\mathrm{DRGW}_{2d}(\Sigma, L)(u,t) = (-1)^{d(k+1-g)}\frac{1}{2}\mathrm{GW}_d^{\mathrm{conn}}(g|k,k)(iu,it),$$

where g is the genus of Σ, $k = c_1(L)[\Sigma]$ is the degree of L, and $\mathrm{GW}_d^{\mathrm{conn}}(g|k,k)$ are the connected invariants defined in [9] for the anti-diagonal action.

As a consequence of the structure result provided by Theorem 3.12, we obtain the local real Gopakumar–Vafa formula (cf. [37, SECTION 5]). The local GV conjecture in the classical setting, proved in [22, PROPOSITION 3.4], states that the connected GW invariants defined in [9] have the following structure:

$$\sum_{d=1}^{\infty} \mathrm{GW}_d^{\mathrm{conn}}(g|g-1,g-1)(u)q^d = \sum_{d=1}^{\infty}\sum_{h} n_{d,h}^{\mathbb{C}}(g)\sum_{k=1}^{\infty}\frac{1}{k}\left(2\sin\left(\frac{ku}{2}\right)\right)^{2h-2} q^{kd}, \tag{3.30}$$

where the coefficients $n_{d,h}^{\mathbb{C}}(g)$, called the local BPS states, satisfy (i) $n_{d,h}^{\mathbb{C}}(g) \in \mathbb{Z}$ and (ii) for each d, $n_{d,h}^{\mathbb{C}}(g) = 0$ for large h.

In the real setting, the local real GV formula takes the following form.

Theorem 3.13 ([15, THEOREM 10.1] (Local real GV formula)). *Fix a genus g symmetric surface Σ and consider the local real Calabi–Yau 3-fold $(L \oplus c^*\bar{L}, c_{tw}) \to \Sigma$. Then the generating function for the connected local RGW invariants has the following structure:*

$$\sum_{d=1}^{\infty} \mathrm{CRGW}_d(\Sigma|L)(u)q^d = \sum_{d=1}^{\infty}\sum_{h=0}^{\infty} n_{d,h}^{\mathbb{R}}(g)\sum_{\substack{k \text{ odd}\\ k>0}}\frac{1}{k}\left(2\sinh\left(\frac{ku}{2}\right)\right)^{h-1} q^{kd}, \tag{3.31}$$

where the coefficients $n_{d,h}^{\mathbb{R}}(g)$ satisfy (i) (integrality) $n_{d,h}^{\mathbb{R}}(g) \in \mathbb{Z}$, (ii) (finiteness) for each d, $n_{d,h}^{\mathbb{R}}(g) = 0$ for large h, and (iii) (parity) $n_{d,h}^{\mathbb{R}}(g) = n_{d,h}^{\mathbb{C}}(g) \mod 2$. Moreover,

(a) *for $g = 0$, $n_{d,h}^{\mathbb{R}}(0) = 1$ when $d = 1$ and $h = 0$ and vanish otherwise.*

(b) *for $g = 1$, $n_{d,h}^{\mathbb{R}}(1) = (-1)^{d-1}$ when $h = 1$ and vanish otherwise.*

(c) *for any $g \geq 0$, $n_{1,h}^{\mathbb{R}}(g) = 1$ when $h = g$ and vanish otherwise.*

The $g = 0$ case of Theorems 3.12, 3.13 give the real Gromov–Witten invariants of the resolved conifold and coincide with the SO/Sp Chern–Simons theory on S^3. This is an instance of the real analogue of the large N-duality [20,28]. Developing a mathematical theory of the real topological vertex [1,26] would allow establishing this correspondence for any toric real Calabi–Yau 3-fold. Furthermore, a relation between Kauffman polynomials and real GW invariants is also expected based on this duality and it would be very interesting to investigate the potential implications of such a relation.

ACKNOWLEDGMENTS

I would like to thank my long-term collaborators Aleksey Zinger and Eleny Ionel the joint works with whom are discussed in this note. I would also like to thank E. Brugallé, A. Chiodo, I. Itenberg, V. Kharlamov, M. Liu, A. Oancea, J. Walcher, D. Zvonkine for many valuable discussions and continued support.

FUNDING

The author is partially supported by ANR grant ANR-18-CE40-0009 and ERC Consolidator Grant ROGW-864919.

REFERENCES

[1] M. Aganagic, A. Klemm, M. Mariño, and C. Vafa, The topological vertex. *Comm. Math. Phys.* **254** (2005), no. 2, 425–478.

[2] A. Alexeevski and S. Natanzon, Noncommutative two-dimensional topological field theories and Hurwitz numbers for real algebraic curves. *Selecta Math. (N.S.)* **12** (2006), 307–377.

[3] V. Bouchard, B. Florea, and M. Mariño, Counting higher genus curves with crosscaps in Calabi–Yau orientifolds. *J. High Energy Phys.* **12** (2004).

[4] C. Braun, Moduli spaces of Klein surfaces and related operads. *Algebr. Geom. Topol.* **12** (2012), 1831–1899.

[5] E. Brugallé, Floor diagrams relative to a conic, and GW-W invariants of Del Pezzo surfaces. *Adv. Math.* **279** (2015), 438–500.

[6] E. Brugallé and P. Georgieva, Pencils of quadrics and Gromov–Witten–Welschinger invariants of $\mathbb{CP}^3$. *Math. Ann.* **365** (2016), no. 1, 363–380.

[7] E. Brugallé and G. Mikhalkin, Enumeration of curves via floor diagrams. *C. R. Acad. Sci. Paris* **345** (2007), no. 6, 329–334.

[8] E. Brugallé and G. Mikhalkin, Floor decompositions of tropical curves: the planar case. In *Proceedings of the Gökova Geometry–Topology Conference 2008*, edited by S. Akbulut, T. Önder, and R.J. Stern, pp. 64–90, Int. Press, Somerville, MA, 2009.

[9] J. Bryan and R. Pandharipande, The local Gromov–Witten theory of curves. *J. Amer. Math. Soc.* **21** (2008), 101–136.

[10] X. Chen and A. Zinger, WDVV-type relations for disk Gromov–Witten invariants in dimension 6. *Math. Ann.* **379** (2021), 1231–1313.

[11] C.-H. Cho, Counting real J-holomorphic discs and spheres in dimension four and six. *J. Korean Math. Soc.* **45** (2008), no. 5, 1427–1442.

[12] R. Dijkgraaf and E. Witten, Topological gauge theories and group cohomology. *Comm. Math. Phys.* **129** (1990), 393–429.

[13] M. Farajzadeh Tehrani, Counting genus zero real curves in symplectic manifolds. *Geom. Topol.* **20** (2016), no. 2, 629–695.

[14] D. Freed and F. Quinn, Chern–Simons theory with finite gauge group. *Comm. Math. Phys.* **156** (1993), 435–472.

[15] P. Georgieva and E. Ionel, A Klein TQFT: The local Real Gromov–Witten theory of curves. *Adv. Math.* **391** (2021), 1–70.

[16] P. Georgieva and E. Ionel, Splitting formulas for the local real Gromov–Witten invariants. *J. Symplectic Geom.* (to appear), arXiv:2005.05928.

[17] P. Georgieva and A. Zinger, Enumeration of real curves in $\mathbb{CP}^{2n-1}$ and a WDVV relation for real Gromov–Witten invariants. *Duke Math. J.* **166** (2017), no. 17, 3291–3347.

[18] P. Georgieva and A. Zinger, Real Gromov–Witten theory in all genera and real enumerative geometry: construction. *Ann. of Math.* **188** (2018), 685–752.

[19] P. Georgieva and A. Zinger, Real Gromov–Witten theory in all genera and real enumerative geometry: computation. *J. Differential Geom.* **113** (2019), no. 3, 417–491.

[20] R. Gopakumar and C. Vafa, On the gauge theory/geometry correspondence. *Adv. Theor. Math. Phys.* **3** (1999), 1415–1443.

[21] M. Gromov, Pseudoholomorphic curves in symplectic manifolds. *Invent. Math.* **82** (1985), no. 2, 307–347.

[22] E. Ionel and T. H. Parker, The Gopakumar–Vafa formula for symplectic manifolds. *Ann. of Math.* **187** (2018), 1–64.

[23] I. Itenberg, V. Kharlamov, and E. Shustin, A Caporaso–Harris type formula for Welschinger invariants of real toric Del Pezzo surfaces. *Comment. Math. Helv.* **84** (2009), 87–126.

[24] I. Itenberg, V. Kharlamov, and E. Shustin, Welschinger invariants of real Del Pezzo surfaces of degree ≥ 2. *Int. J. Math.* **26** (2015), no. 08, 1550060.

[25] M. Kontsevich, Intersection theory on the moduli space of curves and the matrix Airy function. *Comm. Math. Phys.* **147** (1992), no. 1, 1–23.

[26] D. Krefl, S. Pasquetti, and J. Walcher, The real topological vertex at work. *Nuclear Phys. B* **833** (2010), no. 3, 153–198.

[27] C.-C. Liu, Moduli of J-holomorphic curves with Lagrangian boundary condition and open Gromov–Witten invariants for an S^1-pair. 2002, arXiv:math/0210257.

[28] M. Mariño, Chern–Simons theory, the $1/N$ expansion, and string theory. *AMS/IP Stud. Adv. Math.* **50** (2011), 243–260.

[29] G. Mikhalkin, Enumerative tropical algebraic geometry in $\mathbb{R}^2$. *J. Amer. Math. Soc.* **18** (2005), no. 2, 313–377.

[30] J. Niu and A. Zinger, Lower bounds for the enumerative geometry of positive-genus real curves. *Adv. Math.* **339** (2018), 191–247.

[31] A. Okounkov, Toda equations for Hurwitz numbers. *Math. Res. Lett.* **7** (2000), 447–453.

[32] A. Okounkov and R. Pandharipande, Gromov–Witten theory, Hurwitz theory, and completed cycles. *Ann. of Math.* **163** (2006), 517–560.

[33] R. Pandharipande, J. Solomon, and J. Walcher, Disk enumeration on the quintic 3-fold. *J. Amer. Math. Soc.* **21** (2008), no. 4, 1169–1209.

[34] E. Shustin, A tropical calculation of the Welschinger invariants of real toric Del Pezzo surfaces. *J. Algebraic Geom.* **15** (2006), 285–322.

[35] J. Solomon, Intersection theory on the moduli space of holomorphic curves with Lagrangian boundary conditions. 2006, arXiv:math/0606429.

[36] J. Solomon and S. Tukachinsky, Relative quantum cohomology. 2019, arXiv:1906.04795.

[37] J. Walcher, Evidence for tadpole cancellation in the topological string. *Commun. Number Theory Phys.* **3** (2009), 111–172.

[38] J.-Y. Welschinger, Invariants of real symplectic 4-manifolds and lower bounds in real enumerative geometry. *Invent. Math.* **162** (2005), no. 1, 195–234.

[39] J.-Y. Welschinger, Spinor states of real rational curves in real algebraic convex 3-manifolds and enumerative invariants. *Duke Math. J.* **127** (2005), no. 1, 89–121.

[40] E. Witten, Two-dimensional gravity and intersection theory on moduli space. *Surv. Differ. Geom.* **1** (1991), 243–310.

PENKA GEORGIEVA

Sorbonne Université and Université de Paris, CNRS, IMJ-PRG, 4 Place Jussieu, 75005 Paris, France, penka.georgieva@imj-prg.fr

This Page Intentionally Left Blank

GAMMA CLASSES AND QUANTUM COHOMOLOGY

HIROSHI IRITANI

ABSTRACT

The $\widehat{\Gamma}$-class is a characteristic class for complex manifolds with transcendental coefficients. It defines an integral structure of quantum cohomology, or more precisely, an integral lattice in the space of flat sections of the quantum connection. We present several conjectures (the $\widehat{\Gamma}$-conjectures) about this structure, particularly focusing on the Riemann–Hilbert problem it poses. We also discuss a conjectural functoriality of quantum cohomology under birational transformations.

MATHEMATICS SUBJECT CLASSIFICATION 2020

Primary 53D45; Secondary 14N35, 53D37, 34M40, 34M50, 14E05

KEYWORDS

Gamma class, quantum cohomology, Riemann–Hilbert problem, birational transformation

INTERNATIONAL CONGRESS OF MATHEMATICIANS 2022 JULY 6–14

Proc. Int. Cong. Math. 2022, Vol. 4, pp. 2552–2574
DOI 10.4171/ICM2022/156

1. GAMMA-INTEGRAL STRUCTURE IN QUANTUM COHOMOLOGY

We briefly review the definition of the $\widehat{\Gamma}$-integral structure in quantum cohomology introduced in [42]. The corresponding rational structure was introduced independently by Katzarkov, Kontsevich, and Pantev [46] in the framework of nc-Hodge structure.

1.1. Gamma class

Let X be an almost complex manifold and let $\delta_1, \dots, \delta_n$ (with $n = \dim_{\mathbb{C}} X$) be the Chern roots of the tangent bundle, so that $c(TX) = (1+\delta_1)\cdots(1+\delta_n)$. The $\widehat{\Gamma}$-*class* $\widehat{\Gamma}_X \in H^*(X;\mathbb{R})$ [42, 46, 52, 53] is the characteristic class defined by

$$\widehat{\Gamma}_X = \Gamma(1+\delta_1)\cdots\Gamma(1+\delta_n)$$

where $\Gamma(1+x) = \int_0^\infty e^{-t} t^x dt$ is Euler's Γ-function. The right-hand side is expanded in symmetric power series in $\delta_1, \dots, \delta_n$ and then expressed in terms of the Chern characters $\mathrm{ch}_k(TX)$ as follows:

$$\widehat{\Gamma}_X = \exp\left(-\gamma c_1(X) + \sum_{k=2}^{\infty}(-1)^k \zeta(k)(k-1)!\,\mathrm{ch}_k(TX)\right),$$

where $\zeta(s) = \sum_{n=1}^{\infty} n^{-s}$ is the Riemann zeta function and $\gamma = \lim_{n\to\infty}(1 + \frac{1}{2} + \cdots + \frac{1}{n} - \log n)$ is the Euler constant. This is a characteristic class with transcendental coefficients.[1] The identity $\Gamma(1-x)\Gamma(1+x) = \pi x/\sin(\pi x)$ shows that the $\widehat{\Gamma}$-class can be thought of as a "square root" of the $\hat{A}$-class, i.e.,

$$\widehat{\Gamma}_X \cdot \widehat{\Gamma}_X^* = (2\pi \mathtt{i})^{\deg/2} \hat{A}_X, \tag{1.1}$$

where $\widehat{\Gamma}_X^* := (-1)^{\deg/2}\widehat{\Gamma}_X$ denotes the dual $\widehat{\Gamma}$-class. We note that $\hat{A}_X$ depends only on the underlying topological manifold whereas $\widehat{\Gamma}_X$ depends on an almost complex structure on it. The identity (1.1) suggests a relationship between the $\widehat{\Gamma}$-class and the Atiyah–Singer index theorem. In fact, we can interpret $\widehat{\Gamma}_X$ as (a regularization of) the inverse S^1-equivariant Euler class of the positive normal bundle $\mathcal{N}_+$ of the set X of constant loops in the free loop space LX (see [53], [26, APPENDIX A]), i.e.,

$$\frac{1}{e_{S^1}(\mathcal{N}_+)} = \frac{1}{\prod_i \prod_{k>0}(\delta_i + kz)} \sim (2\pi)^{-n/2} z^{(n-\deg)/2} z^{c_1(X)} \widehat{\Gamma}_X, \tag{1.2}$$

where z is a generator of the S^1-equivariant cohomology of a point. This is reminiscent of the loop space heuristics of the index theorem by Atiyah and Witten [4], where the $\hat{A}$-class is interpreted as the inverse Euler class $e_{S^1}(\mathcal{N})^{-1}$ of the normal bundle $\mathcal{N}$ itself,

$$\frac{1}{e_{S^1}(\mathcal{N})} = \frac{1}{e_{S^1}(\mathcal{N}_-)e_{S^1}(\mathcal{N}_+)} = \frac{1}{\prod_i \prod_{k\neq 0}(\delta_i + kz)} \sim \left(\frac{z}{2\pi \mathtt{i}}\right)^{n-(\deg/2)} \hat{A}_X.$$

Since $\mathcal{N}_+$ corresponds to infinitesimal (pseudo)holomorphic loops, the $\widehat{\Gamma}$-class can be thought of as the *localization contribution from constant loops* in symplectic Floer theory.

---

1 It is, however, an algebraic (Hodge) class when X is a smooth projective variety.

1.2. Quantum cohomology D-modules

Let X be a smooth projective variety (or a compact symplectic manifold) and let $H^*(X)$ denote the cohomology group with complex coefficients. The quantum cohomology $\mathrm{QH}^*(X) = (H^*(X), \star_\tau)$ of X is a family of supercommutative product structures $\star_\tau$ on $H^*(X)$ parametrized by $\tau \in H^*(X)$. The quantum product $\star_\tau$ is defined by

$$(\alpha \star_\tau \beta, \gamma) = \sum_{d \in H_2(X,\mathbb{Z}), k \geq 0} \langle \alpha, \beta, \gamma, \tau, \dots, \tau \rangle_{0,k+3,d} \frac{Q^d}{k!}$$

for $\alpha, \beta, \gamma \in H^*(X)$. Here $(\alpha, \beta) = \int_X \alpha \cup \beta$ is the Poincaré pairing and $\langle \alpha_1, \dots, \alpha_k \rangle_{0,k,d}$ denotes the genus-zero, k-point, degree d Gromov–Witten invariants. Strictly speaking, we should treat the odd degree part of τ as anticommuting variables and view the parameter space $H^*(X)$ as a supermanifold. For the most part of this paper, we shall restrict the parameter τ and elements of quantum cohomology to the even part of the cohomology group and write $H^*(X)$ for the even part (see Remark 1.2 for the odd part).

In the above formula, we introduced the Novikov variable Q to ensure the adic convergence of $\star_\tau$. The divisor equation shows that, if we decompose $\tau = \sigma + \tau'$ with $\sigma \in H^2(X)$ and $\tau' \in H^{\neq 2}(X)$,

$$(\alpha \star_\tau \beta, \gamma) = \sum_{d \in H_2(X,\mathbb{Z}), k \geq 0} \left\langle \alpha, \beta, \gamma, \tau', \dots, \tau' \right\rangle_{0,k+3,d} \frac{e^{\langle \sigma, d \rangle} Q^d}{k!}.$$

Thus the quantum product can be expanded in a power series in τ' and e^σ and approaches the cup product in the following *large-radius limit*:

$$\tau' \to 0, \quad e^{\langle \sigma, d \rangle} \to 0 \text{ for all effective classes } d \neq 0. \tag{1.3}$$

Hereafter we shall always specialize the Novikov variable Q to 1 and assume that $\star_\tau|_{Q=1}$ (which we shall write as $\star_\tau$) is convergent in a neighborhood U of the large radius limit.

The quantum cohomology defines the structure of a Frobenius manifold **[20]** on the convergence domain $U \subset H^*(X)$. Specifically, it defines a meromorphic flat connection ∇ on the trivial bundle $F = H^*(X) \times (U \times \mathbb{C}) \to (U \times \mathbb{C})$, called the *quantum connection* or the *Dubrovin connection*. It is defined by the formulae

$$\begin{aligned} \nabla_{\partial/\partial \tau^i} &= \frac{\partial}{\partial \tau^i} + \frac{1}{z}(\phi_i \star_\tau), \\ \nabla_{z\partial/\partial z} &= z\frac{\partial}{\partial z} - \frac{1}{z}(E \star_\tau) + \mu, \end{aligned}$$

where $(\tau, z) \in U \times \mathbb{C}$ denotes a point on the base and $\{\tau^i\}$ are linear coordinates dual to a homogeneous basis $\{\phi_i\}$ of $H^*(X)$ so that $\tau = \sum_i \tau^i \phi_i$. The section $E \in \mathcal{O}(F)$ is the *Euler vector field* given by

$$E = c_1(X) + \sum_i \left(1 - \frac{\deg \phi_i}{2}\right) \tau^i \phi_i$$

and $\mu \in \mathrm{End}(H^*(X))$ is the *grading operator* defined by $\mu(\phi_i) = (\frac{\deg \phi_i}{2} - \frac{n}{2})\phi_i$. The connection ∇ has poles of order two along $z = 0$ and is possibly *irregular singular* there. On the other hand, it has logarithmic poles (and is therefore *regular singular*) along $z = \infty$. The connection ∇ is compatible with the Poincaré pairing in the sense that the z-sesquilinear pairing

$$(-1)^* \mathcal{O}(F) \otimes \mathcal{O}(F) \to \mathcal{O}, \quad s_1 \otimes s_2 \mapsto (s_1, s_2) = \int_X s_1 \cup s_2 \tag{1.4}$$

is flat for ∇, where $(-1)\colon U \times \mathbb{C} \to U \times \mathbb{C}$ is the map sending (τ, z) to $(\tau, -z)$.

The *quantum (cohomology) D-module* is the tuple $\mathrm{QDM}(X) = (\mathcal{O}(F), \nabla, (\cdot, \cdot))$ consisting of the locally free sheaf $\mathcal{O}(F)$ over $U \times \mathbb{C}$, the quantum connection ∇ and the pairing in (1.4). The D-module approach to quantum cohomology has been proposed by Givental **[28]** and Guest **[34]**.

1.3. Gamma-integral structure

The $\widehat{\Gamma}$-integral structure is an integral lattice in the space of flat sections for the quantum connection ∇. We have a fundamental solution for ∇-flat sections of the form $L(\tau, z) z^{-\mu} z^{c_1(X)}$ with an $\mathrm{End}(H^*(X))$-valued function $L(\tau, z)$ uniquely characterized by the following asymptotic condition at the large radius limit (1.3),

$$L(\tau, z) = \big(\mathrm{id} + O(e^{\sigma}, \tau')\big) e^{-\sigma/z}.$$

Here $z^{-\mu} z^{c_1(X)} = \exp(-\mu \log z) \exp(c_1(X) \log z)$ is an $\mathrm{End}(H^*(X))$-valued function on the universal cover of $\mathbb{C}^\times$. Given a basis $\{\phi_i\}$ of $H^*(X)$, $\{L(\tau, z) z^{-\mu} z^{c_1(X)} \phi_i\}$ gives a basis of ∇-flat sections. Explicitly, $L(\tau, z)$ is given in terms of gravitational descendants as follows (see **[19, 29]**):

$$L(\tau, z)\phi_i = e^{-\sigma/z}\phi_i - \sum_j \sum_{(d,k) \neq (0,0)} \left\langle \frac{e^{-\sigma/z}\phi_i}{z + \psi}, \tau', \dots, \tau', \phi^j \right\rangle_{0,k+2,d} \frac{\phi_j}{k!} e^{\langle \sigma, d \rangle}, \tag{1.5}$$

where ψ denotes the universal cotangent class at the first marking, $1/(z + \psi)$ should be expanded as $\sum_{k \geq 0} z^{-k-1}(-\psi)^k$, and $\{\phi_j\}$, $\{\phi^j\}$ are mutually dual bases of $H^*(X)$ such that $\int_X \phi_i \cup \phi^j = \delta_i^j$.

Let $\mathcal{S}(X)$ denote the $\mathbb{C}$-vector space of multivalued flat sections of (F, ∇) over $U \times \mathbb{C}^\times$, i.e., flat sections over the universal cover $U \times \widetilde{\mathbb{C}^\times}$. Let $K(X) = K^0_{\mathrm{top}}(X)$ denote the K-group of topological complex vector bundles and define a map $\mathfrak{s}\colon K(X) \to \mathcal{S}(X)$ as

$$\mathfrak{s}(V)(\tau, z) = L(\tau, z) z^{-\mu} z^{c_1(X)} \big((2\pi)^{-n/2} \widehat{\Gamma}_X (2\pi \mathtt{i})^{\deg/2} \mathrm{ch}(V)\big). \tag{1.6}$$

The factor $(2\pi)^{-n/2} z^{-\mu} z^{c_1(X)} \widehat{\Gamma}_X$ also appears in (1.2) as a regularization of $e_{S^1}(\mathcal{N}_+)^{-1}$. The *$\widehat{\Gamma}$-integral structure* is the integral lattice of $\mathcal{S}(X)$ given as the image of the map $\mathfrak{s}$.

By the compatibility between ∇ and the Poincaré pairing, we have a nondegenerate (not necessarily symmetric or antisymmetric) pairing $[\cdot, \cdot)\colon \mathcal{S}(X) \otimes \mathcal{S}(X) \to \mathbb{C}$ defined by

$$[s_1, s_2) = \big(s_1(\tau, e^{-\pi \mathtt{i}} z), s_2(\tau, z)\big) \tag{1.7}$$

for $s_1, s_2 \in \mathcal{S}(X)$. The property (1.1) of the $\widehat{\Gamma}$-class and the Atiyah–Singer index theorem (or Hirzebruch–Riemann–Roch theorem) show that $\mathfrak{s}$ respects the pairing

$$[\mathfrak{s}(V_1), \mathfrak{s}(V_2)) = \chi(V_1, V_2),$$

where $\chi(V_1, V_2) \in \mathbb{Z}$ is the K-theoretic push-forward of $V_1^\vee \otimes V_2$ to a point (the index of a Dirac operator; it is $\sum_{i\ge 0}(-1)^i \dim \operatorname{Ext}^i(V_1, V_2)$ if V_1, V_2 are holomorphic vector bundles). The $\widehat{\Gamma}$-integral structure is monodromy-invariant in the sense that

$$\mathfrak{s}(V)\big(\tau - 2\pi\mathrm{i}c_1(L), z\big) = \mathfrak{s}(V \otimes L)(\tau, z),$$
$$\mathfrak{s}(V)(\tau, e^{-2\pi\mathrm{i}}z) = \mathfrak{s}\big(V \otimes \omega_X[n]\big)(\tau, z),$$

where L is a (topological) line bundle on X and $\omega_X[n] = (-1)^n\omega_X$ is the canonical line bundle ω_X shifted by n, corresponding to the Serre functor of the derived category.

Remark 1.1. The $\widehat{\Gamma}$-integral structure can be defined more generally for orbifolds [42].

Remark 1.2. We can generalize the $\widehat{\Gamma}$-integral structure including the odd part of the quantum cohomology, using $K^*(X) = K^0_{\mathrm{top}}(X) \oplus K^1_{\mathrm{top}}(X)$ instead of $K^0_{\mathrm{top}}(X)$. We still restrict the parameter τ to lie in the even part, but consider flat sections taking values in the full cohomology group. The formula (1.6) makes sense for all $V \in K^*(X)$ when we choose a square root $\sqrt{2\pi\mathrm{i}}$ and use the Chern character ch: $K^*(X) \to H^*(X)$ of Atiyah–Hirzebruch [5]. The resulting map $\mathfrak{s}_X\colon K^*(X)/\mathrm{tors} \to \mathcal{S}(X)$ then has the advantage that it is natural with respect to the Cartesian product, i.e., $\mathfrak{s}_{X\times Y} = \mathfrak{s}_X \otimes \mathfrak{s}_Y$ (under the Künneth isomorphism). Interestingly, we have

$$[\mathfrak{s}(\alpha_1), \mathfrak{s}(\alpha_2)) = -\mathrm{i}\chi(\alpha_1, \alpha_2) \in \mathrm{i}\mathbb{Z} \qquad \text{for } \alpha_1, \alpha_2 \in K^1(X),$$
$$\mathfrak{s}(\alpha)(\tau, e^{-2\pi\mathrm{i}}z) = (-1)^{\deg\alpha}\mathfrak{s}\big(\alpha \otimes \omega_X[n]\big)(\tau, z) \qquad \text{for } \alpha \in K^*(X),$$

where $\chi(\alpha_1, \alpha_2) \in \mathbb{Z}$ is the K-theoretic push-forward of $\alpha_1^\vee \cdot \alpha_2$ to a point as before; the dual element $\alpha_1^\vee$ here is defined via the isomorphism $K^1(X) \cong K^{-1}(X) \cong \tilde{K}^0(S^1 \wedge X^+)$ (see [5]) and the usual duality in K^0.

Remark 1.3 (Mirror symmetry). The $\widehat{\Gamma}$-integral structure had (implicitly) appeared for a long time in the study of mirror symmetry before it was defined in [42, 46]. Under mirror symmetry of Calabi–Yau manifolds, the quantum differential equation corresponds to the Picard–Fuchs differential equations satisfied by periods of the mirror family, and we can partially see the $\widehat{\Gamma}$-class in the asymptotics of periods near the large-complex structure limit. Libgober [52] introduced the (inverse) $\widehat{\Gamma}$-class based on the observation of Hosono et al. [40] that certain combinations of Chern numbers and ζ values appear in solutions of the mirror Picard–Fuchs equations. Hosono [39] stated a conjecture equating periods of mirrors of complete intersections with explicit hypergeometric series and the $\widehat{\Gamma}$-class is hidden in the series. We also refer the reader to [9, 38, 59] for related works. It has been checked in a number of cases that the $\widehat{\Gamma}$-integral structure corresponds to a natural integral structure on the mirror side [42, 44]. Regarding the compatibility with mirror symmetry, an approach based on the SYZ picture and tropical geometry has been proposed in [1] recently.

2. GAMMA CONJECTURES

In this section we review the $\widehat{\Gamma}$-conjectures I, II discussed by Galkin, Golyshev, and the author [26], and their generalization by Sanda and Shamoto [58]. The $\widehat{\Gamma}$-conjectures can be understood as the compatibility between the Betti (real, rational, or integral) structure and the Stokes structure, discussed by Hertling and Sevenheck [37] in the context of TERP structure and by Katzarkov, Kontevich, and Pantev [46] in the context of nc-Hodge structure. The Gamma conjecture II also refines Dubrovin's conjecture [20].

2.1. Gamma conjecture I

The $\widehat{\Gamma}$-conjecture I is specifically about quantum cohomology of Fano manifolds. It roughly speaking says that we can know the topology (the $\widehat{\Gamma}$-class) of a Fano manifold by counting rational curves on it. In view of (1.1), we may view it as a "square root" of the index theorem.

Let X be a Fano manifold and let $J_X(\tau, z)$ be the (small) J-function defined as

$$J_X(\tau, z) = e^{\tau/z}\Big(1 + \sum_i \sum_{d\in H_2(X,\mathbb{Z}),d\neq 0} e^{\langle\tau,d\rangle}\Big\langle \frac{\phi^i}{z(z-\psi)}\Big\rangle_{0,1,d}\phi_i\Big)$$

where $\tau \in H^2(X)$. This is a cohomology-valued function which is convergent for all $(\tau, z) \in H^2(X) \times \mathbb{C}^\times$ (this follows from the Fano assumption). We can also write this as $J_X(\tau, z) = L(\tau, z)^{-1}1$ using the fundamental solution L in (1.5); hence $J_X(\tau, z)$ gives a solution of the quantum D-module along the τ-direction.

Conjecture 2.1 ($\widehat{\Gamma}$-conjecture I). *For a Fano manifold X, we have the equality*

$$[\widehat{\Gamma}_X] = \lim_{t\to+\infty}\big[J_X\big(c_1(X)\log t, 1\big)\big]$$

in the projective space $\mathbb{P}(H^*(X))$ of cohomology.

This has been proved for the projective spaces, type A Grassmannians [26, 30] and Fano threefolds of Picard rank one [31]. The $\widehat{\Gamma}$-conjecture I for Fano toric manifolds or complete intersections in them follows if these spaces satisfy certain conditions related to Conjecture $\mathcal{O}$ [27]. The $\widehat{\Gamma}$-conjecture I is also compatible with taking hyperplane sections, i.e., if a Fano manifold X satisfies the $\widehat{\Gamma}$-conjecture I and if $Y \subset X$ is a hypersurface in the linear system $|L|$ with L proportional to $-K_X$, Y satisfies the $\widehat{\Gamma}$-conjecture I [27, THEOREM 8.3].

Example 2.2. The J-function of $\mathbb{P}^n$ is given by

$$J_{\mathbb{P}^n}\big(c_1(\mathbb{P}^n)\log t, z\big) = \sum_{d=0}^{\infty}\frac{t^{(n+1)(d+p/z)}}{\prod_{k=1}^{d}(p+kz)^{n+1}},$$

where p is the hyperplane class. Setting $z = 1$ and fixing $t > 0$, we find that the dth summand

$$\frac{t^{(n+1)(d+p)}}{\prod_{k=1}^{d}(p+k)^{n+1}} = \left(\frac{t^d(t/d)^p}{d!}\right)^{n+1}\left(\frac{e^{(\log d-(1+\frac{1}{2}+\cdots+\frac{1}{d}))p}}{\prod_{k=1}^{d}((1+\frac{p}{k})e^{-p/k})}\right)^{n+1}$$

has a strong peak approximately when d is close to t. We can guess from this that the limit of $\mathbb{C}J_{\mathbb{P}^n}(c_1(\mathbb{P}^n)\log t, 1)$ in the projective space should be the line generated by

$$\lim_{d\to\infty}\left(\frac{e^{(\log d-(1+\frac{1}{2}+\cdots+\frac{1}{d}))p}}{\prod_{k=1}^{d}((1+\frac{p}{k})e^{-p/k})}\right)^{n+1} = \Gamma(1+p)^{n+1} = \widehat{\Gamma}_{\mathbb{P}^n}.$$

Remark 2.3. In Givental's heuristic calculation of the J-function [28], the dth summand $\prod_{k=1}^{d}(p+kz)^{-n-1}$ of $J_{\mathbb{P}^n}(\tau,z)$ appears as the localization contribution from constant loops in the polynomial loop space (quasimaps' space) of degree d, and hence it can be viewed as the degree-d truncation of $e_{S^1}(\mathcal{N}_+)^{-1}$ appearing in Section 1.1. In view of the loop space interpretation of the $\widehat{\Gamma}$-class, this gives a geometric explanation for the $\widehat{\Gamma}$-conjecture I in this case. In general, the degree d term of $J_X(\tau,z)$ arises from a localization contribution of an integral over the graph space $G_d = \overline{M}_{0,0}(X\times\mathbb{P}^1,(d,1))$, the moduli space of genus-zero stable maps to $X\times\mathbb{P}^1$ of degree $(d,1)$, equipped with the $\mathbb{C}^\times$-action induced by the $\mathbb{C}^\times$-action on $\mathbb{P}^1$. Let $G_d^\circ \subset G_d$ denote the open subset consisting of stable maps which are genuine graphs near $\infty\in\mathbb{P}^1$, i.e., do not contain components contained in $X\times\{\infty\}$. Then G_d° is preserved by the $\mathbb{C}^\times$-action and has $F_d = \overline{M}_{0,1}(X,d)$ as the fixed locus. Writing $\mathrm{ev}_\infty\colon G_d^\circ\to X$ for the evaluation map at $\infty\in\mathbb{P}^1$, we have

$$\int_{G_d^\circ}\mathrm{ev}_\infty^*\alpha = \int_{[F_d]_{\mathrm{vir}}}\frac{\mathrm{ev}_1^*\alpha}{z(z-\psi)} = (\text{degree } d \text{ term of } J_X,\alpha),$$

where we defined the integral over the improper space G_d° using (virtual) equivariant localization. In the case where $X=\mathbb{P}^n$, Givental gave a birational morphism from G_d° to the polynomial loop space and justified his heuristic calculation (see [29, MAIN LEMMA]). Therefore the $\widehat{\Gamma}$-conjecture I can be viewed as the statement that G_d° *approximates the (positive) loop space of X as d* $\to\infty$ in a suitable sense.

Remark 2.4. A discrete version of the limit in Conjecture 2.1 had been also studied (before the formulation of the $\widehat{\Gamma}$-conjecture) and called *Apéry limits*, in view of the connection to Apéry's proof of the irrationality of $\zeta(2)$ and $\zeta(3)$, see Galkin [25] and Golyshev [30].

2.2. Gamma conjecture I in terms of flat sections

We restate $\widehat{\Gamma}$-conjecture I in terms of ∇-flat sections over the z-plane, in order to explain the relationship with $\widehat{\Gamma}$-conjecture II in the following section. We start with Conjecture $\mathcal{O}$.

Conjecture 2.5 (Conjecture $\mathcal{O}$). *Let X be a Fano manifold and let T denote the maximal norm of the eigenvalues of the quantum multiplication $(E\star_0)=(c_1(X)\star_0)$ at $\tau=0$. Then T is a simple eigenvalue of $(c_1(X)\star_0)$, that is, an eigenvalue whose multiplicity in the characteristic polynomial is one.*

Here we omitted part (2) of Conjecture $\mathcal{O}$ in [26, DEFINITION 3.1.1] since we do not need it. Conjecture $\mathcal{O}$ is a consequence of the Perron–Frobenius theorem if $(c_1(X)\star_0)$ is represented by an irreducible nonnegative matrix. Cheong and Li [11] proved Conjecture $\mathcal{O}$ for homogeneous spaces G/P using the Perron–Frobenius theorem.

Eigenvalues of $(c_1(X)\star_0)$ are closely related to asymptotics of flat sections for $\nabla|_{\tau=0}$ as $z\to 0$. The flatness equation reads

$$\left(z\frac{\partial}{\partial z}-\frac{1}{z}c_1(X)\star_0+\mu\right)s(z)=0.$$

For each eigenvector Ψ of $(c_1(X)\star_0)$ with eigenvalue u, we expect that there should be a flat section $s(z)$ with asymptotics $\sim e^{-u/z}\Psi$ as $z \to 0$. Define a vector space $\mathcal{A}$ as

$$\mathcal{A} = \left\{ s\colon \mathbb{R}_{>0} \to H^*(X) \;\middle|\; \begin{array}{l} s(z) \text{ is flat for } \nabla_{z\partial_z}|_{\tau=0} \\ e^{T/z}s(z) \text{ is at most of polynomial growth as } z \to +0 \end{array} \right\}.$$

Assuming Conjecture $\mathcal{O}$ for X, we can prove that $\mathcal{A}$ is one-dimensional and that, for any $s(z) \in \mathcal{A}$, $e^{T/z}s(z)$ converges to a T-eigenvector of $(c_1(X)\star_0)$ as $z \to +0$. The original formulation of the $\widehat{\Gamma}$-conjecture I in **[26]** was as follows.

Conjecture 2.6 ($\widehat{\Gamma}$-conjecture I: another form). *The space $\mathcal{A}$ is generated by $\mathfrak{s}(\mathcal{O})|_{\tau=0}$.*

It is equivalent to Conjecture 2.1 in Section 2.1 under Conjecture $\mathcal{O}$ **[26, COROLLARY 3.6.9]** and can be viewed as a dual formulation. We have a gauge equivalence between the connections $\nabla|_{\tau=0}$ and $\nabla|_{z=1,\tau=c_1(X)\log t}$, that is, $z^{\mu}(\nabla|_{\tau=0})z^{-\mu} = \nabla|_{z=1,\tau=c_1(X)\log t}$ under the identification $t = z^{-1}$. Thus, flat sections over $\{\tau = 0\} \times \mathbb{C}$ and the solution $J_X(c_1(X)\log t, 1)$ are dual to each other. While the space $\mathcal{A}$ consists of flat sections with most rapid decay ($\sim e^{-T/z}$), the $t \to +\infty$ limit of the J-function detects its most rapidly growing component. In fact, we can show that the J-function has the following asymptotics:

$$J_X\big(c_1(X)\log t, 1\big) = Ct^{-n/2}e^{Tt}\big(\widehat{\Gamma}_X + O(t^{-1})\big) \quad \text{as } t \to +\infty$$

for some $C \in \mathbb{C}$, under Conjecture $\mathcal{O}$ and $\widehat{\Gamma}$-conjecture I (see **[27, PROPOSITION 3.8]**).

2.3. Gamma conjecture II

In this section we assume that the quantum product $\star_\tau$ is semisimple[2] at some $\tau = \tau_0 \in H^*(X)$, i.e., $(H^*(X), \star_{\tau_0})$ is isomorphic to the direct sum of $\mathbb{C}$ as a ring. We do not need to assume that X is Fano. Let $\psi_1, \dots, \psi_N \in H^*(X)$ denote an idempotent basis such that $\psi_i \star_\tau \psi_j = \delta_{ij}\psi_i$ and let $u_1, \dots, u_N \in \mathbb{C}$ be the eigenvalues of $(E\star_\tau)$ such that $E \star_\tau \psi_i = u_i\psi_i$; here ψ_i and u_i are analytic functions of τ defined in a neighborhood of $\tau = \tau_0$. The functions $\{u_i\}$ give a local coordinate system near τ_0 called the *canonical coordinates* **[19, 21]**. We write $\Psi_i = (\psi_i, \psi_i)^{-1/2}\psi_i$ for the normalized idempotent basis, which is unique up to sign. Choose a phase $\phi \in \mathbb{R}$ such that $e^{i\phi} \notin \mathbb{R}_{>0}(u_{i,0} - u_{j,0})$ for all i, j, where $u_{i,0}$ is the value of u_i at τ_0; such a phase ϕ is said to be *admissible*. We have a basis $(y_1^\phi(\tau, z), \dots, y_N^\phi(\tau, z))$ of ∇-flat sections defined in a neighborhood of $\tau = \tau_0$ and $\arg z = \phi$ with the following property:

$$e^{u_i/z}y_i^\phi(\tau, z) \to \Psi_i \quad \text{as } z \to 0 \text{ along the angular sector } |\arg z - \phi| < \pi + \varepsilon$$

for some $\varepsilon > 0$, see **[26, PROPOSITION 2.5.1]**.

Conjecture 2.7 ($\widehat{\Gamma}$-conjecture II: a topological form). *Suppose that the quantum product $\star_\tau$ of X is semisimple at some $\tau_0 \in H^*(X)$ and let $\phi \in \mathbb{R}$ be an admissible phase for the*

---

2 Under the semisimplicity assumption, X has no odd cohomology classes and, if, moreover, X is a smooth projective variety, $H^*(X)$ is necessarily of Hodge–Tate type, i.e., $H^{p,q}(X) = 0$ for $p \neq q$, see **[36]**.

eigenvalues of $(E\star_{\tau_0})$. *There exist* K*-classes* $\mathcal{E}_1^\phi, \dots, \mathcal{E}_N^\phi \in K(X)$ *such that* $y_i^\phi(\tau, z) = \mathfrak{s}(\mathcal{E}_i^\phi)(\tau, z)$ *in a neighborhood of* $\tau = \tau_0$ *and* $\arg z = \phi$.

This refines part (3) of Dubrovin's conjecture [20, CONJECTURE 4.2.2] concerning the central connection matrix. It has been proved for type A Grassmannians [26], Fano toric manifolds [23] and quadric hypersurfaces [41]. The $\widehat{\Gamma}$-conjecture I can be viewed as a special case of the $\widehat{\Gamma}$-conjecture II when $\tau = 0$ and $\phi = 0$.

The flat sections $y_i^\phi(\tau, z)$ depend on the choice of a phase ϕ (or, more precisely, on a chamber of admissible phases) whereas their asymptotic expansions as $z \to 0$ do not. This is the so-called Stokes phenomena. The *Stokes matrix* $S = (S_{ij})$ is a transition matrix between the flat sections associated with opposite directions: it is given by $y_j^\phi(\tau, z) = \sum_{i=1}^N y_i^{\phi+\pi}(\tau, z) S_{ij}$ with $\arg z = \phi + \frac{\pi}{2}$. It can be given in terms of the bilinear form in (1.7) and then as the Euler matrix of $\{\mathcal{E}_i^\phi\}$,

$$S_{ij} = (y_i^\phi, y_j^\phi] = \chi(\mathcal{E}_i^\phi, \mathcal{E}_j^\phi).$$

This corresponds to part (2) of Dubrovin's conjecture saying that the Stokes matrix is integral and is given by the Euler pairing. If follows from the fact that the asymptotics $y_i^\phi \sim e^{-u_i/z}\Psi_i$ holds over a sector of angle $> \pi$ that the Stokes matrix is upper-triangular

$$S_{ij} = \chi(\mathcal{E}_i^\phi, \mathcal{E}_j^\phi) = \begin{cases} 1 & \text{if } i = j; \\ 0 & \text{if } \Im(e^{-\mathrm{i}\phi}u_i) \le \Im(e^{-\mathrm{i}\phi}u_j) \text{ and } i \ne j. \end{cases}$$

Remark 2.8. In [26], the $\widehat{\Gamma}$-conjecture II was stated for Fano manifolds which have semisimple quantum cohomology and full exceptional collections in $D^b(X)$. It is, moreover, conjectured that $\{\mathcal{E}_i^\phi\}$ should lift to a full exceptional collection. We drop these assumptions/conclusions to emphasize a topological nature of the $\widehat{\Gamma}$-conjecture.

Remark 2.9. Dubrovin [22] also formulated a conjecture similar to the $\widehat{\Gamma}$-conjecture II. See Cotti, Dubrovin, and Guzzetti [18] for the formulation.

Example 2.10. For $X = \mathbb{P}^n$, the corresponding exceptional collection is $\{\mathcal{O}, \mathcal{O}(1), \dots, \mathcal{O}(n)\}$ at some τ [26]. The collection at $\tau = 0$ is given explicitly in [18].

Remark 2.11. Suppose that X is Fano and is mirror to a Landau–Ginzburg model $f: Y \to \mathbb{C}$. It is expected that the idempotent ψ_i corresponds to a nondegenerate critical point c_i of f such that the corresponding eigenvalue u_i equals $f(c_i)$. The critical point c_i can associate a Lefschetz thimble $\mathfrak{L}_i^\phi$ extending in the direction of $e^{\mathrm{i}\phi}$, which gives an exceptional object in the Fukaya–Seidel category of (Y, f). The object in $D^b(X)$ corresponding to $\mathfrak{L}_i^\phi$ under homological mirror symmetry should give the class $\mathcal{E}_i^\phi$.

Remark 2.12. The $\widehat{\Gamma}$-conjecture II concerns the connection problem between flat sections y_i^ϕ characterized by the asymptotics at the irregular singular point $z = 0$ and flat sections $\mathfrak{s}(V)$ normalized at the regular singular point $z = \infty$. The connection matrix of flat sections (with respect to a fixed basis) is called the *central connection matrix* by Dubrovin [20]; in the formalism of the $\widehat{\Gamma}$-integral structure, it corresponds to the basis $\{\mathcal{E}_i^\phi\}$ of the K-group. As discussed by Dubrovin, when τ and ϕ vary, the basis $\{y_i^\phi\}$ of flat sections (and hence $\{\mathcal{E}_i^\phi\}$) changes discontinuously by the action of the braid group in N strands. Suppose that we ordered flat sections $\{y_i^\phi\}$ in such a way that $\Im(e^{-\mathrm{i}\phi}u_1) \geq \Im(e^{-\mathrm{i}\phi}u_2) \geq \cdots \geq \Im(e^{-\mathrm{i}\phi}u_N)$. The braid group action is generated by the following right mutations and their inverses (which are the actions of simple braids):

$$(\mathcal{E}_1,\ldots,\mathcal{E}_i,\mathcal{E}_{i+1},\ldots,\mathcal{E}_N) \mapsto \big(\mathcal{E}_1,\ldots,\mathcal{E}_{i+1},\mathcal{E}_i-\chi(\mathcal{E}_i,\mathcal{E}_{i+1})\mathcal{E}_{i+1},\ldots,\mathcal{E}_N\big),$$

where we suppressed ϕ to simplify the notation. This transformation happens when the eigenvalue u_i crosses behind u_{i+1} towards direction $e^{\mathrm{i}\phi}$. As Dubrovin observed, this is consistent with mutations of exceptional collections in the derived category [8].

2.4. Conjecture of Sanda and Shamoto

Sanda and Shamoto [58] proposed a generalization of the $\widehat{\Gamma}$-conjecture II to the case where quantum cohomology is not necessarily semisimple and called it *Dubrovin-type conjecture*. Their formulation involves derived category of coherent sheaves and Hochschild homology, but here we give a topological formulation that has been proposed by Sergey Galkin [24]. This formulation makes sense for any compact symplectic manifolds.

We fix a parameter $\tau \in H^*(X)$ in the convergence domain of the quantum product. We consider the restriction $\mathrm{QDM}(X)_\tau$ of the quantum D-module from Section 1.2 to $\{\tau\}\times\mathbb{C}$ and write $\overline{\mathrm{QDM}}(X)_\tau = \mathrm{QDM}(X)_{\tau,0}\otimes_{\mathbb{C}\{z\}}\mathbb{C}[\![z]\!]$ for the restriction to the formal neighborhood of $z=0$, where $\mathrm{QDM}(X)_{\tau,0}$ denotes the germ of $\mathrm{QDM}(X)_\tau$ at $z=0$. We say that the quantum connection at τ is of *exponential type*[3] if we have the following formal decomposition (see [37, LEMMA 8.2]):

$$\widehat{\Phi}\colon \overline{\mathrm{QDM}}(X)_\tau \cong \bigoplus_{u\in \mathsf{C}} (e^{u/z}\otimes \mathcal{R}_u)\otimes_{\mathbb{C}\{z\}}\mathbb{C}[\![z]\!], \tag{2.1}$$

where we disregard the pairing on $\overline{\mathrm{QDM}}(X)_\tau$ momentarily, C denotes the set of distinct eigenvalues of $(E\star_\tau)$, $e^{u/z}$ denotes the rank-one connection $(\mathbb{C}\{z\}, d+d(u/z))$ and $\mathcal{R}_u$ is a free $\mathbb{C}\{z\}$-module equipped with a regular singular connection (whose pole order at $z=0$ is at most two). In this decomposition, each "regular singular piece" $\mathcal{R}_u$ is unique up to isomorphism. This decomposition is automatically orthogonal with respect to the Poincaré pairing (1.4) and hence each piece $\mathcal{R}_u$ inherits a non-degenerate z-sesquilinear pairing $(\cdot,\cdot)_u\colon (-1)^*\mathcal{R}_u\otimes\mathcal{R}_u\to\mathbb{C}\{z\}$. Hereafter we assume that the quantum connection is of exponential type: this assumption is natural from a mirror symmetry point of view.

We choose an *admissible direction* $e^{\mathrm{i}\phi}$ for C, that is, an element $e^{\mathrm{i}\phi}\in S^1$ satisfying $e^{\mathrm{i}\phi}\notin\mathbb{R}_{>0}(u-u')$ for any $u,u'\in\mathsf{C}$. By the Hukuhara–Turrittin theorem (see [37, LEMMA 8.3]), the above formal decomposition (2.1) lifts uniquely to an analytic decomposition

$$\Phi_I\colon \mathrm{QDM}(X)_\tau|_I \cong \bigoplus_{u\in\mathsf{C}} e^{u/z}\otimes\mathcal{R}_u|_I$$

3 We follow the terminology in [46]; it was called "*require no ramification*" in [37].

over a sector of the form $I = \{z \in \mathbb{C}^\times : |\arg z - \phi| < \frac{\pi}{2} + \varepsilon\}$ for some $\varepsilon > 0$. Here we mean by "lifts" that the map Φ_I admits, when expressed in terms of local holomorphic frames of $\mathrm{QDM}(X)_\tau$ and $\mathcal{R}_u$ around $z = 0$, an asymptotic expansion as $z \to 0$ along the sector I and that the expansion coincides with $\widehat{\Phi}$.

Let $\mathcal{S}_I$ denote the space of ∇-flat sections over the angular sector $\{\tau\} \times I$: it can be identified with $\mathcal{S}(X)$ from Section 1.1 once we specify a lift of the sector I to the universal cover of $\mathbb{C}^\times$. The analytic decomposition Φ_I induces a decomposition of $\mathcal{S}_I$,

$$\mathcal{S}_I = \bigoplus_{u \in \mathsf{C}} \mathcal{V}_u, \tag{2.2}$$

where $\mathcal{V}_u$ can be identified with the space of flat sections of $\mathcal{R}_u$ over I. Since the analytic decomposition Φ_I is valid over a sector of angle greater than π, it follows easily that the decomposition (2.2) is semiorthogonal in the sense that

$$[\mathcal{V}_u, \mathcal{V}_{u'}) = 0 \quad \text{if } \Im(e^{-\mathrm{i}\phi}u) < \Im(e^{-\mathrm{i}\phi}u'),$$

where $[\cdot,\cdot)$ is the pairing on $\mathcal{S}_I \cong \mathcal{S}(X)$ introduced in (1.7). The data of the vector space $\mathcal{S}_I$ equipped with the pairing $[\cdot,\cdot)$ and the semiorthogonal decomposition (SOD) (2.2) constitute a *mutation system* in the sense of [58, Definition 2.30]. In what follows, we ignore the torsion part of the K-group and write $K(X)$ for $K(X)/\mathrm{tors}$.

Conjecture 2.13 ([24, 58]). *Suppose that the quantum connection is of exponential type at $\tau \in H^*(X)$. With notation as above, the SOD (2.2) is induced from a decomposition of the topological K-group lattice, i.e., there exists a decomposition*

$$K(X) = \bigoplus_{u \in \mathsf{C}} V_u^\phi \tag{2.3}$$

such that $\mathcal{V}_u = \mathfrak{s}(V_u^\phi) \otimes \mathbb{C}$, where we identify $\mathcal{S}_I$ with $\mathcal{S}(X)$ by choosing a lift $\phi \in \mathbb{R}$ of the direction $e^{\mathrm{i}\phi} \in I$. (A different choice of the lift ϕ changes V_u^ϕ by monodromy, i.e., $V_u^{\phi+2\pi} = V_u^\phi \otimes \omega_X[n]$.)

When this conjecture holds, the lattices $\{V_u^\phi\}$ are semiorthogonal with respect to the Euler pairing, i.e., $\chi(V_u^\phi, V_{u'}^\phi) = 0$ for $\Im(e^{-\mathrm{i}\phi}u) < \Im(e^{-\mathrm{i}\phi}u')$ and therefore the Euler pairing on each V_u^ϕ is necessarily unimodular (because the Euler pairing on $K(X)$ is unimodular by Poincaré duality). In the semisimple case, we must have $V_u^\phi \cong \mathbb{Z}$ and a generator $\mathcal{E}$ of V_u^ϕ must satisfy $\chi(\mathcal{E}, \mathcal{E}) = \pm 1$; the $\widehat{\Gamma}$-conjecture II (Conjecture 2.7) additionally asserts that $\chi(\mathcal{E}, \mathcal{E}) = 1$ (this point does not follow from Conjecture 2.13).

Remark 2.14. The original formulation in [58] assumes that X is a smooth Fano variety and claims also that the semiorthogonal decomposition (SOD) (2.3) arises from an SOD of the derived category of coherent sheaves. We note that an SOD of the derived category induces an SOD of the topological K-group, since projections to the SOD summands are given by Fourier–Mukai kernels in $D^b(X \times X)$ and these kernels induce projections in the topological K-group (see the discussion in [33, §4] in the context of algebraic K-theory).

Example 2.15 ([58]). Sanda and Shamoto proved their conjecture for Fano complete intersections in the projective spaces of Fano index greater than 1. Let X be a degree d Fano hypersurface in $\mathbb{P}^n$, with $n - d > 0$. The set of eigenvalues of the quantum multiplication $(E\star_0) = (c_1(X)\star_0)$ is $\{0\} \cup \{T\zeta : \zeta^{n+1-d} = 1\}$ where $T = (n+1-d) \cdot d^{d/(n+1-d)}$. The multiplicity of $T\zeta$ is one and that of 0 is the dimension of the primitive cohomology plus $d-1$. In this case, the decomposition (2.2) at $\tau = 0$ arises from (up to mutation) the following SOD of the derived category:

$$D^b(X) = \langle \mathcal{A}, \mathcal{O}, \mathcal{O}(1), \dots, \mathcal{O}(n-d) \rangle,$$

where $\mathcal{O}, \dots, \mathcal{O}(n-d)$ are exceptional objects corresponding to simple eigenvalues $T\zeta$ and $\mathcal{A}$ is the right orthogonal of $\langle \mathcal{O}, \mathcal{O}(1), \dots, \mathcal{O}(n-d) \rangle$ corresponding to the eigenvalue 0.

The following problem naturally arises:

Problem 2.16. Understand a geometric meaning of each regular singular piece $\mathcal{R}_u$ and the corresponding unimodular lattice V_u^ϕ predicted in Conjecture 2.13.

In the semisimple case, each regular singular piece is the quantum connection of a point and the K-class $\mathcal{E}_i^\phi$ in the $\widehat{\Gamma}$-conjecture II (Conjecture 2.7) corresponds to a generator of $K^0(\mathrm{pt}) \cong \mathbb{Z}$. The subcategory $\mathcal{A}$ in Example 2.15 is equivalent to the category of graded matrix factorizations of a degree d polynomial $F(x_0, \dots, x_n)$ defining the hypersurface [56]: it is known to be a fractional Calabi–Yau category (in the sense that a power of the Serre functor equals the shift functor).

2.5. Monodromy data and Riemann–Hilbert problem

Let us assume that X satisfies Conjecture 2.13. In this section we explain how the SOD (2.3) encodes the irregular monodromy (Stokes) data, following [37, §8] and [58]. We also formulate a Riemann–Hilbert problem that reconstructs quantum cohomology from the SOD (2.3), formal data (2.1) and certain additional data.

Monodromy. The monodromy transformation $T: s(z) \mapsto s(e^{2\pi \mathrm{i}} z)$ on $\mathcal{S}_I$ is determined from the pairing $[\cdot, \cdot)$ as

$$[Ts_1, s_2) = \big(s_1(e^{\pi \mathrm{i}} z), s_2(z)\big) = \big(s_2(e^{-\pi \mathrm{i}} z), s_1(z)\big) = [s_2, s_1).$$

The restriction $[\cdot, \cdot)_u$ of the pairing $[\cdot, \cdot)$ to $\mathcal{V}_u$ is nondegenerate and is induced from the pairing $(\cdot, \cdot)_u$ on $\mathcal{R}_u$. The monodromy transformation $T_u: \mathcal{V}_u \to \mathcal{V}_u$ on flat sections of $\mathcal{R}_u$ is likewise determined by $[T_u s_1, s_2)_u = [s_2, s_1)_u$.

Stokes data. Let $\mathcal{S}_{-I}$ denote the space of ∇-flat sections over the opposite sector $\{\tau\} \times (-I)$. The Poincaré pairing $(\cdot, \cdot): \mathcal{S}_{-I} \times \mathcal{S}_I \to \mathbb{C}$ identifies $\mathcal{S}_{-I}$ with the dual space of $\mathcal{S}_I$ and the decomposition $\mathcal{S}_{-I} = \bigoplus_{u \in \mathsf{C}} \mathcal{V}'_u$ associated with the sector $-I$ is dual to that for I, i.e., $(\mathcal{V}'_{u'}, \mathcal{V}_u) = 0$ for $u \neq u'$. The Stokes data are given by the analytic continuation maps $S^\pm: \mathcal{S}_I \to \mathcal{S}_{-I}$, $s(z) \mapsto s(e^{\pm\pi \mathrm{i}} z)$. By the very definition of the pairing $[\cdot, \cdot)$, they are determined from $[\cdot, \cdot)$ as

$$(S^+ s_1, s_2) = [s_2, s_1), \quad (S^- s_1, s_2) = [s_1, s_2).$$

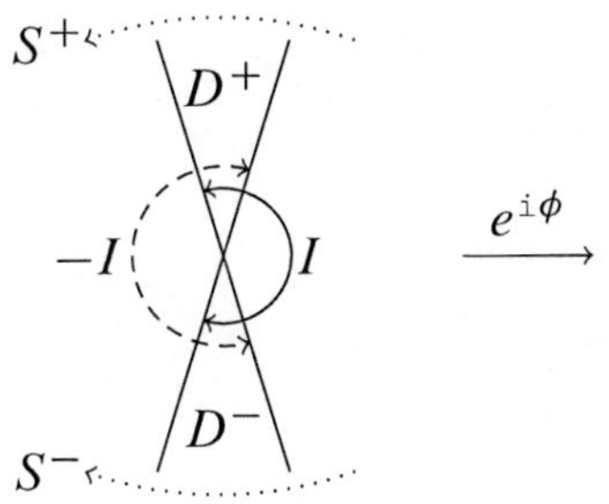

FIGURE 1
The angular sectors $\pm I$ and the paths of analytic continuations used to define the Stokes maps $S^\pm$.

Then we have $T = (S^-)^{-1}S^+$. The Stokes maps $S^\pm$ are upper (or lower) triangular in the sense that $S^+(\mathcal{V}_u) \subset \bigoplus_{\Im(e^{-\mathrm{i}\phi}u') \geq \Im(e^{-\mathrm{i}\phi}u)} \mathcal{V}'_{u'}$ and $S^-(\mathcal{V}_u) \subset \bigoplus_{\Im(e^{-\mathrm{i}\phi}u') \leq \Im(e^{-\mathrm{i}\phi}u)} \mathcal{V}'_{u'}$. They can be used to glue the connections over the opposite sectors

$$\bigoplus_{u\in\mathsf{C}} e^{u/z} \otimes \mathcal{R}_u|_{-I} \quad \text{and} \quad \bigoplus_{u\in\mathsf{C}} e^{u/z} \otimes \mathcal{R}_u|_I$$

along the two overlapping domains $D^\pm = I \cap (-I) \cap \{\pm\Im(ze^{-\mathrm{i}\phi}) > 0\}$ (see Figure 1). Hence the Stokes data reconstruct an analytic germ of the quantum connection at $z = 0$ from the formal data $\{\mathcal{R}_u\}_{u\in\mathsf{C}}$.

Riemann–Hilbert problem. The global quantum connection over $\mathbb{P}^1$ can be reconstructed by gluing the germ of the connection at $z = 0$ with a connection around $z = \infty$ via the $\widehat{\Gamma}$-integral structure. The quantum connection around $z = \infty$ is gauge-equivalent, via $L(\tau, z)$, to the connection

$$\nabla^{(\infty)}_{z\partial_z} = z\frac{\partial}{\partial z} - \frac{c_1(X)}{z} + \mu$$

on the trivial bundle $F_\infty = H^*(X) \times (\mathbb{P}^1 \setminus \{0\}) \to \mathbb{P}^1 \setminus \{0\}$. We identify the space of $\nabla^{(\infty)}$-flat sections with the K-group via the framing $\Psi_\infty \colon K(X) \to H^*(X) \otimes \mathcal{O}_{\widetilde{\mathbb{C}^\times}}$ (cf. (1.6)) given by

$$\Psi_\infty(\alpha) := (2\pi)^{-n/2} z^{-\mu} z^{c_1(X)} \widehat{\Gamma}_X (2\pi\mathrm{i})^{\deg/2} \operatorname{ch}(\alpha). \tag{2.4}$$

We glue the bundle $(F_\infty, \nabla^{(\infty)})$ with the germ around $z = 0$ by identifying the flat section $\Psi_\infty(\alpha)$ with $\alpha \in V_u^\phi$ with the flat section in $\mathcal{V}_u \cong \Gamma(I, \mathcal{R}_u)^\nabla$ corresponding to α (here we need an identification $V_u^\phi \cong \mathcal{V}_u$). This gives us a global vector bundle $\hat{F} \to \mathbb{P}^1$ with a meromorphic connection $\widehat{\nabla}$. The glued bundle $\hat{F}$ must be trivial (although it is not a priori clear); the trivialization of F_∞ at $z = \infty$ induces a trivialization $\hat{F} \cong H^*(X) \times \mathbb{P}^1$. The pair $(\hat{F}, \widehat{\nabla})$ is identified with the quantum connection at τ.

More explicitly, this reconstruction procedure can be described as the following Riemann–Hilbert problem for functions $Y_\pm = (\Phi_{\pm I})^{-1}$ (over the sectors $\pm I$) and $Y_\infty = L(\tau, z)$ (around $z = \infty$). This is an extension of the Riemann–Hilbert problem described by Dubrovin [19], [21, LECTURE 4] in the semisimple case.

Problem 2.17. Suppose that we are given the following data C, $\{\mathcal{R}_u\}_{u\in\mathsf{C}}$, $e^{\mathrm{i}\phi}$, I, ϕ, $(K(X), \chi)$, $K(X) = \bigoplus_{u\in\mathsf{C}} V_u$, Ψ_u, Ψ_∞:

- a subset C of $\mathbb{C}$;
- a finite free $\mathbb{C}\{z\}$-module $\mathcal{R}_u$ with a regular singular connection for each $u \in \mathsf{C}$;
- an admissible direction $e^{\mathrm{i}\phi}$ for C and a sector $I = \{z \in \mathbb{C}^\times : |\arg z - \phi| < \frac{\pi}{2} + \varepsilon\}$ centered around it;
- a unimodular lattice $(K(X), \chi)$ of rank $\dim H^*(X)$, a lift $\phi \in \mathbb{R}$ of $e^{\mathrm{i}\phi}$ and an SOD $K(X) = \bigoplus_{u\in\mathsf{C}} V_u$;
- a framing $\Psi_u\colon V_u \to \Gamma(I, \mathcal{R}_u)^\nabla$ for each $u \in \mathsf{C}$ such that Ψ_u induces an isomorphism over $\mathbb{C}$ and intertwines the transformation $T_u \in \mathrm{End}(V_u)$ given by $\chi(T_u\alpha, \beta) = \chi(\beta, \alpha)$ with the monodromy $s(z) \mapsto s(e^{2\pi\mathrm{i}}z)$ on $\Gamma(I, \mathcal{R}_u)^\nabla$;
- the "$\widehat{\Gamma}$-integral" framing $\Psi_\infty\colon K(X) \to H^*(X) \otimes \mathcal{O}_{\widetilde{\mathbb{C}^\times}}$ given in (2.4), which satisfies $\Psi_\infty(e^{2\pi\mathrm{i}}z) = \Psi_\infty(z) \circ T$ with $T \in \mathrm{End}(K(X))$ given by $\chi(T\alpha, \beta) = \chi(\beta, \alpha)$.

We define Stokes maps $S^\pm\colon K(X) \to K(X)^\vee$ by $\langle S^+\alpha, \beta\rangle = \chi(\beta, \alpha)$, $\langle S^-\alpha, \beta\rangle = \chi(\alpha, \beta)$ and a framing $\Psi_{-,u}\colon V_u^\vee \to \Gamma(-I, \mathcal{R}_u)^\nabla$ over the opposite sector $-I$ by

$$\Psi_{-,u}\big(\chi(\alpha, \cdot)\big) := \text{clockwise analytic continuation of } \Psi_u(\alpha) \text{ through } D^-$$

for $\alpha \in V_u$. We set

$$\Psi := \bigoplus_{u\in\mathsf{C}} \Psi_u\colon K(X) = \bigoplus_{u\in\mathsf{C}} V_u \to \bigoplus_{u\in\mathsf{C}} \Gamma(I, \mathcal{R}_u)^\nabla,$$
$$\Psi_- := \bigoplus_{u\in\mathsf{C}} \Psi_{-,u}\colon K(X)^\vee = \bigoplus_{u\in\mathsf{C}} V_u^\vee \to \bigoplus_{u\in\mathsf{C}} \Gamma(-I, \mathcal{R}_u)^\nabla.$$

The problem is to find (matrix-valued) holomorphic functions

$$Y_\infty \in \mathrm{GL}\big(H^*(X)\big) \otimes \mathcal{O}_{\mathbb{P}^1\setminus\{0\}}, \quad Y_\pm\colon \bigoplus_{u\in\mathsf{C}} \mathcal{R}_u|_{\pm I} \to H^*(X) \otimes \mathcal{O}_{\pm I}$$

such that

$$Y_\infty|_{z=\infty} = \mathrm{id}, \quad Y_\pm \to Y_0 \quad \text{as } z \to 0 \text{ along the sector } \pm I$$

for an invertible operator $Y_0\colon \bigoplus_{u\in\mathsf{C}} \mathcal{R}_u|_{z=0} \to H^*(X)$ and that

$$\begin{aligned} Y_+\Psi e^{-U/z} &= Y_\infty\Psi_\infty && \text{over } I,\\ Y_-\Psi_- e^{-U^\vee/z} S^\pm &= Y_+\Psi e^{-U/z} && \text{over } D^\pm, \end{aligned}$$

where $D^\pm$ is as before, the determination of Ψ_∞ over I is given by $|\arg z - \phi| < \frac{\pi}{2} + \varepsilon$, $U := \bigoplus_{u\in\mathsf{C}} u\,\mathrm{id}_{V_u} \in \mathrm{End}(K(X))$ and $U^\vee := \bigoplus_{u\in\mathsf{C}} u\,\mathrm{id}_{V_u^\vee} \in \mathrm{End}(K(X)^\vee)$.

A solution $(Y_\pm, Y_\infty)$ to this problem is unique if exists. The solution Y_∞ gives the fundamental solution $L(\tau, z)$ and hence recovers the quantum connection. It is interesting to note that we reconstruct not only the connection but also the fundamental solution $L(\tau, z)$ (called *calibration* in the theory of Frobenius manifolds): this implies that the value of the parameter τ can be reconstructed by the asymptotics $L(\tau, z)^{-1}1 = 1 + \tau z^{-1} + O(z^{-2})$ if we know the unit class 1.

Remark 2.18. The additional data we need here (other than those we already mentioned) is the framing Ψ_u for each regular singular piece. In the semisimple case, we have $\mathcal{R}_u \cong (\mathbb{C}\{z\}, d)$ and $V_u \cong \mathbb{Z}$, so there is essentially a unique choice for Ψ_u. A natural candidate for Ψ_u could be given by answering Problem 2.16. See Section 3.3 for the example where we have a natural candidate for the framing.

Remark 2.19. If we include odd classes, the monodromy transformation T on $\mathcal{S}_I$ is given by $(-1)^{\deg \alpha}[T\alpha, \beta) = [\beta, \alpha)$; the Stokes maps $S^{\pm}: \mathcal{S}_I \to \mathcal{S}_{-I}$ are given by $(S^+\alpha, \beta) = (-1)^{\deg \alpha}[\beta, \alpha)$, $(S^-\alpha, \beta) = [\alpha, \beta)$. Problem 2.17 can be also modified accordingly, using the fact that the pairing $[\cdot, \cdot)$ on $\mathcal{S}_I$ corresponds to $-\mathtt{i}\chi(\cdot, \cdot)$ on $K^1(X)$ (see Remark 1.2).

3. FUNCTORIALITY OF QUANTUM COHOMOLOGY

In this section, we discuss a conjectural functoriality of quantum cohomology under birational transformations. Roughly speaking, we expect that the relationship between quantum cohomology is induced from a natural map between K-groups via the $\widehat{\Gamma}$-integral structure. Let X_1, X_2 be smooth projective varieties and let $\varphi: X_1 \dashrightarrow X_2$ be a birational map. Suppose that φ fits into the following commutative diagram:

$$\begin{array}{ccccc} & & \hat{X} & & \\ & {\scriptstyle p_1} \swarrow & & \searrow {\scriptstyle p_2} & \\ X_1 & & \overset{\varphi}{\dashrightarrow} & & X_2, \end{array} \tag{3.1}$$

where p_1, p_2 are projective birational morphisms. We say that φ is *crepant* (or K-equivalent) if $p_1^* K_{X_1} = p_2^* K_{X_2}$ and *discrepant* otherwise. We allow X_i to be smooth Deligne–Mumford stacks (with projective coarse moduli spaces) so that we can include crepant resolutions of orbifolds in the following discussion.

3.1. Crepant transformation

Suppose that $\varphi: X_1 \dashrightarrow X_2$ is crepant. In this case it can be shown that $H^*(X_1) \cong H^*(X_2)$ as graded vector spaces by Kontsevich's motivic integration (see, e.g., [60]). A famous conjecture of Yongbin Ruan [57] says that the quantum cohomologies of X_1 and X_2 become isomorphic after analytic continuation. This problem has been studied by many people, see, e.g., [10,48,51,54]. We give a version of the conjecture stated in terms of quantum D-modules and the $\widehat{\Gamma}$-integral structure following [16, CONJECTURE 5.1], [42, §5.5], [17,43].

Conjecture 3.1 (Crepant Transformation Conjecture). *Let $\varphi: X_1 \dashrightarrow X_2$ be a crepant birational map. There exists a map f from an open subset of $H^*(X_1)$ to an open subset of*

$H^*(X_2)$ such that, after analytic continuation, we have an isomorphism of quantum D-modules $\mathrm{QDM}(X_1) \cong f^* \mathrm{QDM}(X_2)$. *Moreover, via the $\widehat{\Gamma}$-integral structure, the isomorphism is induced by an isomorphism $(K(X_1), \chi) \cong (K(X_2), \chi)$ of topological K-group lattices.*

Recall from Section 1.2 that the quantum D-module $\mathrm{QDM}(X_i)$ is the tuple of the cohomology bundle F, the quantum connection and the Poincaré pairing; the isomorphism in the conjecture is required to respect these structures. Conjecture 3.1 was proved[4] for crepant transformations between complete intersections in toric Deligne–Mumford stacks, which are induced from variation of GIT quotients **[15]**. In that case, it is shown that the map $K(X_1) \cong K(X_2)$ between K-groups is given by a Fourier–Mukai transformation that gives rise to the equivalence of derived categories of X_1 and X_2. The calculation needed in this result is an extension of the work of Borisov–Horja **[9]** that relates analytic continuation of hypergeometric solutions to the GKZ system and Fourier–Mukai transformations between toric orbifolds.

Remark 3.2. (1) We can hope that the isomorphism $K(X_1) \cong K(X_2)$ is induced by an equivalence of derived categories. A different derived equivalence can arise from a different choice of paths of analytic continuation.

(2) When Conjecture 3.1 holds, the map f is necessarily a local isomorphism and identifies the F-manifold structure **[35]** of quantum cohomology. In the case of crepant resolutions of orbifolds, it has been observed in **[16]** that f is not necessarily affine-linear unless the orbifold satisfies the hard Lefschetz condition.

3.2. Discrepant transformation

We present a conjectural picture in the discrepant case following **[45]**. In the discrepant case, the ranks of cohomology are different in general and we expect to have an *orthogonal* decomposition of formal quantum D-modules and a *semiorthogonal* decomposition of the $\widehat{\Gamma}$-integral structure. As in Section 2.4, we write $\overline{\mathrm{QDM}}(X) := \mathrm{QDM}(X) \otimes_{\mathcal{O}[z]} \mathcal{O}[\![z]\!]$ for the quantum D-module formalized along $z = 0$. Because of the lack of abundant evidences, we state our picture as problems rather than conjectures.

Problem 3.3 (Formal decomposition). Let $\varphi\colon X_1 \dashrightarrow X_2$ be a birational map fitting into the diagram (3.1) such that $p_1^* K_{X_1} - p_2^* K_{X_2}$ is an effective divisor. Show that there exists a map f from an open subset of $H^*(X_1)$ to an open subset of $H^*(X_2)$ such that we have an orthogonal decomposition

$$\overline{\mathrm{QDM}}(X_1) \cong f^* \overline{\mathrm{QDM}}(X_2) \oplus \mathscr{D},$$

where $\mathscr{D}$ is a locally free $\mathcal{O}[\![z]\!]$-module equipped with a flat meromorphic connection $\nabla^{\mathscr{D}}$ and a $\nabla^{\mathscr{D}}$-flat pairing $(\cdot,\cdot)_{\mathscr{D}}\colon (-)^* \mathscr{D} \otimes \mathscr{D} \to \mathcal{O}[\![z]\!]$.

4 For complete intersections, we restrict to the ambient part of quantum cohomology in **[15]**.

Problem 3.3 has been solved for discrepant birational transformations between toric Deligne–Mumford stacks which arise from a variation of GIT quotients [45]. The proof is based on mirror symmetry for toric stacks [12,13].

Suppose that Problem 3.3 is solved for some $\varphi\colon X_1 \dashrightarrow X_2$, and also suppose (for simplicity) that $\overline{\mathrm{QDM}}(X_1)$ is of exponential type (see Section 2.4) at some $\tau \in H^*(X_1)$ in the domain of the map f. Then $\overline{\mathrm{QDM}}(X_2)_{f(\tau)}$ and $\mathscr{D}|_\tau$ are also of exponential type. We further assume the following: there exist a phase $\phi \in \mathbb{R}$ and real numbers $l_1 > l_2$ such that

- every eigenvalue u of $-\nabla^{\mathscr{D}}_{z^2\partial_z} \in \mathrm{End}(\mathscr{D}|_{z=0,\tau})$ satisfies either $\Im(e^{-\mathrm{i}\phi}u) > l_1$ or $l_2 > \Im(e^{-\mathrm{i}\phi}u)$ and
- every eigenvalue u of $(E^{X_2}\star_{f(\tau)}) \in \mathrm{End}(H^*(X_2))$ satisfies $l_1 > \Im(e^{-\mathrm{i}\phi}u) > l_2$.

Then $\mathscr{D}|_\tau$ decomposes as $\mathscr{D}|_\tau = \mathscr{D}_1 \oplus \mathscr{D}_2$ so that every eigenvalue of $-\nabla^{\mathscr{D}_1}_{z^2\partial_z}$ on $\mathscr{D}_1|_{z=0}$ satisfies $\Im(e^{-\mathrm{i}\phi}u) > l_1$ and that every eigenvalue of $-\nabla^{\mathscr{D}_2}_{z^2\partial_z}$ on $\mathscr{D}_2|_{z=0}$ satisfies $\Im(e^{-\mathrm{i}\phi}u) < l_2$. By varying ϕ a little, we may assume that $e^{\mathrm{i}\phi}$ is admissible for the eigenvalues of $(E^{X_1}\star_\tau)$. As discussed in Section 2.4, by the Hukuhara–Turrittin theorem, the formal decomposition $\overline{\mathrm{QDM}}(X_1)_\tau \cong \mathscr{D}_1 \oplus \overline{\mathrm{QDM}}(X_2)_{f(\tau)} \oplus \mathscr{D}_2$ lifts to an analytic decomposition of connections over a sector of the form $I = \{z \in \mathbb{C}^\times : |\arg z - \phi| < \frac{\pi}{2} + \varepsilon\}$ for some $\varepsilon > 0$

$$\mathrm{QDM}(X_1)_\tau|_I \cong \mathscr{D}_{1,I} \oplus \mathrm{QDM}(X_2)_\tau|_I \oplus \mathscr{D}_{2,I} \tag{3.2}$$

where $\mathscr{D}_{i,I}$ is an analytic connection over the sector I.

Problem 3.4 (Analytic decomposition). Show that the analytic decomposition (3.2) is induced, via the $\widehat{\Gamma}$-integral structures for X_1 and X_2, by an SOD of topological K-groups:

$$K(X_1) \cong K_1 \oplus K(X_2) \oplus K_2 \tag{3.3}$$

such that the associated inclusion $K(X_2) \to K(X_1)$ respects the Euler pairing.

Problem 3.4 has been answered affirmatively when X_1, X_2 are weak-Fano compact toric Deligne–Mumford stacks (satisfying certain mild technical conditions) and $\varphi\colon X_1 \to X_2$ is a weighted blowup (or a root construction) along a toric substack Z [45]. We also showed that the decomposition (3.3) at some τ and ϕ is given by an Orlov-type SOD [55]. We could hope that the SOD (3.3) in K-theory arises from an SOD $D^b(X_1) \cong \langle \mathcal{A}_1, D^b(X_2), \mathcal{A}_2\rangle$ of the derived category; such an SOD in the derived category has been conjectured in [6].

Remark 3.5. There are closely related works by Bayer [7], Acosta–Shoemaker [2, 3] and González–Woodward [32]. A formal decomposition of quantum D-modules under flips similar to our picture has been also proposed by Lee, Lin, and Wang [49, 50].

Remark 3.6. In Problem 3.3, f is necessarily a submersion and the F-manifold of $\mathrm{QH}^*(X_1)$ locally decomposes into the product of the F-manifold of $\mathrm{QH}^*(X_2)$ and that corresponding to $\mathscr{D}$. *Proof.* The map $T_\tau H^*(X_1) \to \overline{\mathrm{QDM}}(X_1)|_{z=0,\tau} = H^*(X_1)$ given by $v \mapsto z\nabla_v 1$ is an isomorphism. If the unit section 1 maps to $(s,t) \in \overline{\mathrm{QDM}}(X_2) \oplus \mathscr{D}$ under the isomorphism, it follows that the map $df(T_\tau H^*(X_1)) \to \overline{\mathrm{QDM}}(X_2)|_{z=0,f(\tau)}$, $w \mapsto z\nabla_w s$

is surjective. This can only happen when $df: T_\tau H^*(X_1) \to T_{f(\tau)} H^*(X_2)$ is surjective. The ring homomorphism $T_\tau H^*(X_1) \hookrightarrow \mathrm{End}(\overline{\mathrm{QDM}}(X_1)|_{z=0,\tau})$, $v \mapsto z\nabla_v$ factors through $T_\tau H^*(X_1) \to T_{f(\tau)} H^*(X_2) \oplus \mathrm{End}(\mathscr{D}|_{z=0,\tau})$ and this gives a decomposition of the ring $(T_\tau H^*(X_1), \star_\tau)$. The flatness of ∇ shows that the decomposition is integrable.

Remark 3.7. For a higher-genus generalization of Conjecture 3.1 and Problem 3.3, we refer the reader to [14, 45].

3.3. Riemann–Hilbert problem for blowups

Let X be a smooth projective variety and let $Z \subset X$ be a smooth subvariety. Let $\varphi: \tilde{X} \to X$ be the blowup of X along Z. The above mentioned results for toric blowups suggest the following conjectural reconstruction algorithm for quantum cohomology of $\tilde{X}$ from quantum cohomology of X and Z. This is similar to the procedure in Section 2.5.

Orlov decomposition. Let c be the codimension of Z in X. By Orlov [55] we have the SOD of the K-group:

$$K(\tilde{X}) = \varphi^* K(X) \oplus K(Z)_0 \oplus \cdots \oplus K(Z)_{c-2} \cong K(X) \oplus K(Z)^{\oplus(c-1)},$$

where $K(Z)_k = j_*(\mathcal{O}(k) \otimes \pi^* K(Z))$ with $j: E \hookrightarrow X$ the inclusion of the exceptional locus and $\pi: E \cong \mathbb{P}(N_{Z/X}) \to Z$ a projective bundle. We shall fix this decomposition. The cohomology of $\tilde{X}$ is isomorphic to $H^*(X) \oplus H^{*-2}(Z) \oplus \cdots \oplus H^{*-2c+2}(Z)$ as graded vector spaces. The cup product structure on $H^*(\tilde{X})$, the $\widehat{\Gamma}$-class and the Chern character for $\tilde{X}$ can be reconstructed from those for X, Z, the push-forward and pull-back maps between $H^*(X)$, $H^*(Z)$ and the Chern classes $c_i(N_{Z/X}) \in H^{2i}(Z)$.

Formal data. We choose parameters $\sigma \in H^*(X)$ and $\rho_0, \ldots, \rho_{c-2} \in H^*(Z)$ and a phase $\phi \in \mathbb{R}$ so that $\Im(e^{-\mathrm{i}\phi} v) > \Im(e^{-\mathrm{i}\phi} u_0) > \cdots > \Im(e^{-\mathrm{i}\phi} u_{c-2})$ for all eigenvalues v of $(E^X \star_\sigma)$ and all eigenvalues u_i of $(E^Z \star_{\rho_i})$. We define $\overline{\mathrm{QDM}} := \overline{\mathrm{QDM}}(X)_\sigma \oplus \overline{\mathrm{QDM}}(Z)_{\rho_0} \oplus \cdots \oplus \overline{\mathrm{QDM}}(Z)_{\rho_{c-2}}$. This will be the formal quantum D-module for $\tilde{X}$.

Gluing. The given formal decomposition for $\overline{\mathrm{QDM}}$ should lift to analytic decompositions over the sectors I and $-I$, with $I = \{z \in \mathbb{C}^\times : |\arg z - \phi| < \frac{\pi}{2} + \varepsilon\}$ for some $\varepsilon > 0$,

$$\Phi_{\pm I}: \mathrm{QDM}|_{\pm I} \cong \mathrm{QDM}(X)_\sigma \oplus \mathrm{QDM}(Z)_{\rho_0} \oplus \cdots \oplus \mathrm{QDM}(Z)_{\rho_{c-2}}|_{\pm I},$$

and the two analytic decompositions should be glued together by the Stokes data induced from Orlov's SOD. Finally, we glue it with the connection near $z = \infty$ via the $\widehat{\Gamma}$-integral structure to get the quantum D-module for $\tilde{X}$.

The reconstruction can be formulated as a Riemann–Hilbert problem for $Y_\pm = (\Phi_{\pm I})^{-1}$ and $Y_\infty = L(\tau, z)$ (a fundamental solution for $\tilde{X}$, see (1.5)) as follows. Define $S^\pm: K(\tilde{X}) \to K(\tilde{X})^\vee$ by $\langle S^+\alpha, \beta\rangle = \chi(\beta, \alpha)$, $\langle S^-\alpha, \beta\rangle = \chi(\alpha, \beta)$ as before; also define $\Psi: K(\tilde{X}) \cong K(X) \oplus K(Z)^{\oplus(c-1)} \to (H^*(X) \oplus H^*(Z)^{\oplus(c-1)}) \otimes \mathcal{O}_I$ as

$$\Psi(\alpha, \beta_0, \ldots, \beta_{c-2}) = \mathfrak{s}_X(\alpha)(\sigma, z) \oplus \mathfrak{s}_Z(\beta_0)(\rho_0, z) \oplus \cdots \oplus \mathfrak{s}_Z(\beta_{c-2})(\rho_{c-2}, z),$$

where ς_X, ς_Z are the maps (1.6) defined for X and Z, respectively, and define $\Psi_-\colon K(\tilde{X})^\vee \cong K(X)^\vee \oplus (K(Z)^\vee)^{\oplus(c-1)} \to (H^*(X) \oplus H^*(Z)^{\oplus(c-1)}) \otimes \mathcal{O}_{-I}$ as

$$\Psi_-\big(\chi(\alpha,\cdot), \chi(\beta_0,\cdot), \ldots, \chi(\beta_{c-2},\cdot)\big) = \text{clockwise analytic continuation of } \Psi(\alpha, \beta_0, \ldots, \beta_{c-2}).$$

Let Ψ_∞ be the map (2.4) with X there replaced with $\tilde{X}$. The problem is to find functions

$$Y_\infty \in \mathrm{GL}\big(H^*(\tilde{X})\big) \otimes \mathcal{O}_{\mathbb{P}^1\setminus\{0\}}, \quad Y_\pm \in \mathrm{Hom}\big(H^*(X) \oplus H^*(Z)^{\oplus(c-1)}, H^*(\tilde{X})\big) \otimes \mathcal{O}_{\pm I}$$

such that $Y_\infty|_{z=\infty} = \mathrm{id}$, $Y_\pm \to Y_0$ as $z \to 0$ along the sector $\pm I$ for an invertible operator Y_0 and that

$$\begin{aligned} Y_+\Psi &= Y_\infty \Psi_\infty && \text{over } I, \\ Y_-\Psi_- S^\pm &= Y_+\Psi && \text{over } D^\pm, \end{aligned}$$

where $D^\pm$ is as before. As discussed in Section 2.5, we can reconstruct the value of the parameter τ for the quantum D-module of $\tilde{X}$ and it becomes a function of $\sigma, \rho_0, \ldots, \rho_{c-2}$; the parameter space locally splits into the product of $H^*(X)$ and $(c-1)$ copies of $H^*(Z)$ as an F-manifold.

Remark 3.8. Recently, Katzarkov, Kontsevich, and Pantev [47] formulated a closely related conjecture for quantum cohomology of blowups and gave a remarkable application to the problem of rationality.

ACKNOWLEDGMENTS

I thank Sergey Galkin for valuable comments on a draft version of the paper and allowing me to present his formulation in this paper.

FUNDING

This work was partially supported by JSPS grant 16H06335, 20K03582 and 21H04994.

REFERENCES

[1] M. Abouzaid, S. Ganatra, H. Iritani, and N. Sheridan, The Gamma and Strominger–Yau–Zaslow conjectures: a tropical approach to periods. *Geom. Topol.* **24** (2020), 2547–2602.

[2] P. Acosta and M. Shoemaker, Gromov–Witten theory of toric birational transformations. 2016, arXiv:1604.03941.

[3] P. Acosta and M. Shoemaker, Quantum cohomology of toric blowups and Landau–Ginzburg correspondences. *Algebr. Geom.* **5** (2018), 239–263.

[4] M. F. Atiyah, Circular symmetry and stationary phase approximation. In *Colloque en l'honneur de Laurent Schwartz (École Polytechnique, Palaiseau, 30 May – 3 June 1983)*, pp. 43–59, Astérisque 131, Société mathématique de France (Paris), 1985.

[5] M. F. Atiyah and F. Hirzebruch, Vector bundles and homogeneous spaces. In *Proc. Symp. Pure Math. v. III*, pp. 7–38, Amer. Math. Soc., Providence, RI, 1961.

[6] M. Ballard, D. Favero, and L. Katzarkov, Variation of geometric invariant theory quotients and derived categories. *J. Reine Angew. Math.* **746** (2019), 235–303.

[7] A. Bayer, Semisimple quantum cohomology and blowups. *Int. Math. Res. Not.* (2004), 2069–2083.

[8] A. Bondal and A. Polishchuk, Homological properties of associative algebras: the method of helices (in Russian). *Izv. Ross. Akad. Nauk Ser. Mat.* **57** (1993), no. 2, 3–50. English translation in *Russian Acad. Sci. Izv. Math.* **42** (1994), no. 2, 219–260.

[9] L. Borisov and R. Horja, Mellin–Barnes integrals as Fourier–Mukai transforms. *Adv. Math.* **207** (2006), no. 2, 876–927.

[10] J. Bryan and T. Graber, The crepant resolution conjecture. In *Algebraic geometry – Seattle 2005. Part 1*, pp. 23–42, Proc. Sympos. Pure Math. 80, Amer. Math. Soc., Providence, RI, 2009.

[11] D. Cheong and C. Li, On the Conjecture $\mathcal{O}$ of GGI for G/P. *Adv. Math.* **306** (2017), 704–721.

[12] T. Coates, A. Corti, H. Iritani, and H. H. Tseng, A Mirror Theorem for Toric Stacks. *Compos. Math.* **151** (2015), 1878–1912.

[13] T. Coates, A. Corti, H. Iritani, and H. H. Tseng, Hodge-theoretic mirror symmetry for toric stacks. *J. Differential Geom.* **114** (2020), 41–115.

[14] T. Coates and H. Iritani, A Fock sheaf for Givental quantization. *Kyoto J. Math.* **58** (2018), 695–864.

[15] T. Coates, H. Iritani, and Y. Jiang, The crepant transformation conjecture for toric complete intersections. *Adv. Math.* **329** (2018), 1002–1087.

[16] T. Coates, H. Iritani, and H. H. Tseng, Wall-crossings in toric Gromov–Witten theory. I. Crepant examples. *Geom. Topol.* **13** (2009), 2675–2744.

[17] T. Coates and Y. Ruan, Quantum cohomology and crepant resolutions: a conjecture. *Ann. Inst. Fourier (Grenoble)* **63** (2013), no. 2, 431–478.

[18] G. Cotti, B. Dubrovi, and D. Guzzetti, Helix structures in quantum cohomology of Fano varieties. 2018, arXiv:1811.09235.

[19] B. Dubrovin, Geometry of 2D topological field theories. In *Integrable systems and quantum groups*, edited by M. Francaviglia et al., pp. 120–348, Lecture Notes in Math. 1620, Springer, Berlin, 1996.

[20] B. Dubrovin, Geometry and analytic theory of Frobenius manifolds. In *Proceedings of the International Congress of Mathematicians, Vol. II (Berlin, 1998), Doc. Math.* Extra Vol. **II** (1998), 315–326.

[21] B. Dubrovin, Painlevé transcendents in two-dimensional topological field theory. In *The Painlevé property*, pp. 287–412, CRM Ser. Math. Phys., Springer, New York, 1999.

[22] B. Dubrovin, Quantum cohomology and isomonodromic deformation, Lecture at *Recent Progress in the Theory of Painlevé Equations: Algebraic, asymptotic and topological aspects*, Strasbourg, November 2013.

[23] B. Fang and P. Zhou, Gamma II for toric varieties from integrals on T-dual branes and homological mirror symmetry. 2019, arXiv:1903.05300.

[24] S. Galkin, Private communication.

[25] S. Galkin, Apéry constants of homogeneous varieties, preprint SFB45. 2008, arXiv:1604.04652.

[26] S. Galkin, V. Golyshev, and H. Iritani, Gamma classes and quantum cohomology of Fano manifolds: Gamma conjectures. *Duke Math. J.* **165** (2016), no. 11, 2005–2077.

[27] S. Galkin and H. Iritani, Gamma conjecture via mirror symmetry. In *Primitive forms and related subjects – Kavli IPMU 2014*, pp. 55–115, Adv. Stud. Pure Math. 83, 2019.

[28] A. Givental, Homological geometry I. Projective hypersurfaces. *Selecta Math.* **1** (1995), 325–345.

[29] A. Givental, Equivariant Gromov–Witten invariants. *Int. Math. Res. Not.* **1996** (1996), no. 13, 613–663.

[30] V. Golyshev, Deresonating a Tate period. 2009, arXiv:0908.1458.

[31] V. Golyshev and D. Zagier, Proof of the gamma conjecture for Fano 3-folds of Picard rank 1. *Izv. Ross. Akad. Nauk Ser. Mat.* **80** (2016), no. 1, 27–54.

[32] E. González and C. Woodward, Quantum cohomology and toric minimal model programs. *Adv. Math.* **353** (2019), 591–646.

[33] S. Gorchinskiy and D. Dmitri, Geometric phantom categories. *Publ. Math. Inst. Hautes Études Sci.* **117** (2013), 329–349.

[34] M. Guest, Quantum cohomology via D-modules. *Topology* **44** (2005), no. 2, 263–281.

[35] C. Hertling and Y. I. Manin, Weak Frobenius manifolds. *Int. Math. Res. Not.* **6** (1999), 277–286.

[36] C. Hertling, Y. I. Manin, and C. Teleman, An update on semisimple quantum cohomology and F-manifolds. *Tr. Mat. Inst. Steklova* **264** (2009), Mnogomernaya Algebraicheskaya Geometriya, 69–76. Translation in *Proc. Steklov Inst. Math.* **264** (2009), no. 1, 62–69.

[37] C. Hertling and C. Sevenheck, Nilpotent orbits of a generalization of Hodge structures. *J. Reine Angew. Math.* **609** (2007), 23–80.

[38] R. Horja, Hypergeometric functions and mirror symmetry in toric varieties. 1999, arXiv:math/9912109.

[39] S. Hosono, Central charges, symplectic forms, and hypergeometric series in local mirror symmetry. In *Mirror symmetry. V*, pp. 405–439, AMS/IP Stud. Adv. Math. 38, Amer. Math. Soc., Providence, RI, 2006.

[40] S. Hosono, A. Klemm, S. Theisen, and S. T. Yau, Mirror symmetry, mirror map and applications to complete intersection Calabi–Yau spaces. *Nuclear Phys. B* **433** (1995), no. 3, 501–552.

[41] X. Hu and H-Z. Ke, Gamma conjecture II for quadrics. 2021, arXiv:2103.15143.

[42] H. Iritani, An integral structure in quantum cohomology and mirror symmetry for toric orbifolds. *Adv. Math.* **222** (2009), no. 3, 1016–1079.

[43] H. Iritani, Ruan's conjecture and integral structures in quantum cohomology. In *New developments in algebraic geometry, integrable systems and mirror symmetry (RIMS, Kyoto, 2008)*, pp. 111–166, Adv. Stud. Pure Math. 59, Math. Soc. Japan, Tokyo, 2010.

[44] H. Iritani, Quantum cohomology and periods. *Ann. Inst. Fourier (Grenoble)* **61** (2011), 2909–2958.

[45] H. Iritani, Global Mirrors and Discrepant Transformations for Toric Deligne–Mumford Stacks. *SIGMA Symmetry Integrability Geom. Methods Appl.* **16** (2020), 032, 111 pages.

[46] L. Katzarkov, M. Kontsevich, and T. Pantev, Hodge theoretic aspects of mirror symmetry. In *From Hodge theory to integrability and TQFT tt^*-geometry*, pp. 87–174, Proc. Sympos. Pure Math. 78, Amer. Math. Soc., Providence, RI, 2008.

[47] M. Kontsevich, Birational invariants from quantum cohomology. Talk at Higher School of Economics on 27 May 2019, as part of *Homological Mirror Symmetry at HSE, 27 May – 1 June*, 2019.

[48] Y. P. Lee, H. W. Lin, and C. L. Wang, Flops, motives, and invariance of quantum rings. *Ann. of Math. (2)* **172** (2010), 243–290.

[49] Y. P. Lee, H. W. Lin, and C. L. Wang, Quantum cohomology under birational maps and transitions. In *String-Math 2015*, pp. 149–168, Proc. Sympos. Pure Math. 96, Amer. Math. Soc., Providence, RI, 2017.

[50] Y. P. Lee, H. W. Lin, and C. L. Wang, Quantum flips I: local model. 2019, arXiv:1912.03012.

[51] A. M. Li and Y. Ruan, Symplectic surgery and Gromov–Witten invariants of Calabi–Yau 3-folds. *Invent. Math.* **145** (2001), 151–218.

[52] A. S. Libgober, Chern classes and the periods of mirrors. *Math. Res. Lett.* **6** (1999), 141–149.

[53] R. Lu, The $\widehat{\Gamma}$-genus and a regularization of an S^1-equivariant Euler class. *J. Phys. A* **41** (2008), no. 42, 425204 (13 pp).

[54] M. McLean, Birational Calabi–Yau manifolds have the same small quantum products. *Ann. of Math.* **191** (2020), no. 2, 439–579.

[55] D. Orlov, Projective bundles, monoidal transformations, and derived categories of coherent sheaves. *Izv. Ross. Akad. Nauk Ser. Mat.* **56** (1992), 852–862. English translation in *Russian Acad. Sci. Izv. Math.*, **41** (1993) no. 1, 133–141.

[56] D. Orlov, Derived categories of coherent sheaves and triangulated categories of singularities. In *Algebra, arithmetic, and geometry: in honor of Yu. I. Manin. Vol. II*, pp. 503–531, Progr. Math. 270, Birkhäuser Boston Inc., Boston, MA, 2009.

[57] Y. Ruan, The cohomology ring of crepant resolutions of orbifolds. In *Gromov–Witten theory of spin curves and orbifolds*, pp. 117–126, Contemp. Math. 403, Amer. Math. Soc., Providence, RI, 2006.

[58] F. Sanda and Y. Shamoto, An analogue of Dubrovin's conjecture. *Ann. Inst. Fourier* **70** (2020), no. 2, 621–682.

[59] C. van Enckevort and D. van Straten, Monodromy calculations of fourth order equations of Calabi–Yau type. In *Mirror symmetry. V*, pp. 539–559, AMS/IP Stud. Adv. Math. 38, Amer. Math. Soc., Providence, RI, 2006.

[60] T. Yasuda, Motivic integration over Deligne–Mumford stacks. *Adv. Math.* **207** (2006), no. 2, 707–761.

HIROSHI IRITANI

Department of Mathematics, Kitashirakawa-Oiwake-cho, Sakyo-ku, Kyoto, Japan, iritani@math.kyoto-u.ac.jp

This Page Intentionally Left Blank

KÄHLER MANIFOLDS WITH CURVATURE BOUNDED BELOW

GANG LIU

ABSTRACT

This is a survey of certain Kähler manifolds with curvature bounded below. The topics include: (1) the uniformization conjecture of Yau, as well as its related problems; (2) compactification of certain Kähler manifolds of nonnegative curvature; and (3) Gromov–Hausdorff limits of Kähler manifolds.

MATHEMATICS SUBJECT CLASSIFICATION 2020

Primary 53C55; Secondary 32Q15

KEYWORDS

Gromov–Hausdorff convergence, Kähler manifolds, uniformization conjecture, compactification

INTERNATIONAL CONGRESS OF MATHEMATICIANS 2022 JULY 6–14

Proc. Int. Cong. Math. 2022, Vol. 4, pp. 2576–2592
DOI 10.4171/ICM2022/6

Published by EMS Press

1. INTRODUCTION

The study of manifolds with a curvature lower bound has a long history in Riemannian geometry. For instance, we have the comparison theorems, Cheeger–Gromoll splitting theorem [5], Cheng–Yau gradient estimate [18], Li–Yau's heat kernel estimate [44], and many other important theorems. These theorems have been the basic tools to study manifolds with curvature lower bound. In 1981, Gromov proposed a fundamental notion, called the Gromov–Hausdorff convergence. Later, Cheeger, Colding, Tian, Naber [6–9, 19–22] developed an important theory studying the limit of under the Gromov–Hausdorff convergence.

In this survey, we shall consider the applications of Gromov–Hausdorff convergence theory to some problems in Kähler geometry. These include: (1) the uniformization conjecture of Yau, as well as its related problems; (2) compactification of certain noncompact Kähler manifolds of nonnegative curvature; and (3) the structure of Gromov–Hausdorff limits of Kähler manifolds.

2. YAU'S UNIFORMIZATION CONJECTURE AND ITS RELATED PROBLEMS

The classical uniformization theorem states that a simply connected Riemann surface is isomorphic to the Riemann sphere $\mathbb{CP}^1$, the Poincare disk $\mathbb{D}^2$, or the complex plane $\mathbb{C}$. A geometric consequence is that a complete orientable Riemannian surface of positive curvature is necessarily conformal to $\mathbb{CP}^1$ or $\mathbb{C}$. An orientable Riemannian surface can be regarded as a Kähler manifold of complex dimension 1. A natural question is to generalize such result to higher dimensional Kähler manifolds. The curvature we adopt here is the so-called (holomorphic) bisectional curvature.

Definition 2.1 ([42, 64]). On a Kähler manifold M^n, we say the bisectional curvature is greater than or equal to K ($BK \geq K$) if

$$\frac{R(X,\overline{X},Y,\overline{Y})}{\|X\|^2\|Y\|^2+|\langle X,\overline{Y}\rangle|^2} \geq K \tag{2.1}$$

for any two nonzero vectors $X, Y \in T^{1,0}M$.

Observe that the equality holds for complex space forms. The bisectional curvature lower bound condition is weaker than the sectional curvature lower bound, while stronger than the Ricci curvature lower bound.

So the question above can be refined as the classification of Kähler manifolds with positive bisectional curvature. In the compact case, the famous Frankel conjecture, solved by Mori [47] and Siu–Yau [57] independently, states that a compact Kähler manifold of positive bisectional curvature is biholomorphic to $\mathbb{CP}^n$ (in fact, Mori proved a stronger result). The noncompact analogue was proposed by Yau [66] in the 1970s; he asked whether or not a complete noncompact Kähler manifold with positive bisectional curvature is biholomorphic to a complex Euclidean space. For this, Yau further asked in [66] (see also [67, PAGE 117]) whether or not the ring of polynomial growth holomorphic functions is finitely generated, and whether or not the dimension of the spaces of holomorphic functions of polynomial growth is bounded from above by the dimension of the corresponding spaces of polynomials on $\mathbb{C}^n$.

On a complete Kähler manifold M, we say a holomorphic function $f \in \mathcal{O}_d(M)$ if there exists some $C > 0$ with $|f(x)| \leq C(1 + d(x, x_0))^d$ for all $x \in M$. Here x_0 is a fixed point on M. Let $\mathcal{O}_P(M) = \bigcup_{d>0} \mathcal{O}_d(M)$. If one wishes to prove the uniformization conjecture by considering $\mathcal{O}_P(M)$, it is important to know when $\mathcal{O}_P(M) \neq \mathbb{C}$. In [49], Ni proposed an interesting conjecture in this direction. Let us summarize the problems in the four conjectures below:

Conjecture 1. *Let M^n be a complete noncompact Kähler manifold with nonnegative bisectional curvature. Then given any $d > 0$, $\dim(\mathcal{O}_d(M)) \leq \dim(\mathcal{O}_d(\mathbb{C}^n))$.*

Conjecture 2. *Let M^n be a complete noncompact Kähler manifold with nonnegative bisectional curvature. Assume M has positive bisectional curvature at one point p. Then the following three conditions are equivalent:*

(1) $\mathcal{O}_P(M) \neq \mathbb{C}$;

(2) *M has maximal volume growth;*

(3) *There exists a constant C independent of r so that $⨍_{B(p,r)} S \leq \frac{C}{r^2}$. Here S is the scalar curvature; $⨍$ means the average.*

Conjecture 3. *Let M^n be a complete noncompact Kähler manifold with nonnegative bisectional curvature. Then the ring $\mathcal{O}_P(M)$ is finitely generated.*

Conjecture 4. *Let M^n be a complete noncompact Kähler manifold with positive bisectional curvature. Then M is biholomorphic to $\mathbb{C}^n$.*

Conjecture 1 was confirmed by Ni [49] with the assumption that M has maximal volume growth. Later, by using Ni's method, Chen–Fu–Le–Zhu [11] removed the extra condition. The key of Ni's method is a monotonicity formula for the heat flow on a Kähler manifold with nonnegative bisectional curvature. In [34], we discovered a logarithmic convexity result on Kähler manifolds with nonnegative holomorphic sectional curvature. This turns out to be very useful in dealing the problems above.

Definition 2.2. Let M be a complete Kähler manifold. We say that M satisfies the three circle theorem if, for any point $p \in M$, $r > 0$, any holomorphic function f on $B(p, r)$, $\log M_f(r)$ is a convex function of $\log r$. In other words, for $r_1 < r_2 < r_3$,

$$\log\left(\frac{r_3}{r_1}\right) \log M_f(r_2) \leq \log\left(\frac{r_3}{r_2}\right) \log M_f(r_1) + \log\left(\frac{r_2}{r_1}\right) \log M_f(r_3). \tag{2.2}$$

Here $M_f(r) = \sup |f(x)|$ for $x \in B(p, r)$.

Theorem 2.1. *Let M be a complete Kähler manifold. Then M satisfies the three circle theorem if and only if the holomorphic sectional curvature is nonnegative.*

The proof is a simple combination of Hessian comparison and a maximum principle argument.

Corollary 2.1. *Let M be a complete Kähler manifold with nonnegative holomorphic sectional curvature. If $f \in \mathcal{O}_d(M)$, then $\frac{M_f(r)}{r^d}$ is nonincreasing.*

Proof of the corollary. We need to show that

$$\frac{M_f(r_1)}{r_1^d} \geq \frac{M_f(r_2)}{r_2^d}$$

for $r_1 \leq r_2$. By rescaling, we may assume $r_1 = 1$. By the assumption, given any $\varepsilon > 0$, there exists a sequence $\lambda_j \to \infty$ such that

$$\log M_f(\lambda_j) \leq \log M_f(1) + (d + \varepsilon) \log \lambda_j.$$

If we take $r_3 = \lambda_j$ sufficiently large, then

$$M_f(r_2) \leq M_f(r_1) r_2^{d+\varepsilon}.$$

The corollary follows letting $\varepsilon \to 0$. ■

Proof of Conjecture 1. Suppose the inequality fails for some $d > 0$. By linear algebra, at any point $p \in M$, there exists a nonzero holomorphic function $f \in \mathcal{O}_d(M)$ such that the vanishing order at p is at least $[d] + 1$ where $[d]$ is the greatest integer less than or equal to d. Therefore

$$\lim_{r \to 0+} \frac{M_f(r)}{r^d} = 0.$$

Corollary 2.1 says that $\frac{M_f(r)}{r^d}$ is nonincreasing. Thus $f \equiv 0$. This is a contradiction. ■

Now we come to Conjecture 2. Ni and Tam have made important contributions. For example, they proved that (1) is equivalent to (3). In **[35, 36]**, we were able to prove the equivalence of (1) and (2).

Theorem 2.2. *Let (M^n, g) be a complete Kähler manifold with nonnegative bisectional curvature. Assume the universal cover does not split as a product and there exists a nonconstant holomorphic function of polynomial growth on M, then M has maximal volume growth.*

Theorem 2.3. *Let (M^n, g) be a complete Kähler manifold with nonnegative bisectional curvature and maximal volume growth. Then there exists a nonconstant holomorphic function of polynomial growth on M.*

Let us sketch the proof of Theorem 2.2. By a result of Ni and Tam, we have that $\dim(\mathcal{O}_d(M)) \geq cd^n$, where c is independent of n. Assume M does not have maximal volume growth. We can look at a tangent cone of M at infinity. According to Cheeger–Colding, the tangent cone has Hausdorff dimension at most $2n - 1$. Then we pick a regular point on the tangent cone. The tangent cone at that regular point would be $\mathbb{R}^k$, where $1 \leq k \leq 2n - 1$. By the three circle theorem, we were able to pass these polynomial growth holomorphic functions to that Euclidean space without changing the growth rate and the linear independence. If $k < n$, then we obtain a contradiction, by counting the dimension of harmonic functions. However, if $k \geq n + 1$, then this will not work. In this situation, we managed to find some partial complex structure on $\mathbb{R}^k$. This gave extra restriction, which sufficed for the proof.

The existence of polynomial growth holomorphic functions is also very interesting. Usually, this is done by finding plurisubharmonic functions of logarithmic growth. Here we took a different approach. The idea is to look at a tangent cone of M at infinity. Essentially, we managed to establish a complex analytic structure on a tangent cone. Then by pulling back those functions to M, we obtained holomorphic functions on a larger and larger domain. Then the three circle theorem ensured that we can take a subsequence and obtain a nontrivial polynomial growth holomorphic function. The tools we used were the Gromov–Hausdorff convergence theory, Hörmander L^2-estimate of $\bar{\partial}$ [30], heat flow technique by Ni and Tam [52, 53], and the three circle theorem.

Now we come to Conjecture 3. In [45], Mok proved the following:

Theorem 2.4 (Mok). *Let M^n be a complete noncompact Kähler manifold with positive bisectional curvature such that for some fixed point $p \in M$,*

(1) *Scalar curvature $\leq \frac{C_0}{d(p,x)^2}$ for some $C_0 > 0$;*

(2) $\mathrm{vol}(B(p,r)) \geq C_1 r^{2n}$ *for some $C_1 > 0$.*

Then M^n is biholomorphic to an affine algebraic variety.

In Mok's proof, the biholomorphism was given by holomorphic functions of polynomial growth. Therefore, $\mathcal{O}_P(M)$ is finitely generated. In the general case, it was proved by Ni [49] that the transcendental dimension of $\mathcal{O}_P(M)$ over $\mathbb{C}$ is at most n. In [36], we confirmed Conjecture 2:

Theorem 2.5. *Let M^n be a complete noncompact Kähler manifold with nonnegative bisectional curvature. Then the ring $\mathcal{O}_P(M)$ is finitely generated.*

During the course of the proof, we obtained a partial result for Conjecture 4:

Theorem 2.6. *Let M^n be a complete noncompact Kähler manifold with nonnegative bisectional curvature. Assume M is of maximal volume growth, then M is biholomorphic to an affine algebraic variety.*

Our idea is based on the resolution of Conjecture 2. Assume M admits a nontrivial polynomial growth holomorphic function. If we cheat a little bit, say the universal cover does not split, then M is necessarily of maximal volume growth. The main point is to prove that there exist finitely many polynomial growth holomorphic functions $(f_1, \ldots, f_k)$ such that the mapping is proper. Again this heavily uses the convergence theory and the three circle theorem.

Now we come to Conjecture 4. So far, this conjecture is still open. However, there has been much important progress due to various authors. In earlier works, Mok–Siu–Yau [48] and Mok [45] considered embedding by using holomorphic functions of polynomial growth. Later, with the Kähler–Ricci flow (Chern–Ricci flow), results were improved significantly. See, for example, [23–25, 29, 31, 50, 55, 56] for related works. In [38], we obtained

Theorem 2.7. *Let M^n be a complete noncompact Kähler manifold with nonnegative bisectional curvature. Assume M has maximal volume growth, then M is biholomorphic to $\mathbb{C}^n$. In fact, we can find n polynomial growth holomorphic functions $f_1, \ldots, f_n$ which serve as the biholomorphism.*

Remark 2.1. Note that Theorem 2.7 was also proved in [43] by M. C. Li and L. F. Tam. The proof is different from ours.

Let f be a polynomial growth holomorphic function on M. We define the degree of f be the infimum of $d > 0$ such that $f \in \mathcal{O}_d(M)$.

Corollary 2.2. *Under the same assumption as above, if f is a nonconstant polynomial growth holomorphic function on M with minimal degree, then $df \neq 0$ at any point.*

Now let us explain the basic strategy to Theorem 2.7. We follow [26, 27, 36, 37] closely. Recall under the same assumption of Theorem 2.7, it was proved in [36] that the manifold is biholomorphic to an affine algebraic variety. How to prove that the affine variety is in fact $\mathbb{C}^n$? If $n = 2$, Ramanujam's result says that an algebraic surface homeomorphic to $\mathbb{R}^4$ is necessarily isomorphic to $\mathbb{C}^2$. Unfortunately, there is no such criterion in higher dimensions. Moreover, the argument in [36] does not provide information about the topology of the manifold.

Consider a tangent cone V of M at infinity. That is, there exists $r_i \to \infty$ so that the sequence $(M_i, p_i, d_i) = (M, p, \frac{d}{r_i})$ converges to V in the pointed Gromov–Hausdorff sense. Cheeger–Colding theory asserts that V is a metric cone. Let r be the distance to the vertex. Then the vector field $-r\frac{\partial}{\partial r}$ retracts V to the vertex. A key new idea is to solve $\bar{\partial}$ equation on the holomorphic tangent bundle. More precisely, we constructed holomorphic vector fields Z_i on $B(p_i, 1)$ so that in a natural sense, $Re Z_i$ converges to $-r\frac{\partial}{\partial r}$. By using some complex-analytic techniques, we managed to prove that the flow generated by $Re Z_i$ contracts a domain containing $B(p_i, \frac{1}{2})$ to a point. Since $B(p_i, \frac{1}{2})$ exhausts M, we see M is, in fact, exhausted by topological balls. Then by Stalling's result, the manifold is diffeomorphic to $\mathbb{R}^{2n}$. As we see before, if $n = 2$, the manifold is biholomorphic to $\mathbb{C}^2$.

Recall that the domain of Z_i exhausts M. However, it seems difficult to glue these Z_i together. A technical reason is that the unique zero point of Z_i might diverge to infinity.

There are two possible ways to get around this difficulty. One is to prove that the tangent cone V is complex-analytically smooth. Eventually, by using results in algebraic geometry, we can prove that if $n \leq 3$, then V is complex-analytically smooth. Then it is relatively easily to prove that M is biholomorphic to $\mathbb{C}^n$. Unfortunately, the algebro-geometric method fails for higher dimensions.

Another approach is to construct a nice *global* holomorphic vector field on M. This is how we prove Theorem 2.7. A key point is to study a linear space Z consisting of holomorphic vector fields on M so that the action (derivative) on any polynomial growth holomorphic function preserves the degree. It turns out that Z has finite dimension. Arguing by contradiction, we managed to prove that in Z there exists a global holomorphic vector field which contracts M to a point. This gives us the desired biholomorphism from M to $\mathbb{C}^n$. A detailed analysis also gives "canonical" holomorphic coordinate on M.

Finally, let us mention that there have been important progress of Chen–Zhu on the uniformization conjecture without assuming maximal volume growth condition. The reader is referred to [25] for the details.

3. COMPACTIFICATION OF CERTAIN KÄHLER MANIFOLDS OF NONNEGATIVE CURVATURE

In this section, we extend some techniques in [34–37] to study the compactification of certain complete Kähler manifolds with nonnegative Ricci curvature.

In [39], we proved the following

Theorem 3.1. *Let (M^n, g) $(n \geq 2)$ be a complete noncompact Kähler manifold with nonnegative Ricci curvature and maximal volume growth. Fix a point $p \in M$ and set $r(x) := d_g(x, p)$, where d_g denotes the distance with respect to g. Then*

(I) *M is biholomorphic to a Zariski open set of a Moishezon manifold if, for some $\varepsilon > 0$, the bisectional curvature $BK \geq -\frac{C}{r^{2+\varepsilon}}$. In fact, on M, the ring of polynomial growth holomorphic functions is finitely generated;*

(II) *If $BK \geq -\frac{C}{r^2}$ and M has a unique tangent cone at infinity, then M is biholomorphic to a Zariski open set of a Moishezon manifold;*

(III) *M is quasiprojective if the Ricci curvature is positive and $|\mathrm{Rm}| \leq \frac{C}{r^2}$.*

Part (II) states that one can relax the decay assumption on BK in part (I) by assuming the existence of a unique tangent cone at infinity in order to reach the same conclusion. It is desirable to remove the uniqueness of the tangent cone in part (II).

Part (I) is a generalization of Theorem 2.6. Part (II) is connected with many previous results. For instance, Bando–Kasue–Nakajima [2], examples of Tian–Yau [60, 61], Tian [59], and more recent results of Conlon and Hein [12–14]. Part (III) generalizes a theorem of Mok [46, MAIN THEOREM]. There, the same result was obtained under an additional assumption that $\int_M \mathrm{Ric}^n < +\infty$. As a consequence of Theorem 3.1, part (II), we obtain

Corollary 3.1. *Let M be a complete noncompact Ricci-flat Kähler manifold with maximal volume growth. Assume the curvature has quadratic decay. Then M is a crepant resolution of a normal affine algebraic variety. Furthermore, there exists a two-step degeneration from that affine variety to the unique metric tangent cone of M at infinity.*

For a Ricci-flat Riemannian manifold with maximal volume growth, having quadratic curvature decay is equivalent to one (and hence all [15]) tangent cone having a smooth link. Indeed, by [15], the metric on M converges to that on the unique tangent cone at a logarithmic rate. Corollary 3.1 mirrors the result of [27] for tangent cones at a point. In that case, the two-step degeneration should comprise a degeneration of the normal affine variety to the "weighted tangent cone" followed by a $\mathbb{C}^*$-equivariant degeneration of the weighted tangent cone to the tangent cone. Intuitively, the weighted tangent cone is obtained by taking the quotient of the ring of polynomial growth holomorphic functions by the homogeneous ideal generated by weighted homogeneous polynomial growth holomorphic functions so that the restriction to the normal affine variety has lower degree (compare the local weighted tangent cone [27, PP. 354]). For Ricci-flat Kähler manifolds with maximal volume growth and quadratic curvature decay, the weighted tangent cone is expected to be distinct from the tangent cone when the metric converges logarithmically to that on the tangent cone, and is expected to coincide with the tangent cone when the metric converges at a polynomial rate to that on the tangent cone. Note that no metric information is contained in the weighted tangent cone.

A conjecture of Yau [67] states that if a complete Ricci-flat Kähler manifold has finite topological type, then it can be compactified complex analytically. Corollary 2.1 supports this conjecture, at least in this very special setting. Another conjecture of Yau [68, QUESTION 71] states that complete noncompact Kähler manifolds with positive Ricci curvature are biholomorphic to a Zariski open set of a compact Kähler manifold. Part (III) of Theorem 2.2 supports this conjecture.

Now we introduce the strategy of the proof of Theorem 2.2. In parts (I) and (II), we consider polynomial growth holomorphic functions. The three circle theorem was replaced by Donaldson–Sun's three circles theorem [27, PROPOSITION 3.7]. Also note, in the setting of [36], polynomial growth holomorphic functions separate points and tangents. This is no longer true in parts (I) and (II), due to the possibility of compact subvarieties. The two step degeneration in Corollary 3.1 follows from the argument in [17, 27].

The statement of part (III) is very similar to parts (I) and (II). However, the argument is very different. We essentially follow the argument of Mok [46]. The strategy is to consider plurianticanonical sections with polynomial growth. The key new result is a uniform multiplicity estimate for plurianticanonical sections. which provides the dimension estimate for polynomial growth plurianticanonical sections, without the extra assumption $\int_M \mathrm{Ric}^n < +\infty$ (compare [46, THEOREM 2.2]).

4. GROMOV–HAUSDORFF LIMITS OF KÄHLER MANIFOLDS

Recall the seminar work of Cheeger [4]:

Theorem 4.1 (Cheeger, 1970). *Given $K, v, d, n > 0$, consider a class of compact Riemannian n-manifolds with $|\sec| \leq K$, $\mathrm{diam} < d$, $\mathrm{vol} > v$. Then such class is precompact in $C^{1,\alpha}$-topology. In other words, given a suquence of manifolds in this class, there exists a sub-*

sequence convergent in Cheeger–Gromov sense to a smooth manifold M (metric is $C^{1,\alpha}$). As a corollary, this class contains only finite diffeomorphism types.

Later, Anderson [1] generalized Cheeger's work in the Ricci curvature setting:

Theorem 4.2 (Anderson, 1990). *Given $C, i, d, n > 0$, consider a class of compact Riemannian n-manifolds with $|\mathrm{Ric}| \leq C$, $\mathrm{inj} > i$, $\mathrm{diam} < d$. Then the previous theorem holds for such class.*

Remark 4.1. 1. Anderson's theorem satisfies the noncollapsing condition; 2. One cannot replace the injectivity radius bound by noncollapsing condition as in Cheeger's theorem. Otherwise, the limit may not be smooth.

In order to obtain precompactness when the limit is not smooth, one has to consider weaker convergence. An important notion is Gromov–Hausdorff distance. This defines a distance between two compact metric spaces. We say a sequence of metric spaces converge in the Gromov–Hausdorff sense, if the Gromov–Hausdorff distance is approaching zero.

Theorem 4.3 (Gromov, [28]). *Given $C, d, n > 0$, consider a class of compact Riemannian n-manifolds with $\mathrm{Ric} \geq -C$, $\mathrm{diam} < d$. Then this class is precompact in the Gromov–Hausdorff sense (note that the Gromov–Hausdorff limit may be far from smooth).*

For noncompact manifolds, one can consider a manifold with a base pointed. Then the notion of pointed-Gromov–Hausdorff convergence makes sense, i.e., first consider the Gromov–Hausdorff convergence in a geodesic ball of fixed radius, then let the radius go to infinity (diagonal sequence).

A basic problem in metric differential geometry is to study the regularity of the Gromov–Hausdorff limit of manifolds with Ricci curvature lower bound (noncollapsed). There are fundamental contributions by Cheeger, Colding, Tian, and Naber. Given a limit space X and a point $p \in X$, we can consider a blow up of X at p. A blow up limit is called a tangent cone at p (note that the tangent cone at p need not be unique).

Definition 4.1. A point $p \in X$ is called regular if a tangent cone is isometric to a Euclidean space $\mathbb{R}^m$. A point is singular, it is not regular.

Theorem 4.4 (Cheeger–Colding, [6–9]). *Given $n, v > 0$, let (X, o) be the pointed-Gromov–Hausdorff limit of a sequence of n-manifolds (M_i, p_i) with $\mathrm{Ric} \geq -(n-1)$ and $\mathrm{vol}(B(p_i, 1)) > v$. Then*

(1) *X is metric length space of Hausdorff dimension n. The Hausdorff measure is equal to the limit of volume element on M_i.*

(2) *The singular set has Hausdorff codimension at least 2 (sharp).*

(3) *Any tangent cone is a metric cone.*

Note the following: (1) regular points are not so "regular." For example, consider a doubled disk. Then all points are regular. Near the boundary of the disk, the metric is only Lipschitz. It is a conjecture that near a regular point on X, the metric is bi-Lipschitz to a Euclidean ball. Currently, the best known regularity is bi-Hölder; (2) Regular set is not necessarily open. In other words, the singular set need not be closed. In fact, it could be dense. This already appears in the real two-dimensional case (the singular set in this case is countable). In higher dimensions, Li–Naber [19] constructed a limit space so that the singular set is given by a fat Cantor set. In other words, the topology of singular set could be very complicated.

In the above, we considered the Gromov–Hausdorff limit of Riemannian manifolds with Ricci curvature lower bound and noncollapsed volume. What if these manifolds are all Kähler? Can we get extra results? Observe all two-dimensional Riemannian manifolds (oriented) are Kähler. So we cannot expect too much from the extra Kähler assumption. For simplicity, let us call the Gromov–Hausdorff limit of Kähler manifolds with Ricci curvature lower bound Kähler–Ricci limit space. The following is the first important result in this direction:

Theorem 4.5 (Cheeger–Colding–Tian, [18]). *Let X be a noncollapsed Kähler–Ricci limit space. Then any tangent cone splits even dimensional Euclidean factor. In other words, the splitting lines must come in pairs.*

Below is a breakthrough result of Donaldson–Sun [26] and Tian [62] (for simplicity, we only listed a part of their result)

Theorem 4.6. *Let X be the Gromov–Hausdorff limit of a sequence of polarized Kähler manifolds M_i with $|\mathrm{Ric}| \le C$ and $\mathrm{diam} < d$, $\mathrm{vol} > v$. Then X is homeomorphic to a normal projective variety.*

Such result is a key to the existence of Kähler–Einstein metrics on Fano manifolds. In a joint work with G. Szekelyhidi [48], we generalized the result above to the case when the Ricci curvature has a lower bound:

Theorem 4.7. *Let X be the Gromov–Hausdorff limit of a sequence of polarized Kähler manifolds M_i with $\mathrm{Ric} \ge -1$ and $\mathrm{diam} < d$, $\mathrm{vol} > v$. Then X is homeomorphic to a normal projective variety.*

The argument follows [26, 62], and a key step is Tian's partial C^0-estimate [63]. During the proof, we also need a recent deep result of Cheeger–Jiang–Naber [19] on the Minkowski content of the singular set.

A basic technical ingredient in Theorem 4.7 is a result on the existence of holomorphic charts in balls that are Gromov–Hausdorf-close to the Euclidean ball. This is an extension of Proposition 1.3 of [36], where the bisectional curvature lower bound was assumed.

Theorem 4.8. *There exists $\varepsilon > 0$, depending on the dimension n with the following property. Suppose that $B(p, \varepsilon^{-1})$ is a relatively compact ball in a (not necessarily complete) Kähler manifold (M^n, p, ω), satisfying $\mathrm{Ric}(\omega) > -\varepsilon\omega$, and*

$$d_{GH}\big(B(p, \varepsilon^{-1}), B_{\mathbb{C}^n}(0, \varepsilon^{-1})\big) < \varepsilon.$$

Then there is a holomorphic chart $F : B(p,1) \to \mathbb{C}^n$ which is a $\Psi(\varepsilon|n)$-Gromov–Hausdorff approximation to its image. In addition, on $B(p,1)$ we can write $\omega = i\partial\bar{\partial}\phi$ with $|\phi - r^2| < \Psi(\varepsilon|n)$, where r is the distance from p.

We give two applications of Theorem 4.8. The first shows that under Gromov–Hausdorff convergence to a smooth Riemannian manifold, the scalar curvature functions converge as measures. Here we state a simple corollary of this.

Corollary 4.1. *Given any $\varepsilon > 0$, there is a $\delta > 0$ depending on ε, n satisfying the following. Let $B(p,1)$ be a relatively compact unit ball in a Kähler manifold (M,ω) satisfying* $\mathrm{Ric} > -1$, *and $d_{GH}(B(p,1), B_{\mathbb{C}^n}(0,1)) < \delta$. Then $|\int_{B(p,\frac{1}{2})} S| < \varepsilon$, where S is the scalar curvature of ω.*

The other application is the following, which was proved previously under the assumption of non-negative bisectional curvature.

Proposition 4.1. *There exists $\varepsilon > 0$ depending on n, so that if M^n is a complete noncompact Kähler manifold with* $\mathrm{Ric} \geq 0$ *and $\lim_{r\to\infty} r^{-2n}\mathrm{vol}(B(p,r)) \geq \omega_{2n} - \varepsilon$, then M is biholomophic to $\mathbb{C}^n$. Here ω_{2n} is the volume of the Euclidean unit ball.*

Remark 4.2. In the Riemannian setting, Perelman [54] first showed that such a manifold must be contractible. Cheeger–Colding [7] proved such manifold is diffeomorphic to the Euclidean space.

Let us briefly mention the strategy to Theorem 4.8. First, we conformally scale the metric away from a compact domain, in order to make the metric complete (note the metric is no longer Kähler). Thanks to a result of Cavalletti–Mondino [16], our assumptions imply that the almost Euclidean isoperimetric inequality holds in some smaller balls. As in [29], there exists a Ricci flow solution $h(t)$ for a definite time $t \in [0,T]$, satisfying $|\mathrm{Rm}| \leq A/t$ for $t \in (0,T]$. In short, after a fixed time, the metric becomes almost Euclidean is smooth sense. The problem is that, the new metric is not Kähler (not even compatible with the complex structure). Now the observation is that the complex structure J is almost compatible with the new metric. Therefore, by the approach of Newlander–Nirenberg theorem [51], we can find a fixed size holomorphic chart near the center.

Now we glue such chart to a a domain of $\mathbb{CP}^n$. In this way, we were able to run the genuine Kähler–Ricci flow. Thanks to a result of Tian–Zhang [65], the flow has a definite existence time. After a short time, the metric has good regularity near p. The potential estimate follows by integrating along the time line. Finally, the desired holomorphic chart followed by solving another $\bar{\partial}$-problem.

Let us now study the structure of the metric singular set in the Kähler setting. Assume in addition that the N_i^m is a sequence of polarized Kähler manifolds. Then, as we saw above, the limit Y is naturally identified with a projective variety. When the metrics along the sequence are Kähler–Einstein, Donaldson–Sun [26] showed that the metric singular set of Y is the same as the complex analytic singular set of the corresponding projective variety.

Let $\mathcal{R}$ stand for the metric regular set. For small $\varepsilon > 0$, denote by $\mathcal{R}_\varepsilon$ the set of points p so that $\omega_m - \lim_{r\to 0} \frac{\mathrm{vol}(B(p,r))}{r^m} < \varepsilon$. Here ω_m is the volume of the unit ball in $\mathbb{R}^m$. Then $\mathcal{R} = \bigcap_{\varepsilon>0} \mathcal{R}_\varepsilon$. Note that $\mathcal{R}_\varepsilon$ is an open set, while in general $\mathcal{R}$ may not be open. Now we state the theorem

Theorem 4.9. *Let (X, d) be a Gromov–Hausdorff limit as in Theorem 4.7. Then for any $\varepsilon > 0$, $X \setminus \mathcal{R}_\varepsilon$ is contained in a finite union of analytic subvarieties of X. Furthermore, the singular set $X \setminus \mathcal{R}$ is equal to a countable union of subvarieties.*

Remark 4.3. Since the singular set could be dense, the countable union in Theorem 4.9 cannot be replaced by a finite union.

Remark 4.4. In view of Li–Naber's example [19], this result shows that the behavior of singularities in the Kähler case is in sharp contrast with the Riemannian case (see also Theorem 4.10 without polarization). On the one hand, the metric singularities in the Kähler case might seem flexible, since one can perturb Kähler potentials locally. On the other hand, analytic sets are very rigid, and so in particular Theorem 4.9 implies the following: if we perturb the Kähler metric inside a holomorphic chart and assume that the geometric conditions are preserved, then the metric singular set can change by at most a countable set of points.

Let us explain why Theorem 4.9 should be true. The key is the following characterization of the metric singular set:

Proposition 4.2. *A point p is regular in the metric sense if and only if it is complex-analytically regular and the Lelong number for* Ric *vanishes at p.*

Here Ric is the positive $(1,1)$-current on the limit space, regarded as the limit of Ricci form on manifolds. With this in hand, the argument goes as follows: If a point is metric singular, then either it is holomorphic singular point, or the Lelong number of Ric is positive. In the first case, it belongs to a complex-analytic set; in the second case, according to a theorem of Siu [58], this set is given by a countable union of analytic sets.

Let us also sketch the argument in Proposition 4.2. If a point is metric regular, then by Theorem 4.8, we see it is complex-analytically regular. So without loss of generality, we may assume that the point is complex-analytically regular. We are left to show that a point is metric regular iff the Lelong number of Ric vanishes. Note the Lelong number for $2\pi\,\mathrm{Ric}$ at p is given by

$$\liminf_{x\to p} \frac{\log |dz_1 \wedge dz_2 \wedge \cdots \wedge dz_n|^2(x)}{\log |z(x)|},$$

where $z_1, \ldots, z_n$ is a holomorphic chart near p. If the point is metric regular, then by the estimate of Cheeger–Colding, we can show that the value above vanishes. Now assume the point is metric singular, the key is the following claim:

Claim 4.1. Assume p is not a regular point in the metric sense. Then there exist $\varepsilon > 0$ and $r_0 > 0$ so that for all $r < r_0$, if nonzero holomorphic functions $f_1, \ldots, f_n$ on $B(p, 4r)$ satisfy

$f_j(p) = 0$ and $\int_{B(p,r)} f_j \bar{f_k} = 0$ for $j \neq k$, there exists $1 \leq l \leq n$ so that

$$\frac{\fint_{B(p,2r)} |f_l|^2}{\fint_{B(p,r)} |f_l|^2} \geq 2^{2+10n\varepsilon}.$$

Given such claim, we can show that for all small r, on $B(p,r)$,

$$|dz_1 \wedge dz_2 \wedge \cdots \wedge dz_n| \leq 2Cr^{\varepsilon}.$$

Then we obtained the desired lower bound of Lelong number by using the following

Claim 4.2. Let p be a complex-analytically regular point on X. Let $(z_1, \ldots, z_n)$ be a holomorphic chart near p. Assume $z_j(p) = 0$ for all j. Then there exists $\alpha = \alpha(n, v, d) > 0$, $C > 0$, $c > 0$ so that $cr(q)^{\alpha} \leq |z(q)| \leq Cr(q)$ for all q sufficiently close to p. Here r is the distance function to p.

Let us also mention a parallel result under the bisectional curvature lower bound.

Theorem 4.10. *Let (X, p) be the pointed Gromov–Hausdorff limit of complete Kähler manifolds (M_i^n, p_i) with bisectional curvature lower bound -1 and $\mathrm{vol}(B(p_i, 1)) \geq v > 0$. Then X is homeomorphic to a normal complex-analytic space. The metric singular set $X \setminus \mathcal{R}$ is exactly given by a countable union of complex-analytic sets, and for any $\varepsilon > 0$, each compact subset of $X \setminus \mathcal{R}_\varepsilon$ is contained in a finite union of subvarieties.*

Remark 4.5. B. Wilking and R. Bamler [3], M. Lee and L. Tam [32] proved that the limit space is, in fact, complex-analytically smooth. Also J. Lott [33] has developed some theory on the limit of Kähler manifolds with bisectional curvature lower bound.

Now we come to tangent cones of noncollapsed Kähler–Ricci limit space. Recall that, according to Cheeger–Colding, such tangent cones must be metric cones. Also according to Cheeger–Colding–Tian, such tangent cones must split off even dimensional lines. Jointly with G. Szekelyhidi [41], we proved the following:

Theorem 4.11. *Every tangent cone of Z is homeomorphic to a normal affine algebraic variety such that under a suitable embedding into $\mathbf{C}^N$, the homothetic action on the tangent cone extends to a linear torus action.*

Theorem 4.11 was shown previously by Donaldson–Sun [27] under the assumptions that the ω_i are curvature forms of line bundles $L_i \to M_i$, and $|\mathrm{Ric}(\omega_i)| < 1$. An important application of their result is that in their setting the holomorphic spectrum of the tangent cones is rigid, which in turn they used to show the uniqueness of tangent cones. While we are not able to show uniqueness, our result does imply the rigidity of the holomorphic spectrum under two-sided Ricci curvature bounds, even when the (M_i, ω_i) are not polarized.

More precisely, recall that for a Kähler cone $C(Y)$, possibly with singularities, the holomorphic spectrum is defined by

$$\mathcal{S} = \{\deg(f) : f \text{ is homogeneous and holomorphic on } C(Y)\} \subset \mathbf{R}.$$

We then have

Corollary 4.2. *Suppose that we have two sided bounds* $|\mathrm{Ric}(\omega_i)| < 1$ *along the sequence above. Then for any* $q \in Z$ *the holomorphic spectrum of every tangent cone at* q *is the same.*

As in [27], the rigidity of the holomorphic spectrum follows from the fact that the space of tangent cones at each point is connected, and the holomorphic spectrum consists of algebraic numbers. Note that these results hold in particular for tangent cones at infinity of Calabi–Yau manifolds with Euclidean volume growth.

Corollary 4.3. *Let* M *be a complete noncompact Ricci flat Kähler manifold of maximal volume growth. Then the asymptotic volume ratio is an algebraic number.*

ACKNOWLEDGMENTS

The author would like to thank his former advisor Jiaping Wang and former mentor John Lott for guidance and support over the years.

FUNDING

This work was partially supported by NSFC No. 12071140, Program of Shanghai Academic/Technology Research Leader 20XD1401500, the Science and Technology Commission of Shanghai Municipality No. 18dz2271000, as well as the Xplore Prize by Tencent.

REFERENCES

[1] M. Anderson, Convergence and rigidity of manifolds under Ricci curvature bounds. *Invent. Math.* **102** (1990), no. 2, 429–445.

[2] S. Bando, A. Kasue, and H. Nakajima, On a construction of coordinates at infinity on manifolds with fast curvature decay and maximal volume growth. *Invent. Math.* **97** (1989), 313-349.

[3] R. Bamler and B. Wilking, The Ricci flow under almost non-negative curvature conditions. *Invent. Math.* **217** (2019), no. 1, 95–126.

[4] J. Cheeger, Finiteness theorems for Riemannian manifolds. *Amer. J. Math.* **92** (1970), 61–74.

[5] J. Cheeger and D. Gromoll, The splitting theorem for manifolds of nonnegative Ricci curvature. *J. Differential Geom.* **6** (1971), no. 6, 119–128.

[6] J. Cheeger and T. Colding, Lower bounds on Ricci curvature and the almost rigidity of warped products. *Ann. of Math. (2)* **144** (1996), no. 1, 189–237.

[7] J. Cheeger and T. Colding, On the structure of spaces with Ricci curvature bounded below. I. *J. Differential Geom.* **46** (1997), no. 3, 406–480.

[8] J. Cheeger and T. Colding, On the structure of spaces with Ricci curvature bounded below. II. *J. Differential Geom.* **54** (2000), no. 1, 13–35.

[9] J. Cheeger and T. Colding, On the structure of spaces with Ricci curvature bounded below. III. *J. Differential Geom.* **54** (2000), no. 1, 37–74.

[10] J. Cheeger, T. Colding, and G. Tian, On the singularities of spaces with bounded Ricci curvature. *Geom. Funct. Anal.* **12** (2002), 873–914.

[11] B.-L, Chen, X.-Y. Fu, Y. Le, and X.-P. Zhu, Sharp dimension estimates for holomorphic function and rigidity. *Trans. Amer. Math. Soc.* **358** (2006), no. 4, 1435–1454.

[12] R. J. Conlon and H.-J. Hein, Asymptotically conical Calabi–Yau manifolds, I. *Duke Math. J.* **162** (2013), 2855–2902.

[13] R. J. Conlon and H.-J. Hein, Asymptotically conical Calabi–Yau metrics on quasi-projective varieties. *Geom. Funct. Anal.* **25** (2015), 517–552.

[14] R. J. Conlon and H.-J. Hein, Asymptotically conical Calabi–Yau manifolds III, arXiv:1405.7140.

[15] T. Colding and B. Minicozzi. On uniqueness of tangent cones for Einstein manifolds. *Invent. Math.* **196** (2014), no. 3, 515–588.

[16] F. Cavelletti and A. Mondino, Almost Euclidean isoperimetric inequalities in spaces satisfying local Ricci curvature lower bounds. *Int. Math. Res. Not. IMRN* **2020** (2020), no. 5, 1481–1510.

[17] X. X. Chen, S. Sun, and B. Wang, Kähler Ricci flow, Kähler–Einstein metric and K-stability. *Geom. Topol.* **22** (2018), 3145–3173.

[18] S. Y. Cheng and S. T. Yau, Differential equations on Riemannian manifolds and their geometric applications. *Comm. Pure Appl. Math.* **28** (1975), no. 3, 333–354.

[19] J. Cheeger, W. Jiang, and A. Naber, Rectifiability of singular sets in spaces with Ricci curvature bounded below. *Ann. of Math. (2)* **193** (2021), 407–538.

[20] J. Cheeger and A. Naber, Lower bounds on Ricci curvature and quantitative behavior of singular sets. *Invent. Math.* **191** (2013), no. 2, 321–339.

[21] T. Colding and A. Naber, Sharp Hölder continuity of tangent cones for spaces with a lower Ricci curvature bound and applications. *Ann. of Math. (2)* **176** (2012), no. 2, 1173–1229.

[22] W. Jiang and A. A. Naber, L^2 curvature bounds on manifolds with bounded Ricci curvature. *Ann. of Math. (2)* **193** (2021), 107–222.

[23] A. Chau and L.-F. Tam, On the complex structure of Kähler manifolds with non-negative curvature. *J. Differential Geom.* **73** (2006), 107–222.

[24] B. L. Chen, S. H. Tang, and X. P. Zhu, A uniformization theorem of complete noncompact Kähler surfaces with positive bisectional curvature. *J. Differential Geom.* **67** (2004), 519–570.

[25] B. L. Chen and X.-P. Zhu, Yau's uniformization conjecture with nonmaximal volume growth. *Acta Math. Sci.* **38B** (2018), no. 5, 1468–1484.

[26] S. K. Donaldson and S. Sun, Gromov–Hausdorff limits of Kähler manifolds and algebraic geometry. *Acta Math.* **213** (2014), 63–106.

[27] S. K. Donaldson and S. Sun, Gromov–Hausdorff limits of Kähler manifolds and algebraic geometry, II. *J. Differential Geom.* **107** (2017), no. 2, 327–371.

[28] M. Gromov, *Metric structures for Riemannian and non-Riemannian spaces*, Progr. Math., 152, Birkhäuser Boston, Inc., Boston, MA, xx+585 pp, 1999.

[29] F. He, Existence and applications of Ricci flow via pseudolocality, arXiv:1610.01735.

[30] L. Hormander, *An introduction to complex analysis in several variables*, 3rd edition, North Holland, 1990.

[31] S. C. Huang and L.-F. Tam, Kähler–Ricci flow with unbounded curvature. *Amer. J. Math.* **140** (February 2018), no. 1, 189–220.

[32] M.-C. Lee and L.-F. Tam, Kähler manifolds with almost non-negative curvature. *Geom. Topol.* **25** (2021) 1979-2015.

[33] J. Lott, Comparison geometry of holomorphic bisectional curvature for Kaehler manifolds and limit spaces, arXiv:2005.02906.

[34] G. Liu, Three circle theorem and dimension estimate for holomorphic functions on Kähler manifolds. *Duke Math. J.* **15** (2016), 2899–2919.

[35] G. Liu, On the volume growth of Kähler manifolds with nonnegative bisectional curvature. *J. Differential Geom.* **102** (2016), no 3, 485–500.

[36] G. Liu, Gromov–Hausdorff limits of Kähler manifolds and the finite generation conjecture. *Ann. of Math. (2)* **184** (2016), no. 3, 775–815.

[37] G. Liu, Gromov–Hausdorff limits of Kähler manifolds with bisectional curvature lower bound I. *Comm. Pure Appl. Math.* **71** (2018), 267–303.

[38] G. Liu, On Yau's uniformization conjecture. *Camb. J. Math.* **7** (2019), no. 1–2, 33–70.

[39] G. Liu, Compactification of certain noncompact Kähler manifolds with nonnegative Ricci curvature. *Adv. Math.* **382** (14 May 2021).

[40] G. Liu and G. Székelyhidi, Gromov–Hausdorff limits of Kähler manifolds with Ricci curvature bounded below, arXiv:1804.08567.

[41] G. Liu and G. Székelyhidi, Gromov–Hausdorff limits of Kähler manifolds with Ricci curvature bounded below II. *Comm. Pure Appl. Math.* **74** (2021), 909–931.

[42] P. Li and J. Wang, Comparison theorem for Kähler manifolds and positivity of spectrum. *J. Differential Geom.* **69** (2005), 43–74.

[43] M. C. Lee and L. F. Tam, Chern-Ricci flows on noncompact complex manifolds. *J. Differential Geom.* **115** (2020), no. 3, 529-564.

[44] P. Li and S. T. Yau, On the parabolic kernel of the Schrödinger operator. *Acta Math.* **156** (1986), 139–168.

[45] N. Mok, An embedding theorem of complete Kähler manifolds with positive bisectional curvature onto affine algebraic varieties. *Bull. Soc. Math. France* **112** (1984), 197–258.

[46] N. Mok, An embedding theorem of complete Kähler manifolds of positive Ricci curvature onto quasi-projective varieties. *Math. Ann.* **286** (1990), 373–408.

[47] S. Mori, Projective manifolds with ample tangent bundle. *Ann. of Math. (2)* **110** (1979), 593–606.

[48] N. Mok, Y. T. Siu, and S. T. Yau, The Poincare–Lelong equation on complete Kähler manifolds. *Compos. Math.* **44** (1981), 183–281.

[49] L. Ni, A monotonicity formula on complete Kähler manifolds with nonnegative bisectional curvature. *J. Amer. Math. Soc.* **17** (2004), no. 4, 909–946.

[50] L. Ni, Ancient solutions to Kähler–Ricci flow. *Math. Res. Lett.* **12** (2005), 633–654.

[51] A. Newlander and L. Nirenberg, Complex analytic coordinates in almost complex manifolds. *Ann. of Math. (2)* **65** (1957), 391–404.

[52] L, Ni and L.- F. Tam, Plurisubharmonic functions and the structure of complete Kähler manifolds with nonnegative curvature. *J. Differential Geom.* **64** (2003), 457–624.

[53] L, Ni and L.- F. Tam, Poincare–Lelong equation via the Hodge–Laplace heat equation. *Compos. Math.* **149** (2013), no. 11, 1856–1870.

[54] G. Perelman, Manifolds of positive Ricci curvature with almost maximal volume. *J. Amer. Math. Soc.* **7** (1994), no. 2, 299–305.

[55] W. X. Shi, *Ricci deformation of the metric on complete noncompact Kähler manifolds*, PhD thesis, Harvard University, 1990.

[56] W. X. Shi, Ricci flow and the uniformization on complete noncompact Kähler manifolds. *J. Differential Geom.* **45** (1997), 94–220.

[57] Y. T. Siu and S. T. Yau, Compact Kähler manifolds of positive bisectional curvature. *Invent. Math.* **59** (1980), 189–204.

[58] Y.-T. Siu, Analyticity of sets associated to Lelong numbers and the extension of closed positive currents. *Invent. Math.* **27** (1974), 53–156

[59] G. Tian, Compactness theorems for Kähler–Einstein manifolds of dimension 3 and up. *J. Differential Geom.* **35** (1992), 535–558.

[60] G. Tian and S. T. Yau, Complete Kähler manifolds with zero Ricci curvature. I. *J. Amer. Math. Soc.* **3** (1990), 579–609.

[61] G. Tian and S. T. Yau, Complete Kähler manifolds with zero Ricci curvature. II. *Invent. Math.* **106** (1991), 27–60.

[62] G. Tian, Partial C^0-estimate for Kähler–Einstein metrics. *Commun. Math. Stat.* **1** (2013), no. 2, 105–113.

[63] G. Tian, Kähler-Einstein metrics on algebraic manifolds. In *Proc. of Int. Congress of Math.* Kyoto, 1990.

[64] L. F. Tam and C. Yu, Some comparison theorems for Kähler manifolds. *Manuscripta Math.* **137** (2012), no. 3–4, 483–495.

[65] G. Tian and Z. Zhang, On the Kähler–Ricci flow on projective manifolds of general type. *Chin. Ann. Math. Ser. B* **27** (2006), 179–192.

[66] S. T. Yau, Open problems in geometry. *Lectures Differ. Geom.*, by Schoen and Yau, **1** (1994), 365–404.

[67] S. T. Yau, Nonlinear analysis and geometry. *Enseign. Math.* **33** (1987), 109–158.

[68] S. T. Yau, *S. S. Chern: A great geometer of the Twentieth Century*, International Press, 1992.

GANG LIU

Department of Mathematics, East China Normal University, No. 500, Dong Chuan Road, Shanghai, 200241, P.R.China, gliu@math.ecnu.edu.cn

This Page Intentionally Left Blank

GROUPS ACTING AT INFINITY

KATHRYN MANN

ABSTRACT

An action of a group on a topological space is *rigid* if small perturbations to the action have little meaningful influence on its global dynamics. Many examples of rigid actions come from geometric considerations. This introductory survey describes the idea of "looking to infinity" as a source both of rigid examples and proofs of rigidity, starting with some early history then passing quickly to recent developments in topological rigidity of group actions. The examples considered include actions of hyperbolic manifold groups on the visual boundary of their universal cover, automorphism groups of surface groups, boundary actions of hyperbolic groups in the sense of Gromov, and group actions derived from Anosov flows on 3-manifolds.

MATHEMATICS SUBJECT CLASSIFICATION 2020

Primary 37B25; Secondary 20F67, 57M60, 57S25, 37D40

KEYWORDS

Hyperbolic groups, Rigidity, Gromov boundary, Anosov flows

INTERNATIONAL CONGRESS OF MATHEMATICIANS 2022 JULY 6–14

Proc. Int. Cong. Math. 2022, Vol. 4, pp. 2594–2614
DOI 10.4171/ICM2022/47

Published by EMS Press

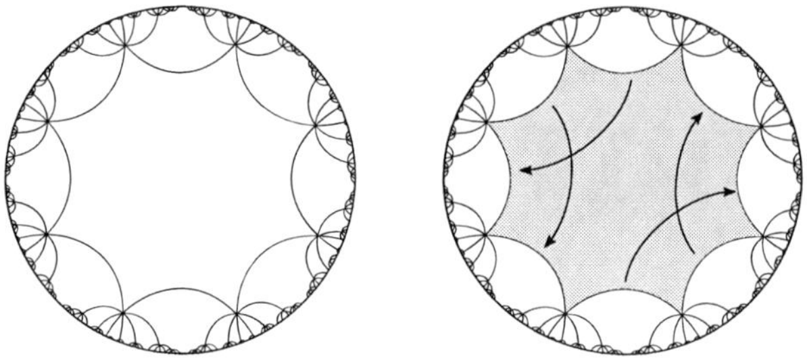

FIGURE 1
A regular tiling of the hyperbolic plane by octagons with geodesic sides. The symmetries of this tiling is a *hyperbolic group* with boundary S^1. Identifying sides of the octagonal domain by translations indicated on the figure gives a genus 2 surface with hyperbolic structure.

1. INTRODUCTION

The boundary at infinity. This survey concerns some recent developments in topological rigidity of group actions that come from "looking to infinity."

To start, let us take a short tour of a familiar object, the Poincaré ball model of the hyperbolic space. Recall that hyperbolic n-space, $\mathbb{H}^n$, is the unique complete, simply connected manifold of constant curvature -1. We can visualize it via the Poincaré ball model, the open unit ball in $\mathbb{R}^n$ equipped with the Riemannian metric $ds^2 = \frac{4\|d\mathbf{x}\|^2}{(1-\|\mathbf{x}\|^2)^2}$. In this model, infinite geodesics are the Euclidean lines and half-circles that meet the boundary of the ball orthogonally (see Figure 1 for an illustration when $n = 2$). Isometries of $\mathbb{H}^n$ preserve geodesics and totally geodesic half-spaces, from which one can deduce that they induce homeomorphisms of the sphere bounding the Poincaré ball. But what is this sphere of "points at infinity"?

Let us try to describe the boundary from the perspective of someone inside $\mathbb{H}^n$. Standing at a point in $\mathbb{H}^n$, your field of vision is a sphere, each line of sight a geodesic ray based at your eye. Each ray in this S^{n-1}-family tends to a unique point on the boundary sphere. To avoid privileging any one point as the eye of an observer, we might broaden our definition of the "sphere at infinity" as follows. In a geodesic metric space (X, d_X), define an equivalence relation on unit speed geodesic rays by declaring that two such rays, say α, $\gamma : [0, \infty) \to X$, are equivalent if there exists a constant D such that $d_X(\alpha(t), \gamma(t)) < D$ holds for all t. Applying this to $\mathbb{H}^n$, the equivalence classes of geodesic rays are in a one-to-one correspondence with points on the boundary sphere of the Poincaré ball, and also to visual directions based at a given point.

The beauty of this definition, beyond being independent of a basepoint, is that we did not need to model hyperbolic space as a ball in $\mathbb{R}^n$ to compactify it. Thus, it lends itself to other spaces with negative—or even coarse analogs of negative—curvature. Applying this definition to any proper geodesic metric space X that is δ-*hyperbolic* in the sense of Gromov (we will return to this later) gives an "ideal boundary" denoted $\partial_\infty X$, which is a compact space when equipped with the quotient topology induced from the compact-open topology on all geodesic rays. Informally, two boundary points are close if they can be represented by rays that stay a bounded distance from each other for a long time. This topology can be naturally extended to $X \cup \partial_\infty X$, compactifying X. The isometries of X preserve geodesics, and preserve the equivalence relation of staying a bounded distance apart, so extend to homeomorphisms of the boundary with this topology. This is the starting point for our story of groups acting on boundary spaces.

Rigidity. A group Γ acting on a space X is *locally rigid* if the only possible small deformations of the action are by trivial procedures—constructions that do not meaningfully change the global dynamics of the action. Of course, what "trivial procedure" and "meaningfully change" mean depends on the context, but a standard interpretation is that there should be either a surjective map, a self-homeomorphism, a diffeomorphism, or an isometry of X (semi)conjugating the perturbed action of Γ back to the original. That is, if $\rho : \Gamma \to \mathrm{Homeo}(X)$ is the original action, and ρ' a perturbation, then *rigid* means that there is a map $h : X \to X$ satisfying $h \circ \rho'(\gamma) = \rho(\gamma) \circ h$ for all $\gamma \in \Gamma$. If invertible, h is a genuine conjugacy; if merely surjective, it is called a *semiconjugacy*. One can strengthen this notion: *globally* rigid means that any deformation, no matter how big or small, is still (semi)conjugate to the original; and one can even do away with the idea of deforming continuously: *strong rigidity* typically means that any other action of the group is either semiconjugate to the original or essentially trivial.

The first major examples of local rigidity to attract significant attention were lattices in linear groups. A *lattice* in a group G is a discrete subgroup Γ such that G/Γ has finite volume, such as $\mathrm{SL}(n, \mathbb{Z})$ in $\mathrm{SL}(n, \mathbb{R})$; it is *cocompact* if the quotient is compact. In 1960, as a means of showing the surprising fact that discrete, cocompact subgroups of $\mathrm{SL}(n, \mathbb{R})$ are conjugate to arithmetic groups, Selberg showed that the inclusion of such a lattice into $\mathrm{SL}(n, \mathbb{R})$ is locally rigid in the sense that any nearby representation is conjugate by an element of $\mathrm{SL}(n, \mathbb{R})$. This, together with concurrent work of Calabi, Weil, Calabi–Vesentini and others, was the birth of rigidity theory. At the time, the major techniques were algebraic (Selberg's work hinged on traces of group elements) and differential geometric. But an important new idea of Selberg, picked up by Mostow, was that of *escape to infinity*, an idea that would prove extremely fruitful to others. Mostow writes:

> *"Upon analyzing Selberg's proof of his rigidity theorem, the key relation shows its force as the elements [considered here in an abelian subgroup] go to infinity... It seemed to me desirable to exploit relations at infinity not only on abelian subfamilies, but among all elements of Γ near infinity."* [35]

What does this mean? Mostow's first success was with lattices in $\mathrm{Isom}(H^n) \cong O(n,1)/\pm 1$, so I will center our discussion here. Given isomorphic cocompact lattices Γ and Γ', one can build a map $f : \mathbb{H}^n \to \mathbb{H}^n$, equivariant with respect to their actions by isometries. Mostow's idea was to ask whether this equivariant map of $\mathbb{H}^n$ extends to a map on the boundary at infinity (the answer is "yes, always") and to study the *regularity* of the extension (a harder answer: "it is necessarily a conformal map, induced by conjugation by an isometry"). This led to the proof of

Theorem 1.1 (Mostow rigidity for hyperbolic manifolds [34]). Suppose M and N are compact, hyperbolic manifolds of dimension at least 3. If M and N are diffeomorphic, then they are isometric.

Shortly afterwards, Margulis noticed that one needs only assume M and N have isomorphic fundamental groups.

Theorem 1.2 (Mostow rigidity, algebraic reformulation [30]). Let M and N be compact, hyperbolic manifolds of dimension at least 3. If $\pi_1(M) \cong \pi_1(N)$, then M and N are isometric. Equivalently, any two isomorphic cocompact lattices in $O(n, 1)$ are conjugate provided $n \geq 3$.

To the reader meeting this theorem for the first time, I wish to emphasize the following dramatic consequence: for hyperbolic manifolds, *every metric invariant is actually a topological invariant.* Volume? Length of a shortest closed geodesic? Theorem 1.2 says these quantities are completely determined by the algebraic structure of the fundamental group of the manifold.

Mostow later extended his rigidity result to lattices in other semisimple Lie groups, the first step again being to construct boundary maps. There are many excellent surveys on these results and their influence in the work of Margulis, Zimmer, and many others; I recommend those of Fisher and Spatzier [12, 36] as a starting point for the curious reader. Here we will take a different tack, skipping ahead to some very recent developments on rigidity of boundary actions. These will occur in settings where, for reasons of low dimension or low regularity, many of the dynamical techniques descended from Mostow and his contemporaries fall short, but the essence of the idea to look at the boundary prevails.

2. SURFACE GROUPS ACTING ON THE CIRCLE

Hyperbolic structures on surfaces. The attentive reader will have noticed the hypothesis "dimension at least 3" in Mostow's theorem. Indeed, this is necessary, as there is a continuum of nonisometric hyperbolic structures on any surface of genus $g \geq 2$, a fact that has been known (depending on how you count) at least since the time of Riemann or Poincaré. These correspond to the continuum of discrete, faithful representations, called *Fuchsian representations*, of the fundamental group of the surface into $\mathrm{Isom}(\mathbb{H}^2)$.

The way in which Mostow's original strategy fails in dimension 2 is quite subtle, having to do with the fact that there is no perfect analog of a quasiconformal map in dimension 1. Agard's survey "Mostow rigidity on the line" [1] contains a nice discussion of what goes right and wrong.[1] That said, for a fixed genus g, each boundary action corresponding to

1 There are many other proofs of Mostow rigidity, all falling short for surfaces in different ways. A proof using Gromov's simplicial volume relies on the fact that an ideal simplex of maximal volume in the hyperbolic n-space is *regular*—a meaningful distinction when $n \geq 3$, but in $\mathbb{H}^2$ all ideal triangles are isometric. Besson–Courtois–Gallot [5] have a strengthening of Mostow's theorem with a different endgame to the proof hinging on an inequality involving the determinant of an $n \times n$ derivative matrix, which simply fails for $n = 2$.

a hyperbolic structure on a genus g surface is conjugate by a *homeomorphism* of the circle.[2] This suggests that one might recover a notion of rigidity by weakening the regularity of the maps in question. Reducing regularity is actually quite a natural consideration: leaving the realm of hyperbolic structures and instead considering the boundary at infinity of the universal cover of the surface with an arbitrary Riemannian metric, or on a Cayley graph for the fundamental group of the surface, one retains only a *topological* circle with an action of the fundamental group by homeomorphisms. These actions are all conjugate to each other, and also all to any Fuchsian action, by a homeomorphism of the circle.

Rigidity at infinity for surfaces. This brings us to a beautiful theorem of Matsumoto on boundary rigidity for surfaces. To state it, I will first make precise the notion of rigidity described before. A group action of Γ on X is a homomorphism $\Gamma \to \mathrm{Homeo}(X)$. The *space of actions* $\mathrm{Hom}(\Gamma, \mathrm{Homeo}(X))$ is the set of all actions equipped with the compact-open topology: informally, two actions are close if, for every element γ in a large finite subset of Γ, the homeomorphisms given by the actions of γ are pointwise close on a large compact subset of X. *Global rigidity* is a statement about the homogeneity of a connected component of $\mathrm{Hom}(\Gamma, \mathrm{Homeo}(X))$; Matsumoto's version is as follows:

Theorem 2.1 (Matsumoto's rigidity **[31]**). Suppose that Σ is a surface of genus at least 2, and $\rho : \pi_1(\Sigma) \to \mathrm{Homeo}(S^1)$ lies in the same connected component as a Fuchsian representation ρ_0. Then there is a continuous, monotone, degree 1 map $h : S^1 \to S^1$ such that $h \circ \rho(\gamma) = \rho_0(\gamma) \circ h$ holds for all $\gamma \in \pi_1(\Sigma)$.

"Monotone" here means that h weakly preserves the cyclic ordering of points on S^1. A triple (x_1, x_2, x_3) of distinct points in S^1 has a *positive* or *negative* orientation depending on whether you proceed anticlockwise or clockwise around the circle as your read the points in order. To *weakly* preserve this order, a map sends each positively oriented triple either to a positively oriented triple or to a degenerate one where two or more points coincide. Thus, h may collapse an interval to a point, but does not double back on itself.

This weakening of the notion of conjugacy in the theorem statement is necessary, as it is easy to construct small, nonconjugate perturbations of almost any action of a countable group on the circle using a "blow up" trick. Enumerate your group Γ, fix a point x and some small $\varepsilon > 0$, and replace the image of x under the action of the nth element of Γ with a closed interval of length $\frac{\varepsilon}{2^n}$. The result is a circle whose circumference is increased by ε. Extend the original action of Γ over the inserted intervals by declaring that g takes the interval corresponding to $f(x)$ to that of $gf(x)$ by the unique affine map. This procedure gives an action by homeomorphisms, and the map $h : S^1 \to S^1$ collapsing each inserted interval to a point is a semiconjugacy.

Group cohomology. A remarkable theorem of Ghys **[14]** says that the equivalence relation on group actions generated by semiconjugacy, in the sense defined above using a mono-

2 This is again a subtle point: one can show that, if such a conjugating homeomorphism is differentiable with nonzero derivative at a single point, then it is real analytic, induced from an isometry of $\mathbb{H}^2$, and the surfaces are isometric.

tone maps, is actually a *cohomological phenomenon*. As is well known, the inclusion of the rotation subgroup $\mathrm{SO}(2) \cong S^1$ into the group $\mathrm{Homeo}_+(S^1)$ of orientation-preserving homeomorphisms of the circle is a homotopy equivalence, thus the cohomology of the classifying space for $\mathrm{Homeo}_+(S^1)$ is generated by an *Euler class* in degree 2. It is less well known (following from a deep theorem of Thurston) that the cohomology of $\mathrm{Homeo}_+(S^1)$ as a *discrete group* is also generated by the Euler class: the map from the discrete $\mathrm{Homeo}_+(S^1)$ to $\mathrm{Homeo}_+(S^1)$ with the usual topology induces an isomorphism on cohomology.

Given an action $\rho : \Gamma \to \mathrm{Homeo}_+(S^1)$ of a group by orientation-preserving homeomorphisms of the circle, one can pull back the discrete, integral Euler class to an element of $H^2(\Gamma; \mathbb{Z})$. Typically, remembering only the Euler class and forgetting the action results in a great loss of information. For instance, if Γ is a surface group, then $H^2(\Gamma; \mathbb{Z}) \cong \mathbb{Z}$, but there are uncountably many distinct actions of Γ on the circle, even up to semiconjugacy. However, the Euler class is a *bounded* cocycle in the sense of Gromov, and the second *bounded cohomology* of a surface group is infinite dimensional. Ghys showed, remarkably, that this pullback of the Euler class in bounded cohomology determines the action up to semiconjugay. The rigidity statement of Matsumoto quoted above is, in fact, a consequence of a stronger theorem, in which he shows that there is a unique, maximal (meaning pairing maximally with the fundamental class of the surface) bounded second cohomology class, corresponding to the semiconjugacy class of a Fuchsian representation.

Rigidity of geometric actions. The Fuchsian surface groups acting on the circle in Matsumoto's rigidity theorem are lattices in $\mathrm{PSL}(2, \mathbb{R})$, the group of orientation-preserving isometries of $\mathbb{H}^2$. With this in mind, we make the following definition:

Definition 2.2. An action of a group Γ on a manifold X is *geometric* if it is obtained by the embedding of Γ as a cocompact lattice in a connected Lie group G acting transitively on X.

This definition is modeled after the idea of a geometry from Klein's *Erlangen program*: a connected Lie group acting transitively on a space with compact point stabilizers. When X is the circle, we can easily list all geometric actions of groups. The connected Lie groups acting transitively on S^1 are the rotation group SO(2), the projective linear group $\mathrm{PSL}(2, \mathbb{R})$ acting on $\mathbb{R}P^1 = S^1$, and finite cyclic covers (i.e., central extensions by finite cyclic groups) of $\mathrm{PSL}(2, \mathbb{R})$, which act naturally on finite covers of S^1—conveniently, also topological circles (see [15]). Any cocompact lattice in one of these groups is, up to finite index, the fundamental group of a surface of genus g for some $g \geq 2$, and these are all obtained by lifting Fuchsian surface groups to a finite cyclic cover. Provided that the degree of the cover divides the Euler characteristic of the surface, such a lift exists and gives a lattice in the corresponding cyclic extension of $\mathrm{PSL}(2, \mathbb{R})$.

Matsumoto's theorem gives global rigidity for geometric examples where the ambient Lie group is $\mathrm{PSL}(2, \mathbb{R})$. With Maxime Wolff, we showed this for all geometric surface groups:

Theorem 2.3 (Rigid $\Leftrightarrow$ geometric on the circle [24,29]). An action of a surface group on S^1 is globally rigid if and only if it is geometric.

The direction *geometric implies rigid* consists in a careful study of geometric representations using the Poincaré rotation number and techniques of Calegari and Walker; a detailed expository account is given in [25] and an alternative proof given later by Matsumoto, using a Markov partition and Maskit combination theorem style of argument, in [32]. The statement that *rigid implies geometric* is much more difficult: handed an action of $\pi_1(\Sigma)$ on the circle, with no knowledge about it except that you cannot deform it, you need to reconstruct the ambient Lie group and lattice surface group.

The proof in [29] is rather technical, but, under a major simplifying assumption one can produce a much easier proof in which the strategy of "reconstruction" is evident. This is carried out in the relative short paper [27]. The simplifying assumption eliminates the possibility that $\rho(\pi_1(\Sigma))$ lies in one of the nontrivial covers of $\mathrm{PSL}(2, \mathbb{R})$. As a consequence of this assumption, one can deform the representation so that some simple closed curve a on the surface has its action $\rho(a)$ conjugate to the boundary action of a hyperbolic isometry of $\mathrm{PSL}(2, \mathbb{R})$. This means that the action of $\rho(a)$ on the circle has exactly two fixed points, one attracting and one repelling.

Having found one such simple closed curve, we find many, and then start the reconstruction: we show that the arrangement of attracting and repelling points of the action of these simple closed curves on the circle agrees with the intersection pattern of lifts of geodesic representatives of these curves on the surface. Under a Fuchsian representation, the axes of translation of elements are exactly such lifts of curves, and this allows us to build the desired semiconjugacy to the Fuchsian class. Figure 1 gives a representative visual: the geodesics shown there are all represented by simple closed curves; it is the cyclic order of their endpoints that we recover.

Open questions. For any real linear algebraic group G and finitely presented group $\Gamma = \langle S \mid R \rangle$, the space of representations $\mathrm{Hom}(\Gamma, G)$ is a real algebraic variety, a subspace of $G^{|S|}$ cut out by the finitely many relations from R. Thus $\mathrm{Hom}(\Gamma, G)$ has finitely many connected components. Goldman [17, THEOREM A] gives a precise count in the case where Γ is the fundamental group of a closed surface, and G the k-fold covering group of $\mathrm{PSL}(2, \mathbb{R})$. Geometric representations exist precisely when k divides $2g - 2$; if k is larger than $2g - 2$, then there is only a single connected component: every representation is deformable to the trivial representation. Using this, one can show that for a given genus g, *up to deformation, there are only finitely many representations of* $\pi_1(\Sigma_g)$ *into a Lie subgroup of* S^1. By contrast, we do not know:

Question 2.4. Does the space $\mathrm{Hom}(\pi_1(\Sigma_g), \mathrm{Homeo}_+(S^1))$ have finitely many connected components?

Question 2.5. Can every action of a surface group on the circle be deformed into one with image in a Lie group?

In fact, the situation is even messier than this. "Deformation" suggests movement along a *path* of representations, while my definition of rigidity referred to homogeneity of *connected components*. Alas, we do not know if path components and connected components of $\mathrm{Hom}(\pi_1(\Sigma_g), \mathrm{Homeo}_+(S^1))$ agree.

At the time of framing Definition 2.2, I was very enthusiastic about pursuing the theme "rigid $\leftrightarrow$ geometric" to see to what extent it might play out in settings beyond group actions on the circle. In retrospect, that seems too optimistic: geometric examples are excellent candidates for rigidity, but I do not think they are precisely all the rigid examples of actions in a general setting. This is not to deter anyone in their attempt to prove such a theorem—perhaps for the right modification of the definition of geometric (maximal dimension Lie group might be a good start), and the right definition of rigid, such a statement is true.

Our next topic should provide evidence for continued optimism, as we look at the first obvious generalization of surface groups acting on the circle—the boundary actions of fundamental groups of hyperbolic manifolds of higher dimension. In fact, one does not need $\mathbb{H}^n$ here (and so we will bid a temporary farewell to lattices) but only a Riemannian metric of negative curvature.

3. MANIFOLD GROUPS ACTING ON BOUNDARY SPHERES

We described in the introduction how to compactify certain spaces by equivalence classes of geodesic rays. When the space in question is the universal cover of a closed hyperbolic or negatively curved surface, Matusmoto's Theorem 2.1 says that the induced boundary action of the fundamental group is rigid. While his proof does not apply for manifolds of higher dimension (there is no bounded Euler class and no cyclic order of points at infinity for groups acting on higher dimensional spheres), Bowden and I recently proved an analogous local result in generality:

Theorem 3.1 ([6]). Let M be a compact, orientable n-manifold with negative curvature, and $\rho_0 : \pi_1(M) \to \mathrm{Homeo}(S^{n-1})$ the boundary action. There exists a neighborhood U of ρ_0 in $\mathrm{Hom}(\pi_1(M), \mathrm{Homeo}(S^{n-1}))$ consisting of representations which are *topological factors* of ρ_0 in the sense of classical dynamics: for each representation $\rho \in U$, there is a continuous, surjective map $h : S^{n-1} \to S^{n-1}$ satisfying $h \circ \rho(\gamma) = \rho_0(\gamma) \circ h$ for all $\gamma \in \pi_1(M)$.

The key idea of the proof is to encode the dynamics of the action ρ_0 and a perturbation ρ in *foliated spaces*, and promote *topologically stable properties* (such as transversality) to the desired dynamical stability. We describe a few details of the strategy now.

From group actions to foliated spaces. Given an arbitrary manifold M and an action ρ of $\pi_1(M)$ by homeomorphisms on a space F, the associated *foliated F-bundle over M* is the quotient of $\tilde{M} \times F$ by the diagonal action of $\pi_1(M)$ via deck transformations on $\tilde{M}$ and via ρ on F. This quotient space is an F-bundle over M, and since the action is diagonal, the "horizontal foliation" of $\tilde{M} \times F$ by leaves of the form $\tilde{M} \times \{*\}$ descends to a topological foliation on this bundle, topologically transverse to the fibers. If M and F have smooth structures (which we will always take to be the case) and the action is by diffeomorphisms, then the result is a smooth bundle with smooth foliation, transverse to the fibers. The idea to

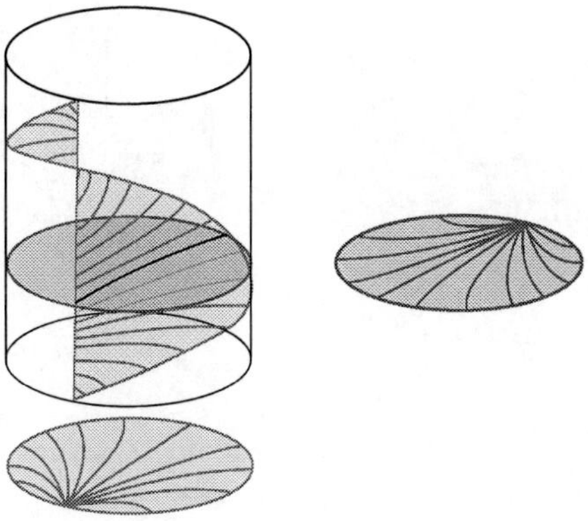

FIGURE 2
$UT(\mathbb{H}^2)$ is trivialized as $\mathbb{H}^2 \times S^1$, where the point on S^1 (height) is given by the endpoint at infinity of a geodesic ray with given tangent vector. Leaves of the weak-stable foliation are horizontal, weak-unstable leaves intersect them transversely along geodesics.

encode actions or representations in foliated spaces is not new—for instance, as one example of a use quite close to our theme, Goldman's 1980 thesis [16] presents the idea of a geometric structure as a section of a foliated bundle (following Kulkarni, Sullivan, and Thurston), using this perspective to understand representations of surface groups into $\mathrm{PSL}(2,\mathbb{R})$.

We are interested in the special case where M and ρ are as in Theorem 3.1. When $\rho = \rho_0$ is the action on the boundary at infinity, the associated foliated sphere bundle over M is isomorphic to the unit tangent bundle of M. This gives additional structure which we exploit in the proof.

Foliations on the unit tangent bundle. The unit tangent bundle $UT(M)$ of a negatively curved manifold M has a natural foliation $\mathcal{F}$ transverse to its sphere fibers. We can describe this by looking at the unit tangent bundle of the universal cover $\tilde{M}$ and using the boundary at infinity. For each point $z \in \partial_\infty \tilde{M}$, let $L_z \subset UT\tilde{M}$ be the set of all unit tangent vectors to geodesic rays that represent z. The leaf L_z is homeomorphic to $\tilde{M}$ (there is exactly one tangent vector in a given direction at each point), so the sets L_z partition $UT\tilde{M}$ into n-dimensional hyperplanes. Following classical work of Anosov, these sets are actually smooth, embedded submanifolds. The action of $\pi_1(M)$ on $UT\tilde{M}$ sends leaves to leaves, so this foliation descends to one on $UT(M)$, and the bundle isomorphism between the foliated sphere bundle associated to the boundary action can be chosen to naturally identify this foliation on $UT(M)$, called the *weak-stable foliation* with the "horizontal" foliation on the foliated bundle. See Figure 2 left for an illustration when $\tilde{M} = \mathbb{H}^2$. Of course, one could equally well make the opposite choice of considering the set of tangent vectors v that *emmanate* from a common point at infinity. This gives the *weak-unstable* foliation, which is also transverse to the fibers, and transverse to leaves of the weak-stable foliation.

Maps between bundles. The proof of Theorem 3.1 starts with the construction of a particularly well behaved map between the foliated bundles associated with nearby actions, as illustrarted in Figure 3. Given a perturbation ρ of the standard boundary action ρ_0, one

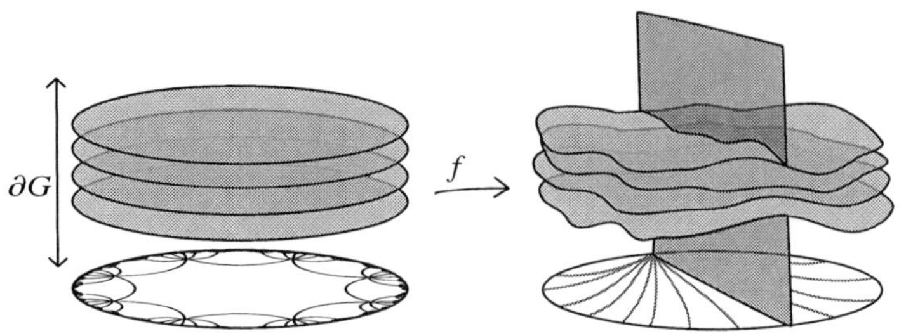

FIGURE 3
An equivariant map locally close to the identity on $\tilde{M} \times \partial G$ captures a perturbation of the action of $G = \pi_1(M)$ on ∂G. The images of horizontal leaves $\tilde{M} \times \{x\}$ intersect the leaves of the stable foliation of geodesic flow (in red) along *quasigeodesics* allowing us to use large-scale metric stability to prove dynamical stability. For general groups, one needs a more complicated space and a substitute for the stable foliation.

builds, by hand, a $\pi_1(M)$-equivariant map $\tilde{M} \times S^{n-1} \to UT\tilde{M}$ where on the left we have an action via ρ on the sphere factor, and on the right the standard action on the unit tangent bundle. Although ρ does not act by diffeomorphisms (only homeomorphisms), by sacrificing injectivity we can design this equivariant map to send each horizontal leaf $\tilde{M} \times \{p\}$ to a C^1-embedded submanifold of $UT\tilde{M}$ that stays C^1-close to leaves of the weak-stable foliation over large compact sets (i.e., its tangent distribution is uniformly close to the weak-stable distribution). This means that the images of horizontal leaves remain *transverse* to the leaves of the weak-*unstable* foliation on $UT\tilde{M}$. We show that any two such leaves that intersect do so along a path that is uniformly close to a geodesic in $UT\tilde{M}$.

Now the boundary at infinity makes another appearance. Using some deep results on the dynamics of the action of $\pi_1(M)$ on its boundary (including the remarkable *convergence group property*, which we will, alas, not have space to discuss here), we show that the near-geodesics on the image of each "horizontal" leaf $\tilde{M} \times \{p\}$ cut out by the weak-stable leaves all share a common endpoint at infinity, depending only on the parameter p. This association of such a point p to this common endpoint at infinity gives a map $S^{n-1} \to S^{n-1}$. And this map, it turns out, gives the desired semiconjugacy of the actions.

4. COARSE HYPERBOLICITY: FROM SPACES TO GROUPS

Following Gromov, a geodesic metric space X is called *δ-hyperbolic* (for some $\delta \geq 0$) if every geodesic triangle T in X has the property that each side of T lies in the metric δ-neighborhood of the union of the other two sides. For example, any tree is 0-hyperbolic, and it is a pleasant exercise in hyperbolic trigonometry to show that $\mathbb{H}^n$, with its metric of constant curvature -1, is δ-hyperbolic for the constant $\delta = \ln(1 + \sqrt{2})$.

This definition also works for groups: a finitely generated group is *hyperbolic* if its Cayley graph is δ-hyperbolic for some δ. While the constant δ depends on the generating set, the notion *hyperbolic for some δ* does not. Indeed, "δ-hyperbolic for some δ" is a metric invariant up to *quasiisometry*, the relation shared between Cayley graphs with different generating sets. This concept is the center of Gromov's highly influential essay *Hyperbolic groups*, and the setting in which one can compactify a space by a boundary at infinity with the method given in the introduction—although Gromov attributes this idea as "essentially due to Mostow and Margulis" due to its appearance in the theorems we quoted earlier [18, 0.3B].

A finitely generated group acts naturally on its Cayley graph by auotmorphisms, hence by isometries when edges are taken to have unit length. Thus, it induces an action by homeomorphisms on the boundary. One can therefore ask:

Question 4.1. Let Γ be a finitely generated hyperbolic group. Is the action of Γ on its boundary (locally) rigid, up to semiconjugacy?

Rather than repeatedly writing "locally rigid up to semiconjugacy," we borrow terminology from classical dynamics, traditionally applied to actions of $\mathbb{Z}$, but just as valid for group actions. An action ρ_0 of Γ on a space X is *topologically stable* if, for any sufficiently nearby action ρ, there is a surjective, continuous map h satisfying $h \circ \rho(\gamma) = \rho_0(\gamma) \circ h$ for all $\gamma \in \Gamma$. Typically, one requires h to depend continuously on the perturbation, being close to the identity if ρ is sufficiently close to ρ_0. Thus, the statement of Theorem 3.1 is simply an assertion of topological stability for manifold fundamental groups.

Boundaries of groups. While the examples of boundaries at infinity we have looked at so far have been spheres, the topology of the boundary of a group is typically quite complicated, both locally and globally (see [20]). In some cases, the topology is so complicated that a positive answer to Question 4.1 follows from the structure of the boundary itself. Kapovich and Kleiner [22] have examples of groups where the only automorphisms of their boundary come from left multiplication. From this one can easily deduce rigidity: the action of the group on its boundary is an isolated point in $\mathrm{Hom}(\Gamma, \mathrm{Homeo}(\partial_\infty \Gamma))$. For free groups, whose boundary is a Cantor set, one can also use the "ping-pong" dynamics of the action to prove local rigidity using a relatively hands-on argument.

At the other end of the spectrum are groups with sphere boundary. In contrast to Kapovich–Kleiner's boundaries, homeomorphisms of the sphere are very easy to perturb, each having an infinite-dimensional family of deformations, making Question 4.1 particularly interesting. It is in this context that Manning and I recently proved an analog to Theorem 3.1:

Theorem 4.2 (Rigidity for sphere boundary actions [26]). For any Gromov hyperbolic group Γ with sphere boundary, the natural action of Γ on $\partial_\infty \Gamma$ is topologically stable.

In proving this, we remove all the differential topological machinery (such as transversality and the regularity of weak-stable foliations) from the proof of Theorem 3.1, replacing it with *coarse metric* machinery. The fundamental starting point is the stability of *quasigeodesics*.

Quasigeodesic stability. A *quasigeodesic* is a map γ from $\mathbb{R}$ into a metric space (X, d_X) such that, for some constants K, C, the bounds

$$\frac{1}{K}d_X(\gamma(t), \gamma(s)) - C \leq |t-s| \leq K d_X(\gamma(t), \gamma(s)) + C$$

hold for all $t, s \in \mathbb{R}$. More generally, a (K, C) *quasiisometric embedding* of a metric space (Y, d_Y) into (X, d_X) is a map $\gamma : Y \to X$ that satisfies the above bounds for all points t, s in Y, with $|t-s|$ replaced by the distance $d_Y(t, s)$. Such a map is called a *quasiisometry* if it has the additional property of being *coarsely surjective*, meaning that each point of Y lies a uniformly bounded distance away from some point in the image.

The idea of quasigeodesic stability comes from work of H. M. Morse in the early 1920s. In [33], he considers the following question: suppose we take a closed surface Σ of genus $g \geq 2$ equipped with an arbitrary Riemannian metric, and a homeomorphism $f : \Sigma \to \Sigma_{\text{hyp}}$ identifying it with some fixed hyperbolic genus g surface Σ_{hyp}. What do length-minimizing geodesic paths on $\widetilde{\Sigma}$ look like under the lifted map $\tilde{f} : \widetilde{\Sigma} \to \widetilde{\Sigma}_{\text{hyp}}$? Do they share properties with genuine hyperbolic geodesics, such as tending in each direction to a unique point on the boundary of the disc in the Poincaré model?

Morse's map $\tilde{f}$ is an example of a quasiisometry, and the image of a length-minimizing geodesic under $\tilde{f}$ is a quasigeodesic. In answering the question above, Morse proves what is now known as the *Morse lemma* on quasigeodesic stability. In its modern, more general form, this lemma states:

Lemma 4.3 (Morse lemma). There exists a constant $B = B(K, C, \delta)$ such that for any δ-hyperbolic metric space X, every (K, C) quasigeodesic $\gamma : \mathbb{R} \to X$ lies in the B-neighborhood of a unique geodesic, and hence every quasigeodesic ray defines a unique point on $\partial_\infty(X)$.

In addition, quasigeodesics satisfy a *local-to-global* principle: a map which is a (K, C) quasigeodesic embedding when restricted to all sufficiently long segments is, in fact, globally a quasigeodesic. This allows us to pursue the broad strategy used in proving Theorem 3.1 in this coarse setting. We first translate a perturbation of an action into a nice map between foliated metric spaces, then show that images of leaves intersect leaves in (a preferred section of) the target foliated space along quasigeodesics. Of course, having no smooth manifold or universal cover on hand makes the strategy nontrivial to even set up, and much harder to execute!

Related results and open questions. Much existing work on the dynamics of groups acting on their boundaries relies on *local expansivity*: for each point of the boundary, there is an element of the group that contracts this neighborhood a uniform amount (thus, one has uniform *expansion* under the inverse, hence the name).[3] Sullivan [37] used this property to demonstrate a structural stability result for Kleinian groups acting on the boundary sphere of $\mathbb{H}^3$: a C^1-small perturbation of such an action has an invariant set on which the action is conjugate

3 There are many related definitions of expansivity, here I am roughly following Sullivan.

to the original action of the group on its limit set. This was recently improved and generalized by Kapovich–Kim–Lee [21] to prove structural stability for a much broader class of expansive actions, including boundary actions of hyperbolic groups, under perturbations which preserve a generalized expansivity property. Lipschitz-small perturbations are one example to which their theory applies, however, general continuous perturbations do not preserve local expansivity and so are not covered by this strategy. Sullivan's method involves a dynamical "coding" of points by sequences of group elements, suggesting a connection to classical symbolic dynamics. This is no coincidence, and we now know a number of rigidity results in this direction (see [7]). However, the following general problem remains open:

Question 4.4. Is the action of every hyperbolic group Γ on its boundary topologically stable? What techniques apply to intermediate cases between the boundary sphere case and Kapovich–Kleiner's rigid examples? Which examples exhibit a rigidity property stronger than topological stability, and what phenomena are responsible for this behavior?

As mentioned before, group boundaries can be topologically complex. One way the sphere plays an essential role in the proof of Theorem 4.2 is that its homeomorphism group is locally contractible, and there is no obvious substitute for this property in other settings. An interesting first case to attack might be the Menger curve, this being the boundary of a *random* group in the standard density model.

We note also that it is unknown which topological spaces occur as boundaries of groups. Thus, an approach to Question 4.4 either has to be restricted to families of understood examples, or avoid explicitly describing the boundary altogether.

5. AUTOMORPHISM GROUPS ACTING AT INFINITY

An automorphism of a finitely generated group Γ defines a quasiisometry of the Cayley graph of Γ, and therefore extends to a homeomorphism of $\partial_\infty \Gamma$. The inner automorphism defined by conjugation by γ is a bounded distance (with bound given by the word length of γ) from the map induced by left-multiplication by γ, so $\mathrm{Inn}(\Gamma) \cong \Gamma$ agrees with the actions we have already discussed and considering the action of $\mathrm{Aut}(\Gamma)$ is a natural next step.

Enlarging $\Gamma \cong \mathrm{Inn}(\Gamma)$ to $\mathrm{Aut}(\Gamma)$ is most interesting when the outer automorphism group of Γ is large. Many hyperbolic groups have trivial or finite outer automorphism group, the most basic case where $\mathrm{Out}(\Gamma)$ is infinite is when Γ is the fundamental group of a surface.[4] This case is particularly interesting to low dimensional topologists due to its relationship with mapping class groups.

Mapping class groups. A *mapping class* is an equivalence class of homeomorphism up to isotopy. Let $\mathrm{MCG}_\pm(\Sigma) := \pi_0(\mathrm{Homeo}(\Sigma))$ denote the group of all mapping classes of

---

4 Following the work of Paulin and Rips, Levitt showed that one-ended hyperbolic groups have infinite outer automorphism group if and only if they split, as an HNN extension or an amalgam of groups with finite center, over a virtually cyclic subgroup with infinite center, so in some sense resemble the surface groups we will discuss.

a surface Σ. One may also consider the subgroup of homeomorphisms fixing a basepoint, in which case $\mathrm{MCG}_{\pm}(\Sigma, x)$ denotes the group of homeomorphisms fixing x up to isotopy preserving x. The subscript $\pm$ here indicates that we consider both orientation preserving and reversing homeomorphisms, the mapping class groups denoted $\mathrm{MCG}(\Sigma, x)$ and $\mathrm{MCG}(\Sigma)$, respectively, are the index two subgroups of orientation-preserving elements.

For a surface Σ (which we continue to assume is of genus at least two, so that its fundamental group is hyperbolic), the *Dehn–Nielsen–Baer theorem* is the statement that the exact sequence

$$\mathrm{Inn}\big(\pi_1(\Sigma)\big) \to \mathrm{Aut}\big(\pi_1(\Sigma)\big) \to \mathrm{Out}\big(\pi_1(\Sigma)\big)$$

is isomorphic, term by term, to the Birman exact sequence

$$\pi_1(\Sigma) \to \mathrm{MCG}_{\pm}(\Sigma, x) \to \mathrm{MCG}_{\pm}(\Sigma).$$

This isomorphism has a particularly nice geometric description, using boundary actions and the identification of $\widetilde{\Sigma}$ with $\mathbb{H}^2$ coming from a choice of hyperbolic structure on Σ. Choose a lift $\tilde{x}$ of the point x on Σ, so each $f \in \mathrm{Homeo}(\Sigma)$ fixing x has a unique lift $\tilde{f}$ to $\mathbb{H}^2$ that fixes $\tilde{x}$. Since Σ is compact and f is continuous, this lift is a quasiisometry of $\mathbb{H}^2$ so induces a continuous map on the boundary circle. If f and g represent the same element of $\mathrm{MCG}_{\pm}(\Sigma, x)$, lifting an isotopy preserving $\tilde{x}$ will move all points on $\widetilde{\Sigma}$ a uniformly bounded distance, so will not change boundary homeomorphism. Thus, considering the action of lifts on the boundary gives a well-defined map from $\mathrm{MCG}_{\pm}(\Sigma, x)$ to homeomorphisms of S^1, agreeing with the action of $\mathrm{Aut}(\pi_1(\Sigma))$ under the identification above.

In his problem list on mapping class groups **[8, QUESTION 6.2]**, Farb asks whether these actions are rigid:

Question 5.1 (Farb). Is every faithful action of $\mathrm{MCG}(\Sigma, x)$ on S^1 by homeomorphisms necessarily semiconjugate to the standard action on the boundary?

Question 5.1 asks for a much stronger form of rigidity than exhibited by the action of the fundamental group $\pi_1(\Sigma)$. There are many distinct semiconjugacy classes of faithful actions of $\pi_1(\Sigma)$ on the circle; in fact, one can even take these to have image in $\mathrm{PSL}(2, \mathbb{R})$ (see, e.g., **[23]** for a detailed discussion and references). Despite this, Farb's question is actually quite reasonable because of *torsion*. The mapping class group of a surface group contains many finite-order elements, and any action of a finite cyclic group on the circle is conjugate to an action by rotations. Thus, the presence of torsion is suggestive, though no guarantee, of rigidity.

Unexpectedly, something even stronger than what Farbs asks for is true—the hypothesis "faithful" is not needed, but only *nontrivial*.

Theorem 5.2 (Mapping class rigidity **[28]**). For any surface Σ of genus at least 2, every nontrivial action of $\mathrm{MCG}(\Sigma, x)$ on the circle is (up to choice of orientation) semiconjugate to the standard boundary action.

The proof, as expected, makes use of torsion, but in a perhaps unexpected way: we study the action of *orbifold fundamental groups* that contain $\pi_1(\Sigma)$. The definition of Euler number for surface group actions on S^1 that we introduced in Section 2 can be extended to actions of orbifold fundamental groups in a natural way so that it is multiplicative under covers, like Euler characteristic. We use geometric and topological arguments (relying on torsion) to show that nontrivial actions of $\mathrm{MCG}(\Sigma, x)$ have a maximal Euler number, and use Matsumoto's theorem to prove rigidity.

Since torsion plays a critical role in this argument, we do not know if a similar result holds for all finite-index subgroups of $\mathrm{MCG}(\Sigma, x)$. It would be interesting to see another approach to mapping class rigidity, relying more on group structure and less on the geometry and topology of the surface, perhaps towards a general theory for rigidity of actions of automorphism groups of other hyperbolic groups.

6. ANOSOV FLOWS ON 3-MANIFOLDS

We conclude this survey by describing a different method of producing actions at infinity, this one coming from the orbit space of an Anosov flows on a 3-manifold. A flow ϕ_t on a Riemannian manifold M is called *Anosov* if there is a ϕ_t-invariant global splitting of the tangent bundle TM as a direct sum

$$TM = X \oplus E^{ss} \oplus E^{uu},$$

where X is the direction of the flow, E^{ss} is the "stable distribution" consisting of vectors whose length is uniformly contracted by the flow, and E^{uu} is the "unstable distribution" consisting of vectors that are uniformly expanded (or, more precisely, uniformly contracted when the direction of the flow is reversed). *Contracted* has a specific meaning: there are positive constants c and $\lambda > 0$ such that the length of the pushforward of a tangent vector $(\phi_t)_*(v)$ under the time t map of the flow is bounded above by $ce^{-\lambda t}||v||$ for all $t > 0$. On a compact manifold, one may always find a Riemannian metric on M *adapted* to the flow for which one may take the multiplicative constant $c = 1$.

It is a classical result that the two distributions $E^s := X \oplus E^{ss}$ and $E^u := X \oplus E^{uu}$ are integrable, meaning they are everywhere tangent to a foliation. We will restrict our attention in this section to 3-manifolds, thus E^{ss}, E^{uu}, and X are all one-dimensional, and the foliations $\mathcal{F}^s$ and $\mathcal{F}^u$ tangent to $X \oplus E^{ss}$ and $X \oplus E^{uu}$ are 2-dimensional transverse foliations that meet along the orbits of the flow.

The two most basic examples of Anosov flows in dimension 3, and indeed the starting points for the construction of all other known examples, are suspensions of linear maps on tori and geodesic flows on surfaces.

Example 6.1 (Linear Anosov maps on tori). Consider a transformation $A \in \mathrm{SL}(2, \mathbb{Z})$ with $\mathrm{trace}(A) > 2$, or equivalently, two distinct, real eigenvalues of norm not equal to 1. Since A preserves the integer lattice in $\mathbb{R}^2$, it descends to a self-diffeomorphism $\bar{A}$ of the square torus $T^2 := \mathbb{R}^2/\mathbb{Z}^2$. Let M be the *mapping torus* of $\bar{A}$, the quotient of $T^2 \times \mathbb{R}$ under the relation $(x, 0) \sim (\bar{A}^n(x), n)$ for $n \in \mathbb{Z}$. The straight line flow $(x, s) \mapsto (x, s + t)$ on $T^2 \times \mathbb{R}$ descends to a flow ϕ_t on M. Each eigendirection of A defines a 1-dimensional $\bar{A}$-invariant line field on M, one of which is uniformly contracted by ϕ_t flow and one uniformly expanded, giving the desired Anosov property.

The reason Anosov flows are such interesting examples in dynamics is that they simultaneously exhibit global stability and local chaos. *Local chaos* means that nearby points have vastly different trajectories. This is already apparent in Example 6.1, for instance, the origin on $T^2 = \mathbb{R}^2/\mathbb{Z}^2$ is fixed by $\bar{A}$ so is a periodic orbit of ϕ_t, but arbitrarily nearby points have infinite trajectories that intersect the torus $T^2 \times \{0\}$ along a dense set. Moreover, this interspersing of periodic and infinite trajectories happens everywhere: since the induced action of elements of $\mathrm{SL}(2, \mathbb{Z})$ preserve the finite set of points of the form $\{(p/q, r/q) : 0 \leq p, r < q\}$ (for any fixed q) on the fundamental domain $[0, 1]^2$ for $\mathbb{R}^2/\mathbb{Z}^2$, any such point eventually returns to itself under iterates of $\bar{A}^n$, giving a closed orbit for ϕ_t.

By contrast, the global picture of the flow is overall stable. Replacing A by a nearby nonlinear diffeomorphism of T^2 and doing the same construction gives a new flow on the same topological space, which turns out always to be *conjugate* to the flow just discussed. In fact, here one can even replace A by any map with the same action on homology of T^2 [13, 19].

The second building block for Anosov flows is geodesic flow in negative curvature. The simplest such examples come from hyperbolic surfaces.

Example 6.2 (Geodesic flow). Let Σ be a surface equipped with a metric of constant curvature -1, and let $M = UT(\Sigma)$ be its unit tangent bundle. *Geodesic flow* is the map that, at time t, sends a unit tangent vector $v \in UT(\Sigma)$ to the vector tangent to the line $\{\exp(sv) : s \in \mathbb{R}\}$ at the point $\exp(tv)$. One can compute explicitly using the identification of $UT(\Sigma)$ with $\mathrm{PSL}(2, \mathbb{R})$ that this flow is Anosov. The weak-stable and weak-unstable foliations from the flow (lifted to $\widetilde{\Sigma}$) are exactly those shown in Figure 2. Leaves of $\mathcal{F}^s$ consist of tangent vectors to geodesics with a common forward endpoint, and $\mathcal{F}^u$ those with a common negative endpoint at infinity.

Classifying flows. Having given two families of examples, we now embark on the ambitious program to understand all Anosov flows in dimension 3. For this, one must answer questions of:

- *Existence.* Which 3-manifolds support an Anosov flow? What techniques can be used to construct families of examples?
- *Abundance.* If M^3 supports one Anosov flow, can it support many dynamically distinct ones?
- *Classification.* What invariants can be used to distinguish distinct flows?

The existence problem has a long history, but is still not completely solved. The work of Palmeira and Verjovsky from the 1970s shows that if M supports an Anosov flow, then $\tilde{M}$ is homeomorphic to $\mathbb{R}^3$, thus M must be irreducible. An irreducible 3-manifold admits a decomposition along tori into geometric pieces, so the next question to ask is which kinds of pieces M may have. Margulis [2, APPENDIX] showed that $\pi_1(M)$ is large in the sense that it has *exponential growth*: for any fixed generating set, the number of reduced words of length r in the group grows exponentially in r. This rules out, for instance, geometric manifolds with a Euclidean structure. There are a number of constructions—Dehn surgery and other gluing techniques—that produce examples on geometric manifolds with exponential growth, or manifolds with a nontrivial torus decomposition, and a few other known special constraints, but we still lack a complete picture.

The approach to classification that I wish to discuss here is the purely topological one, namely, classifying flows *up to orbit equivalence*. Two flows ϕ_t and ψ_t on a manifold M are called *orbit equivalent* if there is a self-homeomorphism of M taking orbits of ϕ_t to orbits of ψ_t, in other words, the 1-dimensional foliations on M by flowlines are homeomorphic.

The remainder of this survey is devoted to describing a very recent rigidity result (Theorem 6.3 and its generalization) saying that Anosov flows can be distinguished up to this equivalence by the set of homotopy classes of loops represented by closed orbits. The proof of this result comes, again, from looking to a boundary at infinity. This time, the boundary is that of the *orbit space* of the flow.

$\mathbb{R}$-covered flows and orbit spaces. A first topological invariant to distinguish flows comes from the global transverse structure $\mathcal{F}^s$. Lifting $\mathcal{F}^s$ to a foliation $\tilde{\mathcal{F}}^s$ on $\tilde{M}$ gives a foliation of $\mathbb{R}^3$ by planes. Collapsing each plane to a point produces a 1-dimensional manifold called the *leaf space* of $\tilde{\mathcal{F}}^s$. The leaf space is either non-Hausdorff, or homeomorphic to $\mathbb{R}$; in the latter case we say the flow is $\mathbb{R}$*-covered*. This terminology does not privilege $\mathcal{F}^s$, by [3, 10] the leaf space of $\tilde{\mathcal{F}}^s$ is Hausdorff if and only if that of the lifted unstable foliation $\tilde{\mathcal{F}}^u$ is as well. These two cases ($\mathbb{R}$ covered and non-Hausdorff leaf space) lend themselves to different techniques for classification. We discuss the $\mathbb{R}$ covered case first.

While being $\mathbb{R}$-covered may seem like a restrictive hypothesis, there are, in fact, many diverse examples. Both Examples 6.1 and 6.2 are $\mathbb{R}$-covered, and many more can be produced by modifying these manifolds using Dehn surgery. A given manifold may support *arbitrarily many* inequivalent $\mathbb{R}$-covered flows even if geometric; Bowden and Mann [6] give constructions of such on closed hyperbolic manifolds.

Our understanding of $\mathbb{R}$-covered flows is due largely to the work of Barbot and Fenley, starting with the work in [3, 10]. They consider the *orbit space* $\mathcal{O}$ of the flow, the quotient of $\tilde{M}$ obtained by collapsing each flowline to a point. Any $\mathbb{R}$-covered flow that is *not* obtained from a hyperbolic toral automorphism such as in Example 6.1 has the remarkable property that its orbit space is homeomorphic to an infinite diagonal strip in the plane, as shown on the left of Figure 4, with $\tilde{\mathcal{F}}^s$ and $\tilde{\mathcal{F}}^u$ being the vertical and horizontal foliations. Such flows are called *skew*. Since the topology of the foliations on this orbit space does not distinguish flows, any classification theorem must rest on a new algebraic or topological invariant. In recent work, Barthelmé and I show that the *spectrum of periodic orbits* (the free homotopy classes of loops represented by periodic orbits of the flow) does the job:

Theorem 6.3 (Spectral rigidity for flows [4]). Suppose ϕ_t and ψ_t are $\mathbb{R}$-covered Anosov flows on a compact 3-manifold M. The conjugacy classes in $\pi_1(M)$ represented by the free

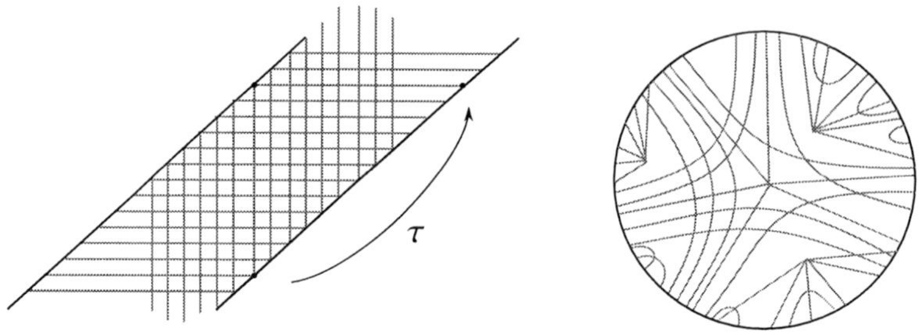

FIGURE 4
The orbit space of a skew flow (left) and a schematic of that of a pseudo-Anosov flow (right). Nonseparated stable and unstable leaves "meeting at infinity" define a shift τ commuting with the action of $\pi_1(M)$ on the skew picture.

homotopy classes of closed orbits for ϕ_t and ψ_t agree if and only if the flows are orbit equivalent via a map isotopic to the identity.

The action on $\mathcal{O}$ at infinity. The interesting case in Theorem 6.3 is for skew flows, as the mapping torus of a linear Anosov map admits only the obvious suspension flow and its inverse. To solve the problem for skew flows, we look to the boundary at infinity—the compactification of $\mathcal{O}$ by the lines in the diagonal strip model. The action of $\pi_1(M)$ on $\tilde{M}$ descends to $\mathcal{O}$ and extends to an action by homeomorphisms on each boundary line, commuting with a translation of the line that comes from the structure of the lifted foliations. The dynamics of individual elements acting on the line at infinity are also well understood. Up to passing to the index-two subgroup of elements preserving orientation, it is an example of what we call a *hyperbolic-like action*.

Definition 6.4 ([4]). An action of a group G on the line is *hyperbolic-like* if it commutes with the translation $x \mapsto x + 1$, and every nontrivial element either acts freely, or has precisely two fixed points in each unit interval, one attracting and one repelling.

In the tradition of classical theorems of Hölder and Solodov, which promote information about the dynamics of individual homeomorphisms of the line to a global conclusion about the structure of a group action, we prove a general result on hyperbolic-like actions on the line.

Theorem 6.5 (Hyperbolic actions are determined by fixed spectra [4]). Given two faithful, minimal, hyperbolic-like actions of a group G on $\mathbb{R}$, if the sets of elements acting with fixed points for each action agree, then the two actions are conjugate by a homeomorphism.

The strategy of the proof for Theorem 6.5 is to recover the linear order of fixed points of elements (and hence reconstruct a dense subset of the line) from the algebraic data of the set of elements with fixed points. This is not far in spirit from the "reconstruction" strategies described in Section 2. Theorem 6.5 is really the heart of the proof of Theorem 6.3, what remains is to promote the conjugacy of actions at infinity to an honest orbit equivalence, a technique already used by Barbot.

Non-$\mathbb{R}$-covered flows and pseudo-Anosov flows. In the case where the leaf spaces of $\tilde{F}^s$ and $\tilde{F}^u$ are non-Hausdorff, one can leverage the topology of these foliations to get information about the flow, a perspective fruitfully exploited by Fenley in [11]. It turns out that the same family of techniques also applies to a strictly broader class of *pseudo-Anosov* flows—topological flows with expanding/contracting behavior as in the Anosov case, but where $\mathcal{F}^s$ and $\mathcal{F}^u$ are allowed to *branch* in a specified way along a discrete set of periodic orbits.

The orbit space of such a flow is a topological plane with two transverse, 1-dimensional, possibly singular foliations, as cartooned in Figure 4 (right). In [9], Fenley gives a natural construction of a compactification of the orbit space of any pseudo-Anosov flow by a boundary circle so that the compactified space is homeomorphic to a disk and the natural action of the fundamental group of the manifold by homeomorphisms of $\mathcal{O}$ extends to the boundary. In the work in preparation with Barthelmé and Frankel, we use this boundary circle and the induced action of $\pi_1(M)$ to prove spectral rigidity for all transitive, non-$\mathbb{R}$-covered Anosov and pseudo-Anosov flows on compact 3-manifolds. Combined with Theorem 6.3, this gives a full spectral rigidity result in the Anosov and pseudo-Anosov setting: provided the flow is transitive, if the conjugacy classes in $\pi_1(M)$ represented by the free homotopy classes of closed orbits for two flows ϕ_t and ψ_t agree, then the flows are orbit equivalent via a map isotopic to the identity.

Although this work gives one answer to the classification problem, many open questions remain, especially regarding existence and abundance. Of particular interest to me is the interplay between geometry of a manifold and topology of the leaf spaces of such flows, hyperbolic manifolds being a particularly interesting example. Which hyperbolic 3-manifolds admit Anosov flows? Does the complexity of the manifold bound the number of distinct flows it may admit? May a hyperbolic manifold admit infinitely many inequivalent Anosov flows?

ACKNOWLEDGMENTS

I thank Jason Manning and Thomas Barthelmé for their helpful comments.

FUNDING

K. Mann was partially supported by NSF grant DMS 1844516, and a Sloan fellowship.

REFERENCES

[1] S. Agard, Mostow rigidity on the line: a survey. In *Holomorphic functions and moduli, Vol. II (Berkeley, CA, 1986)*, pp. 1–12, Math. Sci. Res. Inst. Publ. 11, Springer, New York, 1988.

[2] D. V. Anosov and J. G. Sinaĭ, Certain smooth ergodic systems. *Uspekhi Mat. Nauk* **22** (1967), no. 5 (137), 107–172.

[3] T. Barbot, Caractérisation des flots d'Anosov en dimension 3 par leurs feuilletages faibles. *Ergodic Theory Dynam. Systems* **15** (1995), no. 2, 247–270.

[4] T. Barthelmé and K. Mann, Orbit equivalences of ℝ-covered Anosov flows and applications. 2021, arXiv:2012.11811.

[5] G. Besson, G. Courtois, and S. Gallot, Minimal entropy and Mostow's rigidity theorems. *Ergodic Theory Dynam. Systems* **16** (1996), no. 4, 623–649.

[6] J. Bowden and K. Mann, C^0 stability of boundary actions and inequivalent Anosov flows. 2019, arXiv:1909.02324.

[7] M. Coornaert and A. Papadopoulos, *Symbolic dynamics and hyperbolic groups*. Lecture Notes in Math. 1539, Springer, Berlin, 1993.

[8] B. Farb, Some problems on mapping class groups and moduli space. In *Problems on mapping class groups and related topics*, pp. 11–55, Proc. Sympos. Pure Math. 74, Amer. Math. Soc., Providence, RI, 2006.

[9] S. Fenley, Ideal boundaries of pseudo-Anosov flows and uniform convergence groups with connections and applications to large scale geometry. *Geom. Topol.* **16** (2012), no. 1, 1–110.

[10] S. R. Fenley, Anosov flows in 3-manifolds. *Ann. of Math. (2)* **139** (1994), no. 1, 79–115.

[11] S. R. Fenley, The structure of branching in Anosov flows of 3-manifolds. *Comment. Math. Helv.* **73** (1998), no. 2, 259–297.

[12] D. Fisher, Groups acting on manifolds: around the Zimmer program. In *Geometry, rigidity, and group actions*, pp. 72–157, Chicago Lectures in Math., Univ. Chicago Press, Chicago, IL, 2011.

[13] J. Franks, Anosov diffeomorphisms. In *Global analysis (Proc. Sympos. Pure Math., Vol. XIV, Berkeley, CA, 1968)*, pp. 61–93, Amer. Math. Soc., Providence, RI, 1970.

[14] E. Ghys, Groupes d'homéomorphismes du cercle et cohomologie bornée. In *The Lefschetz centennial conference, Part III (Mexico City, 1984)*, pp. 81–106, Contemp. Math. 58, Amer. Math. Soc., Providence, RI, 1987.

[15] E. Ghys, Groups acting on the circle. *Enseign. Math. (2)* **47** (2001), no. 3–4, 329–407.

[16] W. M. Goldman, *Discontinuous Groups and the Euler Class*. Ph.D. Thesis, ProQuest LLC, Ann Arbor, MI, University of California, Berkeley 1980.

[17] W. M. Goldman, Topological components of spaces of representations. *Invent. Math.* **93** (1988), no. 3, 557–607.

[18] M. Gromov, Hyperbolic groups. In *Essays in group theory*, pp. 75–263, Math. Sci. Res. Inst. Publ. 8, Springer, New York, 1987.

[19] M. Handel, Global shadowing of pseudo-Anosov homeomorphisms. *Ergodic Theory Dynam. Systems* **5** (1985), no. 3, 373–377.

[20] I. Kapovich and N. Benakli, Boundaries of hyperbolic groups. In *Combinatorial and geometric group theory (New York, 2000/Hoboken, NJ, 2001)*, pp. 39–93, Contemp. Math. 296, Amer. Math. Soc., Providence, RI, 2002.

[21] M. Kapovich, S. Kim, and J. Lee, Structural stability of meandering-hyperbolic group actions. 2019, arXiv:1904.06921.

[22] M. Kapovich and B. Kleiner, Hyperbolic groups with low-dimensional boundary. *Ann. Sci. Éc. Norm. Supér. (4)* **33** (2000), no. 5, 647–669.

[23] S-h. Kim, T. Koberda, and M. Mj, *Flexibility of group actions on the circle*. Lecture Notes in Math. 2231, Springer, Cham, 2019.

[24] K. Mann, Spaces of surface group representations. *Invent. Math.* **201** (2015), no. 2, 669–710.

[25] K. Mann, Rigidity and flexibility of group actions on the circle. In *Handbook of group actions. Vol. IV*, pp. 705–752, Adv. Lect. Math. (ALM) 41, Int. Press, Somerville, MA, 2018.

[26] K. Mann and J. Manning, Stability for hyperbolic groups acting on boundary spheres. 2021, arXiv:2104.01269.

[27] K. Mann and M. Wolff, A characterization of Fuchsian actions by topological rigidity. *Pacific J. Math.* **302** (2019), no. 1, 181–200.

[28] K. Mann and M. Wolff, Rigidity of mapping class group actions on S^1. *Geom. Topol.* **24** (2020), no. 3, 1211–1223.

[29] K. Mann and M. Wolff, Rigidity and geometricity for surface group actions on the circle. 2018, arXiv:1710.04902.

[30] G. A. Margulis, The isometry of closed manifolds of constant negative curvature with the same fundamental group. *Dokl. Akad. Nauk SSSR* **192** (1970), 736–737.

[31] S. Matsumoto, Some remarks on foliated S^1 bundles. *Invent. Math.* **90** (1987), no. 2, 343–358.

[32] S. Matsumoto, Basic partitions and combinations of group actions on the circle: a new approach to a theorem of Kathryn Mann. *Enseign. Math.* **62** (2016), no. 1–2, 15–47.

[33] H. M. Morse, A fundamental class of geodesics on any closed surface of genus greater than one. *Trans. Amer. Math. Soc.* **26** (1924), no. 1, 25–60.

[34] G. D. Mostow, On the rigidity of hyperbolic space forms under quasiconformal mappings. *Proc. Natl. Acad. Sci. USA* **57** (1967), 211–215.

[35] G. D. Mostow, Selberg's work on the arithmeticity of lattices and its ramifications. In *Number theory, trace formulas and discrete groups (Oslo, 1987)*, pp. 169–183, Academic Press, Boston, MA, 1989.

[36] R. J. Spatzier, An invitation to rigidity theory. In *Modern dynamical systems and applications*, pp. 211–231, Cambridge Univ. Press, Cambridge, 2004.

[37] D. Sullivan, Quasiconformal homeomorphisms and dynamics. II. Structural stability implies hyperbolicity for Kleinian groups. *Acta Math.* **155** (1985), no. 3–4, 243–260.

KATHRYN MANN

Department of Mathematics, Cornell University, Ithaca, NY 14850, USA, k.mann@cornell.edu

This Page Intentionally Left Blank

FLOER COHOMOLOGY, SINGULARITIES, AND BIRATIONAL GEOMETRY

MARK MCLEAN

ABSTRACT

We explain a few recent results concerning the application of various Floer theories to topics in algebraic geometry, including singularity theory and birational geometry. We will also state conjectures and open problems related to these results. We start out with a purely dynamical interpretation of the minimal discrepancy of an isolated singularity and explain how Floer theory fits into this story. Using similar ideas, we show how one can prove part of the cohomological McKay correspondence by computing a Floer cohomology group in two different ways. Finally, we illustrate how Hamiltonian Floer cohomology can be used to prove that birational Calabi–Yau manifolds have the same small quantum cohomology algebras, and we speculate how this might extend to orbifolds.

MATHEMATICS SUBJECT CLASSIFICATION 2020

Primary 53D40; Secondary 53D42, 14E16, 14N35, 14J17, 14E05, 14E18

INTERNATIONAL CONGRESS OF MATHEMATICIANS
2022 JULY 6–14

Proc. Int. Cong. Math. 2022, Vol. 4, pp. 2616–2637
DOI 10.4171/ICM2022/13

1. INTRODUCTION

Objects in algebraic geometry provide a rich source of symplectic and contact manifolds. Smooth projective varieties $X \subset \mathbb{C}P^N$, for instance, have symplectic structures ω_X given by restricting the standard Fubini–Study form on $\mathbb{C}P^N$. Links of isolated singularities admit natural contact structures, given by a complex hyperplane distribution.

What can such structures tell us about the underlying algebraic variety? Typically, a large amount of data is lost when one forgets everything except the symplectic or contact structure. For instance, if we have a smooth family of projective varieties in $\mathbb{C}P^N$, a Moser argument tells us that they are all symplectomorphic. On the other hand, many properties are retained such as *uniruledness*, which is the property that a rational curve passes through every point. This was shown by Kollár and Ruan in [47, PROPOSITION G].

An important tool in symplectic geometry which can help us understand this question better is *Floer (co)homology*. In order to understand what Floer (co)homology is, it is best to first understand its finite-dimensional counterpart, namely *Morse homology*. Let us suppose that we have a generic Morse function f on a closed Riemannian manifold M. Such a function naturally decomposes M into cells, one for each critical point (Figure 1). Hence one can use f to compute the cellular homology of M. The underlying chain complex is generated as a $\mathbb{Z}$-module by critical points of f and the differential is a matrix with respect to this basis of critical points whose (p, q)-entry is the number (counted with sign) of gradient flowlines of f connecting p and q. The homology of this chain complex is called *Morse homology*.

Floer homology is an infinite-dimensional version of Morse homology. There are many different kinds of Floer homology groups. For instance, the infinite-dimensional version of the manifold M above could be the free loopspace $C^\infty(S^1, X)$ of a symplectic manifold (X, ω) with $\omega = d\theta$ satisfying an additional "convexity" condition at infinity. The infinite-dimensional version of the Morse function f could be an action functional

$$\mathcal{A} : C^\infty(S^1, X) \to \mathbb{R}, \quad \mathcal{A}(\gamma) := -\int_{S^1} \gamma^*\theta - \int_0^{2\pi} H\big(\vartheta, \gamma(e^{2\pi i\vartheta})\big)\,d\vartheta \tag{1.1}$$

where $H : \mathbb{R}/\mathbb{Z} \times X \to \mathbb{R}$ is a time-dependent Hamiltonian, which also has a particular form near infinity. The generators of the chain complex for this Floer cohomology group as

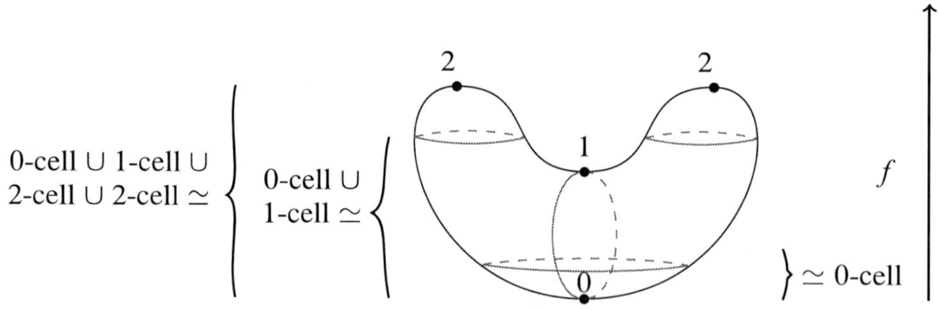

FIGURE 1
Cell decomposition from Morse function f.

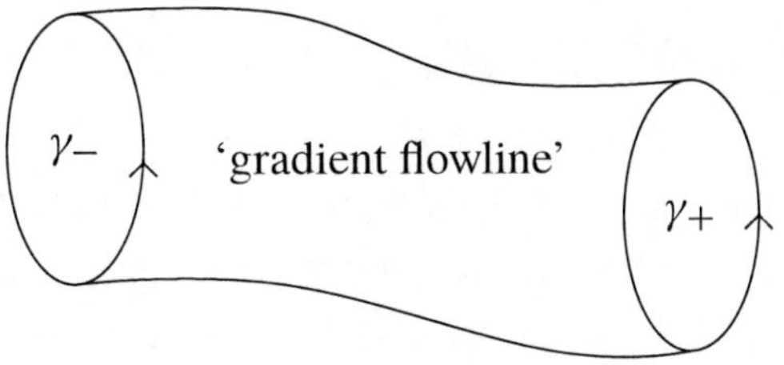

FIGURE 2
Floer differential.

a $\mathbb{Z}$-module are 1-periodic orbits of H, which are the critical points of $\mathcal{A}$. The differential is a matrix whose (γ_-, γ_+)-entry is the number of cylinders mapping to X connecting γ_- and γ_+ satisfying a certain PDE which represents the "gradient flowlines" of $\mathcal{A}$ (Figure 2). See **[40]** for a survey of some of these ideas.

In this article we will demonstrate how certain Floer (co)homology groups can be used to understand the following things:

(1) The minimal discrepancy of isolated singularities (Section 2);

(2) The Cohomological McKay correspondence (Section 3);

(3) Quantum cohomology of birational Calabi–Yau manifolds (Section 4).

2. MINIMAL DISCREPANCY OF ISOLATED SINGULARITIES

Let $A \subset \mathbb{C}^N$ be an irreducible affine variety of complex dimension n with at most one singularity at $0 \in \mathbb{C}^N$. In other words, $A \subset \mathbb{C}^N$ is cut out by a finite number of polynomial equations whose Jacobian matrix has constant rank along $A - 0$. The *link* of A at 0 is the manifold given by the intersection of A with the ε-sphere $S_\varepsilon := \{|z| = \varepsilon\} \subset \mathbb{C}^N$ where $\varepsilon > 0$ is any sufficiently small number. The link admits a contact structure $\xi_A := TL_A \cap iTL_A$ where $i : T\mathbb{C}^N \to T\mathbb{C}^N$ is multiplication by $i = \sqrt{-1}$ so long as $\varepsilon > 0$ is small enough.

Let us give two examples of such links. The first example is the smooth case, $A = \mathbb{C}^n$. Here the link $(L_{\mathbb{C}^n}, \xi_{\mathbb{C}^n})$ is contactomorphic to the $(2n-1)$-dimensional sphere S^{2n-1} with contact structure $\ker(\sum_{k=1}^n x_k dy_k - y_k dx_k)$ where $x_1 + iy_1, \dots, x_n + iy_n$ are the standard complex coordinates on $\mathbb{C}^n$. The second example is the nondegenerate hypersurface singularity $A = \{\sum_{k=1}^{n+1} z_k^2 = 0\}$ inside $\mathbb{C}^{n+1}$ whose link is contactomorphic to the unit cotangent bundle of the n-sphere.

Suppose that $A' \subset \mathbb{C}^{N'}$ is another irreducible affine variety with at most one singularity at 0. If there are neighborhoods $U \subset A$, $U' \subset A'$ of 0 together with a homeomorphism $\phi : U \to U'$ sending $0 \in A$ to $0 \in A'$ so that ϕ is a biholomorphism from $U - 0$ to $U' - 0$, then the links (L_A, ξ_A) and $(L_{A'}, \xi_{A'})$ are contactomorphic (**[56]**).

Question 2.1. Conversely, suppose that (L_A, ξ_A) is contactomorphic to $(L_{A'}, \xi_{A'})$. What does A and A' have in common?

The following definition is inspired by a similar notion in Heegaard's thesis (see [22, PAGE 89]). We say that A is *topologically smooth at* 0 if L_A is diffeomorphic to a sphere. Mumford in [39] showed that if A is of dimension 2, normal, and topologically smooth then A is smooth at 0. Such a fact is false in higher dimensions. For instance, the three-dimensional singularity $\{x^2 + y^2 + z^2 + w^3 = 0\} \subset \mathbb{C}^4$ is not smooth, but it is normal and topologically smooth (see [7]). However, the link of such a singularity is not contactomorphic to the link of $\mathbb{C}^3$ (see [55] for a direct proof). Seidel in [50] conjectured the following:

Conjecture 2.2. *If A is normal and (L_A, ξ_A) is contactomorphic to $(L_{\mathbb{C}^n}, \xi_{\mathbb{C}^n})$ then A is smooth at* 0.

Theorem 2.3 ([34, COROLLAY 1.2]). *Conjecture* 2.2 *is true in dimension* 3.

In fact, we proved a stronger result, which we will see has some connections with birational geometry. First of all, we need some definitions. Let (C, ξ) be a cooriented contact manifold. A 1-form $\alpha \in \Omega^1(C)$ is *compatible* with ξ if $\ker(\alpha) = \xi$ and α respects the coorientation of ξ. The restriction $d\alpha|_\xi$ is a symplectic structure on ξ. Therefore since the natural inclusion map from the unitary group to the linear symplectomorphism group is a homotopy equivalence, we get that the structure group of ξ naturally lifts to the unitary group and hence we can define its first Chern class $c_1(\xi)$. We say that our singularity $0 \in A$ is *numerically $\mathbb{Q}$-Gorenstein* if $c_1(\xi_A)$ vanishes in $H^2(L_A; \mathbb{Q})$.

We will now define the minimal discrepancy of such a singularity. This is an important invariant in the minimal model program (see [51]). Let X be a complex n-manifold with boundary and suppose that the natural map $H^2(X, \partial X; \mathbb{Q}) \to H^2(X; \mathbb{Q})$ is injective. Suppose also that $c_1(TX|_{\partial X})$ vanishes inside $H^2(\partial X; \mathbb{Q})$. Then, we can define the *relative first Chern class* $c_1(X, \partial X) \in H^2(X, \partial X; \mathbb{Q})$ as follows. Consider the long exact sequence

$$H^1(\partial X; \mathbb{Q}) \xrightarrow{0} H^2(X; \partial X; \mathbb{Q}) \to H^2(X; \mathbb{Q}) \xrightarrow{\beta} H^2(\partial X; \mathbb{Q}). \tag{2.1}$$

The vanishing condition on $c_1(TX|_{\partial X})$ implies that $\beta(c_1(X)) = 0$ and so $c_1(X)$ lifts uniquely to a class $c_1(X, \partial X; \mathbb{Q}) \in H^2(X; \partial X; \mathbb{Q})$ which we call the *relative first Chern class of X*.

Now let $\pi : \tilde{A} \to A$ be a *resolution* of A at 0. In other words, a proper morphism from a smooth variety $\tilde{A}$ which is an isomorphism onto its image away from $0 \in A$ so that $\pi^{-1}(0)$ is a union of transversally intersecting connected complex hypersurfaces $E := \bigcup_{i \in S} E_i$. The hypersurfaces $(E_i)_{i \in S}$ are the *irreducible exceptional divisors* of our resolution. Resolutions always exist according to Hironaka [23]. If A is smooth at 0 then we require $\tilde{A} \neq A$. Let $B_\varepsilon \subset \mathbb{C}^N$ be the closed ε-ball. Then $\tilde{A}_\varepsilon := \pi^{-1}(B_\varepsilon)$ deformation retracts onto E for ε small enough and so $H^2(\tilde{A}_\varepsilon; \mathbb{Q})$ is generated freely by the Poincaré duals $\mathrm{PD}(E_i)$, $i \in S$ of $(E_i)_{i \in S}$. Such a fact combined with the negativity lemma can be used to show that the natural map $H^2(\tilde{A}_\varepsilon; \partial \tilde{A}_\varepsilon; \mathbb{Q}) \to H^2(\tilde{A}_\varepsilon; \mathbb{Q})$ is injective (see [34, LEMMA 3.2]). Also $\xi_A \oplus \mathbb{C}$ is isomorphic to $TA|_{L_A}$ and so $c_1(\xi_A) = c_1(TA|_{L_A})$.

Now suppose that our singularity is numerically $\mathbb{Q}$-Gorenstein. Then by the discussion above, $\tilde{A}_\varepsilon$ has a relative first Chern class which is a sum $\sum_{i \in S} a_i \mathrm{PD}(E_i)$ for some unique rational numbers $(a_i)_{i \in S}$. The *discrepancy* of E_i is defined to be a_i for each $i \in S$. We define the *minimal discrepancy* $\mathrm{md}(A) \in \mathbb{Q}$ of $0 \in A$ to be $a := \min_{i \in S} a_i$ if $a \geq -1$ and $-\infty$ otherwise.

We will give a dynamical interpretation of the minimal discrepancy using ξ_A. The *Reeb vector field* associated to a contact form α compatible with ξ_A is the unique vector field R_α in the kernel of $d\alpha$ satisfying $\alpha(R_\alpha) = 1$. The dynamics of the flow of such a vector field can change drastically depending on the choice of contact form compatible with ξ_A. A *Reeb orbit of* α of period $L > 0$ is a periodic flowline $\gamma : \mathbb{R}/L\mathbb{Z} \to L_A$ of this vector field. If (L_A, ξ_A) is $\mathbb{Q}$-Gorenstein and satisfies $H^1(L_A; \mathbb{Q}) = 0$, one can associate an index to this orbit γ called the *Conley–Zehnder index* $\mathrm{CZ}(\gamma) \in \mathbb{Q}$ (see [34, DEFINITION 4.2]). Very roughly, this index "counts" the number of times the Reeb flow "wraps" around γ. We define the *lower SFT index* to be

$$\mathrm{CZ}(\gamma) - \frac{1}{2} \dim \ker(D\phi_L|_{(\xi|_{\gamma(0)})} - \mathrm{id}) + (n-3), \tag{2.2}$$

where $\phi_t : L_A \to L_A$, $t \in \mathbb{R}$ is the flow of R_α. We define the *minimal SFT index* of α to be $\mathrm{mi}(\alpha) := \inf_\gamma \mathrm{lSFT}(\gamma)$ where the infimum is taken over all Reeb orbits γ of α. We define the *highest minimal SFT index* of (L_A, ξ_A) to be $\mathrm{hmi}(L_A, \xi_A) := \sup_\alpha \mathrm{mi}(\alpha)$ where the supremum is taken over all contact forms α compatible with ξ_A. By construction, this is an invariant of (L_A, ξ_A) up to coorientation preserving contactomorphism.

Theorem 2.4 ([34, THEOREM 1.1]). *Let $0 \in A$ be normal and numerically $\mathbb{Q}$-Gorenstein. Suppose $H^1(L_A; \mathbb{Q}) = 0$. Then*

- *if* $\mathrm{md}(A, 0) \geq 0$ *then* $\mathrm{hmi}(L_A, \xi_A) = 2\mathrm{md}(A, 0)$, *and*
- *if* $\mathrm{md}(A, 0) < 0$ *then* $\mathrm{hmi}(L_A, \xi_A) < 0$.

Seidel's conjecture follows immediately from Theorem 2.4 above and the conjecture below due to the fact that $\mathrm{md}(\mathbb{C}^n, 0) = n - 1$.

Conjecture 2.5 ([52, CONJECTURE 2]). *Suppose A is normal and numerically $\mathbb{Q}$-Gorenstein with* $\mathrm{md}(A, 0) = n - 1$ *then A is smooth at* 0.

This conjecture is true when $n = 3$ by [45, MAIN THEOREM (I)] combined with minimal discrepancy calculations from [33] and [25], as well as [5, COROLLARY 5.17]. Therefore we have a proof of Theorem 2.3.

There are two parts to the proof of Theorem 2.4. The first part gives an upper bound of $\mathrm{md}(A, 0)$, and the second part gives a lower bound. It is easier to prove the upper bound since one only needs to find an explicit contact form α compatible with ξ_A satisfying $\mathrm{md}(A, 0) \leq \mathrm{mi}(\alpha)$. In order to construct such a contact form, one starts with a resolution $\pi : \tilde{A} \to A$ as above. Since $\pi^{-1}(0)$ is a transverse intersection of complex hypersurfaces, one can deform the link $\pi^{-1}(S_\varepsilon)$ through contact hypersurfaces so that it is compatible with these hypersurfaces in some sense. The periodic orbits of the corresponding Reeb flow

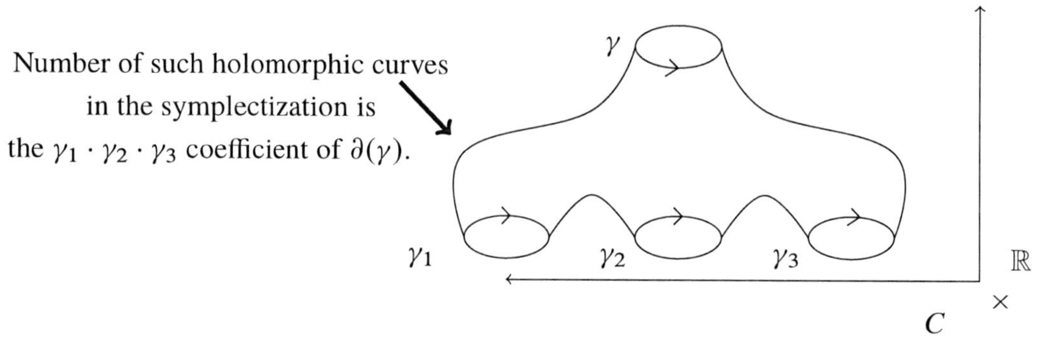

FIGURE 3
Full contact homology differential.

"wrap" around the divisors $(E_i)_{i \in S}$. One can explicitly compute all of their Conley–Zehnder indices, giving our result (see **[34, THEOREM 5.23]**).

In the paper **[34]**, we used pseudoholomorphic curve techniques to give the lower bound for $\mathrm{md}(A, 0)$ (see **[34, SECTIONS 6,7]**). However, this lower bound conjecturally can also be proven using a Floer homology group, called *full contact homology*. We will give a brief sketch of this idea in the case where $\mathrm{md}(A, 0) \geq 0$.

Very roughly, *full contact homology* $\mathrm{CH}_*(C, \xi)$ of a $(2n-1)$-contact manifold (C, ξ) is defined in the following way (see **[2,14,24,42]**). The chain complex is the free super-commutative algebra over $\mathbb{Q}$ generated by Reeb orbits of a generic compatible contact form λ and graded by Conley–Zehnder index plus $(n-3)$. We now put an appropriate translation-invariant almost-complex structure on the symplectization $(\mathbb{R} \times C, d(e^t \lambda))$ of (C, ξ). The differential is the unique $\mathbb{Q}$-linear differential on this algebra satisfying the Leibniz rule and whose $(\gamma, \prod_{i=1}^k \gamma_i)$ coefficient is a count of genus-zero holomorphic curves in $\mathbb{R} \times C$ up to translation "limiting" to the corresponding Reeb orbits γ, $(\gamma_i)_{i=1}^k$ of λ (Figure 3). Full contact homology does not depend on the choice of a compatible contact form.

We have the following conjectural spectral sequence computing $\mathrm{CH}_*(L_A, \xi_A)$. To set up this spectral sequence, we need some preliminary definitions. For each $I \subset S$, let $E_I := \bigcap_{i \in I} E_i$ where $(E_i)_{i \in S}$ are the irreducible exceptional divisors of our resolution as above. Define $E_I^o := E_I - \bigcup_{I \subsetneq I'} E_{I'}$ and let NE_I^o be its normal bundle in $\tilde{A}$ for each $I \subset S$. For each tuple $(k_i)_{i \in I}$ of integers, there is a $U(1)$ action on NE_I^o preserving the fibers so that $\beta \in U(1)$ sends a point $(x_i)_{i \in I} \in NE_I^o = \bigoplus_{i \in S-I}((TE_{I-i}|_{E_I^o})/TE_I^o)$ to $(\beta^{k_i} x_i)_{i \in I}$. Let $NE_I^{/(k_i)_{i \in I}}$ be the quotient of $NE_I^o - E_I^o$ by this action. Suppose our resolution $\tilde{A}$ admits a Kähler form ω with an integral lift. Then one can construct a line bundle with curvature a positive multiple of $-2\pi i \omega$ together with a meromorphic section s so that the divisor associated to s is equal to $-\sum_{i \in S} w_i E_i$ for some positive integers $(w_i)_{i \in S}$.

Conjecture 2.6. *Define*

$$A_{p,q} \equiv \bigoplus_{\{(k_i) \in \mathbb{N}_{\geq 0}^S : \sum_i k_i w_i = p\}} H_{p+q-2\sum_i k_i a_i}\big(NE_{I_{(k_i)_{i \in I}}}^{/(k_i)}; \mathbb{Q}\big) \tag{2.3}$$

where $I_{(k_i)} \equiv \{i \in S : k_i \neq 0\}$. *Then there is a spectral sequence converging to* $\mathrm{CH}_*(L_A, \xi_A)$ *with* E^1 *page equal to the free supercommutative algebra generated by the bigraded vector space* $A_{*,*}$, *i.e.,*

$$E^1_{*,*} = \bigoplus_{n \geq 0} \mathrm{Sym}^n_{\mathbb{Q}}(A_{*,*}). \tag{2.4}$$

This spectral sequence is very similar to those in [19] and [35], and we expect the method of proof for that above to be similar in spirit.

Now let us continue with the proof of the lower bound for $\mathrm{md}(A, 0)$ in the case where $\mathrm{md}(A, 0) \geq 0$. Consider the smallest value of p for which the entry $E^1_{p,2\mathrm{md}(A,0)-p}$ in the spectral sequence above is nonzero. Then for degree reasons, this entry cannot be killed. Hence full contact homology is nonzero in degree $2\mathrm{md}(A, 0)$. This implies that there is a Reeb orbit of lower SFT index $2\mathrm{md}(A, 0)$ for any generic contact form compatible with ξ_A and hence $\mathrm{hmi}(L_A, \xi_A) \leq 2\mathrm{md}(A, 0)$, giving us our lower bound.

It would be interesting to know if there are other properties of the singularity $0 \in A$ captured by full contact homology. Full contact homology is typically very hard to compute since one has to compute the differential by solving a PDE with asymptotic boundary conditions. However, the following definition and conjecture might be of help.

Definition 2.7. Let $\mathbb{D}_z(\delta) \subset \mathbb{C}$ be the closed disk of radius $\delta > 0$ centered at $z \in \mathbb{C}$. Define $\mathbb{D}(\delta) := \mathbb{D}_0(\delta)$ and $\mathbb{D} := \mathbb{D}(1)$. Define the *short arc space* $\mathrm{Arc}^o(A)$ to be the space of holomorphic maps $u : \mathbb{D} \to A$ whose boundary is disjoint from 0 equipped with the C^∞ topology coming from the embedding in $A \subset \mathbb{C}^N$ (see [27, DEFINITION 2(2)]). Let $\mathrm{Arc}^*(A)$ be the disjoint union $\bigsqcup_{m \in \mathbb{N}_{\geq 0}} (\mathrm{Arc}^o(A))^m$. For each $w \in \mathbb{R}$, we define $\mathrm{Arc}^*_{\leq w}(A)$ to be the subspace of those tuples $(u_i)_{i=1}^m$ of arcs for which the sum of the degrees of $u_i^*(\sum_{j \in S} w_j E_j)$, $i = 1, \ldots, m$ is $\leq w$, as well as the case $m = 0$.

For each $l \in \mathbb{N}$, we define $\widetilde{\mathrm{Jet}}^l(A)$ to be the set of l-jets of holomorphic maps $\phi : \mathbb{D}(\delta) \to A$, $\delta > 0$ satisfying $u^{-1}(0) = 0$. We let $\overline{\mathrm{Jet}}^l(A) := \widetilde{\mathrm{Jet}}^l(A)/S^1$ where the S^1 action rotates the arcs. For each $l \in \mathbb{N}$, define $S\overline{\mathrm{Jet}}^{l!}(A) := \{\emptyset\} \sqcup \bigsqcup_{j=1}^l (\overline{\mathrm{Jet}}^{l!/j}(A))^j / S_j$ where S_j is the permutation group on j elements. For $w \in \mathbb{R}$ and $l > w$, we define the map $\pi_{l,w} : \mathrm{Arc}^*_{\leq w}(A) \to S\overline{\mathrm{Jet}}^{l!}(A)$ as follows: Let $(u_i)_{i=1}^m$ be an element of $\mathrm{Arc}^*(A)$ and for each $j = 1, \ldots, m$ let $u_i^{-1}(0) = \{z_1^i, \ldots, z_{j_i}^i\} \subset \mathbb{D}$, $j_i \in \mathbb{N}$. Let $\delta > 0$ be very small. Then the collection of arcs $\phi|_{\mathbb{D}_{z_k^j}(\delta)}$, $k = 1, \ldots, j_i$, $i = 1, \ldots, m$ defines an element of $S\overline{\mathrm{Jet}}^{l!}(A)$. If $m = 0$ then this corresponds to having no arcs, and we map this to $\{\emptyset\}$.

For each $w \in \mathbb{R}$, $l > w$, we define $S\mathrm{Jet}^{l!}_{\leq w}(A)$ to be the image of $\pi_{l,w}$ equipped with the finest topology making $\pi_{l,w}$ continuous (this can be different from the usual jet space topology, [27, EXAMPLE 4]). For each $w \in \mathbb{R}$, there is a natural integration map

$$H_c^*\big(S\mathrm{Jet}^{l!}_{\leq w}(A)\big) \to H_c^{*-n(l!-k!)}\big(S\mathrm{Jet}^{k!}_{\leq w}(A)\big) \tag{2.5}$$

for each $k \leq l$ sufficiently large. Define $H_c^*(S\mathrm{Jet}(A))$ to be

$$\varinjlim_{w \in \mathbb{R}} \varprojlim_{l} H_c^{*+n(l!+1)}\big(S\mathrm{Jet}^{l!}(A)\big).$$

Conjecture 2.8. *We have a natural isomorphism* $\mathrm{CH}_*(L_A, \xi_A) \cong H_c^*(S\mathrm{Jet}(A))$.

The parameter w should, very roughly, correspond to the natural action filtration on full contact homology. Note that for many examples these groups can be nontrivial in both positive and negative degrees. The following conjecture provides evidence for Conjecture 2.8.

Conjecture 2.9. *The same spectral sequence from Conjecture* 2.6 *computes* $H_c^*(S\mathrm{Jet}(A))$.

We hope that the same methods from [8] can be used to prove Conjecture 2.9. The filtration associated to this spectral sequence should come from the parameter w above. In order to prove Conjecture 2.8, one needs to write down an enhanced "PSS" map (see [43]) and show that it respects both spectral sequences (Conjectures 2.6 and 2.9) (or, more precisely, the action filtration and the filtration coming from w). A simpler version of this map is described later on in the next section.

3. COHOMOLOGICAL MCKAY CORRESPONDENCE

Quotient singularities $\mathbb{C}^n/G$, where $G \subset \mathrm{SU}(n)$ is a finite group, are natural examples of singularities to study from a Floer-theoretic perspective. One reason for this is that they are homogeneous, and this ensures that the link has a compatible contact 1-form with nice Reeb dynamics. In this section, we will show how Floer theory can shed light on the *cohomological McKay correspondence* [44].

Definition 3.1. A *crepant resolution* of $\mathbb{C}^n/G$ is a resolution $\pi : Y \to \mathbb{C}^n/G$ satisfying $c_1(Y) = 0$.

Let us consider the following open problem.

Conjecture 3.2 (Cohomological McKay correspondence over $\mathbb{K}$, [46, CONJECTURE 1.1]). *Let* $\mathbb{K}$ *be a field. There is a natural basis of* $H^*(Y;\mathbb{K})$ *consisting of irreducible representations of* G. *In particular, its dimension is the number of conjugacy classes* $|\mathrm{Conj}(G)|$ *of* G.

By blowing up an existing resolution, one can construct new resolutions of the same singularity whose cohomology has arbitrarily large rank. However, such resolutions are typically not crepant. In dimension 3 it was shown that crepant resolutions always exist (see [6, THEOREM 1.2]). However, in dimension 4 there are examples which do not admit any crepant resolution (see [11, EXAMPLE 2.28]). Batyrev [4] showed that when $\mathbb{K} = \mathbb{Q}$, the rank of $H^*(Y;\mathbb{K})$ is the number of conjugacy classes of G. However, he did not give a natural basis for this group.

Theorem 3.3 ([37, THEOREMS 1.4 AND 1.5]). *Suppose that* G *acts freely away from* $0 \in \mathbb{C}^n$ *and suppose* $\mathbb{K}$ *is a field whose characteristic does not divide* $|G|$. *Let* Y *be a quasiprojective crepant resolution. Then there is a Floer cohomology group* $\mathrm{SH}_+^*(Y;\Lambda_{\mathbb{K}})$ *defined over a field* $\Lambda_{\mathbb{K}}$ *of the same characteristic as* $\mathbb{K}$ *satisfying the following properties:*

(1) $\mathrm{SH}^*_+(Y;\Lambda_{\mathbb{K}})$ *is naturally isomorphic to* $H^*(Y;\Lambda_{\mathbb{K}})$ *(up to a shift in degree) and*

(2) $\mathrm{SH}^*_+(Y;\Lambda_{\mathbb{K}})$ *has rank equal to* $|\mathrm{Conj}(G)|$.

Corollary 3.4. *Let* $\mathbb{K}$ *be a field whose characteristic does not divide* $|G|$. *If* G *acts freely away from* 0 *then the rank of* $H^*(Y;\mathbb{K})$ *is equal to the number of conjugacy classes of* G.

The field $\Lambda_{\mathbb{K}}$ is called the *Novikov field* over $\mathbb{K}$ and is defined as the power series ring

$$\Lambda_{\mathbb{K}} = \left\{ \sum_{i \in \mathbb{N}} a_i t^{r_i} \;\middle|\; a_i \in \mathbb{K},\ r_i \in \mathbb{R},\ \forall\, i \in \mathbb{N},\ r_i \to \infty \text{ as } i \to \infty \right\}. \tag{3.1}$$

Let us now explain how the Floer group $\mathrm{SH}^*_+(Y;\Lambda_{\mathbb{K}})$, called *positive symplectic cohomology*, is constructed. In order to do this, we need to define *Hamiltonian Floer cohomology* first. Let $H = (H_t)_{t\in[0,1]}$ be a generic time-dependent Hamiltonian on a symplectic manifold (X,ω) and let us assume $c_1(X) = 0$. The chain complex for Hamiltonian Floer cohomology $\mathrm{HF}^*(H;\Lambda_{\mathbb{K}})$ is freely generated over $\Lambda_{\mathbb{K}}$ by 1-periodic orbits $\gamma : \mathbb{R}/\mathbb{Z} \to X$ of H. This is graded by a version of the Conley–Zehnder index. The differential is a matrix with respect to this basis of 1-periodic orbits whose (γ_-,γ_+) entry is a count of cylinders $u : \mathbb{R}\times\mathbb{R}/\mathbb{Z} \to X$ joining γ_- and γ_+ satisfying a Cauchy–Riemann-like PDE $\partial_s u + J_t(\partial_t u + X_{H_t}) = 0$ where $(J_t)_{t\in\mathbb{R}/\mathbb{Z}}$ is a family of almost complex structures on X (these are called *Floer trajectories*). The count is weighted by energy, which is a particular integral over this cylinder, and this is why we need the Novikov ring $\Lambda_{\mathbb{K}}$. If we did not do this, then the count might be infinite. Hamiltonian Floer cohomology was originally developed by Floer in [15]. The book [1] provides a very good introduction to Hamiltonian Floer cohomology.

Symplectic cohomology is a Hamiltonian Floer cohomology group that is usually defined for noncompact symplectic manifolds satisfying certain convexity properties at infinity. Very roughly, these are Hamiltonian Floer groups associated to Hamiltonians that tend to infinity very rapidly as one travels to infinity in the symplectic manifold. The fact that the symplectic manifold is noncompact can create problems such as infinite counts or the differential not squaring to zero since Floer trajectories can escape to infinity. However, in nice cases, one can define symplectic cohomology. There are many different versions of symplectic cohomology (e.g., [9, 10, 16, 21, 57–59]). Two good surveys of symplectic cohomology are contained in [40] and [49].

Now let us define $\mathrm{SH}^*_+(Y;\Lambda_{\mathbb{K}})$ for our crepant resolution Y. Since Y is crepant, we have that π is an isomorphism onto its image away from the exceptional locus $\pi^{-1}(0)$, and so we have a natural identification $Y - \pi^{-1}(0) = (\mathbb{C}^n - 0)/G$. Since Y is quasiprojective, we can put a natural symplectic structure ω_Y on Y which coincides with the standard linear symplectic structure on $\mathbb{C}^n/G$ away from a small neighborhood of $\pi^{-1}(0)$ (see [37, SECTION 2.5]). We now let H be a Hamiltonian on Y which is equal to $|z|^4$ near infinity (or some other rapidly increasing function of $|z|$). Then we define $\mathrm{SH}^*(Y;\Lambda_{\mathbb{K}}) := \mathrm{HF}^*(H;\Lambda_{\mathbb{K}})$. This group is a symplectomorphism invariant of (Y,ω_Y). There is a natural map $H^*(Y;\Lambda_{\mathbb{K}}) \to \mathrm{SH}^*(Y;\Lambda_{\mathbb{K}})$ and the cone of the corresponding chain map is called the *positive symplectic* *cohomology* $\mathrm{SH}^*_+(Y;\Lambda_{\mathbb{K}})$, which is the key Floer group from Theorem 3.3. Strictly speaking, the Hamiltonian H cannot be *exactly* $|z|^4$ near infinity since it needs to be generic, and so in reality it is a very small generic perturbation of such a function near infinity. A standard argument ensures that the definition does not depend on the specific choice of Hamiltonian H.

Let us explain very roughly how to prove parts (1) and (2) of Theorem 3.3. Let us start with part (1), which states that $\mathrm{SH}^*_+(Y;\Lambda_{\mathbb{K}}) \cong H^*(Y;\Lambda_{\mathbb{K}})$. By the definitions above, it is sufficient to show $\mathrm{SH}^*(Y;\Lambda_{\mathbb{K}}) = 0$. Consider the natural $U(1)$-action on $\mathbb{C}^n$ given by sending a vector $z \in \mathbb{C}^n$ to $e^{2\pi i\vartheta}z$ for each $\vartheta \in \mathbb{R}/\mathbb{Z}$. Such an action lifts to a $U(1)$-action on Y (see [37, LEMMA 3.4]). For appropriate ω_Y, one can show that this $U(1)$-action is the flow of a Hamiltonian $K : Y \to \mathbb{R}$. Now the key point is that we can deform H in a compact region of Y so that it is equal to a large multiple of K near the exceptional locus and is a rapidly increasing function of $|z|$ away from this locus, with high derivatives. This forces the Conley–Zehnder indices of all the orbits to be very large, since the linearized flow near each orbit "spins" extremely fast. Hence $\mathrm{SH}^*(Y;\Lambda_{\mathbb{K}})$ vanishes since the chain complex can be made to vanish at any given degree.

The proof of part (2) of Theorem 3.3 is a spectral sequence argument. One first deforms H in a compact region of Y so that it is C^2-small near the exceptional locus and is a generic perturbation of a function of $|z|$ elsewhere. The generators of $\mathrm{HF}^*(H;\Lambda_{\mathbb{K}})$ away from the exceptional locus correspond to Reeb orbits of an appropriate contact form on the link S^{2n-1}_ε/G of $\mathbb{C}^n/G$ where S^{2n-1}_ε is the sphere of radius $\varepsilon > 0$. Since our link S^{2n-1}_ε is simply connected, we have have a natural bijection $\pi_0(\mathcal{L}(S^{2n-1}/G)) \cong \mathrm{Conj}(G)$ where $\mathcal{L}(S^{2n-1})/G$ is the free loop space of our link. Hence the chain complex computing $\mathrm{SH}^*_+(Y;\Lambda_{\mathbb{K}})$ splits as a direct sum of groups indexed by conjugacy classes of G. However, the Floer differential might not respect this direct sum structure.

The contact form on our link S^{2n-1}_ε/G is the radial one $\alpha = \frac{1}{2}\sum_{i=1}^n r_i^2\,d\vartheta_i$ where (r_i,ϑ_i), $i = 1,\dots,n$ are polar coordinates on each factor of $\mathbb{C}^n$. The Reeb flow of this contact form is the same as the flow of the $U(1)$-action on S^{2n-1}_ε/G up to scaling. Hence one can compute the generators of the chain complex for $\mathrm{SH}^*_+(Y)$ using this $U(1)$-action.

The orbits come in families associated to the eigenspaces of each matrix element $g \in G \subset \mathrm{SU}(n)$ and the cohomology of these families give us the E_1 page of a spectral sequence computing $\mathrm{SH}^*_+(Y;\Lambda_{\mathbb{K}})$. Now, instead of computing $\mathrm{SH}^*_+(Y;\Lambda_{\mathbb{K}})$, one must first compute a variant $\mathrm{SH}^*_{S^1,+}(Y;\Lambda_{\mathbb{K}})$ since in this case the spectral sequence degenerates. One can then show that $\mathrm{SH}^*_+(Y;\Lambda_{\mathbb{K}})$ has rank $|\mathrm{Conj}(G)|$. This ends the sketch of the proof of Theorem 3.3.

The proof of part (1) of Theorem 3.3 naturally identifies $\mathrm{SH}^*_+(Y;\Lambda_{\mathbb{K}})$ with $H^*(Y;\Lambda_{\mathbb{K}})$. However, the proof of part (2) does not produce a similar natural identification.

Open Problem 3.5. Does the natural grading by conjugacy classes of G of the chain complex computing $\mathrm{SH}^*_+(Y;\Lambda_{\mathbb{K}})$ also get respected by the differential?

If the answer to this problem is yes, then we get a natural basis of $H^*(Y; \Lambda_{\mathbb{K}})$ by elements of $\mathrm{Conj}(G)$. Another issue is that we require that the characteristic of $\mathbb{K}$ does not divide $|G|$. We do not know how to show that the spectral sequence computing $\mathrm{SH}^*_{S^1,+}(Y; \Lambda_{\mathbb{K}})$ degenerates when the characteristic of $\mathbb{K}$ divides $|G|$.

Open Problem 3.6. What does the spectral sequence look like for $\mathrm{SH}^*_{S^1,+}(Y; \Lambda_{\mathbb{K}})$ when the characteristic $\mathbb{K}$ divides $|G|$?

Our work [37] was partly inspired by [27, SECTION 6], which tries to understand the cohomological McKay correspondence using arc spaces. Recall in Definition 2.7 that, for an isolated singularity A, we defined the *short arc space* $\mathrm{Arc}^o(A)$. Consider the subspace $\mathrm{ShArc}(A) \subset \mathrm{Arc}^o(A)$ of those arcs $u : \mathbb{D} \to \mathbb{C}^n/G$ satisfying $u^{-1}(0) = \{0\}$. Kollár and Némethi in [27, COROLLARY 32] show that the "irreducible components" of $\mathrm{ShArc}(\mathbb{C}^n/G)$ are in natural 1–1 correspondence with $\mathrm{Conj}(G)$. This correspondence is given by the boundary of each short arc $u : \mathbb{D} \to \mathbb{C}^n/G$, viewed as an element of

$$\pi_0\big(\mathcal{L}(\mathbb{C}^n - 0)/G\big) = \pi_0\big(\mathcal{L}(S^{2n-1}_\varepsilon/G)\big) = \mathrm{Conj}(G). \tag{3.2}$$

One way of connecting $\mathrm{ShArc}(\mathbb{C}^n/G)$ with $\mathrm{SH}^*(Y; \Lambda_{\mathbb{K}})$ might be through the *PSS map* (see [43]). The *PSS map* is a natural map from $\mathrm{SH}^*_+(Y; \Lambda_{\mathbb{K}})$ to $H_*(\mathrm{ShArc}(\mathbb{C}^n/G); \Lambda_{\mathbb{K}})$ given by sending an orbit γ to a "cycle" swept out by the moduli space of maps $u : \mathbb{C} \to Y$ so that $u(re^{2\pi i\vartheta})$ converges to $\gamma(\vartheta)$ as $r \to \infty$ and where the "cycle" is swept out by 0 (Figure 4).

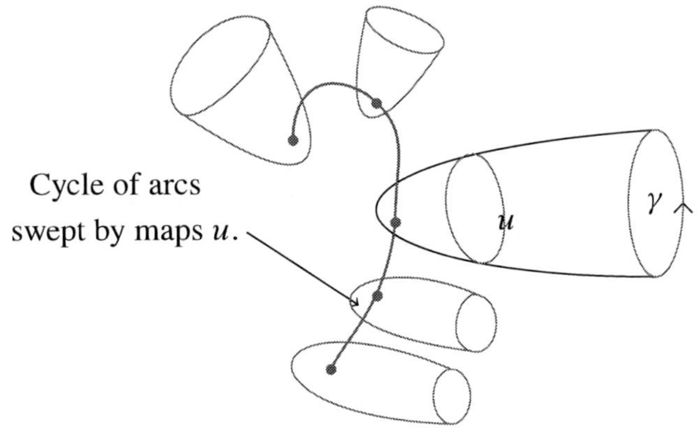

FIGURE 4
Arc space PSS map.

The Floer-theoretic methods used to prove Theorem 3.3 work very well if G acts freely away from 0.

Open Problem 3.7. Is there a way of using the Floer-theoretic methods above to deal with the case where G does not necessarily act freely away from 0?

The ideas of Section 4 below might be of use when we are dealing with this problem (see Open Problem 4.7 below).

4. QUANTUM COHOMOLOGY OF BIRATIONAL CALABI–YAU MANIFOLDS

Recall that two algebraic varieties are *birational* to each other if they have isomorphic dense Zariski-open subsets. The *minimal model program* in algebraic geometry, very roughly, is concerned with finding the "smallest" varieties in their birational equivalence class (minimal models). These minimal models are not necessarily unique. Calabi–Yau manifolds are examples of such minimal models. Therefore it is very natural to ask what properties birational Calabi–Yau manifolds have in common. For our purposes, we will say that a *Calabi–Yau manifold* is a smooth projective variety with trivial first Chern class.

Batyrev showed in [3] that any two birational Calabi–Yau manifolds have the same Betti numbers. In fact, by using ideas in [12, 28] combined with [20], or by [61, COROLLARY 1.6], they have the same integral cohomology groups. However, the methods used do not produce an explicit isomorphism between these groups. Also the cup product structures might not agree (see [17, EXAMPLE 7.7]).

There is a deformed version of the cup product called the quantum cup product. Let us define this. We will fix a field $\mathbb{K}$ and a Calabi–Yau manifold X with a Kähler form ω admitting an integral lift. We define the Novikov ring

$$\Lambda_{\mathbb{K}}^{\omega} = \left\{ \sum_{i \in \mathbb{N}} a_i t^{\beta_i} \;\middle|\; a_i \in \mathbb{K},\ \beta_i \in H_2(X;\mathbb{Z}),\ \omega(\beta_i) \to \infty \text{ as } i \to \infty \right\}. \tag{4.1}$$

Let $A, B, C \in H^*(X;\mathbb{K})$ be cohomology classes whose degrees sum up to $2n$, where n is the complex dimension of X, and let $a, b, c \in C_*(X;\mathbb{K})$ be cycles representing the corresponding Poincaré duals of A, B, C. For each $\beta \in H_2(X;\mathbb{Z})$, we define the *Gromov–Witten invariant* $\mathrm{GW}_{0,3}^{X,\beta}(A, B, C) \in \mathbb{Z}$ to be the "count" of holomorphic maps $u : \mathbb{P}^1 \to X$ representing β so that $u(0)$ maps to a, $u(1)$ maps to b, and $u(\infty)$ maps to c. Technically, in order for this count to make sense, one needs to perturb the complex structure on X to a generic domain-dependent family of almost complex structures and count these curves with sign. Now let $A_1, \dots, A_k \in H^*(X;\mathbb{K})$ be a basis of homogeneous elements and let $\hat{A}_1, \dots, \hat{A}_k \in H^*(X;\mathbb{K})$ be the dual elements with respect to the pairing $(\eta, \nu) \to \int_X \eta \cup \nu$. We define *small quantum cohomology* to be the unique $\Lambda_{\mathbb{K}}^{\omega}$-algebra $\mathrm{QH}^*(X;\Lambda_{\mathbb{K}}^{\omega})$ which is isomorphic as a graded $\Lambda_{\mathbb{K}}^{\omega}$-module to $H^*(X;\Lambda_{\mathbb{K}}^{\omega})$ and whose product $\star_X$ satisfies

$$A_i \star_X A_j = \sum_{\beta \in H_2(X;\mathbb{Z})} \sum_{l=1}^{k} \mathrm{GW}_{0,3}^{X,\beta}(A_i, A_j, A_l)\hat{A}_l t^{\beta}. \tag{4.2}$$

One should think of this product as the cup product which has additional "correction" terms coming from counts of nonconstant holomorphic maps (Figure 5). For instance, if there were no nonconstant genus zero holomorphic maps (e.g., when X is an abelian variety) then this would be equal to the cup product.

Big quantum cohomology is also a deformation of the cup product which is more general than small quantum cohomology. Its definition involves counts of genus-zero curves passing through arbitrarily many cycles.

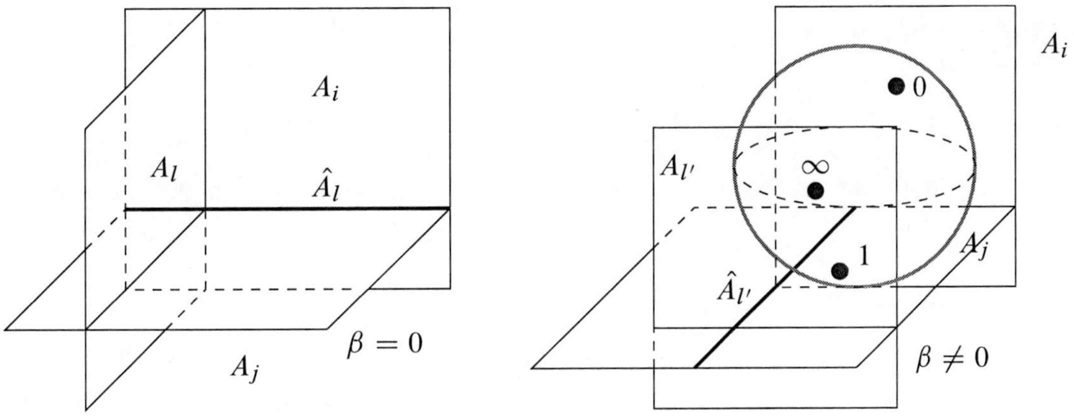

FIGURE 5
Terms in small quantum product. Cycles Poincaré dual to their respective cohomology classes are illustrated.

Conjecture 4.1 (Morrison [38] and Ruan [48]). *Any two birational Calabi–Yau manifolds have isomorphic (small or big) quantum cohomology rings up to analytic continuation.*

This conjecture was proven in dimension 3 in [32]. It was shown in [29–31] that if both Calabi–Yau manifolds are related by a sequence of birational transformations called *ordinary flops* then the conjecture above is true for big quantum cohomology (and hence also for small quantum cohomology). Wang in [60, SECTION 4.3, CONJECTURE IV] conjectured that all such Calabi–Yau manifolds, after deformation, are related by these operations, and so this would imply Conjecture 4.1. The method of proof in the papers [29–32] above is given by degenerating the Calabi–Yau manifold in a particular way and looking at Gromov–Witten invariants on this degeneration. We will describe a completely different approach to Conjecture 4.1 above using Floer theory, and in particular a modified version of symplectic cohomology.

Let X and $\check{X}$ be birational Calabi–Yau manifolds and let ω and $\check{\omega}$ be Kähler forms on X and $\check{X}$, respectively, admitting integral lifts. We get two Novikov rings $\Lambda_{\mathbb{K}}^{\omega}$ and $\Lambda_{\mathbb{K}}^{\check{\omega}}$ defined as in equation (4.1). By [26, LEMMA 4.2], there are natural identifications $H_2(X;\mathbb{Z}) \cong H_2(\hat{X};\mathbb{Z})$, due to the fact that the region in which the birational transform is not an isomorphism has complex codimension ≥ 2. Hence from now on, we will not distinguish between these groups, and so we can define the intersection of both Novikov rings $\Lambda_{\mathbb{K}}^{\omega,\check{\omega}} := \Lambda_{\mathbb{K}}^{\omega} \cap \Lambda_{\mathbb{K}}^{\check{\omega}}$. More explicitly,

$$\Lambda_{\mathbb{K}}^{\omega,\check{\omega}} = \left\{ \sum_{i \in \mathbb{N}} a_i t^{\beta_i} \;\middle|\; a_i \in \mathbb{K},\ \beta_i \in H_2(X;\mathbb{Z}),\ \min\big(\omega(\beta_i), \check{\omega}(\beta_i)\big) \to \infty \text{ as } i \to \infty \right\}. \tag{4.3}$$

The following theorem essentially proves Conjecture 4.1 for small quantum cohomology algebras.

Theorem 4.2 ([36, THEOREM 1.2]). *There exists a graded $\Lambda_{\mathbb{K}}^{\omega,\check{\omega}}$-algebra Z together with algebra isomorphisms*

$$Z \otimes_{\Lambda_{\mathbb{K}}^{\omega,\check{\omega}}} \Lambda_{\mathbb{K}}^{\omega} \cong \mathrm{QH}^*(X; \Lambda_{\mathbb{K}}^{\omega}), \quad Z \otimes_{\Lambda_{\mathbb{K}}^{\omega,\check{\omega}}} \Lambda_{\mathbb{K}}^{\check{\omega}} \cong \mathrm{QH}^*(\check{X}; \Lambda_{\mathbb{K}}^{\check{\omega}}). \tag{4.4}$$

The downside of this theorem is that the algebra Z is unknown in general, as is the isomorphisms in (4.4).

4.1. Example

We will now illustrate Theorem 4.2 with an example (see [38, SECTION 7.3]). Suppose that X and $\check{X}$ are connected Calabi–Yau 3-folds and that there exists a disjoint union of connected genus 0 curves $C_1, \dots, C_k$ in X and $\check{C}_1, \dots, \check{C}_k$ in $\check{X}$ together with a class $\Gamma \in H_2(X; \mathbb{Z})$ so that

- $[C_j] = \Gamma \in H_2(X; \mathbb{Z})$ and $[\check{C}_j] = -\Gamma \in H_2(\check{X}; \mathbb{Z})$ for each j and all connected genus-zero curves mapping to X or $\check{X}$, representing a multiple of Γ, have image equal to one of these curves,
- the normal bundle of C_j and $\check{C}_j$ is $\mathcal{O}(-1) \oplus \mathcal{O}(-1)$ for each j, and
- X and $\check{X}$ are related by an Atiyah flop along all of these curves.

Very roughly, an *Atiyah flop* along C_j removes C_j and glues it back in with the two $\mathcal{O}(-1)$ factors of its normal bundle swapped. One can think of an Atiyah flop as a kind of 0-surgery along the "knot" C_j. Since $H_2(X; \mathbb{Z})$ is naturally identified with $H_2(\check{X}; \mathbb{Z})$, we have by Poincaré duality a natural identification $H^k(X; \mathbb{Z}) = H^k(\check{X}; \mathbb{Z})$ where k is even. Hence from now on we will identify these cohomology groups. Let $\hat{A}_0, \dots, \hat{A}_l \in H^4(X; \mathbb{Q})$ be a basis so that $\hat{A}_0$ is Poincaré dual to Γ and let $A_0, \dots, A_l \in H^2(X; \mathbb{Q})$ be the dual basis with respect to the pairing $(\alpha, \beta) \to \int_X \alpha \cup \beta$. The algebra Z from Theorem 4.2 is equal to $H^*(X; \Lambda_{\mathbb{Q}}^{\omega,\check{\omega}})$ as a $\Lambda^{\omega,\check{\omega}}$-module, and its product $\star_Z$ is the unique $\Lambda^{\omega,\check{\omega}}$-bilinear map satisfying

$$A_i \star_Z A_j = A_i \cup_X A_j + k\delta_{0i}\delta_{0j}\hat{A}_0 t^{\Gamma} + \sum_{k=0}^{l} \sum_{\beta \notin \mathbb{Z}\Gamma} \mathrm{GW}_{0,3}^{X,\beta}(A_i, A_j, A_k)\hat{A}_k t^{\beta} \tag{4.5}$$

for each $i, j \in \{0, \dots, l\}$. By replacing the class A_0 in (4.5) with $\frac{1}{1-t^{\Gamma}} A_0$ and $-\frac{t^{-\Gamma}}{1-t^{-\Gamma}} A_0$, and the class $\hat{A}_0$ with $(1 - t^{\Gamma})\hat{A}_0$ and $-\frac{1-t^{-\Gamma}}{t^{-\Gamma}} \hat{A}_0$, we get the respective isomorphisms (4.4).

4.2. Symplectic cohomology of compact subsets

The main tool in the proof of Theorem 4.2 is a version of symplectic cohomology, which is very similar to definitions of symplectic cohomology in [21, 57, 58].

Definition 4.3. Let (M, ω) be a closed symplectic manifold and let $K \subset M$ be a compact subset. Then we define *symplectic cohomology of $K \subset M$* to be

$$\mathrm{SH}^*(K \subset M) := \varinjlim_{a} \varprojlim_{b} \varinjlim_{H|_K < 0} \mathrm{HF}^*_{[a,b]}(H) \tag{4.6}$$

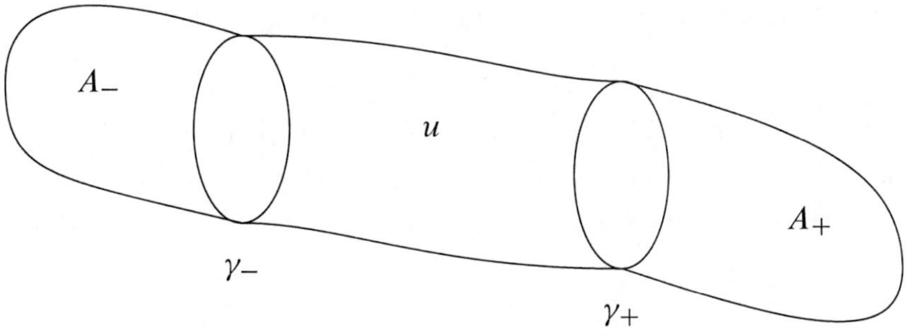

FIGURE 6
Capped Floer trajectory.

where $\mathrm{HF}^*_{[a,b]}(H)$ is a Hamiltonian Floer cohomology group which is defined in the same way as in Section 3, with a few differences:

(1) The chain complex is freely generated over $\mathbb{K}$ by pairs (γ, A), called *capped orbits of action in* $[a, b]$, where $\gamma : \mathbb{R}/\mathbb{Z} \to M$ is a 1-periodic orbit of H and $A \in H_2(M, \gamma; \mathbb{Z})$ is a homology cycle with boundary γ (called a *capping*);

(2) The *action* of (γ, A) is $-\int_A \omega - \int_{\mathbb{R}/\mathbb{Z}} H(\gamma(t))dt$;

(3) The differential only counts cylinders u connecting (γ_-, A_-) and (γ_+, A_+), so that when one caps off each end of u by A_- and A_+, respectively, one gets a null-homologous sphere (Figure 6).

Also the limits are taken with respect to the ordering $\leq$.

The group $\mathrm{SH}^*(K \subset M)$ naturally has a $\Lambda^\omega_{\mathbb{K}}$-module structure induced by the natural $H_2(X; \mathbb{Z})$-action on capped orbits given by adding these classes to the cappings A. It also has a natural "pair of pants" product (see **[43]**) making it a $\Lambda^\omega_{\mathbb{K}}$ algebra. The maps $\mathrm{HF}^*(H_1) \to \mathrm{HF}^*(H_2)$ with $H_1 \leq H_2$ in equation (4.6) above are defined by counting cylinders in a similar way to the differential. As demonstrated in **[57]**, one should really take the direct and inverse limits in equation (4.6) at the chain level in some appropriate homotopy-theoretic sense before taking homology, but we will not do this here for simplicity.

Symplectic cohomology seems to be quite useful when K is a *Liouville subdomain* of (M, ω). A *Liouville subdomain* is a codimension-0 submanifold $K \subset M$ satisfying $\omega|_K = d\theta$ for some 1-form θ with the property that the ω-dual X_θ of θ points outwards along ∂K. One should think of the last condition as a "convexity" condition. Symplectic cohomology satisfies the following properties:

(1) If $c_1(M) = 0$ and K is a Liouville subdomain satisfying a certain "index boundedness" property then $\mathrm{SH}^*(K \subset M)$ only depends on the isotopy class of K. In other words, if we have a smooth family of index-bounded Liouville domains

then the corresponding symplectic cohomology groups are naturally isomorphic.

(2) If $K - M$ is *stably displaceable* then $\mathrm{SH}^*(K \subset M) = \mathrm{SH}^*(M \subset M)$ (a set $P \subset M$ is *stably displaceable* if $\phi(P \times S^1) \cap P \times S^1 = \emptyset$ for some Hamiltonian symplectomorphism ϕ of $M \times T^*S^1$).

(3) $\mathrm{SH}^*(M \subset M) \cong \mathrm{QH}^*(M; \Lambda_{\mathbb{K}}^{\omega})$.

(4) If $c_1(M) = 0$ and K is a Liouville domain satisfying this "index boundedness" property then $\mathrm{SH}^*(K \subset M)$ can be computed using Hamiltonians that are constant outside a neighborhood of K and where these constant orbits do not contribute the chain complex.

4.3. Idea of proof

Here we will give an idea of the proof of Theorem 4.2. Let $\phi : X \dashrightarrow \check{X}$ be our birational isomorphism between Calabi–Yau manifolds X and $\check{X}$, and let ω and $\check{\omega}$ be Kähler forms on X and $\check{X}$, respectively, which admit integral lifts. Choose Zariski-dense affine subvarieties $A \subset X$ and $\check{A} \subset \check{X}$ so that ϕ maps A isomorphically to $\check{A}$. We can modify $\check{\omega}$ so that $\omega|_Q = \phi^*(\check{\omega})|_Q$ for an arbitrarily large compact subset Q of A.

Now one of the key observations is that codimension ≥ 1 subvarieties of Kähler manifolds are stably displaceable (see **[36, SECTION 6.3]**). Combining this with the fact that $c_1(X) = c_1(\check{X}) = 0$, one can choose $\check{\omega}$ very carefully so that there exists a smooth family D_t, $t \in [0,1]$ of Liouville subdomains in $(\check{A}, \check{\omega})$ satisfying an index boundedness property so that $\phi^{-1}(D_0) \subset Q$, $X - \phi^{-1}(D_0)$ is stably displaceable in (X, ω) and $\check{X} - D_1$ is stably displaceable in $(\check{X}, \check{\omega})$. Strictly speaking, such a family of Liouville subdomains $(D_t)_{t \in [0,1]}$ is not constructed in the paper **[36]**, and something slightly more complicated is done instead (see **[36, SECTION 7]**). However, we will assume $(D_t)_{t \in [0,1]}$ exists for simplicity.

We will now explain how to construct the $\Lambda_{\mathbb{K}}^{\omega,\check{\omega}}$-algebra Z in the statement of Theorem 4.2. By property (4), we can find chain complexes computing $\mathrm{SH}^*(X - \phi^{-1}(D_0) \subset X)$ and $\mathrm{SH}^*(\check{X} - D_0)$ involving Hamiltonians which are constant outside a small neighborhood of $\phi^{-1}(D_0)$ (resp. D_0) and such that the constant orbits do not contribute to the chain complex. Now, the regions $V_X \subset X$, $V_{\check{X}} \subset X$ where ϕ and its inverse are ill-defined are of complex codimension at least 2 and hence, by a genericity argument, one can ensure that the Floer trajectories map to the domain or image of ϕ only (Figure 7). This means that we can define these Hamiltonian Floer groups over $\Lambda_{\mathbb{K}}^{\omega,\check{\omega}}$ giving us a new "symplectic cohomology" group Z associated to $\phi^{-1}(D_0) \subset X$. As a result, we can show

$$Z \otimes_{\Lambda_{\mathbb{K}}^{\omega,\check{\omega}}} \Lambda_{\mathbb{K}}^{\omega} \cong \mathrm{SH}^*\big(X - \phi^{-1}(D_0) \subset X\big), \quad Z \otimes_{\Lambda_{\mathbb{K}}^{\omega,\check{\omega}}} \Lambda_{\mathbb{K}}^{\check{\omega}} \cong \mathrm{SH}^*\big(\check{X} - D_0 \subset \check{X}\big). \tag{4.7}$$

The following two equations also hold:

$$\mathrm{SH}^*\big(X - \phi^{-1}(D_0)\big) \overset{(2)}{=} \mathrm{SH}^*(X \subset X) \overset{(3)}{=} \mathrm{QH}^*\big(X; \Lambda_{\mathbb{K}}^{\check{\omega}}\big), \tag{4.8}$$

$$\mathrm{SH}^*(\check{X} - D_0 \subset \check{X}) \overset{(1)}{=} \mathrm{SH}^*(\check{X} - D_1 \subset X) \overset{(2)}{=} \mathrm{SH}^*(\check{X} \subset \check{X}) \overset{(3)}{=} \mathrm{QH}^*\big(\check{X}; \Lambda_{\mathbb{K}}^{\omega}\big). \tag{4.9}$$

Our result now follows from equations (4.7)–(4.9).

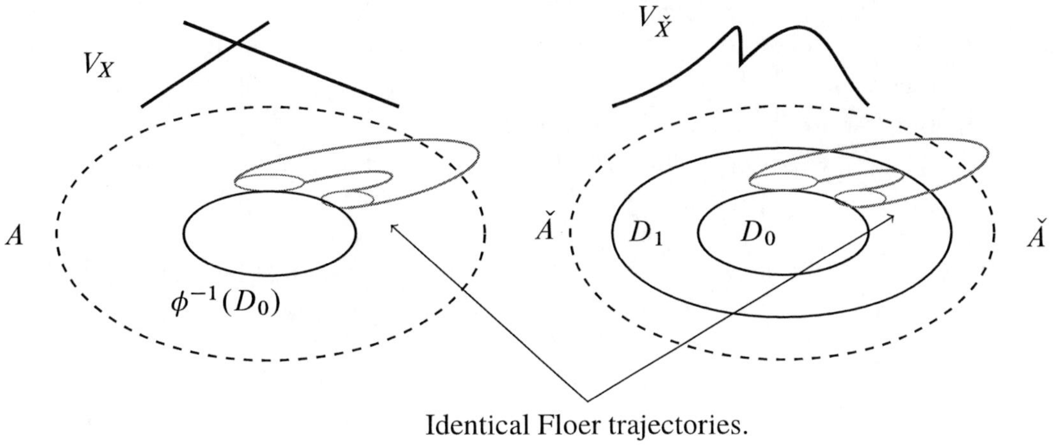

FIGURE 7
Floer trajectories avoiding V_X and $V_{\hat{X}}$.

4.4. Further directions

One of the problems with Theorem 4.2 is that the isomorphisms are not explicit. Let $\Delta \subset X \times \check{X}$ be the closure of the graph of the birational isomorphism ϕ. Then we have a push–pull map

$$\Psi_\Delta : H^*(X;\mathbb{K}) \to H^*(\check{X};\mathbb{K}), \quad \Psi_\Delta(\alpha) := \mathrm{PD}\big((\mathrm{pr}_{\check{X}})_*(\Delta \cap \mathrm{pr}_X^* \alpha)\big) \tag{4.10}$$

where pr_X and $\mathrm{pr}_{\check{X}}$ are the natural projection maps from $X \times \check{X}$ to X and $\check{X}$, respectively, and PD is Poincaré duality.

Conjecture 4.4 ([**60, SECTION 4.3, CONJECTURE I**]). *If $\mathbb{K} = \mathbb{Q}$, then we can identify the quantum cohomology groups of X and $\check{X}$ using the equivalence Ψ_Δ.*

Since the regions V_X and $V_{\check{X}}$ where ϕ and its inverse are ill-defined have complex codimension 2, it should be possible to show that the above conjecture is true if we restrict ourselves to the subalgebra of $H^*(X;\Lambda_X^\omega)$ and $H^*(\check{X};\Lambda_{\check{X}}^\omega)$ generated by elements of degree $0, 1, 2$ and $2n-2$, $2n-1$, and $2n$. Motivated by the fact that symplectic cohomology could, in principle, be computed by relative Gromov–Witten invariants ([**49, REMARK 8.3**], [**13, 19**]), it would be interesting to investigate (over any field $\mathbb{K}$) whether the equivalences (4.4) can be realized in some way by counts of curves in $X \times \check{X}$. This leads us to the following very difficult open problem:

Open Problem 4.5. Can one produce a purely algebraic proof of Theorem 4.2 using relative Gromov–Witten invariants motivated by themes in [**13**] or [**19**]?

Trying to understand what is going on in dimension 3 could be of use here. There should be a version of symplectic cohomology of $M \subset M$ which is defined using *bulk deformed* Hamiltonian Floer cohomology (see [**18, 54**]). This is naturally isomorphic to big quantum cohomology. However, the methods of Section 4.3 do not work using this bulk deformed version of Hamiltonian Floer cohomology due to the fact that the definition involves both cycles and orbits. There should be a version of Hamiltonian Floer cohomology which only uses orbits and Riemann surfaces satisfying the perturbed Floer equation joining them so that the associated symplectic cohomology group of $M \subset M$ is isomorphic to big quantum cohomology (see [41]). However, such a construction requires an additional choice of a "trivialization of a circle action."

Open Problem 4.6. Can one use the techniques above to prove that birational Calabi–Yau manifolds have the "same" big quantum cohomology groups (maybe, up to some additional choices).

The article [53] gave a potential definition for Hamiltonian Floer cohomology in the setting of orbifolds. This leads to the following open problem:

Open Problem 4.7. Suppose that X and $\check{X}$ are birational Calabi–Yau orbifolds. Can one use the techniques in the previous section to relate the quantum cohomology of X and $\check{X}$.

An example of such a birational transform is a crepant resolution as in Section 3. This problem has an additional serious difficulty which is that the birational transform might be ill-defined on a codimension 1 region. This means that the analogue of equation (4.7) does not hold. However, there might be additional genus Gromov–Witten invariants counting curves mapping to the locus where ϕ and ϕ^{-1} are ill-defined which might correct for this.

ACKNOWLEDGMENTS

We thank Stony Brook math community for their support over the years and Mohammed Abouzaid and Ivan Smith for helpful advice.

FUNDING

This work was partially supported by NSF grant DMS-1811861.

REFERENCES

[1] M. Audin and M. Damian, *Morse theory and floer homology*. Universitext, Springer, London; EDP Sciences, Les Ulis, 2014.

[2] E. Bao and K. Honda, Semi-global Kuranishi charts and the definition of contact homology. 2015, arXiv:1512.00580.

[3] V. V. Batyrev, Birational Calabi–Yau n-folds have equal Betti numbers. In *New trends in algebraic geometry (Warwick, 1996)*, edited by K. Hulek, M. Reid, C. Peters, and F. Catanese, pp. 1–11, London Math. Soc. Lecture Note Ser. 264, Cambridge Univ. Press, Cambridge, 1999.

[4] V. V. Batyrev, Non-Archimedean integrals and stringy Euler numbers of log-terminal pairs. *J. Eur. Math. Soc. (JEMS)* **1** (1999), no. 1, 5–33.

[5] S. Boucksom, T. de Fernex, C. Favre, and S. Urbinati, Valuation spaces and multiplier ideals on singular varieties. In *Recent advances in algebraic geometry*, edited by C. Hacon, M. Mustaţă, and M. Popa, pp. 29–51, London Math. Soc. Lecture Note Ser. 417, Cambridge Univ. Press, Cambridge, 2015.

[6] T. Bridgeland, A. King, and M. Reid, The McKay correspondence as an equivalence of derived categories. *J. Amer. Math. Soc.* **14** (2001), no. 3, 535–554.

[7] E. Brieskorn, Beispiele zur Differentialtopologie von Singularitäten. *Invent. Math.* **2** (1966), no. 1, 1–14.

[8] N. Budur, J. Fernández de Bobadilla, Q. T. Lê, and H. D. Nguyen, Cohomology of contact loci. 2019, arXiv:1911.08213.

[9] K. Cieliebak, A. Floer, and H. Hofer, Symplectic homology. II. A general construction. *Math. Z.* **218** (1995), no. 1, 103–122.

[10] K. Cieliebak and A. Oancea, Symplectic homology and the Eilenberg–Steenrod axioms. *Algebr. Geom. Topol.* **18** (2018), no. 4, 1953–2130.

[11] A. Craw, *The McKay correspondence and representations of the McKay quiver*. PhD thesis, University of Warwick, 2001.

[12] J. Denef and J. Loeser, Geometry on arc spaces of algebraic varieties. In *European Congress of Mathematics, Vol. I (Barcelona, 2000)*, edited by A. Chambert-Loir, J. H. Lu, M. Ruzhansky, and Y. Tschinkel, pp. 327–348, Progr. Math. 201, Birkhäuser, Basel, 2001.

[13] L. Diogo and S. Lisi, Symplectic homology of complements of smooth divisors. *J. Topol.* **12** (2019), no. 3, 967–1030.

[14] Y. Eliashberg, A. Givental, and H. Hofer, Introduction to symplectic field theory. In *Visions in mathematics, GAFA 2000 special volume, part II*, edited by N. Alon, J. Bourgain, A. Connes, M. Gromov, and V. Milman, pp. 560–673, Modern Birkhäuser Classics. Birkhäuser, Basel, 2000.

[15] A. Floer, Morse theory for fixed points of symplectic diffeomorphisms. *Bull. Amer. Math. Soc. (N.S.)* **16** (1987), no. 2, 279–281.

[16] A. Floer and H. Hofer, Symplectic homology. I. Open sets in $\mathbb{C}^n$. *Math. Z.* **215** (1994), no. 1, 37–88.

[17] R. Friedman, On threefolds with trivial canonical bundle. In *Complex geometry and Lie theory (Sundance, UT, 1989)*, edited by J. A. Carlson, C. H. Clemens, and D. R. Morrison, pp. 103–134, Proc. Sympos. Pure Math. 53, Amer. Math. Soc., Providence, RI, 1991.

[18] K. Fukaya, Y. G. Oh, H. Ohta, and K. Ono, *Spectral invariants with bulk, quasi-morphisms and Lagrangian Floer theory*. Mem. Amer. Math. Soc. 260, American Mathematical Society, 2019.

[19] S. Ganatra and D. Pomerleano, Symplectic cohomology rings of affine varieties in the topological limit. *Geom. Funct. Anal.* **30** (1993), no. 2, 334–345.

[20] H. Gillet and C. Soulé, Descent, motives and K-theory. *J. Reine Angew. Math.* **478** (1996), 127–176.

[21] Y. Groman, Floer theory and reduced cohomology on open manifolds. 2015, arXiv:1510.04265.

[22] P. Heegaard, *Forstudier til en topologisk teori for de algebraiske fladers sammenhaeng*. Bojesen, 1898.

[23] H. Hironaka, Resolution of singularities of an algebraic variety over a field of characteristic zero. *Ann. of Math.* **79** (1964), no. 2, 109–203.

[24] S. Ishikawa, Construction of general symplectic field theory. 2018, arXiv:1807.09455.

[25] Y. Kawamata, The minimal discrepancy of a 3-fold terminal singularity, Appendix to Vyacheslav Shokurov. 3-fold log flips. *Russian Acad. Sci. Izv. Math* **40** (1993), no. 3, 93–202.

[26] Y. Kawamata, D-equivalence and K-equivalence. *J. Differential Geom.* **61** (2002), no. 1, 147–171.

[27] J. Kollár and A. Némethi, Holomorphic arcs on singularities. *Invent. Math.* **200** (2015), no. 1, 97–147.

[28] M. Kontsevich, *Grothendieck ring of motives and related rings*. Lecture at Orsay, 1995.

[29] Y. P. Lee, H. W. Lin, F. Qu, and C. L. Wang, Invariance of quantum rings under ordinary flops III: a quantum splitting principle. *Camb. J. Math.* **4** (2016), no. 3, 333–401.

[30] Y. P. Lee, H. W. Lin, and C. L. Wang, Invariance of quantum rings under ordinary flops I: quantum corrections and reduction to local models. *Algebr. Geom.* **3** (2016), no. 5, 578–614.

[31] Y. P. Lee, H. W. Lin, and C. L. Wang, Invariance of quantum rings under ordinary flops II: a quantum Leray–Hirsch theorem. *Algebr. Geom.* **3** (2016), no. 5, 615–653.

[32] A. M. Li and Y. Ruan, Symplectic surgery and Gromov–Witten invariants of Calabi–Yau 3-folds. *Invent. Math.* **145** (2001), no. 1, 151–218.

[33] D. Markushevich, Minimal discrepancy for a terminal cDV singularity is 1. *J. Math. Sci. Univ. Tokyo* **3** (1996), 445–456.

[34] M. McLean, Reeb orbits and the minimal discrepancy of an isolated singularity. *Invent. Math.* **204** (2016), no. 2, 505–594.

[35] M. McLean, Floer cohomology, multiplicity and the log canonical threshold. *Geom. Topol.* **23** (2019), no. 2, 957–1056.

[36] M. McLean, Birational Calabi–Yau manifolds have the same small quantum products. *Ann. of Math. (2)* **191** (2020), no. 2, 439–579.

[37] M. McLean and A. Ritter, The McKay correspondence via Floer theory. 2018, arXiv:1802.01534.

[38] D. Morrison, Beyond the Kähler cone. In *Proceedings of the Hirzebruch 65 Conference on Algebraic Geometry (Ramat Gan, 1993)*, edited by M. Teicher and F. Hirzebruch, pp. 361–376, Israel Math. Conf. Proc. 9, Bar-Ilan Univ., Ramat Gan, 1996.

[39] D. Mumford, The topology of normal singularities of an algebraic surface and a criterion for simplicity. *Publ. Math. Inst. Hautes Études Sci.* **9** (1961), no. 1, 5–22.

[40] A. Oancea, A survey of Floer homology for manifolds with contact type boundary or symplectic homology. In *Symplectic geometry and Floer homology. A survey of the Floer homology for manifolds with contact type boundary or symplectic homology*, edited by F. Laudenbach and A. Oancea, pp. 51–91, Ensaios Mat. 7, Soc. Brasil. Mat., Rio de Janeiro, 2004.

[41] A. Oancea and D. Vaintrob, The Deligne–Mumford operad as a trivialization of the circle action. 2020, arXiv:2002.10734.

[42] J. Pardon, Contact homology and virtual fundamental cycles. *J. Amer. Math. Soc.* **32** (2019), no. 3, 825–919.

[43] S. Piunikhin, D. Salamon, and M. Schwarz, Symplectic Floer–Donaldson theory and quantum cohomology. In *Contact and symplectic geometry (Cambridge, 1994)*, edited by C. B. Thomas, pp. 171–200, Publications of the Newton Institute. 8, Camb. Univ. Press, 1996.

[44] M. Reid, La correspondance de McKay. *Astérisque* **276** (2002), 53–72.

[45] M. Reid, Minimal models of canonical 3-folds. In *Algebraic varieties and analytic varieties (Tokyo, 1981)*, edited by S. Iitaka, pp. 131–180, Adv. Stud. Pure Math. 1, North-Holland, Amsterdam, 1983.

[46] M. Reid, McKay correspondence, 1997, arXiv:alg-geom/9702016.

[47] R. Ruan, Virtual neighborhoods and pseudo-holomorphic curves. *Turkish J. Math.* **23** (1999), no. 1, 161–231.

[48] Y. Ruan, Surgery, quantum cohomology and birational geometry. In *Northern California symplectic geometry seminar*, edited by Y. Eliashberg, D. Fuchs, T. Ratiu, and A. Weinstein, pp. 183–198, Amer. Math. Soc. Transl. Ser. 2 196, Amer. Math. Soc., Providence, RI, 1999.

[49] P. Seidel, A biased view of symplectic cohomology. In *Current developments in mathematics, 2006*, edited by D. Jerison, B. Mazur, T. Mrowka, W. Schmid, R. P. Stanley, and S. T. Yau, pp. 211–253, Current Dev. Math., Int. Press, Somerville, MA, 2006.

[50] P. Seidel, Symplectic cohomology graduate course at MIT, 2007.

[51] V. V. Shokurov, Problems about Fano varieties. In *Birational geometry of algebraic varieties, open problems*, edited by S. Mori, J. Kollár, S. Mukai, and Y. Miyaoka, pp. 30–32, the 23rd International Symposium, Division of Mathematics, the Taniguchi Foundation, Katata, 1988.

[52] V. V. Shokurov, Letters of a bi-rationalist. IV. Geometry of log flips. In *Algebraic geometry*, edited by M. C. Beltrametti, F. Catanese, C. Ciliberto, A. Lanteri, and C. Pedrini, pp. 313–328, De Gruyter, New York, 2008.

[53] M. Thaddeus, Floer cohomology with gerbes. In *Enumerative invariants in algebraic geometry and string theory*, edited by M. Marino, M. Thaddeus, and R. Vakil, pp. 105–141, Lecture Notes in Math. 1947, Springer, Berlin, 2008.

[54] M. Usher, Deformed Hamiltonian Floer theory, capacity estimates and Calabi quasimorphisms. *Geom. Topol.* **15** (2011), no. 3, 1313–1417.

[55] I. Ustilovsky, Infinitely many contact structures on S^{4m+1}. *Int. Math. Res. Not.* **14** (1999), 781–791.

[56] A. N. Varčenko, Contact structures and isolated singularities. *Vestnik Moskov. Univ. Ser. I Mat. Mekh.* **101** (1980), no. 2, 18–21.

[57] U. Varolgunes, Mayer–Vietoris property for relative symplectic cohomology. *Geom. Topol.* **25** (2021), no. 2, 547–642.

[58] S. Venkatesh, Rabinowitz Floer homology and mirror symmetry. *J. Topol.* **11** (2018), no. 1, 144–179.

[59] C. Viterbo, Functors and computations in Floer homology with applications. I. *Geom. Funct. Anal.* **9** (1999), no. 5, 985–1033.

[60] C. L. Wang, K-equivalence in birational geometry. In *Second International Congress of Chinese Mathematicians*, edited by C. S. Lin and S. T. Yau, pp. 199–216, New Stud. Adv. Math. 4, Int. Press, Somerville, MA, 2004.

[61] M. Zargar, Integration of Voevodsky motives. 2017, arXiv:1711.02015.

MARK MCLEAN

Stony Brook Mathematics Department, Stony Brook University, 100 Nicolls Rd, Stony Brook, NY 11794, USA, markmclean@math.stonybrook.edu

SURFACES VIA SPINORS AND SOLITON EQUATIONS

ISKANDER A. TAIMANOV

ABSTRACT

This article surveys the Weierstrass representation of surfaces in the three- and four-dimensional spaces, with an emphasis on its relation to the Willmore functional. We also describe an application of this representation to constructing a new type of solutions to the Davey–Stewartson II equation. They have regular initial data, gain one-point singularities at certain moments of time, and extend to smooth solutions for the remaining times.

MATHEMATICS SUBJECT CLASSIFICATION 2020

Primary 53A05; Secondary 35B38, 35Q51, 53C422

KEYWORDS

Surfaces in the Euclidean spaces, Weierstrass (spinor) representation of surfaces, two-dimensional Dirac operator, Willmore functional, Davey–Stewartson equation

ICM INTERNATIONAL CONGRESS OF MATHEMATICIANS 2022 JULY 6–14

Proc. Int. Cong. Math. 2022, Vol. 4, pp. 2638–2654
DOI 10.4171/ICM2022/92

1. THE WEIERSTRASS (SPINOR) REPRESENTATION OF SURFACES IN THE THREE-SPACE

The Weierstrass representation for minimal surfaces in the three-space is as follows: for any pair of holomorphic functions ψ_1 and $\bar{\psi}_2$ defined in a domain $U \subset \mathbb{C}$ in the complex plane, the formulae

$$
\begin{aligned}
x^1(P) &= \frac{i}{2} \int \left[(\psi_1^2 + \bar{\psi}_2^2) dz + (\bar{\psi}_1^2 + \psi_2^2) d\bar{z} \right] + x^1(P_0), \\
x^2(P) &= \frac{1}{2} \int \left[(-\psi_1^2 + \bar{\psi}_2^2) dz + (-\bar{\psi}_1^2 + \psi_2^2) d\bar{z} \right] + x^2(P_0), \\
x^3(P) &= \int [\psi_1 \bar{\psi}_2 dz + \bar{\psi}_1 \psi_2 d\bar{z}] + x^3(P_0)
\end{aligned}
\tag{1.1}
$$

determine a minimal surface in $\mathbb{R}^3$. Here we assume that U is simply-connected or the integrals over cycles in U vanish, and the integrals are taken along a path from a fixed point $P_0 \in U$ to P. Moreover, every minimal surface admits such a representation. Weierstrass used another data, namely $f = \bar{\psi}_2^2$ and $g = \frac{\psi_1}{\bar{\psi}_2}$. However, for the generalization of this representation, it is worth to consider ψ_1 and ψ_2 and treat this pair as a solution of the Dirac equation

$$
D\psi = 0, \quad \psi = \begin{pmatrix} \psi_1 \\ \psi_2 \end{pmatrix}, \tag{1.2}
$$

for a two-dimensional Dirac operator of the form

$$
D = \begin{pmatrix} 0 & \partial \\ -\bar{\partial} & 0 \end{pmatrix} + \begin{pmatrix} U & 0 \\ 0 & U \end{pmatrix}, \quad U = \bar{U},
$$

where a real-valued potential U vanishes for minimal surfaces. Now the Weierstrass representation generalizes as follows:

Theorem 1.1 ([16]). *For every solution ψ of (1.2), the formulae (1.1) define a surface in $\mathbb{R}^3$ for which z is a conformal parameter, the induced metric takes the form*

$$
ds^2 = e^{2\alpha} dz d\bar{z}, \quad e^\alpha = |\psi_1|^2 + |\psi_2|^2,
$$

and the potential U of the Dirac operator equals

$$
U = \frac{He^\alpha}{2},
$$

where H is the mean curvature.

Theorem 1.2 ([26]). *Every surface in $\mathbb{R}^3$ (with a fixed conformal parameter z on it) admits such a representation even globally. Therewith ψ is a section of a spinor bundle over the surface, the form $U^2 dx \wedge dy$ is globally defined and its integral over the surface is proportional to the Willmore functional*

$$
\mathcal{W} = \int H^2 d\mu = 4 \int U^2 dx \wedge dy,
$$

where $d\mu$ is the induced area form of the surface.

Hence, being considered for the Dirac operators with general real-valued potentials, the formulae (1.1) define the Weierstrass (spinor) representation of general surfaces in $\mathbb{R}^3$.

Theorem 1.1 was derived from the similar formulae in the book by Eisenhart [9, PROBLEM 35.4] where instead of (1.2) the following condition is used:

$$L\psi_1 = L\bar{\psi}_2 = 0, \quad L = \partial\bar{\partial} - \frac{\partial \log U}{U}\bar{\partial} + U^2.$$

Here D naturally arises as the "square root" of the Schrödinger operator L. The representation based on the Dirac operator provides many more opportunities because its potential has no singularities and the operator has good spectral properties. In the advanced problems of his textbook, Eisenhart frequently proposed to prove results from various articles, and we cannot exclude that these formulae might be traced to some earlier publication. It appears that this local representation is equivalent to another one derived in [14], where the Dirac operator was not used either.

In [16] the Weierstrass representation was used for introducing the deformations of surfaces admitting such a representation. The operator D generates a hierarchy of solution equations of the form

$$\frac{\partial D}{\partial t_n} = [D, A_n] - B_n D,$$

where A_n and B_n are matrix differential operators such that the principal term of A_n takes the form

$$A_n = \begin{pmatrix} \partial^{2n+1} + \bar{\partial}^{2n+1} & 0 \\ 0 & \partial^{2n+1} + \bar{\partial}^{2n+1} \end{pmatrix} + \cdots .$$

This evolution preserves the zero energy level of D deforming the corresponding eigenfunctions

$$\frac{\partial \psi}{\partial t} + A\psi = 0 \qquad (1.3)$$

and $D\psi_0 = 0$ for the initial data $\psi_0 = \psi|_{t=t_0}$, then $D\psi = 0$ for all $t \geq t_0$.

For $n = 1$, we have the modified Novikov–Veselov (mNV) equation [5]

$$U_t = \left(U_{zzz} + 3U_z V + \frac{3}{2}UV_z\right) + \left(U_{\bar{z}\bar{z}\bar{z}} + 3U_{\bar{z}}\bar{V} + \frac{3}{2}U\bar{V}_{\bar{z}}\right)$$

where

$$V_{\bar{z}} = (U^2)_z.$$

In the case when $U|_{t=0}$ depends only on x, we have $U = U(x, t)$ and the mNV equation reduces to the modified Korteweg–de Vries equation $U_t = \frac{1}{4}U_{xxx} + 6U_x U^2$ (here $V = U^2$). In the same manner, the original Novikov–Veselov equation

$$U_t = U_{zzz} + U_{\bar{z}\bar{z}\bar{z}} + (VU)_z + (\bar{V}U)_{\bar{z}}, \quad V_{\bar{z}} = 3U_z$$

generalizes the Korteweg–de Vries equation.

The mNV deformation introduced in [16] is as follows: let a surface be induced by ψ via (1.1) and consider solutions U and ψ of the mNV equation and (1.3) with given initial data. Then for any moment of time, we have a spinor ψ that determines the deformed surface. In fact, we have infinitely many deformations defined up to translations by $(x^1(P_0,t), x^2(P_0,t), x^3(P_0,t))$. This is some family of the mNV deformations of the surface.

Theorem 1.3 ([26]). *The mNV deformations evolve tori into tori and preserve their conformal classes and the values of the Willmore functional.*

Theorems 1.2 and 1.3 hint at the relation of this representation to the Willmore functional. Formulae (1.1) give immersions of the universal covers of surfaces and there are no compact minimal surfaces without boundary in $\mathbb{R}^3$. Hence the infima for the Willmore functional for various conformal classes of closed surfaces show how much stress must be applied for converting an immersion of the universal cover into an immersion of a closed surface. In Section 2 we briefly expose how the Weierstrass representation was applied to studying the conformal geometry of surfaces.

In Section 4, in contrast to Section 2 where analysis was applied to geometry, we discuss the recent applications of geometry to analysis. We show how to construct exact solutions to the Davey–Stewartson II equation. Therewith, geometry of surfaces helps in finding a new scenario for creating singularities of solutions with regular initial data.

It would be interesting to apply the Weierstrass representation to other problems of the surface theory (bending, existence of umbilics, etc.). In particular, if some conjecture appears false, then methods of integrable systems can help in constructing an explicit counterexample (see, for instance, [1]).

2. SPECTRAL CHARACTERISTICS OF D AND CONFORMAL GEOMETRY OF SURFACES

The Willmore conjecture which states that the minimum of the Willmore functional among tori in $\mathbb{R}^3$ is attained at the Clifford torus was proved in [19] by means of the geometric measure theory and calculus of variations.

In the mid-1990s we proposed an approach to proving it using Theorem 1.3 and the integrable systems theory. This approach was not implemented, but we think it is worth to be briefly exposed here.

It was conjectured in [26] that

a nonstationary torus (with respect to the mNV flow and up to translations) cannot be a local minimum of the Willmore functional.

Otherwise, by Theorem 1.3, the minimum of the Willmore functional would contain an infinite family of tori invariant under the mNV flow and this would be very unlikely. By the general philosophy of integrable systems, the stationary solution to the mNV equation has the simplest possible spectral curve [27].

Since the flow preserves the conformal classes of tori, the same conjecture has to be valid for tori of every fixed conformal class.

For two-dimensional differential operators with periodic coefficients, the spectral curve (on the zero level energy) parameterizes its Floquet eigenfunctions [8]. In our case a Floquet eigenfunction ψ of the operator D with the eigenvalue (or the energy) E is a formal solution to the equation

$$D\psi = E\psi$$

which satisfies the periodicity conditions

$$\psi(q+\gamma_j) = e^{2\pi i (k,\gamma_j)}\psi(z,\bar{z}), \quad j = 1,2,$$

where γ_1 and γ_2 generate the lattice of periods Λ of the potential U and $(k,\gamma) = k_1\gamma_j^1 + k_2\gamma_j^2$ is the inner product. The quantities $k_1, k_2 \in \mathbb{C}$ are called the quasimomenta of ψ and $\mu(\gamma_j) = e^{2\pi i (k,\gamma_j)}$ are Floquet multipliers. All possible triples (k_1, k_2, E) for which Floquet functions exist form an analytic subset $Q(U)$ in $\mathbb{C}^3$, invariant under the dual lattice $\Lambda^* \subset \mathbb{R}^2 \subset \mathbb{C}^2$ acting on the quasimomenta. We proved that for the two-dimensional operators $\Delta + U$ and $\partial_y - \partial_x^2 + U$ in 1985. However, this paper was unpublished, although referred in [18] and was exposed in [30]. Now we define the spectral curve as the complex curve

$$\Gamma = (Q \cap \{E = 0\})/\Lambda^*$$

and consider it up to biholomorphic equivalence, making the definition independent on the choice of a basis for Λ. The curve is an invariant of the mNV flow, it is naturally completed by a couple of points at infinity, which compactify it in the case of finite genus. The Floquet functions are glued into a meromorphic section over Γ. The above rough definition must be detailed for singular spectral curves. In general, the space of Floquet functions corresponding to a point from Γ is one-dimensional and the multiple points have to be normalized in such a manner that for the resulting curve Γ_ψ to every point there corresponds a one-dimensional space, there is a meromorphic section ψ of this bundle, and every Floquet function is a linear combination of sections at different points (see the definition of Γ_ψ in [30]). The spinor ψ generating a torus via (1.1) has the Floquet multipliers equal to ± 1.

The spectral curve defined for D is a particular case of the general spectral curves which play a fundamental role in integrable systems. They are the first integrals of the system (that was first showed for the Korteweg–de Vries equation in [22]. The particular case of them are the spectral curves of constant mean curvature tori which are always of finite genus [13, 25]. In general, this spectral curve is of infinite genus. For finite genera cases, solutions to the integrable systems are expressed in terms of theta functions on spectral curves. In our case all Floquet functions are reconstructed from certain data related to Γ_ψ and the value of the Willmore functional is also determined by them [27]. We conjectured that

> *for tori in $\mathbb{R}^3$, the curve Γ, i.e., the set of the multipliers $\mu(\gamma_j)$, is conformally invariant (as is the Willmore functional).*

Since this is evident for translations and rotations, one was left to prove the same for the Möbius inversion, which was accomplished in [12].

For the Clifford torus parameterized by x, y such that $0 \leq x, y \leq 2\pi$, the potential U of its Weierstrass representation is

$$U(x) = \frac{\sin x}{2\sqrt{2}(\sin x - \sqrt{2})}$$

and its spectral curve Γ_ψ is $\mathbb{C}P^1$ with two pairs of glued points.

For differential operators on surfaces of higher genera, the analog of Floquet–Bloch theory is unknown. It would be interesting to find it, if it exists, for the Dirac operator D.

For spheres, there are no analogs of the Floquet functions and the zero energy level of D just consists of the kernel Ker D.

We notice that there is an antiinvolution

$$\begin{pmatrix} \psi_1 \\ \psi_2 \end{pmatrix} \xrightarrow{\sigma} \begin{pmatrix} -\bar{\psi}_2 \\ \bar{\psi}_1 \end{pmatrix}, \quad \sigma^2 = -1, \tag{2.1}$$

acting on Ker D. This implies that the dimension of the kernel over $\mathbb{C}$ is always even.

We say that a sphere in $\mathbb{R}^3$ admits a spinor representation with a one-dimensional potential if after removing a certain pair of points we obtain the cylinder $\mathbb{R} \times S^1$ for which the potential of the representation depends on x only, i.e., $U = U(x)$. These are, for instance, spheres of revolution. By using the inverse scattering transform of one-dimensional Dirac operators on the line, we proved

Theorem 2.1 ([28]). *For spheres with a one-dimensional potential, we have*

$$\mathcal{W} = 4 \int U^2 dx \wedge dy \geq 4\pi N^2, \tag{2.2}$$

where $\dim_{\mathbb{C}}$ Ker $D = 2N$, *and the equalities are achieved at the soliton potentials*

$$U_N(x) = \frac{N}{2\cosh x}.$$

We call the spheres that correspond to these potentials soliton spheres, and it appears that they have very interesting geometrical properties [6]. In [28] we conjectured that

inequality (2.2) holds for all spheres.

Soon after the preprint of [28] appeared, Friedrich showed that this conjecture implies the following statement:

Given an eigenvalue λ of the Dirac operator D on a two-dimensional spin-manifold homeomorphic to the two-sphere,

$$\lambda^2 \mathrm{Area}(M) \geq \pi m^2(\lambda), \tag{2.3}$$

where $m(\lambda)$ is the multiplicity of λ.

For $m(\lambda) = 2$, inequality (2.3) was already proved by Bär [2].

The arguments by Friedrich were as follows. On a spin-manifold of dimension 2 with the metric $e^{2\alpha} dz d\bar{z}$, the Dirac operator (on the spin-manifold) takes the form

$$D = 2e^{-3\alpha/2} \begin{pmatrix} 0 & \partial \\ -\bar{\partial} & 0 \end{pmatrix} e^{\alpha/2},$$

and the equation

$$D\varphi = \lambda\varphi$$

is rewritten as

$$\left[\begin{pmatrix} 0 & \partial \\ -\bar{\partial} & 0 \end{pmatrix} - \frac{\lambda e^{\alpha}}{2}\right]\psi = 0,$$

where $\psi = e^{\alpha/2}\varphi$, and if λ is constant, then (2.2) implies (2.3). Moreover, if $\lambda = H$, then this is exactly the Dirac equation (1.2) (the sign of the mean curvature can be changed without any loss) and, since $e^{\alpha} = |\psi|^2$, we have $|\varphi| = 1$. Therefore the Weierstrass representation is rewritten in terms of solutions of the Dirac equation

$$D\varphi = H\varphi$$

of constant length, $|\varphi| = 1$ [11, THEOREM 13].

This embedding of the Weierstrass representation into the general framework of Dirac operators on spin-manifolds appears very fruitful: it led to its generalization, the spinorial representation of immersions of manifolds, which are not necessarily two-dimensional, into certain homogeneous spaces (see [3] and references therein).

The Weierstrass representation for surfaces in $\mathbb{R}^3$ was generalized for surfaces in three-dimensional Lie groups with left-invariant metrics in [4]. It helped establish some facts on constant mean curvature surfaces in these groups.

It would be interesting, at least as a test problem, to find a discretization of the Weierstrass representation by means of discrete complex analysis. In [36] that was done for the generalizations of the representation for time-like surfaces in $\mathbb{R}^{2,1}$, $\mathbb{R}^{3,1}$, and $\mathbb{R}^{2,2}$. But in these cases complex analysis is not involved because the principal term of the Dirac operator D has the form $\begin{pmatrix} 0 & \partial_\xi \\ \partial_\eta & 0 \end{pmatrix}$ where ξ and η are isotropic coordinates.

The conjectured inequality (2.2) was finally proved with its generalizations for surfaces of higher genera:

Theorem 2.2 ([10]). *For a closed oriented surface of genus g immersed into $\mathbb{R}^3$ via* (1.1) *and* (1.2), *we have*

$$\int U^2 dx \wedge dy \geq \begin{cases} \pi N^2, & \text{for } g = 0, \\ \begin{cases} \frac{\pi N^2}{4} & \text{for } N \text{ even}, \\ \frac{\pi(N^2-1)}{4} & \text{for } N \text{ odd}, \end{cases} & \text{for } g = 1, \\ \frac{\pi}{4g}(N^2 - g^2), & \text{for } g > 1, \end{cases}$$

where $\dim_{\mathbb{C}} \operatorname{Ker} D = 2N$.

3. SURFACES IN THE FOUR-SPACE AND THE DAVEY–STEWARTSON EQUATION

Theorem 2.2 was derived from the Plücker formula in the quaternionic algebraic geometry [10].

The Weierstrass representation allows applying to surface theory other branches of mathematics. In Section 2 we discuss an approach based on the spectral theory of the Dirac operator. The quaternionic algebraic geometry applies algebro-geometrical methods by considering solutions of the Dirac equation as "holomorphic" sections of spinor bundles. It starts with treating the symmetry (2.1) as a multiplication by an imaginary unit j and considering $\operatorname{Ker} D$ as a linear space over quaternions $\mathbb{H}$ [24]. Therewith one may consider the Dirac operator of the more general form

$$D = \begin{pmatrix} 0 & \partial \\ -\bar{\partial} & 0 \end{pmatrix} + \begin{pmatrix} U & 0 \\ 0 & \bar{U} \end{pmatrix} \tag{3.1}$$

whose kernel is also invariant under (2.1).

For that we identify $\mathbb{C}^2$ with $\mathbb{H}$ as follows:

$$(z_1, z_2) \to z_1 + jz_2 = \begin{pmatrix} z_1 & -\bar{z}_2 \\ z_2 & \bar{z}_1 \end{pmatrix}$$

and consider the two matrix operators

$$\bar{\partial} = \begin{pmatrix} \bar{\partial} & 0 \\ 0 & \partial \end{pmatrix}, \quad jU = j\begin{pmatrix} U & 0 \\ 0 & \bar{U} \end{pmatrix} = \begin{pmatrix} 0 & -\bar{U} \\ U & 0 \end{pmatrix},$$

where $j \in \mathbb{H}$ is the imaginary unit for which we have $j^2 = -1$, $zj = j\bar{z}$, and $\bar{\partial} j = j\partial$. Then the Dirac equation $D\psi = 0$ takes the form

$$(\bar{\partial} + jU)(\psi_1 + j\psi_2) = (\bar{\partial}\psi_1 - \bar{U}\psi_2) + j(\partial\psi_2 + U\psi_1) = 0.$$

Since ψ_1 and $\bar{\psi}_2$ are sections of the same bundle E, we rewrite the Dirac equation as

$$(\bar{\partial} + jU)(\psi_1 + \bar{\psi}_2 j) = 0$$

and treat $E \oplus E$ as a quaternionic line bundle whose sections are of the form $\psi_1 + \bar{\psi}_2 j$. The symmetry (2.1) induces some quaternion linear endomorphism J of E such that $J^2 = -1$, $\psi_1 + \bar{\psi}_2 j \to (\psi_1 + \bar{\psi}_2 j)j = -\bar{\psi}_2 + \psi_1 j$, and this J defines for any quaternion fiber a canonical splitting into $\mathbb{C} \oplus \mathbb{C}$ (in our case this is a splitting into ψ_1 and $\bar{\psi}_2$) and such a bundle is called a "complex quaternionic line bundle." The kernel of $D = \bar{\partial} + jU$ is invariant under the right-side multiplications by constant quaternions and hence is a linear space over $\mathbb{H}$.

The "quaternionic" analog of the classical the Plücker formula established in [10] implies (2.2) and (2.3).

By using the analogy with complex algebraic geometry, other interesting results were obtained, in particular on Bäcklund transformations and special classes of surfaces. Moreover, this approach offers another opportunity: in its framework the Weierstrass representation was also extended to surfaces in $\mathbb{R}^4$ and therewith $\mathbb{R}^4$ was naturally identified with $\mathbb{H}$. In the coordinate language, the representation was written down in [17] and is as follows.

Let D be of the form (3.1) and introduce the formally conjugate operator

$$D^\vee = \begin{pmatrix} 0 & \partial \\ -\bar\partial & 0 \end{pmatrix} + \begin{pmatrix} \bar U & 0 \\ 0 & U \end{pmatrix}.$$

Theorem 3.1 ([17]). *If ψ and φ satisfy the equations*

$$D\psi = 0, \quad D^\vee\varphi = 0. \tag{3.2}$$

then the formulae

$$\begin{aligned} x^k(P) &= x^k(P_0) + \int (x^k_z dz + \bar x^k_z d\bar z), \quad k = 1,2,3,4, \\ x^1_z &= \frac{i}{2}(\bar\varphi_2\bar\psi_2 + \varphi_1\psi_1), \quad x^2_z = \frac{1}{2}(\bar\varphi_2\bar\psi_2 - \varphi_1\psi_1), \\ x^3_z &= \frac{1}{2}(\bar\varphi_2\psi_1 + \varphi_1\bar\psi_2), \quad x^4_z = \frac{i}{2}(\bar\varphi_2\psi_1 - \varphi_1\bar\psi_2), \end{aligned} \tag{3.3}$$

define the surface in $\mathbb{R}^4$ for which the induced metric is given by $e^{2\alpha}dzd\bar z = (|\psi_1|^2 + |\psi_2|^2)(|\varphi_1|^2 + |\varphi_2|^2)dzd\bar z$ and $|U| = \frac{|\mathbf{H}|e^\alpha}{2}$ with $\mathbf{H}$ being the mean curvature vector.

For $U = \bar U$ and $\psi = \varphi$, this representation reduces to (1.1).

The converse is also true but there is a difference with surfaces in $\mathbb{R}^3$ for which a choice of a parameter z defines ψ uniquely up to multiplication by ± 1.

Theorem 3.2 ([29]). *Every oriented surface (with a given conformal parameter) has representation* (3.3). *The spinors ψ and φ are defined up to the gauge transformations*

$$\psi_1 \to e^h\psi_1, \quad \psi_2 \to e^{\bar h}\psi_2, \quad \varphi_1 \to e^{-h}\varphi_1, \quad \varphi_2 \to e^{-\bar h}\varphi_2, \quad U \to e^{\bar h - h}U,$$

where h is holomorphic. For every torus, the potential U may be taken doubly periodic.

Let us explain the appearance of these gauge transformations and, at the same time, why the dimensions 3 and 4 are distinguished by the existence of such spinor representations.

The Grassmannian $\tilde G_{n,2}$ of oriented two-planes in $\mathbb{R}^n$ is diffeomorphic to the quadric Q:

$$z_1^2 + \cdots + z_n^2 = 0, \quad (z_1 : \cdots : z_n) \in Q_n \subset \mathbb{C}P^{n-1}.$$

To every oriented plane with an positively oriented orthonormal basis $e_1 = (x_1, \dots, x_n)$, $e_2 = (y_1, \dots, y_n)$ there corresponds the point $(z_1 : \cdots : z_n)$, $z_k = x_k + iy_k$, $k = 1, \dots, n$, of this quadric. Given a surface $(X^1(z,\bar z), \dots, X^n(z,\bar z))$ in $\mathbb{R}^n$ with a conformal parameter z, we define the Gauss map as

$$z \to \left(\frac{\partial X^1}{\partial z} : \cdots : \frac{\partial X^n}{\partial z}\right) \in Q_n.$$

It is straightforward to derive that the image of the Gauss map lies in the quadric from the conformality of z. For $n = 3$, the quadric Q_3 is diffeomorphic to $\mathbb{C}^1$ and its rational parameterization is

$$z_1 = \frac{i}{2}(a^2 - b^2), \quad z_2 = \frac{1}{2}(b^2 - a^2), \quad z_3 = ab, \quad (a : b) \in \mathbb{C}P^1,$$

and the spinor ψ is reconstructed from the Gauss map as $\psi_1 = a$, $\bar{\psi}_2 = b$. For $n = 4$, we have the diffeomorphic Segre mapping

$$\mathbb{C}P^1 \times \mathbb{C}P^1 \to Q_4$$

of the form $z_1 = \frac{i}{2}(a_1 b_1 + a_2 b_2)$, $z_2 = \frac{1}{2}(a_2 b_2 - a_1 b_1)$, $z_3 = \frac{1}{2}(a_1 b_2 - a_2 b_1)$, $z_4 = \frac{i}{2}(a_2 b_1 - a_1 b_2)$, $(a_1 : a_2) \in \mathbb{C}P^1$, $(b_1 : b_2) \in \mathbb{C}P^1$, the spinors take the form $\varphi = (a_1, \bar{a}_2)$, $\psi = (b_1, \bar{b}_2)$ and are reconstructed up to the gauge transformations. Since they have to satisfy (3.2), h has to be holomorphic. For $n > 4$, the quadrics Q_n have no such rational parameterizations.

The operators D and $D^\vee$ enter the representation of the Davey–Stewartson (DS) equations via compatibility of linear systems. That led to introducing the DS deformations of surfaces, the four-dimensional analog of the mNV deformations [17].

We consider one of such deformations for which we proved that it transforms tori into tori and preserves the Willmore functional $4\int |U|^2 dx \wedge dy$ [29]. It has the form

$$U_t = i\big(U_{zz} + U_{\bar{z}\bar{z}} + (V + \bar{V})U\big), \quad V_{\bar{z}} = 2\big(|U|^2\big)_z \tag{3.4}$$

and is the compatibility condition for the linear problems

$$D\psi = 0, \quad \partial_t \psi = A\psi$$

where

$$A = i\begin{pmatrix} -\partial^2 - V & \bar{U}\bar{\partial} - \bar{U}_{\bar{z}} \\ U\partial - U_z & \bar{\partial}^2 + \bar{V} \end{pmatrix}.$$

It is also the compatibility condition for the system

$$D^\vee \varphi = 0, \quad \varphi_t = A^\vee \varphi,$$

where

$$A^\vee = -i\begin{pmatrix} -\partial^2 - V & U\bar{\partial} - U_{\bar{z}} \\ \bar{U}\partial - \bar{U}_z & \bar{\partial}^2 + \bar{V} \end{pmatrix}.$$

This equation is called the Davey–Stewartson II (DSII) equation.

The evolution of ψ and φ gives us a deformation of the Gauss map of surfaces (3.3) which are at every moment of time defined up to a translation depending on the temporal variable.

4. THE MOUTARD TRANSFORMATION FOR THE DAVEY–STEWARTSON II EQUATION AND ITS APPLICATIONS

The Moutard transformation was introduced in 1876 in projective differential geometry for the equation

$$f_{xy} + Uf = 0.$$

Given a solution f_0 of this equation, the transformation constructs another equation of this form with a different potential $\tilde{U}$ such that to every solution of the first equation there corresponds a solution of the new one and this is done by an explicit analytical formula. One of the problems to which the transformation was applied is an explicit construction of an immersion of the hyperbolic plane into $\mathbb{R}^3$ which, by Hilbert's theorem, appeared to be impossible. Later, the one-dimensional version, the Darboux transformation, was constructed and has found many important applications in mathematical physics.

Recently, the version for the elliptic equation $f_{z\bar{z}} + Uf = 0$ was applied, for instance, to constructing in terms of explicit analytical formulae

(1) blowing up solutions of the Novikov–Veselov equation with regular and fast decaying initial data [34],

(2) two-dimensional von Neumann–Wigner potentials with multiple positive eigenvalues [21].

We recall that a potential of the Schrödinger operator on $\mathbb{R}^n$ is called von Neumann–Wigner if it has a positive eigenvalue.

Here we construct a Moutard-type transformation for (3.2) and extend it to a transformation of solutions of the DSII equation.

Extend spinors ψ and φ to $\mathbb{H}$-valued functions, i.e.,

$$\Psi = \begin{pmatrix} \psi_1 & -\bar{\psi}_2 \\ \psi_2 & \bar{\psi}_1 \end{pmatrix}, \quad \Phi = \begin{pmatrix} \varphi_1 & -\bar{\varphi}_2 \\ \varphi_2 & \bar{\varphi}_1 \end{pmatrix}$$

and put

$$\omega(\Phi, \Psi) = -\frac{i}{2}(\Phi^\top \sigma_3 \Psi + \Phi^\top \Psi)dz - \frac{i}{2}(\Phi^\top \sigma_3 \Psi - \Phi^\top \Psi)d\bar{z},$$

where $X \to X^\top$ is the conjugation of X, and $\sigma_3 = \left(\begin{smallmatrix} 1 & 0 \\ 0 & -1 \end{smallmatrix}\right)$ is the Pauli matrix. If Ψ and Φ satisfy the Dirac equations (3.2) then $\omega(\Phi, \Psi)$ and $\omega(\Psi, \Phi)$ are closed forms. Denote, for brevity, $\Gamma = \left(\begin{smallmatrix} 0 & 1 \\ -1 & 0 \end{smallmatrix}\right)$. The $\mathbb{H}$-valued function

$$\begin{aligned} S(\Phi, \Psi)(z, \bar{z}) &= \Gamma \int \omega(\Phi, \Psi) \\ &= \int \left[i \begin{pmatrix} \psi_1\bar{\varphi}_2 & -\bar{\psi}_2\bar{\varphi}_2 \\ \psi_1\varphi_1 & -\bar{\psi}_2\varphi_1 \end{pmatrix} dz + i \begin{pmatrix} \psi_2\bar{\varphi}_1 & \bar{\psi}_1\bar{\varphi}_1 \\ -\psi_2\varphi_2 & -\bar{\psi}_1\varphi_2 \end{pmatrix} d\bar{z} \right] \\ &= \int d \begin{pmatrix} ix^3 + x^4 & -x^1 - ix^2 \\ x^1 - ix^2 & -ix^3 + x^4 \end{pmatrix} \end{aligned}$$

defines a surface in $\mathbb{R}^4 = \mathbb{H}$ with z as the conformal parameter (3.3). Hence we identify S with a surface in $\mathbb{R}^4$.

Let us define the $\mathbb{H}$-valued function

$$K(\Phi, \Psi) = \Psi S^{-1}(\Phi, \Psi)\Gamma\Phi^\top\Gamma^{-1} = \begin{pmatrix} i\bar{W} & a \\ -\bar{a} & -iW \end{pmatrix}. \tag{4.1}$$

The following theorem gives a Moutard-type transformation for D.

Theorem 4.1 ([20]). *Given Ψ_0 and Φ_0, the solutions of* (3.2), *for every pair Ψ and Φ of solutions of the same equations, the $\mathbb{H}$-valued functions*

$$\widetilde{\Psi} = \Psi - \Psi_0 S^{-1}(\Phi_0, \Psi_0)S(\Phi_0, \Psi), \quad \widetilde{\Phi} = \Phi - \Phi_0 S^{-1}(\Psi_0, \Phi_0)S(\Psi_0, \Phi)$$

satisfy the Dirac equations

$$\tilde{D}\widetilde{\Psi} = 0, \quad \tilde{D}^\vee \widetilde{\Phi} = 0$$

for the Dirac operators with the potential

$$\tilde{U} = U + W, \tag{4.2}$$

where W is defined by (4.1) *for $K(\Phi_0, \Psi_0)$. Here $S(\Psi_0, \Phi_0)$ is normalized by the condition*

$$\Gamma S^{-1}(\Phi_0, \Psi_0)\Gamma = \left(S^{-1}(\Psi_0, \Phi_0)\right)^\top.$$

The potential $\tilde{U}$ is the potential of the Weierstrass representation of the surface S^{-1} with z being a conformal parameter. The surface S^{-1} is obtained from S by composition of the inversion centered at the origin and the reflection $(x_1, x_2, x_3, x_4) \to (-x_1, -x_2, -x_3, x_4)$.

For $U = \bar{U}$ and $\Psi = \Phi$, this transformation reduces to the transformation of Dirac operators with real-valued potentials given in [35] in different form. In [32] it was related to the Weierstrass representation of surfaces in $\mathbb{R}^3$ by proving that it corresponds to the Möbius inversion $S \to S^{-1}$. This gives another proof of the conformal invariance of the Floquet multipliers by explicitly describing the transformations of the Floquet functions. Theorem 4.1 implies its analog for tori in $\mathbb{R}^4$. However, in this case the curve Γ_ψ is not preserved by the Möbius inversions. For instance, for the Clifford torus in the unit sphere $S^3 \subset \mathbb{R}^4$, the spectral curve Γ_ψ of its Möbius inversion centered at some point is $\mathbb{C}P^1$ except for the case when the surface lies in a plane, in which case it is $\mathbb{C}P^1$ with a pair of double points [20].

Let us replace $K(\Phi, \Psi)$ in (4.1) with

$$S(\Phi, \Psi)(z, \bar{z}, t) = \Gamma \int \omega(\Phi, \Psi) + \Gamma \int \omega_1(\Phi, \Psi),$$

where

$$\omega_1(\Phi, \Psi) = \left(\left[\Phi_z^\top \begin{pmatrix} 1 & 0 \\ 0 & 0 \end{pmatrix} + \Phi_{\bar{z}}^\top \begin{pmatrix} 0 & 0 \\ 0 & 1 \end{pmatrix}\right]\Psi \right.$$
$$\left. - \Phi^\top \left[\begin{pmatrix} 1 & 0 \\ 0 & 0 \end{pmatrix}\Psi_z + \begin{pmatrix} 0 & 0 \\ 0 & 1 \end{pmatrix}\Psi_{\bar{z}}\right]\right) dt.$$

We have

Theorem 4.2 ([33]). *If U solves the Davey–Stewartson II equation* (3.4) *and Ψ and Φ satisfy the equations $D\Psi = 0$, $\Psi_t = A\Psi$, $D^\vee \Phi = 0$, and $\Phi_t = A^\vee \Phi$, then the Moutard transformation* (4.2) *of U gives the solution $\tilde{U}$ of the DSII equation*

$$\tilde{U}_t = i\left(\tilde{U}_{zz} + \tilde{U}_{\bar{z}\bar{z}} + 2(\tilde{V} + \bar{\tilde{V}})\tilde{U}\right), \quad \tilde{V}_{\bar{z}} = \left(|\tilde{U}|^2\right)_z$$

with

$$\tilde{V} = V + 2ia_z$$

where a is given by (4.1).

The geometrical meaning of this transformation is as follows: for every fixed t, the spinors Ψ and Φ determine some surface $S(t)$ in $\mathbb{R}^4$ and U is the potential of such a representation. The surfaces $S(t)$ evolve via the DSII equation. We invert every such surface and obtain the t-parameter family of surfaces $\tilde{S}(t) = S^{-1}(t)$ which evolve via the DSII equation. Starting with a family of smooth surfaces and the corresponding smooth potentials U, we may construct singular solutions of the DSII equation: when $S(t)$ passes through the origin, the function $\tilde{U}$ loses continuity or regularity because the origin is mapped into the infinity by the inversion.

One of the simplest applications of Theorem 4.2 consists in constructing exact solutions from holomorphic functions. In this case we start from the trivial solution $U = V = 0$ for which Ψ and Φ are defined by holomorphic data. For instance, we have

Theorem 4.3 ([33]). *Let $f(z,t)$ be a function which is holomorphic in z and satisfies the equation*

$$\frac{\partial f}{\partial t} = i\frac{\partial^2 f}{\partial z^2}.$$

Then

$$U = \frac{i(zf' - f)}{|z|^2 + |f|^2}, \qquad V = 2ia_z,$$

where

$$a = -\frac{i(\bar{z} + f')\bar{f}}{|z|^2 + |f|^2},$$

satisfy the Davey–Stewartson II equation.

Geometrically, we have the deformation of graphs $w = f(z,t)$ which are minimal surfaces in $\mathbb{R}^4 = \mathbb{C}^2$. Whenever $f(z,t)$ vanishes at $z = 0$, the graph passes through the origin and the solution $\tilde{U}$ loses continuity or regularity. Hence the Weierstrass representation visualizes the creation of singularity and gives a method for finding such solutions.

We already applied this idea to constructing a solution with a one-point singularity for the modified Novikov–Veselov equation by using the Enneper surface [31]. However, in contrast to the mNV equation, the DSII has an important physical meaning.

In the variables

$$X = 2y, \quad Y = 2x.$$

the Davey–Stewartson II equation takes the form known in mathematical physics, namely

$$\begin{aligned} iU_t - U_{XX} + U_{YY} &= -4|U|^2U + 8\varphi_X U, \\ \Delta\varphi = \frac{\partial^2\varphi}{\partial X^2} + \frac{\partial^2\varphi}{\partial Y^2} &= \frac{\partial}{\partial X}|U|^2, \end{aligned} \tag{4.3}$$

where $\mathrm{Re}\, V = 2|U|^2 - 4\varphi_X$, $\varphi_X = \frac{\partial\varphi}{\partial X}$ [7]. This version of the DSII equation is called focusing.

Ozawa constructed a blow-up solution to (4.3) with the initial data

$$U(X,Y,0) = \frac{e^{-ib(4a)^{-1}(X^2-Y^2)}}{a(1 + ((X/a)^2 + (Y/a)^2)/2)}$$

and showed that, for constants a and b such that $ab < 0$, we have

$$\|U\|^2 \to 2\pi \cdot \delta \quad \text{as } t \to T = -a/b$$

in $\mathcal{S}'$ where $\|U\|^2 = \int_{\mathbb{R}^2} |U|^2\,dx\,dy$ is the squared L_2-norm of U and δ is the Dirac distribution centered at the origin **[23]**. We remark that $\|U\|^2 = 2\pi$ and the solution extends to $T > -a/b$ and gains regularity. In **[15]** it is conjectured for this equation that the blow-up in all cases is self-similar and the time-dependent scaling is as in the Ozawa solution. This conjecture is based on numerical results.

Let us consider the simplest examples of the solutions given by Theorem 4.3. We denote by c a constant which may take arbitrary complex values, and by r we denote $|z|$, $z \in \mathbb{C}$.

Next we consider

(1) $f = z^2 + 2it + c$,

$$\begin{aligned} U &= \frac{i(z^2 - 2it - c)}{|z|^2 + |z^2 + 2it + c|^2}, \\ V &= \frac{4(\bar{z}^2 - 2it + \bar{c})}{|z|^2 + |z^2 + 2it + c|^2} - \frac{2(2z(\bar{z}^2 - 2it + \bar{c}) + \bar{z})^2}{(|z|^2 + |z^2 + 2it + c|^2)^2}, \end{aligned} \tag{4.4}$$

and $|U| = O(\frac{1}{r^2})$ as $r \to \infty$. If c is not purely imaginary, then the solution is always smooth. If $c = i\tau$, $\tau \in \mathbb{R}$, then for $t = -\frac{\tau}{2}$, U has singularity at $z = 0$ of the type

$$U \sim ie^{2i\phi} \quad \text{as } r \to 0, \text{ where } z = re^{i\phi}.$$

We remark that $U \in L_2(\mathbb{R}^2)$ for all t and c. Since a small variation of c removes singularities, they are unstable.

(2) $f = z^4 + 12itz^2 - 12t^2 + c$,

$$U = \frac{i(3z^4 + 12itz^2 + 12t^2 - c)}{|z|^2 + |z^4 + 12itz^2 - 12t^2 + c|^2}. \tag{4.5}$$

This solution becomes singular for $c = 12t^2$ which is possible if and only if c is real-valued and positive. In this case it has singularities $U \sim -12te^{2i\phi}$ at $z = 0$ for $t = \pm\sqrt{c/12}$.

The solution to the mNV equation constructed in **[31]** is real-valued and regular except for the time T_{sing} when it has singularity at the origin of the form $U \sim -\cos 2\phi$.

We remark that $\|U\|^2$ is the first integral of the system. For (4.4), it is always equal to 2π except for the time T_{sing} when the solution becomes singular. For $t = T_{\text{sing}}$, it is equal to π. Analogously, for (4.5) it is equal to 4π for t such that U is nonsingular, and is equal to 3π for $t = T_{\text{sing}}$. The multiplicity of the value of this functional to π in both cases is explained by that the surfaces $\tilde{S}$ are immersed Willmore spheres (with singularities for singular moments of time).

By taking polynomials of higher degree for f, we can construct such singular solutions for which the regular initial data have any polynomial decay.

Are there another physically relevant wave equations that admit solutions with such singularities?

FUNDING

This work was partially supported by the Mathematical Center in Akademgorodok under the agreement No. 075-15-2019-1675 with the Ministry of Science and Higher Education of the Russian Federation.

REFERENCES

[1] U. Abresch, Constant mean curvature tori in terms of elliptic functions. *J. Reine Angew. Math.* **374** (1987), 169–192.

[2] C. Bär, Lower eigenvalue estimates for Dirac operators. *Math. Ann.* **293** (1992), no. 1, 39–46.

[3] P. Bayard, Spinorial representation of submanifolds in $SL(n,\mathbb{C})/SU(n)$. *Adv. Appl. Clifford Algebr.* **29** (2019), 51, 38 pp.

[4] D. A. Berdinskii and I. A. Taimanov, Surfaces in three-dimensional Lie groups. *Sib. Math. J.* **46** (2005), no. 6, 1005–1019.

[5] L. V. Bogdanov, Veselov–Novikov equation as a natural two-dimensional generalization of the Korteweg–de Vries equation. *Theoret. and Math. Phys.* **70** (1987), no. 2, 309–314.

[6] C. Bohle and P. Peters, Soliton spheres. *Trans. Amer. Math. Soc.* **363** (2011), no. 10, 5419–5463.

[7] A. Davey and K. Stewartson, On three-dimensional packets of surface waves. *Proc. R. Soc. Lond. A* **338** (1974), 101–110.

[8] B. A. Dubrovin, I. M. Krichever, and S. P. Novikov, The Schrödinger equation in a periodic field and Riemann surfaces. *Sov. Math., Dokl.* **17** (1976), 947–951.

[9] L. P. Eisenhart, *A treatise on the differential geometry of curves and surfaces.* Allyn and Bacon, Boston, 1909.

[10] D. Ferus, K. Leschke, F. Pedit, and U. Pinkall, Quaternionic holomorphic geometry: Plücker formula, Dirac eigenvalue estimates and energy estimates of harmonic 2-tori. *Invent. Math.* **146** (2001), no. 3, 505–593.

[11] T. Friedrich, On the spinor representation of surfaces in Euclidean 3-space. *J. Geom. Phys.* **28** (1998), no. 1–2, 143–157.

[12] P. G. Grinevich and M. U. Schmidt, Conformal invariant functionals of immersions of tori into $\mathbb{R}^3$. *J. Geom. Phys.* **26** (1997), no. 1–2, 51–78.

[13] N. J. Hitchin, Harmonic maps from a 2-torus to the 3-sphere. *J. Differential Geom.* **31** (1990), no. 3, 627–710.

[14] K. Kenmotsu, Weierstrass formula for surfaces of prescribed mean curvature. *Math. Ann.* **245** (1979), no. 2, 89–99.

[15] C. Klein and N. Stoilov, Numerical study of blow-up mechanisms for Davey–Stewartson II systems. *Stud. Appl. Math.* **141** (2018), no. 1, 89–112.

[16] B. G. Konopelchenko, Induced surfaces and their integrable dynamics. *Stud. Appl. Math.* **96** (1996), no. 1, 9–51.

[17] B. G. Konopelchenko, Weierstrass representations for surfaces in 4D spaces and their integrable deformations via DS hierarchy. *Ann. Global Anal. Geom.* **18** (2000), no. 1, 61–74.

[18] I. M. Krichever, Spectral theory of two-dimensional periodic operators and its applications. *Russian Math. Surveys* **44** (1989), no. 2, 145–225.

[19] F. C. Marques and A. Neves, Min–max theory and the Willmore conjecture. *Ann. of Math. (2)* **179** (2014), no. 2, 683–782.

[20] R. M. Matuev and I. A. Taimanov, The Moutard transformation of two-dimensional Dirac operators and the conformal geometry of surfaces in four-space. *Math. Notes* **100** (2016), no. 5–6, 835–846.

[21] R. G. Novikov, I. A. Taimanov, and S. P. Tsarev, Two-dimensional von Neumann–Wigner potentials with a multiple positive eigenvalue. *Funct. Anal. Appl.* **48** (2014), no. 4, 295–297.

[22] S. P. Novikov, A periodic problem for the Korteweg–de Vries equation. I. *Funct. Anal. Appl.* **8** (1974), no. 3, 236–246.

[23] T. Ozawa, Exact blow-up solutions to the Cauchy problem for the Davey–Stewartson systems. *Proc. R. Soc. Lond. Ser. A* **436** (1992), no. 1897, 345–349.

[24] F. Pedit and U. Pinkall, Quaternionic analysis on Riemann surfaces and differential geometry. Proceedings of the International Congress of Mathematicians, Vol. II (Berlin, 1998). *Doc. Math.* **Extra Vol. II** (1998), 389–400.

[25] U. Pinkall and I. Sterling, On the classification of constant mean curvature tori. *Ann. of Math. (2)* **130** (1989), no. 2, 407–451.

[26] I. A. Taimanov, Modified Novikov–Veselov equation and differential geometry of surfaces. In *Solitons, geometry, and topology: on the crossroad*, pp. 133–151, Amer. Math. Soc. Transl. Ser. 2 179, Adv. Math. Sci. 33, Amer. Math. Soc., Providence, RI, 1997.

[27] I. A. Taimanov, The Weierstrass representation of closed surfaces in $\mathbb{R}^3$. *Funct. Anal. Appl.* **32** (1998), no. 4, 258–267.

[28] I. A. Taimanov, The Weierstrass representation of spheres in $\mathbb{R}^3$, the Willmore numbers, and soliton spheres. *Proc. Steklov Inst. Math.* **225** (1999), 322–343.

[29] I. A. Taimanov, Surfaces in the four-space and the Davey–Stewartson equations. *J. Geom. Phys.* **56** (2006), no. 8, 1235–1256.

[30] I. A. Taimanov, The two-dimensional Dirac operator and the theory of surfaces. *Russian Math. Surveys* **61** (2006), no. 1, 79–159.

[31] I. A. Taimanov, Blowing up solutions of the modified Novikov–Veselov equation and minimal surfaces. *Theoret. and Math. Phys.* **182** (2015), no. 2, 173–181.

[32] I. A. Taimanov, The Moutard transformation of two-dimensional Dirac operators and the Möbius geometry. *Math. Notes* **97** (2015), no. 5–6, 124–135.

[33] I. A. Taimanov, The Moutard transformation for the Davey–Stewartson II equation and its geometrical meaning. *Math. Notes* **110** (2021), no. 5–6, 754–766.

[34] I. A. Taimanov and S. P. Tsarev, Two-dimensional rational solitons and their blow-up via the Moutard transformation. *Theoret. and Math. Phys.* **157** (2008), no. 2, 1525–1541.

[35] D. Yu, Q. P. Liu, and S. Wang, Darboux transformation for the modified Veselov–Novikov equation. *J. of Physics A* **35** (2001), no. 16, 3779–3785.

[36] D. V. Zakharov, The Weierstrass representation of discrete isotropic surfaces in $\mathbb{R}^{2,1}$, $\mathbb{R}^{3,1}$, and $\mathbb{R}^{2,2}$. *Funct. Anal. Appl.* **45** (2011), no. 1, 25–32.

ISKANDER A. TAIMANOV

Sobolev Institute of Mathematics, 630090 Novosibirsk, Russia, and Department of Mathematics and Mechanics, Novosibirsk State University, 630090 Novosibirsk, Russia, taimanov@math.nsc.ru

This Page Intentionally Left Blank

ENTROPY IN MEAN CURVATURE FLOW

LU WANG

ABSTRACT

The entropy of a hypersurface is defined by the supremum over all Gaussian integrals with varying centers and scales, thus invariant under rigid motions and dilations. It measures geometric complexity and is motivated by the study of mean curvature flow. We will survey recent progress on conjectures of Colding–Ilmanen–Minicozzi–White concerning the sharp lower bound on entropy for hypersurfaces, as well as their extensions.

MATHEMATICS SUBJECT CLASSIFICATION 2020

Primary 53E10; Secondary 35K93, 49R05

KEYWORDS

Entropy, mean curvature flow, stability

INTERNATIONAL CONGRESS OF MATHEMATICIANS 2022 JULY 6–14

Proc. Int. Cong. Math. 2022, Vol. 4, pp. 2656–2676
DOI 10.4171/ICM2022/116

Published by EMS Press

1. INTRODUCTION

In the trailblazing work [34], Colding and Minicozzi define a notion of entropy for hypersurfaces which is given by the supremum over all Gaussian integrals with varying centers and scales (cf. [67]). It is a geometric quantity that measures complexity and is invariant under rigid motions and dilations. In this survey, we discuss recent results on geometric properties of hypersurfaces with low entropy.

Entropy is motivated by the study of mean curvature flow which is a natural analogue of the heat equation in extrinsic curvature flows. Any hypersurface evolves under mean curvature flow in the direction of steepest descent for area, and the flow in general may become singular even before its vanishing. By Huisken's monotonicity formula [53], the entropy is decreasing under mean curvature flow. Thus, the entropy at all future singularities for the flow is bounded from above by that of the initial hypersurface.

By the work of Huisken [53] and Ilmanen [57], all possible blowups at a given singularity for a mean curvature flow are modeled by self-shrinkers which are hypersurfaces that flow in a self-similarly shrinking manner. Despite the abundance of self-shrinkers (see [63,70,71,80]), Colding and Minicozzi [34] study the properties of entropy and prove a striking result that spheres, generalized cylinders, and hyperplanes are the only stable self-shrinkers under mean curvature flow.

Inspired in part by the dynamic approach to mean curvature flow of [34], Colding, Ilmanen, Minicozzi, and White [33] employ a perturbative argument and singularity analysis for mean curvature flow to show that the round sphere minimizes entropy among all closed (i.e., compact without boundary) self-shrinkers. They further conjecture that in dimension less than 7, the round sphere indeed minimizes entropy among all nonflat self-shrinkers and so does it among all closed hypersurfaces.

After reviewing basic properties of entropy in Section 2, we discuss, in Sections 3 and 4, recent progress towards the above conjectures of Colding–Ilmanen–Minicozzi–White, with an emphasis on joint work with Bernstein [10,11]. We conclude our discussion in Section 5 to explain various stability results for round spheres under small perturbation of entropy.

2. ENTROPY FOR HYPERSURFACES

In this section, we discuss related background on the Colding–Minicozzi entropy for hypersurfaces, with an emphasis on its connection with mean curvature flow.

2.1. Basic properties for entropy

Follwing Colding and Minicozzi [34] (cf. [67]), define the entropy for a hypersurface $\Sigma \subset \mathbb{R}^{n+1}$ by

$$\lambda[\Sigma] = \sup_{\mathbf{x}_0 \in \mathbb{R}^{n+1}, t_0 > 0} (4\pi t_0)^{-\frac{n}{2}} \int_\Sigma e^{-\frac{|\mathbf{x}-\mathbf{x}_0|^2}{4t_0}} \, d\mathcal{H}^n \tag{2.1}$$

where $\mathcal{H}^n$ is the n-dimensional Hausdorff measure on $\mathbb{R}^{n+1}$.

It is readily checked that $\lambda[\Sigma \times \mathbb{R}^k] = \lambda[\Sigma]$ and, for $\rho > 0$ and $\mathbf{y} \in \mathbb{R}^{n+1}$,

$$\lambda[\rho\Sigma + \mathbf{y}] = \lambda[\Sigma],$$

where $\rho\Sigma + \mathbf{y}$ is a hypersurface given by

$$\rho\Sigma + \mathbf{y} = \{\mathbf{z} \in \mathbb{R}^{n+1} \mid \mathbf{z} = \rho\mathbf{x} + \mathbf{y} \text{ for some } \mathbf{x} \in \Sigma\}.$$

A direct calculation gives $\lambda[\mathbb{R}^n] = 1$. Moreover, Stone [76] computes:

$$2 > \lambda[\mathbb{S}^1] > \frac{3}{2} > \lambda[\mathbb{S}^2] > \cdots > \lambda[\mathbb{S}^n] > \lambda[\mathbb{S}^{n+1}] > \cdots \to \sqrt{2}. \tag{2.2}$$

The definition of entropy can be extended in a straightforward manner to measures and varifolds on a Euclidean space. There are also interesting studies of analogues of the Colding–Minicozzi entropy in noncompact Riemannian manifolds under certain curvature and volume conditions by Sun [78] and in hyperbolic space by Bernstein [9] (see also [90]).

2.2. Mean curvature flow

A one-parameter family of hypersurfaces $\Sigma_t \subset \mathbb{R}^{n+1}$ is a *mean curvature flow* if, for $\mathbf{x} \in \Sigma_t$,

$$\left(\frac{\partial \mathbf{x}}{\partial t}\right)^{\perp} = \mathbf{H}_{\Sigma_t}, \tag{2.3}$$

where the superscript $\perp$ means the projection to the unit normal $\mathbf{n}_{\Sigma_t}$ on Σ_t, and $\mathbf{H}_{\Sigma_t}$ is the mean curvature given by

$$\mathbf{H}_{\Sigma_t} = -H_{\Sigma_t}\mathbf{n}_{\Sigma_t} = -\operatorname{div}_{\Sigma_t}(\mathbf{n}_{\Sigma_t})\mathbf{n}_{\Sigma_t}.$$

Not only is mean curvature flow a beautiful subject in its own right, it also models various physical phenomena and has potential applications in numerous scientific fields, such as biology, computer imaging, and material sciences (see, e.g., [31, 62, 66, 68]).

By the avoidance principle (see [21, 6.3] and [45, CHAP. 4]), the mean curvature flow starting from any given closed hypersurface becomes singular in finite time. A central topic in the study of mean curvature flow is to understand the asymptotic behavior of the flow near singularities. Namely, suppose $\{\Sigma_t\}_{t\in[0,T)}$ is a mean curvature flow with $T > 0$ the first singular time. Let $\mathbf{x}_i \in \Sigma_{t_i}$ with $\mathbf{x}_i \to \mathbf{x}_0$ and $t_i \to T$ be such that the second fundamental form $|A_{\Sigma_{t_i}}(\mathbf{x}_i)| \to \infty$. If we define

$$\Gamma_s = \frac{1}{\sqrt{T-t}}\Sigma_t \quad \text{and} \quad s = -\log(T-t),$$

then the family $\{\Gamma_s\}_{s\geq -\log T}$ satisfies, for $\mathbf{y} \in \Gamma_s$,

$$\left(\frac{\partial \mathbf{y}}{\partial s}\right)^{\perp} = \mathbf{H}_{\Gamma_s} + \frac{\mathbf{y}^{\perp}}{2}, \tag{2.4}$$

which is called the *rescaled mean curvature flow* associated to the flow $\{\Sigma_t\}_{t\in[0,T)}$. Thus, characterizing the limits of Γ_s as $s \to \infty$ plays a fundamental role in the study of singularity formation for mean curvature flow.

Observe that the rescaled mean curvature flow $\{\Gamma_s\}_{s\geq -\log T}$ satisfies

$$\frac{d}{ds}\left((4\pi)^{-\frac{n}{2}}\int_{\Gamma_s} e^{-\frac{|\mathbf{y}|^2}{4}}\,d\mathcal{H}^n\right) = -(4\pi)^{-\frac{n}{2}}\int_{\Gamma_s}\left|\mathbf{H}_{\Gamma_s}+\frac{\mathbf{y}^\perp}{2}\right|^2 e^{-\frac{|\mathbf{y}|^2}{4}}\,d\mathcal{H}^n, \tag{2.5}$$

so it is the (negative) gradient flow of the Gaussian surface area

$$F[\Gamma] = (4\pi)^{-\frac{n}{2}}\int_{\Gamma} e^{-\frac{|\mathbf{y}|^2}{4}}\,d\mathcal{H}^n. \tag{2.6}$$

Notice that rewinding the change of variables in (2.5) gives exactly the monotonicity formula discovered by Huisken [53]. And it follows that the entropy $\lambda[\Sigma_t]$ is decreasing in t. Moreover, sending $s \to \infty$, up to passing to a subsequence, Γ_s converges weakly to a critical point, Γ_0, of the functional F that satisfies the Euler–Lagrange equation

$$\mathbf{H}_{\Gamma_0} + \frac{\mathbf{y}^\perp}{2} = \mathbf{0}, \tag{2.7}$$

and thus $\lambda[\Gamma_0] \leq \lambda[\Sigma_0]$ (see [53, 57]). A hypersurface satisfying (2.7) is also called a *self-shrinker*. Observe that $\{\sqrt{-t}\,\Gamma_0\}_{t<0}$ is a Brakke flow [21] that satisfies (2.3) weakly, and we call it a *tangent flow* at $(\mathbf{x}_0, T)$. One may also consider a blowup sequence $\rho_i(\Sigma_{t_i} - \mathbf{x}_i)$ with $\rho_i \to \infty$, and, by Brakke's compactness [21] (see also [55, SECT. 7]), the limit is also a Brakke flow, called a *limit flow* at $(\mathbf{x}_0, T)$.

There is a wild zoo of examples of self-shrinkers (see [4, 63, 64, 70, 71, 80]). However, a long-standing conjecture of Huisken [58, #8] (and of Angenent–Chopp–Ilmanen [5] in $\mathbb{R}^3$) asserts that starting with a generic closed hypersurface, the mean curvature flow develops only spherical and cylindrical singularities. Recently, Colding and Minicozzi have pioneered a number of innovative techniques about entropy and made important progress towards Huisken's conjecture (see [34–41]). Among them, the most relevant to this article is the following result.

Theorem 2.1 ([34, THEOREM 0.12]). *The only smooth embedded entropy-stable self-shrinkers with polynomial volume growth are round spheres, generalized cylinders and hyperplanes.*

Here "entropy-stable" means that there are no perturbations of the self-shrinkers to decrease the entropy. An easy consequence of Theorem 2.1 is that any singularities for mean curvature flow that are not spheres or generalized cylinders may be perturbed away in an appropriate sense. Moreover, it is possible to perturb the initial data to avoid certain unstable singularities for mean curvature flow (see Chodosh–Choi–Mantoulidis–Schulze [27, 28], Sun [77] and Sun–Xue [81, 82]).

Without the smoothness assumption, Colding and Minicozzi show that in dimension less than 7, Theorem 2.1 still holds for oriented F-stationary integral varifolds that have singular sets with locally finite codimension-2 Hausdorff measure [34, THEOREM 0.14]. Furthermore, in [91], Zhu utilizes an α-structural hypothesis in minimal surface theory and extends this result to higher dimensions. Notice that the hypothesis on the size of singular set is expected to hold for any self-shrinkers arising in mean curvature flow [57, PAGE 8].

2.3. Conjectures on the sharp lower entropy bound for hypersurfaces

The dynamic perspective of [34] suggests the following two closely related conjectures of Colding, Ilmanen, Minicozzi, and White (cf. [33, CONJECTURES 0.9 AND 0.10]).

Conjecture 2.2. *For $n \leq 6$, there is an $\varepsilon_0 = \varepsilon_0(n) > 0$ so that if $\Sigma \subset \mathbb{R}^{n+1}$ is a nonflat self-shrinker not equal to the round sphere, then $\lambda[\Sigma] \geq \lambda[\mathbb{S}^n] + \varepsilon_0$.*

Conjecture 2.3. *For $n \leq 6$, if $\Sigma \subset \mathbb{R}^{n+1}$ is a closed hypersurface, then $\lambda[\Sigma] \geq \lambda[\mathbb{S}^n]$.*

Both conjectures are known to be true with $n = 1$. Indeed, Conjecture 2.2 follows directly from Abresch–Langer's classification of self-shrinking planar curves [1]. And by work of Gage–Hamilton [50] and Grayson [51], every closed embedded curve in plane evolves under mean curvature flow to a round point which, together with the monotonicity of entropy, proves Conjecture 2.3.

As remarked before, the mean curvature flow starting from any given closed hypersurface becomes singular in finite time and the self-shrinker modeling the singularity of the flow has lower entropy. Thus, Conjecture 2.3 would follow from Conjecture 2.2. Despite Theorem 2.1, one of the difficulties to prove Conjecture 2.2 is that if one perturbs a noncompact self-shrinker, a priori it may flow smoothly without developing singularities.

At last, it may be interesting to think of Conjecture 2.2 as an analogue, in the Gaussian setting, of the question on the sharp lower bound on density for minimal cones (see Ilmanen–White [60] and Marques–Neves [69]).

3. SHARP LOWER BOUND ON ENTROPY FOR SELF-SHRINKERS

In this section, we discuss recent progress towards Conjecture 2.2. First, Brakke's local regularity [21, 6.1] implies that $\mathbb{R}^n$ has the least entropy of all self-shrinkers and, moreover, there is a gap to the next lowest (see also White [88]). As such, Conjecture 2.2 concerns the sharp lower entropy bound with a gap for all *nonflat* self-shrinkers.

Observe that if an immersed hypersurface has entropy strictly less than 2, then it must be embedded. Thus, we always assume embeddedness for the remainder of this section. Moreover, by the Frankel property (see [85, THEOREM 7.4] and [61, THEOREM C]), any embedded self-shrinker is connected.

3.1. Closed self-shrinkers with low entropy

In [33], Colding, Ilmanen, Minicozzi, and White initiate the study of Conjecture 2.2 and prove the following result.

Theorem 3.1 ([33]). *Given n, there exists $\varepsilon = \varepsilon(n) > 0$ so that if $\Sigma \subset \mathbb{R}^{n+1}$ is a closed self-shrinker not equal to the round sphere, then*

$$\lambda[\Sigma] \geq \lambda[\mathbb{S}^n] + \varepsilon. \tag{3.1}$$

Moreover, if

$$\lambda[\Sigma] \leq \min\left\{\lambda[\mathbb{S}^{n-1}], \frac{3}{2}\right\}, \tag{3.2}$$

then Σ *is diffeomorphic to* $\mathbb{S}^n$. *(If* $n > 2$, *then* $\lambda[\mathbb{S}^{n-1}] < \frac{3}{2}$ *and the minimum is unnecessary.)*

Outline of the proof. By Abresch–Langer [1], the theorem is vacuously true with $n = 1$; thus, assume $n \geq 2$ below. We also assume $\lambda[\Sigma] \leq \min\{\lambda[\mathbb{S}^n], 3/2\}$, as otherwise the theorem follows from inequality (2.2). As Σ is closed and not round, it follows from Colding–Minicozzi's classification of stable self-shrinkers, Theorem 2.1, that Σ is entropy unstable. Thus, there is a nearby hypersurface $\tilde{\Sigma}$ with the following properties:

(1) $\lambda[\tilde{\Sigma}] < \lambda[\Sigma]$;

(2) $\tilde{\Sigma}$ is inside of Σ, i.e., the compact region of $\mathbb{R}^{n+1}$ bounded by Σ contains $\tilde{\Sigma}$;

(3) $H_{\tilde{\Sigma}} - \frac{1}{2}\mathbf{x} \cdot \mathbf{n}_{\tilde{\Sigma}} > 0$ (with a suitable choice of the unit normal of $\tilde{\Sigma}$).

(See [34, COROLLARY 5.15, THEOREM 4.30, AND THEOREM 0.15].)

Next, one may use a Simon-type equation and the parabolic maximum principle to show that, starting from $\tilde{\Sigma}$, the rescaled mean curvature flow, i.e., a family of hypersurfaces $\tilde{\Sigma}_t \subset \mathbb{R}^{n+1}$ flowing by equation (2.4), preserves property (2) and bounds the second fundamental form $A_{\tilde{\Sigma}_t}$ by

$$|A_{\tilde{\Sigma}_t}|^2 \leq Ce^{-2t}\left|H_{\tilde{\Sigma}_t} - \frac{1}{2}\mathbf{x} \cdot \mathbf{n}_{\tilde{\Sigma}_t}\right|^2 \tag{3.3}$$

for some constant C depending on $\tilde{\Sigma}$. As $\tilde{\Sigma}_t$ becomes singular in finite time, a (subsequential) limit of blowups of the rescaled flow $\tilde{\Sigma}_t$ at the singularity is given by a (possibly singular) self-shrinker Γ. More crucially, estimate (3.3) gives

$$|A| \leq CH \quad \text{on the regular part of } \Gamma. \tag{3.4}$$

Appealing to the monotonicity of entropy, property (1) and the entropy bound of Σ gives that

$$\lambda[\Gamma] \leq \lambda[\tilde{\Sigma}] < \lambda[\Sigma] \leq \min\left\{\lambda[\mathbb{S}^{n-1}], \frac{3}{2}\right\}.$$

Thus, by Allard's regularity (see [3] or [73]) and estimate (3.4), Γ is a smooth embedded mean-convex self-shrinker. Thus, the classification of mean-convex self-shrinkers of Huisken [53] and of Colding–Minicozzi [34, THEOREM 0.17] implies Γ is of the form $\mathbb{S}^k \times \mathbb{R}^{n-k}$. Furthermore, the entropy bound of Γ ensures Γ is the round sphere. Thus, It follows that $\tilde{\Sigma}_t$ flows smoothly until it vanishes in a round point. Hence, as by construction $\tilde{\Sigma}$ can be chosen to be sufficiently close, in the C^∞ topology, to Σ, it follows that $\lambda[\Sigma] > \lambda[\tilde{\Sigma}] \geq \lambda[\mathbb{S}^n]$ and Σ is diffeomorphic to $\mathbb{S}^n$.

Finally, to see that there is a gap, one argues by contradiction. Suppose there is a sequence of closed self-shrinkers Σ^i that are not round with entropy converging to $\lambda[\mathbb{S}^n]$. Like before, perturbing these self-shrinkers and then applying rescaled mean curvature flow to the perturbations gives a sequence of flows $\tilde{\Sigma}^i_t$ with entropy less than or equal to $\lambda[\Sigma^i]$ and developing a spherical singularity in finite time. By the monotonicity of entropy, rescaling the $\tilde{\Sigma}^i_t$ about the spherical singularity creates a new sequence of rescaled mean curvature flows converging to the static sphere. This contradicts that, by Huisken [53], for i large Σ^i has negative curvature at some point. ■

Remark 3.2. In the proof of Theorem 3.1, the number $\frac{3}{2}$ in the minimum of (3.2) is only used to rule out the possibility of triple junctions arising in the rescaled mean curvature flow. However, by the orientability and results on mod 2 flat chains [89], the second part of Theorem 3.1 still holds under the weaker assumption that $\lambda[\Sigma] \leq \lambda[\mathbb{S}^{n-1}]$.

3.2. Noncompact self-shrinkers with low entropy

The arguments for Theorem 3.1 fail on noncompact self-shrinkers because perturbing a noncompact self-shrinker and applying rescaled mean curvature flow to the perturbation a priori may yield a rescaled mean curvature flow that has no singularities in finite time. To overcome this issue, it is needed to combine ideas from the proof of Theorem 3.1 and [11, 14].

A starting point is to understand the asymptotic structure of noncompact self-shrinkers. It is shown in [83] that any noncompact self-shrinker in $\mathbb{R}^3$ of finite genus is smoothly asymptotic (at infinity) to a cone or a cylinder (see also [79, APPENDIX A]). Assuming the noncompact self-shrinker has entropy bounded by that of the circle instead, a stronger result is true (cf. [10, PROPOSITION 4.5]).

Lemma 3.3. *If $\Sigma \subset \mathbb{R}^3$ is a noncompact self-shrinker with $\lambda[\Sigma] \leq \lambda[\mathbb{S}^1]$, then one of the following is true:*

(1) *Σ is isometric to a cylinder.*

(2) *There is a regular cone $\mathcal{C} \subset \mathbb{R}^3$ so that Σ is smoothly asymptotic to $\mathcal{C}$, i.e., as $\rho \to 0^+$, the $\rho\Sigma$ converges to $\mathcal{C}$ in $C^\infty_{loc}(\mathbb{R}^3 \setminus \{\mathbf{0}\})$. In particular, the curvature of Σ is quadratically decaying at infinity.*

Here a regular cone is a proper subset of $\mathbb{R}^{n+1}$ that is invariant under dilations and the link of the cone is a smooth embedded codimension-one submanifold of $\mathbb{S}^n$.

Proof. By definition, there is a cone $\mathcal{C} \subset \mathbb{R}^3$ so that as $\rho \to 0^+$, the $\rho\Sigma$ converges, in the Hausdorff distance, to $\mathcal{C}$. Fix any $\mathbf{y} \in \mathcal{C} \setminus \{\mathbf{0}\}$. Observe that $\sqrt{-t}\,\Sigma$, $t < 0$, is a mean curvature flow converging to $\mathcal{C}$ at time 0. Thus, as $\lambda[\Sigma] \leq \lambda[\mathbb{S}^1] < 2$, it follows from White's stratification theorem [87] that any tangent flow at $(\mathbf{y}, 0)$ is a multiplicity-one self-shrinker of the form $\Gamma \times \mathbb{R}$. Furthermore, by Abresch–Langer's classification of self-shrinking planar curves [1], $\Gamma = \mathbb{R}$ or $\mathbb{S}^1$. If $\Gamma = \mathbb{S}^1$, then Huisken's monotonicity gives Σ splits off a line and, thus, is isometric to a cylinder. If $\Gamma = \mathbb{R}$, then Brakke's local regularity [21] (see also White [88]) implies the flow is regular near $(\mathbf{y}, 0)$. As $\mathbf{y}$ is arbitrary, the second item follows. ■

Next, it is shown in [11] that there is a topological restriction on asymptotically conical self-shrinkers with entropy less than or equal to that of the round cylinder. This is the key to the proof of Conjecture 2.2 with $n = 2$.

Theorem 3.4 ([11]). *For $n \geq 2$, let $\Sigma \subset \mathbb{R}^{n+1}$ be a self-shrinker that is smoothly asymptotic to a regular cone $\mathcal{C}$. If $\lambda[\Sigma] \leq \lambda[\mathbb{S}^{n-1}]$, then the link of the asymptotic cone $\mathcal{C}$ separates $\mathbb{S}^n$ into two connected components both diffeomorphic to Σ. As a consequence, the link is connected.*

Outline of the proof. The arguments below may be thought of a natural analog, in the asymptotically conical setting, of the arguments in the proof of Theorem 3.1. However, there is an essential difference: while it is exploited there that the flow of a closed hypersurface must form a singularity in finite time, it is shown below that the flow of an asymptotically conical hypersurface with small entropy must exist without singularities for long-time and the flow eventually becomes star-shaped.

As the theorem is trivially true for hyperplanes, without loss of generality assume $\Sigma \neq \mathbb{R}^n$. By Theorem 2.1, Σ is entropy unstable, so there are two nearby hypersurfaces $\tilde{\Sigma}^{\pm} \subset \mathbb{R}^{n+1}$ such that

(1) $\lambda[\tilde{\Sigma}^{\pm}] < \lambda[\Sigma] \leq \lambda[\mathbb{S}^{n-1}] = \lambda[\mathbb{S}^{n-1} \times \mathbb{R}]$;

(2) $\tilde{\Sigma}^{+}$ lies in one side of Σ while $\tilde{\Sigma}^{-}$ lies on the other side of Σ;

(3) $H - \frac{1}{2}\mathbf{x} \cdot \mathbf{n} > K(1 + |\mathbf{x}|^2)^{\mu}$ on $\tilde{\Sigma}^{\pm}$ (with respective to the correct orientation) for constants $K > 0$ and $\mu < -1$ both depending on Σ;

(4) $\tilde{\Sigma}^{\pm}$ are both smoothly asymptotic to the cone $\mathcal{C}$.

In the proof of Theorem 3.1, it is convenient to think of self-shrinkers as static points for the rescaled mean curvature flow and show the sign of $H - \frac{1}{2}\mathbf{x} \cdot \mathbf{n}$ is preserved under the flow. Here it is crucial to instead study *shrinker mean curvature relative to the space-time point $X_0 = (\mathbf{x}_0, t_0)$ and at time t* (see also [75])

$$S^{X_0,t} = 2(t_0 - t)H - (\mathbf{x} - \mathbf{x}_0) \cdot \mathbf{n}. \tag{3.5}$$

By the parabolic maximum principle, the sign of shrinker mean curvature is preserved under the mean curvature flow starting with $\tilde{\Sigma}^{\pm}$ at time -1. Notice that shrinker mean curvature makes senses for mean curvature flows which start close to a self-shrinker but that persist up to (and beyond) the singular time of the self-shrinker, which is the key to this proof.

Arguing similarly as in the proof of Theorem 3.1 and invoking property (1) give that the flows $\tilde{\Sigma}^{\pm}_t$ starting with $\tilde{\Sigma}^{\pm}$ at time -1 both exist smoothly for long-time and become star-shaped at time 0. As $\tilde{\Sigma}^{-}_0$ and $\tilde{\Sigma}^{+}_0$ lie on different sides of $\mathcal{C}$ and are both smoothly asymptotic to $\mathcal{C}$, it follows that the link of $\mathcal{C}$ divides $\mathbb{S}^n$ into two components ω^{+} and ω^{-} over which $\tilde{\Sigma}^{+}_0$ and $\tilde{\Sigma}^{-}_0$ are radial graphs, respectively. Thus, $\tilde{\Sigma}^{\pm}_0$ and $\omega^{\pm}$ are diffeomorphic. Hence, by construction, $\tilde{\Sigma}^{-}_0$ and $\tilde{\Sigma}^{+}_0$ are both diffeomorphic to Σ and so are ω^{-} and ω^{+}. ■

Moreover, by the arguments in the proof of Theorem 3.4 and standard topological facts, Theorem 3.4 can be further refined.

Theorem 3.5 ([14, THEOREM 1.2]). *For $n \geq 2$, let $\Sigma \subset \mathbb{R}^{n+1}$ is a self-shrinker smoothly asymptotic to a regular cone $\mathcal{C}$. If $\lambda[\Sigma] \leq \lambda[\mathbb{S}^{n-1}]$, then Σ is contractible and the link of the asymptotic cone $\mathcal{C}$ is a homology $(n-1)$-sphere.*

Immediately, the classification of surfaces and Alexander's theorem [2] gives the following consequence.

Corollary 3.6. *For $2 \leq n \leq 3$, let $\Sigma \subset \mathbb{R}^{n+1}$ be a self-shrinker smoothly asymptotic to a regular cone. If $\lambda[\Sigma] \leq \lambda[\mathbb{S}^{n-1}]$, then Σ is diffeomorphic to $\mathbb{R}^n$.*

We now explain why Conjecture 2.2 is true with $n = 2$. Let $\Sigma \subset \mathbb{R}^3$ be a self-shrinker with $\lambda[\Sigma] \leq \lambda[\mathbb{S}^1]$. If Σ is closed, then, by Theorem 3.1 and remark (3.2), Σ is diffeomorphic to $\mathbb{S}^2$. If Σ is noncompact, then Lemma 3.3 implies that it is either a cylinder or smoothly asymptotic to a regular cone. In the latter case, Corollary 3.6 implies Σ is diffeomorphic to $\mathbb{R}^2$. Thus, by Brendle's classification for genus-zero self-shrinkers [22], Σ is a round sphere, a cylinder or a plane. Hence, it follows that the round sphere has the lowest entropy among all nonflat self-shrinkers in $\mathbb{R}^3$ and cylinder has the second lowest. In particular, this proves Conjecture 2.2 with $n = 2$ and $\varepsilon_0(2) = \lambda[\mathbb{S}^1] - \lambda[\mathbb{S}^2] > 0$.

Furthermore, there is a gap to the third lowest. To see this, suppose there is a sequence of self-shrinkers $\Sigma^i \subset \mathbb{R}^3$ so that $\lambda[\mathbb{S}^1] < \lambda[\Sigma^i] < \lambda[\mathbb{S}^1] + i^{-1}$. Thus, up to passing to a subsequence, the Σ^i converges smoothly to a self-shrinker Σ' with $\lambda[\Sigma'] = \lambda[\mathbb{S}^1]$. By the preceding discussions, Σ' is a cylinder. In particular, Σ' has positive mean curvature. The nature of convergence ensures that, for large i, Σ^i also has positive mean curvature in a large compact set. Hence, by the cylinder rigidity of Colding–Ilmanen–Minicozzi [32], for i large Σ^i is a cylinder, contradicting the entropy bound of Σ^i. Hence, we arrive at the following gap result.

Corollary 3.7 ([11, COROLLARY 1.2]). *There is a $\delta > 0$ so that if $\Sigma \subset \mathbb{R}^3$ is a self-shrinker not equal to a round sphere, a cylinder or a plane, then $\lambda[\Sigma] \geq \lambda[\mathbb{S}^1] + \delta$.*

4. SHARP LOWER BOUND ON ENTROPY FOR CLOSED HYPERSURFACES

In this section, we discuss a complete resolution of Conjecture 2.3, which asserts that round spheres have the least entropy of closed hypersurfaces of dimension less than 7. By definition (see (2.1)) $\lambda[\Sigma] \geq 1$ for any hypersurface $\Sigma \subset \mathbb{R}^{n+1}$. In fact, Chen [25] shows that $\lambda[\Sigma] = 1$ if and only if Σ is a hyperplane. Though a hyperplane can be approximated by closed hypersurfaces, Conjecture 2.3 claims that the entropy of any closed hypersurface is strictly larger than 1.

In [10], Bernstein and the author use a weak mean curvature flow and analyze terminal singularities to confirm Conjecture 2.3 (compare Ketover–Zhou [65]) After that, Zhu [91] further elaborates on the Colding–Minicozzi classification of stable self-shrinkers (see Theorem 2.1) to extend Conjecture 2.3 to all dimensions.

Theorem 4.1 ([10,91]; cf. [65]). *If $\Sigma \subset \mathbb{R}^{n+1}$ is a closed hypersurface, then $\lambda[\Sigma] \geq \lambda[\mathbb{S}^n]$ with equality if and only if Σ is a round sphere.*

Despite the discussion of Section 2.3, the proof of Theorem 4.1 that we explain below is independent of Conjecture 2.2. We give necessary background on some key ingredients that may be of independent interest and then sketch the proof of Theorem 4.1.

4.1. Weak mean curvature flow

Among various notions of weak mean curvature flow, the most relevant to this article are the Brakke flow and level set flow. Following Ilmanen [55] (cf. [21]), a Brakke flow is a one-parameter family of Radon measures on $\mathbb{R}^{n+1}$ which satisfy equation (2.3) in a certain weak form. The Brakke flow ensures that the mass of the measures decreases along the flow. A Brakke flow is *integral* if, at almost all times, the flow is an integer rectifiable Radon measure. Thinking of hypersurfaces $\Sigma \subset \mathbb{R}^{n+1}$ as measures $\mathcal{H}^n \lfloor \Sigma$, any smooth mean curvature flow is an integral Brakke flow.

Motivated by work of Osher–Sethian [72] in numerical analysis, the theory of level set flows has been established independently by Chen–Giga–Goto [26] and Evans–Spruck [46–49] (cf. [54,55]). A level set flow is a family of hypersurfaces obtained in the following way. First, embed a hypersurface $\Sigma \subset \mathbb{R}^{n+1}$ as the 0-level set of a Lipschitz function on $\mathbb{R}^{n+1}$. Then evolving the function in the way that, intuitively, every level set of the function flows by mean curvature yields a family of Lipschitz functions on $\mathbb{R}^{n+1}$. The level set flow of Σ is given by the 0-level set of the family of functions. It is shown, for instance, in [46], that the level set flow is well defined in the sense that it is independent of the choice of initial functions and coincides with the smooth mean curvature flow as long as the latter exists.

For the purposes of this article, Brakke flows have two important properties. The first is that Huisken's monotonicity formula [53] also holds for Brakke flows (see [57] and [86]). The second is the powerful regularity of Brakke [21] for such flows. A major technical difficulty in using Brakke flows is that there is a great deal of nonuniqueness as, by construction, Brakke flows are allowed to vanish instantaneously. On the other hand, the level set flow satisfies a strong maximum principle and thus is unique. In [55], Ilmanen uses an elliptic regularization procedure to construct a multiplicity-one Brakke flow that is supported on any given nonfattening level set flow (cf. Evans–Spruck [49]). Here a level set flow is *nonfattening* if the flow does not develop nonempty interiors. Observe the nonfattening condition is generic. Thus, it suffices to consider the closure (under Brakke's compactness) of the set of integral Brakke flows constructed by the elliptic regularization procedure.

4.2. Noncollapsed self-shrinkers and Brakke flows

An important notion of being noncollapsed for self-shrinkers and, more generally, for flows is introduced in [10] and is used to ensure nonvanishing. A self-shrinking measure on

$\mathbb{R}^{n+1}$ is an integer n-rectifiable Radon measure μ on $\mathbb{R}^{n+1}$ such that the associated varifold is F-stationary.

Definition 4.2. A self-shrinking measure μ on $\mathbb{R}^{n+1}$ is *noncollapsed* if there are $\mathbf{y} \in \mathbb{R}^{n+1}$ and $R > 4\sqrt{n}$ so that

(1) $\operatorname{spt}(\mu)$ is regular (i.e., smooth properly embedded) in the (open) ball $B_R(\mathbf{y})$;

(2) $\operatorname{spt}(\mu)$ separates $B_R(\mathbf{y}) \subset \mathbb{R}^{n+1}$ into two connected components Ω_+ and Ω_- containing closed balls $\bar{B}_{2\sqrt{n}}(\mathbf{x}_+)$ and $\bar{B}_{2\sqrt{n}}(\mathbf{x}_-)$, respectively.

The measure μ is *strongly noncollapsed* if $\mu \times \mu_{\mathbb{R}^k}$ is noncollapsed for all $k \geq 0$, where $\mu_{\mathbb{R}^k}$ is the k-dimensional Hausdorff measure on $\mathbb{R}^k$.

For instance, if μ is a noncompact self-shrinking measure on $\mathbb{R}^3$ with $\lambda[\mu] < 3/2$, then μ is strongly noncollapsed (cf. Lemma 3.3). On the other hand, the avoidance principle implies compact self-shrinking measures on $\mathbb{R}^{n+1}$ are all collapsed.

In an analogous way, define (strongly) noncollapsed Brakke flows as follows.

Definition 4.3. An integral Brakke flow $\mathcal{K} = \{\mu_t\}_{t \geq t_0}$ in $\mathbb{R}^{n+1}$ is *noncollapsed at time* τ if there are $(\mathbf{y}, s) \in \mathbb{R}^{n+1} \times (t_0, \tau)$, $R > 4\sqrt{n(\tau - t_0)}$, and $0 < \varepsilon < \min\{\tau - s, s - t_0\}$ so that

(1) $\mathcal{K}$ is regular in $B_R(\mathbf{y}) \times (s - \varepsilon, s + \varepsilon)$;

(2) $\operatorname{spt}(\mu_s)$ separates $B_R(\mathbf{y}) \subset \mathbb{R}^{n+1}$ into two connected components Ω_+ and Ω_- containing closed balls $\bar{B}_{2\sqrt{n(\tau-s)}}(\mathbf{x}_+)$ and $\bar{B}_{2\sqrt{n(\tau-s)}}(\mathbf{x}_-)$, respectively.

The Brakke flow $\mathcal{K}$ is *strongly noncollapsed at time* τ if $\{\mu_t \times \mu_{\mathbb{R}^k}\}_{t \geq t_0}$ is noncollapsed at time τ for all $k \geq 0$.

Note that if μ is a self-shrinking measure that is (strongly) noncollapsed, then the associated Brakke flow is (strongly) noncollapsed at time 0. A key observation is that being (strongly) noncollapsed at a time is an open condition for integral Brakke flows. Thus, given an integral Brakke flow $\mathcal{K}$ with finite entropy, if a tangent flow of $\mathcal{K}$ at $(\mathbf{y}, \tau)$ is (strongly) noncollapsed at time 0, then $\mathcal{K}$ is (strongly) noncollapsed at time τ.

There is a general structural result for self-shrinking measures with entropy less than that of a round sphere. It follows from an inductive argument and White's stratification theorem [87].

Proposition 4.4 ([10, PROPOSITION 4.12]). *For $n \geq 2$, if μ is a self-shrinking measure on $\mathbb{R}^{n+1}$ with $\lambda[\mu] < \lambda[\mathbb{S}^n]$, then one of the following holds:*

(1) *μ has compact support.*

(2) *μ is strongly noncollapsed.*

(3) *There is a self-shrinking measure ν on $\mathbb{R}^{n+1}$ so that $\lambda[\nu] \leq \lambda[\mu]$ and $\nu = \hat{\nu} \times \mu_{\mathbb{R}^{n-k}}$ for $\hat{\nu}$ a compact self-shrinking measure and $1 \leq k \leq n-1$.*

4.3. Outline of the proof of Theorem 4.1

By work of Gage–Hamilton [50] and Grayson [51], the claim is true with $n = 1$. To that end, assume $n \geq 2$. Argue by contradiction, then suppose $\lambda[\Sigma] < \lambda[\mathbb{S}^n]$. Let $\mathcal{K} = \{\mu_t\}_{t\geq 0}$ be the integral Brakke flow in $\mathbb{R}^{n+1}$ with $\mu_0 = \mathcal{H}^n \lfloor \Sigma$. By the spheres comparison and avoidance principle, the extinction time of $\mathcal{K}$, $T_0(\mathcal{K})$ satisfies

$$0 < T_0(\mathcal{K}) = \sup\{t \geq 0 \mid \operatorname{spt}(\mu_t) \neq \emptyset\} < \infty.$$

Appealing to Definition 4.3 and the avoidance principle implies that $\mathcal{K}$ is collapsed at time $T_0(\mathcal{K})$. As being noncollapsed is an open condition for Brakke flows, any tangent flow of $\mathcal{K}$ at time $T_0(\mathcal{K})$ is collapsed at time 0. Recall from Section 2.2 that tangent flows are given by self-shrinking measures. Thus, it is enough to show, for all $1 \leq k \leq n$, the set, $\mathcal{CSM}_k(\lambda[\mathbb{S}^n])$, of all compact self-shrinking measures on $\mathbb{R}^{k+1}$ that have entropy less than $\lambda[\mathbb{S}^n]$ is an empty set, because it would follow from Proposition 4.4 that all tangent flows are strongly noncollapsed at time 0, giving a contradiction.

On $\mathbb{R}^2$, all self-shrinking measures with entropy less than $3/2$ are smooth embedded and thus have been classified by Abresch–Langer [1]. Thus, by a direct computation, the claim is true with $k = 1$. Suppose, inductively, the claim holds for all $1 \leq k \leq l-1$. Assume $l < n+1$, as otherwise we are done. Argue by contradiction, then suppose the claim were false for $k = l$. Take an entropy minimizing sequence of compact self-shrinking measures ν_i on $\mathbb{R}^{l+1}$ with $\lambda[\nu_i] < \lambda[\mathbb{S}^n]$. Then, up to passing to a subsequence, the ν_i converges to a self-shrinking measure ν_0 with $\lambda[\nu_0] < \lambda[\mathbb{S}^n]$. As any compact self-shrinking measure is collapsed, the ν_i are all collapsed. By the openness of being noncollapsed, ν_0 is also collapsed. Thus, by the inductive hypothesis and Proposition 4.4, ν_0 has compact support. Our construction and the entropy bound ensure ν_0 is entropy stable and has a singular set of codimension at least 2. Hence, appealing to Colding–Minicozzi [34, THEOREM 0.14] when $l \leq 6$ while to Zhu [91, THEOREM 1.2] when $l \geq 7$ gives $\nu_0 = \mathcal{H}^l \lfloor \mathbb{S}^l$, contradicting $\lambda[\nu_0] < \lambda[\mathbb{S}^l]$.

It remains only to characterize the equality case. Suppose $\lambda[\Sigma] = \lambda[\mathbb{S}^n]$. If Σ is not (modulo translations and dilations) a self-shrinker, then applying mean curvature flow to Σ for short time yields a closed hypersurface $\tilde{\Sigma}$ with $\lambda[\tilde{\Sigma}] < \lambda[\mathbb{S}^n]$. This contradicts what we have just shown. Thus, modulo translations and dilations, Σ is a self-shrinker. Moreover, Σ is entropy stable, as otherwise one finds a perturbation of Σ so that the perturbation is a closed hypersurface with strictly less entropy, giving a contradiction. Thus, the classification of entropy stable self-shrinkers, Theorem 2.1, implies Σ is a round sphere.

5. STABILITY FOR THE ENTROPY INEQUALITY

We continue to discuss a natural follow-up question to Theorem 4.1 that whether a closed hypersurface with entropy sufficiently close to the lowest is itself close to a round sphere. There are various perspectives of this question. For instance, Wang [84] proves a forward analogue of Brakke's clearing-out lemma [21] and establish an explicit relationship between a certain normalized Hausdorff distance of a surface to a round sphere and the difference between their entropy (cf. [12]). It is also interesting to approach this question from a topological viewpoint. Indeed, an immediate application of mean curvature flow and Corollary 3.7 is that any closed surface in $\mathbb{R}^3$ with entropy less than or equal to that of a round cylinder has genus zero. In particular, such a surface is isotopic to $\mathbb{S}^2$.

A conditional isotopic stability result is also true in general dimensions. To state the hypotheses, we follow relevant notions of [14]. For $\Lambda > 0$, let $\mathcal{RMC}_n^*(\Lambda)$ be the set of nonflat regular minimal cones $\mathcal{C} \subset \mathbf{R}^{n+1}$ with $\lambda[\mathcal{C}] < \Lambda$, and let $\mathcal{S}_n^*(\Lambda)$ be the set of nonflat self-shrinkers $\Sigma \subset \mathbb{R}^{n+1}$ with $\lambda[\Sigma] < \Lambda$. The first hypothesis is

$$\mathcal{RMC}_k^*(\Lambda) = \emptyset \quad \text{for all } 3 \le k \le n. \qquad (\star_{n,\Lambda})$$

As all regular minimal cones in $\mathbb{R}^2$ consist of unions of rays, $\mathcal{RMC}_1^*(\Lambda) = \emptyset$. Similarly, as geodesics in $\mathbb{S}^2$ are great circles, $\mathcal{RMC}_2^*(\Lambda) = \emptyset$. The second hypothesis is

$$\mathcal{S}_{n-1}^*(\Lambda) = \emptyset. \qquad (\star\star_{n,\Lambda})$$

As round cylinders are nonflat self-shrinkers, $(\star\star_{n,\Lambda})$ holds only if $\Lambda \le \lambda[\mathbb{S}^{n-1}]$. Bernstein and the author [17] and Chodosh–Choi–Mantoulidis–Schulze [27] employ different strategies to prove the following conditional result in general dimensions.

Theorem 5.1 ([17, THEOREM 1.3]; cf. [27, THEOREM 10.1]). *Fix $n \ge 3$ and $\Lambda \le \lambda[\mathbb{S}^{n-1}]$. If $(\star_{n,\Lambda})$ and $(\star\star_{n,\Lambda})$ both hold and Σ is a closed connected hypersurface in $\mathbb{R}^{n+1}$ with $\lambda[\Sigma] \le \Lambda$, then Σ is smoothly isotopic to $\mathbb{S}^n$.*

Remark 5.2. By Marques–Neves' proof of the Willmore conjecture (see [69, THEOREM B]) $\mathcal{RMC}_3^*(\lambda[\mathbb{S}^2]) = \emptyset$. And Corollary 3.7 ensures $\mathcal{S}_2^*(\lambda[\mathbb{S}^2]) = \emptyset$. Thus, $(\star_{n,\Lambda})$ and $(\star\star_{n,\Lambda})$ are both fulfilled with $n = 3$ and $\Lambda = \lambda[\mathbb{S}^2]$.

5.1. Overview of the proof of Theorem 5.1

The basic idea is, again, to apply mean curvature flow to Σ and then analyze the behavior of the flow near singularities. Let $\mathcal{K}$ be the Brakke flow starting at Σ. If, modulo translations and dilations, Σ is a self-shrinker, then the claim follows from Theorem 3.1. Otherwise, the entropy is strictly decreasing under the flow. Then it is shown in [14, SECT. 3] that, by hypotheses $(\star_{n,\Lambda})$ and $(\star\star_{n,\Lambda})$, appealing to Allard's regularity [3] and White's stratification theorem [87] implies that any singularities for $\mathcal{K}$ are modeled by smooth embedded multiplicity-one self-shrinkers, which are either compact and, by Theorem 3.1, smoothly isotopic to $\mathbb{S}^n$, or noncompact asymptotically conical.

Due to the lack of a classification for self-shrinkers of simple topology in general dimensions, it seems very difficult to rule out the possibility of these asymptotically conical singularities for $\mathcal{K}$. However, as suggested by Theorem 2.1, such singularities are unstable under mean curvature flow and are expected to be perturbed away in an appropriate sense. This strategy has been carried out in [27]. Namely, it is shown that the ancient mean curvature flow that lies on one-side of a given asymptotically conical self-shrinker exists uniquely for long-time. As an application, perturbing Σ and applying mean curvature flow to the perturbation gives a mean curvature flow that is smooth until it disappears in a round point. The claim follows immediately from this.

The strategy of [17] is distinct from that of [27] and relies on the study of self-expanding solutions to mean curvature flow [13,15–20]. There are two key ingredients. One is an application of a forward analogue of Huisken's monotonicity formula for flows emerging from a conical singularity [15] (see also Section 5.2) to show that taking a second blowup gives self-expanding flows. Another is a topological uniqueness for self-expanders asymptotic to a given cone with entropy less than that of a round cylinder [16] (see also Section 5.3). Thus, combining these with a suitable bubble-tree blowup argument implies $\mathcal{K}$ is smooth at almost all times and stay in the same isotopic class whenever it is smooth. Hence, as near its extinction point $\mathcal{K}$ is isotopic to the shrinking spheres, it follows that Σ is isotopic to $\mathbb{S}^n$.

5.2. Forward monotonicity formula for flows coming out of cones

Huisken's monotonicity formula implies any tangent flows are backwardly self-shrinking. On the other hand, it is unknown that whether tangent flows forward in time are self-expanding or not. Nonetheless, suppose $\mathcal{T} = \{\mu_t\}_{t\in\mathbb{R}}$ is a tangent flow with $\mu_0 = \mathcal{H}^n \lfloor \mathcal{C}$ for $\mathcal{C}$ a regular cone in $\mathbb{R}^{n+1}$. Let $\mathcal{T}' = \{\mu'_t\}_{t\in\mathbb{R}}$ be a tangent flow of $\mathcal{T}$ at $(\mathbf{0}, 0)$. Thus, $\mathcal{T}'$ is self-shrinking for negative times and equal to μ_t for all $t \leq 0$. To that end, we explain the reason for that $\mathcal{T}'$ is self-expanding for positive times.

Consider the (forward) rescaled flow $\{\nu_s\}_{s\in\mathbb{R}}$ associated to $\mathcal{T} \lfloor \mathbb{R}^{n+1} \times (0, \infty)$ about $(\mathbf{0}, 0)$. Appealing to [56] and [27, SECT. 8] (cf. [44]) gives, for all $s > 0$, that $\operatorname{spt}(\nu_s)$ is trapped between two self-expanders Γ_- and Γ_+, both smoothly asymptotic to $\mathcal{C}$. Here self-expanders are critical points for the expander energy functional

$$E[\Gamma] = \int_\Gamma e^{\frac{|\mathbf{x}|^2}{4}} \, d\mathcal{H}^n. \tag{5.1}$$

Following a suggestion of Ilmanen [56], define the *relative expander entropy* of ν_s relative to Γ_- by

$$E_{\mathrm{rel}}[\nu_s, \Gamma_-] = \lim_{R\to\infty} \left(\int_{B_R} e^{\frac{|\mathbf{x}|^2}{4}} \, d\nu_s - \int_{\Gamma_-\cap B_R} e^{\frac{|\mathbf{x}|^2}{4}} \, d\mathcal{H}^n \right). \tag{5.2}$$

In [15], Bernstein and the author employ calibration type arguments to show the limit in (5.2) exists and is finite. Prior to that, this relative functional has been studied by Ilmanen–Neves–Schulze [59] in the curve case. Deruelle and Schulze [43] investigate this relative functional in general dimensions and exploit the convergence rate between two self-expanders [8] to show it is well defined and finite for pairs of self-expanders asymptotic to the same cone.

Furthermore, it is shown in [15] that there is a monotonicity formula for the relative expander entropy for flows emerging from a conical singularity. Applying this formula to $\{\nu_s\}_{s\in\mathbb{R}}$ yields that, for $s_1 < s_2$,

$$E_{\mathrm{rel}}[\nu_{s_2}, \Gamma_-] - E_{\mathrm{rel}}[\nu_{s_1}, \Gamma_-] = -\int_{s_1}^{s_2} \int \left| \mathbf{H} - \frac{\mathbf{x}^\perp}{2} \right|^2 e^{\frac{|\mathbf{x}|^2}{4}} \, d\nu_s \, ds. \tag{5.3}$$

As a consequence, given a sequence $s_i \to -\infty$, there is a subsequence s_j so that the ν_{s_j} converges to a critical point of the functional E. This implies that $\mathcal{T}'$ is self-expanding for positive times.

5.3. Topological uniqueness for self-expanders with low entropy

It is illustrated by Angenent–Chopp–Ilmanen [5] that there is an open set of regular cones so that for each cone in the set there are at least two self-expanders asymptotic to the cone (cf. [13]). However, it is proved in [16] that given a regular cone with sufficiently small entropy all self-expanders asymptotic to the cone are in the same isotopic class.

Theorem 5.3 ([16]). *For $0 < \Lambda \leq \lambda[\mathbb{S}^{n-1}]$, let $\mathcal{C} \subset \mathbb{R}^{n+1}$ be a regular cone with $\lambda[\mathcal{C}] < \Lambda$ and assume one of the following holds:*

(1) $2 \leq n \leq 6$ *and* $\Lambda = \lambda[\mathbb{S}^{n-1}]$.

(2) $n \geq 7$ *and* $(\star_{n,\Lambda})$ *holds.*

If $\Gamma_1, \Gamma_2 \subset \mathbb{R}^{n+1}$ are two self-expanders both smoothly asymptotic to $\mathcal{C}$, then Γ_1 and Γ_2 are a.c.-isotopic with fixed cone.

Here two asymptotically conical hypersurfaces are said to be *a.c.-isotopic with fixed cone* if there is an isotopy of hypersurfaces that respects the asymptotically conical behavior and fixes the asymptotic cone.

Outline of the proof of Theorem 5.3. It follows from the main result of [20] that the space of asymptotically conical expanders is an infinite-dimensional smooth Banach manifold. Thus, invoking Smale's version [74] of the Sard theorem gives a residual set $\mathscr{R}$ of regular cones so that for each cone in the set any self-expanders smoothly asymptotic to the cone are nondegenerate in the sense that there are no nontrivial normal Jacobi fields that fix the asymptotic cone. In particular, degenerate asymptotically conical self-expanders can be perturbed with varying asymptotic cones to nondegenerate ones. As such, we focus below on generic cones $\mathcal{C} \in \mathscr{R}$. Our goal is to construct Morse flow lines joining any two self-expanders both smoothly asymptotic to $\mathcal{C}$.

Denote by $\mathcal{ACH}_n(\mathcal{C})$ the space of hypersurfaces in $\mathbb{R}^{n+1}$ that are smoothly asymptotic to the cone $\mathcal{C}$. It is convenient to define an order on $\mathcal{ACH}_n(\mathcal{C})$ as follows. First fix a choice of unit normals $\mathbf{n}_{\mathcal{L}}$ on the link $\mathcal{L}$ of $\mathcal{C}$. We then let $\omega_+ \subset \mathbb{S}^n$ be the open set so that $\partial\omega_+ = \mathcal{L}$ and $\mathbf{n}_{\mathcal{L}}$ points into ω_+. For $\Sigma \in \mathcal{ACH}_n(\mathcal{C})$, let $\Omega_+(\Sigma) \subset \mathbb{R}^{n+1}$ be the open set so that $\partial\Omega_+(\Sigma) = \Sigma$ and the blowdowns of Ω_+ in $\mathbb{S}^n$ converge as sets to ω_+. For $\Sigma_1, \Sigma_2 \in \mathcal{ACH}_n(\mathcal{C})$, we say $\Sigma_1 \preceq \Sigma_2$ provided $\Omega_+(\Sigma_2) \subset \Omega_+(\Sigma_1)$.

Let $\mathcal{ACE}_n(\mathcal{C}) \subset \mathcal{ACH}_n(\mathcal{C})$ be the subset consisting of self-expanders. If $\Gamma \in \mathcal{ACE}_n(\mathcal{C})$ is unstable, then there are two eternal rescaled mean curvature flows that deform Γ to two stable elements $\Gamma_\pm \in \mathcal{ACE}_n(\mathcal{C})$ with $\Gamma_- \preceq \Gamma \preceq \Gamma_+$. Moreover, if $\Gamma' \in \mathcal{ACE}_n(\mathcal{C})$ is stable and $\Gamma' \preceq \Gamma$ (respectively, $\Gamma \preceq \Gamma'$), then $\Gamma' \preceq \Gamma_- \preceq \Gamma$ (respectively, $\Gamma \preceq \Gamma_+ \preceq \Gamma'$). The hypotheses ensure that the eternal flows are smooth. On the other hand, if $\Gamma_0, \Gamma_1 \in \mathcal{ACE}_n(\mathcal{C})$ are (strictly) stable and $\Gamma_0 \preceq \Gamma_1$, then appealing to a min–max construction for relative expander entropy [18] yields an element $\Gamma_2 \in \mathcal{ACE}_n(\mathcal{C})$ with $\Gamma_2 \neq \Gamma_i$ for $i \in \{0, 1\}$ and $\Gamma_0 \preceq \Gamma_2 \preceq \Gamma_1$. Again, the hypotheses guarantee the smoothness of the min–max self-expander. Hence, arguing by induction on the cardinality of the subset consisting of stable elements of $\mathcal{ACE}_n(\mathcal{C})$ (see [19]), it follows that every element of $\mathcal{ACE}_n(\mathcal{C})$ can be deformed via rescaled mean curvature flows through a finite number of intermediate elements of $\mathcal{ACE}_n(\mathcal{C})$ to the lowest (with respect to the order $\preceq$), implying the claim. ∎

6. FURTHER DISCUSSIONS

Instead of assuming low entropy, Hershkovits and White prove a sharp relation between the entropy and topology of closed self-shrinkers for all dimensions [52]. This may be thought of as an extension of Theorem 3.1. Thinking of self-shrinkers as a special class of ancient the mean curvature flow, combining with work of Angenent–Daskalopoulos–Sesum [6,7], Bernstein–Wang [11], and Brendle–Choi [23,24], Choi, Haslhofer. and Hershkovits [29] classify the ancient mean curvature flow in $\mathbb{R}^3$ with entropy less than or equal to that of a cylinder. There is an analogous classification for ancient mean curvature flows in higher dimensions under the assumption that the flows are smoothly asymptotic at time $-\infty$ to a round cylinder [30].

In general, Conjecture 2.2 is wide open, in part because it is unknown whether there is a complete classification for self-shrinkers of dimension at least 3 with simple topology (compare Brendle [22]) It would be also interesting to study analogous questions in higher codimensions. We refer the interested reader to [41] and references therein.

Very recently, Daniels–Holgate [42] combines [29] and [30] with suitable barriers to construct smooth mean curvature flows with surgery that approximate weak mean curvature flows with only spherical and neck-pinch singularities. Together with [28], this implies that any closed hypersurface in $\mathbb{R}^4$ that has entropy less than or equal to $\lambda[\mathbb{S}^1 \times \mathbb{R}^2]$ is smoothly isotopic to $\mathbb{S}^3$, which, together with Theorem 5.1, sheds some light on the smooth Schoenflies conjecture for $\mathbb{R}^4$.

ACKNOWLEDGMENTS

We would like to thank Jacob Bernstein for collaborations on numerous results discussed in this article.

FUNDING

This work was partially supported by NSF grants DMS-2141529 and DMS-2146997.

REFERENCES

[1] U. Abresch and J. Langer, The normalized curve shortening flow and homothetic solutions. *J. Differential Geom.* **23** (1986), no. 2, 175–196.

[2] J. W. Alexander, On the subdivision of 3-space by a polyhedron. *Proc. Natl. Acad. Sci.* **10** (1924), 6–8.

[3] A. K. Allard, On the first variation of a varifold. *Ann. of Math. (2)* **95** (1972), 417–491.

[4] S. Angenent, Shrinking doughnuts. In *Nonlinear diffusion equations and their equilibrium states 3 (Gregynog, 1989)*, pp. 21–38, Progr. Nonlinear Differential Equations Appl. 7, Birkhäuser Boston, Boston, MA, 1992.

[5] S. Angenent, D. L. Chopp, and T. Ilmanen, A computed example of nonuniqueness of mean curvature flow in $\mathbb{R}^3$. *Comm. Partial Differential Equations* **20** (1995), no. 11–12, 1937–1958.

[6] S. Angenent, P. Daskalopoulos, and N. Sesum, Unique asymptotics of ancient convex mean curvature flow solutions. *J. Differential Geom.* **111** (2019), no. 3, 381–455.

[7] S. Angenent, P. Daskalopoulos, and N. Sesum, Uniqueness of two-convex closed ancient solutions to the mean curvature flow. *Ann. of Math. (2)* **192** (2020), no. 2, 353–436.

[8] J. Bernstein, Asymptotic structure of almost eigenfunctions of drift Laplacians on conical ends. *Amer. J. Math.* **142** (2020), no. 6, 1897–1929.

[9] J. Bernstein, Colding Minicozzi entropy in hyperbolic space. *Nonlinear Anal.* **210** (2021), 112401, 16 pp.

[10] J. Bernstein and L. Wang, A sharp lower bound for the entropy of closed hypersurfaces up to dimension six. *Invent. Math.* **206** (2016), no. 3, 601–627.

[11] J. Bernstein and L. Wang, A topological property of asymptotically conical self-shrinkers of small entropy. *Duke Math. J.* **166** (2017), no. 3, 403–435.

[12] J. Bernstein and L. Wang, Hausdorff stability of the round two-sphere under small perturbations of the entropy. *Math. Res. Lett.* **25** (2018), no. 2, 347–365.

[13] J. Bernstein and L. Wang, An integer degree for asymptotically conical self-expanders. 2018, arXiv:1807.06494.

[14] J. Bernstein and L. Wang, Topology of closed hypersurfaces of small entropy. *Geom. Topol.* **22** (2018), no. 2, 1109–1141.

[15] J. Bernstein and L. Wang, Relative expander entropy in the presence of a two-sided obstacle and applications. 2019, arXiv:1906.07863.

[16] J. Bernstein and L. Wang, Topological uniqueness for self-expanders of small entropy. 2019, arXiv:1902.03642.

[17] J. Bernstein and L. Wang, Closed hypersurfaces of low entropy in $\mathbb{R}^4$ are isotopically trivial. 2020, arXiv:2003.13858.

[18] J. Bernstein and L. Wang, A mountain-pass theorem for asymptotically conical self-expanders. 2020, arXiv:2003.13857.

[19] J. Bernstein and L. Wang, Smooth compactness for spaces of asymptotically conical self-expanders of mean curvature flow. *Int. Math. Res. Not. IMRN* **2021** (2021), no. 12, 9016–9044.

[20] J. Bernstein and L. Wang, The space of asymptotically conical self-expanders of mean curvature flow. *Math. Ann.* **380** (2021), no. 1–2, 175–230.

[21] K. A. Brakke, *The motion of a surface by its mean curvature*. Math. Notes 20, Princeton University Press, Princeton, NJ, 1978.

[22] S. Brendle, Embedded self-similar shrinkers of genus 0. *Ann. of Math. (2)* **183** (2016), no. 2, 715–728.

[23] S. Brendle and K. Choi, Uniqueness of convex ancient solutions to mean curvature flow in $\mathbb{R}^3$. *Invent. Math.* **217** (2019), no. 1, 35–76.

[24] S. Brendle and K. Choi, Uniqueness of convex ancient solutions to mean curvature flow in higher dimensions. *Geom. Topol.* **25** (2021), no. 5, 2195–2234.

[25] L. Chen, Rigidity and stability of submanifolds with entropy close to one. 2020, arXiv:2003.07480.

[26] Y. G. Chen, Y. Giga, and S. Goto, Uniqueness and existence of viscosity solutions of generalized mean curvature flow equations. *J. Differential Geom.* **33** (1991), no. 3, 749–786.

[27] K. Choi, O. Chodosh, C. Mantoulidis, and F. Schulze, Mean curvature flow with generic initial data. 2020, arXiv:2003.14344.

[28] K. Choi, O. Chodosh, C. Mantoulidis, and F. Schulze, Mean curvature flow with generic low-entropy initial data. 2021, arXiv:2102.11978.

[29] K. Choi, R. Haslhofer, and O. Hershkovits, Ancient low entropy flows, mean convex neighborhoods, and uniqueness. 2018, arXiv:1810.08467.

[30] K. Choi, R. Haslhofer, O. Hershkovits, and B. White, Ancient asymptotically cylindrical flows and applications. 2019, arXiv:1910.00639.

[31] D. L. Chopp, Computing minimal surfaces via level set curvature flow. *J. Comput. Phys.* **106** (1993), no. 1, 77–91.

[32] T. H. Colding, T. Ilmanen, and W. P. Minicozzi II, Rigidity of generic singularities of mean curvature flow. *Publ. Math. Inst. Hautes Études Sci.* **121** (2015), 363–382.

[33] T. H. Colding, T. Ilmanen, W. P. Minicozzi II, and B. White, The round sphere minimizes entropy among closed self-shrinkers. *J. Differential Geom.* **95** (2013), no. 1, 53–69.

[34] T. H. Colding and W. P. Minicozzi II, Generic mean curvature flow I: generic singularities. *Ann. of Math. (2)* **175** (2012), no. 2, 755–833.

[35] T. H. Colding and W. P. Minicozzi II, Smooth compactness of self-shrinkers. *Comment. Math. Helv.* **87** (2012), no. 2, 463–475.

[36] T. H. Colding and W. P. Minicozzi II, Uniqueness of blowups and Łojasiewicz inequalities. *Ann. of Math. (2)* **182** (2015), no. 1, 221–285.

[37] T. H. Colding and W. P. Minicozzi II, Differentiability of the arrival time. *Comm. Pure Appl. Math.* **69** (2016), no. 12, 2349–2363.

[38] T. H. Colding and W. P. Minicozzi II, The singular set of mean curvature flow with generic singularities. *Invent. Math.* **204** (2016), no. 2, 443–471.

[39] T. H. Colding and W. P. Minicozzi II, Regularity of the level set flow. *Comm. Pure Appl. Math.* **71** (2018), no. 4, 814–824.

[40] T. H. Colding and W. P. Minicozzi II, Dynamics of closed singularities. *Ann. Inst. Fourier (Grenoble)* **69** (2019), no. 7, 2973–3016.

[41] T. H. Colding and W. P. Minicozzi II, Complexity of parabolic systems. *Publ. Math. Inst. Hautes Études Sci.* **132** (2020), 83–135.

[42] J. Daniels-Holgate, Approximation of mean curvature flow with generic singularities by smooth flows with surgery. 2021, arXiv:2104.11647.

[43] A. Deruelle and F. Schulze, Generic uniqueness for expanders with vanishing relative entropy. *Math. Ann.* **377** (2020), no. 3–4, 1095–1127.

[44] Q. Ding, Minimal cones and self-expanding solutions for mean curvature flows. *Math. Ann.* **376** (2020), no. 1–2, 359–405.

[45] K. Ecker, *Regularity theory for mean curvature flow*. Progr. Nonlinear Differential Equations Appl. 57, Birkhäuser Boston, Inc., Boston, MA, 2004.

[46] L. C. Evans and J. Spruck, Motion of level sets by mean curvature. I. *J. Differential Geom.* **33** (1991), no. 3, 635–681.

[47] L. C. Evans and J. Spruck, Motion of level sets by mean curvature. II. *Trans. Amer. Math. Soc.* **330** (1992), no. 1, 321–332.

[48] L. C. Evans and J. Spruck, Motion of level sets by mean curvature. III. *J. Geom. Anal.* **2** (1992), no. 2, 121–150.

[49] L. C. Evans and J. Spruck, Motion of level sets by mean curvature. IV. *J. Geom. Anal.* **5** (1995), no. 1, 77–114.

[50] M. Gage and R. S. Hamilton, The heat equation shrinking convex plane curves. *J. Differential Geom.* **23** (1986), no. 1, 69–96.

[51] M. A. Grayson, Shortening embedded curves. *Ann. of Math. (2)* **129** (1989), no. 1, 71–111.

[52] O. Hershkovits and B. White, Sharp entropy bounds for self-shrinkers in mean curvature flow. *Geom. Topol.* **23** (2019), no. 3, 1611–1619.

[53] G. Huisken, Asymptotic behavior for singularities of the mean curvature flow. *J. Differential Geom.* **31** (1990), no. 1, 285–299.

[54] T. Ilmanen, Generalized flow of sets by mean curvature on a manifold. *Indiana Univ. Math. J.* **41** (1992), no. 3, 671–705.

[55] T. Ilmanen, Elliptic regularization and partial regularity for motion by mean curvature. *Mem. Amer. Math. Soc.* **108** (1994), no. 520, 90 pp.

[56] T. Ilmanen, Lectures on mean curvature flow and related equations. 1995, https://www.math.ethz.ch/~ilmanen/papers/pub.html.

[57] T. Ilmanen, Singularities of mean curvature flow of surfaces. 1995, http://www.math.ethz.ch/~ilmanen/papers/pub.html.

[58] T. Ilmanen, Problems in mean curvature flow. 2003, https://people.math.ethz.ch/~ilmanen/classes/eil03/problems03.ps.

[59] T. Ilmanen, A. Neves, and F. Schulze, On short time existence for the planar network flow. *J. Differential Geom.* **111** (2019), no. 1, 39–89.

[60] T. Ilmanen and B. White, Sharp lower bounds on density for area-minimizing cones. *Camb. J. Math.* **3** (2015), no. 1–2, 1–18.

[61] D. Impera, S. Pigola, and M. Rimoldi, The Frankel property for self-shrinkers from the viewpoint of elliptic PDEs. *J. Reine Angew. Math.* **773** (2021), 1–20.

[62] D. M. Kane, The Gaussian surface area and noise sensitivity of degree-d polynomial threshold functions. *Comput. Complexity* **20** (2011), no. 2, 389–412.
[63] N. Kapouleas, S. J. Kleene, and N. M. Møller, Mean curvature self-shrinkers of high genus: non-compact examples. *J. Reine Angew. Math.* **739** (2018), 1–39.
[64] D. Ketover, Self-shrinking Platonic solids. 2016, arXiv:1602.07271.
[65] D. Ketover and X. Zhou, Entropy of closed surfaces and min-max theory. *J. Differential Geom.* **110** (2018), no. 1, 31–71.
[66] R. D. Macpherson and D. J. Srolovitz, The von Neumann relation generalized to coarsening of three-dimensional microstructures. *Nature* **446** (2007), 1053–1055.
[67] A. Magni and C. Mantegazza, Some remarks on Huisken's monotonicity formula for mean curvature flow. In *Singularities in nonlinear evolution phenomena and applications*, pp. 157–169, CRM Series 9, Ed. Norm, Pisa, 2009.
[68] R. Malladi and J. A. Sethian, Image processing via level set curvature flow. *Proc. Natl. Acad. Sci. USA* **92** (1995), no. 15, 7046–7050.
[69] F. C. Marques and A. Neves, Min-max theory and the Willmore conjecture. *Ann. of Math. (2)* **179** (2014), no. 2, 683–782.
[70] N. M. Møller, Closed self-shrinking surfaces in $\mathbb{R}^3$ via the torus. 2011, arXiv:1111.7318.
[71] X. H. Nguyen, Construction of complete embedded self-similar surfaces under mean curvature flow, Part III. *Duke Math. J.* **163** (2014), no. 11, 2023–2056.
[72] S. Osher and J. A. Sethian, Fronts propagating with curvature-dependent speed: algorithms based on Hamilton–Jacobi formulations. *J. Comput. Phys.* **79** (1988), no. 1, 12–49.
[73] L. Simon, Lectures on geometric measure theory. In *Proceedings of the Centre for Mathematical Analysis, Australian National University 3*, Australian National University, Centre for Mathematical Analysis, Canberra, 1983.
[74] S. Smale, An infinite dimensional version of Sard's theorem. *Amer. J. Math.* **87** (1965), 861–866.
[75] K. Smoczyk, Star-shaped hypersurfaces and the mean curvature flow. *Manuscr. Math.* **95** (1998), no. 2, 225–236.
[76] A. Stone, A density function and the structure of singularities of the mean curvature flow. *Calc. Var. Partial Differential Equations* **2** (1994), no. 4, 443–480.
[77] A. Sun, Local entropy and generic multiplicity one singularities of mean curvature flow of surfaces. 2018, arXiv:1810.08114.
[78] A. Sun, Entropy in a closed manifold and partial regularity of mean curvature flow limit of surfaces. *J. Geom. Anal.* **31** (2021), no. 6, 5619–5635.
[79] A. Sun and Z. Wang, Compactness of self-shrinkers in $\mathbb{R}^3$ with fixed genus. *Adv. Math.* **367** (2020), 107110, 39 pp.
[80] A. Sun, Z. Wang, and X. Zhou, Multiplicity one for min-max theory in compact manifolds with boundary and its applications. 2020, arXiv:2011.04136.
[81] A. Sun and J. Xue, Initial perturbation of the mean curvature flow for asymptotical conical limit shrinker. 2021, arXiv:2107.05066.

[82] A. Sun and J. Xue, Initial perturbation of the mean curvature flow for closed limit shrinker. 2021, arXiv:2104.03101.

[83] L. Wang, Asymptotic structure of self-shrinkers. 2016, arXiv:1610.04904.

[84] S. Wang, Round spheres are Hausdorff stable under small perturbation of entropy. *J. Reine Angew. Math.* **758** (2020), 261–280.

[85] G. Wei and W. Wylie, Comparison geometry for the Bakry–Emery Ricci tensor. *J. Differential Geom.* **83** (2009), no. 2, 377–405.

[86] B. White, Partial regularity of mean-convex hypersurfaces flowing by mean curvature. *Int. Math. Res. Not.* **1994** (1994), no. 4, 185–192.

[87] B. White, Stratification of minimal surfaces, mean curvature flows, and harmonic maps. *J. Reine Angew. Math.* **488** (1997), 1–35.

[88] B. White, A local regularity theorem for mean curvature flow. *Ann. of Math. (2)* **161** (2005), no. 3, 1487–1519.

[89] B. White, Currents and flat chains associated to varifolds, with an application to mean curvature flow. *Duke Math. J.* **148** (2009), no. 1, 41–62.

[90] Y. Zhang, Superconvexity of the heat kernel on hyperbolic space with applications to mean curvature flow. *Proc. Amer. Math. Soc.* **149** (2021), no. 5, 2161–2166.

[91] J. J. Zhu, On the entropy of closed hypersurfaces and singular self-shrinkers. *J. Differential Geom.* **114** (2020), no. 3, 551–593.

LU WANG

Department of Mathematics, Yale University, 10 Hillhouse Avenue, New Haven, CT 06511, USA, lu.wang@yale.edu

This Page Intentionally Left Blank

COMPOSING AND DECOMPOSING SURFACES AND FUNCTIONS

ROBERT J. YOUNG

ABSTRACT

In mathematics, we are often drawn to the simple or elegant, but what lies at the other end of the spectrum? How can we build and study complex objects? How can we break them down? In this note, we will describe some tools for building functions and surfaces with structure at many different scales and, conversely, tools for decomposing complex objects into simple pieces. These methods are based on ideas from geometric measure theory and harmonic analysis, and we will give some applications to quantitative and metric geometry.

MATHEMATICS SUBJECT CLASSIFICATION 2020

Primary 28A75; Secondary 53A05, 51F30

KEYWORDS

Complexity, decompositions, uniform rectifiability

INTERNATIONAL CONGRESS OF MATHEMATICIANS 2022 JULY 6–14

Proc. Int. Cong. Math. 2022, Vol. 4, pp. 2678–2695
DOI 10.4171/ICM2022/19

Published by EMS Press

What makes one object more complex than another? For instance, what makes a high-genus surface more complex than a sphere, or what makes a random graph with n^2 vertices more complex than an $n \times n$ grid? A rough definition of complexity is that complex objects are hard to describe concisely. The *Kolmogorov complexity* of a string of 0's and 1's, for instance, measures the number of bits it takes to describe an algorithm that outputs that bit string. Then, on the one hand, the 1000-bit sequence $0, 1, 0, 1, \dots, 0, 1$ has low Kolmogorov complexity, since it is the output of a simple algorithm. On the other hand, there are at most $2^{k+1} - 1$ possible algorithms that can be described in at most k bits, so a generic string has large complexity: over 99% of the 1000-bit strings have complexity of at least 990 bits.

This highlights one of the properties of complex objects: while nearly all n-bit strings have complexity of at least $0.99n$ bits, it is impossible to construct an example of such a string without using randomness—any explicit deterministic construction of an n-bit string is an algorithm, so its complexity is bounded by the length of the algorithm. Situations like this are not uncommon in combinatorics. Erdős [7], for instance, famously bounded the Ramsey numbers by showing that a random $2^{\frac{s}{2}}$-vertex graph is overwhelmingly likely to have no s-vertex cliques or s-vertex independent sets. Nevertheless, no specific graph is known to have this property, and there is no known way to construct such graphs for large s without using randomness.

This is one reason that random graphs, surfaces, and complexes can behave in strange and unexpected ways. All the familiar examples of graphs, surfaces, and complexes can be constructed algorithmically, so they are simple in the sense of Kolmogorov. Complex objects may be generic, but complex objects can be strange and unexpected to an intuition trained on familiar examples.

One may hope, however, that objects in $\mathbb{R}^n$ may behave in more familiar ways. In this note, we will confirm this intuition by constructing objects in $\mathbb{R}^n$ that are as complex as possible and bounding the complexity of such objects by decomposing them into simpler pieces. In Section 1, we will construct Lipschitz functions, and in Section 2, we will construct closed surfaces.

In both cases, we find that the complexity of these objects is bounded by geometric quantities. A Lipschitz function on the unit interval, for instance, has a graph which is a curve in $\mathbb{R}^n$. One can construct a Lipschitz function by starting with a linear function, then perturbing it repeatedly, but each perturbation increases the length of the graph, so the complexity of the function is ultimately bounded by the length of the graph. Likewise, a surface in $\mathbb{R}^n$ may be complex, but it can often be decomposed into a sum of several pieces. If the pieces of the decomposition are simple and their size is bounded, the decomposition gives an efficient description of the surface and bounds its complexity. Finally, in Section 3, we describe some applications of these techniques to geometric measure theory and metric geometry.

1. HOW TO BUILD A FUNCTION

We start with a simple example. What does a generic 1-Lipschitz function of a single variable look like? (Most of these ideas can be generalized to higher dimensions, but we stick to one dimension for simplicity.) This question turns out to be surprisingly tricky. For example, one possible approach is to discretize; one can construct a Lipschitz function $f\colon[0,1]\to\mathbb{R}$ by choosing some $n\in\mathbb{N}$ and a sequence of bounded i.i.d. random variables $y_1,\dots,y_n$ and defining

$$f\left(\frac{k}{n}\right)=\sum_{i=1}^{k}\frac{y_i}{n}.$$

We extend to all of $[0,1]$ by linear interpolation; as long as $|y_i|\le 1$ for all i, this is 1-Lipschitz. For any finite n, this produces potentially interesting 1-Lipschitz functions, but the central limit theorem implies that as $n\to\infty$, these functions tend toward $g(x)=mx$, where m is the mean of the distribution that the y_i are drawn from.

The problem is that there are no nontrivial scale-invariant models of random 1-Lipschitz functions. By Rademacher's theorem, any Lipschitz function is differentiable almost everywhere, and the same is true for random 1-Lipschitz functions; if $f\colon[0,1]\to\mathbb{R}$ is a random 1-Lipschitz function drawn from some distribution, then for almost every $x\in[0,1]$, there exists a random variable $f'(x)$ with $|f'(x)|\le 1$ such that

$$\mathbf{P}\left[\lim_{h\to 0}\frac{f(x+h)-f(x)}{h}=f'(x)\right]=1. \tag{1.1}$$

For any $r>0$, we can rescale f around x by letting $f_r(h)=r^{-1}(f(x+rh)-f(x))$. Then (1.1) implies that f_r converges almost surely to the random linear function $h\mapsto f'(x)h$. It follows that any scale-invariant distribution on the space of 1-Lipschitz functions must be supported on the space of affine functions.

Instead, we can construct complex Lipschitz functions by combining functions with different scales. The simplest example is a Weierstrass-type construction; if $\phi_k(x)=10^{-k}\sin(10^k x)$, then $f(x)=\frac{1}{L}\sum_{k=0}^{L-1}\phi_k(x)$ is a 1-Lipschitz function whose graph has bumps at the L different scales $1,10^{-1},\dots,10^{-L+1}$. The bumps at scale 10^{-k} have height roughly $\frac{1}{L}$ times their width, so we say that f is $\frac{1}{L}$-bumpy at L different scales. (One can construct a random 1-Lipschitz function with similar complexity by replacing the ϕ_k's by random 1-Lipschitz functions that oscillate with amplitude and wavelength roughly 10^{-k}.)

With a little more care, we can make this function more complex. The key is that

$$f'(x)=\sum_{k=0}^{L-1}\frac{1}{L}\cos(10^k x)$$

and there's little correlation between $\cos(10^k x)$ and $\cos(10^l x)$ when $k\ne l$. That is, $f'(x)$ is a sum of L values between $-\frac{1}{L}$ and $\frac{1}{L}$. At $x=0$, all these values are $\frac{1}{L}$, so $f'(x)=1$, but for a typical $x\in\mathbb{R}$, these values are close to independent, so $f'(x)$ is typically of order $\frac{\sqrt{L}}{L}$, which much smaller than 1. In fact, by the central limit theorem, if L is large and

$$g(x)=\frac{1}{L}\sum_{k=0}^{L^2}10^{-k}\sin(10^k x),$$

then g is $\frac{1}{L}$-bumpy at L^2 different scales and the distribution of $g'(x)$ is close to a Gaussian with variance less than 1. While g is not 1-Lipschitz, it is almost Lipschitz in the sense that

$|g'(x)| < 5$ for all but a tiny fraction of points, and we can make it 5-Lipschitz by changing it on a tiny fraction of its domain. This produces a Lipschitz function that is $\frac{1}{L}$-bumpy at L^2 different scales.

This is just about the bumpiest a Lipschitz function can be. One way to see this is a theorem of Dorronsoro [5]. Let $f: \mathbb{R} \to \mathbb{R}$ be differentiable. For $x \in \mathbb{R}$ and $r > 0$, we define a quantity $\beta_f(x, r)$ that measures how close f is to affine on $(x - r, x + r)$ by

$$\beta_f(x, r) = \frac{1}{r^2} \min_{\lambda} \int_{x-r}^{x+r} \big|f(y) - \lambda(y)\big| \, dy,$$

where λ ranges over all affine functions. This is normalized to be scale invariant; if $g(x) = cf(c^{-1}x)$, then $\beta_g(x, r) = \beta_f(c^{-1}x, c^{-1}r)$, and Dorronsoro's Theorem implies that if f is supported on $[0, 1]$ and satisfies $\|f'\|_2 < \infty$, then

$$\sum_{n=0}^{\infty} \int_0^1 \beta_f(x, 2^{-n})^2 \, dx \frac{dr}{r} \lesssim \|f'\|_2^2. \tag{1.2}$$

It can help to interpret this inequality as an expectation. Let x be a uniformly distributed random point in $[0, 1]$. Then

$$\mathbf{E}_x\left[\sum_{n=0}^{\infty} \beta_f(x, 2^{-n})^2\right] \lesssim \|f'\|_2^2. \tag{1.3}$$

In particular, for any $L > 0$, the expected number of n's such that $\beta_f(x, 2^{-n}) > \frac{1}{L}$ is at most a constant times L^2, meaning that a 1-Lipschitz function is, for the most part, $\frac{1}{L}$-bumpy at a maximum of roughly L^2 different scales.

The exponent 2 in these bounds comes from the Pythagorean Theorem: adding a bump of height $\frac{r}{L}$ to a segment of length r multiplies the length by roughly a factor of $\frac{1}{L^2}$, so covering a curve by such bumps (making it $\frac{1}{L}$-bumpy at scale r) increases its length by roughly a factor of $\frac{1}{L^2}$. If $f: [0, 1] \to \mathbb{R}$ is a 1-Lipschitz function, its graph is a curve of length between 1 and $\sqrt{2}$. The length is minimized when f is constant, and is larger when f is bumpier. Making f to be $\frac{1}{L}$-bumpy at a scale increases the length of the graph by roughly $\frac{1}{L^2}$, so the bound on the length of the graph implies that f can be $\frac{1}{L}$-bumpy at no more than roughly L^2 different scales. Going further in this direction leads to similar results for rectifiable curves, like Jones's Traveling Salesman Theorem [13].

2. HOW TO BUILD A SURFACE

Now we turn our attention to surfaces. There are many ways to construct complicated closed surfaces embedded or immersed in $\mathbb{R}^n$. One can, for instance, construct codimension-1 surfaces by embedding a k-complex X in $\mathbb{R}^n$ and letting Σ be the boundary of a regular neighborhood of X; one can construct self-similar surfaces inductively, like the Koch snowflake or the Menger sponge; or one can use general position arguments or the Whitney Embedding Theorem to embed arbitrary k-manifolds in $\mathbb{R}^n$.

Although these surfaces can be complex, they can still be decomposed into simple pieces. For example, if $X \subset \mathbb{R}^n$ is an embedded simplicial complex, then its regular neighborhood R can be decomposed into neighborhoods of individual simplices R_δ. The fundamental

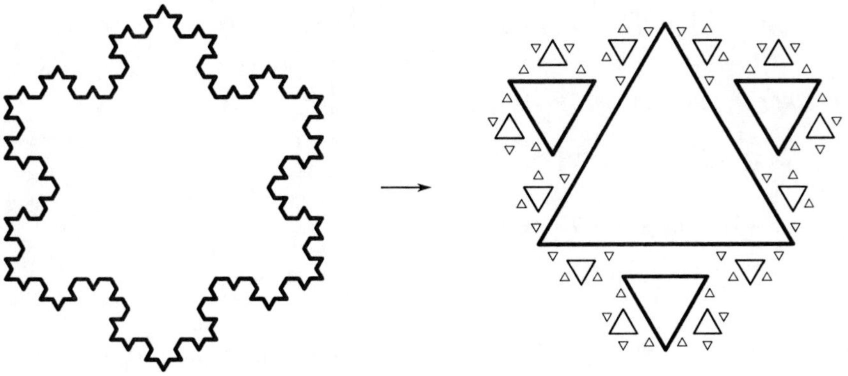

FIGURE 1
A stage in the construction of the Koch snowflake can be decomposed into triangles, and the total length of the triangles is bounded by the length of the original curve.

class $[R]$ of R is the sum $[R] = \sum_\delta [R_\delta]$ of the fundamental classes of the pieces, and the boundary $A = [\partial R] = \partial[R]$ can be written $A = \sum_\delta \partial[R_\delta]$. Likewise, any step in the construction of the Koch snowflake or the Menger sponge can be written as a sum of the boundaries of equilateral triangles or cubes, as in Figure 1. These decompositions are efficient in the sense that the total area of the pieces is bounded by a multiple of the area of the original surface.

In this section, we will argue that arbitrary surfaces in $\mathbb{R}^N$ cannot be too much more complex than Lipschitz functions. That is, given a surface $\Sigma \subset \mathbb{R}^N$, written as a k-cycle $M = [\Sigma] \in C_k(\mathbb{R}^n; \mathbb{Z}_2)$ with coefficients in $\mathbb{Z}_2$, we can write M as a sum $M = \sum_i A_i$ of k-cycles such that each of the A_i's can be approximated by graphs of Lipschitz functions (Lipschitz graphs) with bounded total volume. Furthermore, this decomposition is efficient, i.e., the total area of the A_i's is bounded by a multiple of the area of Σ.

In the following, we take $C_k(X; A)$ to be the set of singular Lipschitz k-chains in X with coefficient group A, i.e., formal sums $M = \sum_i a_i[\delta_i]$ where $a_i \in A \setminus \{0\}$ and $\delta_i \colon \Delta^k \to X$ are distinct Lipschitz k-simplices. When $A = \mathbb{Z}$ or $\mathbb{R}$, we define mass $M = \sum_i |a_i| \operatorname{vol}^k \delta_i$, where $\operatorname{vol}^k \delta_i$ is the k-dimensional Hausdorff measure of δ_i, counted with multiplicity. When $A = \mathbb{Z}_2$, we define mass $M = \sum_i \operatorname{vol}^k \delta_i$.

2.1. Decomposing into cubes

We first use cellular approximation and the Federer–Fleming Deformation Theorem to decompose cycles into sums of boundaries of cubes.

Let $0 < t < 1$ and let τ_t be the grid of side length t in $\mathbb{R}^n$. By cellular approximation, any chain in $\mathbb{R}^n$ can be approximated by a cellular chain in τ_t. The following special case of the Federer–Fleming Deformation Theorem makes this approximation quantitative.

Theorem 2.1 ([6, 8]). *There is a $c_n > 0$ with the following property. Let $t > 0$. Let $T \in C_k(\mathbb{R}^n; A)$ be a singular Lipschitz k-chain over a coefficient group $A = \mathbb{Z}, \mathbb{R}$, or $\mathbb{Z}_\nu$*

(i.e., a formal sum of Lipschitz maps $\Delta^k \to \mathbb{R}^n$ such that $\partial T = 0$). Suppose that $\partial T \in C_{k-1}(\tau_t; A)$ is a cellular chain. Then there are a cellular k-chain $P \in C_k(\tau_t; A)$ and a singular Lipschitz $(k+1)$-chain Q such that:

(1) $\text{mass } P \leq c_n \text{ mass } T$,

(2) $\text{mass } Q \leq c_n t \text{ mass } T$, *and*

(3) $\partial Q = P - T$.

In particular, $\partial P - \partial T = \partial^2 Q = 0$, so P and T have the same boundary.

Furthermore, if T is supported in a subcomplex $K \subset \tau_t$, then P is supported in the same subcomplex.

Let $T \in C_k(\tau_1)$ be a cellular k-cycle and let $M = \text{mass } T$. We will decompose T by constructing a sequence of approximations of T. Let $P_0 = T$, and for each $i \geq 1$, let $P_i \in C_k(\tau_{2^i})$ be a cellular approximation of T in τ_{2^i} as in Theorem 2.1.

On the one hand, P_i is a sum of k-cells of τ_{2^i}, so mass P_i is a multiple of 2^{ki}. On the other hand, $\text{mass } P_i \leq c_n \text{ mass } T$ for all i. Therefore, if $2^{ki} > c_n \text{ mass } T$, then $P_i = 0$. Let i_0 be the smallest integer such that $2^{ki_0} > c_n \text{ mass } T$. Then $P_{i_0} = 0$, so we can decompose T as

$$T = \sum_{i=0}^{i_0-1} (P_i - P_{i+1}).$$

For each i, $P_i - P_{i+1}$ is a cellular cycle in τ_{2^i}, so there is some cellular chain $R_i \in C_{k+1}(\tau_{2^i})$ such that $\partial R_i = P_i - P_{i+1}$. We can use Theorem 2.1 to find R_i. Let Q_i be a $(k+1)$-chain as in Theorem 2.1 so that $\partial Q_i = P_i - T$ and $\text{mass } Q_i \leq 2^i c_n \text{ mass } T$. Then $\partial(Q_i - Q_{i+1}) = P_i - P_{i+1} \in C_k(\tau_{2^i})$. That is, $Q_i - Q_{i+1}$ has cellular boundary, so we can apply Theorem 2.1 again to approximate it by a cellular chain $R_i \in C_{k+1}(\tau_{2^i})$ such that

$$\text{mass } R_i \leq c_n(\text{mass } Q_i + \text{mass } Q_{i+1}) \leq c_n(2^i c_n \text{ mass } T + 2^{i+1} c_n \text{ mass } T) \lesssim 2^i \text{ mass } T$$

and

$$\partial R_i = \partial(Q_i - Q_{i+1}) = P_i - P_{i+1}.$$

We write R_i as a sum $R_i = \sum_j a_{i,j} R_{i,j}$, where the $R_{i,j}$'s are $(k+1)$-cells of τ_{2^i} and $\text{mass } R_i = \sum_j |a_{i,j}| 2^{(k+1)i}$. Then

$$T = \sum_{i=0}^{i_0-1} \partial R_i = \sum_{i=0}^{i_0-1} \sum_j a_{i,j} \partial R_{i,j},$$

decomposes T as a sum of boundaries of cubes.

The number and size of the pieces of the decomposition is bounded in terms of the mass of T. For each $0 \leq i \leq i_0$, the total mass of the boundaries of the cubes is bounded by

$$\sum_j |a_{i,j}| \text{ mass } \partial R_{i,j} \approx \sum_j |a_{i,j}| 2^{ik} = 2^{-i} \text{ mass } R_i \lesssim \text{mass } T,$$

so, since $2^{ki_0} \approx c_n \operatorname{mass} T$, we have

$$\sum_{i=0}^{i_0-1} \sum_j |a_{i,j}| \operatorname{mass}(\partial R_{i,j}) \lesssim i_0 \operatorname{mass}(T) \lesssim \operatorname{mass}(T) \log \operatorname{mass}(T).$$

This decomposition and similar decompositions are useful for studying isoperimetric inequalities. Recall that the isoperimetric inequality in $\mathbb{R}^n$ implies that for any k-cycle T, there is a $(k+1)$-chain S such that $\partial S = T$ and $\operatorname{mass} S \lesssim (\operatorname{mass} T)^{\frac{k+1}{k}}$. If R_i are as above, then $S = \sum_{i=0}^{i_0-1} R_i$ satisfies $\partial S = T$, and since $2^{ki_0} \approx c_n \operatorname{mass} T$,

$$\operatorname{mass}(S) \leq \sum_{i=0}^{i_0-1} \operatorname{mass} R_i \lesssim \sum_{i=0}^{i_0-1} 2^i \operatorname{mass} T \leq 2^{i_0} \operatorname{mass}(T) \lesssim \operatorname{mass}(T)^{\frac{k+1}{k}}.$$

More generally, this decomposition is useful for studying higher-dimensional versions of the *Dehn function* of a group or space, which measure the difficulty of filling a k-cycle in a space by a $(k+1)$-chain. In many cases (see, for instance, [20]), one can use a version of the Federer–Fleming Deformation Theorem to decompose an arbitrary k-cycle T into a sum of scalings of simple pieces $T = \sum_i S_i$ and construct a filling of T by adding together fillings of the S_i's.

2.2. An inductive strategy

One difficulty with this decomposition is that the total volume of the pieces grows when τ_1 is replaced by a finer grid. That is, the decomposition above writes a cycle $T \in C_k(\tau_1)$ as a sum of boundaries of cubes $T = \sum_{i=0}^{M-1} \sum_j \partial R_{i,j}$, where $2^{kM} \approx \operatorname{mass} T$ and $\sum_j \operatorname{mass}(\partial R_{i,j}) \approx \operatorname{mass} T$ for all i; the total mass of the $\partial R_{i,j}$'s is at most $M \operatorname{mass} T$.

Let $m < 0$ and let $T \in C_k(\tau_{2^m})$. By applying the same decomposition to a rescaling of T, we can write $T = \sum_{i=m}^{M-1} \sum_j \partial R_{i,j}$, where each $R_{i,j}$ is a cube of side length 2^i and $2^{kM} \approx \operatorname{mass} T$. Unfortunately, the total mass now satisfies

$$\sum_{i=m}^{M-1} \sum_j \partial R_{i,j} \lesssim (M - m) \operatorname{mass} T,$$

and even if $\operatorname{mass} T$ is fixed, this will get larger as $m \to -\infty$.

Ideally, given $m < 0$ and a cycle $T \in C_k(\tau_{2^m})$, we would like a decomposition $T = \sum_i K_i$ such that $\sum_i \operatorname{mass} K_i \lesssim \operatorname{mass} T$, with constant independent of m; such a decomposition opens up applications in geometric measure theory. In the rest of this section, we will pursue such decompositions.

One such decomposition appears in Wenger's proof of Gromov's Filling Inequality [19]. Gromov's Filling Inequality states the following.

Theorem 2.2 ([12]). *Let $k > 0$. There is a $c > 0$ such that for any Banach space X and any k-cycle $A \in C_k(X)$, there is a $(k+1)$-chain $D \in C_{k+1}(X)$ such that $\partial D = A$ and $\operatorname{mass} D \leq c(\operatorname{mass} A)^{\frac{k+1}{k}}$.*

The methods used in the previous section cannot be used to prove Theorem 2.2, because the constants in the Federer–Fleming Deformation Theorem depend on the dimension of the ambient space. Nevertheless, it is straightforward to prove Theorem 2.2 when A is *round*, i.e., when $\operatorname{diam}\operatorname{supp} A \lesssim (\operatorname{mass} A)^{\frac{1}{k}}$.

Lemma 2.3 (Cone-type inequality). *Let $k > 0$. There is a $c > 0$ such that for any Banach space X and any k-cycle $A \in C_k(X)$, there is a $(k+1)$-chain $D \in C_{k+1}(X)$ such that $\partial D = A$ and* $\operatorname{mass} D \leq c(\operatorname{mass} A)(\operatorname{diam}\operatorname{supp} A)$.

Proof. We translate so that $0 \in \operatorname{supp} A$. Suppose that $A = \sum_i a_i[\delta_i]$ for some $a_i \in \mathbb{Z}$ and some Lipschitz simplices $\delta_i : \Delta^k \to X$. Let $\overline{\delta_i} : \Delta^k \times [0,1] \to X$, $\overline{\delta_i}(x,t) = t\delta_i(x)$. Then $D = \sum_i a_i[\overline{\delta_i}]$ satisfies the desired properties. ■

Difficulties arise, however, when A is not round, for example, when A is a long skinny cylinder. Wenger proves Theorem 2.2 by decomposing an arbitrary cycle A into a sum of round cycles and applying the cone-type inequality to each piece, and in this section, we will describe a version of his strategy, with some details simplified for the sake of brevity. Any inaccuracies and oversimplifications are entirely our fault.

We say that a k-chain T is *c-round* if $\operatorname{mass} T \geq c(\operatorname{diam}\operatorname{supp} T)^k$. Let $c > 0$ be small. The key idea of Wenger's decomposition is that if A is not c-round, then we can "cut" an open ball B out of A so that the cut-off piece is round. That is, we can decompose $A = A' + M$ so that M is a round cycle with $\operatorname{supp} M \subset \bar{B}$ and $\operatorname{supp} A' \subset X \setminus B$.

We find B by noting that since A is not c-round, there are $x \in \operatorname{supp} A$ and $0 < r < \operatorname{diam}\operatorname{supp} A$ such that $\operatorname{mass}(A \cap B(x,r)) \in [\frac{c}{2}r^k, cr^k]$. We let $B = B(x,r)$ and let $K = A \setminus B$ be the restriction of A to $X \setminus B$, so that $\operatorname{supp} K = (\operatorname{supp} A) \setminus B$ and $\partial K \in C_{k-1}(\partial B)$. Let $L \in C_k(\partial B)$ be a chain such that $\partial L = \partial K$ and let $A' = K - L$ so that A' is a cycle.

If c is sufficiently small and if x and r are chosen carefully, we can arrange that $\operatorname{mass} L \leq \frac{c}{4}r^k$,

$$\operatorname{mass} A' = \operatorname{mass} A - \operatorname{mass}(A \cap B) + \operatorname{mass} L \leq \operatorname{mass} A - \frac{c}{4}r^k,$$

and

$$\operatorname{mass}(A - A') = \operatorname{mass}(A \cap B) + \operatorname{mass} L \leq 2cr^k.$$

Let $M = A - A'$. Then

$$\operatorname{diam}\operatorname{supp}(M) \leq \operatorname{diam} B \leq 2r \leq 2c^{-\frac{1}{k}}(\operatorname{mass} M)^{\frac{1}{k}},$$

so M is $(2^{-k}c)$-round. Furthermore,

$$\operatorname{mass} M \lesssim \operatorname{mass} A - \operatorname{mass} A'. \tag{2.1}$$

Equation (2.1) is important because it lets us use this construction in an inductive argument. Let $c > 0$ be as in the argument above and let $A_0 = A$. If we have constructed A_i and A_i is not c-round, then there is a ball B_i and a decomposition $A_i = A_{i+1} + M_i$ where M_i is a round cycle with $\operatorname{supp} M_i \subset \bar{B}_i$ and $\operatorname{mass} M_i \lesssim \operatorname{mass} A_i - \operatorname{mass} A_{i+1}$. Otherwise, if A_i is c-round, we terminate the construction, letting $A_{i+1} = 0$ and $M_i = A_i$.

If this construction terminates with some $A_n = 0$, we have $A = \sum_{i=0}^{n-1} M_i$ and

$$\sum_{i=0}^{n-1} \operatorname{mass} M_i \lesssim \sum_{i=0}^{n-1} (\operatorname{mass} A_i - \operatorname{mass} A_{i+1}) = \operatorname{mass} A_0 - \operatorname{mass} A_n = \operatorname{mass} A.$$

Otherwise, for any $n > 0$, $A = A_n + \sum_{i=0}^{n-1} M_i$, and

$$\operatorname{mass} A_n + \sum_{i=0}^{n-1} \operatorname{mass} M_i \lesssim \operatorname{mass} A_n + \sum_{i=0}^{n-1} (\operatorname{mass} A_i - \operatorname{mass} A_{i+1}) = \operatorname{mass} A.$$

In general, we cannot guarantee termination, but we can choose the B_i's so that $\lim_i \operatorname{mass} A_i = 0$ and $\operatorname{diam} \operatorname{supp} A_i < 2 \operatorname{diam} \operatorname{supp} A$ for all i. Consequently, for any $\varepsilon > 0$, there is an efficient decomposition $A = \sum_{i=0}^{n-1} M_i$ or $A = A_n + \sum_{i=0}^{n-1} M_i$ such that each of the M_i's is round and $\operatorname{mass} A_n < \varepsilon$ (if the decomposition has an A_n term). By applying Lemma 2.3 to each summand, one constructs $D_0, \ldots, D_n$ such that $\partial D_i = M_i$ and $\partial D_n = A_n$. Let $D = \sum D_i$. Then

$$\operatorname{mass} D \lesssim \varepsilon \operatorname{diam} \operatorname{supp} A + \sum_{i=0}^{n-1} (\operatorname{mass} M_i)^{\frac{k+1}{k}},$$

and if ε is sufficiently small,

$$\operatorname{mass} D \lesssim (\operatorname{mass} A)^{\frac{k+1}{k}},$$

as desired.

2.3. Quasiminimizers and uniform rectifiability

Wenger's proof of Gromov's Filling Inequality suggests a general strategy for constructing efficient decompositions inductively:

(1) Let $A \in C_k(\mathbb{R}^n)$ be a k-cycle. Let $A_0 = A$.

(2) Suppose by induction that we have constructed a cycle A_i. Find a region $U_i \subset \mathbb{R}^n$ and a cycle A_{i+1} such that A_i and A_{i+1} are the same outside U_i and

$$\operatorname{mass}(A_i - A_{i+1}) \lesssim \operatorname{mass} A_i - \operatorname{mass} A_{i+1}.$$

Let $M_i = A_i - A_{i+1}$.

(3) Repeat this process until there is some m such that $A_m = 0$ or $\operatorname{mass} A_m$ is as small as desired.

(4) Then $A = A_m + \sum_{i=0}^{m} M_i$, and

$$\operatorname{mass} A_m + \sum_{i=0}^{m} \operatorname{mass} M_i \lesssim \operatorname{mass} A_m + \sum_{i=0}^{m} (\operatorname{mass} A_i - \operatorname{mass} A_{i+1}) \lesssim \operatorname{mass} A.$$

In fact, this is the strategy behind the following theorem, which efficiently decomposes an arbitrary cellular mod-2 cycle as a sum of uniformly rectifiable pieces.

Theorem 2.4 ([21]). *Let $n > 0$. There is a $c > 0$ with the following property. Let $t > 0$ and let τ_t be the grid of side length t in $\mathbb{R}^n$. Any mod-2 cycle $A \in C_d(\tau_t; \mathbb{Z}_2)$ can be written as a sum $A = \sum_{i=0}^{m} M_i$ of mod-2 d-cycles $M_i \in C_d(E_i; \mathbb{Z}_2)$, where each E_i is a d-dimensional, c-uniformly rectifiable subcomplex of τ_t and $\sum |E_i| \lesssim \operatorname{mass} A$.*

Uniform rectifiability is a property defined and studied by David and Semmes [3]. It has many definitions; we will present a definition that uses a function $\beta_E(x, r)$ that measures the "bumpiness" of a set $E \subset \mathbb{R}^n$, similar to the function $\beta_f(x, r)$ in Section 1. Let $c > 1$, let $d > 0$ be an integer, and let $|\cdot|$ denote Hausdorff d-measure. A set $E \subset \mathbb{R}^n$ is *Ahlfors c-regular* if

$$c^{-1} r^d < \left|E \cap B(x, r)\right| < c r^d \quad \text{for all } x \in E \text{ and } 0 < r < \operatorname{diam} E. \tag{2.2}$$

For $x \in E$ and $r > 0$, let

$$\beta_E(x, r) = \frac{1}{r^{d+1}} \min_P \int_{E \cap B(x,r)} d(y, P) \, dy,$$

where P ranges over all d-planes in $\mathbb{R}^n$ and dy represents Hausdorff d-measure. Then $\beta_E(x, r)$ measures how well $E \cap B(x, r)$ can be approximated by a plane. It is scale-invariant in the sense that if $cE = \{cy \mid y \in E\}$, then $\beta_E(x, r) = \beta_{cE}(cx, cr)$.

For any $c > 1$, we say that $E \subset \mathbb{R}^n$ is c-uniformly rectifiable if E is Ahlfors c-regular and if it satisfies the following inequality based on (1.2). For every $x \in E$ and $r > 0$,

$$\frac{1}{r^d} \int_{E \cap B(x,r)} \int_0^s \beta_E(y, s)^2 \frac{ds}{s} \, dy \leq c. \tag{2.3}$$

That is, a uniformly rectifiable set is an Ahlfors regular set which is no bumpier than a Lipschitz function. The prototypical example of a uniformly rectifiable set is the graph of a Lipschitz function, or a *Lipschitz graph*.

The power of uniform rectifiability is that this condition is equivalent to a variety of other conditions on E. The work discussed in the next section, for instance, uses the fact that a uniformly rectifiable set admits a *corona decomposition*. Defining such a decomposition rigorously is rather technical, and we point interested readers to [3], but briefly, if E admits a corona decomposition, then there is a collection $\mathcal{C}$ of Lipschitz graphs with uniformly bounded Lipschitz constants that approximates E efficiently at most points and most scales. That is, on the one hand, for almost every $x \in E$, there is a finite set of "bad scales" $S_x \subset \mathbb{Z}$ such that for all $i \in \mathbb{Z} \setminus S_x$, either $2^{-i} > \operatorname{diam} E$ or the intersection $B(x, 2^{-i}) \cap E$ can be approximated by an element of $\mathcal{C}$. Furthermore, as x ranges over E, the average size of S_x is bounded. On the other hand, the set $\mathcal{C}$ is not too big; in particular, the total measure of the elements of $\mathcal{C}$ is comparable to the measure of E.

Conversely, if a set is not uniformly rectifiable, then it must be complex—it must be far from a plane at many scales. Fractals are typical examples; if E is self-similar but not a plane, then $\beta_E(x, r) > \varepsilon$ for most x's and r's, so $\sum_{i=0}^{\infty} \beta_E(x, r2^{-i})^2$ is typically infinite.

Fractals also provide examples of sets that are rectifiable but not uniformly rectifiable. One such set is based on the four-corners Cantor set. For each $i \geq 0$, let K_i be the ith step in the construction of the four-corners Cantor set, so that $K_0 = \partial[0, 1]^2$ is the boundary

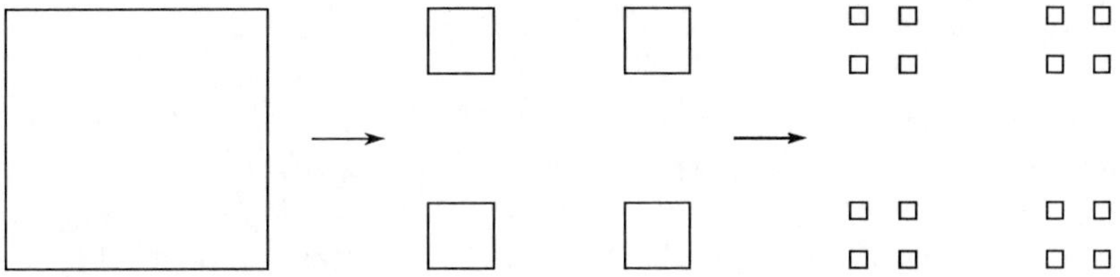

FIGURE 2
Three stages in the construction of the four-corners Cantor set. The stages of the construction are uniformly Ahlfors regular, but there is no c such that they are all c-uniformly rectifiable.

of the unit square, and each set K_{i+1} is obtained by replacing each square in K_i by four squares of one fourth the side length (Figure 2). Then for any $x \in K_i$ and any $j \le i$, the intersection $K_i \cap B(x, 4^{-j})$ is not close to a line, so $\beta_{K_i}(x, 4^{-j+1}) \gtrsim 1$. Therefore, for any $y \in K_i$, we have

$$\int_{4^{-i}}^{1} \beta_{K_1}(y,s)^2 \frac{\mathrm{d}s}{s} \approx i.$$

That is, the increasing complexity of the K_i's implies that there is no $c > 0$ such that all of the K_i's are c-uniformly rectifiable. Nevertheless, for each i, the fundamental class of K_i is a cellular cycle in $\tau_{4^{-i}}$, so Theorem 2.4 implies that it can be written as a sum of uniformly rectifiable pieces. In this case, K_i is a sum of the fundamental classes of 4^i disjoint squares, each of side length 4^{-i}.

The key tools for constructing the decomposition in Theorem 2.4 are the inductive strategy in Section 2.2 and a result of David and Semmes [4], which states that quasiminimizing sets are uniformly rectifiable. A set is quasiminimizing if compactly-supported deformations cannot decrease its area by too much. To state this rigorously, let $U \subset \mathbb{R}^n$ be a bounded open set. We say that a continuous map $f: \mathbb{R}^n \to \mathbb{R}^n$ is a *deformation supported in* U if $f(U) \subset U$ and $f(x) = x$ for all $x \notin U$. Let $k > 1$. A set $S \subset \mathbb{R}^n$ such that $|S \cap B(0,\rho)| < \infty$ for all $\rho > 0$ is said to be (k,r)-*quasiminimizing* if for every open set $U \subset \mathbb{R}^n$ with $\operatorname{diam} U < r$, every deformation f supported in U satisfies

$$\left|f(S \cap U)\right| \ge \frac{1}{k}|S \cap U|.$$

For example, Lipschitz graphs are quasiminimizing sets. Let $\alpha: \mathbb{R}^{n-1} \to \mathbb{R}$ be a 1-Lipschitz function, let $\psi: \mathbb{R}^{n-1} \to \mathbb{R}^n$, $\psi(x) = (x, \alpha(x))$, and let $S = \psi(\mathbb{R}^n)$ be the graph of α. Let $\pi: \mathbb{R}^{n-1} \times \mathbb{R} \to \mathbb{R}^{n-1}$ be the projection $\pi(x,y) = x$. On the one hand, π is distance-decreasing, so $|\pi(E)| \le |E|$ for all $E \subset \mathbb{R}^n$. On the other hand, for any $E \subset S$, we have $E = \psi(\pi(E))$, and since ψ is 2-Lipschitz, we have

$$\left|\pi(E)\right| \le |E| \le 2^n \left|\pi(E)\right|.$$

For any deformation f supported in U, we have $\pi(f(U \cap S)) \supset \pi(U \cap S)$, so

$$\left|f(U \cap S)\right| \ge \left|\pi\left(f(U \cap S)\right)\right| \ge \left|\pi(U \cap S)\right| \ge \frac{1}{2^n}|U \cap S|.$$

That is, S is $(2^n, \infty)$-quasiminimizing.

David and Semmes proved the following theorem.

Theorem 2.5 ([4]). *Let $n, d > 0$ and $k > 1$. There is a $c > 0$ such that if $S \subset \mathbb{R}^n$, $r \geq \frac{1}{10} \operatorname{diam} S$, and S is (k, r)-quasiminimizing, then S is c-uniformly rectifiable.*

The proof of Theorem 2.4 uses Theorem 2.5 to implement the inductive strategy. Let $k > 10$ and let c be as in Theorem 2.5. Let $A_0 = A \in C_d(\mathbb{R}^n; \mathbb{Z}_2)$ be a mod-2 cycle. If $i \geq 0$ and the support $S_i = \operatorname{supp} A_i$ is c-uniformly rectifiable, we let $A_{i+1} = 0$ and terminate the construction. Otherwise, Theorem 2.5 implies that S_i is not a quasiminimizer. That is, there is a deformation $f_i: \mathbb{R}^n \to \mathbb{R}^n$, supported on some bounded set U_i with $\operatorname{diam} U_i < \frac{1}{10} \operatorname{diam} S_i$ such that $|f_i(S_i \cap U_i)| < \frac{1}{k}|S_i \cap U_i|$. If we choose the deformation carefully, we can ensure that $f_i(S_i)$ is a union of d-cells of τ_t, so that the push-forward $A_{i+1} = (f_i)_\sharp(A_i)$ is a cellular chain. We let $M_i = A_i - A_{i+1}$.

Since each A_i is a mod-2 cellular chain, we have $\operatorname{mass} A_i = |\operatorname{supp} A_i| = |S_i|$. Each step of this construction decreases the measure of S_i and thus decreases the mass of A_i. In fact,

$$\operatorname{mass} A_i - \operatorname{mass} A_{i+1} = |S_i \cap U_i| - \left|f_i(S_i \cap U_i)\right| \geq \frac{9}{10}|S_i \cap U_i| > 0$$

and

$$\operatorname{mass} M_i \leq |S_i \cap U_i| + \left|f_i(S_i \cap U_i)\right| \leq 2|S_i \cap U_i| \leq 3(\operatorname{mass} A_i - \operatorname{mass} A_{i+1}).$$

The number of cells in A_i is an integer and decreases with i, so this guarantees that the process terminates, i.e., $A_m = 0$ for some m. Then $A = \sum_{i=0}^{m-1} M_i$ and

$$\sum_{i=0}^{m-1} \operatorname{mass} M_i \leq \sum_{i=0}^{m-1} 3(\operatorname{mass} A_i - \operatorname{mass} A_{i+1}) \leq 3 \operatorname{mass} A.$$

That is, this is an efficient decomposition of A. Choosing the f_i so that the M_i's are supported on uniformly rectifiable sets, however, is more difficult, and we point interested readers to the full proof in [21].

Thus, while surfaces in $\mathbb{R}^n$ can be complex, their complexity is bounded by their volume, and complex surfaces can be efficiently decomposed into pieces that are not too much bumpier than Lipschitz graphs.

3. APPLICATIONS

In this section, we will describe some ways to apply the decompositions described in the previous section to bound the topology of cycles and currents in $\mathbb{R}^n$ and to bound embeddings of nilpotent groups into Banach spaces.

3.1. Geometric measure theory and quantifying the topology of embedded submanifolds

Theorem 2.4 can be used to bound the topological complexity of an arbitrary cycle in $\mathbb{R}^n$. Specifically, it can be used to quantify how difficult it is to orient a mod-2 cycle, that is, to lift it to a cycle with integer coefficients.

Given a cycle $A \in C_d(\mathbb{R}^n; \mathbb{Z}_2)$, a *pseudoorientation* of A is a cycle $P \in C_d(\mathbb{R}^n)$ such that $P \equiv A \pmod 2$. In general, a cycle A has many pseudoorientations. We define the *nonorientability* of A to be the infimal mass of a pseudoorientation, i.e.,

$$\mathrm{NO}(A) = \inf\{\text{mass } P \mid P \text{ is a pseudoorientation of } A\}.$$

The nonorientability of a surface measures how difficult it is to cut that surface into orientable pieces. That is, let $\Sigma \subset \mathbb{R}^4$ be an arbitrary nonorientable surface and let $A = [\Sigma]$. Let Γ be a graph (a 1-dimensional simplicial complex) embedded in Σ such that each connected component of $\Sigma \setminus \Gamma$ is orientable. (For instance, let Σ be a Klein bottle and let Γ be a simple closed curve that cuts Σ into a cylinder.) Let $C_1, \dots, C_k$ be these components. Orient them arbitrarily and let $C = \sum_{i=1}^k [C_i]$. Let $G = \partial C = \sum_{i=1}^k \partial[C_i]$.

Each edge e of Γ is in the boundary of two components C_i and C_j, so e occurs exactly twice in the sum $\sum_{i=1}^k \partial[C_i]$. If the orientations of C_i and C_j agree on e, then the occurrences of e cancel out and it has coefficient 0 in G. If the orientations disagree, then e has coefficient ± 2. Since all coefficients of G are even, $G/2$ is a 1-cycle with integer coefficients, so there is some D such that $\partial D = G/2$. By the isoperimetric inequality, we can choose D such that mass $D \le (\text{mass } G)^2 \le \ell(\Gamma)^2$, where $\ell(\Gamma)$ is the total length of the edges of Γ. Let $P = C - 2D$; then $P \equiv C \equiv A \pmod 2$, so P is a pseudoorientation of A and

$$\mathrm{NO}(A) \le \text{mass } P \le \text{mass } A + 2\ell(\Gamma)^2.$$

In [21], Theorem 2.4 was used to prove the following.

Theorem 3.1 ([21]). *Let $n > 0$ and $0 < d < n$. There is a $c > 0$ with the following property. Let $t > 0$ and let τ_t be the grid of side length t in $\mathbb{R}^n$. Then for any mod-2 cycle $A \in C_d(\tau_t; \mathbb{Z}_2)$, $\mathrm{NO}(A) \le c$ mass A.*

By Theorem 2.4, $A = \sum_i M_i$ where each M_i is a cycle supported on a uniformly rectifiable set E_i. Nonorientability is subadditive, so it suffices to use the uniform rectifiability of the E_i's to bound $\mathrm{NO}(M_i)$.

The proof relies on approximating the uniformly rectifiable set E_i by Lipschitz graphs. To give a brief sketch, since E_i is uniformly rectifiable, it has a corona decomposition, i.e., a collection of Lipschitz graphs $\mathcal{C}$ such that for most points $x \in E_i$ and most scales $0 < r < \operatorname{diam} E_i$, the intersection $B(x,r) \cap E_i$ is close to one of the Lipschitz graphs $\Lambda \in \mathcal{C}$. Then the restriction of M_i to $B(x,r)$ can be approximated by a chain in Λ, and since Λ is orientable, that chain inherits an orientation from Λ. If every intersection $B(x,r) \cap E_i$ could be approximated by oriented surfaces and all of the orientations agreed, then it would be easy to lift M_i to a cycle with integer coefficients. Difficulties only arise where E_i is complex—from choices of x and r such that $B(x,r) \cap E_i$ cannot be approximated by a Lipschitz graph or when elements of $\mathcal{C}$ don't have consistent orientations—but the uniform rectifiability of E_i bounds how much complexity it can have.

3.2. Metric geometry and embeddings of nilpotent groups

In a more recent work, these tools have been applied to study embeddings into Banach spaces, especially the Heisenberg groups. The Heisenberg groups are the simplest family of nonabelian nilpotent groups, and their noncommutativity makes them difficult to embed into Banach spaces in a way that preserves their metrics.

An introduction to the Heisenberg group can be found, for instance, in [1], but briefly, the integer Heisenberg group $H_n^{\mathbb{Z}}$ is the group generated by $X_1, \dots, X_n, Y_1, \dots, Y_n$, and Z such that $[X_i, Y_i] = X_i Y_i X_i^{-1} Y_i^{-1} = Z$ and all other pairs of generators commute. This is isomorphic to a group of upper-triangular $(n+2) \times (n+2)$ matrices of the form

$$\begin{pmatrix} 1 & x_1 & \cdots & x_n & z \\ & \ddots & & & y_1 \\ & & \ddots & & \vdots \\ & & & \ddots & y_n \\ & & & & 1 \end{pmatrix} \tag{3.1}$$

where the x_i's, y_i's, and z are all integers. Here, $X_i \in H_n$ is identified with the matrix of form (3.1) with all coefficients zero except that $x_i = 1$ and likewise for Y_i and Z. The integer Heisenberg group is a lattice in the nilpotent Lie group obtained by taking the matrices of the form (3.1) with real coefficients; we call this Lie group H_n and write elements of H_n as points $(x_1, \dots, x_n, y_1, \dots, y_n, z) \in \mathbb{R}^{2n+1}$.

A key feature of the Heisenberg group is the difference between the vertical direction Z and the horizontal directions X_i and Y_i. For any n, one can calculate that $[X_i^n, Y_i^n] = Z^{n^2}$. That is, quadratically large powers of Z can be written as products of linearly many X_i's and Y_i's. Let $\langle Z \rangle$ be the z-axis in H_n; in the terminology of geometric group theory, $\langle Z \rangle$ is a quadratically distorted subgroup. Left-invariant metrics on H_n inherit this quadratic distortion; we will equip H_n with the left-invariant metric such that

$$d\big(\mathbf{0}, (x_1, \dots, x_n, y_1, \dots, y_n, z)\big) = \max\big\{|x_i|, |y_i|, \sqrt{|z|}\big\}.$$

Then, for $t \neq 0$, the map

$$s_t(x_1, \dots, x_n, y_1, \dots, y_n, z) = (tx_1, \dots, tx_n, ty_1, \dots, ty_n, t^2 z),$$

which scales horizontal directions by t and scales the vertical direction by t^2, is an automorphism of H_n that scales the metric by a factor of t.

A natural question is how well H_n can be embedded into different spaces, especially Banach spaces. This has applications in theoretical computer science, where the accuracy of some algorithms depends on how well certain metric spaces embed in L_1 (see, for instance, [11, 15]). In this section, we will consider embeddings of H_n into L_p spaces.

For any $n > 1$ and any $p \in [1, \infty)$, H_n, equipped with its subriemannian metric, does not embed in L_p by a bi-Lipschitz map. When $p > 1$, this follows from a version of Pansu's differentiability theorem [18], which states that Lipschitz maps from H_n to L_p can be locally approximated by homomorphisms from H_n to L_p. Since L_p is abelian and $Z = [X_1, Y_1]$, any such homomorphism must send each *vertical line* (coset of $\langle Z \rangle$) to a point; thus, Lipschitz maps from H_n to L_p cannot be bi-Lipschitz.

This differentiability fails, however, for maps to L_1; Lipschitz maps need not be differentiable anywhere. Instead, Cheeger and Kleiner [2] proved that H_n does not embed in L_1 by showing that the behavior of maps from H_n to L_1 depends on the structure of surfaces in H_n.

To give a rough idea of this argument, suppose that M is a measure space and that $f: H_n \to L_1(M)$ is a Lipschitz function. We claim that f is not bi-Lipschitz. For almost every $m \in M$, the coordinate function $f_m: H_n \to \mathbb{R}$ has bounded variation [2]. Almost every sublevel set of a BV function has finite perimeter. (A finite perimeter set is a set that can be approximated by sets whose boundary has uniformly bounded Hausdorff measure.) If $E = f_m^{-1}((-\infty, t])$ is such a sublevel set, then an analogue of De Giorgi's theorem in H_n implies that the reduced boundary $\partial^* E$ has a approximate tangent plane at almost every $x \in \partial^* E$ [9]. Furthermore, this tangent plane is a *vertical plane* (a union of vertical lines); vertical planes are the only scale-invariant codimension-1 subgroups of H_n. Thus, when $r > 0$ is small, the intersection $B(x,r) \cap E$ is close to the intersection of $B(x,r)$ with a half-space bounded by a vertical plane.

In fact, for almost every $m \in M$ and $x \in H_n$, there is an $r > 0$ such that $f_m|_{B(x,r)}$ is close to a function $\bar{f}_m$ whose sublevel sets are half-spaces bounded by vertical planes. We call $\bar{f}_m$ a *vertical function*. By Fubini's theorem, there is some $x \in H_n$ and some $r > 0$ such that $f_m|_{B(x,r)}$ is close to a vertical function $\bar{f}_m$ for all but a small fraction of $m \in M$; then $f|_{B(x,r)}$ is close to a map $\bar{f}$ such that every coordinate function of $\bar{f}$ is vertical. Then $\bar{f}$ sends vertical lines in $B(x,r)$ to a point, so f is not bi-Lipschitz on $B(x,r)$.

This argument links the metric properties of f to the shape of the sublevel sets $E_{m,y} = f_m^{-1}((-\infty, y])$. For example, if the $E_{m,y}$'s are all smooth at some scale ρ (i.e., if $B(x,\rho) \cap E_{m,y}$ is close to a vertical half-space for all m, t, and $x \in \partial E_{m,y}$), then f collapses vertical line segments at scale ρ. Conversely, if f is c-bi-Lipschitz at scales between R and R' (i.e., there are c, R, and R' such that

$$cd(p,q) \leq \left\| f(p) - f(q) \right\|_1 \leq d(p,q)$$

for all $p, q \in H_n$ such that $d(p,q) \in [R, R']$), then the $E_{m,y}$'s must be bumpy at scales between R and R'; in the language of Sections 1 and 2, the $E_{m,y}$'s must be roughly c-bumpy at roughly $\log \frac{R'}{R}$ different scales.

The techniques of Sections 1 and 2, however, bound how bumpy a surface can be. Given a set $U \subset \mathbb{R}^n$ of finite perimeter, we can approximate U by a set U^t which is a union of cells of τ_t. As $t \to 0$, this approximation converges to U, and since U has finite perimeter, the $(n-1)$-volume of the boundary $|\partial U^t|$ stays bounded. The boundary ∂U^t can be viewed as an $(n-1)$-cycle, so by Theorem 2.4, ∂U^t is a sum of cycles supported on uniformly rectifiable sets. Each of these pieces is no bumpier than a Lipschitz function; they all satisfy (2.3).

The results of [16] extend this to the Heisenberg group H_n and prove that the boundary ∂U of a set $U \subset H_n$ of finite perimeter can be decomposed as a sum of cycles supported on sets $E \subset H_n$. Uniform rectifiability is not as well studied in H_n as in $\mathbb{R}^n$, and it is not known which definitions of uniform rectifiability generalize to H_n, but [16] shows that the E's admit intrinsic corona decompositions. These are collections of intrinsic Lipschitz graphs that approximate E at most points and scales. (Intrinsic Lipschitz graphs were introduced in [10] as an analogue of Lipschitz graphs.) When $n \geq 2$, intrinsic Lipschitz graphs in H_n are about as bumpy as Lipschitz graphs in $\mathbb{R}^n$. That is, like Lipschitz graphs, an intrinsic Lipschitz graph can be $\frac{1}{L}$-bumpy at no more than roughly L^2 scales. This corresponds to the following bound:

Theorem 3.2 ([16]). *For any $n \geq 2$ there is a $c > 0$ such that for sufficiently large $R > 1$, there are no 1-Lipschitz maps $f: H_n \to L_1$ that are $c(\log R)^{-\frac{1}{2}}$-bi-Lipschitz at scales between 1 and R. Furthermore, this is sharp; there is a $c' > 0$ such that for every sufficiently large R, there is a 1-Lipschitz map $f: H_n \to L_1$ that is $c'(\log R)^{-\frac{1}{2}}$-bi-Lipschitz at scales between 1 and R.*

Intrinsic Lipschitz graphs in H_1, however, can be much bumpier than Lipschitz graphs in $\mathbb{R}^n$. In [17], it is shown that intrinsic Lipschitz graphs in H_1 can be $\frac{1}{L}$-bumpy at up to roughly L^4 scales, but no more than that. This matches the bound for Lipschitz curves proved in [14] and corresponds to the following bound.

Theorem 3.3 ([17]). *There is a $c > 0$ such that for sufficiently large $R > 1$, there are no 1-Lipschitz maps $f: H_1 \to L_1$ that are $c(\log R)^{-\frac{1}{4}}$-bi-Lipschitz at scales between 1 and R. Furthermore, there is a $c' > 0$ such that for every sufficiently large R, there is a 1-Lipschitz map $f: H_1 \to L_1$ that is $c'(\log R)^{-\frac{1}{4}}$-bi-Lipschitz at scales between 1 and R.*

This leads to the following consequence:

Theorem 3.4 ([17]). *There is a metric space M that has a bi-Lipschitz embedding into L_1 and L_4, but not L_p for $1 < p < 4$.*

In short, the extent to which f preserves the metric on H_n depends on the bumpiness/complexity of the sublevel sets $E_{m,y}$, and the maximum possible complexity of a finite-perimeter subset $E \subset H_n$ depends on the ambient dimension n.

4. CONCLUSION

Some possible next questions include:

- How can other topological properties of a manifold be quantified? How do they depend on the complexity of the manifold?
- These decompositions bound the complexity of manifolds embedded in $\mathbb{R}^n$, suggesting that manifolds embedded in $\mathbb{R}^n$ are less complex than abstract manifolds. How does that affect their geometry?

- How does the maximum possible complexity of a manifold embedded in a space X depend on the geometry of X?

More generally, these tools suggest a link between complexity and geometry. If a manifold is simple in the sense that it embeds in $\mathbb{R}^n$, then it is simple in the sense that it can be decomposed into simple pieces. In a way, any object that can be drawn on a piece of paper embeds, more or less, in a low-dimensional Euclidean space; conversely, a key property of very complex objects like random graphs and arithmetic manifolds is that they are hard to embed, hard to decompose, and hard to visualize. Perhaps the objects whose shapes we can imagine are the objects simple enough to fit in our imagination. How can we better understand the shape of objects on the far end of the complexity spectrum?

ACKNOWLEDGMENTS

We would like to thank Larry Guth, Assaf Naor, Shmuel Weinberger, and Stefan Wenger for their conversations and collaborations on geometry, analysis, and complexity that influenced this paper, and to thank the Institute of Advanced Study for its hospitality during part of the preparation of this paper.

FUNDING

This work was partially supported by the National Science Foundation under Grant Nos. 2005609 and 1926686.

REFERENCES

[1] L. Capogna, D. Danielli, S. D. Pauls, and J. T. Tyson, *An introduction to the Heisenberg group and the sub-Riemannian isoperimetric problem*. Progr. Math. 259, Birkhäuser, Basel, 2007.

[2] J. Cheeger and B. Kleiner, Differentiating maps into L^1, and the geometry of BV functions. *Ann. of Math. (2)* **171** (2010), no. 2, 1347–1385.

[3] G. David and S. Semmes, Singular integrals and rectifiable sets in $\mathbf{R}^n$: beyond Lipschitz graphs. *Astérisque* **193** (1991), 1–145.

[4] G. David and S. Semmes, Uniform rectifiability and quasiminimizing sets of arbitrary codimension. *Mem. Amer. Math. Soc.* **144** (2000), no. 687.

[5] J. R. Dorronsoro, A characterization of potential spaces. *Proc. Amer. Math. Soc.* **95** (1985), no. 1, 21–31.

[6] D. B. A. Epstein, J. W. Cannon, D. F. Holt, S. V. F. Levy, M. S. Paterson, and W. P. Thurston, *Word processing in groups*. Jones and Bartlett Publishers, Boston, MA, 1992.

[7] P. Erdős, Some remarks on the theory of graphs. *Bull. Amer. Math. Soc.* **53** (1947), 292–294.

[8] H. Federer and W. H. Fleming, Normal and integral currents. *Ann. of Math. (2)* **72** (1960), 458–520.

[9] B. Franchi, R. Serapioni, and F. Serra Cassano, Rectifiability and perimeter in the Heisenberg group. *Math. Ann.* **321** (2001), no. 3, 479–531.

[10] B. Franchi, R. Serapioni, and F. Serra Cassano, Intrinsic Lipschitz graphs in Heisenberg groups. *J. Nonlinear Convex Anal.* **7** (2006), no. 3, 423–441.

[11] M. X. Goemans, Semidefinite programming in combinatorial optimization. *Math. Program.* **79** (1997), no. 1, 143–161.

[12] M. Gromov, Filling Riemannian manifolds. *J. Differential Geom.* **18** (1983), no. 1, 1–147.

[13] P. W. Jones, Rectifiable sets and the traveling salesman problem. *Invent. Math.* **102** (1990), no. 1, 1–15.

[14] S. Li and R. Schul, The traveling salesman problem in the Heisenberg group: upper bounding curvature. *Trans. Amer. Math. Soc.* **368** (2016), no. 7, 4585–4620.

[15] N. Linial, Finite metric-spaces—combinatorics, geometry and algorithms. In *Proceedings of the International Congress of Mathematicians, Vol. III (Beijing, 2002)*, pp. 573–586, Higher Ed. Press, Beijing, 2002.

[16] A. Naor and R. Young, Vertical perimeter versus horizontal perimeter. *Ann. of Math. (2)* **188** (2018), no. 1, 171–279.

[17] A. Naor and R. Young, Foliated corona decompositions. *Acta Math.* (to appear), (2021).

[18] P. Pansu, Métriques de Carnot–Carathéodory et quasiisométries des espaces symétriques de rang un. *Ann. of Math. (2)* **129** (1989), no. 1, 1–60.

[19] S. Wenger, A short proof of Gromov's filling inequality. *Proc. Amer. Math. Soc.* **136** (2008), no. 8, 2937–2941.

[20] R. Young, Filling inequalities for nilpotent groups through approximations. *Groups Geom. Dyn.* **7** (2013), no. 4, 977–1011.

[21] R. Young, Quantitative nonorientability of embedded cycles. *Duke Math. J.* **167** (2018), no. 1, 41–108.

ROBERT J. YOUNG

New York University, Courant Institute of Mathematical Sciences, 251 Mercer Street, New York, NY 10012, USA, ryoung@cims.nyu.edu

MEAN CURVATURE AND VARIATIONAL THEORY

XIN ZHOU

ABSTRACT

In this article, we survey recent progress on the variational theory related to mean curvature. We will discuss the Morse theory of minimal hypersurfaces with an emphasis on the Multiplicity One Conjecture, generic spatial distributions of minimal hypersurfaces, variational theory for constant mean curvature (CMC) surfaces, and variational theory for minimal surfaces with free boundary.

MATHEMATICS SUBJECT CLASSIFICATION 2020

Primary 53C42; Secondary 53A10, 49Q05, 49J35, 58E12, 58E20

KEYWORDS

Minimal hypersurfaces, constant mean curvature hypersurfaces, hypersurfaces with prescribed mean curvature, min–max theory, Multiplicity One Conjecture, equidistribution and scarring

INTERNATIONAL CONGRESS OF MATHEMATICIANS 2022 JULY 6–14

Proc. Int. Cong. Math. 2022, Vol. 4, pp. 2696–2717
DOI 10.4171/ICM2022/27

1. INTRODUCTION

In geometry, a large class of canonical objects are submanifolds which are stationary with respect to variations of length, area, or volume, possibly under various constraints. The condition of being stationary is tightly linked to a geometric quantity called the mean curvature function. The most notable examples are minimal surfaces, constant mean curvature surfaces, and more generally surfaces with prescribed mean curvature (PMC), where the mean curvature function respectively vanishes, is equal to a constant, or is prescribed by an ambient function. Such objects have been studied extensively by mathematicians for more than two centuries since the work of Lagrange on minimal surfaces in 1762, and various different methods have been developed, including but not limited to complex analysis, calculus of variations, partial differential equations, and geometric measure theory. In addition to their intrinsic beauty, profound applications of such canonical submanifolds have been found and led to solutions of many fundamental problems in other fields like topology, analysis, and physics. We refer to [15, 63, 67] for more discussions on historical backgrounds.

In the calculus of variations or variational theory, which we will focus on in this article, such surfaces are viewed as critical points of certain area- or volume-related functionals. In the past ten years, the variational-theoretic approach has enjoyed spectacular development, and deep new results have been proved on the existence of minimal, CMC, and PMC surfaces. In particular, the famous Yau's conjecture on the existence of infinitely many closed minimal surfaces was confirmed by combining the works of Marques–Neves [58] and Song [79], and general existence of closed CMC and PMC hypersurfaces was established by the author with Zhu [95, 96], and with Cheng [12]. Moreover, surprising new connections between these surfaces have been discovered, leading to the resolution of the Multiplicity One Conjecture for minimal hypersurfaces by the author [94]. In this article, we will provide a survey of these results, as well as some discussion of open problems along this direction.

1.1. Minimal surfaces

We start with a discussion of variational constructions of minimal surfaces in 3-dimensional spaces and, more generally, minimal hypersurfaces when the ambient space has dimension higher than 3. Minimal surfaces are mathematical models of soap films, where the surface tension tends to minimize the area. By the first variation formula of area, the mean curvature of such surfaces has to vanish. In general, minimal surfaces are defined as surfaces with vanishing mean curvature or, equivalently, stationary points of the area functional. The problem of finding area-minimizing surfaces with a given boundary in the 3-dimensional Euclidean space was raised by Lagrange, and later named after Joseph Plateau who systematically experimented with soap films in the 19th century. The Plateau's problem was solved independently by Douglas and Radó in 1930 using mapping methods. Since then, there have been various attempts to generalize this existence result to the case of higher-dimensional submanifolds and in Euclidean or Riemannian spaces of higher co-dimensions. In particular, this led to the development of geometric measure theory (GMT) by many outstanding mathematicians. By combining the works of Federer, Fleming, De Giorgi, Almgren, and Simons

[5, 21, 25, 26, 77], it is known that an area-minimizing current of codimension one is smoothly embedded outside a singular set of codimension 7. (We also refer to the work of De Lellis, particularly the survey [22], for the regularity of higher-codimensional area-minimizing currents.)

Besides the Plateau's problem, it is also natural to consider the existence of closed minimal surfaces in closed Riemannian manifolds. When the ambient space has rich topology, area-minimizing surfaces can be produced using either the mapping approach or geometric measure theory. For instance, when the ambient space M^n contains an incompressible surface $f : S_g \to M$ where S_g is a genus-g surface, Schoen–Yau [75] and Sacks–Uhlenbeck [72] proved the existence of an area-minimizing surface in its conjugacy class, by minimizing first the Dirichlet energy $E(f) = \int_{S_g} |\nabla f|^2$ and then within the Teichmüller space of conformal structures. In [62], Meeks–Simon–Yau proved the existence of embedded minimal surfaces by minimizing area within a nontrivial isotopy class in 3-manifolds. More generally, if there exists a nontrivial element $c \neq 0$ in the homology group $H_{n-1}(M^n, \mathbb{Z})$, by GMT there always exists an area-minimizing integral current $\Sigma \in c$, whose support is smoothly embedded outside a codimension 7 singular set.

The problem of finding closed minimal surfaces in general is more interesting and significantly harder. Inspired by earlier works on finding closed geodesics (one-dimensional minimal submanifolds) on 2-dimensional spheres [9, 53], Almgren [2, 3] initiated a program aiming at finding closed minimal submanifolds in closed Riemannian manifolds of any dimension and codimension. He designed a very general min–max theory applicable to families of integral cycles and showed the existence of a nontrivial stationary integral k-dimensional varifold in any closed M^n for $1 \leq k \leq n$. Later on, in a seminal work [68], Pitts further improved Almgren's theory and proved that the support of a min–max varifold is smoothly embedded in the codimension-one case ($k = n - 1$) when $3 \leq n \leq 6$, by using the famous curvature estimates for stable minimal hypersurfaces by Schoen–Simon–Yau [74]. Schoen–Simon [73] then extended the curvature estimates and hence obtained the regularity for codimension-one min–max varifolds in higher dimensions $n \geq 7$, allowing singular sets of codimension 7. Combining all the results above, the first general existence theorem is:

Theorem 1.1. *Every closed Riemannian manifold (M^n, g) of dimension $n \geq 3$ contains a nontrivial integral $(n-1)$-dimensional stationary varifold V whose support is a smoothly embedded minimal hypersurface outside a singular set of codimension 7. If $3 \leq n \leq 7$, the support of V is a smooth, closed, embedded, minimal hypersurface Σ.*

We also note that when M^n has nontrivial higher homotopy groups, Sacks–Uhlenbeck [72] produced branched, immersed, minimal 2-spheres by developing another min–max theory using perturbation arguments and classical Morse theory on Banach manifolds. Recently, in their proof of the finite-time extinction of the Ricci flow, Colding–Minicozzi [17] found a new proof of Sacks–Uhlenbeck's result by their harmonic replacements method. See also for the works of the author [92, 93] and Rivière [69] for min–max constructions for higher genus minimal surfaces.

Motivated by the these results and the existence theory of closed geodesics, S. T. Yau formulated a famous conjecture in [90] asserting that every closed 3-manifold admits infinitely many distinct smooth, closed, immersed minimal surfaces. One peak of the recent developments on minimal hypersurface theory is the resolution of this conjecture by Marques–Neves [58] and Song [79]. Around the same time, a Morse theory for the area functional has been established [55, 57, 59, 94], and several striking results concerning the spatial distribution of these minimal hypersurfaces were proved [40, 60, 81]. All of these results were obtained by applying the Almgren–Pitts min–max theory to families of cycles of multiple parameters (deeply influenced by the solution of the Willmore Conjecture by Marques–Neves [56]). We will postpone detailed discussions to Sections 2 and 3.

A central difficulty in obtaining the aforementioned results is the a priori existence of integer multiplicity of the min–max varifolds. That is, the min–max varifolds may be represented by integer multiples of embedded minimal hypersurfaces. Therefore, applications of the min–max theory to higher-parameter families of integral cycles may just result in multiple covers of the minimal hypersurfaces associated with lower-parameter families. Motivated by classical Morse theory, Marques–Neves [59] formulated the Multiplicity One Conjecture:

Conjecture 1.2. *For smooth generic metrics on M^n when $3 \leq n \leq 7$, min–max varifolds are represented by multiplicity-one embedded minimal hypersurfaces.*

This conjecture was confirmed by the author in [94] using new ideas which were inspired by the investigation of the existence theory of CMC/PMC hypersurfaces [95, 96], to be discussed below. We will provide a sketch of proof of this conjecture in Section 2. Finally, we also note that the counterpart of the Multiplicity One Conjecture in the phase transition setting was proved by Chodosh–Mantoulidis [14] in 3 dimensions; see also [30, 35].

1.2. CMC and PMC surfaces

Surfaces with constant mean curvature (CMC) are mathematical models of soap bubbles. In the ideal situation, surface tension tends to minimize the surface area while the volume of enclosed air is fixed. Such surfaces must then be stationary points of the area subject to a volume constraint, and hence must have constant mean curvature by the first variation formula. CMC surfaces form a classical topic in differential geometry, and play an essential role in many areas, such as isoperimetric problems, interface theory for polymers, and general relativity. The classification of CMC surfaces in $\mathbb{R}^3$ and other homogeneous 3-manifolds has been a classical problem since the seminal work of Aleksandrov [1], and we refer to the survey paper [64] for this direction. In this article, we will focus on the existence theory of CMC surfaces in general manifolds.

We start with a brief and nonexhaustive review of several previous existence results that are closely related to our main results. The existence of CMC surfaces in $\mathbb{R}^3$ with prescribed mean curvature and Plateau boundary conditions was initiated by Heinz [38] and Hildebrandt [39]. The Rellich conjecture, which asserts the existence of at least two solutions to the CMC Plateau problem, was solved later by Brezis–Coron [11] and Struwe [83]. For the existence of closed CMC hypersurfaces, it is well known that the boundaries of isoperimetric regions are smoothly embedded CMC hypersurfaces (up to a singular set of codimension 7); see [4,66]. By perturbation arguments, one can generate foliations by closed CMC hypersurfaces from a given nondegenerate closed minimal hypersurface, or near minimal submanifolds of strictly lower dimensions (see, for instance, the works of Ye, Mahmoudi–Mazzeo–Pacard [54,91]). The gluing constructions pioneered by Kapouleas produced many important examples of complete or compact CMC surfaces in Euclidean spaces [10,43]. In addition, there is the degree theory approach by Rosenberg–Smith [70]. However, these works left open the fundamental problem of finding closed hypersurfaces with arbitrary prescribed constant mean curvature in general manifolds.

In [95], Zhu and the author settled this problem by establishing the following general existence theory.

Theorem 1.3 ([95]). *Let M^n be a closed Riemannian manifold of dimension $3 \leq n \leq 7$. Given any $c \in \mathbb{R}$, there exists a nontrivial, smooth, closed, almost embedded hypersurface Σ of constant mean curvature c.*

Remark 1.4. A smooth almost embedded hypersurface is a smooth immersion where near any self-intersection point, the hypersurface decomposes into sheets which may touch but not cross. Such hypersurfaces are Alexandrov embedded.

We proved this result by establishing a min–max theory for the area functional with a volume term added, extending the Almgren–Pitts theory to the more general CMC setting. A sketch of proof will be provided in Section 4.

We note that no control on topology of the CMC hypersurfaces in Theorem 1.3 was known due to the use of integral currents as the total variation space. In contrast, we note that using a variant of the Almgren–Pitts theory, Simon–Smith [78] proved the existence of an embedded minimal 2-sphere in any Riemannian 3-sphere. Their work has been generalized to an arbitrary closed 3-manifold M by Colding–De Lellis [16] using sweepouts associated with Heegaard splittings, and the genus of the min–max surface is known to be bounded by the Heegaard genus of M [23,44]. On the other hand, the min–max theory based on the harmonic mapping approach [17,69,72,92,93] naturally produces branched immersed minimal surfaces with controlled genus. With these contrasts in mind, it is tempting to search for closed CMC surfaces with both prescribed mean curvature and controlled genus (bounded by the Heegaard genus) in 3-manifolds. In particular, we note a conjecture by Rosenberg–Smith [70, PAGE 3]: "*for any $H \geq 0$ and any metric g on S^3 of positive sectional curvature, there exists an embedding of S^2 to S^3 of constant mean curvature H*". However, by the works of Torralbo [85] and Meeks–Mira–Pérez–Ros [65], it is known that in certain positively curved homogenous 3-spheres, there are mean curvature values for which the associated immersed CMC 2-spheres must have self-intersections. Motivated by the mapping approach for minimal surfaces, it is natural to modify *embedding* to *branched immersion* in the Rosenberg–Smith conjecture.

In [12], Cheng and the author solved this modified conjecture with a newly devised min–max theory using the mapping approach.

Theorem 1.5 ([12]). *Given a Riemannian 3-sphere (S^3, g) with nonnegative Ricci curvature, for every constant H, there exists a nontrivial branched immersed 2-sphere with constant mean curvature H.*

Remark 1.6. In [12], we also proved that a branched immersed H-CMC 2-sphere exists in (S^3, g) whenever $\mathrm{Ric}_g > -\frac{H^2}{2} g$, or for almost all H (with respect to the Lebesgue measure) without curvature assumptions on g.

A hypersurface Σ^{n-1} in M^n has prescribed mean curvature (PMC) by some function $h : M \to \mathbb{R}$ if its mean curvature is everywhere identical to the value of h. PMC hypersurfaces are natural generalizations of CMC hypersurfaces and are models for capillary surfaces; see [27, §1.6]. The local existence theory for PMC hypersurfaces is quite well understood in the case of Plateau boundary conditions and in the graphical case; see [96] for references. On the other hand, the global theory or the existence for closed PMC hypersurfaces had been largely open except for constant prescription functions. The global existence problem for closed PMC surfaces in closed three manifolds is a conjecture of Yau in the 1980s (by personal communication, see also [90, PROBLEM 59] for a version of his conjecture in $\mathbb{R}^3$).

In [96], Zhu and the author extended our CMC min–max theory developed in [95] to nonconstant prescription functions. In particular, we solved the existence problem for closed PMC hypersurfaces for a generic class of smooth prescription functions.

Theorem 1.7 ([96]). *Let M^n be a closed Riemannian manifold of dimension $3 \leq n \leq 7$. There is an open dense set (in the smooth topology) $\mathcal{S} \subset C^\infty(M)$ of prescription functions h for which there exists a nontrivial, smooth, closed, almost embedded hypersurface Σ of prescribed mean curvature h. That is, $H_\Sigma = h|_\Sigma$.*

In both Theorems 1.3 and 1.7, the min–max theory was devised only for one-parameter families. These results were later generalized to multiparameter min–max constructions together with Morse index upper bounds by the author in [94]. The PMC min–max Theorem 1.7 and its generalizations in [94] had played an essential role in the proof of the Multiplicity One Conjecture by the author in [94].

Finally, we also note the phase transition approach to the PMC existence problem by Bellettini–Wickramasekera [8] for nonnegative Lipschitz prescribing functions.

2. VARIATIONAL THEORY FOR AREA AND THE MULTIPLICITY ONE CONJECTURE

In this part, we introduce the recently developed Morse theory for the area functional and a sketch of proof of the Multiplicity One Conjecture. For simplicity, in what follows, we will denote by n the dimension of a hypersurface Σ^n, and by $(n+1)$ the dimension of the ambient manifold M^{n+1}.

The principle behind Morse theory is to relate the topology of a given total space to all the critical points of a functional defined therein in a generic scenario. We choose the total space for the area functional to be the space of mod-2 n-cycles, denoted by $\mathcal{Z}_n(M,\mathbb{Z}_2)$, which can roughly be regarded as the boundaries of open sets with finite n-dimensional Hausdorff measure. In [2], Almgren calculated all the homotopy groups of $\mathcal{Z}_n(M,\mathbb{Z}_2)$, and proved the following:

Theorem 2.1. *$\mathcal{Z}_n(M,\mathbb{Z}_2)$ is weakly homotopic to $\mathbb{RP}^\infty$.*

Here $\mathbb{RP}^\infty$ denotes the infinite-dimensional real projective space. This fact implies that the $\mathbb{Z}_2$-cohomological ring of $\mathcal{Z}_n(M,\mathbb{Z}_2)$ is a polynomial ring whose generator we denote by $\bar{\lambda}$. That is, $\mathcal{H}^*(\mathcal{Z}_n(M,\mathbb{Z}_2),\mathbb{Z}_2)=\mathbb{Z}_2[\bar{\lambda}]$. Motivated by the topological structures, Gromov [31,32], Guth [36], and Marques–Neves [58] introduced the notion of the volume spectrum for the area functional in $\mathcal{Z}_n(M,\mathbb{Z}_2)$ as a nonlinear version of the Laplacian spectrum. Below, we let X be any finite-dimensional parameter space, for instance, a cubical complex.

Definition 2.2 (Volume spectrum). Given $k\in\mathbb{N}$, a continuous map $\Phi: X\to\mathcal{Z}_n(M,\mathbb{Z}_2)$ is called a *k-sweepout* if $\Phi^*(\bar{\lambda}^k)\neq 0$ in $H^k(X,\mathbb{Z}_2)$. The *kth volume spectrum*, or the *k-width*, is just the min–max value

$$\omega_k(M)=\inf_{\Phi:k\text{-sweepout}}\ \sup_{x\in\mathrm{dmn}(\Phi)}\ \mathrm{Area}\big(\Phi(x)\big),$$

where $\mathrm{dmn}(\Phi)$ stands for the domain of Φ.

It was proved that the sequence $\{\omega_k(M)\}$ grows sublinearly at the rate of $k^{\frac{1}{n+1}}$ as $k\to\infty$ [31,32,36,58]. Moreover, the sequence satisfies a Weyl Law.

Theorem 2.3 (Liokumovich–Marques–Neves [52]). *There exists a universal constant $a(n)>0$ such that for any compact Riemannian manifold M^{n+1},*

$$\lim_{k\to\infty}\omega_k(M)k^{-\frac{1}{n+1}}=a(n)\,\mathrm{Vol}(M)^{\frac{n}{n+1}}.$$

Note that the Almgren–Pitts min–max theory works for families of cycles within a homotopy class, while the definition of the volume spectrum concerns all families via the cohomological condition. To link them together, Marques–Neves systematically studied the Morse index for minimal hypersurfaces produced by the Almgren–Pitts theory [57]. In particular, they proved the following version of the min–max theorem.

Theorem 2.4. *Let M^{n+1} be a closed Riemannian manifold with $3\le n+1\le 7$. For each $k\in\mathbb{N}$, there exists a disjoint collection of connected, closed, smoothly embedded minimal hypersurfaces $\{\Sigma_i^k: i=1,\dots,l_k\}$ with integer multiplicities $\{m_i^k: i=1,\dots,l_k\}\subset\mathbb{N}$ such that*

$$\omega_k(M)=\sum_{i=1}^{l_k}m_i^k\cdot\mathrm{Area}\big(\Sigma_i^k\big)\quad\textit{and}\quad\sum_{i=1}^{l_k}\mathrm{Ind}\big(\Sigma_i^k\big)\le k.$$

Here Ind(Σ) *stands for the Morse index of* Σ, *which is the number of negative eigenvalues of the second variation of area.*

The possible existence of multiplicities greater than 1 formed a major obstacle in applications of the Almgren–Pitts theory since the 1980s. In addition to the possible repeated occurrence of minimal hypersurfaces when applying Theorem 2.4 to $\{\omega_k\}_{k\in\mathbb{N}}$, min–max varifolds with higher multiplicities cannot fit into the program of Marques–Neves [59] to obtain Morse index lower bounds; (see also [55]). The following famous conjecture was formulated by Marques–Neves [59].

Conjecture 2.5 (Multiplicity One Conjecture). *For a bumpy metric on* M^{n+1}, $3 \leq n+1 \leq 7$, *there exists a collection* $\{\Sigma_i^k\}$ *as in Theorem* 2.4 *such that every component* Σ_i^k *is two-sided and of multiplicity one.*

Remark 2.6. A hypersurface is *two-sided* if its normal bundle is trivial. A Riemannian metric is *bumpy* if every closed immersed minimal hypersurface is a nondegenerate critical point of the area functional. White proved that the set of bumpy metrics is generic in the sense of Baire [88, 89].

This conjecture was confirmed by the author in [94].

Theorem 2.7. *Conjecture* 2.5 *is true.*

Theorem 2.7, together with the program on Morse index lower bounds developed by Marques–Neves [59], implies that for bumpy metrics, there exists a closed minimal hypersurface of Morse index k and area $\omega_k(M)$ for each $k \in \mathbb{N}$. The above works together established a satisfactory global Morse theory for the area functional. Recently, Marques–Montezuma–Neves proved Morse inequalities for the area functional [55], and hence established a local Morse theory as well.

By the convergence theorems for minimal hypersurfaces of Sharp [76], the same conclusions in Theorem 2.7 hold true for metrics with a positive Ricci curvature, as well as the following results concerning the multiplicity and Morse index of min–max minimal hypersurfaces for general metrics.

Theorem 2.8 ([94]). *In Theorem* 2.4, *every component* Σ_j^k *which is not weakly stable is two-sided with* $m_j^k = 1$; *and* $\sum_{\Sigma_j^k:\ \textit{two-sided}} \mathrm{Ind}(\Sigma_j^k) \leq k$.

Remark 2.9. A closed minimal hypersurface Σ is weakly stable if the second variation of area at Σ is nonnegative definite with a nontrivial kernel. The results in Theorem 2.8 have been partially generalized to dimensions $n+1 > 7$ by Li [58].

Sketch of proof of Theorem 2.7. The key idea of our proof in [94] is to approximate the area functional by the weighted $\mathcal{A}^h$-functional used in the PMC min–max theory [96]. Here $\mathcal{A}^h$ is defined for Caccioppoli sets Ω by $\mathcal{A}^h(\Omega) = \mathrm{Area}(\partial\Omega) - \int_\Omega h\, dM$, where $h \in C^\infty(M)$. A smooth critical point of $\mathcal{A}^h$ is a Caccioppoli set Ω whose boundary is a smooth hypersurface $\Sigma = \partial\Omega$ and has mean curvature (with respect to the outward unit normal) given by h

restricted to Σ. There are two crucial parts in the proof. First, we show that, given a bumpy metric, the volume spectrum $\omega_k(M)$ can be realized by the area of some minimal hypersurfaces coming from relative min–max constructions using sweepouts of boundaries. Next, we observe that, still assuming bumpiness, if one approximates Area by a sequence $\{\mathcal{A}^{\varepsilon_k h}\}_{k\in\mathbb{N}}$ where $\varepsilon_k \to 0$, and if $h: M \to \mathbb{R}$ is carefully chosen, then the limit min–max minimal hypersurfaces (of min–max PMC hypersurfaces associated with $\mathcal{A}^{\varepsilon_k h}$) are all two-sided and have multiplicity one.

Part 1. Given a bumpy metric, for each $k \in \mathbb{N}$, by [57] there exists a free homotopy class Π of maps $\Phi : X \to \mathcal{Z}_n(M, \mathbb{Z}_2)$, where X is a fixed k-dimensional parameter space such that the min–max value $\mathbf{L} = \inf_{\Phi\in\Pi} \max_{x\in X} \mathrm{Area}(\Phi(x))$ equals $\omega_k(M)$. Choose $\Phi_0 \in \Pi$ so that $\max_{x\in X} \mathrm{Area}(\Phi_0(x))$ is very close to $\mathbf{L}$. Since the space of Caccioppoli sets $\mathcal{C}(M)$ forms a double cover of $\mathcal{Z}_n(M, \mathbb{Z}_2)$ via the boundary map $\partial : \Omega \to [\partial\Omega]$ (see [59]), we can lift Φ_0 to $\widetilde{\Phi}_0 : \tilde{X} \to \mathcal{C}(M)$, where $\pi : \tilde{X} \to X$ is also a double cover. Next let Y be the subset of $x \in X$ where $\Phi_0(x)$ is ε-close to the set $\mathcal{S}$ of closed embedded minimal hypersurfaces Σ with Area $\leq \mathbf{L} + 1$ and Ind $\leq k$, and let $Z = \overline{X \setminus Y}$. As $\mathcal{S}$ is a finite set by [76], Y is topologically trivial, and hence $\tilde{Y} = \pi^{-1}(Y)$ is a disjoint union of two homeomorphic copies of Y, that is, $\tilde{Y} = Y^+ \bigsqcup Y^-$ with $Y \simeq Y^+ \simeq Y^-$. On the other hand, since no element in $\Phi_0(Z)$ is close to being regular, we can deform $\Phi_0|_Z$ based on Pitts's combinatorial argument [68, 4.10], so that

$$\max_{x\in Z} \mathrm{Area}\big(\Phi_0(x)\big) < \mathbf{L}. \tag{2.1}$$

Now consider the $(\tilde{X}, \tilde{Z})$-relative homotopy class of maps generated by $\widetilde{\Phi}_0$: $\widetilde{\Pi} = \{\Psi : \tilde{X} \to \mathcal{C}(M) : \Psi|_{\tilde{Z}} = \widetilde{\Phi}_0|_{\tilde{Z}}\}$.

Lemma 2.10 ([94, LEMMA 5.8]). *The min–max value* $\tilde{\mathbf{L}}$ *of* $\widetilde{\Pi}$ *satisfies*

$$\tilde{\mathbf{L}} := \inf_{\Psi\in\widetilde{\Pi}} \max_{x\in\tilde{X}} \mathrm{Area}\big(\partial\Psi(x)\big) \geq \mathbf{L} = \omega_k(M).$$

Hence by (2.1), *we have the nontriviality condition* $\tilde{\mathbf{L}} > \max_{x\in Z} \mathrm{Area}(\Phi_0(x))$.

Proof. If the conclusion were false, then since

$$\max_{x\in\tilde{Z}} \mathrm{Area}(\partial\widetilde{\Phi}_0(x)) = \max_{x\in Z} \mathrm{Area}(\Phi_0(x)) < \mathbf{L},$$

one could deform $\widetilde{\Phi}_0$ on $\tilde{Y}$ so that the maximum area is less than $\mathbf{L}$. However, as Y^+ and Y^- are disjoint, the deformations on Y^+ (or on Y^-) can be passed to the quotient to give deformations of $\Phi_0|_Y$ in $\mathcal{Z}_n(M, \mathbb{Z}_2)$. As all the maps are fixed on Z, we then obtain deformations of Φ_0 after which the maximum area is less than $\mathbf{L}$, which is a contradiction. ∎

Part 2: The main conclusion follows from the result below.

Theorem 2.11 ([94, THEOREM 4.1]). *In the above notation, if g is bumpy,* $\tilde{\mathbf{L}}$ *can be realized as the area of a multiplicity one, closed, embedded, two-sided, minimal hypersurface.*

To derive Theorem 2.7, first note that by the choice of Φ_0, we know that $\tilde{\mathbf{L}}$ is very close to $\mathbf{L}$. By the bumpiness of g, the values of $\tilde{\mathbf{L}}$ should stabilize to $\mathbf{L}$ when they are close enough.

Proof of Theorem 2.11. To simplify notions, we will drop all the tildes in this part. Given a smooth function $h : M \to \mathbb{R}$ and $\varepsilon > 0$, we can approximate $\mathbf{L}$ by the min–max values for the $\mathcal{A}^{\varepsilon h}$-functional, $\mathbf{L}^{\varepsilon h} = \inf_{\Psi \in \Pi} \max_{x \in X} \mathcal{A}^{\varepsilon h}(\partial \Psi(x))$, that is, $\mathbf{L}^{\varepsilon h} \to \mathbf{L}$ as $\varepsilon \to 0$. Note that we require $\Psi|_Z = \Phi_0|_Z$ for all $\Psi \in \Pi$. By the fact $\mathbf{L} > \max_{x \in Z} \text{Area}(\partial \Phi_0(x))$, and that the term $\varepsilon \int_\Omega h dM$ in $\mathcal{A}^{\varepsilon h}(\Omega)$ is uniformly small, we have, for ε small enough,

$$\mathbf{L}^{\varepsilon h} > \max_{x \in Z} \mathcal{A}^{\varepsilon h}\big(\Psi(x)\big). \tag{2.2}$$

For a generic choice of h, applying the multiparameter PMC min–max theory [94, THEOREM 1.7] (based on the one-parameter version in [96]), we obtain some $\Omega_\varepsilon \in \mathcal{C}(M)$ such that: (1) $\Sigma_\varepsilon = \partial \Omega_\varepsilon$ is an almost embedded hypersurface; (2) the mean curvature (with respect to outward unit normal) $H_{\Sigma_\varepsilon} = \varepsilon h|_{\Sigma_\varepsilon}$; (3) $\mathcal{A}^{\varepsilon h}(\Omega_\varepsilon) = \mathbf{L}^{\varepsilon h}$; and (4) the Morse index (with respect to $\mathcal{A}^{\varepsilon h}$) $\text{Ind}(\Sigma_\varepsilon) \leq k$.

Letting $\varepsilon \to 0$, by (2)–(4) and [94, THEOREM 2.6], up to taking a subsequence, Σ_ε converge locally smoothly away from a finite set $\mathcal{W}$ to a closed embedded minimal hypersurface Σ_0 (assumed to be connected without loss of generality) with an integer multiplicity $m \in \mathbb{N}$. Therefore, $\mathbf{L} = m\,\text{Area}(\Sigma_0)$, and it suffices to prove that Σ_0 is two-sided (which we skip here) and $m = 1$.

The convergence implies that Σ_ε locally decomposes as an m-sheeted graph over $\Sigma_0 \setminus \mathcal{W}$, with graphing functions: $u_\varepsilon^1 \leq u_\varepsilon^2 \leq \cdots \leq u_\varepsilon^m$. And by (1), the outward unit normal of Ω_ε will alternate orientations along these sheets. The proof proceeds depending on whether m is odd or even.

Claim 1. *If $m \geq 3$ is odd, then Σ is degenerate, hence a contradiction.*

Proof. Since m is odd, the top and bottom sheets have the same orientation, so by subtracting the PMC equations of the two sheets, we have $L(u_\varepsilon^m - u_\varepsilon^1) + o(u_\varepsilon^m - u_\varepsilon^1) = \varepsilon(h(x, u_\varepsilon^m) - h(x, u_\varepsilon^1)) = o(u_\varepsilon^m - u_\varepsilon^1)$, where L is the Jacobi operator associated with the second variation of Σ. After renormalizations, the height differences $u_\varepsilon^m - u_\varepsilon^1$ will converge subsequentially to a positive Jacobi field of $\Sigma \setminus \mathcal{W}$, which extends to Σ by a standard trick. ■

Claim 2. *If m is even, there exists a solution of $L\varphi = 2h|_{\Sigma_0}$ which does not change sign.*

Proof. Now the top and bottom sheets have opposite orientations. Thus $L(u_\varepsilon^m - u_\varepsilon^1) + o(u_\varepsilon^m - u_\varepsilon^1) = \pm\varepsilon(h(x, u_\varepsilon^1) + h(x, u_\varepsilon^m))$. Using the renormalization procedure again and noting that $u_\varepsilon^m - u_\varepsilon^1 > 0$, we get either a positive Jacobi field (which cannot happen) or a positive function φ satisfying $L\varphi = 2h|_{\Sigma_0}$ or $L\varphi = -2h|_{\Sigma_0}$. ■

The following key lemma says that Claim 2 cannot hold for a suitably chosen h. Hence the proof of Theorem 2.11 is complete.

Lemma 2.12. *For a suitably chosen h, the solutions of* $L\varphi = 2h|_\Sigma$ *on a closed embedded minimal hypersurface* Σ *with* Area $\leq C$ *and* Ind $\leq k$ *must change sign.*

Proof. By [76], the set of minimal hypersurfaces with Area $\leq C$ and Ind $\leq k$ is finite, which we denote by $\{\Sigma_1, \Sigma_2, \dots, \Sigma_N\}$. Take pairwise disjoint neighborhoods $U_j^{\pm} \subset \Sigma_j$ and a smooth function f defined on $\bigcup U_j^{\pm}$ with compact support such that (1) $f|_{U_j^+}$ is nonnegative and is positive at some point; (2) $f|_{U_j^-}$ is nonpositive and is negative at some point. Next extend Lf to some $h_0 \in C^\infty(M)$ and take a generic h as close to h_0 as we want. Then any solution φ of $L\varphi = 2h|_{\Sigma_j}$ would be close to $2f$ for each Σ_j, and hence must change sign. ■

3. GENERIC DENSENESS, EQUIDISTRIBUTION, AND SCARRING

En route to the proof of Yau's conjecture and the establishment of a Morse theory for the area functional, we have also observed several striking results on the spatial distribution of closed minimal hypersurfaces for generic smooth metrics, which we introduce in this part. There is an intimate analog between closed minimal hypersurfaces and L^2-density of Laplace eigenfunctions regarding their spatial distributions. For instance, both exhibit equidistribution and scarring phenomena. We refer to [81] for a survey of this analogy.

Using the Weyl Law for the volume spectrum (Theorem 2.3), Irie–Marques–Neves [40] obtained a very surprising generic density result for closed minimal hypersurfaces, and hence settled Yau's conjecture in generic case. (See [49] by Li for the generalization to higher dimensions.)

Theorem 3.1 (Irie–Marques–Neves [40]). *Let* M^{n+1} *be a closed manifold with* $3 \leq n+1 \leq 7$. *Then for a* C^∞*-generic Riemannian metric, the union of all closed, smoothly embedded minimal hypersurfaces is dense in* M.

This result was later quantified by Marques–Neves–Song [60] to prove the following generic equidistribution result for closed minimal hypersurfaces.

Theorem 3.2 (Marques–Neves–Song [60]). *Let* M^{n+1} *be a closed manifold with* $3 \leq n+1 \leq 7$. *Then for a* C^∞*-generic Riemannian metric, there exists a sequence of closed, smoothly embedded minimal hypersurfaces* $\{\Sigma_j\}_{j\in\mathbb{N}}$ *that is equidistributed in* M. *That is,* $\forall f \in C^\infty(M)$,

$$\lim_{q\to\infty} \frac{1}{\sum_{j=1}^q \mathrm{Area}(\Sigma_j)} \sum_{j=1}^q \int_{\Sigma_j} f\, d\Sigma_j = \frac{1}{\mathrm{Vol}(M,g)} \int_M f\, dM.$$

The key idea behind these results is that, after bumping up the metric in a neighborhood U of a point $p \in M$ (for instance, by conformal changes), the min–max theory necessarily yields a closed minimal hypersurface passing through U according to the Weyl Law.

In his proof of Yau's conjecture for general nongeneric metrics, Song [79] introduced a localized version of the volume spectrum $\{\widetilde{\omega}_k\}_{k\in\mathbb{N}}$, called the *cylindrical volume spectrum* and defined as the volume spectrum of certain noncompact manifolds with Lipschitz metrics obtained by gluing infinite cylinders to compact manifolds with stable minimal boundaries. In contrast to the sublinear growth of the standard volume spectrum, $\{\widetilde{\omega}_k\}_{k\in\mathbb{N}}$ grows linearly by [79]. By extending the ideas in [60] to the cylindrical volume spectrum, Song and the author [81] obtained a generic scarring result. Namely, we showed that generically there exist closed embedded minimal hypersurfaces with large area and Morse index, which accumulate surrounding any stable minimal hypersurface in a quantitative way. Such a phenomenon is called *scarring*.

Theorem 3.3 (Generic scarring, [81]). *Let M^{n+1} be a closed manifold with $3 \leq n+1 \leq 7$. For a C^∞-generic metric, we have: for any connected, closed, embedded, 2-sided, minimal hypersurface S in M which is stable, there is a sequence $\{\Sigma_k\}$ of closed, embedded, minimal hypersurfaces, such that*

$$(1)\ \Sigma_k \cap S = \emptyset; \quad (2)\ \lim_{k\to\infty} \|\Sigma_k\| = \infty; \quad (3)\ \lim_{k\to\infty} \operatorname{Ind}(\Sigma_k)\|\Sigma_k\|^{-1} = \|S\|^{-1};$$

$$(4)\ \mathbf{F}\left(\frac{[S]}{\|S\|}, \frac{[\Sigma_k]}{\|\Sigma_k\|}\right) \leq 1/\log\big(\|\Sigma_k\|\big).$$

Here $[\Sigma]$ is the varifold associated to Σ, $\|\Sigma\|$ is its area, and $\mathbf{F}$ is the varifold distance.

In dimension $n+1=3$, we also explored the 3-manifold topology to find stable minimal surfaces and showed that generic scarring happens for all closed 3-manifolds but the spherical quotients.

4. MIN–MAX THEORY FOR CMC SURFACES

In this part, we present a sketch of the proof of the CMC existence Theorem 1.3, focusing for simplicity on the one-parameter min–max construction. The proof of the PMC Theorem 1.7, which we omit here, shared several key ideas with the CMC case, with additional challenges including the correct choice of prescribing functions [96, PROPOSITION 0.2], and a more complicated gluing scheme.

Sketch of proof of Theorem 1.3. Fixing a closed manifold (M^{n+1}, g) and a number $c > 0$, a given Morse function $f : M \to [0,1]$ generates a continuous map $\Phi_0 : [0,1] \to \mathcal{C}(M)$ by $\Phi_0(x) = \{f(p) < x\}$ with $\Phi_0(0) = \emptyset$ and $\Phi_0(1) = [M]$. The min–max value of $\mathcal{A}^c$ associated with the relative homotopy class $\Pi = \{\Phi : [0,1] \to \mathcal{C}(M), \Phi|_{\{0,1\}} = \Phi_0|_{\{0,1\}}\}$ is

$$\mathbf{L}^c = \inf_{\Phi\in\Pi} \max_{x\in[0,1]} \mathcal{A}^c\big(\Phi(x)\big),$$

where

$$\mathcal{A}^c(\Omega) = \operatorname{Area}(\partial\Omega) - c\operatorname{Vol}_g(\Omega).$$

Using the isoperimetric inequalities for small volumes, we have

Theorem 4.1 ([95, THEOREM 3.9]). $\mathbf{L}^c > 0$.

Note that $\max\{\mathcal{A}^c(\emptyset), \mathcal{A}^c(M)\} = 0$. This directly implies that $\mathbf{L}^c$ can be realized by some nontrivial weak limit. In the multiparameter cases, one needs to assume that $\mathbf{L}^c$ is strictly greater than the values assumed on the relative boundary; see (2.2).

For an arbitrary critical sequence $\{\Phi_i\} \subset \Pi$, that is, if $\max_{x\in[0,1]} \mathcal{A}^c(\Phi_i(x)) \to \mathbf{L}^c$, we define the *critical set* as the collection of all varifold limits:

$$\mathbf{C}(\{\Phi_i\}) = \Big\{V : V = \lim_{i_j\to\infty} \big|\partial\Phi_{i_j}(x)\big| \text{ as varifolds, where } \mathcal{A}^c\big(\Phi_{i_j}(x_j)\big) = \mathbf{L}^c\Big\}.$$

By a tightening argument adapted from that of Almgren–Pitts, we can homotopically deform $\{\Phi_i\}$ to a new critical sequence **[95, §4]**, denoted still by $\{\Phi_i\}$ by abuse of notation, such that

Lemma 4.2. *Every element in* $\mathbf{C}(\{\Phi_i\})$ *has c-bounded first variation.*

Note that this is the first key novel idea in comparison with the minimal case **[68]**: the $\mathcal{A}^c$-functional is not defined for general varifolds, so we cannot show that every element in $\mathbf{C}(\{\Phi_i\})$ is a stationary point of $\mathcal{A}^c$ as in **[68]**. Nevertheless, having c-bounded first variation provides enough control on elements of $\mathbf{C}(\{\Phi_i\})$ to proceed.

We can then adapt the Almgren–Pitts combinatorial argument to show that at least one element $V \in \mathbf{C}(\{\Phi_i\})$ satisfies an "*almost-minimizing*" property. Heuristically, V is almost-minimizing in an open set $U \subset M$ if it can be approximated by the boundaries of a sequence $\{\Omega_i\} \subset \mathcal{C}(M)$ such that, if we deform Ω_i in U without increasing the $\mathcal{A}^c$-value by δ_i in the process, then we are not allowed to decrease the $\mathcal{A}^c$-value by ε_i at the end. Here $\delta_i, \varepsilon_i \to 0$ as $i \to \infty$. That is, writing the deformation as $\{\Omega_i^t\}_{t\in[0,1]}$,

$$\mathcal{A}^c\big(\Omega_i^t\big) \le \mathcal{A}^c(\Omega_i) + \delta_i, \forall t \in [0,1] \Rightarrow \mathcal{A}^c\big(\Omega_i^1\big) \ge \mathcal{A}^c(\Omega_i) - \varepsilon_i.$$

Using this property, we can construct replacements V^* of V inside any $K \subset\subset U$, satisfying:

Proposition 4.3 (**[95, Proposition 5.8]**). (1) *V^* is the same as V outside K;* (2) $-c\,\mathrm{Vol}(K) \le \|V^*\| - \|V\| \le c\,\mathrm{Vol}(K)$*;* (3) *$V^*$ is the limit of boundaries $\partial\Omega_i^*$ which locally minimize $\mathcal{A}^c$ in the interior of K.*

Note that (2) and (3) form two main differences of the CMC case compared to the minimal case **[68]**. In (2), the mass may change due to the volume term in $\mathcal{A}^c$, but luckily the errors for mass change converge to zero in any blowup procedure. Moreover, in (3) we gain more regularity. In fact, $\partial\Omega_i^*$ are stable CMC hypersurfaces, and hence form a compact family in the smooth topology by curvature estimates, and the limit can still be represented as a boundary due to the one-sided maximum principle satisfied by CMC **[95, Lemma 2.7]**. That is, two embedded CMC hypersurfaces which do not cross each other and have opposite orientations must either be disjoint or touch on at most a codimension-one subset. We point out that in the minimal case, the replacement V^* is smoothly embedded inside K, but may have integer multiplicities. This phenomenon forms the key mechanism for separating minimal sheets in the PMC approximation used in **[94]**.

To obtain the regularity of V, we showed, heuristically, that V coincides with V^*. One key step is to prove that two such replacements V^* and V^{**} glue together as a smooth almost embedded CMC hypersurface along a particularly chosen interface. This amounts to showing that the unit normal vectors match modulo standard regularity theory of elliptic PDEs. Since the CMC equation is not homogeneous, we need to make sure that the orientations of V^* and V^{**} match at a gluing point. Fortunately, this can be justified using the boundary structures. Another challenge is to glue near a self-touching point of V^* and V^{**}. We observed that a blowup of V^* satisfies all the requirements for the existence of good replacements in the minimal case, and hence must be an embedded minimal hypersurface. This fact, together with our particular gluing configurations, implies that all blowups appearing in the gluing procedure are planes. The matching of normal vectors then follows in a standard way. ■

5. MINIMAL SURFACES WITH FREE BOUNDARY AND APPLICATIONS

All the aforementioned results have their counterparts in compact manifolds M with smooth boundary ∂M. The variational problem in $(M, \partial M)$ concerns submanifolds $\Sigma \subset M$ with boundary $\partial \Sigma$ (possibly empty) constrained to lie in ∂M, that is, $\partial \Sigma \subset \partial M$. Critical points of the area functional for this type of variational problems are minimal submanifolds Σ with boundary $\partial \Sigma$ meeting ∂M orthogonally, usually called *minimal submanifolds with free boundary*. Other than earlier works of Gergonne in 1816 and H. A. Schwarz in 1890, Courant was the first mathematician who studied systematically the free boundary problems for minimal surfaces; see [19, CHAPTER VI]. We refer to [48] for a brief historical account of this topic. The study of free boundary minimal surfaces was recently revived by the seminal work of Fraser–Schoen [29], where they revealed a deep connection between extremal Steklov eigenvalue problems and free boundary minimal surface theory in the unit ball. We refer to [47] for a nice survey on this connection and on various constructions of examples in the unit ball.

In this part, we will focus on the codimension-one case, namely *minimal hypersurfaces with free boundary*, abbreviated as FBMHs, in general compact manifolds. We refer to [28,51] for higher-codimensional cases. Parallel to the proof of Yau's conjecture and the development of Morse theory in closed manifolds, we have also witnessed fruitful results in the free boundary setting. In his original proposal, Almgren [2] already included compact manifolds with boundary $(M, \partial M)$. He considered the space of relative cycles $\mathcal{Z}_{\text{rel}}(M, \partial M, G)$, where $G = \mathbb{Z}$ or $\mathbb{Z}_2$. (Those are integral currents or flat chains in M with boundary supported on ∂M.) The min–max procedure was expected to produce smoothly embedded FBMHs in dimensions between 3 and 7. The works by Grüter, Jost [33,41] in the 1980s and by De Lellis–Ramic [24] recently confirmed this regularity result with an additional convexity assumption on ∂M. Without assuming any boundary convexity, Li [46] first attempted this problem in dimension 3, and a general existence and regularity result was later completely established by Li and the author [48] in dimensions between 3 and 7, hence finishing the first step of Almgren's program in the free boundary setting.

A subtle difficulty present in the nonconvex boundary case is the possible touching of the interior of an FBMH with ∂M, usually called the touching phenomenon. In [48], we proved that the min–max varifold is smoothly embedded even if it has nontrivial support on

∂M. That is, the min–max FBMH may touch ∂M along an arbitrary set. This has been further developed by Guang, Li, Wang, and the author [34] to obtain Morse index upper bounds, as well as a generic density result. Very recently, Wang [86] solved the free boundary version of Yau's conjecture based on ideas in [79], thereby proving the existence of infinitely many FBMHs in an arbitrary compact $(M, \partial M)$. With Sun and Wang, we [84] proved the free boundary version of the Multiplicity One Conjecture based on [94]. As an essential tool, we also established the free boundary min–max theory for CMC/PMC hypersurfaces in [84].

Variational theory in ambient manifolds with boundary has potential applications to constructing minimal hypersurfaces in noncompact or singular spaces. The idea is to exhaust those spaces by compact domains with smooth boundary and then take the limit of the free boundary FBMHs constructed therein. In [84], we found one such application to Gaussian spaces, and constructed minimal surfaces in such spaces with arbitrarily large Gaussian area; see also [45]. Note that those minimal surfaces are self-shrinking solutions of the mean curvature flow; see [18]. We hope to see more applications of this idea in the future, for instance, in compact spaces with singularities.

6. FURTHER DISCUSSIONS

We have seen many celebrated results of the min–max theory in closed manifolds with bumpy metrics or with metrics of positive Ricci curvature. However, there still remain many interesting open problems for general metrics, besides those that can be proved by approximations. The solution of Yau's conjecture by Song [79] is almost the only general result about an arbitrary metric in this field. Since contradiction arguments were used in [79] (see also the refined proof in [80] using the notion of saddle point minimal hypersurfaces), it is tempting to find a direct construction of infinitely many closed minimal hypersurfaces by variational methods. This would require a better understanding of multiplicities for nonbumpy metrics. In particular, it would be interesting to know for a general metric whether there exist infinitely many $k \in \mathbb{N}$ such that the min–max minimal hypersurfaces associated with ω_k have multiplicity one. On the other hand, in an ongoing work joint with Wang [87], we exhibit the first nontrivial examples of nonbumpy metrics on $\mathbb{S}^3$ under which the min–max varifolds associated with the second width ω_2 must have multiplicity two. We conjecture that for any $k \in \mathbb{N}$, there exist nonbumpy metrics under which the k-width must be realized by minimal hypersurfaces with higher multiplicity. Upon finishing this survey, we learned that in an ongoing work [82], Stevens and Sun proved a nice dichotomy result for a closed manifold with an arbitrary metric, that is, there exist either closed minimal hypersurfaces with arbitrarily large area, or uncountably many closed minimal hypersurfaces. It would be interesting to know when the second situation happens. It is also natural to ask to what extent equidistribution and scarring of closed minimal hypersurfaces hold for general metrics, where one may first search for a sequence of closed minimal hypersurfaces whose average measures converge to a limit measure with positive density everywhere. Enlightened by the Quantum Unique Ergodicity Conjecture [71], it would be desirable to show that for generic metrics the sequence of min–max minimal hypersurfaces associated with the volume spectrum individually equidistribute (or even just to find a sequence of closed minimal hypersurfaces which individually equidistribute); see [60]. Another interesting question is whether the generic scarring phenomenon can occur surrounding unstable (for instance, index-one) minimal hypersurfaces in general; see [81]. Finally, it would be very interesting to know to what extent the volume spectrum reflects the ambient geometry.

Compared to minimal (hyper)surfaces, the existence theory of CMC (hyper)surfaces, particularly for multiple solutions, is still largely open. For instance, it would be very interesting to know whether the Simon–Smith [78] min–max constructions work for the CMC setting with small prescribed mean curvatures in an arbitrary 3-sphere. If so, the multiplicity-one result for the Simon–Smith min–max could be proved using the ideas in [94], and this will shed light on another famous conjecture of Yau which asserts the existence of four distinct minimal spheres [90, PROBLEM 89]; see also [37, 42]. The existence of multiple closed CMC hypersurfaces with prescribed mean curvature is a very interesting and natural problem (compare with Yau's conjecture on minimal hypersurfaces). Recently, there was some nice progress on this problem presented by Dey [20] and Mazurowski [61] based on [95]. Motivated by a well-known conjecture of Arnold [7, 1981-9] which asserts the existence of at least two distinct closed curves of any prescribed constant geodesic curvature on an arbitrary Riemannian 2-sphere (see [13,97] for more discussions), it is tempting to conjecture that every closed manifold (M^{n+1}, g), $3 \leq n+1 \leq 7$, contains at least two distinct closed CMC hypersurfaces with mean curvature c for any $c > 0$. On a related note, it would be interesting to extend the Rellich conjecture mentioned earlier to higher-dimensional Euclidean spaces using the theory in [95]. Also, since the Euclidean spaces contain closed embedded CMC hypersurfaces of any prescribed curvature, it is natural to conjecture that any asymptotically flat manifold of low dimension contains at least one closed CMC hypersurface for any prescribed curvature. (Note that there is an extensive literature on stable CMC hypersurfaces in those spaces which we do not go into here.) Finally, we note that the equation satisfied by marginally outer-trapped surfaces (MOTS) in general relativity is also of prescribed mean curvature type; see [6]. Even though the MOTS equation is not variational, it is still an interesting question whether one can construct them using a min–max scheme.

ACKNOWLEDGMENTS

The author would like to thank Da Rong Cheng, Qiang Guang, Martin Li, Longzhi Lin, Antoine Song, Ao Sun, Zhichao Wang, and Jonathan Zhu for collaborations on many results discussed in this survey, and also for helpful discussions and comments.

FUNDING

This work was partially supported by NSF grant DMS-1811293, DMS-1945178, and an Alfred P. Sloan Research Fellowship.

REFERENCES

[1] A. D. Aleksandrov, Uniqueness theorems for surfaces in the large. I. *Amer. Math. Soc. Transl. Ser. 2* **21** (1962), 341–354.

[2] F. Almgren, The homotopy groups of the integral cycle groups. *Topology* **1** (1962), 257–299.

[3] F. Almgren, *The theory of varifolds*. Mimeographed notes. Princeton, 1965.

[4] F. Almgren, Existence and regularity almost everywhere of solutions to elliptic variational problems with constraints. *Mem. Amer. Math. Soc.* **4** (1976), no. 165, viii+199.

[5] F. J. Jr. Almgren, Some interior regularity theorems for minimal surfaces and an extension of Bernstein's theorem. *Ann. of Math. (2)* **84** (1966), 277–292.

[6] L. Andersson, M. Eichmair, and J. Metzger, Jang's equation and its applications to marginally trapped surfaces. In *Complex analysis and dynamical systems IV. Part 2*, pp. 13–45, Contemp. Math. 554, Amer. Math. Soc., Providence, RI, 2011.

[7] V. Arnold, *Arnold's problems*. Springer, Berlin; PHASIS, Moscow, 2004.

[8] C. Bellettini and N. Wickramasekera, The inhomogeneous Allen–Cahn equation and the existence of prescribed-mean-curvature hypersurfaces. 2020, arXiv:2010.05847.

[9] G. D. Birkhoff, Dynamical systems with two degrees of freedom. *Trans. Amer. Math. Soc.* **18** (1917), no. 2, 199–300.

[10] C. Breiner and N. Kapouleas, Complete constant mean curvature hypersurfaces in euclidean space of dimension four or higher. *Amer. J. Math.* **143** (2021), no. 4, 1161–1259.

[11] H. Brezis and J.-M. Coron, Multiple solutions of H-systems and Rellich's conjecture. *Comm. Pure Appl. Math.* **37** (1984), no. 2, 149–187.

[12] D. R. Cheng and X. Zhou, Existence of constant mean curvature 2-spheres in Riemannian 3-spheres. *Comm. Pure Appl. Math.* (to appear), arXiv:2011.04136.

[13] D. R. Cheng and X. Zhou, Existence of curves with constant geodesic curvature in a riemannian 2-sphere. *Trans. Amer. Math. Soc.* **374** (2021), no. 12, 9007–9028.

[14] O. Chodosh and C. Mantoulidis, Minimal surfaces and the Allen-Cahn equation on 3-manifolds: index, multiplicity, and curvature estimates. *Ann. of Math. (2)* **191** (2020), no. 1, 213–328.

[15] F. Codá Marques, Minimal surfaces: variational theory and applications. In *Proceedings of the International Congress of Mathematicians—Seoul 2014. Vol. 1*, pp. 283–310, Kyung Moon Sa, Seoul, 2014.

[16] T. Colding and C. De Lellis, The min-max construction of minimal surfaces. In *Surveys in differential geometry, Vol. VIII (Boston, MA, 2002)*, pp. 75–107, Surv. Differ. Geom. VIII, Int. Press, Somerville, MA, 2003.

[17] T. Colding and W. Minicozzi, II, Width and finite extinction time of Ricci flow. *Geom. Topol.* **12** (2008), no. 5, 2537–2586.

[18] T. Colding and W. Minicozzi, II, Generic mean curvature flow I: generic singularities. *Ann. of Math. (2)* **175** (2012), no. 2, 755–833.

[19] R. Courant, *Dirichlet's principle, conformal mapping, and minimal surfaces.* Springer, New York–Heidelberg, 1977.

[20] A. Dey, Existence of multiple closed cmc hypersurfaces with small mean curvature. *J. Differential Geom.* (to appear), arXiv:1910.00989.

[21] E. De Giorgi, *Frontiere orientate di misura minima.* Editrice Tecnico Scientifica, Pisa, 1961.

[22] C. De Lellis, The size of the singular set of area-minimizing currents. In *Surveys in differential geometry 2016. Advances in geometry and mathematical physics,* pp. 1–83, Surv. Differ. Geom. 21, Int. Press, Somerville, MA, 2016.

[23] C. De Lellis and F. Pellandini, Genus bounds for minimal surfaces arising from min-max constructions. *J. Reine Angew. Math.* **644** (2010), 47–99.

[24] C. De Lellis and J. Ramic, Min-max theory for minimal hypersurfaces with boundary. *Ann. Inst. Fourier (Grenoble)* **68** (2018), no. 5, 1909–1986.

[25] H. Federer, The singular sets of area minimizing rectifiable currents with codimension one and of area minimizing flat chains modulo two with arbitrary codimension. *Bull. Amer. Math. Soc.* **76** (1970), 767–771.

[26] H. Federer and W. H. Fleming, Normal and integral currents. *Ann. of Math. (2)* **72** (1960), 458–520.

[27] R. Finn, *Equilibrium capillary surfaces.* Grundlehren Math. Wiss. 284, Springer, New York, 1986.

[28] A. Fraser, On the free boundary variational problem for minimal disks. *Comm. Pure Appl. Math.* **53** (2000), no. 8, 931–971.

[29] A. Fraser and R. Schoen, Sharp eigenvalue bounds and minimal surfaces in the ball. *Invent. Math.* **203** (2016), no. 3, 823–890.

[30] P. Gaspar and M. A. M. Guaraco, The Allen-Cahn equation on closed manifolds. *Calc. Var. Partial Differential Equations* **57** (2018), no. 4, Paper No. 101, 42.

[31] M. Gromov, Dimension, nonlinear spectra and width. In *Geometric aspects of functional analysis (1986/87)*, pp. 132–184, Lecture Notes in Math. 1317, Springer, Berlin, 1988.

[32] M. Gromov, Isoperimetry of waists and concentration of maps. *Geom. Funct. Anal.* **13** (2003), no. 1, 178–215.

[33] M. Grüter and J. Jost, On embedded minimal disks in convex bodies. *Ann. Inst. H. Poincaré Anal. Non Linéaire* **3** (1986), no. 5, 345–390.

[34] Q. Guang, M. M.-c. Li, Z. Wang, and X. Zhou, Min-max theory for free boundary minimal hypersurfaces II: general Morse index bounds and applications. *Math. Ann.* **379** (2021), no. 3–4, 1395–1424.

[35] M. A. M. Guaraco, Min-max for phase transitions and the existence of embedded minimal hypersurfaces. *J. Differential Geom.* **108** (2018), no. 1, 91–133.

[36] L. Guth, Minimax problems related to cup powers and Steenrod squares. *Geom. Funct. Anal.* **18** (2009), no. 6, 1917–1987.

[37] R. Haslhofer and D. Ketover, Minimal 2-spheres in 3-spheres. *Duke Math. J.* **168** (2019), no. 10, 1929–1975.

[38] E. Heinz, über die Existenz einer Fläche konstanter mittlerer Krümmung bei vorgegebener Berandung. *Math. Ann.* **127** (1954), 258–287.

[39] S. Hildebrandt, On the Plateau problem for surfaces of constant mean curvature. *Comm. Pure Appl. Math.* **23** (1970), 97–114.

[40] K. Irie, F. C. Marques, and A. Neves, Density of minimal hypersurfaces for generic metrics. *Ann. of Math. (2)* **187** (2018), no. 3, 963–972.

[41] J. Jost, Existence results for embedded minimal surfaces of controlled topological type. II. *Ann. Sc. Norm. Super. Pisa Cl. Sci. (4)* **13** (1986), no. 3, 401–426.

[42] J. Jost, Embedded minimal surfaces in manifolds diffeomorphic to the three-dimensional ball or sphere. *J. Differential Geom.* **30** (1989), no. 2, 555–577.

[43] N. Kapouleas, Complete constant mean curvature surfaces in Euclidean three-space. *Ann. of Math. (2)* **131** (1990), no. 2, 239–330.

[44] D. Ketover, Genus bounds for min-max minimal surfaces. *J. Differential Geom.* **112** (2019), no. 3, 555–590.

[45] D. Ketover and X. Zhou, Entropy of closed surfaces and min-max theory. *J. Differential Geom.* **110** (2018), no. 1, 31–71.

[46] M. M.-c. Li, A general existence theorem for embedded minimal surfaces with free boundary. *Comm. Pure Appl. Math.* **68** (2015), no. 2, 286–331.

[47] M. M.-c. Li, Free boundary minimal surfaces in the unit ball: recent advances and open questions. In *Proceedings of the International Consortium of Chinese Mathematicians 2017*, pp. 401–435, Int. Press, Boston, MA, 2020.

[48] M. M.-c. Li and X. Zhou, Min-max theory for free boundary minimal hypersurfaces I-regularity theory. *J. Differential Geom.* **118** (2021), no. 3, 487–553.

[49] Y. Li, Existence of infinitely many minimal hypersurfaces in higher-dimensional closed manifolds with generic metrics. *J. Differential Geom.* (to appear), arXiv:1901.08440.

[50] Y. Li, An improved Morse index bound of min-max minimal hypersurfaces. *Geom. Topol.* (to appear), arXiv:2007.14506.

[51] L. Lin, A. Sun, and X. Zhou, Min-max minimal disks with free boundary in Riemannian manifolds. *Geom. Topol.* **24** (2020), no. 1, 471–532.

[52] Y. Liokumovich, F. Marques, and A. Neves, Weyl law for the volume spectrum. *Ann. of Math. (2)* **187** (2018), no. 3, 933–961.

[53] L. Lyusternik and L. Snirelman, Topological methods in variational problems and their application to the differential geometry of surfaces. *Uspekhi Mat. Nauk* **2** (1947), no. 1(17), 166–217.

[54] F. Mahmoudi, R. Mazzeo, and F. Pacard, Constant mean curvature hypersurfaces condensing on a submanifold. *Geom. Funct. Anal.* **16** (2006), no. 4, 924–958.

[55] F. C. Marques, R. Montezuma, and A. Neves, Morse inequalities for the area functional. *J. Differential Geom.* (to appear), arXiv:2003.01301.

[56] F. C. Marques and A. Neves, Min-max theory and the Willmore conjecture. *Ann. of Math. (2)* **179** (2014), no. 2, 683–782.

[57] F. C. Marques and A. Neves, Morse index and multiplicity of min-max minimal hypersurfaces. *Cambridge J. Math.* **4** (2016), no. 4, 463–511.

[58] F. C. Marques and A. Neves, Existence of infinitely many minimal hypersurfaces in positive Ricci curvature. *Invent. Math.* **209** (2017), no. 2, 577–616.

[59] F. C. Marques and A. Neves, Morse index of multiplicity one min-max minimal hypersurfaces. *Adv. Math.* **378** (2021), 107527, 58.

[60] F. C. Marques, A. Neves, and A. Song, Equidistribution of minimal hypersurfaces for generic metrics. *Invent. Math.* **216** (2019), no. 2, 421–443.

[61] L. Mazurowski, CMC doublings of minimal surfaces via min-max. 2020, arXiv:2010.01094.

[62] W. Meeks, III, L. Simon, and S. T. Yau, Embedded minimal surfaces, exotic spheres, and manifolds with positive Ricci curvature. *Ann. of Math. (2)* **116** (1982), no. 3, 621–659.

[63] W. H. Meeks, III and J. Pérez, The classical theory of minimal surfaces. *Bull. Amer. Math. Soc. (N.S.)* **48** (2011), no. 3, 325–407.

[64] W. H. Meeks, III, J. Pérez, and G. Tinaglia, Constant mean curvature surfaces. 2016, arXiv:1605.02512.

[65] W. H. Meeks, P. Mira, J. Pérez, and A. Ros, Constant mean curvature spheres in homogeneous three-spheres. *J. Differential Geom.* (to appear), arXiv:1308.2612.

[66] F. Morgan, Regularity of isoperimetric hypersurfaces in Riemannian manifolds. *Trans. Amer. Math. Soc.* **355** (2003), no. 12, 5041–5052.

[67] A. Neves, New applications of min-max theory. In *Proceedings of the International Congress of Mathematicians—Seoul 2014. Vol. II*, pp. 939–957, Kyung Moon Sa, Seoul, 2014.

[68] J. Pitts, *Existence and regularity of minimal surfaces on Riemannian manifolds*. Math. Notes 27, Princeton University Press, Princeton, N.J., 1981.

[69] T. Rivière, A viscosity method in the min-max theory of minimal surfaces. *Publ. Math. Inst. Hautes Études Sci.* **126** (2017), 177–246.

[70] H. Rosenberg and G. Smith, Degree theory of immersed hypersurfaces. *Mem. Amer. Math. Soc.* **265** (2020), no. 1290, v+62.

[71] Z. Rudnick and P. Sarnak, The behaviour of eigenstates of arithmetic hyperbolic manifolds. *Comm. Math. Phys.* **161** (1994), no. 1, 195–213.

[72] J. Sacks and K. Uhlenbeck, The existence of minimal immersions of 2-spheres. *Ann. of Math. (2)* **113** (1981), no. 1, 1–24.

[73] R. Schoen and L. Simon, Regularity of stable minimal hypersurfaces. *Comm. Pure Appl. Math.* **34** (1981), no. 6, 741–797.

[74] R. Schoen, L. Simon, and S. T. Yau, Curvature estimates for minimal hypersurfaces. *Acta Math.* **134** (1975), no. 3–4, 275–288.

[75] R. Schoen and S. T. Yau, Existence of incompressible minimal surfaces and the topology of three-dimensional manifolds with nonnegative scalar curvature. *Ann. of Math. (2)* **110** (1979), no. 1, 127–142.

[76] B. Sharp, Compactness of minimal hypersurfaces with bounded index. *J. Differential Geom.* **106** (2017), no. 2, 317–339.

[77] J. Simons, Minimal varieties in riemannian manifolds. *Ann. of Math.* **88** (1968), no. 2, 62–105.

[78] F. R. Smith, *On the existence of embedded minimal 2-spheres in the 3-sphere, endowed with an arbitrary riemannian metric*. Ph.D. thesis, Australian National University, superwisor: Leon Simon, 1982.

[79] A. Song, Existence of infinitely many minimal hypersurfaces in closed manifolds. 2018, arXiv:1806.08816v1.

[80] A. Song, A dichotomy for minimal hypersurfaces in manifolds thick at infinity. *Ann. Sci. Éc. Norm. Supér.* (to appear), arXiv:1902.06767.

[81] A. Song and X. Zhou, Generic scarring for minimal hypersurfaces along stable hypersurfaces. *Geom. Funct. Anal.* **31** (2021), no. 4, 948–980.

[82] J. T. Stevens and A. Sun, Existence of minimal hypersurfaces with arbitrarily large area in full generality. 2021, arXiv:2112.02389.

[83] M. Struwe, Nonuniqueness in the Plateau problem for surfaces of constant mean curvature. *Arch. Ration. Mech. Anal.* **93** (1986), no. 2, 135–157.

[84] A. Sun, Z. Wang, and X. Zhou, Multiplicity one for min-max theory in compact manifolds with boundary and its applications. 2020, arXiv:2011.04136.

[85] F. Torralbo, Rotationally invariant constant mean curvature surfaces in homogeneous 3-manifolds. *Differential Geom. Appl.* **28** (2010), no. 5, 593–607.

[86] Z. Wang, Existence of infinitely many free boundary minimal hypersurfaces. *J. Differential Geom.* (to appear), arXiv:2001.04674.

[87] Z. Wang and X. Zhou, Min-max minimal hypersurfaces with multiplicity two. In preparation, 2021.

[88] B. White, The space of minimal submanifolds for varying Riemannian metrics. *Indiana Univ. Math. J.* **40** (1991), no. 1, 161–200.

[89] B. White, On the bumpy metrics theorem for minimal submanifolds. *Amer. J. Math.* **139** (2017), no. 4, 1149–1155.

[90] S. T. Yau, Problem section. In *Seminar on Differential Geometry*, pp. 669–706, Ann. of Math. Stud. 102, Princeton Univ. Press, Princeton, N.J., 1982.

[91] R. Ye, Foliation by constant mean curvature spheres. *Pacific J. Math.* **147** (1991), no. 2, 381–396.

[92] X. Zhou, On the existence of min-max minimal torus. *J. Geom. Anal.* **20** (2010), no. 4, 1026–1055.

[93] X. Zhou, On the existence of min-max minimal surface of genus $g \geq 2$. *Commun. Contemp. Math.* **19** (2017), no. 4, 1750041, 36.

[94] X. Zhou, On the Multiplicity One Conjecture in min-max theory. *Ann. of Math. (2)* **192** (2020), no. 3, 767–820.

X. ZHOU

[95] X. Zhou and J. Zhu, Min-max theory for constant mean curvature hypersurfaces. *Invent. Math.* **218** (2019), 441–490.

[96] X. Zhou and J. Zhu, Existence of hypersurfaces with prescribed mean curvature I—generic min-max. *Cambridge J. Math.* **8** (2020), no. 2, 311–362.

[97] X. Zhou and J. Zhu, Min-max theory for networks of constant geodesic curvature. *Adv. Math.* **361** (2020), 106941, 16.

XIN ZHOU

Department of Mathematics, Cornell University, Ithaca, NY 14853, USA, and Department of Mathematics, University of California Santa Barbara, Santa Barbara, CA 93106, USA, xinzhou@cornell.edu

KÄHLER–RICCI FLOW ON FANO MANIFOLDS

XIAOHUA ZHU

ABSTRACT

This is an expository paper. We will discuss some recent development in Kähler–Ricci flow on Fano manifolds.

MATHEMATICS SUBJECT CLASSIFICATION 2020

Primary 53C25; Secondary 53C55, 32Q20, 32Q10, 58J05

KEYWORDS

Fano manifold, Kähler–Ricci flow, Kähler–Ricci soliton

INTERNATIONAL CONGRESS OF MATHEMATICIANS 2022 JULY 6–14

Proc. Int. Cong. Math. 2022, Vol. 4, pp. 2718–2737
DOI 10.4171/ICM2022/63

Published by EMS Press

1. INTRODUCTION

Ricci flow was introduced by Hamilton in the early 1980s [27], and it preserves the Kählerian structure. The Kähler–Ricci (abbreciated as KR) flow is simply the Ricci flow restricted to Kähler metrics. If M is a Fano manifold, that is, a compact Kähler manifold with positive first Chern class $c_1(M) > 0$, we usually consider the following normalized KR-flow:

$$\frac{\partial \omega(t)}{\partial t} = -\operatorname{Ric}\big(\omega(t)\big) + \omega(t), \quad \omega(0) = \omega_0, \tag{1.1}$$

where ω_0 and $\omega(t)$ denote the Kähler forms of a given Kähler metric g_0 and the solutions of Ricci flow with initial metric g_0 in $2\pi c_1(M)$, respectively.[1] Then the flow preserves the Kähler class, i.e., $[\omega(t)] = 2\pi c_1(M)$ for all t. In particular, the flow preserves the volume of $\omega(t)$.

We may write solutions of (1.1) as

$$\omega(t) = \omega_t = \omega_0 + \sqrt{-1}\partial\bar{\partial}\varphi_t > 0$$

for some Kähler potential $\varphi = \varphi_t$. Let h be a Ricci potential of the background metric ω_0 such that

$$\operatorname{Ric}(\omega_0) - \omega_0 = \sqrt{-1}\partial\bar{\partial}h.$$

In 1985, Cao [11] first reduced (1.1) to solving a parabolic complex Monge–Ampère (MA) equation in the space of Kähler potentials as follows:

$$\frac{\partial \varphi}{\partial t} = \log \frac{\omega_\varphi^n}{\omega_0^n} + \varphi - h. \tag{1.2}$$

By using the maximum principle, he proved that (1.2) has a global solution ω_t for all $t \geq 0$. Thus the main interest in (1.1) is to study the limit behavior of $\omega(t)$, as well as φ_t of (1.2). In particular, if φ_t has a smooth limit, $\omega(t)$ will converge to a Kähler–Eintein (KE) metric. Hence, (1.1) also provides an approach to study KE-metrics on a Fano manifold. Compared to the continuity method used by Yau [68] and Aubin [4], Cao's argument also gives a variant proof via KR-flow of the existence of KE-metrics on a compact Kähler manifold with negative or trivial first Chern class.

In the one-dimensional case, i.e., $M = \mathbb{S}^2$, Hamilton proved the convergence of (1.1) to a round sphere under the assumption of positive curvature of ω_0 [28]. Later, Chow removed the Hamilton's condition [18, 19]. But both proofs depend on the uniformization theorem. An independent proof for the convergence of (1.1) on $\mathbb{S}^2$ was given by Chen–Lu–Tian [13]. As a consequence, they gave a proof of the uniformization theorem by using the Ricci flow.

Motivated by the Frankel conjecture, there are many influential works published for KR-flow on $\mathbb{C}P^n$ under the assumption of positive (or nonnegative) bisectional curvature, for instance, see [6, 16, 26, 41], among other references. In particular, Chen–Sun–Tian gave a proof of the Frankel conjecture by employing the Ricci flow [14].

---

1 For simplicity, we will denote a Kähler metric by its Kähler form thereafter.

Because there are some well-known obstructions for KE-manifolds (cf. [25, 40]), a Fano manifold may not admit a KE-metric, in general. Thus, the solutions φ_t of (1.2) may develop a singularity. It makes the investigation more complicated, when studying the limit behavior of the flow (1.1). In this paper, we will introduce some basic tools, as well as some recent developments of the KR-flow, including Perelman's fundamental estimates for KR-flow, the smooth convergence of KR-flow, the progress on Hamilton–Tian conjecture and the KR-flow on G-manifolds with singular limits.

2. KÄHLER–RICCI SOLITONS

A special class of solutions of (1.1) are related to KR-solitons. A KR-soliton on a Fano manifold M is a pair (X, ω), where X is a holomorphic vector field (HVF) on M and $\omega \in 2\pi c_1(M)$ is a Kähler metric on M such that

$$\mathrm{Ric}(\omega) - \omega = L_X(\omega), \tag{2.1}$$

where L_X denotes the Lie derivative along X. If $X = 0$, the KR-soliton becomes a KE-metric.

In 1985, Bando–Mabuchi proved the following uniqueness result for KE-metrics on a Fano manifold [7].

Theorem 2.1 (Bando–Mabuchi). *For any two KE-metrics ω and ω' on a Fano manifold M, there is a $\sigma \in \mathrm{Aut}(M)$ such that*

$$\omega' = \sigma^* \omega,$$

where $\mathrm{Aut}(M)$ is the group of holomorphism transformations of M.

Bando–Mabuchi's uniqueness theorem was generalized to KR-solitons by Tian and the author in 2000 [58, 59]: a KR-soliton on a compact complex manifold, if it exists, must be unique modulo $\mathrm{Aut}(M)$. Furthermore, X lies in the center of Lie algebra of a reductive part $\mathrm{Aut}_r(M)$ of $\mathrm{Aut}(M)$. We call such an X a soliton HVF, which is also unique modulo $\mathrm{Aut}(M)$. In fact, it is determined by the modified Futaki invariant, regardless of the existence of KR-solitons [59].

An important class of examples of KR-solitons were found in toric manifolds by Wang–Zhu in 2004. They solved (2.1) for a soliton HVF and torus-invariant metrics on a Fano toric manifold by using the technique of real MA-equations.

Let $\sigma_t = \exp\{t\,\mathrm{Re}(X)\}$ be a 1-PS in $\mathrm{Aut}(M)$. Then it is easy to see that the induced metrics by σ_t from a KR-soliton ω,

$$\omega(t) = \sigma_t^* \omega = \omega + \sqrt{-1}\partial\bar{\partial}\varphi_t, \tag{2.2}$$

are solutions of (1.1), as well as φ_t are solutions of (1.2). In particular, a KE-metric is a static solution of (1.1).

Note that φ_t in (2.2) is not uniformly bounded. Thus, we usually study the convergence of (1.1) or (1.2) in the sense of geometric metrics modulo holomorphism or diffeomorphism transformations; see [14, 16, 55, 56, 60, 61, 65, 66, 72, 73], etc.

3. PERELMAN'S ESTIMATES

There is a fundamental estimate for (1.1) established by Perelman in 2003 [43].

Lemma 3.1 (Perelman). *Let h_t be a Ricci potential of ω_t in* (1.1). *Then there are constants $c > 0$ and $C > 0$ depending only on the initial metric ω_0 such that the following are true:*

(1) $\operatorname{diam}(M, \omega_t) \le C$, $\operatorname{vol}(B_r(p), \omega_t) \ge c r^{2n}$;

(2) *For any $t \in (0, \infty)$, there is a constant c_t such that $h_t = -\dot{\phi}_t + c_t$ satisfies*

$$\|h_t\|_{C^0(M)} \le C, \quad \|\nabla h_t\|_{\omega_t} \le C, \quad \|\Delta h_t\|_{C^0(M)} \le C. \tag{3.1}$$

Perelamn's proof of Lemma 3.1 depends on the W-functional and the argument in proving noncollapsing of Ricci flow in the pioneering paper of his solution of the Poincaré conjecture [42]. A detailed proof of Lemma 3.1 can be found in a paper by Sesum–Tian [45]. We note that h_t is a Ricci potential of ω_t. Thus (3.1) means that the Ricci potential is uniformly bounded along the KR-flow (1.1) as well as the scalar curvature.

For Kähler metrics g in $2\pi c_1(M)$ on an n-dimensional Fano manifold M, Perelman's W-functional can be defined with a pair (g, f) by (cf. [62])

$$W(g, f) = (2\pi)^{-n} \int_M \big[R(g) + |\nabla f|^2 + f\big] e^{-f} \omega_g^n, \tag{3.2}$$

where f is a real smooth function normalized by

$$\int_M e^{-f} \omega_g^n = \int_M \omega_g^n = V. \tag{3.3}$$

Then Perelman's entropy $\lambda(g)$ is defined by

$$\lambda(g) = \inf_f \big\{W(g, f) \mid (g, f) \text{ satisfies (3.3)}\big\}.$$

The number $\lambda(g)$ can be attained by some f (cf. [44]). In fact, such an f is a solution of the equation

$$2\Delta f + f - |Df|^2 + R = \lambda(g). \tag{3.4}$$

In particular, $f = \theta_X$ if $\omega_g = \omega_{KS}$, where θ_X is a potential of soliton HVF X associated to the KR-soliton ω_{KS} [61]. As a consequence, one can further prove that the minimizer of $W(g, \cdot)$ is uniquely associated to g near a KR-soliton (cf. [47]).

The first variation of $\lambda(\omega_g)$ with $\omega_g \in 2\pi c_1(M)$ has been computed [61, 62],

$$\delta\lambda(\omega_g) = -(2\pi)^{-n} \int_M \langle \operatorname{Ric}(g) - g + \operatorname{Hess} f, \delta g \rangle e^{-f} \omega_g^n.$$

Then it is easy to see that $\lambda(\omega_t)$ is monotonic along the flow (1.1). Thus the smooth limit of ω_t in Cheeger–Gromov topology should be a KR-soliton. In particular, if the curvature of ω_t is uniformly bounded, then by the regularity of Ricci flow [46]. together with the noncollapse property in Lemma 3.1(1), there exists a sequence of ω_t which converges smoothly to a KR-soliton in Cheeger–Gromov topology.

Since the scalar curvature of ω_t is uniformly bounded along (1.1) by Lemma 3.1(2), the monotonicity of the entropy $\lambda(\omega_t)$ implies a uniform log Sobolev inequality associated to ω_t. It was proved by Zhang that this log Sobolev inequality is equivalent to the following Sobolev inequality [70]:

$$\left(\int_M |\psi|^{\frac{2n}{n-1}} \omega_t^n\right)^{\frac{n-1}{n}} \leq C_s\left(\int_M |\nabla \psi|_{\omega_t}^2 \omega_t^n + \int_M |\psi|_{\omega_t}^2 \omega_t^n\right), \quad \forall \psi \in C^\infty(M), \quad (3.5)$$

where C_s is a uniform constant independent of t.

4. SMOOTH CONVERGENCE

As we know, the smooth limit of KR-flow (1.1) should be a KR-soliton. Thus it is natural to study the convergence of KR-flow on a Fano manifold which admits a KR-soliton. In 2002, Perelman first announced the convergence result on KE-manifolds in his distinguished paper [42] (see the last paragraph in the introduction part). In 2007, Tian and the author gave a proof of Perelman's result with discrete $\mathrm{Aut}(M)$ [60]. The proof is based on an inequality of Moser–Trudinger type established by Tian in his seminal work on KE-mertrics [51]. Then in 2013, we avoided using the Moser–Trudinger inequality and so gave a complete proof of Perelman's result for the convergence of KR-flow on KE-manifolds [61]. In the general case of KR-solitons, we have proved the following convergence result in [56].

Theorem 4.1. *Let (M, ω_{KS}) be a Fano manifold which admits a KR-soliton $\omega_{KS} \in 2\pi c_1(M)$ with respect to an HVF X. Let K_X be a compact 1-PS in* $\mathrm{Aut}(M)$ *generalized by* $\mathrm{Im}(X)$. *Then for any K_X-invariant initial metric $\omega_0 \in 2\pi c_1(M)$, the flow* (1.1) *converges to ω_{KS} exponentially modulo* $\mathrm{Aut}(M)$.

Proof of Theorem 4.1 is reduced to solving the following modified KR-flow equation:

$$\frac{\partial \omega(t)}{\partial t} = -\mathrm{Ric}(\omega(t)) + \omega(t) + L_X \omega(t), \quad \omega(0) = \omega_0. \quad (4.1)$$

Analogous to (1.1), (4.1) is equivalent to

$$\frac{\partial \varphi}{\partial t} = \log \frac{\omega_\varphi^n}{\omega_0^n} + \varphi + X(\varphi) + \theta_X - h, \quad (4.2)$$

where θ_X is a potential of X associated to ω_0. We will deform the K_X-invariant initial metric ω_0 from ω_{KS} by a path ω^τ ($\tau \in [0,1]$), for example, $\omega^\tau = \tau\omega_0 + (1-\tau)\omega_{KS}$, to prove the convergence of φ_t in (4.2) for any initial ω^τ.

The first step is to prove the convergence of Kähler potentials φ_t in (4.2) for an initial metric ω_0 very close to ω_{KS}. This is related to the stability problem of KR-flow (4.1). We use a contradiction argument employing the fact of uniqueness of KR-solitons with the help of the regularity of (4.2) on $M \times [-1+t, 1+t]$ for any t. In a subsequent paper [73], the author actually proved that φ_t is convergent to a Kähler potential exponentially (without any holomorphism transformation). Moreover, the K_X-invariance condition for ω_0 can be removed. But we do not know whether the convergence is still with an exponential rate after holomorphism transformation, in general.

By the first step, we see that there is $\tau_0 \leq 1$ such that the flow (4.1) is convergent for the initial metric ω^τ with any $\tau < \tau_0$. It remains to prove that the convergence still holds for ω^{τ_0}. We can also use a contradiction argument. A key estimate is to show that the energy level $L(\omega^{\tau_0})$ of the flow for the initial metric ω^{τ_0} satisfies

$$L(\omega^{\tau_0}) = \lim_{t\to\infty} \lambda(\omega_t^{\tau_o}) = \lambda(\omega_{KS}). \tag{4.3}$$

In fact, we prove

Proposition 4.2. *Suppose that M is a Fano manifold which admits a KR-soliton (ω_{KS}, X). Then for any K_X-invariant initial metric ω of KR-flow* (1.1), *it holds that*

$$L(\omega) = (2\pi)^{-n}\big(nV - N_X\big(c_1(M)\big)\big), \tag{4.4}$$

where

$$N_X\big(c_1(M)\big) = \int_M \theta_X(\omega)e^{\theta_X(\omega)}\omega^n \tag{4.5}$$

is a holomorphic invariant for HVFs, which is independent of $\omega \in 2\pi c_1(M)$, and $\theta_X(\omega)$ is a potential of X associated to ω with a normalization

$$\int_M e^{\theta_X(\omega)}\omega^n = \int_M \omega^n = V. \tag{4.6}$$

The proof of Proposition 4.2 depends on an estimate of the asymptotic behavior of minimizer f_t in (3.4) for metric ω_t of the flow (1.1) via the Sobolev inequality (3.5) [56, LEMMA 3.3, PROPOSITION 4.2].

Remark 4.3. (1) For a K_X-invariant initial metric ω_0, φ_t in (4.2) is convergent to a Kähler potential exponentially as in the first step (without any holomorphism transformation).

(2) In case of Fano toric manifolds, the author proved the convergence of (1.2) for a torus-invariant initial metric ω_0 after torus transformations by using the technique of the real MA-equation [72]. As a consequence, the result gives an alternative proof of Wang–Zhu's theorem [67] for the existence of KR-solitons on toric manifolds via the Ricci flow.

5. H-INVARIANT

The invariant $N_X(c_1(M))$ in (4.5) can be defined for any $Y \in \eta_r(M)$ as follows (cf. [56]):

$$H(Y) = F_Y(Y) + N_Y\big(c_1(M)\big), \tag{5.1}$$

where

$$F_Y(Y) = \int_M Y(h_\omega - \theta_Y(\omega))e^{\theta_Y(\omega)}\omega^n$$

and

$$N_Y\big(c_1(M)\big) = \int_M \theta_Y(\omega)e^{\theta_Y(\omega)}\omega^n.$$

Here ω is chosen as a K-invariant metric in $2\pi c_1(M)$ and $\mathrm{Im}(Y)$ generates a compact 1-PS in $\mathrm{Aut}_r(M)$ so that $\theta_Y(\omega)$ is a real potential of HVF Y. Note that $F_X(\cdot)$ is just the modified Futaki-invariant and so $F_X(X) = 0$ [59]. Thus,

$$H(X) = N_X\big(c_1(M)\big).$$

Moreover, $H(Y)$ can be also written as (cf. [56, (5.3)])

$$H(Y) = \int_M \theta_Y(\omega) e^{h_\omega} \omega^n, \tag{5.2}$$

where the Ricci potential h_ω is normalized by

$$\int_M e^{h_\omega} \omega^n = \int_M \omega^n = V$$

and is $\theta_Y(\omega)$ given in (4.6). Thus if we do not care about the normalization of $\theta_Y(\omega)$,

$$H(Y) = \int_M \theta_Y(\omega) e^{h_\omega} \omega^n - V \ln\left[\frac{1}{V}\int_M e^{\theta_Y(\omega)} \omega^n\right], \quad \forall Y \in \eta_r(M). \tag{5.3}$$

In the above formula, $(-H(\cdot))$ is usually called an H-invariant in the literature [9, 22, 29, 30], which can be defined for any special degeneration induced by C^*-actions on M as the generalized Futaki-invariant of Ding–Tian [23].

By calculating the H-invariant for the special degeneration arising from the KR-flow (1.1) with the help of Hamilton–Tian conjecture [51] (also see the next section), Dervan–Székelyhidi recently proved (4.4) for any $\omega \in 2\pi c_1(M)$ [22]. As a consequence, they removed the assumption for a K_X-invariant initial metric ω_0 in Theorem 4.1 as follows.

Theorem 5.1 (Dervan–Székelyhidi). *Let (M, ω_{KS}) be a Fano manifold which admits a KR-soliton ω_{KS}. Then for any initial metric $\omega_0 \in 2\pi c_1(M)$, the flow* (1.1) *converges smoothly to ω_{KS} in the sense of Kähler potentials modulo* $\mathrm{Aut}(M)$.

There are other applications of H-invariant to the uniqueness of limit of KR-flow and the optimal degeneration of Fano manifolds; we refer the reader to recent papers [9, 29, 35, 65].

Remark 5.2. As we see, Dervan–Székelyhid's proof of Theorem 5.1 depends on the Hamilton–Tian conjecture. It would be interesting to give a direct proof without using the conjecture as done for Theorem 4.1 in [56].

6. A NEW APPROACH TO THE HAMILTON–TIAN CONJECTURE

In [51], Tian proposed the following conjecture (a folklore conjecture of Hamilton–Tian (HT-conjecture) [5, 17, 42, 57]):

Any sequence of $(M, \omega(t))$ contains a subsequence converging to a length space $(M_\infty, \omega_\infty)$ in the Gromov–Hausdorff topology and $(M_\infty, \omega_\infty)$ is a smooth KR-soliton outside a closed subset S, called the singular set, of codimension at least 4. Moreover, this subsequence of $(M, \omega(t))$ converges locally to the regular part of $(M_\infty, \omega_\infty)$ in the Cheeger–Gromov topology.

The HT-conjecture asserts the existence of a singular limit of (1.1) with local regularity. The above Theorem 4.1 (and Theorem 5.1) confirms the conjecture for the Fano manifold admitting a KR-soliton. In this case, the convergence of flow ω_t is smooth in the sense of Kähler potentials, in particular, the curvature of ω_t is uniformly bounded. However, it has been found in some special Fano manifolds with large symmetric group action that the curvature of ω_t cannot stay uniformly bounded [37] (also see the next section). In other words, there are examples of Fano manifolds on which the KR-flow develops singularities of type II. Thus, in general, there is no smooth limit of a KR-flow.

The Gromov–Hausdorff convergence part in the HT-conjecture follows from Perelman's noncollapse result and Zhang's upper volume estimate [71]. There is significant progress on this conjecture, first by Tian and Zhang in dimension less than 4 [57], then by Chen–Wang [17] and Bamler [5] in higher dimensions. In fact, by using the tools of geometric measure theory, Bamler proved a generalized version of the conjecture for a Ricci flow with uniformly bounded scalar curvature. His result can be also regarded as a version of Ricci flow for Cheeger–Colding compactness theorem [12] for Riemannian metrics with a bounded Ricci curvature.

In this section, we discuss an alternative proof of the HT-conjecture in a joint work with Wang [66]. Precisely, we prove

Theorem 6.1. *For any sequence of (M, ω_t) of (1.1), there is a subsequence $t_i \to \infty$ and a $\mathbb{Q}$-Fano variety $\tilde{M}_\infty$ with klt singularities such that ω_{t_i} is locally C^∞-convergent to a KR-soliton ω_∞ on $\mathrm{Reg}(\tilde{M}_\infty)$ in the Cheeger–Gromov topology. Moreover, ω_∞ can be extended to a singular KR-soliton on $\tilde{M}_\infty$ with a L^∞-bounded Kähler potential ψ_∞ and the completion of $(\mathrm{Reg}(\tilde{M}_\infty), \omega_\infty)$ is isometric to the global limit $(M_\infty, \omega'_\infty)$ of ω_{t_i} in the Gromov–Hausdorff topology. In addition, if ω_∞ is a singular KE-metrics, ψ_∞ is continuous and M_∞ is homeomorphic to $\tilde{M}_\infty$ which has Hausdorff codimension of singularities of $(M_\infty, \omega'_\infty)$ equal to at least 4.*

Compared to the proofs by blowing-up arguments in the two long papers [17] and [5], our proof of Theorem 6.1 is purely analytic by using the technique of the complex MA-equation. In Theorem 6.1, we also obtain a structure of $\mathbb{Q}$-Fano variety with klt singularities for the Gromov–Hausdorff limit in the HT-conjecture.

The proof of Theorem 6.1 is based on a recent result of Liu–Székelyhidi on Tian's partical C^0-estimate for polarized Kähler metrics with Ricci curvature bounded below [39]. In a paper of Zhang [69], it has been observed that Liu–Székelyhidi's result can be applied to prove a partial C^0-estimate for a sequence of Kähler metrics raised from the flow ω_t of (1.1). We note that the HT-conjecture also implies a partial C^0-estimate for the flow (1.1) (cf. [15, 57]). Thus we actually prove that the HT-conjecture and the partial C^0-estimate for the KR-flow are equivalent.

Let (M, L, ω) be a polarized manifold such that ω is a Kähler metric in $2\pi c_1(L)$. Choose a Hermitian metric h on L such that $\mathrm{R}(h) = \omega$. Then for any positive integer l, we have an L^2-inner product on $H^0(M, L^l, \omega)$,

$$(s_1, s_2) = \int_M \langle s_1, s_2\rangle_{h^{\otimes l}} \omega^n, \quad \forall s_1, s_2 \in H^0(M, L^l, \omega). \tag{6.1}$$

Thus for any orthonormal basis $\{s^\alpha\}$ $(0 \le \alpha \le N = N(l))$ of $H^0(M, L^l, \omega)$, we define the Bergman kernel by (cf. [48])

$$\rho_l(M, \omega)(x) = \sum_{i=0}^{N} |s^\alpha|^2_{h^{\otimes l}}(x), \quad \forall x \in M, \tag{6.2}$$

which is independent of choice of the basis $\{s^\alpha\}$.

The following fundamental result was proved by Liu–Székelyhidi [39].

Lemma 6.2 (Liu–Székelyhidi). *Given $n, D, v > 0$, there is a positive integer l and a real number $b > 0$ with the following property: Suppose that (M, L, ω) is a polarized Kähler manifold with $\omega \in 2\pi c_1(L)$ such that*

$$\mathrm{Ric}(\omega) \ge -\omega, \quad \mathrm{vol}(M, \omega) \ge v, \quad \mathrm{diam}(M, \omega) \le D. \tag{6.3}$$

Then for any $x \in M$, one has

$$\rho_l(M, \omega)(x) \ge b. \tag{6.4}$$

An inequality like (6.4) was called a partial C^0-estimate by Tian [49,50,52,53], which plays a critical role in his proof of YTD-conjecture [54]. The upper bound of $\rho_l(M, \omega)$ can be also obtained by using the standard Moser iteration (for example, see [31, LEMMA 3.2]).

By (6.4), we can write ω as a metric with bounded Kähler potential using the Fubini–Study metric as the background metric. In fact, if the orthonormal basis $\{s^\alpha\}$ $(0 \le \alpha \le N)$ defines an embedding Φ, then we have

$$\omega = \Phi^*\left(\frac{1}{l}\omega_{FS}\right) - \frac{1}{l}\sqrt{-1}\partial\bar{\partial} \log \rho_l(M, \omega).$$

By the gradient estimate for s^α (cf. [24,49]) and the lower bound (6.4) for $\rho_l(M, \omega)$, it holds that

$$\Phi^*\left(\frac{1}{l}\omega_{FS}\right) \le C(n, D, v)\omega. \tag{6.5}$$

This is because

$$\Phi^*(\omega_{FS}) = \sqrt{-1}\frac{\sum_{\alpha=0}^N \langle \nabla s^\alpha, \nabla s^\alpha\rangle}{\rho_l(M, \omega)} - \sqrt{-1}\frac{(\sum_{\alpha=0}^N \langle \nabla s^\alpha, s^\alpha\rangle)(\sum_{\alpha=0}^N \langle s^\alpha, \nabla s^\alpha\rangle)}{\rho_l^2(M, \omega)}.$$

As in [69], we modify metric ω_t to η_t so that (6.3) is satisfied by solving the following MA-equation:

$$(\eta_t)^n = (\omega_t + \sqrt{-1}\partial\bar{\partial}\kappa_t)^n = e^{h_t}\omega_t^n, \quad \sup_M \kappa_t = 0, \tag{6.6}$$

where h_t is a uniformly bounded Ricci potential of ω_t chosen as in Lemma 3.1. By Yau's solution to Calabi's problem [68], (6.6) can be solved, and by Moser iteration (cf. [58]) we have

$$|\kappa_t| \le C\left(\|h_t\|_{C^0(M)}, \omega_0\right) \le A. \tag{6.7}$$

By (6.7), each orthonormal basis of $H^0(M, K_M^{-l}, \omega_t)$ is comparable to any one of $H^0(M, K_M^{-l}, \eta_t)$. Thus by Lemma 6.2, we prove [66, PROPOSITION 2.7],

Proposition 6.3. *Given any sequence ω_{t_i} $(i \to \infty)$ of flow ω_t, there is a subsequence of ω_{t_i}, which is still denoted as ω_{t_i}, such that (6.4) holds for ω_{t_i} and the embedding of M by an orthonormal basis $\{s_{t_i}^\alpha\}$ of $H^0(M, K_M^{-l}, \omega_{t_i})$ in $\mathbb{C}P^N$ converges to a normal variety $\tilde{M}_\infty$.*

Denoting the embedding of $\{s_{t_i}^\alpha\}$ by Φ_i, we have $\tilde{M}_i = \Phi_i(M)$ converging to a normal variety $\tilde{M}_\infty$ by Proposition 6.3. Thus

$$(\Phi_i^{-1})^* \omega_{t_i} = \frac{1}{l}\omega_{FS} + \sqrt{-1}\partial\bar{\partial}\varphi_i, \tag{6.8}$$

where $\phi_i = -\frac{1}{l}(\Phi_i^{-1})^*(\log \rho_l(M, \omega_i))$ satisfies

$$|\varphi_i| \le C. \tag{6.9}$$

Moreover, by the gradient estimate for $s_{t_i}^\alpha$ [31, LEMMA 3.1],

$$\|\nabla s_{t_i}^\alpha\|_{\omega_{t_i}} \le C_s(\|h_{t_i}\|_{C^0(M)}, C_s, n) l^{\frac{n}{2}+1}, \tag{6.10}$$

where C_s is the Sobolev constant in (3.5). Thus as in (6.5), we get

$$\frac{1}{l}\omega_{FS}|_{\tilde{M}_i} \le C(\Phi_i^{-1})^* \omega_{t_i}. \tag{6.11}$$

By (6.9) and (6.11), we can derive a local $C^{k,\alpha}$-estimate for ϕ_i via the parabolic equation (1.2). In fact, we may choose exhausting open sets $\Omega_\gamma \subset \tilde{M}_\infty$. Then by Proposition 6.3, there are diffeomorphisms $\Psi_\gamma^i : \Omega_\gamma \to \tilde{M}_i$ such that the curvature of $\omega_{FS}|_{\tilde{\Omega}_\gamma^i}$ is C^k-uniformly bounded independently of i, where $\tilde{\Omega}_\gamma^i = \Psi_\gamma^i(\Omega_\gamma)$. For simplicity, we let $\tilde{\omega}_i = \frac{1}{l}\omega_{FS}|_{\tilde{M}_i}$.

The following key estimate was obtained in [66, LEMMA 3.1].

Lemma 6.4. *There exist constants A, C_γ, A_γ such that*

$$|\varphi_i| \le A \quad \text{in } \tilde{M}_i, \tag{6.12}$$

$$C_\gamma^{-1}\tilde{\omega}_i \le (\Phi_i^{-1})^* \omega_{t_i} \le C_\gamma \tilde{\omega}_i \quad \text{in } \tilde{\Omega}_\gamma^i, \tag{6.13}$$

$$\|\varphi_i\|_{C^{k,\alpha}(\tilde{\Omega}_\gamma^i)} \le A_\gamma. \tag{6.14}$$

The estimates (6.12)–(6.14) in Lemma 6.4 can be extended to Kähler potentials ϕ_i^s of metrics ω_{t_i+s} associated to the background $\hat{\omega}_i$, where $s \in [-1, 1]$ and ϕ_i^s satisfies

$$(\Phi_i^{-1})^* \omega_{t_i} = \hat{\omega}_i + \sqrt{-1}\partial\bar{\partial}\varphi_i^s.$$

By Lemma 6.4, we see that ω_{t_i} (by taking a subsequence) is locally C^∞-convergent to a KR-soliton ω_∞ on $\mathrm{Reg}(\tilde{M}_\infty)$ in the Cheeger–Gromov topology, which can be extended to a singular KR-soliton on $\tilde{M}_\infty$ with a L^∞-bounded Kähler potential ψ_∞ in the sense of full MA-measure [8]. In the case of KE-metrics ω_∞, one can further show that the local limit of η_{t_i} on $\mathrm{Reg}(\tilde{M}_\infty)$ associated to ω_{t_i} in (6.6) is just ω_∞. Thus in this case, we can actually prove that the Gromov–Hausdorff limit $(M_\infty, \omega'_\infty)$ is homeomorphic to $\tilde{M}_\infty$ and the Hausdorff codimension of singularities of $(M_\infty, \omega'_\infty)$ is at least 4. The $\mathbb{Q}$-Fano structure of $\tilde{M}_\infty$ with klt singularities can be proved as in [8,31,54].

Remark 6.5. The uniqueness of $\mathbb{Q}$-Fano structure of $\tilde{M}_\infty$ in Theorem 6.1 is independent both of the sequence and initial metric $\omega_0 \in 2\pi c_1(M)$. We refer the reader to recent papers [15,29,65].

7. KR-FLOW ON G-MANIFOLDS

In this section, we discuss the KR-flow on a Fano G-manifold, which develops singularities of type II [37]. In this joint paper, Li, Tian, and the author prove

Theorem 7.1. *Let G be a complex reductive Lie group and (M, J) be a Fano G-manifold. Suppose that (M, J) does not admit any KR-soliton. Then any solution of KR-flow on (M, J) with an initial metric $\omega_0 \in 2\pi c_1(M, J)$ is of type II.*

Here by a G-manifold we mean a *(biequivariant) compactification of G* which admits a holomorphic $G \times G$-action and has an open and dense orbit isomorphic to G as a $G \times G$-homogeneous space [1–3]. Clearly, toric manifolds form a special class of G-manifolds with G being the torus group.

A criterion theorem for the existence of KE-metrics on Fano G-manifolds was established by Delcroix [20] several years ago.

Theorem 7.2 (Delcroix). *Let M be a Fano G-manifold with associated moment polytope P. Let P_+ be the positive part of P defined by a positive roots' system $\Phi_+ = \{\alpha\}$ of G. Then M admits a KE-metric if and only if the barycenter of P_+ with respect to Φ_+ satisfies*

$$\mathrm{bar}(P_+) \in 2\rho + \Xi, \tag{7.1}$$

where Ξ is the relative interior of the cone generated by Φ_+ and $\rho = \frac{1}{2}\sum_{\alpha\in\Phi_+}\alpha$.

Delcroix's proof obtains a prior C^0-estimate for a class of real MA-equations on the positive cone $\mathfrak{a}_+ \subset \mathfrak{a} = \mathbb{R}^r$ (r is the rank of G, i.e., the dimension of a maximal torus in G) defined by Φ_+ as done for toric Fano manifolds in [67]. Later, Li, Zhou, and the author gave another proof of Delcroix's theorem by verifying the properness of K-energy and also generalized the theorem to the case of KR-solitons [38].

By Theorem 7.2, it is possible to classify all Fano G-manifolds which admit a KE-metric or KR-solitons. For example, for the rank $r = 2$, there are two $SO_4(\mathbb{C})$-manifolds and one $Sp_4(\mathbb{C})$-manifold which cannot admit any KR-solitons, see [20,21,37,74]. Thus Theorem 7.1 provides a class of examples of Ricci flow with singularities of type II on Fano manifolds.

By the HT-conjecture and the uniqueness result in [29,65], we may assume that the initial metric ω_0 in Theorem 7.1 is $K \times K$-invariant. Here K is a maximal compact subgroup of G. The proof of Theorem 7.1 includes two main steps by using a contradiction argument under the assumption of uniformly bounded curvature: first, proving that the

Cheeger–Gromov limit (M_∞, J_∞) of any sequence of $K \times K$-invariant metrics on (M, J) is still a G-manifold; second, showing that the complex structure J_∞ of limit will not jump from J. It is useful to mention that the above two steps work for any sequence of $K \times K$-invariant metrics in $2\pi c_1(M, J)$ and can be also generalized to any sequence of $K \times K$-invariant metrics with uniformly bounded curvature on a polarized G-manifold. For example, as an application one can establish an analogue of Theorem 7.1 for Calabi's flows [10] with singularities of type II on polarized G-manifolds.

7.1. A direct proof of Theorem 7.1 for a sequence of ω_t

In the following, we present a more direct proof of Theorem 7.1 from a paper jointly with Tian [63]. We turn to prove

Theorem 7.3. *Let ω_i $(= \omega_{t_i}, t_i \to \infty)$ be a sequence of (M, ω_t, J) in the flow of* (1.1) *with a $K \times K$-invariant initial metric $\omega_0 \in 2\pi c_1(M, J)$. Suppose that the curvature of ω_i is uniformly bounded. Then ω_i converges to a KR-soliton in the sense of Kähler potentials on (M, J). In particular, (M, J) admits a KR-soliton.*

We need to recall some notation and facts proved in [37]. Let $\{E_1, \ldots, E_n\}$ be a basis of the Lie algebra $\mathfrak{g}$. Then the right (left) action of G induces a space of $\mathrm{span}\{e_1, \ldots, e_n\}$ of HVFs with $\mathrm{Im}(e_a) \in \mathfrak{k}$ on M, where $\mathfrak{k}$ is the Lie algebra of K. By the partial C^0-estimate as in Section 6, there is a sequence of Kodaira embeddings $\Phi_i : M \to \mathbb{C}P^N$ induced by an orthonormal basis $\{s^i_\alpha\}$ in $H^0(K_M^{-m}, \omega_i)$ such that the image $\Phi_i(M) = \hat{M}_i$ converges to the image $\Phi_\infty(M_\infty) = \hat{M}_\infty$ in the topology of complex submanifolds, where $\Phi_\infty : M_\infty \to \mathbb{C}P^N$ is the Kodaira embedding induced by an orthonormal basis $\{s^\infty_\alpha\}$ in $H^0(K_{M_\infty}^{-m}, \omega_\infty)$.

Let $\mathrm{span}\{\hat{e}^i_1, \ldots, \hat{e}^i_n\}$ be a space of HVFs on $\hat{M}_i$ induced by Φ_i. It has been proved in [37, (4.6)] that for each a, $(\hat{e}^i_a, \hat{\omega}_i)$ converges to an HVF $(\hat{e}^\infty_\alpha, \hat{\omega}_\infty)$ on $\hat{M}_\infty$ in the sense of [37, DEFINITION 3.1], where $\hat{\omega}_i = \frac{1}{m}\omega_{FS}|_{\hat{M}_i}$ and $\hat{\omega}_\infty = \frac{1}{m}\omega_{FS}|_{\hat{M}_\infty}$. Moreover, the basis $\{\hat{e}^\infty_1, \ldots, \hat{e}^\infty_n\}$ induces an effective G-action on $\hat{M}_\infty$.

Let $T^{\mathbb{C}}$ be a torus subgroup of G acting on $\hat{M}_\infty$ with a basis $\{X_1, \ldots, X_r\}$ of $\mathfrak{a} = J\mathfrak{k} \cap \mathfrak{t}^{\mathbb{C}}$. Then it can be regarded as a subgroup of the maximal torus group $\tilde{T}^{\mathbb{C}}$ in $\mathbb{C}P^N$. Let $\tilde{W}_1, \ldots, \tilde{W}_{N+1}$ be the $N+1$ hyperplanes in $\mathbb{C}P^N$ where $\tilde{T}^{\mathbb{C}}$ does not act freely. Thus for any induced HVF $\tilde{X}$ of $X \in \mathfrak{t}^{\mathbb{C}}$ on $\hat{M}_\infty$, one has

$$\{\hat{x} \in \hat{M}_\infty \mid \tilde{X}(\hat{x}) = 0\} \subset \bigcup_\alpha \tilde{W}_\alpha. \tag{7.2}$$

Let $\mathcal{O}$ denote an open dense G-orbit in M. Since M has finitely many $G \times G$-orbits [1,2], there are basis points $x_\delta \in M \backslash \mathcal{O}$, $\delta = 1, \ldots, k$, such that

$$M = \mathcal{O} \bigcup_\delta (G \times G)x_\delta. \tag{7.3}$$

Note that the closure of each $G \times G$-orbit $(G \times G)x_\delta$ is a smooth algebraic variety whose dimension is less than n. Then, up to a subsequence, the closure of $\Phi_i((G \times G)x_\delta)$ converges to an algebraic limit in $\mathbb{C}P^N$. As a consequence, $\Phi_i(M\setminus\mathcal{O})$ has an algebraic limit $D\hat{M}_\infty$ in $\hat{M}_\infty \subset \mathbb{C}P^N$.

Let $\hat{\mathcal{O}}_\infty = \hat{M}_\infty \setminus D\hat{M}_\infty$ be an open set in $\hat{M}_\infty$. We set

$$\hat{\mathcal{O}}^0_\infty = \hat{\mathcal{O}}_\infty \setminus \Big(\bigcup_\alpha \tilde{W}_\alpha\Big) \quad \text{and} \quad \mathcal{O}^0_\infty = \Phi^{-1}_\infty(\hat{\mathcal{O}}^0_\infty) \subset M_\infty. \tag{7.4}$$

Note that

$$e^\infty_a = (\Phi^{-1}_\infty)_* \hat{e}^\infty_a, \quad a = 1, \dots, n.$$

Thus

$$\bar{X}(x_\infty) \neq 0, \quad \forall x_\infty \in \mathcal{O}^0_\infty, \quad \bar{X} \in \operatorname{span}\{e^\infty_1, \dots, e^\infty_r\}. \tag{7.5}$$

Fix a point $\hat{x}_\infty \in \hat{\mathcal{O}}_\infty$. We choose $\hat{x}_i \in \hat{M} \to \hat{x}_\infty$ and let $x_i = \Phi_i^{-1}(\hat{x}_i) \in \mathcal{O}_\infty$. Then $x_i \to x_\infty \in M_\infty$ in the Gromov–Hausdorff topology. Let $x_i' = (x_i^1, \dots, x_i^r) \in \mathfrak{a}$ be the real part of local partial coordinates z_i in [37, SECTION 2.1]. Without loss of generality, we may assume $x_i' \in \mathfrak{a}_+$ since the metric ω_i is $K \times K$-invariant. By the argument in the proof of [37, LEMMA 4.4], we have actually proved the following key lemma.

Lemma 7.4. *Suppose that the center of $\mathfrak{g}$ is zero. Then there is an absolute constant A such that*

$$|x_i'|^2 = |x_i^1|^2 + \cdots + |x_i^r|^2 \leq A. \tag{7.6}$$

Proof of Theorem 7.3. Let ψ, ψ^i be Weyl-invariant convex functions on $\mathfrak{a}$ associated to the background metric $\omega_0 = \sqrt{-1}\partial\bar{\partial}\psi$ and $K \times K$-invariant metrics ω_i, respectively. It suffices to prove that

$$|\varphi_i| = |\psi^i - \psi| \leq C \tag{7.7}$$

for some absolute constant C [56].

We first consider Case 1: G is semisimple. Then by Lemma 7.4, (7.6) holds. Thus, by [37, (4.18)], there is a small Euclidean ball B_ε in $\mathfrak{a}$ such that

$$\det(\psi^i_{ab})(x') \geq \delta_0, \quad \forall x' \in B_\varepsilon \cap \mathfrak{a}_+. \tag{7.8}$$

Moreover, as in the proof of [37, (4.23)], there is an open set $B' \subset B_\varepsilon \cap \mathfrak{a}_+$ such that for any $\alpha \in \Phi_+$, we have

$$\langle \alpha, \nabla\psi^i(x') \rangle \geq c_0, \quad \forall x' \in B'. \tag{7.9}$$

Thus, by the metric matrix (2.3) in [37, LEMMA 2.1], we get

$$a_0\omega \leq \omega_i \leq \frac{1}{a_0}\omega \quad \text{in } \Delta'_\varepsilon, \tag{7.10}$$

where a_0 is a small absolute constant and

$$\Delta'_\varepsilon = \{z \in \Delta_\varepsilon \mid x_z' \in B'\} \tag{7.11}$$

is an open set of $\Delta_\varepsilon = \{z = (z^1, \dots, z^n) \mid |z^l - x_i^l| < \varepsilon\}$.

We claim that there are a point $z_0^i \in \Delta'_\varepsilon$ and an absolute constant C_1 such that

$$\varphi_i(z_0^i) \geq \sup_M \varphi_i - C_1 \quad \text{and} \quad -\varphi_i(z_0^i) \geq \sup_M(-\varphi_i) - C_1. \tag{7.12}$$

By the Green formula, we have

$$\sup_M \varphi_i \leq \frac{1}{V}\int_M \varphi_i \omega_i^n + C_2. \tag{7.13}$$

On the other hand, by the Sobolev inequality (3.5) and the Green function estimate [7], there is an absolute constant A_1 such that the Green function $G_t(\cdot,\cdot)$ associated to ω_t satisfies

$$\int_M G_t(x,\cdot)\omega_t^n = 0 \quad \text{and} \quad G_t(x,\cdot) \geq A_1.$$

Thus as in (7.13), we also have

$$\sup_M(-\varphi_i) \leq \frac{1}{V}\int_M (-\varphi_i)\omega_i^n + C_3, \tag{7.14}$$

where C_3 is an absolute constant.

For any small number $\delta \ll 1$, we let $M_\delta = \frac{C_2 V}{\delta}$ and $M'_\delta = \frac{C_3 V}{\delta}$. Set

$$E_\delta = \Big\{z \in \Delta'_\varepsilon \mid \varphi_i(z) \leq \sup_M \varphi_i - M_\delta\Big\}$$

and

$$E_\delta^i = \Big\{z \in \Delta'_\varepsilon \mid \big(-\varphi_i(z)\big) \leq \sup_M(-\varphi_i) - M'_\delta\Big\}.$$

Then, by (7.13) and (7.14), it is easy to see that

$$\text{meas}_\omega(E_\delta) \leq \delta \quad \text{and} \quad \text{mess}_{\omega_i}(E_\delta^i) \leq \delta.$$

By (7.10), it follows that

$$\text{meas}_\omega(E_\delta),\ \text{meas}_\omega(E_\delta^i) \to 0 \quad \text{as } \delta \to 0.$$

Note that $\text{meas}_\omega(\Delta'_\varepsilon)$ is strictly positive. Hence (7.12) must be true.

By (7.12), we get

$$\text{osc}_M\, \varphi_i \leq \sup_M \varphi_i + \sup_M(-\varphi_i) + C \leq C_4.$$

Recall that $\omega_s = \omega_0 + \sqrt{-1}\partial\bar\partial\varphi_s$ satisfies the following complex MA-equation:

$$\frac{\partial\varphi_s}{\partial s} = \log\frac{(\omega_0 + \sqrt{-1}\partial\bar\partial\varphi_s)^n}{\omega_0^n} + \varphi_s - h, \quad s \in [-1+t_i, t_i+1]. \tag{7.15}$$

Then by Lemma 3.1, we may assume that φ_s satisfies (cf. [60])

$$|\varphi_s| \leq C \quad \text{and} \quad \left|\frac{\partial\varphi_s}{\partial s}\right| \leq C, \quad \forall s \in [-1+t_i, t_i+1].$$

Thus by the regularity, we get

$$\|\varphi_i\|_{C^{k,\alpha}} \leq C. \tag{7.16}$$

As a consequence, ω_i converges to a KR-soliton on (M, J) [42, 60].

Next we deal with the general Case 2: The center $\mathfrak{g}_c$ of $\mathfrak{g}$ is not zero. Let $\mathfrak{a}_c \subseteq \mathfrak{a}$ be the real part of $\mathfrak{g}_c$. By the above case 1 and the argument in the proof of [37, LEMMA 4.4] (Case 1 on page 14), we may assume that $|x_i'| \to \infty$ as $i \to \infty$ and

$$\langle \alpha, x_i' \rangle \leq A, \quad \forall \alpha \in \Phi_+, \tag{7.17}$$

for some absolute constant A, where Φ_+ is a positive roots' system. Write x_i' as

$$x_i' = x_i^0 + x_i'',$$

where $x_i^0 \in \mathfrak{a}_c$. Then

$$\langle \alpha, x_i' \rangle = \langle \alpha, x_i'' \rangle \leq A. \tag{7.18}$$

Let

$$\tilde{\psi}^i(x) = \psi^i(x + x_i^0).$$

Then $\tilde{\psi}^i(x)$ is still a Weyl-invariant convex function on $\mathfrak{a}$ and it still satisfies equation (7.15). Moreover, by (7.18), x_i'' is uniformly bounded if $\mathfrak{a}_c \subsetneq \mathfrak{a}$. Thus we can argue as in Case 1 for $\tilde{\psi}^i$ such that (7.16) holds, while the ε-cube Δ_ε of dimension n centered at x_i in (7.11) is replaced by

$$\tilde{\Delta}_\varepsilon = \{z = (z^1, \ldots, z^n) \mid |z^l - \tilde{x}_i^l| < \varepsilon\}$$

in the local coordinates $\{z^l\}_{l=1,\ldots,n}$, where $x_i'' = (\tilde{x}_i^l, \ldots, \tilde{x}_i^l)$. As a consequence,

$$\omega_i = \tilde{\omega}_i = \sqrt{-1}\partial\bar{\partial}\tilde{\psi}^i$$

converges to a KR-soliton in the sense of Kähler potentials on (M, J). The proof is complete. ■

7.2. Examples by Li–Li

By Theorems 7.1 and 6.1 for the HT-conjecture, the limit of KR-flow should be a singular KR-soliton on a $\mathbb{Q}$-Fano variety $\tilde{M}_\infty$ if a Fano G-manifold M does not admit a KR-soliton. It is interesting to study the degenerate structure of $\tilde{M}_\infty$ from M. Recently, Y. Li and Z. Li classified $\tilde{M}_\infty$ for the case of $G = \mathrm{SO}_4(\mathbb{C})$ in [35].

It is known that there are three possible Fano compactifications of $\mathrm{SO}_4(\mathbb{C})$ of dimension 6 (cf. [21]). By Theorem 7.2, one of the compactifications admits a KE-metric and the other two cannot admit any KE-metric (cf. [74]). Since $\mathrm{SO}_4(\mathbb{C})$ is semisimple, the latter two also cannot admit any KR-soliton. Thus by Theorem 7.1, the limit $\tilde{M}_\infty$ of a flow on these two manifolds should be a singular Q-Fano variety.

By studying the minimizer of H-invariant (see Section 5) for $G \times G$-equivariant special degenerations on a Fano G-manifold, Y. Li and Z. Li proved that the minimizer can be attained by a $G \times G$-equivalent special degeneration with a center fiber of $G \times G$-spherical variety [35]. In the case of $G = \mathrm{SO}_4(\mathbb{C})$, they further showed that the two $G \times G$-spherical varieties corresponding to the two non-KE-manifolds above are both relatively modified K-polystable. By Han–Li's uniqueness result for the minimizers of H-invariant [29] and the fact that the Perelman's entropy (see Section 3) attains the maximum along the KR-flow [15, 22, 65], these two spherical varieties should be limits $\tilde{M}_\infty$ of KR-flows.

Remark 7.5. (1) Examples in [36] show that the singular limit $\tilde{M}_\infty$ of KR-flow on a G-manifold cannot be a $\mathbb{Q}$-Fano variety of G-compactification, in general. We expect it is a $G \times G$-spherical variety as in the above case of $G = \mathrm{SO}_4(\mathbb{C})$.

(2) To the best of the author's knowledge, there is no known example of Fano manifold M with discrete $\mathrm{Aut}(M)$ on which the solution of a KR-flow is of type II. In fact, we do not know whether there is a Fano manifold with discrete $\mathrm{Aut}(M)$ on which the limit of a KR-flow is a singular KE-metric. In the latter, M must be K-semistable (cf. [34, 65]).

ACKNOWLEDGMENTS

The author would like to thank Professor Gang Tian for his guidance and encouragement while studying Kähler geometry in the past years. Most results in this paper were achieved in collaboration with him. The author also thanks his collaborators, B. Zhou, F. Wang, Y. Li, and others for their contributions.

FUNDING

This work is partially supported by National Key R&D Program of China SQ2020YFA07-0059 and BJSF Grant Z180004.

REFERENCES

[1] V. Alexeev and M. Brion, Stable reductive varieties I: Affine varieties. *Invent. Math.* **157** (2004), 227–274.

[2] V. Alexeev and M. Brion, Stable reductive varieties II: Projective case. *Adv. Math.* **184** (2004), 382–408.

[3] V. Alexeev and L. Katzarkov, On K-stability of reductive varieties. *Geom. Funct. Anal.* **15** (2005), 297–310.

[4] T. Aubin, Equations du type Monge–Ampère sur les variétés kählériennes compactes. *Bull. Sci. Math.* **102** (1978), 63–95.

[5] R. Bamler, Convergence of Ricci flows with bounded scalar curvature. *Ann. of Math.* **188** (2018), 753–831.

[6] S. Bando, On three dimensional compact Kähler manifolds of nonnegative bisectional curvature. *J. Differential Geom.* **19** (1984), 283–297.

[7] S. Bando and T. Mabuchi, In *Uniqueness of Kähler–Einstein metrics modulo connected group actions, Sendai 1985*, pp. 11–40, Adv. Stud. Pure Math. 10, 1987.

[8] R. Berman, S. Boucksom, P. Essydieux, V. Guedj, and A. Zeriahi, Kähler–Einstein metrics and the Kähler–Ricci flow on log Fano varieties. *J. Reine Angew. Math.* **751** (2019), 27–89.

[9] H. Blum, Y. Liu, C. Xu, and Z. Zhuang, The existence of the Kähler–Ricci soliton degeneration. 2021, arXiv:2103.15278.

[10] E. Calabi, Extremal metrics. In *Seminar on Differ. Geom., no. 16*, pp. 259–290, Ann. of Math. Stud. 16, University of Princeton, 1982.

[11] H. D. Cao, Deformation of Kähler metrics to Kähler–Einstein metrics on compact Kähler manifolds. *Invent. Math.* **81** (1985), 359–372.

[12] J. Cheeger and T. Colding, On the structure of spaces with Ricci curvature bounded below I. *J. Differential Geom.* **45** (1997), 406–480.

[13] X. Chen, P. Lu, and G. Tian, A note on uniformization of Riemann surfaces by Ricci flow. *Proc. Amer. Math. Soc.* **134** (2006), no. 11, 3391–3393.

[14] X. Chen, S. Sun, and G. Tian, A note on Kähler–Ricci soliton. *Int. Math. Res. Not. IMRN* **17** (2009), 3328–3336.

[15] X. Chen, S. Sun, and B. Wang, Kähler–Ricci flow, Kähler–Einstein metric, and K-stability. *Geom. Topol.* **22** (2018), 3145–3173.

[16] X. Chen and G. Tian, Ricci flow on Kähler–Einstein surfaces. *Invent. Math.* **147** (2002), no. 3, 487–544.

[17] X. Chen and B. Wang, Space of Ricci flows (II). Part A: Moduli of singular Calabi–Yau spaces. *Forum Math. Sigma* **5** (2017), e32 103 pp. Space of Ricci flows (II). Part B: Weak compactness of the flows. *J. Differential Geom.* **116** (2020), 1–123.

[18] B. Chow, On the entropy estimate for the Ricci flow on compact 2-orbifolds. *J. Differential Geom.* **33** (1991), 597–600.

[19] B. Chow, The Ricci flow on the 2-sphere. *J. Differential Geom.* **33** (1991), 325–334.

[20] T. Delcroix, Kähler–Einstein metrics on group compactifications. *Geom. Funct. Anal.* **27** (2017), 78–129.

[21] T. Delcroix, K-Stability of Fano spherical varieties. *Ann. Sci. Éc. Norm. Supér.* **53** (2020), no. 4, 615–662.

[22] R. Dervan and G. Székelyhidi, Kähler–Ricci flow and optimal degenerations. *J. Differential Geom.* **116** (2020), no. 1, 187–203.

[23] W. Ding and G. Tian, Kähler–Einstein metrics and the generalized Futaki invariants. *Invent. Math.* **110** (1992), 315–335.

[24] S. Donaldson and S. Sun, Gromov–Hausdorff limits of Kähler manifolds and algebraic geometry. *Acta Math.* **213** (2014), no. 1, 63–106.

[25] A. Futaki, An obstruction to the existence of Einstein–Kähler metrics. *Invent. Math.* **73** (1983), 437–443.

[26] H. Gu, A new proof of Mok's generalized Frankel Conjecture theorem. *Proc. Amer. Math. Soc.* **137** (2009), no. 3, 1063–1068.

[27] R. S. Hamilton, Three-manifolds with positive Ricci curvature. *J. Differential Geom.* **17** (1982), 255–306.

[28] R. S. Hamilton, The Ricci flow on surfaces, Contemp. Math. 71, AMS, Providence, RI, 1988.

[29] J. Han and C. Li, Algebraic uniqueness of Kähler–Ricci flow limits and optimal degenerations of Fano varieties. 2020, arXiv:2009.01010v1.

[30] W. He, Kähler–Ricci soliton and H-functional. *Asian J. Math.* **20** (2016), 645–664.

[31] W. Jiang, F. Wang, and X. H. Zhu, Bergman Kernels for a sequence of almost Kähler–Ricci solitons. *Ann. Inst. Fourier (Grenoble)* **67** (2017), 1279–1320.

[32] K. Kodaira, *Complex manifolds and deformation of complex structures*. Springer, Berlin, 2005, x+465 pp. ISBN: 3-540-22614-1.

[33] M. Kuranishi, New proof for the existence of locally complete families of complex structures. In *Proc. Conf. Complex Analysis*, pp. 142–154, 1964, Springer, Berlin, 1965.

[34] C. Li, Yau–Tian–Donaldson correspondence for K-semistable Fano manifolds. *J. Reine Angew. Math.* **733** (2017), 55–85.

[35] Y. Li and Z. Li, Semistable degenerations of Q-Fano compactifications. 2021, arXiv:2103.06439v4.

[36] Y. Li, G. Tian, and X. H. Zhu, Singular Kähler–Einstein metrics on $\mathbb{Q}$-Fano compactifications of a Lie group. 2020, arXiv:2001.11320.

[37] Y. Li, G. Tian, and X. H. Zhu, Singular limits of Kähler–Ricci flow on Fano G-manifolds. 2020, arXiv:1807.09167v4.

[38] Y. Li, B. Zhou, and X. H. Zhu, K-energy on polarized compactifications of Lie groups. *J. Funct. Anal.* **275** (2018), 1023–1072.

[39] G. Liu and G. Székelyhidi, Gromov–Hausdorff limit of Kähler manifolds with Ricci bounded below. 2020, arXiv:1804.03084v2.

[40] Y. Matsushima, Sur la structure du group d'homeomorphismes analytiques d'une certaine varietie Kaehlerinne. *Nagoya Math. J.* **11** (1957), 145–150.

[41] N. Mok, The uniformization theorem for compact Kähler manifolds of nonnegative bisectional curvature. *J. Differential Geom.* **27** (1988), 179–214.

[42] G. Perelman, The entropy formula for the Ricci flow and its geometric applications. 2002, arXiv:math/0211159.

[43] G. Perelman, unpublished, 2003.

[44] O. Rothaus, Logarithmic Sobolev inequality and the spectrum of Schördinger operators. *J. Funct. Anal.* **42** (1981), 110–120.

[45] N. Sesum and G. Tian, Bounding scalar curvature and diameter along the Kähler–Ricci flow (after Perelman). *J. Inst. Math. Jussieu* **7** (2008), 575–587.

[46] W. X. Shi, Ricci deformation of the metric on complete noncompact Riemannian manifolds. *J. Differential Geom.* **30** (1989), 223–301.

[47] S. Sun and Y. Wang, On the Kähler–Ricci flow near a Kähler–Einstein metric. *J. Reine Angew. Math.* **699** (2015), 143–158.

[48] G. Tian, On a set of polarized Kahler metrics on algebraic manifolds. *J. Differential Geom.* **32** (1990), 99–130.

[49] G. Tian, On Calabi's conjecture for complex surfaces. *Invent. Math.* **101** (1990), 101–172.

[50] G. Tian, Kähler–Einstein on algebraic manifolds. In *Proceedings of the International Congress of Mathematicians, Vol. I, II (Kyoto, 1990)*, pp. 587–598, Math. Soc. Japan, Tokyo, 1991.

[51] G. Tian, Kähler–Einstein metrics with positive scalar curvature. *Invent. Math.* **130** (1997), 1–37.

[52] G. Tian, Existence of Einstein metrics on Fano manifolds. In *Jeff Cheeger anniversary volume: metric and differential geometry*, pp. 119–162, Progr. Math. 297, 2012.

[53] G. Tian, Partial C^0-estimates for Kähler–Einstein metrics. *Commun. Math. Stat.* **1** (2013), 105–113.

[54] G. Tian, K-stability and Kähler–Einstein metrics. *Comm. Pure Appl. Math.* **68** (2015), 1085–1156.

[55] G. Tian, L. Zhang, and X. H. Zhu, Kähler–Ricci flow for deformed complex structures. 2021, arXiv:2107.12680.

[56] G. Tian, S. Zhang, Z. Zhang, and X. H. Zhu, Perelman's entropy and Kähler–Ricci flow an a Fano manifold. *Trans. Amer. Math. Soc.* **365** (2013), 6669–6695.

[57] G. Tian and Z. L. Zhang, Regularity of Kähler–Ricci flows on Fano manifolds. *Acta Math.* **216** (2016), 127–176.

[58] G. Tian and X. H. Zhu, Uniqueness of Kähler–Ricci solitons. *Acta Math.* **184** (2000), 271–305.

[59] G. Tian and X. H. Zhu, A new holomorphic invariant and uniqueness of Kähler–Ricci solitons. *Comment. Math. Helv.* **77** (2002), 297–325.

[60] G. Tian and X. H. Zhu, Convergence of the Kähler–Ricci flow. *J. Amer. Math. Sci.* **17** (2007), 675–699.

[61] G. Tian and X. H. Zhu, Convergence of the Kähler–Ricci flow on Fano manifolds. *J. Reine Angew. Math.* **678** (2013), 223–245.

[62] G. Tian and X. H. Zhu, Perelman's W-functional and stability of Kähler–Ricci flows. *Progr. Math.* **2** (2018), no. 1, 1–14. arXiv:0801.3504v2.

[63] G. Tian and X. H. Zhu, Kähler–Ricci flow on Fano G-manifolds. Preprint, 2020.

[64] F. Wang, B. Zhou, and X. H. Zhu, Modified Futaki invariant and equivariant Riemann–Roch formula. *Adv. Math.* **289** (2016), 1205–1235.

[65] F. Wang and X. H. Zhu, Uniformly strong convergence of Kähler–Ricci flows on a Fano manifold. 2020, arXiv:2009.10354.

[66] F. Wang and X. H. Zhu, Tian's partial C^0-estimate implies Hamilton–Tian's conjecture. *Adv. Math.* **381** (2021), 1–29.

[67] X. Wang and X. H. Zhu, Kähler–Ricci solitons on toric manifolds with positive first Chern class. *Adv. Math.* **188** (2004), 87–103.

[68] S. T. Yau, On the Ricci curvature of a compact Kähler manifold and the complex Monge–Ampère equation, I. *Comm. Pure Appl. Math.* **31** (1978), 339–411.

[69] K. Zhang, Some refinements of the partial C^0-estimate. *Anal. PDE* **14** (2021), 2307–2326.

[70] Q. Zhang, A uniform Sobolev inequality under Ricci flow. *Int. Math. Res. Not.* **17** (2007), 1–17.

[71] Q. Zhang, Bounds on volume growth of geodesic balls under Ricci flow. *Math. Res. Lett.* **19** (2012), 245–253.

[72] X. H. Zhu, Kähler–Ricci flow on a toric manifold with positive first Chern class. In *Differential geometry*, pp. 323–336, Adv. Lect. Math. (ALM) 22, Int. Press, Somerville, MA, 2012, arXiv:math/0703486.

[73] X. H. Zhu, Stability on Kähler–Ricci flow on a compact Kähler manifold with positive first Chern class. *Math. Ann.* **356** (2013), 1425–1454.

[74] X. H. Zhu, Kähler–Einstein metrics on toric manifolds and G-manifolds. *Progr. Math.* **330** (2020), 545–585.

XIAOHUA ZHU

School of Mathematical Sciences, Peking University, Beijing 100871, China, xhzhu@math.pku.edu.cn

6. TOPOLOGY

HOMOLOGY COBORDISM, KNOT CONCORDANCE, AND HEEGAARD FLOER HOMOLOGY

JENNIFER HOM

ABSTRACT

We review some recent results in knot concordance and homology cobordism. The proofs rely on various forms of Heegaard Floer homology. We also discuss related open problems.

MATHEMATICS SUBJECT CLASSIFICATION 2020

Primary 57Q60; Secondary 57K18, 57R58

KEYWORDS

Knot concordance, homology cobordism, Heegaard Floer homology, 3-manifolds

INTERNATIONAL CONGRESS OF MATHEMATICIANS 2022 JULY 6–14

Proc. Int. Cong. Math. 2022, Vol. 4, pp. 2740–2766
DOI 10.4171/ICM2022/45

Published by EMS Press

1. INTRODUCTION

Many interesting phenomena occur in three and four dimensions that do not occur in higher dimensions. Indeed, the Poincaré conjecture in dimensions five and higher was proved by Smale [83] in 1961 using techniques from surgery theory, while the Poincaré conjecture in dimension three remained unsolved for another 40 years until Perelman's proof of Thurston's geometrization conjecture [74–76]. In 1982, Freedman [19] proved the topological 4-dimensional Poincaré conjecture, while the smooth 4-dimensional Poincaré conjecture remains open. As another example of how dimension four is special, by work of Freedman [19] and Donaldson [12], $\mathbb{R}^n$ admits smooth structures that are not diffeomorphic to the standard one only when $n = 4$.[1]

The 3-dimensional homology cobordism group and the knot concordance group are fundamental structures in low-dimensional topology. The former played a key role in Manolescu's disproof of the high dimensional triangulation conjecture [55], while the latter has the potential to shed light on the smooth 4-dimensional Poincaré conjecture (see, for example, [18, 58]).

Our goal is to review some recent applications of Heegaard Floer theory to homology cobordism and knot concordance, and to discuss the power and limitations of these tools to address major open questions in the field.

1.1. Homology cobordism

Two closed, oriented 3-manifolds Y_0, Y_1 are *homology cobordant* if there exists a smooth, compact, oriented 4-manifold W such that $\partial W = -Y_0 \sqcup Y_1$ and the inclusions $\iota : Y_i \to W$ induce isomorphisms

$$\iota_* : H_*(Y_i; \mathbb{Z}) \to H_*(W; \mathbb{Z})$$

for $i = 0, 1$. The key point is that, on the level of homology, W looks like a product. The *3-dimensional homology cobordism group* $\Theta^3_{\mathbb{Z}}$ consists of integer homology 3-spheres modulo homology cobordism, under the operation induced by connected sum. A homology sphere Y represents the identity in $\Theta^3_{\mathbb{Z}}$ if and only if Y bounds a homology 4-ball, and the inverse of $[Y]$ in $\Theta^3_{\mathbb{Z}}$ is $[-Y]$, where $-Y$ denotes Y with the opposite orientation. The Rokhlin invariant [80] gives a surjective homomorphism

$$\mu : \Theta^3_{\mathbb{Z}} \to \mathbb{Z}/\mathbb{Z}2,$$

showing that $\Theta^3_{\mathbb{Z}}$ is nontrivial. Manolescu [55] showed that if $\mu(Y) = 1$, then Y is not of order two in $\Theta^3_{\mathbb{Z}}$. By work of Galewski–Stern [22] and Matumoto [59], this leads to a disproof of the triangulation conjecture in dimensions ≥ 5. See [56, 57] for an overview of this work. The triangulation conjecture is also false in dimension four, by work of Casson; see [3].

Fintushel–Stern [14] used gauge theory to show that $\Theta^3_{\mathbb{Z}}$ is infinite, and Furuta [21] and Fintushel–Stern [15] improved this result to show that $\Theta^3_{\mathbb{Z}}$ contains a subgroup isomorphic to $\mathbb{Z}^\infty$. Frøyshov [20] used Yang–Mills theory to define a surjective homomorphism

1 Smale, Perelman, Freedman, and Donaldson all won Fields medals for their work discussed here; Perelman declined the award.

$\Theta^3_{\mathbb{Z}} \to \mathbb{Z}$, showing that $\Theta^3_{\mathbb{Z}}$ has a direct summand isomorphic to $\mathbb{Z}$. (This is stronger than having a $\mathbb{Z}$ subgroup, since, for example, $\mathbb{Z}$ is a subgroup of $\mathbb{Q}$ but not a summand.) In joint work with Dai, Stoffregen, and Truong, we use Hendricks–Manolescu's involutive Heegaard Floer homology [32] to prove the following:

Theorem 1.1 ([9]). *The homology cobordism group $\Theta^3_{\mathbb{Z}}$ contains a direct summand isomorphic to $\mathbb{Z}^\infty$.*

Fundamental questions about the structure of $\Theta^3_{\mathbb{Z}}$ remain open:

Question 1.2. Does $\Theta^3_{\mathbb{Z}}$ contain any torsion? Modulo torsion, is $\Theta^3_{\mathbb{Z}}$ free abelian?

If there is any torsion in $\Theta^3_{\mathbb{Z}}$, two-torsion seems the most likely. Indeed, any integer homology sphere Y admitting an orientation-reversing self-diffeomorphism is of order at most two in $\Theta^3_{\mathbb{Z}}$. However, as far as we are aware, all known examples of such Y bound integer homology balls, and hence are trivial in $\Theta^3_{\mathbb{Z}}$.

In a different direction, it is natural to ask which types of manifolds can represent a given class $[Y] \in \Theta^3_{\mathbb{Z}}$. The first answers to this question were in the positive. Livingston [52] showed that every class in $\Theta^3_{\mathbb{Z}}$ can be represented by an irreducible integer homology sphere and Myers [61] improved this to show that every class has a hyperbolic representative.

In the negative direction, Frøyshov (in an unpublished work), F. Lin [49], and Stoffregen [84] showed that there are classes in $\Theta^3_{\mathbb{Z}}$ that do not admit Seifert fibered representatives. Nozaki–Sato–Taniguchi [63] improved this result to show that there are classes that do not admit a Seifert fibered representative or a representative that is surgery on a knot in S^3. However, none of these results were sufficient to obstruct $\Theta^3_{\mathbb{Z}}$ from being generated by Seifert fibered spaces. In joint work with Hendricks, Stoffregen, and Zemke, we prove the following:

Theorem 1.3 ([30, 31]). *The homology cobordism group $\Theta^3_{\mathbb{Z}}$ is not generated by Seifert fibered spaces. More specifically, let Θ_{SF} denote the subgroup generated by Seifert fibered spaces. The quotient $\Theta^3_{\mathbb{Z}}/\Theta_{SF}$ is infinitely generated.*

In light of the aforementioned Nozaki–Sato–Taniguchi result, it is natural to ask:

Question 1.4. Do surgeries on knots in S^3 generate $\Theta^3_{\mathbb{Z}}$?

The expectation is that surgeries on knots in S^3 are not sufficiently generic to generate $\Theta^3_{\mathbb{Z}}$, but such a result seems beyond the capabilities of current tools.

1.2. Knot concordance

Two knots $K_0, K_1 \subset S^3$ are *concordant* if there exists a smooth, properly embedded annulus A in $S^3 \times [0,1]$ such that $K_i = A \cap (S^3 \times \{i\})$ for $i = 0, 1$. The *knot concordance group* $\mathcal{C}$ consists of knots in S^3 modulo concordance, under the operation induced by connected sum. The inverse of $[K]$ in $\mathcal{C}$ is given by $[-K]$, where $-K$ denotes the reverse of the mirror image of K. A knot $K \subset S^3 = \partial B^4$ is *slice* if it bounds a smoothly embedded disk in B^4. Fox–Milnor [17] and Murasugi [60] showed that $\mathcal{C}$ is nontrivial, and J. Levine [48] used the Seifert form to define a surjective homomorphism $\mathcal{C} \to \mathbb{Z}^\infty \oplus (\mathbb{Z}/2\mathbb{Z})^\infty \oplus (\mathbb{Z}/4\mathbb{Z})^\infty$, demonstrating that $\mathcal{C}$ is, in fact, highly nontrivial. In higher odd dimensions (that is, knotted

S^{2n+1} in S^{2n+3}, $n \geq 1$), Levine's homomorphism is an isomorphism, while in the classical dimension, the kernel is nontrivial [6]. See [53] for a survey of knot concordance.

We can consider various generalizations of the knot concordance group. For example, rather than considering annuli in $S^3 \times [0, 1]$, we may consider annuli in homology cobordisms. Two knots $K_0 \subset Y_0$ and $K_1 \subset Y_1$ are *homology concordant* if they cobound a smooth, properly embedded annulus in a homology cobordism between Y_0 and Y_1.

Let $\mathcal{C}_{\mathbb{Z}}$ denote the group of knots in S^3, modulo homology concordance. A knot $K \subset S^3$ represents the identity in $\mathcal{C}_{\mathbb{Z}}$ if and only if K bounds a smoothly embedded disk in some homology 4-ball. Since B^4 is of course a homology 4-ball, there is naturally a surjection from $\mathcal{C}$ to $\mathcal{C}_{\mathbb{Z}}$. A natural question is whether or not this map is injective; in other words,

Question 1.5. If a knot $K \subset S^3$ bounds a disk in a homology 4-ball, must K also bound a disk in B^4?

One reason why the question above is challenging is that many obstructions to a knot K bounding a disk in B^4 also obstruct K from bounding a disk in a homology 4-ball.

An even more difficult question is the following:

Question 1.6. If a knot $K \subset S^3$ bounds a disk in a homotopy 4-ball, must K also bound a disk in B^4?

Recall the smooth 4-dimensional Poincaré conjecture which by work of Freedman [19] may be stated as follows:

Conjecture 1.7 (Smooth 4-dimensional Poincaré conjecture). *If a smooth 4-manifold X is homeomorphic to S^4, then X is actually diffeomorphic to S^4.*

A negative answer to Question 1.6 provides one possible strategy for disproving Conjecture 1.7. Indeed, by Freedman [19], any homotopy 4-sphere is homeomorphic to S^4. Now suppose we found a homotopy 4-sphere X and a knot $K \subset S^3 = \partial(X \setminus \mathring{B}^4)$ such that K bounds a smoothly embedded disk in $X \setminus \mathring{B}^4$. If we could obstruct K from bounding a smoothly embedded disk in B^4, then it follows that X cannot be diffeomorphic to S^4. This approach was attempted in [18, 58], but has yet to lead to a disproof of the conjecture.

We now return to the group $\mathcal{C}_{\mathbb{Z}}$. This group is naturally a subgroup of $\hat{\mathcal{C}}_{\mathbb{Z}}$, the group of manifold–knot pairs (Y, K), where Y is a homology sphere bounding a homology ball and K is a knot in Y, modulo homology concordance. One can ask whether the injection from $\mathcal{C}_{\mathbb{Z}}$ to $\hat{\mathcal{C}}_{\mathbb{Z}}$ is a surjection. Adam Levine [47] answered this question in the negative, showing that there exist knots in a homology null-bordant Y (in fact, his example bounds a contractible 4-manifold) that are not concordant to any knot in S^3. Expanding on this result, in joint work with Levine and Lidman, we prove the following:

Theorem 1.8 ([39]). *The subgroup $\mathcal{C}_{\mathbb{Z}} \subset \hat{\mathcal{C}}_{\mathbb{Z}}$ is of infinite index. More specifically,*

(1) *the quotient $\hat{\mathcal{C}}_{\mathbb{Z}}/\mathcal{C}_{\mathbb{Z}}$ is infinitely generated, and*

(2) *the quotient $\hat{\mathcal{C}}_{\mathbb{Z}}/\mathcal{C}_{\mathbb{Z}}$ contains a subgroup isomorphic to $\mathbb{Z}$.*

This result demonstrates the vast difference between knots in S^3 and knots in arbitrary homology spheres bounding homology balls, up to concordance. The examples in the infinite generation part of the theorem bound contractible 4-manifolds; it is unknown whether the examples in the $\mathbb{Z}$ subgroup do. Zhou [90] proved that the quotient $\hat{\mathcal{C}}_{\mathbb{Z}}/\mathcal{C}_{\mathbb{Z}}$ has a subgroup isomorphic to $\mathbb{Z}^\infty$. (It is unknown whether his examples bound contractible 4-manifolds.) The forthcoming joint work with Dai, Stoffregen, and Truong [10] improves Zhou's result to a $\mathbb{Z}^\infty$-summand.

One can also consider concordance in more general 4-manifolds. Let R be a ring. Two closed, oriented, connected 3-manifolds Y_0, Y_1 are *R-homology cobordant* if there exists a smooth, compact, oriented 4-manifold W such that $\partial W = -Y_0 \sqcup Y_1$ and the inclusions $\iota : Y_i \to W$ induce isomorphisms

$$\iota_* : H_*(Y_i; R) \to H_*(W; R)$$

for $i = 0, 1$. We have already discussed the case $R = \mathbb{Z}$. The rational homology cobordism group $\Theta^3_{\mathbb{Q}}$ contains elements of order two, for example, $[\mathbb{R}P^3]$; in contrast, as asked in Question 1.2, it remains open whether there is any torsion in the integer homology cobordism group $\Theta^3_{\mathbb{Z}}$.

We can consider concordances in other R-homology cobordisms, such as $\mathbb{Q}$-homology cobordisms. A knot $K \subset S^3$ is *rationally slice* if it is $\mathbb{Q}$-homology concordant to the unknot, or equivalently, if K bounds a smoothly embedded disk in a rational homology 4-ball.

Let $\mathcal{C}_{\mathbb{Q}S}$ denote the subgroup of $\mathcal{C}$ consisting of rationally slice knots. Cochran, based on work of Fintushel–Stern [13], showed that the figure-eight knot is rationally slice. Hence $\mathbb{Z}/2\mathbb{Z}$ is a subgroup of $\mathcal{C}_{\mathbb{Q}S}$, since the figure-eight is negative amphichiral and not slice. Cha [7] extended this result to show that $\mathcal{C}_{\mathbb{Q}S}$ has a subgroup isomorphic to $(\mathbb{Z}/2\mathbb{Z})^\infty$. A natural question to ask is whether $\mathcal{C}_{\mathbb{Q}S}$ contains elements of infinite order (see, for example, [86, PROBLEM 1.11]). Joint work with Kang, Park, and Stoffregen uses the involutive knot Floer package of Hendricks–Manolescu [32] to prove:

Theorem 1.9 ([38]). *The group of rationally slice knots $\mathcal{C}_{\mathbb{Q}S}$ contains a subgroup isomorphic to $\mathbb{Z}^\infty$.*

The figure-eight is slice in a rational homology 4-ball W with $H_1(W; \mathbb{Z}) = \mathbb{Z}/2\mathbb{Z}$ (see, for example, [2, SECTION 3]), as are Cha's examples [7].

Question 1.10. Does there exist a knot $K \subset S^3$ that is not slice in B^4 but is slice in a rational homology 4-ball W with $|H_1(W; \mathbb{Z})|$ odd?

Compare this question to Question 1.5, which asks whether there is a knot $K \subset S^3$ that is not slice in B^4 but is slice in a integer homology 4-ball W. Indeed, both Questions 1.5 and 1.10 can be viewed as incremental steps towards Question 1.6, a negative answer to which would in turn disprove the smooth 4-dimensional Poincaré conjecture.

1.3. Ribbon concordance

We conclude the introduction with a discussion of ribbon knots, ribbon concordances, and ribbon homology cobordisms. A knot $K \subset S^3$ is *ribbon* if it bounds an immersed disk in S^3 with only ribbon singularities. A *ribbon singularity* is a closed arc consisting of intersection points of the disk with itself such that the preimage of this arc is two disjoint arcs in the disk with one arc a_1 being contained entirely in the interior of the disk and the other arc a_2 having its endpoints on the boundary of the disk; see Figure 1. Note that ribbon knots are slice, since ribbon singularities can be resolved in the 4-ball (namely, by pushing the arc a_1 farther into the 4-ball).

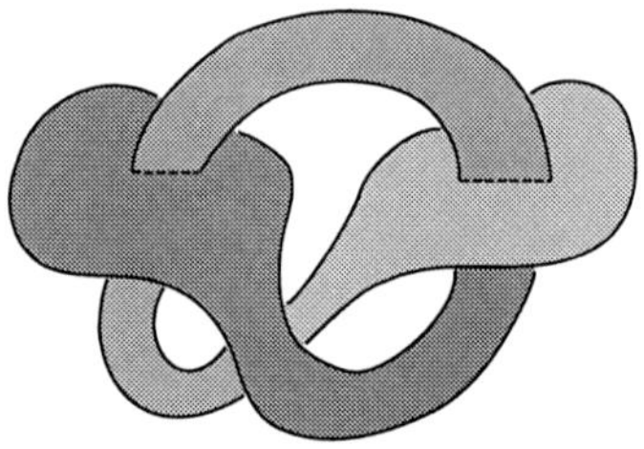

FIGURE 1
An example of a ribbon disk.

Conjecture 1.11 (Slice-ribbon conjecture **[16]**). *Every slice knot is ribbon.*

The slice-ribbon conjecture is true for two-bridge knots **[51]** and many infinite families of pretzel knots **[25, 45]**. On the other hand, potential counterexamples exist; see, for example, **[1, 23]**.

Equivalently, a ribbon knot can be defined as a knot $K \subset S^3$ that bounds a *ribbon disk* in B^4, that is, a smoothly embedded disk $D \subset B^4$ such that the radial Morse function on B^4 restricted to D has no interior local maxima. There is a *ribbon concordance* from K_0 to K_1 if there is a concordance from K_0 to K_1 in $S^3 \times [0, 1]$ with no interior local maxima (with respect to the natural height function on $S^3 \times [0, 1]$) or, equivalently, if projection to $[0, 1]$ is Morse with only index 0 and index 1 critical points. Note that, unlike ordinary concordance, ribbon concordance is not symmetric (and that the convention regarding the direction of the ribbon concordance varies in the literature).

Conjecture 1.12 (**[24]**). *Ribbon concordance is a partial order. That is, if there exist a ribbon concordance from K_0 to K_1 and a ribbon concordance from K_1 to K_0, then $K_0 = K_1$.*

Gordon **[24]** proved that Conjecture 1.12 holds for fibered knots and two-bridge knots, and more generally for the class of knots generated by such knots under the operations of connected sum and cabling. He also proved that if S is a ribbon concordance from K_0 to K_1, then $\pi_1(S^3 \setminus \nu(K_0)) \to \pi_1(X)$ is injective and $\pi_1(S^3 \setminus \nu(K_1)) \to \pi_1(X)$ is surjective, where X denotes the exterior of S in $S^3 \times [0, 1]$.

The notion of a ribbon concordance can also be generalized to homology cobordisms. A *ribbon cobordism* between two 3-manifolds is a cobordism admitting a handle decomposition with only 1- and 2-handles. Observe that the complement of a ribbon concordance from one knot to another is naturally a ribbon cobordism between their knot complements. Daemi, Lidman, Vela-Vick, and Wong proposed the following 3-manifold analog of Conjecture 1.12:

Conjecture 1.13 ([8]). *Ribbon $\mathbb{Q}$-homology cobordism is a partial order on closed, oriented, connected 3-manifolds. That is, if there exist a ribbon $\mathbb{Q}$-homology cobordism from Y_0 to Y_1 and a ribbon $\mathbb{Q}$-homology cobordism from Y_1 to Y_0, then Y_0 and Y_1 are diffeomorphic.*

The results in [8] provide evidence in support of Conjecture 1.13.

1.4. Organization

The remainder of this article is devoted to discussing applications of Heegaard Floer homology to the theorems and problems discussed above. We will describe various Heegaard Floer chain complexes associated to 3-manifolds and knots inside of them. As we progress, our chain complexes will have more and more structure; we will sketch how this additional structure leads to the results described in the introduction.

In Section 2, we discuss properties of the Heegaard Floer 3-manifold invariant of [66, 70] and applications to homology cobordism. In Section 3, we move on to knot Floer homology [69, 78] and applications to concordance. With the advent of involutive Heegaard Floer homology [32], these invariants can be endowed with additional structure; in Section 4, we describe involutive Heegaard Floer homology, and in Section 5, we delve into involutive knot Floer homology. Lastly, in Section 6, we discuss the potential (or lack thereof) for Heegaard Floer homology to answer the questions posed in Section 1.

For an introduction to many of the tools described in this article, we refer the reader to Sections 1–3 of [34].

2. HEEGAARD FLOER HOMOLOGY: THE 3-MANIFOLD INVARIANT

In this section, we consider the Heegaard Floer 3-manifold invariant and maps induced by 4-dimensional cobordisms. For expository overviews of Heegaard Floer homology, see, for example, [40, 72], and [34, SECTION 2].

2.1. Properties and examples

Given a closed, oriented 3-manifold Y, its Heegaard Floer homology $\mathrm{HF}^-(Y)$ is a finitely generated, graded module over $\mathbb{F}[U]$, where $\mathbb{F} = \mathbb{Z}/2\mathbb{Z}$ and U is a formal variable in degree -2. (There are other flavors, $\mathrm{HF}^+(Y)$, $\widehat{\mathrm{HF}}(Y)$ of Heegaard Floer homology, but for the purposes of this article, we will focus on the minus version.)

More precisely, every closed, oriented 3-manifold Y can be described as a union of two handlebodies; such a decomposition is called a *Heegaard splitting*. In turn, a Heegaard splitting can be described via a *Heegaard diagram* $\mathcal{H}$, consisting of a closed, oriented surface Σ of genus g, together with g α-circles and g β-circles, which describe how the handlebodies fill in the surface on either side. (These circles are required to satisfy a certain homological condition.) For technical reasons, we also fix a basepoint z in the complement of the α- and β-circles. Any two diagrams representing the same 3-manifold can be related by a sequence of *Heegaard moves*, as described in [72, SECTION 2.6]; see also [34, SECTION 1].

From this data, Ozsváth–Szabó [70] construct $CF^-(\mathcal{H})$, a free, finitely generated, graded chain complex over $\mathbb{F}[U]$. The variable U keeps track of the basepoint z. The chain homotopy type of $CF^-(\mathcal{H})$ is an invariant of Y, that is, it does not depend on the choice of Heegaard diagram, or on any other choices made in the construction. We often write $CF^-(Y)$ to denote this chain homotopy class, or a representative thereof. Juhász–Thurston–Zemke [41] prove something even stronger: Heegaard Floer homology is natural, in the sense that it assigns a concrete module, rather than an isomorphism class of modules, to a 3-manifold.

Example 2.1. The Heegaard Floer homology of S^3 is $HF^-(S^3) = \mathbb{F}_{(0)}[U]$, where the subscript (0) denotes that $1 \in \mathbb{F}[U]$ is in grading 0. (This can easily be computed using the definition of the Heegaard Floer chain complex.)

Example 2.2. The Heegaard Floer homology of the Brieskorn homology sphere $\Sigma(2,3,7)$ is $HF^-(\Sigma(2,3,7)) = \mathbb{F}_{(0)}[U] \oplus \mathbb{F}_{(0)}$. (This is not so easy to compute directly from the definition of Heegaard Floer homology; however, it is a straightforward consequence of some of the formal properties of Heegaard Floer homology.)

Example 2.3. The Heegaard Floer homology of the Brieskorn homology sphere $\Sigma(2,3,5)$ is $HF^-(\Sigma(2,3,5)) = \mathbb{F}_{(-2)}[U]$. (This can be computed using some of the formal properties of Heegaard Floer homology.)

Remark 2.4. In this article, we take the slightly unconventional grading convention above, which simplifies the formula for gradings in, for example, HF^- of connected sums. Many other sources use the convention that $HF^-(S^3) = \mathbb{F}_{(-2)}[U]$, which simplifies calculations in HF^+.

Remark 2.5. For rational homology spheres, the gradings in Heegaard Floer homology take values in $\mathbb{Q}$. For integer homology spheres, the gradings take values in $2\mathbb{Z}$. For 3-manifolds Y with $H_1(Y;\mathbb{Z})$ infinite, the gradings are slightly more complicated; see [65, SECTION 4.2].

Since the degree of U is -2, any homogenously graded polynomial in $\mathbb{F}[U]$ is of the form U^n for some $n \in \mathbb{N}$. Thus, by the fundamental theorem of finitely generated graded modules over a PID, we have that $HF^-(Y)$ is of the form

$$\bigoplus_{i=1}^{N} \mathbb{F}_{(d_i)}[U] \oplus \bigoplus_{j=1}^{M} \mathbb{F}_{(c_j)}[U]/U^{n_j}\mathbb{F}[U],$$

for Y a rational homology sphere; that is, $HF^-(Y)$ is a direct sum of a free part and a U-torsion part. Ozsváth–Szabó [66, THEOREM 10.1] show that when Y is an integer homology sphere, $N = 1$, that is, $HF^-(Y)$ is of the form

$$HF^-(Y) = \mathbb{F}_{(d)}[U] \oplus \bigoplus_{j=1}^{M} \mathbb{F}_{(c_j)}[U]/U^{n_j}\mathbb{F}[U].$$

The number d above is called the *d-invariant* of Y, denoted $d(Y)$. More generally, when Y is a rational homology sphere with $|H_1(Y;\mathbb{Z})| = k$, there are exactly k free summands in $\mathrm{HF}^-(Y)$, in which case one obtains a k-tuple of d-invariants of Y.

Heegaard Floer homology satisfies a Künneth-type formula under connected sums [66, THEOREM 1.5]. That is, $\mathrm{CF}^-(Y_1 \# Y_2)$ is chain homotopy equivalent to

$$\mathrm{CF}^-(Y_1) \otimes_{\mathbb{F}[U]} \mathrm{CF}^-(Y_2).$$

In particular, if Y_1 and Y_2 are integer homology spheres, then the d-invariant is additive under connected sum. (An analogous statement also holds for more general 3-manifolds.)

2.2. Cobordism maps

Heegaard Floer homology is a $(3+1)$-dimensional topological quantum field theory (TQFT) [71, THEOREM 1.1]. That is, to a 3-manifold, Heegaard Floer homology associates a module, and to a 4-manifold cobordism W from Y_0 to Y_1, it associates a chain map

$$F_W : \mathrm{CF}^-(Y_0) \to \mathrm{CF}^-(Y_1).$$

When W is a homology cobordism, F_W induces an isomorphism

$$(F_W \otimes \mathrm{id})_* : U^{-1}\mathrm{HF}^-(Y_0) \to U^{-1}\mathrm{HF}^-(Y_1),$$

where $U^{-1}\mathrm{HF}^-(Y) = H_*(\mathrm{CF}^-(Y) \otimes_{\mathbb{F}[U]} \mathbb{F}[U, U^{-1}])$ [65, PROOF OF THEOREM 9.1]. A straightforward algebra calculation then implies that d-invariants are invariants of homology cobordism. In particular, we have a homomorphism

$$d : \Theta^3_{\mathbb{Z}} \to 2\mathbb{Z},$$

and this homomorphism is surjective, since $d(\Sigma(2,3,5)) = -2$.

In light of the discussion above, and motivated by the desire to study the homology cobordism group $\Theta^3_{\mathbb{Z}}$, one could define an equivalence relation $\sim$, called *local equivalence*, on Heegaard Floer chain complexes, where $\mathrm{CF}^-(Y_0) \sim \mathrm{CF}^-(Y_1)$ if there exist $\mathbb{F}[U]$-module chain maps

$$f : \mathrm{CF}^-(Y_0) \to \mathrm{CF}^-(Y_1) \quad \text{and} \quad g : \mathrm{CF}^-(Y_1) \to \mathrm{CF}^-(Y_0),$$

inducing isomorphisms on $U^{-1}\mathrm{HF}^-(Y_i)$. We can now consider the group

$$\mathfrak{D} = \{\mathrm{CF}^-(Y) \mid Y \text{ an integer homology sphere}\}/\sim$$

under the operation induced by tensor product. This construction yields a homomorphism

$$\Theta^3_{\mathbb{Z}} \to \mathfrak{D}$$

obtained by sending $[Y]$ to $[\mathrm{CF}^-(Y)]$. However, it turns out that $\mathfrak{D}$ is isomorphic to $\mathbb{Z}$, with the isomorphism being given by $[\mathrm{CF}^-(Y)] \mapsto d(Y)/2$.

2.3. Ribbon homology cobordisms

Ribbon homology cobordisms induce particularly nice maps on Heegaard Floer homology:

Theorem 2.6 ([8, THEOREM 1.19]). *Let W be a ribbon homology cobordism from Y_0 to Y_1. Then*

$$F_W : \mathrm{HF}^-(Y_0) \to \mathrm{HF}^-(Y_1)$$

is injective, and includes $\mathrm{HF}^-(Y_0)$ *into* $\mathrm{HF}^-(Y_1)$ *as a direct summand.*

The proof of Theorem 2.6 relies on considering the *double* $D(W)$ of W, formed by gluing W to $-W$ along Y_1; they also prove that the analogous statement holds for ribbon $\mathbb{Z}/2\mathbb{Z}$-homology cobordisms. Their approach was inspired by [88].

3. KNOT FLOER HOMOLOGY

In this section, we will discuss the Heegaard Floer knot invariant, maps induced by concordances, and various concordance invariants arising from the knot Floer complex. For expository overviews of knot Floer homology, see [72, SECTION 10], [37, 54], [29, SECTION 2], and [34, SECTION 3]. Note that [34] uses the same notation and conventions used here (i.e., viewing the knot Floer complex as a module over a two-variable polynomial ring), while the others use a different but equivalent formulation in terms of filtered chain complexes.

3.1. Properties and examples

For simplicity, we will focus on knots in integer homology spheres. Let K be a knot in an integer homology sphere Y. We can describe the pair (Y, K) via a *doubly pointed Heegaard diagram* $\mathcal{H}$, which consists of a Heegaard diagram for Y with an extra basepoint w. The knot K is the union of two arcs, specified by connecting the basepoint w to the basepoint z in the complement of the α-arcs, pushed slightly into one handlebody, and connecting z to w in the complement of the β-arcs, pushed slightly into the other handlebody. See, for example, [34, SECTION 1] for more details.

From this data, Ozsváth–Szabo [70] and independently J. Rasmussen [78] construct a chain complex $\mathrm{CFK}(\mathcal{H})$. One way of constructing this chain complex (see, for example, [89, SECTION 1.5]) is as a free, finitely generated, bigraded chain complex over $\mathbb{F}[U, V]$, the second formal variable V corresponding to the second basepoint w. As one would hope, the chain homotopy type of $\mathrm{CFK}(\mathcal{H})$ is an invariant of the pair (Y, K), and does not depend on the choice of diagram, or on any of the other choices made in the construction. We often write $\mathrm{CFK}(Y, K)$, or simply $\mathrm{CFK}(K)$ when $Y = S^3$, to denote this chain homotopy class, or a representative thereof. Moreover, like the 3-manifold version, knot Floer homology is natural [41].

Example 3.1. The knot Floer complex of the unknot in S^3 is generated over $\mathbb{F}[U, V]$ by a single generator x in bigrading $(0, 0)$ with trivial differential. (This can be computed directly from the definition of the knot Floer chain complex.)

Example 3.2. The knot Floer complex of the right-handed trefoil is generated over $\mathbb{F}[U, V]$ by a, b, and c with the following differentials and bigradings:

	∂	gr
a	0	$(0, -2)$
b	$Ua + Vc$	$(-1, -1)$
c	0	$(-2, 0)$

(This can be computed directly from the definition of the knot Floer chain complex.)

There is not a simple characterization of finitely-generated, graded modules over $\mathbb{F}[U, V]$, since $\mathbb{F}[U, V]$ is not a PID. One way to obtain a module over a PID is to set $V = 0$ on the chain level. (There is a symmetry between U and V, so one could instead choose to set $U = 0$.) Taking homology of the resulting chain complex, we obtain a version of knot Floer homology, namely $\mathrm{HFK}^-(Y, K)$ given by

$$\mathrm{HFK}^-(Y, K) = H_*\big(\mathrm{CFK}(Y, K)/(V = 0)\big).$$

If we prefer an even simpler algebraic structure, we can set both U and V equal to zero on the chain level. Taking homology of the resulting chain complex, we obtain another version of knot Floer homology, denoted $\widehat{\mathrm{HFK}}(Y, K)$, namely

$$\widehat{\mathrm{HFK}}(Y, K) = H_*\big(\mathrm{CFK}(Y, K)/(U = V = 0)\big).$$

Using a suitable renormalized bigrading, this is the version of knot Floer homology whose graded Euler characteristic is the Alexander polynomial.

Like the 3-manifold invariant, the knot Floer complex satisfies a Künneth-type formula under connected sums [66, THEOREM 1.5]. That is, $\mathrm{CFK}(Y_1 \# Y_2, K_1 \# K_2)$ is chain homotopy equivalent to

$$\mathrm{CFK}(Y_1, K_1) \otimes_{\mathbb{F}[U,V]} \mathrm{CFK}(Y_2, K_2).$$

3.2. Maps induced by concordances

Knot Floer homology also behaves nicely under cobordisms. Consider a cobordism (W, S) from (Y_0, K_0) to (Y_1, K_1), that is, W is a 4-manifold cobordism from Y_1 to Y_2 and $S \subset W$ a properly embedded connected surface with boundary $-K_0 \sqcup K_1$. The pair (W, S) induces a module homomorphism

$$F_{W,S} : \mathrm{CFK}(Y_0, K_0) \to \mathrm{CFK}(Y_1, K_1).$$

When W is a homology cobordism and S is an annnulus, $F_{W,S}$ induces an isomorphism

$$(F_{W,S} \otimes \mathrm{id})_* : (U, V)^{-1}\,\mathrm{HFK}(Y_0, K_0) \xrightarrow{\cong} (U, V)^{-1}\,\mathrm{HFK}(Y_1, K_1),$$

where $(U, V)^{-1}\,\mathrm{HFK}(Y_0, K_0) = H_*(\mathrm{CFK}(Y, K) \otimes_{\mathbb{F}[U,V]} \mathbb{F}[U, U^{-1}, V, V^{-1}]$ [89, THEOREM 1.7]. When $W = S^3 \times [0, 1]$, we may simply write F_S instead of $F_{W,S}$.

Using this additional structure, we define an equivalence relation on these chain complexes that is well suited to studying the knot concordance group, and more generally, concordances in homology cobordisms.

Definition 3.3. Two knot Floer complexes, $\mathrm{CFK}(Y_0, K_0)$ and $\mathrm{CFK}(Y_1, K_1)$, are *locally equivalent*, denoted $\mathrm{CFK}(Y_0, K_0) \sim \mathrm{CFK}(Y_1, K_1)$, if there exist $\mathbb{F}[U, V]$-module chain maps

$$f : \mathrm{CFK}(Y_0, K_0) \to \mathrm{CFK}(Y_1, K_1) \quad \text{and} \quad g : \mathrm{CFK}(Y_1, K_1) \to \mathrm{CFK}(Y_0, K_0),$$

inducing isomorphisms on $(U, V)^{-1}\,\mathrm{HFK}(Y_i, K_i)$.

Remark 3.4. In the literature, this equivalence relation is also referred to as ν^+-equivalence [43] and stable equivalence [37].

As in the 3-manifold case above, we can now consider the group

$$\mathfrak{C} = \{\mathrm{CFK}(Y, K) \mid Y \text{ a } \mathbb{Z}HS^3 \text{ bounding a } \mathbb{Z}HB^4,\ K \text{ a knot in } Y\}/\sim$$

under the operation induced by tensor product, giving us a homomorphism

$$\hat{\mathcal{C}}_{\mathbb{Z}} \to \mathfrak{C}$$

obtained by sending $[(Y, K)]$ to $[\mathrm{CFK}(Y, K)]$. (We require Y to bound a $\mathbb{Z}HB^4$ to parallel the definition of $\hat{\mathcal{C}}_{\mathbb{Z}}$.) By precomposing with the map $\mathcal{C} \to \hat{\mathcal{C}}_{\mathbb{Z}}$, we obtain a homomorphism $\mathcal{C} \to \mathfrak{C}$.

The group $\mathfrak{C}$ is not easy to study. One way to obtain a simpler algebraic structure is to set $V = 0$, in which case we can run the analogous construction, that is, we can consider the group

$$\mathfrak{C}' = \{\mathrm{CFK}(Y, K)/(V = 0) \mid Y \text{ a } \mathbb{Z}HS^3 \text{ bounding a } \mathbb{Z}HB^4,\ K \text{ a knot in } Y\}/\sim$$

where $\mathrm{CFK}(Y_0, K_0)/(V = 0) \sim \mathrm{CFK}(Y_1, K_1)/(V = 0)$ if there exist $\mathbb{F}[U]$-module chain maps

$$f : \mathrm{CFK}(Y_0, K_0)/(V = 0) \to \mathrm{CFK}(Y_1, K_1)/(V = 0),$$
$$g : \mathrm{CFK}(Y_1, K_1)/(V = 0) \to \mathrm{CFK}(Y_0, K_0)/(V = 0),$$

inducing isomorphisms on $U^{-1}\,\mathrm{HFK}^-(Y_i, K_i)$; we call this equivalence relation *local equivalence mod V*. As in the 3-manifold case, this group $\mathfrak{C}'$ is isomorphic to $\mathbb{Z}$; indeed, up to renormalization, this construction yields the Ozsváth–Szabó τ-invariant [67] (see also [73, APPENDIX A]).

In a case of mathematical Goldilocks, working over the full ring $\mathbb{F}[U, V]$ yields a group that is too complicated to study, while working over the ring $\mathbb{F}[U] = \mathbb{F}[U, V]/(V = 0)$ yields a group that is too simple. Somewhat miraculously, it turns out that working over the ring $\mathbb{F}[U, V]/(UV = 0)$ is just right, at least for knots in S^3. The main idea is that although $\mathbb{F}[U, V]/(UV = 0)$ has zero-divisors, it is somehow closer to being a PID than $\mathbb{F}[U, V]$ is. Furthermore, for knots in S^3, the local equivalence group mod UV is totally ordered, as we now describe.

Definition 3.5. We say that

$$\mathrm{CFK}(K_1) \leq \mathrm{CFK}(K_2)$$

if there exists an $\mathbb{F}[U,V]/(UV=0)$-module chain map

$$f : \mathrm{CFK}(K_1) \to \mathrm{CFK}(K_2)$$

such that f induces an isomorphism on $H_*(\mathrm{CFK}(K_i)/U)/V$-torsion. If

$$\mathrm{CFK}(K_1) \leq \mathrm{CFK}(K_2) \quad \text{and} \quad \mathrm{CFK}(K_2) \leq \mathrm{CFK}(K_1),$$

then we say that $\mathrm{CFK}(K_1)$ and $\mathrm{CFK}(K_2)$ are *locally equivalent mod UV*.

Remark 3.6. Note that since U and V are both zero-divisors in $\mathbb{F}[U,V]/(UV)$, we cannot invert them directly. Also note that requiring f to induce an isomorphism on the module $H_*(\mathrm{CFK}(K_i)/U)/V$-torsion is ever so slightly stronger than requiring f to induce an isomorphism on $V^{-1}H_*(\mathrm{CFK}(K_i)/U)$.

This is the same total order as that induced by ε in [36]. Indeed, one way to define the $\{-1,0,1\}$-valued concordance invariant ε [35] (see also [11, SECTION 3]) is as follows:

- $\varepsilon(K) \leq 0$ if and only if $\mathrm{CFK}(K) \leq \mathbb{F}[U,V]$,
- $\varepsilon(K) \geq 0$ if and only if $\mathrm{CFK}(K) \geq \mathbb{F}[U,V]$.

In particular, $\varepsilon(K) = 0$ if and only if $\mathbb{F}[U,V] \leq \mathrm{CFK}(K) \leq \mathbb{F}[U,V]$.

For knots in S^3, we are able to characterize their knot Floer complexes, up to local equivalence mod UV:

Theorem 3.7 ([11, THEOREM 1.3]). *Let K be a knot in S^3. The knot Floer complex of K is locally equivalent mod UV to a standard complex, which can be represented by a finite sequence of nonzero integers. Moreover, if we endow the integers with the following unusual order:*

$$-1 <^! -2 <^! -3 <^! \cdots <^! 0 <^! \cdots <^! 3 <^! 2 <^! 1,$$

then local equivalence classes mod UV are ordered lexicographically with respect to their standard representatives.

See Section 4 of [11] for the definition of a standard complex. This characterization of knot Floer complexes up to local equivalence mod UV is a key step in the definition of the linearly independent family of concordance homomorphisms

$$\varphi_i : \mathcal{C} \to \mathbb{Z}, \quad i \in \mathbb{N}$$

from [11]. The main idea is that $\varphi_i(K)$ is the signed count of the number of times that i appears in the sequence of integers parametrizing the local equivalence class mod UV of $\mathrm{CFK}(K)$. (For symmetry reasons, we actually only consider every other term in the sequence; see [11, SECTION 7] for more details.)

For knots in S^3, we have the following relationships between τ, ε, and φ_i:

Theorem 3.8 ([11, PROPOSITION 1.2], [36, PROPOSITION 3.2]). *Let K be a knot in S^3.*

(1) *If $\varepsilon(K) = 0$, then $\varphi_i(K) = 0$ for all i.*

(2) *The invariant τ is equal to*

$$\tau(K) = \sum_{i=1}^{\infty} i\varphi_i(K).$$

In particular, if $\varepsilon(K) = 0$, then $\tau(K) = 0$.

The invariant ε can be generalized to a homology concordance invariant [39, SECTION 4], which behaves like a sign under connected sum, in the sense that

- if $\varepsilon(Y_1, K_1) = \varepsilon(Y_2, K_2)$, then $\varepsilon(Y_1 \# Y_2, K_1 \# K_2) = \varepsilon(Y_1, K_1)$, and
- if $\varepsilon(Y_1, K_1) = 0$, then $\varepsilon(Y_1 \# Y_2, K_1 \# K_2) = \varepsilon(Y_2, K_2)$.

The proof of Theorem 1.8(2) relies on using the filtered mapping cone of [28] to produce a manifold–knot pair (Y, K) with $\varepsilon(Y, K) = 0$ and $\tau(Y, K) = 1$. By Theorem 3.8, we know that if a knot K in a manifold Y is homology concordant to any knot J in S^3 and $\varepsilon(Y, K) = 0$, then $\tau(Y, K) = 0$. Hence $K \subset Y$ is not homology concordant to any knot in S^3. Moreover, since τ is a concordance homomorphism, it follows that any nonzero multiple of (Y, K) has $\varepsilon = 0$ and τ nonzero, hence cannot be homology concordant to any knot in S^3.

Remark 3.9. Note that $\varepsilon(K)$, $\tau(K)$, $\varphi_i(K)$, and, more generally, the local equivalence class of $\mathrm{CFK}(K)$ are all invariant under concordances in rational homology cobordisms. In particular, they all vanish for rationally slice knots. Thus, in order to study $\mathcal{C}_{\mathbb{Q}S}$, the group of rationally slice knots, we will need additional structure, as discussed in Section 5.

3.3. Ribbon concordances

As in the case for 3-manifolds, ribbon concordances induce particularly nice maps on the knot Floer complex:

Theorem 3.10 ([88, THEOREM 1.7]). *Let S be a ribbon concordance from K_0 to K_1 in $S^3 \times [0,1]$, and let S' denote the concordance obtained by reversing S. Then*

$$F_{S'} \circ F_S : \mathrm{CFK}(K_0) \to \mathrm{CFK}(K_0)$$

is chain homotopic to the identity, via an $\mathbb{F}[U,V]$-equivariant chain homotopy. In particular, if S is a ribbon concordance from K_0 to K_1, then $\widehat{\mathrm{HFK}}(K_0)$ is a direct summand of $\widehat{\mathrm{HFK}}(K_1)$ and $\mathrm{HFK}^-(K_0)$ is a direct summand of $\mathrm{HFK}^-(K_1)$.

Since knot Floer homology detects the knot genus $g(K)$ [68], an immediate consequence of the above theorem is that if there is a ribbon concordance from K_0 to K_1 then $g(K_0) \leq g(K_1)$ [88, THEOREM 1.3].

4. INVOLUTIVE HEEGAARD FLOER HOMOLOGY

We would like to use the Heegaard Floer package to study homology cobordism. In light of Theorem 3.7, we see that a richer algebraic structure, namely, chain complexes over a more complicated ring than $\mathbb{F}[U]$, can give us richer invariants. Fortunately for us, Hendricks and Manolescu [32] endowed the Heegaard Floer chain complex with the additional structure of a homotopy involution ι. Very roughly, this additional data lets us think of the Heegaard Floer chain complex as a module over (a quotient of) a two-variable polynomial ring, allowing us to employ the techniques used in the proof of Theorem 3.7 to define an infinite family of $\mathbb{Z}$-valued homology cobordism homomorphisms. These homomorphisms lead to the proof of Theorem 1.1. Furthermore, the characterization of such chain complexes up to a suitable notion of local equivalence is a key ingredient in the proof of Theorem 1.3.

4.1. Properties and examples

Recall that in the construction of Heegaard Floer homology, we specify our 3-manifold Y via a pointed Heegaard diagram $\mathcal{H} = (\Sigma, \boldsymbol{\alpha}, \boldsymbol{\beta}, z)$, where Σ is a closed, oriented surface of genus g, and $\boldsymbol{\alpha}$ and $\boldsymbol{\beta}$ are each a collection of g disjoint embedded circles in Σ. Reversing the orientation of Σ reverses the orientation of Y, as does reversing the roles of the α- and β-circles. In particular, the Heegaard diagram $\overline{\mathcal{H}} = (-\Sigma, \boldsymbol{\beta}, \boldsymbol{\alpha}, z)$ describes the same manifold as $\mathcal{H}$, namely Y. Thus, there is a sequence of Heegaard moves taking $\overline{\mathcal{H}}$ to $\mathcal{H}$, inducing an $\mathbb{F}[U]$-equivariant chain map

$$\Phi_{\overline{\mathcal{H}},\mathcal{H}} : \mathrm{CF}^-(\overline{\mathcal{H}}) \to \mathrm{CF}^-(\mathcal{H});$$

this chain map is well defined since Heegaard Floer homology is natural [41]. There is also a canonical $\mathbb{F}[U]$-equivariant chain complex isomorphism

$$\eta : \mathrm{CF}^-(\mathcal{H}) \to \mathrm{CF}^-(\overline{\mathcal{H}})$$

given by the "obvious" identification of the generators of the two chain complexes. Hendricks and Manolescu show that $\iota = \Phi_{\overline{\mathcal{H}},\mathcal{H}} \circ \eta$ is a homotopy involution (that is, $\iota^2 \simeq \mathrm{id}$) and prove that for a homology sphere Y, the chain homotopy type of the pair $(\mathrm{CF}^-(\mathcal{H}), \iota)$ is an invariant of Y [32, PROOF OF PROPOSOTION 2.7]; we will write $(\mathrm{CF}^-(Y), \iota)$, called the *$\iota$-complex of Y*, to denote a representative of this equivalence class. (An analogous statement holds for a general 3-manifold equipped with a self-conjugate spin^c-structure; for ease of exposition, we have chosen to focus on homology spheres to eliminate the need to discuss spin^c-structures.)

Example 4.1. The ι-complex of S^3 is $(\mathbb{F}[U], \mathrm{id})$. (The map ι is uniquely determined by the fact that $\iota^2 \simeq \mathrm{id}$.)

Example 4.2. The ι-complex of $\Sigma(2, 3, 7)$ is generated over $\mathbb{F}[U]$ by a, b, and c with ∂, ι, and gradings as follows:

	∂	ι	gr
a	0	c	0
b	$Ua + Uc$	b	-1
c	0	a	0

(The map ι can be computed by realizing $-\Sigma(2,3,7)$ as $+1$-surgery on the left-handed trefoil. See [32, SECTION 6.8].)

We have the following Künneth-type formula for ι-complexes of connected sums [33, THEOREM 1.1]:

$$\big(\mathrm{CF}^-(Y_1\#Y_2), \iota_{1\#2}\big) \simeq \big(\mathrm{CF}^-(Y_1) \otimes_{\mathbb{F}[U]} \mathrm{CF}^-(Y_2), \iota_1 \otimes \iota_2\big),$$

where ι_1 (respectively ι_2) denotes the homotopy involution on Y_1 (respectively Y_2) and $\iota_{1\#2}$ denotes the homotopy involution on $Y_1\#Y_2$.

4.2. Cobordism maps

The map ι behaves nicely with respect to cobordism maps. For expositional simplicity, we will focus on homology cobordisms. (One can also consider general cobordisms with a conjugacy class of spin^c-structures.) A homology cobordism W from Y_0 to Y_1 induces a chain map

$$F_W : \mathrm{CF}^-(Y_1) \to \mathrm{CF}^-(Y_2),$$

which commutes with ι, up to homotopy [32, PROOF OF PROPOSITION 4.9]:

$$F_W \circ \iota_1 \simeq \iota_2 \circ F_W.$$

In particular, we can define a refined version of local equivalence as follows:

Definition 4.3. Two ι-complexes (C_1, ι_1) and (C_2, ι_2) are *ι-locally equivalent*, denoted $(C_1, \iota_1) \sim (C_2, \iota_2)$, if there exist $\mathbb{F}[U]$-module chain maps

$$f : \mathrm{CF}^-(Y_0) \to \mathrm{CF}^-(Y_1) \quad \text{and} \quad g : \mathrm{CF}^-(Y_1) \to \mathrm{CF}^-(Y_0),$$

inducing isomorphisms on $U^{-1}\,\mathrm{HF}^-(Y_i)$, such that

$$f \circ \iota_1 \simeq \iota_2 \circ f \quad \text{and} \quad g \circ \iota_2 \simeq \iota_1 \circ g.$$

We can now consider the group

$$\mathfrak{I} = \big\{\big(\mathrm{CF}^-(Y), \iota\big) \mid Y \text{ an integer homology sphere}\big\}/\sim$$

under the operation induced by tensor product. This construction yields a homomorphism

$$\Theta^3_{\mathbb{Z}} \to \mathfrak{I}$$

obtained by sending $[Y]$ to $[(\mathrm{CF}^-(Y), \iota)]$. The additional requirement that the local equivalences homotopy commute with ι makes this group more interesting than before. However, this group is almost too interesting, in the sense that it is very difficult to understand.

As in the knot case above, a certain algebraic simplification allows us to characterize elements in this group, up to an ever so slightly weaker notion of equivalence. The main idea is to append "mod U" to every statement in Definition 4.3 involving ι.

Definition 4.4. Two ι-complexes (C_1, ι_1) and (C_2, ι_2) are *almost ι-locally equivalent*, if there exist $\mathbb{F}[U]$-module chain maps

$$f : \mathrm{CF}^-(Y_0) \to \mathrm{CF}^-(Y_1) \quad \text{and} \quad g : \mathrm{CF}^-(Y_1) \to \mathrm{CF}^-(Y_0),$$

inducing isomorphisms on $U^{-1}\,\mathrm{HF}^-(Y_i)$, such that

$$f \circ \iota_1 \simeq \iota_2 \circ f \bmod U \quad \text{and} \quad g \circ \iota_2 \simeq \iota_1 \circ g \bmod U.$$

Similarly, we can relax the definition of an ι-complex so as to only require that $\iota^2 \simeq \mathrm{id} \bmod U$; we will call such a complex an *almost ι-complex.*[2] Almost ι-local equivalence classes of almost ι-complexes are totally ordered, with

$$(C_1, \iota_1) \leq (C_2, \iota_2),$$

if there exists an $\mathbb{F}[U]$-module chain map $f : C_1 \to C_2$, inducing an isomorphism on $H_*(C_i)/U$-torsion, such that $f \circ \iota_1 \simeq \iota_2 \circ f \bmod U$.

Theorem 4.5 ([9, THEOREM 6.2]). *Every ι-complex is almost ι-locally equivalent to a standard complex, which can be represented by a finite sequence of the form $(a_i, b_i)_{i=1}^n$ where $a_i \in \{\pm 1\}$ and $b_i \in \mathbb{Z} \setminus \{0\}$. Moreover, if we endow the integers with the following unusual order:*

$$-1 <^! -2 <^! -3 <^! \cdots <^! 0 <^! \cdots <^! 3 <^! 2 <^! 1,$$

then almost ι-local equivalence classes are ordered lexicographically with respect to their standard representatives.

The astute reader may notice that Theorem 4.5 looks very similar to Theorem 3.7. Indeed, the idea is that ι-complexes can roughly be thought of as chain complexes over $\mathbb{F}[U, Q]/(Q^2)$, which are then very similar to chain complexes over $\mathbb{F}[U, V]$. Analogously, we can define a linearly independent family of homology cobordism homomorphisms

$$\phi_i : \Theta^3_{\mathbb{Z}} \to \mathbb{Z}, \quad i \in \mathbb{N}.$$

These homomorphisms can be used to show that the Brieskorn homology spheres $\Sigma(2j+1, 4j+1, 4j+3)$ span a free infinite rank subgroup of $\Theta^3_{\mathbb{Z}}$, proving Theorem 1.1. Rostovtsev [81] gives an alternate proof of Theorem 4.5 and extends our result to define an additional, linearly independent integer-valued homology cobordism homomorphism.

Let $\hat{\mathfrak{I}}$ denote the group of almost ι-complexes up to almost ι-local equivalence. We have the homomorphism

$$\hat{h} : \Theta^3_{\mathbb{Z}} \to \hat{\mathfrak{I}}$$

2 Here, we use the word "almost" to denote that any statement regarding ι should be taken mod U. In the next section, we use the word "almost" to denote that any statement regarding ι_K should be taken mod (U, V).

defined by sending $[Y]$ to $[(\mathrm{CF}^-(Y), \iota)]$. We now describe the main ideas behind the proof of Theorem 1.3, which states that Seifert fibered spaces do not generate $\Theta^3_{\mathbb{Z}}$. Let Θ_{SF} denote the subgroup of $\Theta^3_{\mathbb{Z}}$ generated by Seifert fibered spaces. In [9, SECTION 8.1], we determine $\hat{h}(\Theta_{SF})$, the image of Θ_{SF} in $\hat{\mathfrak{I}}$. Recall that elements of $\hat{\mathfrak{I}}$ are finite sequences. Elements in $\hat{h}(\Theta_{SF})$ are exactly the sequences satisfying a certain monotonicity condition on their terms; see [9, THEOREM 8.1] for the precise statement. We then use the involutive surgery formula [30, THEOREM 1.6] to determine the ι-complexes of surgeries on a family of connected sums of torus knots and iterated cables. The sequences associated to these surgeries do not satisfy the monotonicity condition of $\hat{h}(\Theta_{SF})$. A similar statement applies to linear combinations of these surgeries, giving Theorem 1.3.

5. INVOLUTIVE KNOT FLOER HOMOLOGY

In the previous section, we put additional structure, namely the homotopy involution ι, on the Heegaard Floer chain complex $\mathrm{CF}^-(Y)$. In this section, we put additional structure, namely a skew-graded, skew-equivariant (i.e., interchanges the actions of U and V) chain map ι_K, on the knot Floer chain complex $\mathrm{CFK}(K)$. The map ι_K is not a homotopy involution; rather, Hendricks–Manolescu [32, SECTION 6.2] show that ι_K squares to be homotopic to the Sarkar map [82], which is induced by moving the basepoints w and z once around the knot K. The map ι_K is the additional structure alluded to in Remark 3.9.

5.1. Properties and examples

The map ι_K is defined in a similar way to ι. Consider a doubly-pointed Heegaard diagram $\mathcal{H} = (\Sigma, \boldsymbol{\alpha}, \boldsymbol{\beta}, z, w)$ for a knot K in an integer homology sphere Y. (With minor modifications, these constructions also work for null-homologous knots in any 3-manifold.) The doubly-pointed Heegaard diagram $\overline{\mathcal{H}} = (-\Sigma, \boldsymbol{\beta}, \boldsymbol{\alpha}, w, z)$ also describes $K \subset Y$, and thus there is a sequence of Heegaard moves from $\overline{\mathcal{H}}$ to $\mathcal{H}$, inducing a $\mathbb{F}[U,V]$-equivariant chain map

$$\Phi_{\overline{\mathcal{H}},\mathcal{H}} : \mathrm{CFK}(\overline{\mathcal{H}}) \to \mathrm{CFK}(\mathcal{H}).$$

(Care should be taken with regards to the basepoints; see [32, SECTION 6.2] for details.) There is also a canonical skew-equivariant isomorphism

$$\eta_K : \mathrm{CFK}(\mathcal{H}) \to \mathrm{CFK}(\overline{\mathcal{H}}),$$

given by the "obvious" identification of the generators of the two chain complexes. Then ι_K is defined to be $\Phi_{\overline{\mathcal{H}},\mathcal{H}} \circ \eta_K$. The chain homotopy type of the pair $(\mathrm{CFK}(\mathcal{H}), \iota_K)$ is an invariant of the knot K in Y [32, PROPOSITION 6.3]. As usual, we write $(\mathrm{CFK}(K), \iota_K)$ to denote a representative of this chain homotopy equivalence class, and we call the pair $(\mathrm{CFK}(K), \iota_K)$ an *ι_K-complex.*

Example 5.1. The ι_K-complex of the unknot in S^3 is $(\mathbb{F}[U,V], \mathrm{id})$. (The map ι_K is uniquely determined by the fact that it squares to be homotopic to the Sarkar map.)

Example 5.2. The ι_K-complex of the right-handed trefoil is described by

	∂	ι_K	gr
a	0	c	$(0,-2)$
b	$Ua+Vc$	b	$(-1,-1)$
c	0	a	$(-2,0)$

(The map ι_K is uniquely determined by the fact that it is skew-graded and squares to be homotopic to the Sarkar map.)

There is a Künneth-type formula for ι_K-complexes [87, THEOREM 1.1],

$$\begin{aligned}&\big(\mathrm{CFK}(Y_1\#Y_2, K_1\#K_2), \iota_{K_1\#K_2}\big)\\ &\quad\simeq \big(\mathrm{CFK}(K_1)\otimes \mathrm{CFK}(K_2), \iota_{K_1}\otimes\iota_{K_2} + (\Phi\otimes\Psi)\circ(\iota_{K_1}\otimes\iota_{K_2})\big).\end{aligned}$$

See [87, SECTION 4.2] for the definitions of Φ and Ψ; for an expository overview, see [38, SECTION 2].

5.2. Maps induced by concordances

Let (W,S) be a $\mathbb{Z}/2\mathbb{Z}$-homology cobordism from (Y_0,K_0) to (Y_1,K_1), that is, W is a $\mathbb{Z}/2\mathbb{Z}$-homology cobordism from Y_0 to Y_1 and S is a concordance from K_0 to K_1. (More generally, one can consider spin cobordisms; see [32].) As one would hope, the module homomorphism $F_{W,S}$ induced by (W,S) behaves nicely with respect to ι_K [87, THEOREM 1.3] in the sense that

$$F_{W,S}\circ\iota_{K_0}\simeq\iota_{K_1}\circ F_{W,S}.$$

Based on previous sections, it may now be apparent to the reader what we do next. We jump straight to the definition of almost ι_K-local equivalence; the definition of ι_K-local equivalence can be obtained by striking out both instances of "mod (U,V)" in the definition below.

Definition 5.3. Two ι_K-complexes (C_1,ι_1) and (C_2,ι_2) are *almost ι_K-locally equivalent* if there exist $\mathbb{F}[U,V]$-module chain maps

$$f:\mathrm{CFK}(Y_0,K_0)\to\mathrm{CFK}(Y_1,K_1)\quad\text{and}\quad g:\mathrm{CFK}(Y_1,K_1)\to\mathrm{CFK}(Y_0,K_0),$$

inducing isomorphisms on $(U,V)^{-1}\,\mathrm{HFK}(Y_i,K_i)$, such that

$$f\circ\iota_{K_0}\eqsim\iota_{K_1}\circ f \bmod (U,V)\quad\text{and}\quad g\circ\iota_{K_1}\eqsim\iota_{K_0}\circ g \bmod (U,V),$$

where $\eqsim$ denotes skew-equivariant homotopy equivalence.

We can now consider $\hat{\mathfrak{I}}_K$, the group of ι_K-complexes modulo almost ι_K-local equivalence, with the operation induced by tensor product. Note that this group has 2-torsion, generated by, for example, the figure-eight knot, and hence this group is not totally ordered. In particular, there are rationally slice knots, such as the figure-eight, with nontrivial image in $\hat{\mathfrak{I}}_K$; this is possible because the (almost) local equivalence class of a knot is an invariant of concordances in $\mathbb{Z}/2\mathbb{Z}$-homology cobordisms, rather than $\mathbb{Q}$-homology cobordisms, as is the case in the noninvolutive setting. Thus, this toolkit is particularly well equipped for studying rationally slice knots.

The proof of Theorem 1.9 relies on finding a linearly independent family of rationally slice knots. The knots under consideration are K_n, the $(2n+1,1)$-cable of the figure-eight; these knots are rationally slice because the figure-eight is rationally slice, and thus its $(2n+1,1)$-cable is rationally concordant (i.e., concordant in a rational homology cobordism) to the $(2n+1,1)$-torus knot, which is the unknot.

We compute the almost ι_K-local equivalence class of $(\mathrm{CFK}(K_n), \iota_K)$ using bordered Floer homology [26, 50], in particular, its applications to cables [27, 77], together with formal properties of ι_K, such as the fact that it squares to be homotopic to the Sarkar map. With this computation in hand, we observe that there is certain structure in $(\mathrm{CFK}(K_n), \iota_K)$ (roughly, a particular $\mathbb{F}[U]/U^n$ summand in $\mathrm{HFK}^-(K_n)$) which, using the formula for connected sums of ι_K-complexes and properties of almost ι_K-local equivalences, allows us to determine that the K_n are linearly independent.

6. WHAT NEXT?

As we have demonstrated, the Heegaard Floer package can answer a range of questions in low-dimensional topology. Do these techniques have the potential to answer any of the open questions from Section 1?

Question 1.2 asks whether $\Theta^3_{\mathbb{Z}}$ contains any torsion. The most likely torsion is order two, generated by a homology sphere admitting an orientation-reversing self-diffeomorphism. There are many constructions for building homology spheres with orientation-reversing self-diffeomorphisms (for example, the double-branched cover of an amphichiral knot with determinant one, or the splice of a knot complement with that of its mirror), but so far, there has been no success in obstructing such an example from being homology cobordant to S^3. One can consider an algebraic version of the local equivalence group $\mathfrak{I}$, by considering all ι-complexes (not just those known to be realized by a 3-manifold) modulo local equivalence. This algebraic group is known to have two-torsion; the difficulty lies in finding a 3-manifold that realizes such an algebraic example. At present, computations of ι-complexes are limited to certain special families of manifolds (e.g., Seifert fibered spaces, surgeries on knots in S^3); we hope to improve this shortcoming in the future.

Question 1.4 asks whether surgeries on knots in S^3 generate $\Theta^3_{\mathbb{Z}}$. This seems like a hard question to answer with Heegaard Floer homology, as the question about which ι-complexes can be realized by surgery on a knot in S^3 then reduces to the question of which ι_K-complexes can be realized by knots in S^3. Even without the additional structure of ι_K, this is a difficult question; for some partial answers, see [29], as well as more recent progress in [4, 44, 62].

As for Questions 1.5 and 1.6, which ask for knots that are not slice in B^4 but are slice in a homology B^4 or a homotopy B^4, respectively, it seems unlikely that Heegaard

Floer homology will be able to provide an answer. Indeed, with the current technology, if the Heegaard Floer package obstructs a knot from being slice in B^4, then it also obstructs the knot from being slice in a homology or homotopy B^4. There are other invariants which may be able to shed light on this question. For example, at present, it remains open whether or not the Rasmussen s-invariant [79], defined using the Lee [46] deformation of Khovanov homology [42] (see [5] for an expository overview), vanishes for knots that are slice in a homology or homotopy B^4.

We now turn to Question 1.10, which asks whether there exists a knot $K \subset S^3$ that is not slice in B^4 but is slice in a rational homology 4-ball W with $|H_1(W;\mathbb{Z})|$ odd. It seems unlikely that Heegaard Floer homology, in its present form, can address this condition. Note that involutive Heegaard Floer homology gives obstructions to being slice in a $\mathbb{Z}/2\mathbb{Z}$-homology 4-ball; if W is a rational homology 4-ball with $|H_1(W;\mathbb{Z})|$ odd, then it is a $\mathbb{Z}/2\mathbb{Z}$-homology ball. However, recall that prior to the advent of involutive Heegaard Floer homology, there was no way to use Heegaard Floer homology to obstruct a knot (such as the figure-eight) from being slice in any rational homology 4-ball. Perhaps there is some other additional structure that we can add to the Heegaard Floer package, yielding new obstructions. Alternatively, it remains possible that the s-invariant may have something to say about this question.

Conjecture 1.12 posits that ribbon concordance is a partial order. Zemke [88, THEOREM 1.7] proved that if there is a ribbon concordance from K_0 to K_1, then $\widehat{\mathrm{HFK}}(K_0)$ injects into $\widehat{\mathrm{HFK}}(K_1)$. Thus, if there is also a ribbon concordance from K_1 to K_0, then $\widehat{\mathrm{HFK}}(K_0) \cong \widehat{\mathrm{HFK}}(K_1)$. Note that there are infinite families of knots with the same knot Floer homology [29, THEOREM 1]. However, as far as the author knows, there are no known ribbon concordances between distinct knots in any of those families. Further investigation is needed before we rule out knot Floer homology as a tool for resolving Conjecture 1.12.

Closely related is Conjecture 1.13, which posits that ribbon $\mathbb{Q}$-homology cobordism is a partial order on 3-manifolds. There is a ribbon homology cobordism from S^3 to $Y \# -Y$ for any homology sphere Y. Taking $Y = \Sigma(2,3,5)$, and noting that $\mathrm{HF}^-(\Sigma(2,3,5)\# -\Sigma(2,3,5)) \cong \mathrm{HF}^-(S^3) \cong \mathbb{F}_{(0)}[U]$, we see that we have two distinct 3-manifolds with the same Heegaard Floer homology and a ribbon homology cobordism in one direction. (As alluded to above, we do not know of an analogous example in the knot case.) However, since $\Sigma(2,3,5)\# -\Sigma(2,3,5)$ does not bound a simply-connected homology 4-ball [85, PROPOSITION 1.7], it follows that there is no ribbon homology cobordism from $\Sigma(2,3,5)\# -\Sigma(2,3,5)$ to S^3 (for if there were, we could glue a 4-ball to the S^3 end and obtain a simply-connected homology ball with boundary $\Sigma(2,3,5)\# -\Sigma(2,3,5)$). We refer the reader to [8] for further evidence, some of it coming from various Floer homologies, in support of Conjecture 1.13.

As we have seen, advances in Heegaard Floer homology have answered many questions about homology cobordism and knot concordance. These successes were not immediate; they began in 2003, when Ozsváth–Szabó [65, 67] defined the homomorphisms

$$d : \Theta^3_{\mathbb{Z}} \to 2\mathbb{Z} \quad \text{and} \quad \tau : \mathcal{C} \to \mathbb{Z}.$$

The next major step in extracting concordance information from the knot Floer complex was the definition of ε in 2014 [36], which in turn led to two infinite families of concordance homomorphisms

$$\Upsilon_t : \mathcal{C} \to \mathbb{R},\ t \in [0,2] \quad \text{and} \quad \varphi_i : \mathcal{C} \to \mathbb{Z},\ i \in \mathbb{N},$$

the former defined by Ozsváth–Stipsicz–Szabó [64], and the latter by Dai, Stoffregen, Truong, and the author [11]. The algebraic framework necessary to define Υ_t and φ_i existed since the inception of knot Floer homology in the early 2000s, yet it took over a decade for anyone to exploit this structure to define these new homomorphisms. Concurrent with these developments was the advent of Hendricks–Manolescu's involutive Heegaard Floer homology [32], which put new, more refined structure on the Heegaard Floer and knot Floer homology packages, yielding new homology cobordism homomorphisms and new rational concordance obstructions. We look forward to seeing whether the Heegaard Floer package in its present form can be further mined for new applications, to refining the structure on these invariants even more to prove new theorems, and to developing new, unanticipated tools for resolving the questions and conjectures that we have posed here.

ACKNOWLEDGMENTS

I would like to thank my collaborators Irving Dai, Kristen Hendricks, Sungkyung Kang, Adam Levine, Tye Lidman, JungHwan Park, Matthew Stoffregen, Linh Truong, and Ian Zemke for the privilege of working with them and for their patience when working with me. The questions posed in this article are generally well known in the field; I learned of most of them through problems sessions and discussions over the years. I am grateful to all of my colleagues who have generously given their time and energy to conference organization and conversation. I would also like to thank Robert Lipshitz, Chuck Livingston, Ciprian Manolescu, and András Stipsicz for helpful comments on an earlier draft.

FUNDING

The author was partially supported by NSF grants DMS-1552285 and DMS-2104144.

REFERENCES

[1] T. Abe and K. Tagami, Fibered knots with the same 0-surgery and the slice-ribbon conjecture. *Math. Res. Lett.* **23** (2016), no. 2, 303–323.

[2] S. Akbulut and K. Larson, Brieskorn spheres bounding rational balls. *Proc. Amer. Math. Soc.* **146** (2018), no. 4, 1817–1824.

[3] S. Akbulut and J. D. McCarthy, *Casson's invariant for oriented homology* 3-*spheres*. Mathematical Notes 36, Princeton University Press, Princeton, NJ, 1990.

[4] J. Baldwin and D. S. Vela-Vick, A note on the knot Floer homology of fibered knots. *Algebr. Geom. Topol.* **18** (2018), no. 6, 3669–3690.

[5] D. Bar-Natan, On Khovanov's categorification of the Jones polynomial. *Algebr. Geom. Topol.* **2** (2002), 337–370.

[6] A. J. Casson and C. M. Gordon, Cobordism of classical knots. In *à la recherche de la topologie perdue*, pp. 181–199, Progr. Math. 62, Birkhäuser Boston, Boston, MA, 1986.

[7] J. C. Cha, The structure of the rational concordance group of knots. *Mem. Amer. Math. Soc.* **189** (2007), no. 885, x+95 pp.

[8] A. Daemi, T. Lidman, D. S. Vela-Vick, and C.-M. M. Wong, Ribbon homology cobordisms. 2019, arXiv:1904.09721.

[9] I. Dai, J. Hom, M. Stoffregen, and L. Truong, An infinite-rank summand of the homology cobordism group. 2018, arXiv:1810.06145.

[10] I. Dai, J. Hom, M. Stoffregen, and L. Truong, Homology concordance and knot Floer homology. 2021, arXiv:2110.14803.

[11] I. Dai, J. Hom, M. Stoffregen, and L. Truong, More concordance homomorphisms from knot Floer homology. *Geom. Topol.* **25** (2021), no. 1, 275–338.

[12] S. K. Donaldson, Self-dual connections and the topology of smooth 4-manifolds. *Bull. Amer. Math. Soc. (N. S.)* **8** (1983), no. 1, 81–83.

[13] R. Fintushel and R. J. Stern, A μ-invariant one homology 3-sphere that bounds an orientable rational ball. In *Four-manifold theory (Durham, NH, 1982)*, pp. 265–268, Contemp. Math. 35, Amer. Math. Soc., Providence, RI, 1984.

[14] R. Fintushel and R. J. Stern, Pseudofree orbifolds. *Ann. of Math. (2)* **122** (1985), no. 2, 335–364.

[15] R. Fintushel and R. J. Stern, Instanton homology of Seifert fibred homology three spheres. *Proc. London Math. Soc. (3)* **61** (1990), no. 1, 109–137.

[16] R. H. Fox, Some problems in knot theory. In *Topology of 3-manifolds and related topics (Proc. The Univ. of Georgia Institute, 1961)*, pp. 168–176, Prentice-Hall, Englewood Cliffs, NJ, 1962.

[17] R. H. Fox and J. W. Milnor, Singularities of 2-spheres in 4-space and cobordism of knots. *Osaka Math. J.* **3** (1966), 257–267.

[18] M. Freedman, R. Gompf, S. Morrison, and K. Walker, Man and machine thinking about the smooth 4-dimensional Poincaré conjecture. *Quantum Topol.* **1** (2010), no. 2, 171–208.

[19] M. H. Freedman, The topology of four-dimensional manifolds. *J. Differential Geometry* **17** (1982), no. 3, 357–453.

[20] K. A. Frøyshov, Equivariant aspects of Yang–Mills Floer theory. *Topology* **41** (2002), no. 3, 525–552.

[21] M. Furuta, Homology cobordism group of homology 3-spheres. *Inventiones mathematicae* **100** (1990), no. 1, 339–355.

[22] D. E. Galewski and R. J. Stern, Classification of simplicial triangulations of topological manifolds. *Ann. of Math. (2)* **111** (1980), no. 1, 11–34.

[23] R. E. Gompf, M. Scharlemann, and A. Thompson, Fibered knots and potential counterexamples to the property 2R and slice-ribbon conjectures. *Geom. Topol.* **14** (2010), no. 4, 2305–2347.

[24] C. Gordon, Ribbon concordance of knots in the 3-sphere. *Math. Ann.* **257** (1981), 157–170.

[25] J. Greene and S. Jabuka, The slice-ribbon conjecture for 3-stranded pretzel knots. *Amer. J. Math.* **133** (2011), no. 3, 555–580.

[26] J. Hanselman, J. Rasmussen, and L. Watson, Bordered Floer homology for manifolds with torus boundary via immersed curves. 2016, arXiv:1604.03466.

[27] J. Hanselman and L. Watson, Cabling in terms of immersed curves. 2019, arXiv:1908.04397.

[28] M. Hedden and A. S. Levine, A surgery formula for knot Floer homology. 2019, arXiv:1901.02488.

[29] M. Hedden and L. Watson, On the geography and botany of knot Floer homology. *Selecta Math. (N. S.)* **24** (2018), no. 2, 997–1037.

[30] K. Hendricks, J. Hom, M. Stoffregen, and I. Zemke, Surgery exact triangles in involutive Heegaard Floer homology. 2020, arXiv:2011.00113.

[31] K. Hendricks, J. Hom, M. Stoffregen, and I. Zemke, On the quotient of the homology cobordism group by Seifert spaces. 2021, arXiv:2103.04363.

[32] K. Hendricks and C. Manolescu, Involutive Heegaard Floer homology. *Duke Math. J.* **166** (2017), no. 7, 1211–1299.

[33] K. Hendricks, C. Manolescu, and I. Zemke, A connected sum formula for involutive Heegaard Floer homology. *Selecta Math. (N. S.)* **24** (2018), no. 2, 1183–1245.

[34] J. Hom, Lecture notes on Heegaard Floer homology. 2020, arXiv:2008.01836.

[35] J. Hom, Bordered Heegaard Floer homology and the tau-invariant of cable knots. *J. Topol.* **7** (2014), no. 2, 287–326.

[36] J. Hom, The knot Floer complex and the smooth concordance group. *Comment. Math. Helv.* **89** (2014), no. 3, 537–570.

[37] J. Hom, A survey on Heegaard Floer homology and concordance. *J. Knot Theory Ramifications* **26** (2017), no. 2, 1740015, 24.

[38] J. Hom, S. Kang, J. Park, and M. Stoffregen, Linear independence of rationally slice knots. 2020, arXiv:2011.07659.

[39] J. Hom, A. Levine, and T. Lidman, Knot concordance in homology cobordisms. 2018, arXiv:1801.07770.

[40] A. Juhász, A survey of Heegaard Floer homology. In *New ideas in low dimensional topology*, pp. 237–296, Ser. Knots Everything 56, World Sci. Publ., Hackensack, NJ, 2015.

[41] A. Juhász, D. Thurston, and I. Zemke, Naturality and mapping class groups in Heegaard Floer homology. 2012, arXiv:1210.4996.

[42] M. Khovanov, A categorification of the Jones polynomial. *Duke Math. J.* **101** (2000), no. 3, 359–426.

[43] M. H. Kim and K. Park, An infinite-rank summand of knots with trivial Alexander polynomial. *J. Symplectic Geom.* **16** (2018), no. 6, 1749–1771.

[44] D. Krcatovich, The reduced knot Floer complex. *Topology Appl.* **194** (2015), 171–201.

[45] A. G. Lecuona, On the slice-ribbon conjecture for pretzel knots. *Algebr. Geom. Topol.* **15** (2015), no. 4, 2133–2173.

[46] E. S. Lee, An endomorphism of the Khovanov invariant. *Adv. Math.* **197** (2005), no. 2, 554–586.

[47] A. S. Levine, Nonsurjective satellite operators and piecewise-linear concordance. *Forum Math. Sigma* **4** (2016), e34, 47 pp.

[48] J. Levine, Knot cobordism groups in codimension two. *Comment. Math. Helv.* **44** (1969), 229–244.

[49] F. Lin, The surgery exact triangle in Pin(2)-monopole Floer homology. *Algebr. Geom. Topol.* **17** (2017), no. 5, 2915–2960.

[50] R. Lipshitz, P. S. Ozsvath, and D. P. Thurston, Bordered Heegaard Floer homology. *Mem. Amer. Math. Soc.* **254** (2018), no. 1216, viii+279.

[51] P. Lisca, Lens spaces, rational balls and the ribbon conjecture. *Geom. Topol.* **11** (2007), 429–472.

[52] C. Livingston, Homology cobordisms of 3-manifolds, knot concordances, and prime knots. *Pacific Journal of Mathematics* **94** (1981), no. 1, 193–206.

[53] C. Livingston, A survey of classical knot concordance. In *Handbook of knot theory*, pp. 319–347, Elsevier B. V., Amsterdam, 2005.

[54] C. Manolescu, An introduction to knot Floer homology. In *Physics and mathematics of link homology*, pp. 99–135, Contemp. Math. 680, Amer. Math. Soc., Providence, RI, 2016.

[55] C. Manolescu, Pin(2)-equivariant Seiberg–Witten Floer homology and the triangulation conjecture. *J. Amer. Math. Soc.* **29** (2016), no. 1, 147–176.

[56] C. Manolescu, Homology cobordism and triangulations. In *Proceedings of the International Congress of Mathematicians—Rio de Janeiro 2018. Vol. II. Invited lectures*, pp. 1175–1191, World Sci. Publ., Hackensack, NJ, 2018.

[57] C. Manolescu, Lectures on the triangulation conjecture. In *Proceedings of the Gökova Geometry–Topology Conference 2015*, pp. 1–38, Gökova Geometry/Topology Conference (GGT), Gökova, 2016.

[58] C. Manolescu and L. Piccirillo, From zero surgeries to candidates for exotic definite four-manifolds. 2021, arXiv:2102.04391.

[59] T. Matumoto, Triangulation of manifolds. In *Algebraic and geometric topology (Proc. Sympos. Pure Math., Stanford Univ., Stanford, CA, 1976), Part 2*, pp. 3–6, Proc. Sympos. Pure Math. XXXII, Amer. Math. Soc., Providence, RI, 1976.

[60] K. Murasugi, On a certain numerical invariant of link types. *Trans. Amer. Math. Soc.* **117** (1965), 387–422.

[61] R. Myers, Homology cobordisms, link concordances, and hyperbolic 3-manifolds. *Transactions of the American Mathematical Society* **278** (1983), no. 1, 271–288.

[62] Y. Ni, The next-to-top term in knot Floer homology. 2021, arXiv:2104.14687.

[63] Y. Nozaki, K. Sato, and M. Taniguchi, Filtered instanton Floer homology and the homology cobordism group. 2019, arXiv:1905.04001.

[64] P. S. Ozsváth, A. I. Stipsicz, and Z. Szabó, Concordance homomorphisms from knot Floer homology. *Adv. Math.* **315** (2017), 366–426.

[65] P. S. Ozsváth and Z. Szabó, Absolutely graded Floer homologies and intersection forms for four-manifolds with boundary. *Adv. Math.* **173** (2003), no. 2, 179–261.

[66] P. S. Ozsváth and Z. Szabó, Holomorphic disks and three-manifold invariants: properties and applications. *Ann. of Math. (2)* **159** (2004), no. 3, 1159–1245.

[67] P. Ozsváth and Z. Szabó, Knot Floer homology and the four-ball genus. *Geom. Topol.* **7** (2003), 615–639.

[68] P. Ozsváth and Z. Szabó, Holomorphic disks and genus bounds. *Geom. Topol.* **8** (2004), 311–334.

[69] P. Ozsváth and Z. Szabó, Holomorphic disks and knot invariants. *Adv. Math.* **186** (2004), no. 1, 58–116.

[70] P. Ozsváth and Z. Szabó, Holomorphic disks and topological invariants for closed three-manifolds. *Ann. of Math. (2)* **159** (2004), no. 3, 1027–1158.

[71] P. Ozsváth and Z. Szabó, Holomorphic triangles and invariants for smooth four-manifolds. *Adv. Math.* **202** (2006), no. 2, 326–400.

[72] P. Ozsváth and Z. Szabó, An introduction to Heegaard Floer homology. In *Floer homology, gauge theory, and low-dimensional topology*, pp. 3–27, Clay Math. Proc. 5, Amer. Math. Soc., Providence, RI, 2006.

[73] P. Ozsváth, Z. Szabó, and D. Thurston, Legendrian knots, transverse knots and combinatorial Floer homology. *Geom. Topol.* **12** (2008), no. 2, 941–980.

[74] G. Perelman, The entropy formula for the Ricci flow and its geometric applications. 2002, arXiv:math/0211159.

[75] G. Perelman, Finite extinction time for the solutions to the Ricci flow on certain three-manifolds. 2003, arXiv:math/0307245.

[76] G. Perelman, Ricci flow with surgery on three-manifolds. 2003, arXiv:math/0303109.

[77] I. Petkova, Cables of thin knots and bordered Heegaard Floer homology. *Quantum Topol.* **4** (2013), no. 4, 377–409.

[78] J. Rasmussen, *Floer homology and knot complements*. Ph.D. thesis, Harvard University, 2003. arXiv:math/0306378.

[79] J. Rasmussen, Khovanov homology and the slice genus. *Invent. Math.* **182** (2010), no. 2, 419–447.

[80] V. A. Rohlin, New results in the theory of four-dimensional manifolds. *Doklady Akad. Nauk SSSR (N. S.)* **84** (1952), 221–224.

[81] D. Rostovtsev, Almost ι-complexes as immersed curves. 2020, arXiv:2012.07189.

[82] S. Sarkar, Moving basepoints and the induced automorphisms of link Floer homology. *Algebr. Geom. Topol.* **15** (2015), no. 5, 2479–2515.

[83] S. Smale, Generalized Poincaré's conjecture in dimensions greater than four. *Ann. of Math. (2)* **74** (1961), 391–406.

[84] M. Stoffregen, Manolescu invariants of connected sums. *Proceedings of the London Mathematical Society* **115** (2017), no. 5, 1072–1117.

[85] C. H. Taubes, Gauge theory on asymptotically periodic 4-manifolds. *J. Differential Geom.* **25** (1987), no. 3, 363–430.

[86] M. Usher (ed.), Low-dimensional and symplectic topology, Proceedings of Symposia in Pure Mathematics 82, American Mathematical Society, Providence, RI, 2011.

[87] I. Zemke, Connected sums and involutive knot Floer homology. *Proc. Lond. Math. Soc. (3)* **119** (2019), no. 1, 214–265.

[88] I. Zemke, Knot Floer homology obstructs ribbon concordance. *Ann. of Math. (2)* **190** (2019), no. 3, 931–947.

[89] I. Zemke, Link cobordisms and absolute gradings on link Floer homology. *Quantum Topol.* **10** (2019), no. 2, 207–323.

[90] H. Zhou, Homology concordance and an infinite rank subgroup. 2020, arXiv:2009.05145.

JENNIFER HOM

School of Mathematics, Georgia Institute of Technology, Atlanta, GA 30332, USA, hom@math.gatech.edu

This Page Intentionally Left Blank

STABLE HOMOTOPY GROUPS OF SPHERES AND MOTIVIC HOMOTOPY THEORY

DANIEL C. ISAKSEN, GUOZHEN WANG, AND ZHOULI XU

Dedicated to Mark Mahowald

ABSTRACT

We consider the problem of computing the stable homotopy groups of spheres, including applications and history. We describe a new technique that yields streamlined computations through dimension 61 and gives new computations through dimension 90 with very few exceptions. We discuss questions and conjectures for further study, including a new approach to the computation of motivic stable homotopy groups over arbitrary base fields. We provide complete charts for the Adams spectral sequence through dimension 90.

MATHEMATICS SUBJECT CLASSIFICATION 2020

Primary 55Q45; Secondary 55T15, 14F42, 18E30, 57R55, 13P20

KEYWORDS

Stable homotopy groups of spheres, Adams spectral sequences, motivic homotopy theory, smooth structures on spheres, stable infinity categories, derived categories, t-structure

INTERNATIONAL CONGRESS OF MATHEMATICIANS 2022 JULY 6–14

Proc. Int. Cong. Math. 2022, Vol. 4, pp. 2768–2790
DOI 10.4171/ICM2022/32

1. INTRODUCTION

The computation of stable homotopy groups of spheres is one of the most fundamental and important problems in topology. It has connections to many topics in topology, such as the cobordism theory of framed manifolds, the classification of smooth structures on spheres, obstruction theory, the theory of topological modular forms, algebraic K-theory, and equivariant homotopy theory.

Consider the set of homotopy classes of continuous based maps $S^{n+k} \to S^k$ between spheres of dimensions $n + k$ and k. This set admits a natural group structure. By the Freudenthal Suspension Theorem [12], this group only depends on k when $n > k + 1$. This group is called the nth stable homotopy group of spheres, or the nth stable stem, and is denoted by π_n. If $n < 0$, then π_n is the zero group. Moreover, π_0 is isomorphic to the group of integers. Serre's finiteness theorem [40] tells us that π_n is a finite abelian group for $n > 0$.

Despite their simple definition, which was available 80 years ago, the stable homotopy groups are notoriously hard to compute. All known methods only give a complete answer through a range, and then reach an obstacle that can only be surmounted by the introduction of a new method. **Mahowald's Uncertainty Principles** attempt to quantify the inherent difficulty of the problem. Despite its difficulty, many mathematicians have made significant progress. We will briefly review the history and Mahowald's Uncertainty Principles in Section 3.

Recently, the authors have developed a new method [14] using motivic homotopy theory. Using this new method, we have already greatly improved our knowledge of stable stems [19,20], and we have ongoing computations into even higher dimensions. Our method is currently the most effective and less prone to human error, partly due to the fact that it relies more heavily on machine computation than previous methods.

The original purpose of motivic homotopy theory was to apply abstract homotopy theory to problems in number theory and algebraic geometry. In contrast, our work has reversed the information flow and applied motivic homotopy theory to discover new phenomena in classical topology.

2. SMOOTH STRUCTURES ON SPHERES

The work of Kervaire and Milnor [22] on the classification of smooth structures on spheres in dimensions at least 5 is an important example of an application of stable stem computations. Let Θ_n be the group of h-cobordism classes of homotopy n-spheres. This group classifies the differential structures on S^n for $n \geq 5$. Kervaire and Milnor [22] reduced the computation of the group Θ_n to the computation of the stable homotopy group π_n and the Kervaire invariant problem. The latter was resolved by Hill, Hopkins, and Ravenel [16] in all dimensions except for 126. In particular, Kervaire and Milnor observed that the spheres in dimensions 5, 6, and 12 have unique smooth structures.

We restate the following conjecture from [47], which is based on the current knowledge of stable stems and a problem proposed by Milnor [28].

Conjecture 2.1. *In dimensions greater than 4, the only spheres with unique smooth structures are S^5, S^6, S^{12}, S^{56}, and S^{61}.*

Uniqueness in dimension 56 is due to the first author [18], and uniqueness in dimension 61 is due to the second and third authors [47].

Conjecture 2.1 is equivalent to the claim that the group Θ_n is not of order 1 for dimensions greater than 61. This conjecture has been confirmed in all odd dimensions by the second and the third authors [47] based on the work of Hill, Hopkins, and Ravenel [16].

Theorem 2.2 ([47, COROLLARY 1.13]). *The only odd-dimensional spheres with unique smooth structures are S^1, S^3, S^5, and S^{61}.*

For even dimensions, Conjecture 2.1 has been confirmed for over half of all even dimensions by Behrens, Hill, Hopkins, Mahowald, and Quigley [6,7].

3. HISTORY AND MAHOWALD'S UNCERTAINTY PRINCIPLES

We review the history of computing stable stems. See [46] for a survey of classical methods and also Section 2 of [47].

After the geometric computation of the first three stems, Serre [39] computed π_n for $n \leq 8$ using the cohomology of Eilenberg–MacLane spaces and the Serre spectral sequence. Using the EHP sequence and higher compositions such as Toda brackets, Toda [41] computed a large range of unstable homotopy groups of spheres and obtained π_n for $n \leq 19$.

Since π_n is finite abelian, it can be reconstructed from its p-primary components for each prime p. History has demonstrated the effectiveness of this approach. The standard approach to computing stable stems at each prime is to use Adams-type spectral sequences that converge from algebra to homotopy. To identify the algebraic E_2-pages, one needs auxiliary algebraic spectral sequences that converge from simpler algebra to more complicated algebra. For any spectral sequence, difficulties arise in computing differentials and in solving extension problems. Typically, a variety of complementary methods are required to compute a spectral sequence. One method may compute some types of differentials and extension problems efficiently but leave other types unanswered. To obtain complete computations, one must be *eclectic*, applying and combining different methodologies. Even so, combining all known methods, there are eventually some problems that cannot currently be solved.

In fact, we have the following principle, first named by Ravenel [15].

The First Mahowald Uncertainty Principle. Any spectral sequence converging to the homotopy groups of spheres with an E_2-page that can be named using homological algebra will be infinitely far from the actual answer.

The first principle essentially says that the computation of stable stems is not an algebraic problem—there are infinitely many nonzero differentials that must be resolved in such a spectral sequence. Based on experience of learning from Mark Mahowald, the third author [48] named the second principle:

The Second Mahowald Uncertainty Principle. Any method that computes nontrivial differentials in such a spectral sequence will leave infinitely many differentials undecided.

At odd primes, the state-of-the-art is computed by Ravenel in [36] with the Adams–Novikov spectral sequence [33] and the chromatic spectral sequence, which are based on complex cobordism and formal groups. As the prime grows, so does the range of computation, since the spectral sequences become sparser. For example, for $p = 3$ and $p = 5$, we have complete knowledge up to around 100 and 1000 stems, respectively [36]. These ranges are both approximately equal to $p^3(2p-2)$.

At the prime 2, the classical Adams spectral sequence [1] is still the most efficient method. May [26] constructed the May spectral sequence at all primes, which converges to the E_2-page of the Adams spectral sequence, and he computed π_n for $n \leq 28$. Using higher structure such as the interactions between Massey products and Toda brackets, Mahowald (with Barratt, Bruner, Jones, and Tangora) [3, 4, 8, 25] computed π_n for $n \leq 47$. We also mention [23, 24], which take an entirely different approach. However, the computations in [23, 24] are now known to contain several errors.

More recently, in 2014, the first author [18] gave a thorough accounting of the Adams spectral sequence up to dimension 59 with the exception of only one differential, but then he reached an obstacle as predicted by Mahowald's Uncertainty Principles. The new idea was to compare classical computations with the motivic Adams spectral sequence. The exception was later proved by the third author [21] based on the first author's computations.

In 2016, with tremendous efforts, the second and third authors [47] bypassed the above obstacle by computing two more stable stems using the $\mathbb{R}P^\infty$-method. In particular, the second and third authors proved that π_{61} is the zero group; Theorem 2.2 is a consequence. The $\mathbb{R}P^\infty$-method is useful for finding specific, particularly difficult, Adams differentials and is not designed to study all differentials systematically.

A major breakthrough occurred in 2017 and the next few years. A new method [14] allowed the authors [19, 20] to recompute most Adams differentials up to dimension 61 very easily and to extend computations to dimension 90 with only a few exceptions. For example, the hardest differential $d_3(D_3) = B_3$ proved in [47] is now an immediate consequence of this new method; it comes immediately from the output of a computer program. Our new method is discussed in the next section.

Further computations into higher dimensions are still ongoing. We have not yet reached an insurmountable obstacle that will require a new method to resolve.

4. MOTIVIC HOMOTOPY THEORY AND ALGEBRAICITY OF THE COFIBER OF τ

Morel and Voevodsky [30, 31] developed motivic homotopy theory in the mid-1990s in order to import homotopical techniques into algebraic geometry. This program found great success in Voevodsky's resolutions of the Milnor Conjecture [42] and the Bloch–Kato Conjecture [43].

There is a cellular subcategory of the motivic stable homotopy category that is generated by two types of spheres: the simplicial sphere $S^{1,0}$, and the multiplicative group $\mathbb{G}_m = \mathbb{A}^1 - 0$, denoted by $S^{1,1}$. After p-completion, there is a stable map $\tau : S^{0,-1} \to S^{0,0}$ over $\mathbb{C}$ that induces a nonzero map on mod p motivic homology. We denote by $S^{0,0}/\tau$ the cofiber of τ.

One may view the p-completed $\mathbb{C}$-motivic stable homotopy category as a deformation of the p-completed classical stable homotopy category, with this element τ as a parameter. Dugger and the first author [11] identified the generic fiber "$\tau = 1$" with the p-completed classical stable homotopy category. Gheorghe and the second and third authors [14] identified the special fiber "$\tau = 0$" with a purely algebraic category.

Theorem 4.1 ([14]). *At each prime p, there is an equivalence*

$$S^{0,0}/\tau\text{-}\mathbf{Mod} \simeq \mathcal{D}(\mathrm{BP}_{2*}\mathrm{BP}\text{-}\mathbf{Comod})$$

of stable ∞-categories, equipped with t-structures, between the category of cellular module spectra over $S^{0,0}/\tau$ and Hovey's [17] derived category of $\mathrm{BP}_{2*}\mathrm{BP}$*-comodules.*

The right-hand side is also known as the derived category of quasicoherent sheaves on the moduli stack of formal groups over $\mathbb{Z}_p$-algebras, which is foundational to chromatic homotopy theory [29, 35].

The first author [18] observed that the homotopy groups of $S^{0,0}/\tau$ are isomorphic to the classical Adams–Novikov E_2-page $\mathrm{Ext}^{*,*}_{\mathrm{BP}_*\mathrm{BP}}(\mathrm{BP}_*, \mathrm{BP}_*)$. In 2017, the second author [45] made a computer program that computes the algebraic Novikov spectral sequence, which converges to the Adams–Novikov E_2-page, in a large range. The computer data aligned with the motivic Adams spectral sequence for $S^{0,0}/\tau$ obtained by the first author. This discovery motivated the following theorem by Gheorghe and the second and third authors [14], which is crucial to the computation of classical and $\mathbb{C}$-motivic stable homotopy groups of spheres.

Theorem 4.2 ([14]). *The tri-graded motivic Adams spectral sequence for $S^{0,0}/\tau$ is isomorphic to the algebraic Novikov spectral sequence for* BP_{2*} [27, 33]*:*

$$\begin{array}{ccc}
\mathrm{Ext}^{a,2w-s+a,w}_{A^{\mathrm{mot}}}(\mathbb{F}_p[\tau], \mathbb{F}_p) & \xrightarrow{\quad\cong\quad} & \mathrm{Ext}^{s,2w}_{\mathrm{BP}_{2*}\mathrm{BP}/I}(\mathbb{F}_p, I^{a-s}/I^{a-s+1}) \\
\Big\Downarrow \text{Motivic Adams SS} & & \Big\Downarrow \text{Algebraic Novikov SS} \\
\pi_{2w-s,w}(S^{0,0}/\tau) & \xrightarrow{\quad\cong\quad} & \mathrm{Ext}^{s,2w}_{\mathrm{BP}_{2*}\mathrm{BP}}(\mathrm{BP}_{2*}, \mathrm{BP}_{2*}).
\end{array}$$

Here A^{mot} is the motivic Steenrod algebra, and $\mathbb{F}_p[\tau]$ is the mod p motivic cohomology of $S^{0,0}$.

There is a Betti realization functor $\mathbf{Re}$ from the motivic stable homotopy category over $\mathbb{C}$ to the classical stable homotopy category, which extends the functor that sends a complex algebraic variety to its $\mathbb{C}$-points. We have $\mathbf{Re}(S^{n,w}) \simeq S^n$ and $\mathbf{Re}(\tau) = 1$.

The naturality of Adams spectral sequences then gives us a zigzag diagram

$$\begin{array}{ccccc}
\mathrm{Ext}_A^{s,t}(\mathbb{F}_p,\mathbb{F}_p) & \xleftarrow{\mathrm{Re}} & \mathrm{Ext}_{A^{\mathrm{mot}}}^{s,t,w}(\mathbb{F}_p[\tau],\mathbb{F}_p[\tau]) & \longrightarrow & \mathrm{Ext}_{A^{\mathrm{mot}}}^{s,t,w}(\mathbb{F}_p[\tau],\mathbb{F}_p) \\
\Downarrow \text{Adams SS} & & \Downarrow \text{Motivic Adams SS} & & \Downarrow \text{Motivic Adams SS} \\
\pi_{t-s}S^0 & \xleftarrow{\mathrm{Re}} & \pi_{t-s,w}S^{0,0} & \longrightarrow & \pi_{t-s,w}S^{0,0}/\tau
\end{array}$$

of spectral sequences. Here the left map is given by the Betti realization functor, and the right map is induced by the quotient map $S^{0,0} \to S^{0,0}/\tau$. This diagram of spectral sequences is very powerful. The differentials on the right side are purely algebraic by Theorem 4.2 and can be obtained by the output of a computer program!

In fact, this method obtains all differentials up to the 45-stem with essentially only one exception. Consistent with the Second Mahowald Uncertainty Principle, more and more exceptions occur in higher dimensions. See Appendix A of **[14]** for more details.

In practice, our method can be summarized in the following steps:

(1) Compute the $\mathbb{C}$-motivic Adams E_2-page with a machine in a large range.

(2) Compute the algebraic Novikov spectral sequence with a machine in a large range, including all differentials and multiplicative structure, and use Theorem 4.2 to identify it with the motivic Adams spectral sequence for $S^{0,0}/\tau$.

(3) Use the cofiber sequence

$$S^{0,-1} \xrightarrow{\tau} S^{0,0} \to S^{0,0}/\tau \to S^{1,-1}$$

and naturality of Adams spectral sequences to pull back and push forward Adams differentials for $S^{0,0}/\tau$ to Adams differentials for the motivic sphere.

(4) Apply a variety of ad hoc arguments to deduce additional Adams differentials for the motivic sphere.

(5) Invert τ to obtain the classical Adams spectral sequence and the classical stable homotopy groups.

The machine-generated data that we use in steps (1) and (2) are available at **[44]**.

5. RESULTS AND ADAMS CHARTS

Our computational results of the classical Adams spectral sequence are best summarized in charts, which we include at the end of this article. The charts are displayed in pieces so that they fit onto individual pages. For tables that describe the stable homotopy groups π_n for $n \leq 90$, see **[19, 20]**.

The first eight charts (Figures 1–8) represent the Adams E_2-page. The dimension is on the horizontal axis, and Adams filtration is on the vertical axis. Each dot represents a copy of $\mathbb{F}_2$. Dark gray vertical lines and lines of slope 1 and 1/3 represent multiplications by h_0, h_1, and h_2, respectively. Light gray lines of slope $-r$ represent Adams d_r differentials.

Nearly all of the differentials through dimension 90 have been computed. The only exceptions are that $d_9(x_{85,6} + h_0^3 c_3)$ or $d_{10}(h_1 f_2)$ might equal $M\Delta h_1 d_0$ in the 84-stem.

The last three charts (Figures 9–11) represent the Adams E_∞-page. The dark gray lines represent a multiplicative structure that is inherited from the E_2-page. Light gray lines represent a multiplicative structure that is hidden by the Adams filtration. Beyond the 70-stem, there remain some unresolved ν extensions that are not shown on the chart. Beyond the 80-stem, there remain unresolved 2 extensions and η extensions that are not shown.

The 2-primary part of π_n can be read from this chart. The vertical column in dimension n represents the associated graded object of the Adams filtration of π_n. The presence of k dots in the nth column means that π_n has order 2^k.

The vertical lines determine the group structure of π_n. Each vertical line represents a nontrivial extension of abelian groups. Therefore, a sequence of m dots connected by vertical lines represents a copy of $\mathbb{Z}/2^m$ inside of π_n. For example, the 2-primary part of π_{23} is $\mathbb{Z}/2 \oplus \mathbb{Z}/8 \oplus \mathbb{Z}/16$.

In stems beyond 30, a regular pattern emerges along the top of the E_∞-page that is distinct from the much more complicated and irregular pattern below. This regular pattern represents the v_1-periodic part. We omit this pattern starting from the high 40s.

6. DEFORMATIONS OF STABLE HOMOTOPY THEORY

One interpretation of Theorem 4.1 is that the $\mathbb{C}$-motivic cellular stable homotopy category is a deformation of classical stable homotopy category. Although our work is heavily motivated by motivic homotopy theory, it is not logically dependent because of purely topological constructions [13, 34] of this cellular subcategory.

There are other deformations of classical stable homotopy theory that are also computationally useful, such as $H\mathbb{F}_2$-synthetic stable homotopy theory [9]. Beyond the 90-stem, $H\mathbb{F}_2$-synthetic stable homotopy theory has provided additional information to our method, and can be viewed as one more tool in the "ad hoc" step (4) of Section 4.

Lately, we have begun to study $H\mathbb{F}_2$-synthetic $\mathbb{C}$-motivic stable homotopy theory. This can be viewed as a deformation of a deformation of classical stable homotopy theory. On the other hand, one could also perform the deformations in the other order by considering BP-synthetic, $H\mathbb{F}_2$-synthetic stable homotopy theory. We believe that these double deformations are equivalent, and we propose the name "bimotivic homotopy theory" for this triply-graded stable homotopy theory.

7. THE CHOW t-STRUCTURE

Over an arbitrary based field k, the story is more complicated than just a deformation—it becomes the Postnikov–Whitehead tower associated to the Chow t-structure.

In [2], Bachmann, Kong, and the second and third authors defined the Chow t-structure on the motivic stable homotopy category $\mathrm{SH}(k)$ over any base field k. Its non-negative part $\mathrm{SH}(k)_{c\geq 0}$ is generated by Thom spectra $\mathrm{Th}(\xi)$ associated to K-theory points $\xi \in K(X)$ on smooth and proper schemes X. We implicitly invert the exponential characteristic of k and denote by $E \mapsto E_{c=i}$ the truncations with respect to the Chow t-structure.

Theorem 7.1 ([2]). *Let $E \in \mathrm{SH}(k)$. Then there is a canonical isomorphism*

$$\pi_{2w-s,w} E_{c=i} \cong \mathrm{Ext}^{s,2w}_{\mathrm{MU}_{2*}\mathrm{MU}}(\mathrm{MU}_{2*}, \mathrm{MGL}_{2*+i,*}E).$$

Here MGL is the algebraic cobordism spectrum. Theorem 7.1 generalizes the isomorphism on the abutments in Theorem 4.2 over $\mathbb{C}$ to an arbitrary base field and is an integral statement.

Moreover, the heart of the Chow t-structure $\mathrm{SH}(k)^{c\heartsuit}$ can be described as a category of enriched presheaves (see, e.g., [37, SECTION 3.5]) over the category of pure MGL-motives $\mathrm{PM}_{\mathrm{MGL}}(k)$ [2, DEFINITION 1.4].

Theorem 7.2 ([2]). *The Chow heart $\mathrm{SH}(k)^{c\heartsuit}$ is equivalent to the category of enriched presheaves on $\mathrm{PM}_{\mathrm{MGL}}(k)$ with values in $\mathrm{MU}_{2*}\mathrm{MU}$-comodules.*

Restricting to the subcategory of cellular objects, the Chow heart can be identified as the abelian category of $\mathrm{MU}_{2*}\mathrm{MU}$-comodules. The category of cellular objects over $(S^{0,0})_{c=0}$ is equivalent to Hovey's [17] derived category of $\mathrm{MU}_{2*}\mathrm{MU}$-comodules.

Theorem 7.3 ([2]). *There are equivalences of stable ∞-categories*

$$\mathrm{SH}(k)^{\mathrm{cell},c\heartsuit} \simeq \mathrm{MU}_{2*}\mathrm{MU}\text{-}\mathbf{Comod},$$
$$(S^{0,0})_{c=0}\text{-}\mathbf{Mod}^{\mathrm{cell}} \simeq \mathcal{D}(\mathrm{MU}_{2*}\mathrm{MU}\text{-}\mathbf{Comod}).$$

Theorem 7.3 allows us to identify the motivic Adams spectral sequence of $(S^{0,0})_{c=i}$ as an algebraic Novikov spectral sequence, which can be computed by a machine. We anticipate that Adams differentials for the k-motivic sphere can be computed through the Postikov–Whitehead tower associated to the Chow t-structure (see [2] for more details).

It would be interesting to compare our approach with methods developed in [38].

8. FURTHER QUESTIONS AND CONJECTURES

We include a few questions and conjectures for future study.

The orders of individual p-primary stable homotopy groups do not follow a clear pattern. However, an empirically observed pattern emerges if we consider the cumulative size of the groups.

Conjecture 8.1 (Stable stems growth conjecture). *Let $f(n)$ be the product of the orders of the p-primary stable homotopy groups in dimensions 1 through n. Then $\log_p f(n) = O(n^2)$.*

The ring spectrum of topological modular forms *tmf* is very useful for computing Adams differentials for the sphere spectrum, since *tmf* detects many classes above a line of slope $1/6$ on the Adams chart. Starting in the high 60s, the Mahowald operator $Ma = \langle g_2, h_0^3, a \rangle$ organizes many more classes just below this line, where a is detected by *tmf*.

Question 8.2 (Mahowald operator detection question). *Does there exist a ring spectrum whose Adams spectral sequence is completely computable such that its E_2-page detects $M^n a$ for all $n > 0$ and all classes a that are detected by tmf?*

Baues, Jibladze, and Nassau [5, 32] described how consideration of the secondary Steenrod algebra leads to the computation of Adams differentials. Recently, Chua [10] has used these ideas to obtain machine-generated values of the Adams d_2-differentials. This allows us to take the Adams E_3-page as given by machine.

Question 8.3 (Automated Adams differential computation question). *Are there effective algorithms that can compute all Adams d_3 or even d_4-differentials in a given range?*

Question 8.4 (Automated Adams–Novikov differential computation question). *Are there effective algorithms that can compute all Adams–Novikov d_3-differentials in a given range?*

Within any Adams filtration on the E_2-page, there is an operation Sq^0 that doubles the internal degree t. The following Conjecture 8.5 is due to Minami.

Conjecture 8.5 (New doomsday conjecture). *For any Sq^0-family*

$$\{x, Sq^0 x, \ldots, (Sq^0)^n x, \ldots\}$$

in the Adams spectral sequence, only finitely many classes survive to the E_∞-page.

Conjecture 8.6 (Stable length conjecture). *Nonzero Adams differentials supported by any Sq^0-family a_n are of the form $d_r(a_n) = c \cdot b_n$ when n is large enough, where b_n is an Sq^0-family and c is a fixed element in* Ext.

In Adams filtrations 1 and 2, the New Doomsday Conjecture is essentially equivalent to the Hopf invariant one problem and the Kervaire invariant one problem, respectively.

ACKNOWLEDGMENTS

We thank Mark Behrens, Robert Bruner, Houhong Fan, Paul Goerss, Jesper Grodal, Lars Hesselholt, Mike Hopkins, Peter May, Haynes Miller, Christian Nassau, Doug Ravenel, and John Rognes for their support and encouragement throughout this project.

FUNDING

The first author was supported by NSF grant DMS-1904241. The second author was supported by grant NSFC-11801082 and by the Shanghai Rising-Star Program. The third author was supported by NSF grant DMS-2105462. Many of the associated machine computations were performed on the Wayne State University Grid high performance computing cluster.

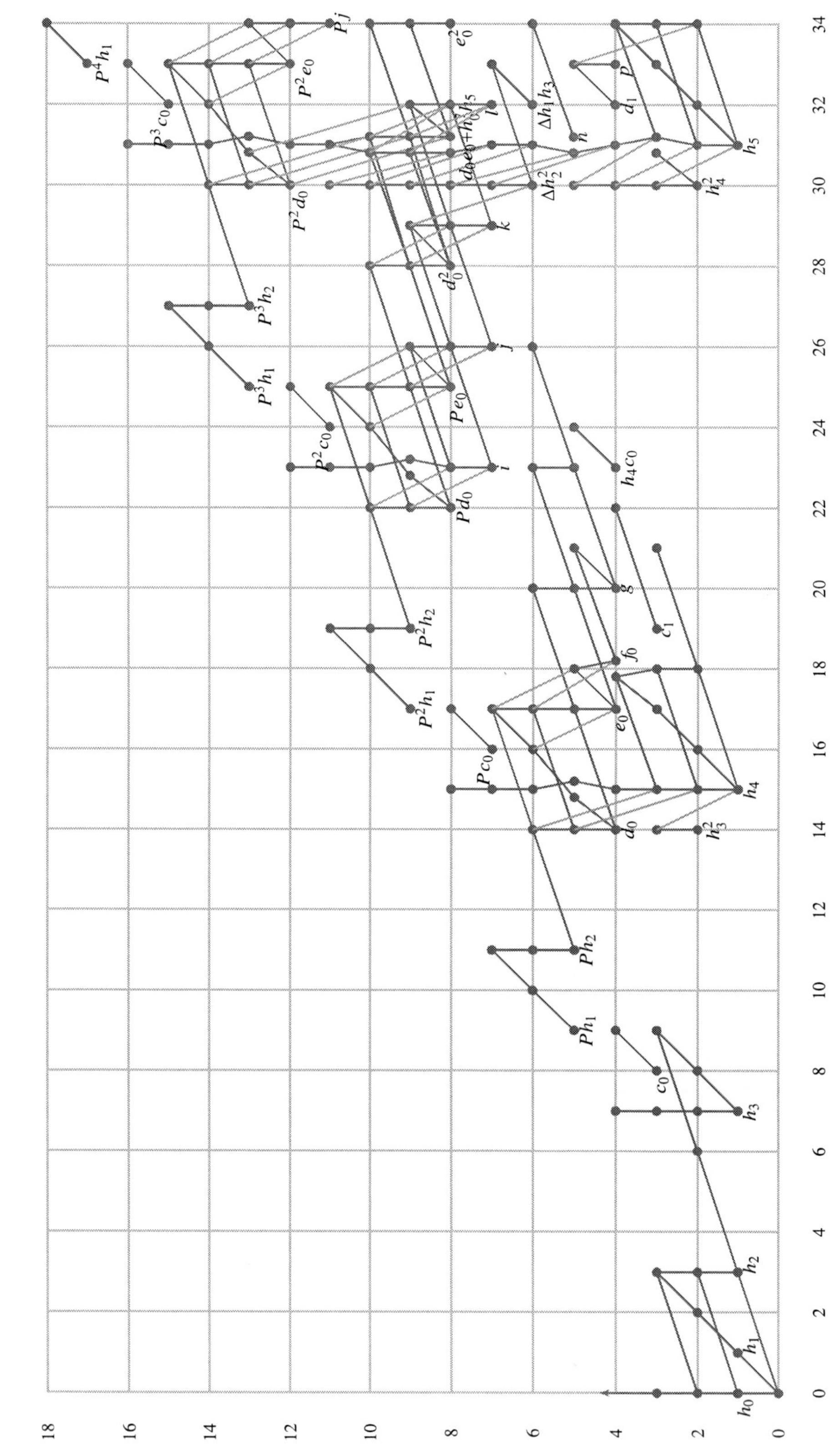

FIGURE 1
The Adams E_2-page in dimensions 0–34

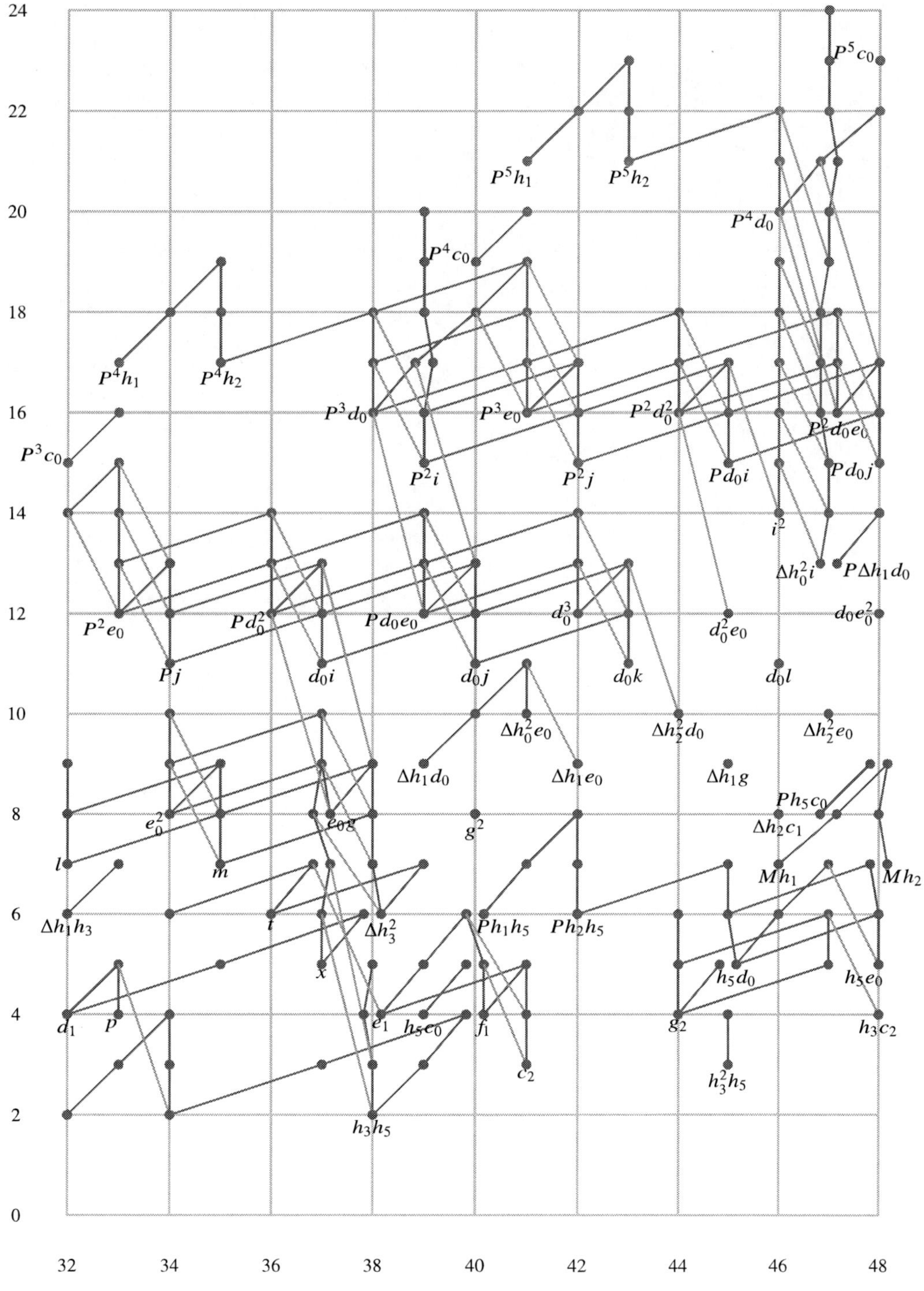

FIGURE 2
The Adams E_2-page in dimensions 32–48

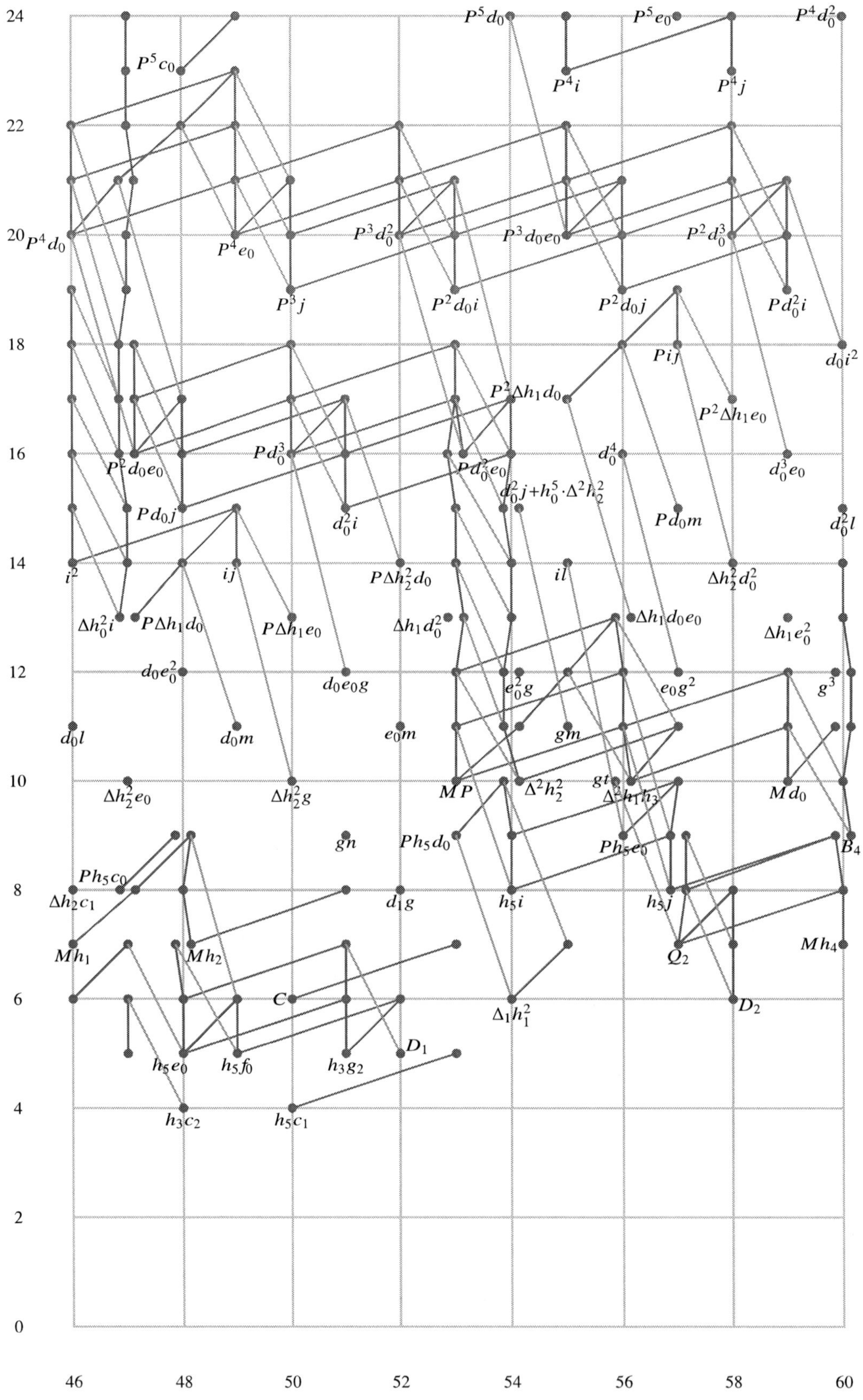

FIGURE 3
The Adams E_2-page in dimensions 46–60

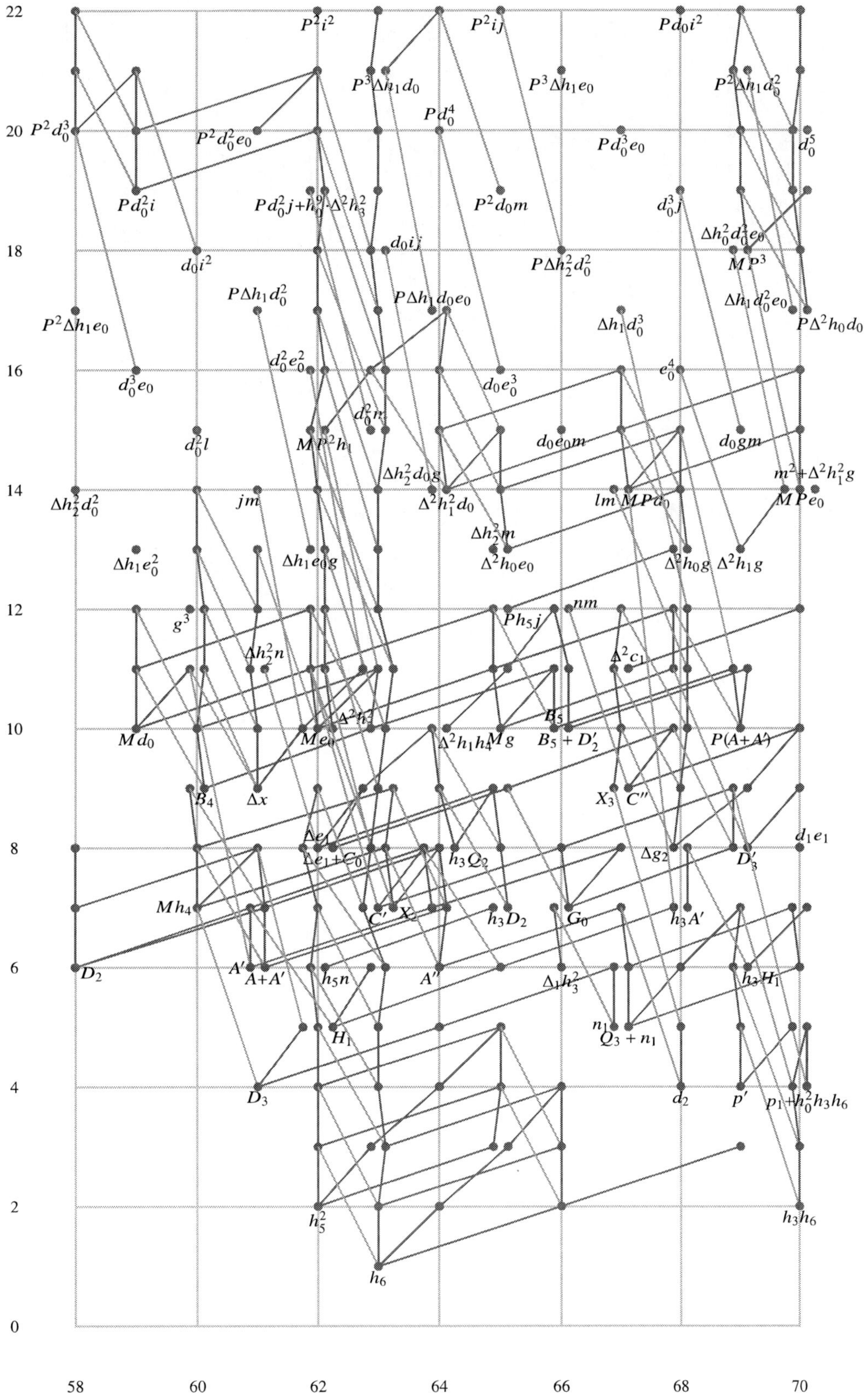

FIGURE 4

The Adams E_2-page in dimensions 58–70

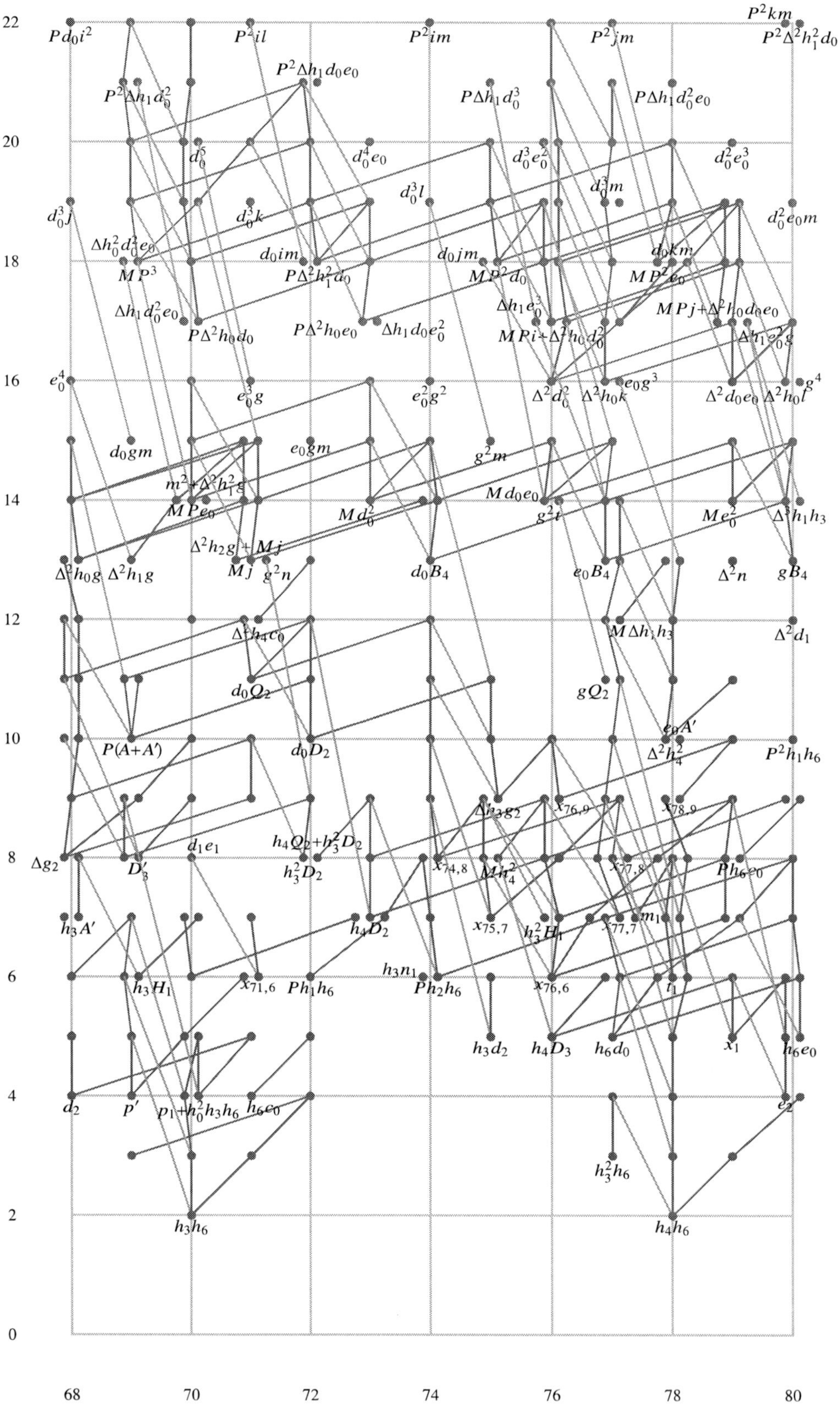

FIGURE 5

The Adams E_2-page in dimensions 68–80

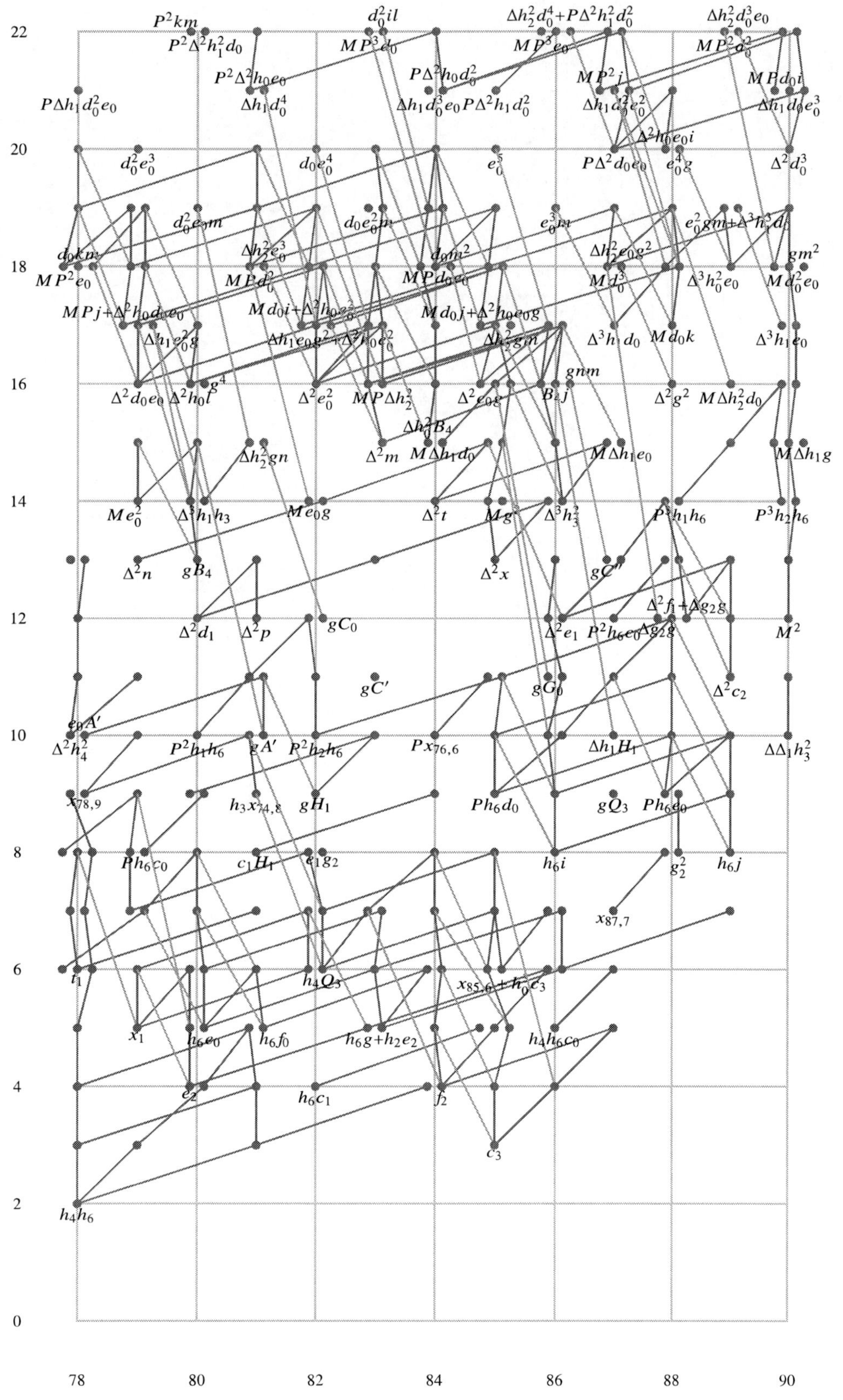

FIGURE 6
The Adams E_2-page in dimensions 78–90

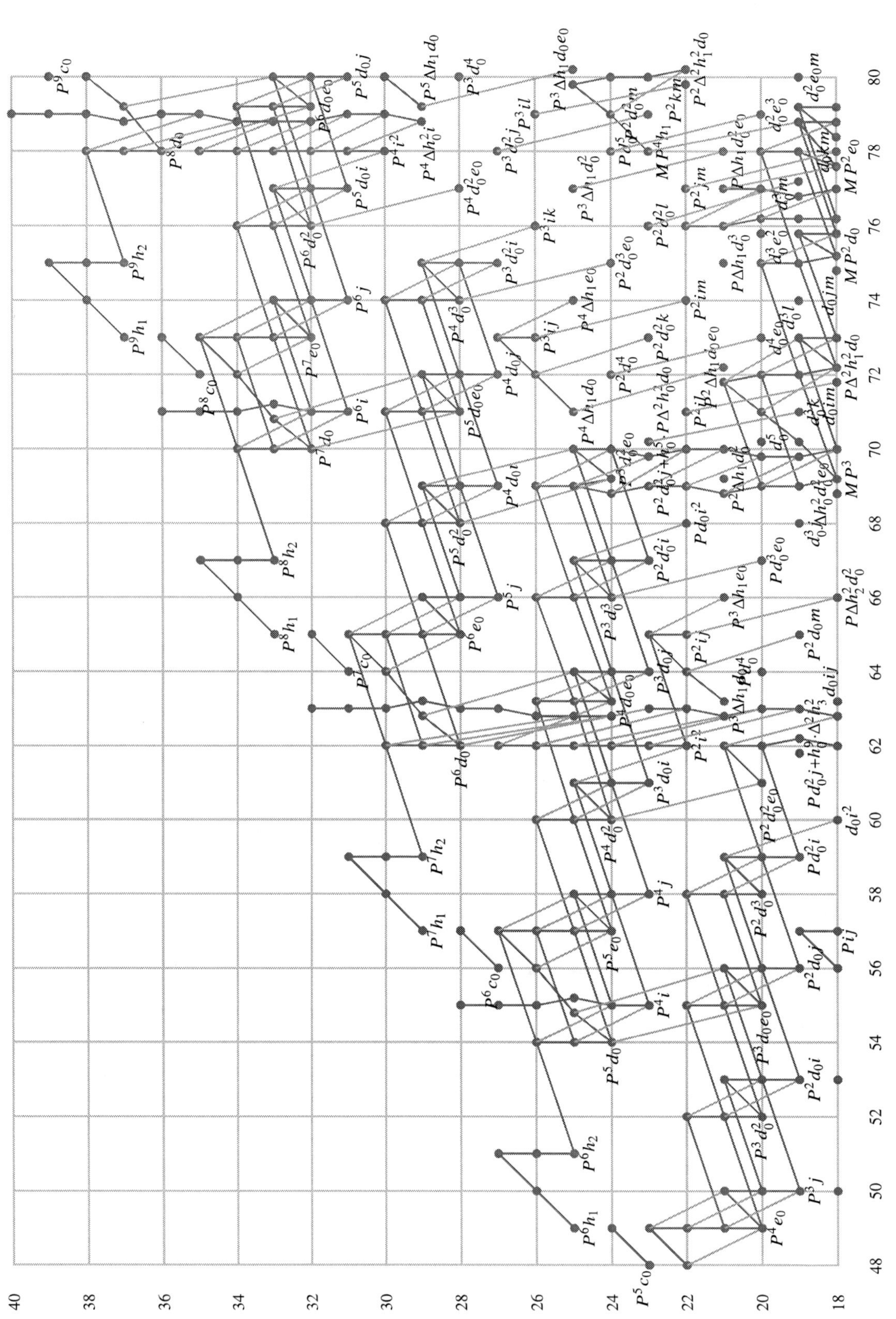

FIGURE 7

The Adams E_2-page in dimensions 48–80 in high filtration

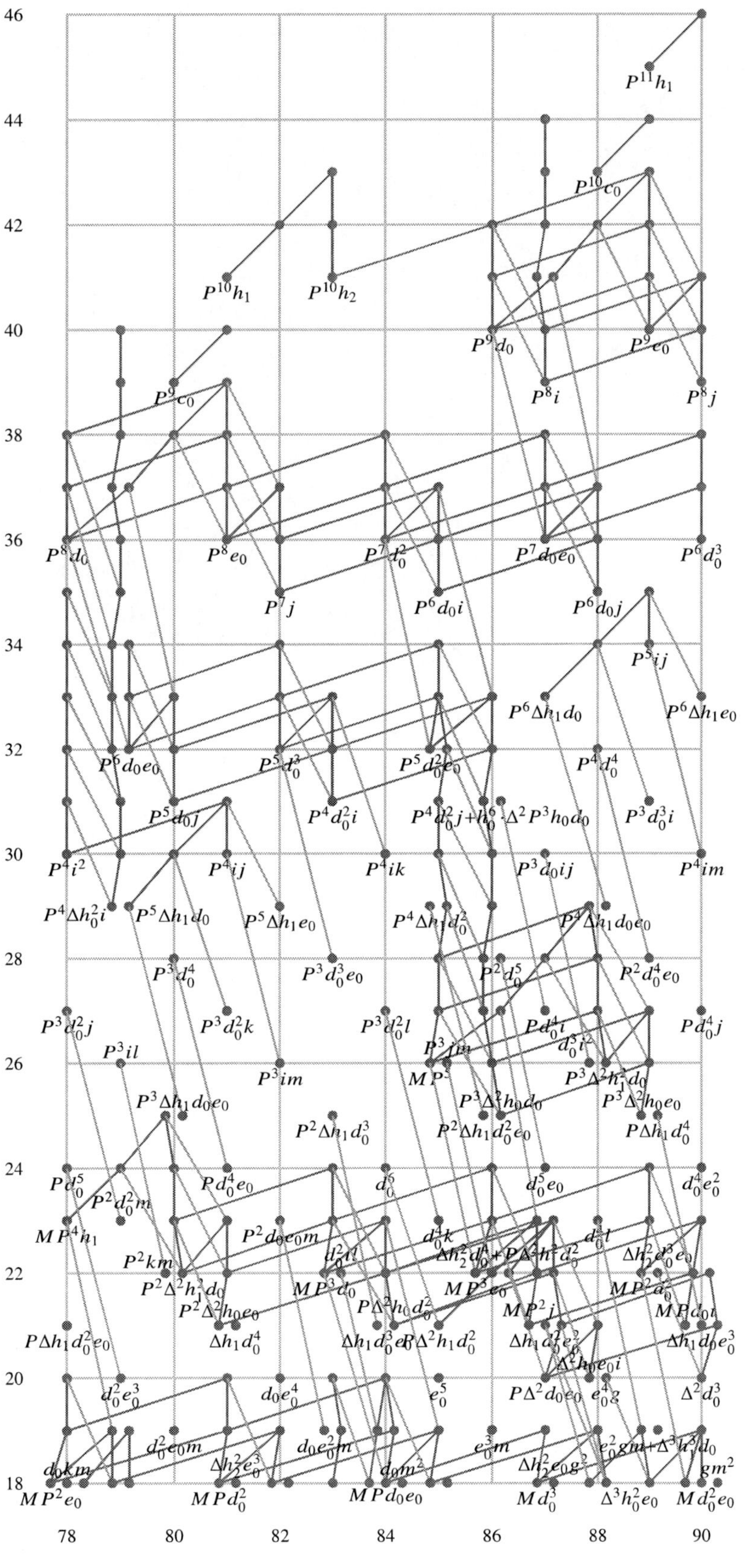

FIGURE 8

The Adams E_2-page in dimensions 78–90 in high filtration

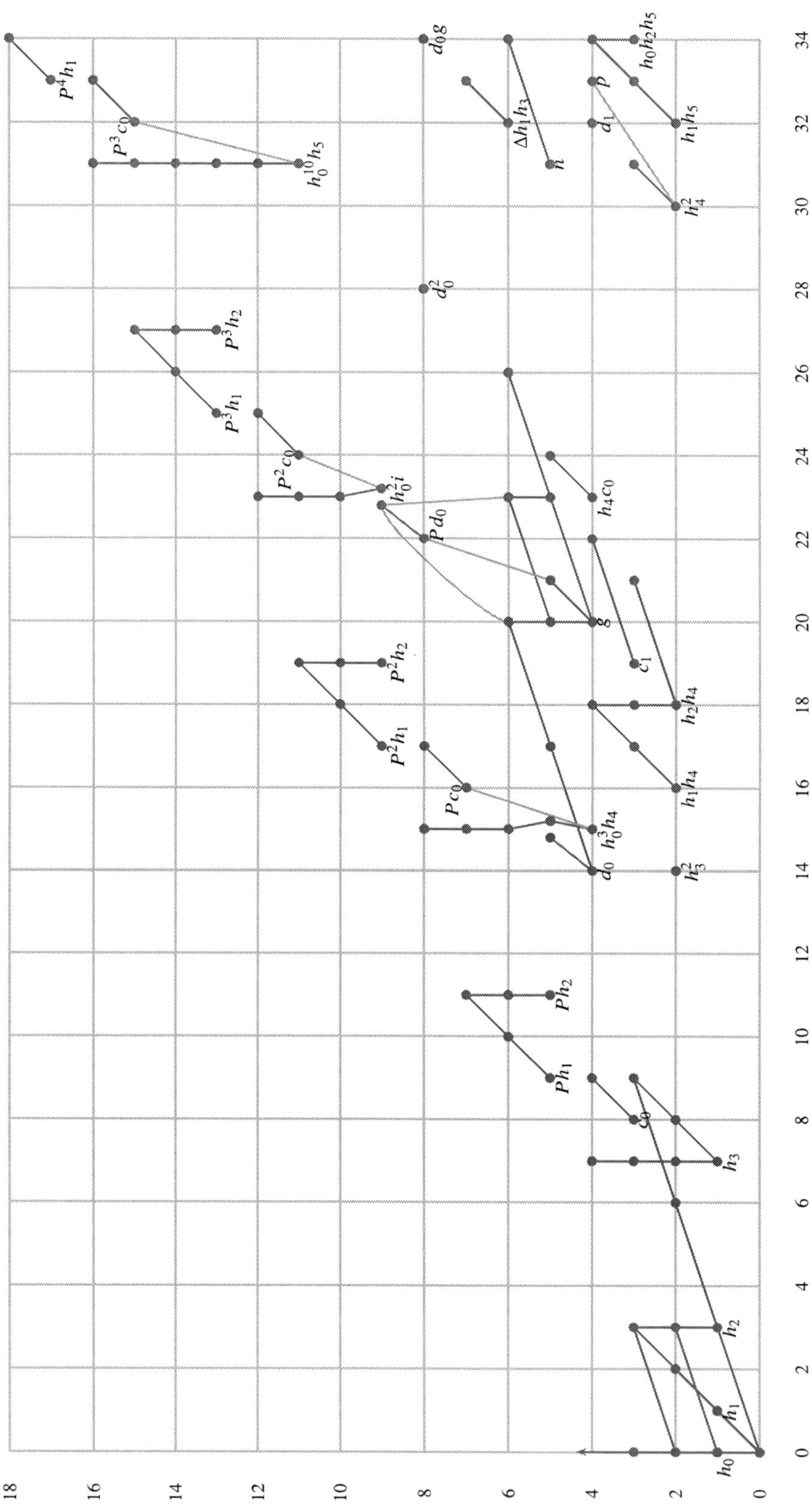

FIGURE 9

The Adams E_∞-page in dimensions 0–34

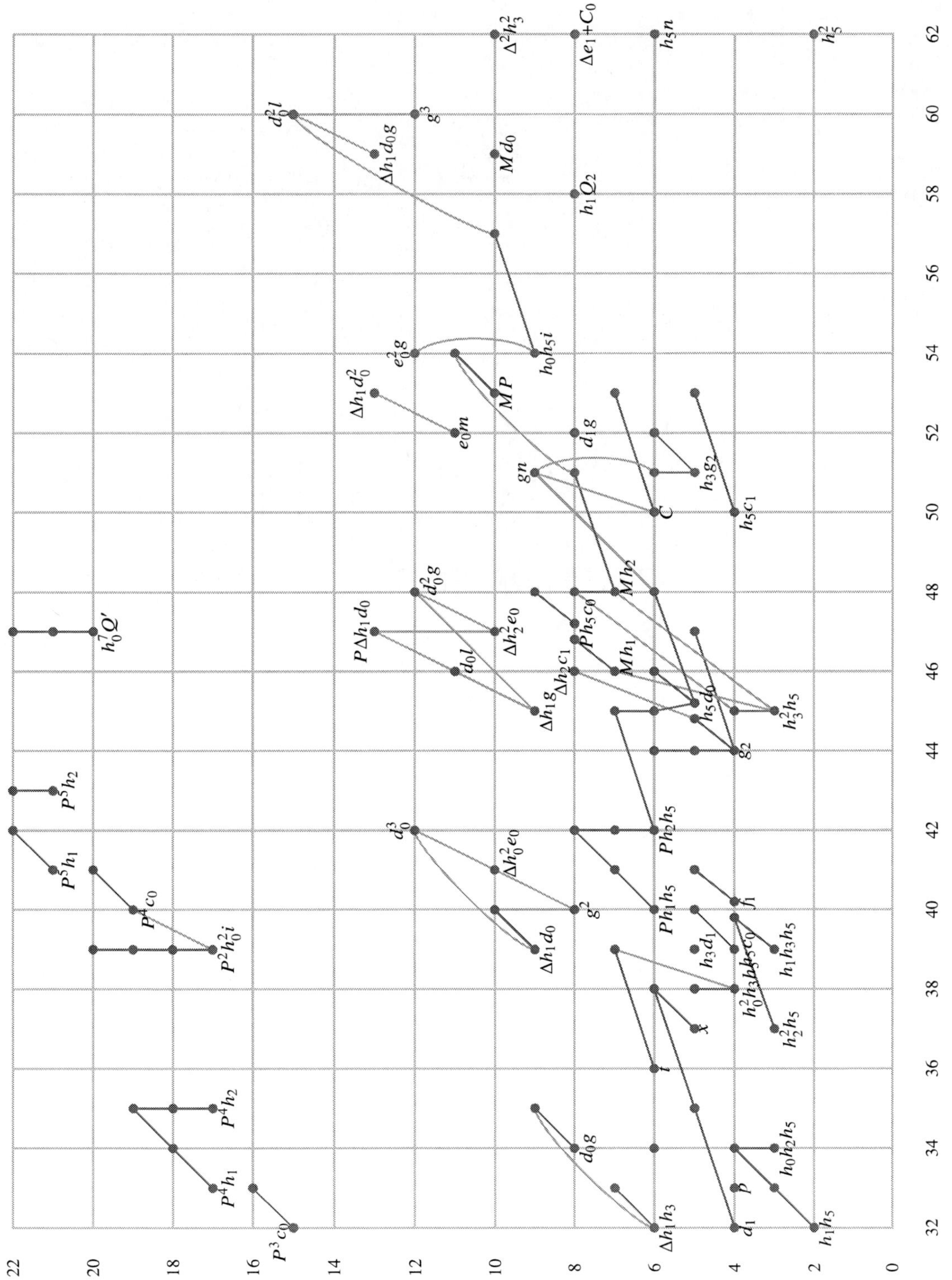

FIGURE 10
The Adams E_∞-page in dimensions 32–62

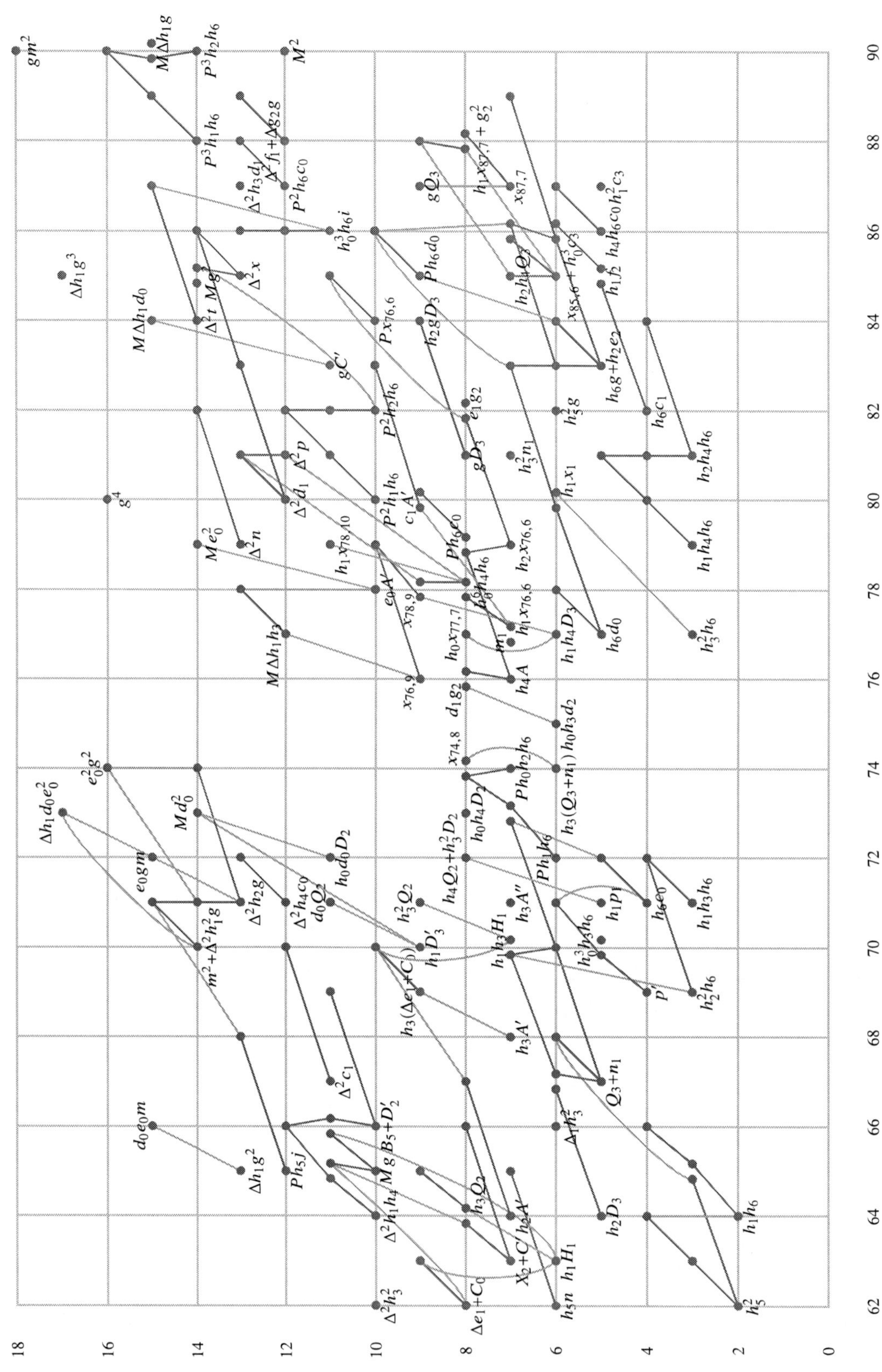

FIGURE 11
The Adams E_∞-page in dimensions 62–90

REFERENCES

[1] J. F. Adams, On the structure and applications of the Steenrod algebra. *Comment. Math. Helv.* **32** (1958), 180–214.

[2] T. Bachmann, H. J. Kong, G. Wang, and Z. Xu, The Chow t-structure on the ∞-category of motivic spectra. 2020, arXiv:2012.02687. To appear in *Ann. of Math.*

[3] M. G. Barratt, J. D. S. Jones, and M. E. Mahowald, Relations amongst Toda brackets and the Kervaire invariant in dimension 62. *J. Lond. Math. Soc. (2)* **30** (1984), no. 3, 533–550.

[4] M. G. Barratt, M. E. Mahowald, and M. C. Tangora, Some differentials in the Adams spectral sequence. II. *Topology* **9** (1970), 309–316.

[5] H.-J. Baues and M. Jibladze, Dualization of the Hopf algebra of secondary cohomology operations and the Adams spectral sequence. *J. K-Theory* **7** (2011), no. 2, 203–347.

[6] M. Behrens, M. Hill, M. J. Hopkins, and M. Mahowald, Detecting exotic spheres in low dimensions using coker J. *J. Lond. Math. Soc. (2)* **101** (2020), no. 3, 1173–1218.

[7] M. Behrens, M. Mahowald, and J. D. Quigley, The 2-primary Hurewicz image of *tmf*. 2020, arXiv:2011.08956.

[8] R. Bruner, A new differential in the Adams spectral sequence. *Topology* **23** (1984), no. 3, 271–276.

[9] R. Burklund, An extension in the adams spectral sequence in dimension 54. *Bull. Lond. Math. Soc.* **53** (2021), 404–407.

[10] D. Chua, Adams differentials via the secondary Steenrod algebra. 2021, arXiv:2105.07628.

[11] D. Dugger and D. C. Isaksen, The motivic Adams spectral sequence. *Geom. Topol.* **14** (2010), no. 2, 967–1014.

[12] H. Freudenthal, Über die Klassen der Sphärenabbildungen I. Große Dimensionen. *Compos. Math.* **5** (1938), 299–314.

[13] B. Gheorghe, D. C. Isaksen, A. Krause, and N. Ricka, $\mathbb{C}$-motivic modular forms. To appear in *J. Eur. Math. Soc. (JEMS).*

[14] B. Gheorghe, G. Wang, and Z. Xu, The special fiber of the motivic deformation of the stable homotopy category is algebraic. *Acta Math.* **226** (2021), no. 2, 319–407.

[15] P. Goerss, The Adams–Novikov spectral sequence and the homotopy groups of spheres, 2007, https://sites.math.northwestern.edu/~pgoerss/papers/stras1.pdf.

[16] M. A. Hill, M. J. Hopkins, and D. C. Ravenel, On the nonexistence of elements of Kervaire invariant one. *Ann. of Math. (2)* **184** (2016), no. 1, 1–262.

[17] M. Hovey, Homotopy theory of comodules over a Hopf algebroid. In *Homotopy theory: relations with algebraic geometry, group cohomology, and algebraic K-theory*, pp. 261–304, Contemp. Math. 346, Amer. Math. Soc., Providence, RI, 2004.

[18] D. C. Isaksen and S. stems. *Mem. Amer. Math. Soc.* **262** (2019), no. 1269, viii+159.

[19] D. C. Isaksen, G. Wang, and Z. Xu, More stable stems. 2020, arXiv:2001.04511.

[20] D. C. Isaksen, G. Wang, and Z. Xu, Stable homotopy groups of spheres. *Proc. Natl. Acad. Sci.* **117** (2020), no. 40, 24757–24763.

[21] D. C. Isaksen and Z. Xu, Motivic stable homotopy and the stable 51 and 52 stems. *Topology Appl.* **190** (2015), 31–34.

[22] M. A. Kervaire and J. W. Milnor, Groups of homotopy spheres. I. *Ann. of Math. (2)* **77** (1963), 504–537.

[23] S. O. Kochman, *Stable homotopy groups of spheres. A computer-assisted approach*. Lecture Notes in Math. 1423, Springer, Berlin, 1990.

[24] S. O. Kochman and M. E. Mahowald, On the computation of stable stems. In *The Čech centennial (Boston, MA, 1993)*, pp. 299–316, Contemp. Math. 181, Amer. Math. Soc., Providence, RI, 1995.

[25] M. Mahowald and M. Tangora, Some differentials in the Adams spectral sequence. *Topology* **6** (1967), 349–369.

[26] J. P. May, *The cohomology of restricted Lie algebras and of Hopf algebras: application to the Steenrod algebra*. Ph.D. Thesis, ProQuest LLC, Ann Arbor, MI, 1964.

[27] H. R. Miller, *Some algebraic aspects of the Adams–Novikov spectral sequence*. Ph.D. Thesis, ProQuest LLC, Ann Arbor, MI, 1975.

[28] J. Milnor, Differential topology forty-six years later. *Notices Amer. Math. Soc.* **58** (2011), no. 6, 804–809.

[29] J. Morava, Noetherian localisations of categories of cobordism comodules. *Ann. of Math. (2)* **121** (1985), no. 1, 1–39.

[30] F. Morel, *Théorie homotopique des schémas*. Astérisque (1999), no. 256, vi+119.

[31] F. Morel and V. Voevodsky, $\boldsymbol{A}^1$-homotopy theory of schemes. *Publ. Math. Inst. Hautes Études Sci.* **90** (1999), no. 90, 45–143.

[32] C. Nassau, On the secondary Steenrod algebra. *New York J. Math.* **18** (2012), 679–705.

[33] S. P. Novikov, Methods of algebraic topology from the point of view of cobordism theory. *Izv. Ross. Akad. Nauk Ser. Mat.* **31** (1967), 855–951.

[34] P. Pstrągowski, Synthetic spectra and the cellular motivic category. 2018, arXiv:1803.01804.

[35] D. Quillen, On the formal group laws of unoriented and complex cobordism theory. *Bull. Amer. Math. Soc.* **75** (1969), 1293–1298.

[36] D. C. Ravenel, *Complex cobordism and stable homotopy groups of spheres*. Pure Appl. Math. 121, Academic Press, Inc., Orlando, FL, 1986.

[37] E. Riehl, *Categorical homotopy theory*. New Math. Monogr. 24, Cambridge University Press, 2014.

[38] O. Röndigs, M. Spitzweck, and P. A. Østvær, The first stable homotopy groups of motivic spheres. *Ann. of Math. (2)* **189** (2019), no. 1, 1–74.

[39] J.-P. Serre, Homologie singulière des espaces fibrés. Applications. *Ann. of Math. (2)* **54** (1951), 425–505.

[40] J.-P. Serre, Groupes d'homotopie et classes de groupes abéliens. *Ann. of Math. (2)* **58** (1953), 258–294.

[41] H. Toda, *Composition methods in homotopy groups of spheres*. Ann. of Math. Stud. 49, Princeton University Press, Princeton, NJ, 1962.

[42] V. Voevodsky, Motivic cohomology with $\mathbf{Z}/2$-coefficients. *Publ. Math. Inst. Hautes Études Sci.* **98** (2003), 59–104.

[43] V. Voevodsky, On motivic cohomology with $\mathbf{Z}/l$-coefficients. *Ann. of Math. (2)* **174** (2011), no. 1, 401–438.

[44] G. Wang, https://github.com/pouiyter/morestablestems.

[45] G. Wang, Computations of the Adams–Novikov E_2-term. *Chin. Ann. Math. Ser. B* **42** (2021), no. 4, 551–560.

[46] G. Wang and Z. Xu, A survey of computations of homotopy groups of spheres and cobordisms, 2010, https://sites.google.com/view/xuzhouli/research.

[47] G. Wang and Z. Xu, The triviality of the 61-stem in the stable homotopy groups of spheres. *Ann. of Math. (2)* **186** (2017), no. 2, 501–580.

[48] Z. Xu, Conference talk "computing stable homotopy groups of spheres" at homotopy theory: tools and applications, University of Illinois at Urbana-Champaign, 2017.

DANIEL C. ISAKSEN

Department of Mathematics, Wayne State University, Detroit, MI 48202, USA, isaksen@wayne.edu

GUOZHEN WANG

Shanghai Center for Mathematical Sciences, Fudan University, Shanghai 200433, China, wangguozhen@fudan.edu.cn

ZHOULI XU

Department of Mathematics, University of California San Diego, La Jolla, CA 92093, USA, xuzhouli@ucsd.edu

This Page Intentionally Left Blank

SURFACE AUTOMORPHISMS AND FINITE COVERS

YI LIU

ABSTRACT

This article surveys recent progress on virtual properties of surface automorphisms.

MATHEMATICS SUBJECT CLASSIFICATION 2020

Primary 57K32; Secondary 20F65, 37C25

KEYWORDS

Pseudo-Anosov mapping, fixed point theory, 3-manifold, profinite group

ICM INTERNATIONAL CONGRESS OF MATHEMATICIANS 2022 JULY 6–14

Proc. Int. Cong. Math. 2022, Vol. 4, pp. 2792–2805
DOI 10.4171/ICM2022/70

Published by EMS Press

1. INTRODUCTION

Self-homeomorphisms of a topological space can be studied through their mapping tori. This very basic observation connects surface automorphisms with 3-manifold theory. In this survey, we focus on recent applications of virtual properties of 3-manifold groups to surface automorphisms and their lifts to finite covers. We collect results and techniques in that direction. We mention some currently open questions, most of which are reformulated from more general 3-manifold versions. We review necessary background to make our exposition accessible to non-expert readers.

Throughout this survey, a *surface* S refers to a connected compact orientable 2-manifold, possibly with boundary, and a *surface automorphism* (S, f) refers to an orientation-preserving self-homeomorphism $f: S \to S$. A *covering* between surface automorphisms $(S', f') \to (S, f)$ refers to an (unramified) covering projection $\kappa: S' \to S$ which is equivariant with respect to the pair of automorphisms (that is, $f \circ \kappa = \kappa \circ f'$). By saying that a covering $(S', f') \to (S, f)$ is finite, regular, characteristic, or so on, we mean that the referred property holds for $S' \to S$.

2. SURFACE AUTOMORPHISMS AFTER NIELSEN AND THURSTON

We recall some aspects about surface automorphisms that have been well developed since the mid-1970s. Our summary puts an emphasis on characterizing dynamical properties of a surface automorphism in terms of the fundamental group of its mapping torus. To this end, we denote the *mapping torus* of any surface automorphism (S, f) as

$$M_f = \frac{S \times \mathbb{R}}{(x, r+1) \sim (f(x), r)},$$

which is topologically a connected compact orientable 3-manifold, with boundary a possibly empty disjoint union of tori. Note that we follow the dynamical convention. It makes sure that translation along the $\mathbb{R}$-factor $S \times \mathbb{R} \to S \times \mathbb{R}: (x, r) \mapsto (x, r+t)$ descends to the (forward) *suspension flow* $\theta_t: M_f \to M_f$, which is a continuous family of self-homeomorphisms parametrized by $t \in \mathbb{R}$, such that θ_0 is the identity. We denote by $\phi_f \in H^1(M_f; \mathbb{Z})$ the distinguished cohomology class homotopically represented by the natural projection $M_f \to \mathbb{R}/\mathbb{Z}$.

2.1. Classification of mapping classes

For surfaces of positive or zero Euler characteristic, the isotopy classes of their automorphisms are easy to describe. When S is a sphere or a disk, any automorphism f of S is isotopic to the identity. When S is an annulus, parametrized as $\mathbb{R}/\mathbb{Z} \times [-1, 1]$, any automorphism f of S is isotopic to either the identity or the involution $(x + \mathbb{Z}, y) \mapsto (-x + \mathbb{Z}, -y)$. When S is a torus, parametrized as $(\mathbb{R} \times \mathbb{R})/(\mathbb{Z} \times \mathbb{Z})$, any automorphism f of S is isotopic to a unique linear automorphism represented by a matrix in $\mathrm{SL}(2, \mathbb{Z})$.

In general, the Nielsen–Thurston classification asserts that the isotopy class of any surface automorphism falls into one of three types: periodic, reducible, or pseudo-Anosov. The above description with torus automorphisms provides a prototype of the classification, and the three types correspond to the representing matrix in $\mathrm{SL}(2, \mathbb{Z})$ being elliptic/central, parabolic/central, or hyperbolic, respectively (as a fractional linear transformation on the upper-half complex plane). In general, a surface automorphism is said to be *periodic* if it has finite order under iteration, or *reducible* if it preserves a union of mutually disjoint, essential simple closed curves on the surface. A *pseudo-Anosov* automorphism refers to a surface automorphism (S, f) such that f preserves a pair of (transversely) measured foliations on the interior of S, and rescales the measures by some factors $\lambda > 1$ and λ^{-1}, respectively. Unlike an Anosov torus automorphism, the foliations in the pseudo-Anosov case are allowed to have prong singularities (having prong number ≥ 3 at points in the interior, or ≥ 1 at the ends as punctures). We also require the pair of foliations to be transverse to each other except at a common finite set of singular points. See [8, EXPOSÉ 1].

When S has negative Euler characteristic, an automorphism f of S is periodic up to isotopy if and only if the mapping torus M_f supports the $\mathbf{H}^2 \times \mathbb{R}$ geometry (and hence also the $\widetilde{\mathrm{SL}_2}(\mathbb{R})$ geometry with $\partial S \neq \emptyset$); f is reducible up to isotopy if and only if M_f has nontrivial geometric decomposition; f is pseudo-Anosov up to isotopy if and only if M_f supports the 3-dimensional hyperbolic geometry $\mathbf{H}^3$. Moreover, in the reducible case, a collection of curves on S for reducing f can be obtained by intersecting any essential torus or Klein bottle in M_f (minimally up to isotopy) with the distinguished fiber $S \times \{0\}$. See [3, CHAPTER 1].

2.2. Periodic orbit classes and indices

Given a surface automorphism, one could freely ask if there are any fixed points. Nielsen's fixed point theory is more than to answer yes or no. The general theory applies to continuous self-maps of compact connected simplicial complexes. Instead of considering individual fixed points, which may disappear or duplicate under homotopy, one will consider abstract fixed point classes, and distinguish finitely many essential ones from the others. Then the essential fixed point classes will depend only on the homotopy class of the self-map, and each of them will guarantee at least one distinct fixed point. Below we follow the mapping torus approach as suggested by B. Jiang in the survey [13]; see also [11] for more detail.

Let (S, f) be a surface automorphism. Denote by $\mathrm{Fix}(f) \subset S$ the set of fixed points. For any $x \in \mathrm{Fix}(f)$, we obtain a 1-periodic trajectory (of the suspension flow) $\gamma_x: \mathbb{R}/\mathbb{Z} \to M_f$ (coming from the line $x \times \mathbb{R}$ in $S \times \mathbb{R}$). We say that $x, y \in \mathrm{Fix}(f)$ are *of the same* fixed point class if γ_x and γ_y are freely homotopic in M_f. More abstractly, a *fixed point class* of f can be defined as a free homotopy loop γ in M_f, such that $\phi_f(\gamma) = \langle \phi_f, [\gamma] \rangle$ equals 1. Every fixed point class $\mathbf{p}$ has a well-defined *fixed point index* $\mathrm{ind}(f; \mathbf{p}) \in \mathbb{Z}$, which can be described as follows.

Note that $\mathrm{Fix}(f) \subset S$ is a union of mutually disjoint isolated connected closed subsets, with only finitely many components (since S is compact). The subset $\mathrm{Fix}(f; \mathbf{p}) \subset \mathrm{Fix}(f)$ of fixed point class $\mathbf{p}$ is a subunion of those components. If $\mathrm{Fix}(f; \mathbf{p})$ is empty, $\mathrm{ind}(f; \mathbf{p})$ equals zero. Otherwise, take (a smooth structure of S and) a smooth homotopy perturbation $\tilde{f}$ of f, supported in an open neighborhood U of $\mathrm{Fix}(f; \mathbf{p})$ away from the rest of $\mathrm{Fix}(f)$; make sure that $\tilde{f}$ has only nondegenerate fixed points in U (that is, for any $x \in \mathrm{Fix}(\tilde{f}) \cap U$, the tangent map $\mathrm{d}\tilde{f}|_x: T_x S \to T_x S$ has no eigenvalue 1). Then $\mathrm{ind}(f; \mathbf{p})$

can be calculated as the sum of the sign $+1$ or -1 of $\det(\mathrm{id} - \mathrm{d}\tilde{f}|_x)$, where x ranges over $\mathrm{Fix}(\tilde{f}) \cap U$.

A fixed point class is said to be *essential* if its index is nonzero. In particular, every essential fixed point class is represented by a fixed point of f. If S is closed and if f is pseudo-Anosov, every fixed point represents a distinct essential fixed point class. In this case, there are simple rules for telling the fixed point index. When a fixed point x is a k-prong singularity (of either of the invariant foliations), its index equals $1-k$ if f preserves every prong at x, otherwise its index equals 1; when x is not singular, its index equals -1 or 1 according as f preserves or reverses an orientation of the leaf through x. For general surface automorphisms, it is also possible to characterize all the essential fixed point classes and their index, in terms of normal forms in the Nielsen–Thurston classification [14].

For any positive integer $m \in \mathbb{N}$, an *m-periodic orbit class* of f can be defined as a free homotopy loop γ in M_f with $\phi_f(\gamma) = m$. The *m-index* of the m-periodic class is defined as the sum of the fixed point indices of γ' with respect to f^m, where γ' ranges over all the free homotopically distinct lifts of γ to the m-cyclic cover M_{f^m}. (The indices of different lifts are actually equal, so the summation only counts one value with suitable multiplicity.) Finally, *essential* m-periodic orbit classes are those of nonzero m-index.

2.3. Homological directions

Let (S, f) be a surface automorphism. As every periodic orbit class $\mathbf{p}$ is free-homotopically represented by a periodic trajectory in M_f, its homology class is a well-defined element $[\mathbf{p}] \in H_1(M_f; \mathbb{Z})$. Passing to real coefficients, there is a unique minimal convex cone in $H_1(M_f; \mathbb{R})$ (formed by linear rays emanating from the origin) that contains all the homology classes of the essential periodic trajectories. Fried shows that this cone is polyhedral. In other words, it is the convex hull of finitely many extreme rays. If f is pseudo-Anosov, this cone has codimension zero in $H_1(M_f; \mathbb{R})$. The directions of rays in the cone are called the (essential) *homological directions* of the suspension flow. Fried's cone of homological directions is exactly dual to the Thurston's fibered cone that contains $\phi_f \in H^1(M_f; \mathbb{R})$, as we elaborate below, based on Fried's exposition [8, EXPOSÉ 14].

We recall that the Thurston norm is defined for any compact connected orientable 3-manifold N as a seminorm on the real linear space $H^1(N; \mathbb{R})$. It is nondegenerate if N supports the 3-dimensional hyperbolic geometry (of finite volume). It is characterized by the property that for any integral cohomology class $\phi \in H^1(N; \mathbb{Z})$, the Thurston norm of ϕ is the minimum of the complexity among all properly embedded oriented surfaces $(S, \partial S) \subset (N, \partial N)$ homologous to the Poincaré–Lefschetz dual of ϕ in $H_2(N, \partial N; \mathbb{Z})$. Here, the complexity of S refers to $\sum_{i=1}^{k} \max(0, -\chi(S_i))$, where $S_1, \dots, S_k$ enumerate the connected components of S.

The unit ball of the Thurston norm of N is a (possibly noncompact) convex polyhedron of codimension zero in $H^1(N; \mathbb{R})$, central symmetric about the origin. Its dual is a (possibly positive codimensional) compact convex polyhedron in $H_1(N; \mathbb{R}) \cong \mathrm{Hom}_{\mathbb{R}}(H^1(N; \mathbb{R}), \mathbb{R})$. Moreover, if $\phi \in H^1(N; \mathbb{Z})$ is fibered (that is, homotopically represented by a bundle projection onto the circle with surface fibers), Thurston shows that ϕ

is contained in the cone over an open codimension-one face the Thurston norm unit ball, in which the integral classes are all fibered. Such cones are called the *fibered cones* of N, over the *fibered faces* of the Thurston norm unit ball in $H^1(N;\mathbb{R})$. The fibered faces are dual to a collection of vertices in the dual of the Thurston norm unit ball in $H_1(N;\mathbb{R})$, which we may reasonably call the *flow vertices*.

For the mapping torus M_f of a surface automorphism (S,f), we find a distinguished fibered cone in $H^1(M_f;\mathbb{R})$ that contains the distinguished cohomology class ϕ_f. When S has negative Euler characteristic, the corresponding flow vertex can be figured out as $-e_f/2$ in $H^2(M_f,\partial M_f;\mathbb{R}) \cong H_1(M_f;\mathbb{R})$, where $e_f \in H^2(M_f,\partial M_f;\mathbb{Z})$ denotes the relative Euler class of the (oriented, ∂-transverse) vertical tangent bundle of M_f with respect to its fibering over $\mathbb{R}/\mathbb{Z}$, oriented compatibly to make $\phi_f(e_f) = \chi(S)$. We can naturally identify the tangent space of $H_1(M_f;\mathbb{R})$ at the opposite vertex $e_f/2$ as $H_1(M_f;\mathbb{R})$. Then Fried's cone of homological directions consists exactly of those tangent vectors at $e_f/2$ pointing into (the corner of) the polytope dual to the Thurston norm unit ball.

2.4. Various zeta functions

For any surface automorphism (S,f), the *Nielsen zeta function* is a useful tool for analyzing the iteration dynamics. It can be defined by the following expression:

$$\zeta_{N,f}(t) = \exp\left(\sum_{m=1}^{\infty} \frac{N(f^m)}{m} \cdot t^m\right)$$

where $N(f^m)$ denotes the number of essential fixed points of f^m, called the *Nielsen number* of f^m. When f is pseudo-Anosov with stretch factor $\lambda > 1$, the Nielsen numbers $N(f^m)$ grow exponentially as

$$\overline{\lim_{m\to\infty}}\, N(f^m)^{1/m} = \lambda.$$

More generally, the above limit superior is equal to the maximum stretch factor among the pseudo-Anosov components in the Nielsen–Thurston decomposition, or 1 if all the components are periodic. In other characterizations, the logarithm of that value is known to be the mapping-class topological entropy of f. In particular, $\zeta_{N,f}(t)$ converges absolutely as a complex analytic function in t in a neighborhood of 0. It is known that $\zeta_{N,f}(t)$ is a radical of a rational function in t near 0.

The *Lefschetz zeta function* $\zeta_{L,f}(t)$ of (S,f) is defined using the Lefschetz numbers $L(f^m)$ instead of the Nielsen numbers $N(f^m)$. This makes $\zeta_{L,f}(t)$ easier to calculate than $\zeta_{N,f}(t)$. Indeed, recall that $L(f^m)$ is equal to the alternating sum of the traces of f_*^m on $H_*(S;\mathbb{Q})$. It follows that $\zeta_{L,f}(t)$ is equal to $t^{-\chi(S)}$ divided by the alternating product of the characteristic polynomials of f_* on $H_*(S;\mathbb{Q})$. The resulting form can be recognized as (a representative of) the Reidemeister torsion of M_f with respect to ϕ_f. This is an instance of a general connection between twisted Lefschetz zeta functions and twisted Reidemeister torsions.

Let R be a commutative domain that contains $\mathbb{Z}$. Suppose that $\rho\colon \pi_1(M_f) \to \mathrm{GL}(n,R)$ is a linear representation over R^n. The *twisted Lefschetz zeta function* of (S,f)

with respect to ρ is defined as

$$\zeta^{\rho}_{L,f}(t) = \exp\left(\sum_{m=1}^{\infty} \frac{\sum_{\mathbf{p}} \chi_\rho(\mathbf{p}) \cdot \mathrm{ind}_m(f;\mathbf{p})}{m} \cdot t^m\right),$$

where $\mathbf{p}$ ranges over all the m-periodic orbit classes; being a free homotopy loop, $\mathbf{p}$ also represents a conjugacy class in $\pi_1(M_f)$, so the character χ_ρ of ρ can be evaluated at $\mathbf{p}$, as the trace of ρ evaluated at any group element in that conjugacy class; the notation $\mathrm{ind}_m(f;\mathbf{p})$ stands for the m-index of $\mathbf{p}$ with respect to f (so the summation over $\mathbf{p}$ is essentially finite); finally, the whole expression is understood as a formal power series over the field of fractions $F(R)$ in an indeterminate t, with $\exp(z) = \sum_{n=0}^{\infty} z^n/n!$. In particular, $\zeta^{\rho}_{L,f}(t)$ is invariant under homotopy of f and conjugation of ρ. On the other hand, we simply recall that the *twisted Reidemeister torsion* $\tau^{\rho,\phi_f}_{M_f}(t)$ of M_f with respect to ϕ_f and ρ is well defined as an element in $F(R[t,t^{-1}]) = F(R)(t)$ up to units of $R[t,t^{-1}]$ (that is, up to factors in the multiplicative subgroup $(R[t,t^{-1}])^\times = R^\times \times t^{\mathbb{Z}}$); see **[9]**. Under the above assumptions, $\zeta^{\rho}_{L,f}(t)$ agrees with the power series expansion in t of a unique rational function over $F(R)$, and the identity

$$\tau^{\rho,\phi_f}_{M_f}(t) \doteq \zeta^{\rho}_{L,f}(t)$$

holds up to units of $R[t,t^{-1}]$; see **[12]** (and also **[17, LEMMA 8.2]**).

3. VIRTUAL HOMOLOGICAL EIGENVALUES

Let (S,f) be a surface automorphism. Since $H_1(S;\mathbb{Z})$ is a finitely generated free abelian group, the induced linear automorphism $f_*: H_1(S;\mathbb{Z}) \to H_1(S;\mathbb{Z})$ has a characteristic polynomial, which we denote as

$$\Delta_f(t) = \det_{\mathbb{Z}[t]}(t \cdot \mathrm{id} - f_*).$$

This is a monic polynomial over $\mathbb{Z}$ with the property $\Delta_f(1) = \pm 1$. If S has genus g and h boundary components, $\Delta_f(t)$ factorizes as the product of a reciprocal polynomial of degree $2g$ and cyclotomic factors of total degree $\max(0, h-1)$, because f preserves ∂S and descends to an automorphism of the closed surface obtained by filling ∂S with disks. Moreover, $\Delta_f(t)$ can be recognized as the (first) Alexander polynomial of M_f with respect to ϕ_f, the latter being well-defined in $\mathbb{Z}[t,t^{-1}]$ up to units.

A *homological eigenvalue* of a surface automorphism (S,f) refers to a complex root of the polynomial $\Delta_f(t)$, and a *virtual homological eigenvalue* of (S,f) refers to a complex root of the polynomial $\Delta_{f'}(t)$ where $(S',f') \to (S,f)$ is some finite covering. We are interested in a general question as to which complex values may occur as virtual homological eigenvalues of a given surface automorphism. Moreover, how do they reflect the dynamical complexity of its isotopy class?

We start with the following well-known, simple observation.

Theorem 3.1. *If a surface automorphism has no pseudo-Anosov type components in its Nielsen–Thurston decomposition, then its virtual homological eigenvalues are all roots of unity.*

Proof. The condition implies that some finite iterate of the given surface automorphism is isotopic to a finite product of left- or right-hand Dehn twists along mutually disjoint simple closed curves. Then the characteristic polynomial of that iterate is a power of $t - 1$. The conclusion follows because the condition also holds for any finite covering of the given surface automorphism. ∎

There are many pseudo-Anosov automorphisms on any surface of negative Euler characteristic, such that the induced homological action is trivial. However, when (S, f) is a pseudo-Anosov automorphism with transversely orientable invariant foliations, the stretch factor $\lambda > 1$ must occur as a homological eigenvalue. Moreover, if every singularity of the invariant foliations is formed with an even number of prongs, one may achieve the transverse orientability condition by passing to a covering (S', f') of degree at most 2, so λ still occurs as a virtual homological eigenvalue. Note that the same trick does not apply when there are singularities of odd prong numbers, because they locally obstruct the transverse orientability, and they lift locally homeomorphically to any covering.

The above facts lead to the first part of the following theorem. The second part is truly surprising, known as the gap theorem due to C. T. McMullen [22]. It is proved by comparing the Teichmüller metric on the Teichmüller space and the Kobayashi metric on Siegel spaces associated to finite covers.

Theorem 3.2. *Let (S, f) be a pseudo-Anosov automorphism with stretch factor $\lambda > 1$.*

(1) *If the invariant foliations of (S, f) have no singularities of odd prong numbers, then λ is a virtual homological eigenvalue of (S, f).*

(2) *Otherwise, there exists some constant $1 < r < \lambda$, depending only on (S, f), such that every virtual homological eigenvalue μ of (S, f) satisfies $|\mu| < r$.*

McMullen conjectured the converse of Theorem 3.1. The converse has been proved by the author [18], as the following theorem. The proof relies on the virtual specialization of hyperbolic 3-manifold groups.

Theorem 3.3. *If a surface automorphism has a pseudo-Anosov type component in its Nielsen–Thurston decomposition, then it has a virtual homological eigenvalue outside the complex unit circle.*

Remark 3.4. An analogous conjecture for outer automorphisms of finitely generated free groups is proved by A. Hadari [10]. The desired finite-index normal subgroup therein is constructed using nilpotent quotients. Hadari's result also implies Theorem 3.3 for surfaces with nonempty boundary.

An effective version of Theorem 3.3 is yet unknown. We pose the following question, as analogous to the Kojima–McShane inequality regarding pseudo-Anosov stretch factors [15].

Question 3.5. Suppose that (S, f) is an automorphism of a surface of negative Euler characteristic. Does the following inequality hold as μ ranges over all virtual homological eigenvalues of (S, f):

$$\sup_{\mu} \log|\mu| \geq \frac{1}{3\pi \cdot |\chi(S)|} \cdot \mathrm{Vol}(M_f)?$$

Here, $\mathrm{Vol}(M_f)$ denotes the Gromov norm of M_f times the volume of a regular ideal hyperbolic tetrahedron.

We mention another upper-bound estimate regarding the distribution of virtual homological eigenvalues. It is a quick consequence of a theorem due to T. T. Q. Lê [16]. Recall that the (multiplicative) Mahler measure of a nonzero complex polynomial $P(t) = c \cdot \prod_{j=1}^{d}(t - \xi_j)$ refers to the positive value $\mathbb{M}(P) = |c| \cdot \prod_{j=1}^{d} \max(1, |\xi_j|)$. In particular, $\mathbb{M}(P) \geq 1$ holds for any nonzero $P(t) \in \mathbb{Z}[t]$.

Theorem 3.6. *Suppose that* $\cdots \to (S'_n, f'_n) \to \cdots \to (S'_1, f'_1)$ *is a cofinal tower of surface automorphisms which are regular finite coverings of* (S, f)*. Then*

$$\overline{\lim_{n\to\infty}} \frac{\log \mathbb{M}(\Delta_n)}{[S'_n : S]} \leq \frac{1}{6\pi} \cdot \mathrm{Vol}(M_f),$$

where Δ_n *denotes the characteristic polynomial of* f'_{n*} *on* $H_1(S'_n; \mathbb{Z})$.

Proof. Note that $\mathbb{M}(\Delta_{g^m}) = \mathbb{M}(\Delta_g)^m$ for any $g = f'_n$ and any $m \in \mathbb{N}$. We can find some sequence $m_n \in \mathbb{N}$, such that the mapping tori M''_n of $(f'_n)^{m_n}$ form a cofinal tower of regular finite coverings over M_f. Then apply [16, THEOREM 1.1]. ■

The estimate in Theorem 3.6 would become an equality if the homological torsion growth conjecture could be proved to that generality [4, 20] (see also [16, CONJECTURE 1.3]). That would also imply a positive answer to Question 3.5.

4. DETERMINING PROPERTIES USING FINITE QUOTIENT ACTIONS

Let (S, f) be a surface automorphism. Fix a base point of S for speaking of $\pi_1(S)$. Using any path from the base point to its image under f, we can construct an automorphism of $\pi_1(S)$. Different choices of the path only affect the construction by inner automorphisms of $\pi_1(S)$. Therefore, for any characteristic subgroup K of $\pi_1(S)$, (S, f) induces a well-defined outer automorphism of the quotient group $\pi_1(S)/K$, which we denote as $[f]_K \in \mathrm{Out}(\pi_1(S)/K)$.

Theorem 4.1. *Let* (S, f_A) *and* (S, f_B) *be automorphisms of a closed surface. If* $[f_A]_K$ *is conjugate to* $[f_B]_K$ *in* $\mathrm{Out}(\pi_1(S)/K)$ *for every characteristic finite index subgroup* K *of* $\pi_1(S)$*, then* f_A *and* f_B *are of identical type in the Nielsen–Thurston classification.*

Remark 4.2. (1) Theorem 4.1 is a consequence of a theorem due to H. Wilton and P. A. Zalesskii [28]. They prove that the profinite group completion detects the geometric decomposition of any finitely generated 3-manifold group. In fact, their result implies that f_A and f_B as in Theorem 4.1 have isomorphic Nielsen–Thurston decomposition graph decorated with vertex types (as being periodic or pseudo-Anosov). See **[17, SECTION 12]** for an exposition.

(2) The condition in Theorem 4.1 defines an equivalence relation on the set of automorphisms, which passes to an equivalence relation on the mapping class group $\mathrm{Mod}(S)$. Equivalent mapping classes in this sense are said to be *procongruently conjugate* (thinking of any K as a "principal congruence subgroup" in $\pi_1(S)$ by analogy). Any procongruent conjugacy class in $\mathrm{Mod}(S)$ is a disjoint union of conjugacy classes in $\mathrm{Mod}(S)$. Being procongruently conjugate is equivalent as having conjugate image under the natural homomorphism $\mathrm{Mod}(S) \to \mathrm{Out}(\widehat{\pi})$, where $\widehat{\pi}$ denotes the profinite completion of $\pi = \pi_1(S)$. The homomorphism naturally factors through the profinite completion of $\mathrm{Out}(\pi)$, which is very different from $\mathrm{Out}(\widehat{\pi})$ in general. See **[17, SECTION 3]** for detailed discussion.

The following theorems are proved in **[17]**.

Theorem 4.3. *Let (S, f_A) and (S, f_B) be pseudo-Anosov automorphisms of a closed surface of genus ≥ 2. If $[f_A]_K$ is conjugate to $[f_B]_K$ in $\mathrm{Out}(\pi_1(S)/K)$ for every characteristic finite index subgroup K of $\pi_1(S)$, then f_A and f_B have identical stretch factor and their invariant foliations have identical number of singularities of each prong number. In fact, f_A and f_B have identical number of fixed points of each index.*

Theorem 4.4. *Let (S, f_B) be an automorphism of a closed surface. Then there exists a finite collection $\mathcal{B}$ of automorphisms of S with the following property. If (S, f_A) is any automorphism, such that $[f_A]_K$ is conjugate to $[f_B]_K$ in $\mathrm{Out}(\pi_1(S)/K)$ for every characteristic finite index subgroup K of $\pi_1(S)$, then f_A is isotopic to a topological conjugate of some $f_b \in \mathcal{B}$.*

Remark 4.5. Theorem 4.4 is equivalent to saying that every procongruent conjugacy class in $\mathrm{Mod}(S)$ is the disjoint union of finitely many conjugacy classes.

In the pseudo-Anosov case, the finiteness follows immediately from Theorem 4.3 (and Theorem 4.1), together with the well-known finiteness of pseudo-Anosov automorphisms with uniformly bounded stretch factor. See also **[19]** for a more recent finiteness result regarding profinite completions of finite-volume hyperbolic 3-manifold groups.

Example 4.6. Let S be the torus $(\mathbb{R} \times \mathbb{R})/(\mathbb{Z} \times \mathbb{Z})$. The mapping class group $\mathrm{Mod}(S)$ can be identified with $\mathrm{SL}(2, \mathbb{Z})$. In 1972, P. F. Stebe **[26]** discovered a pair of matrices

$$\begin{bmatrix} 188 & 275 \\ 121 & 177 \end{bmatrix} \quad \text{and} \quad \begin{bmatrix} 188 & 11 \\ 3025 & 177 \end{bmatrix}$$

which are not conjugate in $\mathrm{SL}(2, \mathbb{Z})$, or in $\mathrm{GL}(2, \mathbb{Z})$, but are conjugate in $\mathrm{GL}(2, \mathbb{Z}/N\mathbb{Z})$ for any natural number N.

The above example shows that the finiteness in Theorem 4.4 cannot be improved to uniqueness, in general. Nevertheless, we pose the following question:

Question 4.7. Let S be a surface of negative Euler characteristic. If (S, f_A) and (S, f_B) are pseudo-Anosov automorphisms, such that $[f_A]_K$ and $[f_B]_K$ are conjugate in $\mathrm{Out}(\pi_1(S)/K)$ for every characteristic finite index subgroup K of $\pi_1(S)$, is it true that $[f_A]$ and $[f_B]$ are conjugate in $\mathrm{Out}(\pi_1(S))$?

For a one-holed torus, M. R. Bridson, A. W. Reid, and H. Wilton have answered Question 4.7 affirmatively [6]. More generally, if one could prove that finite-volume hyperbolic 3-manifold groups are profinitely rigid among 3-manifold groups, a positive answer to Question 4.7 should follow from the $\widehat{\mathbb{Z}}^{\times}$-regularity of profinite isomorphisms as in [19]. See [25] for a survey of the profinite rigidity problem, and [5] for some recent evidence in finite-volume hyperbolic 3-manifold groups.

Question 4.8. Input a pair of surface automorphisms (S, f_A) and (S, f_B). Is there an algorithm to certify the statement that for all characteristic finite index subgroups K of $\pi_1(S)$, $[f_A]_K$ and $[f_B]_K$ are conjugate in $\mathrm{Out}(\pi_1(S)/K)$?

Question 4.9. Let (S, f) be a pseudo-Anosov automorphism on a closed surface of genus ≥ 2. Is it possible to characterize the Heegaard Floer homology $\mathrm{HF}^+(M_f)$ [23, 24] in terms of $[f] \in \mathrm{Out}(\widehat{\pi})$, where $\widehat{\pi}$ denotes the profinite completion of $\pi = \pi_1(S)$?

5. MISCELLANEOUS ON FIBERED CONES

Let (S, f) be a surface automorphism. For any regular finite cover M' of the mapping torus M_f, the pullback ϕ' of the distinguished cohomology class ϕ_f lies in the interior of a unique fibered cone in $H^1(M'; \mathbb{R})$, which we simply refer to as the *distinguished fibered cone* of M'. The induced action of deck transformations on $H^1(M'; \mathbb{R})$ fixes ϕ', and hence preserves the distinguished fibered cone.

Theorem 5.1. *If (S, f) is a pseudo-Anosov automorphism on a surface of negative Euler characteristic, then for any natural number n, there exists a finite regular cover M' of M_f, such that the distinguished fibered cone of M' has at least n distinct deck transformation orbits of codimension-one faces.*

Remark 5.2.

(1) Theorem 5.1 is a key ingredient in the proof of Theorem 3.3. For the case with $\partial S = \emptyset$, see [18, PROBLEM 1.5] for an outline in dual terms of cones of homological directions; see also [19, SECTION 6.2] for a more detailed proof. The case with $\partial S \neq \emptyset$ can be derived easily using a well-known hyperbolic Dehn filling trick.

(2) By virtual specialization, every finite-volume hyperbolic 3-manifold is virtually fibered, and has unbounded virtual first Betti numbers [2]. Moreover, the virtual numbers of fibered cones are unbounded [1]. Theorem 5.1 shows that any fibered cone can virtually become as complicated as you want.

Question 5.3. Let (S, f) be a pseudo-Anosov automorphism of a surface of negative Euler characteristic. For any primitive periodic trajectory γ of $\pi_1(M_f)$, does there always exist a regular finite cover M' of M_f such that the homological direction along $[\gamma']$ is extreme on the distinguished cone of homological directions of M'? Here, γ' denotes any preimage component of γ in M'', and $[\gamma']$ denotes its homology class in $H_1(M;\mathbb{R})$.

In other words, is every primitive periodic trajectory covered by some virtual periodic trajectory in an extreme virtual homological direction?

Question 5.4. Let N be an orientable finite-volume hyperbolic 3-manifold. For any embedded closed geodesic γ, does there always exist a finite cover N' of N with a fibered class ϕ', such that some preimage component γ' of γ is freely homotopic to a periodic trajectory of the pseudo-Anosov suspension flow on N' dual to ϕ'?

In other words, is every primitive closed geodesic covered by some (essential) virtual periodic trajectory, with respect to some virtual fibering?

Question 5.5. Let N be an orientable finite-volume hyperbolic 3-manifold with cusps. In similar brief words, is every peripheral slope covered by some virtual slope of degeneracy, with respect to some virtual fibering?

Any cohomology class $\psi \in H^1(M_f;\mathbb{Z})$ in the distinguished fibered cone is homotopically represented by a bundle projection $M_f \to \mathbb{R}/\mathbb{Z}$. The monodromy of that bundle defines a surface automorphism whose mapping torus is homeomorphic to M_f. If f is pseudo-Anosov, then the surface automorphism (S^ψ, f^ψ) associated to ψ is also pseudo-Anosov, and its suspension flow is isotopic to θ_t up to parametrization.

Fried showed that the stretch factor $\lambda: \psi \mapsto \lambda(f^\psi)$ extends to a continuous function on the distinguished fibered cone of ϕ_f valued in $(1, +\infty)$, such that $\lambda(r\psi) = \lambda(\psi)^r$ holds for any $r > 0$. Moreover, restricted to the corresponding fibered face of the Thurston norm unit ball, $1/\log\lambda$ is a strictly concave function, and converges zero as ψ tends to the boundary. McMullen introduced a Teichmüller polynomial in the group ring $\Theta \in \mathbb{Z}H$, where H denotes the free abelianization of $\pi_1(M_f)$. One may think of Θ as a multivariable Laurent polynomial by fixing a basis of H, and Θ can be characterized by the property that $\lambda(\psi)$ is the maximum modulus among the zeros of the ψ-specialization of Θ (that is, $\sum_{h\in H} a_h t^{\psi(h)}$ in $\mathbb{Z}[t,t^{-1}]$, denoting $\Theta = \sum_{h\in H} a_h h$ in $\mathbb{Z}H$ and $\psi: H \to \mathbb{Z}$ in $H^1(M_f;\mathbb{Z})$). With the Teichmüller polynomial, McMullen reproved the above properties of Fried's stretch factor function λ, and went on to ask if the unique minimum of λ on the distinguished fibered face is achieved at a rational point (that is, a point in $H^1(M_f;\mathbb{Q})$) [21].

H. Sun exhibits examples where the rationality holds, and other more generic examples where the rationality fails [27]. The following theorem summarizes some properties of the stretch factor function as discovered in [27].

Theorem 5.6. *If (S, f) is a pseudo-Anosov automorphism, then the stretch factor minimizing point ψ_0 on the distinguished fibered face of M_f is either rational or transcendental. Moreover, for any finite cover M' of M_f, then the stretch factor minimizing point on the distinguished fibered face of M' is the pullback of ψ_0 divided by $[M' : M_f]$.*

As is implied by Theorem 5.6, the rationality/transcendence of the stretch factor minimizing point is a property that depends only the fibered cone, and is a commensurability class invariant with respect to cone in a certain sense.

We record the following question as suggested in Sun [27].

Question 5.7. Let N be a finite-volume hyperbolic 3-manifold. Does there always exist a finite cover N' of N, such that N' has a fibered face on which the stretch factor function is minimized at a transcendental point?

In [7], D. Calegari, H. Sun, and S. Wang initiate a systematic study of commensurability relations of surface automorphisms. A pair of surface automorphisms (S_A, f_A) and (S_B, f_B) are said to be *commensurable* if (S_A, f_A^k) and (S_B, f_B^l) admit a common finite covering surface automorphism for some natural numbers nonzero integers k and l. This is equivalent to saying that the mapping tori M_A of (S_A, f_A) and M_B of (S_B, f_B) admit a common finite cover such that the distinguished cohomology classes ϕ_A and ϕ_B are pulled back to rationally commensurable cohomology classes.

With natural extension of the terminology to 2-orbifold automorphisms, Calegari, Sun, and Wang prove the following theorem.

Theorem 5.8. *The commensurability class of any pseudo-Anosov surface automorphism has a unique (possibly orbifold) minimal member.*

Remark 5.9. Theorem 5.8 contrasts the well-known fact that the commensurability class of any arithmetic hyperbolic 3-manifold has infinitely many orbifold minimal members.

FUNDING

This work was partially supported by NSFC grant 11925101 and 2020YFA0712800.

REFERENCES

[1] I. Agol, Criteria for virtual fibering. *J. Topol.* **1** (2008), 269–284.

[2] I. Agol, The virtual Haken conjecture. With an appendix by I. Agol, D. Groves, and J. Manning. *Doc. Math.* **18** (2013), 1045–1087.

[3] M. Aschenbrenner, S. Friedl, and H. Wilton, *3-manifold groups*. EMS Ser. Lect. Math., 2015.

[4] N. Bergeron and A. Venkatesh, The asymptotic growth of torsion homology for arithmetic groups. *J. Inst. Math. Jussieu* **12** (2013), no. 2, 391–447.

[5] M. R. Bridson, D. B. McReynolds, A. W. Reid, and R. Spitler, Absolute profinite rigidity and hyperbolic geometry. *Ann. of Math. (2)* **192** (2020), 679–719.

[6] M. R. Bridson, A. W. Reid, and H. Wilton, Profinite rigidity and surface bundles over the circle. *Bull. Lond. Math. Soc.* **49** (2017), 831–841.

[7] D. Calegari, H. Sun, and S. Wang, On fibered commensurability. *Pacific J. Math.* **250** (2011), 287–317.

[8] A. Fathi, F. Laudenbach, and V. Poénaru, *Thurston's work on surfaces*. Translated from the 1979 French original by D. M. Kim and D. Margalit. Princeton University Press, Princeton, NJ, 2012.

[9] S. Friedl and S. Vidussi, A survey of twisted Alexander polynomials. In *The Mathematics of Knots*, pp. 45–94, Contrib. Math. Comput. Sci. 1, Springer, Heidelberg, 2011.

[10] A. Hadari, Homological eigenvalues of lifts of pseudo-Anosov mapping classes to finite covers. *Geom. Topol.* **24** (2020), 1717–1750.

[11] B. Jiang, *Lectures on Nielsen fixed point theory*. Contemp. Math. 14, Amer. Math. Soc., Province, RI, 1983.

[12] B. Jiang, Estimation of the number of periodic orbits. *Pacific J. Math.* **172** (1996), 151–185.

[13] B. Jiang, A primer of Nielsen fixed point theory. In *Handbook of topological fixed point theory*, pp. 617–645, Springer, Dordrecht, 2005.

[14] B. Jiang and J. Guo, Fixed points of surface diffeomorphisms. *Pacific J. Math.* **160** (1993), 67–89.

[15] S. Kojima and G. McShane, Normalized entropy versus volume for pseudo-Anosovs. *Geom. Topol.* **22** (2018), 2403–2426.

[16] T. T. Q. Lê, Growth of homology torsion in finite coverings and hyperbolic volume. *Ann. Inst. Fourier (Grenoble)* **68** (2018), 611–645.

[17] Y. Liu, Mapping classes are almost determined by their finite quotient actions. 2019, arXiv:1906.03602v2.

[18] Y. Liu, Virtual homological spectral radii for automorphisms of surfaces. *J. Amer. Math. Soc.* **33** (2020), 1167–1227.

[19] Y. Liu, Finite-volume hyperbolic 3-manifolds are almost determined by their finite quotient groups. 2021, arXiv:1906.03602v1.

[20] W. Lück, Approximating L^2-invariants and homology growth. *Geom. Funct. Anal.* **23** (2013), 622–663.

[21] C. T. McMullen, Polynomial invariants for fibered 3-manifolds and Teichmüller geodesics for foliations. *Ann. Sci. Éc. Norm. Supér. (4)* **33** (2000), 519–560.

[22] C. T. McMullen, Entropy on Riemann surfaces and the Jacobians of finite covers. *Comment. Math. Helv.* **88** (2013), 953–964.

[23] P. S. Ozsváth and Z. Szabó, Holomorphic disks and three-manifold invariants: Properties and applications. *Ann. of Math.* **159** (2004), 1159–1245.

[24] P. S. Ozsváth and Z. Szabó, Holomorphic disks and topological invariants for closed three-manifolds. *Ann. of Math.* **159** (2004), 1027–1158.

[25] A. W. Reid, Profinite rigidity. In *Proceedings of the International Congress of Mathematicians— Rio de Janeiro 2018. Vol. II. Invited lectures*, pp. 1193–1216, World Sci. Publ., Hackensack, NJ, 2018.

[26] P. F. Stebe, Conjugacy separability of groups of integer matrices. *Proc. Amer. Math. Soc.* **32** (1972), 1–7.

[27] H. Sun, A transcendental invariant of pseudo-Anosov maps. *J. Topol.* **8** (2015), 711–743.

[28] H. Wilton and P. A. Zalesskii, Profinite detection of 3-manifold decompositions. *Compos. Math.* **155** (2019), 246–259.

YI LIU

Beijing International Center for Mathematical Research, Peking University, 5 Yiheyuan Road, Haidian District, Beijing 100871, China, liuyi@bicmr.pku.edu.cn

HOMOTOPY PATTERNS IN GROUP THEORY

ROMAN MIKHAILOV

ABSTRACT

This is a survey. The main subject of this survey is the homotopical or homological nature of certain structures which appear in classical problems about groups, Lie rings and group rings. It is well known that the (generalized) dimension subgroups have complicated combinatorial theories. In this paper we show that, in certain cases, the complexity of these theories is based on homotopy theory. The derived functors of nonadditive functors, homotopy groups of spheres, group homology, etc., appear naturally in problems formulated in purely group-theoretical terms. The variety of structures appearing in the considered context is very rich. In order to illustrate it, we present this survey as a trip passing through examples having a similar nature.

MATHEMATICS SUBJECT CLASSIFICATION 2020

Primary 20A10; Secondary 20J05, 20J99, 20F18, 55Q40, 18G99

KEYWORDS

Group ring, Lie ring, dimension subgroup, lower central series, derived functor, homotopy groups of spheres, fr-codes, homotopy pattern, group homolog

ICM INTERNATIONAL CONGRESS OF MATHEMATICIANS 2022 JULY 6–14

Proc. Int. Cong. Math. 2022, Vol. 4, pp. 2806–2825
DOI 10.4171/ICM2022/21

Published by EMS Press

1. INTRODUCTION

It would be nice to discuss the title first. What will we mean by "group theory"? Obviously, not a collection of funny stories on various constructions of groups with exotic properties, as well as not a classification of groups from the point of view of certain structure theory. In this paper, the groups will be the main category we will work in. In some cases it will be changed by Lie rings over integers. So, for us the "group theory" will mean just a (mostly functorial) life inside the category of groups.

We will not define the *homotopy pattern* as a concept. The term *pattern* itself is multi-valued. It is used in different philosophical contexts and usually is understood intuitively. One can try to define it after reading this paper. Somebody can understand it as a *system of signs* of homotopy origin or a *collection of relations* that comes from the homotopy theory.

The main claim, or "formula," which we state here is the following abstract relation:

$$(\Phi) \qquad \frac{\text{Intersection of subsctructures}}{\text{Obvious part}} \supset \text{Homotopy pattern.}$$

We will take some structure like a group, group ring, Lie ring or universal enveloping algebra, consider certain substructures, their intersection, take its quotient by some obvious part and see that, in many cases, this quotient contains elements of homotopical or homological nature. An *obvious part* is not defined in a unified way, usually it is a maximal substructure of the intersection defined using a given type of operations (for example, it is not an intersection itself). As a rule, the *obvious part* is a more explicit construction than the *intersection of substructures*. One can consider the left-hand site of the formula (Φ) as "implicity modulo explicity," or define a hierarchy of the explicity and take the quotients of its terms.

It is a time to stop saying general words and do some math. Let G be a group, $\mathbb{Z}[G]$ its integral group ring, and $\mathbf{b}$ a two-sided ideal of $\mathbb{Z}[G]$. The problem of identification of the subgroup

$$D(G, \mathbf{b}) := G \cap (1 + \mathbf{b}) = \langle g \in G \mid g - 1 \in \mathbf{b} \rangle$$

is a fundamental problem in the theory of groups and group rings. It is often the case that a certain normal subgroup $N(G, \mathbf{b})$ of G is easily seen to be contained in $D(G, \mathbf{b})$ and explicitly defined in terms of G only, without using the group ring (and is the largest subgroup with such a property). The computation of the quotient $D(G, \mathbf{b})/N(G, \mathbf{b})$ usually becomes a challenging problem. This case will be the first example of our formula (Φ). Various choices of the ideal $\mathbf{b}$ lead to the derived functors inside the quotients $D(G, \mathbf{b})/N(G, \mathbf{b})$. We will discuss the derived functors and their appearance in this context in Section 2.

The main examples of the above type are classical dimension subgroups. Let $\mathbf{g}$ be the augmentation ideal of $\mathbb{Z}[G]$. The subgroups $D_n(G) := G \cap (1 + \mathbf{g}^n)$, $n \geq 1$ are known as *dimension subgroups*. It is easy to see that, for any G and $n \geq 1$, the dimension subgroups contain the terms of lower central series of G: $\gamma_n(G) \subseteq D_n(G)$. The lower central series are defined inductively as follows: $\gamma_1(G) := G$, $\gamma_{n+1}(G) := [\gamma_n(G), G] = \langle [x, y], x \in \gamma_n(G), y \in G \rangle^G$, $n \geq 1$. Is it true that $D_n(G) = \gamma_n(G)$? This problem was open for many years (see Section 5 for some history). Our formula (Φ) states that the dimension quotients D_n/γ_n may contain certain elements of homotopical nature. This is exactly the case. Section 5 is about it. In particular, we will show in details in Section 5 how an element $\mathbb{Z}/3 \subset \pi_6(S^2)$ is related to the 3-torsion in dimension quotients for Lie rings.

The formula (Φ), which presents not a rigorous statement but a feeling, goes through the paper. In Section 3 we will see that the homotopy groups of certain spaces can be described as intersections of subgroups in a group modulo a natural commutator subgroup. In particular, the homotopy groups of the 2-sphere are given in this way. This is the well-known Wu formula and its variations. Section 4 is about combinatorics of Wu-type formulas. One can see examples of *homotopy patterns* as combinations of brackets in groups and Lie rings. As we mentioned already, Section 5 shows how these combinations can be applied to the classical dimension problem.

Section 6 is a bit isolated, however, its subject fits smoothly in a general context of the paper. Section 6 is about the method of construction of functors via (derived) limits. We show how to present certain derived functors, group homology, the forth dimension quotient via limits over the category of group presentations. So the theory of limits becomes a unified theory for different functors discussed in this paper. At the end of Section 6, we briefly discuss the so-called **fr**-language, a combinatorial-linguistic game which can be used in the study of functors.

On October 2, 2021, my friend and teacher Inder Bir Singh Passi passed away. For all those 19 years that we were in contact, I was deeply touched by his delicacy, kindness, and empathy for other people. In 2002, he invited me to visit India for the first time. That visit changed my life. I dedicate this text to his memory.

2. DERIVED FUNCTORS

The derived functors in the sense of Dold–Puppe [10] are defined as follows. For an abelian group A and an endofunctor F on the category of abelian groups, the derived functor of F is given as

$$L_i F(A, n) = \pi_i\big(FKP_*[n]\big), \quad i \geq 0, n \geq 0,$$

where $P_* \to A$ is a projective resolution of A, and K is the Dold–Kan transform, inverse to the Moore normalization functor from simplicial abelian groups to chain complexes. For simplicity, we write $L_i F(A) := L_i F(A, 0)$.

Derived functors appear naturally in the theory of Eilenberg–MacLane spaces, Moore spaces and general homotopy theory. For example, for an abelian group A, homology $H_*(A)$ can be filtred in a way that the graded pieces are derived functors of the exterior powers $L_i \Lambda^j(A)$ [6]. In particular, there exist the following natural exact sequences:

$$0 \to \Lambda^3(A) \to H_3(A) \to L_1\Lambda^2(A) \to 0,$$
$$0 \to \Lambda^4(A) \to H_4(A) \to L_1\Lambda^3(A) \to 0.$$

These exact sequences split as sequences of abelian groups but do not split naturally. This is a common situation, the homology and homotopy functors usually present nontrivial gluing of derived functors of different type.

As a rule, the derived functors have a complicated structure. Here we will describe a couple of them, in a sense the simplest ones, and show how they appear in the context of group rings. First, let us recall the well-known description (see, for example, MacLane [25]), of the derived functor of the tensor square $L_1 \otimes^2 (A) = \mathrm{Tor}(A, A)$. Given an abelian group A, the group $\mathrm{Tor}(A, A)$ is generated by the n-linear expressions $\tau_h(a_1, a_2)$ (where all a_i belong to the subgroup ${}_hA := \{a \in A \mid ha = 0\}$, $h > 0$), subject to the so-called slide relations

$$\tau_{hk}(a_1, a_2) = \tau_h(ka_1, a_2), \tag{1}$$

for all i whenever $hka_1 = 0$ and $ha_2 = 0$, and an analogous relation, where the roles of a_1, a_2 are interchanged.

The natural projection of the tensor square to the symmetric square $\otimes^2 \to \mathsf{S}^2$ induces a natural epimorphism $L_1 \otimes^2 (A) \to L_1\mathsf{S}^2(A)$ which maps the generator $\tau_h(a_1, a_2)$ of $L_1 \otimes^2 (A) = \mathrm{Tor}(A, A)$ to the generator $\beta_h(a_1, a_2)$ of $L_1\mathsf{S}^2(A)$ so that the kernel of this map is generated by the elements $\tau_h(a, a), a \in {}_hA$.

The functor $L_1\mathsf{S}^2$ appears as a quadratic piece of the homology of Eilenberg–MacLane spaces $H_5K(-, 2)$:

$$0 \to L_1\mathsf{S}^2(A) \to H_5K(A, 2) \to \mathrm{Tor}(A, \mathbb{Z}/2) \to 0.$$

It is shown by Jean in [22] that the first derived functor of the symmetric cube can be described as follows:

$$L_1\mathsf{S}^3(A) \simeq \big(L_1\mathsf{S}^2(A) \otimes A\big) / \operatorname{Jac}_S, \tag{2}$$

where Jac_S is the subgroup generated by elements of the form (Jacobi-type elements)

$$\beta_h(x_1, x_2) \otimes x_3 + \beta_h(x_1, x_3) \otimes x_2 + \beta_h(x_2, x_3) \otimes x_1$$

with $x_i \in {}_hA$.

Recall one more property of the functor $L_1\mathsf{S}^2$. Suppose that an abelian group A is presented as a quotient $A = Q/U$, for a free abelian Q and its subgroup U. Then $\mathsf{S}^2(A)$, $L_1\mathsf{S}^2(A)$ are naturally isomorphic to the zeroth and first homology of the Kozsul-type complex [1]

$$\Lambda^2(U) \to U \otimes Q \to \mathsf{S}^2(Q).$$

The maps in this sequence are natural and can be easily recognized. Now consider the following diagram with exact columns:

$$\begin{array}{ccccccc}
 & & \Lambda^2(U) & \rightarrowtail & U \otimes Q & \longrightarrow & \mathsf{S}^2(Q) \\
 & & \downarrow & & \downarrow & & \| \\
 & & \Lambda^2(Q) & \rightarrowtail & Q \otimes Q & \longrightarrow & \mathsf{S}^2(Q) \\
 & & \downarrow & & \downarrow & & \\
L_1\mathsf{S}^2(A) & \rightarrowtail & \Lambda^2(Q)/\Lambda^2(U) & \longrightarrow & Q/U \otimes Q & \twoheadrightarrow & \mathsf{S}^2(A)
\end{array} \tag{3}$$

All vertical maps in this diagram are obvious, the middle horizontal sequence is the classical Kozsul short exact sequence. The lower sequence is exact.

Now we return to the theory of nonabelian groups. Let F be a free group, R its normal subgroup, $\mathbf{f} = (F-1)\mathbb{Z}[F]$, $\mathbf{r} = (R-1)\mathbb{Z}[F]$, as always. Let us describe the generalized dimension subgroup $F \cap (1+\mathbf{rf}+\mathbf{f}^3)$. Obviously, this subgroup contains $\gamma_2(R)\gamma_3(F)$, however, this is not a complete description. In order to find the remaining part, denote $Q := F_{ab}$, $U := R/R \cap [F, F]$. Observe now that there are natural vertical isomorphisms

$$\begin{array}{ccc} \Lambda^2(Q)/\Lambda^2(U) & \longrightarrow & Q/U \otimes Q \\ \Vert & & \Vert \\ \frac{\gamma_2(F)}{\gamma_2(R)\gamma_3(F)} & \longrightarrow & \frac{\mathbf{f}^2}{\mathbf{rf}+\mathbf{f}^3} \end{array}$$

and the lower map is induced by $g \mapsto g-1$. That is, the needed generalized dimension quotient can be described as (see [16] for another proof of this statement)

$$\frac{F \cap (1+\mathbf{rf}+\mathbf{f}^3)}{\gamma_2(R)\gamma_3(F)} = L_1\mathsf{S}^2(G_{ab}), \tag{4}$$

where $G = F/R$. It is easy to lift the element from $L_1\mathsf{S}^2(F/R\gamma_2(F))$ to $F \cap (1+\mathbf{rf}+\mathbf{f}^3)$. Given $\beta_h(a_1, a_2)$, denote by f_1, f_2 the preimages of a_1, a_2 in F, then $f_1^h, f_2^h \in R\gamma_2(F)$. Now the needed image is given as $[f_1, f_2]^h$.

The identification (4) is an example of a situation described in our formula (Φ). The subgroup $\gamma_2(R)\gamma_3(F)$ is the maximal obvious subgroup of the intersection $F \cap (1+\mathbf{rf}+\mathbf{f}^3)$. In general, for two ideals $\mathbf{a}, \mathbf{b}$, the maximal obvious part of $D(F, \mathbf{a}+\mathbf{b})$ is the product $D(F, \mathbf{a})D(F, \mathbf{b})$. In the case above, $D(F, \mathbf{rf}) = \gamma_2(R)$ and $D(F, \mathbf{f}^3) = \gamma_3(F)$. For the situation of arbitrary ideals $\mathbf{a}$ and $\mathbf{b}$, the quotient

$$\frac{D(F, \mathbf{a}+\mathbf{b})}{D(F, \mathbf{a})D(F, \mathbf{b})}$$

shows how these ideals are "linked" in $\mathbb{Z}[F]$ and in many cases has a homological description, what agrees with our formula (Φ).

Here is one more example related to the functor $L_1\mathsf{S}^2$.

Recall the so-called Fox subgroup problem (see [13, p. 557]; [4, Problem 13]; [14]). It asks for the identification of the normal subgroup $F(n, R) := F \cap (1+\mathbf{rf}^n)$ for a free group F and its normal subgroup R. A solution to this problem has been given by I. A. Yunus [37] and Narain Gupta [14, Chapter III]. It turns out that, while $F(1, R) = \gamma_2(R)$ and $F(2, R) = [R \cap \gamma_2(F), R \cap \gamma_2(F)]\gamma_3(R)$, the identification of $F(n, R), n \geq 3$, is given as an isolator of a subgroup. For instance, $F(3, R) = \sqrt{G(3, R)}$, where

$$G(3, R) := \gamma_2\big(R \cap \gamma_3(F)\big)\big[[R \cap \gamma_2(F), R], R \cap \gamma_2(F)\big]\gamma_4(R).$$

It is shown in [31] that there is a natural isomorphism

$$\frac{F(3, R)}{G(3, R)} \simeq L_1\mathsf{S}^2\left(\frac{R \cap \gamma_2(F)}{\gamma_2(R)(R \cap \gamma_3(F))}\right).$$

As in the simplest example, the derived functor $L_1\mathsf{S}^2$ can be lifted to the generalized dimension subgroup. This gives a complete description of the third Fox subgroup, $F(3,R) = G(3,R)W$, where W is a subgroup of F, generated by elements

$$[x,y]^m[x,s_y]^{-1}[y,s_x]$$

with

$$\begin{aligned} x^m &= r_x s_x, \quad r_x \in R \cap \gamma_3(F), \quad s_x \in \gamma_2(R), \\ y^m &= r_y s_y, \quad r_y \in R \cap \gamma_3(F), \quad s_y \in \gamma_2(R). \end{aligned}$$

Higher generalizations of this description are of interest. For instance, the complete description of the fourth Fox subgroup $F \cap (1 + \mathbf{rf}^4)$ should be related to the derived functors of certain cubical functors.

The above examples lie at the tip of an iceberg, and present the simplest illustration of a deep relation between generalized dimension subgroups and derived functors. The derived functors of high degree polynomial functors, as well as subgroups defined by ideals, have complicated structure. In many cases the same kind of tricks as above, like diagram chasing together with group-theoretical identifications, lead to surprising connections.

Here are some other examples of descriptions of the generalized dimension subgroups which use the derived functors (we assume that $G = F/R$):

$$\frac{F \cap (1 + \mathbf{r}^2\mathbf{f} + \mathbf{f}^4)}{\gamma_3(R)\gamma_4(F)} = L_2\mathsf{L}^3_s(G_{ab}),$$

$$\frac{F \cap (1 + \mathbf{f}(F'-1) + \mathbf{rf}^3 + \mathbf{f}^4)}{[\gamma_2(R),F]\gamma_4(F)} = L_1\mathsf{S}^3(G_{ab}),$$

$$\frac{F \cap (1 + \mathbf{frf} + \mathbf{f}^4)}{[\gamma_2(R),F]\gamma_4(F)} = L_1\mathsf{S}^3(G_{ab}) \quad \text{(provided } G_{ab} \text{ is 2-torsion-free).}$$

For an abelian group A, the third super-Lie functor $\mathsf{L}^3_s(A)$ is generated by brackets $\{a,b,c\}$, $a,b,c \in A$, which are additive in each variable with the following relations: $\{a,b,c\} = \{b,a,c\}$, $\{a,b,c\} + \{c,a,b\} + \{b,c,a\} = 0$. The derived functors of higher super-Lie functors appear in the description of the subgroup $F \cap (1 + \mathbf{r}^n\mathbf{f} + \mathbf{f}^{n+2})$ for $n \geq 2$.

Some exotic examples of the same nature can be found in **[30]**. In particular, there are the following descriptions of the generalized dimension subgroups:

$$\frac{F \cap (1 + \mathbf{rfr} + \mathbf{sr})}{\gamma_2(S)\gamma_3(R)} = L_1\mathsf{S}^2\big(H_2(G)\big),$$

$$\frac{F \cap (1 + \mathbf{s}^2\mathbf{r} + \mathbf{r}^2\mathbf{fr})}{\gamma_3(S)\gamma_4(R)} = L_2\mathsf{L}^3_s\big(H_2(G)\big),$$

where $S = [F,R]$, $\mathbf{s} = (S-1)\mathbb{Z}[F]$ and $H_2(G)$ the second integral group homology.

3. HOMOTOPY PUSHOUTS

In this section we will see that not only derived functors but also the homotopy groups of certain spaces can be presented in group-theoretical terms. We start with the simplest case. Let G be a group, and R, S its normal subgroups. Consider a homotopy pushout

$$\begin{array}{ccc} K(G,1) & \longrightarrow & K(G/R,1) \\ \downarrow & & \downarrow \\ K(G/S,1) & \longrightarrow & X \end{array}$$

where the maps between classifying spaces $K(-,1)$ are induced by natural epimorphisms $G \to G/R$, $G \to G/S$. Classical van Kampen theorem implies that $\pi_1(X) = G/RS$. In [8], the second homotopy group of X is described as follows:

$$\pi_2(X) \simeq \frac{R \cap S}{[R,S]}.$$

This result gives a topological interpretation of the difference between the intersection and the commutator of a pair of subgroups. (For a pair R, S of normal subgroups of a free group F, $[R,R] \cap [S,S] \subseteq [R,S]$. One can prove this as an exercise.)

A fact that the second homotopy group is related to a certain group-theoretical construction is not surprising. The theory of (non)aspherical group presentations, identity sequences, etc., is about the properties of the second homotopy group of a standard complex constructed from a given group presentation. Next we will show how to extend the above result for the case of three normal subgroups.

Let R, S, T be normal subgroups of G. Define an analog of the commutator subgroup as

$$[\![R,S,T]\!] := [R \cap S, T][S \cap T, R][T \cap R, S].$$

Consider a homotopy pushout

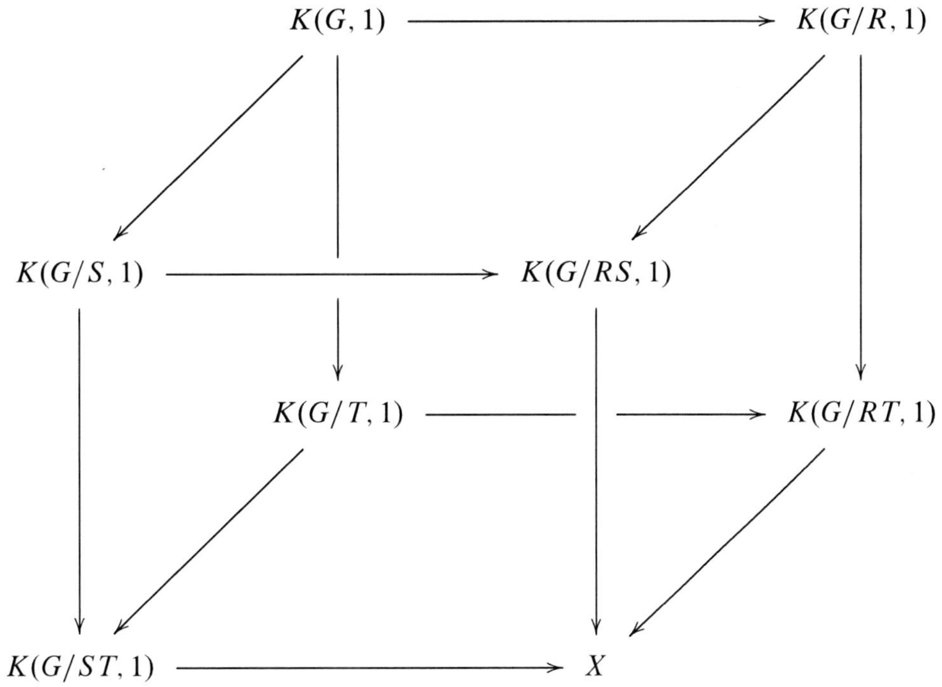

The lower homotopy groups of X are described as follows (see [11]):

$$\pi_1(X) \simeq G/RST,$$
$$\pi_2(X) \simeq \frac{RS \cap RT}{R(S \cap T)},$$
$$\pi_3(X) \simeq \frac{R \cap S \cap T}{[\![R, S, T]\!]}.$$

At first glance, it might seem that $\frac{RS \cap RT}{R(S \cap T)}$ is not symmetric in R, S, T and that the subgroup R plays a special role. However, the homotopy pushout X is symmetric and the above description of π_2 provides a proof of the following isomorphisms:

$$\frac{RS \cap RT}{R(S \cap T)} \simeq \frac{SR \cap ST}{S(R \cap T)} \simeq \frac{TR \cap TS}{T(R \cap S)}.$$

The above description shows that the third homotopy group appears quite naturally in the context of group theory. Next we will show how to get a purely group-theoretic result using given homotopy identifications.

For a triple of normal subgroup R, S, T of G, consider the corresponding ideals in $\mathbb{Z}[G]$: $\mathbf{r} := (R-1)\mathbb{Z}[G]$, $\mathbf{s} := (S-1)\mathbb{Z}[G]$, and $\mathbf{t} := (T-1)\mathbb{Z}[G]$. An obvious ring-theoretic analog of the subgroup $[\![R, S, T]\!]$ is the following ideal:

$$(\!(\mathbf{r}, \mathbf{s}, \mathbf{t})\!) := \mathbf{r}(\mathbf{s} \cap \mathbf{t}) + (\mathbf{s} \cap \mathbf{t})\mathbf{r} + \mathbf{s}(\mathbf{t} \cap \mathbf{r}) + (\mathbf{t} \cap \mathbf{r})\mathbf{s} + \mathbf{t}(\mathbf{r} \cap \mathbf{s}) + (\mathbf{r} \cap \mathbf{s})\mathbf{t}.$$

It is easy to check that, for any $w \in [\![R, S, T]\!]$, $w - 1 \in (\!(\mathbf{r}, \mathbf{s}, \mathbf{t})\!)$ and then one asks about the structure of the quotient

$$\frac{G \cap (1 + (\!(\mathbf{r}, \mathbf{s}, \mathbf{t})\!))}{[\![R, S, T]\!]}.$$

It is shown in [21] that there exists the following commutative diagram:

$$\begin{array}{ccc} \frac{R \cap S \cap T}{[\![R,S,T]\!]} & \longrightarrow & \frac{\mathbf{r} \cap \mathbf{s} \cap \mathbf{t}}{(\!(\mathbf{r},\mathbf{s},\mathbf{t})\!)} \\ \| & & \downarrow \\ \pi_2(\Omega X) & \longrightarrow & H_2(\Omega X) \end{array}$$

Here X is the homotopy pushout described above, ΩX is the loop space, the lower horizontal map is the Hurewicz homomorphism, while the upper is the natural map induced by $g \mapsto g - 1$. For a connected space Y, the kernel of the second Hurewicz homomorphism $\pi_2(Y) \to H_2(Y)$ is a 2-torsion group [21]. As a consequence, we get the following theorem from [21].

Theorem 3.1. *For any group G and its normal subgroups R, S, T, the quotient*

$$\frac{G \cap (1 + (\!(\mathbf{r}, \mathbf{s}, \mathbf{t})\!))}{[\![R, S, T]\!]}$$

is an abelian 2-torsion group.

The author does not know how to prove this results without using the homotopy theory.

What about the higher homotopy groups? Two normal subgroups give a possibility to model π_2 for a certain space, three normal subgroups correspond to π_3. Is it true that n normal subgroups of a group allow constructing a space with a group-theoretical description of its π_n? The answer is "yes," however, under certain conditions. For a group G and its normal subgroups $R_1, \dots, R_n$, $n \geq 2$, construct the homotopy pushout of the n-dimensional cubical diagram with $2^n - 1$ classifying spaces $K(G/\prod_{i \in I} R_i)$, $I \subset \{1, \dots, n\}$. There exist the so-called connectivity conditions on the collection of subgroups R_i, which imply that (see [11])

$$\pi_n(\text{homotopy pushout}) \simeq \frac{R_1 \cap \dots \cap R_n}{[\![R_1, \dots, R_n]\!]},$$

where

$$[\![R_1, \dots, R_n]\!] := \prod_{I \cup J = \{1,\dots,n\}, I \cap J = \emptyset} \Big[\bigcap_{i \in I} R_i, \bigcap_{j \in J} R_j\Big].$$

The main example we will consider here is the Wu formula for homotopy groups of S^2. Let $F = F(x_1, \dots, x_n)$ be a free group of rank $n \geq 2$. Consider the following normal subgroups of F: $R_i = \langle x_i \rangle^F$, $i = 1, \dots, n$, $R_{n+1} = \langle x_1 x_2 \dots x_n \rangle^F$. The homotopy pushout of a diagram of the corresponding $2^n - 1$ classifying spaces is S^2. Therefore we get the following:

$$\pi_{n+1}(S^2) \simeq \frac{R_1 \cap \dots \cap R_{n+1}}{[\![R_1, \dots, R_{n+1}]\!]}.$$

This is a version of the Wu formula, proved in [36] using simplicial methods. In this particular case, the above commutator subgroup equals to the symmetric commutator subgroup

$$[R_1, \dots, R_{n+1}]_S := \prod_{\sigma \in \Sigma_{n+1}} \big[\dots[R_{\sigma(1)}, R_{\sigma(2)}], \dots, R_{\sigma(n+1)}\big].$$

Here Σ_{n+1} is the group of $(n+1)$-permutations. That is, the homotopy groups of S^2 can be presented as

$$\pi_{n+1}(S^2) \simeq \frac{R_1 \cap \dots \cap R_{n+1}}{[R_1, \dots, R_{n+1}]_S}.$$

It turns out that the above intersection modulo the symmetric commutator coincides with the center of the quotient of the free group F modulo the symmetric commutator, that is,

$$\pi_{n+1}(S^2) \simeq Z\big(F/[R_1, \dots, R_{n+1}]_S\big).$$

A generalization of this construction to the higher spheres, as well as Moore spaces, is given in [32]. For any $n, k > 3$, a finitely generated group $G_{n,k}$ given by explicit generators and relations is constructed such that $\pi_n(S^k) \simeq Z(G_{n,k})$. The group $G_{n,k}$ is defined in [32] as a certain quotient of the amalgamated square of the pure braid group on n strands.

Should we mention that the presentation of homotopy groups in this section follows the idea of our formula (Φ)? It is always an intersection of some subgroups modulo an obvious part. In this way, one can reflect on the difference between explicity of terms like $RS \cap RT$ and $R(S \cap T)$.

4. WU-TYPE FORMULAS

What are the first questions that come to mind when you look at Wu formula for homotopy groups of S^2? How to get some results on $\pi_*(S^2)$ using group theory? Can one find any new element from $\pi_*(S^2)$ using its group-theoretic presentation? How to present generators of the known elements of the homotopy groups in terms of the free group? Here we will briefly discuss the latter question.

We know that $\pi_3(S^2) = \mathbb{Z}$, $\pi_4(S^2) = \mathbb{Z}/2$, $\pi_5(S^2) = \mathbb{Z}/2$, $\pi_6(S^2) = \mathbb{Z}/12, \dots$ all homotopy groups in degree > 1 of S^2 are nonzero (see [20]). In general, the sequence of finite abelian groups $\pi_n(S^2)$, $n \geq 4$, is one of the most mysterious objects in math, it is difficult to speculate how far we are from its understanding. It is a strange luck that we can realize this extremely complicated sequence as a series of simply formulated subquotients of free groups.

Looking at the Wu formula, one can also ask the following. What about other algebraic systems, different from groups? Clearly, the same type of quotients can be considered for associative rings, Lie rings, etc. As a rule, the answers will be easier. In the associative case, this is almost obvious. The case of Lie rings (over $\mathbb{Z}$) is interesting and meaningful. Let $L_n = L(y_1, \dots, y_n)$, $n \geq 2$ be a free Lie ring over $\mathbb{Z}$. Consider its ideals $I_i = (y_i)^L$, $i = 1, \dots, n$, $I_{n+1} = (y_1 + \dots + y_n)^L$. Define the Lie analog of the symmetric commutator of ideas

$$[I_1, \dots, I_{n+1}]_S := \prod_{\sigma \in \Sigma_{n+1}} \big[\dots [I_{\sigma(1)}, I_{\sigma(2)}], \dots, I_{\sigma(n+1)}\big].$$

There is the following isomorphism:

$$\frac{I_1 \cap \dots \cap I_{n+1}}{[I_1, \dots, I_{n+1}]_S} \simeq \bigoplus_{i \geq 1} E^1_{i,n}, \tag{5}$$

where $E^1_{*,*}$ is the first page of the Curtis spectral sequence, defined via derived functors as

$$E^1_{i,j} = L_j \mathcal{L}^i(\mathbb{Z}, 1).$$

Here $\mathcal{L}^i$ is the ith Lie functor (see [7] for the discussion of this spectral sequence and properties of derived functors). The values of $E^1_{*,*}$ are known and can be described in terms of Lambda-algebra (see [5, 23]).

Now let us return to the problem of describing the generators in terms of free groups or Lie rings. It is clear that the generators can be chosen in different ways, since we work modulo the symmetric commutators. However, we try to find those with a simple form. Here are the results in low dimensions (see [2]).

Case $n = 2$. In this case, the generators are given as commutators $[x_1, x_2]$ in the group, as well as $[y_1, y_2]$ the Lie ring case (here we will write the group-expressions in terms of x's and Lie ring expressions in terms of y's). Indeed, $[x_1, x_2] \in R_1 \cap R_2 \cap R_3 \setminus [R_1, R_2, R_3]_S$. In this case, $R_1 \cap R_2 \cap R_3 = \gamma_2(F)$, $[R_1, R_2, R_3] = \gamma_3(F)$ and $\gamma_2(F)/\gamma_3(F) \simeq \mathbb{Z} \simeq \pi_3(S^2)$.

Case $n = 3$. In this case, $\pi_4(S^2) = \mathbb{Z}/2$, $E^1_{4,3} = \mathbb{Z}/2$, $E^1_{i,3} = 0$, $i \neq 4$. The generators of these $\mathbb{Z}/2$-terms are given by

$$[[x_1, x_2], [x_1, x_2x_3]], \quad [[y_1, y_2], [y_1, y_3]].$$

It can be easily checked that these terms lie in $R_1 \cap \cdots \cap R_4$ and $I_1 \cap \cdots \cap I_4$, respectively. Since we work in low dimensions, it can be proved directly that they do not belong to the symmetric commutators.

Case $n = 4$. This case is just a suspension of the previous one, $\pi_5(S^2) = \mathbb{Z}/2$, $E^1_{8,4} = \mathbb{Z}/2$, $E^1_{i,4} = 0$, $i \geq 8$. The generators are given as

$$[[[x_1, x_2], [x_1, x_2x_3]], [[x_1, x_2], [x_1, x_2x_3x_4]]]$$

and

$$[[[y_1, y_2], [y_1, y_3]], [[y_1, y_2]], [y_1, y_4]]].$$

The fact that these elements lie in the intersection of subgroups R_i or ideals I_i is obvious. To prove that they do not lie in the symmetric commutators, we need some homotopy theory. It can be done using simplicial methods, by realizing these elements as cycles in the Milnor construction $F[S^1]$ and its Lie analog.

Case $n = 5$. This case is already complicated, $\pi_6(s^2) = \mathbb{Z}/12$, $E^1_{6,5} = \mathbb{Z}/3$, $E^1_{8,5} = E^1_{16,5} = \mathbb{Z}/2$, $E^1_{i,5} = 0$, $i \neq 6, 8, 16$. This horizontal line is the first place in the spectral sequence where one can see a nontrivial gluing of the E^∞-term: two cells in degrees 8 and 16 with values $\mathbb{Z}/2$ are glued into $\mathbb{Z}/4 \subset \pi_6(S^2)$. It is easy to write down the generator of $E^1_{16,5}$, since it comes as a suspension of the previously written element $E^1_{8,4}$, namely

$$[[[[y_1, y_2], [y_1, y_3]], [[y_1, y_2]], [y_1, y_4]]], [[[y_1, y_2], [y_1, y_3]], [[y_1, y_2]], [y_1, y_5]]]].$$

The term $E^1_{6,5} = \mathbb{Z}/3$ is generated by the element (see [2] for the proof using simplicial methods and derived functors)

$$\begin{aligned} \alpha_3 := {} & [[[y_1, y_5], [y_2, y_5]], [y_3, y_4]] - [[[y_1, y_5], [y_3, y_5]], [y_2, y_4]] \\ & + [[[y_1, y_5], [y_4, y_5]], [y_2, y_3]] + [[[y_2, y_5], [y_3, y_5]], [y_1, y_4]] \\ & - [[[y_2, y_5], [y_4, y_5]], [y_1, y_3]] + [[[y_3, y_5], [y_4, y_5]], [y_1, y_2]]. \end{aligned}$$

The term $E^1_{8,5}$ and the group-case liftings are much more complicated (see [2]). For example, the 3-torsion from $\pi_6(S^2)$ can be written as a product of 14 commutators in a free group of weight ≥ 6.

The element α_3 corresponds to the Serre element $\mathbb{Z}/3 \subset \pi_6(S^2)$. Analogous picture takes place for all primes. For an odd prime p, the Serre element of order p appears in the homotopy group $\pi_{2p}(S^2)$. These elements can be easily seen from the structure of the first page of the spectral sequence $E^1_{2p,2p-1} = \mathbb{Z}/p$ (these terms are labeled as λ_1 in the language of Lambda-algebras). It turns out that the $E^1_{2p,2p-1}$-term is isolated from other $\mathbb{Z}/p$-torsion terms of the spectral sequence, hence $E^1_{2p,2p-1} = E^\infty_{2p,2p-1}$ (in the theory of spectral sequences, such arguments are sometimes called *lacunary reasons*). Let y_i for $i = 1, \dots, 2p-1$ be free generators of a free Lie algebra and consider the following element:

$$\alpha_p = \sum_{\substack{\rho \in \Sigma_{2p-2} \text{ a } 2^{p-1}\text{-shuffle} \\ \rho(1)<\rho(3)<\cdots<\rho(2p-5)}} (-1)^\rho \big[[y_{\rho(0)}, y_{2p-2}], [y_{\rho(1)}, y_{2p-2}], [y_{\rho(2)}, y_{\rho(3)}], \dots, [y_{\rho(2p-4)}, y_{\rho(2p-3)}]\big];$$

the sum is taken over all permutations $(\rho(0), \dots, \rho(2p-3)) \in \Sigma_{2p-2}$ satisfying $\rho(0) < \rho(1), \dots, \rho(2p-4) < \rho(2p-3)$, as well as $\rho(1) < \rho(3) < \cdots < \rho(2p-5)$. Here we use the left-normalized notation, i.e., $[x, y, z] := [[x, y], z]$. Then α_p presents a generator of $L_{2p-1}\mathcal{L}^{2p}(\mathbb{Z}, 1) = E^1_{2p,2p-1}$ (see [2] for the proof).

Looking at the elements of free groups and Lie rings like $[[x_1, x_2], [x_1, x_2x_3]]$ or α_3, one can get some impression of *homotopy patterns*. Next we will see how they work in the context of dimension subgroups.

5. CLASSICAL DIMENSION SUBGROUPS

Is it true that, for any group G and $n \geq 1$, $D_n(G) = \gamma_n(G)$? This question is known as the *dimension problem* and has a long history. For a detailed discussion of this problem, we refer to [14, 28, 33]. The first results in this direction are due to Magnus and Witt. They proved that, for a free group F, the dimension subgroups coincide with the lower central series [26, 35]. Incorrect solutions of the dimension problem appeared more than once, see [9, 24] and [27, THEOREM 5.15(I)]. The first example of a group with $D_4(G) \neq \gamma_4(G)$ is due to Ilya Rips [34]. The group constructed in [34] has order 2^{38} and it seems that this is the smallest finite group with the property $D_4 \neq \gamma_4$. The next point we have to mention regarding the history of the question is the series of works of Narain Gupta. In order to describe the dimension subgroups and solve the dimension problem completely, Gupta spent about 20 years developing a special calculus. As a final result, he published the paper [15], where he claims that the dimension property holds for all groups of odd order. In particular, it follows from his claim that, for an odd prime p, it is not possible to construct a group G and $n \geq 1$ with $D_n(G)/\gamma_n(G) \supseteq \mathbb{Z}/p$. In fact, Gupta claimed even more, namely that, for any group G, the dimension quotients $D_*(G)/\gamma_*(G)$ are just $\mathbb{Z}/2$-vector spaces. The last statement was written in the unpublished manuscript of Gupta, which was available from 1990s to the experts in the area.

The proofs given in the mentioned papers of Gupta are extremely complicated. During many years the author, together with I. B. S. Passi, tried to understand these proofs. It became clear already about 10 years ago that they contain gaps, however, it was not easy to find counterexamples to the main statements. Finally, the following result was proved in [2]:

Theorem 5.1. *For any prime p, there exists a group G and integer n, such that $D_n(G)/\gamma_n(G)$ contains $\mathbb{Z}/p$ as a subgroup.*

Among other things, a small finite group G with $D_7(G) \neq \gamma_7(G)$ is constructed in [2]. The needed statement that $D_7(G) \neq \gamma_7(G)$ is checked using GAP. The order of G is

3^{494} and this is a 3-group without dimension property of the smallest order the authors were able to construct.

In [2], both categories are considered, groups and Lie rings. What often happens is that the computations for Lie rings are simpler. Here we will give detailed examples for the case of Lie rings. For a Lie ring over integers L, consider its universal enveloping algebra $U(L)$. The algebra $U(L)$ admits the augmentation ideal ω. The Lie ring L embeds in $U(L)$, and the dimension subgroups of L are defined as $\delta_n(L) := L \cap \omega^n$. The lower central series term $\gamma_n(L)$ lies in $\delta_n(L)$, and almost all main statements of the theory of dimension subgroups can be extended from groups to Lie rings, with simpler proofs (see [3]). In particular, for any L, $\delta_n(L) = \gamma_n(L)$, $n = 1, 2, 3$, and there exists a Lie analog of Rips example such that $\delta_4/\gamma_4 = \mathbb{Z}/2$. The following result also is from [2].

Theorem 5.2. *For any prime p, there exist a Lie ring A and integer n such that the abelian group $\delta_n(A)/\gamma_n(A)$ contains $\mathbb{Z}/p$ as a subgroup.*

In a Lie ring presentation, we introduce the following notation: for $d \in \mathbb{N}$, when we write a generator $y^{(d)}$ of degree d, we mean a list of generators $y_1, \dots, y_d$; and when $y^{(d)}$ is written for the left-normed iterated commutator, then $y^{(d)} := [y_1, \dots, y_d]$. Thus, for example, "$\langle y_1^{(2)}, y_2^{(3)} \mid [y_1, y_2]\rangle$" is shorthand for "$\langle y_{1,1}, y_{1,2}, y_{2,1}, y_{2,2}, y_{2,3} \mid [[y_{1,1}, y_{1,2}], [y_{2,1}, y_{2,2}, y_{2,3}]]\rangle$." The following is proved in [2]. Given an integer $s \geq 3$, there are integers $e, c_0, \dots, c_s$ and $n = c_0 + \cdots + c_s$ such that, for the Lie ring

$$A = \langle y_0 \dots, y_s, z_0^{(c_0)}, \dots, z_s^{(c_s)} \mid y_0 + \cdots + y_s = 0, e^{c_i} y_i = z_i \text{ for } i = 0, \dots, s\rangle,$$

there exists a natural embedding

$$\bigoplus_i E_{i,s}^1 \hookrightarrow \delta_n(A)/\gamma_n(A).$$

To illustrate how it works, we first rewrite formula (5) as follows. Take L to be a free Lie ring with generators $y_0, \dots, y_s$ and one relation $y_0 + \cdots + y_s = 0$. Set $I_i = (y_i)^L$. In this notation,

$$\frac{I_0 \cap \cdots \cap I_s}{[I_0, \dots, I_s]_S} \simeq \bigoplus_{i \geq 1} E_{i,s}^1.$$

Consider the universal enveloping algebra $U(L)$, the corresponding ideals $J_i = y_i U(L)$ in $U(L)$, and their symmetric product

$$(J_0, \dots, J_s)_S = \sum_{\rho \in \Sigma_{s+1}} J_{\rho(0)} \cdots J_{\rho(s)}.$$

The natural map $L \to U(L)$ induces

$$\begin{array}{ccc} \frac{I_0 \cap \cdots \cap I_s}{[I_0, \dots, I_s]_S} & \longrightarrow & \frac{J_0 \cap \cdots \cap J_s}{(J_0, \dots, J_s)_S} \\ \Vert & & \Vert \\ \bigoplus_{i \geq 1} E_{i,s}^1 & \longrightarrow & H_s(U(L[S^1])) \end{array}$$

Here $U(L[S^1])$ is the universal enveloping algebra of the simplicial Lie ring $L[S^1]$, it has infinite cyclic homology groups in all dimensions. At the same time, $E^1_{i,j}$-terms of the lower central series spectral sequence for S^2 are finite for all $j \geq 3$. It follows that the map is 0. Therefore, for $s \geq 3$ we have $I_0 \cap \cdots \cap I_s \leq L \cap (J_0, \dots, J_s)_S$ when considered in the universal enveloping algebra. See [2] for details.

Next we add the relations $e^{c_i} y_i = z_i$, for a fixed choose of $e, c_i, i = 0, \dots, s$. Recall that the element z_i is of degree c_i. These relations imply that, for any $w \in I_0 \cap \cdots \cap I_s$, the image of $e^n w$ in A, lies in δ_n. An element w is our *homotopy pattern*, we see that, if we write it in the universal enveloping algebra $U(L)$, it lies in the symmetric product $(J_0, \dots, J_n)_S$. At the same time, it does not lie in the symmetric product $[I_0, \dots, I_s]_S$. Now the relations $e^{c_i} y_i = z_i$ guarantee that $e^n w$ lies in the nth augmentation power of the universal enveloping algebra. There is no obvious reason for the element $e^n w$ to lie in the nth term of the lower central series of A. The detailed analysis shows that one can choose the constants e, c_i, such that $e^n w$ will not lie in $\gamma_n(A)$ (see [2]). We can call the described method the *bloating of a homotopy pattern.*

In particular, we can take $e = p, s = 2p - 1$ and realize the p-torsion elements α_p described in the previous section inside dimension quotients of Lie rings.

In some cases it is possible to find c_i's sufficiently small. The following two examples are from [2], they are checked with computer assistance. Consider the Lie ring

$$A = \langle y_0, y_1, y_2, y_3, z_0^{(1)}, z_1^{(2)}, z_2^{(2)}, z_3^{(2)} \mid$$
$$y_0 + y_1 + y_2 + y_3 = 0, y_0 = 2^6 z_0, 2^6 y_1 = 2^5 z_1, 2^5 y_2 = 2^3 z_2, 2^3 y_3 = z_3 \rangle$$

and the element $\omega = [[y_0, y_1], [y_0, y_2]]$. In that Lie algebra, we have $\omega \in \delta_7(A) \setminus \gamma_7(A)$ and $2\omega \in \gamma_7(A)$. So, our *homotopy pattern,* which generated $L_3\mathcal{L}^4(\mathbb{Z}, 1)$ and $\pi_4(S^2)$, makes the difference between δ_7 and γ_7.

For $p = 3$, the situation is similar and the construction can be simplified. Consider the following Lie ring:

$$\begin{aligned} A = \langle y_{ij}, z_{ij}^{(i+j+1)} \text{ for } 0 \leq i < j \leq 5 \mid & \, y_{01} + y_{02} + y_{03} + y_{04} + y_{05} = 0, \\ & -y_{01} + y_{12} + y_{13} + y_{14} + y_{15} = 0, \\ & -y_{02} - y_{12} + y_{23} + y_{24} + y_{25} = 0, \\ & -y_{03} - y_{13} - y_{23} + y_{34} + y_{35} = 0, \\ & -y_{04} - y_{14} - y_{24} - y_{34} + y_{45} = 0, \\ & 3^{i+j} y_{ij} = z_{ij} \text{ for } 0 \leq i < j \leq 5 \rangle. \end{aligned}$$

Then the element $\omega = 3^{15}([y_{04}, y_{14}, y_{23}] - [y_{04}, y_{24}, y_{13}] + [y_{04}, y_{34}, y_{12}] + [y_{14}, y_{24}, y_{03}] - [y_{14}, y_{34}, y_{02}] + [y_{24}, y_{34}, y_{01}])$ belongs to $\delta_{18}(A) \setminus \gamma_{18}(A)$ with $3\omega \in \gamma_{18}(A)$. One can easily recognize our *homotopy pattern* α_3 rewritten in the way $[y_i, y_j] \rightsquigarrow y_{i-1,j-1}$.

The situation for groups is similar. We start with a free group

$$F = \langle x_0, \dots, x_s \mid x_0 \dots x_s = 1 \rangle, \quad R_i = \langle x_i \rangle^F,$$

and take an element q from the intersection $R_0 \cap \cdots \cap R_s \setminus [R_0, \ldots, R_s]_S$. Since, for $s \geq 3$, the Hurewicz homomorphism $\pi_{s+1}(S^2) = \pi_s(\Omega S^2) \to H_s(\Omega S^2)$ is the zero map, the element $q - 1$, written in the group ring $\mathbb{Z}[F]$, belongs to the symmetric product of ideals $(r_0, \ldots, r_s)_S$. Next we use a *bloating of a homotopy pattern* applied to the group-case. The difference between Lie ring and group cases is that, in the case of groups, we cannot transfer exponents in a free way, $[x^e, y] \neq [x, y^e]$. That is, the process has to be accomplished more carefully. Again we refer to [2] for details. The $\mathbb{Z}/p$-torsion terms in the dimension quotients can be realized, for example, using Serre elements $\mathbb{Z}/p \subset \pi_{2p}(S^2)$. The experiments with GAP show that Rips-type examples with $D_4 \neq \gamma_4$ also can be constructed using the described method, by bloating of a homotopy pattern $[[x_1, x_2], [x_1, x_2x_3]]$ which corresponds to the homotopy element $\mathbb{Z}/2 = \pi_4(S^2)$.

6. LIMITS. SPECULATIVE FUNCTOR THEORY

"All Rajayoga, for instance, depends on this perception and experience that our inner elements, combinations, functions, forces, can be separated or dissolved, can be new-combined and set to novel and formerly impossible workings or can be transformed and resolved into a new general synthesis by fixed internal processes." (Sri Aurobindo, "The Synthesis of Yoga")

A standard way to define a group or Lie algebra is combinatorial, i.e., via generators and relations. In order to construct an algebraic objects with complicated properties, one can play with generators and relations. Take, for example, two symbols a and b and one relation, namely

$$\langle a, b \mid a^{-1}b^2ab^{-1}a^3b^{-1} = 1\rangle.$$

It turns out that the resulting group has interesting and nonobvious properties. When we consider some *functor*, we usually mean that this functor comes from a natural consideration and is not constructed in a speculative way. There are no obvious combinatorial games which give a possibility to define functors in terms of generators and relations. However, there is one nonobvious way, and this section will be about it. We will see that the characters we discussed before, like derived functors, dimension quotients and group homology, will appear in those functorial constructions.

Let G be a group. By $\mathsf{Pres}(G)$ we denote the *category of presentations* of G with the objects being free groups F together with epimorphisms to G. Morphisms are group homomorphisms over G. For a functor $\mathcal{F} : \mathsf{Pres}(G) \to \mathsf{Ab}$ from the category $\mathsf{Pres}(G)$ to the category of abelian groups, one can consider the (higher) limits $\varprojlim^i \mathcal{F}$, $i \geq 0$, over the category of presentations. That is, we fix our group G, consider free presentations $R \hookrightarrow F \twoheadrightarrow G$ and make functorial (on F, R) constructions $\mathcal{F}(F, R)$. The limits $\varprojlim^i \mathcal{F}(F, R)$, $i \geq 0$, will depend only on G and, moreover, present functors from the category of all groups to abelian groups.

The category $\mathrm{Pres}(G)$ is strongly connected and has pairwise coproducts. The limit $\varprojlim F = \varprojlim^0 F$ has the following properties. For any $c \in \mathrm{Pres}(G)$,

$$\varprojlim \mathcal{F} = \{x \in \mathcal{F}(c) \mid \forall c' \in \mathrm{Pres}(G), \varphi, \psi : c \to c', \mathcal{F}(\varphi)(x) = \mathcal{F}(\psi)(x)\}. \quad (6)$$

Moreover, it is known [31] that the limit of a functor from a strongly connected category with pairwise coproducts is equal to the equalizer

$$\varprojlim \mathcal{F} \cong \mathrm{eq}(\mathcal{F}(c) \rightrightarrows \mathcal{F}(c \sqcup c))$$

for any $c \in \mathrm{Pres}(G)$. In particular, this equalizer does not depend on c.

The standard and one of the simplest examples is the following. For a group $G = F/R$, the Hopf formula for the second homology is very useful: $H_2(G) = \frac{R \cap \gamma_2(F)}{[F,R]}$. The second homology can be presented as a limit as well:

$$H_2(G) = \varprojlim R/[F, R].$$

That is, the quotient $\frac{R \cap \gamma_2(F)}{[F,R]}$ is the maximal subgroup of $R/[F, R]$, which depends on G only.

The limits $\varprojlim^i \mathcal{F}$ are studied in the series of papers [12, 17, 19, 29, 31]. Next we will give examples which illustrate the variety and complexity of functors which can be obtained playing with limits.

The group homology can be presented as follows (see [12, 17]):

$$\varprojlim (R_{ab}^{\otimes n})_F = H_{2n}(G), \quad n \geq 1,$$
$$\varprojlim^1 (R_{ab}^{\otimes n})_F = H_{2n-1}(G).$$

Here the tensor powers of the relation modules R_{ab} are considered with diagonal action of F. The derived functors $L_1\mathsf{S}^2$ and $L_1\mathsf{S}^3$, which we discussed in Section 2, can be described via limits (see [29, 30])

$$\varprojlim \frac{\gamma_2(F)}{\gamma_2(R)\gamma_3(F)} = L_1\mathsf{S}^2(G_{ab}),$$
$$\varprojlim \frac{\gamma_3(F)}{[\gamma_2(R), F]\gamma_4(F)} = L_1\mathsf{S}^3(G_{ab}),$$
$$\varprojlim \frac{\gamma_3(F)}{\gamma_3(R)\gamma_4(F)} = L_2\mathsf{L}_s^3(G_{ab}),$$
$$\varprojlim \frac{\gamma_4(F)}{[\gamma_2(R), F, F]\gamma_2(\gamma_2(F))\gamma_5(F)} = L_1\mathsf{S}^4(G_{ab}),$$
$$\varprojlim \frac{\gamma_2(R)}{\gamma_2([R, F])\gamma_3(R)} = L_1\mathsf{S}^2(H_2(G)).$$

Comparing the first three limits with the results of Section 2, we see that, in certain cases, when the formula (Φ) can be applied, our *homotopy patterns* can be described as limits

$$\varprojlim \frac{\text{Whole structure}}{\text{Obvious part of an intersection}} = \text{Homotopy pattern}.$$

Observe that a simple deformation of the considered functor $\mathsf{Pres} \to \mathsf{Ab}$ may change the limits completely. For example,

$$\varprojlim \frac{\gamma_3(F)}{[\gamma_2(F), R]\gamma_4(F)} = 0.$$

Let us mention a couple of exotic examples. If a group G does not have a 2-torsion, then [17]

$$\varprojlim \gamma_2(R)/[\gamma_2(R), F] = H_4(G; \mathbb{Z}/2),$$

$$\varprojlim{}^1 \gamma_2(R)/[\gamma_2(R), F] = H_3(G; \mathbb{Z}/2).$$

Recall the Fox quotient $\frac{F(3,R)}{G(3,R)}$ from Section 2. This quotient depends on F and R, not only on G. The limit of this quotient is computed in [31] as

$$\varprojlim \frac{F(3, R)}{G(3, R)} = L_1\mathsf{S}^2(L_1\mathsf{S}^2(G_{ab})).$$

The next result shows how to present the fourth dimension quotient via limits (see [31] for the proof)

Theorem 6.1. *There is a natural short exact sequence*

$$\varprojlim \frac{R \cap \gamma_2(F)}{\gamma_2(R)(R \cap \gamma_4(F))} \hookrightarrow \varprojlim \frac{\gamma_2(F)}{\gamma_2(R)\gamma_4(F)} \twoheadrightarrow \frac{D_4(G)}{\gamma_4(G)}.$$

Theorem 6.1 shows how to describe the fourth dimension quotient as a functor without using the group ring. The limits in Theorem 6.1 are just equalizers (6), one can compute them for simple examples. The author thanks L. Bartholdi for computing these limits for Rips-type examples using a computer. These computations show that the short exact sequence from Theorem 6.1 does not split.

We finish the paper by briefly reviewing the so-called **fr**-*language* (see [18,19]). The ideals $\mathbf{f} = (F-1)\mathbb{Z}[F]$, $\mathbf{r} = (R-1)\mathbb{Z}[F]$ define functors $\mathsf{Pres}(G) \to \mathsf{Ab}$. Moreover, all possible products of ideals $\mathbf{f}, \mathbf{r}$, their sums and intersections define functors $\mathsf{Pres}(G) \to \mathsf{Ab}$ as well, and we can ask how to describe their limits. One can take any *sentence* of symbols $\mathbf{f}, \mathbf{r}$ like $\mathbf{rr} + \mathbf{ffr} + \mathbf{frf} + \mathbf{rff}$ or $\mathbf{rrrf} + \mathbf{frrr}$ and consider their $\varprojlim^i$ as functors. The author does not know any unified method to describe the limits for a given **fr**-sentence. For any particular case, there are some special tricks, based on homological algebra or group theory. Sometimes the results are surprising. For example, $\varprojlim^2$ of two mentioned sequences are the following:

$$\varprojlim{}^2(\mathbf{rr} + \mathbf{ffr} + \mathbf{frf} + \mathbf{rff}) = G_{ab} \otimes G_{ab},$$

$$\varprojlim{}^2(\mathbf{rrrf} + \mathbf{frrr}) = H_5(G).$$

Here are some more examples of computations, which show that the variety of functors which can be presented as limits of **fr**-sentences is rich enough:

$$\varprojlim{}^1(\mathbf{rff} + \mathbf{frr}) = \mathsf{Tor}(H_2(G), G_{ab}),$$

$$\varprojlim{}^1(\mathbf{rr} + \mathbf{frf} + \mathbf{rff}) = H_2(G, G_{ab}),$$

$$\varprojlim{}^1(\mathbf{rr} + \mathbf{frf}) = H_3(G),$$

$$\varprojlim{}^2(\mathbf{rr}+\mathbf{frf}) = \mathbf{g}\otimes_{\mathbb{Z}[G]}\mathbf{g},$$
$$\varprojlim{}^1(\mathbf{rrf}+\mathbf{frr}) = H_4(G),$$
$$\varprojlim{}^1(\mathbf{rr}+\mathbf{fff}) = \mathsf{Tor}(G_{ab}, G_{ab}),$$
$$\varprojlim{}^2(\mathbf{rr}+\mathbf{fff}) = G_{ab}\otimes G_{ab}.$$

The point of this theory (which we also call **fr**-*language*) is that the formal manipulations with codes in two letters may induce deep and unexpected transformations of functors. Simple transformations of **fr**-codes, like changing the symbol **r** by **f** in a certain place, adding a monomial to the **fr**-code, etc., induce natural transformations of (higher) limits determined by these **fr**-codes. For example, the transformation of the **fr**-codes

$$\mathbf{rr}+\mathbf{frf} \rightsquigarrow \mathbf{rr}+\mathbf{frf}+\mathbf{rff}$$

induces the natural transformation of functors

$$H_3(G) = \varprojlim{}^1(\mathbf{rr}+\mathbf{frf}) \rightsquigarrow \varprojlim{}^1(\mathbf{rr}+\mathbf{frf}+\mathbf{frr}) = H_2(G, G_{ab}).$$

Here the map $H_3(G)\to H_2(G, G_{ab})$ is constructed as

$$H_3(G) = H_2(G,\mathbf{g}) \to H_2(G, \mathbf{g}/\mathbf{g}^2) = H_2(G, G_{ab}),$$

where the last map is induced by the natural projection $\mathbf{g}\twoheadrightarrow \mathbf{g}/\mathbf{g}^2 = G_{ab}$.

We end this section with an observation that, in many cases, when the formula (Φ) can be applied to the structures described in terms of **f** and **r**, the *homotopy patterns* can be seen via limits. For example, the well-known description of the $(2n)$th homology $(n\geq 1)$,

$$H_{2n}(G) = \frac{\mathbf{r}^n\cap \mathbf{f}\mathbf{r}^{n-1}\mathbf{f}}{\mathbf{r}^n\mathbf{f}+\mathbf{f}\mathbf{r}^n},$$

represents the formula (Φ). A simple computation (see [18]) shows that

$$\varprojlim{}^1(\mathbf{r}^n\mathbf{f}+\mathbf{f}\mathbf{r}^n) = H_{2n}(G), \quad n>1.$$

The case $n=1$ is an exception: $\varprojlim{}^1(\mathbf{rf}+\mathbf{fr}) = \mathbf{g}\otimes_{\mathbb{Z}[G]}\mathbf{g}\supset H_2(G)$. In such cases, the higher limits give a way to consider derived versions of the *homotopy patterns* as well. It seems that this is a good point to end this survey.

FUNDING

The work is supported by the grant of the Government of the Russian Federation for the state support of scientific research carried out under the supervision of leading scientists, agreement 14.W03.31.0030 dated 15.02.2018.1.

REFERENCES

[1] B. Köck, Computing the homology of Koszul complexes. *Trans. Amer. Math. Soc.* **353** (2001), 3115–3147.

[2] L. Bartholdi and R. Mikhailov, Group and Lie algebra filtrations and homotopy groups of spheres. 2021, arXiv:1805.10894.

[3] L. Bartholdi and I. B. S. Passi, Lie dimension subrings. *Internat. J. Algebra Comput.* **25** (2015), 1301–1325.

[4] J. S. Birman, *Braids, links and mapping class groups*. Ann. Math. Stud. 82, Princeton Univ. Press, 1974.

[5] A. K. Bousfield, E. B. Curtis, D. M. Kan, D. G. Quillen, D. L. Rector, and J. W. Schlesinger, The mod-p lower central series and the Adams spectral sequence. *Topology* **5**, (1966), 331–342.

[6] L. Breen, On the functorial homology of abelian groups. *J. Pure Appl. Algebra* **142** (1999), 199–237.

[7] L. Breen and R. Mikhailov, Derived functors of non-additive functors and homotopy theory. *Algebr. Geom. Topol.* **11** (2011), 327–415.

[8] R. Brown, Coproducts of crossed P-modules: applications to second homotopy groups and to the homology of groups. *Topology* **23** (1984), 337–345.

[9] P. M. Cohn, Generalization of a theorem of Magnus. *Proc. Lond. Math. Soc. (3)* **2** (1952), 297–310.

[10] A. Dold and D. Puppe, Homologie nicht-additiver Funktoren. Anwendungen. *Ann. Inst. Fourier* **11** (1961), 201–312.

[11] G. Ellis and R. Mikhailov, A colimit of classifying spaces. *Adv. Math.* **223** (2010), 2097–2113.

[12] I. Emmanouil and R. Mikhailov, A limit approach to group homology. *J. Algebra* **319** (2008), 1450–1461.

[13] R. Fox, Free differential calculus I: derivation in the free group ring. *Ann. of Math.* **57** (1953), 547–560.

[14] N. Gupta, *Free group rings*. Contemp. Math. 66, American Mathematical Society, 1987.

[15] N. Gupta, The dimension subgroup conjecture holds for odd order groups. *J. Group Theory* **5** (2002), 481–491.

[16] M. Hartl, R. Mikhailov, and I. B. S. Passi, Dimension quotients. *J. Indian Math. Soc., New Ser. Spec. Centenary* (2007), 63–107.

[17] S. O. Ivanov and R. Mikhailov, A higher limit approach to homology theories. *J. Pure Appl. Algebra* **219** (2015), 1915–1939.

[18] S. O. Ivanov and R. Mikhailov, Higher limits, homology theories and frcodes. In *Combinatorial And Toric Homotopy: Introductory Lectures*, pp. 229–261, Lect. Notes Ser. Inst. Math. Sci. 35, World Scientific, 2017.

[19] S. O. Ivanov, R. Mikhailov, and F. Pavutnitsky, Limits, standard complexes and fr-codes. *Sb. Math.* **211** (2020), 1568–1591.

[20] S. O. Ivanov, R. Mikhailov, and J. Wu, On nontriviality of certain homotopy groups of spheres. *Homology, Homotopy Appl.* **18** (2016), 337–344.

[21] S. O. Ivanov, R. Mikhailov, and J. Wu, Homotopy theory and generalized dimension subgroups. *J. Algebra* **484** (2017), 224–246.

[22] F. Jean, *Foncteurs derives de lalgebre symetrique: Application au calcul de certains groupes d'homologie fonctorielle des espaces* $K(B, n)$. Doctoral thesis, University of Paris 13, 2002.

[23] D. Leibowitz, *The* E^1 *term of the lower central series spectral sequence for the homotopy of spaces*. Ph.D. thesis, Brandeis University, 1972.

[24] G. Losey, On dimension subgroups. *Trans. Amer. Math. Soc.* **97** (1960), 474–486.

[25] S. Mac Lane, Decker's sharper Künneth formula. *Lecture Notes in Math.* **1348** (1988), 242–256.

[26] W. Magnus, Über Beziehungen zwischen höheren Kommutatoren. *J. Reine Angew. Math.* **177** (1937), 105–115.

[27] W. Magnus, A. Karrass, and D. Soliter, *Combinatorial group theory: presentations of groups in terms of generators and relations*. Interscience Publishers [John Wiley & Sons, Inc.], New York–London–Sydney, 1966.

[28] R. Mikhailov and I. B. S. Passi, *Lower central and dimension series of groups*. Lecture Notes in Math. 1952, Springer, 2009.

[29] R. Mikhailov and I. B. S. Passi, Generalized dimension subgroups and derived functors. *J. Pure Appl. Algebra* **220** (2016), 2143–2163.

[30] R. Mikhailov and I. B. S. Passi, Free group rings and derived functors. In *Proceedings of the 7th ECM congress*, pp. 407–425, European Mathematical Society, Berlin, Germany, 2018.

[31] R. Mikhailov and I. B. S. Passi, Dimension quotients, Fox subgroups and limits of functors. *Forum Math.* **31** (2019), 385–401.

[32] R. Mikhailov and J. Wu, Combinatorial group theory and the homotopy groups of finite complexes. *Geom. Topol.* **17** (2013), 235–272.

[33] I. B. S. Passi, *Group rings and their augmentation ideals*. Lecture Notes in Math. 715, Springer, Berlin, 1979.

[34] E. Rips, On the fourth integer dimension subgroup. *Israel J. Math.* **12** (1972), 342–346.

[35] E. Witt, Treue Darstellung Liescher Ringe. *J. Reine Angew. Math.* **177** (1937), 152–160.

[36] J. Wu, Combinatorial descriptions of the homotopy groups of certain spaces. *Math. Proc. Cambridge Philos. Soc.* **130** (2001), 489–513.

[37] A. I. Yunus, On a problem of Fox. *Sov. Math., Dokl.* **30** (1984), 346–350.

ROMAN MIKHAILOV

Laboratory of Modern Algebra and Applications, St. Petersburg State University, 14th Line, 29b, Saint Petersburg, 199178, Russia, and St. Petersburg Department of Steklov Mathematical Institute, Fontanka 27, Saint Petersburg, 191023, Russia, romanvm@mi.ras.ru

FROBENIUS HOMOMORPHISMS IN HIGHER ALGEBRA

THOMAS NIKOLAUS

ABSTRACT

We will discuss versions of the Frobenius homomorphism for a ring spectrum R: the *Tate-valued Frobenius* $R \to R^{tC_p}$ and the *Frobenius on topological Hochschild homology* $\mathrm{THH}(R) \to \mathrm{THH}(R)^{tC_p}$. Similar to ordinary algebra, these morphisms play an important role in higher algebra and are related to various concepts in stable homotopy theory and algebraic K-theory. We discuss the notion of *perfectness*, which is to say that these morphisms are equivalences, and relate this notion to the Segal conjecture, the red-shift conjecture, and the classification of spaces by stable data.

MATHEMATICS SUBJECT CLASSIFICATION 2020

Primary 18N70; Secondary 55P43, 19D55, 13A35

KEYWORDS

Higher algebra, Frobenius homomorphism, topological Hochschild homology

ICM INTERNATIONAL CONGRESS OF MATHEMATICIANS 2022 JULY 6–14

Proc. Int. Cong. Math. 2022, Vol. 4, pp. 2826–2854
DOI 10.4171/ICM2022/159

Published by EMS Press

In this survey I would like to give an overview of some of the ideas in higher algebra that have been central to my research over the last few years and which will also be crucial for the next years. Those ideas arose in joint work and discussions with a lot of people and many of the aspects are due to them.

The classical *Frobenius homomorphism* for an ordinary commutative ring R and a prime number p is the ring map

$$\varphi_p : R \to R/pR, \quad r \mapsto [r^p].$$

This homomorphism plays a major role in the analysis and structure theory of algebras and schemes in characteristic p. A very important class of $\mathbb{F}_p$-algebras are the *perfect* ones, i.e., those for which the Frobenius homomorphism is an isomorphism. For example, those algebras admit unique torsion-free lifts to $\mathbb{Z}_p$, called the Witt vectors. There are also mixed characteristic versions of perfect $\mathbb{F}_p$-algebras, called perfectoid rings, which have similar deformations. In this survey I will discuss several incarnations of Frobenius homomorphisms in higher algebra and shed light on the role of perfectness in this setting.

I will use the term "higher algebra" to mean "algebra" over the sphere spectrum $\mathbb{S}$, or said differently, the theory of spectra and ring spectra. The reader unfamiliar with the notion of spectra should just think of the sphere spectrum $\mathbb{S}$ as a stable homotopy-theoretic version of the ring of integers $\mathbb{Z}$. Spectra are modules over $\mathbb{S}$, thus a stable homotopy-theoretic version of abelian groups. Similarly, $\mathbb{E}_\infty$-ring spectra are commutative algebras over $\mathbb{S}$, a homotopy coherent version of commutative rings.

It was Waldhausen's vision to apply arithmetic ideas to ring spectra and develop a theory similar to ordinary algebra. He called this branch "brave new algebra" to express this idea. Nowadays there has been a lot of work by many people towards realizing this vision. Many ideas and concept have been transferred from ordinary to higher algebra, including scheme theory, obstruction theory, K-theory, and Hochschild homology. We view the results and ideas that we present in this survey as a further step in this program.

Specifically, we will define and discuss the following two incarnations of Frobenius homomorphisms for a ring spectrum R:

(1) The *Tate-valued Frobenius*

$$\varphi_p : R \to R^{tC_p}.$$

Here R^{tC_p} is the Tate construction which we discuss in Section 1.

(2) The Frobenius on topological Hochschild homology

$$\varphi_p : \mathrm{THH}(R) \to \mathrm{THH}(R)^{tC_p},$$

where $\mathrm{THH}(R)$ is a spectrum associated with R, called *topological Hochschild homology* which generalizes ordinary Hochschild homology.

We will describe how these morphisms are related to many important aspects of stable homotopy theory such as power operations, the characterization of spaces by stable data, and the computation of algebraic K-theory. A major role for this will be played by the notion of perfectness, which is to say that these maps are equivalences (at least in a range of degrees). We will see how perfectness or ring spectra is related to the *Segal conjecture*, various far reaching generalizations of it and the *red-shift conjecture* in algebraic K-theory.

Higher algebra and algebraic K-theory. Let us first say a view general words about the motivation for higher algebra. Since spectra are the central objects in stable homotopy theory, it is clear that gaining an understanding of these objects is the major goal. For example, the homotopy groups of $\mathbb{S}$ are the stable homotopy groups of spheres, which besides their central role in homotopy theory are also closely related to understanding cobordism classes of manifolds by the Pontryagin–Thom construction. Waldhausen's original motivation was to study spaces of diffeomorphism and h-cobordisms, which he managed to relate to the algebraic K-theory groups of certain ring spectra.

More generally, algebraic K-theory is a topic which is a major motivation to study ring spectra. Recall that for an ordinary ring R, Quillen assigned groups $K_*(R)$ for every $* \in \mathbb{N}$, called the algebraic K-theory groups. Those generalized previously defined groups for $* = 0, 1, 2$ and by now play an important role in many areas of mathematics, from geometric topology to arithmetic. In his ICM address 1974, Quillen conjectured a deep relation of the higher K-theory groups to étale cohomology [42], the so-called Lichtenbaum–Quillen conjecture which has later been proven by Voevodsky [48].

By definition, the K-theory groups are the homotopy groups of a spectrum $K(R)$, and it turns out that studying the spectrum $K(R)$ itself is very often more fruitful than just its homotopy groups. For example, Waldhausen observed that Quillen's conjecture could be restated in terms of chromatic homotopy theory, namely that the map $K(R) \to L_1^f K(R)$ induces a p-adic equivalence in high enough degrees (for a prime p and suitable rings R). Here L_1^f is a certain *chromatic* localization, which is a spectral analogue of localizing a ring at a prime number. There are also higher versions of this localization denoted L_n^f for $n \geq 0$ and Rognes' red-shift problem in algebraic K-theory is about whether for certain ring spectra R the corresponding map $K(R) \to L_n^f K(R)$ is a p-adic equivalence in high degrees, see [44] for a more precise formulation. This question can be seen as a natural higher variant of the Lichtenbaum–Quillen conjecture.

In order to study the K-theory spectrum $K(R)$, Bökstedt invented the theory of *topological Hochschild homology* [9], which is the natural generalization of Hochschild homology to the sphere spectrum. Indeed, he defined for every ring R a spectrum $\mathrm{THH}(R)$. Later Bökstedt–Hsiang–Madsen [10] used the definition of THH to define another spectrum $\mathrm{TC}(R)$, called *topological cyclic homology* which comes with a map $K(R) \to \mathrm{TC}(R)$, called the *cyclotomic trace*. This map is a natural generalization of the Chern-character and is often close to an isomorphism by work of many people (see [13] for a very definitive statement). Therefore, if one has a good understanding of topological cyclic homology, one can use this to compute and understand algebraic K-theory. We will explain how the Frobenius on THH is the key structure needed to gain such an understanding.

Overview of this survey. Generally, our aim is not to give a technically precise account or an exhaustive list of applications and results, but rather explain some fundamental ideas and give the reader an insight into the phenomena that can show up. In particular, we do not talk about the concrete computations that have been done using the technology presented here (see, e.g., [21, 43, 45]) since this would go beyond the scope of this article.

We will motivate and complement all the homotopy theoretic construction in higher algebra by the corresponding statements and constructions in ordinary algebra. Therefore most of the sections contain a short paragraph about the "classical" analogue, presented in a way that makes it easier to understand what happens in the higher case.

In Section 1 we introduce for a spectrum X and an action of a finite group G on X the Tate construction X^{tG}. This is a new spectrum whose homotopy groups are closely related to ordinary Tate cohomology. We explain an important theorem of Lin and Gunawardena, which states that for X the sphere spectrum this Tate spectrum is a completion of the sphere spectrum. In Section 2 we construct the Tate diagonal, which is a certain natural map of spectra $X \to (X \otimes_{\mathbb{S}} \cdots \otimes_{\mathbb{S}} X)^{tC_p}$ akin to the diagonal map of a set. This map is the key needed to define all the Frobenius homomorphisms in higher algebra that we discuss in subsequent sections. By a deep result of Rognes–Lunøe-Nielsen (with a small refinement by the author with P. Scholze) this Tate diagonal is often an equivalence. Section 3 contains the construction of the Tate-valued Frobenius $R \to R^{tC_p}$ for an $\mathbb{E}_\infty$-ring spectrum. We also discuss when it gives rise to an equivalence and the notion of perfectness for $\mathbb{E}_\infty$-rings. In Section 4 we introduce the dual version of the Frobenius for $\mathbb{E}_\infty$-coalgebra spectra C. Magically, this gives an endomorphism $\varphi_p : C \to C$ if C is connective and p-complete. We discuss why this endomorphism is the identity for the suspension spectrum $C := (\Sigma^\infty_+ X)^\wedge_p$ of a space X and conjecture that this precisely characterizes suspension spectra among all spectra. In Section 5 we quickly review topological Hochschild homology and then introduce the Frobenius $\varphi_p : \mathrm{THH}(R) \to \mathrm{THH}(R)^{tC_p}$. We explain how this is gives rise to a cyclotomic structure and how one can define $\mathrm{TC}(R)$ from that. Finally, we elaborate on the role of perfectness of THH, meaning that the Frobenius is an equivalence. Section 6 contains a discussion of cyclotomic spectra with *Frobenius lifts* and the relation between perfectness and boundedness of cyclotomic spectra. We also explain how this relates to the Quillen–Lichtenbaum conjecture following work of Mathew and Hahn–Wilson. In the final Section 7 we explain the Segal conjecture and state a conjecture which would drastically generalize it. This conjecture is the key to prove the conjectures made in Section 4.

All of the section are based on the results and notation introduced in the first two sections. But the reader only interested in the THH-aspects can skip Sections 3 and 4 and jump right into Section 5 and 6. Also Section 7 is independent of Sections 3–6.

1. THE TATE CONSTRUCTION

In this first section we will introduce and study the Tate construction X^{tG} for an action of a finite group G on a spectrum X. The Tate construction is a stable homotopy theoretic version of Tate cohomology, hence the name. We will mostly be interested in the case of the cyclic group C_p with p elements for a prime number p.

The Tate construction in algebra. Let A be an abelian group with an action by a finite group G. In this situation we have a norm map from the coinvariants to the invariants:

$$A_G \to A^G, \quad [a] \mapsto \sum_{g \in G} ga.$$

The zeroth Tate cohomology is the cokernel of this map,

$$\hat{H}^0(G; A) := A^G / A_G.$$

By definition, $\hat{H}^0(G; A)$ comes with a canonical quotient map can : $A^G \to \hat{H}^0(G; A)$.

Example 1.1. Assume that G acts trivially on A. Then we get $H^0(G; A) = A/|G|$. To say this a bit more systematically, we note that for the trivial action of G on A we have a natural map

$$\text{triv} : A \xrightarrow{=} A^G \xrightarrow{\text{can}} H^0(G; A), \quad a \mapsto [a],$$

and this map induces the isomorphism $A/|G| \xrightarrow{\cong} H^0(G; A)$.

The "external product"

$$H^0(G; A) \otimes H^0(G; B) \to H^0(G; A \otimes B), \quad [a] \otimes [b] \mapsto [a \otimes b],$$

makes the zeroth Tate cohomology into a lax symmetric monoidal functor. It follows that if A has a ring structure such that G acts through ring homomorphisms, then also $H^0(G; A)$ is a ring.

Example 1.2. For every abelian group A with G-action, we have that multiplication by $|G|$ is zero on $H^0(G; A)$. To see this, note that A is a module over $\mathbb{Z}$ with trivial G-action and so it follows that $H^0(G; A)$ is a $\mathbb{Z}/|G|$-module.

The Tate construction in higher algebra. Let G be a finite group acting on a spectrum X, that is, a functor $BG \to \text{Sp}$. Then we can form the homotopy orbits and the homotopy fixed points X_{hG} and X^{hG} defined as the colimit and limit of this functor. There is a natural norm map

$$\text{Nm}_G : X_{hG} \to X^{hG}$$

such that the composition $X \to X_{hG} \to X^{hG} \to X$ is given by the sum over all the maps $\cdot g : X \to X$ for $g \in G$. The precise definition requires some coherence technology, see, e.g., [32, SECTION 6.16] or [41, CHAPTER 1]. The *Tate construction* is then defined as the cofiber

$$X^{tG} := \text{cofib}(\text{Nm}_G) = X^{hG}/X_{hG}.$$

Example 1.3. Let HA be the Eilenberg–MacLane spectrum of an ordinary abelian group with a G-action. Then we have that $\pi_0(HA^{tG}) = \hat{H}^0(A; G)$. That is the sense in which the

Tate construction for spectra generalizes the construction from the previous section. We can also describe the other homotopy groups as follows:

$$\pi_*(A) = \begin{cases} \ker(A_G \to A^G), & * = 1, \\ H_{*-1}(G; A), & * > 1, \\ H^{-*}(G; A), & * < 0. \end{cases}$$

This follows immediately from the long exact sequence obtained from the defining cofiber sequence. These are the ordinary *Tate cohomology groups*.

By construction, we have a canonical map

$$\mathrm{can} : X^{hG} \to X^{tG}$$

and if X carries the trivial G-action then we can compose this with the map $X \to X^{hG} = X^{BG}$ induced by the projection $BG \to \mathrm{pt}$ to obtain a natural map

$$\mathrm{triv} : X \to X^{tG}.$$

Theorem 1.4 (Lin [28], Gunawardena [17]). *For the sphere spectrum $\mathbb{S}$ equipped with the trivial action of C_p, the map* $\mathrm{triv} : \mathbb{S} \to \mathbb{S}^{tC_p}$ *is a p-completion, i.e., exhibits $\mathbb{S}^{tC_p}$ as the p-complete sphere $\mathbb{S}_p^\wedge$.*

In my opinion, this result is one of the deepest and most striking results in stable homotopy theory. For example, the fact that $\mathbb{S}^{tC_p}$ is connective is already quite surprising if one thinks about Example 1.3. In fact, there is a convergent spectral sequence

$$\hat{H}^i\big(C_p, \pi_j(\mathbb{S})\big) \Rightarrow \pi_{j-i}(\mathbb{S}^{tC_p}).$$

Every single page of this spectral sequence will be periodic, still the result is connective. Theorem 1.4 also shows that the homotopy groups of the Tate construction X^{tC_p} are in contrast to the algebraic case generally not p-torsion groups, as $\pi_0(\mathbb{S}_p^\wedge) = \mathbb{Z}_p$.

From Theorem 1.4 it follows that for any p-adically finite[1] spectrum X with trivial C_p-action the map triv map also induces an equivalence $X_p^\wedge \to X^{tC_p}$ which should be considered as the higher algebra analogue of Example 1.1. For a general spectrum X, this, however, completely fails as the example of $X = H\mathbb{Z}$ already shows. However, the completeness part is still true very generally and should be considered as the "correct" analogue of Example 1.2.

Proposition 1.5 ([41, LEMMA I.2.9.]). *Let X be a bounded below spectrum with C_p-action. Then X^{tC_p} is p-complete.*

The bounded below assumption is crucial here, without that one can easily find counterexamples, e.g., the complex K-theory spectrum KU with trivial C_p-action for which KU^{tC_p} is rational and nontrivial.

---

1 This means that the p-completion of X is finite over the p-complete sphere. For example, the p-complete sphere itself is an example which is, of course, not finite over $\mathbb{S}$.

Proposition 1.6 ([41, THEOREM I.3.1]). *The functor* $(-)^{tG} : \mathrm{Sp}^{BG} \to \mathrm{Sp}$ *admits a canonical lax symmetric monoidal structure.*

In particular, when applied to ring spectra, the Tate construction produces ring spectra. For example, for an ordinary ring R with G-action, we find that $(HR)^{tG}$ is a ring spectrum, which on homotopy groups induces a graded multiplication. This is the multiplication in ordinary Tate cohomology, whose existence I find quite striking.

2. THE TATE DIAGONAL

The goal of this section is to explain that for every spectrum X and every prime p there is a natural map $\Delta_p : X \to (X \otimes_{\mathbb{S}} \cdots \otimes_{\mathbb{S}} X)^{tC_p}$ which we call the *Tate diagonal*. Here C_p-acts by cyclic permutation on the p-fold tensor product $X \otimes_{\mathbb{S}} \cdots \otimes_{\mathbb{S}} X$. This map Δ_p will be the source of all the Frobenius homomorphisms in higher algebra. As usual we want to explain the algebraic analogue first.

The Tate diagonal in algebra. Let A be an abelian group and p a prime number. We consider the map

$$A \to (A \otimes_{\mathbb{Z}} \cdots \otimes_{\mathbb{Z}} A)^{C_p}, \quad a \mapsto a \otimes \cdots \otimes a, \tag{2.1}$$

where the target has p tensor factors and C_p acts by cyclic permutations. This map is obviously not additive, as it sends $a_0 + a_1$ to

$$(a_0 + a_1) \otimes \cdots \otimes (a_0 + a_1) = \sum_{(i_0,\ldots,i_p)} a_{i_0} \otimes \cdots \otimes a_{i_p},$$

where the sum ranges over all sequences in $\{0, 1\}^p$. We see that the deviation from additivity is exactly an element in the image of the norm, so that the composite

$$\Delta_p : A \to (A \otimes_{\mathbb{Z}} \cdots \otimes_{\mathbb{Z}} A)^{C_p} \xrightarrow{\text{can}} \hat{H}^0\big(C_p; (A \otimes_{\mathbb{Z}} \cdots \otimes_{\mathbb{Z}} A)\big) \tag{2.2}$$

is a group homomorphism. This map is the algebraic version of the Tate diagonal.

Proposition 2.1. *For any abelian group and every prime p, the map Δ_p induces an isomorphism $A/p \to \hat{H}^0(C_p; A^{\otimes p})$.*

Sketch. The target is a p-torsion group by Example 1.2. Therefore we get an induced map $A/p \to \hat{H}^0(C_p; A^{\otimes p})$. One checks by hand that this map is an isomorphism for $A = \mathbb{Z}/n$ and $A = \mathbb{Z}$.

The key fact to observe is that the construction $\hat{H}^0(C_p; A^{\otimes p})$ commutes with direct sums in A. This can be seen by expanding the expressing for a direct sum and using that $\hat{H}^0(C_p; -)$ vanishes on induced C_p-modules. Therefore we immediately deduce the result for all finitely generated abelian groups A. Finally, the functor also commutes with filtered colimits in A so that it follows for all abelian groups. ■

The Tate diagonal in higher algebra. The higher version of the diagonal (2.2) is incarnated as follows. Considers the functor

$$\mathrm{Sp} \to \mathrm{Sp}, \quad X \mapsto (X^{\otimes p})^{tC_p}. \tag{2.3}$$

This functor admits a lax symmetric monoidal structure induced by the lax symmetric monoidal structure of the Tate construction.

Proposition 2.2. *For any finite spectrum X, there is a map of spectra*

$$\Delta_p : X \to (X \otimes_{\mathbb{S}} \cdots \otimes_{\mathbb{S}} X)^{tC_p}.$$

This map natural and symmetric monoidal in X and unique with respect to these properties.

We will refer to Δ_p as the Tate diagonal. The key observation to prove Proposition 2.2 is that the functor (2.3) is exact. This exactness can be reduced to showing that it preserves binary direct sums $X \oplus Y$ which then amounts to a categorification of the argument given in the last section. Finally, one uses that the identity functor $\mathrm{Sp} \to \mathrm{Sp}$ is the initial lax symmetric monoidal endofunctor [39] to deduce Proposition 2.2.

We have the following analogue of Proposition 2.1:

Theorem 2.3 (Lunøe-Nielsen–Rognes [30], Nikolaus–Scholze [41]). *For any bounded below spectrum X and any prime p, the Tate diagonal*

$$X \to (X \otimes \cdots \otimes X)^{tC_p}$$

is a p-completion.

Note that for $X = \mathbb{S}$ this theorem reduces to the Theorem of Lin and Gunawardena (Theorem 1.4). Rognes and Lunøe-Nielsen have proven Theorem 2.3 under the additional assumption that X is of finite type over $\mathbb{F}_p$ by relating it to the algebraic "Singer construction" of Li–Singer and Adams–Gunawardena–Miller. In joint work with P. Scholze, we prove the extension to all bounded below spectra. In fact, we deduce that this follows using formal properties of the Tate construction from the case $X = H\mathbb{F}_p$.

Remark 2.4. For $X = H\mathbb{F}_p$ the Theorem asserts that $(H\mathbb{F}_p \otimes_{\mathbb{S}} \cdots \otimes_{\mathbb{S}} H\mathbb{F}_p)^{tC_p}$ is concentrated in degree 0. One can try proving this directly using the Tate spectral sequence, which for example for $p = 2$ takes the form

$$\hat{H}^i\big(C_2, \mathcal{A}_2^\vee\big) \Rightarrow \pi_{j-i}\big((H\mathbb{F}_2 \otimes_{\mathbb{S}} H\mathbb{F}_2)^{tC_2}\big),$$

where $\mathcal{A}_2^\vee$ is the dual Steenrod algebra which carries a C_2-action by the conjugation χ. The dual Steenrod algebra, as well as the conjugation, have been calculated by Milnor [38], but even the invariants of this action are unknown in general (even though it is easy to calculate them in small degrees with a computer). This means that one cannot completely calculate the first page of the spectral sequence. But even in the range where this is possible, it is in my opinion impossible to determine the differentials. Using Theorem 2.3 and the knowledge that almost everything has to cancel out, one can determine the differentials for the first 20 stems or so. This leads to a wild pattern in which I did not see any regularity.

3. THE TATE-VALUED FROBENIUS

Let R be an $\mathbb{E}_\infty$-ring spectrum and p a fixed prime number. The goal of this section is to discuss a variant of the Frobenius homomorphism for ring spectra, called the *Tate-valued Frobenius*, which will be an $\mathbb{E}_\infty$-map $R \to R^{tC_p}$. Here C_p-acts trivially on R.

Recall that for an ordinary commutative ring R the Frobenius homomorphism is the ring morphism

$$R \to R/p, \quad r \mapsto r^p.$$

Using the algebraic Tate-diagonal (2.1), we can write this map as the composite

$$R \xrightarrow{\Delta_p} \hat{H}^0(C_p; R \otimes_{\mathbb{Z}} \cdots \otimes_{\mathbb{Z}} R) \xrightarrow{m} \hat{H}^0(C_p; R) \xrightarrow{\cong} R/p,$$

where $m : R \otimes_{\mathbb{Z}} \cdots \otimes_{\mathbb{Z}} R \to R$ is the p-fold multiplication map of the ring R, which is C_p-equivariant for the cyclic action on the source and the trivial action on the target.

Definition 3.1 ([41, DEFINITION IV.1.1]). Let R be an $\mathbb{E}_\infty$-ring spectrum. The Tate-valued Frobenius is the composite

$$\varphi_p : R \xrightarrow{\Delta_p} (R \otimes_{\mathbb{S}} \cdots \otimes_{\mathbb{S}} R)^{tC_p} \to R^{tC_p},$$

where Δ_p is the Tate diagonal (cf. Proposition 2.2). The Tate-valued Frobenius is by construction a natural map of $\mathbb{E}_\infty$-ring spectra.

Example 3.2. Let HR be the Eilenberg–MacLane spectrum of an ordinary ring R. Then on π_0 the Tate-valued Frobenius induces the ordinary Frobenius $R \to R/p$.

Example 3.3. One can generalize the last example: if R is a connective ring spectrum, then we have a canonical map $d : \pi_0(R^{tC_p}) \to \pi_0(R)/p$ and the composite

$$\pi_0(R) \xrightarrow{\pi_0(\varphi_p)} \pi_0(R^{tC_p}) \xrightarrow{d} \pi_0(R)/p$$

is the Frobenius of $\pi_0(R)$. Note that in contrast to the discrete case, the map d is in general far from an isomorphism; for example, in the case $R = \mathbb{S}$ of the sphere spectrum, it can be identified with the projection $\mathbb{Z}_p \to \mathbb{F}_p$ (using Theorem 1.4).

The Tate-valued Frobenius is closely related to power operations. For example, if $R = C^*(X, \mathbb{F}_2)$ is the $\mathbb{E}_\infty$-ring of $\mathbb{F}_2$-valued cochains on a space X then we have that $\pi_*(C^*(X, \mathbb{F}_2)^{tC_2}) = H^{-*}(X; \mathbb{F}_2)((t))$ with t in homological degree -1. On homotopy groups the Tate-valued Frobenius induces the map

$$H^*(X; \mathbb{F}_2) \to H^*(X; \mathbb{F}_2)((t)), \quad x \mapsto \sum_{i=0}^{\infty} \mathrm{Sq}^i(x) t^{-i},$$

where $\mathrm{Sq}^i : H^*(X; \mathbb{F}_2) \to H^{*+i}(X; \mathbb{F}_2)$ is the ith Steenrod operation, see [41, PROPOSITION IV.1.16] or [49, SECTION 3.5]. Note that this sum is finite by the instability relation $\mathrm{Sq}^i(x) = 0$ for $i > -|x|$ in homological grading. More generally, for an arbitrary $\mathbb{E}_\infty$-algebra R over $\mathbb{F}_2$, we have that the Frobenius induces the map

$$\pi_*(R) \to \pi_*(R)((t)), \quad x \mapsto \sum_{i=-\infty}^{\infty} Q^i(x) t^i,$$

where $Q^i : \pi_*(R) \to \pi_{*+i}(R)$ is the ith Dyer–Lashof operation. This sum is in general not finite, but a Laurent series since $Q^{-i}(x) = 0$ for $i > -|x|$. In fact, one could take this as the definition of the Dyer–Lashof operations and derive the usual properties from it, see Wilson's paper [49] for a very nice discussion employing this perspective.

One can also get the odd primary Steenrod and Dyer–Lashof operations through this perspective. More generally one can relate power operations for algebras over any base $\mathbb{E}_\infty$-algebra to the Tate-valued Frobenius. See [16] for a very nice discussion of general power operations in the language of ring spectra and the relation to Tate spectra.

The Frobenius as an endomorphism. The main difference of the Tate-valued Frobenius to the ordinary Frobenius homomorphism is that the target is in general not p-torsion but rather p-complete as explained around Proposition 1.5. Recall that by Theorem 1.4 we know that if the underlying spectrum of R is a finite spectrum, then the map

$$\mathrm{triv} : R \to R^{tC_p}$$

is a p-completion. Thus using the inverse to this map we can consider the Tate-valued Frobenius as a map

$$\varphi_p : R \to R_p^\wedge$$

for every p-adically finite ring spectrum R. The finiteness assumption is crucial here and in general the Tate-valued Frobenius does not induce an endomorphism. However, sometimes it does, for example, if R is the dual of a connective coalgebra, as we will see in the next section.

Example 3.4. For the sphere spectrum $\mathbb{S}$, we have that φ_p is the completion map

$$\mathbb{S} \to \mathbb{S}_p^\wedge.$$

This is forced, since it is a map of ring spectra. More generally, let k be a finite field of characteristic p. Then there is an $\mathbb{E}_\infty$-ring $\mathbb{S}_{W(k)}$ called the *ring of spherical Witt vectors* uniquely characterized by the property that it is p-complete, flat over $\mathbb{S}_p^\wedge$ and that $\pi_0(\mathbb{S}_{W(k)})$ is the ordinary ring of Witt vectors $W(k)$. This ring spectrum is finite over the p-complete sphere (see, e.g., [33, EXAMPLE 5.2.7] for a discussion of spherical Witt vectors). Thus the Frobenius is an endomorphism

$$\varphi_p : \mathbb{S}_{W(k)} \to \mathbb{S}_{W(k)}$$

which can be identified with the map induced by the Witt vector Frobenius $F : W(k) \to W(k)$. This follows by an obstruction theoretic argument from Example 3.3 since maps between $\mathbb{S}_{W(k)}$ are uniquely determined by their effect on the modulo p-reduction of π_0.

Definition 3.5. We say that a p-complete and p-adically finite $\mathbb{E}_\infty$-ring spectrum R is *perfect* if the map $\varphi_p : R \to R$ is an equivalence.

Example 3.6. Let X be a finite CW complex. Then we consider the mapping spectrum $R := \mathrm{map}(X, \mathbb{S}_p^\wedge)$ which is an $\mathbb{E}_\infty$-algebra. Since the Frobenius is natural and is the identity on the sphere, it follows that it is also the identity on R. In particular R is perfect.

More generally, we can consider a hypercomplete sheaf of spaces on the étale site of $\mathrm{Spec}(\mathbb{F}_p)$, or said differently, a homotopy type X with a (continuous) action of the profinite group $\mathbb{Z}^\wedge$.[2] We will denote this ∞-category by $\mathcal{S}^{\mathbb{Z}^\wedge}$. Then we can form a twisted version $\mathrm{map}_\sigma(X, \mathbb{S}_p^\wedge)$ of $\mathrm{map}(X, \mathbb{S}_p^\wedge)$ as

$$\mathrm{map}_\sigma\big(X, \mathbb{S}_p^\wedge\big) := \mathrm{map}_{\mathbb{Z}^\wedge}(X, \mathbb{S}_{W(\overline{\mathbb{F}}_p)}).$$

In this finite case this p-complete $\mathbb{E}_\infty$-algebra $\mathrm{map}_\sigma(X, \mathbb{S}_p^\wedge)$ is also perfect. In fact, one can deduce from Mandell's theorem [34] that the assignment

$$\mathcal{S}^{\mathbb{Z}^\wedge} \to \mathrm{Alg}_{\mathbb{E}_\infty}(\mathrm{Sp}), \quad X \mapsto \mathrm{map}_\sigma\big(X, \mathbb{S}_p^\wedge\big)$$

defines an equivalence between the full subcategory of equivariantly p-complete, finite, and simply connected spaces on the left and of perfect $\mathbb{E}_\infty$-algebras R which are p-complete, p-completely finite, and whose reduced $\mathbb{F}_p$-cohomology is simply connected. See [50, SECTION 7] and [31, SECTION 3.5] for similar statements. This should be seen as a classification of a large class of perfect $\mathbb{E}_\infty$-algebras.

4. THE COALGEBRA FROBENIUS

In the last section we have construction the Tate-valued Frobenius $R \to R^{tC_p}$ which could be interpreted as an endomorphism for R finite. In this section we construct a dual morphism for connective coalgebra spectra C which will always be an endomorphism. The construction crucially relies on Theorem 2.3.

The Frobenius homomorphism for ordinary coalgebras. Recall that a cocommutative coalgebra over a field k is a k-module C together with k-linear maps

$$\Delta : C \to C \otimes_k C \qquad \text{(comultiplication)},$$
$$\varepsilon : C \to k \qquad \text{(counit)}$$

satisfying the duals of the axioms of a commutative k-algebra. In fact, one can simply define a coalgebra as an algebra in the opposite category of the category of k-vector spaces.

The dual $C^\vee = \mathrm{Hom}_k(C, k)$ of a coalgebra is always an algebra but the dual $A^\vee$ of an algebra A is in general not a coalgebra unless A is finite dimensional. More precisely, dualization induces a functor

$$(-)^\vee : \mathrm{coAlg}_k^{\mathrm{op}} \to \mathrm{Alg}_k$$

which restricts to an equivalence between finite dimensional coalgebras and algebras.

The fundamental theorem of coalgebras asserts that every coalgebra is the colimit of its finite dimensional subcoalgebras, see [46]. One can use this to show that the category coAlg_k is the ind-category of the category of finite dimensional coalgebras. In particular, it

2 Here continuity is not meant on the naive sense, but in a sense similar to continuous actions of profinite groups on discrete sets. One can make this precise using condensed mathematics.

is opposite equivalent to the category of profinite algebras. From this we can directly deduce the existence of a Frobenius homomorphism for coalgebras over $\mathbb{F}_p$:

Proposition 4.1. *For every cocommutative coalgebra C over $\mathbb{F}_p$ there is a natural map*

$$\varphi_p : C \to C$$

uniquely characterized by the property that its dual $\varphi_p^\vee : C^\vee \to C^\vee$ is the Frobenius homomorphism of the commutative $\mathbb{F}_p$-algebra $C^\vee$.

One can also give a more direct construction of the Frobenius for coalgebras involving the algebraic Tate diagonal. Namely one shows that for every coalgebra over $\mathbb{F}_p$ the diagram

$$\begin{array}{ccc} C & \xrightarrow{\Delta^p} & (C \otimes_{\mathbb{F}_p} \cdots \otimes_{\mathbb{F}_p} C)^{C_p} \\ {\scriptstyle \varphi_p}\downarrow & & \downarrow{\scriptstyle \mathrm{can}} \\ C & \xrightarrow{\Delta_p} & \hat{H}^0(C_p; C \otimes_{\mathbb{F}_p} \cdots \otimes_{\mathbb{F}_p} C) \end{array}$$

commutes, where the upper horizontal map is the p-fold comultiplication of C and the lower horizontal map is the algebraic Tate diagonal. The algebraic Tate diagonal is an isomorphism (Proposition 2.1) so that this diagram defines φ_p, see **[40]** for a discussion along those lines.

The Frobenius for coalgebra spectra. We now want to discuss the higher algebra analogue of the Frobenius for coalgebras. Thus let C be an $\mathbb{E}_\infty$-coalgebra spectrum, that is, an $\mathbb{E}_\infty$-algebra object in the opposite category of spectra. We can form the dual spectrum $C^\vee = \mathrm{map}(C, \mathbb{S})$ and this will be an $\mathbb{E}_\infty$-ring spectrum. As in the algebraic case, we find that this construction induces a functor of ∞-categories

$$(-)^\vee : \mathrm{coAlg}_{\mathbb{E}_\infty}(\mathrm{Sp}) \to \mathrm{Alg}_{\mathbb{E}_\infty}(\mathrm{Sp})$$

which restricts to an equivalence on those full subcategories spanned by objects that are finite over $\mathbb{S}$.

Warning 4.2. The analogue of the fundamental theorem for coalgebras fails over $\mathbb{S}$: not every coalgebra over $\mathbb{S}$ is a filtered colimit of finite coalgebras. Also finite coalgebras are not necessarily compact as objects of $\mathrm{coAlg}_{\mathbb{E}_\infty}(\mathrm{Sp})$ and $\mathrm{coAlg}_{\mathbb{E}_\infty}(\mathrm{Sp})$ is not compactly generated. But it is still a presentable ∞-category and comonadic over the ∞-category of spectra as one can see using the monadicity theorem.

Definition 4.3. For every connective coalgebra $C \in \mathrm{coAlg}_{\mathbb{E}_\infty}(\mathrm{Sp})$, we define a Frobenius morphism $C \to C_p^\wedge$ such that the diagram

$$\begin{array}{ccc} C & \xrightarrow{\Delta^p} & (C \otimes_{\mathbb{S}} \cdots \otimes_{\mathbb{S}} C)^{hC_p} \\ {\scriptstyle \varphi_p}\downarrow & & \downarrow{\scriptstyle \mathrm{can}} \\ C_p^\wedge & \xrightarrow{\Delta_p} & (C \otimes_{\mathbb{S}} \cdots \otimes_{\mathbb{S}} C)^{tC_p} \end{array} \tag{4.1}$$

commutes. Here we use that the lower horizontal map is an equivalence by Theorem 2.3.

A p-complete, connective $\mathbb{E}_\infty$-coalgebra C over $\mathbb{S}_p^\wedge$ is called perfect if the Frobenius $\varphi_p : C \to C$ is an equivalence.

By definition, φ_p is a map of spectra. However, we can consider it as a map $C_p^\wedge \to C_p^\wedge$ and using the monoidal properties of the Tate diagonal one gets the following result:

Proposition 4.4. *The map $\varphi_p : C_p^\wedge \to C_p^\wedge$ canonically refines to a natural map of co-$\mathbb{E}_\infty$-algebras in p-complete spectra. For finite C this morphisms dualizes to the Frobenius of ring spectra as discussed in Section* 3.

Sketch of proof. A coalgebra in $\mathrm{Sp}_p^\wedge$ is essentially the same as a symmetric monoidal functor

$$\mathrm{Fin}^{\mathrm{op}} \to \mathrm{Sp}_p^\wedge,$$

where Fin is the category of finite sets with disjoint union as symmetric monoidal structure. The opposite category $\mathrm{Fin}^{\mathrm{op}}$ can then naturally be considered as a symmetric monoidal category and thus as a symmetric monoidal ∞-category. It is the free symmetric monoidal ∞-category with a cocommutative coalgebra object (given by the singleton).

Thus for a given p-complete coalgebra $C \in \mathrm{Sp}_p^\wedge$, we get an essentially unique symmetric monoidal functor $\underline{C} : \mathrm{Fin}^{\mathrm{op}} \to \mathrm{Sp}_p^\wedge$ which sends the singleton to C. We can now take the diagram (4.1) and replace all the instances of C by the functor $\underline{C}$ to obtain a diagram

$$\begin{array}{ccc}
\underline{C} & \xrightarrow{\Delta^p} & (\underline{C} \otimes_{\mathbb{S}} \cdots \otimes_{\mathbb{S}} \underline{C})^{hC_p} \\
{\scriptstyle \varphi_p} \downarrow & & \downarrow {\scriptstyle \mathrm{can}} \\
\underline{C} & \xrightarrow{\Delta_p} & (\underline{C} \otimes_{\mathbb{S}} \cdots \otimes_{\mathbb{S}} \underline{C})^{tC_p}
\end{array}$$

of functors $\mathrm{Fin}^{\mathrm{op}} \to \mathrm{Sp}_p^\wedge$. The functors are not strong symmetric monoidal, but still lax symmetric monoidal, and all the maps admit the structure of symmetric monoidal transformations. Therefore it follows that the left vertical morphism is also symmetric monoidal which shows the first claim. We skip the argument for the second. ■

We conclude this section by a conjectural further refinement of the Frobenius for coalgebras.

Conjecture 4.5. *The Frobenius refines to an action of the monoidal category* $B\mathbb{N}$ *on the ∞-category* $\mathrm{coAlg}_{\mathbb{E}_\infty}(\mathrm{Sp}_p^\wedge)$ *of p-complete $\mathbb{E}_\infty$-coalgebras.*

Here $B\mathbb{N}$ is the category with a single object and the natural numbers as endomorphisms. This category is itself symmetric monoidal since $\mathbb{N}$ is abelian.

Remark 4.6. An action of $B\mathbb{N}$ is the same as an $\mathbb{E}_2$-map[3]

$$\rho : \mathbb{N} \to Z := Z\big(\mathrm{coAlg}_{\mathbb{E}_\infty}(\mathrm{Sp}_p^\wedge)\big),$$

3 By $\mathbb{E}_2$-monoid we mean $\mathbb{E}_2$-algebra in the ∞-category $\mathcal{S}$ of spaces.

where the target is the center of the ∞-category of coalgebras, by which we mean the $\mathbb{E}_2$-monoid of endomorphisms of the identity functor $\mathrm{id} : \mathrm{coAlg}_{\mathbb{E}_\infty}(\mathrm{Sp}_p^\wedge) \to \mathrm{coAlg}_{\mathbb{E}_\infty}(\mathrm{Sp}_p^\wedge)$. Of course, the map ρ should send the generator of $\mathbb{N}$ to the Frobenius. This determines it as an $\mathbb{E}_1$-map, since $\mathbb{N}$ is the free $\mathbb{E}_1$-space on a single generator. So the conjecture is about a refinement of the $\mathbb{E}_1$-map given by the Frobenius to an $\mathbb{E}_2$-map. Informally, this means that the Frobenius has to coherently commute with itself.

One can try to understand such a question by obstruction theory: the $\mathbb{E}_2$-monoid $\mathbb{N}$ admits a cell structure

$$\mathbb{N} = \varinjlim(M_1 \to M_2 \to M_3 \to M_4 \to \cdots)$$

in which M_1 is the free $\mathbb{E}_2$-monoid on a single generator in degree 0 and each M_{n+1} is obtained from M_n by attaching an $\mathbb{E}_2$-cell of dimension $2n$. This can be seen applying arguments similar to those used in [2, APPENDIX B] and we learned this statement from Achim Krause. After group completion the induced $\mathbb{E}_2$-cell structure on $\mathbb{Z} = \Omega^2\mathbb{C}\mathrm{P}^\infty$ corresponds to the standard cell structure of $\mathbb{C}\mathrm{P}^\infty$.

The Frobenius defines a map $M_1 \to Z$, and we get iteratively for $n \geq 1$ a sequence of obstructions $o_n \in \pi_{2n-1}(Z)$. One could hope that all these homotopy groups vanish, but we have no insight into whether this might be true. However, if we restrict to perfect coalgebras this could hold, at least it does under additional finiteness conditions, by the results outlined at the end of Section 3.

In recent work [50], Allen Yuan has proven a version of this conjecture, namely he shows that it is true when we restrict to the full subcategory spanned by those coalgebras whose underlying spectrum is p-adically finite. Equivalently, this can be then be translated into the dual setting of $\mathbb{E}_\infty$-algebras by Proposition 4.4. Yuan really proves the corresponding statement about the Frobenius of finite $\mathbb{E}_\infty$-algebras. His proof crucially relies on the Segal conjecture, that we will discuss in Section 7 below, specifically Theorem 7.8. In order to apply similar techniques to prove Conjecture 4.5 in its full generality, one would have to prove the version of the Segal conjecture for the norm that we propose as Conjecture 7.10.

Boolean coalgebras. In this section we want to talk about a specific class of coalgebras over the sphere, which we call Boolean coalgebras. The definition is conditional to Conjecture 4.5. We first give the corresponding definition for ordinary coalgebras.

Definition 4.7. Let C be an (ordinary) coalgebra over $\mathbb{F}_p$. Then it is called Boolean (or p-Boolean if we want to emphasize the prime p) if the Frobenius endomorphism $\varphi_p : C \to C$ is the identity.

The name Boolean is motivated by the fact that the dual notion (i.e., algebras over $\mathbb{F}_p$ with $\varphi_p = \mathrm{id}$) is for $p = 2$ the same as a Boolean algebra in the sense of logic.

Proposition 4.8 (coStone duality). *The category of p-Boolean coalgebras is equivalent to the category of sets. The equivalence is given by sending a coalgebra C to the set of coalgebra morphisms from $\mathbb{F}_p$ to C, i.e., the set of grouplike elements in C. Conversely, a set S is sent to the coalgebra $\bigoplus_S \mathbb{F}_p$ with the pointwise comultiplication.*

Now we would like to generalize this to higher algebra, i.e., to give a definition of Boolean coalgebra spectra. The definition of those is conditional to Conjecture 4.5, which was about the existence of an $\mathrm{B}\mathbb{N}$-action on the ∞-category of p-complete coalgebras, which is pointwise given by the Frobenius endomorphism $\varphi_p : C \to C$. Under suitable finiteness conditions, this in fact follows from the results of Yuan.

Definition 4.9. A p-Boolean coalgebra spectrum is a fixed point for the action of $\mathrm{B}\mathbb{N}$ on the ∞-category of $\mathbb{E}_\infty$-coalgebras. A (global) Boolean coalgebra spectrum is an $\mathbb{E}_\infty$-coalgebra C over $\mathbb{S}$ together with a refinement to a p-Boolean coalgebra spectrum for every prime p.

Concretely, a p-Boolean coalgebra is a p-complete Boolean $\mathbb{E}_\infty$-coalgebra C together with a *coherent* equivalence between the Frobenius endomorphism $\varphi_p : C \to C$ and the identity $\mathrm{id} : C \to C$.

Example 4.10. The sphere $\mathbb{S}$ refines uniquely to a Boolean coalgebra. Therefore for every space X the suspension spectrum $\Sigma^\infty_+ X$ also does, as it can be considered as the colimit of the constant X-indexed diagram in Boolean coalgebras with value the sphere. This way we get a unique refinement of the functor Σ^∞_+ through Boolean coalgebras.

Now we conjecture that these Boolean coalgebras can be used to describe spaces algebraically.

Conjecture 4.11. *The functor Σ^∞_+ induces an equivalence between simply connected objects in the ∞-category of spaces and in the ∞-category of Boolean coalgebras (here simply connected relative to the counit map). The inverse is given by sending a Boolean coalgebra C to the mapping space from $\mathbb{S}$ to C.*

This conjecture came up in discussions with Heuts and Klein based on a theorem of Heuts giving an inductive description of spaces using Tate coalgebras [23]. The version for finite simply connected spaces has been implemented by Yuan based on a similar theorem of Mandell [34,35] (using $\mathbb{E}_\infty$-algebras over $\mathbb{Z}$).

Conjecture 4.11 relies on the generalized Segal conjecture that we will explain later (Conjecture 7.10) for two reasons: first, to define the notion of Boolean coalgebras, but also to analyze the adjunction counit of the functors between spaces and Boolean coalgebras. We finally note that one can also make a more general conjecture about a description of p-adic perfect coalgebras in terms of p-complete spaces with a (suitably continuous) action of the profinite integers $\mathbb{Z}^\wedge$. In the finite case, this can again be deduced from Mandell's results [34,35] similar to the result at the end of Section 3. See also the discussion in [50, SECTION 7].

5. THE FROBENIUS ON THH

In this section we will discuss a further instance of the Frobenius operator in higher algebra, namely the Frobenius operator on topological Hochschild homology. We first recall that topological Hochschild homology is a variant of Hochschild homology, namely the variant where one works relative to the sphere spectrum $\mathbb{S}$.

The motivation to consider THH and refined invariants such as TC is that they are good approximations to algebraic K-theory. More precisely, for any ring R there is a natural map $K(R) \to \mathrm{TC}(R)$, called the cyclotomic trace, which is often close to an isomorphism. For example, it is a seminal result of Mathew–Clausen–Morrow [13] based on previous work by McCarthy and many others that for a p-adic ring R the spectrum $\mathrm{TC}(R)$ is after p-adic completion equivalent to étale K-theory. The proof of this result relies crucially on the Frobenius-perspective on cyclotomic spectra that we will explain in this section.

THH and Bökstedt's theorem. Let us review the definition and basic properties of THH here. For more details, see, e.g., [27] or the YouTube lectures based on this document.

Definition 5.1. Let R be a ring spectrum, not necessarily commutative. Then $\mathrm{THH}(R)$ is defined as the relative tensor product

$$\mathrm{THH}(R) = R \otimes_{R \otimes_{\mathbb{S}} R^{\mathrm{op}}} R.$$

Here R^{op} is R equipped with the opposite multiplication so that left $R \otimes_{\mathbb{S}} R^{\mathrm{op}}$-modules are R-bimodules. Then we can view R as a left module over $R \otimes_{\mathbb{S}} R^{\mathrm{op}}$ via its canonical bimodule structure, but also as a right module over $R \otimes_{\mathbb{S}} R^{\mathrm{op}}$ using the flip-involution on $R \otimes_{\mathbb{S}} R^{\mathrm{op}}$.

We could have worked over other bases than the sphere $\mathbb{S}$ by simply taking the tensor product $R \otimes R^{\mathrm{op}}$ over other bases. In particular, if R is an algebra spectrum over the Eilenberg–MacLane spectrum $H\mathbb{Z}$, e.g., if R is itself the Eilenberg–MacLane spectrum of an ordinary ring, then we can form the relative tensor product

$$R \otimes_{R \otimes_{H\mathbb{Z}} R^{\mathrm{op}}} R,$$

and it turns out that this is equivalent to ordinary Hochschild homology $\mathrm{HH}(R)$.[4]

There is a canonical map $R \to \mathrm{THH}(R)$ induced by the inclusion of R into any of the two tensor factors. If R is an $\mathbb{E}_\infty$-ring spectrum then $\mathrm{THH}(R)$ also inherits a natural $\mathbb{E}_\infty$-structure, in particular the homotopy groups $\mathrm{THH}_*(R)$ are a graded commutative ring. The map $R \to \mathrm{THH}(R)$ refines to a map of $\mathbb{E}_\infty$-maps in this case. The most important result about THH is Bökstedts calculation of THH for the finite field $\mathbb{F}_p$, which we implicitly consider as an Eilenberg–MacLane spectrum.

Theorem 5.2 (Bökstedt [9]). *We have that*

$$\mathrm{THH}_*(\mathbb{F}_p) \cong \mathbb{F}_p[x], \quad |x| = 2.$$

This result is central to almost everything that has been done with THH. It is the basis of essentially all the K-theory computations that have been carried out using trace methods, such as the seminal computations of Hesselholt–Madsen [20]. It is used to deduce the very important computations of Bhatt–Morrow–Scholze for perfectoid rings [6]. These

4 Really one gets a derived variant sometimes called Shukla-homology, but we will not distinguish that here.

are the basis of the relation of THH to p-adic Hodge theory and prismatic cohomology (pioneered by Lars Hesselholt). Using Bökstedt's theorem combined with the relation of THH to K-theory, one can even deduce classical Bott periodicity for complex K-theory (see [22, SECTION 1.3.2] for details). There are also computations related to the red-shift conjecture pioneered by Ausoni–Rognes [3] and with a recent breakthrough by Hahn–Wilson [18] which we will describe in Section 6. These result also dependent on the computation of $\mathrm{THH}(\mathbb{F}_p)$.

The way Theorem 5.2 was originally proven is by computing the $\mathbb{F}_p$-homology of the spectrum $\mathrm{THH}(\mathbb{F}_p)$ using a specific spectral sequence:

$$\mathrm{HH}\big(\pi_*(\mathbb{F}_p \otimes_{\mathbb{S}} \mathbb{F}_p)/\mathbb{F}_p\big) \Rightarrow \pi_*\big(\mathrm{THH}(\mathbb{F}_p) \otimes_{\mathbb{S}} \mathbb{F}_p\big).$$

Here $\pi_*(\mathbb{F}_p \otimes_{\mathbb{S}} \mathbb{F}_p) = \mathcal{A}_p^\vee$ is the dual Steenrod algebra, which has been calculated by Milnor [38]. It has a number of polynomial and exterior generators. For the computation of Bökstedt, one needs a bit more information about the dual Steenrod algebra, namely its Dyer–Lashof operations which have been calculated by Steinberger [11, CHAPTER 3, THEOREMS 2.2 AND 2.3]. The interesting thing about these calculations for the dual Steenrod algebra is that if one takes these operations into account, then the homotopy ring of the dual Steenrod algebra is completely generated by a single element in $\pi_1(\mathbb{F}_p \otimes_{\mathbb{S}} \mathbb{F}_p)$. One can in fact combine all these computations into a more conceptual statement as follows (see [26, SECTION 1.1] for a review):

Theorem 5.3 (Milnor, Steinberger, Araki-Kudo, Dyer-Lashof, etc.). *The ring spectrum $\mathbb{F}_p \otimes_{\mathbb{S}} \mathbb{F}_p$ is, as an $\mathbb{E}_2$-algebra over $\mathbb{F}_p$, free on a generator in degree* 1.

This result really is a combination of all the computations above. For example, Milnor's computation of the dual Steenrod algebra can easily be deduced from it. Once one has phrased the statement about the dual Steenrod algebra in this way it is very easy to deduce Bökstedt's theorem from it. But one can also do the opposite: Theorem 5.3 follows formally from Theorem 5.2 using a version of bar–cobar duality, see, e.g., [26] for details.

The Frobenius on THH. Now we describe extra structure on THH, namely a Frobenius homomorphism. Let us note that while we are mostly interested in commutative rings the Frobenius homomorphism even exist in the noncommutative setting. To understand this, let us first understand the corresponding construction in ordinary algebra.

Assume R is an ordinary, not necessarily commutative ring. Then we have

$$R \otimes_{R \otimes_{\mathbb{Z}} R^{\mathrm{op}}} R = R/[R, R],$$

where $[R, R] \subseteq R$ is the subgroup additively generated by commutators $rs - sr$. This quotient $R/[R, R]$ is the algebraic analogue of THH. In fact, it is isomorphic to π_0 of $\mathrm{THH}(R)$.

Observation 5.4. *The map*

$$R/[R, R] \to \big(R/[R, R]\big)/p = R/\big([R, R] + pR\big), \quad [r] \mapsto \big[r^p\big],$$

is a well-defined group homomorphism.

This is the algebraic version of the Frobenius that we will construct on THH now. Before we can discuss the Frobenius on THH, we have to introduce another bit of structure: THH carries a natural action of the circle group $\mathbb{T} = U(1)$. This goes back to Connes and is a homotopy-theoretic incarnation of the Connes' operator. We will take this as a black-box here, see [41] for a careful construction. For an $\mathbb{E}_\infty$-ring R, this $\mathbb{T}$-action on $\mathrm{THH}(R)$ is an action through $\mathbb{E}_\infty$-homomorphisms and it is a result of McClure, Schwänzel, and Vogt that $\mathrm{THH}(R)$ is initial among $\mathbb{E}_\infty$-rings under R with an action of $\mathbb{T}$ [37].

For any action of $\mathbb{T}$, we get an induced C_p-action by restriction to the subgroup $C_p \subseteq \mathbb{T}$ of pth roots. Then we have the following result:

Proposition 5.5. *For any ring spectrum R, there is a map*

$$\varphi_p : \mathrm{THH}(R) \to \mathrm{THH}(R)^{tC_p}$$

called the cyclotomic Frobenius, which is a natural and symmetric monoidal transformation. Moreover, it is $\mathbb{T}$-equivariant where the target carries the residual action by $\mathbb{T}/C_p \cong \mathbb{T}$.

Proof sketch. Let us sketch the construction of φ_p in the case of an $\mathbb{E}_\infty$-ring R. There is a unique extension of the $\mathbb{E}_\infty$-map $R \to \mathrm{THH}(R)$ to a C_p-equivariant $\mathbb{E}_\infty$-map

$$R^{\otimes p} = R \otimes_{\mathbb{S}} \cdots \otimes_{\mathbb{S}} R \to \mathrm{THH}(R)$$

using that the source is the free $\mathbb{E}_\infty$-ring with an action of C_p. In particular, we get an induced map $(R \otimes_{\mathbb{S}} \cdots \otimes_{\mathbb{S}} R)^{tC_p} \to \mathrm{THH}(R)^{tC_p}$. Then the Frobenius φ_p is the unique $\mathbb{T}$-equivariant $\mathbb{E}_\infty$-map rendering the diagram

$$\begin{array}{ccc} R & \longrightarrow & \mathrm{THH}(R) \\ \downarrow{\scriptstyle \Delta_p} & & \downarrow{\scriptstyle \varphi_p} \\ (R \otimes_{\mathbb{S}} \cdots \otimes_{\mathbb{S}} R)^{tC_p} & \longrightarrow & \mathrm{THH}(R)^{tC_p} \end{array}$$

commutative. Such a map exists by the result of McClure, Schwänzel, and Vogt that $\mathrm{THH}(R)$ is the initial $\mathbb{E}_\infty$-ring under R with a $\mathbb{T}$-action. ∎

The map $\mathrm{THH}(R) \to \mathrm{THH}(R)^{tC_p}$ is a refinement of the map of Observation 5.4 in the sense that for an ordinary ring R on π_0 it recovers the map of Observation 5.4.

Definition 5.6. A cyclotomic spectrum is a spectrum X with $\mathbb{T}$-action and for every prime p a $\mathbb{T}$-equivariant map $X \to X^{tC_p}$.

Using this definition we can rephrase Proposition 5.5 as the existence of a natural cyclotomic structure on $\mathrm{THH}(R)$. The main content of the paper [41] is a discussion of the theory of cyclotomic spectra and of topological cyclic homology from this perspective. In particular topological cyclic homology $\mathrm{TC}(R)$ of a connective ring spectrum R can be computed as the mapping spectrum in the stable ∞-category of cyclotomic spectra from $\mathrm{THH}(\mathbb{S})$ to $\mathrm{THH}(R)$. Prior to that, $\mathrm{TC}(R)$ was defined using point set models of $\mathrm{THH}(R)$ and genuine equivariant homotopy theory.

Remark 5.7. For an $\mathbb{E}_\infty$-ring R the Tate-valued Frobenius $\varphi_p : R \to R^{tC_p}$ constructed in Section 3 and the map $\mathrm{THH}(R) \to \mathrm{THH}(R)^{tC_p}$ of Proposition 5.5 are related in the following sense: there is an $\mathbb{E}_\infty$-map $b : \mathrm{THH}(R) \to R$. This can be constructed as the unique $\mathbb{T}$-equivariant map (with trivial action on the target) extending the identity $R \to R$. Then the following diagram commutes:

$$\begin{array}{ccc} R & \longrightarrow & \mathrm{THH}(R) \\ \downarrow{\scriptstyle \varphi_p} & & \downarrow{\scriptstyle \varphi_p} \\ R^{tC_p} & \xleftarrow{b^{tC_p}} & \mathrm{THH}(R)^{tC_p} \end{array}$$

so that the Frobenius on THH refines the Tate-valued Frobenius.

Perfectness of THH. Recall that we have seen that for a p-adically finite spectrum X the map $\mathrm{triv} : X \to X^{tC_p}$ is a p-completion (Theorem 1.4). If X is bounded below then the map $\Delta_p : X \to (X \otimes_{\mathbb{S}} \cdots \otimes_{\mathbb{S}} X)^{tC_p}$ is a p-completion. In this section we will see that in many situations also the cyclotomic Frobenius

$$\mathrm{THH}(R) \to \mathrm{THH}(R)^{tC_p}$$

is an equivalence, at least in large degrees. We capture this in the following definition:

Definition 5.8. We say that a cyclotomic spectrum X is *eventually perfect* (or it satisfies the *Segal conjecture*) if the map

$$X_p^\wedge \to X^{tC_p}$$

is an equivalence in sufficiently large degrees $* \gg 0$.

It turns out that there are many cases of eventually perfect cyclotomic spectra. We try to give a short (and incomplete list) to illustrate this.

Example 5.9 (Bökstedt–Madsen). For $R = \mathbb{F}_p$, the map

$$\varphi_p : \mathrm{THH}(\mathbb{F}_p) \to \mathrm{THH}(\mathbb{F}_p)^{tC_p}$$

is an equivalence in degrees ≥ -1. The map

$$\varphi_p : \mathrm{THH}(\mathbb{Z}_p) \to \mathrm{THH}(\mathbb{Z}_p)^{tC_p}$$

is an equivalence in degrees ≥ 0. This can be considerably generalized to see that $\mathrm{THH}(R)$ is eventually perfect for DVRs of mixed characteristic with perfect residue field [20], for smooth algebras in positive characteristic [19, PROP. 6.6], and for torsion-free excellent noetherian rings R with R/p finitely generated over its pth powers [36]. Finally, Bhatt–Morrow–Scholze show that also for R a perfectoid ring the spectrum $\mathrm{THH}(R)$ is eventually perfect.

Example 5.10 (Rognes–Lunøe-Nielsen). For $R = \mathrm{MU}$, the complex cobordism spectrum, or $R = \mathrm{BP}$, the Brown–Peterson spectrum (which is a retract of p-localized MU), the map

$$\mathrm{THH}(R) \to \mathrm{THH}(R)^{tC_p}$$

is a p-completion [29].

Example 5.11 (Ausoni–Rognes [3] for $n = 1$, Hahn–Wilson [18] for general n). Let F be any finite type $(n+1)$-complex. Then the cyclotomic spectrum $F \otimes_{\mathbb{S}} \mathrm{THH}(\mathrm{BP}\langle n\rangle)$ is eventually perfect, in other words, the map

$$F_*(\mathrm{THH}(\mathrm{BP}\langle n\rangle)) \to F_*(\mathrm{THH}(\mathrm{BP}\langle n\rangle)^{tC_p})$$

is an equivalence for $* \gg 0$.

Remark 5.12. Spectrum $\mathrm{THH}(R)$ is the geometric realization (i.e., colimit) of a simplicial spectrum $\mathrm{THH}_\bullet(R) = R^{\otimes(\bullet+1)}$, the *cyclic Bar complex*. By a p-fold edgewise subdivision we see that this is also the colimit over $\mathrm{sd}_p\mathrm{THH}_\bullet(R) = R^{\otimes p(\bullet+1)}$. One can show that applying the Tate diagonal levelwise extends to a map of simplicial objects $\mathrm{THH}_\bullet(R) \to (\mathrm{sd}_p\mathrm{THH}_\bullet(R))^{tC_p}$. In this picture, the THH-Frobenius φ_p is induced by the map

$$\mathrm{THH}(R) = |\mathrm{THH}_\bullet(R)| \xrightarrow{\Delta_p} |(\mathrm{sd}_p\mathrm{THH}_\bullet(R))^{tC_p}| \xrightarrow{i} |\mathrm{sd}_p\mathrm{THH}_\bullet(R)|^{tC_p} = \mathrm{THH}(R)^{tC_p},$$

where i is the canonical interchange map. For a connective ring spectrum R, the first map Δ_p is a p-adic equivalence by Theorem 2.3. Thus the question whether THH is (eventually) perfect is equivalent to the question whether i is an equivalence (in a range).

6. FROBENIUS LIFTS AND TR

In this section we will talk about Frobenius lifts for cyclotomic spectra. We fix a prime p and focus on this prime. One can also set up a global theory for all primes, but we will not get into this here.

Frobenius lifts in algebra. Recall that for an ordinary commutative ring R a Frobenius lift (for the fixed prime p) is a ring homomorphism $F_p : R \to R$ such that the composite

$$R \xrightarrow{F_p} R \xrightarrow{\mathrm{can}} R/p$$

is the Frobenius homomorphism. Here can is the canonical projection. If R is p-torsion free this precisely captures the notion of a δ-ring (also known as p-typical λ-ring or λ_p-ring). If R has p-torsion this not quite true on the nose. In this case one should define a Frobenius lift as an endomorphism $F_p : R \to R$ *together* with a homotopy between

$$R \xrightarrow{F_p} R \xrightarrow{\mathrm{can}} R \otimes^L_{\mathbb{Z}} \mathbb{F}_p \quad \text{and} \quad R \xrightarrow{\varphi_p} R \otimes^L_{\mathbb{Z}} \mathbb{F}_p$$

considered as maps of animated commutative rings (also known as simplicial commutative rings). A Frobenius lift on R in this sense is then always equivalent to the structure of a δ-ring on R, see [7, REMARK 2.5].

An example of a ring with a Frobenius lift is given by the ring of p-typical Witt vectors $W(R)$ for any ring R. The Frobenius lift $F : W(R) \to W(R)$ is the Witt vector Frobenius. By definition, the Witt vectors come with a reduction map $W(R) \to R$, and we have:

Proposition 6.1. *For any commutative ring R, the ring of p-typical Witt vectors $W(R)$ is the universal ring with Frobenius lift over R, i.e., the cofree δ-ring on R.*

In fact, the Witt vectors have additional structure, namely a map $V : W(R) \to W(R)$ called *Verschiebung*, which is additive but not multiplicative. It instead satisfies the relations

$$FV = p \cdot \mathrm{id} \quad \text{and} \quad V\big(F(a)b\big) = aV(b) \quad \text{for all } a, b \in W(R).$$

Moreover, we have that $W(R)/V = R$ and that $W(R)$ is derived complete with respect to the filtration induced by V, i.e., the derived inverse limit $\mathrm{R}\varprojlim_V W(R)$ vanishes.

One can express this a bit more systematically as follows: an abelian group A with operators $F, V : A \to A$ such that $FV = p \cdot \mathrm{id}$ is called a (p-typical) *Cartier module*. On the category of Cartier modules, there is a tensor product $\boxtimes$ and the Witt vectors form an algebra with respect to that, see [1, SECTION 4.2] for these facts.

Cyclotomic spectra with Frobenius lift. Now we would like to find analogous statements to the statements of the last paragraph for cyclotomic spectra. We fix a prime p throughout this section, and everything will depend on p.

Definition 6.2. Let X be a cyclotomic spectrum. A Frobenius lift is a $\mathbb{T}$-equivariant map $F : X \to X^{hC_p}$ together with an equivalence $\mathrm{can} \circ F \simeq \varphi_p$ of $\mathbb{T}$-equivariant maps $X \to X^{tC_p}$.

Now we have an analogue to the algebraic situation, i.e., the analogue of the Witt vectors:

Proposition 6.3 ([27]). *For every cyclotomic spectrum X, there is a universal cyclotomic spectrum $\mathrm{TR}(X) \to X$ with Frobenius lift. If $X = \mathrm{THH}(R)$ for a connective $\mathbb{E}_\infty$-ring spectrum R, then $\mathrm{TR}(X)$ is also an $\mathbb{E}_\infty$-ring spectrum and the ring $\pi_0\mathrm{TR}(X)$ is canonically isomorphic to the ring $W(\pi_0 R)$ of p-typical Witt vectors.*

This spectrum $\mathrm{TR}(X)$ can be described explicitly by an iterated pullback and has appeared in the theory of cyclotomic spectra much earlier (in fact, it was used to define TC in the first place by Bökstedt–Hsiang–Madsen [10]). By a result of Blumberg–Mandell, the functor $X \mapsto \mathrm{TR}(X)$ is even corepresentable on the stable ∞-category of cyclotomic spectra [8]. We shall not need these facts here, but there is the following abstract characterization:

Proposition 6.4 (Nikolaus–Antieau [1]). *For every cyclotomic spectrum X, there is a $\mathbb{T}$-equivariant map*

$$V : \mathrm{TR}(X)_{hC_p} \to \mathrm{TR}(X)$$

called Verschiebung *such that the cofiber of V is equivalent to X and the composite $F \circ V$: $\mathrm{TR}(X)_{hC_2} \to \mathrm{TR}(X)^{hC_2}$ is canonically identified with the C_2-norm. If X is additionally bounded below, then we have a pullback square of spectra with $\mathbb{T}$-action of the form*

$$\begin{array}{ccc} \mathrm{TR}(X) & \longrightarrow & X \\ \downarrow F & & \downarrow \varphi_p \\ \mathrm{TR}(X)^{hC_p} & \xrightarrow{\mathrm{can}} & \mathrm{TR}(X)^{tC_p} \simeq X^{tC_p}. \end{array} \tag{6.1}$$

From the pullback square (6.1), we see that the cyclotomic spectrum X is eventually perfect precisely if the map $F : \mathrm{TR}(X) \to \mathrm{TR}(X)^{hC_p}$ is an equivalence in high degrees. This will be important in the next section.

Definition 6.5. A *topological Cartier module* is a spectrum M with $\mathbb{T}$-action, $\mathbb{T}$-equivariant maps $V : M_{hC_p} \to M$, $F : M \to M^{hC_p}$ and a $\mathbb{T}$-equivariant equivalence of the composite $F \circ V$ to the C_p-norm.

With this language, Proposition 6.4 can be summarized by saying that for every cyclotomic spectrum X the spectrum $\mathrm{TR}(X)$ is a topological Cartier module. Moreover, one can show that the functor TR induces an equivalence between the ∞-category of bounded below cyclotomic spectra and the ∞-category of bounded below topological Cartier modules which are complete with respect to the Verschiebung, see **[1]**.

Example 6.6. For the spectrum $X = \mathrm{THH}(\mathbb{F}_p)$, one can compute the spectrum $\mathrm{TR}(X)$ and finds that it is given by the Eilenberg–MacLane spectrum $H\mathbb{Z}_p$. The $\mathbb{T}$-action is (necessarily) trivial and the Frobenius is given by the map $F : H\mathbb{Z}_p \to H\mathbb{Z}_p^{hC_p}$ which is the identity on π_0. The Verschiebung is the map $V : (H\mathbb{Z}_p)_{hC_p} \to H\mathbb{Z}_p$ which is multiplication by p on π_0. The pullback then takes the form

$$\begin{array}{ccc} H\mathbb{Z}_p & \longrightarrow & \mathrm{THH}(\mathbb{F}_p) \\ \big\downarrow F & & \big\downarrow \varphi_p \\ H\mathbb{Z}_p^{hC_p} & \xrightarrow{\text{can}} & \mathbb{Z}_p^{tC_p} \end{array} \tag{6.2}$$

and implies that $\mathrm{THH}(\mathbb{F}_p)$ is equivalent to the connective cover of $H\mathbb{Z}_p^{tC_p}$. In fact, one can reverse the logic here: since $\pi_0(\mathrm{TR}(\mathbb{F}_p)) = W(\mathbb{F}_p) = \mathbb{Z}_p$ by Proposition 6.3, we see that Bökstedt's theorem (Theorem 5.2) is equivalent to the assertion that $\mathrm{TR}(\mathbb{F}_p)$ has no higher homotopy groups, i.e., is 0-truncated. As explained around Theorem 5.3, this is then also equivalent to the combination of Milnor's and Steinberger's computations and implies topological Bott periodicity.

One can attempt to prove the truncatedness of $\mathrm{TR}(\mathbb{F}_p)$ directly to give new proofs of all of these theorems. Such a direct proof is given in **[5]** using the theory of polynomial functors and the fact that TR can be evaluated on polynomial functors.

Boundedness of TR. Now we shall relate the perfectness of THH to the red-shift conjecture and computations of algebraic K-theory. This is based on recent work of Mathew and Hahn–Wilson. We first review a bit more of the theory of cyclotomic spectra. The stable ∞-category of cyclotomic spectra carries a t-structure, defined as follows:

Definition 6.7. A cyclotomic spectrum X is called *(cyclotomically) n-connective* if the underlying spectrum is n-connective and *cyclotomically n-truncated* if every map $Y \to X$ where Y is $(n+1)$-connective is nullhomotopic (as a map of cyclotomic spectra). We say that X is *cyclotomically bounded* if it is cyclotomically bounded above and below, i.e., n-connective for some $n \gg 0$ and k-truncated for some $k \gg 0$.

It is not hard to see that this defines a t-structure and as usual the heart consists of those cyclotomic spectra which are cyclotomically 0-connective and 0-truncated.

Remark 6.8. Note that we are a bit loose with the adjective "cyclotomic" for connective things, since this simply means that the underlying spectrum is connective and there is no danger of confusion. For truncated objects, we will be careful though, since this cyclotomic notion of truncatedness is completely different to the truncatedness of the underlying spectrum X. For example, $\mathrm{THH}(\mathbb{F}_p)$ is cyclotomically 0-truncated as one can see from the following theorem together with Example 6.6, but the underlying spectrum is not truncated at all (recall Theorem 5.2).

Theorem 6.9 (Antieau–Nikolaus [1]). *A p-complete cyclotomic spectrum is cyclotomically bounded precisely if the spectrum* $\mathrm{TR}(X)$ *is bounded as a spectrum. In this case X is eventually perfect (see Definition 5.8) and the p-completion of* $\mathrm{TC}(X)$ *is bounded as a spectrum. The heart of the t-structure is equivalent to the abelian category of derived V-complete Cartier modules.*

One reason why we care about the boundedness of TR and TC is its relation to the red shift conjecture in algebraic K-theory, which we want to outline now (following the recent paper [18] of Hahn–Wilson). The first ingredient is the following result, which is a consequence of work of Mahowald–Rezk:

Proposition 6.10 (Hahn–Wilson [18]). *Assume that for a connective ring spectrum R and a type $(n+2)$-complex F the spectrum $F \otimes_{\mathbb{S}} \mathrm{THH}(R)$ is cyclotomically bounded and $\pi_i(R_p^{\wedge})$ is finitely generated for each i. Then the map*

$$\mathrm{TC}(R) \to L_{n+1}^{f} \mathrm{TC}(R)$$

is an equivalence in degrees $* \gg 0$.

In order to use this result, one needs an efficient way of verifying the boundedness of a cyclotomic spectrum which is provided by the following criterion.

Proposition 6.11 (Mathew [36], Hahn–Wilson [18]). *Assume a cyclotomic spectrum X is bounded below, p-power torsion and eventually perfect. Then the following are equivalent:*

(1) *X is cyclotomically bounded, i.e.,* TR *is bounded.*

(2) *The $\mathbb{T}$-spectrum X^{tC_p} is $\mathbb{T}$-nilpotent, i.e., lies in the thick subcategory generated by free $\mathbb{T}$-spectra.*

(3) *For $* \gg 0$ the maps $\pi_*(\mathrm{can}) : \pi_*(X^{hC_{p^k}}) \to \pi_*(X^{tC_{p^k}})$ with $1 \leq k \leq \infty$ are zero.*

Especially condition (3) can be verified in practice. Using this, Hahn–Wilson prove that $\mathrm{THH}(\mathrm{BP}\langle n\rangle)$ for every $\mathbb{E}_3$-form of $\mathrm{BP}\langle n\rangle$ satisfies the assumptions of Proposition 6.10. From this, together they deduce the following groundbreaking result.

Theorem 6.12 (Hahn–Wilson [18]). *For every $\mathbb{E}_3$-form of* $\mathrm{BP}\langle n\rangle$, *the morphisms*

$$\mathrm{TC}(\mathrm{BP}\langle n\rangle) \to L_{n+1}^f(\mathrm{TC}(\mathrm{BP}\langle n\rangle)),$$
$$K(\mathrm{BP}\langle n\rangle) \to L_{n+1}^f(K(\mathrm{BP}\langle n\rangle))$$

are after p-completion equivalences in degrees $* \gg 0$.

Remark 6.13. The thick subcategory of $\mathbb{T}$-nilpotent spectra is in fact a tensor ideal. Therefore Mathew concludes from Proposition 6.11 and the results of Hahn–Wilson that, for a connective $\mathrm{BP}\langle n\rangle$-algebra spectrum R and a finite type $(n+2)$-complex F, the cyclotomic spectrum $F \otimes \mathrm{THH}(R)$ is cyclotomically bounded precisely if it is eventually perfect [18, PROPOSITION 3.3.7].

7. THE SEGAL CONJECTURE

In this last section we want to discuss the Segal conjecture also known as Segal's Burnside ring conjecture. Despite its name, this is a theorem that was proven by Carlsson [12]. We will conjecture a generalization of this theorem.

In order to do this, we first have to discuss a variant of the Tate construction called the *proper Tate construction*. We will first sketch the algebraic counterpart as usual.

The proper Tate construction in algebra. Let A be an abelian group with G-action and $H \subseteq G$ a subgroup of G. Then we have a relative norm map

$$A^H \to A^G, \quad a \mapsto \sum_{[g]\in G/H} ga.$$

We can form the quotient of A^G by all these relative norms

$$\frac{A^G}{\bigoplus_{H \subsetneq G} A^H}, \tag{7.1}$$

where H ranges trough all proper subgroups of H. This quotient equals the zeroth Tate construction if $G = C_p$ for a prime p but differs in general.

Example 7.1. If G has an index n-subgroup then the quotient (7.1) is n-torsion, as one easily verifies. Applying this observation to Sylow subgroups, we see that the quotient (7.1) vanishes if G is not a p-group. If G is a p-group which acts trivially on A then the quotient (7.1) is isomorphic to A/p.

The proper Tate construction in higher algebra. Now we want to make the analogous construction for spectra. Let X be a spectrum with G-action. Then for any subgroup $H \subseteq G$ there is a similar relative norm map

$$\mathrm{Nm}_H^G : X^{hH} \to X^{hG}$$

and these maps Nm_H^G are compatible as H ranges through all subgroups of G. To make this compatibility precise, we consider the orbit category Orb_G consisting of all G-orbits, i.e., transitive G-sets. All of these orbits are of the form G/H for a subgroup $H \subseteq G$. The precise compatibility statement that we need is the following statement, which follows, for example, from the work of Barwick [4]:

Proposition 7.2. *For any spectrum X with G-action, there is a canonical functor*

$$\mathrm{Orb}_G \to \mathrm{Sp}$$

which sends G/H to X^{hH} and a map in the orbit category to a relative norm.

Definition 7.3. For a spectrum X with an action by a finite group G, we define the proper Tate construction $X^{\varphi G}$ as the cofiber

$$\mathrm{colim}_{H \subsetneq G} X^{hH} \to X^{hG},$$

where the colimit is indexed over the full subcategory of Orb_G consisting of the orbits G/H for proper subgroups H of G (which we abusively denote by $H \subsetneq G$).

Example 7.4. For $G = C_p$ with p prime, the proper Tate construction agrees with the Tate construction, i.e., $X^{\varphi C_p} = X^{tC_p}$.

Example 7.5. For $G = C_{p^n}$ with $n > 0$, a cofinality argument shows that the colimit in Definition 7.3 is equivalent to the colimit over the full subcategory of Orb_G spanned by the orbit with p-elements. Thus we find that

$$X^{\varphi(C_{p^n})} \simeq \mathrm{cofib}\big((X^{hC_{p^{n-1}}})_{hC_p} \to X^{hC_{p^n}}\big) \simeq (X^{hC_{p^{n-1}}})^{tC_p},$$

where the action of C_p on $X^{hC_{p^{n-1}}}$ is the "residual" action under the identification $C_p = C_{p^n}/C_{p^{n-1}}$.

Warning 7.6. For any abelian group A with G-action, there is a canonical map

$$A^G/\oplus_{H \subsetneq G} A^H \to \pi_0(HA^{\varphi G}),$$

where HA is the Eilenberg–MacLane spectrum associated with A. In contrast to the case for $G = C_p$, this map is in general not an isomorphism.

Similar to the Tate construction, the proper Tate construction refines to a lax symmetric monoidal functor

$$(-)^{\varphi G} : \mathrm{Sp}^{BG} \to \mathrm{Sp}$$

and, if X is equipped with the trivial G-action, there is a natural map

$$\mathrm{triv} : X \to X^{\varphi G}.$$

Moreover, there are homotopy-theoretic versions of the statements in Example 7.1:

Proposition 7.7 ([41]). *Let X be a spectrum with G action. Then we have $X^{\varphi G} = 0$, unless G is a p-group.*

Now the following deep result is a generalization of the theorem of Lin and Gunawardena (Theorem 1.4) and a higher analogue of the second part of Example 7.1.

Theorem 7.8 (Carlsson [12]). *For any finite p-group G and every finite spectrum X with trivial G-action, the map* $\mathrm{triv} : X \to X^{\varphi G}$ *is a p-completion.*

Remark 7.9. This theorem is equivalent to the Segal conjecture which states that for any group G the canonical map

$$\mathbb{S}^G \to \mathbb{S}^{hG}$$

is a completion. Here $\mathbb{S}^G$ is the spectrum obtained by the group completion of the category of finite G-sets. Its π_0 is given by the Burnside ring and the relevant completion is the completion with respect to the augmentation ideal. In this form the conjecture was inspired by the Atiyah–Segal completion theorem which states that a similar map in complex K-theory is a completion.

Now similar to the Tate diagonal discussed in Section 2, there is a unique natural and symmetric monoidal map $\Delta_G : X \to (X^{\otimes G})^{\varphi G}$ for every finite group G. Note that such a map does not exist if we replace the proper Tate construction $(-)^{\varphi G}$ by the actual Tate construction $(-)^{tG}$.

Conjecture 7.10. *For any finite p-group G and any bounded below spectrum X, the Tate diagonal*

$$X \to (X^{\otimes G})^{\varphi G}$$

is a p-completion.

This conjecture reduces for $X = \mathbb{S}$ to Theorem 7.8. For $G = C_p$, it reduces to Theorem 2.3 above. One can formally deduce the case $G = C_{p^n}$ from this case using the Tate orbit lemma of [41] which implies that for any bounded below spectrum X with C_{p^n}-action we have $X^{\varphi C_{p^n}} = (X^{tC_p})^{\varphi C_{p^{n-1}}}$. The key step to verify Conjecture 7.10 in general is the case of G an elementary abelian p-group. We have been informed that Håkon Bergsaker is very close proving this conjecture using the continuous Adams spectral sequence and an Ext-calculation based on the Singer construction similar to the case of the ordinary Segal conjecture.

Remark 7.11. Conjecture 7.10 is equivalent to the assertion that for any finite group G and any bounded below spectrum X the map

$$N_e^G(X)^G \to (X^{\otimes G})^{hG}$$

from the fixed points of the Hill–Hopkins–Ravenel norm to the homotopy fixed points is a completion (at the augmentation ideal of the Burnside ring). Equivalently, the HHR-norm is Borel complete after this completion. In this language, it becomes clear that this is a direct analogue of the classical Burnside ring conjecture.

A consequence would be that for any finite group G and every connective spectrum X, the spectrum $(X^{\otimes G})^{hG}$ is connective and π_0 is a completion of π_0 of the norm. The latter has an explicit algebraic expression in terms of $\pi_0(X)$. This was described in [24, 47] but remains somewhat inexplicit. In the case that the group G is given by C_{p^n}, this algebraic expression is given by the Witt vectors of $\mathbb{Z}$ with values in the abelian group $\pi_0(X)$ defined and discussed in [14], see also [25] for a similar description. For a general group G, the group π_0 of the norm should similarly be a certain (yet to be defined) version of Dress–Siebeneicher's Witt–Burnside ring [15] for the group G with coefficients in $\pi_0(X)$.

ACKNOWLEDGMENTS

I would like to thank all the coauthors of the work that is presented in this survey for great collaborations: Benjamin Antieau, Clark Barwick, Saul Glasman, Lars Hesselholt, Achim Krause, Akhil Mathew, and Peter Scholze. I would also like to thank Achim Krause for comments on a draft of this note.

FUNDING

The author was funded by the Deutsche Forschungsgemeinschaft (DFG, German Research Foundation) – Project-ID 427320536 – SFB 1442 and under Germany's Excellence Strategy EXC 2044 390685587, Mathematics Münster: Dynamics–Geometry–Structure.

REFERENCES

[1] B. Antieau and T. Nikolaus, Cartier modules and cyclotomic spectra. *J. Amer. Math. Soc.* **34** (2021), no. 1, 1–78.

[2] S. Ariotta, Coherent cochain complexes and Beilinson t-structures, with an appendix by Achim Krause. 2021, arXiv:2109.01017.

[3] C. Ausoni and J. Rognes, Algebraic K-theory of the first Morava K-theory. *J. Eur. Math. Soc. (JEMS)* **14** (2012), no. 4, 1041–1079.

[4] C. Barwick, Spectral Mackey functors and equivariant algebraic K-theory (I). *Adv. Math.* **304** (2017), 646–727.

[5] C. Barwick, S. Glasman, A. Mathew, and T. Nikolaus, K-theory and polynomial functors. 2021, arXiv:2102.00936.

[6] B. Bhatt, M. Morrow, and P. Scholze, Topological Hochschild homology and integral p-adic Hodge theory. *Publ. Math. Inst. Hautes Études Sci.* **129** (2019), 199–310.

[7] B. Bhatt and P. Scholze, Prisms and prismatic cohomology. 2019, arXiv:1905.08229.

[8] A. J. Blumberg and M. A. Mandell, The homotopy theory of cyclotomic spectra. *Geom. Topol.* **19** (2015), no. 6, 3105–3147.

[9] M. Bökstedt, Topological hochschild homology of $\mathbb{Z}$ and $\mathbb{Z}/p$. 1985, preprint, Universität Bielefeld.

[10] M. Bökstedt, W. C. Hsiang, and I. Madsen, The cyclotomic trace and algebraic K-theory of spaces. *Invent. Math.* **111** (1993), no. 3, 465–539.

[11] R. R. Bruner, J. P. May, J. E. McClure, and M. Steinberger, *H_∞ ring spectra and their applications*. Lecture Notes in Math. 1176, Springer, Berlin, 1986.

[12] G. Carlsson, Equivariant stable homotopy and Segal's Burnside ring conjecture. *Ann. of Math. (2)* **120** (1984), no. 2, 189–224.

[13] D. Clausen, A. Mathew, and M. Morrow, K-theory and topological cyclic homology of henselian pairs. *J. Amer. Math. Soc.* **34** (2021), no. 2, 411–473.

[14] E. Dotto, A. Krause, T. Nikolaus, and I. Patchkoria, Witt vectors with coefficients and characteristic polynomials over non-commutative rings. 2020, arXiv:2002.01538.

[15] A. W. M. Dress and C. Siebeneicher, The Burnside ring of profinite groups and the Witt vector construction. *Adv. Math.* **70** (1988), no. 1, 87–132.

[16] S. Glasman and T. Lawson, Stable power operations. 2020, arXiv:2002.02035.

[17] J. Gunawardena, Segal's conjecture for cyclic groups of (odd) prime order. *JT Knight prize essay, Cambridge* **224** (1980).

[18] J. Hahn and D. Wilson, Redshift and multiplication for truncated Brown–Peterson spectra. 2020, arXiv:2012.00864.

[19] L. Hesselholt, Periodic topological cyclic homology and the Hasse–Weil zeta function. 2016.

[20] L. Hesselholt and I. Madsen, On the K-theory of local fields. *Ann. of Math. (2)* **158** (2003), no. 1, 1–113.

[21] L. Hesselholt and T. Nikolaus, Algebraic K-theory of planar cuspidal curves. In *K-theory in algebra, analysis and topology*, pp. 139–148, Contemp. Math. 749, Amer. Math. Soc., Providence, RI, 2020. ©2020

[22] L. Hesselholt and T. Nikolaus, Topological cyclic homology. In *Handbook of homotopy theory*, pp. 619–656, CRC Press/Chapman Hall Handb. Math. Ser., CRC Press, Boca Raton, FL, 2020. ©2020

[23] G. Heuts, *Goodwillie approximations to higher categories*. PhD thesis, Harvard University, 2015, arXiv:1510.03304.

[24] M. A. Hill and M. J. Hopkins, Equivariant symmetric monoidal structures. 2016, arXiv:1610.03114.

[25] M. A. Hill and K. Mazur, An equivariant tensor product on Mackey functors. *J. Pure Appl. Algebra* **223** (2019), no. 12, 5310–5345.

[26] A. Krause and T. Nikolaus, Bökstedt periodicity and quotients of DVRs. 2019, arXiv:1907.03477.

[27] A. Krause and T. Nikolaus, Lectures on topological Hochschild homology and cyclotomic spectra. Available at https://wwwmath.uni-muenster.de/u/nikolaus

[28] W. H. Lin, On conjectures of Mahowald, Segal and Sullivan. *Math. Proc. Cambridge Philos. Soc.* **87** (1980), no. 3, 449–458.

[29] S. Lunøe-Nielsen and J. Rognes, The Segal conjecture for topological Hochschild homology of complex cobordism. *J. Topol.* **4** (2011), no. 3, 591–622.

[30] S. Lunøe-Nielsen and J. Rognes, The topological Singer construction. *Doc. Math.* **17** (2012), 861–909.

[31] J. Lurie, Rational and p-adic homotopy theory. 2011, preprint, available at https://www.math.ias.edu/~lurie/.

[32] J. Lurie, Higher algebra. 2017, preprint, available at https://www.math.ias.edu/~lurie/.

[33] J. Lurie, Elliptic cohomology II: orientations. 2018, preprint, available at https://www.math.ias.edu/~lurie/.

[34] M. A. Mandell, E_∞ algebras and p-adic homotopy theory. *Topology* **40** (2001), no. 1, 43–94.

[35] M. A. Mandell, Cochains and homotopy type. *Publ. Math. Inst. Hautes Études Sci.* **103** (2006), 213–246.

[36] A. Mathew, On $K(1)$-local TR. *Compos. Math.* **157** (2021), no. 5, 1079–1119.

[37] J. McClure, R. Schwänzl, and R. Vogt, $\mathrm{THH}(R) \cong R \otimes S^1$ for E_∞ ring spectra. *J. Pure Appl. Algebra* **121** (1997), no. 2, 137–159.

[38] J. Milnor, The Steenrod algebra and its dual. *Ann. of Math. (2)* **67** (1958), 150–171.

[39] T. Nikolaus, Stable ∞-operads and the multiplicative Yoneda lemma. 2016, arXiv:1608.02901.

[40] T. Nikolaus, Rational and p-adic homotopy theory, available at https://www.math.uni-muenster.de/u/nikolaus.

[41] T. Nikolaus and P. Scholze, On topological cyclic homology. *Acta Math.* **221** (2018), no. 2, 203–409.

[42] D. Quillen, Higher algebraic K-theory. In *Proceedings of the International Congress of Mathematicians (Vancouver, BC, 1974), Vol. 1*, pp. 171–176, 1975.

[43] N. Riggenbach, On the algebraic k-theory of double points. 2020, arXiv:2007.01227.

[44] J. Rognes, Algebraic K-theory of finitely presented ring spectra. 2000, available at https://www.mn.uio.no/math/personer/vit/rognes/papers/red-shift.pdf.

[45] M. Speirs, On the K-theory of truncated polynomial algebras, revisited. *Adv. Math.* **366** (2020), 107083, 18.

[46] M. E. Sweedler, Book Review: Hopf algebras. *Bull. Amer. Math. Soc. (N.S.)* **5** (1981), no. 3, 349–354.

[47] J. Ullman, Symmetric powers and norms of Mackey functors. 2013, arXiv:1304.5648.

[48] V. Voevodsky, On motivic cohomology with $\mathbb{Z}/l$-coefficients. *Ann. of Math. (2)* **174** (2011), no. 1, 401–438.

[49] D. Wilson, Mod 2 power operations revisited. 2019, arXiv:1905.00054.

[50] A. Yuan, Integral models for spaces via the higher Frobenius. 2019, arXiv:1910.00999.

THOMAS NIKOLAUS

FB Mathematik und Informatik, Universität Münster, Einsteinstr. 62, D-48149 Münster, Germany, nikolaus@wwu.de

This Page Intentionally Left Blank

DIFFEOMORPHISMS OF DISCS

OSCAR RANDAL-WILLIAMS

ABSTRACT

I describe what is currently known, for $d \geq 5$, about the rational homotopy type of the group of diffeomorphisms of the d-disc relative to its boundary, and the closely related group of homeomorphisms of d-dimensional Euclidean space.

MATHEMATICS SUBJECT CLASSIFICATION 2020

Primary 57S05; Secondary 55R40, 58D10

KEYWORDS

Diffeomorphism groups, pseudoisotopy, configuration space integrals, graph complexes, operads, characteristic classes, orthogonal calculus, parameterised surgery, embedding calculus, Torelli groups

INTERNATIONAL CONGRESS OF MATHEMATICIANS 2022 JULY 6–14

Proc. Int. Cong. Math. 2022, Vol. 4, pp. 2856–2878
DOI 10.4171/ICM2022/16

Published by EMS Press

1. INTRODUCTION

I will be concerned with the homotopy types of the topological groups of diffeomorphisms $\mathrm{Diff}(\mathbb{R}^d)$ and homeomorphisms $\mathrm{Homeo}(\mathbb{R}^d)$ of Euclidean space, and of diffeomorphisms $\mathrm{Diff}_\partial(D^d)$ and homeomorphisms $\mathrm{Homeo}_\partial(D^d)$ of closed discs fixing the boundary pointwise (equivalently, compactly-supported diffeomorphisms and homeomorphisms of Euclidean space). By scaling outwards, the group $\mathrm{Diff}(\mathbb{R}^d)$ deformation retracts to the subgroup of linear diffeomorphisms, and thence to the subgroup $\mathrm{O}(d)$ of orthogonal diffeomorphisms of $\mathbb{R}^d$. By scaling inwards (the "Alexander trick"), the group $\mathrm{Homeo}_\partial(D^d)$ contracts to a point.

In contrast, the groups $\mathrm{Homeo}(\mathbb{R}^d)$ and $\mathrm{Diff}_\partial(D^d)$ have much more mysterious homotopy types. As long as $d \neq 4$, all four groups are related by smoothing theory **[23, ESSAY V]**, which provides a homotopy equivalence

$$\frac{\mathrm{Homeo}_\partial(D^d)}{\mathrm{Diff}_\partial(D^d)} \simeq \Omega_0^d\left(\frac{\mathrm{Homeo}(\mathbb{R}^d)}{\mathrm{Diff}(\mathbb{R}^d)}\right)$$

and so, incorporating the above and writing $\mathrm{Top}(d) := \mathrm{Homeo}(\mathbb{R}^d)$, provides a homotopy equivalence (the "Morlet equivalence")

$$B\mathrm{Diff}_\partial(D^d) \simeq \Omega_0^d\left(\frac{\mathrm{Top}(d)}{\mathrm{O}(d)}\right).$$

Taking the homotopy type of $\mathrm{O}(d)$ as given, the Morlet equivalence shows that understanding the homotopy types of $\mathrm{Top}(d)$ and $B\mathrm{Diff}_\partial(D^d)$ are more or less equivalent. The latter is independently interesting as it classifies smooth D^d-bundles $\pi : E \to B$ trivialised near the boundary, and this perspective offers a useful way to study it: it is the perspective I usually adopt. For any manifold M of dimension $d \neq 4$, smoothing theory identifies $\frac{\mathrm{Homeo}_\partial(M)}{\mathrm{Diff}_\partial(M)}$ with certain path components of a space of sections $\Gamma_\partial(\mathrm{Fr}(TM) \times_{\mathrm{O}(d)} \frac{\mathrm{Top}(d)}{\mathrm{O}(d)} \to M)$, so these homotopy types furthermore describe the difference between diffeomorphisms and homeomorphisms of all d-manifolds. It is therefore an important goal of geometric topology to investigate these homotopy types.

In this essay I will describe what is known about the *rational* homotopy type of $B\mathrm{Diff}_\partial(D^d)$, and some recent techniques which are allowing us to say more about it. Along the way the influence of Michael Weiss will be seen at every turn, and it is a pleasure to acknowledge and celebrate his many profound contributions to this subject.

2. SOME PHENOMENA

I will first describe the classical approach to calculating $\pi_*(B\mathrm{Diff}_\partial(D^d)) \otimes \mathbb{Q}$, which describes it completely in the so-called pseudoisotopy stable range, and then explain two more recent results which indicate the existence of new phenomena outside of this range: the work of Watanabe on configuration space integrals, and the work of Weiss on unstable topological Pontrjagin classes.

2.1. Pseudoisotopy and algebraic K-theory

The topological group of smooth pseudoisotopies

$$\mathrm{C}(M) := \{f : M \times [0,1] \xrightarrow{\text{diffeo}} M \times [0,1] \mid f \text{ fixes } \partial M \times [0,1] \cup M \times \{0\} \text{ pointwise}\}$$

of a manifold M participates in a fibre sequence

$$\mathrm{Diff}_\partial(M \times [0,1]) \to \mathrm{C}(M) \xrightarrow{f \mapsto f|_{M\times\{1\}}} \mathrm{Diff}_\partial(M), \tag{2.1}$$

and so measures to what extent diffeomorphisms of $M \times [0,1]$ may be represented as loops of diffeomorphisms of M. There is a stabilisation map

$$\sigma_M : \mathrm{C}(M) \to \mathrm{C}(M \times [0,1])$$

morally induced by crossing with the interval, but technically slightly more involved. The (smooth) *pseudoisotopy stable range* $\Phi(d)$ is the minimum of the connectivities of σ_M taken over all manifolds M of dimension $\geq d$, and it is a deep theorem of Igusa **[21]** that $\Phi(d) \geq \min(\frac{d-7}{2}, \frac{d-4}{3})$. The stabilisation $\mathscr{C}(M) := \mathrm{hocolim}_n\, \mathrm{C}(M \times [0,1]^n)$ may be promoted to a homotopy-invariant functor from the category of spaces to the category of infinite loop spaces. The stable parameterised h-cobordism theorem **[42]** relates the functor $\mathscr{C}(-)$ to Waldhausen's **[41]** algebraic K-theory of spaces – or to the K-theory of ring spectra – by the fibration sequence (with cosection) of infinite loop spaces

$$B\mathscr{C}(M) \to Q(M_+) \to \Omega^\infty \mathrm{K}(\mathbb{S}[\Omega M]).$$

In particular, for $M = D^d$ the rational homotopy equivalence $\mathrm{K}(\mathbb{S}) \to \mathrm{K}(\mathbb{Z})$ and Borel's **[3]** calculation

$$K_i(\mathbb{Z}) \otimes \mathbb{Q} = \begin{cases} \mathbb{Q}, & i = 0, 5, 9, 13, 17, 21, \ldots, \\ 0, & \text{else} \end{cases}$$

determines $\pi_*(\mathrm{C}(D^d)) \otimes \mathbb{Q}$ in the pseudoisotopy stable range as being a copy of $\mathbb{Q}$ in each degree $\equiv 3 \bmod 4$.

A further piece of structure on $\mathrm{C}(M)$ is the *pseudoisotopy involution*. Writing τ for the reflection of $[0,1]$ at $\frac{1}{2}$, this is given by

$$f \longmapsto \bar{f} = (f|_{M\times\{1\}} \times [0,1])^{-1} \circ (M \times \tau) \circ f \circ (M \times \tau),$$

and there are compatible involutions on the fibration sequence (2.1) given by inversion on $\mathrm{Diff}_\partial(M)$, and by conjugating by the reflection $M \times \tau$ on $\mathrm{Diff}_\partial(M \times [0,1])$. By analysing this involution, as well as the other involution on $\mathrm{C}(M \times [0,1])$ induced by $\mathrm{C}(M \times \tau)$ and their compatibility with σ_M, it can be shown (see **[22, SECTION 6.5]** for a nice discussion) that

$$\pi_i(B\mathrm{Diff}_\partial(D^{2n})) \otimes \mathbb{Q} = 0,$$

$$\pi_i(B\mathrm{Diff}_\partial(D^{2n+1})) \otimes \mathbb{Q} = \begin{cases} \mathbb{Q}, & i \equiv 0 \bmod 4, \\ 0, & \text{else} \end{cases} \tag{2.2}$$

in the pseudoisotopy stable range. This calculation was first obtained by Farrell and Hsiang **[9]**, though by somewhat different means.

2.2. Configuration space integrals

Kontsevich [25] proposed a method to produce invariants of smooth *vertically framed* D^d-bundles $\pi : E \to B$ trivialised near the boundary, by forming (certain compactifications of) the fibrewise configuration spaces and integrating suitably chosen differential forms along these associated configuration space bundles. The combinatorics of these forms are organised in terms of graphs, and the result is a chain map $\mathrm{GC}_d^2[-d]^\vee \otimes_{\mathbb{Q}} \mathbb{R} \to \Omega^\bullet(B)$ from a certain graph complex to the de Rham complex of the base, so the homology of this graph complex yields invariants of the original bundle. Up to regrading, the chain complexes GC_d^2 only depend on the parity of d, and they split $\mathrm{GC}_d^2 = \bigoplus_g \mathrm{GC}_d^{2,g\text{-loop}}$ as a sum of subcomplexes of fixed loop order. See Willwacher's contribution to the 2018 ICM for an introduction to these objects.

The work of Watanabe. The detailed investigation of this construction has been taken up by Watanabe, firstly in [43–45] for odd-dimensional discs as Kontsevich proposed, and latterly [46, 47] for even-dimensional discs, too. Write $B\mathrm{Diff}_\partial^{\mathrm{fr}}(D^d)$ for the space classifying smooth vertically framed D^d-bundles trivialised near the boundary. (The analogue of the Morlet equivalence in this setting has the form $B\mathrm{Diff}_\partial^{\mathrm{fr}}(D^d) \simeq \Omega^d \mathrm{Top}(d)$—up to a small correction of path-components which I shall ignore—so studying smooth framed disc bundles has an even closer connection to homeomorphisms of Euclidean space.) Kontsevich's construction in particular gives characteristic classes

$$\xi_r \in H^{r\cdot(d-3)}\big(B\mathrm{Diff}_\partial^{\mathrm{fr}}(D^d); \mathcal{A}_r^{(-1)^d}\big),$$

where $\mathcal{A}_r^+$ and $\mathcal{A}_r^-$ are real vector spaces spanned by connected trivalent graphs of loop order $r+1$ (equipped with certain orientation data which I shall neglect), modulo the IHX relation (and modulo certain signs when changing orientation data: these signs depend on the superscript $+$ and $-$). This vector space arises as the lowest nontrivial homology of $\mathrm{GC}_d^{2,(r+1)\text{-loop}} \otimes_{\mathbb{Q}} \mathbb{R}$: the differential is given by summing over splitting vertices, so all trivalent graphs are automatically cycles, and the IHX relation arises from the three ways to split a 4-valent vertex. For small values of r, the dimension of $\mathcal{A}_r^-$ has been calculated to be $1, 1, 1, 2, 2, 3, 4, 5, 6, 8, 9$ for $r = 1, 2, \dots, 11$, and the dimension of $\mathcal{A}_r^+$ has been calculated to be $0, 1, 0, 0, 1, 0, 0, 0, 1$ for $r = 1, 2, \dots, 9$.

Watanabe's results in this direction ([44, THEOREM 3.1] taking into account the improvement in [45], and [46] taking into account the improvement in [47]) is that, as long as $d \geq 4$, the evaluation map

$$\xi_r : \pi_{r\cdot(d-3)}\big(B\mathrm{Diff}_\partial^{\mathrm{fr}}(D^d)\big) \otimes \mathbb{R} \to \mathcal{A}_r^{(-1)^d}$$

is surjective. In fact, his result is somewhat more precise: he constructs for each trivalent graph Γ a more-or-less explicit framed D^d-bundle over a sphere which is sent by the map ξ_r to the class of Γ.

This does not directly tell us about $B\mathrm{Diff}_\partial(D^d)$ because of the framing data, but the difference is easily understood. Forgetting framings defines a homotopy fibre sequence

$$\Omega^d \mathrm{O}(d) \to B\mathrm{Diff}_\partial^{\mathrm{fr}}(D^d) \to B\mathrm{Diff}_\partial(D^d),$$

and it is not hard to calculate

$$\pi_*\big(\Omega^d \mathrm{O}(d)\big) \otimes \mathbb{Q} = \bigoplus_{\substack{k\geq 3,\\ k\equiv 2d+1 \bmod 4}} \mathbb{Q}[d-k].$$

Thus as long as d is even, or d is odd and $r > 1$, one still has the lower bound

$$\dim_{\mathbb{Q}} \pi_{r\cdot(d-3)}\big(B\mathrm{Diff}_\partial(D^d)\big) \otimes \mathbb{Q} \geq \dim_{\mathbb{R}} \mathcal{A}_r^{(-1)^d}.$$

It is worth pausing at this point to emphasise how remarkable it is that Watanabe's results apply for $d = 4$, and imply, for example, that $\pi_2(B\mathrm{Diff}_\partial(D^4)) \otimes \mathbb{Q} \neq 0$.

However, when d is odd—say $d = 2n+1$—and $r = 1$, the composition

$$\mathbb{R} \cong \pi_{2n-2}\big(\Omega^{2n+1}\mathrm{O}(2n+1)\big) \otimes \mathbb{R} \to \pi_{2n-2}\big(B\mathrm{Diff}_\partial^{\mathrm{fr}}(D^{2n+1})\big) \otimes \mathbb{R} \xrightarrow{\xi_1} \mathcal{A}_1^{\mathrm{odd}} \cong \mathbb{R}$$

might be nontrivial, and in fact it is. Watanabe addresses this difficulty in his earlier work **[43]**, by constructing an integral refinement of ξ_1, and playing off this integrality against the non-integrality of the topological Pontrjagin classes: his conclusion is that, as long as certain arithmetic conditions (**[43, COROLLARY 2]**, **[44, COROLLARY 3.5]**) involving Bernoulli numbers and the orders of stable homotopy groups of spheres are satisfied, one may still conclude that $\pi_{2n-2}(B\mathrm{Diff}_\partial(D^{2n+1})) \otimes \mathbb{Q} \neq 0$. He verified this computationally for all odd $n \leq 399$.

Automorphisms of the little discs operads. Configuration space integrals are constructed from (suitable compactifications of) all the ordered configuration spaces $\mathrm{Conf}_k(D^d)$, and all the natural maps between them, applied fibrewise to a (vertically framed) D^d-bundle. There is another way to encode (the homotopy types of) these configuration spaces and the natural maps between them, namely as the little d-discs operad E_d. There is a topological version of the framed little d-discs operad, which means that in a suitable homotopical sense the group $\mathrm{Top}(d)$ acts on E_d, giving a map

$$B\mathrm{Top}(d) \to B\mathrm{hAut}(E_d),$$

where the latter is the classifying space of the E_1-algebra of derived automorphisms of the little d-discs operad. Looping this $(d+1)$ times gives a map

$$B\mathrm{Diff}_\partial^{\mathrm{fr}}(D^d) \simeq \Omega^d \mathrm{Top}(d) \to \Omega^d \mathrm{hAut}(E_d).$$

(This corresponds **[2]** to applying the embedding calculus of Goodwillie and Weiss **[18, 49]** to framed self-embeddings of D^d relative to the boundary, though that point of view is not necessary for this discussion.)

The derived automorphisms of the *rationalised* little d-discs operad $E_d^{\mathbb{Q}}$ have been studied by Fresse, Turchin, and Willwacher **[11]**, who for $d \geq 3$ give an identification $\pi_i(\mathrm{hAut}(E_d^{\mathbb{Q}})) = H_i(\mathrm{GC}_d^2)$. Combined with the above, this gives a map

$$\pi_i\big(B\mathrm{Diff}_\partial^{\mathrm{fr}}(D^d)\big) \otimes \mathbb{Q} \to H_{i+d}(\mathrm{GC}_d^2),$$

and it is difficult to imagine that this is given by anything other than evaluation of Kontsevich's invariant, but as far as I know the connection between this point of view and configuration space integrals has not yet been made precise. Assuming for now that this is so, Watanabe's results show that this map hits those graph homology classes represented by trivalent graphs.

2.3. Pontrjagin–Weiss classes

It follows from the work of Sullivan and of Kirby and Siebenmann that the homotopy fibre Top/O of the map $BO \to B\mathrm{Top}$ has finite homotopy groups, and therefore that

$$H^*(B\mathrm{Top}; \mathbb{Q}) = \mathbb{Q}[p_1, p_2, p_3, \ldots],$$

a polynomial ring on certain classes p_i of degree $4i$ which pull back to the Pontrjagin classes on BO: these are the topological Pontrjagin classes. By pulling back along the stabilisation map $B\mathrm{Top}(d) \to B\mathrm{Top}$, they are defined for all $\mathbb{R}^d$-bundles.

For real vector bundles of dimension $2n$, and so universally in the cohomology of $B\mathrm{O}(2n)$, the definition of Pontrjagin classes in terms of Chern classes of the complexification immediately gives that

$$p_i = 0 \quad \text{for } i > n, \tag{2.3}$$

$$p_n = e^2, \tag{2.4}$$

where e denotes the Euler class. The Euler class only depends on the underlying spherical fibration of the vector bundle, obtained by removing the zero-section. An $\mathbb{R}^d$-bundle also has an associated spherical fibration (by removing any section), and hence also has an Euler class: as both the Euler and Pontrjagin classes are defined on $B\mathrm{Top}(2n)$, one may then ask about the validity of the identities (2.3) and (2.4) there.

The work of Weiss. Reis and Weiss [39] had proposed an elaborate strategy for establishing these identities, but in a spectacular turnaround Weiss [50] then showed that these identities are in fact generally *false* for $\mathbb{R}^d$-bundles. I will comment further on his strategy in Section 4.1, as its philosophy is fundamental to all the results in Section 3.

To say more precisely what Weiss proved, consider the fibration sequence

$$\frac{\mathrm{Top}(2n)}{\mathrm{O}(2n)} \to B\mathrm{O}(2n) \to B\mathrm{Top}(2n).$$

The rational cohomology classes $p_n - e^2$ and p_i for $i > n$ are defined on $B\mathrm{Top}(2n)$ and are canonically trivial on $B\mathrm{O}(2n)$, and hence yield (pre-)transgressed cohomology classes $(p_n - e^2)^\tau$ and p_i^τ on $\frac{\mathrm{Top}(2n)}{\mathrm{O}(2n)}$. Weiss showed [50, SECTION 6] that for many n and $i \geq n$ (he shows that $n \geq 83$ and $i < \frac{9n}{4} - 11$, or $n \geq 59$ and $i < \frac{7n}{4}$ will do) these evaluate nontrivially against $\pi_{4i-1}(\frac{\mathrm{Top}(2n)}{\mathrm{O}(2n)})$: this certainly implies that the corresponding $p_n - e^2$ and p_i are nontrivial in the cohomology of $B\mathrm{Top}(2n)$, but is stronger. Translated to diffeomorphisms groups of discs via the Morlet equivalence, Weiss' result shows that the map

$$\begin{aligned}\pi_{4i-2n-1}(B\mathrm{Diff}_\partial(D^{2n})) &\cong \pi_{4i-2n-1}\left(\Omega_0^{2n}\frac{\mathrm{Top}(2n)}{\mathrm{O}(2n)}\right)\\ &\cong \pi_{4i-1}\left(\frac{\mathrm{Top}(2n)}{\mathrm{O}(2n)}\right) \xrightarrow{(p_n-e^2)^\tau \text{ or } p_i^\tau} \mathbb{Q}\end{aligned}$$

is nontrivial for many n and $i \geq n$. I will call an element of $\pi_{4i-2n-1}(B\mathrm{Diff}_\partial(D^{2n}))$ a *Pontrjagin–Weiss class* if it is detected by such maps.

Odd dimensions. On $BO(2n+1)$ the Pontrjagin classes still satisfy $p_i = 0$ for $i > n$, which is the analogue of (2.3), so there are (pre-)transgressed classes p_i^τ on $\frac{\mathrm{Top}(2n+1)}{\mathrm{O}(2n+1)}$ for $i > n$, and one may ask about their nontriviality on homotopy groups. As these classes pull back to the classes of the same name on $\frac{\mathrm{Top}(2n)}{\mathrm{O}(2n)}$, this nontriviality follows from Weiss' theorem in many cases.

It seems to be less well known that there is also an analogue of (2.4) in odd dimensions. Namely, there is a characteristic class E of S^{2n}-fibrations such that

$$p_n = E \tag{2.5}$$

in $H^{4n}(BO(2n+1);\mathbb{Q})$. It may be defined as follows. Given an S^{2n}-fibration $S^{2n} \to X \xrightarrow{\pi} Y$ with orientation local system $\mathcal{O}$, the self-intersection of the fibrewise diagonal map $\Delta : X \to X \times_Y X$ defines a *fibrewise Euler class* $e^{\mathrm{fw}}(\pi) \in H^{2n}(X;\mathcal{O})$, and then $E(\pi) := \frac{1}{2}\int_\pi e^{\mathrm{fw}}(\pi)^3 \in H^{4n}(Y;\mathbb{Q})$.

As E depends only on the underlying spherical fibration, it is also defined in $H^*(B\mathrm{Top}(2n+1);\mathbb{Q})$, so one can also ask whether the identity (2.5) fails to hold here, or even better whether the cohomology class $(p_n - E)^\tau$ on $\frac{\mathrm{Top}(2n+1)}{\mathrm{O}(2n+1)}$ evaluates nontrivially on $\pi_{4n-1}(\frac{\mathrm{Top}(2n+1)}{\mathrm{O}(2n+1)})$. It can be checked that under $B\mathrm{Top}(2n) \to B\mathrm{Top}(2n+1)$ the class E pulls back to e^2, so $(p_n - E)^\tau$ pulls back to $(p_n - e^2)^\tau$ on $\frac{\mathrm{Top}(2n)}{\mathrm{O}(2n)}$, and hence the nontriviality of $(p_n - E)^\tau$ on homotopy groups in many degrees also follows from Weiss' theorem.

Propagating. Formalising the method used above, the stabilisation maps

$$\begin{array}{ccc} \Omega_0^d \frac{\mathrm{Top}(d-1)}{\mathrm{O}(d-1)} & = & \Omega_0 B\mathrm{Diff}_\partial(D^{d-1}) \\ \downarrow & & \downarrow \\ \Omega_0^d \frac{\mathrm{Top}(d)}{\mathrm{O}(d)} & = & B\mathrm{Diff}_\partial(D^d) \end{array}$$

(known as "Gromoll maps" on the diffeomorphism group side) show that if a Pontrjagin–Weiss class exists on $B\mathrm{Diff}_\partial(D^{d-1})$, and the cohomology class detecting it can be defined on $\frac{\mathrm{Top}(d)}{\mathrm{O}(d)}$, then it survives to $B\mathrm{Diff}_\partial(D^d)$.

Relation to configuration space integrals. Somewhat surprisingly, the map

$$\pi_{2n-2}\big(B\mathrm{Diff}_\partial^{\mathrm{fr}}(D^{2n+1})\big) \cong \pi_{4n}\big(B\mathrm{Top}(2n+1)\big) \xrightarrow{E} \mathbb{Q}$$

can be identified with the simplest Kontsevich invariant ξ_1 (which is that associated to the Θ-graph) as studied by Watanabe in [43]. From this point of view, Watanabe's argument in that paper shows that $p_n \neq E$ in $H^{4n}(B\mathrm{Top}(2n+1);\mathbb{Q})$, so is closely related to Weiss' theorem (but does not imply it!). This is explained in detail in [28, APPENDIX B].

3. THE RATIONAL HOMOTOPY TYPE OF $B\mathrm{Diff}_\partial(D^d)$

The results of the last section give a complete calculation (2.2) of the rational homotopy groups $\pi_*(B\mathrm{Diff}_\partial(D^d)) \otimes \mathbb{Q}$ valid in the pseudoisotopy stable range, but also indicate the existence of various new phenomena outside of this range. These new phenomena start in degrees $\sim d$, and Krannich [26] and I [37] had shown that (2.2) is in fact valid in degrees $\lesssim d$. In this section I present two recent results, obtained in collaboration with Kupers and with Krannich, giving detailed information quite far outside of this range, and I then speculate about what they might be indicating.

3.1. Even-dimensional discs

Kupers and I [32,34] have investigated the rational homotopy type of $B\mathrm{Diff}_\partial(D^{2n})$. The following is the main result of [32], incorporating the improvement from [34, SECTION 7.1].

Theorem 3.1 (Kupers–Randal-Williams). *Let $2n \geq 6$. Then $\pi_j(B\mathrm{Diff}_\partial(D^{2n})) \otimes \mathbb{Q} = 0$ for $j < 2n-1$, and for $j \geq 2n-1$ we have*

$$\pi_j\big(B\mathrm{Diff}_\partial(D^{2n})\big) \otimes \mathbb{Q} = \begin{cases} \mathbb{Q}, & j \equiv 2n-1 \bmod 4,\ j \notin \bigcup_{r\geq 2}[2r(n-2)-1, 2r(n-1)+1], \\ 0, & j \not\equiv 2n-1 \bmod 4,\ j \notin \bigcup_{r\geq 2}[2r(n-2)-1, 2r(n-1)+1], \\ ?, & \textit{otherwise}. \end{cases}$$

The copies of $\mathbb{Q}$ in this theorem are generated by Pontrjagin–Weiss classes in the sense of Section 2.3, and the theorem gives a complete calculation in degrees $\leq 4n-10$, as well as in higher degrees outside of the indicated "bands".

It can be cast in a somewhat stronger form by using the Morlet equivalence $B\mathrm{Diff}_\partial(D^{2n}) \simeq \Omega_0^{2n} \frac{\mathrm{Top}(2n)}{\mathrm{O}(2n)}$ and considering the fibration sequence

$$\Omega^{2n+1}\left(\frac{\mathrm{Top}}{\mathrm{Top}(2n)}\right) \to \Omega^{2n}\left(\frac{\mathrm{Top}(2n)}{\mathrm{O}(2n)}\right) \to \Omega^{2n}\left(\frac{\mathrm{Top}}{\mathrm{O}(2n)}\right).$$

A slight strengthening of the theorem is then that $\pi_*(\Omega^{2n+1}\frac{\mathrm{Top}}{\mathrm{Top}(2n)}) \otimes \mathbb{Q}$ is supported in degrees $\bigcup_{r\geq 2}[2r(n-2)-1, 2r(n-1)+1]$; the rational homotopy groups of $\Omega^{2n}\frac{\mathrm{Top}}{\mathrm{O}(2n)}$ are $\mathbb{Q}$ in every degree $\equiv 2n-1 \bmod 4$, and the right-hand map detects the Pontrjagin–Weiss classes.

The result can also be given a little more structure by using the involution on $B\mathrm{Diff}_\partial(D^{2n}) \simeq \Omega_0^{2n}\frac{\mathrm{Top}(2n)}{\mathrm{O}(2n)}$ induced by conjugation by a reflection of the disc. The terms in the fibration sequence above have compatible involutions, which on $\pi_*(\Omega_0^{2n}\frac{\mathrm{Top}}{\mathrm{O}(2n)}) \otimes \mathbb{Q}$ acts as (-1), and on $\pi_*(\Omega^{2n+1}\frac{\mathrm{Top}}{\mathrm{Top}(2n)}) \otimes \mathbb{Q}$ acts as $(-1)^r$ in the band of degrees $[2r(n-2)-1, 2r(n-1)+1]$ (when such bands overlap this should be regarded as inconclusive). This implies the existence of Pontrjagin–Weiss classes outside of degrees $\bigcup_{r\geq 2, r \text{ odd}}[2r(n-2)-1, 2r(n-1)+1]$.

Finally, as explained in Section 2.3, Pontrjagin–Weiss classes can be propagated from smaller discs to larger ones. The conclusion of this discussion is depicted in Figure 1. It seems likely that all possible Pontrjagin–Weiss classes already exist in $\pi_*(B\mathrm{Diff}_\partial(D^6))$.

3.2. Odd-dimensional discs

Krannich and I [28] have investigated the rational homotopy type of $B\mathrm{Diff}_\partial(D^{2n+1})$.

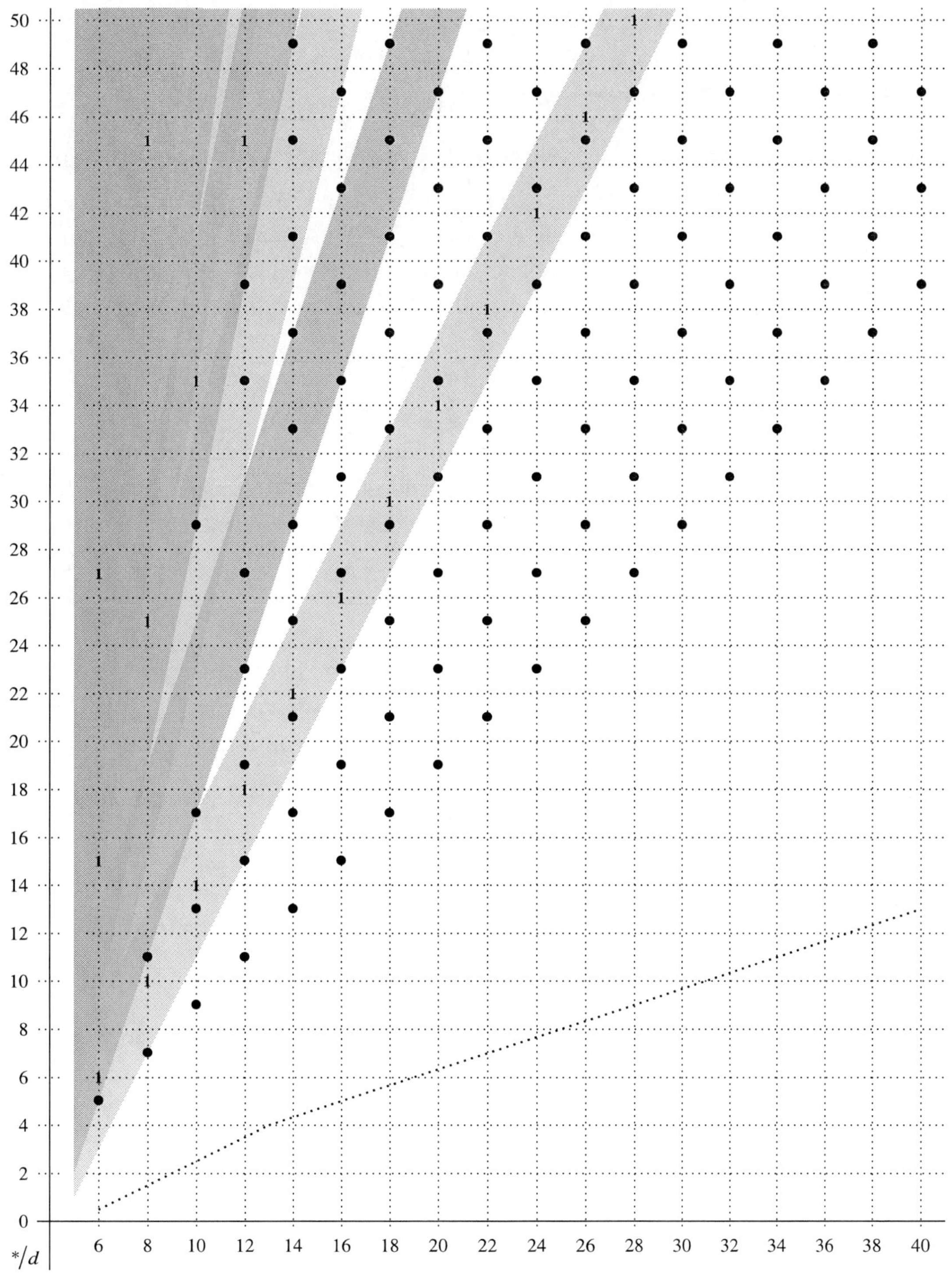

FIGURE 1

Rational homotopy groups of $B\mathrm{Diff}_\partial(D^{2n})$. The calculation is complete in the unshaded region, and • denotes Pontrjagin–Weiss classes. The lightly shaded bands are those on which the reflection acts as $+1$, and Pontrjagin–Weiss classes are still present in these; the darkly-shaded bands are those where the reflection acts as -1. Existing copies of • have been propagated downwards along lines of slope -1 as in Section 2.3. The numbers denote Watanabe's lower bounds on these groups. The dotted line indicates the Igusa stable range.

Theorem 3.2 (Krannich–Randal-Williams). *For degrees $j \leq 3n-8$, we have*

$$\pi_j\big(B\mathrm{Diff}_\partial(D^{2n+1})\big) \otimes \mathbb{Q} = K_{j+1}(\mathbb{Z}) \otimes \mathbb{Q} \oplus \begin{cases} \mathbb{Q}, & j \equiv 2n-2 \bmod 4,\ j \geq 2n-2, \\ 0, & \text{otherwise}. \end{cases}$$

The first term is the rational algebraic K-theory of the integers, extending the classes discussed in Section 2.1. The second term consists of Pontrjagin–Weiss classes. As discussed in Section 2.3, the lowest of these—in degree $(2n-2)$—corresponds to the configuration space integral associated to the Θ-graph, and so accounts for the class in this degree found for odd $n \leq 399$ by Watanabe [43], and show that such classes exist for all n. The conclusion of this discussion is depicted in Figure 2.

In proving Theorem 3.2, Krannich and I were only attempting to calculate within the indicated range, and with the method we used it is not clear how to establish the "band" pattern in higher degrees for odd-dimensional discs, too. But it does seem feasible that the method used to prove Theorem 3.1 could be adapted to the odd-dimensional case (though there are significant hurdles) and I think it very likely that the "band" pattern occurs in this case, too.

3.3. Outlook and speculation

These two theorems have sufficient detail that one is tempted to propose a structural description of $\pi_*(B\mathrm{Diff}_\partial(D^d)) \otimes \mathbb{Q}$. In fact, it seems better to describe $\pi_*(B\mathrm{Top}(d)) \otimes \mathbb{Q}$. Summarising the structural features of the above results, $\pi_*(B\mathrm{Top}(d)) \otimes \mathbb{Q}$ has

(i) classes corresponding to Pontrjagin classes, i.e. detected by $B\mathrm{Top}(d) \to B\mathrm{Top}$, in degrees ≥ 0,

(ii) classes corresponding to $K_{*>0}(\mathbb{Z}) \otimes \mathbb{Q}$ if d is odd, in degrees $\gtrsim d$, and a class corresponding to $K_0(\mathbb{Z}) \otimes \mathbb{Q}$ in degree d if d is even (detected by the Euler class),

(iii) classes supported in bands of degrees around $k \cdot d$ for each $k \geq 2$ (at least for d even, but lets suppose that this also occurs for d odd).

Orthogonal calculus. This behaviour could be explained by Weiss' theory of orthogonal calculus [48], a calculus of functors for continuous functors $\mathsf{F} : \mathcal{J} \to \mathcal{T}op$ defined on the category $\mathcal{J}$ of real inner product spaces and their isometric embeddings. It may be applied to the functor $\mathsf{Bt} : V \mapsto B\mathrm{Top}(V)$, where it provides a tower of Taylor approximations

$$\begin{array}{ccccc} & & \vdots & & \\ & & T_2\mathsf{Bt}(V) & \longleftarrow & \Omega^\infty\big(S^{2\cdot V} \wedge \Theta\mathsf{Bt}^{(2)}\big)_{hO(2)} \\ & & \downarrow & & \\ & & T_1\mathsf{Bt}(V) & \longleftarrow & \Omega^\infty\big(S^{1\cdot V} \wedge \Theta\mathsf{Bt}^{(1)}\big)_{hO(1)} \\ & & \downarrow & & \\ \mathsf{Bt}(V) & \longrightarrow & T_0\mathsf{Bt}(V) & = & B\mathrm{Top}, \end{array}$$

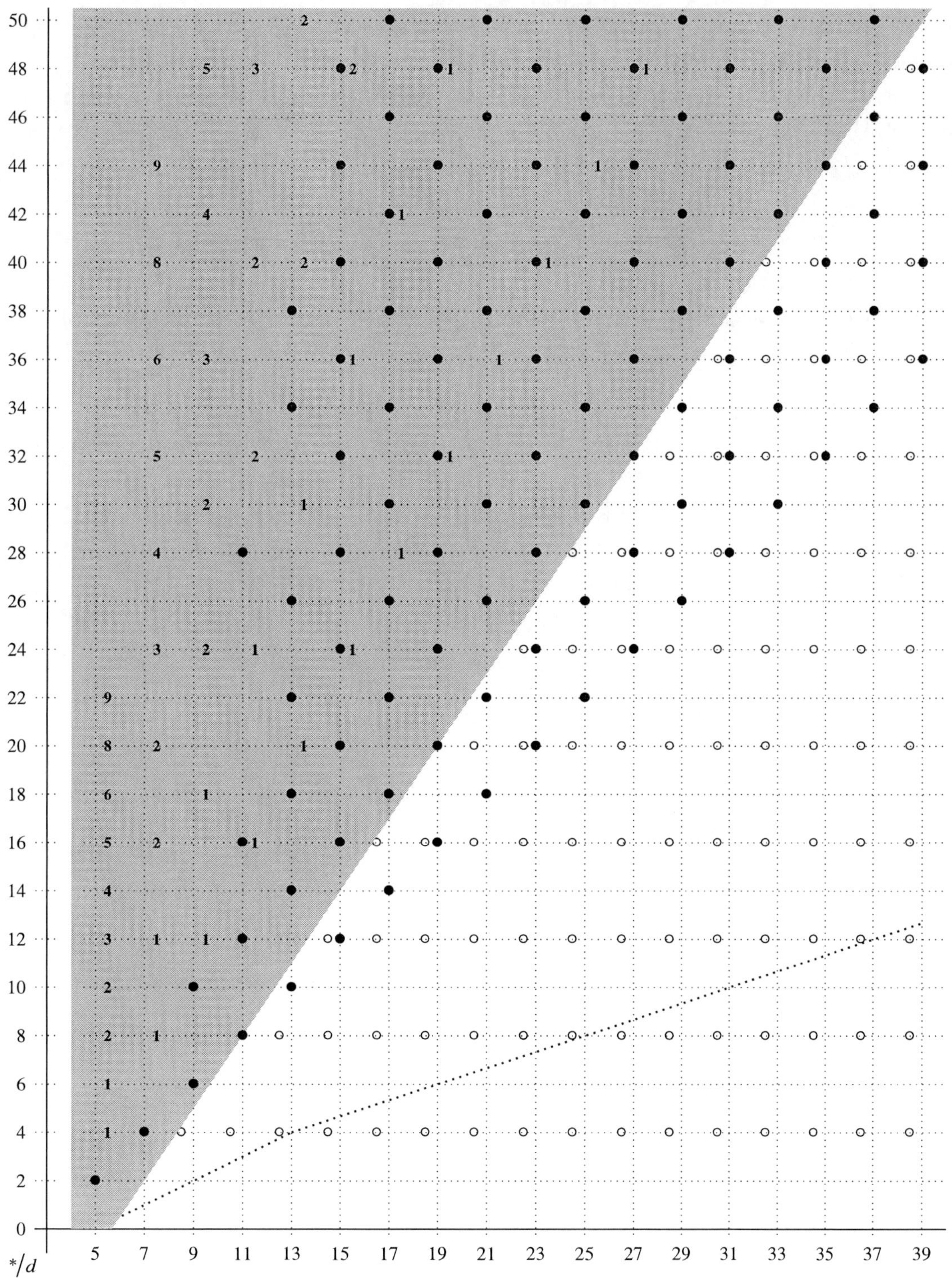

FIGURE 2

Rational homotopy groups of $B\mathrm{Diff}_\partial(D^{2n+1})$. The calculation is complete in the unshaded region: • denotes Pontrjagin–Weiss classes, and ∘ denotes algebraic K-theory classes. In the shaded region we have indicated existing Pontrjagin–Weiss classes, and the numbers denote Watanabe's lower bounds on these groups.

whose kth layer is described in terms of an O(k)-spectrum $\Theta\mathsf{Bt}^{(k)}$, the *kth derivative*, and the 1-point compactifications $S^{k\cdot V}$ of the vector spaces $\mathbb{R}^k \otimes V$. The zeroth Taylor approximation is the stabilisation of the functor, in this case $B\mathrm{Top}$. It is a theorem of Waldhausen that $\Theta\mathsf{Bt}^{(1)}$ is $\mathrm{A}(*) = \mathrm{K}(\mathbb{S})$, with a certain O(1)-action, and, in view of the rational equivalence $\mathrm{K}(\mathbb{S}) \to \mathrm{K}(\mathbb{Z})$, points (i) and (ii) above can be accounted for by the first Taylor approximation. Point (iii) would then be accounted for if

(i) the Taylor tower converges (rationally), and

(ii) for each $k \geq 2$ the homotopy orbits $(\Theta\mathsf{Bt}^{(k)})_{h\mathrm{SO}(k)}$ of the derivative spectra have finitely-many nontrivial (rational) homotopy groups.

In this worldview, the (finitely-many) rational homotopy groups of $(\Theta\mathsf{Bt}^{(k)})_{h\mathrm{SO}(k)}$ correspond to the rational homotopy classes of $B\mathrm{Top}(d)$ in the kth band which are not detected by Pontrjagin classes or algebraic K-theory: more precisely, the residual $\mathrm{O}(k)/\mathrm{SO}(k)$-action splits $\pi_*((\Theta\mathsf{Bt}^{(k)})_{h\mathrm{SO}(k)}) \otimes \mathbb{Q}$ into eigenspaces, and the $(-1)^d$-eigenspace provides the kth band of $B\mathrm{Top}(d)$. In particular, this worldview predicts that the homotopy groups in the kth band depend only on the parity of d.

By Theorems 3.1 and 3.2, this property does indeed hold for the second band, and Krannich and I **[28]** have used this to investigate the second derivative $\Theta\mathsf{Bt}^{(2)}$, establishing a rational equivalence

$$\Theta\mathsf{Bt}^{(2)} \simeq_{\mathbb{Q}} \mathrm{map}(S^1_+, \mathbb{S}^{-1})$$

of O(2)-spectra, where O(2) acts in the usual way on S^1. Reis and Weiss **[39]** had earlier shown that $\mathrm{map}(S^1_+, \mathbb{S}^{-1})$ is the second derivative of the orthogonal functor $\mathsf{Bg}(V) := B\mathrm{hAut}(S(V))$, the classifying space of the monoid of homotopy automorphisms of the unit sphere in the inner product space V, and the natural map $\mathsf{Bt} \to \mathsf{Bg}$ (in fact, zigzag) induces an equivalence on rationalised second derivatives.

Automorphisms of little discs operads and graph complexes. In Section 2.2 I explained that there is a map

$$B\mathrm{Top}(d) \to B\mathrm{hAut}(E_d) \tag{3.1}$$

corresponding to a derived action of $\mathrm{Top}(d)$ on the little d-discs operad E_d. I mentioned also that the derived automorphisms of the rationalisation $E_d^{\mathbb{Q}}$ have been analysed by Fresse, Turchin, and Willwacher **[11]**, giving an identification $\pi_i(\mathrm{hAut}(E_d^{\mathbb{Q}})) = H_i(\mathrm{GC}_d^2)$ for $d \geq 3$ in terms of a version of Kontsevich's graph complex. There is a loop-order decomposition $\mathrm{GC}_d^2 = \bigoplus_{g\geq 1} \mathrm{GC}_d^{2,g\text{-loop}}$ and, for $g \geq 2$,

$$H_*(\mathrm{GC}_d^{2,g\text{-loop}}) \text{ is supported in degrees } * \in \left[g(d-3)+3, g(d-2)+1\right],$$

and furthermore up to translating degrees this homology depends only on the parity of d. There are some computer calculations of these groups, but they are largely unknown. On the other hand, the 1-loop part is completely known, and is

$$H_*(\mathrm{GC}_d^{2,1\text{-loop}}) = \bigoplus_{\substack{k\geq 1,\\ k\equiv 2d+1 \bmod 4}} \mathbb{Q}[d-k].$$

Writing E_V for the little discs operad modelled on the unit disc in an inner product space V, one can consider orthogonal calculus applied to the functor $\mathsf{Ba} : V \mapsto B\mathrm{hAut}(E_V)$ (or perhaps better $B\mathrm{hAut}(E_V^{\mathbb{Q}})$: there is an important and subtle question if $B\mathrm{hAut}(E_V) \to B\mathrm{hAut}(E_V^{\mathbb{Q}})$ is a rationalisation on universal covers, which I shall elide). Presumably by passing to appropriate models one can upgrade the maps (3.1) to a map $\mathsf{Bt} \to \mathsf{Ba}$ of orthogonal functors. The data above would seem to suggest that Ba enjoys precisely the property described in the last section, namely that $\Theta\mathsf{Ba}^{(1)}$ is rationally equivalent to the $\mathrm{O}(1)$-spectrum $\mathrm{map}(\mathbb{CP}^\infty_+, \mathbb{S}^0)$, where the action is by complex conjugation, and that for $k \geq 2$ the rational homotopy groups of $(\Theta\mathsf{Ba}^{(k)})_{h\mathrm{SO}(k)}$ are supported in degrees $[4-3k, 2-2k]$ and combine the k-loop graph homology for both parities encoded as the $\mathrm{O}(k)/\mathrm{SO}(k)$-eigenspace decomposition. (In particular, this suggests an action of the twisted group ring $H^{-*}(B\mathrm{SO}(k); \mathbb{Q})[\mathrm{O}(k)/\mathrm{SO}(k)]$ on the graded vector space

$$s^{-2k} H_*(\mathrm{GC}_2^{2,k\text{-loop}}) \oplus s^{-3k} H_*(\mathrm{GC}_3^{2,k\text{-loop}}),$$

giving a potentially nontrivial relationship between even and odd graph homology.)

There are two reasons $B\mathrm{Top}(d) \to B\mathrm{hAut}(E_d^{\mathbb{Q}})$ cannot be a rational equivalence:

(i) this map tends to kill the Pontrjagin–Weiss classes (and for d odd the algebraic K-theory classes),

(ii) the 1-loop graph contribution $H_{*+1}(\mathrm{GC}_d^{2,1\text{-loop}}) \subset \pi_*(B\mathrm{hAut}(E_d^{\mathbb{Q}}))$ does not come from $\pi_*(B\mathrm{Top}(d)) \otimes \mathbb{Q}$.

Point (i) concerns the contribution to $B\mathrm{Top}(d) = \mathsf{Bt}(\mathbb{R}^d)$ due to the first Taylor approximation $T_1\mathsf{Bt}(\mathbb{R}^d)$, and, given the degrees in which the 1-loop graphs contribute, point (ii) presumably concerns the contribution to $B\mathrm{hAut}(E_d^{\mathbb{Q}}) = \mathsf{Ba}(\mathbb{R}^d)$ due to the first Taylor approximation $T_1\mathsf{Ba}(\mathbb{R}^d)$. Together these suggest that a better question is to ask about the rational homotopy cartesianness of

$$\begin{array}{ccccc} B\mathrm{Top}(d) & = & \mathsf{Bt}(\mathbb{R}^d) & \longrightarrow & T_1\mathsf{Bt}(\mathbb{R}^d) \\ & & \downarrow & & \downarrow \\ B\mathrm{hAut}(E_d^{\mathbb{Q}}) & = & \mathsf{Ba}(\mathbb{R}^d) & \longrightarrow & T_1\mathsf{Ba}(\mathbb{R}^d). \end{array} \tag{3.2}$$

If this square were rationally homotopy cartesian for all large enough d then in particular the maps on derivatives $\Theta\mathsf{Bt}^{(k)} \to \Theta\mathsf{Ba}^{(k)}$ would be rational equivalences for all $k \geq 2$. As evidence for this, from the proposed description of the rational homotopy groups of $(\Theta\mathsf{Ba}^{(k)})_{h\mathrm{SO}(k)}$ described above, and the calculation of 2-loop graph homology, one can easily deduce that $\Theta\mathsf{Ba}^{(2)} \simeq_{\mathbb{Q}} \mathrm{map}(S^1_+, \mathbb{S}^{-1})$ and that the induced map $\Theta\mathsf{Bt}^{(2)} \to \Theta\mathsf{Ba}^{(2)}$ is indeed a rational equivalence.

When $d = 2n$, rational homotopy cartesianness of (3.2) is equivalent to the map

$$B\mathrm{Diff}^{\mathrm{fr}}_\partial(D^{2n}) \simeq \Omega_0^{2n}\mathrm{Top}(2n) \to \Omega_0^{2n}\left(\mathrm{hAut}(E_{2n}^{\mathbb{Q}}) \times \mathrm{Top}\right)$$

being a rational equivalence, and by comparing certain graphical models arising in [32] with graph complexes arising in the operadic model for embedding calculus it looks like this might be true. Kupers, Willwacher, and I are trying to make this precise.

4. METHODS

I will now explain some of the ideas which go into the proofs of Theorems 3.1 and 3.2, though my goal is to give an overall impression of the methods involved rather than explain how exactly they are combined to prove these two particular results.

4.1. Weiss fibre sequences and the general strategy

Weiss' proof of the existence of Pontrjagin–Weiss classes contains an observation [50, REMARK 2.1.3] which technically does not play a role in his argument but is central to its philosophy. It is more a general principle than a specific formulation, and I shall give it only in a modestly general form: variants of it underlie many recent results about diffeomorphism groups [6,7,24,28,30–32]. Let W be a manifold with boundary ∂W decomposed as $\partial_- W \cup \partial_+ W$ into codimension zero submanifolds with common boundary. Then there is a fibration

$$B\mathrm{Diff}_\partial\big(\partial_- W \times [0,1]\big) \to B\mathrm{Diff}_\partial(W) \to B\mathrm{Emb}^{\cong}_{\partial_+ W}(W, W). \tag{4.1}$$

The rightmost term needs a little explaining: it is the classifying space of the group-like topological monoid $\mathrm{Emb}^{\cong}_{\partial_+ W}(W, W)$ of those self-embeddings of W which are the identity on $\partial_+ W$, and which are isotopic to diffeomorphisms. But crucially these self-embeddings are allowed to send $\partial_- W$ into the interior of W, as indicated in Figure 3.

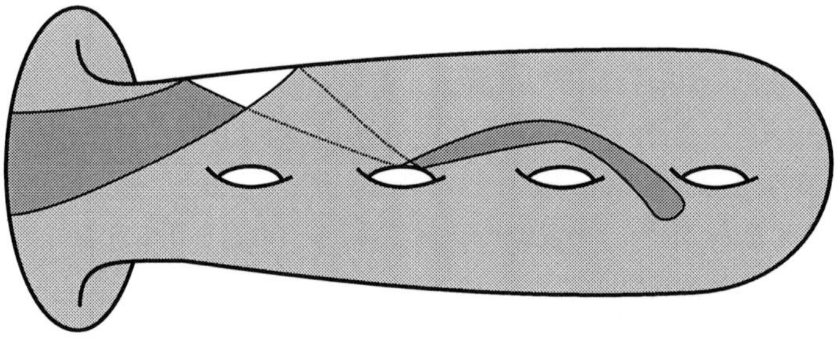

FIGURE 3
A self-embedding of W relative only to $\partial_+ W$.

This is not a technically difficult result (after passing to a different model of the rightmost term, it is a simple consequence of the parameterised isotopy extension theorem). Somewhat more technical is Kupers' theorem [30, SECTION 4] that this fibration sequence deloops—with respect to the evident composition law on $B\mathrm{Diff}_\partial(\partial_- W \times [0,1])$—which is sometimes convenient.

The importance of this fibration sequence is the following strategy which it indicates: to understand $B\mathrm{Diff}_\partial(\partial_- W \times [0,1])$, you can instead try to understand $B\mathrm{Diff}_\partial(W)$ and $B\mathrm{Emb}^{\cong}_{\partial_+ W}(W, W)$, *for any manifold W containing $\partial_- W$ in its boundary*. This is powerful because these two spaces can sometimes be accessed, though by very different methods: the homology of $B\mathrm{Diff}_\partial(W)$ by parameterised surgery theory, and the homotopy of $B\mathrm{Emb}^{\cong}_{\partial_+ W}(W, W)$ by embedding calculus. Let me explain how this strategy may be implemented to study $B\mathrm{Diff}_\partial(D^d)$.

Set-up for discs. If $d = 2n$ then take $W_{g,1} := D^{2n}\#g(S^n \times S^n)$ and $\partial_- W_{g,1} = D^{2n-1} \subset \partial W_{g,1}$, so that the fibration (4.1) takes the form

$$B\mathrm{Diff}_\partial(D^{2n}) \to B\mathrm{Diff}_\partial(W_{g,1}) \to B\mathrm{Emb}^{\cong}_{\partial_+ W_{g,1}}(W_{g,1}, W_{g,1}), \tag{4.2}$$

where we have identified $D^{2n-1} \times [0,1] \approx D^{2n}$.

If $d = 2n+1$ then instead of directly using (4.1), take the handlebody $V_g := \natural g(S^n \times D^{n+1})$, and use a variant of (4.1) allowing most of the boundary of V_g to not be fixed. This takes the form

$$BC(D^{2n}) \to B\mathrm{Diff}_{D^{2n}}(V_g) \to B\mathrm{Emb}^{\cong}_{D^{2n}, W_{g,1}}(V_g, V_g), \tag{4.3}$$

where the left-hand term is given by pseudoisotopies of D^{2n}, the middle term by diffeomorphisms of V_g fixing a disc $D^{2n} \subset \partial V_g$ but allowing the rest of the boundary to move, and the right-hand term is given by self-embeddings of V_g which preserve $\partial V_g \setminus \mathrm{int}(D^{2n}) = W_{g,1}$ setwise, and furthermore preserve a disc $D^{2n} \subset W_{g,1}$ pointwise. By the fibration sequence

$$B\mathrm{Diff}_\partial(D^{2n+1}) \to BC(D^{2n}) \to B\mathrm{Diff}_\partial(D^{2n}) \tag{4.4}$$

from (2.1), given $B\mathrm{Diff}_\partial(D^{2n})$ it is equivalent to get at $B\mathrm{Diff}_\partial(D^{2n+1})$ or $BC(D^{2n})$, and I will explain below why the latter is more accessible.

Parameterised surgery. The reason for the choice of manifold $W_{g,1}$ is that Galatius and I [15] have shown that the maps $B\mathrm{Diff}_\partial(W_{g,1}) \to B\mathrm{Diff}_\partial(W_{g+1,1})$, induced by the evident embeddings $W_{g,1} \hookrightarrow W_{g+1,1}$, are homology isomorphisms in a range of homological degrees tending to infinity with g, as long as $2n \geq 6$, and furthermore [13] that a certain parameterised Pontrjagin–Thom map

$$\operatorname*{hocolim}_{g\to\infty} B\mathrm{Diff}_\partial(W_{g,1}) \to \Omega^\infty_0 \mathrm{MT}\theta^n,$$

to the infinite loop space of a certain Thom spectrum, induces an isomorphism on homology. The rational cohomology of the right-hand side is quite simple: it is a polynomial algebra on certain easily-defined cohomology classes, known as Miller–Morita–Mumford classes.

These results are analogues in high dimensions of Harer's [20] theorem on the stability of the homology of mapping class groups of oriented surfaces, and Madsen and Weiss' [35] theorem on the stable homology of these mapping class groups. In fact, the stability result holds much more generally for all $2n$-manifolds of the form $W\#g(S^n \times S^n)$ with $2n \geq 6$ and W simply-connected (or even with virtually polycyclic fundamental group [12]), and there is an analogous description of the stable homology for any W of any even dimension [14] (including dimension 4). See Galatius' contribution to the 2014 ICM for an overview of this theory.

In odd dimensions the stable homology of the diffeomorphism groups of the analogous manifolds $D^{2n+1}\#g(S^n \times S^{n+1})$ is not yet known, but Botvinnik and Perlmutter [5] have a version for $B\mathrm{Diff}_{D^{2n}}(V_g)$, and Perlmutter [36] has the appropriate stability theorem in this case. This accounts for the use of the modified Weiss fibre sequence (4.3) in odd dimensions, rather than a more obvious analogue of (4.2) involving $D^{2n+1}\#g(S^n \times S^{n+1})$.

Embedding calculus. The difficulty of studying embeddings of one manifold into another depends on the codimension, but this must be counted appropriately. What matters is the *geometric dimension* of the target minus the *handle dimension* of the source. In particular, if W is d-dimensional but can be constructed from $\partial_+ W \times [0,1]$ by attaching handles of index $\leq h$, then self-embeddings of W relative to $\partial_+ W$ have codimension $d-h$. If this codimension is ≥ 3, then the theory of embedding calculus as developed by Goodwillie, Klein, and Weiss [17, 18, 49] can be used to access spaces of self-embeddings of W relative to $\partial_+ W$. This theory provides a tower

$$\begin{array}{ccc} & & \vdots \\ & & BT_3\mathrm{Emb}^{\cong}_{\partial_+ W}(W,W) \\ & \nearrow & \downarrow \\ & & BT_2\mathrm{Emb}^{\cong}_{\partial_+ W}(W,W) \\ & \nearrow & \downarrow \\ B\mathrm{Emb}^{\cong}_{\partial_+ W}(W,W) & \longrightarrow & BT_1\mathrm{Emb}^{\cong}_{\partial_+ W}(W,W), \end{array} \tag{4.5}$$

such that, as long as the codimension (as described above) is ≥ 3, the map

$$B\mathrm{Emb}^{\cong}_{\partial_+ W}(W,W) \to BT_\infty \mathrm{Emb}^{\cong}_{\partial_+ W}(W,W) = \operatorname*{holim}_{k\to\infty} BT_k\mathrm{Emb}^{\cong}_{\partial_+ W}(W,W)$$

is an equivalence. The bottom stage $BT_1\mathrm{Emb}^{\cong}_{\partial_+ W}(W,W)$ is equivalent to the classifying space of the monoid $\mathrm{Bun}^{\cong}_{\partial_+ W}(TW,TW)$ of bundle maps $TW \to TW$ which are the identity over $\partial_+ W$ and which are homotopic to the derivative of a diffeomorphism, and the homotopy fibre of $BT_k\mathrm{Emb}^{\cong}_{\partial_+ W}(W,W) \to BT_{k-1}\mathrm{Emb}^{\cong}_{\partial_+ W}(W,W)$ has a description in terms of a space of sections of a bundle $Z_k \to C_k(W)$ over the configuration space of k unordered points in W, whose fibres are constructed from the configuration spaces of $\leq k$ ordered points in W. Thus in principle the bottom stage and these homotopy fibres are amenable to calculation by homotopical methods.

For the manifold $W_{g,1}$ and $\partial_+ W_{g,1} = D^{2n}$, the codimension in the sense described is n, so the embedding calculus tower converges as long as $2n \geq 6$. For the manifolds V_g, there is a similar tower for $B\mathrm{Emb}^{\cong}_{D^{2n},W_{g,1}}(V_g,V_g)$, which converges for $2n+1 \geq 7$.

The strategy which suggests itself is then to calculate as much as you can about the middle and right-hand terms of (4.2) and (4.3) using these two very different methods, and then use these fibre sequences and (4.4) to deduce things about $B\mathrm{Diff}_\partial(D^d)$. This is a very attractive picture, but for getting explicit answers there is a serious

Difficulty. Parameterised surgery fundamentally gets at the *homology* groups of diffeomorphism groups, whereas embedding calculus, at least if applied in the most classical way, naturally allows one to get at the *homotopy* groups of embedding spaces.

4.2. Qualitative results

One situation in which this Difficulty is not so serious is if one wishes to obtain qualitative results about $B\mathrm{Diff}_\partial(\partial_- W \times [0,1])$, for example, that its homology or homotopy groups lie in a given Serre class. This was pioneered by Kupers [30], to prove that the homotopy (or equivalently, homology) groups of $B\mathrm{Diff}_\partial(D^d)$ are finitely-generated for $d \neq 4, 5, 7$. A slight variant of his line of reasoning is as follows.

Firstly, for D^{2n} consider the Weiss fibre sequence (4.2), which may be delooped. Using [13, 15], as long as $2n \geq 6$, the homology of $B\mathrm{Diff}_\partial(W_{g,1})$ is easily seen to be finitely-generated in degrees $* \leq \frac{g-3}{2}$, so it suffices to show that the homology of

$$X := B\mathrm{Emb}^{\cong}_{\partial_+ W_{g,1}}(W_{g,1}, W_{g,1})$$

is finitely-generated, too. Using the embedding calculus tower (4.5), it is not difficult to show that the higher homotopy groups of X are all finitely-generated, and hence to deduce that the homology of the universal cover $\widetilde{X}$ is finitely-generated. It remains to study the spectral sequence for the fibration

$$\widetilde{X} \to X \to B\pi_1(X)$$

and the crucial point here is that the group

$$\pi_1(X) = \pi_0\big(\mathrm{Emb}^{\cong}_{\partial_+ W_{g,1}}(W_{g,1}, W_{g,1})\big) \cong \pi_0\big(\mathrm{Diff}_\partial(W_{g,1})\big)/\pi_0\big(\mathrm{Diff}_\partial(D^{2n})\big)$$

enjoys Wall's finiteness property (F_∞). In this case it is clear by Kreck's [29] calculation of the group $\pi_0(\mathrm{Diff}_\partial(W_{g,1}))$, but as a general principle it follows from Sullivan's theorem [40] that mapping class groups of simply-connected manifolds of dimension ≥ 5 are commensurable (up to finite kernel, see [27]) to arithmetic groups.

Secondly, for D^{2n+1} it suffices, given the above, to prove finite-generation of the homology of $BC(D^{2n})$, so consider the Weiss fibre sequence (4.3), which may also be delooped. Using [5, 36], as long as $2n+1 \geq 9$, the homology of $B\mathrm{Diff}_{D^{2n}}(V_g)$ is finitely-generated in a stable range, and embedding calculus considerations as above show that the homology of $B\mathrm{Emb}^{\cong}_{D^{2n}, W_{g,1}}(V_g, V_g)$ is finitely-generated, too.

When working modulo a Serre class, one can sometimes also determine the lowest nontrivial term modulo that class. Bustamante and I [7] have used the above strategy with a Weiss fibre sequence for the manifolds $X_g := S^1 \times D^{2n-1} \# g(S^n \times S^n)$ to analyse $\pi_*(B\mathrm{Diff}_\partial(S^1 \times D^{2n-1}))_{(p)}$ for $2n \geq 6$ modulo the Serre class of finitely-generated $\mathbb{Z}_{(p)}$-modules, where we show that it vanishes in degrees $* < \min(2p-3, n-2)$ and is $\bigoplus^\infty \mathbb{Z}/p$ in degree $2p-3$ as long as $2p-3 < n-2$. This was known in the pseudoisotopy stable range using algebraic K-theory methods [19], but our work gives a rather different perspective on this infinitely-generated subgroup.

Turning the Weiss fibre sequence around. A further point of view on the Weiss fibre sequence is that, assuming $B\mathrm{Emb}^{\cong}_{\partial_+ W}(W, W)$ may be understood using embedding calculus, it reduces questions about $B\mathrm{Diff}_\partial(W)$ for a whole class of manifolds W to questions about the single space $B\mathrm{Diff}_\partial(\partial_- W \times [0,1])$. As the minimal choice of $\partial_- W$ is $\partial_- W = D^{d-1}$, this gives another reason to be particularly interested in $B\mathrm{Diff}_\partial(D^d)$.

Kupers [30] exploits this point of view to show—given the homological, and so also homotopical, finite generation of $B\mathrm{Diff}_\partial(D^d)$ discussed above—that $B\mathrm{Diff}_\partial(W)$ has finitely-generated higher homotopy groups for *any* closed 2-connected manifold W of dimension $d \neq 4, 5, 7$. More recently Bustamante, Krannich, and Kupers [6] have extended this to any closed manifold of dimension $2n \geq 6$ with finite fundamental group.

4.3. Quantitative results

To obtain quantitative results one must confront the Difficulty. The most obvious way to do this—in the case of discs—is to try to make Kupers' method from the last section quantitative, by trying to calculate the homology of the right-hand terms of (4.2) and (4.3).

This is the strategy I pursued with Krannich in [27] to prove Theorem 3.2, and (though for $S^1 \times D^{2n-1}$ and not for discs) with Bustamante in [7]. An important preliminary simplification is to consider the Weiss fibre sequence with framings (or similar): for example, in the framed version

$$B\mathrm{Diff}^{\mathrm{fr}}_\partial(D^{2n}) \to B\mathrm{Diff}^{\mathrm{fr}}_\partial(W_{g,1}) \to B\mathrm{Emb}^{\mathrm{fr},\cong}_{\partial_+ W_{g,1}}(W_{g,1}, W_{g,1}) \tag{4.6}$$

of the sequence (4.2), $B\mathrm{Diff}^{\mathrm{fr}}_\partial(D^{2n})$ differs from $B\mathrm{Diff}_\partial(D^{2n})$ by a copy of $\Omega^{2n}\mathrm{O}(2n)$, whose rational homotopy groups are completely understood, but [13, 15] shows that the rational homology of $B\mathrm{Diff}^{\mathrm{fr}}_\partial(W_{g,1})$ is *trivial* in the stable range, which is far simpler than the rational homology of $B\mathrm{Diff}_\partial(W_{g,1})$. Luckily, the effect on the homology of the self-embeddings term is also beneficial. The way we calculate the homology of the (framed) self-embedding spaces in these papers is not in fact using embedding calculus as I have been advertising, but rather using disjunction theory (which is in any case the fuel which makes embedding calculus work [17, 18], but using it directly is sometimes more convenient), in the form of Morlet's lemma of disjunction in [7] and Goodwillie's multi-relative disjunction lemma [16] in [27]. The nature of the calculations involved makes it hard to say anything very general about them, so I shall not try to.

Instead, I should like to discuss an alternative strategy, which is what Kupers and I do in [32] and prepare for in the companion papers [31, 33, 34], and is what leads to the proof of Theorem 3.1. There we adopt the view that embedding calculus is very well suited to calculating—or estimating—the rational *homotopy* groups of $B\mathrm{Emb}^{\cong}_{\partial_+ W_{g,1}}(W_{g,1}, W_{g,1})$, and so we propose to calculate—or estimate—the rational *homotopy* groups of $B\mathrm{Diff}_\partial(W_{g,1})$ (in fact, we consider the framed version $B\mathrm{Diff}^{\mathrm{fr}}_\partial(W_{g,1})$, but again the difference on rational homotopy groups is very mild). Describing these is an interesting problem in its own right, especially in view of Berglund and Madsen's [1] calculation of the rational homotopy and stable cohomology of the groups of block diffeomorphisms and of homotopy automorphisms of $W_{g,1}$. In the remainder I will focus on this calculation, and not try to explain exactly how it implies Theorem 3.1.

4.4. Torelli groups

Diffeomorphisms of $W_{g,1}$ induce automorphisms of $H_n(W_{g,1};\mathbb{Z})$ which preserve the intersection form, giving a homomorphism

$$\alpha_g : \mathrm{Diff}_\partial(W_{g,1}) \to G_g := \begin{cases} \mathrm{O}_{g,g}(\mathbb{Z}), & n \text{ even}, \\ \mathrm{Sp}_{2g}(\mathbb{Z}), & n \text{ odd}. \end{cases}$$

This is surjective if n is even or $n = 1, 3, 7$, but for other odd values of n has image a certain finite-index subgroup G'_g. By analogy with the case $2n = 2$, the kernel of α_g is called the *Torelli group* and denoted $\mathrm{Tor}_\partial(W_{g,1})$, so that there is a fibration sequence

$$B\mathrm{Tor}_\partial(W_{g,1}) \to B\mathrm{Diff}_\partial(W_{g,1}) \to BG'_g. \tag{4.7}$$

Now the fundamental group of $B\mathrm{Diff}_\partial(W_{g,1})$ is quite complicated, as it surjects onto the arithmetic group G'_g, so although the results of [13, 15] describe the rational cohomology of this space, there is no reason to think that this has much to do with its rational higher homotopy groups. However, by (4.7) these higher homotopy groups are the same as those of $B\mathrm{Tor}_\partial(W_{g,1})$, and, as long as $2n \geq 6$, the Weiss fibre sequence can be used (in "qualitative mode") to prove that the space $B\mathrm{Tor}_\partial(W_{g,1})$ is nilpotent [31, THEOREM C]. Thus $B\mathrm{Tor}_\partial(W_{g,1})$ has a meaningful rationalisation, and its rational homotopy and cohomology groups are closely related (in the sense that there are spectral sequences computing each from the other). On the other hand, passing to the infinite covering space $B\mathrm{Tor}_\partial(W_{g,1})$ of $B\mathrm{Diff}_\partial(W_{g,1})$ has an unknown effect on cohomology, so the problem is now to understand the rational cohomology of $B\mathrm{Tor}_\partial(W_{g,1})$.

Cohomology of Torelli groups. The fibration (4.7) provides a representation of the arithmetic group G'_g on the rational vector spaces $H^i(B\mathrm{Tor}_\partial(W_{g,1}); \mathbb{Q})$, and, as long as $2n \geq 6$ and $g \geq 2$, a further application of the Weiss fibre sequence in "qualitative mode" shows [31, THEOREM A] that these are *algebraic representations* of G'_g, i.e. they extend to representations of the ambient algebraic group $\mathbf{G}_g \in \{\mathbf{O}_{g,g}, \mathbf{Sp}_{2g}\}$. As $\mathbf{G}_g$-representations are semisimple, and the irreducibles are classified in terms of Young diagrams and are all summands of tensor powers of the defining $\mathbf{G}_g$-representation $H := H_n(W_{g,1}; \mathbb{Q})$, the G'_g-representation $H^i(B\mathrm{Tor}_\partial(W_{g,1}); \mathbb{Q})$ may be determined in terms of the vector spaces

$$\left[H^i(B\mathrm{Tor}_\partial(W_{g,1}); \mathbb{Q}) \otimes H^{\otimes S}\right]^{G'_g} \tag{4.8}$$

for all finite sets S, and the structure maps between them given by applying permutations and contractions $H \otimes H \to \mathbb{Q}$. On the other hand, the vector spaces (4.8) are related, by the Serre spectral sequence for (4.7), to the cohomology groups

$$H^*\left(B\mathrm{Diff}_\partial(W_{g,1}); \mathcal{H}^{\otimes S}\right) \tag{4.9}$$

with coefficients in the Sth tensor power of the local system $\mathcal{H}$ on $B\mathrm{Diff}_\partial(W_{g,1})$ provided by the G'_g-representation H. Using work of Borel [4] this Serre spectral sequence can be shown to degenerate: Ebert and I [8] introduced this strategy, but as the results of [13,15] only describe the cohomology of $B\mathrm{Diff}_\partial(W_{g,1})$ with constant coefficients, we were only able to use it to determine $[H^i(B\mathrm{Tor}_\partial(W_{g,1}); \mathbb{Q})]^{G'_g}$ in a stable range. However, the results of [13,15] also apply to $B\mathrm{Diff}^\theta_\partial(W_{g,1})$ for quite arbitrary tangential structures θ (such as framings, but also including "maps to a space Y"), and exploiting the functoriality of the result with respect to θ (this kind of argument originates in [38]) it is possible to calculate (4.9) in a stable range, and hence by the strategy outlined here to calculate $H^*(B\mathrm{Tor}_\partial(W_{g,1}); \mathbb{Q})$ in a stable range of degrees, as a $\mathbb{Q}$-algebra and as a G'_g-representation. This is done in [33].

A similar strategy can be applied to $B\mathrm{Diff}_\partial^{\mathrm{fr}}(W_{g,1})$, though by a subtlety in the proof the argument above does not directly calculate the cohomology of $B\mathrm{Tor}_\partial^{\mathrm{fr}}(W_{g,1})$. Instead, there is a certain fibration sequence

$$X_1(g) \to B\mathrm{Tor}_\partial^{\mathrm{fr}}(W_{g,1}) \to X_0$$

with X_0 a loop space having rational cohomology $\Lambda_{\mathbb{Q}}[\bar{\sigma}_{4j-2n-1} \mid j > n/2]$. The $\bar{\sigma}_{4j-2n-1}$ are secondary characteristic classes associated to the fact that the family signature vanishes for two different reasons on $B\mathrm{Tor}_\partial^{\mathrm{fr}}(W_{g,1})$: because of the framing, and because of the triviality of the action on $H^n(W_{g,1};\mathbb{Z})$. The analogue of the argument above leads to the following description of the cohomology of the nilpotent space $X_1(g)$. For $r \geq 3$ and $v_1, \ldots, v_r \in H^n(W_{g,1};\mathbb{Q})$, there are defined *twisted Miller–Morita–Mumford classes*

$$\kappa_1(v_1 \otimes \cdots \otimes v_r) \in H^{(r-2)n}(X_1(g);\mathbb{Q}),$$

which satisfy:

(i) linearity in each v_i,

(ii) $\kappa_1(v_{\sigma(1)} \otimes \cdots \otimes v_{\sigma(r)}) = \mathrm{sign}(\sigma)^n \cdot \kappa_1(v_1 \otimes \cdots \otimes v_r)$,

(iii) $\sum_i \kappa_1(v_1 \otimes \cdots \otimes v_{k-1} \otimes a_i) \smile \kappa_1(a_i^\# \otimes v_k \otimes \cdots \otimes v_r) = \kappa_1(v_1 \otimes \cdots \otimes v_r)$,

(iv) $\sum_i \kappa_1(v_1 \otimes \cdots \otimes v_r \otimes a_i \otimes a_i^\#) = 0$,

where $\sum_i a_i \otimes a_i^\# \in H^n(W_{g,1};\mathbb{Q})^{\otimes 2}$ is dual to the intersection form. The framed analogue of the group G'_g acts on $\kappa_1(v_1 \otimes \cdots \otimes v_r)$ by its evident action on the $v_i \in H^n(W_{g,1};\mathbb{Q})$. Kupers and I [32] show that, in a range of degrees tending to infinity with g, the cohomology algebra of $X_1(g)$ is generated by the classes $\kappa_1(v_1 \otimes \cdots \otimes v_r)$ and subject only to the relations (i)–(iv).

Homotopy of Torelli groups. As the rational cohomology of $X_1(g)$ is supported in degrees divisible by n in a stable range, it follows formally that its rational homotopy groups are supported in degrees $\bigcup_{r\geq 1}[r(n-1)+1, rn]$ in this stable range, so exhibit a band pattern.

But it turns out that we can do a lot better. It is not hard to see that the above data in fact presents a quadratic algebra, generated by the elements $\kappa_1(v_1 \otimes v_2 \otimes v_3)$ of degree n (modulo (iv)), and it is then tempting to ask whether this quadratic algebra is Koszul. Kupers and I [34] prove that $H^*(X_1(g);\mathbb{Q})$ is indeed Koszul in a stable range of degrees (this was simultaneously proved by Felder, Naef, and Willwacher [10]), so it follows that in this range $\pi_*(X_1(g)) \otimes \mathbb{Q}$ is in fact supported in degrees of the form $r(n-1)+1$, and is furthermore given by the quadratic dual Lie algebra. Up to a few extension questions, this calculates $\pi_*(B\mathrm{Diff}_\partial(W_{g,1})) \otimes \mathbb{Q}$ in a stable range.

ACKNOWLEDGEMENTS

It is a pleasure to thank those with whom I have collaborated on the ideas described here: Mauricio Bustamante, Johannes Ebert, Søren Galatius, Manuel Krannich, and Alexander Kupers.

FUNDING

The author was supported by the ERC under the European Union's Horizon 2020 research and innovation programme (grant agreement No. 756444), and by a Philip Leverhulme Prize from the Leverhulme Trust.

REFERENCES

[1] A. Berglund and I. Madsen, Rational homotopy theory of automorphisms of manifolds. *Acta Math.* **224** (2020), no. 1, 67–185.

[2] P. Boavida de Brito and M. Weiss, Spaces of smooth embeddings and configuration categories. *J. Topol.* **11** (2018), no. 1, 65–143.

[3] A. Borel, Stable real cohomology of arithmetic groups. *Ann. Sci. Éc. Norm. Supér. (4)* **7** (1974), 235–272.

[4] A. Borel, Stable real cohomology of arithmetic groups. II. In *Manifolds and Lie groups (Notre Dame, Ind., 1980)*, pp. 21–55, Progr. Math. 14, Birkhäuser, Boston, Mass., 1981.

[5] B. Botvinnik and N. Perlmutter, Stable moduli spaces of high-dimensional handlebodies. *J. Topol.* **10** (2017), no. 1, 101–163.

[6] M. Bustamante, M. Krannich, and A. Kupers, Finiteness properties of automorphism spaces of manifolds with finite fundamental group. 2021, arXiv:2103.13468.

[7] M. Bustamante and O. Randal-Williams, On automorphisms of high-dimensional solid tori. 2020, arXiv:2010.10887.

[8] J. Ebert and O. Randal-Williams, Torelli spaces of high-dimensional manifolds. *J. Topol.* **8** (2015), no. 1, 38–64.

[9] F. T. Farrell and W.-C. Hsiang, On the rational homotopy groups of the diffeomorphism groups of discs, spheres and aspherical manifolds. In *Algebraic and geometric topology, Stanford 1976*, pp. 325–337, Proc. Sympos. Pure Math. XXXII, Amer. Math. Soc., Providence, RI, 1976.

[10] M. Felder, F. Naef, and T. Willwacher, Stable cohomology of graph complexes. 2021, arXiv:2106.12826.

[11] B. Fresse, V. Turchin, and T. Willwacher, The rational homotopy of mapping spaces of E_n operads. 2017, arXiv:1703.06123.

[12] N. Friedrich, Homological stability of automorphism groups of quadratic modules and manifolds. *Doc. Math.* **22** (2017), 1729–1774.

[13] S. Galatius and O. Randal-Williams, Stable moduli spaces of high-dimensional manifolds. *Acta Math.* **212** (2014), no. 2, 257–377.

[14] S. Galatius and O. Randal-Williams, Homological stability for moduli spaces of high dimensional manifolds. II. *Ann. of Math. (2)* **186** (2017), no. 1, 127–204.

[15] S. Galatius and O. Randal-Williams, Homological stability for moduli spaces of high dimensional manifolds. I. *J. Amer. Math. Soc.* **31** (2018), no. 1, 215–264.

[16] T. G. Goodwillie, A multiple disjunction lemma for smooth concordance embeddings. *Mem. Amer. Math. Soc.* **86** (1990), no. 431, viii+317.

[17] T. G. Goodwillie and J. R. Klein, Multiple disjunction for spaces of smooth embeddings. *J. Topol.* **8** (2015), no. 3, 651–674.

[18] T. G. Goodwillie and M. Weiss, Embeddings from the point of view of immersion theory. II. *Geom. Topol.* **3** (1999), 103–118 (electronic).

[19] J. Grunewald, J. R. Klein, and T. Macko, Operations on the A-theoretic nil-terms. *J. Topol.* **1** (2008), no. 2, 317–341.

[20] J. L. Harer, Stability of the homology of the mapping class groups of orientable surfaces. *Ann. of Math. (2)* **121** (1985), no. 2, 215–249.

[21] K. Igusa, The stability theorem for smooth pseudoisotopies. *K-Theory* **2** (1988), no. 1–2, 1–355.

[22] K. Igusa, *Higher Franz–Reidemeister torsion*. AMS/IP Stud. Adv. Math. 31, Amer. Math. Soc., Providence, RI/International Press, Somerville, MA, 2002.

[23] R. C. Kirby and L. C. Siebenmann, *Foundational essays on topological manifolds, smoothings, and triangulations*. Ann. of Math. Stud. 88, Princeton University Press, Princeton, NJ/University of Tokyo Press, Tokyo, 1977.

[24] B. Knudsen and A. Kupers, Embedding calculus and smooth structures. 2020, arXiv:2006.03109.

[25] M. Kontsevich, Feynman diagrams and low-dimensional topology. In *First European Congress of Mathematics, Vol. II (Paris, 1992)*, pp. 97–121, Progr. Math. 120, Birkhäuser, Basel, 1994.

[26] M. Krannich, A homological approach to pseudoisotopy theory. I. *Invent. Math.* (to appear), arXiv:2002.04647.

[27] M. Krannich and O. Randal-Williams, Mapping class groups of simply connected high-dimensional manifolds need not be arithmetic. *C. R. Math. Acad. Sci. Paris* **358** (2020), no. 4, 469–473.

[28] M. Krannich and O. Randal-Williams, Diffeomorphisms of discs and the second Weiss derivative of $B\mathrm{Top}(-)$. 2021, arXiv:2109.03500.

[29] M. Kreck, Isotopy classes of diffeomorphisms of $(k-1)$-connected almost-parallelizable $2k$-manifolds. In *Algebraic topology, Aarhus 1978*, pp. 643–663, Lecture Notes in Math. 763, Springer, Berlin, 1979.

[30] A. Kupers, Some finiteness results for groups of automorphisms of manifolds. *Geom. Topol.* **23** (2019), 2277–2333 (electronic).

[31] A. Kupers and O. Randal-Williams, The cohomology of Torelli groups is algebraic. *Forum Math. Sigma* **8** (2020), e64.

[32] A. Kupers and O. Randal-Williams, On diffeomorphisms of even-dimensional discs. 2020, arXiv:2007.13884.

[33] A. Kupers and O. Randal-Williams, On the cohomology of Torelli groups. *Forum Math. Pi* **8** (2020), e7.

[34] A. Kupers and O. Randal-Williams, On the Torelli Lie algebra. 2021, arXiv:2106.16010.

[35] I. Madsen and M. Weiss, The stable moduli space of Riemann surfaces: Mumford's conjecture. *Ann. of Math. (2)* **165** (2007), no. 3, 843–941.

[36] N. Perlmutter, Homological stability for diffeomorphism groups of high-dimensional handlebodies. *Algebr. Geom. Topol.* **18** (2018), no. 5, 2769–2820.

[37] O. Randal-Williams, An upper bound for the pseudoisotopy stable range. *Math. Ann.* **368** (2017), no. 3–4, 1081–1094.

[38] O. Randal-Williams, Cohomology of automorphism groups of free groups with twisted coefficients. *Selecta Math. (N.S.)* **24** (2018), no. 2, 1453–1478.

[39] R. Reis and M. Weiss, Rational Pontryagin classes and functor calculus. *J. Eur. Math. Soc. (JEMS)* **18** (2016), no. 8, 1769–1811.

[40] D. Sullivan, Infinitesimal computations in topology. *Publ. Math. Inst. Hautes Études Sci.* **47** (1977), 269–331.

[41] F. Waldhausen, Algebraic K-theory of spaces. In *Algebraic and geometric topology (New Brunswick, NJ, 1983)*, pp. 318–419, Lecture Notes in Math. 1126, Springer, Berlin, 1985.

[42] F. Waldhausen, B. Jahren, and J. Rognes, *Spaces of PL manifolds and categories of simple maps*. Ann. of Math. Stud. 186, Princeton University Press, Princeton, NJ, 2013.

[43] T. Watanabe, On Kontsevich's characteristic classes for higher dimensional sphere bundles. I. The simplest class. *Math. Z.* **262** (2009), no. 3, 683–712.

[44] T. Watanabe, On Kontsevich's characteristic classes for higher-dimensional sphere bundles. II. Higher classes. *J. Topol.* **2** (2009), no. 3, 624–660.

[45] T. Watanabe, Erratum to: On Kontsevich's characteristic classes for higher-dimensional sphere bundles. II. Higher classes. 2018, https://www.math.kyoto-u.ac.jp/~tadayuki.watanabe/kon2-erratum.pdf.

[46] T. Watanabe, Some exotic nontrivial elements of the rational homotopy groups of $\mathrm{Diff}(S^4)$. 2018, arXiv:1812.02448.

[47] T. Watanabe, Addendum to: Some exotic nontrivial elements of the rational homotopy groups of $\mathrm{Diff}(S^4)$ (homological interpretation). 2021, arXiv:2109.01609.

[48] M. Weiss, Orthogonal calculus. *Trans. Amer. Math. Soc.* **347** (1995), no. 10, 3743–3796.

[49] M. Weiss, Embeddings from the point of view of immersion theory. I. *Geom. Topol.* **3** (1999), 67–101 (electronic).

[50] M. Weiss, Dalian notes on Pontryagin classes. *Geom. Topol.* (to appear), arXiv:1507.00153.

OSCAR RANDAL-WILLIAMS

Centre for Mathematical Sciences, Wilberforce Road, Cambridge CB3 0WB, UK, or257@cam.ac.uk

This Page Intentionally Left Blank

International Congress of Mathematicians (ICM 2022)

Sections 5-8
Volume 4 of 7

Online
6-14 July 2022

Part B

Editors:

Dmitry Beliaev
Stanislav Smirnov

ISBN: 978-1-7138-9871-9

ICM INTERNATIONAL CONGRESS OF MATHEMATICIANS 2022 JULY 6—14

SECTIONS 5–8

EDITED BY D. BELIAEV AND S. SMIRNOV

INTERNATIONAL MATHEMATICAL UNION IMU

EMS PRESS

Editors

Dmitry Beliaev
Mathematical Institute
University of Oxford
Andrew Wiles Building
Radcliffe Observatory Quarter
Woodstock Road
Oxford OX2 6GG, UK

Email: belyaev@maths.ox.ac.uk

Stanislav Smirnov
Section de mathématiques
Université de Genève
rue du Conseil-Général 7–9
1205 Genève, Switzerland

Email: stanislav.smirnov@unige.ch

2020 Mathematics Subject Classification: 00B25

ISBN 978-3-98547-058-7, eISBN 978-3-98547-558-2, DOI 10.4171/ICM2022

Volume 1. Prize Lectures
ISBN 978-3-98547-059-4, eISBN 978-3-98547-559-9, DOI 10.4171/ICM2022-1
Volume 2. Plenary Lectures
ISBN 978-3-98547-060-0, eISBN 978-3-98547-560-5, DOI 10.4171/ICM2022-2
Volume 3. Sections 1–4
ISBN 978-3-98547-061-7, eISBN 978-3-98547-561-2, DOI 10.4171/ICM2022-3
⟶ **Volume 4. Sections 5–8**
ISBN 978-3-98547-062-4, eISBN 978-3-98547-562-9, DOI 10.4171/ICM2022-4
Volume 5. Sections 9–11
ISBN 978-3-98547-063-1, eISBN 978-3-98547-563-6, DOI 10.4171/ICM2022-5
Volume 6. Sections 12–14
ISBN 978-3-98547-064-8, eISBN 978-3-98547-564-3, DOI 10.4171/ICM2022-6
Volume 7. Sections 15–20
ISBN 978-3-98547-065-5, eISBN 978-3-98547-565-0, DOI 10.4171/ICM2022-7

This edition printed by Curran Associates, Inc.
ISBN 978-1-7138-9871-9

e-Book available: https://ems.press/books/standalone/276

No changes have been made to the content of these proceedings. There may be changes to pagination and minor adjustments for aesthetics.

Bibliographic information published by the Deutsche Nationalbibliothek
The Deutsche Nationalbibliothek lists this publication in the Deutsche Nationalbibliografie; detailed bibliographic data are available on the Internet at http://dnb.dnb.de.

Published by EMS Press, an imprint of the

European Mathematical Society – EMS – Publishing House GmbH
Institut für Mathematik
Technische Universität Berlin
Straße des 17. Juni 136
10623 Berlin, Germany

https://ems.press

CONTENTS

VOLUME 1

PRIZE LECTURES

POPULAR SCIENTIFIC EXPOSITIONS

SUMMARIES OF PRIZE WINNERS' WORK

VOLUME 2

SPECIAL PLENARY LECTURES

PLENARY LECTURES

VOLUME 3

1. LOGIC

2. ALGEBRA

3. NUMBER THEORY – SPECIAL LECTURE

3. NUMBER THEORY

4. ALGEBRAIC AND COMPLEX GEOMETRY – SPECIAL LECTURE

4. ALGEBRAIC AND COMPLEX GEOMETRY

VOLUME 4

5. GEOMETRY – SPECIAL LECTURES

5. GEOMETRY

6. TOPOLOGY

7. LIE THEORY AND GENERALIZATIONS

8. ANALYSIS – SPECIAL LECTURE

8. ANALYSIS

VOLUME 5

9. DYNAMICS

10. PARTIAL DIFFERENTIAL EQUATIONS

11. MATHEMATICAL PHYSICS – SPECIAL LECTURE

11. MATHEMATICAL PHYSICS

VOLUME 6

12. PROBABILITY – SPECIAL LECTURE

12. PROBABILITY

13. COMBINATORICS – SPECIAL LECTURE

13. COMBINATORICS

14. MATHEMATICS OF COMPUTER SCIENCE – SPECIAL LECTURES

14. MATHEMATICS OF COMPUTER SCIENCE

VOLUME 7

15. NUMERICAL ANALYSIS AND SCIENTIFIC COMPUTING

18. STOCHASTIC AND DIFFERENTIAL MODELLING

19. MATHEMATICAL EDUCATION AND POPULARIZATION OF MATHEMATICS

20. HISTORY OF MATHEMATICS

FLOER HOMOLOGY OF 3-MANIFOLDS WITH TORUS BOUNDARY

JACOB RASMUSSEN

ABSTRACT

Manifolds with torus boundary have played a special role in the study of Floer homology for 3-manifolds since the early days of the subject. In joint work with Jonathan Hanselman and Liam Watson, we defined a geometrical Heegaard Floer invariant for 3-manifolds with torus boundary. The invariant is a reformulation of the bordered Floer homology of Lipshitz, Ozsváth, and Thurston, and takes the form of a collection of immersed closed curves (possibly decorated with local systems) in a covering space of the punctured torus. We briefly discuss the construction of the invariant and some applications to the L-space conjecture of Boyer–Gordon–Watson and Juhász. We then describe a generalization to manifolds with sutured boundary, and some applications to the study of satellite knots.

MATHEMATICS SUBJECT CLASSIFICATION 2020

Primary 57R58; Secondary 57K18, 57K31, 57K10

KEYWORDS

Heegaard Floer homology, three-manifold, toroidal boundary, satellite, knot

ICM INTERNATIONAL CONGRESS OF MATHEMATICIANS 2022 JULY 6–14

Proc. Int. Cong. Math. 2022, Vol. 4, pp. 2880–2902
DOI 10.4171/ICM2022/103

1. INTRODUCTION

In the seminal papers [16,17], Andreas Floer created two branches of the theory which now bear his name. The first branch is concerned with symplectic geometry, and provides invariants of symplectomorphisms and Lagrangian submanifolds. The second branch, with which we will be concerned, provides invariants of 3-manifolds. In addition to Floer's work, which uses SU_2 instantons, there are many different approaches to defining Floer homology for 3-manifolds, including the monopole Floer homology of Kronheimer and Mrowka [35] or Hutchings' theory of embedded contact homology [29]. We will mainly use the Heegaard Floer homology of Ozsváth and Szabó [44] which is easy to compute and relatively simple to work with from a technical standpoint. Regardless of their definition, all Floer theories assign an abelian group to a closed connected oriented 3-manifold Y. The key question we will be concerned with is:

"What is the Floer homology of a 3-manifold M with $\partial M \simeq T^2$?"

Here our main criterion for defining the Floer homology is that if $Y = M_1 \cup_{T^2} M_2$, we should be able to recover the Floer homology of Y from the Floer homologies of M_1 and M_2.

1.1. The view from 1992

It is illuminating to consider the situation for Floer's original instanton homology, as it was understood 30 years ago. Let $I^*(Y)$ be the homology of a chain complex $\mathrm{CI}^*(Y)$, which is generated by irreducible flat SU_2 connections on Y if things are nice enough. By considering the holonomy representation, we see that the set of flat connections is in bijection with the SU_2 character variety of Y:

$$X_{SU_2}(Y) = \{\rho : \pi_1(Y) \to SU_2\}/\, SU_2$$

where SU_2 acts on the set of representations by conjugation, $(A \cdot \rho)(x) = A\rho(x)A^{-1}$. A representation ρ is *reducible* if its image is contained in an abelian subgroup of SU_2; otherwise, it is *irreducible*. Any reducible representation can be factored as $\pi_1(Y) \to H_1(Y) \to S^1 \subset SU_2$, so if $H_1(Y) = 0$, the unique irreducible representation is the trivial one.

If $\partial M \simeq T^2$, we can likewise consider the character variety $X_{SU_2}(M)$. The inclusion $i_* : \pi_1(\partial M) \to \pi_1(M)$ induces a map $i^* : X_{SU_2}(M) \to X_{SU_2}(\partial M)$. Since $\pi_1(T^2) \simeq \mathbb{Z}^2$ is abelian, every $\rho : \pi_1(T^2) \to SU_2$ is reducible. Any two 1-parameter subgroups in SU_2 are conjugate, and the stabilizer of a fixed 1-parameter subgroup is the Weyl group $W = \mathbb{Z}/2$. It follows that

$$X_{SU_2}(T^2) = \{\rho : \mathbb{Z}^2 \to S^1\}/W = T^2/(\mathbb{Z}/2)$$

is the *pillowcase orbifold* $T^2/(x \sim -x)$. Figure 1 shows the image $i^*(X_{SU_2}(M))$ for two simple 3-manifolds, namely $M = S^1 \times D^2$ and $M = M_{T_{2,3}}$ – the exterior of the right-hand trefoil knot in S^3.

If $Y = M_1 \cup_{T^2} M_2$, the inclusions $i_{j*} : \pi_1(T^2) \to \pi_2(M_j)$ induce maps $i_j^* : X_{SU_2}(M_j) \to X_{SU_2}(T^2)$. The simplest example of such a gluing is a Dehn filling, where $M_2 = S^1 \times D^2$. In this case, it is easy to see:

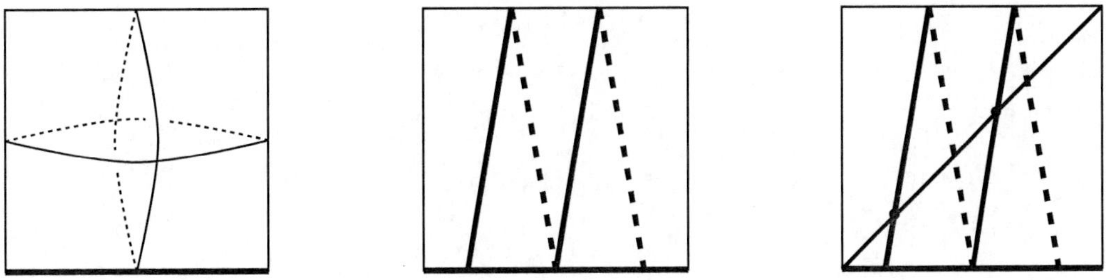

FIGURE 1
The pillowcase orbifold $X_{\mathrm{SU}_2}(T^2)$ is shown on the left. The solid line at the bottom is $i^*(X_{\mathrm{SU}_2}(S^1 \times D^2))$. The middle figure shows the image of $X_{\mathrm{SU}_2}(M_{T_{2,3}})$, which consists of a reducible part (the line at the bottom of the figure) and an irreducible part (the line segment of slope 6). The figure on the right shows the intersection between the two character varieties corresponding to $+1$ surgery on the trefoil.

Lemma 1.1. *If* $Y = M_1 \cup_{T^2} (S^1 \times D^2)$, *then* $X(Y)$ *is naturally identified with the fiber product* $X_{\mathrm{SU}_2}(M_1) \times_{X_{\mathrm{SU}_2}(T^2)} X_{\mathrm{SU}_2}(S^1 \times D^2)$.

The Poincaré sphere is the result of $+1$ surgery on $T_{2,3}$. The corresponding fiber product is illustrated on the right-hand side of Figure 1. There are 3 intersection points (circled) between $X_{\mathrm{SU}_2}(M_{T_{2,3}})$ and $X_{\mathrm{SU}_2}(S^1 \times D^2)$, which tells us that $X_{\mathrm{SU}_2}(P)$ consists of two irreducible characters and a single reducible character.

It is tempting to consider the image $i^*(X_{\mathrm{SU}_2}(M)) \subset X_{\mathrm{SU}_2}(\partial M)$ as a proxy for the Floer homology of M. However, a closer consideration of this picture reveals many difficulties:

- How should the reducible flat connections be treated? If Y is a homology sphere, the only reducible connection is the trivial one, which we can afford to ignore. As soon as $H_1(Y) \neq 0$, this is no longer feasible.
- If instead of taking $M_2 = S^1 \times D^2$ we use another 3-manifold, the fiber product in Lemma 1.1 becomes more complicated. Each intersection between irreducible points in $i_1^*(X_{\mathrm{SU}_2}(M_1))$ and $i_2^*(X_{\mathrm{SU}_2}(M_2))$ gives an entire circle of irreducible flat connections in $X_{\mathrm{SU}_2}(Y)$.
- Perhaps most importantly, $X_{\mathrm{SU}_2}(Y)$ is only the set of generators for $\mathrm{CI}^*(Y)$. To compute the homology, we must understand the differential, which involves counting solutions to the SU_2 ASD equation on $Y \times \mathbb{R}$. *A priori,* there is no reason to believe that the character variety should tell us anything about this.

Despite these problems, there were reasons for optimism as well. Indeed, $\mathrm{CI}^*(Y)$ was $\mathbb{Z}/8$ graded, and the reduction of this grading to $\mathbb{Z}/2$ agreed with the sign of intersection in the fiber product. Hence the two irreducible generators for $\mathrm{CI}^*(P)$ have the same $\mathbb{Z}/2$ grading, and there were no differentials in the chain complex. Fintushel and Stern [15] showed that the same was true for any Seifert fibered space.

The second, very potent reason was Floer's exact triangle, which related the Floer homologies of different Dehn fillings of M. Suppose $K \subset Y_\infty$ is a null-homologous knot, and let Y_0 and Y_1 be the manifolds obtained by 0 and 1 surgery on K. Then we have:

Theorem 1.2 ([8, 18]). *There is a long exact sequence*

$$\cdots \to I^*(Y_\infty) \to I^*(Y_0) \to I^{*-1}(Y_1) \to I^{*-1}(Y_\infty) \to \cdots .$$

The original motivation for this theorem was the relation between character varieties shown in Figure 2, and the fact that it holds suggests that our naive idea for thinking about the Floer homology of M in terms of the picture provided by the character variety might have something to it after all.

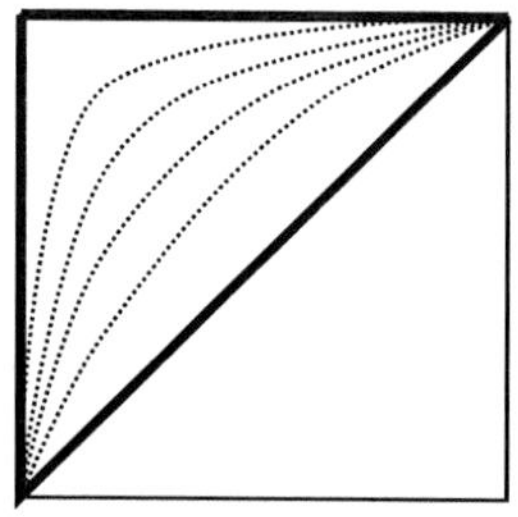

FIGURE 2
The pillowcase with curves corresponding to $+1$, 0, and ∞ surgery slopes. The $+1$ curve can be continuously deformed to the union of the other two, producing a chain complex which computes $\mathrm{HI}^*(Y_1)$ but whose generators are the union of the generators of $\mathrm{CI}^*(Y_0)$ and $\mathrm{CI}^*(Y_\infty)$.

1.2. The modern perspective

In 30 years, we have made a lot of progress. Many of the technical difficulties associated with instanton theory have been simplified or elided, first by the appearance of Seiberg–Witten theory [53] and then by the development of Heegaard Floer homology [44]. The problems associated with reducibles (such as the first two points above) have been addressed in several ways: by working with appropriate equivariant versions of the theory, as in [35]; by restricting to sectors in which reducibles do not appear [36]; or by dividing out by the based gauge group (or something similar) rather than the full gauge group [34].

We will focus on the Heegaard Floer invariant $\widehat{\mathrm{HF}}$, which, roughly speaking, corresponds to the monopole Floer invariant obtained by using the based gauge group instead of the full gauge group. If (Y, z) is a closed, connected, oriented, pointed 3-manifold, $\widehat{\mathrm{HF}}(Y, z)$ is a finite-dimensional vector space over the field $\mathbb{F} = \mathbb{Z}/2$. Some parts of the theory are also known to work with $\mathbb{Z}$ coefficients, but we will stick to $\mathbb{Z}/2$ coefficients throughout. If $z, z' \in Y$, there is a diffeomorphism $\psi : Y \to Y$ with $\psi(z) = z'$, so $\widehat{\mathrm{HF}}(Y, z) \simeq \widehat{\mathrm{HF}}(Y, z')$, but this isomorphism is not canonical, as was first observed by Juhász [32]. (The point z corresponds to the point used to define the based gauge group in Seiberg–Witten theory.)

Suppose that $Y = M_1 \cup_\Sigma M_2$, where Σ is a connected surface containing z. In this situation, the yoga of extended TQFTs suggests that to Σ we should associate an additive category $\mathcal{A}(\Sigma)$, that M_1 and M_2 should determine objects $\mathcal{A}(M_i)$ of $\mathcal{A}(\Sigma)$, and that we should have

$$\widehat{\mathrm{HF}}(Y, z) \cong \mathrm{Hom}\big(\mathcal{A}(M_1), \mathcal{A}(M_2)\big).$$

This picture was realized by Lipshitz, Ozsváth, and Thurston [39] in their seminal work on *bordered Floer homology*. They described the category $\mathcal{A}(\Sigma)$ in terms of algebraic objects which they called Type D and Type A structures. Their work was given a beautiful geometrical interpretation by Auroux [2,3] who showed that $\mathcal{A}(\Sigma)$ can be interpreted as the *partially wrapped Fukaya category* of the symmetric product $\mathrm{Sym}^g(\Sigma - z)$.

Auroux's result is very important from a philosophical standpoint, but its practical applications are limited by the difficulties of working with the Fukaya category of $\mathrm{Sym}^g(\Sigma - z)$. Naively, the objects of the Fukaya category are Lagrangian submanifolds, but in reality one must also consider arbitrary mapping cones built up out of Lagrangians. The algebra required to do this is essentially the same as that of Type A and Type D structures invented by Lipshitz, Ozsváth, and Thurston.

The one exception to this rule is the case $g = 1$, where we can give a simple geometric description of the (compactly supported) Fukaya category of $T^2 - z$. We say that a curve in $T^2 - z$ is *nice* if it is an immersed closed curve which is unobstructed, in the sense that it bounds no monogon in $T^2 - z$. Then we can formulate a version of the (compactly supported) Fukaya category, which we denote by $\mathcal{F}(T^2 - z)$. The objects of $\mathcal{F}(T^2 - z)$ are finite unions of nice curves equipped with local systems, and the group $\mathrm{Hom}(\gamma_1, \gamma_2)$ can be computed combinatorially. In particular, if γ_1 and γ_2 are primitive nonisotopic curves in $T^2 - z$, $\mathrm{Hom}(\gamma_1, \gamma_2)$ is determined by the minimal geometric intersection number $i(\gamma_1, \gamma_2)$.

Let $\overline{\mathcal{F}}(T^2 - z)$ be the set of isomorphism classes of objects in $\mathcal{F}(T^2 - z)$. Together with Hanselman and Watson, we proved the following theorem, which realizes the geometrical hope expressed in the previous section in the context of $\widehat{\mathrm{HF}}$:

Theorem 1.3 ([21,22]). *If (M, z) is a closed, connected, oriented, and pointed 3-manifold with $z \in \partial M \simeq T^2$, there is a well-defined invariant $\widehat{\mathrm{HF}}(M, z) \in \overline{\mathcal{F}}(\partial M - z)$, which satisfies*

$$\widehat{\mathrm{HF}}(M_1 \cup_{T^2} M_2, z) \simeq \mathrm{Hom}\big(\widehat{\mathrm{HF}}(M_1, z), \widehat{\mathrm{HF}}(M_2, z)\big).$$

In this context, the fact that $\widehat{\mathrm{HF}}$ satisfies an exact triangle analogous to that of Theorem 1.2 (proved by Ozsváth and Szabó in [43]) is a consequence of the fact that the lines of slope 0, 1, and ∞ in T^2 form an exact triangle in the Fukaya category. This argument is due to Lipshitz, Ozsváth, and Thurston [39].

The invariant $\widehat{\mathrm{HF}}(M)$ can be effectively computed in many examples; for example, by the work of F. Ye [54], it is known for all but 9 of the 286 orientable 1-cusped hyperbolic manifolds in the SnapPy census of hyperbolic 3-manifolds built from 5 or fewer tetrahedra [11,27].

Some examples of $\widehat{\mathrm{HF}}$ for simple 3-manifolds are shown in Figure 3. Each curve in the figure lives in an infinite cylinder obtained by identifying the dashed lines on the left- and right-hand sides. To pass to the invariant in $T^2 - z$, we divide out by the obvious $\mathbb{Z}$ action. Part (d) shows the pairing between $\widehat{\mathrm{HF}}(M_T)$ and $\widehat{\mathrm{HF}}(S^1 \times D^2)$ corresponding to $+1$ surgery on the trefoil. There is a single intersection point, which matches the fact that $\widehat{\mathrm{HF}}(P) \cong \mathbb{Z}/2$.

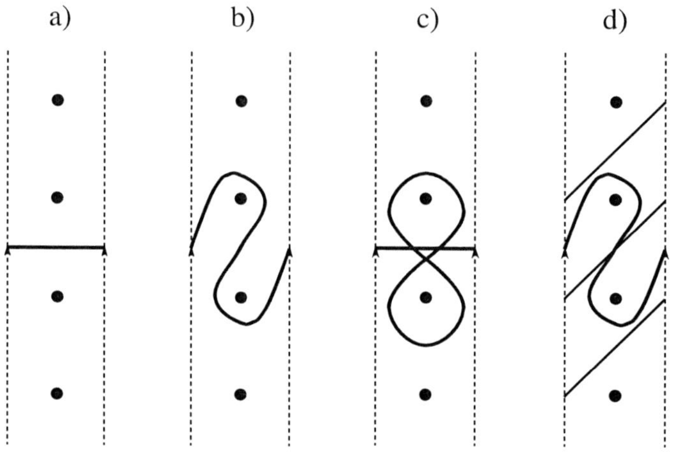

FIGURE 3

Heegaard Floer invariant $\widehat{\mathrm{HF}}$ of some simple manifolds, including (a) $S^1 \times D^2$, (b) the exterior of the right-hand trefoil, and (c) the exterior of the figure-eight knot. The dots indicate punctures.

We close this section with a question about instanton Floer homology. Kronheimer and Mrowka have defined **[36]** an invariant $I^{\#}(Y)$ which is an instanton analog of $\widehat{\mathrm{HF}}(Y)$ and is conjectured to be isomorphic to it. It is thus very natural to ask:

Question 1.4. Is there an instanton analog of the invariant $\widehat{\mathrm{HF}}(M)$, and, if so, can it be related directly to $X_{\mathrm{SU}_2}(M)$?

It is probably too much to ask for a direct relation with the character variety in every case, but one might still hope for it in some simple examples. (See **[22]** for something along these lines using the Seiberg–Witten moduli space.)

In the sections that follow, we will briefly explain how the bordered Floer homology of a manifold with torus boundary can be reinterpreted to define the invariant $\widehat{\mathrm{HF}}(M)$, describe some applications of the theorem, and discuss generalizations and further directions.

2. CONSTRUCTION AND PROPERTIES OF THE INVARIANT

2.1. The Fukaya category

We begin with a brief and imprecise account of the Fukaya category. For more careful discussions, we refer the reader to **[4, 51, 52]**. Suppose that (M, ω) is an exact symplectic manifold. Taken naively, objects of the Fukaya category $\mathcal{F}(M)$ are Lagrangian submanifolds $L_i \subset M$, and $\mathrm{Hom}(L_1, L_2) = \mathrm{HF}(L_1, L_2)$ is Lagrangian Floer homology—the other sort of homology invented by Floer. The Floer chain complex is generated by intersections between L_1 and L_2, and the differential is given by counting J-holomorphic disks with respect to a compatible almost-complex structure. This is an oversimplification for many reasons: first, $\mathcal{F}(M)$ is an A_∞-category, with higher morphisms given by counts of holomorphic polygons with higher numbers of sides; and second, $\mathcal{F}(M)$ is triangulated, so a typical object is actually a *twisted complex*—an iterated mapping cone built out of geometric Lagrangians.

All this extra structure may seem daunting to the newcomer, but it has its advantages. Although there are usually infinitely many different Lagrangians in M, in many cases it is possible to show that every object of $\mathcal{F}(M)$ is isomorphic to a twisted complex built up out of a finite number of Lagrangians $L_1, \ldots, L_n$. In this case the L_i are said to generate the Fukaya category. This is easiest to arrange in the case where both M and the L_i are noncompact. In fact, M should be a Liouville manifold, so that near infinity it looks like the symplectization of a contact manifold N. In this situation, we need to be more careful about what is meant by $\mathrm{Hom}(L_i, L_j)$. The correct answer turns out to be the *wrapped Floer homology*, in which we replace L_i by its image under a flow determined by the Reeb flow on N. More generally, we can consider the *partially wrapped Floer homology* [4] in which the flow is stopped on some $X \subset N$.

If $L_1, \ldots, L_n$ generate, we define $\mathcal{L} = \bigoplus_i L_i$, and consider the A_∞ algebra $\mathcal{A} = \mathrm{End}(\mathcal{L})$. If L is an object of $\mathcal{F}(M)$, we can consider $\mathcal{M}_L = \mathrm{Hom}(\mathcal{L}, L)$, which is an A_∞ module over $\mathcal{A}$. By the Yoneda embedding lemma, L and $\mathcal{M}_L$ carry the same information [2].

2.2. Bordered Floer homology

Next, we discuss the work of Lipshitz, Ozsváth, and Thurston [39].

Definition 2.1. Let Σ be a closed, connected, and oriented surface. A *parametrization* $\mathcal{P}$ of Σ is a minimal handle decomposition of Σ, together with a choice of basepoint z on the boundary of the 2-handle. A *bordered 3-manifold* $(M, \mathcal{P})$ is a compact, connected, and oriented 3-manifold M, together with a parametrization $\mathcal{P}$ of ∂M.

Up to isotopy, $\mathcal{P}$ is specified by the position of the 2-handle and the cocores of its 1-handles. These form a system of disjoint arcs $\alpha_1, \ldots, \alpha_{2g} \subset \Sigma$ with ends on the boundary of the 2-handle. To a parametrized surface $(\Sigma, \mathcal{P})$, Lipshitz, Ozsváth, and Thurston associate an explicit A_∞ algebra $\mathcal{A}(\mathcal{P})$. (In fact, $\mathcal{A}(\mathcal{P})$ is a dga: $\mu_i = 0$ for all $i > 2$.) They also define the notions of Type D and Type A structures over $\mathcal{A}(\mathcal{P})$. In the language of Section 2.1, a (bounded) Type D structure is essentially a twisted complex over the category determined by $\mathcal{A}(P)$. The Type A structure corresponding to a Type D structure $\mathcal{D}$ is essentially $\mathrm{Hom}(\mathcal{A}, \mathcal{D})$. Thus the relation between Type D and Type A structures is the same as the relation between L and $\mathcal{M}_L$ in the Fukaya category. The main theorem of bordered Floer homology is:

Theorem 2.2 ([38, 39]). *A bordered 3-manifold $(M, \mathcal{P})$ determines a Type D structure $\widehat{\mathrm{CFD}}(M, \mathcal{P})$ over $\mathcal{A}(\mathcal{P})$ which is well defined up to quasiisomorphism. If $Y = M_1 \cup_\Sigma M_2$ and $\mathcal{P}$ is a parametrization of Σ, then $\widehat{\mathrm{HF}}(Y) \cong \mathrm{Hom}(\widehat{\mathrm{CFD}}(M_1, \mathcal{P}), \widehat{\mathrm{CFD}}(M_2, \mathcal{P}))$.*

Suppose Σ is a parametrized surface, and let $\Sigma_0 \subset \Sigma$ be the complement of the 2-handle. If $I = \{i_1, \dots, i_g\}$ is a g-element subset of $\{1, \dots, 2g\}$, we define L_I to be the image of $\alpha_{i_i} \times \cdots \times \alpha_{i_g}$ in $\mathrm{Sym}^g \Sigma_0$. The L_I are noncompact Lagrangian submanifolds of $M = \mathrm{Sym}^g \Sigma_0$. In addition, the point $z \in \partial\Sigma_0$ determines a stop X_z for M. Auroux proved:

Theorem 2.3 ([2, 3]). *The L_I generate $\mathcal{F}_{X_z}(\mathrm{Sym}^g(\Sigma_0))$, and $\mathcal{A}(\mathcal{P}) \cong \mathrm{End}\, \mathcal{L}$, where $\mathcal{L} = \oplus L_I$.*

Hence $\widehat{\mathrm{CFD}}(M, \mathcal{P})$ determines an object of $\mathcal{F}_{X_z}(\mathrm{Sym}^g(\Sigma_0))$. A priori, this object is neither compact nor geometric—it is a twisted complex built up out of noncompact Lagrangians.

2.3. The torus

Up to isotopy, the torus T^2 has a unique parametrization, $\mathcal{P}$, as shown in Figure 4. The corresponding algebra $\mathcal{A}(T^2) = \mathcal{A}(\mathcal{P})$ is a quotient of the quiver algebra generated by the quiver below by the quadratic relations $\rho_2\rho_1 = \rho_3\rho_2 = 0$. Geometrically speaking, the arrows in the quiver correspond to the labeled arcs in on the boundary of the punctured torus, as shown in Figure 4. Composition is given by concatenation (when possible) and is 0 otherwise. We write $\rho_1\rho_2 = \rho_{12}$, etc.

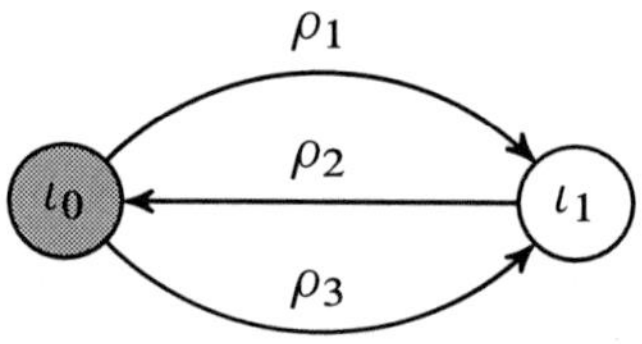

A Type D structure over $\mathcal{A}(T^2)$ can be represented by a decorated graph, whose vertices are labeled by idempotents of $\mathcal{A}$ (we use $\bullet$ for L_0, and $\circ$ for L_1) and whose edges are labeled by morphisms. The labels on the edges determine the differential D in the twisted chain complex, which must satisfy $D^2 = 0$. Here is an example of a twisted complex built out of three objects—one copy of L_0 and two of L_1:

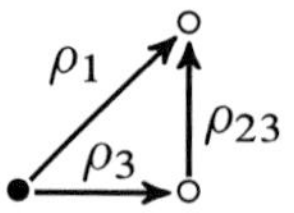

The key step in the proof of Theorem 1.3 is an algebraic structure theorem, which shows that every Type D structure over $\mathcal{A}(T^2)$ is homotopy equivalent to that with a particularly nice form.

Definition 2.4. A Type D structure over is a *loop* if its underlying graph (forgetting labels and orientations) is a cycle. More generally, a Type D structure with graph G is a *loop with*

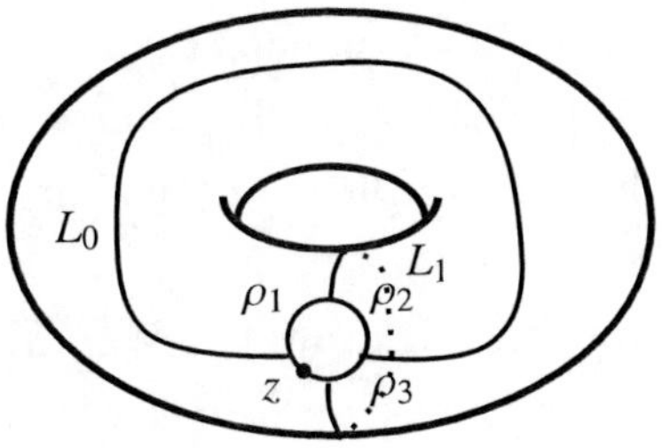

FIGURE 4
Parametrization of T^2, showing the 2-handle (large circle), α arcs (labeled L_0 and L_1), and basepoint.

a local system of dimension k if there is a loop L and a map $\pi : G \to L$ which preserves the labels edges and vertices and is a k-to-1 covering map away from one edge of L.

The torus algebra is a quotient of a slightly larger algebra $\widetilde{\mathcal{A}}$, which is obtained by adding in a generator ρ_0 corresponding to the arc that runs over the basepoint z, and setting any word that contains two copies of ρ_0 to 0, as well as the usual quadratic relations. An important result due to Lipshitz, Ozsváth, and Thurston is that if $\partial M = T^2$, then $\widehat{\mathrm{CFD}}(M)$ is extendable, that is, we can add in additional arrows labeled by elements of $\widetilde{\mathcal{A}}$ so that

$$D^2 = \sum_{j=0}^{3} \rho_j \rho_{j+1} \rho_{j+2} \rho_{j+3},$$

where the subscripts are to be interpreted modulo 4. The main technical result of [21] is:

Theorem 2.5. *An extendable Type D structure over* $\mathcal{A}$ *is homotopy equivalent to a disjoint union of loops with local systems.*

The first theorem of this type was proved by Haiden, Katzarkov, and Kontsevich [19], who showed that any twisted complex over $\mathcal{A}(T^2)$ (or more generally, the algebra associated to the Fukaya category of a higher genus surface) is a direct sum of loops with local systems and chains. The key role that such loops play in the study of bordered Floer homology was first observed by Hanselman and Watson in [23]. In [21] we give an effective algorithm for reducing an arbitrary extendable Type D structure to a disjoint union of loops. Alternately, one can appeal to [19], and then use the fact that the Type D structure is extendable to rule out the presence of any chains.

The final step in the proof of Theorem 1.3 is to associate a geometric loop $\gamma_{\mathcal{D}}$ (a closed curve in $T^2 \setminus z$) to a loop-type Type D structure $\mathcal{D}$, and show that

$$\mathrm{Hom}(\mathcal{D}_1, \mathcal{D}_2) \cong \mathrm{HF}(\gamma_{\mathcal{D}_1}, \gamma_{\mathcal{D}_2}).$$

Here the left-hand side is Hom in the category of Type D structures, and the right-hand side is an appropriately formulated version of Floer homology in $T^2 - z$; $\Gamma_{\mathcal{D}}$ is constructed by taking a straight line segment for each object in the loop, and joining the ends of consecutive objects according to the label on the arrow that joins them. There are two ways to do this.

In [19], the authors use the noncompact Lagrangians L_0, L_1, and join them by arcs along the boundary. In [21] we take a dual approach, using the compact arcs coming from the cores of the 1-handles and joining them by curves along the boundary of the 0-handle.

2.4. Spin^c structures and the Alexander polynomial

In this section, we review three basic properties of $\widehat{\mathrm{HF}}$ for closed 3-manifolds, and explain their generalization to manifolds with torus boundary. First, it is well known that $\widehat{\mathrm{HF}}(Y)$ can be decomposed according to the set of Spin^c structures on Y:

$$\widehat{\mathrm{HF}}(Y) = \bigoplus_{\mathfrak{s}\in \mathrm{Spin}^c(Y)} \widehat{\mathrm{HF}}(Y, \mathfrak{s}).$$

The same statement is true when M has torus boundary:

$$\widehat{\mathrm{HF}}(M) = \bigoplus_{\mathfrak{s}\in \mathrm{Spin}^c(M)} \widehat{\mathrm{HF}}(M, \mathfrak{s}),$$

where now the direct sum is taken in the Fukaya category, where it is given by disjoint union of curves. More interestingly, each summand $\widehat{\mathrm{HF}}(M, \mathfrak{s})$ can be lifted to a covering space of $\partial M - z$. To be precise, let $\overline{T}_M$ be the covering space of $T_M = \partial M$ whose fundamental group is the kernel of the composite map $\pi_1(\partial M) \to H_1(\partial M) \to H_1(M)$. Let $T_M^\circ = \partial M - z$, and define $\overline{T}_M^\circ$ to be its preimage in $\overline{T}_M$. It is shown in [21] that $\widehat{\mathrm{HF}}(M, \mathfrak{s})$ lifts to $\overline{T}_M^\circ$. For example, if $H_1(M) \cong \mathbb{Z}$, $\overline{T}_M^\circ$ is a punctured cylinder like those shown in Figure 3. There is a unique $\mathfrak{s} \in \mathrm{Spin}^c(M, \partial M)$, and the curves shown in Figure 3 are $\widehat{\mathrm{HF}}(M, \mathfrak{s})$ for their respective M's.

Suppose $Y = M_1 \cup_{T^2} M_2$. By considering the pairing of the lifted curves for $\widehat{\mathrm{HF}}(M_1)$ and $\widehat{\mathrm{HF}}(M_2)$, together with the action of the deck group, one can recover the Spin^c decomposition on $\widehat{\mathrm{HF}}(Y)$. Figure 5(b) illustrates this computation for 0 surgery on the torus knot $T(2,5)$.

Second, $\widehat{\mathrm{HF}}(Y)$ carries a natural $\mathbb{Z}/2$ grading. For manifolds with torus boundary, we have the following analog:

Proposition 2.6. *If M is a 3-manifold with torus boundary, then there is natural orientation on $\widehat{\mathrm{HF}}(M)$. If $Y = M_1 \cup_{T^2} M_2$ and x is a generator of $\widehat{\mathrm{HF}}(Y)$ corresponding to an intersection point of $\widehat{\mathrm{HF}}(M_1)$ and $\widehat{\mathrm{HF}}(M_2)$, then the $\mathbb{Z}/2$ grading of $x \in \widehat{\mathrm{HF}}(M_1) \cap \widehat{\mathrm{HF}}(M_2)$ is given by the sign of intersection of $\widehat{\mathrm{HF}}(M_1)$ and $\widehat{\mathrm{HF}}(M_2)$ at x.*

Finally, it is well known (going back to Casson [1]) that the Euler characteristic of Floer homology is related to the Alexander polynomial. We describe this relation in our context, restricting to the case where $H_1(M) = \mathbb{Z}$ for simplicity. Let $\pi : \overline{T}_M \to \partial M$ be the projection. The set $\pi^{-1}(z)$ can be naturally identified with $\mathbb{Z}$ by the action of the deck group. The space $\overline{T}_M$ has two ends: a positive end to which the z_n converge as $n \to \infty$ and a negative end to which the z_n converge as $n \to -\infty$. Let η_n be a path from z_n to the negative end, and define a_n to be the signed intersection number of η_n with $\widehat{\mathrm{HF}}(M)$. (Since $\widehat{\mathrm{HF}}(M)$ is compact, $z_n = 0$ for $n \ll 0$.) Then we have:

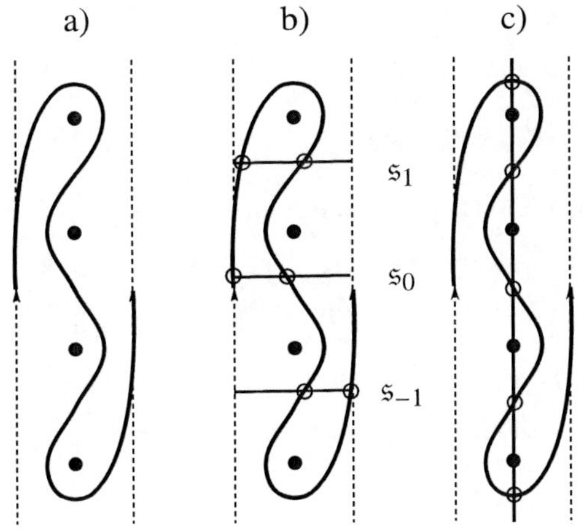

FIGURE 5
Some computations with the (2, 5) torus knot: (a) $\widehat{\mathrm{HF}}(M_{T(2,5)})$, (b) $\widehat{\mathrm{HF}}$ of 0-surgery on $T(2,5)$ has dimension 2 in each of the Spin^c structures $\mathfrak{s}_{-1}, \mathfrak{s}_0$, and $\mathfrak{s}_1$, and (c) $\widehat{\mathrm{HFK}}(T(2,5), i)$ has dimension 1 for $i = -2, -1, 0, 1, 2$.

Proposition 2.7 ([21]). *If $\Delta(M) \in \mathbb{Z}[t^{\pm 1}]$ is the Alexander polynomial, then*

$$\frac{\Delta_M}{1-t} \sim \sum_{n \in \mathbb{Z}} a_n t^n.$$

Here both sides are to be interpreted as Laurent series, and $\sim$ indicates equality up to multiplication by some power of t. The quantity on the left-hand side is the *Milnor torsion* of M. For example, by referring to Figure 3, one can easily compute that

$$\frac{\Delta(T)}{1-t} \sim t^{-1} + t + t^3 + t^4 + t^5 + \cdots,$$

corresponding to the well-known fact that $\Delta(T) = t^{-1} - 1 + t$.

2.5. Knot Floer homology

The definition of Heegaard Floer homology for closed 3-manifolds can be generalized to give an invariant of a pair $K \subset Y$, where K is a knot in Y. This invariant is called knot Floer homology (written $\widehat{\mathrm{HFK}}(K)$), and was discovered by Ozsváth and Szabó [42] and independently by the author [48]. Two basic properties of knot Floer homology are:

- If $b_1(Y) = 0$, $\widehat{\mathrm{HFK}}(K)$ splits as a direct sum $\widehat{\mathrm{HFK}}(K) = \bigoplus_{i \in \mathbb{Z}} \widehat{\mathrm{HFK}}(K, i)$. The grading i is called the *Alexander grading* and satisfies

$$\sum_i \chi\big(\widehat{\mathrm{HFK}}(K, i)\big) \cdot t^i \sim \Delta_K(t).$$

- If $b_1(Y) = 0$, there are two spectral sequences with E_1 term $\widehat{\mathrm{HFK}}(K)$ which converge to $\widehat{\mathrm{HF}}(Y)$. In one sequence, the differentials decrease the Alexander grading, while in the other they increase it.

If $\partial M = T^2$ and α is a simple closed curve on ∂M, we can form the Dehn filling $M(\alpha) = M \cup_\phi S^1 \times D^2$, where $\phi_*([\partial D^2]) = [\alpha]$. We consider the *core knot*

$$K_\alpha = S^1 \times 0 \subset S^1 \times D^2 \subset M(\alpha)$$

whose complement is again M. Let L'_α be the (noncompact) Lagrangian in $\partial M \setminus z$ which consists of a line of slope α passing through z. Then we have, the following result, which is essentially due to Lipshitz, Ozsváth, and Thurston:

Proposition 2.8. $\widehat{\mathrm{HFK}}(K_\alpha) \simeq \mathrm{HF}(\widehat{\mathrm{HF}}(M), L'_\alpha)$. *Conversely, if* $K \subset S^3$ *and we understand both sets of differentials on* $\widehat{\mathrm{HFK}}(K)$, *we can reconstruct* $\widehat{\mathrm{HF}}(M_K)$.

Since $\widehat{\mathrm{HF}}(M)$ is compact, the pairing $\mathrm{HF}(\widehat{\mathrm{HF}}(M), L'_\alpha)$ can still be computed by taking the minimal number of intersections between the two curves. By way of comparison, if L_α is the compact line of slope α in ∂M, then $\mathrm{HF}(\widehat{\mathrm{HF}}(M), L'_\alpha) \cong \widehat{\mathrm{HF}}(M(\alpha))$.

As in the previous section, we can understand the Alexander grading by passing to the lift $\widehat{\mathrm{HF}}(M, \mathfrak{s}) \subset \overline{T}^{\circ}_M$. For simplicity, we again restrict to the case where $H_1(M) \cong \mathbb{Z}$. The Lagrangian L'_α is homeomorphic to an open interval, so the set of lifts to $\overline{T}_M$ can be labeled as $L_{\alpha,i}$ for $i \in \mathbb{Z}$. With an appropriate choice of labeling,

$$\widehat{\mathrm{HFK}}(K_\alpha, i) \cong \mathrm{HF}\big(\widehat{\mathrm{HF}}(M), L_{\alpha,i}\big).$$

From this perspective, the spectral sequences from $\widehat{\mathrm{HFK}}(K_\alpha)$ to $\widehat{\mathrm{HF}}(M(\alpha))$ arise from the fact that if we push $\bigcup_i L'_{\alpha,i}$ off of the preimage of z, we get $\pi^{-1}(L_\alpha)$. The fact that there are two such sequences corresponds that we can push either to the left or to the right. From a more algebraic point of view, as an object of the Fukaya category, $\pi^{-1}(L_\alpha)$ is isomorphic to the filtered complex

$$\cdots \xrightarrow{\rho_{12}} L'_{\alpha,2} \xrightarrow{\rho_{12}} L'_{\alpha,1} \xrightarrow{\rho_{12}} L'_{\alpha,0} \xrightarrow{\rho_{12}} L'_{\alpha,-1} \xrightarrow{\rho_{12}} L'_{\alpha,-2} \xrightarrow{\rho_{12}} \cdots$$

whose associated grading is $\bigoplus_{i \in \mathbb{Z}} L'_{\alpha,i}$.

3. FLOER SIMPLE MANIFOLDS AND THE L-SPACE GLUING THEOREM

3.1. Floer simple manifolds

We say that Y is an *L-space* if Y is a rational homology sphere and $\dim \widehat{\mathrm{HF}}(Y, \mathfrak{s}) = 1$ for each $\mathfrak{s} \in \mathrm{Spin}^c(Y)$. Since $\chi(\widehat{\mathrm{HF}}(Y, \mathfrak{s}) = 1$, this is as small as it can be, and L-spaces are the closed manifolds with the simplest possible Floer homology. In this section, we discuss the analogous notion for manifolds with torus boundary.

If M is such a manifold, let $\mathrm{Sl}(M)$ be the set of possible Dehn filling slopes on M. Then $\mathrm{Sl}(M)$ is naturally identified with the projective space on $H_1(\partial M; \mathbb{Z})$. By choosing a basis of $H_1(M; \mathbb{Z})$, we can identify $\mathrm{Sl}(M)$ with the rational projective space $\mathbb{QP}^1$. Let

$$\mathcal{L}(M) = \{\alpha \in \mathrm{Sl}(M) \mid M(\alpha) \text{ is an L-space}\}$$

be the set of L-space Dehn filling slopes of M. With S. Rasmussen, we proved

Theorem 3.1 ([49]). *If $\partial M \cong T^2$ and $b_1(M) = 1$, $\mathcal{L}(M)$ is one of the following:*

- *the empty set,*
- *a single point*
- *a closed interval with rational endpoints in $\mathbb{QP}^1$, or*
- *$\mathbb{QP}^1 \setminus [\ell]$ where ℓ is the rational longitude.*

Definition 3.2. Manifold M is *Floer simple* if $\mathcal{L}(M)$ contains more than one element. In this case we call $\mathcal{L}(M)$ the *L-space interval* and write $\mathcal{L}^\circ(M)$ for its interior.

If M is Floer simple, $\widehat{\mathrm{HF}}(M)$ is easy to describe. If $\alpha \in Sl(M)$, let $n_\alpha = \alpha \cdot \ell$ and consider the map $p_\alpha : T_M \to \mathbb{R}/(n_\alpha)$ given by $p_\alpha(x) = \alpha \cdot x$. For $\mathfrak{s} \in \mathrm{Spin}^c(M)$, let $\gamma_{M,\mathfrak{s}}$ be curve obtained by pulling $\widehat{\mathrm{HF}}(M, \mathfrak{s})$ tight. Then we have:

Proposition 3.3 ([21]). *Manifold M is Floer simple with $\alpha \in \mathcal{L}^\circ(M)$ if and only if p_α maps $\gamma_{M,\mathfrak{s}}$ bijectively to $\mathbb{R}/(n_\alpha)$ bijectively for all $\mathfrak{s} \in \mathrm{Spin}^c(M)$.*

If M is Seifert fibered with $b_1 = 1$, then M is Floer simple, and the class of the fiber slope is in $\mathcal{L}^\circ(Y)$. But many hyperbolic 3-manifolds are Floer simple as well. In [12], Dunfield studied Burton's census [9] of all 59,068 1-cusped hyperbolic 3-manifolds which have $b_1 = 1$ and admit an ideal triangulation with 9 or fewer tetrahedra. He found that 50,598 of them were Floer simple and 8,352 were not, leaving only 118 where he was unable to decide. It is natural to ask if the condition of being Floer simple has any geometrical meaning. By applying the fibration detection theorem of Ni [41], it is easy to see

Proposition 3.4. *If M is a Floer simple manifold with $H_1(M) \cong \mathbb{Z}$, then M fibers over the circle.*

Conversely, we could ask if there is a geometric characterization of a the monodromy of a fibered Floer simple manifold. To make the question precise, write $\mathrm{Mod}_{g,1}$ for the mapping class group of a genus g surface with one puncture. For $\phi \in \mathrm{Mod}_{g,1}$, let M_ϕ be the mapping torus of ϕ, and define $\mathcal{FS}_g = \{\phi \in \mathrm{Mod}_{g,1} \mid M_\phi \text{ is Floer simple}\}$.

Question 3.5. Describe $\mathcal{FS}_g$ as a subset of $\mathrm{Mod}_{g,1}$.

When $g = 1$, $\mathrm{Mod}_{1,1} \cong \mathrm{SL}_2(\mathbb{Z})$. Using Baldwin's work on the Floer homology of genus-one fibered manifolds [5], it is not difficult to see that

$$\mathcal{FS}_1 = \{A \in \mathrm{SL}_2(\mathbb{Z}) \mid \operatorname{tr} A \le 1\}.$$

In other words, $A \in \mathrm{SL}_2(\mathbb{Z})$ is a Floer simple monodromy if A is elliptic or negatively hyperbolic or parabolic, but not if A is positively parabolic or hyperbolic. For $g > 1$, virtually nothing is known. It would be interesting to know if $g = 1$ is typical (in the sense that $\mathcal{FS}_g$ forms a large subset of $\mathrm{Mod}_{g,1}$) or atypical (in the sense that $\mathcal{FS}_g$ is relatively sparse.)

In another direction, if M is Floer simple, $\mathcal{L}(M)$ forms a distinguished interval in the circle of slopes. If we know a single point of $\mathcal{L}^\circ(M)$, the entire interval can be determined from the Turaev torsion of M [49], but from a purely homological perspective, it is difficult to say anything about $\mathcal{L}(M)$. We have a distinguished slope given by the homological longitude ℓ which is not contained in $\mathcal{L}(M)$, but otherwise $\text{Sl}(M)$ looks quite homogenous. If M is Seifert fibered, then the fiber slope is a distinguished element of $\mathcal{L}(M)$. When M is a 1-cusped hyperbolic manifold, there is also a lot more geometry available: the hyperbolic metric on M naturally induces a flat metric (the cusp metric) on ∂M or, equivalently, on $H_1(\partial M)$, and we can talk about shortest geodesics, length of curves, the degeneracy slope, etc.

Question 3.6. If M is hyperbolic, can $\mathcal{L}(M)$ be related to the geometry of the induced metric on the cusp?

In this direction, there is an interesting unpublished observation of T. Brown, who pointed out that for many (but not all) manifolds in Burton's census, $\ell^{\perp}$ (the slope orthogonal to the homological longitude with respect to the cusp metric) is contained in $\mathcal{L}(M)$.

3.2. L-space gluings

Our work on the immersed curve picture for Floer homology arose out of earlier attempts [23,49] to understand the L-space conjecture of Boyer–Gordon–Watson and Juhász. This conjecture posits a surprising relation between Heegaard Floer homology and the fundamental group. To be precise, we say that a nontrivial group G is left orderable if there is a total order $<$ on G satisfying $gx < gy$ whenever $x < y$. (By convention, the trivial group is not left orderable.)

Conjecture 3.7 (The L-space conjecture [7,31]). *If Y is prime, the following statements are equivalent:*

- *Y is an not an L-space,*
- *$\pi_1(Y)$ is left-orderable,*
- *Y admits a coorientable taut foliation.*

The notion of the L-space interval, along with similar sets for fibered and non-left orderable fillings, was first introduced by Boyer and Clay [6], who used it to study the L-space conjecture for graph manifolds. Building on their work, we proved

Theorem 3.8 ([20,50]). *The L-space conjecture holds for graph manifolds.*

There are two independent proofs of this theorem—one of them by S. Rasmussen [50], and the other by Hansel, Watson, and Rasmussen.[2] Based on their work, we conjectured the following *L-space Gluing Theorem*, which was proved in [21].

Theorem 3.9. *$Y = M_1 \cup_{T^2} M_2$ is an L-space if and only if $\mathcal{L}^{\circ}(M_1) \cup \mathcal{L}^{\circ}(M_2) = \text{Sl}(T^2)$. (In particular, if Y is an L-space, both M_1 and M_2 must be Floer simple.)*

When M_1 and M_2 are Floer simple, partial results in this direction were obtained in [23, 49], but the proof of the full result relies in an essential way on Theorem 1.3. As a corollary, we were able to reprove the following result of Eftekhary:

Corollary 3.10 ([14]). *An L-space homology sphere cannot contain an incompressible torus.*

4. LINKS, SATELLITES, AND SUTURES

The results of Theorem 1.3 can be extended without much difficulty to describe a broad class of manifolds with sutured boundary. We describe an application of this method to the knot Floer homology of satellite knots.

4.1. Sutured manifolds

Sutured Floer homology is a very important variant of Heegaard Floer homology introduced by Juhasz [30]. A *balanced sutured manifold* is a compact, oriented 3-manifold with boundary, together with a multicurve γ (the suture) which divides ∂M into two parts R_+ and R_- such that (a) $\chi(R_+) = \chi(R_-)$ and (b) each component of ∂M contains at least one component of γ. If (M, γ) is a balanced sutured manifold, its sutured Floer homology $\mathrm{SFH}(M, \gamma)$ is a vector space over $\mathbb{Z}/2$.

Both $\widehat{\mathrm{HF}}$ and $\widehat{\mathrm{HFK}}$ appear as special cases of SFH. If (Y, z) is a connected, pointed 3-manifold, we define $Y(1) \subset Y$ to be the complement of a small ball in centered at z, and let $\gamma \subset \partial Y$ be a simple closed curve on ∂Y. Then $\mathrm{SFH}(Y(1), \gamma) \cong \widehat{\mathrm{HF}}(Y)$. More generally, if $Y(n)$ is the analogous manifold where we remove n balls, we have

$$\mathrm{SFH}\big(Y(n), \gamma\big) \cong \widehat{\mathrm{HF}}(Y) \otimes H^*(S^1)^{\otimes n}.$$

Similarly, if $K \subset Y$ has meridian μ, we let $M_K \subset Y$ be the exterior of K and define γ_μ to be two parallel copies of μ in ∂M_K. Then $\mathrm{SFH}(M_K, \gamma_\mu) \cong \widehat{\mathrm{HFK}}(K)$.

Zarev [55] extended the framework of bordered Floer homology to the class of *bordered sutured manifolds*. Such a manifold consists of a compact oriented 3-manifold M together with a decomposition $\partial M = F \cup R$, where F is the bordered (or glueable) part of the boundary, and R is sutured, that is, it is equipped with a multicurve γ which divides R into two parts R_+ and R_-. We do not impose a condition on the Euler characteristic of $R_\pm$, but do require that each boundary component of R intersects both R_+ and R_-.

Following Auroux, Zarev's bordered sutured Floer homology can be interpreted as defining an object in a partially wrapped Fukaya category of $\mathrm{Sym}^k(F)$ for some k, where the set of stops used to define the wrapping is determined by the set of intersections of R_+ with ∂R. The usual bordered Floer homology corresponds to the special case where R is a small disk centered at z which is divided in half by a single arc.

Suppose that F is a once punctured torus. When R_+ intersects ∂F in a single interval and $\chi(R_+) = \chi(R_-)$, the bordered sutured Floer homology can be interpreted as an object of the partially wrapped Fukaya category of (a covering space of) F. Rather than discussing this situation in generality, we will focus on a special case. Suppose that $\partial M \simeq T^2_a \cup T^2_b$, and give M the bordered sutured where $R = R_\mu$ consists of all of T^2_b, equipped with a pair of parallel sutures of slope μ, together with a disk in T^2_a which is divided in half by a single suture. Then we have

Proposition 4.1. *The pair* (M, R_μ) *determines* $\widehat{\mathrm{HF}}(M, R_\mu)$*, which is a compactly supported object of* $\mathcal{F}(T^2_a - z)$*. As in the closed case,* $\widehat{\mathrm{HF}}(M, R_\mu)$ *can be represented by a union of immersed closed curves equipped with local systems.*

Zarev's gluing theorem [55] then implies that if M' is a manifold with torus boundary and $\phi : T^2_a \to \partial M$ is an orientation reversing diffeomorphism, then

$$\mathrm{HF}\big(\widehat{\mathrm{HF}}(M'), \widehat{\mathrm{HF}}(M, R_\mu)\big) \cong \mathrm{SFH}(M' \cup_\phi M, \gamma_\mu) \otimes H^*(S^1). \tag{4.1}$$

The extra factor of $H^*(S^1)$ appears because the sutured manifold obtained by gluing (M, R_μ) and (M', R_z) together has *three* boundary components—there is an extra bubble in the middle coming from the two sutured disks. To get $M' \cup_\phi M$ without the bubble, we would need to use a bordered sutured manifold $(M(\eta), R_{\eta,\mu})$ which is constructed by choosing a framed path η from a point on γ_μ to z, removing a tubular neighborhood of η, and using the framing to extend the sutures over the boundary of the tubular neighborhood. The difference between (M, R_μ) and $(M(\eta), R_{\eta,\mu})$ is illustrated in Figure 6.

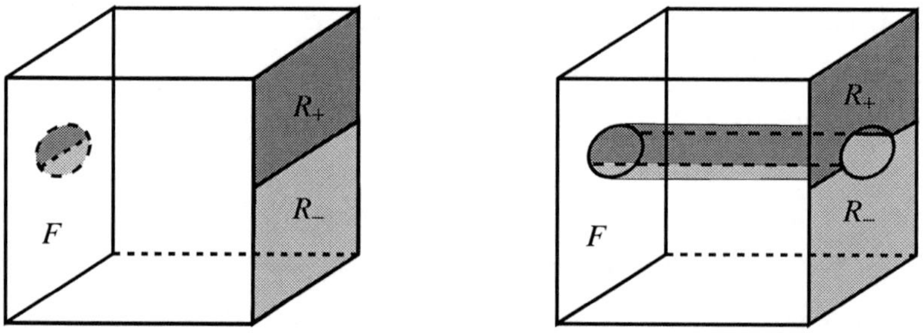

FIGURE 6
The bordered sutured manifolds (M, R_μ) (on the left) and $(M(\eta), R_{\eta,\mu})$ (on the right). The left- and right-hand faces of the cubes are part of the boundary of M. All the other faces are in the interior of M.

4.2. Link complements

We now specialize to the situation where $M = M_L$ is the exterior of a 2-component link $L \subset S^3$, and $\mu = \mu_2$ is the meridian of the second component of L. (We label the meridians and longitude of L_i by μ_i, λ_i.) Let $(M_L, \gamma_{\mu_1,\mu_2})$ be the sutured manifold with meridinal sutures on both boundary components. Then

$$\mathrm{SFH}(M_L, \gamma_{\mu_1,\mu_2}) = \widehat{\mathrm{HFL}}(L)$$

is the link Floer homology defined by Ozsváth and Szabó in [45]. In the same way that we can compute $\widehat{\mathrm{HF}}(M_K)$ from $\widehat{\mathrm{HFK}}(K)$ (and the differentials on it) when K is a knot in S^3, we can compute $\widehat{\mathrm{HF}}(M_L, R_{\mu_2})$ from $\widehat{\mathrm{HFL}}(L)$ (and the differentials on it) when L_2 is the unknot in S^3.

We will describe some examples, but before doing so, we pause to discuss Spin^c structures and lifts. As before, the Floer homology decomposes as

$$\widehat{\mathrm{HF}}(M_L, R_\mu) = \bigoplus_{\mathfrak{s} \in \mathrm{Spin}^c(M_L, R_\mu)} \widehat{\mathrm{HF}}(M_L, R_\mu, \mathfrak{s}).$$

Here $\mathrm{Spin}^c(M_L, R_\mu)$ is an affine set modeled on $\operatorname{coker} i_{a*}$, where $i_{a*} : \pi_1(F) \to H_1(M_L)$. It is easy to see that $\operatorname{coker} i_{a*} \cong \mathbb{Z}/n$, where n is the linking number of L. Also as before, $\widehat{\mathrm{HF}}(M_L, R_\mu, \mathfrak{s})$ lifts to the covering space $\overline{T}^\circ_M$ of $T^\circ_M := F$ given by $\pi_1(\overline{T}_M) = \ker i_{a*}$. Note that $\overline{T}^\circ_M$ is also determined by the linking number: it is the universal abelian cover of T°_M if $n \neq 0$, but an infinite punctured cylinder if $n = 0$.

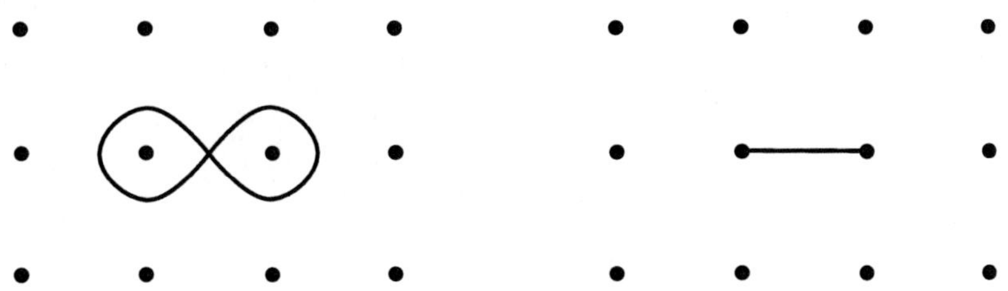

FIGURE 7

Curves when L is the Hopf link: $\widehat{\mathrm{HF}}(M_L, R_\mu)$ (left) and $\widehat{\mathrm{HF}}(M_L(\eta), R_{\eta,\mu})$ (right).

Example 4.2. If L is the Hopf link, then $M_L = T^2 \times I$. The linking number is 1, so there is a unique Spin^c structure and $\overline{T}_M$ is the universal abelian cover of T_M; $\widehat{\mathrm{HF}}(M_L, R_\mu) = \gamma_1$ is shown on the left-hand side of Figure 7. In this case there is a canonical choice of the path η, namely $z \times I$. With this choice, $\widehat{\mathrm{HF}}(M_L(\eta), R_{\eta,\mu}) = \gamma_2$ is shown on the right. The vector defined by the line segment is $\lambda_1 = \mu_2$ (the homology class of the suture.) Note that γ_1 is obtained by "inflating" γ_2 to form a figure-eight. It is not hard to see that $\mathrm{HF}(\gamma, \gamma_1) = HF(\gamma, \gamma_2) \otimes H^*(S^1)$ for any closed curve γ, as predicted by equation (4.1).

Example 4.3. If L is the $(2, 4)$ torus link, the linking number is 2, and there are 2 distinct Spin^c structures. In each case $\widehat{\mathrm{HF}}(M_L, R_\mu, \mathfrak{s})$ consists of a single figure-eight obtained by inflating a line segment. In one Spin^c structure the segment represents the vector μ_1, and in the other it represents the vector $\mu_1 + \lambda_1$.

Example 4.4. If L is the positive Whitehead link, the linking number is 0, so we have Spin^c structures $\mathfrak{s}_i$ for $i \in \mathbb{Z}$. In $\mathfrak{s}_{\pm 1}$, we have a single figure-eight representing μ_1, while in $\mathfrak{s}_0$, we have two figure-eights representing $\mu_1 + \lambda_1$ and μ_1, respectively.

More generally, the we can make the same calculations when L is a 2-bridge link. In this case, both components of L are unknots (so we are in the situation where can compute

$\widehat{\mathrm{HF}}(M, R_\mu)$ from $\widehat{\mathrm{HFL}}(L)$), and L is alternating, so $\widehat{\mathrm{HFL}}(L)$ can be computed by a theorem of Ozsváth and Szabó. We deduce:

Theorem 4.5. *If M is the complement of a 2-bridge link L, then $\widehat{\mathrm{HF}}(M_L, R_\mu)$ is a collection of figure-eights determined by the multivariable Alexander polynomial, signature, and linking number of L.*

We expect that in this case there should be a natural choice of curve η for which $\widehat{\mathrm{HF}}(M_L(\eta), R_{\eta,\mu})$ is a collection of line segments which are the "cores" of the figure-eights in $\widehat{\mathrm{HF}}(M, R_\mu)$.

4.3. Satellites

Suppose $L \subset S^3$ is a 2-component link, where L_1 is the unknot. If $C \subset S^3$ is a knot, we choose $\phi : \partial M_{L_1} \to \partial M_C$ with $\phi_*(l_1) = \mu_C$ and $\phi_*(\mu_1) = \lambda_C$ Then $M_C \cup_\phi M_{L_1} \cong S^3$, so the image of L_2 in this union is knot in S^3. It is called the *satellite knot* $C(P)$, where C is the *companion* and $P := L_2$ is the *pattern.*

There is a well-known formula for the Alexander polynomial of a satellite,

$$\Delta_{C(P)}(t) = \Delta_C(t^n)\Delta_P(t),$$

where n is the winding number of P (its homology class in the solid torus) and $\Delta_P(t)$ is the single-variable Alexander polynomial of $P \subset S^3$. It is thus very natural to ask whether there is a formula for the knot Floer homology of $C(P)$.

The knot Floer homology of satellites has been studied extensively, starting with the work of Eftekhary [13] and Hedden [25, 26], and including important contributions by Hom [28] and Levine [37]. More recently, Chen [10] gave a very interesting method for computing $\widehat{\mathrm{HFK}}(K(P))$ when P is a component of a 2-bridge link.

The method described above gives an alternate approach to the same problem. From equation (4.1) above, it is clear that to compute $\widehat{\mathrm{HFK}}(C(P))$, it suffices to understand $\widehat{\mathrm{HF}}(M_L, R_\mu)$. Hence if L is a 2-bridge link, Theorem 4.5 implies that there is a formula for $\widehat{\mathrm{HFK}}$ of the satellite, in the sense that there is a finite set of slopes $\alpha_i \in \mathrm{Sl}(M_K)$ such that

$$\widehat{\mathrm{HFK}}\big(C(P)\big) \cong \bigoplus_i \widehat{\mathrm{HFK}}(K_{\alpha_i})).$$

The slopes α_i are determined by the multivariable Alexander polynomial, signature, and linking number of L.

Chen's method also makes use of the curve invariant $\widehat{\mathrm{HF}}(M_K)$, but in a rather different way. (For example, he is able to compute $\tau(C(P))$, which the method above does not allow us to do.) It would be interesting to understand how the two approaches are related.

Ideally, one would like to compute the full curve invariant $\widehat{\mathrm{HF}}(M_{C(P)}$ rather than just the knot Floer homology. It is unknown how to do this in general, but Hanselman and Watson have given a very beautiful description of how to do this for cables [24].

5. FURTHER DEVELOPMENTS AND QUESTIONS

5.1. Tangles

There are other situations in which Zarev's bordered Floer sutured homology gives an invariant which lives in the Fukaya category of a surface. One of the most interesting invariants corresponds to the case of 2-strand tangles. This situation has been studied by Zibrowius, who proved

Theorem 5.1 ([58]). *If T is a 2-strand tangle in B^3, then there is a well-defined invariant $\widehat{\mathrm{HFT}}(T)$ which takes the form of a collection of immersed closed curves with local systems in the 4-punctured sphere. If $L = T_1 \cup T_2$, where T_1 and T_2 are such tangles, then $\widehat{\mathrm{HFL}}(L)$ can be computed by pairing the curves $\widehat{\mathrm{HFT}}(T_1)$ and $\widehat{\mathrm{HFT}}(T_2)$.*

Other tangle invariants analogous to bordered Floer homology have been developed by Ozsváth and Szabó [46] and Petkova and Vertesi [47].

Unlike the case of a manifold with torus boundary (where relatively few restrictions on the form of the curve invariant are known), Zibrowius was able to prove some very strong constraints on the form of the curves that appear in $\widehat{\mathrm{HFT}}(T)$. This enabled him to answer a long-standing question about the effect of mutation on the total dimension of knot Floer homology.

Theorem 5.2 ([57]). *If K_1 and K_2 are mutant knots, then* $\dim \widehat{\mathrm{HFK}}(K_1) = \dim \widehat{\mathrm{HFK}}(K_2)$.

More recently, Kotelskiy, Watson, and Zibrowius have introduced some similar interpretations of the Khovanov homology of a 4-ended tangle T [33]. At the level of polynomials, the Jones polynomial of a 4-ended tangle is not so different from its $\mathfrak{sl}(n)$/HOMFLY-PT polynomial. (Both live in 2-dimensional vector spaces.) Hence it is natural to ask:

Question 5.3. Can the $\mathfrak{sl}(n)$ homology of a 4-ended tangle be interpreted as a curve invariant?

5.2. Cobordisms and extended TQFTs

Although we have not discussed it here, $\widehat{\mathrm{HF}}$ fits into the structure of a (relative) $3+1$ dimensional TQFT, as established by Zemke [56]. A cobordism $(W, \eta) : (Y_0, z_0) \to (Y_1, z_1)$ induces a map $F_{W,\eta} : \widehat{\mathrm{HF}}(Y_0, z_0) \to \widehat{\mathrm{HF}}(Y_1, z_1)$. It is an important foundational problem to show that the structure of bordered Floer homology can be extended to give a (pointed) extended TQFT, so that we associate a category $\mathcal{A}(\Sigma, z)$ to a pointed surface Σ, an object of that category $\mathcal{A}(M, z)$ to a pointed 3-manifold M with $\partial M \cong \Sigma$, and a morphism $\mathcal{A}(M_0, z) \to \mathcal{A}(M_1, z)$ to a cobordism with corners $W : M_0 \to M_1$. The lower-dimensional parts of this structure have already been established by Lipshitz–Ozsváth–Thurston, and it is not difficult to understand what the cobordism maps should be. The real work is in showing that they are well defined and satisfy an appropriate gluing theorem.

5.3. HF^-

For closed 3-manifolds, $\widehat{HF}$ is part of a larger package that also includes the equivariant homologies HF^+ and HF^-. One might hope to understand what these invariants mean for a manifold with torus boundary. Lipshitz, Ozsváth, and Thurston are in the process of developing a bordered theory for HF^- (see [40] for a first installment), and it will be interesting to see whether and how this can be interpreted in terms of curves and the Fukaya category. Some ideas for knot Floer homology have already been developed by Hanselman.

ACKNOWLEDGMENTS

The author would like to thank Steven Boyer, Nathan Dunfield, Cameron Gordon, Jonathan Hanselman, Peter Kronheimer, Yanki Lekili, Robert Lipshitz, Tom Mrowka, Peter Ozsváth, Sarah Rasmussen, Ivan Smith, Zoltán Szabó, Liam Watson, and Claudius Zibrowius for many helpful conversations on this subject over the years, and his family (Sarah most of all) for their love and support.

FUNDING

Part of this work was supported by EPSRC grant EP/M000648/1.

REFERENCES

[1] S. Akbulut and J. D. McCarthy, *Casson's invariant for oriented homology 3-spheres. An exposition.* Math. Notes 36, Princeton University Press, Princeton, NJ, 1990.

[2] D. Auroux, Fukaya categories and bordered Heegaard-Floer homology. In *Proceedings of the International Congress of Mathematicians. Volume II*, pp. 917–941, Hindustan Book Agency, New Delhi, 2010.

[3] D. Auroux, Fukaya categories of symmetric products and bordered Heegaard-Floer homology. *J. Gökova Geom. Topol. GGT* **4** (2010), 1–54.

[4] D. Auroux, A beginner's introduction to Fukaya categories. In *Contact and symplectic topology*, pp. 85–136, Bolyai Soc. Math. Stud. 26, János Bolyai Math. Soc., Budapest, 2014.

[5] J. A. Baldwin, Tight contact structures and genus one fibered knots. *Algebr. Geom. Topol.* **7** (2007), 701–735.

[6] S. Boyer and A. Clay, Foliations, orders, representations, L-spaces and graph manifolds. *Adv. Math.* **310** (2017), 159–234.

[7] S. Boyer, C. M. Gordon, and L. Watson, On L-spaces and left-orderable fundamental groups. *Math. Ann.* **356** (2013), no. 4, 1213–1245.

[8] P. J. Braam and S. K. Donaldson, Floer's work on instanton homology, knots and surgery. In *The Floer memorial volume*, pp. 195–256, Progr. Math. 133, Birkhäuser, Basel, 1995.

[9] B. A. Burton, The cusped hyperbolic census is complete. 2014, arXiv:1405.2695.

[10] W. Chen, Knot Floer homology of satellite knots with (1, 1)-patterns. 2019, arXiv:1912.07914.

[11] M. Culler, N. M. Dunfield, M. Goerner, and J. R. Weeks, SnapPy, a computer program for studying the geometry and topology of 3-manifolds. http://snappy.computop.org.

[12] N. M. Dunfield, Floer homology, group orderability, and taut foliations of hyperbolic 3-manifolds. *Geom. Topol.* **24** (2020), no. 4, 2075–2125.

[13] E. Eftekhary, Longitude Floer homology and the Whitehead double. *Algebr. Geom. Topol.* **5** (2005), 1389–1418.

[14] E. Eftekhary, Bordered Floer homology and existence of incompressible tori in homology spheres. *Compos. Math.* **154** (2018), no. 6, 1222–1268.

[15] R. Fintushel and R. J. Stern, Instanton homology of Seifert fibred homology three spheres. *Proc. Lond. Math. Soc. (3)* **61** (1990), no. 1, 109–137.

[16] A. Floer, An instanton-invariant for 3-manifolds. *Comm. Math. Phys.* **118** (1988), no. 2, 215–240.

[17] A. Floer, Morse theory for Lagrangian intersections. *J. Differential Geom.* **28** (1988), no. 3, 513–547.

[18] A. Floer, Instanton homology, surgery, and knots. In *Geometry of low-dimensional manifolds, 1 (Durham, 1989)*, pp. 97–114, London Math. Soc. Lecture Note Ser., 150, Cambridge Univ. Press, Cambridge, 1990.

[19] F. Haiden, L. Katzarkov, and M. Kontsevich, Flat surfaces and stability structures. *Publ. Math. Inst. Hautes Études Sci.* **126** (2017), 247–318.

[20] J. Hanselman, J. Rasmussen, S. D. Rasmussen, and L. Watson, L-spaces, taut foliations, and graph manifolds. *Compos. Math.* **156** (2020), no. 3, 604–612.

[21] J. Hanselman, J. Rasmussen, and L. Watson, Bordered Floer homology for manifolds with torus boundary via immersed curves. 2016, arXiv:1604.03466.

[22] J. Hanselman, J. Rasmussen, and L. Watson, Heegaard Floer homology for manifolds with torus boundary: properties and examples. 2018, arXiv:1810.10355.

[23] J. Hanselman and L. Watson, A calculus for bordered Floer homology. 2015, arXiv:1508.05445.

[24] J. Hanselman and L. Watson, Cabling in terms of immersed curves. 2019, arXiv:1908.04397.

[25] M. Hedden, On knot Floer homology and cabling. *Algebr. Geom. Topol.* **5** (2005), 1197–1222.

[26] M. Hedden, On knot Floer homology and cabling. II. *Int. Math. Res. Not. IMRN* **12** (2009), 2248–2274.

[27] M. Hildebrand, and J.Weeks. A computer generated census of cusped hyperbolic 3-manifolds. In *Computers and mathematics (Cambridge, MA, 1989)*, pp. 53–59, Springer, New York, 1989.

[28] J. Hom, Bordered Heegaard Floer homology and the tau-invariant of cable knots. *J. Topol.* **7** (2014), no. 2, 287–326.

[29] M. Hutchings, An index inequality for embedded pseudoholomorphic curves in symplectizations. *J. Eur. Math. Soc. (JEMS)* **4** (2002), no. 4, 313–361.

[30] A. Juhász, Holomorphic discs and sutured manifolds. *Algebr. Geom. Topol.* **6** (2006), 1429–1457.

[31] A. Juhász, A survey of Heegaard Floer homology. In *New ideas in low dimensional topology*, pp. 237–296, Ser. Knots Everything 56, World Sci. Publ., Hackensack, NJ, 2015.

[32] A. Juhász, D. P. Thurston, and I. Zemke, Naturality and mapping class groups in Heegaard Floer homology. *Mem. Amer. Math. Soc.* **273** (2021), no. 1338.

[33] A. Kotelskiy, L. Watson, and C. Zibrowius, Immersed curves in Khovanov homology. 2019, arXiv:1910.14584.

[34] P. B. Kronheimer and T. S. Mrowka, Knot homology groups from instantons. *J. Topol.* **4** (2011), no. 4, 835–918.

[35] P. Kronheimer and T. Mrowka, *Monopoles and three-manifolds*. New Math. Monogr. 10, Cambridge University Press, Cambridge, 2007.

[36] P. Kronheimer and T. Mrowka, Knots, sutures, and excision. *J. Differential Geom.* **84** (2010), no. 2, 301–364.

[37] A. S. Levine, Nonsurjective satellite operators and piecewise-linear concordance. *Forum Math. Sigma* **4:Paper No. e34** (2016), 47.

[38] R. Lipshitz, P. S. Ozsváth, and D. P. Thurston, Bimodules in bordered Heegaard Floer homology. *Geom. Topol.* **19** (2015), no. 2, 525–724.

[39] R. Lipshitz, P. S. Ozsvath, and D. P. Thurston, Bordered Heegaard Floer homology. *Mem. Amer. Math. Soc.* **254** (2018), no. 1216, viii+279 pp.

[40] R. Lipshitz, P. Ozsváth, and D. Thurston, A bordered HF^- algebra for the torus. 2021, arXiv:2108.12488.

[41] Y. Ni, Knot Floer homology detects fibred knots. *Invent. Math.* **170** (2007), no. 3, 577–608.

[42] P. Ozsváth and Z. Szabó, Holomorphic disks and knot invariants. *Adv. Math.* **186** (2004), no. 1, 58–116.

[43] P. Ozsváth and Z. Szabó, Holomorphic disks and three-manifold invariants: properties and applications. *Ann. of Math. (2)* **159** (2004), no. 3, 1159–1245.

[44] P. Ozsváth and Z. Szabó, Holomorphic disks and topological invariants for closed three-manifolds. *Ann. of Math. (2)* **159** (2004), no. 3, 1027–1158.

[45] P. Ozsváth and Z. Szabó, Holomorphic disks, link invariants and the multivariable Alexander polynomial. *Algebr. Geom. Topol.* **8** (2008), no. 2, 615–692.

[46] P. Ozsváth and Z. Szabó, Kauffman states, bordered algebras, and a bigraded knot invariant. *Adv. Math.* **328** (2018), 1088–1198.

[47] I. Petkova and V. Vértesi, Combinatorial tangle Floer homology. *Geom. Topol.* **20** (2016), no. 6, 3219–3332.

[48] J. A. Rasmussen, *Floer homology and knot complements. ProQuest LLC.* Thesis (PhD), Harvard University, Ann Arbor, MI, 2003.

[49] J. Rasmussen and S. D. Rasmussen, Floer simple manifolds and L-space intervals. *Adv. Math.* **322** (2017), 738–805.

[50] S. D. Rasmussen, L-space intervals for graph manifolds and cables. *Compos. Math.* **153** (2017), no. 5, 1008–1049.

[51] P. Seidel, *Fukaya categories and Picard–Lefschetz theory*. Zur. Lect. Adv. Math., European Mathematical Society (EMS), Zürich, 2008.

[52] I. Smith, A symplectic prolegomenon. *Bull. Amer. Math. Soc. (N.S.)* **52** (2015), no. 3, 415–464.

[53] E. Witten, Monopoles and four-manifolds. *Math. Res. Lett.* **1** (1994), no. 6, 769–796.

[54] F. Ye, Constrained knots in lens spaces. 2020, arXiv:2007.04237.

[55] R. Zarev, *Bordered Sutured Floer Homology. ProQuest LLC*. Thesis (PhD), Columbia University, Ann Arbor, MI, 2011.

[56] I. Zemke, A graph TQFT for hat Heegaard Floer homology. 2015, arXiv:1503.05846.

[57] C. Zibrowius, On symmetries of peculiar modules; or, δ-graded link Floer homology is mutation invariant. 2019, arXiv:1909.04267.

[58] C. Zibrowius, Peculiar modules for 4-ended tangles. *J. Topol.* **13** (2020), no. 1, 77–158.

JACOB RASMUSSEN

Department of Pure Mathematics and Mathematical Statistics, University of Cambridge, Cambridge, United Kingdom, jar60@cam.ac.uk

This Page Intentionally Left Blank

HOMOLOGICAL STABILITY: A TOOL FOR COMPUTATIONS

NATHALIE WAHL

ABSTRACT

Homological stability has shown itself to be a powerful tool for the computation of homology of families of groups such as general linear groups, mapping class groups, or automorphisms of free groups. We survey here tools and techniques for proving homological stability theorems and for computing the stable homology, and illustrate the method through the computation of the homology of Higman–Thompson groups.

MATHEMATICS SUBJECT CLASSIFICATION 2020

20J05

KEYWORDS

Homology of groups, buildings for stability, stable homology

INTERNATIONAL CONGRESS OF MATHEMATICIANS 2022 JULY 6–14

Proc. Int. Cong. Math. 2022, Vol. 4, pp. 2904–2927
DOI 10.4171/ICM2022/137

Published by EMS Press

1. INTRODUCTION

Homology is an invariant that comes in many flavors. We will here mostly be concerned with group homology, but the story we will tell can be told in other contexts as well. Like many invariants, while easy to define, homology is often difficult to compute. What homological stability has shown to us over the years is that in some situations, it is easier to compute infinitely many homology groups at once than computing a single one by itself. We will in this paper illustrate this through examples, and try to give the reader a sense of how to do homological computations using stability methods, and a sense of when such methods are likely to work.

Many mathematical objects come in families. We will here be interested in families of groups like the symmetric groups Σ_n, the braid group B_n, the general linear groups $\mathrm{GL}_n(R)$ over a ring R, or automorphism groups $\mathrm{Aut}(F_n)$ of free groups. These examples are more than collections of groups: they all have an additional structure in the form of maps

$$\begin{aligned}
\oplus &: \Sigma_n \times \Sigma_m \to \Sigma_{n+m}, \\
\oplus &: B_n \times B_m \to B_{n+m}, \\
\oplus &: \mathrm{GL}_n(R) \times \mathrm{GL}_m(R) \to \mathrm{GL}_{n+m}(R), \\
\oplus &: \mathrm{Aut}(F_n) \times \mathrm{Aut}(F_m) \to \mathrm{Aut}(F_{n+m}),
\end{aligned}$$

by "block sum" of permutations, braids or matrices, or juxtaposition of automorphisms. Another important flavor of example for us will be families of mapping class groups of surfaces or 3-manifolds with sum $\oplus$ an appropriate boundary connected sum.

Taking $m = 1$ in the above and evaluating the maps $\oplus$ at the identity element in $\Sigma_1, B_1, \mathrm{GL}_1(R)$, or $\mathrm{Aut}(F_1)$ gives sequences of groups

$$\begin{aligned}
&\Sigma_1 \to \Sigma_2 \to \Sigma_3 \to \cdots, \\
&B_1 \to B_2 \to B_3 \to \cdots, \\
&\mathrm{GL}_1(R) \to \mathrm{GL}_2(R) \to \mathrm{GL}_3(R) \to \cdots, \\
&\mathrm{Aut}(F_1) \to \mathrm{Aut}(F_2) \to \mathrm{Aut}(F_3) \to \cdots.
\end{aligned}$$

We are here interested in the following property of such sequences of groups:

Definition 1.1. A sequence of groups $G_1 \to G_2 \to \cdots$ satisfies *homological stability* if the associated sequence of homology groups

$$H_i(G_1) \to H_i(G_2) \to H_i(G_3) \to \cdots \tag{1.1}$$

is eventually constant for each i, that is, if $H_i(G_n) \xrightarrow{\cong} H_i(G_{n+1})$ for n large enough with respect to i.

Unless explicitly otherwise stated, homology here means homology with integral coefficients, $H_*(-) = H_*(-;\mathbb{Z})$. Typical stability bounds are linear, of the form $n \geq ki + a$, for k the *slope* of stability.

Definition 1.1 clearly makes sense in other contexts, with the groups and group homology replaced by some other type of object and associated homology theory. Much of what we will present here is known to generalize to sequences of spaces, and, to some level, sequences of algebras. We will focus here on the case of groups for simplicity, and because it is already very rich.

All the examples of families of groups mentioned above are known to satisfy homological stability. In the 1960s, Nakaoka computed the homology of the symmetric groups Σ_n in [53] and observed that

$$H_i(\Sigma_n) \xrightarrow{\cong} H_i(\Sigma_{n+1}) \quad \text{for } i \leq \frac{n}{2}.$$

Arnold computed in [1] that the same holds for the homology of the braid groups, and around the same time Quillen, interested in algebraic K-theory [59] (see also [66]), showed, for example, that, for a field $\mathbb{F} \neq \mathbb{F}_2$,

$$H_i(\mathrm{GL}_n(\mathbb{F})) \xrightarrow{\cong} H_i(\mathrm{GL}_{n+1}(\mathbb{F})) \quad \text{for } i \leq n.$$

Harer showed in the 1980s that also mapping class groups of surfaces satisfy homological stability [29], a result that was extended to nonorientable surfaces by the author [71]. For the automorphisms of free groups, the first proof goes back to Hatcher [30], while Hatcher and the author proved a very general stability theorem for mapping class groups of 3-manifolds [34]: if M, N are any orientable 3-manifolds such that $\partial M \neq \emptyset$, then the map $\pi_0 \operatorname{Diff}(M \# N^{\# n}) \to \pi_0 \operatorname{Diff}(M \# N^{\# n+1})$ extending diffeomorphisms by the identity on the added summand N, induces an isomorphism on H_i for $i \leq \frac{n-2}{2}$. And many more stability results for families of groups are known!

Quillen's stability argument. Quillen devised a strategy for proving homological stability using a spectral sequence associated to the action of the groups on appropriate spaces: To apply Quillen's strategy to a family of groups $\{G_n\}_{n\geq 0}$, one needs to find a family of simplicial objects $\{W_n\}_{n\in\mathbb{N}}$, with G_n acting on W_n, satisfying (roughly) the following:

(1) the action is transitive on vertices, and has "manageable" sets of orbits of p–simplices for every p;

(2) the stabilizer of a p-simplex is a previous group in the sequence, e.g., G_{n-p-1};

(3) each W_n is highly connected.

From this data, one can construct a spectral sequence with E^1-term

$$E^1_{p,q} = \bigoplus_{\sigma_p} H_q(\operatorname{Stab}(\sigma_p); \mathbb{Z}_\sigma),$$

where the sum runs over representatives of the orbits of p-simplices σ_p in W_n. The spectral sequence, together with conditions (1)–(3) and a few minor additional assumptions, allows then for an inductive argument. (See Section 2.2 for some more details.)

Variants and extensions of this strategy have been applied in a variety of contexts. In addition to the examples already mentioned, stability has been shown using this strategy for $\mathrm{GL}_n(R)$ for many rings R [44, 69], for other classical groups like symplectic groups, orthogonal groups, unitary groups, see, e.g., [9, 52, 64, 70], and many other groups. The strategy was also adapted to prove homological stability for moduli spaces of manifolds and configuration spaces [26, 56, 62], or for certain families of algebras [6, 37]. So many stability theorems have been proved using this method that it is difficult to mention them all.

Stable homology. Let

$$G_\infty := \bigcup_{n=1}^{\infty} G_n = \operatorname*{colim}_{n\to\infty} G_n$$

be the limit of the sequence of groups. Homological stability can be reformulated as saying that the map $G_n \to G_\infty$ induces an isomorphism

$$H_i(G_n) \xrightarrow{\cong} H_i(G_\infty)$$

in an increasing range of degrees $i \le b(n)$ for $b(n)$ a bound depending on n. The limit homology $H_*(G_\infty)$ is called the *stable homology*. The power of homological stability comes from the fact that often the stable homology is easier to compute because it most often belongs to the world of spectra, where methods of homotopy theory come into play. We give here known stable computations for the examples of families of groups described above.

The Barratt–Priddy–Quillen theorem identifies the stable homology $H_*(\Sigma_\infty)$ of the symmetric groups with that of the basepoint component $\Omega_0^\infty \mathbb{S}$ of the infinite loop space of the sphere spectrum $\mathbb{S}$. Galatius showed that the same holds for the stable homology of automorphisms of free groups. Combining these results with the best known homological stability ranges gives

Theorem 1.2 ([2, 20, 31, 53]). *For all* $i \le \frac{n}{2}$, $H_i(\Sigma_n) \cong H_i(\Omega_0^\infty \mathbb{S}) \cong H_i(\mathrm{Aut}(F_{n+3}))$.

A direct consequence is that the stable rational homology of $\mathrm{Aut}(F_n)$ is trivial. The result also gives that the inclusion $\Sigma_n \hookrightarrow \mathrm{Aut}(F_n)$ induces a homology isomorphism in the range $i \le \frac{n-3}{2}$, a fact we only know through the above stable homology computation.

For the braid groups B_n, the corresponding result is

Theorem 1.3 ([1, 12]). *For all* $i \le \frac{n}{2}$, $H_i(B_n) \cong H_i(\Omega_0^2 S^2)$.

F. Cohen completely computed homology of the right-hand side, see [12, PAPER III, APP A], yielding a full computation of the stable homology of the braid groups.

For $\mathrm{GL}_n(R)$, the relevant spectrum is the K-theory spectrum, and here the flow of information has gone the other way around compared to the above examples: In the case where $R = \mathbb{F}_{p^r}$ is a finite field, the homology $H_*(\mathrm{GL}(\mathbb{F}_{p^r}); \mathbb{F}_\ell)$ was completely computed by Quillen for any prime $\ell \neq p$, a computation he used to deduce information about the K-theory spectrum [60]. When $\ell = p$, only the stable homology is fully known:

Theorem 1.4 ([21, 59, 60, 66]). [1] $H_i(\mathrm{GL}_n(\mathbb{F}_{p^r}); \mathbb{F}_p) = 0$ *for all* $i \le n + r(p-1) - 3$ *if* $p^r \neq 2$, *and for all* $i < \frac{2n-2}{3}$ *if* $p^r = 2$.

A similar result holds for symplectic, orthogonal, and unitary groups, see [18, 66].

---

1 Note that the paper [21] uses a different stability method than Quillen's, see Section 3.

For mapping class groups of surfaces, the stable homology was computed by Madsen and Weiss, in a breakthrough work that lead to much progress in manifold topology: Denoting by $\Sigma_{g,b}$ an orientable surface of genus g with b boundary components, and by $S_{h,b}$ a nonorientable surface of genus h with b boundary components, and combining the Madsen–Weiss theorem with the best known ranges for homological stability, as well as the unoriented versions of these theorems, we have

Theorem 1.5 ([4,46,62,71]).

$$H_i(\pi_0 \operatorname{Diff}(\Sigma_{g,b})) \cong H_i(\Omega_0^\infty \mathrm{MTSO}(2)), \quad i \le \frac{2g-2}{3},$$
$$H_i(\pi_0 \operatorname{Diff}(S_{h,b})) \cong H_i(\Omega_0^\infty \mathrm{MTO}(2)), \quad i \le \frac{h-3}{3}.$$

Here MTSO(2) or MTO(2) are the Thom spectra of the orthogonal bundle to the universal bundle over the Grassmannian of oriented or nonoriented 2-planes in $\mathbb{R}^\infty$, respectively. A direct consequence is that the stable rational homology of these groups are the polynomial algebras $\mathbb{Q}[\kappa_1, \kappa_2, \dots]$ and $\mathbb{Q}[\zeta_1, \zeta_2, \dots]$, respectively, where $|\kappa_i| = 2i$ and $|\zeta_i| = 4i$. In the oriented case, this rational computation had been a conjecture of Mumford. This result was generalized to higher dimensional manifolds of even dimension by Galatius and Randal-Williams [25,26] and to odd-dimensional handlebodies by Botvinnik and Perlmutter [5,57]. This has since been used to compute, e.g., homotopy groups of the diffeomorphisms of discs, or give a totally new approach to pseudoisotopy theory [40,42].

In Section 4, we will explain such a theorem for the Higman–Thompson groups, see Theorem 4.1, which computes, as a corollary, the full homology of Thompson's group V. And we will explain in Section 3 why one should not be surprised to see double or infinite loop spaces in the above statements.

Content of the paper. In this article, we want to address the following questions:

(1) When can one expect that homological stability holds?

(2) How does one find appropriate G_n-space for Quillen's stability argument?

(3) How does one compute the stable homology?

Let us though make clear from the start that we will, of course, not give full answers to any of the three questions.

A priori one only needs a sequence of groups $G_1 \to G_2 \to \cdots$ to talk about homological stability. Following the article [63] and its generalization [39], Sections 2 of the present paper shows that having the additional data of a "block sum," as exhibited above for the groups Σ_n, B_n, $\mathrm{GL}_n(R)$, or $\mathrm{Aut}(F_n)$, is enough input to run Quillen's argument in the following sense: in Section 2.1, we construct a canonical *space of destabilizations* W_n when the sum operation is braided, and Theorem 2.9 in Section 2.2 states that homological stability holds whenever these spaces are sufficiently connected. In Section 2.3, we explain how homological stability with *abelian* and *polynomial* coefficients automatically also holds under the same assumption, see Theorems 2.12 and 2.13.

In Section 3 we will see that the braiding forces the stable homology, through the "group completion theorem," to be that of a double loop space, or an infinite loop space when the braiding is a symmetry.

In Section 4, we will then explain how all these ideas were used in [68] to show that the homology of Thompson's group V is trivial, via a stability theorem and stable computation for the more general Higman–Thompson groups.

The article ends with a short section addressing the wider perspective.

2. A GENERAL FRAMEWORK FOR QUILLEN'S STABILITY ARGUMENT

In this section, we describe a framework in which Quillen's strategy for proving homological stability can always be implemented.

Recall that a *groupoid* $\mathcal{G}$ is a category whose morphisms are all invertible. A *monoidal groupoid* is a groupoid $\mathcal{G}$ equipped with a sum

$$\oplus : \mathcal{G} \times \mathcal{G} \to \mathcal{G}$$

that is associative and unital. It is *braided* if it is in addition equipped with an isomorphism $\sigma_{A,B} : A \oplus B \to B \oplus A$ for every pair of objects, satisfying the braid identity $\sigma_{A,B}\sigma_{A,C}\sigma_{B,C} = \sigma_{B,C}\sigma_{A,C}\sigma_{A,B} : A \oplus B \oplus C \to C \oplus B \oplus A$ and such that

$$\begin{array}{ccc} A \oplus B & \xrightarrow{\sigma_{A,B}} & B \oplus A \\ {\scriptstyle f\oplus g}\downarrow & & \downarrow{\scriptstyle g\oplus f} \\ C \oplus D & \xrightarrow{\sigma_{C,D}} & D \oplus C \end{array}$$

commutes whenever it is defined, see, e.g., [45]. The groupoid is *symmetric monoidal* if the braiding squares to the identity.

There are many examples of braided and symmetric monoidal groupoids. Standard examples of interest to us are the groupoid of sets with disjoint union, the groupoid of R–modules with direct sum, the groupoid of groups with free or direct product, the groupoid of vector spaces equipped with a symplectic or Hermitian form with the direct sum, or the groupoid of manifolds of a given dimension with an appropriate connected sum operation. Each of these examples are actually the groupoid of isomorphisms in a braided or symmetric monoidal category. For us only the isomorphisms will play a role.

From groups to groupoids. If we start with a family of groups $\{G_n\}_{n\in\mathbb{N}}$ and defined $\mathcal{G} = \coprod_n G_n$ to be the groupoid with objects the formal sums $X^{\oplus n}$ for $n \in \mathbb{N}$ of a generating object X, and only nontrivial morphisms $G_n := \mathrm{Aut}_{\mathcal{G}}(X^{\oplus n})$, then a monoidal structure on $\mathcal{G}$, extending the sum in $\mathbb{N}$, is the data of an associative "sum" operation $G_n \times G_m \xrightarrow{\oplus} G_{n+m}$, and a braiding is the data of a homomorphism $\phi : B_n \to G_n$ from the braid group, such that the block braid $b_{n,m}$ satisfies that $\phi(b_{n,m})(g \oplus g')\phi(b_{n,m})^{-1} = g' \oplus g$ for each $g \in G_n$ and $g' \in G_m$ (see Figure 1). The groupoid is symmetric precisely if the homomorphism ϕ factors through the symmetric group Σ_n.

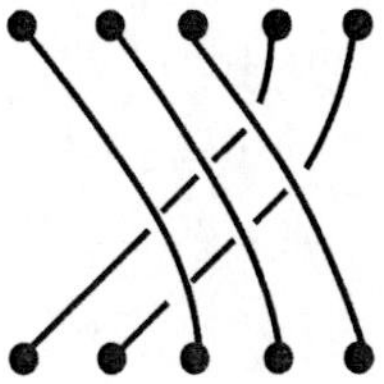

FIGURE 1
Block braid $b_{3,2}$.

For example, applying this construction to the symmetric groups $\{\Sigma_n\}_{n\in\mathbb{N}}$ with the block sum of permutations yields the following: the objects are the natural numbers, where we can think of n as representing a set $[n]$ with n elements, and the automorphism group of $[n]$ is Σ_n. As $[n+m] \cong [n] \sqcup [m]$, we see that the monoidal sum corresponds to the disjoint union of sets. The resulting groupoid is a skeleton of the groupoid of finite sets of Example 2.1 below. If we instead start with the general linear groups $\{\mathrm{GL}_n(R)\}_{n\in\mathbb{N}}$, we can think of the object n in the resulting groupoid as representing $R^n = R^{\oplus n}$, whose automorphism group is $\mathrm{GL}_n(R)$. The monoidal product then correspond to the direct sum of R-modules, yielding a subcategory of the category of R-modules of Example 2.2 below.

From groupoids to groups. If we start instead with a monoidal groupoid $\mathcal{G} = (\mathcal{G}, \oplus)$, for any two objects A, X in $\mathcal{G}$, we get a sequence of groups $G_1 \to G_2 \to \cdots$ by setting

$$G_n = \mathrm{Aut}_{\mathcal{G}}\big(A \oplus X^{\oplus n}\big)$$

with

$$G_n = \mathrm{Aut}_{\mathcal{G}}\big(A \oplus X^{\oplus n}\big) \xrightarrow{\_\oplus \mathrm{id}_X} G_{n+1} = \mathrm{Aut}_{\mathcal{G}}\big(A \oplus X^{\oplus n} \oplus X\big).$$

We think of G_n as the automorphism group of "A stabilized n times in the X direction."

Example 2.1. Let $\mathcal{G} = \mathrm{Sets}^{\mathrm{iso}}$ denote the groupoid of finite sets and bijections, with monoidal structure $\oplus = \sqcup$ given by disjoint union. It is a symmetric monoidal groupoid with the symmetry the standard bijection $A \sqcup B \xrightarrow{\cong} B \sqcup A$. Taking $A = \emptyset$ and $X = \{*\}$ in the above yields $G_n = \Sigma_n$ the symmetric group on n letters, with $\Sigma_n \to \Sigma_{n+1}$ the standard inclusion as the subgroup of permutations fixing the last element.

Example 2.2. Let R be a ring and let $\mathcal{G} = R$–Mod denote the groupoid of R-modules and their isomorphisms, with monoidal product the direct sum $\oplus$ of modules. This is again a symmetric monoidal groupoid with symmetry given by the standard isomorphism $M \oplus N \xrightarrow{\cong} N \oplus M$. Taking $A = 0$ and $X = R$, we get $G_n = \mathrm{GL}_n(R)$, the automorphism group of the module $R^{\oplus n}$, with the map $\mathrm{GL}_n(R) \to \mathrm{GL}_{n+1}(R)$ adding a 1 in the bottom corner of the matrix. If we take A to be any R-module, the group $G_n = \mathrm{GL}(A \oplus R^{\oplus n})$ is the automorphism group of the module A stabilized n times.

Example 2.3. Let $\mathcal{G} = \mathrm{Groups}^{\mathrm{iso}}$ be the groupoid of groups with free product as monoidal structure. This is again a symmetric monoidal groupoid with symmetry the natural isomorphism $G * H \to H * G$. If we take $A = \langle e \rangle$ to be the trivial group and $X = \mathbb{Z}$, we get

$G_n = \mathrm{Aut}(F_n)$, the already considered automorphism group of the free group F_n. For $A = H$ and $X = G$ any group, the group $G_n = \mathrm{Aut}(H * G * \cdots * G)$ is the automorphism group H free product with n copies of G, whose stability is studied in [13, 34].

Modules over monoidal groupoids. Let $\mathcal{G} = (\mathcal{G}, \oplus)$ be a monoidal groupoid. A category $\mathcal{C}$ is a *right module over* $\mathcal{G}$ if $\mathcal{C}$ is equipped with a unital and associative action

$$\mathcal{C} \times \mathcal{G} \xrightarrow{\oplus} \mathcal{C}$$

of $\mathcal{G}$. (See [39, SECT. 7.1].) Taking $A \in \mathcal{C}$ and $X \in \mathcal{G}$, we can again define $G_n = \mathrm{Aut}_{\mathcal{C}}(A \oplus X^{\oplus n})$, and this yields just as above a sequence of groups $G_1 \to G_2 \to \cdots$ with the map $\_ \oplus \mathrm{id}_X : G_n \to G_{n+1}$ adding the identity on the extra copy of X.

The sequence of groups G_n obtained above from a monoidal groupoid $\mathcal{G}$ only is the special case when $\mathcal{C} = \mathcal{G}$, considering $\mathcal{G}$ as a module over itself. Most of our examples will be of that form, but there are examples from, e.g., manifolds [26, 34], or Coxeter groups [36], that require the more general setup of a module over a groupoid. (See also [39] for examples in the context of homological stability for topological spaces.)

2.1. The space of destabilizations

Recall from the introduction that to apply Quillen's strategy for proving homological stability, one needs for each n a simplicial object W_n on which the group G_n acts, with appropriate transitivity, stabilizer, and connectivity properties. The spaces W_n used in homological stability are most typically one of three types: simplicial complexes, (semi)simplicial sets, or posets. We will here only discuss spaces of the first two types.

Ad hoc simplicial objects W_n associated to families of groups G_n have been defined in very many situations to prove stability statements; in fact, most homological stability theorems for families of groups have been so far proved using Quillen's strategy. Following [63] and its generalization [39], we construct here the smallest such semisimplicial set W_n for any family of groups of the form $G_n = \mathrm{Aut}_{\mathcal{C}}(A \oplus X^{\oplus n})$ arising as above from the action of a *braided* monoidal groupoid $\mathcal{G}$ on a groupoid $\mathcal{C}$; the definition of the face maps in W_n will use the braiding of $\mathcal{G}$. We also define an associated simplicial complex S_n.

Fix $\mathcal{C}$ a module over a braided monoidal groupoid $\mathcal{G}$, with A an object of $\mathcal{C}$, and X an object of $\mathcal{G}$ as above.

Definition 2.4 ([63, DEF. 2.1], [39, DEF. 7.5]). The *space of destabilizations* $W_n(A, X)_\bullet$ is the semisimplicial set with set of p-simplices

$$W_n(A,X)_p = \{(B, f) \mid B \in \mathrm{Ob}(\mathcal{C}) \text{ and } f : B \oplus X^{\oplus p+1} \to A \oplus X^{\oplus n} \text{ in } \mathcal{C}\}/_\sim$$

where $(B, f) \sim (B', f')$ if there exists an isomorphism $g : B \to B'$ in $\mathcal{C}$ satisfying that $f = f' \circ (g \oplus \mathrm{id}_{X^{\oplus p+1}})$. The face map $d_i : W_n(A, X)_p \to W_n(A, X)_{p-1}$ is defined by $d_i[B, f] = [B \oplus X, d_i f]$ for

$$d_i f : B \oplus X \oplus X^p \xrightarrow{\mathrm{id}_B \oplus b^{-1}_{X^{\oplus i}, X} \oplus \mathrm{id}_{X^{\oplus p-i}}} B \oplus X^{\oplus i} \oplus X \oplus X^{\oplus p-i} \xrightarrow{f} A \oplus X^{\oplus n},$$

for $b^{-1}_{X^{\oplus i}, X} : X \oplus X^{\oplus i} \to X^{\oplus i} \oplus X$ coming from the braiding in $\mathcal{G}$.

The group $G_n = \mathrm{Aut}_{\mathcal{C}}(A \oplus X^{\oplus n})$ acts on $W_n(A, X)_\bullet$ by postcomposition. The following holds for the action:

(local cancelation) If $Y \oplus X^{\oplus p+1} \cong A \oplus X^{\oplus n} \Longrightarrow Y \cong A \oplus X^{\oplus n-p-1}$,

then G_n acts transitively on $W_n(A, X)_p$. (2.1)

(injectivity) If the stabilization $G_{n-p-1} \to G_n$ taking f to $f \oplus \mathrm{id}_{X^{p+1}}$ is injective,

then there is an isomorphism $G_{n-p-1} \cong \mathrm{Stab}(\sigma_p)$ for of any p-simplex σ_p. (2.2)

A direct consequence is that, under these two mild conditions on the $\mathcal{G}$-module $\mathcal{C}$, the set of p-simplices $W_n(A, X)_p$ of the space of destabilizations is isomorphic to G_n/G_{n-p-1}. As we will see in Section 4 in an example, local cancelation can actually be forced by changing the definition of $\mathcal{C}$ and $\mathcal{G}$, declaring in particular $A \oplus X^{\oplus n}$ and $A \oplus X^{\oplus m}$ for $n \neq m$ to be nonisomorphic. If the second condition is not satisfied, $W_n(A, X)$ needs to be replaced by a semisimplicial space in Quillen's argument, see [39, SECT. 7.3].

Remark 2.5. In the case of a groupoid $\mathcal{G} = \mathcal{C}$ acting on itself, the set of p-simplices of $W_n(A, X)$ can be interpreted as the set of morphisms from $X^{\oplus p+1}$ to $A \oplus X^{\oplus n}$ in a category $\langle \mathcal{G}, \mathcal{G} \rangle$ constructed from the action, see Appendix A. The face maps are then given by precomposition with standard morphisms $X^{\oplus p} \to X^{\oplus p+1}$ in that category.

From $W_n(A, X)$, one can also define a simplicial complex $S_n(A, X)$ as follows:

Definition 2.6. Let $S_n(A, X)$ be the simplicial complex with the same vertices as $W_n(A, X)$. A set of vertices $\{x_0. \dots, x_p\}$ spans a p-simplex in $S_n(A, X)$ if and only if they are the vertices of a p-simplex of $W_n(A, X)$.

We will see in Section 2.2 that it is often equivalent, and more convenient, to work with $S_n(A, X)$ instead of $W_n(A, X)$ for connectivity questions.

Example 2.7. As in Example 2.1, consider $(\mathcal{G}, \oplus) = (\mathrm{Sets}^{\mathrm{iso}}, \sqcup)$ the symmetric monoidal groupoid of finite sets, seen as a module over itself, with $A = \emptyset$ and $X = \{*\}$, giving $G_n = \Sigma_n$ the symmetric group. A p-simplex $[B, f]$ of $W_n(\emptyset, \{*\})$ is determined by the restriction of the bijection $f : B \sqcup [p+1] \to [n]$ to $[p+1]$. So a p-simplex of $W_n(\emptyset, \{*\})$ is an ordered tuple of $p+1$ elements of $[n] = \{1, \dots, n\}$. The ith boundary map forgets the $(i+1)$st element. This semisimplicial set is known as the *complex of injective words*, and it is known to be $(n-2)$-connected [17] (see also [63, SECT. 5.1]). The simplicial complex $S_n(\emptyset, \{*\})$ has the same vertices as $W_n(\emptyset, \{*\})$, namely the elements of $[n]$, and $p+1$ such elements form a simplex in $S_n(\emptyset, \{*\})$ precisely when there exist an injective word in these letters, i.e., if they are distinct. Hence $S_n(\emptyset, \{*\})$ identifies with the $(n-1)$-simplex Δ^{n-1}.

Example 2.8. Let $(\mathcal{G}, \oplus) = (R\text{–Mod}, \oplus)$ be the symmetric monoidal groupoid of R-modules acting on itself, with $A = 0$ and $X = R$, giving $G_n = \mathrm{GL}_n(R)$ as in Example 2.2. A p-simplex $[B, f]$ in $W_n(A, X)$, with $f : B \oplus R^{p+1} \xrightarrow{\cong} R^n$, is determined by the pair $(f(B) < R^n, f|_{R^{p+1}} : R^{p+1} \hookrightarrow R^n)$. The simplicial complex $S_n(A, X)$ thus has vertices pairs (H, f) with $H < R^n$ and $f : R \hookrightarrow R^n$ so that $R^n = H \oplus f(R)$, and vertices $((H_0, f_0), \dots, (H_p, f_p))$ form a p-simplex if together the maps $(f_0 \oplus \cdots \oplus f_p) : R^{p+1} \to R^n$ form an injective map with a complement H such that each $H_i = H \oplus \bigoplus_{j \neq i} f_j(R)$. This complex is very closely related to complexes studied by van der Kallen [69] and Charney [10] and is $\frac{n-a}{2}$-connected for a a constant depending on the stable rank of R (see [63, LEMMA 5.10]). The fact that simplices are not just split injective homomorphisms, but rather split homomorphisms *with a choice of complement* H, makes the simplicial complex more intricate to study, but it forces the stabilizer of a p-simplex to be exactly $\mathrm{GL}(H)$, instead of an affine version of the group, which would be the case if complements had not been chosen.

The simplicial complex $S_n(A, X)$ has appeared in the literature in many examples long before it was defined in the above generality. Here are a few additional examples: for the automorphisms of free groups $\mathrm{Aut}(F_n)$, it is essentially the complex of split factorizations of Hatcher and Vogtmann [32] (see [63, SECT. 5.2.1]), for mapping class groups of surfaces with genus stabilization, this identifies with the tethered arc complex of the same authors [33] (see [63, SECT. 5.6.3]), while the poset of simplices of $W_n(A, X)$ in the case of unitary groups already appeared in [52] (see [63, SECT. 5.4]).

2.2. Homological stability

Let $\mathcal{C}$ be a module over a braided monoidal groupoid $\mathcal{G}$ as above, with A and X objects of $\mathcal{C}$ and $\mathcal{G}$, respectively. We have so far associated a sequence of groups $G_n = \mathrm{Aut}_{\mathcal{C}}(A \oplus X^{\oplus n})$ to this data, together with a collection of associated G_n-spaces $W_n = W_n(A, X)$ and $S_n = S_n(A, X)$. We will now use this as an input for Quillen's strategy for proving homological stability for the groups G_n. It turns out that W_n is best suited for the spectral sequence argument.

The spectral sequence in Quillen's argument is obtained as follows. Let $E_\bullet G_n$ be a free resolution of $\mathbb{Z}$ as a $\mathbb{Z}G_n$-module, and let $\tilde{C}_*(W_n)$ denote the augmented cellular complex of W_n. Tensoring these two objects together, we get a first quadrant double complex

$$C_{\bullet,*} = E_\bullet G_n \otimes_{G_n} \tilde{C}_*(W_n).$$

Filtering $C_{\bullet,*}$ in the first direction gives a spectral sequence whose E^1-page is trivial in a range under the assumption that W_n is highly connected, from which it follows that the spectral sequence coming from filtering in the second direction must converge to zero in a range. By transitivity of the action and Shapiro's lemma, this latter sequence has E^1-term

$$E^1_{p,q} = H_q\big(\mathrm{Stab}(\sigma_p)\big) \cong H_q(G_{n-p-1})$$

under the local cancelation and injectivity assumption of Section 2.1, where there are no twisted coefficients because the stabilizer of a p-simplex in W_n always fixes the simplex pointwise. This spectral sequence allows for an inductive argument. This argument has been written in full details many places, see [63, THM. 3.1] for the case where W_n is precisely the complex of destabilization considered here, or, e.g., [34, THM. 5.1] for a version adaptable to more general simplicial objects W_n.

Theorem 2.9 ([63, THM. 3.1]). *Let $G_n = \mathrm{Aut}_{\mathcal{C}}(A \oplus X^{\oplus n})$ for $\mathcal{C}, \mathcal{G}, A$ and X as above satisfying* (2.1) *and* (2.2), *and assume that for all $n \geq 0$, there is a $k \geq 2$ such that the space $W_n(A, X)$ is $\frac{n-2}{k}$-connected. Then the stabilization map*

$$H_i(G_n) \to H_i(G_{n+1})$$

is an isomorphism for $i \leq \frac{n-1}{k}$ and a surjection for $i \leq \frac{n}{k}$.

Remark 2.10. The paper [63] has two additional assumptions on $\mathcal{G}$: it should have no zero divisors and the unit has no nontrivial isomorphisms, but, as pointed out by Krannich in [39, SECT. 7.3], these two assumptions are not actually necessary. Indeed, these assumptions ensure that $\mathrm{Aut}_{\mathcal{G}}(A \oplus X^{\oplus n}) \cong \mathrm{Aut}_{U\mathcal{G}}(A \oplus X^{\oplus n})$ for $U\mathcal{G} = \langle \mathcal{G}, \mathcal{G} \rangle$ a certain category associated to the groupoid $\mathcal{G}$ (see Section A), but in fact stability just holds for both groups whether they are equal or not, with the same proof. The paper [63] also only formulates the result for the case of $\mathcal{G}$ acting on itself, but the proof generalizes with no significant change, as noted in [39].

Remark 2.11 (Stability slope). The slope k of stability given by the theorem depends on the slope of connectivity of the spaces $W_n(A, X)$, though with the constrain that the best possible slope is slope 2. This last restriction is due to the structure of the spectral sequence. To obtain a better slope than slope 2 with the spectral sequence described here, one needs additional information about the groups or differentials appearing in the spectral sequence; such better slopes do not follow from a direct inductive argument.

It is an open question whether stability holds if and only if the spaces $W_n(A, X)$ are highly connected, see [63, CONJ. C].

Connectivity of buildings. Stability can only be proved using the above argument under the condition that the spaces W_n (the above defined spaces of destabilizations or some other appropriate buildings) are highly connected. This is a place where work that depends on the groups in question comes in. Under mild conditions, the connectivity of $W_n(A, X)$ is controlled by that of the associated simplicial complex $S_n(A, X)$, and $S_n(A, X)$ will also typically be (weakly) Cohen–Macaulay, a very useful property in connectivity arguments, see [63, SECT. 2.1].

There are a few general useful facts and tricks that are good to know when working on connectivity questions for such simplicial complexes or semisimplicial sets, see, e.g., [33, SECT. 2], [15, SECT. 2,4,5], or [34, SECT. 3], for expositions of tools and techniques. For an example of how such arguments look like, the survey paper [72] gives a proof of high connectivity of simplicial complexes of arcs relevant for the stability of the mapping class groups of surfaces, assembling tricks and techniques from the literature.

2.3. Twisted coefficients

Homological stability is also often considered in the context of homology with twisted coefficients: Given a sequence of groups $G_1 \to G_2 \to \cdots$, and a sequence of modules $M_1 \to M_2 \to \cdots$ such that G_n acts on M_n and the map $M_n \to M_{n+1}$ is equivariant with respect to the map $G_n \to G_{n+1}$, one can ask whether the resulting sequence

$$H_i(G_1, M_1) \to H_i(G_2, M_2) \to H_i(G_3, M_3) \to \cdots$$

stabilizes. We explain briefly here how the same assumptions as Theorem 2.9 yield that stability also holds for certain types of "abelian" and "polynomial"coefficients.

Abelian coefficients. Suppose that M is a G_∞-module. Then we can consider M as a G_n-module via the maps $G_n \to G_\infty = \bigcup_n G_n$. If we let M_n be this module, this gives an example of a compatible family of coefficients for the groups G_n. We say that M is *abelian* if the action of G_∞ factors through its abelianization $H_1(G_\infty)$.

Theorem 2.12 ([63, THM. 3.4]). *Let $G_n = \mathrm{Aut}_{\mathcal{C}}(A \oplus X^{\oplus n})$ be as in Theorem 2.9 and assume that for all $n \geq 0$, there is a $k \geq 3$ such that the space $W_n(A, X)$ is $\frac{n-2}{k}$-connected. Then for any $H_1(G_\infty)$-module M, the stabilization map*

$$H_i(G_n; M) \to H_i(G_{n+1}; M)$$

is an isomorphism for $i \leq \frac{n-k}{k}$ and a surjection for $i \leq \frac{n-k+2}{k}$.

The simplest example of such an abelian coefficient system is $M = \mathbb{Z}H_1(G_\infty)$. Because untwisted homological stability gives that $H_1(G_\infty) \cong H_1(G_n)$ for n large enough, we have that the twisted homology in that case computes the homology of the commutator subgroup. A direct corollary is thus that, under the same hypothesis as Theorem 2.9 (with $k \geq 3$), homological stability also holds for the commutator subgroups G'_n: the stabilization map also induces isomorphisms

$$H_i(G'_n) \xrightarrow{\cong} H_i(G'_{n+1})$$

for $i \leq \frac{n-k}{k}$ and a surjection for $i \leq \frac{n-k+2}{k}$. This gives, for example, homological stability for alternating groups (= commutator subgroups of symmetric groups), or special automorphism groups of free groups (= commutator subgroups of $\mathrm{Aut}(F_n)$).

Note that the best possible slope given by the statement is now slope 3. This is optimal as stated because we know from [35, PROP. B] that slope 3 is optimal for alternating groups, despite the fact that the spaces $W_n(A, X)$ in this case are slope 2 connected.

Polynomial coefficients. Twisted coefficients classically used in homological stability have been of "polynomial type," as introduced by Dwyer in [14] in the case of general linear groups. It turns out that polynomiality in the sense of Dwyer makes sense in our current general framework of groups of the form $G_n = \mathrm{Aut}_{\mathcal{C}}(A \oplus X^{\oplus n})$, as we explain now.

To define a coefficient system for the groups G_n, we need the data of a module M_n over G_n for each n, and a map $M_n \to M_{n+1}$ compatible with the actions. We will here encode this data in a functor from a category built from the $\mathcal{G}$-module $\mathcal{C}$, in similar fashion as the spaces $W_n(A, X)$ were build from $\mathcal{G}$ and $\mathcal{C}^2$: Let $\mathcal{C}_{A,X}$ be the category with objects $A \oplus X^{\oplus n}$ and morphisms from $A \oplus X^{\oplus m}$ to $A \oplus X^{\oplus n}$ empty unless $m \leq n$, in which case

2 This is again an example of a bracket construction for categories, as described in Section A.

a morphism is an equivalence class of maps $f : A \oplus X^{\oplus n} \to A \oplus X^{\oplus n}$ in $\mathcal{C}$, with $f \sim f'$ if there is an isomorphism $g : X^{\oplus n-m} \to X^{\oplus n-m}$ in $\mathcal{G}$ such that $f = f' \circ (\mathrm{id} \oplus g)$.

A functor $M : \mathcal{C}_{A,X} \to R$–Mod defines a coefficient system in the above sense, by setting $M_n := M(A \oplus X^{\oplus n})$. Because of the equivalence relation in the definition of the morphisms in $\mathcal{C}_{A,X}$, such coefficient systems have the particularity that $\mathrm{Aut}_{\mathcal{G}}(X^{\oplus m})$ acts trivially on the image of the map $M_n \to M_{n+m}$; they are in fact characterized by this property [63, PROP. 4.2].

Using the braiding of $\mathcal{G}$, we can define a functor

$$\Sigma_X : \mathcal{C}_{A,X} \to \mathcal{C}_{A,X}$$

that adds a copy of X "to the left," taking $A \oplus X^{\oplus n}$ to $A \oplus X^{\oplus n+1}$, and a morphism f to the composition $(\mathrm{id}_A \oplus b_{X,X^{\oplus n}}) \circ (f \oplus \mathrm{id}_X) \circ (\mathrm{id}_A \oplus b^{-1}_{X,X^{\oplus n}})$. This functor comes with a natural transformation $\sigma_X : \mathrm{id} \Rightarrow \Sigma_X$ (see [63, 4.2]). For $M : \mathcal{C}_{A,X} \to R$–Mod, we define its suspension

$$\Sigma M = M \circ \Sigma_X : \mathcal{C}_{A,X} \to R\text{–Mod}.$$

It comes with a natural transformation $M \to \Sigma M$ induced by σ_X.

A *finite degree coefficient system* is defined inductively as follows: the trivial coefficient system $M \equiv 0$ is by definition of degree -1, and a coefficient system M is of *degree* r if the natural transformation $M \to \Sigma M$ has trivial kernel, and cokernel of degree $r-1$ [63, DEF. 4.10]. For example, constant coefficient systems are of degree 0, and all finitely presented FI-modules are coefficient systems of finite degree for the symmetric groups [63, PROP. 4.18]. The Burau representation of the braid group is an example of a coefficient system of degree 1 [63, EX. 4.15].

Theorem 2.13 ([63, THM. A]). *Under the same hypothesis as Theorem* 2.9, *if* $\{M_n\}_{n\in\mathbb{N}}$ *is a polynomial coefficient system of degree* r, *then*

$$H_i(G_n; M_n) \to H_i(G_{n+1}; M_{n+1})$$

is an isomorphism for $i \leq \frac{n}{k} - r - 1$ *and a surjection for* $i \leq \frac{n}{k} - r$.

3. GROUP COMPLETION AND THE STABLE HOMOLOGY

The fact that the groupoid $\mathcal{G}$ is braided or symmetric monoidal has direct implications for the stable homology of the groups $G_n = \mathrm{Aut}_{\mathcal{C}}(A \oplus X^{\oplus n})$ we have been considering here. We briefly discuss here the case of automorphism groups $G_n = \mathrm{Aut}_{\mathcal{G}}(X^{\oplus n})$ in $\mathcal{G}$, and refer to [63, SECT. 3.2] for more details, and for some words about the case $G_n = \mathrm{Aut}_{\mathcal{G}}(A \oplus X^{\oplus n})$.

E_n-algebras. A (topological) E_n-algebra is an algebra over the little n-disc operad. When $n = 1$, such an object goes also under the name A_∞-algebra; it is a space with a multiplication that is associative "up to all higher homotopies". When $n \geq 2$, the multiplication is in addition homotopy commutative, with "more and more" homotopies as n grows, all the way to an E_∞-algebra that is commutative up to all higher homotopies. In particular, any E_n-algebra is a topological monoid, that is homotopy commutative whenever $n \geq 2$. (See, e.g., [3, 48].)

These algebraic structures are relevant for us for the following reason: the geometric realization $|\mathcal{G}|$ of the nerve of a monoidal, braided monoidal, or symmetric monoidal category $\mathcal{G}$ is respectively an E_1-, E_2-, or E_∞-algebra, see, e.g., [47], [19, SECT. 8]. When $\mathcal{C}$ is a module over a braided monoidal groupoid $\mathcal{G}$, then $|\mathcal{C}|$ is an *E_1-module over the E_2-algebra* $|\mathcal{G}|$ in the sense of [39].

The primary example of an E_n-algebra is the n-fold loop space $\Omega^n X = \mathrm{Maps}_*(S^n, X)$ of a space X. For $n = \infty$, an ∞-loop space is an n-fold loop space $Y = \Omega^n X_n$ for every n, where the spaces X_n together form a *spectrum* $\mathbb{X}$. Loops have the particularity that they possess homotopy inverses with respect to concatenation, which is the monoid structure underlying their E_n-algebra structure. The *recognition principle* for iterated loop spaces says that, after "group completion," i.e., after adding homotopy inverses, any E_n-algebra is an n-fold loop space [48] (see also [3, 65]). Explicitly, the group completion of a topological monoid $(M, \oplus)$ is the space $\Omega B_\oplus M$, where $B_\oplus$ denotes the bar construction, a simplicial space constructed from M and the sum $\oplus$. The *group completion theorem* states that, if $(M, \oplus)$ is homotopy commutative, then $H_*(\Omega B_\oplus M) \cong H_*(M_\infty)$ for M_∞ an appropriate "limit" space defined from M, see [50, 61].

Applying this to the realization $|\mathcal{G}|$ of a braided monoidal groupoid, we get that its group completion $\Omega B_\oplus |\mathcal{G}|$ is a double loop space $\Omega^2 X$, or an infinite loop space $\Omega^\infty \mathbb{X}$ if $\mathcal{G}$ was actually symmetric. For $\mathcal{G}$ of the form $\mathcal{G} = \coprod_{n \geq 0} G_n$ with $G_n = \mathrm{Aut}_{\mathcal{G}}(X^{\oplus n})$, the limit space $|\mathcal{G}|_\infty$ identifies with $\mathbb{Z} \times BG_\infty$ for $G_\infty = \bigcup_n G_n = \mathrm{colim}(G_0 \to G_1 \to G_2 \to \cdots)$, and the group completion theorem thus takes the form $H_*(\Omega B_\oplus |\mathcal{G}|) \cong H_*(\mathbb{Z} \times BG_\infty)$. Equivalently, it gives that the stable homology of the groups G_n has the following form:

$$H_*(G_\infty) \cong H_*(\Omega_0 B_\oplus |\mathcal{G}|) \cong \begin{cases} H_*(\Omega_0^2 X) & \text{if } \mathcal{G} \text{ is braided,} \\ H_*(\Omega_0^\infty \mathbb{X}) & \text{if } \mathcal{G} \text{ is symmetric,} \end{cases}$$

for some space X, respectively spectrum $\mathbb{X}$, just as in the examples we have seen so far, namely Theorems 1.2–1.5. The work in identifying the stable homology of a family of groups thus comes down, through these classical results, to the question of identifying certain double or, most often, infinite loop spaces arising as classifying spaces of groupoids. Considered very broadly, this is the subject of K-theory. In Section 4, we sketch one such computation.

"Higher" stability and the E_k-splitting complex. The stabilization maps we study here only use a very small part of the E_2- or E_∞-structures we have at hand: taking the sum $\oplus X$ just uses part of the underlying E_1-module structure. The space of destabilizations $W_n(A, X)$ associated to the E_1-module $|\mathcal{C}|$ over the E_2-algebra $|\mathcal{G}|$ and the elements $A \in |\mathcal{C}|$ and $X \in |\mathcal{G}|$, can be thought of as a form of resolution of the space $\coprod_n BG_n$ as E_1-module generated by A over the E_2-algebra generated by X.

One can ask whether there are interesting "higher" stabilization maps, summing for example with higher dimensional homology classes, or whether the whole E_k-structure can tell us more about the homology of the family of groups. The answer is yes, and is the subject of the body of work [21–24] (see also [51] in the context of representation stability). Considering the full E_k-structure has turned out to be powerful, and these papers manage to go further than with the classical arguments, including to obtain information about the homology past the stable range. (See also the related paper [38].) The authors define E_k-*splitting complexes* that resolve the full E_k-structure. For a relationship between the connectivity of the spaces $W_n(A, X)$ defined here and that of the E_1-splitting complex, see [38, THM. 13.2].

4. HIGMAN–THOMSON GROUPS

Sometimes homological stability is useful in unexpected situations, as turned out to be the case in the study of the homology of Thompson's group V. Thompson's groups come in three flavors: $F < T < V$ where F is a subgroup of the piecewise-linear homeomorphisms of the interval, T a subgroup of those of the circle, and V of the homeomorphisms of the Cantor set. The homology of F and T was computed in the 1980s by Brown–Geoghegan and Ghys–Sergiescu in [8, 27]. Brown proved a few years later that the rational homology of Thompson's group V was trivial, and conjectured that it was also integrally trivial [7]. Brown's conjecture was proved 25 years later by Szymik and the author in [68] using the following unexpected strategy:

(1) $V = V_1$ is part of a family of groups $V_1 \to V_2 \to V_3 \to \cdots$ that satisfies homological stability;

(2) The homology $H_*(V)$ is entirely stable, i.e., $H_*(V) \cong H_*(V_\infty)$;

(3) The stable homology identifies with that of a trivial infinite loop space.

In fact, we will see below that each group V_n in the sequence is isomorphic to V, but the maps $V_n \to V_{n+1}$ are only isomorphisms after passing to homology. The strategy works more generally to compute the homology of the Higman–Thompson groups, so we describe it now in more details in that context.

The Higman–Thompson group $V_{k,n}$ is the group of self-maps of a disjoint union of n intervals $I^{\sqcup n}$ obtained by choosing k-ary subdivisions of the source and target, subdividing the interval into k equal sized subintervals and repeating on some of the intervals thus obtained, and matching the resulting subintervals by a chosen bijection. (See [68, SECT. 1.2], and Figure 2 for an example when $k = 2$ and $n = 1$.) Thompson's group $V = V_{2,1}$ is the group obtained this way from binary subdivisions of a single interval. Fixing some $k \geq 2$, we can think of $V_{k,n}$ as the automorphism group of an object $X^{\oplus n} = I^{\sqcup n}$ in a groupoid $\mathcal{V}_k = \bigsqcup_{n \geq 0} V_{k,n}$, just as we have considered in this paper. Juxtaposition of intervals induces maps

$$V_{k,n} \times V_{k,m} \to V_{k,n+m}$$

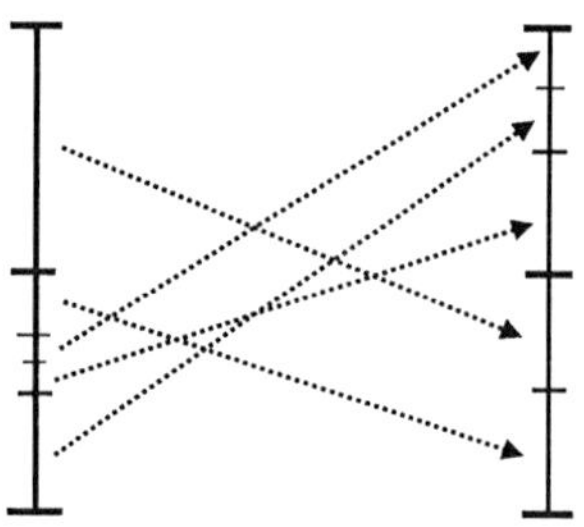

FIGURE 2

An element of Thompson's group $V = V_{2,1}$ obtained from a binary subdivision of the source and target interval, and a choice of permutation of the subintervals.

that make the groupoid $\mathcal{V}_k$ symmetric monoidal, with the symmetries coming for block permutations of the intervals. Hence we can try to apply the stability machine described in the present paper to prove homological stability for the groups $\{V_{k,n}\}_{n\geq 0}$.

Note that there are group isomorphisms $V_{k,n} \cong V_{k,n+(k-1)}$ induced by subdividing an interval into k subintervals, but these isomorphisms are not encoded in the groupoid $\mathcal{V}_k$. For the purpose of homological stability, it is convenient to have a rank function, that is, to know what "n" is at all times. Ignoring these isomorphisms also gives, by construction, the local cancelation property (2.1) which was necessary for the transitivity of the action on the associated complex of destabilization W_n. So from the point of view of the groups, the objects I and $I^{\sqcup k}$ are isomorphic, but we will consciously suppress that information in the first part of our argument.

Let $W_n = W_n(0, I)$ be the space of destabilizations associated to the symmetric monoidal groupoid $\mathcal{V}_k$ (acting on itself) and the objects 0 and I, and let $S_n = S_n(0, I)$ be its associated simplicial complex, as defined in Section 2.1. The group $V_{k,n}$ can be defined as the automorphism group of an object called the free *Cantor algebra* $C_k(n)$ of arity k on n generators (see **[68, DEF. 1.1]**), and a p-simplex in W_n corresponds to an embedding $C_k(p+1) \hookrightarrow C_k(n)$ with complement isomorphic to $C_k(n-p-1)$. It is shown in **[68, COR. 3.4]** that S_n, and hence also W_n (by **[63, THM. 2.10]**), is at least $(n-3)$-connected for all $n \geq 2$. The complex S_n has dimension $n-1$ and the idea of the proof of connectivity is to work with its $(n-2)$-skeleton, as simplices that are not maximal correspond to embeddings that have a complement of rank at least 1, i.e., at least as big as $C_k(1)$. But there are isomorphisms $C_k(1) \cong C_k(1+(k-1)) \cong C_k(1+2(k-1)) \cong \cdots$ so that in practice, a nontrivial complement is actually a complement that is "as large as one likes," which is useful for coning off simplices.

Applying Theorem 2.9, we immediately get that the stabilization map $V_{k,n} \to V_{k,n+1}$ that adds the identity on the new interval, induces an isomorphism $H_i(V_{k,n}) \xrightarrow{\cong} H_i(V_{k,n+1})$ in a range increasing with n. Coupling this with the fact that the isomorphisms $V_{k,n} \cong V_{k,n+(k-1)} \cong V_{k,n+2(k-1)} \cong \cdots$ can be chosen compatibly with the stabilization maps, we get that the rank n can be assumed as large as one like, so that the isomorphism $H_i(V_{k,n}) \xrightarrow{\cong} H_i(V_{k,n+1})$ actually holds without any bound.

It remains to compute the stable homology. From the results described in Section 3, given that $\mathcal{V}_k$ is a symmetric monoidal groupoid, we know that the stable homology of the groups is that of an infinite loop space. Now here it turns out to be more convenient to do the computation using a different symmetric monoidal groupoid whose group completion also yields the stable homology of the groups $V_{k,n}$, namely the groupoid $\overline{\mathcal{V}}_k$ where we now remember the isomorphism $I \to I^{\sqcup k}$ or, equivalently, the isomorphisms of Cantor algebras $C_k(n) \cong C_k(n + (k - 1))$. Theorem 5.4 of [68] says that

$$H_*(V_{k,\infty}) \cong H_*(\Omega_0 B_\oplus |\overline{\mathcal{V}}_k|).$$

As $\overline{\mathcal{V}}_k$ is symmetric monoidal, we again have that its group completion is an infinite loop space and what remains is to find out what the corresponding spectrum is.

So now we are in the world of symmetric monoidal categories, and the idea is simply to find a symmetric monoidal category that is equivalent to $\overline{\mathcal{V}}_k$ as symmetric monoidal category, and hence group completes to the same infinite loop space, but whose associated spectrum is easier to recognize. Our search was guided by the following observation: the category $\overline{\mathcal{V}}_k$ resembles the category of finite sets and isomorphisms, to which one has declared one extra isomorphism, namely that $[1]$ is now isomorphic to $[k]$, or, after group completion, $[0]$ is isomorphic to $[k-1]$. As already mentioned in the introduction, the spectrum associated to the category of finite sets (or, equivalently, to the symmetric groups) is the sphere spectrum $\mathbb{S}$. In homotopy theory, we trivialize by taking cofibers, and the cofiber of the map $\mathbb{S} \xrightarrow{(k-1)} \mathbb{S}$ multiplying by $(k-1)$ is a well-known spectrum $\mathbb{M}_{k-1}$ called the Moore spectrum. Making this idea precise, formulating it on the level of symmetric monoidal categories, and combining it with the homological stability result described above, led to the following result

Theorem 4.1 ([68]). *There are isomorphisms*

$$H_*(V_{n,k}) \cong H_*\big(\Omega_0^\infty \mathbb{M}_{k-1}\big).$$

Specializing to the case $k = 2$ *yields that* $H_*(V) = 0$ *for* $* > 0$ *as the spectrum* $\mathbb{M}_1 = \text{cofiber}(\mathbb{S} \xrightarrow{\text{id}} \mathbb{S})$ *is trivial.*

Note that the homology of $\Omega_0^\infty \mathbb{M}_{k-1}$ for $k \geq 3$ is tractable, and we have many tools available to compute it. For example, it is immediate that the rational homology is trivial, but also that the integral homology is not. We confirm, for instance, in [68, SECT. 6] that $H_1(V_{k,n}) = \mathbb{Z}/2$ for k odd and show that the first nontrivial homology group in the k even case is $H_{2p-3}(V_{k,n}) = \mathbb{Z}/p$ for p the smallest prime dividing $k-1$.

5. PERSPECTIVES

Many stability results have been proved over the past decades, and one is left to wonder how far homological stability methods can reach. We have highlighted here the idea that braidings seem to be relevant. This is, however, neither a necessary nor a sufficient condition. We give here some examples that tests the limits of stability, as well as a hint to the wider context homological stability can be considered in.

No braiding = no stability? Such a statement is not going to ever be literally true, but here are some standard types of examples that are good to have in mind: The full braid groups $G_n = B_n$ satisfy homological stability, but not the pure braid groups $K_n = \ker(B_n \to \Sigma_n)$. Likewise, the general linear groups $G_n = \mathrm{GL}_n(R)$ satisfy stability for many rings R but not the congruence subgroups $K_n = \mathrm{GL}_n(R, I) = \ker(\mathrm{GL}_n(R) \to \mathrm{GL}_n(I))$. There are in fact many examples of that form with a family of groups $K_n < G_n$ with the groupoid $\mathcal{G} = \bigsqcup G_n$ braided monoidal while the groupoid $\mathcal{K} = \bigsqcup K_n$ is monoidal but not braided, and with the family G_n stabilizing but not the family K_n. It turns out that such families $\{K_n\}_{n\geq 0}$ often satisfy instead a form of *representation stability* in the sense of [11], see also [16, 55].

Braiding $\not\Rightarrow$ stability. There are very few examples of braided monoidal categories where we know that homological stability for the associated groups G_n does not hold. One such example, constructed by Patzt [54], is the following: consider the category of sets, but using the product $\times$ instead of the disjoint union as monoidal structure. This is a symmetric monoidal groupoid, and if we pick $A = [1]$ and $X = [2]$, we get $G_n = \Sigma_{2^n}$ is the symmetric group on 2^n elements. The resulting space $W_n(A, X)$ is, however, disconnected in this case! And indeed, even though the symmetric groups satisfy homological stability, the stabilization maps in this case do not induce isomorphisms; the induced map on first homology is instead the zero map. So the existence of a braiding does not imply stability, which in hindsight is probably not surprising.

There are in addition plenty of examples where we have a braided monoidal groupoid at hand but we do not know that stability holds. For example, the category of R-modules over any ring R is symmetric monoidal, but stability for the groups $\mathrm{GL}_n(R)$ is essentially only known under the condition that the ring has finite Bass stable range [69]. But examples of rings for which we know that stability for $\mathrm{GL}_n(R)$ does not hold are surprisingly rare; see [41] for one example of a ring for which $H_1(\mathrm{GL}_n(R))$ does not stabilize. For mapping class groups or diffeomorphism groups of manifolds, we essentially know stability in full generality in dimension 2 and 3, but in higher dimension, homological stability for the classifying spaces of diffeomorphism groups is only known for stabilization by connected sums with certain products $S^p \times S^q$ [5, 26]. Similarly, homological stability for the automorphism groups of vector spaces equipped with a form (symplectic, unitary, or orthogonal groups), is mostly known in the particular case of stabilizing with the hyperbolic form, see, e.g., [66]. In the other cases, we just do not know the connectivity of the complex of destabilizations.

Homological stability in other contexts. We have already mentioned a number of stability results for sequences of spaces. The most classical examples are configuration spaces, going back to the work of McDuff, Segal and F. Cohen in the 1970s [12, 49, 65]. In other contexts, examples seem to be more rare so far, but there is currently a growing interest in stability in the homology of families of algebras, see, e.g., [6, 37, 67], and there exist, e.g., some results for bounded cohomology of groups [43]. These results are of a very similar flavor as what we have described in the present paper.

A. ADDING COMPLEMENTS CATEGORICALLY

The semisimplicial sets $W_n(A, X)$ of Section 2.1 and the categories $\mathcal{C}_{A,X}$ used to define polynomial coefficients in Section 2.3, were constructed using equivalence classes of maps in the groupoid $\mathcal{C}$. Both these constructions are related to a categorical construction, first considered by Quillen in the context of K-theory [28, p. 219]. We recall this construction here and give a few examples. The resulting categories will be natural "homes" of the spaces $W_n(A, X)$, and for the polynomial twisted coefficients, which gives some insights.

Let $\mathcal{M}$ be a category, that is, a left module over a monoidal groupoid $(\mathcal{G}, \oplus)$. We define a category $\langle \mathcal{G}, \mathcal{M} \rangle$ as follows: $\langle \mathcal{G}, \mathcal{M} \rangle$ has the same objects as $\mathcal{M}$, and morphisms from A to B are defined as equivalence classes of pairs (X, f) with X an object of $\mathcal{G}$ and $f : X \oplus A \to B$ a morphism of $\mathcal{M}$, where $(X, f) \sim (X', f)$ if there is a commuting diagram

$$\begin{array}{ccc} X \oplus A & \xrightarrow{\ f\ } & B \\ {\scriptstyle g\oplus \mathrm{id}}\downarrow & \nearrow{\scriptstyle f'} & \\ X' \oplus A & & \end{array}$$

in $\mathcal{M}$. (If $\mathcal{C}$ is a right module instead, a category $\langle \mathcal{C}, \mathcal{G} \rangle$ is defined analogously.) When $\mathcal{M}$ is a groupoid, as will be the case in our examples, the maps f are isomorphisms and the object X can be thought of as a choice of complement for A inside B.

We will here only consider the case where $\mathcal{M} = \mathcal{G}$ is a monoidal groupoid acting on itself, and denote by $U\mathcal{G} = \langle \mathcal{G}, \mathcal{G} \rangle$ the resulting category.

Example A.1. Let $(\mathcal{G}, \oplus) = (\mathrm{Sets}^{\mathrm{iso}}, \sqcup)$ be the monoidal groupoid of finite sets and bijections of Example 2.1, with the monoidal structure induced by disjoint unions. Then $U\mathcal{G} = \mathrm{FI}$ is the category of finite sets and injections. Indeed, any injection $f : A \hookrightarrow B$ has, up to isomorphism, a unique complement $X = B \backslash f(A)$.

Example A.2. Let $(\mathcal{G}, \oplus) = (R\text{–Mod}, \oplus)$ be the groupoid of R-modules and isomorphisms of Example 2.2, with the monoidal structure given by direct sum. Then $U\mathcal{G}$ is closely related to the category sometimes called VIC, with the same objects as R–Mod and with morphisms from M to N given by pairs (H, f) with $f : M \to N$ a split injective homomorphism and H a choice of complement in N of the image, $N = H \oplus f(M)$ (see, e.g., [58]).

If the monoidal groupoid $(\mathcal{G}, \oplus)$ is braided, one can define a monoidal structure on $U\mathcal{G}$ as follows: on objects the monoidal structure $\oplus$ is that of $\mathcal{G}$, and for $[X, f]$ a morphism from A to B and $[Y, g]$ a morphism from C to D, we set

$$[X, f] \oplus [Y, g] = \big[X \oplus Y, f \oplus g \circ \mathrm{id}_X \oplus b_{A,Y}^{-1} \oplus \mathrm{id}_C\big] : A \oplus C \to B \oplus D$$

where we use the braiding to switch A and Y in $X \oplus Y \oplus A \oplus C$ to be able to apply the morphism $f \oplus g$. The category $U\mathcal{G}$ is not in general braided (see the next example), though it is symmetric when $(\mathcal{G}, \oplus)$ is a symmetric monoidal groupoid, see [63, PROP. 1.8].

Example A.3. The braid groups B_n form together a groupoid $\mathcal{B} = \bigsqcup_n B_n$, that is, the free braided monoidal groupoid on one element, where the monoidal structure comes from the juxtaposition of braids. The category $U\mathcal{B}$ can be described in terms of braids with free ends: a morphism from m to n for $m \leq n$ in $U\mathcal{B}$ is an equivalence class of braid in B_n where the braid has $n - m$ free ends that can freely pass under, but not over, any other strand, see [63, SECT. 1.2]. (It can alternatively be defined in terms of embeddings of punctured discs, see [63, SECT. 5.6.2].) The category $U\mathcal{B}$ is not braided monoidal, but only *prebraided* in the sense of [63, DEF. 1.5].

Remark A.4. The forgetful map $B_n \to \Sigma_n$ from the braid groups to the symmetric groups induces a map $U\mathcal{B} \to \mathrm{FI} = U(\mathrm{Sets}^{\mathrm{iso}})$. Because $\mathcal{B}$ is the free braided monoidal category on one object, it encodes all the structure we have when we picked objects A and X in the groupoids $\mathcal{C}$ and $\mathcal{G}$ in Section 2. As pointed out in [39, REMARK 2.8], the reason we can construct a semisimplicial set $W_n(A, X)$ comes from the following: A semisimplicial set is a functor $\Delta_{\mathrm{inj}}^{\mathrm{op}} \to \mathrm{Sets}$ for Δ_{inj} the category of finite ordered sets and ordered injections. One can consider Δ_{inj} as a subcategory of the category FI of finite sets and injections. Now while the forgetful map $U\mathcal{B} \to \mathrm{FI}$ does not admit a splitting, it does admit a partial splitting, in the form of a functor $\Delta_{\mathrm{inj}} \to U\mathcal{B}$, and this partial splitting is what rules the semisimplicial structure of the space of destabilization $W_n(A, X)$.

ACKNOWLEDGMENTS

My interest in homological stability originates in the work of Tillmann and Madsen–Weiss on the moduli space of Riemann surfaces. My thanks goes to them, as well as Ruth Charney, Bill Dwyer, John Harer, Nikolai Ivanov, Dan Quillen, and Willem van der Kallen, for the beautiful papers that inspired much of the research presented here, and, of course, also to my stability-collaborators Giovanni Gandini, Allen Hatcher, Oscar Randal-Williams, David Sprehn, Markus Szymik, and Karen Vogtmann. Finally, I would like to thank Søren Galatius, Manuel Krannich, and Oscar Randal-Williams for thoughtful comments on earlier versions of this paper.

FUNDING

This work was partially supported by the Danish National Research Foundation through the Copenhagen Centre for Geometry and Topology (DNRF151), and the European Research Council (ERC) under the European Union's Horizon 2020 research and innovation programme (grant agreement No. 772960).

REFERENCES

[1] V. I. Arnold, On some topological invariants of algebraic functions. *Tr. Mosk. Mat. Obŝ.* **21** (1970), 27–46.

[2] M. Barratt and S. Priddy, On the homology of non-connected monoids and their associated groups. *Comment. Math. Helv.* **47** (1972), 1–14.

[3] J. M. Boardman and R. M. Vogt, *Homotopy invariant algebraic structures on topological spaces*. Lecture Notes in Math. 347, Springer, Berlin–New York, 1973.

[4] S. K. Boldsen, Improved homological stability for the mapping class group with integral or twisted coefficients. *Math. Z.* **270** (2012), no. 1–2, 297–329.

[5] B. Botvinnik and N. Perlmutter, Stable moduli spaces of high-dimensional handlebodies. *J. Topol.* **10** (2017), no. 1, 101–163.

[6] R. Boyd and R. Hepworth, On the homology of the Temperley–Lieb algebras. 2020, arXiv:2006.04256.

[7] K. S. Brown, The geometry of finitely presented infinite simple groups. In *Algorithms and classification in combinatorial group theory (Berkeley, CA, 1989)*, pp. 121–136, Math. Sci. Res. Inst. Publ. 23, Springer, New York, 1992.

[8] K. S. Brown and R. Geoghegan, An infinite-dimensional torsion-free FP_∞ group. *Invent. Math.* **77** (1984), no. 2, 367–381.

[9] J.-L. Cathelineau, Homology stability for orthogonal groups over algebraically closed fields. *Ann. Sci. Éc. Norm. Supér. (4)* **40** (2007), no. 3, 487–517.

[10] R. Charney, On the problem of homology stability for congruence subgroups. *Comm. Algebra* **12** (1984), no. 17–18, 2081–2123.

[11] T. Church, J. S. Ellenberg, and B. Farb, FI-modules and stability for representations of symmetric groups. *Duke Math. J.* **164** (2015), no. 9, 1833–1910.

[12] F. R. Cohen, T. J. Lada, and J. P. May, *The homology of iterated loop spaces*. Lecture Notes in Math. 533, Springer, Berlin–New York, 1976.

[13] G. Collinet, A. Djament, and J. T. Griffin, Stabilité homologique pour les groupes d'automorphismes des produits libres. *Int. Math. Res. Not. IMRN* **19** (2013), 4451–4476.

[14] W. G. Dwyer, Twisted homological stability for general linear groups. *Ann. of Math. (2)* **111** (1980), no. 2, 239–251.

[15] J. Ebert and O. Randal-Williams, Semisimplicial spaces. *Algebr. Geom. Topol.* **19** (2019), no. 4, 2099–2150.

[16] B. Farb, Representation stability. In *Proceedings of the International Congress of Mathematicians—Seoul 2014. Vol. II*, pp. 1173–1196, Kyung Moon Sa, Seoul, 2014.

[17] F. D. Farmer, Cellular homology for posets. *Math. Jpn.* **23** (1978/1979), no. 6, 607–613.

[18] Z. Fiedorowicz and S. Priddy, *Homology of classical groups over finite fields and their associated infinite loop spaces*. Lecture Notes in Math. 674, Springer, Berlin, 1978.

[19] Z. Fiedorowicz, M. Stelzer, and R. M. Vogt, Homotopy colimits of algebras over cat-operads and iterated loop spaces. *Adv. Math.* **248** (2013), 1089–1155.

[20] S. Galatius, Stable homology of automorphism groups of free groups. *Ann. of Math. (2)* **173** (2011), no. 2, 705–768.

[21] S. Galatius, A. Kupers, and O. Randal-Williams, E_∞-cells and general linear groups of finite fields. 2018, arXiv:1810.11931.

[22] S. Galatius, A. Kupers, and O. Randal-Williams, Cellular E_k-algebras. 2018, arXiv:1805.07184.

[23] S. Galatius, A. Kupers, and O. Randal-Williams, E_2-cells and mapping class groups. *Publ. Math. Inst. Hautes Études Sci.* **130** (2019), 1–61.

[24] S. Galatius, A. Kupers, and O. Randal-Williams, E_∞-cells and general linear groups of infinite fields. 2020, arXiv:2005.05620.

[25] S. Galatius and O. Randal-Williams, Stable moduli spaces of high-dimensional manifolds. *Acta Math.* **212** (2014), no. 2, 257–377.

[26] S. Galatius and O. Randal-Williams, Homological stability for moduli spaces of high dimensional manifolds. I. *J. Amer. Math. Soc.* **31** (2018), no. 1, 215–264.

[27] E. Ghys and V. Sergiescu, Sur un groupe remarquable de difféomorphismes du cercle. *Comment. Math. Helv.* **62** (1987), no. 2, 185–239.

[28] D. Grayson, Higher algebraic K-theory. II (after Daniel Quillen). In *Algebraic K-theory (Proc. Conf., Northwestern Univ., Evanston, IL, 1976)*, pp. 217–240, Lecture Notes in Math. 551, Springer, Berlin, 1976.

[29] J. L. Harer, Stability of the homology of the mapping class groups of orientable surfaces. *Ann. of Math. (2)* **121** (1985), no. 2, 215–249.

[30] A. Hatcher, Homological stability for automorphism groups of free groups. *Comment. Math. Helv.* **70** (1995), no. 1, 39–62.

[31] A. Hatcher and K. Vogtmann, Cerf theory for graphs. *J. Lond. Math. Soc. (2)* **58** (1998), no. 3, 633–655.

[32] A. Hatcher and K. Vogtmann, Rational homology of $\mathrm{Aut}(F_n)$. *Math. Res. Lett.* **5** (1998), no. 6, 759–780.

[33] A. Hatcher and K. Vogtmann, Tethers and homology stability for surfaces. *Algebr. Geom. Topol.* **17** (2017), no. 3, 1871–1916.

[34] A. Hatcher and N. Wahl, Stabilization for mapping class groups of 3-manifolds. *Duke Math. J.* **155** (2010), no. 2, 205–269.

[35] J.-C. Hausmann, Manifolds with a given homology and fundamental group. *Comment. Math. Helv.* **53** (1978), no. 1, 113–134.

[36] R. Hepworth, Homological stability for families of Coxeter groups. *Algebr. Geom. Topol.* **16** (2016), no. 5, 2779–2811.

[37] R. Hepworth, Homological stability for Iwahori–Hecke algebras. 2020, arXiv:2006.04252.

[38] R. Hepworth, On the edge of the stable range. *Math. Ann.* **377** (2020), no. 1–2, 123–181.

[39] M. Krannich, Homological stability of topological moduli spaces. *Geom. Topol.* **23** (2019), no. 5, 2397–2474.

[40] M. Krannich, A homological approach to pseudoisotopy theory. I. 2021, arXiv:2002.04647.

[41] A. Kupers, Rings whose general linear groups do not exhibit homological stability. 2021, https://www.utsc.utoronto.ca/people/kupers/wp-content/uploads/sites/50/instability.pdf.

[42] A. Kupers and O. Randal-Williams, On diffeomorphisms of even-dimensional discs. 2020, arXiv:2007.13884.

[43] C. D. la Cruz Mengual and T. Hartnick, Stability in bounded cohomology for classical groups, I: The symplectic case. 2019, arXiv:1902.01383.

[44] H. Maazen, Homology stability for the general linear group. 1979, http://www.staff.science.uu.nl/~kalle101/maazen1979.pdf.

[45] S. Mac Lane, *Categories for the working mathematician. Second edn.* Grad. Texts in Math. 5, Springer, New York, 1998.

[46] I. Madsen and M. S. Weiss, The stable moduli space of Riemann surfaces: Mumford's conjecture. *Ann. of Math.* **165** (2007), no. 3, 843–941.

[47] J. P. May, E_∞ spaces, group completions, and permutative categories. In *New developments in topology (Proc. Sympos. Algebraic Topology, Oxford, 1972)*, pp. 61–93, London Math. Soc. Lecture Note Ser. 11, Cambridge Univ. Press, Oxford, 1972.

[48] J. P. May, *The geometry of iterated loop spaces*. Lecture Notes in Math. 271, Springer, Berlin–New York, 1972.

[49] D. McDuff, Configuration spaces of positive and negative particles. *Topology* **14** (1975), 91–107.

[50] D. McDuff and G. Segal, Homology fibrations and the "group-completion" theorem. *Invent. Math.* **31** (1975/1976), no. 3, 279–284.

[51] J. Miller and J. C. H. Wilson, Higher-order representation stability and ordered configuration spaces of manifolds. *Geom. Topol.* **23** (2019), no. 5, 2519–2591.

[52] B. Mirzaii and W. van der Kallen, Homology stability for unitary groups. *Doc. Math.* **7** (2002), 143–166 (electronic).

[53] M. Nakaoka, Decomposition theorem for homology groups of symmetric groups. *Ann. of Math. (2)* **71** (1960), 16–42.

[54] P. Patzt, Counterexample to homological stability. 2015, unpublished note.

[55] P. Patzt, Central stability homology. *Math. Z.* **295** (2020), no. 3–4, 877–916.

[56] N. Perlmutter, Homological stability for the moduli spaces of products of spheres. *Trans. Amer. Math. Soc.* **368** (2016), no. 7, 5197–5228.

[57] N. Perlmutter, Homological stability for diffeomorphism groups of high-dimensional handlebodies. *Algebr. Geom. Topol.* **18** (2018), no. 5, 2769–2820.

[58] A. Putman and S. V. Sam, Representation stability and finite linear groups. *Duke Math. J.* **166** (2017), no. 13, 2521–2598.

[59] D. Quillen, Notes 1971, August 8, 1971, http://www.claymath.org/publications/quillen-notebooks.

[60] D. Quillen, On the cohomology and K-theory of the general linear groups over a finite field. *Ann. of Math. (2)* **96** (1972), 552–586.

[61] O. Randal-Williams, 'Group-completion', local coefficient systems and perfection. *Q. J. Math.* **64** (2013), no. 3, 795–803.

[62] O. Randal-Williams, Resolutions of moduli spaces and homological stability. *J. Eur. Math. Soc. (JEMS)* **18** (2016), no. 1, 1–81.

[63] O. Randal-Williams and N. Wahl, Homological stability for automorphism groups. *Adv. Math.* **318** (2017), 534–626.

[64] C.-H. Sah, Homology of classical Lie groups made discrete. I. Stability theorems and Schur multipliers. *Comment. Math. Helv.* **61** (1986), no. 2, 308–347.

[65] G. Segal, Configuration-spaces and iterated loop-spaces. *Invent. Math.* **21** (1973), 213–221.

[66] D. Sprehn and N. Wahl, Homological stability for classical groups. *Trans. Amer. Math. Soc.* **373** (2020), no. 7, 4807–4861.

[67] R. Sroka, *Patterns in the homology of algebras: Vanishing, stability, and higher structures*. Ph.D. thesis, 2021, http://web.math.ku.dk/noter/filer/phd21rs.pdf.

[68] M. Szymik and N. Wahl, The homology of the Higman–Thompson groups. *Invent. Math.* **216** (2019), no. 2, 445–518.

[69] W. van der Kallen, Homology stability for linear groups. *Invent. Math.* **60** (1980), no. 3, 269–295.

[70] K. Vogtmann, Homology stability for $O_{n,n}$. *Comm. Algebra* **7** (1979), no. 1, 9–38.

[71] N. Wahl, Homological stability for the mapping class groups of non-orientable surfaces. *Invent. Math.* **171** (2008), no. 2, 389–424.

[72] N. Wahl, Homological stability for mapping class groups of surfaces. In *Handbook of moduli. Vol. III*, pp. 547–583, Adv. Lect. Math. (ALM) 26, Int. Press, Somerville, MA, 2013.

NATHALIE WAHL

Dept. of Mathematics, University of Copenhagen, Universitetsparken 5, 2100 Copenhagen, Denmark, wahl@math.ku.dk

This Page Intentionally Left Blank

7. LIE THEORY AND GENERALIZATIONS

PBW DEGENERATIONS, QUIVER GRASSMANNIANS, AND TORIC VARIETIES

EVGENY FEIGIN

Dedicated to the memory of Ernest Borisovich Vinberg

ABSTRACT

We present a review on the recently discovered link between the Lie theory, the theory of quiver Grassmannians, and various degenerations of flag varieties. Our starting point is the induced Poincaré–Birkhoff–Witt filtration on the highest weight representations and the corresponding PBW degenerate flag varieties.

MATHEMATICS SUBJECT CLASSIFICATION 2020

Primary 14M15; Secondary 16G20, 17B10, 14M25, 14D06

KEYWORDS

PBW filtration, flag varieties, degenerations, quiver Grassmannians, toric varieties

ICM INTERNATIONAL CONGRESS OF MATHEMATICIANS 2022 JULY 6–14

Proc. Int. Cong. Math. 2022, Vol. 4, pp. 2930–2946
DOI 10.4171/ICM2022/81

Published by EMS Press

1. INTRODUCTION

The celebrated Poincaré–Birkhoff–Witt theorem claims that there exists a filtration on the universal enveloping of a Lie algebra such that the associated graded algebra is isomorphic to the symmetric algebra. The PBW filtration on the universal enveloping algebra of a nilpotent subalgebra of a simple Lie algebra induces a filtration on the representation space of a highest weight module. The natural problem is to study this filtration and the corresponding graded space. Quite unexpectedly, the problem turned out to be related to numerous representation-theoretic, algebro-geometric, and combinatorial questions. Our goal is to give an overview of the whole story and to describe various links between different parts of the picture. The main objects of study are monomial bases, convex polytopes, flag and Schubert varieties, their degenerations, quiver Grassmannians, and toric varieties.

The paper is organized as follows. In Section 2 we collect representation-theoretic results of algebraic nature. Section 3 is devoted to the geometric representation theory. In Section 4 we discuss combinatorics emerging from the cellular decomposition of the PBW degenerate flag varieties. In Section 5 we describe the link between the Lie theory and the theory of quiver Grassmannians. Finally, Section 6 treats toric degenerations.

Throughout the paper we work over the field of complex numbers.

2. REPRESENTATION THEORY: ALGEBRA

Let $\mathfrak{g}$ be a simple Lie algebra with the set R^+ of positive roots. Let α_i, ω_i, $i = 1, \dots, n-1$ be the simple roots and the fundamental weights. Let $\mathfrak{g} = \mathfrak{n}^+ \oplus \mathfrak{h} \oplus \mathfrak{n}^-$ be the Cartan decomposition. For $\alpha \in R^+$, we denote by $e_\alpha \in \mathfrak{n}^+$ and $f_\alpha \in \mathfrak{n}^-$ the corresponding Chevalley generators. We denote by P^+ the set of dominant integral weights.

Consider the PBW filtration on the universal enveloping algebra $\mathrm{U}(\mathfrak{n}^-)$:

$$\mathrm{U}(\mathfrak{n}^-)_s = \operatorname{span}\{x_1 \cdots x_l : x_i \in \mathfrak{n}^-, l \le s\},$$

for example, $\mathrm{U}(\mathfrak{n}^-)_0 = \mathbb{C} \cdot 1$.

For a dominant integral weight $\lambda = m_1\omega_1 + \cdots + m_{n-1}\omega_{n-1}$, let V_λ be the corresponding irreducible highest weight $\mathfrak{g}$-module with a highest weight vector v_λ. Since $V_\lambda = \mathrm{U}(\mathfrak{n}^-)v_\lambda$, we have an increasing filtration $(V_\lambda)_s$ on V_λ,

$$(V_\lambda)_s = \mathrm{U}(\mathfrak{n}^-)_s v_\lambda.$$

We call this filtration the PBW filtration and study the associated graded space $V_\lambda^a = gr\, V_\lambda$.

Let us consider an example for fundamental weights in type A. Let V_{ω_1} be the vector representation of $\mathfrak{sl}_n$ with a basis $w_1, \dots, w_n$ and consider $V_{\omega_k} \simeq \Lambda^k V_{\omega_1}$ for $k = 1, \dots, n-1$. Then $(V_{\omega_k})_s$ is spanned by the wedge products $w_{i_1} \wedge \cdots \wedge w_{i_k}$ such that the number of indices a with $i_a > k$ is at most s.

The following holds true [60]:

(1) The action of $\mathrm{U}(\mathfrak{n}^-)$ on V_λ induces the structure of an $S(\mathfrak{n}^-)$-module on V_λ^a and $V_\lambda^a = S(\mathfrak{n}^-)v_\lambda$.

(2) The action of $\mathrm{U}(\mathfrak{n}^+)$ on V_λ induces the structure of a $U(\mathfrak{n}^+)$-module on V_λ^a.

Our aims are to describe V_λ^a as an $S(\mathfrak{n}^-)$-module and to find a basis of V_λ^a. We present the answer in type A. For similar results in other types, see [12, 61, 62, 77, 78, 99].

The positive roots in type A_{n-1} are of the form $\alpha_{i,j} = \alpha_i + \cdots + \alpha_j$ with $1 \le i \le j \le n-1$. Recall that a Dyck path is a sequence $\mathbf{p} = (\beta(0), \beta(1), \ldots, \beta(k))$ of positive roots of $\mathfrak{sl}_n$ satisfying the following conditions: if $k = 0$, then $\mathbf{p}$ is of the form $\mathbf{p} = (\alpha_i)$ for some simple root α_i, and if $k \ge 1$, then the first and last elements are simple roots, and if $\beta(s) = \alpha_{p,q}$, then $\beta(s+1) = \alpha_{p,q+1}$ or $\beta(s+1) = \alpha_{p+1,q}$.

Here is an example of a path for $\mathfrak{sl}_6$:

$$(\alpha_2, \alpha_2 + \alpha_3, \alpha_2 + \alpha_3 + \alpha_4, \alpha_3 + \alpha_4, \alpha_4, \alpha_4 + \alpha_5, \alpha_5).$$

For a multiexponent $\mathbf{s} = \{s_\beta\}_{\beta > 0}$, $s_\beta \in \mathbb{Z}_{\ge 0}$, let $f^{\mathbf{s}} = \prod_{\beta \in R^+} f_\beta^{s_\beta} \in S(\mathfrak{n}^-)$. For an integral dominant $\mathfrak{sl}_n$-weight $\lambda = \sum_{i=1}^{n-1} m_i \omega_i$, let $S(\lambda)$ be the set of all multiexponents $\mathbf{s} = (s_\beta)_{\beta \in R^+} \in \mathbb{Z}_{\ge 0}^{R^+}$ such that for all Dyck paths $\mathbf{p} = (\beta(0), \ldots, \beta(k))$,

$$s_{\beta(0)} + s_{\beta(1)} + \cdots + s_{\beta(k)} \le m_i + m_{i+1} + \cdots + m_j, \tag{2.1}$$

where $\beta(0) = \alpha_i$ and $\beta(k) = \alpha_j$.

The polytopes in $\mathbb{R}_{\ge}^{R^+}$ defined by inequalities (2.1) are referred to as the FFLV polytopes. For their combinatorial properties and connection to the Gelfand–Tsetlin polytopes [75], see [6, 44, 47, 67]. The following theorem holds true [60].

Theorem 2.1. *The vectors $f^{\mathbf{s}} v_\lambda$, $\mathbf{s} \in S(\lambda)$, form a basis of V_λ^a. In addition, $S(\lambda) + S(\mu) = S(\lambda + \mu)$.*

We note that Theorem 2.1 implies that the elements $f^{\mathbf{s}} v_\lambda$, $\mathbf{s} \in S(\lambda)$ form a basis of the classical representation V_λ provided an order of factors is fixed in each monomial $f^{\mathbf{s}}$ (see [121]).

Let us describe the Lie algebra $\mathfrak{g}^a$ acting on V_λ^a. As a vector space, $\mathfrak{g}^a$ is isomorphic to $\mathfrak{g}$. The Borel $\mathfrak{b} \subset \mathfrak{g}^a$ is a subalgebra, the nilpotent subalgebra $\mathfrak{n}^- \subset \mathfrak{g}^a$ is an abelian ideal, and $\mathfrak{b}$ acts on the space $\mathfrak{n}^-$ as on the quotient $\mathfrak{g}/\mathfrak{b}$. Then for any $\lambda \in P^+$ the structure of the $\mathfrak{g}$-module on V_λ induces the structures of $\mathfrak{g}^a$ module on V_λ^a.

Note that $V_\lambda^a = S(\mathfrak{n}^-)v_\lambda$ is a cyclic $S(\mathfrak{n}^-)$-module, so we can write $V_\lambda^a \simeq S(\mathfrak{n}^-)/I(\lambda)$, for some ideal $I(\lambda) \subset S(\mathfrak{n}^-)$.

The following theorem holds in types A and C [60]:

Theorem 2.2. $I(\lambda) = S(\mathfrak{n}^-)(\mathrm{U}(\mathfrak{n}^+) \circ \operatorname{span}\{f_\alpha^{(\lambda,\alpha^\vee)+1}, \alpha \in R^+\})$.

This theorem should be understood as a commutative analogue of the well-known description of V_λ as the quotient

$$V_\lambda \simeq \mathrm{U}(\mathfrak{n}^-)/\langle f_\alpha^{(\lambda,\alpha^\vee)+1}, \alpha \in R^+ \rangle$$

(see, for example, [74, 86]).

The proof of the theorems above is based on the following claim available in types A and C [60–62].

Theorem 2.3. *Let $\lambda, \mu \in P^+$. Then*

$$V^a_{\lambda+\mu} \simeq \mathrm{U}(\mathfrak{g}^a)(v_\lambda \otimes v_\mu) \hookrightarrow V^a_\lambda \otimes V^a_\mu$$

as $\mathfrak{g}^a$-modules.

The algebraic and representation theoretic properties of the PBW filtration and the $\mathfrak{g}^a$ action in more general settings are considered in [10–12,32,48,51,52,56,65,70,77,78,99,106, 107].

3. REPRESENTATION THEORY: GEOMETRY

Let G be a simple simply-connected Lie group with the Lie algebra $\mathfrak{g}$. Let $B \subset G$ be a Borel subgroup with the Lie algebra $\mathfrak{b}$. Each space V_λ, $\lambda \in P^+$ is equipped with the natural structure of a G-module. Therefore G acts on the projectivization $\mathbb{P}(V_\lambda)$. The (generalized) flag variety $\mathcal{F}_\lambda \hookrightarrow \mathbb{P}(V_\lambda)$ is defined as the G-orbit of the line $\mathbb{C}v_\lambda$ (see [73,94]). Each variety $\mathcal{F}_\lambda$ is isomorphic to the quotient of G by the parabolic subgroup leaving the point $\mathbb{C}v_\lambda \in \mathbb{P}(V_\lambda)$ invariant. In particular, for $G = \mathrm{SL}_n$ and a fundamental weight $\lambda = \omega_k$ the flag variety $\mathcal{F}_\lambda$ is isomorphic to the Grassmannian $\mathrm{Gr}_k(n)$. For a regular weight λ, the flag variety $\mathcal{F}_\lambda$ sits inside $\prod_{k=1}^{n-1} \mathrm{Gr}_k(n)$ and consists of chains of embedded subspaces. In what follows, we mostly consider the case $G = \mathrm{SL}_n$ and regular λ, the general type A case can be treated similarly (see [53,54,56,57]). We denote the complete type A_{n-1} flag variety by $\mathcal{F}_n$ (it is known to be independent of a regular weight λ). The variety $\mathcal{F}_n$ admits Plücker embedding into the product of projective spaces $\prod_{k=1}^{n-1} \mathbb{P}(\Lambda^k(\mathbb{C}^n))$. The homogeneous coordinate ring (also known as the Plücker algebra) is a quotient of the polynomial algebra in Plücker variables X_I, $I \subset [n]$ by the quadratic Plücker ideal.

Recall the Lie algebra $\mathfrak{g}^a$ acting on V^a_λ. We now describe the corresponding Lie group G^a. Let $M = \dim \mathfrak{n}$ and let $\mathbb{G}_a$ be the additive group of the field $\mathbb{C}$. The Lie group G^a is a semidirect product $\mathbb{G}_a^M \rtimes B$ of the normal subgroup $\mathbb{G}_a^M$ and the Borel subgroup B. The action by conjugation of B on $\mathbb{G}_a^M$ is induced from the B-action on $(\mathfrak{n}^-)^a \simeq \mathfrak{g}/\mathfrak{b}$.

We now define the degenerate flag varieties $\mathcal{F}^a_\lambda$ [54]. Let $[v_\lambda] \in \mathbb{P}(V^a_\lambda)$ be the line $\mathbb{C}v_\lambda$.

Definition 3.1. The variety $\mathcal{F}^a_\lambda \hookrightarrow \mathbb{P}(V^a_\lambda)$ is the closure of the G^a-orbit of $[v_\lambda]$,

$$\mathcal{F}^a_\lambda = \overline{G^a[v_\lambda]} = \overline{\mathbb{G}_a^M[v_\lambda]} \hookrightarrow \mathbb{P}(V^a_\lambda).$$

We note that the orbit $G[v_\lambda] \hookrightarrow \mathbb{P}(V_\lambda)$ coincides with its closure, but the orbit $G^a[v_\lambda]$ does not; in fact, $\mathcal{F}^a_\lambda$ is the so-called $\mathbb{G}_a^M$-variety, see [7,8,82]. Theorem 2.3 implies that in types A and C the varieties $\mathcal{F}^a_\lambda$ depend only on the regularity class of λ, i.e., $\mathcal{F}^a_\lambda$ is isomorphic to $\mathcal{F}^a_\mu$ if and only if the sets of fundamental weights showing up in λ and μ coincide (see [102] for the study of a similar question for Schubert varieties).

In types A and C, we have rather explicit description of the degenerate flag varieties [53,58]. In particular, for $\mathfrak{g} = \mathfrak{sl}_n$ one has $\mathcal{F}^a_{\omega_k} \simeq \mathrm{Gr}_k(n)$. To describe the PBW degenerate flag varieties in type A, we introduce the following notation: let W be an n-dimensional vector space with a basis $w_1, \dots, w_n$. Let us denote by $\mathrm{pr}_k : W \to W$ the projection along w_k. We denote the regular PBW degenerate flag variety by $\mathcal{F}^a_n$. The following theorem holds [53,54] (we use the shorthand notation $[n] = \{1, \dots, n\}$).

Theorem 3.2. *One has*

$$\mathcal{F}^a_n \simeq \big\{(V_1, \dots, V_{n-1}) : V_k \in \mathrm{Gr}_k(W), k \in [n]; \mathrm{pr}_{k+1} V_k \subset V_{k+1}, k \in [n-1]\big\}.$$

Using this description, one proves the following theorem [29–31] (see also [96]).

Theorem 3.3. *The variety $\mathcal{F}^a_n$ is isomorphic to a Schubert variety for the group* SL_{2n-1}.

The symplectic PBW degenerations are described in [58] (see also [16]).

For a partition $\lambda = (\lambda_1 \geq \dots \geq \lambda_{n-1} \geq 0)$, we denote by Y_λ the corresponding Young diagram. Recall that the classical SL_n flag variety admits an embedding to the product of Grassmannians. The corresponding homogeneous coordinate ring (the Plücker algebra) is generated by the Plücker variables X_I, $I \subset [n]$ and is known to be isomorphic to the direct sum $\bigoplus_{\lambda \in P^+} V^*_\lambda$ (see [73]). There is a one-to-one bijection between the Plücker variables and columns filled with numbers from $[n]$ (the numbers increase from top to bottom). Then the semistandard Young tableaux provide a basis of the homogeneous coordinate ring of SL_n/B (one takes the product of Plücker variables, corresponding to the columns of a tableau). Similar result holds true in the PBW degenerate situation.

We denote by μ_j the length of the jth column of a diagram.

Definition 3.4. A semistandard PBW-tableau of shape λ is a filling $T_{i,j}$ of the Young diagram Y_λ with numbers $1, \dots, n$. The number $T_{i,j} \in \{1, \dots, n\}$ is attached to the box (i, j). The filling $T_{i,j}$ has to satisfy the following properties:

(1) if $T_{i,j} \leq \mu_j$, then $T_{i,j} = i$;

(2) if $i_1 < i_2$ and $T_{i_1,j} \neq i_1$, then $T_{i_1,j} > T_{i_2,j}$;

(3) for any $j > 1$ and any i, there exists $i_1 \geq i$ such that $T_{i_1,j-1} \geq T_{i,j}$.

One can show that the number of shape λ semistandard PBW-tableaux is equal to $\dim V_\lambda$. Moreover, the following theorem holds [54] (see also [63,79]).

Theorem 3.5. *The homogeneous coordinate ring of $\mathcal{F}^a_n$ (also known as the PBW degenerate Plücker algebra) is isomorphic to the direct sum of dual PBW degenerate modules $(V^a_\lambda)^*$, $\lambda \in P^+$. The ideal of relations is quadratic and is generated by degenerate Plücker relations. The PBW semistandard tableaux parametrize a basis in the coordinate ring.*

Certain infinite-dimensional analogues of the results described above are obtained in [59,108]. However, this direction has not been seriously pursued so far.

4. TOPOLOGY AND COMBINATORICS

In this section we describe a cellular decomposition of the type A complete PBW degenerate flag varieties $\mathcal{F}_n^a$ (see [16, 53, 58] for a more general picture).

Let us fix an n-dimensional vector space W with a basis $w_1, \dots, w_n$. Let $\mathbf{I} = (I_1, \dots, I_{n-1})$ be a collection of subsets of the set $[n]$ such that $|I_k| = k$. We denote by $p_\mathbf{I} \in \prod_{k=1}^{n-1} \mathrm{Gr}_k(W)$ a point in the product of Grassmann varieties such that the kth component is equal to the linear span of w_i with $i \in I_k$. Theorem 3.2 implies that $p_\mathbf{I} \in \mathcal{F}_n^a$ if and only if

$$I_k \subset I_{k+1} \cup \{k+1\} \quad \text{for all } k = 1, \dots, n-2. \tag{4.1}$$

The following theorem is proved in [53].

Theorem 4.1. *The G^a orbits of the points $p_\mathbf{I}$ provide a cellular decomposition of $\mathcal{F}_n^a$.*

A natural problem is to compute the Euler characteristic and Poincaré polynomial of $\mathcal{F}_n^a$. The answer is given in terms of the normalized median Genocchi numbers and the Dellac configurations.

The normalized median Genocchi numbers h_n, $n = 0, 1, 2, \dots$ form a sequence which starts with $1, 1, 2, 7, 38, 295$ [1]. The earliest definition was given by Dellac in [36] (see also [13, 15, 17, 40–42, 55, 80, 93, 120, 122]). Consider a rectangle with n columns and $2n$ rows. It contains $n \times 2n$ boxes labeled by pairs (l, j), where $l = 1, \dots, n$ is the number of a column and $j = 1, \dots, 2n$ is the number of a row. A Dellac configuration D is a subset of boxes, subject to the following conditions:

(1) each column contains exactly two boxes from D,

(2) each row contains exactly one box from D,

(3) if the (l, j)th box is in D, then $l \leq j \leq n + l$.

Let DC_n be the set of such configurations. Then the number of elements in DC_n is equal to h_n.

We list all Dellac's configurations for $n = 3$.

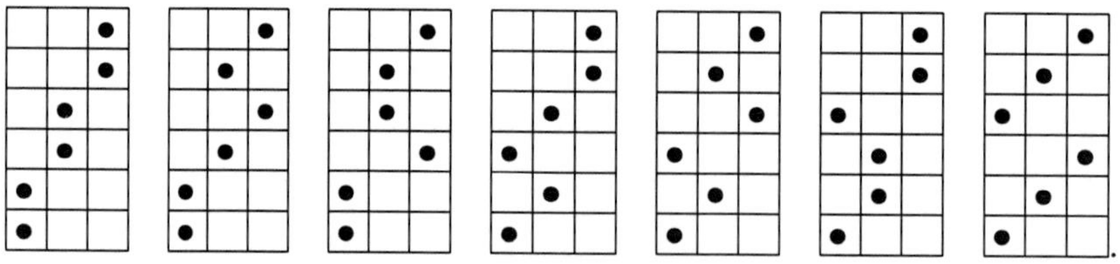

The importance of the median Genocchi numbers comes from the following theorem [53].

Theorem 4.2. *The number of collections $\mathbf{I}$ subject to conditions* (4.1) *is equal to the normalized median Genocchi number h_n. The Euler characteristic of $\mathcal{F}_n^a$ is equal to h_n.*

An explicit formula for the numbers h_n is available (see [25]), namely

$$h_n = \sum_{f_0,\dots,f_n \geq 0} \prod_{k=1}^{n} \binom{1+f_{k-1}}{f_k} \prod_{k=0}^{n-1} \binom{1+f_{k+1}}{f_k} \tag{4.2}$$

with $f_0 = f_n = 0$.

In order to compute the Poincaré polynomial of $\mathcal{F}_n^a$, we define a length $l(D)$ of a Dellac configuration D as the number of pairs (l_1, j_1), (l_2, j_2) such that the boxes (l_1, j_1) and (l_2, j_2) are both in D and $l_1 < l_2$, $j_1 > j_2$. This definition resembles the definition of the length of a permutation. We note that in the classical case the complex dimension of the cell attached to a permutation σ in a flag variety is equal to the number of pairs $j_1 < j_2$ such that $\sigma(j_1) > \sigma(j_2)$, which equals the length of σ. One has [53]:

Theorem 4.3. *The complex dimension of the cell in $\mathcal{F}_n^a$ containing a point p_{I} is equal to $l(D)$. Thus the Poincaré polynomial $h_n(q) = P_{\mathcal{F}_n^a}(q)$ is given by $h_n(q) = \sum_{D \in DC_n} q^{l(D)}$.*

The first four polynomials $h_n(q)$ are as follows:

$$\begin{aligned} &h_1(q) = 1, \quad h_2(q) = 1 + q, \\ &h_3(q) = 1 + 2q + 3q^2 + q^3, \\ &h_4(q) = 1 + 3q + 7q^2 + 10q^3 + 10q^4 + 6q^5 + q^6. \end{aligned}$$

The following (fermionic type) formula for the polynomials $h_n(q)$ is obtained in [25] using the geometry of quiver Grassmannians:

$$h_n(q) = \sum_{f_1,\dots,f_{n-1} \geq 0} q^{\sum_{k=1}^{n-1}(k-f_k)(1-f_k+f_{k+1})} \prod_{k=1}^{n} \binom{1+f_{k-1}}{f_k}_q \prod_{k=0}^{n-1} \binom{1+f_{k+1}}{f_k}_q \tag{4.3}$$

(we assume $f_0 = f_n = 0$). The formula is given in terms of the q-binomial coefficients

$$\binom{m}{n}_q = \frac{m_q!}{n_q!(m-n)_q!}, \quad m_q! = \prod_{i=1}^{m} \frac{1-q^i}{1-q}.$$

5. QUIVER GRASSMANNIANS

Theorem 3.2 provides a link between the PBW degenerate flag varieties and quiver Grassmannians. Let Q be a quiver with the set of vertices Q_0 and the set of arrows Q_1. For two vectors $\mathbf{e}, \mathbf{d} \in \mathbb{Z}^{Q_0}$, we denote by $\langle \mathbf{e}, \mathbf{d} \rangle$ the value of the Euler from of the quiver. For a Q module M and a dimension vector $\mathbf{e} \in \mathbb{Z}_{\geq 0}^{Q_0}$, we denote by $\mathrm{Gr}_{\mathbf{e}}(M)$ the quiver Grassmannian consisting of $\mathbf{e}$-dimensional subrepresentations of M. For more details on the quiver representation theory, see [9, 34, 35, 113]. The general theory of quiver Grassmannians can be found in [21] (see also [2, 3, 20, 22, 85, 97, 109, 110]).

Now let Q be an equioriented type A_{n-1} quiver. We label the vertices by the numbers from 1 to n. Then the set Q_1 consists of arrows $i \to i+1$, $i \in [n-1]$. The indecomposable representations of Q are labeled by pairs $1 \leq i \leq j \leq n$; the representation $U_{i,j}$ is supported on vertices from i to j and is one-dimensional at each vertex. The projective indecomposable representations are given by $P_k = U_{k,n}$ and the injective indecomposables are $I_k = U_{1,k}$. In particular, the path algebra A of Q is isomorphic to the direct sum $\bigoplus_{k=1}^{n-1} P_k$ and the dual A^* is the direct sum $\bigoplus_{k=1}^{n-1} I_k$ of all indecomposable injectives.

By the very definition, the classical complete flag variety SL_n / B is isomorphic to the quiver Grassmannian $\mathrm{Gr}_{\dim A}(P_1^{\oplus n})$. The following observation was made in [25]:

$$\mathcal{F}_n^a \simeq \mathrm{Gr}_{\dim A}(A \oplus A^*). \tag{5.1}$$

The realization (5.1) provides additional tools for the study of algebro-geometric and combinatorial properties of the degenerate flag varieties (see [25–27,29]). In particular, one recovers and generalizes [27,28] the Bott–Samelson type construction for the resolution of singularities of $\mathcal{F}_n^a$ [57] (see also [89,112] for further generalizations). The resolution is constructed as a quiver Grassmannian for a larger quiver attached to Q.

Since the degenerate flag varieties have many nice properties, it is natural to study the quiver Grassmannians $\mathrm{Gr}_{\dim P}(P \oplus I)$ for arbitrary projective representation P and an injective representation I and a Dynkin quiver Q (the so-called well-behaved quiver Grassmannians). We summarize the main properties of these quiver Grassmannians in the following theorem (see [25,26]).

Theorem 5.1. *Let P and I be a projective and an injective representations of a Dynkin quiver Q. Then the quiver Grassmannian $X = \mathrm{Gr}_{\dim P}(P \oplus I)$ has the following properties:*

(1) $\dim X = \langle \dim P, \dim I \rangle$,

(2) *X is irreducible and normal,*

(3) *X is locally a complete intersection,*

(4) *there exists an algebraic group $G \subset \mathrm{Aut}(P \oplus I)$ acting on X with finitely many orbits.*

For a dimension vector $\mathbf{d} \in \mathbb{Z}_{\geq 0}^{Q_0}$, let $R_{\mathbf{d}}$ be the variety of Q-representations of dimension $\mathbf{d}$. The group $GL_{\mathbf{d}} = \prod_{i \in Q_0} GL_{d_i}$ acts on $R_{\mathbf{d}}$ by base change and the orbits are parameterized by the isoclasses of $\mathbf{d}$-dimensional representations of Q. The closure of orbits induces the degeneration order on the set of isoclasses. Fixing a dimension vector $\mathbf{e}$, we obtain a family $\mathrm{Gr}_{\mathbf{e}}(\mathbf{d})$ of $\mathbf{e}$-dimensional quiver Grassmannians over the representation space $R_{\mathbf{d}}$ (the so-called universal quiver Grassmannian). Let us denote the projection map $\mathrm{Gr}_{\mathbf{e}}(\mathbf{d}) \to R_{\mathbf{d}}$ by $p_{\mathbf{e},\mathbf{d}}$.

We are interested in the case when Q is the equioriented type A_{n-1} quiver, $\mathbf{d} = (n, \dots, n)$ and $\mathbf{e} = (1, 2, \dots, n-1)$. Then both the classical and the PBW degenerate flag varieties are isomorphic to the fibers of $p_{\mathbf{e},\mathbf{d}}$. It is thus natural to ask about the properties of the whole family. The $GL_{\mathbf{d}}$ orbits on $R_{\mathbf{d}}$ are parametrized by the tuples $\mathbf{r}$ of ranks $r_{i,j}$ of the compositions of the maps between the ith and jth vertices. We define three rank tuples $\mathbf{r}^0$, $\mathbf{r}^1$, and $\mathbf{r}^2$ by

$$r_{i,j}^0 = n + 1, \quad r_{i,j}^1 = n + 1 - (j - i), \quad r_{i,j}^2 = n - (j - i).$$

Then the corresponding representations of Q are given by $M^0 = P_1^{\oplus n}$, $M^1 = A \oplus A^*$, and

$$M^2 = \bigoplus_{k=1}^{n-1} P_k \oplus \bigoplus_{k=1}^{n-2} I_k \oplus S,$$

where S is the direct sum of all simple modules of Q. One has $\mathrm{SL}_n / B \simeq \mathrm{Gr}_{\mathbf{e}}(M^0)$, $\mathcal{F}_n^a \simeq \mathrm{Gr}_{\mathbf{e}}(M^1)$. In [23] we prove the following theorem:

Theorem 5.2. (a) *The quiver Grassmannian $p_{\mathbf{e},\mathbf{d}}^{-1}(M^2)$ is of expected dimension $n(n-1)/2$. It is reducible and the number of irreducible components is equal to the nth Catalan number.*

(b) *The flat irreducible locus of* $\mathrm{Gr}_{\mathbf{e}}(\mathbf{d})$ *consists of the fibers $p_{\mathbf{e},\mathbf{d}}^{-1}(M)$ such that M degenerates to M^1.*

(c) *The flat locus of* $\mathrm{Gr}_{\mathbf{e}}(\mathbf{d})$ *consists of the fibers $p_{\mathbf{e},\mathbf{d}}^{-1}(M)$ such that M degenerates to M^2.*

The case of partial flag varieties is considered in [24].

6. TORIC DEGENERATIONS

As explained in Section 5, the degeneration of the classical flag variety into the PBW degenerate flag variety can be considered within a family of quiver Grassmannians over the representation space of the quiver. In particular, the study of other degenerations (intermediate and deeper ones) leads to the new and interesting results and examples. Yet another direction is to make a connection between the PBW degeneration and toric degenerations [33] of flag varieties (the latter attracted a lot of attention in the last two decades, see [4, 5, 14, 18, 19, 45, 46, 64, 81, 95]). One of the most famous examples is a flat degeneration of $\mathcal{F}_n$ into the toric variety with the Newton polytope being the Gelfand–Tsetlin polytope [75, 76, 92, 119]. We are able to prove the following theorem.

Theorem 6.1. *The complete flag variety $\mathcal{F}_n = \mathrm{SL}_n / B$ admits a flat degeneration to the toric variety corresponding to a FFLV polytope with regular highest weight. This degeneration factors through the PBW degeneration.*

The GT and FFLV polytopes are identified with the order and chain polytopes of a certain poset (see [6, 47, 100, 103, 118]). Several proofs of Theorem 6.1 are available. Essentially, there are three different approaches:

(1) via the representation space V_λ,

(2) via the Gröbner theory for the Plücker ideal,

(3) via the SAGBI theory for the Plücker algebra.

The first approach is utilized in [43, 48, 63]. The approach is similar to the PBW degeneration construction: instead of attaching degree one to each Chevalley generator, one uses a weight system, attaching weight $a_{i,j}$ to the generators $f_{\alpha_{i,j}}$ for all positive roots. For certain weight systems, one gets a filtration on the universal enveloping algebra, which leads to a filtered (and then graded) representation space and degenerate flag variety. The following theorem holds (see [48,63]).

Theorem 6.2. *Consider the PBW filtration with the weight system* $a_{i,j} = (j-i+1)(n-j)$. *Then*

(1) *in the associated graded space the nonzero monomials in* f_α *form a basis,*

(2) *the associated graded space is acted upon by the symmetric algebra* $S(\mathfrak{n}^-)$ *and the degenerate flag variety is a* $\mathbb{G}_a^{n(n-1)/2}$ *variety,*

(3) *the corresponding degenerate flag variety is toric with the Newton polytope being the FFLV polytope.*

Instead of working with the representation space, one may start with the algebraic variety $\mathcal{F}_n$ from the very beginning. As an intermediate step one considers the theory of Newton–Okounkov (NO) bodies [88,105]. The connection between the NO bodies and toric degenerations is used in many papers, see, e.g., [5,49,81,87]. The following holds true.

Theorem 6.3. *The toric variety attached to the FFLV polytope can be constructed as a Newton–Okounkov body for certain valuations. The valuations are obtained via Lie theory* [63] *or geometrically* [71,72,90,91].

Recall the Plücker coordinates X_I, the quadratic Plücker ideal defining the flag variety $\mathcal{F}_n$ inside the product of the projectivized fundamental representations and the Plücker algebra (the quotient by the Plücker ideal). There are two general constructions leading to the degenerations of algebraic varieties: Gröbner theory for the defining ideals [83,104] (see also [43,116,117] for the tropical version) and the SAGBI (subalgebra analogues of the Gröbner bases for ideals) theory [111] (see also [37,83,84]). The former construction works with the defining ideals, attaching certain degrees to the variables, and the latter deals with the quotient algebras, using certain monomial orders. In our setting the following claims hold (see [43,101]).

Theorem 6.4. *There exists a maximal cone in the Gröbner fan of the Plücker ideal such that a general point corresponds to the monomial ideal defined by the PBW semistandard tableaux. There exists a monomial order on the set of Plücker variables such that the monomials in Plücker variables corresponding to the PBW semistandard tableaux form a SAGBI basis of the Plücker algebra.*

Let us close with the remark that it would be very interesting to construct and study toric degenerations for affine flag varieties [94] and semiinfinite flag varieties [50,68]. The first steps in this direction were made in [114,115]. From the representation theory point of view, this would lead to new constructions of bases and character formulas for the integrable representations of affine algebras and global Weyl and Demazure modules for current algebras [38,39,66,98].

ACKNOWLEDGMENTS

I dedicate this review to the memory of Ernest Borisovich Vinberg, who passed away in 2020. I am indebted to him for sharing his ideas on monomial bases in irreducible representations of simple Lie algebras. I am grateful to Giovanni Cerulli Irelli, Xin Fang, Michael Finkelberg, Ghislain Fourier, Peter Littelmann, Igor Makhlin and Markus Reineke for fruitful collaboration.

FUNDING

This work was partially funded within the framework of the HSE University Basic Research Program.

REFERENCES

[1] The On-Line Encyclopedia of Integer Sequences, A000366, https://oeis.org/.

[2] S. Abeasis and A. Del, Fra, Degenerations for the representations of quiver of type A_m. *J. Algebra* **93** (1985), no. 2, 376–412.

[3] S. Abeasis and A. Del Fra, Degenerations for the representations of an equioriented quiver of type A_m. *Boll. Unione Mat. Ital. Suppl.* **2** (1984), 81–172.

[4] V. Alexeev and M. Brion, Toric degenerations of spherical varieties. *Selecta Math. (N.S.)* **10** (2004), no. 4, 453–478.

[5] D. Anderson, Okounkov bodies and toric degenerations. *Math. Ann.* **356** (2013), 1183–1202.

[6] F. Ardila, T. Bliem, and D. Salazar, Gelfand–Tsetlin polytopes and Feigin–Fourier–Littelmann–Vinberg polytopes as marked poset polytopes. *J. Combin. Theory Ser. A* **118** (2011), no. 8, 2454–2462.

[7] I. Arzhantsev, Flag varieties as equivariant compactifications of $\mathbb{G}_a^n$. *Proc. Amer. Math. Soc.* **139** (2011), no. 3, 783–786.

[8] I. Arzhantsev and E. Sharoiko, Hassett–Tschinkel correspondence: modality and projective hypersurfaces. *J. Algebra* **348** (2011), no. 1, 217–232.

[9] I. Assem, D. Simson, and A. Skowronski, *Elements of the representation theory of associative algebras, Vol. 1. Techniques of representation theory*. London Math. Soc. Stud. Texts 65, Cambridge University Press, Cambridge, 2006.

[10] T. Backhaus and C. Desczyk, PBW filtration: Feigin–Fourier–Littelmann modules via Hasse diagrams. *J. Lie Theory* **25** (2015), no. 3, 818–856.

[11] T. Backhaus, X. Fang, and G. Fourier, Degree cones and monomial bases of Lie algebras and quantum groups. *Glasg. Math. J.* **59** (2017), no. 3, 595–621.

[12] T. Backhaus and D. Kus, The PBW filtration and convex polytopes in type B. *J. Pure Appl. Algebra* **223** (2019), no. 1, 245–276.

[13] D. Barsky, Congruences pour les nombres de Genocchi de 2e espèce. *Groupe Étude Anal. Ultramétr., 8e année* (1980/81), no. 34, 13 pp.

[14] V. Batyrev, I. Ciocan-Fontanine, B. Kim, and D. Van Straten, Mirror symmetry and toric degenerations of partial flag manifolds. *Acta Math.* **184** (2000), no. 1, 1–39.

[15] A. Bigeni, Combinatorial study of Dellac configurations and q-extended normalized median Genocchi numbers. *Electron. J. Combin.* **21** (2014), no. 2, 2.32, 27 pp.

[16] A. Bigeni and E. Feigin, Symmetric Dellac configurations and symplectic/orthogonal flag varieties. *Linear Algebra Appl.* **573** (2019), 54–79.

[17] A. Bigeni and E. Feigin, Symmetric Dellac configurations. *J. Integer Seq.* **23** (2020), no. 4, 20.4.6, 32 pp.

[18] L. Bossinger, X. Fang, G. Fourier, M. Hering, and M. Lanini, Toric degenerations of $Gr(2, n)$ and $Gr(3, 6)$ via plabic graphs. *Ann. Comb.* **22** (2018), no. 3, 491–512.

[19] P. Caldero, Toric degenerations of Schubert varieties. *Transform. Groups* **7** (2002), no. 1, 51–60.

[20] P. Caldero and M. Reineke, On the quiver Grassmannian in the acyclic case. *J. Pure Appl. Algebra* **212** (2008), no. 11, 2369–2380.

[21] G. Cerulli Irelli, *Three lectures on quiver Grassmannians*. Contemp. Math. 758, American mathematical Society, 2020.

[22] G. Cerulli Irelli, F. Esposito, H. Franzen, and M. Reineke, Cellular decomposition and algebraicity of cohomology for quiver Grassmannians. *Adv. Math.* **379** (2021), 107544, 47 pp.

[23] G. Cerulli Irelli, X. Fang, E. Feigin, G. Fourier, and M. Reineke, Linear degeneration of flag varieties. *Math. Z.* **287** (2017), no. 1–2, 615–654.

[24] G. Cerulli Irelli, X. Fang, E. Feigin, G. Fourier, and M. Reineke, Linear degenerations of flag varieties: partial flags, defining equations, and group actions. *Math. Z.* **296** (2020), no. 1–2, 453–477.

[25] G. Cerulli Irelli, E. Feigin, and M. Reineke, Quiver Grassmannians and degenerate flag varieties. *Algebra Number Theory* **6** (2012), no. 1, 165–194.

[26] G. Cerulli Irelli, E. Feigin, and M. Reineke, Degenerate flag varieties: moment graphs and Schröder numbers. *J. Algebraic Combin.* **38** (2013), no. 1.

[27] G. Cerulli Irelli, E. Feigin, and M. Reineke, Desingularization of quiver Grassmannians for Dynkin quivers. *Adv. Math.* **245** (2013), 182–207.

[28] G. Cerulli Irelli, E. Feigin, and M. Reineke, Homological approach to the Hernandez–Leclerc construction and quiver varieties. *Represent. Theory* **18** (2014), 1–14.

[29] G. Cerulli Irelli, E. Feigin, and M. Reineke, Schubert Quiver Grassmannians. *Algebr. Represent. Theory* **20** (2017), no. 1, 147–161.

[30] G. Cerulli Irelli and M. Lanini, Degenerate flag varieties of type A and C are Schubert varieties. *Int. Math. Res. Not.* **15** (2015), 6353–6374.

[31] G. Cerulli Irelli, M. Lanini, and P. Littelmann, Degenerate flag varieties and Schubert varieties: a characteristic free approach. *Pacific J. Math.* **284** (2016), no. 2, 283–308.

[32] I. Cherednik and E. Feigin, Extremal part of the PBW-filtration and E-polynomials. *Adv. Math.* **282** (2015), 220–264.

[33] D. Cox, J. Little, and H. Schenck, *Toric varieties*. Grad. Stud. Math. 124, American Mathematical Society, Providence, RI, 2011.

[34] W. Crawley-Boevey, Lectures on representations of quivers. Preprint, 1992. https://www.math.uni-bielefeld.de/~wcrawley/quivlecs.pdf.

[35] W. Crawley-Boevey, More lectures on representations of quivers. Preprint, 1992. https://www.math.uni-bielefeld.de/~wcrawley/morequivlecs.pdf.

[36] H. Dellac, Problem 1735. *L'Intermédiaire des Math.* **7** (1900), 9–10.

[37] C. De Concini, D. Eisenbud, and C. Procesi, *Hodge algebras*. Astérisque 91, Société Mathématique de France, Paris, 1982, 87 pp.

[38] I. Dumanski and E. Feigin, Reduced arc schemes for Veronese embeddings and global Demazure modules. 2019, arXiv:1912.07988.

[39] I. Dumanski, E. Feigin, and M. Finkelberg, Beilinson–Drinfeld Schubert varieties and global Demazure modules. *Forum Math. Sigma* **9** (2021), 1–25.

[40] D. Dumont, Interprétations combinatoires des nombres de Genocchi. *Duke Math. J.* **41** (1974), 305–318.

[41] D. Dumont and A. Randrianarivony, Dérangements et nombres de Genocchi. *Discrete Math.* **132** (1994), 37–49.

[42] D. Dumont and G. Viennot, A combinatorial interpretation of the Seidel generation of Genocchi numbers. *Discrete Math.* **6** (1980), 77–87.

[43] X. Fang, E. Feigin, G. Fourier, and I. Makhlin, Weighted PBW degenerations and tropical flag varieties. *Commun. Contemp. Math.* **21** (2019), no. 1, 1850016, 27 pp.

[44] X. Fang and G. Fourier, Marked chain-order polytopes. *European J. Combin.* **58** (2016), 267–282.

[45] X. Fang, G. Fourier, and P. Littelmann, Essential bases and toric degenerations arising from birational sequences. *Adv. Math.* **312** (2017), 107–149.

[46] X. Fang, G. Fourier, and P. Littelmann, On toric degenerations of flag varieties. In *Representation theory – current trends and perspectives*, pp. 187–232, EMS Ser. Congr. Rep., Eur. Math. Soc., Zürich, 2017.

[47] X. Fang, G. Fourier, J.-P. Litza, and C. Pegel, A Continuous Family of Marked Poset Polytopes. *SIAM J. Discrete Math.* **34** (2020), no. 1, 611–639.

[48] X. Fang, G. Fourier, and M. Reineke, PBW-Filtration on quantum groups of type A_n. *J. Algebra* **449** (2016), 321–345.

[49] X. Fang and P. Littelmann, From standard monomial theory to semi-toric degenerations via Newton–Okounkov bodies. *Trans. Moscow Math. Soc.* **78** (2017), 275–297.

[50] B. Feigin and E. Frenkel, Affine Kac–Moody algebras and semi-infinite flag manifolds. *Comm. Math. Phys.* **128** (1990), 161–189.

[51] E. Feigin, The PBW filtration, Demazure modules and toroidal current algebras. *SIGMA* **4** (2008), 070, 21 pp.

[52] E. Feigin, The PBW filtration. *Represent. Theory* **13** (2009), 165–181.

[53] E. Feigin, Degenerate flag varieties and the median Genocchi numbers. *Math. Res. Lett.* **18** (2011), no. 6, 1–16.

[54] E. Feigin, $\mathbb{G}_a^M$ degeneration of flag varieties. *Selecta Math.* **18** (2012), no. 3, 513–537.

[55] E. Feigin, The median Genocchi numbers, Q-analogues and continued fractions. *European J. Combin.* **33** (2012), 1913–1918.

[56] E. Feigin, Degenerate SL_n: representations and flag varieties. *Funct. Anal. Appl.* **48** (2014), no. 1, 59–71.

[57] E. Feigin and M. Finkelberg, Degenerate flag varieties of type A: Frobenius splitting and BWB theorem. *Math. Z.* **275** (2013), no. 1–2, 55–77.

[58] E. Feigin, M. Finkelberg, and P. Littelmann, Symplectic degenerate flag varieties. *Canad. J. Math.* **66** (2014), no. 6, 1250–1286.

[59] E. Feigin, M. Finkelberg, and M. Reineke, Degenerate affine Grassmannians and loop quivers. *Kyoto J. Math.* **57** (2017), no. 2, 445–474.

[60] E. Feigin, G. Fourier, and P. Littelmann, PBW filtration and bases for irreducible modules in type A_n. *Transform. Groups* **16** (2011), no. 1, 71–89.

[61] E. Feigin, G. Fourier, and P. Littelmann, PBW filtration and bases for symplectic Lie algebras. *Int. Math. Res. Not. IMRN* **24** (2011), 5760–5784.

[62] E. Feigin, G. Fourier, and P. Littelmann, PBW-filtration over $\mathbb{Z}$ and compatible bases for $V_{\mathbb{Z}}(\lambda)$ in type A_n and C_n. In *Symmetries, integrable systems and representations*, pp. 35–63, Springer Proc. Math. Stat. 40, Springer, Heidelberg, 2013.

[63] E. Feigin, G. Fourier, and P. Littelmann, Favourable modules: filtrations, polytopes, Newton–Okounkov bodies and flat degenerations. *Transform. Groups* **22** (2017), no. 2, 321–352.

[64] E. Feigin, M. Lanini, and A. Pütz, Totally nonnegative Grassmannians, Grassmann necklaces and quiver Grassmannians. 2021, arXiv:2108.10236.

[65] E. Feigin and I. Makedonskyi, Nonsymmetric Macdonald polynomials, Demazure modules and PBW filtration. *J. Combin. Theory Ser. A* (2015), 60–84.

[66] E. Feigin and I. Makedonskyi, Semi-infinite Plücker relations and Weyl modules. *Int. Math. Res. Not.* **14** (2020), 4357–4394.

[67] E. Feigin and I. Makhlin, Vertices of FFLV polytopes. *J. Algebraic Combin.* **45** (2017), no. 4, 1083–1110.

[68] M. Finkelberg and I. Mirković, Semi-infinite flags. I. Case of global curve $\mathbb{P}^1$. Differential topology, infinite-dimensional Lie algebras, and applications. In *Differential topology, infinite-dimensional Lie algebras, and applications*, pp. 81–112, Amer. Math. Soc. Transl. Ser. 2 194, Amer. Math. Soc., Providence, RI, 1999.

[69] S. Fomin and A. Zelevinsky, Cluster algebras I: foundations. *J. Acad. Mark. Sci.* **15** (2002), no. 2, 497–529.

[70] G. Fourier and D. Kus, PBW degenerations of Lie superalgebras and their typical representations. *J. Lie Theory* **31** (2021), no. 2, 313–334.

[71] N. Fujita, Newton–Okounkov polytopes of flag varieties and marked chain-order polytopes. 2021, arXiv:2104.09929.

[72] N. Fujita and A. Higashitani, Newton–Okounkov bodies of flag varieties and combinatorial mutations. *Int. Math. Res. Not. IMRN* **12** (2021), 9567–9607.

[73] W. Fulton, *Young tableaux, with applications to representation theory and geometry*. Cambridge University Press, 1997.

[74] W. Fulton and J. Harris, *Representation theory. A first course*. Undergrad. Texts Math. Read. Math. 129, Springer, New York, 1991.

[75] I. Gelfand and M. Tsetlin, Finite dimensional representations of the group of unimodular matrices. *Dokl. Akad. Nauk USSR* **71** (1950), no. 5, 825–828.

[76] N. Gonciulea and V. Lakshmibai, Degenerations of flag and Schubert varieties to toric varieties. *Transform. Groups* **1** (1996), no. 3, 215–248.

[77] A. Gornitskii, Essential signatures and canonical bases of irreducible representations of the group G_2. *Math. Notes* **97** (2015), no. 1, 30–41.

[78] A. Gornitskii, Essential signatures and monomial bases for B_n and D_n. *J. Lie Theory* **29** (2019), no. 1, 277–302.

[79] C. Hague, Degenerate coordinate rings of flag varieties and Frobenius splitting. *Selecta Math. (N.S.)* **20** (2014), no. 3, 823–838.

[80] G.-N. Han and J. Zeng, On a q-sequence that generalizes the median Genocchi numbers. *Ann. Sci. Math. Québec* **23** (1999), 63–72.

[81] M. Harada and K. Kaveh, Integrable systems, toric degenerations and Okounkov bodies. *Invent. Math.* **202** (2015), no. 3, 927–985.

[82] B. Hassett and Yu. Tschinkel, Geometry of equivariant compactifications of $\mathbb{G}_a^n$. *Int. Math. Res. Not.* **20** (1999), 1211–1230.

[83] J. Herzog and T. Hibi, *Monomial ideals*. Grad. Texts in Math. 260, Springer, London, 2011.

[84] T. Hibi, Distributive lattices, affine semigroup rings and algebras with straightening laws. In *Commutative algebra and combinatorics*, pp. 93–109 Adv. Stud. Pure Math. 11, North-Holland, Amsterdam, 1987.

[85] A. Hubery, Irreducible components of quiver Grassmannians. *Trans. Amer. Math. Soc.* **369** (2017), no. 2, 1395–1458.

[86] A. Joseph, On the Demazure character formula. *Ann. Sci. Éc. Norm. Supér.* (1985), 389–419.

[87] K. Kaveh, Crystal bases and Newton–Okounkov bodies. *Duke Math. J.* **164** (2015), 2461–2506.

[88] K. Kaveh and A. G. Khovanskii, Newton–Okounkov bodies, semigroups of integral points, graded algebras and intersection theory. *Ann. of Math.* **176** (2012), no. 2, 925–978.

[89] B. Keller and S. Scherotzke, Desingularizations of quiver Grassmannians via graded quiver varieties. 2013, arXiv:1305.7502.

[90] V. Kiritchenko, Newton–Okounkov polytopes of flag varieties. *Transform. Groups* **22** (2017), no. 2, 387–402.

[91] V. Kiritchenko, Newton–Okounkov polytopes of flag varieties for classical groups. *Arnold Math. J.* **5** (2019), no. 2–3, 355–371.

[92] M. Kogan and E. Miller, Toric degeneration of Schubert varieties and Gelfand–Tsetlin polytopes. *Adv. in Math.* **193** (2015), 1–17.

[93] G. Kreweras, Sur les permutations comptées par les nombres de Genocchi de 1-ière et 2-ième espèce. *European J. Combin.* **18** (1997), 49–58.

[94] S. Kumar, *Kac–Moody groups, their flag varieties and representation theory.* Progr. Math. 204, Birkhäuser Boston, Inc., Boston, MA, 2002.

[95] V. Lakshmibai, Degenerations of flag varieties to toric varieties. *C. R. Acad. Sci. Paris* **321** (1995), 1229–1234.

[96] M. Lanini and E. Strickland, Cohomology of the flag variety under PBW degenerations. *Transform. Groups* **24** (2019), no. 3, 835–844.

[97] O. Lorscheid and T. Weist, Plücker relations for quiver Grassmannians. *Algebr. Represent. Theory* **22** (2019), no. 1, 211–218.

[98] I. Makedonskyi, Semi-infinite Plücker relations and arcs over toric degeneration. 2020, arXiv:2006.04172.

[99] I. Makhlin, FFLV-type monomial bases for type B. *Algebraic Combin.* **2** (2019), no. 2, 305–322.

[100] I. Makhlin, Gelfand–Tsetlin degenerations of representations and flag varieties. *Transform. Groups* (2020). DOI 10.1007/s00031-020-09622-z.

[101] I. Makhlin, Gröbner fans of Hibi ideals, generalized Hibi ideals and flag varieties. 2020, arXiv:2003.02916.

[102] I. Makhlin, PBW degenerate Schubert varieties: Cartan components and counterexamples. *Algebr. Represent. Theory* **23** (2020), no. 6, 2315–2330.

[103] A. Molev and O. Yakimova, Monomial bases an branching rules. *Transform. Groups* **26** (2021), 995–1024.

[104] T. Mora and L. Robbiano, The Gröbner fan of an ideal. *J. Symbolic Comput.* **6** (1988), 183–208.

[105] A. Okounkov, Multiplicities and Newton polytopes. In *Kirillov's seminar on representation theory*, pp. 231–244, Amer. Math. Soc. Transl. Ser. 2 181, Amer. Math. Soc., Providence, RI, 1998.

[106] D. Panyushev and O. Yakimova, A remarkable contraction of semi-simple Lie algebras. *Ann. Inst. Fourier (Grenoble)* **62** (2012), no. 6, 2053–2068.

[107] D. Panyushev and O. Yakimova, Parabolic contractions of semi-simple Lie algebras and their invariants. *Selecta Math.* **19** (2013), no. 3, 699–717.

[108] A. Pütz, Degenerate affine flag varieties and quiver Grassmannians. *Algebr. Represent. Theory* (2020). DOI 10.1007/s10468-020-10012-y.

[109] M. Reineke, Every projective variety is a quiver Grassmannian. *Algebr. Represent. Theory* **16** (2013), 1313–1314.

[110] C. M. Ringel, Quiver Grassmannians for wild acyclic quivers. *Proc. Amer. Math. Soc.* **146** (2018), no. 5, 1873–1877.

[111] L. Robbiano and M. Sweedler, Subalgebra Bases. In *Proc. Commutative Algebra (Salvador, 1988)*, pp. 61–87, Lecture Notes in Math. 1430, Springer, Berlin, 1990.

[112] S. Scherotzke, Desingularization of Quiver Grassmannians via Nakajima Categories. *Algebr. Represent. Theory* **20** (2017), 231–243.

[113] R. Schiffler, *Quiver representations*. CMS Books Math., Springer, 2014.

[114] F. Sottile, Real rational curves in Grassmannians. *J. Amer. Math. Soc.* **13** (2000), 333–341.

[115] F. Sottile and B. Sturmfels, A sagbi basis for the quantum Grassmannian. *J. Pure Appl. Algebra* **158** (2001), no. 2–3, 347–366.

[116] D. Speyer and B. Sturmfels, The tropical Grassmannian. *Adv. Geom.* **4** (2003), 389–411.

[117] D. Speyer and L. Williams, The tropical totally positive Grassmannian. *J. Algebraic Combin.* **22** (2005), no. 2, 189–210.

[118] R. P. Stanley, Two poset polytopes. *Discrete Comput. Geom.* **1** (1986), no. 1, 9–23.

[119] B. Sturmfels, *Algorithms in invariant theory*. Texts Monogr. Symbol. Comput., Springer, Vienna, 1993.

[120] G. Viennot, Interprétations combinatoires des nombres d'Euler et de Genocchi. In *Seminar on Number Theory*, Univ. Bordeaux I, Talence, 1981/1982, no. 11, 94 pp.

[121] E. Vinberg, *On some canonical bases of representation spaces of simple Lie algebras, conference talk*, Bielefeld, 2005.

[122] J. Zeng and J. Zhou, A q-analog of the Seidel generation of Genocchi numbers. *European J. Combin.* (2006), 364–381.

EVGENY FEIGIN

HSE University, Faculty of Mathematics, Usacheva 6, Moscow, 119048, Russia, and Skolkovo Institute of Science and Technology, Center for Advanced Studies, Bolshoy Boulevard 30, bld. 1, Moscow 121205, Russia, evgfeig@gmail.com

This Page Intentionally Left Blank

REPRESENTATIONS OF REDUCTIVE GROUPS OVER LOCAL FIELDS

TASHO KALETHA

ABSTRACT

We discuss progress towards the classification of irreducible admissible representations of reductive groups over non-archimedean local fields and the local Langlands correspondence.

MATHEMATICS SUBJECT CLASSIFICATION 2020

Primary 22E50; Secondary 11S37, 11F70

KEYWORDS

Representation, Harish-Chandra character, local field, Langlands correspondence

INTERNATIONAL CONGRESS OF MATHEMATICIANS 2022 JULY 6–14

Proc. Int. Cong. Math. 2022, Vol. 4, pp. 2948–2975
DOI 10.4171/ICM2022/115

Published by EMS Press

Let F be a local field, i.e., a finite extension of the field $\mathbb{R}$ of real numbers, or the field $\mathbb{Q}_p$ of p-adic numbers, or the field $\mathbb{F}_p((t))$ of Laurent series over a finite field. Let G be a connected reductive F-group. Motivated by the theory of automorphic forms, the study of irreducible admissible representations of the topological group $G(F)$ with complex coefficients has been an active area of research since the pioneering work of Bargmann on $\mathrm{SL}_2(\mathbb{R})$. Some of the main problems in this area are:

(1) the classification of irreducible admissible representations,

(2) the determination of their character functions.

In addition, motivated by Langlands' conjectures, one can add

(3) the relation with representations of the Galois/Weil group of F,

(4) the proof of character identities stemming from endoscopy and more general functoriality, and

(5) the appropriate normalization of intertwining operators for parabolic induction.

In this note I would like to discuss progress towards some of these questions. There are, of course, many other interesting questions such as, for example, the classification of unitary representations, about which I will not say anything.

1. CLASSIFICATION OF IRREDUCIBLE REPRESENTATIONS AND CHARACTERS

1.1. The archimedean case

The archimedean case, where F is a finite extension of $\mathbb{R}$, thus equal to $\mathbb{R}$ or $\mathbb{C}$, has been largely resolved by the work of Harish-Chandra, Langlands, Shelstad, and others. I will discuss it briefly, because it will serve as a useful guide to the non-archimedean case.

One of Harish-Chandra's fundamental contributions was the introduction of the notion of a discrete series representation, i.e., those unitary representations whose matrix coefficients are square-integrable modulo center, and the classification of such representations. Slightly more generally, one considers essentially discrete series representations—those which become discrete series after tensor product with a character of $G(F)$. His theorem can roughly be stated as follows (this is a reformulation due to Langlands [57] and is easily seen to be equivalent to the original formulation).

Theorem 1.1. *The set of isomorphism classes of essentially discrete series representations of $G(F)$ is in a natural bijection with the set of $G(F)$-conjugacy classes of triples (S, B, θ), where $S \subset G$ is an elliptic (i.e., anisotropic modulo center) maximal torus, B is a Borel subgroup of $G_{\bar{F}}$ containing S, and θ is a character of $S(F)$ whose differential is B-dominant.*

The group G may fail to have an elliptic maximal torus. For example, when $F = \mathbb{C}$, such a torus never exists, unless G itself is a torus. Even when $F = \mathbb{R}$ an elliptic maximal torus may not exist as, for example, in the case of SL_n for $n > 2$. When an elliptic maximal torus $S \subset G$ exists, it is unique up to $G(\mathbb{R})$-conjugacy. The corresponding maximal torus S_{sc} of the simply connected cover of the derived subgroup of G is anisotropic, and the restriction of θ to it is an algebraic character, i.e., an element of the coweight lattice $X^*(S_{sc})$ of the absolute root system of G relative to S, so it makes sense to ask that $d\theta$ be B-dominant. Moreover, when this differential is a regular element of the weight lattice, it uniquely determines B, so the classifying datum is just a $G(\mathbb{R})$-conjugacy class of pairs (S, θ). We might be tempted to call the corresponding essentially discrete series representation "regular," a notion that will find its analog in the non-archimedean case.

The essentially discrete series representation $\pi_{(S,B,\theta)}$ associated to the triple (S, B, θ) by the above theorem can be specified by its character function. This uses another fundamental result of Harish-Chandra (valid for an arbitrary local field F of characteristic zero, and extended to local fields of positive characteristic by **[16]** under some assumptions): the fact that the character distribution

$$f \mapsto \operatorname{tr} \pi(f), \quad \pi(f)v = \int_{G(F)} f(g)\pi(g)v\,dg$$

of an admissible representation π of $G(F)$ is representable by a locally integrable function $\Theta_\pi : G(F) \to \mathbb{C}$. Just like in the case of finite groups, this function determines π up to equivalence. In fact, when $F = \mathbb{R}$ and π is essentially discrete series, already the restriction of Θ_π to $S(\mathbb{R})$ determines π. More precisely, we have

Theorem 1.2. *$\pi_{(S,B,\theta)}$ is the unique essentially discrete series representations such that for all $s \in S(F) \cap G(F)_{\text{reg}}$,*

$$\Theta_{\pi_{(S,B,\theta)}}(s) = (-1)^{q(G)} \sum_{w \in N(S,G)(\mathbb{R})/S(\mathbb{R})} \frac{\theta(s^w)}{\prod_{\alpha>0}(1-\alpha(s^w)^{-1})},$$

where $q(G)$ is half of the dimension of the symmetric space of $G(\mathbb{R})$ and $\alpha > 0$ indicates the product over all those absolute roots with respect to S that are positive with respect to B.

The classification of (essentially) discrete series representations of $G(F)$ is a key step in the classification of all irreducible admissible representations of $G(F)$. The next step is the classification of the irreducible tempered representations, i.e., those unitary representations whose matrix coefficients are almost square-integrable modulo center. Harish-Chandra has shown that these are precisely the irreducible constituents of parabolically induced discrete series representations of Levi subgroups of G. Moreover, the theory of the R-group due again to Harish-Chandra provides a description of the various irreducible constituents of such a parabolic induction, and hence a full classification of the tempered representations, provided one has suitably normalized the intertwining operators. What the right normalization is has been conjectured by Langlands for any local field. For archimedean local fields Arthur **[6, §3]** proved Langlands' conjecture, while for non-archimedean local fields of characteristic zero Arthur **[6, §4]** proved abstractly that a normalization exists, without being able to prove that it is provided by Langlands' formula.

The final step is the Langlands classification theorem, which states that every irreducible admissible representation is equivalent to one of the form $j_P^G(\sigma \otimes \nu)$. Here P is a parabolic subgroup of G; we denote by M the Levi quotient of P and by A_M the maximal split torus in the center of M; σ is a tempered representation of $M(F)$ and ν is an element of $X^*(A_M) \otimes \mathbb{R}$ that lies in the acute open cone associated to P, and which is identified with a character of $M(F)$ using the exponential map; and j_P^G is the unique irreducible quotient of the parabolic induction $i_P^G(\sigma \otimes \nu)$. It is known that two such representations are equivalent if and only if their Langlands data (P, σ, ν) are $G(F)$-conjugate.

1.2. The non-archimedean case

Consider now a finite extension F of $\mathbb{Q}_p$ or $\mathbb{F}_p((t))$. Harish-Chandra's classification of tempered representations in terms of discrete series representations of Levi subgroups by means of the theory of the R-group, and Langlands' classification theorem, continue to hold, with minor modifications to their statements and proofs, cf. [72, CH. VII]. This is an instance of Harish-Chandra's "Lefschetz principle," which is the philosophy that the representation theory of real and p-adic groups (and even the automorphic representations of adele groups) exhibit parallel behavior, despite the stark differences in the fine structure of these groups. But, unlike in the archimedean case, our understanding of the discrete series representations in the non-archimedean case is less developed, and the classification of these representations is at this moment incomplete.

A special subclass of the discrete series is made out of the supercuspidal representations, which are those whose matrix coefficients have compact support modulo the center. Real reductive groups do not have such representations, except for the trivial case of tori. The last 30 years have seen a significant improvement of our understanding of supercuspidal representations, beginning with the work of Moy–Prasad [66,67] and Morris [64,65] in the case of depth zero, the constructions of general depth supercuspidal representations due to Adler [4] and Yu [88], and the exhaustion results of Kim [49] and Fintzen [24]. These works rely crucially on the filtrations of the topological group $G(F)$ coming from Bruhat–Tits theory and its extensions by Moy–Prasad (an example is the filtration of the compact group $\mathrm{SL}_2(\mathbb{Z}_p)$ by congruence subgroups), and as such are very much a p-adic phenomenon with no clear analog in the archimedean case.

The most comprehensive construction, due to Yu, produces supercuspidal representations out of what is nowadays customarily called "Yu-data," rather complicated structures consisting among other things of a tower of twisted Levi subgroups, a depth-zero supercuspidal representation of the smallest subgroup, and a sequence of characters of each subgroup subject to a genericity condition. Fintzen's result shows that all supercuspidal representations arise from this construction when G is tamely ramified and p does not divide the order of the absolute Weyl group of G. Work of Hakim–Murnaghan [31] defines an explicit equivalence relation on the set of Yu-data which describes when two data produce the same representation. These results amount to a classification of all supercuspidal representations of $G(F)$ under the given conditions on G, in terms of equivalence classes of Yu-data. However, a simpler classification may be desirable. In fact, with Harish-Chandra's Lefschetz principle and his work for real groups in mind, we would ideally like a classification in terms of objects close to $G(F)$-conjugacy classes of pairs (S, θ) consisting of an elliptic maximal torus $S \subset G$ and a character θ of it.

Moy–Prasad introduce the notion of *depth* of a representation and show that an irreducible depth-zero supercuspidal representation always arises via compact induction from a maximal open and compact-mod-center subgroup $G(F)_x$—the stabilizer of a vertex in the Bruhat–Tits building—of an irreducible representation σ of $G(F)_x$ with the following very special property: $G(F)_x$ has a natural quotient that is the group $\mathsf{G}_x(k_F)$ of k_F-points of a usually disconnected reductive k_F-group G_x, and σ is required to factor through this quotient and moreover the restriction to the identity component $\mathsf{G}_x^\circ(k_F)$ must contain a cuspidal representation of this finite group of Lie type; here k_F is the residue field of F. In this way, the representation theory of finite groups of Lie type (including disconnected ones) is reflected in the representation theory of reductive p-adic groups. Given a connected reductive k_F-group G and a pair (S, θ) of a maximal torus $\mathsf{S} \subset \mathsf{G}$ and a character θ of $\mathsf{S}(k_F)$ (customarily taking ℓ-adic values), the construction of Deligne–Lusztig [19] assigns a virtual representation $R_{\mathsf{S},\theta}$ of $\mathsf{G}(k_F)$. In general, this virtual representation is not an actual representation (even up to sign), but quite often it is. More precisely, Deligne–Lusztig define the notions of a character θ to be "nonsingular" and in "general position," which are dual to the notions of a semisimple element in a connected reductive group to be "regular" and "strongly regular". They show [19, REMARK 9.15.1] that $R_{\mathsf{S},\theta}$ is an actual representation (up to a well-understood sign) whenever θ is nonsingular (originally under a certain affineness assumption, which was later shown to always hold by He [34]). Moreover, Deligne–Lusztig show that the representation $\pm R_{\mathsf{S},\theta}$ is irreducible when θ is in general position, and cuspidal if S is elliptic.

These results are both encouraging for our quest to parameterize supercuspidal representations in terms of pairs (S, θ), but also cautioning us that there will be supercuspidal representations that do not obey such a parameterization. More precisely, in [45, §3] I define the notion of a "regular" supercuspidal representation, which is one that arises from Yu's construction and for which the depth-zero part of the Yu-datum comes from, via the results of Moy–Prasad and Deligne–Lusztig, a character in general position, and then prove the following classification.

Theorem 1.3 ([45, COR. 3.7.10]). *Assume that G splits over a tame extension of F and p does not divide the order of the Weyl group of G. The set of isomorphism classes of regular supercuspidal representations of $G(F)$ is in a natural bijection with the set of $G(F)$-conjugacy classes of pairs (S, θ), where S is an elliptic maximal torus that splits over a tame extension, and θ is a regular character of $S(F)$.*

Remark 1.4.

(1) The condition on p can be weakened; I have opted here for the one which is easiest to state.

(2) I have not explicated here the definition of a "regular" character θ, but the main point is that it is an explicit Lie-theoretic condition, essentially amounting to the

stabilizer in $N_G(S)(F)/S(F)$ of the restriction θ^0 of θ to the Iwahori subgroup of $S(F)$ being trivial. For details, cf. [45, DEF. 3.7.5].

(3) In the definition of regular supercuspidal representation, "general position" should be taken with respect to the p-adic group G, which is slightly stronger than taking it with respect to the finite group of Lie type G_x°.

(4) One of the useful properties of this theorem is that it does not reference the fine structure of the topological group $G(F)$ coming from the p-adic field F, such as the various filtrations coming from Bruhat–Tits and Moy–Prasad theory.

(5) Continuing the previous point, this theorem is, in fact, rather analogous to Theorem 1.1 restricted to regular discrete series representations (in the sense of the previous subsection). In this way, it establishes the Harish-Chandra Lefschetz principle among a wide class of discrete series representations, setting up a parallel between the regular discrete series representations of real reductive groups and the regular supercuspidal representations of p-adic reductive groups.

(6) One important difference between the real and p-adic cases is that while in the real case S is unique up to $G(F)$-conjugacy, in the non-archimedean case there usually are (finitely) many different $G(F)$-conjugacy classes (in fact, even isomorphism classes) of elliptic tamely ramified maximal tori of G. Moreover, in the non-archimedean case, elliptic maximal tori always exist, and supercuspidal representations always exist, cf. [56].

There is also an analog of Theorem 1.2 which we will discuss in a moment, but before doing so we briefly consider going beyond the "regular" case. We can impose on θ the p-adic analog of Deligne–Lusztig's "nonsingularity" condition, which is weaker than the condition of being regular. The arguments involved in the proof of Theorem 1.3 still apply and produce a supercuspidal representation $\pi_{(S,\theta)}$, which may, however, be reducible, in fact, a direct sum of finitely many irreducible supercuspidal representations.

The irreducible representations obtained this way, i.e., the irreducible constituents of $\pi_{(S,\theta)}$ for all possible pairs (S,θ), can be characterized in the same way as the regular ones, but where we replace the "general position" requirement with a "nonsingularity" requirement [47, DEF. 3.1.1]. We can thus call these supercuspidal representations "non-singular" (although a better term might be "semisimple," as a contrast to the concept of a unipotent supercuspidal representation). It may not be clear at first sight why this class of representations is interesting, beyond being a generalization of the class of regular supercuspidal representations. The main interest in them comes from the fact that these are precisely those supercuspidal representations whose Langlands parameters are "supercuspidal," i.e., discrete with trivial monodromy, at least when p does not divide the order of the Weyl group and according to the construction of [47]; we will discuss this point in the next section.

The classification of all irreducible nonsingular supercuspidal representations reduces, via the analog of Theorem 1.3, to the study of the internal structure of the representations $\pi_{(S,\theta)}$, i.e., its decomposition into irreducible factors and their multiplicities. The situation is made subtle by the fact that an irreducible constituent of $\pi_{(S,\theta)}$ may occur with multiplicity greater than 1, which is a phenomenon that does not occur for connected reductive groups over finite fields. The study of the internal structure of $\pi_{(S,\theta)}$ reduces to the case of depth-zero, where it relies on geometric intertwining operators acting on Deligne–Lusztig induction in the setting of disconnected groups, building on the work of Bonnafé–Dat–Rouquier [10]. These operators must be suitably normalized. Using square brackets to denote sets of irreducible constituents, the following classification result is proved in [47, §3].

Theorem 1.5 ([47, FACT 2.4.11, PROPOSITION 3.2.4, PROPOSITION 3.4.6, COROLLARY 3.4.7]). *Assume that G splits over a tame extension of F and p does not divide the order of the Weyl group of G.*

(1) *The set of normalizations of the geometric intertwining operators is a nonempty torsor under the Pontryagin dual of the finite abelian group $N_G(S)(F)_{\theta^0}/S(F)$ (cf. Remark 1.4(2)).*

(2) *Any such normalization provides a multiplicity-preserving bijection*

$$[\pi_{(S,\theta)}] \leftrightarrow [\mathrm{Irr}_\theta(N_G(S)(F)_\theta],$$

where on the right we have those irreducible representations of $N_G(S)(F)_\theta$ whose restriction to $S(F)$ is θ-isotypic.

The existence of normalized intertwining operators, which is part of the first point, is formally analogous to Arthur's result [6, §4] on the existence of suitable normalization of standard intertwining operators between parabolically induced representations. As in Arthur's situation, I have not been able to provide a specific normalization. In fact, at the moment there is not even a conjectural expectation of what a good normalization might look like. The fact that the decomposition of the supercuspidal representation $\pi_{(S,\theta)}$ is formally analogous to the decomposition of a parabolic induction (via standard intertwining operators and the R-group) is quite intriguing.

The above theorem implies that the set of isomorphism classes of nonsingular supercuspidal representations of $G(F)$ is in bijection with the set of $G(F)$-conjugacy classes of triples (S,θ,ρ), where S is an elliptic maximal torus that splits over a tame extension, θ is a nonsingular character of $S(F)$, and ρ is an irreducible representation of $N_G(S)(F)_\theta$ whose restriction to $S(F)$ is θ-isotypic. The bijection is at the moment not completely natural, due the lack of natural normalization of the intertwining operators.

We now come to the non-archimedean analog of Theorem 1.2. It is based on work of Adler, DeBacker, Reeder, and Spice [5, 17, 18, 84, 85], and is ultimately formulated in [26]. First, we state a simpler version.

Theorem 1.6 ([26, PROPOSITION 4.3.2]). *Let $\pi_{(S,\theta)}$ be the (possibly reducible) supercuspidal representation associated to a pair (S,θ) of a tame elliptic maximal torus and a nonsingular* *character. Let $s \in S(F) \cap G(F)_{\mathrm{reg}}$ be topologically semisimple modulo center. The value of $\Theta_{\pi(S,\theta)}$ at s is given by*

$$e(G)\varepsilon_L\big(X^*(T_G)_{\mathbb{C}} - X^*(S)_{\mathbb{C}}, \Lambda\big)D(s)^{-\frac{1}{2}} \sum_{w \in N(S,G)(F)/S(F)} \Delta_{II}^{\mathrm{abs}}[a, \chi''](^w s)\theta(^w s).$$

To briefly explain the notation, $D(s) = |\prod_\alpha (1 - \alpha(s))|$ is the usual Weyl discriminant, the product being taken over all absolute roots of S, $e(G)$ is the Kottwitz sign of G as in [51], T_G is the minimal Levi subgroup of the quasisplit inner form of G, Λ is an arbitrarily chosen nontrivial character of the additive group of the base field F, ε_L is the root number of the given virtual Artin representation of degree 0, and $\Delta_{II}^{\mathrm{abs}}$ is the function of $S(F)$ given by the formula

$$\Delta_{II}^{\mathrm{abs}}(s) = \prod_\alpha \chi''_\alpha\left(\frac{\alpha(s) - 1}{a_\alpha}\right).$$

The product runs over the Γ-orbits of absolute roots of S that are symmetric, i.e., invariant under multiplication by -1, and Γ is the absolute Galois group of F. If α represents such an orbit, we can associate the subgroups $\Gamma_\alpha \subset \Gamma_{\pm\alpha} \subset \Gamma$ and the corresponding field extensions $F_\alpha/F_{\pm\alpha}/F$. Then $a_\alpha \in F_\alpha^\times$ and $\chi''_\alpha : F_\alpha^\times \to \mathbb{C}^\times$ are computed explicitly in terms of θ, and a_α depends moreover on Λ. We refer to [26, §4] for the precise formulas. The first main takeaway is this:

All constituents of this formula make sense for $F = \mathbb{R}$, and with this interpretation this formula recovers Harish-Chandra's formula from Theorem 1.2.

There is, however, a key difference: Theorem 1.6 applies only to very special elements of $S(F) \cap G(F)_{\mathrm{reg}}$—those that are topologically semisimple modulo center. It may happen that there are no such elements at all! So the *values* of the given function may not uniquely characterize the representation $\pi_{(S,\theta)}$. There is a sense in which the *formula* itself does characterize it, but such a statement may be met with skepticism by some colleagues, and in any event the question remains as to characterizing $\pi_{(S,\theta)}$ by its character function. There are two approaches to this problem. One, taken by Chan–Oi in [14], is to extend the validity of this formula to some more general elements of $S(F) \cap G(F)_{\mathrm{reg}}$ and prove that the resulting values are enough to characterize the representation; so far this has been successful under additional assumptions on (S, θ), including the assumption that S is unramified. One can hope that such methods can be generalized to yield the validity of Theorem 1.6 for all elements of $S(F)$ whose topologically semisimple modulo center part is regular. The other approach, taken by [26, §4], is to establish a more general character formula, valid for all elements of $G(F)_{\mathrm{reg}}$ and all (S, θ), but under stricter conditions on F, as follows.

Theorem 1.7 ([26, THEOREM 4.3.5]). *Assume F has characteristic zero and p does not divide the order of the Weyl group of G and is larger than $(2 + e)n$, where e is the ramification degree of $F/\mathbb{Q}_p$ and n is the smallest dimension of a faithful algebraic representation of G. For any $\gamma \in G(F)_{\mathrm{reg}}$ with topological Jordan decomposition modulo center $\gamma = \gamma_0 \cdot \gamma_{0+}$,* *the value of* $\Theta_{\pi(S,\theta)}$ *at* γ *is given by*

$$e(G)e(J)\varepsilon_L\big(X^*(T_G)_{\mathbb{C}} - X^*(T_J)_{\mathbb{C}}, \Lambda\big)D(\gamma)^{-\frac{1}{2}}$$
$$\times \sum_{\substack{g \in S(F)\backslash G(F)/J(F) \\ {}^g\gamma_0 \in S(F)}} \Delta_{II}^{\mathrm{abs}}\big[a, \chi''\big]({}^g\gamma_0)\theta({}^g\gamma_0)\hat{O}^J_{X^g}(\log \gamma_{0+}).$$

To explain the new notation, J is the identity component of the centralizer of γ_0 in G, X is any element of $\mathrm{Lie}^*(S)(F)$ such that $\theta(\exp(Y)) = \Lambda(\langle X, Y\rangle)$ for all $Y \in \mathrm{Lie}(S)(F)_{0+}$, and $\hat{O}^J_{X^g}$ is the function on $\mathrm{Lie}(J)(F)$ representing the Fourier transform of the orbital integral on $\mathrm{Lie}^*(J)(F)$ at X^g. In $\Delta_{II}^{\mathrm{abs}}$ we now drop those roots trivial on ${}^g\gamma_0$.

The reason we impose the stricter conditions on F is so that the exponential map converges on the set of topologically nilpotent elements of $\mathrm{Lie}(G)(F)$, cf. [17, APP. A]. This leads naturally to the following question:

Question 1.8. Is there a formulation of the above character formula in which the function $\hat{O}^J_{X^g}(\log \gamma_{0+})$ is replaced by another function on the set of topologically unipotent modulo center elements of $J(F)$, which does not involve the logarithm map, but is still conjugation-invariant. Such a function can be seen as a p-adic analog of Lusztig's Green functions, and this formulation would be valid also in positive characteristic and with possibly weaker conditions on p, as it avoids the use of the logarithm.

It should be noted that the results of [85] are formulated with weaker conditions on F, but use a pseudologarithm map that may not have good equivariance properties.

Remark 1.9. For Theorems 1.6 and 1.7, one must use the twisted Yu construction of [26], which is a modification of the original Yu construction whose purpose is to remedy an error in [88] that goes back to [30]. That error invalidates some results of [88], rendering invalid Yu's proof that the construction produces irreducible supercuspidal representations. It was shown by Fintzen in [25] that despite the error, the original Yu construction does produce irreducible supercuspidal representations. Nonetheless, the error introduces problems that lead to the appearance of auxiliary sign characters in the character formula (the characters $\varepsilon^{\mathrm{ram}}$ and $\varepsilon_{f,\mathrm{ram}}$ of [45, COROLLARIES 4.8.2, 4.10.1], as well as the character of [46, PROPOSITION 5.27]), which make some applications of the resulting formula very difficult. In addition, the arguments of [85] can only be carried out for the twisted Yu construction.

1.3. Double covers of tori

In the archimedean setting, Adams and Vogan [2] have shown that Theorem 1.1, as well as the local Langlands correspondence that will be our next topic, are more naturally formulated if instead of characters θ of an elliptic maximal torus $S(\mathbb{R})$ one uses genuine characters of a certain double cover $S(\mathbb{R})_\rho$. This double cover is obtained by choosing a Borel $\mathbb{C}$-subgroup B of G that contains S and considering the algebraic double cover S_ρ obtained as the pullback of the diagram $S \xrightarrow{2\rho} \mathbb{C}^\times \xleftarrow{2} \mathbb{C}^\times$, where 2ρ is the sum of the B-positive absolute roots and 2 denotes the squaring map. Then $S(\mathbb{R})_\rho$ is defined as the preimage of

$S(\mathbb{R})$ under the isogeny $S_\rho(\mathbb{C}) \to S(\mathbb{C})$. Note that there is a canonical character $\rho : S_\rho \to \mathbb{G}_m$. The choice of B is immaterial, as one can take the limit over all possible choices.

One can then reformulate Theorems 1.1 and 1.2 as the statement that there is a bijection between the set of discrete series representations of $G(\mathbb{R})$ and the set of $G(\mathbb{R})$-conjugacy classes of pairs $(S, \tilde{\theta})$, where $S \subset G$ is an elliptic maximal torus and $\tilde{\theta}$ is a genuine character of the double cover $S(\mathbb{R})_\rho$ such that $d\tilde{\theta}$ is regular. Note that we are not restricting here to what we called "regular" discrete series representations, i.e., this formulation of the theorem covers all discrete series representations. Moreover, the representation $\pi_{(S,\tilde{\theta})}$ is the unique one whose Harish-Chandra character function evaluated at an element $s \in S(\mathbb{R}) \cap G(\mathbb{R})_{\text{reg}}$ has the form

$$(-1)^{q(G)} \frac{\sum_{w \in N(S,G)(\mathbb{R})/S(\mathbb{R})} \operatorname{sgn}(w) \tilde{\theta}(\tilde{s}^w)}{\prod_{\alpha>0} (\alpha^{\frac{1}{2}}(\tilde{s}) - \alpha^{-\frac{1}{2}}(\tilde{s}))}, \tag{1.1}$$

where $\alpha > 0$ runs over all absolute roots that are positive with respect to the Weyl chamber determined by the regular element $d\tilde{\theta}$, and $\tilde{s} \in S(\mathbb{R})_\rho$ is any lift of s. Both numerator and denominator are well-defined genuine functions on $S(\mathbb{R})_\rho$, and their quotient descends to $S(\mathbb{R})$.

The double cover of Adams–Vogan generalizes to all local fields, but the generalization takes a different form than the original definition, in that it is of Galois-theoretic rather than algebraic nature. Without going into technical details, for which we refer to **[46]**, we just mention that for any local field F, a connected reductive F-group G, and a maximal torus $S \subset G$, there exists a double cover $S(F)_\pm$ whose elements can be represented by tuples $(s, (\delta_\alpha))$ with $s \in S(F)$ and $\delta_\alpha \in F_\alpha^\times$ for every symmetric $\alpha \in R(S, G)$ such that $\delta_{\sigma(\alpha)} = \sigma(\delta_\alpha)$ for all $\sigma \in \Gamma$ and $\delta_\alpha / \delta_{-\alpha} = \alpha(s)$. When $F = \mathbb{R}$ and the torus S is elliptic, $S(\mathbb{R})_\pm$ is canonically identified with $S(\mathbb{R})_\rho$.

Theorems 1.3 and 1.6 take the following shape in terms of this double cover: There is a natural bijection between the set of $G(F)$-conjugacy classes of regular supercuspidal representations and the set of $G(F)$-conjugacy classes of pairs $(S, \tilde{\theta})$, where S is a tame elliptic maximal torus and $\tilde{\theta}$ is a regular genuine character of the double cover $S(F)_\pm$. For any $s \in S(F) \cap G(F)_{\text{reg}}$ that is topologically semisimple modulo center, $\Theta_{\pi_{(S,\tilde{\theta})}}$ takes the value

$$e(G)\varepsilon_L\big(X^*(T_G)_{\mathbb{C}} - X^*(S)_{\mathbb{C}}, \Lambda\big) D(s)^{-\frac{1}{2}} \sum_{w \in N_G(S)(F)/S(F)} a_S(\tilde{s}^w) \tilde{\theta}(\tilde{s}^w),$$

where $a_S : S(F)_\pm \to \{\pm 1\}$ is the genuine function sending $\tilde{s} = (s, (\delta_\alpha)) \in S(F)_\pm$ to

$$\prod_\alpha \kappa_\alpha\left(\frac{\delta_\alpha - \delta_{-\alpha}}{a_\alpha}\right),$$

the product runs over the set of Γ-orbits of symmetric elements in $R(S, G)$, and $\kappa_\alpha : F_{\pm\alpha}^\times \to \{\pm 1\}$ is the quadratic character associated to the extension $F_\alpha / F_{\pm\alpha}$. One can also formulate Theorem 1.7 in terms of $S(F)_\pm$; we skip this for now, but will formulate an analogous formula when discussing the local Langlands correspondence.

The advantage of using $S(F)_{\pm}$ is that it removes the somewhat mysterious characters χ''_{α} used in Theorem 1.6 (or rather, it clarifies their role as mediating between characters of $S(F)$ and genuine characters of $S(F)_{\pm}$). Unfortunately, unlike in the archimedean case, this formulation does not allow the parameterization in terms of pairs $(S, \tilde{\theta})$ to be extended beyond the case of regular supercuspidal representations.

2. THE LOCAL LANGLANDS CORRESPONDENCE

2.1. The basic version

Let F be a local field and G be a connected reductive F-group. Let ${}^L G = \widehat{G} \rtimes \Gamma$ be the L-group of G and let L_F be the local Langlands group of F, i.e., the Weil group W_F when F is archimedean, or the group $W_F \times \mathrm{SL}_2(\mathbb{C})$ when F is non-archimedean. The basic version of the local Langlands conjecture states that there exists a surjective finite-to-one map from the set of equivalence classes of irreducible admissible representations of $G(F)$ to the set of $\widehat{G}$-conjugacy classes of relevant L-parameters $\varphi : L_F \to {}^L G$. The fiber over φ is called an *L-packet*, denoted by $\Pi_{\varphi}(G)$.

There are reduction steps on the side of L-parameters that are parallel to the reduction steps "admissible" $\to$ "tempered" $\to$ "discrete" in the classification of irreducible admissible representations, but amount to simple exercises. The step "admissible" $\to$ "tempered" produces from an arbitrary Langlands parameter φ a triple (P, φ_M, ν) consisting of a parabolic subgroup P of G, a tempered parameter φ_M for the Levi quotient M of P, and an element ν of $X^*(A_M)_{\mathbb{R}}$ that lies in the P-positive open cone, cf. [81] for the non-archimedean case. The L-packet $\Pi_{\varphi}(G)$ then consists of the representations $j_P^G(\sigma \otimes \nu)$ for any $\sigma \in \Pi_{\varphi_M}(M)$. The step "tempered" $\to$ "discrete" is even simpler, and just records the Levi subgroup M of G so that a given tempered L-parameter factors through ${}^L M$ and through no smaller Levi subgroup. The L-packet $\Pi_{\varphi}(G)$ consists of the irreducible constituents of the parabolic induction $i_P^G(\sigma)$ for any $\sigma \in \Pi_{\varphi}(M)$.

This reduces the construction of the correspondence to the case of discrete parameters, i.e., those that do not factor through any proper Levi subgroup, and essentially discrete series representations. At this point it becomes clear that the conjecture, as stated so far, is almost vacuous: nothing prevents us from randomly assigning discrete series representations to discrete parameters. This raises the following fundamental question, raised on various occasions by M. Harris, K. Buzzard, and others, which is so far unresolved in full generality:

Question 2.1. Find a list of properties that uniquely characterize the local Langlands correspondence.

As discussed above, it is enough to answer this question for discrete parameters. While eventually a compatibility with a given global correspondence would be a key requirement, at the moment this is not feasible, and we seek a purely local characterization.

A number of expected properties have already been formulated, for example, compatibility with central and cocentral characters and homomorphisms with abelian kernel and cokernel [11, 10.3(1),(2),(5)], the strong tempered L-packet conjecture (a strengthening of [76, CONJECTURE 9.4] stating that each tempered L-packet on a quasisplit group contains a unique member that is generic with respect to a fixed Whittaker datum), the formal degree conjecture [38], and the contragredient conjecture [3, 39]. These are, however, not enough to pin down the correspondence uniquely. The following property is also expected:

Conjecture 2.2. *Each discrete*[1] *series L-packet is atomically stable, i.e., there exists a linear combination of the Harish-Chandra characters of its members that is a stable distribution, and no proper subset of the L-packet has this property.*

It is expected that Conjecture 2.2 uniquely characterizes the partition of the set of equivalence classes of irreducible discrete series representations of $G(F)$ into L-packets. However, it does not determine the matching between L-packets and L-parameters.

In the case of GL_N, stability is a vacuous condition and L-packets are singletons. On the other hand, Henniart has found [35, 37] a list of conditions that uniquely determine the local Langlands correspondence for GL_N when F is non-archimedean. Besides the already listed conditions regarding central and cocentral characters and contragredient, what is needed is equality of L- and ε-factors of pairs, which on the Galois side are the Artin factors of the tensor product of the two Galois representations, and on the automorphic side are given by Rankin–Selberg integrals. While analogous factors can be defined for some other groups as well, such as classical groups, it is unfortunately not yet known how to define them for general reductive groups intrinsically. For some interesting ideas in this direction, see [13]. Another approach to characterizing the local Langlands correspondence for non-archimedean F was pioneered by Scholze in [73] for the group GL_N, and extended to a certain list of other groups in [63] based on [74].

One way to characterize the assignment $\varphi \mapsto \Pi_\varphi(G)$ would be to associate to each φ a stably invariant distribution $S\Theta_\varphi^G$ that would be the stable character of the corresponding L-packet (unique up to nonzero scalar multiple by Conjecture 2.2). In the archimedean case, where the local correspondence has been constructed by Langlands [57], this stable distribution (in fact function) can be described most conceptually using the double covers discussed in Section 1.3. According to Adams–Vogan [2], this description is as follows. The existence of a discrete parameter φ easily implies the existence of an elliptic maximal torus $S \subset G$. There is an L-group ${}^L S_\pm$ associated to the double cover $S(\mathbb{R})_\pm$. One key property of the double cover is that there is a canonical $\widehat{G}$-conjugacy class of L-embeddings ${}^L S_\pm \to {}^L G$ (this is not the case for the L-group ${}^L S$ of the torus S itself), cf. [46, §4.1]. It is again easy to see that φ factors through this L-embedding, and thus leads canonically to a genuine character $\tilde{\theta}$ of $S(\mathbb{R})_\pm$, well defined up to $W_G(S)(\mathbb{R})$. The stable character associated to φ is

1 Conjecture 2.2 is expected more generally for tempered L-packets, but not for nontempered L-packets; the latter need to be enlarged to Arthur packets, or more generally ABV packets, in order to provide stable distributions, cf. [1, 8].

uniquely characterized by its restriction to $S(\mathbb{R})$, where it takes the form

$$S\Theta_{\varphi}^{G}(s) = (-1)^{q(G^*)} \cdot \frac{\sum_{w \in W_G(S)(\mathbb{R})} \operatorname{sgn}(w)\tilde{\theta}(\tilde{s}^w)}{\prod_{\alpha>0}(\alpha^{\frac{1}{2}}(\tilde{s}) - \alpha^{-\frac{1}{2}}(\tilde{s}))}. \tag{2.1}$$

Here again $\alpha > 0$ means $\langle \alpha^\vee, d\tilde{\theta}\rangle > 0$. We denote by $W_G(S) = N_G(S)/S$ the absolute Weyl group, and by G^* the quasisplit inner form of G. Note the very close relationship to (1.1). It may appear odd that we insist on the constant $(-1)^{q(G^*)}$ even though the entire function is supposed to be well defined only up to a constant; we will see in the next subsection that in fact this function is conjecturally well defined "on the nose," and not just up to a constant.

Turning to a non-archimedean base field F, we can use these ideas of Adams–Vogan and our experience from Theorems 1.6 and 1.7 to formulate the following conjectures describing the stable character associated to a *supercuspidal* parameter φ, i.e., a discrete parameter that is trivial on the subgroup $\mathrm{SL}_2(\mathbb{C})$ of L_F. It is easy to see that, when G is tame and p does not divide the order of the Weyl group of G, such a parameter determines a stable conjugacy class of elliptic maximal tori $S \subset G$, and that φ factors through the canonical embedding ${}^LS_{\pm} \to {}^LG$, thereby providing a genuine character $\tilde{\theta}$ of $S(F)_{\pm}$.

Conjecture 2.3. *Let $\gamma \in G(F)$ be regular semisimple and topologically semisimple modulo center. Then $S\Theta_{\varphi}^{G}(\gamma)$ is zero unless γ lies in the image of an admissible embedding $S \to G$, in which case (after identifying S with that image), we have*

$$S\Theta_{\varphi}^{G}(\gamma) = \varepsilon_L\big(X^*(T_G)_{\mathbb{C}} - X^*(S)_{\mathbb{C}}\big)D(\gamma)^{-\frac{1}{2}} \sum_{w \in W_G(S)(F)} [a_S \cdot \tilde{\theta}](\gamma^w). \tag{2.2}$$

Note the strong similarity between (2.1) and (2.2). In fact, this is more than just a similarity:

Formula (2.2) *makes sense for any local field F, and recovers formula* (2.1) *when $F = \mathbb{R}$. Therefore, it gives a conjectural description of the stable character associated to a discrete Langlands parameter $\varphi : W_F \to {}^LG$ that is uniform for any local field F.*

Of course, this conjecture only applies to discrete parameters that, in the non-archimedean case, have trivial restriction to $\mathrm{SL}_2(\mathbb{C})$, and in addition p is prime to $|W_G(S)|$. This conjecture was the guiding principle behind the constructions of [45,47]. One drawback it has is that in the non-archimedean case there may not be enough topologically semisimple elements of $S(F)$ to fully determine the function $S\Theta_{\varphi}^{G}$. This is not an issue in the setting of [45, 47] because we are not using the values of the function, but rather the *entire formula*, which carries more information. Nonetheless, a more complete solution is desirable. It is conceivable that the ideas of [14] might lead to a stronger version of this formula. Another approach is to allow arbitrary regular semisimple elements of $G(F)$, in the vein of Theorem 1.7. This leads to the following conjecture.

Conjecture 2.4 ([46, §4.4]). *For any strongly regular semisimple $\gamma \in G(F)$ with topological Jordan decomposition modulo center $\gamma = \gamma_0 \cdot \gamma_{0+}$,*

$$S\Theta_\varphi^G(\gamma) = e(J)\varepsilon_L\big(X^*(T_G)_{\mathbb{C}} - X^*(T_J)_{\mathbb{C}}\big)D(\gamma)^{-\frac{1}{2}} \sum_{j:S\to J} [a_S \cdot \tilde{\theta}](\gamma_0^j) \cdot \widehat{SO}^J_{jX}\big(\log(\gamma_{0+})\big),$$

assuming F has characteristic zero and $p \geq (2+e)n$.

The notation is the same as that in Theorem 1.7, except now we are using the stable orbital integral at ${}^jX$ instead of the usual orbital integral, and the sum runs over the set of stable classes of admissible embeddings $S \to J$.

Again, the reason we require the characteristic of F to be zero and p to be very large is to ensure the convergence of exp on $\mathrm{Lie}(G)(F)_{0+}$, in particular on $\mathrm{Lie}(J)(F)_{0+}$. A positive resolution to Question 1.8 would weaken this requirement.

The most general constructions of the non-archimedean basic local Langlands correspondence are given by Genestier–Lafforgue [29] in positive characteristic and Fargues–Scholze [22] for arbitrary non-archimedean local fields. These constructions only produce semisimplified parameters, but, at least in positive characteristic, recent work of Gan–Harris–Sawin [27] (based on arguments of Gan–Lomelí [28] in the case of classical groups) provides a unique enrichment of such a semisimplified parameter to a full Langlands parameter when the representation in question is supercuspidal. Earlier constructions include [33, 36, 59] for GL_N and [9] for quasisplit symplectic and orthogonal groups in characteristic zero, and [32] for generic supercuspidal representations of the exceptional group G_2.

At the moment there are many open questions regarding the constructions [22, 29], such as

(1) Is the map $\pi \mapsto \varphi$ surjective?

(2) Are the resulting L-packets always finite?

(3) Does the construction of Fargues–Scholze specialize to that of Genestier–Lafforgue when F has positive characteristic?

(4) Do Conjectures 2.2 and 2.4 hold?

(5) Does the formal degree conjecture hold?

On the other hand, [47] gives an *explicit* construction of the correspondence under the assumption that G is tame and p does not divide the order of the Weyl group, and φ is a supercuspidal parameter, building on prior work [17, 40, 71]. This setting is more restrictive than that of [22] or [29], but in return provides much more knowledge about the resulting correspondence. For example, we know that

(1) The map $\pi \mapsto \varphi$ has as domain all nonsingular supercuspidal representations, and as image all supercuspidal parameters.

(2) The map $\pi \mapsto \varphi$ is compatible with central and cocentral characters.

(3) The resulting L-packets are always finite and, in fact, have the desired internal structure (see the next section).

(4) Both Conjectures 2.2 and 2.4 hold [26, §4.4].

(5) The formal degree conjecture holds, as shown by Schwein [75] and Ohara [68].

The question whether the constructions of [22] and [47] agree is equivalent (in the setting of F having characteristic zero and p being sufficiently large) to the question of whether [22] satisfies Conjecture 2.4. A strong indication that they agree would be given if Conjecture 2.3 could be proved instead; the latter can also be pursued for [29], since it allows the characteristic of F to be positive.

The opposite setting of that of supercuspidal parameters and nonsingular (i.e., semisimple) supercuspidal representations is that of unipotent supercuspidal representations, and more generally arbitrary unipotent representations. Much progress has been made on the local Langlands correspondence for these representations via detailed study of affine Hecke algebras and formal degrees [23, 61, 62, 70, 82].

2.2. The refined version

For many applications, such as the Gan–Gross–Prasad conjecture, or the multiplicity formula for discrete automorphic representations, the basic version of the local Langlands conjecture is insufficient because it describes packets of representations rather than individual representations. The refined local Langlands conjecture remedies this by enhancing the notion of a Langlands parameter to allow the description of individual irreducible admissible representations. In fact, already the statement of the Hiraga–Ichino–Ikeda conjecture requires the refined correspondence, a point that we glossed over in the previous subsection.

As pointed out by Vogan [87], the refined conjecture requires a rigidification of the concept of inner forms of reductive groups. In the archimedean case, a good rigidification was obtained by Adams–Barbasch–Vogan [1]. In the non-archimedean case, different ways of rigidification have led to different versions of the refined conjecture. We only give a brief summary, referring the reader to [41] for more details.

The set of equivalence classes of inner forms of a connected reductive group G is $H^1(F, G_{\mathrm{ad}})$, where $G_{\mathrm{ad}} = G/Z(G)$ is the adjoint group of G. In every inner class there is a unique quasisplit form, and we normalize things by taking G to be that form. A rigidification of the notion of an inner form can be achieved by choosing a Galois gerbe $\mathcal{E}$. When F has characteristic zero such a gerbe can be understood, following Langlands–Rapoport [58], as an extension $1 \to u(\overline{F}) \to \mathcal{E} \to \Gamma \to 1$ of the absolute Galois group Γ of F by an algebraic or proalgebraic group u. Following Kottwitz [54], one then considers the set $H^1_{\mathrm{bas}}(\mathcal{E}, G)$ of cohomology classes of $\mathcal{E}$ with values in $G(\overline{F})$ whose restriction to u factors through the center $Z(G)$ of G. There is a natural embedding $H^1(F, G) \to H^1_{\mathrm{bas}}(\mathcal{E}, G)$ and a natural map

$$H^1_{\mathrm{bas}}(\mathcal{E}, G) \to H^1(F, G_{\mathrm{ad}}). \tag{2.3}$$

A rigidification of an inner form of G is the choice of an element of $H^1_{\mathrm{bas}}(\mathcal{E}, G)$ that lifts the class of that inner form.

Vogan's notion of pure inner forms comes from the trivial Galois gerbe $\mathcal{E}^{\text{triv}} = \Gamma$, for which $u = \{1\}$. Kottwitz's theory of isocrystals with G-structure **[53]** employs the gerbe $\mathcal{E}^{\text{iso}}$ of the Tannakian category of F-isocrystals. The problem with $\mathcal{E}^{\text{triv}}$ and $\mathcal{E}^{\text{iso}}$ is that in general the map (2.3) is not surjective, so not all inner forms can be rigidified. For $\mathcal{E}^{\text{iso}}$, that map is surjective when $Z(G)$ is connected. In **[42]** I define a gerbe $\mathcal{E}^{\text{rig}}$ for which the map (2.3) is always surjective. It turns out that, when $F = \mathbb{R}$, this gerbe recovers the notion of strong real forms introduced by **[1]**. A key property of the gerbes $\mathcal{E}^{\text{iso}}$ and $\mathcal{E}^{\text{rig}}$ is that they satisfy a generalization of Tate–Nakayama duality.

When F has positive characteristic the simplified concept of a Galois gerbe as an extension of the absolute Galois group becomes inadequate, due to the possible nonsmoothness of u. Despite this difficulty, Peter Dillery **[20]** has found a way to construct a suitable analog of $\mathcal{E}^{\text{rig}}$. In fact, his construction works uniformly for all non-archimedean local fields and recovers $\mathcal{E}^{\text{rig}}$ when F has characteristic zero. Therefore, we now have a satisfactory definition of $\mathcal{E}^{\text{rig}}$ for any local field.

The refined local Langlands conjecture parameterizes all irreducible admissible representations of all inner forms of G at once. More precisely, one considers tuples (G', ξ, z, π), where G' is a connected reductive F-group, $\xi : G_{F^s} \to G'_{F^s}$ is an isomorphism of F^s-groups, where F^s is a fixed separable closure of F, $z \in Z^1_{\text{bas}}(\mathcal{E}^{\text{rig}}, G)$, and $\xi^{-1}\sigma(\xi) = \text{Ad}(\bar{z}_\sigma)$, where $\bar{z}_\sigma$ is the image of z in $Z^1(F, G_{\text{ad}})$ (since G_{ad} is smooth we can interpret the latter as étale, i.e., Galois, cohomology), and finally π is an irreducible admissible representation of $G'(F)$. An isomorphism $(G_1, \xi_1, z_1, \pi_1) \to (G_2, \xi_2, z_2, \pi_2)$ is a pair (g, f) with $g \in G(F^s)$ and $f : G_1 \to G_2$ an isomorphism of F-groups such that $f \circ \xi_1 = \xi_2 \circ \text{Ad}(g)$ and $z_2(e) = g z_1(e) \sigma_e(g)^{-1}$; here $\sigma_e \in \Gamma$ is the image of $e \in \mathcal{E}^{\text{rig}}$. The key property of the set $Z^1_{\text{bas}}(\mathcal{E}^{\text{rig}}, G)$ is that if we fix the triple (G', ξ, z), then two tuples (G', ξ, z, π_1) and (G', ξ, z, π_2) are isomorphic if and only if the representations π_1 and π_2 of $G'(F)$ are equivalent.

Assuming the validity of the basic local Langlands conjecture, we can define for each Langlands parameter $\varphi : L_F \to {}^L G$ the "compound L-packet" Π_φ as the set of isomorphism classes of tuples (G', ξ, z, π), for all possible (G', ξ, z) and all $\pi \in \Pi_\varphi(G')$. We further let S_φ be the centralizer of φ in $\widehat{G}$, and S_φ^+ and $Z([\widehat{\bar{G}}]^+)$ be the preimages of S_φ and $Z(\widehat{G})^\Gamma$ in the universal cover $\widehat{\bar{G}}$ of $\widehat{G}$. The refined conjecture, inspired by the work of Adams-Barbasch-Vogan and Vogan, is then the following.

Conjecture 2.5 (**[42, §5.4]**). *Fix a Whittaker datum $\mathfrak{w}$ for G.*

(1) *There exists a bijection $\iota_{\varphi,\mathfrak{w}}$ that is the top map in the following commutative diagram*

$$\begin{array}{ccc} \Pi_\varphi & \xrightarrow{\iota_{\varphi,\mathfrak{w}}} & \text{Irr}(\pi_0(S_\varphi^+)) \\ \downarrow & & \downarrow \\ H^1_{\text{bas}}(\mathcal{E}^{\text{rig}}, G) & \longrightarrow & \pi_0(Z(\widehat{\bar{G}})^+)^* \end{array}$$

in which the left map sends the isomorphism class of (G', ξ, z, π) to the class of z, the right map is the central character map, and the bottom map is generalized Tate–Nakayama duality. This bijection relates the unique $\mathfrak{w}$-generic constituent of Π_φ to the trivial representation of $\pi_0(S_\varphi^+)$.

Given a semisimple element $\dot{s} \in S_\varphi^+$, a rigid inner twist (G', ξ, z), and a tempered *parameter φ, define the virtual character*

$$\Theta_{\varphi,\mathfrak{w},\dot{s}}^{G',\xi,z} = e(G') \sum_{\pi \in \Pi_\varphi(G')} \operatorname{tr}\big(\iota_{\varphi,\mathfrak{w}}(G', \xi, z, \pi)(\dot{s})\big) \cdot \Theta_\pi.$$

(2) *The distribution $\Theta_{\varphi,\mathfrak{w},1}^{G',\xi,z}$ is stable and independent of $\mathfrak{w}$ and z.*

(3) *For a general $\dot{s} \in S_\varphi^+$, the distribution $\Theta_{\varphi,\mathfrak{w},\dot{s}}^{G',\xi,z}$ is the endoscopic lift of the distribution $\Theta_{\varphi,*,\dot{s}}^{H,\mathrm{id},1}$ for the endoscopic datum $(H, \dot{s})$ associated to φ, with respect to the transfer factor normalized via $\mathfrak{w}$ and z as in* [42, (5.10)].

Remark 2.6. (1) Point (3) specifies the bijection $\iota_{\varphi,\mathfrak{w}}$ uniquely, provided such a bijection exists.

(2) The distribution in (2) is what we referred to as $S\Theta_\varphi^{G'}$ in the previous subsection. Note that the inner twist ξ is used to identify ${}^LG'$ with ${}^LG$, so if we use ${}^LG$ as codomain for Langlands parameters, we should write $S\Theta_\varphi^{G',\xi}$ to indicate that this distribution depends also on ξ, not just G'.

(3) The fiber over $[z] \in H^1_{\mathrm{bas}}(\mathcal{E}^{\mathrm{rig}}, G)$ of the left vertical map in (1) is by definition the L-packet $\Pi_\varphi(G_{\bar{z}})$ of the inner form $G_{\bar{z}}$ of G associated to the image $[\bar{z}] \in H^1(F, G_{\mathrm{ad}})$ of z. Note that there can be two distinct $[z_1], [z_2] \in H^1(\mathcal{E}^{\mathrm{rig}}, G)$ mapping to $[\bar{z}]$. This leads to the appearance of the same L-packet $\Pi_\varphi(G_{\bar{z}})$ multiple times in the compound L-packet Π_φ. This "overcounting" is the spectral incarnation of the rigidification of the notion of inner forms.

(4) When F is non-archimedean, the bottom map in (1) is bijective. This means that the L-packet $\Pi_\varphi(G_{\bar{z}})$ for the individual group $G_{\bar{z}}$ is described precisely by the set $\mathrm{Irr}(\pi_0(S_\varphi^+), [z])$ of irreducible representations of $\pi_0(S_\varphi^+)$ that transform under $\pi_0(Z(\hat{\bar{G}})^+)$ via the character corresponding to $[z]$. As just discussed, there can be multiple $[z]$ lifting $[\bar{z}]$, leading to multiple ways to parameterize $\Pi_\varphi(G_{\bar{z}})$. Thankfully, it is a rather straightforward matter to relate these two parameterizations of $\Pi_\varphi(G_{\bar{z}})$, as we will discuss in the next subsection.

(5) When $F = \mathbb{R}$, the bottom map in (1) need not be injective or surjective. These failures are well understood, cf [42, §3.4, §4, PROPOSITION 5.3]. The nonsurjectivity implies that, for some $\eta \in \pi_0(Z(\hat{\bar{G}})^+)^*$, the set $\mathrm{Irr}(\pi_0(S_\varphi^+), \eta)$ does not parameterize any L-packet. The noninjectivity implies that for some $\eta \in \pi_0(Z(\hat{\bar{G}})^+)^*$, the set $\mathrm{Irr}(\pi_0(S_\varphi^+), \eta)$ parameterizes a union of L-packets over certain rigid inner forms. The set of rigid inner forms appearing in such a union consists of

exactly those inner forms of G that are related to each other by Galois 1-cocycles valued in the simply connected covers of their adjoint groups; this set is termed a *K-group* by Arthur in [7, §1].

(6) An analogous conjecture can be stated with $\mathcal{E}^{\text{iso}}$ or $\mathcal{E}^{\text{triv}}$ in place of $\mathcal{E}^{\text{rig}}$. One has to then replace $\pi_0(Z(\widehat{\bar{G}})^+)^*$ by $X^*(Z(\widehat{G})^\Gamma)$ or $\pi_0(Z(\widehat{G})^\Gamma)^*$, respectively, and $\pi_0(S_\varphi^+)$ by $S_\varphi^\natural = S_\varphi/(S_\varphi \cap \widehat{G}_{\text{der}})^\circ$ or $\pi_0(S_\varphi)$, respectively. There is a commutative diagram

$$\begin{array}{ccccc} \pi_0(S_\varphi) & \longleftarrow & \pi_0(S_\varphi^+) & \longrightarrow & S_\varphi^\natural \\ \uparrow & & \uparrow & & \uparrow \\ \pi_0(Z(\widehat{G})^\Gamma) & \longleftarrow & \pi_0(Z(\widehat{\bar{G}})^+) & \longrightarrow & Z(\widehat{G})^\Gamma \end{array}$$

that relates the three versions of the conjecture. Given $\xi_1 \in \pi_0(S_\varphi)^*$ and $\xi_2 \in X^*(Z(\widehat{G})^\Gamma)$ with common image $\xi \in \pi_0(Z(\widehat{\bar{G}})^+)^*$, this diagram induces bijections $\mathrm{Irr}(\pi_0(S_\varphi), \xi_1) = \mathrm{Irr}(\pi_0(S_\varphi^+), \xi) = \mathrm{Irr}(S_\varphi^\natural, \xi_2)$. Therefore, the three versions of the conjecture differ only in the amount of inner forms of G that they can reach, and the way in which these inner forms are overcounted by rigidification. For a more thorough comparison, cf. [44, §4.2].

When $F = \mathbb{R}$, Conjecture 2.5 was established by Shelstad in a series of papers, especially [77–80], cf. also [42, §5.6]. A geometric approach to Conjecture 2.5 was developed in [1], which also produces and parameterizes Arthur packets, not just L-packets.

For non-archimedean local fields of characteristic zero, the restriction of Conjecture 2.5 to quasisplit classical groups was proved by Arthur [9].

For more general groups over non-archimedean local fields, we have the following.

Theorem 2.7 ([26, §4.4]). *Under the assumptions of Theorem 1.7, Conjecture 2.5 holds for the construction of regular supercuspidal L-packets of* [45], *and more generally for all supercuspidal L-packets constructed* [47], *but in that larger generality point (3) is proved only for certain elements $\dot{s}$ (the remaining elements $\dot{s}$ are work in progress).*

It is at the moment not known if the constructions of [29] or [22] satisfy Conjecture 2.5. This would follow from Theorem 2.7 once Conjecture 2.4 is established for these constructions, via comparison with [47], at least under the assumptions under which that theorem and conjecture are formulated. We note that [22] works with the concept of inner forms rigidified via $\mathcal{E}^{\text{iso}}$ because isocrystals with G-structure appear naturally in the theory of the Fargues–Fontaine curve, but the relationship between the two versions of the refined correspondence is well understood as per Remark 2.6(6).

The refined local Langlands conjecture can be combined with a global version of $\mathcal{E}^{\text{rig}}$, developed in [43] over number fields, and in [21] over global function fields, to obtain a precise formulation of the conjectural multiplicity formula for discrete automorphic representations originally due to Kottwitz [52, (12.3)], cf. [43, §4.5], [21, §5.4]. Special cases of this formula have been proved by O. Taïbi, cf. [86]. One can also obtain a global multiplicity conjecture using the global version of $\mathcal{E}^{\text{iso}}$ defined by Kottwitz in [50], under the assumption that the global group G has connected center and satisfies the Hasse principle. Under those conditions, the two global multiplicity formulas coming from $\mathcal{E}^{\text{iso}}$ and $\mathcal{E}^{\text{rig}}$ are equivalent, cf. [48].

Another setting in which information about the refined local Langlands conjecture, more precisely part (1) of Conjecture 2.5, has been obtained is that of unipotent representations, cf. [23, 61, 62, 70, 82].

2.3. Compatibility properties

The basic version of the local Langlands correspondence is expected to satisfy many compatibility properties, cf. [11, §10]. The refined version of the local Langlands correspondence allows one to formulate more precise compatibility properties. Some of them turn out to be formal consequences of the refined conjecture, while others have to be proved independently.

Among the simplest such properties are those regarding the dependence of the parameterizing bijection $\iota_{\varphi,\mathfrak{w}}$ on the Whittaker datum $\mathfrak{w}$ and on the element of $H^1_{\text{bas}}(\mathcal{E}^{\text{rig}}, G)$ lifting the class of $H^1(F, G_{\text{ad}})$ that describes the relevant inner form.

2.3.1. Whittaker data

If another Whittaker datum $\mathfrak{w}'$ is chosen, there exists an element $g \in G_{\text{ad}}(F)$ such that $\mathfrak{w}' = g\mathfrak{w}g^{-1}$. Denote by $(\mathfrak{w}', \mathfrak{w})$ the image of g under the connecting homomorphism $H^0(F, G_{\text{ad}}) \to H^1(F, Z(G_{\text{sc}}))$. Local Tate duality identifies $H^1(F, Z(G_{\text{sc}}))$ with the dual of $H^1(F, Z(\widehat{G}_{\text{sc}})) = H^1(L_F, Z(\widehat{G}_{\text{sc}}))$. A Langlands parameter $\varphi : L_F \to {}^LG$ induces an action of L_F on $\widehat{G}$ by conjugation via φ, which coincides with the usual action of L_F on $Z(\widehat{G}_{\text{sc}})$. Via the connecting homomorphism $H^0(\varphi(L_F), \widehat{G}_{\text{ad}})) \to H^1(L_F, Z(\widehat{G}_{\text{sc}}))$ in the resulting long exact cohomology sequence, we can pull back the character $(\mathfrak{w}', \mathfrak{w})$ to the group $S_\varphi/Z(\widehat{G})^\Gamma = S_\varphi^+/Z(\widehat{\bar{G}})^+$. The following result is stated in [39, THEOREM 4.3] for real groups or quasisplit classical p-adic groups, but the proof actually shows that it is a formal consequence of Conjecture 2.5:

Theorem 2.8. *The validity of Conjecture* 2.5 *implies*

$$\iota_{\varphi,\mathfrak{w}'}(\dot{\pi}) = \iota_{\varphi,\mathfrak{w}}(\dot{\pi}) \otimes (\mathfrak{w}', \mathfrak{w}) \quad \forall \dot{\pi} \in \Pi_\varphi.$$

2.3.2. Rigidifying data

The question of changing the rigidifying element can be resolved in an analogous way. Consider an element $\bar{z} \in Z^1(F, G_{\text{ad}})$ leading to the inner form $G_{\bar{z}}$. Let $z_1, z_2 \in Z^1_{\text{bas}}(\mathcal{E}^{\text{rig}}, G)$ both lift $\bar{z}$. Then $z_2z_1^{-1} \in Z^1(\mathcal{E}^{\text{rig}}, Z(G))$, and we denote by $[z_2z_1^{-1}]$ its class. Given a Langlands parameter $\varphi : L_F \to {}^LG$ and $\pi \in \Pi_\varphi(G_{\bar{z}})$, we can consider the elements $\dot{\pi}_1 = (G_{\bar{z}}, \xi, z_1, \pi)$ and $\dot{\pi}_2 = (G_{\bar{z}}, \xi, z_2, \pi)$ of Π_φ, where ξ denotes the identification of G_{F^s} with $G_{\bar{z},F^s}$. These elements both describe the representation π of the group $G_{\bar{z}}(F)$, but are distinct elements of the compound packet Π_φ. This is the overcounting phenomenon mentioned in the previous subsection. To relate these two "reflections" of π, consider the exact sequence

$$1 \to \pi_1(\widehat{G}) \to \widehat{\widehat{G}} \to \widehat{G} \to 1$$

equipped with the action of L_F via conjugation by φ. It leads to the differential $d : S_\varphi^+ \to Z^1(F, \pi_1(\widehat{G}))$ that factors through $\pi_0(S_\varphi^+)$. We denote by $-d$ the composition of this differential with the inversion automorphism of the abelian group $\pi_1(\widehat{G})$. It is shown in [44, §§6.1,6.2] that local Tate duality generalizes to a duality between $H^1(\mathcal{E}^{\text{rig}}, Z(G))$ and $Z^1(F, \pi_1(\widehat{G}))$. The element $[z_2 z_1^{-1}]$ thus becomes a character of $Z^1(F, \pi_1(\widehat{G}))$, which can be pulled back to $\pi_0(S_\varphi^+)$ by $-d$.

Theorem 2.9 ([44, §6.3]). *The validity of Conjecture 2.5 implies*

$$\iota_{\varphi,\mathfrak{w}}(\dot{\pi}_2) = \iota_{\varphi,\mathfrak{w}}(\dot{\pi}_1) \otimes (-d)^*\big([z_2 z_1^{-1}]\big).$$

Remark 2.10. The cohomological constructions in Theorem 2.8 and 2.9 are very closely related: the composition of the differential $d : S_\varphi^+ \to Z^1(F, \pi_1(\widehat{G}))$ with the natural projection $Z^1(F, \pi_1(\widehat{G})) \to H^1(F, \pi_1(\widehat{G}))$ and the natural map $\pi_1(\widehat{G}) \to \pi_1(\widehat{G}_{\text{ad}}) = Z(\widehat{G}_{\text{sc}})$ equals the composition of the natural projection $S_\varphi^+ \to H^0(\varphi(L_F), \widehat{G}_{\text{ad}})$ with the connecting homomorphism $H^0(\varphi(L_F), \widehat{G}_{\text{ad}}) \to H^1(L_F, Z(\widehat{G}_{\text{sc}})) = H^1(F, Z(\widehat{G}_{\text{sc}}))$. This is embodied in the commutative diagrams [44, (6.1),(6.2),(6.6)].

2.3.3. Contragredients

We now discuss the compatibility of the local Langlands conjecture with respect to taking contragredients. To state it, we write $\widehat{C}$ for the Chevalley involution of $\widehat{G}$, well defined up to conjugation, and by ${}^LC$ its extension to an automorphism of ${}^LG$. We write $\pi^\vee$ to denote the contragredient of a representation π, and given $\dot{\pi} = (G', \xi, z, \pi)$ we write $\dot{\pi}^\vee = (G', \xi, z, \pi^\vee)$.

Conjecture 2.11. *Assume Conjecture 2.5. Let $\varphi : L_F \to {}^LG$ be a Langlands parameter with compound L-packet Π_φ. Then*

(1) $\Pi_{{}^LC \circ \varphi} = \{\dot{\pi}^\vee \mid \dot{\pi} \in \Pi_\varphi\}$,

(1′) $S\Theta^{G',\xi}_{{}^LC \circ \varphi}(\delta) = S\Theta^{G',\xi}_{\varphi}(\delta^{-1})$ *whenever φ is tempered,*

(2) $\iota_{{}^LC \circ \varphi, \mathfrak{w}^{-1}}(\dot{\pi}^\vee) = (\iota_{\varphi,\mathfrak{w}}(\pi) \circ \widehat{C}^{-1})^\vee$.

Part (1) can be stated for L-packets on an individual group, and hence without assuming Conjecture 2.5. In this form it was formulated by Adams–Vogan in [3], where it was also proved for $F = \mathbb{R}$. Wen-Wei Li proved that statement of (1) in [60] for the semisimplified basic correspondence of [29]. Part (1′) is a variation of part (1)—it implies part (1) via linear independence of characters, provided one assumes Conjecture 2.2. Over $F = \mathbb{R}$ parts (1) and (1′) are, in fact, equivalent, since Conjectures 2.2 and 2.5 are known and $S\Theta^{G',\xi}_\varphi$ is simply the sum of Θ_π over all $\pi \in \Pi_\varphi(G')$.

Part (2) was formulated by D. Prasad, cf. [69, CONJECTURE 2]. It was proved in [39, THEOREM 5.9] that Conjecture 2.5 for G, together with part (1′) for all endoscopic groups of G, implies part (2) for G. It was furthermore proved in [39, THEOREM 5.8] that for p-adic fields part (1′) holds for quasisplit symplectic and orthogonal groups; the same argument also applies to quasisplit unitary groups.

2.3.4. Automorphisms

The next compatibility we discuss is with respect to automorphisms. Initially, one may be interested in a particular connected reductive F-group G' and wonder how an F-automorphism θ' of G' respects the (basic or refined) local Langlands correspondence. We will see, however, that this is a special case of the following more general consideration.

Let as before G be a quasisplit connected reductive F-group. Let $\theta \in \mathrm{Aut}_F(G)$. Then θ acts on $Z^1_{\mathrm{bas}}(\mathcal{E}^{\mathrm{rig}}, G)$ via its action on G, and furthermore acts on the set of tuples (G', ξ, z, π) by the rule $\theta(G', \xi, z, \pi) = (G', \xi \circ \theta^{-1}, \theta(z), \pi)$. This induces an action of $\mathrm{Aut}_F(G)$ on the set of isomorphism classes of such tuples. The subgroup $\mathrm{Int}_F(G)$ of inner F-automorphisms of G does *not* act trivially, but it follows from Theorems 2.8 and 2.9 and Remark 2.10 that $\iota_{\varphi,\theta(\mathfrak{w})}(\theta\dot{\pi}) = \iota_{\varphi,\mathfrak{w}}(\dot{\pi})$ for $\theta \in \mathrm{Int}_F(G)$ and $\dot{\pi} \in \Pi_\varphi$. Note this implies that θ preserves the compound L-packet Π_φ.

On the other hand, the group $\mathrm{Aut}_F(\widehat{G})$ of Γ-equivariant automorphisms of $\widehat{G}$ acts on the set of refined Langlands parameters by the rule $\theta(\varphi, \rho) = (\theta \circ \varphi, \rho \circ \theta^{-1})$, and the subgroup $\mathrm{Int}_F(\widehat{G})$ acts trivially on the $\widehat{G}$-conjugacy classes of such parameters by definition. Recall that the exact sequences of Γ-modules $1 \to \mathrm{Int}_{F^s}(G) \to \mathrm{Aut}_{F^s}(G) \to \mathrm{Out}_{F^s}(G) \to 1$ and $1 \to \mathrm{Int}(\widehat{G}) \to \mathrm{Aut}(\widehat{G}) \to \mathrm{Out}(\widehat{G}) \to 1$ are split, and the choice of an F-pinning of G resp. of a Γ-stable pinning of $\widehat{G}$ determine Γ-equivariant splittings of these sequences. Recall finally that there is a natural Γ-equivariant identification $\mathrm{Out}(G) = \mathrm{Out}(\widehat{G})$, via which we obtain the identification $\mathrm{Aut}_F(G)/\mathrm{Int}_F(G) = \mathrm{Out}_F(G) = \mathrm{Out}_F(\widehat{G}) = \mathrm{Aut}_F(\widehat{G})/\mathrm{Int}_F(\widehat{G})$.

To state the following conjecture, it is convenient to write $\iota(\mathfrak{w}, \dot{\pi}) = (\varphi, \rho)$, where φ is the unique Langlands parameter with $\dot{\pi} \in \Pi_\varphi$, and $\rho = \iota_{\varphi,\mathfrak{w}}(\dot{\pi})$.

Conjecture 2.12. *Assume Conjecture 2.5. Let $\varphi : L_F \to {}^LG$ be a Langlands parameter with compound L-packet Π_φ. For $\theta \in \mathrm{Out}_F(G) = \mathrm{Out}_F(\widehat{G})$,*

(1) $\Pi_{\theta\circ\varphi} = \{\theta\dot{\pi} \mid \dot{\pi} \in \Pi_\varphi\}$.

(2) $\iota(\theta(\mathfrak{w}), \theta(\dot{\pi})) = \theta(\iota(\mathfrak{w}, \dot{\pi}))$.

Again part (1) can be formulated for L-packets on an individual group, and hence without assuming Conjecture 2.5. In that form it appears as [3, LEMMA 6.18] when $F = \mathbb{R}$.

Note the formal similarity between Conjectures 2.11 and 2.12. If we let θ be the element of $\mathrm{Out}_F(G) = \mathrm{Out}_F(\widehat{G})$ that corresponds to the Chevalley involutions and set $\iota(\mathfrak{w}, \dot{\pi}) = (\varphi, \rho)$, then Conjecture 2.11(2) states $\iota(\mathfrak{w}^{-1}, \dot{\pi}^\vee) = (\theta(\varphi), \theta(\rho)^\vee)$, while Conjecture 2.12(2) states $\iota(\theta(\mathfrak{w}), \theta(\dot{\pi})) = (\theta(\varphi), \theta(\rho))$. We are free to lift θ to an element of $\mathrm{Aut}_F(G)$ any way we like; if we take it to be the involution $\iota_{G,\mathcal{P}}$ defined by D. Prasad in [69, DEFINITION 1] with respect to a pinning related to the Whittaker datum $\mathfrak{w}$ as in [55, §5.3], then $\theta(\mathfrak{w}) = \mathfrak{w}^{-1}$. This shows that [69, CONJECTURE 1] follows from Conjectures 2.11 and 2.12.

Let us now discuss how Conjecture 2.12 gives information about the compatibility of the refined local Langlands correspondence with F-automorphisms of a fixed group G'. We realize G' as a rigid inner twist $(\xi, z) : G \to G'$ of its quasisplit inner form G. Given $\theta' \in \mathrm{Aut}_F(G')$, the automorphism $\xi^{-1}\theta'\xi$ of G need not respect the F-structure. However, it does respect the F^s-structure and, moreover, the difference $(\xi^{-1}\theta'\xi)^{-1} \circ \sigma(\xi^{-1}\theta'\xi)$ is an inner automorphism of G for each $\sigma \in \Gamma$. In other words, the image of $\xi^{-1}\theta'\xi$ in the group $\mathrm{Out}(G) = \mathrm{Aut}(G)/\mathrm{Int}(G)$ is an F-point. Let $\theta \in \mathrm{Aut}_F(G)$ be a lift of that F-point that preserves the Whittaker datum $\mathfrak{w}$. Then $\xi^{-1}\theta'\xi = \mathrm{Ad}(g) \circ \theta$ for some $g \in G(F^s)$. Since the automorphism $\xi^{-1}\theta'\xi$ of G commutes with the twisted action $\mathrm{Ad}(\bar{z}_\sigma)\sigma$ on $G(F^s)$ for all $\sigma \in \Gamma$, we see that the equality $\theta(\bar{z}_\sigma) = g^{-1}\bar{z}_\sigma\sigma(g)$ holds in $Z^1(F, G_{\mathrm{ad}})$. Then $y_e := (g^{-1}z_e\sigma_e(g)) \cdot \theta(z_e)^{-1} \in Z^1(\mathcal{E}^{\mathrm{rig}}, Z(G))$ and we see that (g, θ') is an isomorphism $(G', \xi \circ \theta^{-1}, y \cdot \theta(z), \pi) \to (G', \xi, z, \pi \circ \theta'^{-1})$. The class $[y]$ of y is uniquely determined by θ' and (ξ, z). From Theorem 2.9, we obtain

Corollary 2.13. *Assume Conjectures 2.5 and 2.12. Then*

$$\iota(G', \xi, z, \pi \circ \theta'^{-1}) = \theta\big(\iota(G', \xi, z, \pi)\big) \otimes (-d)^*([y]),$$

where the tensor product affects the second component ρ of the refined parameter (φ, ρ).

The class $[y]$ is trivial if and only if g can be chosen so that (g, θ') is an isomorphism $(G', \xi \circ \theta^{-1}, \theta(z), \pi) \to (G', \xi, z, \pi \circ \theta'^{-1})$. This is not always possible: the simplest example is when θ' is an inner automorphism of G' that comes from an element of $G'_{\mathrm{ad}}(F)$ which does not lift to $G'(F)$. In fact, this example is very useful and leads to the following application of Corollary 2.13.

Corollary 2.14. *Assume Conjectures 2.5 and 2.12. The action of $G'_{\mathrm{ad}}(F)$ on the set of representations of $G'(F)$ preserves each L-packet. More precisely, if $\theta' = \mathrm{Ad}(\bar{g})$ for some $\bar{g} \in G'_{\mathrm{ad}}(F)$, then*

$$\iota(G', \xi, z, \pi \circ \theta'^{-1}) = \iota(G', \xi, z, \pi) \otimes (-d)^*\big([g^{-1}z_e\sigma_e(g)z_e^{-1}]\big),$$

for any lift $g \in G(F^s)$ of $\xi^{-1}(\bar{g})$.

2.3.5. Homomorphism with abelian kernel and cokernel

Let $f' : G'_1 \to G'_2$ be a homomorphism of connected reductive F-groups with abelian kernel and cokernel, and ${}^L f : {}^L G'_2 \to {}^L G'_1$ be the corresponding L-homomorphism. For a Langlands parameter $\varphi_2 : L_F \to {}^L G'_2$ we can consider the composed parameter $\varphi_1 := {}^L f \circ \varphi_2$ and the corresponding L-packets $\Pi_{\varphi_2}(G'_2)$ and $\Pi_{\varphi_1}(G'_1)$ provided by the basic local Langlands conjecture.

It is asserted in [11, §10] that for any $\pi_2 \in \Pi_{\varphi_2}(G'_2)$ the representation $\pi_2 \circ f'$ of $G'_1(F)$ is a direct sum of finitely many members of $\Pi_{\varphi_1}(G'_1)$. The refined local Langlands correspondence allows us to formulate a more precise expectation, namely about the multiplicity

$$m(\pi_1, \pi_2) = \dim \mathrm{Hom}\big(\pi_1, \pi_2 \circ f'\big) = \dim \mathrm{Hom}\big(\pi_2 \circ f', \pi_1\big)$$

for each $\pi_1 \in \Pi_{\varphi_1}(G_1')$. To that end, note that ${}^L f$ induces a map $\widehat{f} : S_{\varphi_2}^+ \to S_{\varphi_1}^+$ via which we can define for any $\rho_1 \in \mathrm{Irr}(\pi_0(S_{\varphi_1}))$ and $\rho_2 \in \mathrm{Irr}(\pi_0(S_{\varphi_2}))$ the number

$$m(\rho_1, \rho_2) = \dim \mathrm{Hom}(\rho_2, \rho_1 \circ \widehat{f}) = \dim \mathrm{Hom}(\rho_1 \circ \widehat{f}, \rho_2).$$

We can realize G_i' as a rigid inner twist $(\xi_i, z_i) : G_i \to G_i'$ of its quasisplit inner form G_i in such a way that $f = \xi_2^{-1} \circ f' \circ \xi_1$ is a homomorphism of F-groups $G_1 \to G_2$ and $z_2 = f(z_1)$.

Conjecture 2.15. *Let* $\rho_i = \iota_{\varphi_i,\mathfrak{w}}(G_i', \xi_i, z_i, \pi_i)$. *Then*

$$m(\pi_1, \pi_2) = m(\rho_2, \rho_1).$$

This conjecture has been stated by Solleveld as [83, CONJECTURE 2], where it has been proven in some cases. A weaker form has been stated by Choiy in [15], where it has been proved under certain working hypotheses numbered 4.1, 4.3, 4.6, 4.11, and 4.15. Among these, 4.1, 4.3, and 4.6 amount to Conjecture 2.5, 4.11 is Corollary 2.14, and 4.15 is a certain dual version to 4.11 that also appears plausible. A direct verification of the basic version of this conjecture stated in [11, §10] in the setting of [47] has been announced by Bourgeois–Mezo in [12], again assuming F has characteristic zero.

ACKNOWLEDGMENTS

We thank Michael Harris, Robert Kottwitz, and Gopal Prasad, for helpful remarks during the preparation of this article.

FUNDING

This work was partially supported by NSF Grant DMS-1801687 and a Simons Fellowship.

REFERENCES

[1] J. Adams, D. Barbasch, and D. A. Jr. Vogan, *The Langlands classification and irreducible characters for real reductive groups*. Progr. Math. 104, Birkhäuser Boston, Inc., Boston, MA, 1992.

[2] J. Adams and D. A. Jr. Vogan, L-groups, projective representations, and the Langlands classification. *Amer. J. Math.* **114** (1992), no. 1, 45–138.

[3] J. Adams and D. A. Jr. Vogan, Contragredient representations and characterizing the local Langlands correspondence. *Amer. J. Math.* **138** (2016), no. 3, 657–682.

[4] J. D. Adler, Refined anisotropic K-types and supercuspidal representations. *Pacific J. Math.* **185** (1998), no. 1, 1–32.

[5] J. D. Adler and L. Spice, Supercuspidal characters of reductive p-adic groups. *Amer. J. Math.* **131** (2009), no. 4, 1137–1210.

[6] J. Arthur, Intertwining operators and residues. I. Weighted characters. *J. Funct. Anal.* **84** (1989), no. 1, 19–84.

[7] J. Arthur, On the transfer of distributions: weighted orbital integrals. *Duke Math. J.* **99** (1999), no. 2, 209–283.

[8] J. Arthur, Unipotent automorphic representations: conjectures. *Astérisque* **171–172** (1989), 13–71.

[9] J. Arthur, *The endoscopic classification of representations*. Amer. Math. Soc. Colloq. Publ. 61, American Mathematical Society, Providence, RI, 2013.

[10] C. Bonnafé, J.-F. Dat, and R. Rouquier, Derived categories and Deligne–Lusztig varieties II. *Ann. of Math. (2)* **185** (2017), no. 2, 609–670.

[11] A. Borel, Automorphic L-functions. In *Automorphic forms, representations and L-functions (Proc. Sympos. Pure Math., Oregon State Univ., Corvallis, Ore., 1977), Part 2*, pp. 27–61, Proc. Sympos. Pure Math. XXXIII, Amer. Math. Soc., Providence, RI, 1979.

[12] A. Bourgeois and P. Mezo, Functoriality for supercuspidal L-packets. 2021, arXiv:2109.09552.

[13] A. Bouthier, B. C. Ngô, and Y. Sakellaridis, On the formal arc space of a reductive monoid. *Amer. J. Math.* **138** (2016), no. 1, 81–108.

[14] C. Chan and M. Oi, Geometric L-packets of Howe-unramified toral supercuspidal representations. 2021, arXiv:2105.06341.

[15] K. Choiy, On multiplicity in restriction of tempered representations of p-adic groups. *Math. Z.* **291** (2019), no. 1–2, 449–471.

[16] R. Cluckers, J. Gordon, and I. Halupczok, Local integrability results in harmonic analysis on reductive groups in large positive characteristic. *Ann. Sci. Éc. Norm. Supér. (4)* **47** (2014), no. 6, 1163–1195.

[17] S. DeBacker and M. Reeder, Depth-zero supercuspidal L-packets and their stability. *Ann. of Math. (2)* **169** (2009), no. 3, 795–901.

[18] S. DeBacker and L. Spice, Stability of character sums for positive-depth, supercuspidal representations. *J. Reine Angew. Math.* **742** (2018), 47–78.

[19] P. Deligne and G. Lusztig, Representations of reductive groups over finite fields. *Ann. of Math. (2)* **103** (1976), no. 1, 103–161.

[20] P. Dillery, Rigid inner forms over local function fields. 2020, arXiv:2008.04472.

[21] P. Dillery, Rigid inner forms over global function fields. 2021, arXiv:2110.10820.

[22] L. Fargues and P. Scholze, Geometrization of the local Langlands correspondence. 2021, arXiv:2102.13459.

[23] Y. Feng, E. Opdam, and M. Solleveld, Supercuspidal unipotent representations: L-packets and formal degrees. *J. Éc. Polytech. Math.* **7** (2020), 1133–1193.

[24] J. Fintzen, Types for tame p-adic groups. *Ann. of Math. (2)* **193** (2021), no. 1, 303–346.

[25] J. Fintzen, On the construction of tame supercuspidal representations. 2019, arXiv:1908.09819.

[26] J. Fintzen, T. Kaletha, and L. Spice, A twisted Yu construction, Harish-Chandra characters, and endoscopy. 2021, arXiv:2106.09120.

[27] W. T. Gan, M. Harris, and W. Sawin, Local parameters of supercuspidal representations. 2021, arXiv:2109.07737.

[28] W. T. Gan and L. Lomelí, Globalization of supercuspidal representations over function fields and applications. *J. Eur. Math. Soc. (JEMS)* **20** (2018), no. 11, 2813–2858.

[29] A. Genestier and V. Lafforgue, Chtoucas restreints pour les groupes réductifs et paramétrisation de Langlands locale. 2017, arXiv:1709.00978.

[30] P. Gérardin, Weil representations associated to finite fields. *J. Algebra* **46** (1977), no. 1, 54–101.

[31] J. Hakim and F. Murnaghan, Distinguished tame supercuspidal representations. *Int. Math. Res. Pap.* **2** (2008), rpn005, 166 pp.

[32] M. Harris, C. B. Khare, and J. A. Thorne, A local langlands parameterization for generic supercuspidal representations of p-adic G_2. 2019, arXiv:1909.05933.

[33] M. Harris and R. Taylor, *The geometry and cohomology of some simple Shimura varieties*. Ann. of Math. Stud. 151, Princeton University Press, Princeton, NJ, 2001.

[34] X. He, On the affineness of Deligne–Lusztig varieties. *J. Algebra* **320** (2008), no. 3, 1207–1219.

[35] G. Henniart, Caractérisation de la correspondance de Langlands locale par les facteurs ϵ de paires. *Invent. Math.* **113** (1993), no. 2, 339–350.

[36] G. Henniart, Une preuve simple des conjectures de Langlands pour $\mathrm{GL}(n)$ sur un corps p-adique. *Invent. Math.* **139** (2000), no. 2, 439–455.

[37] G. Henniart, Une caractérisation de la correspondance de langlands locale pour $\mathrm{GL}(n)$. *Bull. Soc. Math. France* **130** (2002), no. 4, 587–602.

[38] K. Hiraga, A. Ichino, and T. Ikeda, Correction to: "Formal degrees and adjoint γ-factors" [*J. Amer. Math. Soc.* **21** (2008), no. 1, 283–304; mr2350057]. *J. Amer. Math. Soc.* **21** (2008), no. 4, 1211–1213.

[39] T. Kaletha, Genericity and contragredience in the local Langlands correspondence. *Algebra Number Theory* **7** (2013), no. 10, 2447–2474.

[40] T. Kaletha, Epipelagic L-packets and rectifying characters. *Invent. Math.* **202** (2015), no. 1, 1–89.

[41] T. Kaletha, The local Langlands conjectures for non-quasi-split groups. In *Families of automorphic forms and the trace formula*, pp. 217–257, Springer, 2016.

[42] T. Kaletha, Rigid inner forms of real and p-adic groups. *Ann. of Math. (2)* **184** (2016), no. 2, 559–632.

[43] T. Kaletha, Global rigid inner forms and multiplicities of discrete automorphic representations. *Invent. Math.* **213** (2018), no. 1, 271–369.

[44] T. Kaletha, Rigid inner forms vs isocrystals. *J. Eur. Math. Soc. (JEMS)* **20** (2018), no. 1, 61–101.

[45] T. Kaletha, Regular supercuspidal representations. *J. Amer. Math. Soc.* **32** (2019), no. 4, 1071–1170.

[46] T. Kaletha, On L-embeddings and double covers of tori over local fields. 2019, arXiv:1907.05173.

[47] T. Kaletha, Supercuspidal L-packets. 2019, arXiv:1912.03274.

[48] T. Kaletha and O. Taïbi, Global rigid inner forms vs isocrystals. 2018, arXiv:1812.11373.

[49] J.-L. Kim, Supercuspidal representations: an exhaustion theorem. *J. Amer. Math. Soc.* **20** (2007), no. 2, 273–320.

[50] R. E. Kottwitz, $B(G)$ for all local and global fields. 2014, arXiv:1401.5728.

[51] R. E. Kottwitz, Sign changes in harmonic analysis on reductive groups. *Trans. Amer. Math. Soc.* **278** (1983), no. 1, 289–297.

[52] R. E. Kottwitz, Stable trace formula: cuspidal tempered terms. *Duke Math. J.* **51** (1984), no. 3, 611–650.

[53] R. E. Kottwitz, Isocrystals with additional structure. *Compos. Math.* **56** (1985), no. 2, 201–220.

[54] R. E. Kottwitz, Isocrystals with additional structure. II. *Compos. Math.* **109** (1997), no. 3, 255–339.

[55] R. E. Kottwitz and D. Shelstad, Foundations of twisted endoscopy. *Astérisque* **255** (1999), vi+190 pp.

[56] A. Kret, Existence of cuspidal representations of p-adic reductive groups. 2012, arXiv:1205.2771.

[57] R. P. Langlands, On the classification of irreducible representations of real algebraic groups. In *Representation theory and harmonic analysis on semisimple Lie groups*, pp. 101–170, Math. Surveys Monogr. 31, Amer. Math. Soc., Providence, RI, 1989.

[58] R. P. Langlands and M. Rapoport, Shimuravarietäten und Gerben. *J. Reine Angew. Math.* **378** (1987), 113–220.

[59] G. Laumon, M. Rapoport, and U. Stuhler, $\mathscr{D}$-elliptic sheaves and the Langlands correspondence. *Invent. Math.* **113** (1993), no. 2, 217–338.

[60] W.-W. Li, Contragredient representations over local fields of positive characteristic. *Algebra Number Theory* **13** (2019), no. 5, 1197–1242.

[61] G. Lusztig, Classification of unipotent representations of simple p-adic groups. *Int. Math. Res. Not.* **11** (1995), 517–589.

[62] G. Lusztig, Classification of unipotent representations of simple p-adic groups. II. *Represent. Theory* **6** (2002), 243–289.

[63] A. B. Meli and A. Youcis, An approach to the characterization of the local Langlands correspondence. 2020, arXiv:2003.11484.

[64] L. Morris, P-cuspidal representations. *Proc. Lond. Math. Soc. (3)* **57** (1988), no. 2, 329–356.

[65] L. Morris, P-cuspidal representations of level one. *Proc. Lond. Math. Soc. (3)* **58** (1989), no. 3, 550–558.

[66] A. Moy and G. Prasad, Unrefined minimal K-types for p-adic groups. *Invent. Math.* **116** (1994), no. 1–3, 393–408.

[67] A. Moy and G. Prasad, Jacquet functors and unrefined minimal K-types. *Comment. Math. Helv.* **71** (1996), no. 1, 98–121.

[68] K. Ohara, On the formal degree conjecture for non-singular supercuspidal representations. 2021, arXiv:2106.00878.
[69] D. Prasad, Generalizing the MVW involution, and the contragredient. *Trans. Amer. Math. Soc.* **372** (2019), no. 1, 615–633.
[70] M. Reeder, Formal degrees and L-packets of unipotent discrete series representations of exceptional p-adic groups. *J. Reine Angew. Math.* **520** (2000), 37–93.
[71] M. Reeder, Supercuspidal L-packets of positive depth and twisted Coxeter elements. *J. Reine Angew. Math.* **620** (2008), 1–33.
[72] D. Renard, *Représentations des groupes réductifs p-adiques*. Cours Spéc. 17. Société Mathématique de France, Paris, 2010.
[73] P. Scholze, The local Langlands correspondence for GL_n over p-adic fields. *Invent. Math.* **192** (2013), no. 3, 663–715.
[74] P. Scholze and S. W. Shin, On the cohomology of compact unitary group Shimura varieties at ramified split places. *J. Amer. Math. Soc.* **26** (2013), no. 1, 261–294.
[75] D. Schwein, Formal degree of regular supercuspidals. 2021, arXiv:2101.00658.
[76] F. Shahidi, A proof of Langlands' conjecture on Plancherel measures; complementary series for p-adic groups. *Ann. of Math. (2)* **132** (1990), no. 2, 273–330.
[77] D. Shelstad, L-indistinguishability for real groups. *Math. Ann.* **259** (1982), no. 3, 385–430.
[78] D. Shelstad, Tempered endoscopy for real groups. I. Geometric transfer with canonical factors. In *Representation theory of real reductive Lie groups*, pp. 215–246, Contemp. Math. 472, Amer. Math. Soc., Providence, RI, 2008.
[79] D. Shelstad, Tempered endoscopy for real groups. III. Inversion of transfer and L-packet structure. *Represent. Theory* **12** (2008), 369–402.
[80] D. Shelstad, Tempered endoscopy for real groups. II. Spectral transfer factors. In *Automorphic forms and the Langlands program*, pp. 236–276, Adv. Lect. Math. (ALM) 9, Int. Press, Somerville, MA, 2010.
[81] A. Silberger and E.-W. Zink, Langlands classification for L-parameters. *J. Algebra* **511** (2018), 299–357.
[82] M. Solleveld, On unipotent representations of ramified p-adic groups. 2019, arXiv:1912.08451.
[83] M. Solleveld, Langlands parameters, functoriality and Hecke algebras. *Pacific J. Math.* **304** (2020), no. 1, 209–302. DOI 10.2140/pjm.2020.304.209.
[84] L. Spice, Explicit asymptotic expansions for tame supercuspidal characters. *Compos. Math.* **154** (2018), no. 11, 2305–2378.
[85] L. Spice, Explicit asymptotic expansions in p-adic harmonic analysis II. 2021, arXiv:2108.12935.
[86] O. Taïbi, Arthur's multiplicity formula for certain inner forms of special orthogonal and symplectic groups. *J. Eur. Math. Soc. (JEMS)* **21** (2019), no. 3, 839–871.

[87] D. A. Jr. Vogan, The local Langlands conjecture. In *Representation theory of groups and algebras*, pp. 305–379, Contemp. Math. 145, Amer. Math. Soc., Providence, RI, 1993.

[88] J.-K. Yu, Construction of tame supercuspidal representations. *J. Amer. Math. Soc.* **14** (2001), no. 3, 579–622 (electronic).

TASHO KALETHA

University of Michigan, 530 Church Street, Ann Arbor, MI 48109, USA,
kaletha@umich.edu

PERFECT BASES IN REPRESENTATION THEORY: THREE MOUNTAINS AND THEIR SPRINGS

JOEL KAMNITZER

ABSTRACT

In order to give a combinatorial descriptions of tensor product multiplicites for semisimple groups, it is useful to find bases for representations which are compatible with the actions of Chevalley generators of the Lie algebra. There are three known examples of such bases, each of which flows from geometric or algebraic mountain. Remarkably, each mountain gives the same combinatorial shadow: the crystal $B(\infty)$ and the Mirković–Vilonen polytopes. In order to distinguish between the three bases, we introduce measures supported on these polytopes. We also report on the interaction of these bases with the cluster structure on the coordinate ring of the maximal unipotent subgroup.

MATHEMATICS SUBJECT CLASSIFICATION 2020

Primary 22E46; Secondary 14M15, 16G20, 13F60, 05E10

KEYWORDS

Canonical basis, crystals, affine Grassmannian, quiver varieties

INTERNATIONAL CONGRESS OF MATHEMATICIANS 2022 JULY 6–14

Proc. Int. Cong. Math. 2022, Vol. 4, pp. 2976–2996
DOI 10.4171/ICM2022/132

1. REPRESENTATIONS AND THEIR BASES

1.1. Semisimple Lie algebras and their representations

Let G be a complex semisimple group. The representation theory of G is very well understood. The irreducible G-representations are labeled by dominant weights, and every representation is a direct sum of these irreducible representations. For $\lambda \in P_+$, the irreducible representation $V(\lambda)$ admits a decomposition into eigenspaces $V(\lambda)_\mu$ for the action of T. These eigenspaces are called weight spaces and their dimensions are called *weight multiplicities*.

The tensor product of two irreducible representations decomposes into a direct sum of irreducible representations with *tensor product multiplicities* $c^\nu_{\lambda\mu}$,

$$V(\lambda) \otimes V(\mu) \cong \bigoplus_{\nu \in P_+} V(\nu)^{\oplus c^\nu_{\lambda\mu}}.$$

Problem 1.1. Determine combinatorial formulae for weight multiplicities and tensor product multiplicities.

Weight and tensor product multiplicities are closely related by the following construction. Let $C^\nu_{\lambda\mu} = \mathrm{Hom}(V(\nu), V(\lambda) \otimes V(\mu))$, a vector space whose dimension is $c^\nu_{\lambda\mu}$.

Proposition 1.2. *There is an injective map* $C^\nu_{\lambda\mu} \to V(\lambda)_{\nu-\mu}$ *with image* $\bigcap_{i\in I} \ker e_i^{\alpha_i^\vee(\lambda)+1}$.

Here we use the Chevalley presentation of $\mathfrak{g}$, with generators $e_i, f_i, \alpha_i^\vee$, for $i \in I$.

1.2. Good and perfect bases

Problem 1.1 was first solved by Littelmann [34] and Berenstein–Zelevinsky [9], following an approach first proposed by Gel'fand–Zelevinsky [21]. They suggested finding weight bases for each $V(\lambda)$ which restrict to bases of tensor product multiplicity spaces.

Let V be a G-representation. A *weight basis* for V is a basis consisting of weight vectors. A weight basis B for $V(\lambda)$ is called *good*, if for each $i \in I$, it is compatible with the filtration of $V(\lambda)$ given by the kernels of powers of e_i. From Proposition 1.2, it follows that a good basis restricts to a basis of each tensor product multiplicity space.

A slight strengthening of the notion of good basis was proposed by Berenstein–Kazhdan [8]. One might imagine that we could find a basis for a representation such that each e_i takes each basis vector to another basis vector (or 0). However, this is not always possible (see Example 1.4). So instead we will demand that each e_i permutes the basis up to lower order terms.

To formulate this, we define a map $\varepsilon_i : V \to \mathbb{N}$ giving the nilpotence degree of e_i on a vector $v \in V$; more precisely, $\varepsilon_i(v) = \max\{n \in \mathbb{N} : e_i^n b \neq 0\}$.

A good basis B of V is called *perfect*, if for each $i \in I$, and $b \in B$, either $e_i b = 0$ or there exists $\tilde{e}_i(b) \in B$ such that

$$e_i b = \varepsilon_i(b)\, \tilde{e}_i(b) + v \quad \text{for some } v \in \ker e_i^{\varepsilon_i(b)-1}.$$

In other words (up to a predictable scalar) $e_i b$ equals $\tilde{e}_i(b)$ modulo a vector with lower nilpotence degree. Note that this definition only requires V to be a representation of the Borel subalgebra $\mathfrak{b}$.

Example 1.3. To understand these scalars and gain some intuition, it is instructive to consider the case of $\mathfrak{g} = \mathfrak{sl}_2$. In this case, $P_+ = \mathbb{N}$ and $V(n) = \mathbb{C}[x, y]_n$, the space of homogenous polynomials of degree n. The Chevalley generator e acts by $y\partial_x$ (on both the left and right) and the unique perfect basis (up to a scalar) is $\{x^n, x^{n-1}y, \ldots, y^n\}$.

Note that $y\partial_x(x^k y^{n-k}) = kx^{k-1}y^{n-k+1}$ and $\varepsilon(x^k y^{n-k}) = k$. In this case there is no lower order term.

Example 1.4. The simplest irreducible representation where lower order terms occur is the adjoint representation of $\mathfrak{sl}_3$. In this representation, $V = \mathfrak{sl}_3$ with the action given by matrix commutator. If we assume that B contains the highest weight vector

$$\begin{pmatrix} 0 & 0 & 1 \\ 0 & 0 & 0 \\ 0 & 0 & 0 \end{pmatrix},$$

then it is easy to see that the perfect basis condition forces B to contain

$$\begin{pmatrix} 0 & 0 & 1 \\ 0 & 0 & 0 \\ 0 & 0 & 0 \end{pmatrix}, \begin{pmatrix} 0 & -1 & 0 \\ 0 & 0 & 0 \\ 0 & 0 & 0 \end{pmatrix}, \begin{pmatrix} 0 & 0 & 0 \\ 0 & 0 & -1 \\ 0 & 0 & 0 \end{pmatrix}, \begin{pmatrix} 0 & 0 & 0 \\ 1 & 0 & 0 \\ 0 & 0 & 0 \end{pmatrix}, \begin{pmatrix} 0 & 0 & 0 \\ 0 & 0 & 0 \\ 0 & 1 & 0 \end{pmatrix}, \begin{pmatrix} 0 & 0 & 0 \\ 0 & 0 & 0 \\ -1 & 0 & 0 \end{pmatrix}.$$

The choice of basis for the diagonal matrices is more interesting. The requirement that B be compatible with the kernels of e_1, e_2 forces B to contain matrices of the form

$$\begin{pmatrix} -a & 0 & 0 \\ 0 & -a & 0 \\ 0 & 0 & 2a \end{pmatrix}, \begin{pmatrix} 2b & 0 & 0 \\ 0 & -b & 0 \\ 0 & 0 & -b \end{pmatrix}$$

for some nonzero a, b. We are forced to take $b = 1/3$ and similarly $a = 1/3$, since

$$e_1 \begin{pmatrix} 0 & 0 & 0 \\ 1 & 0 & 0 \\ 0 & 0 & 0 \end{pmatrix} = \begin{pmatrix} 1 & 0 & 0 \\ 0 & -1 & 0 \\ 0 & 0 & 0 \end{pmatrix} = 2\begin{pmatrix} 2/3 & 0 & 0 \\ 0 & -1/3 & 0 \\ 0 & 0 & -1/3 \end{pmatrix} + \begin{pmatrix} -1/3 & 0 & 0 \\ 0 & -1/3 & 0 \\ 0 & 0 & 2/3 \end{pmatrix},$$

where the second term is of lower nilpotence degree (since it lies in the kernel of e_1).

1.3. Perfect bases and crystals

Any perfect basis gives rise to a combinatorial structure called a crystal. Crystals were first introduced by Kashiwara [30] as the $q = 0$ limit of a basis for a representation of a quantum group. However, we prefer to view them as recording the leading order behavior of e_i acting on a perfect basis.

A *crystal* is a finite set B, along with a map $\mathrm{wt} : B \to P$, and for each $i \in I$, a partially defined map $\tilde{e}_i : B \to B$. If B is a perfect basis, then it automatically acquires a crystal structure. The following result of Berenstein–Kazhdan [8, THEOREM 5.37] shows that this combinatorial structure depends only on the representation.

Theorem 1.5. *Let V be a representation and let B, B' be two perfect bases. Then there exists a bijection $B \cong B'$ which is an isomorphism of crystals.*

Because of this theorem, we may speak of *the* crystal of a representation. In particular, the crystal of $V(\lambda)$ is denoted $B(\lambda)$. Many different explicit combinatorial realizations of $B(\lambda)$ are possible. In this talk, we will focus on MV polytopes Section 2.3.

1.4. Biperfect bases

Rather than looking at each irreducible representation individually, we can study them all at once, using the following trick. The maximal unipotent subgroup N has left and right actions of N and thus the coordinate ring $\mathbb{C}[N]$ has left and right actions of $\mathfrak{n}$ by differential operators. For each $i \in I$, we write $e_i : \mathbb{C}[N] \to \mathbb{C}[N]$ for the left action and $e_i^* : \mathbb{C}[N] \to \mathbb{C}[N]$ for the right action.

For each $\lambda \in P_+$, choose a highest weight vector $v_\lambda \in V(\lambda)$ and let $v_\lambda^* : V(\lambda) \to \mathbb{C}$ be a dual linear form. We define an N-equivariant map

$$\Psi_\lambda : V(\lambda) \to \mathbb{C}[N], \quad \Psi_\lambda(v)(g) = v_\lambda^*(gv).$$

This linear map is injective and its image is

$$\operatorname{im}\Psi_\lambda = \bigcap_{i\in I} \ker(e_i^*)^{\alpha_i^\vee(\lambda)+1} \subset \mathbb{C}[N]. \tag{1.1}$$

Thus a basis for $\mathbb{C}[N]$ compatible with the kernels of all powers of e_i^* gives a basis for each $V(\lambda)$. Conversely, a collection of bases for each $V(\lambda)$ can sometimes glue together to give a basis for $\mathbb{C}[N]$.

A basis B of $\mathbb{C}[N]$ is called *biperfect* if it contains 1, and it is perfect with respect to both the left and right actions of $\mathfrak{n}$. Thus, B will have two families of crystal operators, written $\tilde{e}_i, \tilde{e}_i^*$ and two families of maps $\varepsilon_i, \varepsilon_i^* : B \to \mathbb{N}$ (but only one weight map).

From Proposition 1.2 and (1.1), we immediately deduce the following corollary, which can be regarded as a generalization (from the canonical basis to arbitrary biperfect bases) of [9, COROLLARY 3.4].

Corollary 1.6. *Let B be a biperfect basis of $\mathbb{C}[N]$.*

(1) *For any $\lambda \in P_+$, the set $\{b \in B : \varepsilon_i^*(b) \le \alpha_i^\vee(\lambda)\}$ restricts via Ψ_λ to a perfect basis for $V(\lambda)$.*

(2) *For any $\lambda, \mu, \nu \in P_+$, the set*

$$\{b \in B : \mathrm{wt}(b) = \nu - \mu - \lambda \text{ and } \forall i \in I,\ \varepsilon_i(b) \le \alpha_i^\vee(\mu),\ \varepsilon_i^*(b) \le \alpha_i^\vee(\lambda)\}$$

restricts to a basis for $C_{\lambda\mu}^\nu$.

Thus we can solve Problem 1.1 by understanding well the bicrystal structure on B.

1.5. The bicrystal $B(\infty)$

The Berenstein–Kazhdan result (Theorem 1.5) generalizes to biperfect bases.

Theorem 1.7 ([5, THEOREM 2.4]). *Let B and B' be two biperfect bases of $\mathbb{C}[N]$. Then there is a unique bijection $B \cong B'$ that respects the bicrystal structure.*

The abstract combinatorial crystal underlying any biperfect basis is denoted $B(\infty)$. On $B(\infty)$ we have the Kashiwara involution $* : B(\infty) \to B(\infty)$ exchanging $\tilde{e}_i$ and $\tilde{e}_i^*$.

Remark 1.8. The algebra $\mathbb{C}[N]$ has an involutive automorphism $*$ (coming from the inverse map on N) which exchanges the left and right actions of $\mathfrak{n}$. A *$*$-invariant perfect basis* is a perfect basis which is invariant under $*$. Every example of a biperfect basis that we know is $*$-invariant. The Kashiwara involution $*$ on $B(\infty)$ is the combinatorial manifestation of the involution $*$ on $\mathbb{C}[N]$.

Example 1.9. When $G = \mathbf{SL}_3$, $B(\infty)$ can be drawn in the following way. The action of e_1 is given by right-pointing diagonal arrows and the action of e_2 is given by the left-pointing ones. Each horizontal group of dots have the same weight and the Kashiwara involution flips each such group. We would like to thank Mark Haiman for showing us this drawing many years ago.

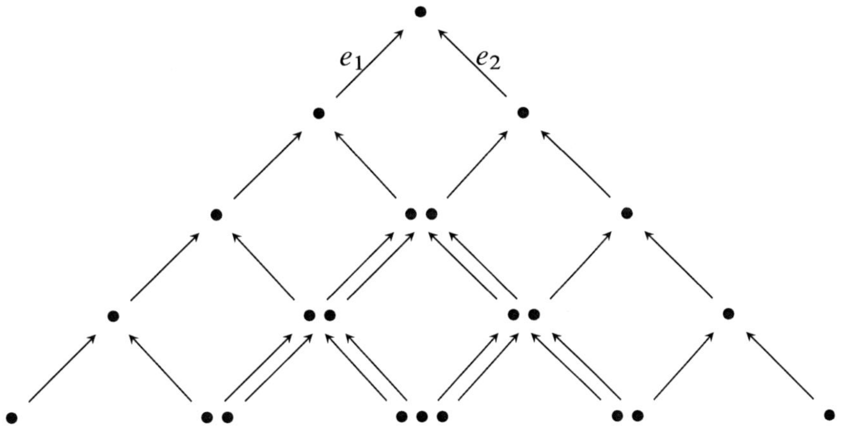

1.6. Biperfect bases in small rank

For small rank groups, it is easy to show the existence and uniqueness of biperfect bases of $\mathbb{C}[N]$ by elementary means.

Theorem 1.10. *For $G = \mathbf{SL}_2, \mathbf{SL}_3, \mathbf{SL}_4$, $\mathbb{C}[N]$ has a unique biperfect basis.*

Example 1.11. Suppose $G = \mathbf{SL}_2$, then $\mathbb{C}[N] = \mathbb{C}[x]$ and $\Psi_n : \mathbb{C}[x,y]_n \to \mathbb{C}[x]$ is the map sending y to 1. The left and right actions of $e \in \mathfrak{n}$ on $\mathbb{C}[x]$ agree and are given by $e = \partial_x$. The unique biperfect basis of $\mathbb{C}[x]$ is $\{1, x, x^2, \dots\}$.

Example 1.12. Suppose $G = \mathbf{SL}_3$, with the standard choice for B, T and N. Then $\mathbb{C}[N] = \mathbb{C}[x, y, z]$ where x, y and z are the three matrix entries of an upper unitriangular matrix

$$\begin{pmatrix} 1 & x & z \\ 0 & 1 & y \\ 0 & 0 & 1 \end{pmatrix} \in N.$$

The unique biperfect basis of $\mathbb{C}[N]$ is

$$B = \{x^a z^b (xy - z)^c : (a, b, c) \in \mathbb{N}^3\} \cup \{y^a z^b (xy - z)^c : (a, b, c) \in \mathbb{N}^3\}.$$

1.7. Three different biperfect bases

For general G, biperfect bases are not unique, nor is it very easy to show their existence.

The first example of a biperfect basis was Lusztig's *dual canonical basis* which is also known as Kashiwara's upper global basis [31, 37]. This is actually a basis for the corresponding quantum deformation of $\mathbb{C}[N]$, but it can be specialized at $q = 1$ to give a biperfect basis.

Another example, when $\mathfrak{g}$ is simply-laced, is Lusztig's *dual semicanonical basis* [38], which is constructed by means of the representation theory of the preprojective algebra.

A third example is the *Mirković–Vilonen basis* [39] coming from the geometry of the affine Grassmannian.

This trichotomy of bases will be the focus of this paper. Each of these bases comes from a complicated algebraic or geometric source. Following Arun Ram, we can imagine three high mountains whose springs give these three bases.

These bases are all different, combining [5, THM. 1.7], [4, PROP. 2.7], and [18, (3)].

Theorem 1.13. *For $G = \mathbf{SL}_6$ in the weight space $2\alpha_1 + 4\alpha_2 + 4\alpha_3 + 4\alpha_4 + 2\alpha_5$, and for $G = \mathbf{SO}_8$ in the weight space $2\alpha_1 + 4\alpha_2 + 2\alpha_3 + 2\alpha_4$, there is a point of $B(\infty)$ whose corresponding dual canonical, dual semicanonical, and MV bases are all different.*

Moreover, in both these examples, we have the following specific situation:

$$d = b + v, \quad c = b + 2v, \tag{1.2}$$

where b, c, d denote the MV, dual semicanonical, and dual canonical basis vectors, all of which define the same point in $B(\infty)$, and v denotes a vector common to all three bases.

Question 1.14. What can we say about the set of all biperfect bases of $\mathbb{C}[N]$ for a fixed G?

2. MIRKOVIĆ–VILONEN BASIS

2.1. MV cycles

Mirković–Vilonen [39] used the geometric Satake correspondence to define the MV basis for irreducible representations of G. This basis is indexed by certain subvarieties in the *affine Grassmannian*, known as MV cycles.

Let $G^\vee$ be the Langlands dual group and let $\mathrm{Gr} = G^\vee((t))/G^\vee[\![t]\!]$ denote the affine Grassmannian of this group. By definition, the coweight lattice of $G^\vee$ coincides with the weight lattice P of G. For each coweight $\mu \in P$, we get a point of $G^\vee((t))$ and hence a point L_μ in Gr. Let $S^\mu_\pm := N^\vee_\pm((t))L_\mu$ denote semiinfinite orbits in Gr, where $N^\vee_\pm$ denote opposite unipotent subgroups in $G^\vee$.

For $\lambda \in P_+$, let $\mathrm{Gr}^\lambda := \overline{G^\vee[\![t]\!]L_\lambda}$ be a *spherical Schubert variety*. This is a finite-dimensional singular projective variety whose geometry is closely related to the irreducible representation $V(\lambda)$. Let $P(\mathrm{Gr})$ denote the category of perverse sheaves on Gr which are constructible with respect to the stratification by $G^\vee[\![t]\!]$ orbits. This is a semisimple category whose simple objects are the intersection cohomology sheaves IC_λ of the spherical Schubert varieties. There is a monoidal structure on $P(\mathrm{Gr})$ by convolution.

The following geometric Satake correspondence was established by Mirković–Vilonen [39], following earlier work by Lusztig [35] and Ginzburg [23].

Theorem 2.1. (1) *There is an equivalence of monoidal categories,* $P(\mathrm{Gr}) \cong \mathrm{Rep}\, G$.

(2) *Under this equivalence, for each* $\lambda \in P_+$, IC_λ *is sent to* $V(\lambda)$.

(3) *Under this equivalence, for each* $\mu \in P$, *the hyperbolic stalk functor* $H^\bullet_{S^\mu_-}(-)$ *matches the functor of taking the* μ*-weight space.*

Combining these statements, we conclude $H_{\mathrm{top}}(\mathrm{Gr}^\lambda \cap S^\mu_-) \cong V(\lambda)_\mu$. The irreducible components of $\mathrm{Gr}^\lambda \cap S^\mu_-$ are called *MV cycles*. Via this theorem, they provide a basis for each $V(\lambda)_\mu$.

2.2. Stable MV cycles

For bases of $\mathbb{C}[N]$, we will be concerned with the intersection of opposite semi-infinite orbits. For any $\nu \in Q_+$, the positive root cone, the irreducible components of $\overline{S^\nu_+ \cap S^0_-}$ are called *stable MV cycles*.

Given an MV cycle $Z \subset \mathrm{Gr}^\lambda \cap S^\mu_-$, we can translate by $t^{-\mu}$ to produce a stable MV cycle $\overline{t^{-\mu}Z}$. This process is the geometric analog of the map $\Psi_\lambda : V(\lambda) \to \mathbb{C}[N]$.

In [5], we combined work of Ginzburg [23] and Mirković–Vilonen [39] to prove the following result, which had been conjectured by Anderson [1].

Theorem 2.2. (1) *The MV bases for each* $V(\lambda)$ *can be collected together to form a biperfect basis for* $\mathbb{C}[N]$, *which is indexed by stable MV cycles.*

(2) *For each* i, *the action of* e_i *on an MV basis vector* b_Z *is given by the intersection of the stable MV cycle* Z *with a hyperplane section.*

(3) *Given two MV cycles* Z_1, Z_2, *the product* $b_{Z_1}b_{Z_2}$ *in* $\mathbb{C}[N]$ *is given by the Beilinson–Drinfeld degeneration of* $Z_1 \times Z_2$.

In particular, the structure constants for the action of e_i *and for the multiplication are nonnegative integers.*

2.3. MV polytopes

For each stable MV cycle Z, we define $\mathrm{Pol}(Z)$ to be its moment map image (for a real Hamiltonian torus action). Equivalently, we have

$$\mathrm{Pol}(Z) = \mathrm{Conv}(\mu : L_\mu \in Z).$$

The polytopes produced this way are called MV polytopes. In [27], we proved the following result.

Theorem 2.3. *The map $Z \mapsto \mathrm{Pol}(Z)$ gives a bijection between the stable MV cycles and the MV polytopes. The MV polytopes are precisely those lattice polytopes whose dual fan is a coarsening of the Weyl fan and whose 2-faces are MV polygons for the appropriate rank 2 groups (which can be described explicitly).*

This theorem was reinterpreted by Goncharov–Shen [24] as the following statement.

Corollary 2.4. *The MV polytopes are in natural bijection with $(G^\vee/B^\vee)(\mathbb{Z}_{\mathrm{trop}})_{\geq}$, the non-negative tropical points of the flag variety.*

Following the historical order, we have described MV polytopes as the moment map images of MV cycles. However, we emphasize that Theorem 2.3 shows that they are purely combinatorial objects. We will see in the next two sections that these same polytopes are naturally obtained from general preprojective algebras modules and simple KLR modules. They are the common shadows from all three mountains.

In [26], we gave an explicit description of the crystal structure on the set of MV polytopes. This provides a convenient combinatorial framework for describing the crystal $B(\infty)$ and is easily connected to many other combinatorial models. In particular, for each reduced word $s_{i_1} \cdots s_{i_m} = w_0$ for the longest element of the Weyl group, Lusztig [36] constructed a bijection $B(\infty) \to \mathbb{N}^m$ using the relation between PBW monomials and the canonical basis. In [26], we showed that the Lusztig datum of $b \in B(\infty)$ is the list of lengths along a path following the edges of the MV polytope $\mathrm{Pol}(b)$, in root directions determined by the reduced word.

Example 2.5. Take $G = \mathbf{SL}_3$. In this case, an MV polytope is a hexagon with all 120° angles, whose "width" A is equal to the maximum of its two "heights" B, C.

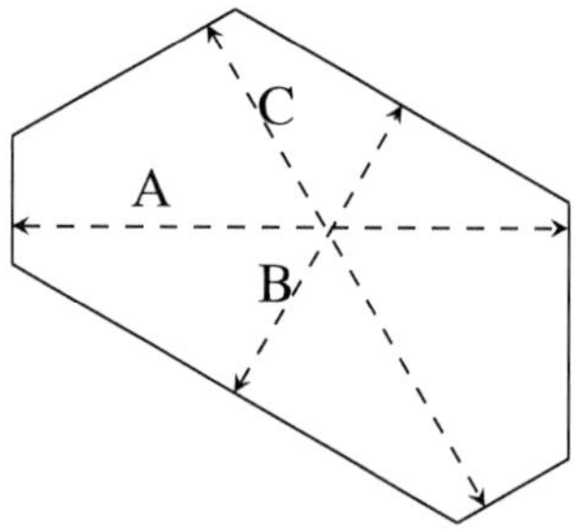

For this polytope, the two Lusztig data are $(3, 2, 1)$ and $(2, 1, 4)$.

3. DUAL SEMICANONICAL BASIS

3.1. Preprojective algebra

Assume for this section that $\mathfrak{g}$ is simply-laced. Let H denote the set of oriented edges of the Dynkin diagram of $\mathfrak{g}$. If $h = (i, j)$, write $\bar{h} = (j, i)$. Fix a map $\tau : H \to \{1, -1\}$ such that for each h, $\tau(h) + \tau(\bar{h}) = 0$ (such a τ corresponds to an orientation of each edge of the Dynkin diagram).

The *preprojective algebra* Λ is the quotient of the path algebra of (I, H) by the relation $\sum_{h\in H} \tau(h)h\bar{h} = 0$. So a Λ-module M consists of vector spaces M_i, for $i \in I$, and linear maps $M_h : M_i \to M_j$ for each $h = (i, j) \in H$, such that

$$\sum_{h\in H} \tau(h) M_h M_{\bar{h}} = 0. \tag{3.1}$$

Given a Λ-module M, we define its dimension vector by

$$\underrightarrow{\dim}\, M = \sum_{i\in I} (\dim M_i)\, \alpha_i.$$

We write S_i for the simple module at vertex i, the unique module with $\underrightarrow{\dim}\, S_i = \alpha_i$.

For each $\nu = \sum_{i\in I} \nu_i \alpha_i \in Q_+$, we consider the affine variety of Λ-module structures on $\bigoplus_{i\in I} \mathbb{C}^{\nu_i}$. More precisely, we define

$$\Lambda(\nu) \subset \bigoplus_{(i,j)\in H} \operatorname{Hom}(\mathbb{C}^{\nu_i}, \mathbb{C}^{\nu_j})$$

to be the subvariety defined by equation (3.1).

3.2. The dual semicanonical basis

Let M be a Λ-module. Following Lusztig [38] and Geiss–Leclerc–Schröer [18, §5], we define an element $\xi_M \in \mathbb{C}[N]$ as follows. First, for each $\underline{i} \in I^p$, we define the projective variety of composition series of type $\underline{i}$,

$$F_{\underline{i}}(M) = \{0 = M^0 \subset M^1 \subset \cdots \subset M^p = M : M^k/M^{k-1} \cong S_{i_k} \text{ for all } k\}$$

and then we define $\xi_M \in \mathbb{C}[N]$ by requiring that

$$\langle e_{i_1} \cdots e_{i_p}, \xi_M \rangle = \chi(F_{\underline{i}}(M))$$

for any $\underline{i} \in I^p$, where χ denotes topological Euler characteristic, and where $\langle \cdot, \cdot \rangle$ denotes the pairing between $U\mathfrak{n}$ and $\mathbb{C}[N]$.

This map $M \mapsto \xi_M$ is constructible and so for any irreducible component $Y \subset \Lambda(\nu)$, we can define $c_Y \in \mathbb{C}[N]_\nu$ by setting $c_Y = \xi_M$, for M a general point in Y.

The following result is due to Lusztig [38].

Theorem 3.1. (1) *For each $\nu \in Q_+$, $\{c_Y \mid Y \in \operatorname{Irr} \Lambda(\nu)\}$ is a basis for $\mathbb{C}[N]_\nu$.*

(2) *Together they form a biperfect basis of $\mathbb{C}[N]$, called the dual semicanonical basis.*

3.3. Polytopes from preprojective algebra modules

In the resulting bicrystal structure on the set $\sqcup_\nu \operatorname{Irr} \Lambda(\nu)$, we have

$$\varepsilon_i(Y) = \dim \operatorname{Hom}_\Lambda(M, S_i), \quad \varepsilon_i^*(Y) = \dim \operatorname{Hom}_\Lambda(S_i, M),$$

where M is a general point of Y.

Remark 3.2. Fix $\lambda \in P_+$. From Corollary 1.6(1), those components Y which satisfy $\varepsilon_i^*(Y) \leq \alpha_i^*(\lambda)$ index a basis for $V(\lambda)$. These same irreducible components form the core of the corresponding Nakajima quiver variety via the correspondence explained in [43, SECTION 4.6].

The bicrystal $\sqcup \operatorname{Irr} \Lambda(\nu)$ is isomorphic to $B(\infty)$ by Theorem 1.7. Thus, an MV polytope is canonically associated to component Y. We can describe this polytope using the module structure on a general point of Y.

Theorem 3.3 ([6, §1.3]). *Let Y be a component of $\Lambda(\nu)$ and M be a general point of Y. The MV polytope of the basis vector c_Y is given by the Harder–Narasimhan polytope of M,*

$$\operatorname{Pol}(M) := \operatorname{Conv}(\underrightarrow{\dim} N : N \subseteq M \textit{ is a submodule}).$$

Example 3.4. Take $\mathfrak{g} = \mathfrak{sl}_3$, and consider $\nu = \alpha_1 + \alpha_2$.

Then $\Lambda(\nu) = \{(a, b) \in \mathbb{C}^2 : ab = 0\}$ where (a, b) corresponds to the Λ-module

$$\mathbb{C} \underset{a}{\overset{b}{\leftrightarrows}} \mathbb{C}.$$

Further, $\Lambda(\nu)$ has two components in this case. If Y is the component given by $b = 0$, then for general $M \in Y$, we have submodules in M of dimension $0, \alpha_1, \alpha_1 + \alpha_2$ and so $\operatorname{Pol}(M)$ is the triangle with these vertices.

4. DUAL CANONICAL BASES

4.1. KLR algebras

Let $\mathfrak{g}$ be an arbitrary semisimple Lie algebra. For each $\nu \in Q_+$, Khovanov–Lauda–Rouquier defined an algebra R_ν. It can either be defined by generators and relations using a modification of the presentation of a Hecke algebra [42], or as an algebra of decorated string diagrams as in [32]. This algebra can also be realized as an Ext algebra of certain perverse sheaves constructed by Lusztig (see [45]).

In this section, we work over $\mathbb{C}$, though it is possible to work over fields of positive characteristic; this produces different bases, called the dual p-canonical bases.

The algebra R_ν contains indempotents $e_{\underline{i}}$ indexed by sequences $(i_1, \ldots, i_p) \in I^p$ such that $\alpha_{i_1} + \cdots + \alpha_{i_p} = \nu$. We write $K(R_\nu)_\mathbb{C}$ for the complexified Grothendieck group of finite-dimensional R_ν modules. The following result is due independently to Rouquier and Khovanov–Lauda.

Theorem 4.1. *For each $\nu \in Q_+$, there is an isomorphism $K(R_\nu)_\mathbb{C} \cong \mathbb{C}[N]_\nu$, written $[L] \mapsto d_L$ such that $\langle e_{\underline{i}}, d_L \rangle = \dim e_{\underline{i}} L$, for any module L and $\underline{i}$ as above.*

The vector space $K(R_\nu)_{\mathbb{C}}$ has a basis given by the simple finite-dimensional R_ν-modules. By [45], the resulting basis for $\mathbb{C}[N]$ coincides with Lusztig's dual canonical basis.

4.2. Polytopes from KLR modules

The bicrystal structure on the set of simple KLR modules was carefully studied by Lauda–Vazirani [33]. By Theorem 1.7, this bicrystal is isomorphic to $B(\infty)$, and thus an MV polytope is canonically associated to each simple KLR module.

On the other hand, these algebras come with nonunital morphisms

$$R_\mu \otimes R_{\nu-\mu} \to R_\nu.$$

We write $e_{\mu,\nu-\mu}$ for the image of the identity under this map. Tingley–Webster [44] used these morphisms to define a polytope associated to each KLR module.

Theorem 4.2. *The MV polytope of a simple R_ν-module L is the character polytope*

$$\operatorname{Pol}(L) := \operatorname{Conv}(\mu : e_{\mu,\nu-\mu} L \neq 0).$$

4.3. Generalizations to affine and Kac–Moody cases

Unlike the MV basis, the dual semicanonical basis and dual canonical basis admit straightforward generalizations to the setting where $\mathfrak{g}$ is a symmetric (resp. symmetrizable) Kac–Moody Lie algebra.

The polytopes $\operatorname{Pol}(M)$ and $\operatorname{Pol}(L)$ associated to a general Λ-module or a simple KLR module admit obvious generalizations in this setting. However, due to higher root multiplicities, the polytope is not enough to characterize the point in $B(\infty)$. Thus, we must enhance the polytope with some extra information. This was carried out in [6] (using Λ-modules) and in [44] (using KLR modules).

In the affine case, this extra information consists of partitions associated to vertical edges of the polytope (vertical edges are those pointing in the imaginary root direction). Moreover, these decorated polytopes are characterized by their 2-faces (as in the finite case, Theorem 2.3) and the new relevant polygons were described combinatorially in [3].

Question 4.3.

(1) Is it possible to give a "tropical" description of affine MV polytopes, similar to Corollary 2.4?

(2) Outside of the affine type, it is possible to give a combinatorial description to the extra information carried on the MV polytope?

(3) Though more complicated, the theory of MV cycles exists for affine Kac–Moody Lie algebras. How can we relate these MV cycles to the affine MV polytopes? In particular, what information about the cycles is encoded in the partitions along vertical edges?

5. COMPARING BIPERFECT BASES

5.1. Change of basis matrix

Let B, B' be two biperfect bases for $\mathbb{C}[N]$. By Theorem 1.7, we obtain bijections $B \to B(\infty) \leftarrow B'$ and thus a bijection between B and B'. Thus, it makes sense to speak of the change of basis matrix between B and B'. In [2], Baumann proved that this matrix is upper unitriangular with respect to a partial order on $B(\infty)$, defined combinatorially using the crystal structure. Many elements of $B(\infty)$ are incomparable using this order. Thus, many off-diagonal elements of the change of basis matrix must vanish. In low rank, the order becomes trivial and gives the proof of Theorem 1.10.

5.2. Measures

We think of an MV polytope as the shadow of a biperfect basis vector. Unfortunately, this shadow is not precise enough to distinguish between different biperfect basis vectors which represent the same element of $B(\infty)$. For this reason, we now introduce a measure supported on the MV polytope.

Consider the vector space $\mathrm{Dist}(\mathfrak{t}^*_{\mathbb{R}})$ of $\mathbb{C}$-valued compactly supported distributions on $\mathfrak{t}^*_{\mathbb{R}}$. It forms an algebra under convolution, using the addition map $\mathfrak{t}^*_{\mathbb{R}} \times \mathfrak{t}^*_{\mathbb{R}} \xrightarrow{+} \mathfrak{t}^*_{\mathbb{R}}$.

Let $\Delta^p := \{(c_0, \dots, c_p) \in \mathbb{R}^{p+1} : \text{ each } c_i \geq 0,\ c_0 + \cdots + c_p = 1\}$ be the standard p-simplex. For $\underline{i} \in I^p$, we define the linear map $\pi_{\underline{i}} : \mathbb{R}^{p+1} \to \mathfrak{t}^*_{\mathbb{R}}$ by

$$\pi_{\underline{i}}(c_0, \dots, c_p) = \sum_{k=0}^{p} c_k(\alpha_{i_1} + \cdots + \alpha_{i_k}).$$

We define the measure $D_{\underline{i}}$ on $\mathfrak{t}^*_{\mathbb{R}}$ by $D_{\underline{i}} := (\pi_{\underline{i}})_*(\delta_{\Delta^p})$, the push-forward of Lebesgue measure on the p-simplex. The measures $D_{\underline{i}}$ satisfy the shuffle identity.

Lemma 5.1 ([5, LEMMA 8.5]). *For $\underline{j} \in I^p, \underline{k} \in I^q$,*

$$D_{\underline{j}} * D_{\underline{k}} = \sum_{\underline{i} \in \underline{j} \sqcup\!\sqcup \underline{k}} D_{\underline{i}},$$

where $\underline{j} \sqcup\!\sqcup \underline{k}$ is the set of all sequences obtained by shuffling $\underline{j}$ and $\underline{k}$.

Elementary considerations involving the coproduct structure on $U\mathfrak{n}$ show that the shuffle identity implies that there is an algebra morphism $D : \mathbb{C}[N] \to \mathrm{Dist}(\mathfrak{t}^*_{\mathbb{R}})$ defined by

$$D(f) = \sum_{\underline{i}} \langle e_{\underline{i}}, f \rangle D_{\underline{i}}.$$

5.3. Fourier transform

For each weight $\beta \in P$, we define e^β to be the function $x \mapsto e^{\langle \beta, x \rangle}$ on $\mathfrak{t}_{\mathbb{C}}$. Given a measure $D(f)$ as above, we can consider its Fourier transform $\mathrm{FT}(D(f))$ which lies in the space of meromorphic functions on $\mathfrak{t}_{\mathbb{C}}$, spanned by these exponentials over the field $\mathbb{C}(\mathfrak{t})$ of rational functions. In this way, we obtain the following result.

Proposition 5.2. *The composition* $\mathrm{FT} \circ D$ *defines an algebra morphism*

$$\mathbb{C}[N] \to \mathbb{C}(\mathfrak{t}) \otimes \mathbb{C}[T].$$

This algebra morphism defines a rational map $\mathfrak{t} \times T \to N$. This map is actually regular on $\mathfrak{t}^{\text{reg}} \times T$, where $\mathfrak{t}^{\text{reg}}$ is the complement of the root hyperplanes in $\mathfrak{t}$.

Theorem 5.3 ([5, THEOREM 8.11]). (1) *For all $x \in \mathfrak{t}^{\text{reg}}$, there exists a unique $n_x \in N$ such that $\operatorname{Ad}_{n_x}(x) = x + e$.*

(2) *The rational map from Proposition 5.2 is given by $(x, t) \mapsto t^{-1} n_x t n_x^{-1}$.*

We now study a simpler invariant $\overline{D}$. For a sequence $\underline{i} = (i_1, \dots, i_p)$, we define

$$\overline{D}_{\underline{i}} = \prod_{k=1}^{p} \frac{1}{\alpha_{i_k} + \cdots + \alpha_{i_p}} \in \mathbb{C}(\mathfrak{t}).$$

Proposition 5.4 ([5, PROPOSITION 8.4 AND LEMMA 8.7]). (1) *These rational functions satisfy the shuffle identity from Lemma 5.1 and thus define an algebra morphism $\overline{D} : \mathbb{C}[N] \to \mathbb{C}(\mathfrak{t})$ by $\overline{D}(f) = \sum_{\underline{i}} \langle e_{\underline{i}}, f \rangle \overline{D}_{\underline{i}}$.*

(2) *For any $f \in \mathbb{C}[N]$, $\overline{D}(f)$ is the coefficient of e^0 in $\operatorname{FT}(D(f))$.*

(3) *The algebra morphism from (1) comes from the morphism of varieties $\mathfrak{t}^{\text{reg}} \to N$ given by $x \mapsto n_x$.*

5.4. Duistermaat–Heckman measure

In symplectic geometry, the Duistermaat–Heckman (DH) measure of a symplectic manifold with a Hamiltonian torus action is defined to be the push-forward of the Liouville measure under the moment map. Brion–Procesi [14] reformulated this notion in algebraic geometry by considering the asymptotics of sections of equivariant line bundles.

Fix a W-invariant bilinear form on $\mathfrak{t}$ (normalized so that short roots have length 1). This leads to a central extension of $G^\vee((t))$ and thus an equivariant line bundle $\mathcal{O}(1)$ on Gr.

Let $Z \subset \mathrm{Gr}$ be an MV cycle. The torus $T^\vee$ acts on the space of sections $\Gamma(Z, \mathcal{O}(n))$. We consider $[\Gamma(Z, \mathcal{O}(n))]$ as a distribution on $\mathfrak{t}^*_{\mathbb{R}}$ by

$$\left[\Gamma\left(Z, \mathcal{O}(n)\right)\right] = \sum_{\mu \in P} \dim \Gamma\left(Z, \mathcal{O}(n)\right)_\mu \delta_\mu.$$

(Implicitly, we use the bilinear form to identify $\mathfrak{t}$ and $\mathfrak{t}^*$.)

Let $\tau_n : \mathfrak{t}^*_{\mathbb{R}} \to \mathfrak{t}^*_{\mathbb{R}}$ be the automorphism given by scaling by $\frac{1}{n}$. The *Duistermaat–Heckman measure* of Z is defined to be the limit

$$\mathrm{DH}(Z) := \lim_{n \to \infty} \frac{1}{n^{\dim Z}} (\tau_n)_* \left[\Gamma\left(Z, \mathcal{O}(n)\right)\right]$$

within the space of distributions on $\mathfrak{t}^*_{\mathbb{R}}$. Note that each $(\tau_n)_*[\Gamma(Z, \mathcal{O}(n))]$ is supported on $\operatorname{Pol}(Z)$, and hence so is $\mathrm{DH}(Z)$.

Via the Fourier transform, $\mathrm{DH}(Z)$ is closely related to the class of Z in the equivariant homology of the affine Grassmannian. The following ideas are not specific to the affine Grassmannian: they apply to any (ind-)projective variety equipped with a torus action having isolated fixed points. First, we have the localization theorem in equivariant homology (see, for example, [13]).

Theorem 5.5. *The inclusion* $\mathrm{Gr}^{T^\vee} \to \mathrm{Gr}$ *induces an isomorphism*

$$H_\bullet^{T^\vee}(\mathrm{Gr}^{T^\vee}) \otimes_{\mathbb{C}[\mathfrak{t}]} \mathbb{C}(\mathfrak{t}) \xrightarrow{\sim} H_\bullet^{T^\vee}(\mathrm{Gr}) \otimes_{\mathbb{C}[\mathfrak{t}]} \mathbb{C}(\mathfrak{t}).$$

Because of this theorem and using $\mathrm{Gr}^{T^\vee} = \{L_\mu : \mu \in P\}$, we can write

$$[Z] = \sum_{\mu \in P} m_\mu(Z)\big[\{L_\mu\}\big]$$

for unique $m_\mu(Z) \in \mathbb{C}(\mathfrak{t})$. Following Brion [13], we call $m_\mu(Z)$ the *equivariant multiplicity* of Z at L_μ. In [5], we proved the following result (again in the general context of projective varieties with torus actions), following ideas of Brion.

Theorem 5.6. *For any MV cycle* Z,

$$\mathrm{FT}\big(\mathrm{DH}(Z)\big) = \sum_{\mu \in P} m_\mu(Z) e^\mu.$$

5.5. DH measures and measures from $\mathbb{C}[N]$

The $T^\vee$-equivariant homology of Gr was computed by Yun–Zhu [46]. They defined a commutative convolution algebra structure on $H_\bullet^{T^\vee}(\mathrm{Gr})$ and described this algebra using the geometric Satake correspondence.

Let $e = \sum e_i$ be a regular nilpotent element. We define the *universal centralizer space* to be

$$C := \big\{(x, b) \in \mathfrak{t} \times B : \mathrm{Ad}_b(x + e) = x + e\big\}.$$

Remark 5.7. For any $x \in \mathfrak{t}$, $x + e$ is regular and has centralizer contained in B. Thus our space C is the base change over $\mathfrak{t} \to \mathfrak{t}/W$ of the usual universal centralizer (often denoted J), as defined in, for example [10, §2.2].

From the definition, we have a map $C \to \mathfrak{t} \times T$ given by $(x, tn) \mapsto (x, t)$. The dual algebra map fits into the following diagram.

Theorem 5.8 ([46, PROP 3.3 AND 5.7]). *There is an isomorphism of algebras* $\theta : \mathbb{C}[C] \to H_\bullet^{T^\vee}(\mathrm{Gr})$ *making the following diagram commute:*

$$\begin{array}{ccc}
\mathbb{C}[\mathfrak{t}] \otimes \mathbb{C}[T] & \xrightarrow{\sim} & H_\bullet^{T^\vee}(\mathrm{Gr}^{T^\vee}) \\
\downarrow & & \downarrow{\scriptstyle \text{Theorem 5.5}} \\
\mathbb{C}[C] & \xrightarrow{\theta} & H_\bullet^{T^\vee}(\mathrm{Gr})
\end{array}$$

Recall the map $D : \mathbb{C}[N] \to \mathrm{Dist}(\mathfrak{t}_{\mathbb{R}}^*)$ defined in §5.2 and the map $\overline{D} : \mathbb{C}[N] \to \mathbb{C}(\mathfrak{t})$ defined in Section 5.3. Combining Theorems 5.3, 5.6, and 5.8, we proved the following in [5].

Corollary 5.9. *For any stable MV cycle* Z,

$$D(b_Z) = \mathrm{DH}(Z), \quad \overline{D}(b_Z) = m_0(Z).$$

This corollary is very useful since the equivariant multiplicity $m_0(Z)$ is easily computed using computer algebra programs. In the appendix of [5] (written with C. Morton-Ferguson and A. Dranowski), we used this approach to establish part of Theorem 1.13.

5.6. Measures from preprojective algebra modules

Let M be a Λ-module of dimension vector ν. By the definition of ξ_M and the map D, we have that

$$D(\xi_M) = \sum_{\underline{i}} \chi(F_{\underline{i}}(M))\, D_{\underline{i}}.$$

In the previous section (Corollary 5.9), we saw that the measure $D(b_Z)$ of an MV basis vector equals the asymptotics of sections of line bundles on Z. In a similar fashion, we will now explain that $D(\xi_M)$ can also be regarded as an asymptotic.

Consider the algebra $\Lambda[t] := \Lambda \otimes_{\mathbb{C}} \mathbb{C}[t]$. We define

$$\mathbb{G}_\mu(M[t]/t^n) = \{N \subset M \otimes \mathbb{C}[t]/t^n : N \text{ is a } \Lambda[t]\text{-submodule, } \underrightarrow{\dim} N = \mu\}.$$

We will record the information of the Euler characteristics of these varieties as an element of $\mathrm{Dist}(\mathfrak{t}^*_{\mathbb{R}})$ by

$$[H^\bullet(\mathbb{G}(M[t]/t^n))] = \sum_{\mu \in Q_+} \chi(\mathbb{G}_\mu(M[t]/t^n))\, \delta_\mu.$$

Theorem 5.10 ([5, THEOREM 11.4 AND LEMMA 12.3]). *For any Λ-module M, we have*

$$D(\xi_M) = \lim_{n\to\infty} \frac{1}{n^{\dim M}} (\tau_n)_* [H^\bullet(\mathbb{G}(M[t]/t^n))].$$

5.7. A conjecture and symplectic duality

Suppose that $Y \in \mathrm{Irr}\, \Lambda(\nu)$, with general point M, and Z is a stable MV cycle, such that $c_Y = b_Z$ (there are many such pairs, conjecturally). Then $D(c_Y) = D(b_Z)$. Via Theorem 5.10 and Corollary 5.9, both sides are the asymptotics of $T^\vee$-representations. So it is natural to expect equality before taking the limit. If we further assume that the odd cohomology of $\mathbb{G}_\mu(M[t]/t^n)$ vanishes, this implies that there is an isomorphism of $T^\vee$-representations,

$$\Gamma(Z, \mathscr{L}^{\otimes n}) \cong H^\bullet(\mathbb{G}(M[t]/t^n)), \quad \text{for all } n \in \mathbb{N}, \tag{5.1}$$

where $T^\vee$ acts on the right-hand side through the decomposition $\mathbb{G}(M[t]/t^n) = \sqcup \mathbb{G}_\mu(M[t]/t^n)$.

The left-hand sides of (5.1) form the components of a graded algebra, so it is natural to search for a similar structure on the right-hand sides. After studying this question for some time, we are pessimistic about finding this algebra structure. On the other hand, $\mathbb{C}[Z] := \bigoplus_n \Gamma(Z, \mathscr{L}^{\otimes n})$ is also a module over $\mathbb{C}[\overline{S^\nu_+ \cap S^0_-}]$. We believe that such a module structure naturally exists for the direct sums of the right-hand side of (5.1).

Conjecture 5.11.
(1) *For any preprojective algebra module M of dimension vector ν, $\bigoplus_{n\in\mathbb{N}} H^\bullet(\mathbb{G}(M[t]/t^n))$ carries the structure of a $C[\overline{S^\nu_+ \cap S^0_-}]$-module.*

(2) *When $b_Z = c_Y$ and M is a general point of Y, then there is an isomorphism (5.1) of $C[\overline{S^\nu_+ \cap S^0_-}]$-modules.*

This conjecture should be a manifestation of the symplectic duality between generalized affine Grassmannian slices and Nakajima quiver varieties. In particular, Braverman–Finkelberg–Nakajima [12] proved that a generalized affine Grassmannian slice is the Coulomb branch associated to the quiver gauge theory defining the corresponding Nakajima quiver variety. In [25], with Hilburn and Weekes, we developed a Springer theory for Coulomb branch algebras and proved a weak form of Conjecture 5.11 for those modules M which come from a representation of the undoubled quiver.

The symplectic singularity viewpoint is also a useful framework for thinking about our three bases. In particular, following the philosophy of Braden–Licata–Proudfoot–Webster [11], the MV cycles and quiver variety components can be categorified using category $\mathcal{O}$ for quantizations of affine Grassmannian slices and quiver varieties, respectively. Moreover, these categories are closely related to categories of modules for KLR algebras [28]. From this perspective, the failures of these bases to agree with the dual canonical basis in Theorem 1.13 can be attributed to the nonirreducibility of the character varieties of simple objects in these categories.

6. CLUSTER STRUCTURES

6.1. Cluster structures on $\mathbb{C}[N]$

Cluster algebras were defined by Fomin–Zelevinsky in order to understand the dual canonical basis of $\mathbb{C}[N]$. A cluster algebra is a commutative algebra A with a distinguished collection of "clusters." Each *cluster* consists of an algebraically independent subset $T = \{x_1, \ldots, x_n\} \subset A$, such that $A \subset \mathbb{C}(x_1, \ldots, x_n)$. We pass from one cluster to another using an "exchange procedure" which removes one of the x_i and replaces it with a certain rational function. A *cluster monomial* is a monomial in the variables in one cluster.

Berenstein–Fomin–Zelevinsky [7] proved that $\mathbb{C}[N]$ carries a cluster algebra structure. Every reduced word $s_{i_1} \cdots s_{i_m} = w_0$ for the longest element of the Weyl group gives us a cluster $T(\underline{i})$ (though these are not all the clusters).

Geiss–Leclerc–Schröer [19] established the following beautiful result explaining how the theory of preprojective algebras provides an additive categorification of the cluster algebra structure on $\mathbb{C}[N]$.

Theorem 6.1. *Assume that $\mathfrak{g}$ is simply-laced.*

(1) *A maximal rigid Λ-module T gives a cluster with cluster variables $x_1 = \xi_{T_1}, \ldots, x_n = \xi_{T_n}$, where T_i are the distinct indecomposable summands of T.*

(2) *Every cluster is of this form and all cluster monomials lie in the dual semicanonical basis.*

(3) *The exchange relations in $\mathbb{C}[N]$ comes from short exact sequences in Λ-mod.*

On the other hand, Kang–Kashiwara–Kim–Oh [29] proved that the categories of R_ν-modules provide a monoidal categorification of the cluster algebra $\mathbb{C}[N]$. In particular, they proved the following result, which was obtained at around the same time by Qin [40].

Theorem 6.2. *Every cluster monomial in $\mathbb{C}[N]$ lies in the dual canonical basis.*

Together, Theorems 6.1 and 6.2 imply that the dual semicanonical and canonical bases contain many common elements, since they both contain all cluster monomials.

6.2. g-vectors

Fix a cluster $T = \{x_1, \dots, x_n\}$ in a cluster algebra A. Let $u \in A$ be a cluster monomial. Fomin–Zelevinsky [17] defined combinatorially the g-vector $g_T(u) \in \mathbb{Z}^n$ of u using a mutation procedure. In [15], Derksen–Weyman–Zelevinsky proved that this g-vector encodes the "leading monomial" appearing in u. In this case, we say u is g-pointed. As we vary the cluster T, the data of these $g_T(u)$ defines a tropical point in the Langlands dual cluster X variety, as studied by Fock–Goncharov [16].

The following observation is due to Genz–Koshevoy–Schumann [22, SECTION 6].

Proposition 6.3. *Let $\underline{i}$ be a reduced word for w_0, giving a cluster $T(\underline{i})$. Let u be a cluster monomial. Then $g_{T(\underline{i})}(u)$ agrees with the $\underline{i}$-Lusztig data of u, up to a simple linear change of coordinates.*

In this way, we see explicitly how the information of all these g-vectors is the same information as the MV polytope (this is also closely related to Corollary 2.4).

In the setting of the dual semicanonical basis, this can be generalized as follows. Let T be a maximal rigid Λ-module and let $M \in \Lambda$-mod. Geiss–Leclerc–Schröer [20] defined $g_T(M) \in \mathbb{Z}^n$ using homological algebra, extending the above notion of g-vector. Moreover, they proved that ξ_M is g-pointed in each cluster.

6.3. Theta basis

A cluster algebra that contains finitely many clusters is called *finite type*. The cluster algebra $\mathbb{C}[N]$ is of finite type only when $G = \mathbf{SL}_2, \mathbf{SL}_3, \mathbf{SL}_4, \mathbf{SL}_5$. When a cluster algebra A is not finite type, then the cluster monomials do not span A. It was a longstanding open problem to extend the set of cluster monomials to a basis for A. This problem was solved by the following remarkable theorem of Gross–Hacking–Keel–Kontsevich.

Theorem 6.4. *Let A be a cluster algebra, satisfying some hypotheses (which hold for $\mathbb{C}[N]$). There is a natural basis for A, called the theta basis, extending the set of cluster monomials. This theta basis is parametrized by the set of tropical points of the Langlands dual cluster X variety. Moreover, each theta basis element is g-pointed in each cluster.*

Combining Theorem 6.4 with Proposition 6.3 or Corollary 2.4, we get a natural parametrization of the theta basis of $\mathbb{C}[N]$ by $B(\infty)$. In Section 1.5, we saw that biperfect bases are parametrized by $B(\infty)$, so it is natural to ask the following.

Question 6.5. Is the theta basis for $\mathbb{C}[N]$ a biperfect basis?

6.4. Cluster structure on the MV basis

From Theorems 6.1 and 6.2 the dual semicanonical and dual canonical bases for $\mathbb{C}[N]$ contain all cluster monomials. In another direction, Qin [41] studied bases which are g-pointed in each cluster and gave a description of the set of all such bases. In particular, he showed that all such bases contain all the cluster monomials.

This motivates the following conjecture.

Conjecture 6.6. *The MV basis for $\mathbb{C}[N]$ contains all cluster monomials. Moreover, its elements are g-pointed in each cluster.*

Note that the conjecture would imply that the MV basis and dual semicanonical basis agree for $\mathbf{SL}_5$ which is not known.

As evidence for this conjecture, let us mention that the first counterexamples (in both $\mathbf{SO}_8$ and $\mathbf{SL}_6$) to the equality of the MV and dual semicanonical bases (see Theorem 1.13) occur for the square of the simplest basis element which is not a cluster monomial.

Baumann–Gaussent–Littelmann proved this conjecture for certain clusters.

Theorem 6.7 ([4, PROP 7.2]). *If a reduced word $\underline{i}$ for w_0 satisfies a certain condition (which holds for all reduced words in small rank), then all cluster monomials in the cluster $T(\underline{i})$ lie in the MV basis.*

At the moment, we are far from Conjecture 6.6, but thinking about this conjecture motivates the following questions.

Question 6.8.

(1) Is there a refinement of the notion of biperfect basis which would imply that such a basis contains all cluster monomials?

(2) What do cluster exchange relations correspond to geometrically? Which collections of MV cycles form clusters?

Finally, we close with the following wild conjecture.

Conjecture 6.9. *The MV and theta bases for $\mathbb{C}[N]$ coincide.*

We have three pieces of weak evidence for this conjecture. First, the way in which the MV, dual canonical, and dual semicanonical bases differ in $\mathbb{C}[N]$ for $\mathbf{SL}_6$ in (1.2) is very reminiscent of the way in which the theta, dual canonical, and generic bases differ for rank 2, affine type cluster algebras. Second, the construction of the MV and theta bases are both related to the geometry of loop spaces. Finally, the trichotomy of bases studied here seems to match the trichotomy of bases for cluster algebras, see [41].

ACKNOWLEDGMENTS

I would like to thank Pierre Baumann and Peter Tingley for many years of fruitful collaboration and friendship. I also thank Alexander Braverman, Michel Brion, Elie Casbi, Anne

Dranowski, Tom Dunlap, Michael Finkelberg, Stephane Gaussent, Justin Hilburn, Allen Knutson, Bernard Leclerc, Peter Littelmann, Calder Morton-Ferguson, Dinakar Muthiah, Ben Webster, Alex Weekes, and Harold Williams for collaboration and useful conversations. Finally, I would like to thank Arun Ram his three mountains analogy.

FUNDING

This work was partially supported by an NSERC Discovery Grant.

REFERENCES

[1] J. E. Anderson, A polytope calculus for semisimple groups. *Duke Math. J.* **116** (2003), 567–588.

[2] P. Baumann, The canonical basis and the quantum Frobenius morphism. 2012, arXiv:1201.0303.

[3] P. Baumann, T. Dunlap, J. Kamnitzer, and P. Tingley, Rank 2 affine MV polytopes. *Represent. Theory* **17** (2013), 442–468.

[4] P. Baumann, S. Gaussent, and P. Littelmann, Bases of tensor products and geometric Satake correspondence. 2020, arXiv:2009.00042.

[5] P. Baumann, J. Kamnitzer, and A. Knutson, The Mirković–Vilonen basis and Duistermaat–Heckman measures. *Acta Math.* **227** (2021), 1–101.

[6] P. Baumann, J. Kamnitzer, and P. Tingley, Affine Mirković–Vilonen polytopes. *Publ. Math. Inst. Hautes Études Sci.* **120** (2014), 113–205.

[7] A. Berenstein, S. Fomin, and A. Zelevinsky, Cluster algebras III: upper bounds and double Bruhat cells. *Duke Math. J.* **126** (2005), 1–52.

[8] A. Berenstein and D. Kazhdan, Geometric and unipotent crystals. II. From unipotent bicrystals to crystal bases. In *Quantum groups (Israel Math. Conf. Proc.)*, pp. 13–88, Contemp. Math. 433, Amer. Math. Soc., 2007.

[9] A. Berenstein and A. Zelevinsky, Tensor product multiplicities, canonical bases and totally positive varieties. *Invent. Math.* **143** (2001), 77–128.

[10] R. Bezrukavnikov, M. Finkelberg, and I. Mirković, Equivariant homology and K-theory of affine Grassmannians and Toda lattices. *Compos. Math.* **141** (2005), 746–768.

[11] T. Braden, T. Licata, N. Proudfoot, and B. Webster, Quantizations of conical symplectic resolutions II: category O and symplectic duality. *Astérisque* **384** (2016), 75–179.

[12] A. Braverman, M. Finkelberg, and H. Nakajima, Coulomb branches of $3d$ $\mathcal{N} = 4$ quiver gauge theories and slices in the affine Grassmannian. *Adv. Theor. Math. Phys.* **23** (2019), no. 1, 75–166.

[13] M. Brion, Equivariant Chow groups for torus actions. *Transform. Groups* **3** (1997), 225–267.

[14] M. Brion and C. Procesi, Action d'un tore dans une variété projective. In *Operator algebras, unitary representations, enveloping algebras, and invariant theory (Paris, 1989)*, pp. 509–539, Progr. Math. 92, Birkhauser, Boston, 1990.

[15] H. Derksen, J. Weyman, and A. Zelevinsky, Quivers with potentials and their representations II: applications to cluster algebras. *J. Amer. Math. Soc.* **23** (2010), no. 3, 749–790.

[16] V. Fock and A. Goncharov, Cluster ensembles, quantization and the dilogarithm. *Ann. Sci. Éc. Norm. Supér.* **42** (2009), no. 6, 865–930.

[17] S. Fomin and A. Zelevinsky, Cluster algebras IV: coefficients. *Compos. Math.* **143** (2007), 112–164.

[18] C. Geiss, B. Leclerc, and J. Schröer, Semicanonical bases and preprojective algebras. *Ann. Sci. Éc. Norm. Supér.* **38** (2005), 193–253.

[19] C. Geiss, B. Leclerc, and J. Schröer, Rigid modules over preprojective algebras. *Invent. Math.* **165** (2006), 589–632.

[20] C. Geiss, B. Leclerc, and J. Schröer, Generic bases for cluster algebras and the Chamber ansatz. *J. Amer. Math. Soc.* **25** (2012), 21–76.

[21] I. Gelfand and A. Zelevinsky, Polytopes in the pattern space and canonical bases for the irreducible representations of gl_3. *Funct. Anal. Appl.* **19** (1985), 72–75.

[22] V. Genz, G. Koshevoy, and B. Schumann, Polyhedral parametrizations of canonical bases and cluster duality. *Adv. Math.* **369** (2020).

[23] V. Ginzburg, Perverse sheaves on a loop group and Langlands duality. 2000, arXiv:alg-geom/9511007.

[24] A. Goncharov and L. Shen, Geometry of canonical bases and mirror symmetry. *Invent. Math.* **202** (2015), no. 2, 487–633.

[25] J. Hilburn, J. Kamnitzer, and A. Weekes, BFN Springer theory. 2020, arXiv:2004.14998.

[26] J. Kamnitzer, The crystal structure on the set of Mirković–Vilonen polytopes. *Adv. Math.* **215** (2007), 66–93.

[27] J. Kamnitzer, Mirković–Vilonen cycles and polytopes. *Ann. of Math.* **171** (2010), 245–294.

[28] J. Kamnitzer, P. Tingley, B. Webster, A. Weekes, and O. Yacobi, On category $\mathcal{O}$ for affine Grassmannian slices and categorified tensor products. *Proc. Lond. Math. Soc.* **119** (2019), no. 5, 1179–1233.

[29] S.-J. Kang, M. Kashiwara, M. Kim, and S-j. Oh, Monoidal categorification of cluster algebras. *J. Amer. Math. Soc.* **31** (2018), 349–426.

[30] M. Kashiwara, Crystalizing the q-analogue of universal enveloping algebras. *Comm. Math. Phys.* **133** (1990), no. 2, 249–260.

[31] M. Kashiwara, Global crystal bases of quantum groups. *Duke Math. J.* **69** (1993), 455–485.

[32] M. Khovanov and A. Lauda, A diagrammatic approach to categorification of quantum groups I. *Represent. Theory* **13** (2009), 309–347.

[33] A. Lauda and M. Vazirani, Crystals from categorified quantum groups. *Adv. Math.* **228** (2011), 803–861.

[34] P. Littelmann, The path model for representations of symmetrizable Kac–Moody algebras. In *Proc. ICM, Zürich, 1994*. Birkhäuser, Basel, 1995.

[35] G. Lusztig, Singularities, character formulas and a q-analog of weight multiplicities. *Astérisque* **101–102** (1983), 208–229.

[36] G. Lusztig, Canonical bases arising from quantized enveloping algebras. *J. Amer. Math. Soc.* **3** (1990), no. 2, 447–498.

[37] G. Lusztig, Canonical bases arising from quantized enveloping algebras II. *Progr. Theoret. Phys. Suppl.* **102** (1990), 175–201.

[38] G. Lusztig, Semicanonical bases arising from enveloping algebras. *Adv. Math.* **151** (2000), 129–139.

[39] I. Mirković and K. Vilonen, Geometric Langlands duality and representations of algebraic groups over commutative rings. *Ann. of Math.* **166** (2007), 95–143.

[40] F. Qin, Triangular bases in quantum cluster algebras and monoidal categorification conjectures. *Duke Math. J.* **166** (2017), no. 12, 2337–2442.

[41] F. Qin, Bases for upper cluster algebras and tropical points. *J. Eur. Math. Soc. (JEMS)* (to appear).

[42] R. Rouquier, 2-Kac–Moody Lie algebras. 2008, arXiv:0812.5023.

[43] Y. Saito, Crystal bases and quiver varieties. *Math. Ann.* **324** (2002), 675–688.

[44] P. Tingley and B. Webster, Mirković–Vilonen polytopes and Khovanov–Lauda–Rouquier algebras. *Compos. Math.* **152** (2016), 1648–1696.

[45] M. Varagnolo and E. Vasserot, Canonical bases and KLR algebras. *J. Reine Angew. Math.* **659** (2011), 67–100.

[46] Z. Yun and X. Zhu, Integral homology of loop groups via Langlands dual groups. *Represent. Theory* **15** (2011), 347–369.

JOEL KAMNITZER

Department of Mathematics, University of Toronto, Toronto, ON, Canada, jkamnitz@math.toronto.edu

This Page Intentionally Left Blank

SPHERICAL VARIETIES, FUNCTORIALITY, AND QUANTIZATION

YIANNIS SAKELLARIDIS

ABSTRACT

We discuss generalizations of the Langlands program, from *reductive groups* to the local and automorphic spectra of *spherical varieties*, and to more general representations arising as "quantizations" of suitable Hamiltonian spaces. To a spherical G-variety X, one associates a *dual group* ${}^L G_X$ and an L-*value* (encoded in a representation of ${}^L G_X$), which conjecturally describe the local and automorphic spectra of the variety. This sets up a problem of functoriality, for any morphism ${}^L G_X \to {}^L G_Y$ of dual groups. We review, and generalize, Langlands' "beyond endoscopy" approach to this problem. Then, we describe the cotangent bundles of quotient stacks of the relative trace formula, and show that transfer operators of functoriality between relative trace formulas in rank 1 can be interpreted as a change of "geometric quantization" for these cotangent stacks.

MATHEMATICS SUBJECT CLASSIFICATION 2020

Primary 11F70; Secondary 11F67, 22E50, 22E55

KEYWORDS

Langlands program, L-functions, periods, spherical varieties, relative trace formula, functoriality, quantization

INTERNATIONAL CONGRESS OF MATHEMATICIANS 2022 JULY 6–14

Proc. Int. Cong. Math. 2022, Vol. 4, pp. 2998–3037
DOI 10.4171/ICM2022/88

Published by EMS Press

1. INTEGRAL REPRESENTATIONS OF L-FUNCTIONS

1.1. Classical periods

1.1.1. In his legendary 1859 paper [66], Riemann proved the functional equation of the zeta function by representing it as the Mellin transform of a theta series

$$\pi^{-\frac{s}{2}}\Gamma\left(\frac{s}{2}\right)\zeta(s) = \int_0^\infty y^{\frac{s}{2}} \sum_{n=1}^\infty e^{-n^2\pi y}\, d^\times y.$$

The proof used the functional equation of the latter with respect to the substitution $y \leftrightarrow y^{-1}$, previously established by Jacobi and based on the Poisson summation formula.

About 90 years later, Iwasawa, in a short announcement [37], and Tate, in his thesis [83], reformulated this integral in the language of the adeles. The new formulation could be directly applied to the generalizations of the zeta function to arbitrary Dirichlet characters (by Dirichlet), or number fields (by Dedekind) and Grössencharacters (by Hecke), and clarified the meaning of the Euler factors of the zeta function, as Mellin transforms of Schwartz functions on the p-adic completions of $\mathbb{Q}$. Namely, we have an identity

$$\zeta_p(s) = \int_{\mathbb{Q}_p^\times} \Phi_p(x)|x|^s\, d^\times x,$$

where, for finite primes p, $\zeta_p(s) = (1-p^{-s})^{-1}$ and $\Phi_p = 1_{\mathbb{Z}_p}$, the characteristic function of the p-adic integers, is what we will call the *basic Schwartz function* on $\mathbb{Q}_p$; the same interpretation extends to the "Archimedean factor" $\zeta_\infty(s) = \pi^{-\frac{s}{2}}\Gamma(\frac{s}{2})$ of the functional equation, with $\mathbb{Q}_\infty = \mathbb{R}$, and Φ_∞ the Gaussian $e^{-\pi x^2}$.

1.1.2. Meanwhile, in 1936–1937, Hecke [34] introduced what is today called the L-function of a modular form, generalized to nonholomorphic automorphic forms by Maass in 1944 [56]. Recast in the adelic language by Jacquet and Langlands in their seminal 1970 work [41], these L-functions, with appropriate Archimedean factors, can be represented as Mellin transforms

$$\int_{k^\times\backslash\mathbb{A}^\times} f\begin{pmatrix} a & \\ & 1\end{pmatrix} |a|^s\, d^\times a,$$

where k denotes a number field, $\mathbb{A}$ its ring of adeles, and f is a cuspidal automorphic form on $\mathrm{GL}_2(k)\backslash\mathrm{GL}_2(\mathbb{A})$.

Shortly after Hecke, Rankin [65] and Selberg [80] discovered an integral representation for the L-function that carries their names, which today is seen as a special case of a Langlands L-function, attached to the tensor product representation

$$\check{G} = \mathrm{GL}_2 \times \mathrm{GL}_2 \xrightarrow{\otimes} \mathrm{GL}_4$$

of the Langlands dual of the group $G = \mathrm{GL}_2 \times \mathrm{GL}_2$. This integral is, on the surface, very different from the Mellin transforms of Riemann and Hecke, as it involves a pair of cusp forms and an Eisenstein series:

$$\int_{\mathrm{GL}_2(\mathbb{A})} f_1(g) f_2(g) E(g,s)\, dg.$$

1.2. The theta series of spherical varieties

1.2.1. The aforementioned works, and their adelic reformulations, led to an explosion of research around L-functions from the 1970s onward, with numerous new integral representations discovered by Godement, Jacquet, Rallis, Piatetski-Shapiro, Gelbart, Shalika, Waldspurger, Ginzburg, Bump, Friedberg, Garrett, and others [19], combining elements from all of the methods above, such as the theta series (from Riemann), the "period integrals" over subgroups (from Hecke), and the Eisenstein series (from Rankin and Selberg).

A uniform approach to many of these methods was proposed in [68]; it relies on the following ingredients:

- A (suitable) *affine spherical variety* X for a group G over a number field k; that is, X is a normal, affine G-variety, with a dense orbit for the Borel subgroup of G. This space is $X = \mathbb{A}^1$ for $G = \mathbb{G}_m$ in Riemann–Iwasawa–Tate theory, $X = \mathrm{GL}_2$ for $G = \mathrm{GL}_1 \times \mathrm{GL}_2$ in Hecke–Jacquet–Langlands theory, and $X = V \times^{\mathrm{GL}_2^{\mathrm{diag}}} (\mathrm{GL}_2 \times \mathrm{GL}_2)$, where V is the standard representation of GL_2, in Rankin–Selberg theory.

- A suitable space of "Schwartz functions" $\mathcal{F}(X(k_v))$ for every completion k_v of k; at almost every place, it contains a distinguished vector $\Phi_{0,v}$, giving rise to a restricted tensor product $\mathcal{F}(X(\mathbb{A})) = \bigotimes'_v \mathcal{F}(X(k_v))$. When X is smooth and k_v is non-Archimedean with ring of integers $\mathfrak{o}_v$, we have $\Phi_{0,v} = 1_{X(\mathfrak{o}_v)}$.

- The X-theta series
$$\Theta : \mathcal{F}\big(X(\mathbb{A})\big) \to C^\infty\big([G]\big),$$
where $[G] = G(k)\backslash G(\mathbb{A})$, given by $\Theta_\Phi(g) := \Theta(\Phi)(g) := \sum_{\gamma \in X(k)} \Phi(\gamma g)$. This generalizes the Jacobi theta series used by Riemann, and many other series of classical analytic number theory, such as Poincaré series (if we allow X to stand for the Whittaker model, which is not just a space but also a nontrivial "line bundle" over it, see § 2.2.5), and Eisenstein series (after we pair a suitable theta series with an automorphic form for some Levi subgroup).

The theta series (for varying inputs Φ) are then integrated against automorphic forms f, and, under some assumptions on the space X, the "period pairing"
$$\langle f, \Theta_\Phi \rangle := \int_{[G]} f(g)\Theta_\Phi(g)\, dg \tag{1.1}$$
is expected to be related to a special value of an L-function of f. This relation will be discussed in Sections 2.3.2–3.1.1.

1.2.2. While it is not the main focus of the present article, it should be mentioned that the main point of the proposal of [68] was to include *singular* affine spherical varieties, in which case the "basic function" of $X(k_v)$ is the "IC function," obtained through the sheaf–function dictionary from the intersection complex of a suitable geometric model of $X(\mathfrak{o}_v)$, see § 3.1.3. This was inspired by the work of Braverman–Kazhdan on the basic affine space [14, 16], which goes back to the geometric Langlands program [13] and ideas of Drinfeld.

The conjecture was refined by Ngô [62] for a class of affine embeddings of reductive groups; the IC function, for non-Archimedean local fields in equal characteristic, was defined in [12], where Ngô's conjecture was proven. In recent joint work with Jonathan Wang [78], we have obtained similar results for the IC function of a broad class of spherical varieties, including a straightforward generalization of the Hecke and Rankin–Selberg integral representations to the Langlands L-function associated to the n-fold tensor product representation of the dual group

$$\check{G} = \underbrace{\mathrm{GL}_2 \times \cdots \times \mathrm{GL}_2}_{n \text{ times}} \xrightarrow{\otimes} \mathrm{GL}_{2^n} .$$

1.2.3. The elephant in the room, of course, is the global functional equation (and meromorphic continuation), which is not available for these L-functions, yet. In favorable cases, it should arise from a Poisson summation formula for a "Fourier transform"

$$\mathcal{F}\big(X(\mathbb{A})\big) \to \mathcal{F}\big(X^*(\mathbb{A})\big),$$

where X^* is the same variety X, with the G-action twisted by a Chevalley involution. Such a Fourier transform and a Poisson summation formula are often available for smooth affine spherical varieties, which are vector bundles over homogeneous spaces, but are quite mysterious in the singular case. For the moment, they are known for spaces of the form $X =$ the affine closure of $[P, P]\backslash G$, where $P \subset G$ is a parabolic, by the work of Braverman–Kazhdan [14,16] (and its refinement [33]). An extension to $X =$ the affine closure of $U_P\backslash G$, where U_P is the unipotent radical of P, would give rise to the functional equation of normalized Eisenstein series, greatly simplifying and generalizing the theory of L-functions obtained through the Langlands–Shahidi method [58,81]. In recent work, Getz and his collaborators [31,32] have proven a Poisson summation formula for a singular space Y which is not directly related to Eisenstein series—the only example of this sort to date, to my knowledge.

In general, this "Fourier" transform may only be available at the level of trace formulas—see [73,88], as well as the discussion of § 6.4 below.

1.2.4. The "period pairing" (1.1) between theta series coming from spherical varieties and automorphic forms is not general enough to include all known integral representations of L-functions. At the very least, we need to replace the Schwartz space of a spherical G-variety by more general *quantizations of Hamiltonian G-spaces*. In the smooth case, those are affine symplectic G-spaces M, equipped with a moment map $M \to \mathfrak{g}^*$, which generalize the cotangent bundle T^*X of a smooth spherical G-variety. The analog of the "spherical" condition for a Hamiltonian G-space M is that it be *coisotropic*, that is, the Poisson algebra $k(M)^G$ of G-invariant rational functions on M be Poisson-commutative.

An example of such a space, that is not the cotangent bundle of a spherical variety, is a symplectic vector space M under the action of a Howe dual pair G; that is, G is, up to central isogeny, equal to a product $G_1 \times G_2$ of two subgroups of $\mathrm{Sp}(M)$, where G_1 is the commutator of G_2, and vice versa. As "quantization" of M we understand the Weil representation of the metaplectic group $\mathrm{Mp}(M)$ associated with an additive character ψ, restricted to (the metaplectic cover of) G. Theta series and the pairing (1.1) still make sense in this setting. More general examples mixing the Weil representation with periods are contained in the influential conjectures of Gan, Gross, and Prasad [27].

In an ongoing work with Ben-Zvi and Venkatesh, we describe a class of coisotropic Hamiltonian spaces M whose "quantizations" in the form of theta series are expected to be related to special values of L-functions, and we demonstrate, by means of known examples, that the L-value associated to such a space gives rise to a *dual Hamiltonian space* $\check{M}$ *for the Langlands dual group*. In the context of the geometric Langlands program, this leads to a hierarchy of conjectures, with connections to mathematical physics. We will encounter one of these conjectures in our discussion of unramified L-factors in § 3.2 below.

1.3. Outline of this paper

In Section 2 we introduce the relative Langlands program, up to the conjectural Euler factorization of the period pairings (1.1).

In Section 3 we discuss the relationship between the local unramified Euler factors and special values of L-functions.

In Section 4 we discuss the "beyond endoscopy" approach to functoriality, generalized to the setting of the relative Langlands program.

Section 5 provides a new interpretation for the transfer operators of functoriality studied in [74], based on the concept of quantization. (Proofs for the results of this section will appear in an expanded version of this article on the arXiv.)

Finally, in Section 6 we discuss interesting research directions for the near future.

1.4. Notation and language

- In general, when a variety is defined over a local field F, and there is no danger of confusion, we will use the same letter to denote its F-points, e.g., "a Schwartz function on X" really means "on $X(F)$."
- For a quasiaffine G-variety over a field F, we will denote by X/G the stack quotient, and by $X /\!\!/ G$ the invariant-theoretic quotient Spec $F[X]^G$.
- A "complex line bundle" on the points of a smooth variety X over a local field F will be
  - when $F = \mathbb{R}$ or $\mathbb{C}$, a complex line bundle on $X(F)$, viewed as a smooth (Nash) manifold;
  - when F is non-Archimedean, a locally constant sheaf of complex vector spaces (l-sheaf) on $X(F)$, for the p-adic (Hausdorff) topology, with 1-dimensional stalks.

  When no confusion arises, we will just say "line bundle" for a complex line bundle; when we want to distinguish it from a line bundle on X in the sense of algebraic geometry, we will say "algebraic line bundle" for the latter.

- An algebraic line bundle L over a smooth F-variety X, where F is a local field, together with a complex number s, give rise to a complex line bundle $|L|^s$ on $X(F)$, by reduction of the corresponding $\mathbb{G}_m$-torsor via the sequence of maps $\mathbb{G}_m(F) \xrightarrow{|\bullet|} \mathbb{R}_+^\times \xrightarrow{x \mapsto x^s} \mathbb{C}^\times$. (When F is non-Archimedean, the absolute value map is discretely-valued, giving rise to the structure of a locally constant sheaf.)

- When $L = \det T^*X$, the line bundle of volume forms on X, the associated complex vector bundle $|L|^s$ is known as the bundle of s-densities. We will, in general, understand the field F as endowed with a Haar measure; this identifies densities, i.e., sections of $|L|$, as measures on $X(F)$. When no confusion arises, we will denote $|dx|$, the density attached to a volume form dx, simply by dx.

- The space of Schwartz functions on the F-points of a variety X, where X is a local field, will be denoted by $\mathcal{F}(X(F))$, the space of Schwartz measures by $\mathcal{S}(X(F))$, and the space of Schwartz half-densities by $\mathcal{D}(X(F))$. These are smooth, compactly supported sections of the corresponding bundles of s-densities, in the non-Archimedean case. In the Archimedean case, they are smooth sections of rapid decay, see [1]. We will also say "test functions/measures," etc., for "Schwartz."

- For an admissible, smooth, complex representation π of a reductive group over a local field, we will denote by $\tilde{\pi}$ its contragredient. When π is unitary, $\tilde{\pi}$ is identified with the complex conjugate $\bar{\pi}$.

- We will generally prefer to replace a hermitian pairing H between functions by the associated bilinear pairing $B(\Phi_1, \Phi_2) = H(\Phi_1, \overline{\Phi_2})$. When H is an inner product, we will sometimes call B, by abuse of language, an "inner product."

- $\mathcal{W}_F$ will denote the Weil group of a local or global field, and $\mathcal{L}_F$ will be the "Langlands group," whose representations should parametrize local and automorphic L-packets. It is the Weil group for Archimedean local fields and global function fields, the Weil–Deligne group for non-Archimedean local fields, and a conjectural extension of the Weil group for number fields.

- We adopt the "Weil group" convention for L-groups of reductive groups, ${}^LG = \check{G} \rtimes \mathcal{W}_F$; the dual group $\check{G}$ is identified with the set of its complex points.

2. THE RELATIVE LANGLANDS CONJECTURES

2.1. The local and global spectrum of a spherical variety

2.1.1. To understand the relationship between period pairings (1.1) and L-functions, one needs to start by understanding the phenomenon of *distinction*, highlighted by the groundbreaking work of Jacquet and his collaborators [38, 42]. A naive formulation of this phenomenon goes as follows:

The local and global spectrum of a spherical G-variety X only contain representations with Langlands parameters in a certain subgroup ${}^LG_X \subset {}^LG$ of the L-group of G.

To make sense of this statement, we need to explain "the local and global spectrum of a G-variety." Then, we need to talk about the L-group ${}^LG_X$. Finally, the statement needs to be corrected, for some "nontempered" varieties X, replacing Langlands parameters by appropriate Arthur parameters.

Let R denote either a local field, or the adelic points of a global field. In order to define the local and global spectrum of a G-variety X (defined over the corresponding field), we will introduce the Plancherel formula and the relative trace formula. These decompose certain distributions—or rather, generalized functions—on the R-points of $X \times X$, invariant under the diagonal action of G. For the purposes of the Langlands program, it turns out to be more natural to think of them as generalized functions on the R-points of the quotient stack $\mathfrak{X} = (X \times X)/G^{\mathrm{diag}}$, which naturally includes "pure inner forms" of the pair (G, X).

2.1.2. Let F be a local field. The space $L^2(X)$ is the Hilbert space completion of the space $\mathcal{D}(X)$ of Schwartz half-densities on $X(F)$, with respect to the L^2-inner product, and furnishes a unitary representation of G. By the Plancherel decomposition, there are a measure μ_X on the unitary dual $\hat{G}$ of G and a measurable family of linear forms

$$J_\pi : \mathcal{D}(X \times X) \to \mathbb{C}$$

such that:

- for μ_X-almost every π, J_π factors as $\mathcal{D}(X \times X) \to \pi \hat{\otimes} \bar{\pi} \to \mathbb{C}$, and
- for all $\Phi \in \mathcal{D}(X \times X)$, we have

$$\int_{X^{\mathrm{diag}}} \Phi = \int_{\hat{G}} J_\pi(\Phi)\mu_X(\pi). \tag{2.1}$$

A linear form satisfying the first property above will be called a *relative character*. The product $J_\bullet \mu_X$, which can be thought of as a measure valued in the space of functionals on $\mathcal{D}(X \times X)$, is uniquely defined. Moreover, the relative characters J_π are invariant under the diagonal action of $G = G(F)$; thus, they factor through the coinvariant space $\mathcal{D}(X \times X)_G =$ the quotient of $\mathcal{D}(X \times X)$ by the (closed, in the Archimedean case) subspace generated by elements of the form $f - g \cdot f$, where $g \cdot f$ denotes the action of $g \in G$ on f by diagonal translation.

Let us assume that X carries a positive G-invariant measure dx, and use it to identify functions, half-densities, and measures on X through the G-equivariant maps $\Phi \mapsto \Phi(dx)^{\frac{1}{2}} \mapsto \Phi dx$ (and similarly on $X \times X$). Then, the coinvariant space $\mathcal{D}(X \times X)_G \simeq \mathcal{S}(X \times X)_G$ is more naturally understood as a subspace of the *Schwartz space of the quotient stack* $\mathfrak{X} := (X \times X)/G$ [71]. This Schwartz space is really a complex of vector spaces, but here we will focus only on its zeroth cohomology, which has the explicit description

$$\mathcal{S}(\mathfrak{X}) = \bigoplus_{\alpha} \mathcal{S}(X^{\alpha} \times X^{\alpha})_{G^{\alpha}}. \tag{2.2}$$

Here, α runs over isomorphism classes of G-torsors (parametrized by the Galois cohomology set $H^1(\Gamma_F, G)$, where Γ_F is the Galois group of a separable closure of F); if R^{α} is a representative of a class α, we let $G^{\alpha} = \mathrm{Aut}_G(R^{\alpha})$, and $X^{\alpha} = X \times^G R^{\alpha}$, a G^{α}-space. In other words, G^{α} is what is called a "pure inner form" of G, and X^{α} can similarly be called a "pure inner form" of X, if its set of F-points is nonempty.

The Plancherel formula (2.1) extends to $\mathcal{S}(\mathfrak{X})$, with a measure $\mu_{\mathfrak{X}}$, on the right-hand side, on the union of the unitary duals of the pure inner forms G^{α}. The support $\Pi_{\mathfrak{X}}$ of this measure (avoiding redundancy—i.e., the support of the canonical linear form-valued measure $J_{\bullet}\mu_{\mathfrak{X}}$) can be called the *local (L^2-)spectrum of the quotient stack* $\mathfrak{X}$.

2.1.3. The global (automorphic) spectrum of X (or rather, again, of the stack $\mathfrak{X} = (X \times X)/G^{\mathrm{diag}}$) can be defined through the relative trace formula of Jacquet. This is a generalization of the Arthur–Selberg trace formula, and an automorphic analog of the local Plancherel formula. Its definition uses the theta series encountered in § 1.2.1, therefore we assume here that X, defined over a global field k, is quasiaffine, so that $X(k)$ is discrete in the adelic points $X(\mathbb{A})$. As before, we write $[G] = G(k)\backslash G(\mathbb{A})$ for the automorphic quotient space.

Roughly speaking, the relative trace formula is the Plancherel formula for $L^2([G])$, applied to the inner product of two theta series for X, i.e., decomposing the functional

$$\begin{aligned}\mathrm{RTF}_X : \mathcal{F}\big(X(\mathbb{A})\big) \otimes \mathcal{F}\big(X(\mathbb{A})\big) \ni \Phi_1 \otimes \Phi_2 &\mapsto \Theta_{\Phi_1} \otimes \Theta_{\Phi_2} \\ &\mapsto \int_{[G]} \Theta_{\Phi_1}(g)\Theta_{\Phi_2}(g)\, dg \in \mathbb{C}.\end{aligned} \tag{2.3}$$

This naive point of view requires some caution:

- The inner product on the right-hand side of (2.3) does not, in general, converge, and needs to be regularized. Depending on X, there may be a *canonical* way to regularize it, described in **[71, §6]**. In many cases of interest, though, notably in the case of the Arthur–Selberg trace formula (where $X = H$, a reductive group, and $G = H \times H$), a canonical regularization is not available, and it takes the mastery of Arthur's work **[6]** to engineer an invariant expression. Such work has not yet been done in the general setting of the relative trace formula.

- We can again choose a G-invariant measure on $X(\mathbb{A})$ (e.g., Tamagawa measure) to identify functions with measures, and understand the $G(\mathbb{A})^{\mathrm{diag}}$-invariant functional (2.3) as a functional on $\mathcal{S}(X \times X(\mathbb{A}))_{G(\mathbb{A})}$. As in the local case, this space is a subspace of the global Schwartz space of the stack $\mathfrak{X} = (X \times X)/G^{\mathrm{diag}}$,

$$\mathcal{S}(\mathfrak{X}(\mathbb{A})) = \bigotimes_{v}{}' \mathcal{S}(\mathfrak{X}(k_v)),$$

and the relative trace formula should be defined as a functional on the bigger space,

$$\mathrm{RTF}_{\mathfrak{X}} = \sum_{\alpha} \mathrm{RTF}_{X^{\alpha}},$$

where now α runs over isomorphism classes of G-torsors over the global field k.

Ignoring the regularization issue, if we could apply the Plancherel formula for $\bigoplus_{\alpha} L^2([G^{\alpha}])$ to the pairing (2.3), we would obtain the *spectral side* of the relative trace formula,

$$\mathrm{RTF}_{\mathfrak{X}} = \int J_{\pi}^{\mathrm{aut}} \mu_{\mathfrak{X}}^{\mathrm{aut}}(\pi), \tag{2.4}$$

where the product $J_{\bullet}^{\mathrm{aut}} \mu_{\mathfrak{X}}^{\mathrm{aut}}$ is a measure on the union $\bigsqcup_{\alpha} \widehat{G^{\alpha}}^{\mathrm{aut}}$ of (L^2-)automorphic spectra of the pure inner forms of G, valued in linear forms on $\mathcal{S}(\mathfrak{X}(\mathbb{A}))$.

The global (automorphic) spectrum of $\mathfrak{X}$ is defined as the support of $J_{\bullet}^{\mathrm{aut}} \mu_{\mathfrak{X}}^{\mathrm{aut}}$. Clearly, this definition is incomplete, as it relies on overcoming the aforementioned issues of regularization, and developing a spectral decomposition for the relative trace formula.

2.2. The Langlands dual group

2.2.1. The local and global spectrum of a spherical variety X are conjecturally governed by the *L-group* ${}^L G_X$ of X. We owe this dual group to the insights developed by Nadler in his thesis [60], and in his joint work with Gaitsgory [25]. They realized that the "little Weyl group" of a spherical G-variety (defined by Brion in [18], and generalizing the little Weyl group of a symmetric space) corresponds to a subgroup $\check{G}_X \subset \check{G}$ of the Langlands dual group of G, and gives rise to a form of the geometric Satake isomorphism for the spherical variety. In [77], it was proposed that this dual group comes equipped with a distinguished morphism

$$\check{G}_X \times \mathrm{SL}_2 \to \check{G} \tag{2.5}$$

that governs the harmonic analysis of X, in a way that will be described below. Since Gaitsgory and Nadler did not fully identify their dual group $\check{G}_X$ (constructed in a Tannakian way), an independent description of a morphism (2.5) was achieved by Knop and Schalke [47]; we can take this as the definition of the dual group, for what follows. Finally, for the purposes of the Langlands program, we need an L-group, in the form of an extension

$$1 \to \check{G}_X \to {}^L G_X \to \mathcal{W}_F \to 1.$$

The correct definition of this L-group, when G is not split, is not completely understood yet, although it is probably within reach. For what follows, we will assume such an L-group, and an extension of the homomorphism (2.5) to the L-groups, in the sense that the conjectures to be stated should hold for an appropriate definition of ${}^L G_X$.

2.2.2. We briefly describe one way to characterize the root datum of the dual group $\check{G}_X$: As in the case of reductive groups, the first step is to describe a canonical maximal torus, which in turn is dual to an "abstract Cartan" group. Let A be the abstract Cartan group of G; it is canonically equal to the reductive quotient of any Borel subgroup of G. Fix such a Borel subgroup B, with unipotent radical N, and let X° be the open B-orbit. On the quotient $X^\circ /\!\!/ N$, B acts through a quotient A_X of A; this is *the Cartan group of* X, and it can be seen to be independent of B, in the sense that any two choices induce canonical tori up to a canonical isomorphism. (These definitions assume that B is defined over the base field, but by Galois descent the Cartan groups A and A_X are defined over the field, even if B is not.)

The quotient $A \to A_X$ gives rise to a morphism of dual tori $\check{A}_X \to \check{A}$, which could have nontrivial (finite) kernel. The image of this morphism is the canonical maximal torus of the Gaitsgory–Nadler dual group $\check{G}_X$. (We caution the reader that in [77] the group $\check{G}_X$ was not necessarily defined as a subgroup of $\check{G}$, and had $\check{A}_X$ as its maximal torus.)

It is slightly harder to define the little Weyl group W_X. Once this is done, the coroots of $\check{G}_X$, which will be called the *normalized spherical roots* of X, are uniquely determined up to multiple, and that multiple is fixed by the following axiom:

> *A normalized spherical root is either a root of G, or the sum of two strongly orthogonal roots, i.e., two roots whose linear span contains no other roots but their multiples.*

2.2.3. For the Weyl group, there are many equivalent definitions. Most relevant to our purposes, when X is defined over a field F in characteristic zero, is the following one, due to Knop [45]: We may assume that X is smooth and that F is algebraically closed (since $\check{G}_X$ only depends on the open G-orbit over the algebraic closure). Consider the cotangent space $M = T^*X$, equipped with the moment map $\mu : M \to \mathfrak{g}^*$. If $\mathfrak{a}^*$ denotes the dual Lie algebra of the Cartan of G, Chevalley's isomorphism identifies the invariant-theoretic quotient $\mathfrak{g}^* /\!\!/ G$ with $\mathfrak{a}^* /\!\!/ W$. The *polarized cotangent bundle*

$$\hat{M} := M \times_{\mathfrak{a}^* /\!\!/ W} \mathfrak{a}^*$$

is not, in general, irreducible. Knop describes a distinguished irreducible component $\hat{M}^\circ$ living over the dual Lie algebra $\mathfrak{a}_X^* \subset \mathfrak{a}^*$ of A_X, and shows that the map $\hat{M}^\circ \to M$ is generically a Galois cover with covering group a subquotient W_X of the Weyl group; this is the little Weyl group of X [45, §6].

For later use, we mention a related result of Knop, still in the homogeneous case. Let $\mathfrak{g}_X^*$ = the normalization of the image of the moment map *in* M (i.e., the spectrum of the integral closure of the image of $F[\mathfrak{g}^*]$ in $F[M]$). The composition $\hat{M}^\circ \to \mathfrak{a}_X^* \to \mathfrak{c}_X^* := \mathfrak{a}_X^* /\!\!/ W_X$ factors through a map $\mu_G : M \to \mathfrak{g}_X^* \to \mathfrak{c}_X^*$, called the *invariant moment map*, and identifies $\mathfrak{c}_X^*$ with the invariant-theoretic quotient $M /\!\!/ G$ [45, KOROLLAR 7.2].

2.2.4. Finally, the restriction of the map (2.5) to SL_2, which we will call the "Arthur-SL_2" of X, is determined by the conjugacy class of parabolics of the form

$$P(X) = \{g \in G \mid X^\circ g = X^\circ\},$$

where X° is the open orbit for a Borel subgroup B. In the quasiaffine case, $P(X)$ is the largest parabolic such that all highest weight vectors in $k[X]$ are $P(X)$-eigenvectors. To this class is canonically associated a standard Levi subgroup $\check{L}(X)$ of $\check{G}$, and the Arthur-SL_2 of X is a principal $\mathrm{SL}_2 \to \check{L}(X)$.

2.2.5. In order to keep the discussion that follows as simple as possible, let us single out a convenient class of spherical varieties. We will say that a spherical G-variety X is *excellent* if it is affine, homogeneous, and the kernel of the map $A \to A_X$ is connected (equivalently, the map $\check{A}_X \to \check{A}$ of dual tori is injective).

We also need to enlarge the class of spherical varieties, in order to include objects such as the Whittaker model. The Whittaker model is the space $N\backslash G$, where G is quasisplit and N is a maximal unipotent subgroup, endowed with a nondegenerate character $\psi : N(F) \to \mathbb{C}^\times$. This character defines, by induction, a complex line bundle L_ψ over $N\backslash G$, and the Whittaker model consists of sections of this line bundle. In the sequel, when we say that Y is "the Whittaker model," we will mean the space $N\backslash G$ together with this line bundle, and we will be using the Schwartz space notation $\mathcal{F}(Y)$, $\mathcal{S}(Y)$, etc., to denote Schwartz sections (resp. measures) valued in this line bundle. For a more general discussion of "Whittaker induction," see [77, §2.6].

2.3. Conjectures

2.3.1. Let X be defined over a local field F, and let $\Pi_{\mathfrak{X}}$ be the set of L^2-distinguished representations of X and its pure inner forms, as in § 2.1.2. We assume, for simplicity, that X carries an invariant measure, to identify measures with half-densities. The "relative local Langlands conjecture" of my work with Venkatesh [77, §16] states:

Conjecture. *Let $\mu_{{}^LG_X}$ be the natural measure on the set of tempered local Langlands parameters into ${}^LG_X$. There is a decomposition of the inner product on $\mathcal{D}(X \times X)$,*

$$\int_{X^{\mathrm{diag}}} \Phi = \int J_\phi(\Phi)\mu_{{}^LG_X}(\phi), \tag{2.6}$$

where the "stable relative characters" J_ϕ are linear combinations of relative characters for representations belonging to Arthur packets with parameter

$$\mathcal{L}_F \times \mathrm{SL}_2 \xrightarrow{\phi\times\mathrm{Id}} {}^LG_X \times \mathrm{SL}_2 \xrightarrow{(2.5)} {}^LG. \tag{2.7}$$

For the "natural measure" on such parameters, see [77, §16]. Comparing with the Plancherel formula (2.1), the conjecture implies that the local L^2-spectrum $\Pi_{\mathfrak{X}}$ of $\mathfrak{X}$ belongs to the union of Arthur packets with parameters of the form (2.7). Developing a Plancherel formula for X in terms of discrete-mod-center spectra of its "boundary degenerations," as in [22,77], reduces the conjecture to discrete spectra.

When G is quasisplit, acts faithfully on X, and the map $\check{A}_X \to \check{A}$ is injective (§ 2.2.2), one would expect the functionals J_ϕ of (2.6), after summing over all pure inner forms of X, to be nonzero. In a broad range of individual cases, including the Gan–Gross–Prasad conjectures [27] and other cases considered by D. Prasad [64] and C. Wan [87], we have much more precise conjectures about how many and which elements in the given Arthur packets are distinguished. A number of cases have been proven by Waldspurger, Mœglin, Beuzart-Plessis, Gan, Ichino, and others [9, 28, 58, 86], and by Mœglin–Renard [57] for symmetric spaces over $\mathbb{R}$.

Besides the question of L^2-distinction, one can ask the question of smooth distinction: Which irreducible representations embed as $\pi \hookrightarrow C^\infty(X)$? The general answer to this question is less understood.

2.3.2. Now, let X be defined over a global field k, and let $\Pi_{\mathfrak{X}}^{\mathrm{aut}}$ be the automorphic spectrum of $\mathfrak{X}$, as in § 2.1.3. We recall that its definition is, in general, conditional on developing the spectral decomposition of the relative trace formula. Nonetheless, one can often restrict to parts of the spectrum where the full relative trace formula is not needed; for example, if π is a discrete automorphic representation where the period pairing (1.1) is absolutely convergent, the corresponding functional-valued measure $J_\bullet^{\mathrm{aut}}\mu_{\mathfrak{X}}^{\mathrm{aut}}$ of (2.4), applied to a test function $\Phi_1 \otimes \Phi_2$ on $(X \times X)(\mathbb{A})$, should have the meaning of

$$J_\pi^{\mathrm{aut}}(\Phi_1 \otimes \Phi_2)\mu_{\mathfrak{X}}^{\mathrm{aut}}(\pi) = \sum_f \left(\int_{[G]} \Theta_{\Phi_1} f\right)\left(\int_{[G]} \Theta_{\Phi_2} \bar{f}\right),$$

where f runs over an orthonormal basis of π.

A landmark in our understanding of these global relative characters was the paper [36] of Ichino and Ikeda, generalizing the formula of Waldspurger [85] to a precise conjectural Euler factorization of J_π^{aut}, in the case of orthogonal Gross–Prasad periods. The conjectures of Ichino and Ikeda gave rise to the realization that there was a general pattern in the Euler factorization of automorphic periods, and were quickly adapted to other cases. Unlike the orthogonal case, which remains open, the conjecture for unitary Gross–Prasad periods has been proven in [8, 10, 90], its analog for Whittaker periods of metaplectic and unitary groups was proven in [53, 54, 59], and there are significant partial results in many other cases.

A generalization of the Ichino–Ikeda conjecture to a wide range of spherical periods (satisfying certain conditions) was proposed in [77]. As in the local case, it lacks the precision of conjectures known in special cases, hence leaving an open problem that should be addressed in the near future. On the other hand, the conjecture of [77] makes clear the connection between the (global) relative trace formula and the (local) Plancherel formula. I will formulate a variant of this conjecture here, using the hypothetical notion of global Arthur parameters, and being a bit vague on choices of measures (see [77, §17] for some hints). Its formulation also relies on Conjecture 3.1.1 below, expressing the local Plancherel density of the basic function $\Phi_{0,v} \in \mathcal{F}(X(k_v))$ at almost every place v in terms of a local L-value $L_X(\phi_v) := L(\phi_v, \rho_X, 0)$, where ϕ_v is a local unramified Langlands parameter into ${}^L G_X$ and $\rho_X : {}^L G_X \to \mathrm{GL}(V_X)$ is a certain representation of the L-group of X.

Conjecture. *There is a decomposition*

$$\mathrm{RTF}_{\mathfrak{X}} = \int J_\phi^{\mathrm{aut}} \mu_X^{\mathrm{aut}}(\phi),$$

where μ_X^{aut} *is a measure on the set of global Arthur parameters which factor as*

$$\mathcal{L}_k \times \mathrm{SL}_2 \xrightarrow{\phi} {}^L G_X \times \mathrm{SL}_2 \xrightarrow{(2.5)} {}^L G$$

(with ϕ *lying over the identity map for the projections to* SL_2*), and* J_ϕ^{aut} *is a sum of relative characters* $\mathcal{S}(\mathfrak{X}(\mathbb{A})) \to \pi \hat{\otimes} \bar{\pi} \to \mathbb{C}$ *for automorphic representations* π *belonging to the corresponding Arthur packet.*

Moreover, when $\mathfrak{X}$ *is stable, the restriction of* $J_\phi^{\mathrm{aut}} \mu_X^{\mathrm{aut}}(\phi)$ *to the* most tempered *Arthur type (*$\phi(\mathrm{SL}_2) = \mathrm{SL}_2$*),* away from the poles of $L_X(\phi|_{\mathcal{L}_k})$ *is equal to*

$$\frac{1}{|S_\phi|} \prod_v' J_{\phi_v} \cdot \mu_{{}^L G_X}(\phi), \tag{2.8}$$

where $\mu_{{}^L G_X}$ *is the natural measure on the set of such parameters,* S_ϕ *is the stabilizer of* ϕ *in* $\check{G}_X$*, and the factors* J_{ϕ_v} *of the Euler product are the local Plancherel relative characters of Conjecture* 2.3.1.

"Stable," here, means that the stabilizers of generic points have trivial Galois cohomology; one can drop this assumption, replacing $\mathrm{RTF}_{\mathfrak{X}}$ by its (properly defined) stable analog. The Euler product of the conjecture needs to be understood, outside of a finite set S of places, as the partial L-value $L_X^S(\phi)/L^S(\phi, \check{\mathfrak{g}}_X, 1)$, according to Conjecture 3.1.1 below. The conjecture can be generalized to other quantizations of suitable Hamiltonian spaces, such as the theta correspondence, where it was shown in **[72]** to follow from a version of the Rallis inner product formula proven in **[29, 89]**. The conjecture is compatible with earlier results and methods for computing period integrals, such as the "unfolding" method **[77, §18]**, or the work of Jacquet and Feigon–Lapid–Offen on unitary periods **[7, 23, 39]**.

3. THE L-VALUE OF A SPHERICAL VARIETY

3.1. Plancherel density of the basic function

3.1.1. It is a very interesting problem to relate the Euler factors of (2.8)—that is, the local Plancherel densities—to special values of local L-functions at every place, including ramified and Archimedean ones. However, we will confine ourselves here to the calculation of the local Plancherel density of the *basic function* $\Phi_0 \in \mathcal{F}(X(F))$, for a local non-Archimedean field F. We assume that G, X are defined over the integers $\mathfrak{o}$ of F, with G reductive, and recall that the basic function is equal to $1_{X(\mathfrak{o})}$, when X is smooth and affine; in general, it is the "IC function," see § 3.1.3 below. We assume that the map $\check{A}_X \to \check{A}$ is injective (§ 2.2.2).

For simplicity of presentation, we will assume that G is split, so that the maximal compact subgroup $\check{A}_X^1 \subset \check{A}_X$ is identified with the group of unramified unitary characters of A_X. The unramified representations appearing in Conjecture 2.3.1 are those obtained by unitary induction of those characters from the parabolic $P(X)$ (§ 2.2.4) through the quotient $P(X) \to A_X$, and the "natural measure" of the conjecture, restricted to unramified parameters (i.e., to $\check{A}^1_X/W_X$), reads

$$\mu_{\check{G}_X}(\phi) = \frac{L(\phi, \check{\mathfrak{g}}_X/\check{\mathfrak{a}}_X, 1)}{L(\phi, \check{\mathfrak{g}}_X/\check{\mathfrak{a}}_X, 0)} d_{\text{Haar}}\phi.$$

Conjecture. *Let X be an affine spherical variety satisfying the conditions above, with a good model over $\mathfrak{o}$, and Φ_0 its basic function. There is a representation $\rho_X : {}^L G_X \to \mathrm{GL}(V_X)$ such that, setting $L_X(\phi) = L(\phi, \rho_X, 0)$, the Plancherel decomposition of Φ_0 reads*

$$\|\Phi_0\|^2 = \int_{\check{A}^1_X/W_X} \frac{L_X(\phi)}{L(\phi, \check{\mathfrak{g}}_X, 1)} \mu_{\check{G}_X}(\phi). \tag{3.1}$$

We refrain from giving details on the precise normalization of Haar measures, or the precise meaning to "good model;" at a minimum, the conjecture should be valid at almost every place for any global model. This, in particular, will identify almost every Euler factor of Conjecture 2.3.2 as a quotient of special values of local L-functions. Note that the "true" point of evaluation of L_X is not 0, but is encoded in ρ_X, which is a representation of the full L-group. Here, this representation would factor through the unramified quotient, and the "true" point of evaluation depends on the action of Frobenius.

The relation between local L-values and Plancherel densities is a fascinating one. On the surface, it is just the outcome of a local integral. For example, when $X(\mathfrak{o}) = H(\mathfrak{o})\backslash G(\mathfrak{o})$, the value $J_\phi(1_{X(\mathfrak{o})})$ is given by the following Ichino–Ikeda local period, in the so-called *strongly tempered* cases where it is convergent:

$$J_\phi(1_{X(\mathfrak{o})}) = \int_H m_\phi(h)dh,$$

where m_ϕ is the zonal spherical function (= unramified matrix coefficient with value 1 at the identity) for the unramified representation with Satake parameter ϕ. This calculation, however, has a conceptual meaning, in terms of both harmonic analysis and geometry. We will only attempt to give a flavor of the richness of the topic here.

3.1.2. The study of the Plancherel density of the basic function is a topic with a long history. The mainstream method for calculating it is the Casselman–Shalika method [20, 21], and it is essentially equivalent to the problem of calculating eigenvectors for the unramified (spherical) Hecke algebra $\mathcal{H}(G(F), G(\mathfrak{o}))$ on the space $C^\infty(X(F))$.

The calculation was related to the structure theory of spherical varieties in [70], for G split. Here, we will formulate the result in the special case when X is an excellent spherical variety (§ 2.2.5) with $\check{G}_X = \check{G}$. Fix a Borel subgroup $B \subset G$, with unipotent radical N. The important geometric invariants determining the L-value are the *colors* of the spherical variety X: those are the B-stable prime divisors on X, over the algebraic closure. For simplicity, we will assume all those divisors to be defined over F. Each such divisor D induces a valuation on the function field $F(X)$, which we restrict to the multiplicative group of nonzero B-eigenfunctions. This gives rise to a homomorphism factoring as

$$F(X)^{(B)} \to \mathcal{X}^\bullet(A_X) \to \mathbb{Z},$$

i.e., an element $\check{v}_D \in X_\bullet(A_X)$, the character group of the dual torus $\check{A}_X$ (which here is equal to $\check{A}$). By [78, COROLLARY 7.3.4], the weights $\check{v}_D$ are all minuscule. Let V_X be the smallest self-dual (algebraic) representation of $\check{G}$ which contains all those weights (with multiplicity, if some of the $\check{v}_D$'s coincide). For an alternative interpretation of this representation, in terms of the structure of the Hamiltonian space T^*X, see Theorem 6.2.2 below. The following was proven in [70] under some assumptions, and in [78] in general:

Theorem. *The Plancherel density of the basic function $1_{X(\mathfrak{o})}$ is given by*

$$J_\phi(1_{X(\mathfrak{o})} \otimes 1_{X(\mathfrak{o})})\mu_{\check{G}}(\phi) = \frac{L(\phi, V_X, \frac{1}{2})}{L(\phi, \check{\mathfrak{g}}, 1)}\mu_{\check{G}}(\phi). \tag{3.2}$$

Note that, for simplicity, we have assumed that $\check{G}_X = \check{G}$. The point $\frac{1}{2}$ of evaluation changes in the general case.

3.1.3. The case of *singular* affine varieties X was undertaken in [78], for the cases with $\check{G}_X = \check{G}$. As mentioned, here one needs to work in a geometric setting, assuming that F is a local field in equal characteristic, $F \simeq \mathbb{F}_q((t))$, with G, X defined over $\mathbb{F}_q$. (There are ad hoc ways to transfer the results to mixed characteristic, but it would be nice to see a direct geometric approach.) The basic function Φ_0 is then defined as the Frobenius trace on the stalks of the intersection complex of finite-dimensional formal models of L^+X, the formal arc space of X [12].

Let us discuss the special case when X is the affine closure Spec $\mathbb{F}_q[X^\bullet]$ of its open G-orbit $X^\bullet$. Colors, here, do not need to be minuscule, but one can still define V_X as before. We have the following generalization of Theorem 3.1.2:

Theorem. *There is a representation V'_X of $\check{A}$, with the same weights as V_X and W-invariant multiplicities, such that the Plancherel density of the basic function Φ_0 is*

$$J_\phi(\Phi_0 \otimes \Phi_0)\mu_{\check{G}}(\phi) = \frac{L(\phi, V'_X, \frac{1}{2})}{L(\phi, \check{\mathfrak{g}}, 1)}\mu_{\check{G}}(\phi). \tag{3.3}$$

Of course, we expect that $V'_X = V_X$. This is automatic in the minuscule case. For example [78, EXAMPLE 1.1.3], there is a family of varieties X_n, $n \in \mathbb{N}$, which gives rise to the generalization of the Hecke and Rankin–Selberg integrals, mentioned in § 1.2.2.

3.2. Derived Satake equivalence for spherical varieties

Ongoing joint work with Ben-Zvi and Venkatesh has revealed deeper relations between periods and L-functions; currently, those can be formulated over function fields and their completions. In the local setting, $F \simeq \mathbb{F}_q((t))$, Conjecture 3.1.1 should be obtained by applying the sheaf–function dictionary to a categorical statement, along the following lines:

We retain the assumptions of the previous subsection, with X and G defined over $\mathbb{F}_q$, and also assume X to be smooth. We denote formal loop and arc spaces by L, resp. L^+, so that $LX(\mathbb{F}_q) = X(F)$, $L^+G(\mathbb{F}_q) = G(\mathfrak{o})$. For appropriate measures, the left-hand side of (3.1), and, more generally, the pairing of two $G(\mathfrak{o})$-invariant functions f, g obtained via the sheaf–function dictionary from objects $\mathscr{F}, \mathscr{G}$ in the bounded derived category

$\mathrm{Shv}(LX/L^+G)$ of constructible ℓ-adic étale sheaves on LX/L^+G can be computed as the (alternating) trace of geometric Frobenius on a derived homomorphism complex:

$$\int_{X(F)} f(x)g(x)dx = \mathrm{tr}\big(\mathrm{Frob}_q, \mathrm{Hom}(\mathscr{F}, D\mathscr{G})^*\big),$$

where $D =$ Verdier dual. The pairing is really a finite sum, and makes sense over $\overline{\mathbb{Q}_\ell}$.

The right-hand side of (3.1), through a simple application of the Weyl integration formula, can be interpreted as the Frobenius trace on

$$\mathbb{C}[V_X]^{\check{G}_X} = \mathbb{C}[\check{M}]^{\check{G}},$$

where V_X is the space of the representation ρ_X, and we have set

$$\check{M} = V_X \times^{\check{G}_X} \check{G}.$$

The empirical observation is that the space $\check{M}$ *has a natural symplectic structure*, and, moreover, that *the assignment* $M = T^*X \to \check{M}$ *is involutive*, although, to make sense of this, one needs to allow for more general coisotropic Hamiltonian spaces, as mentioned in § 1.2.4. For the categorical analog of Conjecture 3.1.1, we need to *shear* the ring $\mathbb{C}[\check{M}]$ into a dg-algebra $\mathbb{C}[\check{M}]^{/\!/}$, with zero differentials, in degrees related to the action of Frobenius in ρ_X.

Conjecture. *Fix an isomorphism* $\mathbb{C} = \overline{\mathbb{Q}_\ell}$. *There is an equivalence of triangulated* $\mathbb{C}$*-linear categories*

$$\mathrm{Shv}(LX/L^+G) \xrightarrow{\sim} D^{/\!/}_{\mathrm{per}}(\check{M}/\check{G}),$$

where $D^{/\!/}_{\mathrm{per}}(\check{M}/\check{G})$ *denotes the full triangulated subcategory, generated by perfect complexes, of the category of* $\check{G}$*-equivariant differential graded* $\mathbb{C}[\check{M}]^{/\!/}$*-modules localized by quasi-isomorphisms.*

This generalizes the derived Satake equivalence of Bezrukavnikov–Finkelberg **[11]**; it should be compatible with it, under the action of $\mathrm{Shv}(L^+G\backslash LG/L^+G)$ on the left, and the moment map $\check{M} \to \check{\mathfrak{g}}^*$ on the right. There is a similar, categorical version of the global Conjecture 2.3.2, for which I defer to the upcoming article.

4. BEYOND ENDOSCOPY

4.1. Relative functoriality

4.1.1. Let X, Y be two spherical varieties (for possibly different groups G, G'), and let r be a morphism of their L-groups, $r : {}^LG_X \to {}^LG_Y$. According to the relative local Langlands conjecture of § 2.3.1, it should give rise to a map

$$\{X\text{-distinguished } L\text{-packets}\} \to \{Y\text{-distinguished } L\text{-packets}\},$$

at least for L-packets distinguished in the L^2-sense.

A basic tenet of Langlands' "beyond endoscopy" proposal **[51]**, generalized to the relative setting, states that the resulting map of stable relative characters $J^X_{\phi_1} \mapsto J^Y_{\phi_2}$ should be realized as the adjoint of a "transfer operator" between spaces of stable test measures,

$$\mathcal{T} : \mathcal{S}(\mathfrak{Y})^{\mathrm{st}} \to \mathcal{S}(\mathfrak{X})^{\mathrm{st}}, \tag{4.1}$$

where $\mathfrak{Y}$ denotes the stack $(Y \times Y)/G'$, and $\mathfrak{X} = (X \times X)/G$. In most cases, one can take "stable" to mean the image of the canonical pushforward map

$$\mathcal{S}(\mathfrak{X}) \to \text{Measures}\big((X \times X) /\!\!/ G\big).$$

When the map $\mathcal{S}(\mathfrak{X}) \to \mathcal{S}(\mathfrak{X})^{\text{st}}$ is an isomorphism (e.g., for the Kuznetsov formula), we will be dropping the exponent "st."

In the group case, this operator has been studied by Langlands [52] and Johnstone [43] when X is a torus and Y is GL_n. Understanding these transfer operators could be considered as the basic problem of functoriality, at least in the local setting.

4.1.2. In the global setting, one would have to find a way to employ these transfer operators in a comparison of relative trace formulas. Langlands' proposal, generalized to our setting, is to extract the part of the automorphic spectrum of $\mathfrak{Y}$ that is in the image of the functorial lift from $\mathfrak{X}$ from the (stable) relative trace formula for $\mathfrak{Y}$ by means of poles of L-functions.

The question of whether it is possible to identify the spectrum of $\mathfrak{X}$ by orders of poles of L-functions has been studied and is known to have a negative answer, in general [5]. Other difficulties with this proposal include the isolation of the tempered part of the spectrum; a lot of hard work has gone into this problem, already for the case of GL_2 [2–4, 24].

4.2. An example: symmetric square lift

4.2.1. Rather than speculating on how to overcome these difficulties, it may be more instructive to look at a variant of the idea, which was applied successfully in the thesis of Venkatesh [84], and to understand what the structure of local transfer operators can tell us about the global problem. Here, $X = T$ is a 1-dimensional torus over a global field k (the kernel of the norm map for a quadratic etale algebra E/k whose quadratic idele class character we will denote by η), and Y is the Whittaker model of the group $G = \mathbb{G}_m \times \mathrm{SL}_2$, so that $\check{G}_Y = \check{G}$. There is a morphism of L-groups $r : {}^LT \to {}^LG$, whose image stabilizes a vector under the product of the standard representation of $\mathbb{G}_m$ with the adjoint representation of PGL_2 (that is, the symmetric-square representation of GL_2, as it factors through $\mathrm{GL}_2 \to \check{G} = \mathbb{G}_m \times \mathrm{PGL}_2 \to \mathrm{GL}_3$). Let $\mathbb{Z}/2$ act on T by inversion. The local transfer operator for this morphism was computed in [76]:

Theorem. *Let G, T as above be defined over a local field F. There is a transfer operator*

$$\mathcal{T} : \mathcal{S}(N, \psi \backslash G / N, \psi) \to \mathcal{S}(T)^{\mathbb{Z}/2},$$

such that the pullback of every unitary character of T is the Kuznetsov relative character of its functorial lift. In natural coordinates (r, t) for $N \backslash G /\!\!/ N \simeq \mathbb{G}_m \times \mathbb{A}^1$, it is given by

$$(da)^{-1} \mathcal{T} f(a) = (dt)^{-1} \lambda(\eta, \psi) \int_r \int_x f\left(r, \frac{t}{x}\right) \eta(xrt) \psi(x)\, dx, \tag{4.2}$$

where $a \in T$ and $t = t(a)$ is its image through an isomorphism $T /\!\!/ (\mathbb{Z}/2) \simeq \mathbb{A}^1 \simeq N \backslash \mathrm{SL}_2 /\!\!/ N$, and $\lambda(\eta, \psi)$ is a constant.

What does this theorem tell us about how to extract from the relative trace formula of Y (that is, the Kuznetsov formula of G) the part of the spectrum that is due to the torus T? Venkatesh [84] performs this extraction in two steps, a Poisson summation formula followed by taking the pole of a zeta integral. As explained in [76, §10], the adelic reformulation of the first step is the Poisson summation formula for the Fourier transform corresponding to the inner integral of (4.2), while the second step is a global version of the Mellin transform represented by the outer integral.

4.2.2. Thus, we see that understanding the local transfer operators can guide our steps for the global "beyond endoscopy" comparisons of trace formulas. Another example of such a comparison is that between the Kuznetsov formula of $G = \mathrm{GL}_2$ and the Selberg trace formula for the same group. The local transfer operator for this comparison was computed in [75, §4], and is given by a simple Fourier tranform (see Theorem 5.3.5). Restricted to holomorphic cusp forms, this global comparison via a Poisson summation formula was performed in the thesis of Zeev Rudnick [67], about 10 years before Langlands' "beyond endoscopy" proposal. A generalization of this comparison to the full Kuznetsov formula, and for GL_n with n arbitrary, is the object of ongoing joint work with Chen Wan.

4.2.3. It has hopefully become clear that understanding the transfer operators is of paramount importance for the problem of functoriality. In [74], I showed that these operators have a very uniform form, for spherical varieties of rank 1. In the remainder of this paper, I would like to propose a reinterpretation of this work, which provides an understanding of those transfer operators as "change of Schrödinger model/geometric quantization" associated to a symplectic group scheme.

5. TRANSFER OPERATORS AND QUANTIZATION

The goal of this section is to recast the transfer operators of functoriality, studied in [74], in the language of quantization. The idea that quantization should have something to do with functoriality is not new; V. Lafforgue suggested it several years ago (private communication), in order to interpret the Rankin–Selberg method, and the functoriality kernels of L. Lafforgue [49]. Here, however, we apply this idea in a different setting: the setting of "beyond endoscopy," and of the quotient stacks showing up in the relative trace formula—the hope being that these operators of functoriality will always exist in this setting, even if they do not exist for the spaces "upstairs."

Geometric quantization was introduced by Kostant and Souriau [48, 82], following the work of Kirillov on the orbit method [44]. Since the notion of quantization for measures on stacks that we need has not been developed yet, we will take a phenomenological approach, with ad hoc definitions that provide the desired reformulation of the results of [74].

5.1. Cotangent space of the RTF stack

5.1.1. The groundbreaking work of Friedrich Knop has shown that, although spherical varieties can be very different from each other, their *cotangent bundles* are quite similar.

This will be the basis of our considerations, when we try to relate cotangent bundles of different quotient stacks of the form $(X \times X)/G$.

For the rest of this paper we will assume, for simplicity, that all groups are split, defined over a field F in characteristic zero. The results of Knop, then, recalled in § 2.2.3, hold verbatim over F. We assume that X is smooth and quasiaffine, and set again $\mu : M = T^*X \to \mathfrak{g}^*$ for the cotangent bundle and its moment map.

The ring of regular functions $F[M]$ has a Poisson structure. Knop has studied the subalgebra $F[M]^G$ of G-invariants; when X is spherical, this subalgebra is Poisson-commutative, and can be naturally identified with the algebra of regular functions on the affine space $\mathfrak{c}_X^*$, defined in § 2.2.3. Hence, regular functions on $\mathfrak{c}_X^*$ pull back to a Poisson-commuting algebra of G-invariant functions ("Hamiltonians") on M. One can ask whether the corresponding Hamiltonian vector fields can be integrated to the action of an abelian group scheme J_X (over $\mathfrak{c}_X^*$) of G-automorphisms of M, and Knop has answered this in the affirmative [46].

5.1.2. More precisely, J_X is "the group scheme of regular centralizers in the split reductive group G_X dual to $\check{G}_X$." When $\check{G}_X$ is adjoint, so that G_X is simply connected, the group scheme J_X has an explicit description as

$$(\mathrm{Res}_{\mathfrak{a}_X^*/\mathfrak{c}_X^*} T^*A_X)^{W_X} \tag{5.1}$$

(see [61, §2.4]), where $\mathrm{Res}_{\mathfrak{a}_X^*/\mathfrak{c}_X^*}$ denotes Weil restriction of scalars from $\mathfrak{a}_X^*$ to $\mathfrak{c}_X^*$.

In general, the group scheme J_X acting on M is an open subgroup scheme of (5.1), which depends not only on the pair (A_X, W_X), but also on the root datum of X. Knop defines a slightly different root datum than ours in [46, §6], giving the maximal possible subgroup scheme acting on M. For our purposes, we will be content with taking $J_X =$ the open subgroup scheme of (5.1) that corresponds to the set of normalized spherical roots, see § 2.2.2. This can be described as the regular centralizer group scheme of G_X, and is the complement of a divisor in (5.1), see [46, THEOREM 7.7], [61, §2.4].

5.1.3. Let us discuss the rank-1 cases. Consider first the case $\check{G}_X = \mathrm{PGL}_2$, so that $A_X = \mathbb{G}_m$, $W_X = \mathbb{Z}/2$, and the normalized spherical root is twice the generator of the character lattice. (The isomorphism $A_X \simeq \mathbb{G}_m$ is canonical, if we require positive roots to correspond to positive powers.) Then, $J_X = J_{\mathrm{SL}_2}$ is given by the restriction of scalars (5.1), which can explicitly be described as follows: Identify $\mathfrak{a}_X^* = \mathfrak{g}_m^*$ with the affine line, with coordinate $\sigma =$ the differential of the identity cocharacter, and set $\xi = \sigma^2$, a coordinate on $\mathfrak{c}_X^*$. We can write

$$J_X = \mathrm{Spec}\, F[t_0, t_1, \xi]/(t_0^2 - \xi t_1^2 - 1),$$

so that the canonical base change map

$$J_X \bullet \mathfrak{a}_X^* \to T^*A_X = T^*\mathbb{G}_m = \mathbb{G}_m \times \mathfrak{g}_m^*,$$

where by $\bullet$ we denote fiber product over $\mathfrak{c}_X^*$, is given by $(t_0, t_1, \sigma) \mapsto (t_0 + \sigma t_1, \sigma)$.

The symplectic form is given by

$$\omega = \frac{dt_1 \wedge d\xi}{2t_0} = \frac{dt_0 \wedge d\xi}{2\xi t_1} = dt_0 \wedge d(t_1^{-1}).$$

It is immediate to check that this is regular and nondegenerate everywhere on J_X.

On the other hand, when the normalized root datum of X is that of PGL_2 (i.e., the normalized spherical root is a generator of the character lattice), the fiber of (5.1) over the nilpotent point $0 \in \mathfrak{c}_X^*$ is isomorphic to $\mathbb{G}_a \times \{\pm 1\}$, but the fiber of J_X is just $\mathbb{G}_a$.

5.1.4. Returning to the general case, the Lie algebra of J_X is canonically isomorphic to the cotangent space of $\mathfrak{c}_X^*$. Thus, the Hamiltonian vector fields associated to $F[\mathfrak{c}_X^*]$ give rise to a homomorphism from $\mathrm{Lie}(J_X)$ to G-invariant vector fields along the fibers of $M \to \mathfrak{g}_X^*$ (notation as in § 2.2.2). Knop has shown [46] that these vector fields integrate to an action of J_X on $M = T^*X$ over $\mathfrak{g}_X^*$, commuting with the action of G.

Moreover, over a dense open subset $\mathring{\mathfrak{c}}_X^* \subset \mathfrak{c}_X^*$, the map $M \to \mathfrak{g}_X^*$ is a J_X-torsor, and the action of J_X arises from the stabilizers of points of $\mathfrak{g}_X^*$, or even of $\mathfrak{g}^*$, in G, i.e., the stabilizer G_z of a generic point $z \in \mathfrak{g}^*$ in the image of the moment map acts transitively on the fiber through a map $G_z \to J_X$.

5.1.5. In order to study the relative trace formula for a stack of the form $\mathfrak{X} = (X \times X)/G$, I propose to use its cotangent stack

$$T^*\mathfrak{X} = (T^*X \times_{\mathfrak{g}^*} T^*X)/G.$$

(Strictly speaking, the fiber product over $\mathfrak{g}^*$ should taken with respect to the moment map and its negative, $(\mu, -\mu)$, so that the quotient above corresponds to symplectic reduction with respect to the diagonal action of G. We apply multiplication by -1 on the fibers of one factor, in order to have fiber product with respect to (μ, μ), which is notationally simpler.)

The fiber product, here, should be taken in the derived sense, turning this quotient into a symplectic derived stack. Although the derived features are likely to be important in the future, for the purposes of the current paper we will ignore them. Then, the fiber product over $\mathfrak{g}^*$ coincides with the fiber product over $\mathfrak{g}_X^*$ over a dense open $\mathring{\mathfrak{c}}_X^* \subset \mathfrak{c}_X^*$, which we will take small enough so that it also has the properties of § 5.1.4.

Knop's theory, now, gives a very satisfactory description of a dense substack. Namely, consider the diagonal embedding $T^*X \hookrightarrow T^*X \times_{\mathfrak{g}^*} T^*X$, which by the action of J_X on the first variable gives rise to

$$J_X \bullet T^*X \to T^*X \times_{\mathfrak{g}^*} T^*X, \tag{5.2}$$

where again by $\bullet$ denotes $\times_{\mathfrak{c}_X^*}$. Given that $T^*X \to \mathfrak{g}^*$ is generically a J_X-torsor over its image, this map is birational into an irreducible component of the right hand side.

Taking quotients by the G-action in (5.2), and using the invariant moment map $T^*X \to \mathfrak{c}_X^*$, we obtain a correspondence

$$J_X \leftarrow (J_X \bullet T^*X)/G \to T^*\mathfrak{X} = (T^*X \times_{\mathfrak{g}^*} T^*X)/G. \tag{5.3}$$

Over the dense open subset $\mathring{\mathfrak{c}}_X^*$, the right arrow is an isomorphism, and the left arrow is an isomorphism if we ignore stabilizers. We would like to think of (5.3) as saying that the symplectic stack $T^*\mathfrak{X}$ is "birational" to the symplectic group scheme J_X. Of course, this is a very naive notion of birationality since, even over a good open dense subset, we are ignoring stacky and derived structures in $T^*\mathfrak{X}$. Nonetheless, even this weak correspondence is quite remarkable, since J_X depends only on the dual group of X. *It is not too far-fetched to imagine that this correspondence plays a key role in functoriality.*

5.2. Rank 1 spherical varieties

5.2.1. We will now specialize to spherical varieties $X = H\backslash G$ with G and H split reductive groups, whose dual group is $\check{G}_X = \mathrm{PGL}_2$ or SL_2. The group scheme J_X was described in § 5.1.3.

In this setting, [74] gave explicit formulas for transfer operators (4.1) between $\mathfrak{Y}$ = the Kuznetsov stack for the group with dual group $\check{G}_X$ (see § 5.2.2 below), and $\mathfrak{X} = (X \times X)/G^{\mathrm{diag}}$. These operators transfer spaces of test measures to each other, but in a number of cases, studied in [30,75,76], properties such as the transfer of characters or the appropriate fundamental lemma are also known; thus, there is enough evidence to believe that these are the "correct" operators of functoriality for these comparisons.

The simplest form of transfer operators appears when the normalized root datum of X is simply connected, i.e., $\check{G}_X = \mathrm{PGL}_2$. For some varieties, this is possible to achieve by passing to a finite cover; the only cases where this can be done lead to the following spaces:

$$X = \mathrm{SO}_{2n-1} \backslash \mathrm{SO}_{2n},\ \mathrm{Spin}_7 \backslash \mathrm{Spin}_8,\ \mathrm{G}_2 \backslash \mathrm{Spin}_7. \tag{5.4}$$

We will call them "cases of type G," because the base case is the group variety $\mathrm{SO}_3 \backslash \mathrm{SO}_4 \simeq \mathrm{SL}_2$. (Here, the $\mathrm{SO}_7 \backslash \mathrm{SO}_8 \simeq \mathrm{Spin}_7 \backslash \mathrm{Spin}_8 \simeq \mathrm{G}_2 \backslash \mathrm{Spin}_7$ as varieties, but with the action twisted by the triality automorphism of Spin_8, for the second, and restricted to the subgroup Spin_7, for the third.)

The remaining cases of spherical varieties satisfying our assumptions are

$$X = \mathrm{GL}_n \backslash \mathrm{PGL}_{n+1},\ \mathrm{SO}_{2n} \backslash \mathrm{SO}_{2n+1},\ \mathrm{Sp}_{2n-2} \times \mathrm{Sp}_2 \backslash \mathrm{Sp}_{2n}\ (\text{with } n \geq 2),$$
$$\mathrm{Spin}_9 \backslash F_4,\ \mathrm{SL}_3 \backslash G_2. \tag{5.5}$$

We will call them "cases of type T," because in the base case $\mathbb{G}_m \backslash \mathrm{PGL}_2$ the stabilizer is a torus. (Again, $\mathrm{SL}_3 \backslash G_2 \simeq \mathrm{SO}_6 \backslash \mathrm{SO}_7$, but with the action restricted to $G_2 \subset \mathrm{SO}_7$.)

In those cases, the normalized root datum of X is that of PGL_2, and, as we will see, it will be necessary to "lift" our description of transfer operators to the root datum of GL_2.

5.2.2. Let X be as above. Let G' be the split reductive group with the same dual group as X, that is, $G' = \mathrm{SL}_2$ for the varieties of (5.4) and $G' = \mathrm{PGL}_2$ for the varieties of (5.5). Let $N \subset G'$ be the upper triangular unipotent subgroup, identified with the additive group $\mathbb{G}_a$, fix a nontrivial character ψ of F, and let Y be the Whittaker model of G' with respect to (N, ψ), § 2.2.5. Let Y^- be the Whittaker model with respect to the inverse character, ψ^{-1}.

We will symbolically write $\mathfrak{Y} = (Y \times Y^-)/G'$ for the "Kuznetsov stack," but we will really treat it not as an abstract quotient stack, but as one equipped with the line bundle defined by the Whittaker character. More precisely, this symbol will only find a rigorous meaning in its Schwartz space $\mathcal{S}(\mathfrak{Y})$, which we define to be the G'^{diag}-coinvariant space of the space $\mathcal{S}(Y \times Y^-)$ of Whittaker Schwartz measures.

Having identified N with $\mathbb{G}_a$, we let $M' = (f + \mathfrak{n}^\perp) \times^N G'$ be the *Whittaker cotangent bundle*, where f is a nilpotent element of $(\mathfrak{g}')^*$ that is equal to the identity functional on $\mathfrak{n} = \mathbb{G}_a$. The corresponding bundle for Y^- is $M'^- = (-f + \mathfrak{n}^\perp) \times^N G'$. Both come equipped with natural moment maps to $\mathfrak{g}'^*$, which we will indiscriminately denote by μ. We now define the *Kuznetsov cotangent stack* as

$$T^*\mathfrak{Y} = \left(M' \times_{\mu, \mathfrak{g}'^*, (-\mu)} M'^-\right)/G' \simeq (M' \times_{\mu, \mathfrak{g}'^*, \mu} M')/G'.$$

Note that the invariant-theoretic quotient $\mathfrak{g}' /\!/ G'$ is canonically identified with $\mathfrak{c}_X^*$, and the group scheme of regular centralizers in G' is canonically identified with J_X. It is well known that M' is a J_X-torsor over the regular subset of $\mathfrak{g}'^*$; in fact, a Kostant section provides a section for this torsor. Therefore, the same considerations that led us to (5.3) hold, but here we have an exact isomorphism

$$J_X \simeq T^*\mathfrak{Y}. \tag{5.6}$$

Our hope, now, is to demonstrate the following idea:

> The cotangent stacks $T^*\mathfrak{X}$ and $T^*\mathfrak{Y}$ being roughly isomorphic to J_X (by (5.3), (5.6)), there is a transfer operator of functoriality
>
> $$\mathcal{T} : \mathcal{S}(\mathfrak{Y}) \to \mathcal{S}(\mathfrak{X})^{\text{st}},$$
>
> corresponding to a "change of geometric quantization" for J_X.

5.2.3. Quantization, of course, is as much of a science as an art, and the reader should not expect a rigorous formulation of this hope in this article. In particular, the type of geometric quantization that we need (suitable for encoding measures on stacks) has not, to my knowledge, been developed, yet. Therefore, the real content of the results that follow is already contained in [74]; but we will dress them up in an ad hoc language of quantization, in order to exhibit some deeper structure that seems to be lying behind them.

We will also assume, from now on, that our base field is $F = \mathbb{R}$, in order to use the language of line bundles with connection, and will write the chosen additive character as $\psi(x) = e^{i\hbar x}$, where $\hbar$ is a nonzero real constant. The final results, contained in Theorems 5.3.5 and 5.4.3, are valid and were proven in [74] over an arbitrary local field in characteristic 0, just by an obvious translation of the formulas. We will only care to describe transfer operators up to an absolute scalar; therefore, we will feel free to choose measures that only modify the result by a scalar, without commenting on those choices.

5.3. Geometric quantization for type G

5.3.1. The process of geometric quantization on a (real) symplectic manifold (M, ω) consists in fixing a (complex) Hermitian vector bundle L, equipped with a connection ∇ whose curvature is $i\hbar\omega$, as well as a Lagrangian foliation $\mathcal{F}$, such that the space of leaves $M/\mathcal{F}$ is a Hausdorff manifold. Then, one attaches to these data the vector space $\mathcal{D}_{\mathcal{F}}(M, L)$ of smooth half-densities on $M/\mathcal{F}$ valued in the space of sections of L over M that are constant along the foliation $\mathcal{F}$ (with respect to the connection). This space has a canonical inner product (namely, the L^2-inner product over $M/\mathcal{F}$), giving rise to a Hilbert space, by completion.

To reformulate the results of [74] in this language, we will now recast the space $\mathcal{S}(\mathfrak{X})^{\mathrm{st}}$ of stable test measures for the relative trace formula as a space of half-densities on the quotient of the group scheme J_X by a Lagrangian foliation, valued in a line bundle L_X.

5.3.2. Fix a rank-1 space X "of type G," i.e., in the list (5.4).

Consider the composition of maps

$$J_X \bullet T^*X \to T^*X \times_{\mathfrak{g}^*} T^*X \to X \times X \tag{5.7}$$

induced from (5.2). There is a natural scaling $\mathbb{G}_m$-action on the left-hand side, under which this composition is invariant, and if we consider the "projectivization" of the space on the left (= remove the zero section in T^*X and divide by $\mathbb{G}_m$), it was shown in [74, §3] that the resulting map

$$\mathbb{P}\big(J_X \bullet T^*X\big) \to X \times X \tag{5.8}$$

is generically an isomorphism. More precisely, in the type-G cases it is an isomorphism over $\mathbb{P}(J_X^\circ \bullet T^*X)$, where $J_X^\circ \subset J_X$ is the complement of the divisor given by the homogeneous equation $t_1 = 0$. (This is a combination of Propositions 3.3.2, 3.5.1 in [74], and the fact that those spaces have an involutive G-automorphism.)

The invariant-theoretic quotient $X \times X \to (X \times X) /\!\!/ G$ is an affine line, and its composition with (5.7) is the map $J_X \bullet T^*X \to J_X \to \mathbb{A}^1$ that remembers only t_0 from the triple (t_0, t_1, ξ) [74, PROPOSITION 3.4.2]. We notice that the level sets of t_0 on J_X° form a Lagrangian foliation; we will call this foliation "vertical," and denote it by $\mathcal{F}_{\mathrm{ver}}$.

5.3.3. Let $d = \dim X$. The short version of the story that follows is that we replace the element $f \in \mathcal{S}(\mathfrak{X})^{\mathrm{st}}$ (a measure in the variable t_0) by

$$f(t_0)(dt_0)^{-\frac{1}{2}}|t_1|^{-\frac{d}{2}+1}, \tag{5.9}$$

obtaining a half-density on $J_X^\circ/\mathcal{F}_{\mathrm{ver}}$ valued in the line bundle L_X whose sections are functions on $J_X^\circ/\mathcal{F}_{\mathrm{ver}}$ multiplied by the factor $|t_1|^{-\frac{d}{2}+1}$. More precisely, L_X will be identified with the trivial line bundle on J_X°, but endowed with a connection

$$\nabla^X = \nabla^0 + d\log|t_1|^{\frac{d}{2}-1} - i\hbar t_1^{-1}dt_0 = \nabla^0 + \left(\frac{d}{2}-1\right)t_1^{-1}dt_1 - i\hbar t_1^{-1}dt_0, \tag{5.10}$$

with curvature $i\hbar\omega$, where ∇^0 is the standard flat connection, so that its parallel sections along the vertical foliation are as described.

Presented this way, this connection is completely unmotivated. In § 5.3.6 below, we will discuss a more natural description of the pair (L_X, ∇^X). Continuing, for now, in this ad hoc fashion, (5.9) defines a map

$$\mathcal{S}(\mathfrak{X})^{\mathrm{st}} \to D_{\mathrm{hor}}(J_X^\circ, L_X), \tag{5.11}$$

where D_{hor} (with regular font) denotes *continuous* (not necessarily Schwartz, or even smooth) "horizontal" half-densities valued in L_X (i.e., half-densities on $J_X^\circ/\mathscr{F}_{\mathrm{ver}}$ valued in the descent of L_X by parallel transport). The image of this map will be denoted by $\mathcal{D}(\mathfrak{X})$.

5.3.4. We consider another Lagrangian foliation $\mathscr{F}_{\mathrm{hor}}$ on J_X°, which we will call "horizontal": its leaves are the level sets of t_1. For the line bundle with connection (L_X, ∇^X), as above, flat sections along horizontal leaves are simply functions of $t_1 \neq 0$, multiplied by the factor $\psi(\frac{t_0}{t_1})$; note that this description is independent of the dimension d used to define ∇^X.

We now propose to think of the space of test measures $\mathcal{S}(\mathfrak{Y})$ for the Kuznetsov formula as a subspace $\mathcal{D}(\mathfrak{Y}) \subset D_{\mathrm{ver}}(J_X^\circ, L_X)$, where D_{ver} denotes continuous "vertical" half-densities (i.e., half-densities on $J_X^\circ/\mathscr{F}_{\mathrm{hor}}$) valued in L_X. First of all, consider the map

$$J_X \bullet M' \xrightarrow{\sim} M' \times_{\mathfrak{g}'^*} M',$$

with notation as in § 5.2.2, where the action of J_X is again on the first copy of M'.

Lemma. *The composition of the map above with $M' \times_{\mathfrak{g}'^*} M' \to Y \times Y \to (Y \times Y) /\!\!/ G$ is the map that only remembers the coordinate t_1 of J_X.*

This is the reason why the foliation $\mathscr{F}_{\mathrm{hor}}$ is relevant to the Kuznetsov formula.

There is a natural pullback from Whittaker functions on $Y \times Y^-$ to scalar-valued functions on $M' \times_{\mathfrak{g}'^*} M'$, as follows: Thinking of elements of $\mathcal{F}(Y)$ (that is, Whittaker functions) as sections of a line bundle L_ψ over $Y = N\backslash G'$ (and similarly for Y^-, just replacing ψ by ψ^{-1}), we note that the line bundle $L_\psi \boxtimes L_{\psi^{-1}}$ is canonically trivial over the diagonal $Y^{\mathrm{diag}} \subset Y \times Y^-$. There is now a unique trivialization of its pullback to $J_X \bullet M'$ which coincides with the canonical one over the diagonal, and is equivariant with respect to the action of $J_X \times G'$. More explicitly, if we use a Kostant section to identify $T^*Y \simeq \mathfrak{c}_X^* \times G'$, and the negative of that section for Y^-, we pull back Whittaker functions to scalar-valued functions on T^*Y via the projection to G', and then restrict to $T^*Y \times_{\mu, \mathfrak{g}'^*, (-\mu)} T^*Y^- \simeq M' \times_{\mathfrak{g}'^*} M'$.

The short version of the story, now, is that we fix a G'-invariant measure on $N\backslash G'$, use it to identify Schwartz (Whittaker) measures on $Y \times Y^-$ with Schwartz (Whittaker) functions, pull them back to scalar-valued functions on $J_X^\circ \bullet M' \simeq J_X^\circ \times G'$, and integrate them against a chosen Haar measure on G'. This gives functions on J_X° that, as can be easily confirmed, correspond to sections of L_X, flat along the leaves of $\mathscr{F}_{\mathrm{hor}}$; further multiplying them by the factor $|t_1|^{\frac{1}{2}} dt_1^{\frac{1}{2}}$ gives rise to an element of $D_{\mathrm{ver}}(J_X^\circ, L_X)$. This descends to an injective map

$$\mathcal{S}(\mathfrak{Y}) \to D_{\mathrm{ver}}(J_X^\circ, L_X), \tag{5.12}$$

whose image will be denoted by $\mathcal{D}(\mathfrak{Y})$. Again, the factor $|t_1|^{\frac{1}{2}} dt_1^{\frac{1}{2}}$ seems unmotivated, and we will attempt to explain it in § 5.3.9 below, after formulating the main theorem.

5.3.5. To recap, we have defined a line bundle L_X on J_X°, endowed with a connection ∇^X with curvature $i\hbar\omega$, "vertical" and "horizontal" foliations $\mathscr{F}_{\mathrm{ver}}$, $\mathscr{F}_{\mathrm{hor}}$ on J_X°, and have identified the spaces $\mathcal{S}(\mathfrak{X})^{\mathrm{st}}$, $\mathcal{S}(\mathfrak{Y})$ of test measures for the corresponding quotients with spaces $\mathcal{D}(\mathfrak{X})$, $\mathfrak{D}(\mathfrak{Y})$ of "horizontal" and "vertical" half-densities for (L_X, J_X°). The main result **[74, THEOREM 1.3.1]** for the transfer operator in this case can now be formulated as follows:

Theorem. *There is an injective operator*

$$\mathcal{T} : \mathcal{D}(\mathfrak{Y}) \to \mathcal{D}(\mathfrak{X})$$

given by integration along the leaves of the vertical foliation:

$$\begin{gathered} D_{\mathrm{ver}}(J_X^\circ, L_X) \dashrightarrow D_{\mathrm{hor}}(J_X^\circ, L_X), \\ \mathcal{T}\varphi(j) = \int_{\mathscr{F}_{\mathrm{ver},j}} T_{j,j'}(\varphi(j'))\left|\omega(j')\right|^{\frac{1}{2}}, \end{gathered} \tag{5.13}$$

where $\mathscr{F}_{\mathrm{ver},j}$ denotes the leaf of $\mathscr{F}_{\mathrm{hor}}$ through the point j, and $T_{j,j'}$ denotes parallel transport from the fiber of L_X over j' to the fiber over j' along this leaf.

Its inverse $\mathcal{T}^{-1}$, valued in an enlargement $\mathcal{D}^-_{L_X}(\mathfrak{Y}) \supset \mathcal{D}(\mathfrak{Y})$ described in **[74, §1.3]**, *is given by integration along the leaves of the horizontal foliation:*

$$D_{\mathrm{hor}}(J_X^\circ, L_X) \dashrightarrow D_{\mathrm{ver}}(J_X^\circ, L_X),$$

Note that a horizontal half-density on J_X°, multiplied by the half-density $|\omega|^{\frac{1}{2}}$, gives rise to a vertical half-density valued in the bundle of densities on the leaves of $\mathscr{F}_{\mathrm{hor}}$; thus, it makes sense to integrate it along these leaves, obtaining a half-density on $J_X^\circ/\mathscr{F}_{\mathrm{hor}}$. This, of course, is completely analogous to the canonical intertwiners for the Schrödinger models quantizing a symplectic vector space **[55]**.

5.3.6. The line bundle L_X, with its connection, admits a more natural description as the dual to *a line bundle of half-densities on the fibers of the invariant moment map* $\mu_G : M = T^*X \to \mathfrak{c}_X^*$. The map (5.11), then, admits a more natural description as descending, up to a choice of invariant measure on X, from a map from Schwartz half-densities on $X \times X$,

$$\mathcal{D}(X \times X) \to D_{\mathrm{hor}}(J_X^\circ, L_X). \tag{5.14}$$

Let us see how this works.

It will be convenient to choose a section s of the invariant moment map μ_G. Such a section exists in the cases $X = H\backslash G$ of (5.4) when G and H are split; it suffices to check the case of $\mathrm{SO}_{2n-1} \backslash \mathrm{SO}_{2n}$, and we refrain from attempting to give an abstract argument. See § 5.4.7 for a further discussion of this issue.

Obviously, the section s has image in the smooth locus of the map μ_G, which implies that the fibers of this map are transversal to the section. If $\mathcal{O}_\xi$ denotes the fiber over $\xi \in \mathfrak{c}_X^*$, let D_X be the algebraic line bundle over $\mathfrak{c}_X^*$ whose fiber over ξ is the determinant of the tangent space of $\mathcal{O}_\xi$ at $s(\xi)$. Let $L_X = |D_X|^{\frac{1}{2}}$, a complex line bundle whose fiber over ξ is dual to the space of Haar half-densities on this tangent space. By pullback, we will also consider L_X as a line bundle over J_X.

For $\xi \neq 0$, the fibers of the invariant moment map μ_G are G-orbits, therefore the tangent space of $\mathcal{O}_\xi$ is $\mathfrak{g}/\mathfrak{g}_{s(\xi)}$, and the fiber of L_X over ξ is the complex line $|\det \mathfrak{g} \otimes \det \mathfrak{g}^*_{s(\xi)}|^{\frac{1}{2}}$. Its dual is the line of invariant half-densities on $\mathcal{O}_\xi$.

There is a natural way to trivialize the bundle L_X, up to a scalar. It uses the fact that the stabilizers $G_{s(\xi)}$, for $\xi \neq 0$, are isomorphic over the algebraic closure. Thus, G-conjugacy gives canonical isomorphisms between the complex line bundles $|\det \mathfrak{g}_{s(\xi)}|$, and this allows us to uniformly fix an invariant measure $d\dot{g}$ on all orbits $\mathcal{O}_\xi$, for $\xi \neq 0$, see **[74, §4]**. Then, by **[74, THEOREM 4.0.3]**:

Proposition. *For a suitable choice of $d\dot{g}$ as above, and the canonical measure dz on T^*X induced by the symplectic form, we have the integration formula*

$$\int_{T^*X} \Phi(z)dz = \int_{\mathfrak{c}^*_X} |\xi|^{\frac{d}{2}-1} \int_{\mathcal{O}_\xi} \Phi(s(\xi)\dot{g})\, d\dot{g}\, d\xi.$$

Hence, the family of Haar half-densities $\xi \mapsto (|\xi|^{\frac{d}{2}-1} d\dot{g})^{\frac{1}{2}}$ on the orbits $\mathcal{O}_\xi$, for $\xi \neq 0$, extends to a nonvanishing half-density on the fiber over 0. We now use this family (depending up to a constant on our choice of $d\dot{g}$) to trivialize L^*_X, hence also L_X, i.e., we have an isomorphism

$$L_X \simeq \underline{\mathbb{C}} \tag{5.15}$$

with the trivial line bundle. Moreover, the proposition above shows that a nonzero element of the fiber of L^*_X over 0 corresponds to a unique half-density on $\mathcal{O}_0$, obtained as the limit of G-invariant half-densities over the fibers $\mathcal{O}_\xi$ with $\xi \neq 0$. Hence, each element in the total space of L^*_X gives rise to a half-density on the corresponding fiber of μ_G.

5.3.7. We can now define the map (5.14). Let $\varphi \in \mathcal{D}(X \times X)$. The product $\varphi \cdot (dt_0)^{-\frac{1}{2}}$ restricts to a half-density on each fiber of the smooth locus of the invariant-theoretic quotient $X \times X \to (X \times X) /\!/ G$. The idea is to integrate this half-density, but for that purpose we need to turn it into a measure. We will do so after pulling it back to J°_X via the maps

$$J^\circ_X \bullet (T^*X \smallsetminus X) \to \mathbb{P}(J^\circ_X \bullet T^*X) \to X \times X,$$

where the second arrow is (5.8), an isomorphism onto its image.

For every $j \in J^\circ_X$ with image $\xi(j) \in \mathfrak{c}^*_X$, the map (5.7) restricts to a map $\{j\} \bullet \mathcal{O}_{\xi(j)} \to X \times X$ that is, by (5.8), an isomorphism onto its image (up to removing the zero section of T^*X, if $\xi(j) = 0$). Thus, the pullback of $\varphi \cdot (dt_0)^{-\frac{1}{2}}$ induces a half-density on $\mathcal{O}_{\xi(j)}$. Multiplying by the half-density corresponding to an element of the fiber of L^*_X over $\xi(j)$ gives rise to a measure, which we can integrate. This way, we get a canonical map

$$\mathcal{D}(X \times X)(dt_0)^{-\frac{1}{2}} \times \underline{L^*_X} \to \mathbb{C},$$

where $\underline{L^*_X}$ denotes the total space of the line bundle L^*_X over J°_X. This corresponds to a map

$$\mathcal{D}(X \times X)(dt_0)^{-\frac{1}{2}} \to \Gamma(J^\circ_X, L_X), \tag{5.16}$$

where the right-hand side denotes (continuous) sections of L_X over J°_X.

5.3.8. The image of $\varphi \cdot (dt_0)^{-\frac{1}{2}}$ under this map has an invariance property: namely, its values at different points j with the same value of t_0 "coincide." To make sense of this, we need to endow L_X with a connection, whose parallel sections along vertical Lagrangians descend to duals of half-densities on the corresponding G-orbits on $X \times X$. To explicate this, consider the integration formula of [74, THEOREM 4.0.2]:

Proposition. *For a function Φ and an invariant measure dx on $X \times X$, we have*

$$\int_{X\times X} \Phi(x)\, dx = \int_{(X\times X)/\!\!/G} |t_0^2 - 1|^{\frac{d}{2}-1} \int_{\mathcal{O}_{t_0}} \Phi(\dot{g})\, d\dot{g}\, dt_0. \tag{5.17}$$

Here, we have denoted by $\mathcal{O}_{t_0}$ the preimage of $t_0 \in (X \times X) /\!\!/ G$ in $X \times X$, using similar notation as for the preimages of points of $\mathfrak{c}_X^*$ in T^*X. The reason is that, as above, for $j \in J_X^\circ$ with $\xi(j) \neq 0$, we can identify the G-orbit $\mathcal{O}_{\xi(j)}$ with the image of $\{j\} \bullet \mathcal{O}_{\xi(j)}$ in $X \times X$, which is equal to $\mathcal{O}_{t_0(j)}$. The measure $d\dot{g}$ in the Proposition, then, is the same measure on $\mathcal{O}_{\xi(j)}$ as that used to trivialize the bundle L_X in § 5.3.6.

The proposition above tells us that, if for every j with $\xi(j) \neq 0$ we multiply $\varphi(dt_0)^{-\frac{1}{2}}$ by the half-density $(|t_0^2(j) - 1|^{\frac{d}{2}-1} d\dot{g})^{\frac{1}{2}}$, the integral will depend only on the function $t_0(j)$ of j. On the other hand, the trivialization (5.15) of L_X uses the half-density $(|\xi|^{\frac{d}{2}-1} d\dot{g})^{\frac{1}{2}}$. The quotient of the two is $|t_1|^{\frac{d}{2}-1}$. We conclude that the map (5.16), composed with the trivialization (5.15), gives rise to *functions f on J_X° such that $|t_1|^{\frac{d}{2}-1} f$ is constant along fibers of* t_0. This explains the definition of ∇^X in (5.10), and completes the construction of the map (5.14).

5.3.9. In a similar way, we define a line bundle L_Y on J_X, pulled back from $\mathfrak{c}_X^*$, as the dual of the line bundle of G'-invariant half-densities on the fibers of $M' = T^*Y \to \mathfrak{c}_X^*$. Here, the fibers are G'-torsors, hence fixing a Haar half-density on G' gives rise to a trivialization

$$L_Y \xrightarrow{\sim} \underline{\mathbb{C}}. \tag{5.18}$$

Through the trivializations (5.15) and (5.18), the line bundles L_X and L_Y are identified.

Recall from § 5.3.4 that the description of "vertical" half-densities for (L_X, ∇^X) is the same in every case (does not depend on the dimension d of X). We can now define a map

$$\mathcal{D}(Y \times Y^-) \to D_{\mathrm{ver}}(J_Y^\circ, L_Y), \tag{5.19}$$

in a completely analogous way to (5.14), using also the scalar-valued pullback from Whittaker functions on $Y \times Y^-$ to scalar-valued functions on $M' \times_{\mathfrak{g}'^*} M'$, described in § 5.3.4. To describe it explicitly, consider the integration formula for functions on Y^2, analogous to (5.17),

$$\int_{Y\times Y} |\Phi(y)|\, dy = \int_{(Y\times Y)/\!\!/G'} |t_1| \int_{G'} |\Phi(\tilde{t_1} g)|\, dg\, dt_1 \tag{5.20}$$

(where $\tilde{t_1}$ denotes any lift of t_1 to $Y \times Y$). Symbolically, we can write the $G' \times G'$-invariant measure dy on $Y \times Y$ as $dg \times |t_1| dt_1$. Similarly, we can write the spaces of Schwartz measures and half-densities as

$$\mathcal{S}(Y \times Y) = \mathcal{F}(Y \times Y) \cdot (dg \times |t_1| dt_1),$$
$$\mathcal{D}(Y \times Y) = \mathcal{F}(Y \times Y) \cdot (dg \times |t_1| dt_1)^{\frac{1}{2}}.$$

Choosing the half-density $(dg)^{\frac{1}{2}}$ to define the trivialization (5.18), the image of a Schwartz half-density $\Phi \cdot (dg \times |t_1| dt_1)^{\frac{1}{2}}$ under (5.19) is precisely the image of the Schwartz measure $\Phi \cdot (dg \times |t_1| dt_1)$ described in the definition of (5.12), giving a natural meaning to that definition.

5.4. Geometric quantization for type T

5.4.1. In the cases (5.5) of "type T," the analog of Theorem 5.3.5 does not directly hold. In turns out, however, that there is a similar interpretation of transfer operators, if we pass from J_X, the group scheme of regular centralizers in PGL_2, to $\tilde{J_X}$ = the group scheme of regular centralizers in GL_2. It lives over $\tilde{\mathfrak{c}_X}^* := \mathfrak{gl}_2^* /\!/ \mathrm{GL}_2$.

If we write $\mathrm{GL}_2 = (\mathrm{SL}_2 \times \mathbb{G}_m)/\mu_2$, use coordinates (t_0, t_1, ξ), as before, for J_{SL_2}, and coordinates (z, τ) for $T^*\mathbb{G}_m = \mathbb{G}_m \times \mathfrak{g}_m^*$, we obtain

$$\tilde{J_X} = \operatorname{Spec} F\big[t_0, t_1, \xi, z^{\pm 1}, \tau\big]^{\mu_2} / (t_0^2 - \xi t_1^2 - 1),$$

where $-1 \in \mu_2$ acts by $(t_0, t_1, \xi, z, \tau) \mapsto (-t_0, -t_1, \xi, -z, \tau)$. The map to $J_X = J_{\mathrm{PGL}_2}$ is then obtained by symplectic reduction modulo $\mathbb{G}_m$, and the symplectic form on $\tilde{J_X}$ reads

$$\omega = dt_0 \wedge d(t_1^{-1}) + d^{\times} z \wedge d\tau,$$

where the notation is $d^{\times} z := d \log z = \frac{dz}{z}$.

5.4.2. To motivate the passage to $\tilde{J_X}$, we should first look at the simple case $X = \mathbb{G}_m \backslash \mathrm{PGL}_2$. This space is a quotient of $\tilde{X} = \mathbb{G}_m \backslash \mathrm{GL}_2$, where $\mathbb{G}_m$ is embedded as the general linear group of a 1-dimensional subspace, and one can think of $\mathcal{S}((X \times X)/G)$ as the $\mathbb{G}_m$-coinvariants of the space $\mathcal{S}((\tilde{X} \times \tilde{X})/\tilde{G})$, where $\mathbb{G}_m$ stands for the center of $\tilde{G} = \mathrm{GL}_2$,

$$\mathcal{S}\big((X \times X)/G\big) = \mathcal{S}\big((\tilde{X} \times \tilde{X})/\tilde{G}\big)_{\mathbb{G}_m}.$$

In this setting, one can easily study a transfer operator

$$\mathcal{T} : \mathcal{S}(\tilde{\mathfrak{Y}}) \to \mathcal{S}(\tilde{\mathfrak{X}}),$$

where $\tilde{\mathfrak{X}} = (\tilde{X} \times \tilde{X})/\tilde{G}$, and $\tilde{\mathfrak{Y}}$ is the "Kuznetsov quotient stack" of GL_2, via the "unfolding" method. The "unfolding" method [77, §9.5] gives rise to an explicit morphism of Schwartz half-densities

$$\mathcal{U} : \mathcal{D}(\tilde{Y}) \to \mathcal{D}(\tilde{X}) \tag{5.21}$$

(where $\tilde{Y}$ denotes the Whittaker model of GL_2), which extends to an L^2-isometry.

We can repeat the earlier constructions to identify the spaces of test measures above with spaces of "vertical," resp. "horizontal," half-densities on $\tilde{J_X}^\circ$ = the complement of $t_1 = 0$, valued in a line bundle L_X (with connection). Here, the corresponding foliations are determined by the following:

Lemma. *The invariant-theoretic quotient* $(\tilde{X} \times \tilde{X}) /\!/ \tilde{G} = \mathbb{G}_m \backslash \mathrm{GL}_2 /\!/ \mathbb{G}_m$ *is a two-dimensional affine space. One can choose the coordinates* (x, y) *so that the resulting map*

$$\tilde{J_X} \to \tilde{J_X} \bullet T^*\tilde{X} \to T^*\tilde{X} \times_{\tilde{\mathfrak{g}}^*} T^*\tilde{X} \to (\tilde{X} \times \tilde{X}) /\!/ \tilde{G}$$

(the first arrow again by choosing a section of $T^*\tilde{X} \to \tilde{\mathfrak{c}_X}^*$*) is given by*

$$(t_0, t_1, \xi, z, \tau) \mapsto \big(x = z^{-1}(t_0 - \tau t_1), y = z(t_0 + \tau t_1)\big). \tag{5.22}$$

The invariant-theoretic quotient for the Kuznetsov formula of GL_2*, composed with the analogous map from* $\tilde{J_X}$*,*

$$\tilde{J_X} \to \tilde{J_X} \bullet T^*\tilde{Y} \to N\backslash\tilde{G} /\!/ N$$

is the map that remembers all even-order monomials in the coordinates t_1 *and* $z^{\pm 1}$*.*

Note that the map $\tilde{J_X} \to (\tilde{X} \times \tilde{X}) /\!/ \tilde{G}$ is smooth, when restricted to $\tilde{J_X}^\circ$ = the complement of the divisor $t_1 = 0$. The "vertical" foliation $\mathscr{F}_{\mathrm{ver}}$ is defined as the set of fibers of this map. The "horizontal" foliation $\mathscr{F}_{\mathrm{hor}}$ on $\tilde{J_X}^\circ$ is defined by the level sets of $(t_1 z, z^2)$.

Remark. The passage to $(X \times X) /\!/ G = \mathbb{A}^1$ is given by the coordinate

$$(x, y) \mapsto c := xy = t_0^2 - \tau^2 t_1^2 = (\xi - \tau^2)t_1^2 + 1. \tag{5.23}$$

5.4.3. Still in the case of $\tilde{X} = \mathbb{G}_m \backslash \mathrm{GL}_2$, defining line bundles L_X, L_Y over $\tilde{J_X}^\circ$ exactly as before, we can repeat the constructions of the maps (5.14) and (5.19) for $\tilde{G}$, to identify test measures as spaces

$$\mathcal{D}(\tilde{\mathfrak{X}}) \hookrightarrow D_{\mathrm{hor}}(\tilde{J_X}^\circ, L_X),$$
$$\mathcal{D}(\tilde{\mathfrak{Y}}) \hookrightarrow D_{\mathrm{ver}}(\tilde{J_X}^\circ, L_Y)$$

of "horizontal," resp. "vertical," half-densities valued in those line bundles.

With the appropriate identification $L_X \simeq L_Y$ over $\tilde{J_X}^\circ$ (which we will present for the general case in § 5.4.4), we can now descend the unfolding map (5.21), applied to $\mathcal{D}(\tilde{Y}) \otimes \mathcal{D}(\tilde{Y}^-)$, to coinvariants for the diagonal action of $\tilde{G}$, obtaining a transfer operator, which can be explicitly described, along the lines of [69, THEOREM 5.4]:

Theorem. *The transfer operator*

$$\tilde{\mathcal{T}} : \mathcal{D}(\tilde{\mathfrak{Y}}) \xrightarrow{\sim} \mathcal{D}(\tilde{\mathfrak{X}})$$

is the operator of integration along the leaves of the vertical foliation:

$$D_{\mathrm{ver}}(\tilde{J_X}^\circ, L_X) \dashrightarrow D_{\mathrm{hor}}(\tilde{J_X}^\circ, L_X).$$

The transfer operator

$$\mathcal{T} : \mathcal{D}(\mathfrak{Y}) \to \mathcal{D}(\mathfrak{X})$$

is the descent of $\tilde{\mathcal{T}}$ *to* $\mathbb{G}_m$*-coinvariants.*

5.4.4. Let, now, X be a general space from the list (5.5). The idea is to generalize the statement of Theorem 5.4.3 for the transfer operator $\mathcal{T}$, *even though the space X does not, in general, admit a cover such as $\tilde{X}$*. (Such a cover exists, more generally, for $X = \mathrm{GL}_n \backslash \mathrm{PGL}_{n+1}$, and can be used to motivate some of the definitions that follow.)

An important feature of the general case is that we will extend the map $J_X \to (X \times X) /\!\!/ G \simeq \mathbb{A}^1$ to a map $\tilde{J_X} \to \mathbb{A}^1$, given by the same coordinate c, (5.23), as in the case of $\mathbb{G}_m \backslash \mathrm{GL}_2$, and will define the vertical and horizontal foliations on $\tilde{J_X}^\circ$ as in § 5.4.2, e.g., the vertical foliation consists of level sets of the pair of functions (x, y) of (5.22). We define the complex line bundle L_X over $\mathfrak{c}_X^*$ as in § 5.3.6, and *extend it to a line bundle L_X on $\tilde{\mathfrak{c}_X}^*$*, by pullback along the map

$$\tilde{\mathfrak{c}_X}^* \ni (\xi, \tau) \mapsto \xi - \tau^2 \in \mathfrak{c}_X^*. \tag{5.24}$$

By pullback from $\tilde{\mathfrak{c}_X}^*$, this also becomes a line bundle over $\tilde{J_X}$. We endow it with the same trivialization (5.15) as before.

Roughly speaking, now, if we fix a Haar measure on $F^\times$, the map (5.24) allows us to pull back elements of $D_{\mathrm{hor}}(J_X^\circ, L_X)$ to $\mathbb{G}_m$-invariant elements of $D_{\mathrm{hor}}(\tilde{J_X}^\circ, L_X)$, thus obtaining maps

$$\mathcal{D}(X \times X) \to D_{\mathrm{hor}}(J_X^\circ, L_X) \to D_{\mathrm{hor}}(\tilde{J_X}^\circ, L_X)^{\mathbb{G}_m}. \tag{5.25}$$

(Fixing a Haar measure on $F^\times$ allows the switch from the coinvariants of Theorem 5.4.3 to invariants.) However, there is one more twist, which is not seen in the cases of $X = \mathrm{GL}_n \backslash \mathrm{PGL}_{n+1}$, but is needed in the general case. Namely, instead of $\mathbb{G}_m$-invariants, one needs *twisted* invariants with respect to a character of $\mathbb{G}_m$ (that is, of $F^\times$).

5.4.5. To introduce this final piece of the puzzle, we recall from [74] that the space $X \times X$ has two closed G-orbits of codimension larger than 1: the diagonal X^{diag} (whose codimension we keep denoting by d), and a second closed G-orbit, whose codimension we will denote by d'. We define a character of $\mathbb{G}_m$ by $\chi_{d'} : z \mapsto |z|^{-\frac{d'}{2}+1}$. We will then understand the space of test densities for $(X \times X)/G$ as a subspace of the $(\mathbb{G}_m, \chi_{d'})$-equivariant elements of $D_{\mathrm{hor}}(\tilde{J_X}^\circ, L_X)$, by composing (5.25) with multiplication by

$$|y|^{-\frac{d'}{2}+1} = |z(t_0 + \tau t_1)|^{-\frac{d'}{2}+1},$$

obtaining a map

$$\mathcal{D}(X \times X) \to D_{\mathrm{hor}}(\tilde{J_X}^\circ, L_X)^{(\mathbb{G}_m, \chi_{d'})}. \tag{5.26}$$

The image $\mathcal{D}(\mathfrak{X})$ is identified, as before, with the space $\mathcal{S}(\mathfrak{X})^{\mathrm{st}}$ of stable test measures, if we fix an invariant measure on $X \times X$. To summarize, the map $\mathcal{S}(\mathfrak{X})^{\mathrm{st}} \to \mathcal{D}(\mathfrak{X})$ takes a measure $f(c)$ to

$$f(c)(dc)^{-\frac{1}{2}}|y|^{-\frac{d'}{2}+1}|t_1|^{-\frac{d}{2}+1}(d^\times z)^{\frac{1}{2}}, \tag{5.27}$$

in the trivialization (5.15), where y, c are given by (5.22) and (5.23).

Remark. The most convincing argument for the relevance of the character $\chi_{d'}$ is [74, PROPOSITION 6.1.5], describing orbital integrals in the neighborhood of $c = xy = 0$ in terms of

$\mathbb{G}_m$-orbital integrals on the (x, y)-plane, twisted by this character. However, a more conceptual understanding of it would be highly desirable.

5.4.6. We also replace half-densities for the Kuznetsov formula of $G' = \mathrm{PGL}_2$ by half-densities for GL_2 with central character $\chi_{d'}$, namely, we define an embedding

$$\mathcal{D}(\mathfrak{Y}) \hookrightarrow D_{\mathrm{ver}}(\tilde{J_X}^{\circ}, L_X)^{(\mathbb{G}_m, \chi_{d'})}$$

simply by multiplying the embedding $\mathcal{D}(\mathfrak{Y}) \hookrightarrow D_{\mathrm{ver}}(J_X^{\circ}, L_Y)$ of § 5.3.9 by the factor $|z|^{-(\frac{d'}{2}-1)}(d^{\times}z)^{\frac{1}{2}}$, and use the same trivialization (5.18) to identify $L_Y \simeq \mathbb{C} \simeq L_X$.

The main result **[74, THEOREM 1.3.1]** for the transfer operator in this case can now be formulated as follows:

Theorem. *There is an injective operator*

$$\mathcal{T} : \mathcal{D}(\mathfrak{Y}) \to \mathcal{D}(\mathfrak{X})$$

given by integration along the leaves of the vertical foliation:

$$D_{\mathrm{ver}}(\tilde{J_X}^{\circ}, L_X) \dashrightarrow D_{\mathrm{hor}}(\tilde{J_X}^{\circ}, L_X).$$

Its inverse $\mathcal{T}^{-1}$, valued in an enlargement $\mathcal{D}^{-}_{L_X}(\mathfrak{Y}) \supset \mathcal{D}(\mathfrak{Y})$ described in **[74, §1.3]**, *is given by integration along the leaves of the horizontal foliation:*

$$D_{\mathrm{hor}}(\tilde{J_X}^{\circ}, L_X) \dashrightarrow D_{\mathrm{ver}}(\tilde{J_X}^{\circ}, L_X),$$

5.4.7. I finish this section with a brief discussion of a case where a section $\mathfrak{c}_X^* \to T^*X$ does not exist. Let $X = T\backslash\mathrm{PGL}_2$, where T is a nonsplit torus, splitting over a quadratic extension E/F. In this case, the transfer operator

$$\mathcal{T} : \mathcal{S}(\mathfrak{Y}) \to \mathcal{S}(\mathfrak{X})$$

was computed in **[69]**, and can be described as follows:

Instead of defining $\tilde{J_X}$ to be the group scheme of regular centralizers in GL_2, define it to be the $\mathrm{Gal}(E/F)$-twist of that, determined by the automorphism $(t_0, t_1, \xi, z, \tau) \mapsto (t_0, t_1, \xi, z^{-1}, -\tau)$; that is, $\tilde{J_X}$ will be isomorphic to the group scheme of regular centralizers in the quasisplit unitary group U_2. The transfer operator, now, it obtained as in Theorem 5.4.3, by descending the operator of integration along the leaves of the corresponding "vertical" foliation on $\tilde{J_X}$ to U_1-coinvariants.

Both the Schwartz space and the descent to U_1-coinvariants, here, need to be understood in a sophisticated, "stacky" way. Namely, the full space $\mathcal{S}(\mathfrak{X})$ includes a "pure inner form" as in (2.2),

$$\mathcal{S}(\mathfrak{X}) = \mathcal{S}(X \times X)_G \oplus \mathcal{S}(X^{\alpha} \times X^{\alpha})_{G^{\alpha}},$$

where $X^{\alpha} \simeq T\backslash G^{\alpha}$, with $G^{\alpha} = PD^{\times}$, the projective multiplicative group of the quaternion division algebra. Similarly, the U_1-coinvariants of $D_{\mathrm{ver}}(\tilde{J_X}^{\circ}, L_X)$, $D_{\mathrm{hor}}(\tilde{J_X}^{\circ}, L_X)$ need to be understood in a stacky way. Explicitly, recall that in the split case the space $D_{\mathrm{hor}}(\tilde{J_X}^{\circ}, L_X)$ was the space of half-densities on the (x, y)-plane (in coordinates (5.22)), with $\mathbb{G}_m$ acting as $z \cdot (x, y) = (z^{-1}x, zy)$. In the nonsplit case, the (x, y)-plane becomes the space $V = \mathrm{Res}_{E/F}\,\mathbb{G}_a$, and instead of U_1-coinvariants of the space $\mathcal{D}(V(F))$, one needs to consider the direct sum

$$\mathcal{D}\big(V(F)\big)_{U_1} \oplus \mathcal{D}\big(V'(F)\big)_{U_1},$$

where V' is the twist of V by the nontrivial U_1-torsor. The same interpretation is needed for "stacky" U_1-coinvariants of the space $D_{\mathrm{ver}}(\tilde{J_X}^{\circ}, L_X)$ (with the coordinates $a = zt_1$, $b = z^{-1}t_1$ for the leaves of the horizontal foliation interchanged by the Galois action), and the operator of "integration along vertical half-densities"—essentially, a Fourier transform from the Galois-twisted (x, y)-plane to the Galois-twisted (a^{-1}, b^{-1})-plane—naturally descends to give the transfer operator $\mathcal{T}$ of [69, THEOREM 5.1].

6. PROBLEMS FOR THE NEAR FUTURE

6.1. The relative Langlands conjectures

The relative Langlands conjectures presented in Sections 2.3.1–2.3.2 do not have the precision of the local conjectures of Gan–Gross–Prasad [27], or the global conjectures of Ichino–Ikeda [36]. Moreover, an extension of those conjectures to Arthur packets not appearing in the L^2-decomposition is required, as in [26].

It is therefore an important problem to refine the existing conjectures. It is also a fascinating one: as always, finding a way to blend several known cases into a uniform theory can lead to new insights about the nature of the problems. The geometric relative Langlands conjectures proposed in joint work with Ben-Zvi and Venkatesh can probably assist in this direction, providing a geometric spectral answer to automorphic problems, which can then be translated to number theory by decategorifying.

6.2. Transfer operators in higher rank

The most important problem "beyond endoscopy," in my view, is to understand transfer operators in higher rank, and for morphisms of L-groups ${}^LG_X \to {}^LG_Y$ that are not isomorphisms. Regarding the latter, despite the traditional emphasis on the Arthur–Selberg trace formula, it might be better, as first observed by Sarnak [79], to try to compare Kuznetsov formulas, which, according to the results of [74], seem to be the "base cases" for every comparison. This can be "explained" by the simple structure (5.6) of the Kuznetsov cotangent stack.

If the ideas discussed in this paper have any merit, understanding transfer operators in terms of "quantization" would involve several steps, including the following:

6.2.1. Develop a theory of "geometric quantization" for derived symplectic stacks, whose output includes the Schwartz spaces of stacks defined in [71]. The cases presented here, and in particular the construction of the maps (5.14) and (5.26), could provide some hints on how to do that, but the various twists involved need to be better understood.

6.2.2. Obtain a better understanding of the structure of coisotropic Hamiltonian spaces and the cotangent stacks appearing in the relative trace formula. The diagram (5.3), arising from the work of Knop, which was interpreted as a "birational" description of $T^*\mathfrak{X}$, does not capture the difference between $T^*\mathfrak{X}$, and the Kuznetsov cotangent stack with the same L-group. This difference seems to be significant for the structure of transfer operators and for spaces of test measures. For example, the enlarged spaces $\mathcal{S}^-_{L_X}(\mathfrak{Y})$ of test measures for the Kuznetsov formula in Theorems 5.3.5 and 5.4.3 should be seen as quantizations of $T^*\mathfrak{X}$, which is strictly larger than the Kuznetsov stack $T^*\mathfrak{Y}$. This difference can also explain "Galois twists" of transfer operators, as in the example of § 5.4.7, where $T^*\mathfrak{X}$ failed to admit an analog of a Kostant section.

A baby case of the idea that embeddings of Hamiltonian spaces correspond to enlarged Schwartz spaces is Iwasawa–Tate theory, where the embedding $T^*\mathbb{G}_m \hookrightarrow T^*\mathbb{A}^1$ corresponds to the enlargement $\mathcal{S}(F^\times) \hookrightarrow \mathcal{S}(F)$. This can be generalized to toric stacks, as described in [63, §5.2]. For a torus A, a collection of coweights $\mu : \mathbb{G}_m^r \to A$ defines a stack $\overline{A}_\mu := \mathbb{A}^r / \ker(\mu)$ with an action of A. The *dual Hamiltonian space* of such a stack is defined as the symplectic $\check{A}$-vector space $\check{M}$ with weights $\{\pm\mu\}$.

To give an example of the descriptions of Hamiltonian spaces envisioned here, in an ongoing joint work with Ben-Zvi and Venkatesh we take a step beyond Knop's theory, modeling the most regular locus of $M = T^*X$, up to codimension 2, on a the analog of a "toric stack" for the group scheme J_X. Our result confirms an observation of V. Lafforgue, shared in private communication several years ago. For example, for spherical varieties X with $\check{G}_X = \check{G}$ and $\check{A}_X = \check{A}$, our description uses the toric stack $\overline{A_X}$ corresponding to the dual Hamiltonian space $\check{M} = V_X$ of the spherical variety, described in terms of its colors in § 3.1.2.

Theorem. *In the setting above, there is an action of W_X on the toric cotangent stack $T^*\overline{A_X}$, and an open dense subset $\mathfrak{c}_X^{*\prime} \subset \mathfrak{c}_X^*$, whose complement has codimension ≥ 2, such that the restriction $M' \subset M$ to the image of $\mathfrak{c}_X^{*\prime}$ under a Kostant section $\kappa : \mathfrak{c}_X^* \to \mathfrak{g}^*$ admits a J_X-equivariant symplectomorphism*

$$M' \simeq (\mathrm{Res}_{\mathfrak{a}_X^*/\mathfrak{c}_X^*} T^*\overline{A_X})^{W_X} \times_{\mathfrak{c}_X^*} \mathfrak{c}_X^{*\prime}.$$

6.2.3. Use cotangent spaces to understand transfer operators. As we saw in the discussion of cases of type T in § 5.4, a "naive" change of geometric quantization on J_X did not give the correct transfer operators; instead, one has to pass to the group scheme $\tilde{J}_X$ associated with the root datum of GL_2, producing a 2-dimensional Fourier transform.

Ongoing joint work with C. Wan, comparing the Kuznetsov to the Arthur–Selberg trace formula for GL_n, suggests that, in higher-rank cases, the transfer operator for a comparison to a Kuznetsov quotient with the same dual group might be given by an r-dimensional integration, where r is, roughly, half the dimension of the nonzero weight spaces of the representation V_X (§ 3.1.1) of the dual group. In view of the discussion of § 6.2.2, this seems to be closely related to the structure of $T^*\mathfrak{X}$. For example, in the setting of Theorem 6.2.2 (such as in the case of Gan–Gross–Prasad periods), one could speculate that the transfer operator to the Kuznetsov formula is somehow a "descent" of a Fourier transform in $\frac{\dim \check{M}}{2}$-dimensions. In this "dream," the following three objects would be closely related: the L-value L_X associated to a spherical variety (encoded in a dual Hamiltonian space $\check{M}$), the fine structure of the Hamiltonian space $M = T^*X$ (whose quantization is the space of test measures on X), and the transfer operator between the relative trace formula of X and the Kuznetsov formula with the same dual group.

6.3. Poisson summation formula

Understanding the local transfer operators should be followed by a global comparison of trace formulas. For example, comparing the relative trace formula for any spherical variety X with the Kuznetsov formula with the same dual group should amount to commutativity of the diagram

$$\begin{array}{ccc} \mathcal{S}^-_{L_X}(\mathfrak{Y}(\mathbb{A})) & \xrightarrow{\mathcal{T}} & \mathcal{S}(\mathfrak{X}(\mathbb{A}))^{\mathrm{st}}, \\ & {\scriptstyle \mathrm{RTF}_{\mathfrak{Y}}} \searrow \quad \swarrow {\scriptstyle \mathrm{RTF}_{\mathfrak{X}}} & \\ & \mathbb{C} & \end{array} \tag{6.1}$$

where $\mathcal{S}^-_{L_X}(\mathfrak{Y}(\mathbb{A}))$ is a suitable enlarged space of test measures for the Kuznetsov formula, related to the L-value L_X of X.

Having a formula for the transfer operator in terms of Fourier transforms (as in Theorems 5.3.5 and 5.4.3) gives hope of employing the Poisson summation formula to establish commutativity of (6.1). However, this is far from straightforward, as the spaces of stable test measures are nonstandard. In Altuğ's work **[2]**, the approximate functional equation was used for the trace formula of $X = \mathrm{GL}_2$, obtaining an expression similar to the Kuznetsov formula (in particular, containing Kloosterman sums), but not quite equal to it.

A different approach was introduced in **[73]**, for the case $X = T\backslash \mathrm{PGL}_2$. It is based on the idea of *deforming* spaces of test measures and transfer operators in analytic families depending on a parameter s (which moves the point of evaluation of L_X), so that in some domain for s the Poisson summation formula is valid. It is likely that this method can be applied more generally, but it requires a better understanding of the idea of deforming spaces of test measures (orbital integrals).

6.4. Hankel transforms

In the recent literature on automorphic forms, the term "Hankel transforms" has been used to describe two distinct conjectural notions:

- The nonlinear Fourier transforms $\mathcal{S}_\rho(G(F)) \to \mathcal{S}_{\rho^*}(G(F))$ between nonstandard spaces of Schwartz functions (or measures) on a reductive group over a local field, adapted to a representation ρ of its dual group, and its dual. These spaces and operators would generalize Fourier transform on the space $\mathcal{S}(\mathrm{Mat}_n(F))$ of Godement–Jacquet theory, for the case $\rho =$ the standard representation of $\check{G} = \mathrm{GL}_n$, and would similarly give rise to the local functional equation for the L-functions associated to ρ. They were introduced by Braverman and Kazhdan **[15, 17]**, and advanced in the work of Ngô **[63]**.

- The descent of such transforms to spaces of test measures for the Arthur–Selberg trace formula, or for the Kuznetsov formula. In the latter case, those would be operators

$$\mathcal{H}_\rho : \mathcal{S}_\rho(\mathfrak{Y}) \to \mathcal{S}_{\rho^*}(\mathfrak{Y})$$

between enlarged spaces of measures for the Kuznetsov formula, such as those encountered in Theorems 5.3.5 and 5.4.3.

The two notions are closely related, but it is the latter that we would like to focus on here. It is natural to ask the question of whether one can describe $\mathcal{H}_\rho$ explicitly, and prove a Poisson summation formula, globally, in the sense that the diagram

$$\begin{array}{ccc} \mathcal{S}_\rho(\mathfrak{Y}(\mathbb{A})) & \xrightarrow{\mathcal{H}_\rho} & \mathcal{S}_{\rho^*}(\mathfrak{Y}(\mathbb{A})) \\ & {\scriptstyle \mathrm{RTF}_{\mathfrak{Y}}} \searrow \quad \swarrow {\scriptstyle \mathrm{RTF}_{\mathfrak{Y}}} & \\ & \mathbb{C} & \end{array} \tag{6.2}$$

should commute. This would lead to an independent proof of the functional equation of the pertinent L-functions.

Such Hankel transforms have been described by Jacquet [40] for $\rho =$ the standard representation of GL_n (the paper [35] is closely related), and by me [76] for $\rho =$ the symmetric square representation of GL_2. It would be interesting to examine if these formulas admit an interpretation in terms of quantization, like the transfer operators in this paper.

It seems counter to the strategy of the Langlands program to seek such a proof of the functional equation, independent of functoriality. On the other hand, the similarity between diagrams (6.1) and (6.2) is enticing. More fundamentally, trace formulas with nonstandard test functions, such as those in Langlands' original "beyond endoscopy" proposal, or the Kuznetsov formula appearing in (6.1), have the L-functions embedded into them. Obtaining the spectral decomposition of those formulas will likely require more than "brute force" analytic number theory, and a Poisson summation formula of the form (6.2) could help resolve the problem. This idea was successfully employed in [73] for a new proof of Waldspurger's formula for toric periods in PGL_2 via a nonstandard comparison of the form (6.1). Therefore, it might be that, in the "beyond endoscopy" program, functoriality and the functional equation of L-functions should be studied hand-in-hand.

ACKNOWLEDGMENTS

My understanding of the subject has been greatly influenced by conversations with David Ben-Zvi, Raphaël Beuzart-Plessis, Ngô Bao Châu, Akshay Venkatesh, Chen Wan, and Jonathan Wang. I thank them all for generously sharing their ideas in our exciting mathematical journeys.

FUNDING

This work was partially supported by NSF grants DMS-1939672 and DMS-2101700, and by a Simons Fellowship in Mathematics.

REFERENCES

[1] A. Aizenbud and D. Gourevitch, Schwartz functions on Nash manifolds. *Int. Math. Res. Not. IMRN* (2008), no. 5, 155, 37 pp.

[2] S. A. Altuğ, Beyond endoscopy via the trace formula: 1. Poisson summation and isolation of special representations. *Compos. Math.* **151** (2015), no. 10, 1791–1820.

[3] S. A. Altuğ, Beyond endoscopy via the trace formula, II: asymptotic expansions of Fourier transforms and bounds towards the Ramanujan conjecture. *Amer. J. Math.* **139** (2017), no. 4, 863–913.

[4] S. A. Altuğ, Beyond endoscopy via the trace formula—III: the standard representation. *J. Inst. Math. Jussieu* **19** (2020), no. 4, 1349–1387.

[5] J. An, J.-K. Yu, and J. Yu, On the dimension datum of a subgroup and its application to isospectral manifolds. *J. Differential Geom.* **94** (2013), no. 1, 59–85.

[6] J. Arthur, The trace formula in invariant form. *Ann. of Math. (2)* **114** (1981), no. 1, 1–74.

[7] R. Beuzart-Plesis, Multiplicities and Plancherel formula for the space of nondegenerate Hermitian matrices. 2021, arXiv:2008.05036.

[8] R. Beuzart-Plesis, P.-H. Chaudouard, and M. Zydor, The global Gan–Gross–Prasad conjecture for unitary groups: the endoscopic case. 2020, arXiv:2007.05601.

[9] R. Beuzart-Plessis, A local trace formula for the Gan–Gross–Prasad conjecture for unitary groups: the Archimedean case. *Astérisque* (2020), no. 418, viii+299 pp.

[10] R. Beuzart-Plessis, Y. Liu, W. Zhang, and X. Zhu, Isolation of cuspidal spectrum, with application to the Gan–Gross–Prasad conjecture. *Ann. of Math. (2)* **194** (2021), no. 2, 519–584.

[11] R. Bezrukavnikov and M. Finkelberg, Equivariant Satake category and Kostant–Whittaker reduction. *Mosc. Math. J.* **8** (2008), no. 1, 39–72, 183.

[12] A. Bouthier, B. C. Ngô, and Y. Sakellaridis, On the formal arc space of a reductive monoid. *Amer. J. Math.* **138** (2016), no. 1, 81–108.

[13] A. Braverman, M. Finkelberg, D. Gaitsgory, and I. Mirković, Intersection cohomology of Drinfeld's compactifications. *Selecta Math. (N.S.)* **8** (2002), no. 3, 381–418.

[14] A. Braverman and D. Kazhdan, On the Schwartz space of the basic affine space. *Selecta Math. (N.S.)* **5** (1999), no. 1, 1–28.

[15] A. Braverman and D. Kazhdan, γ-functions of representations and lifting. In *Visions in Mathematics*, pp. 237–278, GAFA 2000 Special Volume Part I, Birkhäuser, Basel, 2000.

[16] A. Braverman and D. Kazhdan, Normalized intertwining operators and nilpotent elements in the Langlands dual group. *Mosc. Math. J.* **2** (2002), 533–553.

[17] A. Braverman and D. Kazhdan, γ-sheaves on reductive groups. In *Studies in memory of Issai Schur (Chevaleret/Rehovot, 2000)*, pp. 27–47, Progr. Math. 210, Birkhäuser Boston, Boston, MA, 2003.

[18] M. Brion, Vers une généralisation des espaces symétriques. *J. Algebra* **134** (1990), no. 1, 115–143.

[19] D. Bump, The Rankin–Selberg method: an introduction and survey. In *Automorphic representations, L-functions and applications: progress and prospects*, pp. 41–73, Ohio State Univ. Math. Res. Inst. Publ. 11, de Gruyter, Berlin, 2005.

[20] W. Casselman, The unramified principal series of p-adic groups. I. The spherical function. *Compos. Math.* **40** (1980), no. 3, 387–406.

[21] W. Casselman and J. Shalika, The unramified principal series of p-adic groups. II. The Whittaker function. *Compos. Math.* **41** (1980), no. 2, 207–231.

[22] P. Delorme, Neighborhoods at infinity and the Plancherel formula for a reductive p-adic symmetric space. *Math. Ann.* **370** (2018), no. 3–4, 1177–1229.

[23] B. Feigon, E. Lapid, and O. Offen, On representations distinguished by unitary groups. *Publ. Math. Inst. Hautes Études Sci.* **115** (2012), 185–323.

[24] E. Frenkel, R. Langlands, and B. C. Ngô, Formule des traces et fonctorialité: le début d'un programme. *Ann. Sci. Math. Québec* **34** (2010), no. 2, 199–243.

[25] D. Gaitsgory and D. Nadler, Spherical varieties and Langlands duality. *Mosc. Math. J.* **10** (2010), no. 1, 65–137, 271.

[26] W. T. Gan, B. H. Gross, and D. Prasad, Branching laws for classical groups: the non-tempered case. *Compos. Math.* **156** (2020), no. 11, 2298–2367.

[27] W. T. Gan, B. H. Gross, and D. Prasad, Symplectic local root numbers, central critical L values, and restriction problems in the representation theory of classical groups. *Astérisque* **346** (2012), 1–109.

[28] W. T. Gan and A. Ichino, The Gross–Prasad conjecture and local theta correspondence. *Invent. Math.* **206** (2016), no. 3, 705–799.

[29] W. T. Gan, Y. Qiu, and S. Takeda, The regularized Siegel–Weil formula (the second term identity) and the Rallis inner product formula. *Invent. Math.* **198** (2014), no. 3, 739–831.

[30] W. T. Gan and X. Wan, Relative character identities and theta correspondence. In *Relative trace formulas*, pp. 101–186, Springer, Cham, 2021.

[31] J. Getz, C.-H. Hsu, and S. Leslie, Harmonic analysis on certain spherical varieties. 2021, arXiv:2103.10261.

[32] J. R. Getz and B. Liu, A summation formula for triples of quadratic spaces. *Adv. Math.* **347** (2019), 150–191.

[33] J. R. Getz and B. Liu, A refined Poisson summation formula for certain Braverman–Kazhdan spaces. *Sci. China Math.* **64** (2021), no. 6, 1127–1156.

[34] E. Hecke, *Mathematische Werke. Third edn.* Vandenhoeck & Ruprecht, Göttingen, 1983.

[35] P. E. Herman, The functional equation and beyond endoscopy. *Pacific J. Math.* **260** (2012), no. 2, 497–513.

[36] A. Ichino and T. Ikeda, On the periods of automorphic forms on special orthogonal groups and the Gross–Prasad conjecture. *Geom. Funct. Anal.* **19** (2010), no. 5, 1378–1425.

[37] K. Iwasawa, A note on functions. In *Proceedings of the international congress of mathematicians, Cambridge, Mass., 1950, vol. 1*, p. 322, Amer. Math. Soc., Cambridge, Mass, 1950.

[38] H. Jacquet, Automorphic spectrum of symmetric spaces. In *Representation theory and automorphic forms (Edinburgh, 1996)*, pp. 443–455, Proc. Sympos. Pure Math. 61, Amer. Math. Soc., Providence, RI, 1997.

[39] H. Jacquet, Factorization of period integrals. *J. Number Theory* **87** (2001), no. 1, 109–143.

[40] H. Jacquet, Smooth transfer of Kloosterman integrals. *Duke Math. J.* **120** (2003), no. 1, 121–152.

[41] H. Jacquet and K. F. Lai, A relative trace formula. *Compos. Math.* **54** (1985), no. 2, 243–310.

[42] H. Jacquet, K. F. Lai, and S. Rallis, A trace formula for symmetric spaces. *Duke Math. J.* **70** (1993), no. 2, 305–372.

[43] D. Johnstone, A Gelfand–Graev formula and stable transfer factors for $\mathrm{SL}_n(f)$, 2019. arXiv:1611.06291.

[44] A. A. Kirillov, *Lectures on the orbit method*. Grad. Stud. Math. 64, American Mathematical Society, Providence, RI, 2004.

[45] F. Knop, Weylgruppe und Momentabbildung. *Invent. Math.* **99** (1990), no. 1, 1–23.

[46] F. Knop, Automorphisms, root systems, and compactifications of homogeneous varieties. *J. Amer. Math. Soc.* **9** (1996), no. 1, 153–174.

[47] F. Knop and B. Schalke, The dual group of a spherical variety. *Trans. Moscow Math. Soc.* **78** (2017), 187–216.

[48] B. Kostant, Quantization and unitary representations. In *Lectures in modern analysis and applications, III*, pp. 87–208, Lecture Notes in Math. 170, Springer, Berlin, 1970.

[49] L. Lafforgue, Noyaux du transfert automorphe de Langlands et formules de Poisson non linéaires. *Jpn. J. Math.* **9** (2014), no. 1, 1–68.

[50] R. P. Langlands, Euler products, Yale Math. Monogr. 1, Yale University Press, New Haven, CT–London, 1971.

[51] R. P. Langlands, Beyond endoscopy. In *Contributions to automorphic forms, geometry, and number theory*, pp. 611–697, Johns Hopkins Univ. Press, Baltimore, MD, 2004.

[52] R. P. Langlands, Singularités et transfert. *Ann. Sci. Math. Québec* **37** (2013), no. 2, 173–253.

[53] E. Lapid and Z. Mao, On Whittaker–Fourier coefficients of automorphic forms on unitary groups: reduction to a local identity. In *Advances in the theory of automorphic forms and their L-functions*, pp. 295–320, Contemp. Math. 664, Amer. Math. Soc., Providence, RI, 2016.

[54] E. Lapid and Z. Mao, On an analogue of the Ichino–Ikeda conjecture for Whittaker coefficients on the metaplectic group. *Algebra Number Theory* **11** (2017), no. 3, 713–765.

[55] W.-W. Li, *The Weil representation and its character*. Master's thesis, Universiteit Leiden, 2008.

[56] H. Maass, Über eine neue Art von nichtanalytischen automorphen Funktionen und die Bestimmung Dirichletscher Reihen durch Funktionalgleichungen. *Math. Ann.* **121** (1949), 141–183.

[57] C. Mœglin and D. Renard, Séries discrètes des espaces symétriques et paquets d'Arthur. 2019, arXiv:1906.00725.

[58] C. Mœglin and J.-L. Waldspurger, La conjecture locale de Gross–Prasad pour les groupes spéciaux orthogonaux: le cas général. *Astérisque* **347** (2012), 167–216.

[59] K. Morimoto, On a certain local identity for Lapid–Mao's conjecture and formal degree conjecture: even unitary group case. 2019, arXiv:1902.04910.

[60] D. Nadler, Perverse sheaves on real loop Grassmannians. *Invent. Math.* **159** (2005), no. 1, 1–73.

[61] B. C. Ngô, Le lemme fondamental pour les algèbres de Lie. *Publ. Math. Inst. Hautes Études Sci.* **111** (2010), 1–169.

[62] B. C. Ngô, In *On a certain sum of automorphic L-functions. In Automorphic forms and related geometry: assessing the legacy of I. I. Piatetski-Shapiro*, pp. 337–343, Contemp. Math. 614, Amer. Math. Soc., Providence, RI, 2014.

[63] B. C. Ngô, Hankel transform, Langlands functoriality and functional equation of automorphic L-functions. *Jpn. J. Math.* **15** (2020), no. 1, 121–167.

[64] D. Prasad, A 'relative' local Langlands correspondence. 2015, arXiv:1512.04347.

[65] R. A. Rankin, Contributions to the theory of Ramanujan's function $\tau(n)$ and similar arithmetical functions: II. The order of the Fourier coefficients of integral modular forms. *Math. Proc. Cambridge Philos. Soc.* **35** (1939), no. 3, 357–372.

[66] B. Riemann, *Ueber die Anzahl der Primzahlen unter einer gegebenen Grösse*. Monatsberichte der Berliner Akademie, 1859.

[67] Z. Rudnick, *Poincare series*. Ph.D. thesis, Yale University, 1990.

[68] Y. Sakellaridis, Spherical varieties and integral representations of L-functions. *Algebra Number Theory* **6** (2012), no. 4, 611–667.

[69] Y. Sakellaridis, Beyond endoscopy for the relative trace formula I: local theory. In *Automorphic representations and L-functions*, pp. 521–590, Amer. Math. Soc., Providence, RI, 2013.

[70] Y. Sakellaridis, Spherical functions on spherical varieties. *Amer. J. Math.* **135** (2013), no. 5, 1291–1381.

[71] Y. Sakellaridis, The Schwartz space of a smooth semi-algebraic stack. *Selecta Math. (N.S.)* **22** (2016), no. 4, 2401–2490.

[72] Y. Sakellaridis, Plancherel decomposition of Howe duality and Euler factorization of automorphic functionals. In *Representation theory, number theory, and invariant theory*, pp. 545–585, Progr. Math. 323, Springer, Cham, 2017.

[73] Y. Sakellaridis, Beyond endoscopy for the relative trace formula II: global theory. *J. Inst. Math. Jussieu* **18** (2019), no. 2, 347–447.

[74] Y. Sakellaridis, Functorial transfer between relative trace formulas in rank 1. *Duke Math. J.* **170** (2021), no. 2, 279–364.

[75] Y. Sakellaridis, Transfer operators and Hankel transforms between relative trace formulas, I: Character theory. *Adv. Math.* **394** (2022), Paper No. 108010.

[76] Y. Sakellaridis, Transfer operators and Hankel transforms between relative trace formulas, II: Rankin–Selberg theory. *Adv. Math.* **394** (2022), Paper No. 108039.

[77] Y. Sakellaridis and A. Venkatesh, Periods and harmonic analysis on spherical varieties. *Astérisque* **396** (2017), 360.

[78] Y. Sakellaridis and J. Wang, Intersection complexes and unramified L-factors. *J. Amer. Math. Soc.* (2021), published online first.

[79] P. Sarnak, Comments on Robert Langlands' lecture: "Endoscopy and beyond".

[80] A. Selberg, Bemerkungen über eine Dirichletsche Reihe, die mit der Theorie der Modulformen nahe verbunden ist. *Arch. Math. Naturvidensk.* **43** (1940), 47–50.

[81] F. Shahidi, Intertwining operators, L-functions and representation theory. Lecture notes of the eleventh, KAIST mathematics workshop, 1996.

[82] J.-M. Souriau, *Structure des systèmes dynamiques*. Dunod, Paris, 1970.

[83] J. T. Tate, Fourier analysis in number fields, and Hecke's zeta-functions. In *Algebraic Number Theory (Proc. Instructional Conf., Brighton, 1965)*, pp. 305–347, Thompson, Washington, DC, 1967.

[84] A. Venkatesh, "Beyond endoscopy" and special forms on GL(2). *J. Reine Angew. Math.* **577** (2004), 23–80.

[85] J.-L. Waldspurger, Sur les valeurs de certaines fonctions L automorphes en leur centre de symétrie. *Compos. Math.* **54** (1985), no. 2, 173–242.

[86] J.-L. Waldspurger, La conjecture locale de Gross-Prasad pour les représentations tempérées des groupes spéciaux orthogonaux. *Astérisque* **347** (2012), 103–165.

[87] C. Wan, On a multiplicity formula for spherical varieties. *J. Eur. Math. Soc.* (2021), published online first.

[88] H. Xue, Epsilon dichotomy for linear models. *Algebra Number Theory* **15** (2021), no. 1, 173–215.

[89] S. Yamana, L-functions and theta correspondence for classical groups. *Invent. Math.* **196** (2014), no. 3, 651–732.

[90] W. Zhang, Automorphic period and the central value of Rankin–Selberg L-function. *J. Amer. Math. Soc.* **27** (2014), no. 2, 541–612.

YIANNIS SAKELLARIDIS

Johns Hopkins University, 3400 N. Charles St., Baltimore, MD 21218, USA, sakellar@jhu.edu

CATEGORIFICATION AND APPLICATIONS

PENG SHAN

ABSTRACT

We survey some new developments for categorification of quantum groups and their applications in representation theory.

MATHEMATICS SUBJECT CLASSIFICATION 2020

Primary 20C08; Secondary 18N25, 20G42

KEYWORDS

Categorification, quiver Hecke algebra, quantum group, quantized loop algebra, center, cohomology

INTERNATIONAL CONGRESS OF MATHEMATICIANS 2022 JULY 6–14

Proc. Int. Cong. Math. 2022, Vol. 4, pp. 3038–3060
DOI 10.4171/ICM2022/93

Published by EMS Press

1. INTRODUCTION

Categorification, in the broad sense, refers to the realization of a mathematical object as the Grothendieck group of certain category. Lie theory is a domain with many interesting examples of categorifications. An important family among them consists of categorification of quantum groups and their representations.

Let $\mathfrak{g}$ be a Kac–Moody Lie algebra, $\mathbf{U}_v(\mathfrak{g})$ be its quantized enveloping algebra, and $\mathbf{U}_v^+$ be the positive part. The first categorification of $\mathbf{U}_v^+$ was constructed by Lusztig [41, 42] using perverse sheaves on the moduli stack of quiver representations. It reveals some deep structures of $\mathbf{U}_v^+$, including the existence of a remarkable basis called the canonical basis. It is defined as the classes of intersection complexes in the Grothendieck group. Kashiwara [29] also gave a construction of this basis using a different method.

A (naive) categorical $\mathfrak{g}$-action on an exact category $\mathcal{C}$ consists of pairs of exact adjoint endofunctors $\mathcal{E}_i$, $\mathcal{F}_i$ on $\mathcal{C}$ such that the Grothendieck group of $\mathcal{C}$ is a $\mathfrak{g}$-representation with the Chevalley generators in $\mathfrak{g}$ acting by the operators induced by $\mathcal{E}_i$, $\mathcal{F}_i$. A famous example is the categorical action of the affine Lie algebra $\widehat{\mathfrak{sl}}_p$ on the category of representations over a field of characteristic p of the symmetric groups of all ranks. The functors $\mathcal{E}_i$, $\mathcal{F}_i$ are given by some i-restriction and i-induction functors. Categorifical actions also had remarkable applications to representation theory of affine Hecke algebras and their cyclotomic quotients. Ariki [1] proved that the category of modules over affine Hecke algebras of type A at an eth root of unity categorifies the positive half of $\widehat{\mathfrak{sl}}_e$, and the category of modules over cyclotomic Hecke algebras categorifies an integrable irreducible $\widehat{\mathfrak{sl}}_e$-representation, with the classes of simple modules corresponding to the dual canonical basis. This result confirms a conjecture by Lascoux–Leclerc–Thibon [39] and provides character formulae for simple modules over cyclotomic Hecke algebras.

In 2008, a seminal work of Chuang–Rouquier [17] brought some new perspectives on categorifications. They introduced an enhanced notion of $\mathfrak{sl}_2$-categorical action, whose input requires not only exact adjoint endofunctors $\mathcal{E}$, $\mathcal{F}$ as above, but also some natural transformations $x \in \mathrm{End}(\mathcal{E})$, $\tau \in \mathrm{End}(\mathcal{E}^2)$ satisfying the defining relations for nil-affine Hecke algebras. They showed that many previously known examples of categorifications in Lie theory can be enhanced. The enhancement has two important advantages among others. First, it guarantees that the categorification of a simple integrable $\mathfrak{sl}_2$-representation is unique. Second, it provides derived self-equivalences between different blocks, categorifying the Weyl group action on the underlying representation. As an application, Broué's abelian defect group conjecture for symmetric groups was proved.

To extend this powerful theory from $\mathfrak{sl}_2$ to arbitrary Kac–Moody Lie algebra $\mathfrak{g}$, one needs correct substitutions for the nil-affine Hecke algebras. This is provided by a new family of $\mathbb{Z}$-graded algebras introduced by Khovanov–Lauda [33] and independently by Rouquier [53], called quiver Hecke algebras (also known as KLR algebras). The category of graded projective modules over these algebras gives a categorification of $\mathbf{U}_v^+$ in purely algebraic terms. Moreover, by work of Rouquier [55] and Varagnolo–Vasserot [65], it is equivalent to Lusztig's categorification. Integrable simple $\mathfrak{g}$-representations also admit categorifications by representations of cyclotomic quotients of quiver Hecke algebras by Kang–Kashiwara [25]. Moreover, Rouquier [53] proved the unicity of categorifications for these simple representations.

Quiver Hecke algebras have captured a tremendous amount of interest in the last decade and many important progress have been made. We do not intend to give a complete survey of the subject. Instead, we will focus on some new equivalences of categorifications and some recent applications to representation theory. Here is an outline of this report.

In Section 2 we review Lusztig's categorification of $\mathbf{U}_v^+$, quiver Hecke algebra, its realization as extension algebra of some ℓ-adic complexes, the theory of standard modules for quiver Hecke algebras and their relations to PBW basis. We also mention the monoidal categorification of the quantum cluster algebra structure on the quantum coordinate ring $\mathbf{A}_v^+$ by Kang–Kashiwara–Kim–Oh [27].

In Section 3 we discuss categorifications of quantized loop algebras. By the theory of K-theoretical Hall algebras, there is a natural categorification of the positive part of quantized loop algebras in terms of coherent sheaves on the cotangent dg-stacks of quiver representations. For a quiver of finite type, since the loop algebra is part of the corresponding affine Lie algebra, it can also be categorified using representations of quiver Hecke algebras. It is natural to ask whether these two categorifications are equivalent. An equivalence of this kind was given for $\mathfrak{sl}_2$ in [61]. It also gives an interesting comparison between two monoidal categorifications of the quantum cluster algebra structure on a quantum unipotent coordinate ring for $\widehat{\mathfrak{sl}}_2$, one in terms of quiver Hecke algebras as mentioned above and the other in terms of perverse coherent sheaves on affine Grassmannians, constructed by Cautis–Williams [15].

In Section 4 we discuss some recent applications of categorical actions to representation theory, including those of rational double affine Hecke algebras and those of finite reductive groups of classical types. We also discuss how to use categorical actions to construct representations of current algebras on the center of the underlying categories. As an application, we obtain an isomorphism between the center of cyclotomic quiver Hecke algebras and the singular cohomology of Nakajima quiver varieties in finite types. In a parallel context, we get an explicit computation of the cohomology of Gieseker moduli spaces.

2. QUIVER HECKE ALGEBRAS

2.1. Notation

For an exact (resp. triangulated) category $\mathcal{C}$, its Grothendieck group $[\mathcal{C}]$ is the quotient of the free $\mathbb{Z}$-module spanned by the isomorphism classes of objects in $\mathcal{C}$ by the relations $[M] = [M'] + [M'']$ whenever there is a short exact sequence $M' \to M \to M''$(resp. distinguished triangle). Abelian categories are naturally exact categories. Additive categories can be viewed as exact categories with short exact sequences being the split ones.

An exact category $\mathcal{C}$ is graded if it is equipped with an exact autoequivalence $\langle 1\rangle : \mathcal{C} \to \mathcal{C}$. For such a category, the Grothendieck group $[\mathcal{C}]$ is a $\mathbb{Z}[v^{\pm 1}]$-module with $v[M] = [M\langle 1\rangle]$. Here v is a formal variable. In particular, for an object $M \in \mathcal{C}$ and

$a(v) = \sum_r a_r v^r \in \mathbb{N}[v^{\pm 1}]$, we write $a(v)M = \oplus_r M\langle r\rangle^{\oplus a_r}$. We set

$$\mathrm{Hom}^{\bullet}_{\mathcal{C}}(-,-) = \bigoplus_{r\in\mathbb{Z}} \mathrm{Hom}_{\mathcal{C}}(-,-\langle r\rangle).$$

A standard example of graded category is the category of graded vector spaces over a field $\Bbbk$ with $(M\langle 1\rangle)_n = M_{n+1}$. Its Grothendieck group is isomorphic to $\mathbb{Z}[v^{\pm 1}]$, with the class of $M = \bigoplus_{n\in\mathbb{Z}} M_n$ mapping to its graded dimension $\mathrm{gdim}(M) = \sum_{n\in\mathbb{Z}}(\dim_{\Bbbk} M_n)v^n$.

An exact category $\mathcal{C}$ is monoidal if it is equipped with an exact bifunctor $\mathcal{C}\times\mathcal{C}\to\mathcal{C}$ satisfying certain associativity constraints. Such a bifunctor yields a ring structure on the abelian group $[\mathcal{C}]$. If $\mathcal{C}$ is graded monoidal, then $[\mathcal{C}]$ becomes a $\mathbb{Z}[v^{\pm 1}]$-algebra.

Let $\Bbbk$ be a field. For a graded $\Bbbk$-algebra A, let $\mathrm{Mod}(A)$ be the category of graded A-modules. Let $\mathrm{mod}(A)$, $\mathrm{proj}(A)$, $\mathrm{fmod}(A)$ be respectively the full subcategories consisting of finitely generated graded A-modules, finitely generated graded projective A-modules, and graded A-modules which are finite dimensional over $\Bbbk$.

2.2. The quantized enveloping algebra

Let I be a finite set. Fix a Cartan datum with a symmetric generalized Cartan matrix $A = (a_{ij})_{i,j\in I}$, a set of simple roots $\{\alpha_i \mid i \in I\}$, a weight lattice P, and a symmetric bilinear form $P\times P\to\mathbb{Q}$, $(\lambda,\mu)\mapsto\lambda\cdot\mu$ such that $\alpha_i\cdot\alpha_j = a_{ij}$ and $\omega_i\cdot\alpha_j = \delta_{ij}$, where $\{\omega_i \mid i\in I\}$ are the fundamental weights. Let $\mathfrak{g} = \mathfrak{g}_A$ be the associated Kac–Moody Lie algebra. Let $Q = \bigoplus_{i\in I}\mathbb{Z}\alpha_i$ be the root lattice. Set $Q^+ = \bigoplus_{i\in I}\mathbb{N}\alpha_i$ and $P^+ = \bigoplus_{i\in I}\mathbb{N}\omega_i$. Let Φ be the set of roots, and Φ^+ the set of positive roots. Let $\mathfrak{n}\subset\mathfrak{g}$ be the Lie subalgebra spanned by positive root spaces.

For $n\in\mathbb{N}$, $1\leqslant l\leqslant n$, define the following quantum integers in $\mathbb{Z}[v^{\pm 1}]$:

$$[n] = \frac{v^n - v^{-n}}{v - v^{-1}}, \quad [n]^! = \prod_{r=1}^{n}[r], \quad \left[{n \atop l}\right] = \frac{[n]^!}{[l]^![n-l]^!}.$$

The positive part ${}'\mathbf{U}_v^+$ of the quantized enveloping algebra is the unital $\mathbb{Q}(v)$-algebra generated by E_i for $i\in I$, subject to defining relations

$$\sum_{s+r=1-a_{ij}} (-1)^s E_i^{(s)} E_j E_i^{(r)} = 0, \quad \text{for } i\neq j\in I,$$

where $E_i^{(n)} = E_i^n/[n]^!$. It is a Q^+-graded algebra with $\deg(E_i) = \alpha_i$. There is a coproduct

$$\Delta : {}'\mathbf{U}_v^+ \to {}'\mathbf{U}_v^+ \otimes {}'\mathbf{U}_v^+, \quad E_i \mapsto E_i\otimes 1 + 1\otimes E_i, \quad \forall i\in I,$$

such that ${}'\mathbf{U}_v^+$ is a twisted[1] bialgebra. There is a unique nondegenerate symmetric bilinear form $(\cdot,\cdot)_v$ on ${}'\mathbf{U}_v^+$ determined by $(1,1)_v = 1$, $(E_i,E_j)_v = \delta_{i,j}/(1-v^2)$, and $(ab,c)_v = (a\otimes b,\Delta(c))_v$ for $a,b,c\in{}'\mathbf{U}_v^+$, where $(a\otimes b, c\otimes d)_v = (a,c)_v(b,d)_v$ on ${}'\mathbf{U}_v^+\otimes{}'\mathbf{U}_v^+$.

---

1 Here and below, twisted means the multiplication on ${}'\mathbf{U}_v^+\otimes{}'\mathbf{U}_v^+$ is $(a\otimes b)(c\otimes d) = v^{-\deg(b)\cdot\deg(c)}ac\otimes bd$ for homogeneous elements $a,b,c,d\in{}'\mathbf{U}_v^+$.

Let $\mathbf{U}_v^+$ be the $\mathbb{Z}[v^{\pm 1}]$-subalgebra of $'\mathbf{U}_v^+$ generated by $E_i^{(n)}$ for $i \in I, n \geqslant 1$. It is a $\mathbb{Z}[v^{\pm 1}]$-form for $'\mathbf{U}_v^+$, and its specialization at $v = 1$ is an integral form for the universal enveloping algebra of $\mathfrak{n}$. The dual form

$$\mathbf{A}_v^+ = \{x \in {'\mathbf{U}_v^+} \mid (x, y)_v \in \mathbb{Z}[v^{\pm 1}],\ \forall y \in \mathbf{U}_v^+\}$$

is called the *quantum unipotent coordinate ring*. Its specialization at $v = 1$ is an integral form for the coordinate ring of the unipotent group associated with $\mathfrak{n}$.

2.3. Quivers and Ringel's Hall algebra

A quiver $\Gamma = (I, H)$ is an oriented graph with vertices set I and arrows set H. For $h : i \to j \in H$, we write $h' = i, h'' = j$. Let h_{ij} be the total number of arrows from i to j. Assume that Γ has no edge loops, then it determines a symmetric generalized Cartan matrix $A = (a_{ij})_{i,j \in I}$ given by $a_{ii} = 2$ and $a_{ij} = -h_{ij} - h_{ji}$ for $i \neq j$. Write $\mathfrak{g}_\Gamma = \mathfrak{g}_A$. We say Γ is of finite type if $\mathfrak{g}_\Gamma$ is finite dimensional.

Fix a field $\mathbb{F}$. A Γ-representation over $\mathbb{F}$ is a pair (x, V), where $V = \bigoplus_{i \in I} V_i$ is an I-graded $\mathbb{F}$-vector space, and $x = (x_h)_{h \in H}$ is a collection of linear maps $x_h : V_{h'} \to V_{h''}$. Its dimension is the element $\sum_{i \in I} \dim_{\mathbb{F}}(V_i)\alpha_i$ in Q^+. A morphism of representations $(x, V) \to (x', V')$ is a family of linear maps $a_i : V_i \to V_i'$ such that $a_{h''} x_h = x'_h a_{h'}$ for all $h \in H$. Fix $\beta = \sum_{i \in I} d_i \alpha_i \in Q^+$, the space of Γ-representations of dimension β is

$$X_\beta = \bigoplus_{h \in H} \mathrm{Hom}_{\mathbb{F}}(\mathbb{F}^{d_{h'}}, \mathbb{F}^{d_{h''}}).$$

The group $G_\beta = \prod_{i \in I} \mathrm{GL}_{d_i}$ acts on X_β by $(gx)_h = g_{h''} x_h g_{h'}^{-1}$ for all $h \in H, g = (g_i) \in G_\beta$. Two representations are isomorphic if and only if they are in the same G_β-orbits. So the quotient stack $\mathcal{X}_\beta = [X_\beta / G_\beta]$ parametrizes the isomorphism classes of Γ-representations. For each $i \in I$, there is a unique simple Γ-representation S_i of dimension α_i with $x = 0$. All simple Γ-representations are of this form. For a sequence $\nu = (\nu_1, \nu_2, \ldots, \nu_m)$ of elements in Q^+ whose entries sum up to β, let $\tilde{\mathcal{X}}_\nu$ be the moduli stack of flags of Γ-representations $\phi_\bullet = (\phi_1 \subset \cdots \subset \phi_m)$ such that $\phi_r / \phi_{r-1} \in \mathcal{X}_{\nu_r}$ for $1 \leqslant r \leqslant m$. We have a proper morphism

$$f_\nu : \tilde{\mathcal{X}}_\nu \to \mathcal{X}_\beta, \quad \phi_\bullet \mapsto \phi_m. \tag{2.1}$$

Let $\mathcal{X}_\beta^{\mathrm{nil}}$ be the union of the image of f_ν for all possible ν. Representations in $\mathcal{X}_\beta^{\mathrm{nil}}$ are called nilpotent. If Γ has no oriented cycles, then $\mathcal{X}_\beta = \mathcal{X}_\beta^{\mathrm{nil}}$ for all β.

Let $\mathbb{F} = \mathbb{F}_q$ be a finite field. The space $\mathbb{Q}[\mathcal{X}_\beta^{\mathrm{nil}}]$ of $\mathbb{Q}$-valued functions on the finite set $\mathcal{X}_\beta^{\mathrm{nil}}$ is spanned by the characteristic functions 1_ϕ for $\phi \in \mathcal{X}_\beta^{\mathrm{nil}}$. The Hall algebra $\mathbf{H}(\Gamma, q) = \bigoplus_{\beta \in Q^+} \mathbb{Q}[\mathcal{X}_\beta^{\mathrm{nil}}]$ is an associative algebra with the multiplication given by

$$1_{\phi_1} * 1_{\phi_2} = \sum_{\phi \in \mathcal{X}_{\beta+\gamma}} c_\nu \left| f_\nu^{-1}(\phi) \cap p_\nu^{-1}(\phi_1, \phi_2) \right| 1_\phi,$$

for $\phi_1 \in \mathcal{X}_\beta, \phi_2 \in \mathcal{X}_\gamma$. Here $\nu = (\beta, \gamma)$, c_ν is some structural constant and p_ν is the morphism

$$p_\nu : \tilde{\mathcal{X}}_\nu \to \mathcal{X}_\beta \times \mathcal{X}_\gamma, \quad (\phi' \subset \phi) \mapsto (\phi', \phi/\phi'). \tag{2.2}$$

Theorem 2.1 ([52]). *The assignments $E_i \mapsto 1_{S_i}$ for $i \in I$ defines a Q^+-graded algebra embedding*

$$'\mathbf{U}_v^+|_{v=q^{-1/2}} \hookrightarrow \mathbf{H}(\Gamma, q).$$

It is an isomorphism when Γ is a quiver of finite type.

We recall some properties of $\mathcal{X}_\beta$ in two special examples which will be used later.

Example 2.2. Assume Γ is a quiver of finite type. Let $\mathbb{F} = \overline{\mathbb{F}}_q$. By a theorem of Gabriel, there is a bijection between Φ^+ and the set of isomorphism classes of indecomposable Γ-representations, sending α to M_α. There exists a total order on Φ^+ such that $\alpha' \prec \alpha$ if $\mathrm{Hom}_\Gamma(M_\alpha, M_{\alpha'}) = 0$. Fix such an order, then G_β-orbits on X_β are in bijection with descending sequences of elements in Φ^+ whose entries sum up to β. Such sequences are called *Kostant partitions*. We denote by Π_β the set of all Kostant partitions of β.

Example 2.3. Let $\mathbb{F} = \overline{\mathbb{F}}_q$. Consider the quiver $\hat{\Gamma} = (0 \rightrightarrows 1)$ associated with the affine Lie algebra $\widehat{\mathfrak{sl}}_2$. The path algebra of $\hat{\Gamma}$ is isomorphic to the self-extension algebra of the tilting bundle $\mathcal{T} = \mathcal{O}_{\mathbb{P}^1} \oplus \mathcal{O}_{\mathbb{P}^1}(1)$ on the projective line $\mathbb{P}^1$. Thus there is an equivalence of derived categories

$$\mathrm{Ext}^\bullet(\mathcal{T}, -) : D^b \,\mathrm{Coh}(\mathbb{P}^1) \simeq D^b \,\mathrm{Rep}(\hat{\Gamma}).$$

It sends the line bundle $\mathcal{O}_{\mathbb{P}^1}(k)$ for $k \geqslant 0$ to the indecomposable preprojective $\hat{\Gamma}$-representation of dimension $\alpha_1 + k\delta$. Since any vector bundle on $\mathbb{P}^1$ is isomorphic to a direct sum of line bundles, the isomorphism classes of rank r vector bundles are in bijection with the decreasing sequences in $\mathbb{Z}^r$. For $\beta = r\alpha_1 + n\delta$, let

$$\Lambda_\beta = \{(\lambda_1 \geqslant \cdots \geqslant \lambda_r) \in \mathbb{N}^r \mid \lambda_1 + \cdots + \lambda_r = n\}. \tag{2.3}$$

Let Bun_β^+ be the stack of vector bundles parametrized by Λ_β. Let $\mathcal{Y}_\beta$ be the open substack of $\mathcal{X}_\beta$ consisting of preprojective representations. Then the derived equivalence above yields an isomorphism of stacks $\mathrm{Bun}_\beta^+ \simeq \mathcal{Y}_\beta$.

2.4. Lusztig's categorification

Based on Ringel's construction and Grothendieck's sheaf–function correspondence, Lusztig constructed the following categorification of $\mathbf{U}_v^+$. Let $\mathbb{F} = \overline{\mathbb{F}}_q$ and $\Bbbk = \overline{\mathbb{Q}}_\ell$. Let $D_c^b(\mathcal{X}_\beta)$ be the bounded derived category of constructible ℓ-adic sheaves on $\mathcal{X}_\beta$. The category

$$D_c^b(\mathcal{X}) = \bigoplus_\beta D_c^b(\mathcal{X}_\beta)$$

admits a monoidal structure given by the convolution product

$$\mathcal{F}_1 * \mathcal{F}_2 = f_{\nu*} p_\nu^*(\mathcal{F}_1 \boxtimes \mathcal{F}_2)[-\beta \cdot \gamma], \quad \text{for } \mathcal{F}_1 \in D_c^b(\mathcal{X}_\beta),\ \mathcal{F}_2 \in D_c^b(\mathcal{X}_\gamma),$$

where $\nu = (\beta, \gamma)$ and p_ν, f_ν are the morphisms defined in (2.1) and (2.2). For $i \in I$, let $\mathcal{L}_i = \Bbbk_{\mathcal{X}_{\alpha_i}}[-1]$ be the (shifted) constant sheaf on $\mathcal{X}_{\alpha_i}$. For $\nu = (\nu_1, \ldots, \nu_n) \in I^n$, we have

$$\mathcal{L}_{\nu_1} * \cdots * \mathcal{L}_{\nu_n} \simeq f_{\nu*}(\Bbbk_{\tilde{\mathcal{X}}_\nu}[\dim(\tilde{\mathcal{X}}_\nu)]).$$

Denote this complex by $\mathscr{L}_\nu$. Note that by the decomposition theorem, it is a direct sum of intersection complexes on $\mathcal{X}_\beta$ up to some shifts. Let I^β be the subset of I^n consisting of sequences ν whose entries sum up to β. Set $\mathscr{L}_\beta = \bigoplus_{\nu \in I^\beta} \mathscr{L}_\nu$.

Lusztig defined a full additive subcategory $\mathcal{U}_\beta^+$ of $D_c^b(\mathcal{X}_\beta)$ which is generated by the indecomposable summands of $\mathscr{L}_\beta$ and closed under the shifts by [1]. The sum $\mathcal{U}^+ = \bigoplus_{\beta \in Q^+} \mathcal{U}_\beta^+$ is stable under convolution. Hence it is a graded monoidal subcategory of $D_c^b(\mathcal{X})$. We have an isomorphism of $\mathbb{Z}[v^{\pm 1}]$-algebras

$$[\mathcal{U}^+] \simeq \mathbf{U}_v^+$$

with $[\mathscr{L}_i] \mapsto E_i$ for $i \in I$ and $[1] \mapsto v$.

A basis of $[\mathcal{U}^+]$ as a $\mathbb{Z}[v^{\pm 1}]$-module is given by the isomorphism classes of indecomposable objects in $\mathcal{U}^+$ modulo shifts. Let $\mathbf{B}_\beta$ be the set of isomorphism classes of intersection complexes which appear as direct summands of $\mathscr{L}_\beta$ up to some shift. Then $\mathbf{B} = \bigsqcup_{\beta \in Q^+} \mathbf{B}_\beta$ is a basis of $[\mathcal{U}^+]$, called the *canonical basis*.

2.5. Quiver Hecke algebras

Let $\Bbbk$ be a field. For $i, j \in I$, set $Q_{ij}(u, v) = (-1)^{h_{ij}}(u - v)^{-a_{ij}}$ for $i \neq j$ and $Q_{ii} = 0$. For $\beta \in Q^+$ of height n, the symmetric group S_n acts on the set I^n by permutation. The subset I^β is stable under this action. Write $s_k = (k, k+1) \in S_n$ for $1 \leqslant k \leqslant n-1$.

Definition 2.4 ([33, 53]). For $\beta \in Q^+$ of height n, the quiver Hecke algebra R_β is the unital $\Bbbk$-algebra generated by $x_1, \dots, x_n$, $\tau_1, \dots, \tau_{n-1}$ and e_ν for $\nu \in I^\beta$, subject to the following defining relations:

$$x_k x_l = x_l x_k, \quad x_k e_\nu = e_\nu x_k, \quad e_\nu e_{\nu'} = \delta_{\nu,\nu'} e_\nu, \quad \sum_{\nu \in I^\beta} e_\nu = 1,$$

$$\tau_l e_\nu = e_{s_l(\nu)} \tau_l, \quad \tau_k \tau_l = \tau_l \tau_k \quad \text{if } |k - l| > 1, \quad \tau_k^2 e_\nu = Q_{\nu_k, \nu_{k+1}}(x_k, x_{k+1}) e_\nu,$$

$$(\tau_{k+1}\tau_k\tau_{k+1} - \tau_k\tau_{k+1}\tau_k) e_\nu = \delta_{\nu_k, \nu_{k+2}} \frac{Q_{\nu_k,\nu_{k+1}}(x_k, x_{k+1}) - Q_{\nu_{k+2},\nu_{k+1}}(x_{k+2}, x_{k+1})}{x_k - x_{k+2}} e_\nu,$$

$$(\tau_k x_l - x_{s_k(l)} \tau_k) e_\nu = \begin{cases} -e_\nu, & \text{if } l = k, \nu_k = \nu_{k+1}, \\ e_\nu, & \text{if } l = k+1, \nu_k = \nu_{k+1}, \\ 0, & \text{otherwise.} \end{cases}$$

It admits a $\mathbb{Z}$-grading with $\deg(e_\nu) = 0$, $\deg(x_k) = 2$, and $\deg(\tau_l e_\nu) = -a_{\nu_l, \nu_{l+1}}$.

Remark 2.5. Quiver Hecke algebras are also defined for symmetrizable Cartan datum and for more general choices of parameters Q_{ij}. These generalizations are important as they provide new categorification results beyond those with geometrical origins. But to simplify exposition, we will not discuss them in this survey.

Example 2.6. For any $i \in I$, the algebra $R_{n\alpha_i}$ is isomorphic to the nil-affine Hecke algebra NH_n. Consider the polynomial ring $\mathrm{Pol}_n = \Bbbk[x_1, \dots, x_n]$. Let Z_n be the subring of symmetric polynomials. Recall that Pol_n is a free graded Z_n-module of rank $v^{\frac{n(n-1)}{2}}[n]^!$. We have an algebra isomorphism $\rho : \mathrm{NH}_n \to \mathrm{End}_{Z_n}(\mathrm{Pol}_n)$ such that $\rho(x_k)$ is the multiplication by x_k and $\rho(\tau_l) = \frac{s_l - 1}{x_l - x_{l+1}}$ is the Demazure operator. So NH_n is a matrix algebra over Z_n. It has a unique indecomposable self-dual grade projective module $P_n = v^{-\frac{n(n-1)}{2}} \mathrm{Pol}_n$, and $\mathrm{NH}_n \simeq [n]^! P_n$ as graded NH_n-modules.

Example 2.7. Let $\mathbf{H}_{n,q}$ be the affine Hecke algebra of type A. It is generated by $X_1^{\pm 1}, \ldots, X_n^{\pm 1}, T_1, \ldots, T_{n-1}$ subject to usual defining relations. Assume $q \neq 1$. Fix a finite subset I in $\Bbbk^\times$, define a quiver Γ_q with vertices set I and arrows $i \to qi$. Its connected components are either of type A or of affine type A. Let $\mathrm{Mod}(\mathbf{H}_{n,q})_I$ be the category of $\mathbf{H}_{n,q}$-modules over which $X_1, \ldots, X_n$ acts locally finitely with eigenvalues in I. Brundan–Kleshchev **[12]** and Rouquier **[53]** proved that $\mathrm{Mod}(\mathbf{H}_{n,q})_I$ is equivalent to the category of R_n-modules over which $x_1, \ldots, x_n$ act nilpotently, where $R_n = \bigoplus_{|\beta|=n} R_\beta$ is the quiver Hecke algebra for Γ_q.

For $\beta, \gamma \in Q^+$, the element $e_{\beta,\gamma} = \sum_{\nu \in I^\beta, \nu' \in I^\gamma} e_{\nu\nu'}$ is an idempotent in $R_{\beta+\gamma}$. There is a natural algebra embedding $R_\beta \otimes_\Bbbk R_\gamma \to e_{\beta,\gamma} R_{\beta+\gamma} e_{\beta,\gamma}$. For $M \in \mathrm{Mod}(R_\beta)$, $N \in \mathrm{Mod}(R_\gamma)$, the induction

$$M \circ N = R_{\beta+\gamma} e_{\beta,\gamma} \otimes_{R_\beta \otimes_\Bbbk R_\gamma} (M \otimes_\Bbbk N).$$

yields a monoidal structure on $\mathrm{Mod}(R) = \bigoplus_{\beta \in Q^+} \mathrm{Mod}(R_\beta)$. The restriction $\mathrm{Res}_{\beta,\gamma}(-) = e_{\beta,\gamma}(-)$ is right adjoint to the induction. Both functors are exact and preserve the subcategories $\mathrm{mod}(R)$, $\mathrm{proj}(R)$, and $\mathrm{fmod}(R)$. So the Grothendieck groups of these categories become twisted bialgebras.

There are also duality functors $\circledast$ and $\sharp$ on $\mathrm{fmod}(R_\beta)$ and $\mathrm{proj}(R_\beta)$, given respectively by $M^\circledast = \mathrm{Hom}_\Bbbk(M, \Bbbk)$, $P^\sharp = \mathrm{Hom}_{R_\beta}(P, R_\beta)$, both viewed as left R_β-modules via the unique antiinvolution on R_β fixing the generators. They induce involutions on $[\mathrm{fmod}(R)]$ and on $[\mathrm{proj}(R)]$ such that $v \mapsto v^{-1}$.

Theorem 2.8 (**[33]**). *There are unique isomorphisms of twisted $\mathbb{Z}[v^{\pm 1}]$-bialgebras*

$$\begin{aligned} [\mathrm{proj}(R)] &\simeq \mathbf{U}_v^+, \quad [R_{\alpha_i}] \mapsto E_i, \quad i \in I, \\ [\mathrm{fmod}(R)] &\simeq \mathbf{A}_v^+, \end{aligned}$$

such that $([P],[N])_v = \mathrm{gdim}\,\mathrm{Hom}_R(P^\sharp, N)$ *for* $P \in \mathrm{proj}(R)$ *and* $N \in \mathrm{fmod}(R)$.

The following theorem shows that quiver Hecke algebras provide a purely algebraic description of Lusztig's category $\mathcal{U}^+$.

Theorem 2.9 (**[55, 65]**). *There is an isomorphism of graded algebras*

$$R_\beta \simeq \mathrm{Ext}^\bullet_{D^b_c(\mathcal{X}_\beta)}(\mathscr{L}_\beta, \mathscr{L}_\beta)^{\mathrm{op}}. \tag{2.4}$$

The functor $\oplus_\beta \mathrm{Ext}^\bullet_{D^b_c(\mathcal{X}_\beta)}(\mathscr{L}_\beta, -)$ *yields an equivalence of graded monoidal categories*

$$\mathcal{U}^+ \simeq \mathrm{proj}(R),$$

which sends $\mathbf{B}$ *to the classes of indecomposable self-dual projective modules.*

Remark 2.10. This theorem was extended to the setting $\mathbb{F} = \mathbb{C}$ and $\Bbbk$ a field of positive characteristic by Maksimau [44].

Remark 2.11. Kang–Kashiwara–Park [28] introduced quiver Hecke algebras for quivers with edge loops and established the analog of isomorphism (2.4) in this setting.

2.6. Representations of geometric extension algebras

Consider an arbitrary algebraic variety X over $\mathbb{F} = \overline{\mathbb{F}}_q$ equipped with an action of a reductive group G and a G-equivariant proper morphism $f : \tilde{X} \to X$ from a smooth variety $\tilde{X}$. Assume that X, G, f are defined over $\mathbb{F}_q$. Let $\mathcal{X} = [X/G]$ and $\tilde{\mathcal{X}} = [\tilde{X}/G]$. Let $\Bbbk = \overline{\mathbb{Q}}_\ell$. The push-forward of constant sheaf $\mathscr{L}_f = f_* \Bbbk_{\tilde{\mathcal{X}}}[\dim(\tilde{X})]$ is a self-dual semisimple complex in $D^b_c(\mathcal{X})$. The main player of this section is the Yoneda algebra

$$A_f = \operatorname{Ext}^\bullet_{D^b_c(\mathcal{X})}(\mathscr{L}_f, \mathscr{L}_f)^{\mathrm{op}}.$$

Many interesting algebras in representation theory arise in this way. We have just seen that quiver Hecke algebras are of this kind. Historically, Lusztig first gave such a realization for degenerate affine Hecke algebras using the Springer resolutions. There are many other examples, including Schur algebras, degenerate double affine Hecke algebras, and the mathematical definition of Coulomb branch by Braverman–Finkelberg–Nakajima.

In [32], Kato studied homological properties of A_f. Assume G acts on X with finitely many orbits $\{O_\lambda\}_{\lambda \in \Xi}$, and every point in X has a connected stabilizer in G. Then each orbit supports a unique G-equivariant simple perverse sheaf IC_λ given by the intermediate extension of $\Bbbk_{O_\lambda}[\dim O_\lambda]$ to X. By the decomposition theorem, we have

$$\mathscr{L}_f = \bigoplus_{\lambda \in \Xi_f} \mathrm{IC}_\lambda \otimes L_\lambda,$$

where L_λ are self-dual graded vector spaces and $\Xi_f = \{\lambda \in \Xi \mid L_\lambda \neq 0\}$. The set $\{L_\lambda\}_{\lambda \in \Xi_f}$ is a complete collection of nonisomorphic self-dual simple graded A_f-modules. For each λ, the A_f-module $P_\lambda = \operatorname{Ext}^\bullet_{D^b_c(\mathcal{X})}(\mathscr{L}_f, \mathrm{IC}_\lambda)$ is a projective cover of L_λ. Let $j_\lambda : O_\lambda \to X$ be the natural embedding, define the standard module as

$$\Delta_\lambda = \operatorname{Ext}^\bullet_{D^b_c(\mathcal{X})}\Big(\mathscr{L}_f, j_{\lambda *}\big(\Bbbk_{O_\lambda}[\dim O_\lambda]\big)\Big). \tag{2.5}$$

We equip Ξ with the partial order $\prec$ given by the closure relations on the orbits.

Theorem 2.12 ([32]). *Assume that* $\Xi_f = \Xi$ *and that*

(1) *the algebra* A_f *is pure of weight zero,*

(2) *the complex* IC_λ *is pointwise pure for every* $\lambda \in \Xi$.

Then the category $(\mathrm{mod}(A_f), \{\Delta_\lambda\}_{\lambda \in \Xi}, \prec)$ *is a polynomial highest weight category.*

The notion of polynomial highest weight category was introduced by Kleshchev [36]. Being of *polynomial highest weight* means that in $\mathrm{mod}(A_f)$ the projective module P_λ is filtered by standard modules Δ_μ with $\mu \succeq \lambda$ and Δ_λ appears only once as a quotient, and that $\mathrm{End}^\bullet_{A_f}(\Delta_\lambda)$ is a polynomial ring over which Δ_λ is a finitely generated free module, for all $\lambda \in \Xi$. In this case, the algebra A_f has finite global dimension and the following generalized BGG reciprocity holds:

$$[P_\lambda : \Delta_\mu]_v = [\overline{\Delta}_\mu : L_\lambda]_v, \tag{2.6}$$

where $\overline{\Delta}_\mu = \Delta_\mu \otimes_{\mathrm{End}^\bullet_{A_f}(\Delta_\mu)} \Bbbk$ is called a *proper standard module*, and $[- : -]_v$ stands for the graded multiplicities.

Remark 2.13. A version of Theorem 2.12 for $\mathbb{F} = \mathbb{C}$ and $\Bbbk$ a field of positive characteristic was established by McNamara [48], where purity conditions are replaced by parity conditions.

The theorem is applicable to quiver Hecke algebras when Γ is of finite type. Thus $\mathrm{mod}(R_\beta)$ is a polynomial highest weight category in these cases. In fact, for any quiver, the stabilizer of a point in X_β is connected, and the purity assumption (1) is satisfied. However, the finite type assumption is crucial to guarantee that $\Xi_f = \Xi$ is finite. The purity assumption (2) is proved by Lusztig [42] for finite type quiver, but unknown in general. In the affine case, one needs to modify $\mathrm{mod}(R_\beta)$ to get a similar result. Here is an example.

Example 2.14 ([61]). Let $\hat{\Gamma}$ be the Kronecker quiver. Let $\mathcal{Y}_\beta$ be the open substack of preprojective representations in $\mathcal{X}_\beta$ defined in Example 2.3. Recall that the points in $\mathcal{Y}_\beta$ are indexed by the finite set Λ_β. Let $j_\beta : \mathcal{Y}_\beta \to \mathcal{X}_\beta$ be the natural embedding. Set

$$S_\beta = \mathrm{Ext}^\bullet_{D^b_c(\mathcal{Y}_\beta)}\big(j^*_\beta(\mathcal{L}_\beta), j^*_\beta(\mathcal{L}_\beta)\big)^{\mathrm{op}}. \tag{2.7}$$

Then the category $\mathrm{mod}(S_\beta)$ is polynomial highest weight. In this case, the purity assumption (2) is proved using the isomorphism $\mathcal{Y}_\beta \simeq \mathrm{Bun}^+_\beta$ and the fact that Bun^+_β admits an affine paving.

2.7. Standard modules and PBW bases

There is also an algebraic approach for standard modules over quiver Hecke algebras, which works for symmetrizable generalized Cartan matrices as well.

Assume that $\mathfrak{g}_\Gamma$ is of finite type. A convex order on Φ^+ is a total order $\prec$ such that if $\alpha \preceq \beta$ and $\alpha + \beta$ is a root, then $\alpha \preceq \alpha + \beta \preceq \beta$. For each positive root α, and any $n \in \mathbb{N}$, a finitely generated $R_{n\alpha}$-module L is called *semicuspidal* if $\mathrm{Res}_{\lambda,\mu}(L) \neq 0$ implies λ is a sum of roots $\preceq \alpha$ and μ is a sum of roots $\succeq \alpha$. Semicuspidal modules form an abelian subcategory in $\mathrm{mod}(R_{n\alpha})$ which is equivalent to $\mathrm{mod}(\mathrm{NH}_n)$ in Example 2.6. In particular, it contains a unique self-dual simple module $L_{n\alpha}$. Let $\Delta_{n\alpha}$ be its projective cover inside this subcategory of semicuspidal modules. Then $\Delta_\alpha^{\circ n} = [n]^! \Delta_{n\alpha}$ and $L_{n\alpha} = v^{\frac{n(n-1)}{2}} L_\alpha^{\circ n}$. Recall that a Kostant partition π is of the form $(\beta_1^{m_1}, \dots, \beta_k^{m_k})$ with $\beta_1 \succ \cdots \succ \beta_k$. Set

$$\Delta_\pi = \Delta_{m_1\beta_1} \circ \cdots \circ \Delta_{m_k\beta_k}, \quad \overline{\Delta}_\pi = L_{m_1\beta_1} \circ \cdots \circ L_{m_k\beta_k}.$$

Let Π_β be the set of Kostant partitions of β, and equip it with the bilexicographic order.

Theorem 2.15 ([13]). *Assume that* $\mathfrak{g}_\Gamma$ *is of finite type.*

(1) *The category* $\mathrm{mod}(R_\beta)$ *is polynomial highest weight with* $\{\Delta_\pi\}_{\pi\in\Pi_\beta}$ *being the standard modules, and* $\{\overline{\Delta}_\pi\}_{\pi\in\Pi_\beta}$ *being the proper standard modules.*

(2) *For each* π*, the module* $\overline{\Delta}_\pi$ *has a unique simple quotient* L_π*. The set* $\{L_\pi\}_{\pi\in\Pi_\beta}$ *is a complete collection of nonisomorphic self-dual simple graded* R_β*-modules.*

Note that if the convex order on Φ^+ satisfies the property given in Example 2.2, then the standard module Δ_π here coincide with the geometrical one in (2.5).

Remark 2.16. Part (2) gives a new parametrization of simple R_β-modules. It generalizes Zelevinsky's parametrization of simple modules for affine Hecke algebras of type A in terms of multi-segments.

For each choice of a reduced expression for the longest element $w_0 = s_{i_1} \cdots s_{i_N}$ in the Weyl group, we have a convex order $\alpha_1 \succ \cdots \succ \alpha_N$ on Φ^+ with $\alpha_k = s_{i_1} \cdots s_{i_{k-1}}(\alpha_{i_k})$. Lusztig [43] defined the PBW basis for $\mathbf{U}_v^+$ as follows. The root vectors are $E_{\alpha_k} := T_{i_1} \cdots T_{i_{k-1}}(E_{i_k})$ in $\mathbf{U}_v^+$, where T_i are certain braid group operators. The dual root vectors are $E_\alpha^* = (1-v^2)E_\alpha \in \mathbf{A}_v^+$. The PBW basis is $\{E_\pi = E_{\beta_1}^{(m_1)} \cdots E_{\beta_k}^{(m_k)} \mid \pi \in \Pi\}$, and the dual PBW basis for $\mathbf{A}_v^+$ is $\{E_\pi^* = v^{s_\pi} E_{\beta_1}^{*m_1} \cdots E_{\beta_k}^{*m_k} \mid \pi \in \Pi\}$, where $s_\pi = \sum_{r=1}^k m_r(m_r-1)/2$ and $\Pi = \bigsqcup_{\beta\in Q^+} \Pi_\beta$. Under the isomorphisms $[\mathrm{proj}(R)] \simeq \mathbf{U}_v^+$ and $[\mathrm{fmod}(R)] \simeq \mathbf{A}_v^+$, we have $[\Delta_\pi] = E_\pi$, $[\overline{\Delta}_\pi] = E_\pi^*$. In particular, the homological property (2.6) implies that the transfer matrix between the PBW basis and the canonical basis is unitriangular with off-diagonal entries belong to $v\mathbb{N}[v]$. This confirms a conjecture of Lusztig.

This theory has been extended to symmetric affine type by McNamara [49] and Kleshchev–Muth [37]. For a real positive root α, the category of semicuspidal $R_{n\alpha}$-modules is again equivalent to $\mathrm{mod}(\mathrm{NH}_n)$. The new ingredient is a classification of semicuspidal representations for the imaginary roots. Once these representations are constructed, one can proceed as above to define Δ_π, $\overline{\Delta}_\pi$ indexed by (generalized) Kostant partitions. They give categorifications for the PBW basis and the dual PBW basis defined by Beck [3]. There is a similar positivity on the coefficients of the transfer matrix.

An important difference in the affine case is that the category $\mathrm{mod}(R_\beta)$ is no more polynomial highest weight, and its global dimension may be infinite. However, it has an interesting monoidal subcategory $\mathscr{D}_\beta$ with nice properties. Namely, let $\hat{\mathfrak{g}}$ be the affine Lie algebra associated with a finite Lie algebra $\mathfrak{g}$. Let $\hat{\Phi}^{++}$ be the subset of real roots $\alpha + k\delta$ such that $\alpha \in \Phi^+$, $k \geqslant 0$. We can choose a preorder on the set $\hat{\Phi}$ of affine roots such that $\hat{\Phi}^{++} \subset \hat{\Phi}_{\prec\delta}$. Let Π_β^+ be the set of Kostant partitions of β which are supported on $\hat{\Phi}^{++}$. Then the subcategory $\mathscr{D}_\beta$ of $\mathrm{mod}(R_\beta)$ generated by L_π for $\pi \in \Pi_\beta^+$ is a polynomial highest weight category. In particular, it has finite global dimension. So the category of projective objects in $\mathscr{D}_\beta$ and the derived category $D^b(\mathscr{D}_\beta)$ have the same Grothendieck group. Put

$\mathcal{D} = \bigoplus_{\beta \in Q^+} \mathcal{D}_\beta$. We have

$$[D^b(\mathcal{D})] \simeq \mathbf{U}_v(\mathfrak{n}[z]). \tag{2.8}$$

Example 2.17. For the Kronecker quiver in Example 2.3, the closure relation on the orbits in $\mathcal{Y}_\beta$ is compatible with the convex preorder $\alpha_0 \succ \alpha_0 + \delta \succ \cdots \succ \mathbb{Z}\delta \succ \cdots \succ \alpha_1 + \delta \succ \alpha_1$. We have an equivalence of categories $\mathcal{D}_\beta \simeq \mathrm{mod}(S_\beta)$, where S_β is the algebra in (2.7). We have seen in Example 2.14 that this category is polynomial highest weight. The algebraic standard modules again coincide with the geometric ones.

Remark 2.18. The algebra S_β is a semicuspidal algebra for the category $\mathcal{D}_\beta$. The semicuspidal algebra for the imaginary part was studied by Klechshev–Muth [37] and Maksimau–Minets [45].

2.8. Monoidal categorification of quantum cluster algebras

For a symmetric Kac–Moody algebra $\mathfrak{g}$ and any element w in its Weyl group, Geiß–Leclerc–Schröer [23] showed that the quantum unipotent coordinate ring $\mathbf{A}_v(\mathfrak{n}(w))$ is a quantum cluster algebra, where $\mathfrak{n}(w) = \bigoplus_{\alpha \in \Phi^+ \cap w^{-1}(\Phi^-)} \mathfrak{n}_\alpha$. A cluster algebra is a subring of the fraction field of a quantum torus, with some special elements called cluster variables, which are grouped into some overlapping subsets called clusters. The clusters are obtained from an initial one by a combinatorial procedure called mutations. A product of elements inside the same cluster is called a cluster monomial. It was conjectured that the cluster monomials in $\mathbf{A}_v(\mathfrak{n}(w))$ all belong to the dual canonical basis, see Kimura [35]. This conjecture was proved by Kang–Kashiwara–Kim–Oh [27] using a monoidal categorification of $\mathbf{A}_v(\mathfrak{n}(w))$ by modules over quiver Hecke algebras, see Kashiwara's ICM talk [30] for a nice survey on this subject.

A key ingredient in this construction is the study of products of real simple objects. A simple R_β-module L is called *real* if $L \circ L$ is simple. It was shown in [26, 27] that if either M or N is a real simple object, then $\mathrm{Hom}_R(M \circ N, N \circ M) = \Bbbk\mathbf{r}$, where $\mathbf{r} : M \circ N \to N \circ M$ is a nonzero map given by a construction called *renormalized r-matrix*. Moreover, the image of $\mathbf{r}$ is simple, isomorphic to the head of $M \circ N$, and the socle of $N \circ M$ (with grading ignored). In [27] it was shown that given the presence of a quantum cluster structure on the Grothendieck ring of a monoidal category, how renormalized r-matrices reduce the existence of iterated mutations to the existence of one step mutation.

Renormalized r-matrices naturally show up in other contexts, including finite-dimensional representations of quantum affine algebras (which was studied before quiver Hecke algebras), as well as in representations of p-adic groups, see, e.g., [38]. Recently, Cautis–Williams [15] constructed renormalized r-matrices for perverse coherent sheaves on affine Grassmannians, and used them to construct a monoidal categorification of a quantum coordinate ring for $\widehat{\mathfrak{sl}}_2$.

3. COHERENT CATEGORIFICATION OF QUANTIZED LOOP ALGEBRAS

3.1. K-theoretical Hall algebra

We have explained the categorifications of $\mathbf{U}_v^+$ by perverse sheaves on the stack of quiver representations, and its algebraic counterpart by modules over quiver Hecke algebras. Now, we discuss a categorification of the quantized loop algebra via coherent sheaves.

Let $\Gamma = (I, H)$ be a quiver, $\mathfrak{g} = \mathfrak{g}_\Gamma$ and let $\mathfrak{n}$ be the positive part in $\mathfrak{g}$. The loop algebra $\mathfrak{n}[z^{\pm 1}]$ (with z being a formal variable) is a Lie algebra with bracket $[xz^m, yz^n] = [x, y]z^{m+n}$ for $x, y \in \mathfrak{n}$.

Definition 3.1 (Drinfeld). The quantized enveloping algebra $'\tilde{\mathbf{U}}_v^+$ for $\mathfrak{n}[z^{\pm 1}]$ is the $\mathbb{Q}(v)$-algebra generated by $E_{i,n}$ with $i \in I, n \in \mathbb{Z}$, subject to the following defining relations:

(1) for $i, j \in I$, we have $(v^{a_{ij}} z - w)E_i(z)E_j(w) = (z - v^{a_{ij}} w)E_j(z)E_i(w)$,

(2) for $i \neq j$, put $l = 1 - a_{ij}$, we have the Serre relation

$$\mathrm{Sym}_z \sum_{r=0}^{l} (-1)^r \left[\begin{smallmatrix} l \\ r \end{smallmatrix}\right] E_i(z_1) \cdots E_i(z_r) E_j(w) E_i(z_{r+1}) \cdots E_i(z_l) = 0.$$

Here $z, w, z_1, \ldots, z_l$ are variables, $E_i(z) = \sum_{n \in \mathbb{Z}} E_{i,n} z^{-n}$, the operator Sym_z is averaging with respect to the commutator $[a, b]_z = ab - zba$.

Let $\tilde{\mathbf{U}}_v^+$ be the $\mathbb{Z}[v^{\pm 1}]$-subalgebra of $'\tilde{\mathbf{U}}^+$ generated by the quantum divided powers $E_{i,n}^{(r)}$ with $i \in I, n \in \mathbb{Z}, r \geqslant 1$. We explain now its relationship with K-theoretical Hall algebra.

Let $\bar{\Gamma}$ be the quiver obtained from Γ by adding an arrow $\bar{h} : h'' \to h'$ for each $h \in H$. For $\beta \in Q^+$, let $\bar{X}_\beta$ be the space of representations of $\bar{\Gamma}$ of dimension β. Then we have a natural isomorphism $\bar{X}_\beta \simeq T^* X_\beta$. The action of G_β on $\bar{X}_\beta$ is Hamiltonian with the moment map given by

$$\mu_\beta : \bar{X}_\beta \to \mathfrak{g}_\beta, \quad (x_h, x_{\bar{h}})_{h \in H} \mapsto \sum_{h \in H} [x_h, x_{\bar{h}}].$$

We impose $\mathbb{C}^\times$-actions on $\bar{X}_\beta$ and on $\mathfrak{g}_\beta$ by dilations of weight 1 and weight 2, respectively. Set $G_\beta^c = G_\beta \times \mathbb{C}^\times$. Then μ_β is G_β^c-equivariant. The cotangent dg-stack of $\mathcal{X}_\beta$ is the quotient stack

$$T^* \mathcal{X}_\beta = \left[\bar{X}_\beta \times_{\mathfrak{g}_\beta}^R \{0\} / G_\beta^c\right].$$

Here $\bar{X}_\beta \times_{\mathfrak{g}_\beta}^R \{0\}$ is the derived fiber of μ_β at zero. In concrete terms, $\bar{X}_\beta \times_{\mathfrak{g}_\beta}^R \{0\} = \mathrm{Spec}(\mathbf{A}_\beta)$, where $\mathbf{A}_\beta = S(\bar{X}_\beta) \otimes S(\mathfrak{g}_\beta[1]\langle 2 \rangle)$ is a graded dg-algebra with the differential given by the contraction by $\mu_\beta \in S^2(\bar{X}_\beta) \otimes \mathfrak{g}_\beta^* \langle -2 \rangle$. Here $\langle 1 \rangle$ is the degree shift for the internal grading induced by the $\mathbb{C}^\times$-action.

Let $D^b \mathrm{Coh}(T^* \mathcal{X}_\beta)$ be the derived category of coherent sheaves on the dg-stack $T^* \mathcal{X}_\beta$. Equivalently, it is the derived category of graded $\mathbf{A}_\beta \rtimes G_\beta$-modules whose cohomology is finitely generated over $H^0(\mathbf{A}_\beta)$. There is a convolution product on

$$D^b \mathrm{Coh}(T^* \mathcal{X}) = \bigoplus_{\beta \in Q^+} D^b \mathrm{Coh}(T^* \mathcal{X}_\beta),$$

so that it becomes a graded monoidal triangulated category. The $\mathbb{Z}[v^{\pm 1}]$-algebra $[D^b\operatorname{Coh}(T^*\mathcal{X})]$ is called the *K-theoretical Hall algebra*. We also consider a triangulated monoidal subcategory $D^b\operatorname{Coh}(T^*\mathcal{X})_{\mathrm{nil}}$ consisting of complexes with cohomology supported on a closed substack of nilpotent elements.

Example 3.2. Consider $\Gamma = \bullet$, the quiver for $\mathfrak{sl}_2$. Then $\mathfrak{n}[z^{\pm 1}] \simeq \mathbb{C}[z^{\pm 1}]$. Write $Q^+ = \mathbb{N}\alpha$. Since Γ has no arrow, we have $X_{r\alpha} = \bar{X}_{r\alpha} = \{0\}$ and $T^*\mathcal{X}_{r\alpha} = [\{0\} \times^R_{\mathfrak{gl}_r} \{0\}/\operatorname{GL}_r^c]$. Hence $\mathbf{A}_{r\alpha}$ is the exterior algebra $S(\mathfrak{gl}_r[1]\langle 2\rangle)$ with zero differential. By Koszul duality, we have

$$D^b\operatorname{Coh}(T^*\mathcal{X}_{r\alpha}) \simeq D^b\operatorname{Coh}([\mathfrak{gl}_r/\operatorname{GL}_r^c]).$$

For each irreducible GL_r-representation $V(\lambda)$ of highest weight λ, set $\mathcal{O}(\lambda)_{r\alpha} = \mathcal{O}_{\mathfrak{gl}_r} \otimes V(\lambda)$. Let $\omega_1, \ldots, \omega_r$ be the fundamental weights. Then $\mathcal{O}(n\omega_r)_{r\alpha}$ with $r \geqslant 1$ and $n \in \mathbb{Z}$ generate $\bigoplus_{r\geqslant 0} D^b\operatorname{Coh}([\mathfrak{gl}_r/\operatorname{GL}_r^c])$ as a monoidal triangulated category. We have an isomorphism of $\mathbb{Z}[v^{\pm 1}]$-algebras

$$\bigoplus_{r\geqslant 0}[D^b\operatorname{Coh}([\mathfrak{gl}_r/\operatorname{GL}_r^c])] \simeq \tilde{\mathbf{U}}_v^+, \quad [\mathcal{O}(n\omega_r)_{r\alpha}] \mapsto E^{(r)}_{\alpha,n}.$$

Now, assume that Γ is an arbitrary quiver with no edge loop. Then for each $i \in I$, we have $\mathcal{X}_{r\alpha_i} = [\{0\}/\operatorname{GL}_r]$ and the vector bundles $\mathcal{O}(n\omega_r)_{r\alpha_i} \in \operatorname{Coh}(T^*\mathcal{X}_{r\alpha_i})$ as defined above. We have the following theorem.

Theorem 3.3 ([66]). *There is a unique surjective $\mathbb{Z}[v^{\pm 1}]$-algebra homomorphism*

$$\phi : \tilde{\mathbf{U}}_v^+ \to [D^b\operatorname{Coh}(T^*\mathcal{X})_{\mathrm{nil}}], \quad E^{(r)}_{i,n} \mapsto [\mathcal{O}(n\omega_r)_{r\alpha_i}].$$

Moreover, ϕ is an isomorphism if Γ is of finite or affine type except $A_1^{(1)}$. In particular, $D^b\operatorname{Coh}(T^*\mathcal{X})_{\mathrm{nil}}$ gives a categorification of $\tilde{\mathbf{U}}_v^+$.

Remark 3.4. K-theoretical Hall algebras are constructed more generally for quivers with potential by Padurariu [51] using a category of singularities. Conjecturally, they are isomorphic to the positive part of Okounkov–Smirnov quantum affine algebras.

3.2. Equivalence of constructible and coherent categorifications

If $\mathfrak{g}$ is of finite type, its affine Lie algebra $\hat{\mathfrak{g}}$ is a central extension of the loop algebra $\mathfrak{g}[z^{\pm 1}]$. The Kac–Moody positive part $\hat{\mathfrak{n}}$ and the loop algebra $\mathfrak{n}[z^{\pm 1}]$ shares a common Lie subalgebra, which is $\mathfrak{n}[z]$.

Recall that for quiver Hecke algebras R_β of type $\hat{\mathfrak{g}}$, we have introduced the category $\mathcal{D}$ which categorifies $\mathbf{U}_v(\mathfrak{n}[z])$, see (2.8). On the other side, $\tilde{\mathbf{U}}_v^+$ is categorified by coherent sheaves on $T^*\mathcal{X}$, and $\mathbf{U}_v(\mathfrak{n}[z])$ is the subalgebra generated by divided powers $E^{(r)}_{i,n}$ for $i \in I$, $n \geqslant 0$, $r \geqslant 1$. Let $D^b\operatorname{Coh}(T^*\mathcal{X})_+$ be the triangulated subcategory of $D^b\operatorname{Coh}(T^*\mathcal{X})$ generated by $\mathcal{O}(n\omega_r)_{r\alpha_i}$ for $i \in I$, $n \geqslant 0$, $r \geqslant 1$. It also categorifies $\mathbf{U}_v(\mathfrak{n}[z])$. It is natural to ask whether these two categorifications are equivalent.

Question 3.5. Is there an equivalence of triangulated graded monoidal categories

$$D^b(\mathcal{D}) \simeq D^b \operatorname{Coh}(T^* \mathcal{X})_+$$

which induces the identity on the Grothendieck group?

In [61], a version of such an equivalence was given for $\mathfrak{g} = \mathfrak{sl}_2$. On the quiver Hecke side, we consider the category $\mathcal{D}_\beta$ attached to the Kronecker quiver in Example 2.17. The simple objects in this category are parametrized by the finite set Λ_β in (2.3). For $\beta = r\alpha_1 + n\delta$, let $D^b \operatorname{Coh}([\mathfrak{gl}_r / \mathrm{GL}_r^c])_\beta$ be the triangulated subcategory of $D^b \operatorname{Coh}([\mathfrak{gl}_r / \mathrm{GL}_r^c])$ generated by $\mathcal{O}(\lambda)_{r\alpha}$ for $\lambda \in \Lambda_\beta$, see Example 3.2. We conjecture that in this case there is an equivalence of graded monoidal categories

$$D^b(\mathcal{D}_\beta) \simeq D^b \operatorname{Coh}\big([\mathfrak{gl}_r / \mathrm{GL}_r^c]\big)_\beta.$$

Note that both categories can be viewed as categories over $\mathfrak{gl}_r // \mathrm{GL}_r$. The fiber at zero on the coherent side is $D^b \operatorname{Coh}([\mathcal{N}_r / \mathrm{GL}_r])_\beta$, where $\mathcal{N}_r \subset \mathfrak{gl}_r$ is the nilpotent cone. It has a perverse coherent t-structure defined by Arinkin–Berzukavnikov [2], whose heart $\operatorname{PCoh}([\mathcal{N}_r / \mathrm{GL}_r^c])_\beta$ is the category of equivariant perverse coherent sheaves on $\mathcal{N}_r$. The fiber at zero on the quiver Hecke side is a subcategory $\mathcal{D}_\beta^\sharp$ of $\mathcal{D}_\beta$ with the same simple objects as in $\mathcal{D}_\beta$.

Theorem 3.6 ([61]). *For $\beta = r\alpha_1 + n\delta$ with $r \geqslant 1$, there is an equivalence of graded triangulated categories*[2]

$$D^{\mathrm{perf}}(\mathcal{D}_\beta^\sharp) \simeq D^{\mathrm{perf}} \operatorname{Coh}\big([\mathcal{N}_r / \mathrm{GL}_r^c]\big)_\beta,$$

which induces an equivalence of graded abelian categories

$$\mathcal{D}_\beta^\sharp \simeq \operatorname{PCoh}\big([\mathcal{N}_r / \mathrm{GL}_r^c]\big)_\beta.$$

Further, this equivalence is compatible with the proper stratified structures on both sides.

The proof of this theorem uses a derived equivalence between $\mathcal{D}_\beta$ and the category of constructible sheaves on the stack of preprojective representations $\mathcal{Y}_\beta \simeq \mathrm{Bun}_\beta^+$ in Example 2.3, the derived geometric Satake equivalence between $D^b \operatorname{Coh}([\mathfrak{gl}_r / \mathrm{GL}_r^c])$ and the equivariant derived category of constructible sheaves on the affine Grassmannian for GL_r established by Bezrukavnikov–Finkelberg [8], and a version of Radon transform between Bun_β^+ and the affine Grassmannian.

Remark 3.7. This theorem has a similar flavor as the equivalence between two categorifications of affine Hecke algebras established by Bezrukavnikov [6].

Remark 3.8. For $w = (s_0 s_1)^N$ in the affine Weyl group of $\widehat{\mathfrak{sl}}_2$, by [27] the quantum cluster algebra $\mathbf{A}_v(w)$ has a monoidal categorification by a subcategory in $\mathcal{D}$. Cautis–Williams [15] constructed another monoidal categorification using equivariant perverse coherent sheaves on the affine Grassmannian for GL_N. The theorem above combined with a functor of Finkelberg–Fujita [22] yields a faithful functor between these two categorifications, which is expected to be an equivalence.

2 Here "perf" refers to the subcategory of perfect complexes.

4. CATEGORICAL REPRESENTATIONS AND APPLICATIONS

4.1. Categorical representations

The categorified quantum group is a monoidal $\Bbbk$-linear 2-category $\mathcal{U}$ with objects being elements in the weight lattice P, the set of 1-morphisms generated by $\mathcal{E}_i$, $\mathcal{F}_i$ for $i \in I$, and 2-morphisms generated by $x \in \mathrm{End}(\mathcal{E}_i)$, $\tau \in \mathrm{End}(\mathcal{E}_i \mathcal{E}_j)$, $\eta_i : 1 \to \mathcal{F}_i \mathcal{E}_i$, $\varepsilon_i : \mathcal{E}_i \mathcal{F}_i \to 1$, subject to a list of relations. Khovanov–Lauda [34] and Rouquier [53] independently introduced a definition of $\mathcal{U}$, with different sets of generators and relations for 2-morphisms. Brundan [11] proved that they are equivalent.

A categorical $\mathcal{U}$-representation is a 2-functor from $\mathcal{U}$ to the 2-category of $\Bbbk$-linear categories. In concrete terms, it consists of a collection of $\Bbbk$-linear categories $\{\mathcal{C}_\mu\}_{\mu \in P}$ equipped with adjoint functors $\mathcal{E}_i : \mathcal{C}_\mu \to \mathcal{C}_{\mu+\alpha_i}$, $\mathcal{F}_i : \mathcal{C}_{\mu+\alpha_i} \to \mathcal{C}_\mu$, and natural transformations x, τ, ε_i, η_i satisfying the defining relations in $\mathcal{U}$. In this case, we also say that $\mathcal{C} = \bigoplus_{\mu \in P} \mathcal{C}_\mu$ carries a categorical $\mathfrak{g}$-action.

For $\mathfrak{g}$ of type A or affine type A, a *categorical* $\mathfrak{g}$*-action* on an abelian and Artinian category $\mathcal{C}$ is equivalent to the following data (see [53]):

- a decomposition $\mathcal{C} = \bigoplus_{\mu \in P} \mathcal{C}_\mu$,
- a pair of biadjoint endofunctors $\mathcal{E}$, $\mathcal{F}$ on $\mathcal{C}$,
- natural transformations $X \in \mathrm{End}(\mathcal{E})$, $T \in \mathrm{End}(\mathcal{E}^2)$,

such that X acts on $\mathcal{E}$, $\mathcal{F}$ with eigenvalues in I, the generalized eigenfunctors $\mathcal{E}_i$, $\mathcal{F}_i$ for $i \in I$ yield a $\mathfrak{g}$-action on the Grothendieck group $[\mathcal{C}]$ such that $[\mathcal{C}_\mu]$ is the μ-weight space, and X, T satisfy defining relations for affine Hecke algebras.

Many representation categories carrie such actions, including those of symmetric groups, cyclotomic Hecke algebras, the category $\mathcal{O}$ for $\mathfrak{gl}_n$, etc. By Chuang–Rouquier [17], the existence of such a categorical action implies that the categories $\mathcal{C}_\mu$ for μ lying in the same Weyl group orbit are derived equivalent. They also constructed a crystal structure on the set of simple objets in $\mathcal{C}$. In [58], it is proved that the classes of these simple objects form a perfect basis in the Grothendieck group, which has nice unicity properties.

4.2. Minimal categorification

Let $\mathfrak{g}$ be any symmetrizable Kac–Moody algebra. For a dominant weight $\lambda \in P^+$, the irreducible $\mathfrak{g}$-representation of highest weight λ has an integral form $\mathbf{V}_v(\lambda)$, which is a quotient of $\mathbf{U}_v^+$. Let $\mathbf{V}_v^*(\lambda)$ be the dual form.

Definition 4.1. The *cyclotomic quiver Hecke algebra* R_β^λ is the $\mathbb{Z}$-graded algebra defined as the quotient of R_β by the two-sided ideal generated by $\sum_{\nu \in I^\beta} x_1^{\lambda \cdot \alpha_{\nu_1}} e_\nu$.

Kang–Kashiwara [25] proved that $\mathbf{V}_v(\lambda)$ is categorified by $\mathcal{C}_\mu = \mathrm{proj}(R_\beta^\lambda)$ for $\mu = \lambda - \beta$, with $\mathcal{F}_i : \mathrm{proj}(R_\beta^\lambda) \to \mathrm{proj}(R_{\beta+\alpha_i}^\lambda)$ given by $R_{\beta+\alpha_i}^\lambda e_{\beta,i} \otimes_{R_\beta^\lambda} -$, and the adjoint functor given by $\mathcal{E}_i(-) = e_{\beta,i}(-)$ viewed as left R_β^λ-modules. The 2-morphisms x, τ are given by multiplying with the same named generators in R_β. The representation $\mathbf{V}_v^*(\lambda)$ is categorified by $\bigoplus_{\beta \in Q^+} \operatorname{fmod}(R^\lambda_\beta)$. This result generalizes Ariki's [1] categorification theorem for cyclotomic Hecke algebras. Rouquier [53] proved that the categorification of $\mathbf{V}_v(\lambda)$ by $\Bbbk$-linear additive categories is unique.

4.3. Applications to representations of rational double affine Hecke algebras

Cyclotomic rational double affine Hecke algebra (=CRDAHA) is a special family of symplectic reflection algebras introduced by Etingof–Ginzburg [21]. They are associated with complex reflection groups $G(l, 1, n)$ and some parameters. They have a category $\mathcal{O}$, which is a highest weight cover of the category of finitely generated modules over cyclotomic Hecke algebras. This representation category can be viewed as a generalization of the q-Schur algebra, and provides an important example of category $\mathcal{O}$ associated with quantization of symplectic resolutions. The Grothendieck group of these category $\mathcal{O}$ (summed over n) can be naturally identified with the *Fock space* $\mathbf{F}_v$ of level l. The latter is a combinatorial model which gives a concrete realization of integrable $\widehat{\mathfrak{sl}}_e$-representations. Rouquier [54] conjectured that the classes of simple modules in $\mathcal{O}$ correspond to the dual canonical basis in $\mathbf{F}_v$. This yields character formulae for these simple modules in terms of affine Kazhdan–Lusztig polynomials.

In [58], a categorical $\widehat{\mathfrak{sl}}_e$-action on $\mathcal{O}$ was constructed using the induction and restriction functors defined by Bezukavnikov–Etingof [7]. Varagnolo–Vasserot [64] constructed a categorical $\widehat{\mathfrak{sl}}_e$-action on an affine type A parabolic category $\mathcal{O}$, and conjectured it should be equivalent to that for CRDAHA. This conjecture was proved independently by Losev [40] and Rouquier–Shan–Varagnolo–Vasserot [56]. As a consequence, Rouquier's conjecture was confirmed. Further, by [59], the parabolic affine category $\mathcal{O}$ admits a Koszul grading. By the equivalence above, this transfers to a Koszul grading on the category $\mathcal{O}$ of CRDAHA. Moreover, its Koszul dual is the category $\mathcal{O}$ of another CRDAHA. This confirms a conjecture of Chuang–Miyachi [16]. The Koszul duality categorifies the level–rank duality on the Fock space.

On the Fock space, there is also an interesting Heisenberg algebra action. A categorification of this action was constructed in [62] and it was used to prove a conjecture of Etingof [20] on the number of finite dimensional representations of these CRDAHA.

4.4. Applications to representations of finite reductive groups

Categorical actions are also constructed on the category of unipotent representations of classical finite algebraic groups, over a field of characteristic ℓ different from the defining characteristic. For $G = \mathrm{GL}_n(\mathbb{F}_q)$, this was done by Chuang–Rouquier [17]. For finite unitary groups and finite classical groups of type B, C, it was constructed by Dudas–Varagnolo–Vasserot [18, 19]. In all these cases, the functors $\mathcal{E}$ and $\mathcal{F}$ are given by Harish-Chandra restriction and induction functors. The underlying Grothendieck group is a level one Fock space in the case of $\mathrm{GL}_n(\mathbb{F}_q)$, and some explicit level 2 Fock spaces for the other classical types. As a consequence, Broué's abelian defect group conjecture is proved for unipotent ℓ-blocks of these groups at linear prime ℓ.

4.5. Applications to the study of center and cohomology

Let $\mathcal{C}$ be a graded $\mathbb{k}$-linear category. Denote the identity functor by $1_{\mathcal{C}}$. The center of $\mathcal{C}$ is the graded $\mathbb{k}$-algebra $Z^\bullet(\mathcal{C}) = \mathrm{End}(1_{\mathcal{C}})$. Given a pair of biadjoint endofunctors $\mathcal{E}, \mathcal{F}$ and $x \in \mathrm{End}(\mathcal{E})$, Bernstein [5] introduced the following operator:

$$Z_{\mathcal{E}}(x) : Z^\bullet(\mathcal{C}) \to Z^\bullet(\mathcal{C}),$$

$$z \mapsto \left(1_{\mathcal{C}} \xrightarrow{\eta} \mathcal{F} 1_{\mathcal{C}} \mathcal{E} \xrightarrow{\mathcal{F} z x} \mathcal{F} 1_{\mathcal{C}} \mathcal{E} \xrightarrow{\varepsilon'} 1_{\mathcal{C}} \right),$$

where η and ε' are the unit and counit in the biadjunction.

When $\mathcal{C}$ carries a categorical $\mathfrak{g}$-action, it is equipped with a family of biadjoint functors $\mathcal{E}_i, \mathcal{F}_i$ and an endomorphism $x \in \mathrm{End}(\mathcal{E}_i) \simeq \mathrm{End}(\mathcal{F}_i)^{\mathrm{op}}$. So we get a family of operators $x_{i,r}^+ = Z_{\mathcal{F}_i}(x^r)$, $x_{i,r}^- = Z_{\mathcal{E}_i}(x^r)$ for $i \in I$, $r \geqslant 0$. By Beliakova–Habiro–Lauda–Webster [4] and Shan–Varagnolo–Vasserot [60], these operators define an action of the current algebra $L\mathfrak{g}$ on $Z^\bullet(\mathcal{C})$. If $\mathfrak{g}$ is of type ADE, then $L\mathfrak{g} = \mathfrak{g}[z]$ and the operators $x_{i,r}^+, x_{i,r}^-$ correspond to $E_i \otimes z^r$, $F_i \otimes z^r$, respectively.

This construction applied to the minimal categorification in Section 4.2 allows to establish an isomorphism between the center of cyclotomic quiver Hecke algebras and the singular cohomology of quiver varieties. Quiver varieties are a family of complex symplectic varieties $\mathfrak{M}_\beta^\lambda$ introduced by Nakajima [50]. Here $\lambda \in P^+$, $\beta \in Q^+$. Nakajima defined a $\mathfrak{g}$-action on the sum over β of the middle cohomology of $\mathfrak{M}_\beta^\lambda$ with coefficient in $\mathbb{k}$. Varagnolo [63] extended this to an $L\mathfrak{g}$-action on the total cohomology $\oplus_\beta \mathrm{H}^\bullet(\mathfrak{M}_\beta^\lambda)$.

Theorem 4.2 ([4, 60]). *Assume $\mathfrak{g}$ is of type ADE. Fix $\lambda \in P^+$. There is an isomorphism of $L\mathfrak{g}$-modules*

$$\bigoplus_{\beta \in Q^+} Z^\bullet(R_\beta^\lambda) \simeq \bigoplus_{\beta \in Q^+} \mathrm{H}^\bullet(\mathfrak{M}_\beta^\lambda), \tag{4.1}$$

which respects Q^+-grading and intertwines the product on the center and the cup product on the cohomology.

This isomorphism is canonical in the following sense. It is not hard to show that the center of R_β is canonically isomorphic to $\mathrm{H}^\bullet([\mathrm{pt}/G_\beta])$. The quotient map $R_\beta \to R_\beta^\lambda$ induces a map on the center $\kappa_a : Z^\bullet(R_\beta) \to Z^\bullet(R_\beta^\lambda)$, which may not be surjective in general. On the geometrical side, the quiver variety $\mathfrak{M}_\beta^\lambda$ admits an open embedding into $[\mathrm{pt}/G_\beta]$. The pull-back gives the so-called *Kirwan map* $\kappa_g : \mathrm{H}^\bullet([\mathrm{pt}/G_\beta]) \to \mathrm{H}^\bullet(\mathfrak{M}_\beta^\lambda)$. McGerty and Nevins [47] proved that κ_g is surjective for any quiver, including those with edge loops. The isomorphism (4.1) fits into the following diagram:

$$\begin{array}{ccc} Z^\bullet(R_\beta) & \xrightarrow{\sim} & \mathrm{H}^\bullet([\mathrm{pt}/G_\beta]) \\ \downarrow{\scriptstyle \kappa_a} & & \downarrow{\scriptstyle \kappa_g} \\ Z^\bullet(R_\beta^\lambda) & \xrightarrow{(4.1)} & \mathrm{H}^\bullet(\mathfrak{M}_\beta^\lambda). \end{array}$$

Remark 4.3. Quiver varieties carry a symplectic G_λ-action and an additional $\mathbb{C}^\times$-action rescaling the symplectic form. The theorem admits a G_λ-equivariant version by considering cyclotomic quiver Hecke algebras defined over the ring $H_{G_\lambda}^\bullet(\mathrm{pt})$. However, adding

$\mathbb{C}^\times$-equivariance on the geometrical side changes the $L\mathfrak{g}$-action to a Yangian action. It is not known how to realize the Yangian action on the center side.

Remark 4.4. There is also a similar isomorphism between the cocenter $\mathrm{Tr}^\bullet(R^\lambda_\beta)$ of R^λ_β and the Borel–Moore homology of a Langrangian subvariety in $\mathfrak{M}^\lambda_\beta$. Moreover, in [4] it is proved that for $\mathfrak{g}$ of type ADE, the cocenter of the 2-category $\mathcal{U}$ is an idempotent version of $L\mathfrak{g}$. For general type of Kac–Moody algebra $\mathfrak{g}$, it is proved in [60] that $\mathrm{Tr}^\bullet(R^\lambda) = \oplus_\beta \mathrm{Tr}^\bullet(R^\lambda_\beta)$ is always a cyclic $\mathfrak{g}[z]$-module.

There is an interesting variation of this result for the Jordan quiver $\bullet\circlearrowright$.

In this case, the quiver variety $\mathfrak{M}^r_n$ is the Gieseker moduli space parametrizing framed rank r torsion-free sheaves on $\mathbb{P}^2$ with the second Chern class equal to n. It carries an action of $\mathrm{GL}_r \times \mathrm{GL}_2$, where GL_r acts on the framing and GL_2 acts on $\mathbb{P}^2$. Let $G_r = \mathrm{GL}_r \times \mathbb{C}^\times$, with $\mathbb{C}^\times = \mathrm{diag}(t, t^{-1})$ in GL_2. Let $\mathbf{k} = \mathrm{H}_{G_r}(\mathrm{pt}) = \Bbbk[\hbar][y_1, \dots, y_r]^{\mathfrak{S}_r}$ and $\mathbf{k}'$ be its fraction field. Maulik–Okounkov [46] and Schiffmann–Vasserot [57] independently proved that for fixed r, there is an affine W-algebra action on the (localized) equivariant cohomology $\mathbb{M}_r = \oplus_n \mathrm{H}^\bullet_{G_r}(\mathfrak{M}^r_n) \otimes_{\mathbf{k}} \mathbf{k}'$, confirming a version of the AGT conjecture concerning pure $N = 2$ gauge theory for the group SU_r. The quiver Hecke algebra R_n associated with the Jordan quiver is the degenerate affine Hecke algebra over the ring $\mathbf{k}$. Define its cyclotomic quotient $R^r_n = R_n/(x_1 - y_1)\cdots(x_1 - y_r)$. The quotient map induces a morphism $\kappa^r_n : Z^\bullet(R_n) \to Z^\bullet(R^r_n)$, which is only surjective after localization.

Theorem 4.5 ([60]). *Fix $r \geqslant 1$. There is an action of the affine W-algebra on*

$$\bigoplus_{n\in\mathbb{N}} Z^\bullet(R^r_n) \otimes_{\mathbf{k}} \mathbf{k}'$$

constructed using Bernstein operators. The module obtained is isomorphic to $\mathbb{M}_r$. Moreover, there is a ring isomorphism

$$\mathrm{im}(\kappa^r_n) \simeq \mathrm{H}^\bullet_{G_r}(\mathfrak{M}^r_n), \quad \forall n \in \mathbb{N}.$$

In particular, since the ring $\mathrm{im}(\kappa^r_n)$ has a presentation by generators and relations given by Brundan [10], this theorem gives an explicit description for the ring structure on $\mathrm{H}^\bullet_{G_r}(\mathfrak{M}^r_n)$. It also generalizes the results of Göttsche–Soergel [24] and Vasserot [67] for Hilbert scheme of n points on $\mathbb{C}^2$, which is $\mathfrak{M}^1_n$.

Remark 4.6. A similar description for the equivariant cohomology of Calogero–Moser spaces was established in [9].

Remark 4.7. S. Cautis, A. Lauda, A. Licata, and J. Sussan [14] showed that the cocenter of Khovanov's Heisenberg category is a quotient of the W-algebra above.

ACKNOWLEDGMENTS

I benefited greatly from discussions with G. Bellamy, R. Bezrukavnikov, C. Bonnafé, P. Etingof, I. Gordon, B. Leclerc, P. McNamara, S. Riche, R. Rouquier, O. Schiffmann, M. Varagnolo, and E. Vasserot over the years. I would like to thank them for their help.

P. SHAN

REFERENCES

[1] S. Ariki, On the decomposition numbers of the Hecke algebra of $G(m, 1, n)$. *J. Math. Kyoto Univ.* **36** (1996), no. 4, 789–808.

[2] D. Arinkin and R. Bezrukavnikov, Perverse coherent sheaves. *Mosc. Math. J.* **10** (2010), no. 1, 3–29.

[3] J. Beck, Convex bases of PBW type for quantum affine algebras. *Comm. Math. Phys.* **165** (1994), no. 1, 193–199.

[4] A. Beliakova, K. Habiro, A. Lauda, and B. Webster, Current algebras and categorified quantum groups. *J. Lond. Math. Soc. (2)* **95** (2017), no. 1, 248–276.

[5] J. Bernstein, Traces in categories. In *Operator algebras, unitary representations, enveloping algebras, and invariant theory (Paris, 1989)*, pp. 417–423, Progr. Math. 92, Birkhäuser, Boston, MA, 1990.

[6] R. Bezrukavnikov, On two geometric realizations of an affine Hecke algebra. *Publ. Math. Inst. Hautes Études Sci.* **123** (2016), 1–67.

[7] R. Bezrukavnikov and P. Etingof, Parabolic induction and restriction functors for rational Cherednik algebras. *Selecta Math. (N.S.)* **14** (2009), no. 3–4, 397–425.

[8] R. Bezrukavnikov and M. Finkelberg, Equivariant Satake category and Kostant–Whittaker reduction. *Mosc. Math. J.* **8** (2008), no. 1, 39–72.

[9] C. Bonnafé and P. Shan, On the cohomology of Calogero–Moser spaces. *Int. Math. Res. Not. IMRN* **4** (2020), 1091–1111.

[10] J. Brundan, Centers of degenerate cyclotomic Hecke algebras and parabolic category $\mathcal{O}$. *Represent. Theory* **12** (2008), 236–259.

[11] J. Brundan, On the definition of Kac–Moody 2-category. *Math. Ann.* **364** (2016), no. 1–2, 353–372.

[12] J. Brundan and A. Kleshchev, Blocks of cyclotomic Hecke algebras and Khovanov–Lauda algebras. *Invent. Math.* **178** (2009), no. 3, 451–484.

[13] J. Brundan, A. Kleshchev, and P. J. McNamara, Homological properties of finite-type Khovanov–Lauda–Rouquier algebras. *Duke Math. J.* **163** (2014), no. 7, 1353–1404.

[14] S. Cautis, A. Lauda, A. Licata, and J. Sussan, W-Algebras from Heisenberg categories. *J. Inst. Math. Jussieu* **17** (2018), no. 5, 981–1017.

[15] S. Cautis and H. Williams, Cluster theory of the coherent Satake category. *J. Amer. Math. Soc.* **32** (2019), no. 3, 709–778.

[16] J. Chuang and H. Miyachi, Hidden Hecke algebras and Koszul dualities. Preprint, 2011. http://www.math.nagoya-u.ac.jp/~miyachi/preprints/lrkszl21.pdf.

[17] J. Chuang and R. Rouquier, Derived equivalences for symmetric groups and $\mathfrak{sl}_2$-categorification. *Ann. of Math. (2)* **167** (2008), no. 1, 245–298.

[18] O. Dudas, M. Varagnolo, and E. Vasserot, Categorical actions on unipotent representations of finite classical groups. In *Categorification and higher representation theory*, pp. 41–104, Contemp. Math. 683, Amer. Math. Soc., Providence, RI, 2017.

[19] O. Dudas, M. Varagnolo, and E. Vasserot, Categorical actions on unipotent representations of finite unitary groups. *Publ. Math. Inst. Hautes Études Sci.* **129** (2019), 129–197.

[20] P. Etingof, Symplectic reflection algebras and affine Lie algebras. *Mosc. Math. J.* **12** (2012), no. 3, 543–565.

[21] P. Etingof and V. Ginzburg, Symplectic reflection algebras, Calogero–Moser space, and deformed Harish-Chandra homomorphism. *Invent. Math.* **147** (2002), no. 2, 243–348.

[22] M. Finkelberg and R. Fujita, Coherent IC-sheaves on type A_n affine Grassmannians and dual canonical basis of affine type A_1. *Represent. Theory* **25** (2021), 67–89.

[23] C. Geiß, B. Leclerc, and J. Schröer, Cluster structures on quantum coordinate rings. *Selecta Math. (N.S.)* **19** (2013), no. 2, 337–97.

[24] L. Göttsche and W. Soergel, Perverse sheaves and the cohomology of Hilbert schemes of smooth algebraic surfaces. *Math. Ann.* **296** (1993), no. 2, 235–245.

[25] S.-J. Kang and M. Kashiwara, Categorification of highest weight modules via Khovanov–Lauda–Rouquier algebras. *Invent. Math.* **190** (2012), no. 3, 699–742.

[26] S.-J. Kang, M. Kashiwara, M. Kim, and S.-J. Oh, Simplicity of heads and socles of tensor products. *Compos. Math.* **151** (2015), no. 2, 377–396.

[27] S.-J. Kang, M. Kashiwara, M. Kim, and S.-J. Oh, Monoidal categorification of cluster algebras. *J. Amer. Math. Soc.* **31** (2018), no. 2, 349–426.

[28] S.-J. Kang, M. Kashiwara, and E. Park, Geometric realization of Khovanov–Lauda–Rouquier algebras associated with Borcherds–Cartan data. *Proc. Lond. Math. Soc. (3)* **107** (2013), no. 4, 907–931.

[29] M. Kashiwara, On crystal bases of the Q-analogue of universal enveloping algebras. *Duke Math. J.* **63** (1991), no. 2, 465–516.

[30] M. Kashiwara, Crystal bases and categorifications – Chern Medal lecture. In *Proceedings of the International Congress of Mathematicians – Rio de Janeiro 2018. Vol. I. Plenary lectures*, pp. 249–258, World Sci. Publ., Hackensack, NJ, 2018.

[31] S. Kato, Poincaré–Birkhoff–Witt bases and Khovanov–Lauda–Rouquier algebras. *Duke Math. J.* **163** (2014), no. 3, 619–663.

[32] S. Kato, An algebraic study of extension algebras. *Amer. J. Math.* **139** (2017), no. 3, 567–615.

[33] M. Khovanov and A. Lauda, A diagrammatic approach to categorification of quantum groups. I. *Represent. Theory* **13** (2009), 309–347.

[34] M. Khovanov and A. Lauda, A categorification of quantum $\mathfrak{sl}(n)$. *Quantum Topol.* **1** (2010), no. 1, 1–92.

[35] Y. Kimura, Quantum unipotent subgroup and dual canonical basis. *Kyoto J. Math.* **52** (2012), no. 2, 277–331.

[36] A. Kleshchev, Affine highest weight categories and affine quasihereditary algebras. *Proc. Lond. Math. Soc. (3)* **110** (2015), no. 4, 841–882.

[37] A. Kleshchev and R. Muth, Stratifying KLR algebras of affine ADE types. *J. Algebra* **475** (2017), 133–170.

[38] E. Lapid and A. Mínguez, Conjectures and results about parabolic induction of representations of $\mathrm{GL}_n(F)$. *Invent. Math.* **222** (2020), no. 3, 695–747.

[39] A. Lascoux, B. Leclerc, and J.-Y. Thibon, Hecke algebras at roots of unity and crystal bases of quantum affine algebras. *Comm. Math. Phys.* **181** (1996), no. 1, 205–263.

[40] I. Losev, Proof of Varagnolo–Vasserot conjecture on cyclotomic categories $\mathcal{O}$. *Selecta Math. (N.S.)* **22** (2016), no. 2, 631–668.

[41] G. Lusztig, Canonical bases arising from quantized enveloping algebras. *J. Amer. Math. Soc.* **3** (1990), no. 2, 447–498.

[42] G. Lusztig, Quivers, perverse sheaves, and quantized enveloping algebras. *J. Amer. Math. Soc.* **4** (1991), no. 2, 365–421.

[43] G. Lusztig, Introduction to quantized enveloping algebras. In *New developments in Lie theory and their applications (Córdoba, 1989)*, pp. 49–65, Progr. Math. 105, Birkhäuser, Boston, MA, 1992.

[44] R. Maksimau, Canonical basis, KLR algebras and parity sheaves. *J. Algebra* **422** (2015), 563–610.

[45] R. Maksimau and A. Minets, KLR and Schur algebras for curves and semi-cuspidal representations. 2020, arXiv:2010.01419.

[46] D. Maulik and A. Okounkov, Quantum groups and quantum cohomology. *Astérisque* **408** (2019), ix+209 pp.

[47] K. McGerty and T. Nevins, Kirwan surjectivity for quiver varieties. *Invent. Math.* **212** (2018), no. 1, 161–187.

[48] P. J. McNamara, Representation theory of geometric extension algebras. 2017, arXiv:1701.07949.

[49] P. J. McNamara, Representations of Khovanov–Lauda–Rouquier algebras III: symmetric affine type. *Math. Z.* **287** (2017), no. 1–2, 243–286.

[50] H. Nakajima, Instantons on ALE spaces, quiver varieties, and Kac–Moody algebras. *Duke Math. J.* **76** (1994), no. 2, 365–416.

[51] T. Padurariu, *K-theoretic hall algebras for quivers with potential.* PhD thesis, Massachusetts Institute of Technology, 2020.

[52] C. Ringel, Hall algebras and quantum groups. *Invent. Math.* **101** (1990), no. 3, 583–591.

[53] R. Rouquier, 2-Kac–Moody algebras, 2008. arXiv:0812.5023.

[54] R. Rouquier, q-Schur algebras and complex reflection groups. *Mosc. Math. J.* **8** (2008), no. 1, 119–158.

[55] R. Rouquier, Quiver Hecke algebras and 2-Lie algebras. *Algebra Colloq.* **19** (2012), no. 2, 359–410.

[56] R. Rouquier, P. Shan, M. Varagnolo, and E. Vasserot, Categorifications and cyclotomic rational double affine Hecke algebras. *Invent. Math.* **204** (2016), no. 3, 671–786.

[57] O. Schiffmannand and E. Vasserot, Cherednik algebras, W-algebras and the equivariant cohomology of the moduli space of instantons on $\mathbb{A}^2$. *Publ. Math. Inst. Hautes Études Sci.* **118** (2013), 213–342.
[58] P. Shan, Crystals of Fock spaces and cyclotomic rational double affine Hecke algebras. *Ann. Sci. Éc. Norm. Supér. (4)* **44** (2011), no. 1, 147–182.
[59] P. Shan, M. Varagnolo, and E. Vasserot, Koszul duality of affine Kac–Moody algebras and cyclotomic rational double affine Hecke algebras. *Adv. Math.* **262** (2014), 370–435.
[60] P. Shan, M. Varagnolo, and E. Vasserot, On the center of quiver Hecke algebras. *Duke Math. J.* **166** (2017), no. 6, 1005–1101.
[61] P. Shan, M. Varagnolo, and E. Vasserot, Coherent categorification of quantum loop algebras: the $SL(2)$ case. 2019, arXiv:1912.03325.
[62] P. Shan and E. Vasserot, Heisenberg algebras and rational double affine Hecke algebras. *J. Amer. Math. Soc.* **25** (2012), no. 4, 959–1031.
[63] M. Varagnolo, Quiver varieties and Yangians. *Lett. Math. Phys.* **53** (2000), no. 4, 273–283.
[64] M. Varagnolo and E. Vasserot, Cyclotomic double affine Hecke algebras and affine parabolic category $\mathcal{O}$. *Adv. Math.* **225** (2010), no. 3, 1523–1588.
[65] M. Varagnolo and E. Vasserot, Canonical bases and KLR-algebras. *J. Reine Angew. Math.* **659** (2011), 67–100.
[66] M. Varagnolo and E. Vasserot, K-Theoretic Hall algebras, quantum groups and super quantum groups. 2021, arXiv:2011.01203.
[67] E. Vasserot, Sur l'anneau de cohomologie du schéma de Hilbert de $\mathbb{C}^2$. *C. R. Acad. Sci. Paris Sér. I Math.* **332** (2001), no. 1, 7–12.

PENG SHAN

Department of Mathematical Sciences and Yau Mathematical Sciences Centre, Tsinghua University, 100084, Beijing, China, pengshan@tsinghua.edu.cn

This Page Intentionally Left Blank

THETA CORRESPONDENCE AND THE ORBIT METHOD

BINYONG SUN AND CHEN-BO ZHU

ABSTRACT

The theory of theta correspondence, initiated by R. Howe, provides a powerful method of constructing irreducible admissible representations of classical groups over local fields. For archimedean local fields, a principle of great importance is the orbit method introduced by A. A. Kirillov, and it seeks to describe irreducible unitary representations of a Lie group by its coadjoint orbits. In this article, we examine implications of Howe's theory for the orbit method and unitary representation theory, with a focus on a recent work of Barbasch, Ma, and the authors on the construction and classification of special unipotent representations of real classical groups (in the sense of Arthur and Barbasch-Vogan).

MATHEMATICS SUBJECT CLASSIFICATION 2020

Primary 22E46; Secondary 22E45, 22E47

KEYWORDS

Orbit method, special unipotent representation, classical group, theta correspondence, associated cycle

ICM INTERNATIONAL CONGRESS OF MATHEMATICIANS 2022 JULY 6–14

Proc. Int. Cong. Math. 2022, Vol. 4, pp. 3062–3078
DOI 10.4171/ICM2022/23

1. THETA LIFTING: THE BASIC CONSTRUCTION

Classical invariant theory, as expounded by H. Weyl [40], is the study of the polynomial invariants for an arbitrary number of (contravariant or covariant) variables for a standard classical group action. A related theme is the study of the isotypic decomposition of the full tensor algebra for such an action. It is well known that Weyl's approach to classical invariant theory yields in particular a full description of all irreducible rational representations of a classical group. See [16] and [11, 18] for a modern treatment. The theory of theta correspondences, initiated by R. Howe in the 1970s, is a transcendental version and a profound generalization of classical invariant theory [14, 17]. The theory includes both global and local aspects, and has been investigated extensively and by many authors. We will focus on the archimedean local aspect and will thus be concerned with admissible representations of classical Lie groups.

Let W be a finite-dimensional real symplectic vector space with symplectic form $\langle \cdot, \cdot \rangle_W : W \times W \to \mathbb{R}$. Denote by σ the anti-involution of $\mathrm{End}_{\mathbb{R}}(W)$ specified by

$$\langle x \cdot u, v \rangle_W = \langle u, x^\sigma \cdot v \rangle_W, \quad u, v \in W,\ x \in \mathrm{End}_{\mathbb{R}}(W).$$

Then the symplectic group is $\mathrm{Sp}(W) = \{x \in \mathrm{End}_{\mathbb{R}}(W) \mid x^\sigma x = 1\}$. Let (A, A') be a pair of σ-stable semisimple $\mathbb{R}$-subalgebras of $\mathrm{End}_{\mathbb{R}}(W)$ that are mutual centralizers of each other. Put $G := A \cap \mathrm{Sp}(W)$ and $G' := A' \cap \mathrm{Sp}(W)$, which are closed subgroups of $\mathrm{Sp}(W)$. Following Howe [14], the pair of groups (G, G') is called a reductive dual pair in $\mathrm{Sp}(W)$. The dual pair (G, G') is said to be irreducible if the algebra A (or equivalently, A') is either simple or the product of two simple algebras that are exchanged by σ.

Every reductive dual pair is uniquely a product of irreducible dual pairs, and complete classification of irreducible reductive dual pairs has been given by Howe [14, 28], as described in what follows. Let (D, σ_0) be one of the following seven pairs so that D is an $\mathbb{R}$-algebra and σ_0 is an anti-involution of D:

$$(\mathbb{R}, \text{identity map}), \quad (\mathbb{C}, \text{identity map}), \quad (\mathbb{C}, \bar{\ }), \quad (\mathbb{H}, \bar{\ }),$$
$$\big(\mathbb{R} \times \mathbb{R}, (x, y) \mapsto (y, x)\big), \quad \big(\mathbb{C} \times \mathbb{C}, (x, y) \mapsto (y, x)\big), \quad \big(\mathbb{H} \times \mathbb{H}, (x, y) \mapsto (\bar{y}, \bar{x})\big),$$

where $\mathbb{H}$ denotes the algebra of Hamiltonian quaternions, and $\bar{\ }$ indicates the complex or quaternionic conjugation.

Let $\epsilon = \pm 1$. Let V be an ϵ-Hermitian right D-module, namely a free right D-module of finite rank, equipped with a nondegenerate $\mathbb{R}$-bilinear map

$$\langle \cdot, \cdot \rangle_V : V \times V \to \mathrm{D}$$

such that

$$\langle ua, v \rangle_V = \langle u, v \rangle_V a, \quad \langle u, v \rangle_V = \epsilon \big(\langle v, u \rangle_V \big)^{\sigma_0}, \quad \text{for all } u, v \in V,\ a \in \mathrm{D}.$$

This $\mathbb{R}$-bilinear map is called the ϵ-Hermitian form on V. The isometry group $\mathrm{G}(V)$ is a classical Lie group, namely, a real orthogonal group, a real symplectic group, a complex orthogonal group, a complex symplectic group, a unitary group, a quaternionic symplectic group, a quaternionic orthogonal group, a real general linear group, a complex general linear group, or a quaternionic general linear group.

Let V' be an ϵ'-Hermitian right D-module, equipped with the ϵ'-Hermitian form $\langle\cdot,\cdot\rangle_{V'}$, where $\epsilon\epsilon' = -1$. Let $W := \mathrm{Hom}_{\mathrm{D}}(V, V')$, equipped with the symplectic form $\langle\cdot,\cdot\rangle_W$ given by

$$\langle T, S\rangle_W := \mathrm{Tr}_{\mathbb{R}}(T^*S), \qquad T, S \in \mathrm{Hom}_{\mathrm{D}}(V, V'),$$

where $\mathrm{Tr}_{\mathbb{R}}(T^*S)$ is the trace of T^*S as a $\mathbb{R}$-linear transformation, and $T^* \in \mathrm{Hom}_{\mathrm{D}}(V', V)$ is the adjoint of T defined by

$$\langle Tv, v'\rangle_{V'} = \langle v, T^*v'\rangle_V, \quad \text{for all } v \in V, v' \in V'. \tag{1.1}$$

There is a natural homomorphism: $\mathrm{G}(V) \times \mathrm{G}(V') \to \mathrm{Sp}(W)$ given by

$$(g, g') \cdot T = g'Tg^{-1} \quad \text{for } T \in \mathrm{Hom}_{\mathrm{D}}(V, V'),\ g \in G,\ g' \in G'.$$

If both V and V' are nonzero, then $\mathrm{G}(V)$ and $\mathrm{G}(V')$ are both identified with subgroups of $\mathrm{Sp}(W)$, and $(\mathrm{G}(V), \mathrm{G}(V'))$ is an irreducible reductive dual pair in $\mathrm{Sp}(W)$. Moreover, all irreducible reductive dual pairs arise in this way.

Now we return to the general setting so that (G, G') is an arbitrary reductive dual pair in $\mathrm{Sp}(W)$. Write $\mathrm{H}(W) := W \times \mathbb{R}$ for the Heisenberg group with group multiplication

$$(u, t) \cdot (u', t') = (u + u', t + t' + \langle u, u'\rangle_W), \quad u, u' \in W,\ t, t' \in \mathbb{R}.$$

Its center is obviously identified with $\mathbb{R}$. Fix a nontrivial unitary character $\psi : \mathbb{R} \to \mathbb{C}^\times$. Recall the Stone–von Neumann Theorem which asserts that up to isomorphism, there exists a unique irreducible unitary representation of $\mathrm{H}(W)$ with central character ψ.

Let $\tilde{G}$ and $\tilde{G}'$ be a pair of reductive Lie groups together with surjective Lie group homomorphisms $\tilde{G} \to G$ and $\tilde{G}' \to G'$. The group $\tilde{G} \times \tilde{G}'$ acts on the Heisenberg group $\mathrm{H}(W)$ as group automorphisms through its obvious action on W. Using this action, we define the Jacobi group

$$J := (\tilde{G} \times \tilde{G}') \ltimes \mathrm{H}(W).$$

Suppose that J has a unitary representation $\widehat{\omega}$ whose restriction $\widehat{\omega}|_{\mathrm{H}(W)}$ to $\mathrm{H}(W)$ is irreducible with central character ψ. All such representations, if they exist, are isomorphic to each other up to twisting by unitary characters. We fix one such $\widehat{\omega}$ and write ω for the space of smooth vectors of $\widehat{\omega}|_{\mathrm{H}(W)}$, which is J-stable and is a smooth representation of J. We will refer to ω as a smooth oscillator representation.

Remark. A typical pair $(\tilde{G}, \tilde{G}')$ is obtained by taking the inverse image of (G, G') in $\widetilde{\mathrm{Sp}}(W)$, where $\widetilde{\mathrm{Sp}}(W)$ is the real metaplectic group, namely the unique double cover of $\mathrm{Sp}(W)$ that is nonsplit whenever W is nonzero. It is well known that smooth oscillator representations exist in this setting **[14, 39]**. For the related issue of splittings, see **[23]**.

Let π be a Casselman–Wallach representation of $\tilde{G}$, whose contragredient representation is denoted by $\pi^\vee$. (We refer the reader to **[38, CHAPTER 11]** for generalities on Casselman–Wallach representations.) The full theta lift of π is defined to be

$$\Theta_{\tilde{G}}^{\tilde{G}'}(\pi) := (\omega \widehat{\otimes} \pi^\vee)_{\tilde{G}},$$

which is a Casselman–Wallach representation of $\tilde{G}'$. Here and henceforth, $\widehat{\otimes}$ indicates the completed projective tensor product, and a subscript group indicates the Hausdorff coinvariant space. The theta lift $\theta_{\tilde{G}}^{\tilde{G}'}(\pi)$ of π is defined to be the largest semisimple quotient of $\Theta_{\tilde{G}}^{\tilde{G}'}(\pi)$. The following result is one formulation of Howe's duality theorem.

Theorem 1.1 ([17]). *Suppose that π is irreducible. Then $\theta_{\tilde{G}}^{\tilde{G}'}(\pi)$ is irreducible or zero.*

By reversing the role of $\tilde{G}$ and $\tilde{G}'$, Theorem 1.1 implies that the theta lift is injective in the following sense: for any irreducible Casselman–Wallach representations π_1 and π_2 of $\tilde{G}$, if $\theta_{\tilde{G}}^{\tilde{G}'}(\pi_1) \cong \theta_{\tilde{G}}^{\tilde{G}'}(\pi_2) \neq \{0\}$, then $\pi_1 \cong \pi_2$.

2. THETA LIFTING VIA MATRIX COEFFICIENT INTEGRALS AND PRESERVATION OF UNITARITY

Let V be an ϵ-Hermitian right D-module as in Section 1. Fix a maximal compact subgroup K_V of $\mathrm{G}(V)$. Recall that an element $g \in \mathrm{G}(V)$ is said to be hyperbolic if the linear operator $g \otimes 1 : V \otimes_{\mathbb{R}} \mathbb{C} \to V \otimes_{\mathbb{R}} \mathbb{C}$ is diagonalizable and all its eigenvalues are positive real numbers. Denote by Ψ_V the function of $\mathrm{G}(V)$ satisfying the following conditions:

- it is both left and right K_V-invariant;
- for all hyperbolic elements $g \in \mathrm{G}(V)$,

$$\Psi_V(g) = \prod_a \left(\frac{1+a}{2}\right)^{-\frac{1}{2}},$$

  where a runs over all eigenvalues of $g \otimes 1 : V \otimes_{\mathbb{R}} \mathbb{C} \to V \otimes_{\mathbb{R}} \mathbb{C}$, counted with multiplicities.

Note that $0 < \Psi_V(g) \leqslant 1$ for all $g \in \mathrm{G}(V)$.

Denote by Ξ_V the bi-K_V-invariant Harish-Chandra's Ξ function on $\mathrm{G}(V)$. (For a convenient reference, see [37].) Put

$$\nu_V := \mathrm{rank}_{\mathrm{D}}(V) - \frac{2\dim_{\mathbb{R}}\{t \in \mathrm{D} \mid t^{\sigma_0} = \epsilon t\}}{\dim_{\mathbb{R}}(\mathrm{D})}.$$

If $\mathrm{G}(V)$ is noncompact, then ν_V is the smallest real number such that

$$\Psi_V^{\nu_V} \cdot \Xi_V^{-1} \text{ is bounded.}$$

Given $\nu \in \mathbb{R}$, a positive function Ψ on $\mathrm{G}(V)$ is said to be ν-bounded if there is a real number $r > 0$ such that

$$\Psi(kak') \leqslant \left(\log(3 + \mathrm{Tr}_{\mathbb{R}}(a))\right)^r \cdot \Psi_V^{\nu}(a) \cdot \Xi_V(a)$$

for all $k, k' \in K_V$ and all hyperbolic elements $a \in \mathrm{G}(V)$.

In the rest of this section, we assume that $G = \mathrm{G}(V)$, $G' = \mathrm{G}(V')$, $W = \mathrm{Hom}_{\mathrm{D}}(V, V')$, and both V and V' are nonzero so that (G, G') is an irreducible dual pair in $\mathrm{Sp}(W)$. Let $\tilde{G} \to G$, $\tilde{G}' \to G'$, J, $\widehat{\omega}$, and ω be as in Section 1.

Definition 2.1. A Casselman–Wallach representation π of $\tilde{G}$ is said to be ν-bounded if there exist a ν-bounded positive function Ψ on G and continuous seminorms $|\cdot|_\pi$ and $|\cdot|_{\pi^\vee}$ on π and $\pi^\vee$, respectively, such that

$$\left|\langle \tilde{g} \cdot u, v\rangle\right| \leqslant \Psi(g) \cdot |u|_\pi \cdot |v|_{\pi^\vee}$$

for all $u \in \pi$, $v \in \pi^\vee$, and $\tilde{g} \in \tilde{G}$, where g denotes the image of $\tilde{g}$ under the homomorphism $\tilde{G} \to G$.

For a complex vector space E, denote by $\bar{E}$ its complex conjugate. Thus $\bar{E}$ is a complex vector space equipped with a conjugate linear isomorphism $E \to \bar{E}$, $v \mapsto \bar{v}$. In the setting of Section 1, $\bar{\omega}$ is a smooth representation of J in the obvious way, and the inner product on $\widehat{\omega}$ induces a J-invariant continuous bilinear form

$$\langle \cdot, \cdot \rangle : \omega \times \bar{\omega} \to \mathbb{C}.$$

Write Z for the kernel of the homomorphism $\tilde{G} \to G$, and denote by χ_Z the unitary character of Z by which Z acts on ω.

Let π be a Casselman–Wallach representation of $\tilde{G}$. Assume that π is genuine, namely Z acts on π by the character χ_Z.

Definition 2.2. The Casselman–Wallach representation π of $\tilde{G}$ is convergent for $\Theta_{\tilde{G}}^{\tilde{G}'}$ if it is ν-bounded for some $\nu > \nu_V - \mathrm{rank}_{\mathrm{D}}(V')$.

Suppose that π is convergent for $\Theta_{\tilde{G}}^{\tilde{G}'}$. Then the integral

$$\begin{aligned} &\omega \times \pi^\vee \times \bar{\omega} \times \pi \to \mathbb{C}, \\ &(\phi, v', \phi', v) \mapsto \int_G \langle \tilde{g} \cdot \phi, \phi' \rangle \cdot \langle \tilde{g} \cdot v', v\rangle \, dg, \end{aligned} \tag{2.1}$$

is absolutely convergent **[6]** and defines a continuous multilinear map, where dg is a fixed Haar measure on G, and $\tilde{g} \in \tilde{G}$ is an element whose image under the homomorphism $\tilde{G} \to G$ equals g.

The map (2.1) yields a continuous bilinear map

$$(\omega \widehat{\otimes} \pi^\vee) \times (\bar{\omega} \widehat{\otimes} \pi) \to \mathbb{C}. \tag{2.2}$$

Define

$$\bar{\theta}_{\tilde{G}}^{\tilde{G}'}(\pi) := \frac{\omega \widehat{\otimes} \pi^\vee}{\text{the left kernel of (2.2)}}. \tag{2.3}$$

This is a quotient of $\Theta_{\tilde{G}}^{\tilde{G}'}(\pi)$, and hence a Casselman–Wallach representation of $\tilde{G}'$.

Remark. The idea of studying theta lifting by matrix coefficient integrals, as in (2.3), first appeared in Li's work **[25, 26]**.

Definition 2.3. The Casselman–Wallach representation π of $\tilde{G}$ is overconvergent for $\Theta_{\tilde{G}}^{\tilde{G}'}$ if it is ν-bounded for some $\nu > \nu_V^\circ - \mathrm{rank}_{\mathrm{D}}(V')$, where

$$\nu_V^\circ := \begin{cases} \nu_V + 1, & \text{if } G \text{ is a real or complex odd orthogonal group;} \\ \nu_V + \frac{1}{2}, & \text{if } G \text{ is a quaternionic symplectic or quaternionic orthogonal group;} \\ \nu_V, & \text{otherwise.} \end{cases}$$

The idea that one could produce interesting sets of unitary representations from theta lifting is due to Howe [15]. The following result gives a sufficient condition for the preservation of unitarity (see [12, 13, 25, 26] for some earlier results along the same direction).

Theorem 2.4 ([6]). *Assume that* $\mathrm{rank}_{\mathrm{D}}(V') \geq \nu_V^{\circ}$, *and* π *is overconvergent for* $\Theta_{\tilde{G}}^{\tilde{G}'}$. *If* π *is unitarizable, then so is* $\bar{\theta}_{\tilde{G}}^{\tilde{G}'}(\pi)$.

Remark. Given that $\bar{\theta}_{\tilde{G}}^{\tilde{G}'}(\pi)$ is unitarizable, it is clearly a semisimple quotient of $\Theta_{\tilde{G}}^{\tilde{G}'}(\pi)$. If, in addition, π is irreducible and $\bar{\theta}_{\tilde{G}}^{\tilde{G}'}(\pi) \neq \{0\}$, then the fundamental result of Howe implies that $\theta_{\tilde{G}}^{\tilde{G}'}(\pi) = \bar{\theta}_{\tilde{G}}^{\tilde{G}'}(\pi)$ and is irreducible.

Conjecture 2.5. *Suppose that* π *is irreducible and convergent for* $\Theta_{\tilde{G}}^{\tilde{G}'}$. *Then* $\theta_{\tilde{G}}^{\tilde{G}'}(\pi) = \bar{\theta}_{\tilde{G}}^{\tilde{G}'}(\pi)$ *as quotients of* $\Theta_{\tilde{G}}^{\tilde{G}'}(\pi)$.

Remark. When π is not convergent for $\Theta_{\tilde{G}}^{\tilde{G}'}$, by the doubling method and by taking the leading coefficient of the local zeta integral ([32] and [24, SECTION 3]), we may still define a continuous bilinear map as in (2.2), and therefore $\bar{\theta}_{\tilde{G}}^{\tilde{G}'}(\pi)$. We expect that the statement of Conjecture 2.5 remains true for any irreducible π, whether or not it is convergent for $\Theta_{\tilde{G}}^{\tilde{G}'}$. It will be interesting to establish a version of Theorem 2.4 in this more general setting.

3. ALGEBRAIC THETA LIFTING AND BOUND VIA MOMENT MAPS

We continue with the notation of Section 2, and further assume that the homomorphisms $\tilde{G} \to G$ and $\tilde{G}' \to G'$ are finite fold covering maps. We fix a choice of maximal compact subgroups K of G and K' of G', compatible with a given choice of maximal compact subgroup U of $\mathrm{Sp}(W)$. Let $\Omega \subset \omega$ be the Harish-Chandra module associated to U, which is naturally a $(\mathfrak{g} \times \mathfrak{g}', \tilde{K} \times \tilde{K}')$-module. Here and as usual, $\mathfrak{g}$ and $\mathfrak{g}'$ denote the complexified Lie algebras of G and G', respectively, and $\tilde{K} \subset \tilde{G}$ and $\tilde{K}' \subset \tilde{G}'$ are respectively the preimages of K and K'.

Let Π be a $(\mathfrak{g}, \tilde{K})$-module of finite length, whose Harish-Chandra dual is denoted by $\Pi^{\vee}$. The (algebraic) full theta lift of Π is defined to be

$$\Theta_V^{V'}(\Pi) := (\Omega \otimes \Pi^{\vee})_{\mathfrak{g}, \tilde{K}} \quad \text{(the coinvariant space)}.$$

The $(\mathfrak{g}', \tilde{K}')$-module $\Theta_V^{V'}(\Pi)$ is of finite length [17].

We will be concerned with the so-called associated cycles of $\Theta_V^{V'}(\Pi)$.

3.1. The associated cycle map

We recall basic notions from the theory of associated varieties [35]. The theory is a key part of Vogan's formulation of the orbit method for reductive Lie groups [34, 36].

Write $V_{\mathbb{C}} := V \otimes_{\mathbb{R}} \mathbb{C}$, which is a right $\mathrm{D} \otimes_{\mathbb{R}} \mathbb{C}$-module. The $\mathbb{R}$-bilinear map $\langle \cdot, \cdot \rangle_V : V \times V \to \mathrm{D}$ extends to a $\mathbb{C}$-bilinear map $\langle \cdot, \cdot \rangle_{V_{\mathbb{C}}} : V_{\mathbb{C}} \times V_{\mathbb{C}} \to \mathrm{D} \otimes_{\mathbb{R}} \mathbb{C}$. Write $G_{\mathbb{C}}$ for the isometry group of $(V_{\mathbb{C}}, \langle \cdot, \cdot \rangle_{V_{\mathbb{C}}})$, which is a complexification of G. Write $K_{\mathbb{C}}$ and $\tilde{K}_{\mathbb{C}}$ for the complexifications of the compact groups K and $\tilde{K}$, respectively. The space $V'_{\mathbb{C}}$ and the groups $G'_{\mathbb{C}}$, $K'_{\mathbb{C}}$, and $\tilde{K}'_{\mathbb{C}}$ are similarly defined. We identify $\mathfrak{g}$ with its dual space $\mathfrak{g}^*$ by using the trace form

$$\mathfrak{g} \times \mathfrak{g} \to \mathbb{C}, \quad (x, y) \mapsto \text{the trace of the } \mathbb{C}\text{-linear endomorphism } xy : V_{\mathbb{C}} \to V_{\mathbb{C}}.$$

Likewise, $\mathfrak{g}'$ is identified with $\mathfrak{g}'^*$.

Let $\mathrm{Nil}_{G_{\mathbb{C}}}(\mathfrak{g})$ be the set of nilpotent $G_{\mathbb{C}}$-orbits in $\mathfrak{g}$. Suppose that $\mathcal{O} \in \mathrm{Nil}_{G_{\mathbb{C}}}(\mathfrak{g})$. We say that a finite length $(\mathfrak{g}, \tilde{K})$-module Π is *$\mathcal{O}$-bounded* if the associated variety of the annihilator ideal in $\mathcal{U}(\mathfrak{g})$ (the universal enveloping algebra of $\mathfrak{g}$) is contained in the Zariski closure $\overline{\mathcal{O}}$ of $\mathcal{O}$. Denote by

$$\mathfrak{g} = \mathfrak{k} \oplus \mathfrak{p}$$

the complexified Cartan decomposition fixed by our choice of the maximal compact subgroup K of G, and by $\mathrm{Nil}_{K_{\mathbb{C}}}(\mathfrak{p})$ the set of nilpotent $K_{\mathbb{C}}$-orbits in $\mathfrak{p}$. It follows from **[34, THEOREM 8.4]** that Π is $\mathcal{O}$-bounded if and only if its associated variety $\mathrm{AV}(\Pi)$ is contained in $\overline{\mathcal{O}} \cap \mathfrak{p}$. Let $\mathcal{M}_{\mathcal{O}}(\mathfrak{g}, \tilde{K})$ denote the category of $\mathcal{O}$-bounded finite length $(\mathfrak{g}, \tilde{K})$-modules, and write $\mathcal{K}_{\mathcal{O}}(\mathfrak{g}, \tilde{K})$ for its Grothendieck group.

Under the adjoint action of $K_{\mathbb{C}}$, the complex variety $\mathcal{O} \cap \mathfrak{p}$ is a union of finitely many orbits, each of dimension $\frac{\dim_{\mathbb{C}} \mathcal{O}}{2}$. For any $K_{\mathbb{C}}$-orbit $\mathscr{O} \subset \mathcal{O} \cap \mathfrak{p}$, let $\mathcal{K}_{\mathscr{O}}(\tilde{K}_{\mathbb{C}})$ denote the Grothendieck group of the category of $\tilde{K}_{\mathbb{C}}$-equivariant algebraic vector bundles on $\mathscr{O}$, and $\mathcal{K}^+_{\mathscr{O}}(\tilde{K}_{\mathbb{C}})$ the submonoid generated by the $\tilde{K}_{\mathbb{C}}$-equivariant algebraic vector bundles. Taking the isotropy representation at a point $X \in \mathscr{O}$ yields an identification

$$\mathcal{K}_{\mathscr{O}}(\tilde{K}_{\mathbb{C}}) = \mathcal{R}\big((\tilde{K}_{\mathbb{C}})_X\big),$$

where the right-hand side denotes the Grothendieck group of the category of algebraic representations of the stabilizer group $(\tilde{K}_{\mathbb{C}})_X$.

Put

$$\mathcal{K}_{\mathcal{O}}(\tilde{K}_{\mathbb{C}}) := \bigoplus_{\mathscr{O} \text{ is a } K_{\mathbb{C}}\text{-orbit in } \mathcal{O} \cap \mathfrak{p}} \mathcal{K}_{\mathscr{O}}(\tilde{K}_{\mathbb{C}})$$

and

$$\mathcal{K}^+_{\mathcal{O}}(\tilde{K}_{\mathbb{C}}) := \bigoplus_{\mathscr{O} \text{ is a } K_{\mathbb{C}}\text{-orbit in } \mathcal{O} \cap \mathfrak{p}} \mathcal{K}^+_{\mathscr{O}}(\tilde{K}_{\mathbb{C}}).$$

There is a partial order $\preceq$ on $\mathcal{K}_{\mathcal{O}}(\tilde{K}_{\mathbb{C}})$ defined by

$$\mathcal{E}_1 \preceq \mathcal{E}_2 \Leftrightarrow \mathcal{E}_2 - \mathcal{E}_1 \in \mathcal{K}^+_{\mathcal{O}}(\tilde{K}_{\mathbb{C}}), \quad \mathcal{E}_1, \mathcal{E}_2 \in \mathcal{K}_{\mathcal{O}}(\tilde{K}_{\mathbb{C}}).$$

According to Vogan **[34, THEOREM 2.13]**, we have a canonical homomorphism, called the associated cycle map:

$$\mathrm{AC}_{\mathcal{O}} : \mathcal{K}_{\mathcal{O}}(\mathfrak{g}, \tilde{K}) \to \mathcal{K}_{\mathcal{O}}(\tilde{K}_{\mathbb{C}}).$$

For a $(\mathfrak{g}, \tilde{K})$-module of finite length Π which is $\mathcal{O}$-bounded, we call $\mathrm{AC}_{\mathcal{O}}(\Pi)$ the associated cycle of Π. This is a fundamental invariant attached to Π.

3.2. The moment maps

Put $\mathcal{W} = \mathrm{Hom}_{\mathrm{D}\otimes_{\mathbb{R}}\mathbb{C}}(V_{\mathbb{C}}, V'_{\mathbb{C}}) = W \otimes_{\mathbb{R}} \mathbb{C}$. Recall we have the moment maps [10,22]

$$\mathfrak{g} \xleftarrow{\ \mathcal{M}\ } \mathcal{W} \xrightarrow{\ \mathcal{M}'\ } \mathfrak{g}'$$

that are given by

$$\mathcal{M}(\phi) = \phi^*\phi \quad \text{and} \quad \mathcal{M}'(\phi) = \phi\phi^*.$$

Here ϕ^* denotes the adjoint map as in (1.1).

As in [6, SECTION 3], we may find "Cartan transforms" L on $V_{\mathbb{C}}$, L' on $V'_{\mathbb{C}}$, and $\mathcal{L}$ on $\mathcal{W}$ which will induce compatible Cartan involutions on G, G', and $\mathrm{Sp}(W)$, respectively. Then $K_{\mathbb{C}} = G^L_{\mathbb{C}}$ (the centralizer of L) and $K'_{\mathbb{C}} = (G'_{\mathbb{C}})^{L'}$.

We decompose

$$\mathcal{W} = \mathcal{X} \oplus \mathcal{Y} \tag{3.1}$$

where $\mathcal{X}$ and $\mathcal{Y}$ are $\sqrt{-1}$ and $-\sqrt{-1}$ eigenspaces of $\mathcal{L}$, respectively. We have the following two algebraic maps [30]:

$$\begin{array}{ccccc} \mathfrak{p} & \xleftarrow{M=\mathcal{M}|_{\mathcal{X}}} & \mathcal{X} & \xrightarrow{M'=\mathcal{M}'|_{\mathcal{X}}} & \mathfrak{p}', \\ \phi^*\phi & \longleftarrow & \phi & \longmapsto & \phi\phi^*. \end{array}$$

These two maps M and M' are also called the moment maps. They are both $K_{\mathbb{C}} \times K'_{\mathbb{C}}$-equivariant. Here $K'_{\mathbb{C}}$ acts trivially on $\mathfrak{p}$, $K_{\mathbb{C}}$ acts trivially on $\mathfrak{p}'$, and all the other actions are the obvious ones.

Put

$$\mathcal{W}^\circ := \{\phi \in \mathcal{W} \mid \text{the image of } \phi^* \text{ is nondegenerate with respect to } \langle\cdot,\cdot\rangle_{V_{\mathbb{C}}}\}$$

and

$$\mathcal{X}^\circ := \mathcal{X} \cap \mathcal{W}^\circ.$$

Lemma 3.1 ([6]). *Let $\mathcal{O}'$ be a $K'_{\mathbb{C}}$-orbit in $\mathfrak{p}'$. Suppose that $\mathcal{O}'$ is contained in the image of the moment map M'. Then the set*

$$(M')^{-1}(\mathcal{O}') \cap \mathcal{X}^\circ \tag{3.2}$$

is a single $K_{\mathbb{C}} \times K'_{\mathbb{C}}$-orbit. Moreover, for every element ϕ in $(M')^{-1}(\mathcal{O}') \cap \mathcal{X}^\circ$, there is an exact sequence of algebraic groups,

$$1 \to (K_{\mathbb{C}})_\phi \to (K_{\mathbb{C}} \times K'_{\mathbb{C}})_\phi \xrightarrow{\text{the projection to the second factor}} (K'_{\mathbb{C}})_{e'} \to 1,$$

where $e' := M'(\phi) \in \mathcal{O}'$, and a subscript element indicates the stabilizer group of the element.

In the notation of Lemma 3.1, write

$$\nabla(\mathcal{O}') := \text{the image of the set (3.2) under the moment map } M,$$

which is a $K_{\mathbb{C}}$-orbit in $\mathfrak{p}$. This is called the descent of $\mathcal{O}'$. It is an element of $\mathrm{Nil}_{K_{\mathbb{C}}}(\mathfrak{p})$ if $\mathcal{O}' \in \mathrm{Nil}_{K'_{\mathbb{C}}}(\mathfrak{p}')$.

Now suppose that we have a $G'_{\mathbb{C}}$-orbit $\mathcal{O}' \subset \mathfrak{g}'$, which is contained in the image of the moment map $\mathcal{M}'$. Similar to the first assertion of Lemma 3.1, the set

$$(\mathcal{M}')^{-1}(\mathcal{O}') \cap \mathcal{W}^{\circ} \tag{3.3}$$

is a single $G_{\mathbb{C}} \times G'_{\mathbb{C}}$-orbit. Write

$$\nabla(\mathcal{O}') := \text{the image of the set (3.3) under the moment map } \mathcal{M},$$

which is a $G_{\mathbb{C}}$-orbit in $\mathfrak{g}$. This is called the descent of $\mathcal{O}'$. It is an element of $\mathrm{Nil}_{G_{\mathbb{C}}}(\mathfrak{g})$ if $\mathcal{O}' \in \mathrm{Nil}_{G'_{\mathbb{C}}}(\mathfrak{g}')$.

3.3. Geometric theta lift

We are in the setting of Section 3. We assume that the choice of $\mathcal{X}$ in (3.1) is compatible with Ω, as follows. As a module for the Lie algebra $\mathfrak{h}(W)$, Ω is the submodule of ω generated by $\omega^{\mathcal{X}}$ (the invariant space of $\mathcal{X}$, which is one-dimensional). Write $\tilde{U} \to U$ for the double cover of U induced by the metaplectic double cover $\widetilde{\mathrm{Sp}}(W) \to \mathrm{Sp}(W)$. Recall that Ω is naturally an $(\mathfrak{sp}(\mathcal{W}), \tilde{U})$-module. Recall also from [6, SECTION 5.1] that $\tilde{K}_{\mathbb{C}} \times \tilde{K}'_{\mathbb{C}}$ acts on $\omega^{\mathcal{X}}$ by a character, henceforth denoted by ζ.

We are now back in the setting of Lemma 3.1, with $\mathscr{O}' \in \mathrm{Nil}_{K'_{\mathbb{C}}}(\mathfrak{p}')$. Write $\mathscr{O} := \nabla(\mathscr{O}')$, and let $e := M(\phi)$.

Let $\mathcal{E}$ be a $\tilde{K}_{\mathbb{C}}$-equivariant algebraic vector bundle over $\mathscr{O}$. Its fiber $\mathcal{E}_e$ at e is an algebraic representation of the stabilizer group $(\tilde{K}_{\mathbb{C}})_e$, which is the preimage of $(K_{\mathbb{C}})_e$. We also view it as a representation of the group $(\tilde{K}_{\mathbb{C}} \times \tilde{K}'_{\mathbb{C}})_\phi$ via the pull-back through the homomorphism

$$\left(K_{\mathbb{C}} \times K'_{\mathbb{C}}\right)_\phi \xrightarrow{\text{the projection to the first factor}} (K_{\mathbb{C}})_e.$$

We may thus view $\mathcal{E}_e \otimes \zeta$ as a representation of $(\tilde{K}_{\mathbb{C}} \times \tilde{K}'_{\mathbb{C}})_\phi$ and, by taking the coinvariant space $(\mathcal{E}_e \otimes \zeta)_{(\tilde{K}_{\mathbb{C}})_\phi}$, we get an algebraic representation of $(\tilde{K}'_{\mathbb{C}})_{e'}$. Write $\mathcal{E}' := \check{\vartheta}_{\mathscr{O}}^{\mathscr{O}'}(\mathcal{E})$ for the $\tilde{K}'_{\mathbb{C}}$-equivariant algebraic vector bundle over $\mathscr{O}'$ whose fiber at e' equals this coinvariant space representation. In this way, we get an exact functor $\check{\vartheta}_{\mathscr{O}}^{\mathscr{O}'}$ from the category of $\tilde{K}_{\mathbb{C}}$-equivariant algebraic vector bundle over $\mathscr{O}$ to the category of $\tilde{K}'_{\mathbb{C}}$-equivariant algebraic vector bundle over $\mathscr{O}'$. This exact functor induces a homomorphism of the Grothendieck groups:

$$\check{\vartheta}_{\mathscr{O}}^{\mathscr{O}'} : \mathcal{K}_{\mathscr{O}}(\tilde{K}_{\mathbb{C}}) \to \mathcal{K}_{\mathscr{O}'}(\tilde{K}'_{\mathbb{C}}).$$

The above homomorphism is independent of the choice of ϕ in Lemma 3.1.

Now let $\mathcal{O} := \nabla(\mathcal{O}')$, where $\mathcal{O}' \in \mathrm{Nil}_{G'_{\mathbb{C}}}(\mathfrak{g}')$. We define the geometric theta lift to be the homomorphism

$$\check{\vartheta}_{\mathcal{O}}^{\mathcal{O}'} : \mathcal{K}_{\mathcal{O}}(\tilde{K}_{\mathbb{C}}) \to \mathcal{K}_{\mathcal{O}'}(\tilde{K}'_{\mathbb{C}})$$

such that

$$\check{\vartheta}_{\mathcal{O}}^{\mathcal{O}'}(\mathcal{E}) = \sum_{\mathscr{O}' \text{ is a } K'_{\mathbb{C}}\text{-orbit in } \mathcal{O}' \cap \mathfrak{p}',\ \nabla(\mathscr{O}') = \mathscr{O}} \check{\vartheta}_{\mathscr{O}}^{\mathscr{O}'}(\mathcal{E}),$$

for any $K_{\mathbb{C}}$-orbit $\mathscr{O}$ in $\mathcal{O} \cap \mathfrak{p}$, and any $\tilde{K}_{\mathbb{C}}$-equivariant algebraic vector bundle $\mathcal{E}$ over $\mathscr{O}$.

A basic result in algebraic theta lifting is the following theorem.

Theorem 3.2 ([6]). *Suppose that $\mathcal{O} := \nabla(\mathcal{O}')$ and $\mathcal{O}'$ is regular for ∇ (see* [6, DEFINITION 7.6]*). Let Π be an $\mathcal{O}$-bounded $(\mathfrak{g}, \tilde{K})$-module of finite length. Then $\Theta_V^{V'}(\check{\Pi})$ is $\mathcal{O}'$-bounded, and*

$$\mathrm{AC}_{\mathcal{O}'}\big(\Theta_V^{V'}(\check{\Pi})\big) \preceq \check{\vartheta}_{\mathcal{O}}^{\mathcal{O}'}\big(\mathrm{AC}_{\mathcal{O}}(\Pi)\big).$$

Remark. Earlier results on the associated cycles of $\Theta_V^{V'}(\check{\Pi})$ appeared in [27, 29, 31].

4. COMBINATORIAL PARAMETERS FOR SPECIAL UNIPOTENT REPRESENTATIONS

In [33], Vogan proposed that the orbit method (introduced by A. A. Kirillov [19]; see also [21] for an extension in geometric terms) should serve as a unifying principle in the description of the unitary duals of reductive Lie groups. Furthermore, the quantization problem (attaching irreducible unitary representations to coadjoint orbits) should involve three steps in accordance with the Jordan decomposition of the element representing an (co)adjoint orbit, and in the order of the nilpotent step, the elliptic step, and the hyperbolic step. The elliptic and hyperbolic steps are implemented by cohomological and parabolic induction, respectively, and are well understood. The nilpotent step is the most difficult, and is the theory of unipotent representations [34, 35], which is still in development. We refer the reader to [36] for a comprehensive account of Vogan's conception of the orbit method for reductive Lie groups.

We will be concerned with special unipotent representations, which originated in Arthur's work [3, 4] and are defined by Vogan and Barbasch [2, 8]. It will turn out that all special unipotent representations of classical Lie groups can be constructed via iterated theta lifts, supplemented by irreducible unitary parabolic inductions. We take even real orthogonal groups and real symplectic groups as examples, and will construct a (combinatorially defined) parameter set which underlies the special unipotent representations of both groups.

For every Young diagram $\imath$, write $\mathscr{R}_i(\imath)$ and $\mathscr{C}_i(\imath)$ ($i \in \mathbb{N}^+$, the set of positive integers), respectively, for its ith row length and ith column length. Let $\check{\mathcal{O}}$ be a nonempty Young diagram which satisfies the following good parity condition (for type D and C):

$$\text{All nonzero row lengths of } \check{\mathcal{O}} \text{ are odd.} \tag{4.1}$$

Put

$$m := |\check{\mathcal{O}}| := \sum_{i=1}^{\infty} \mathscr{R}_i(\check{\mathcal{O}}) \quad \text{and} \quad l := \mathscr{C}_1(\check{\mathcal{O}}).$$

We define a pair $(\imath_{\check{\mathcal{O}}}, \jmath_{\check{\mathcal{O}}})$ of Young diagrams such that the nonzero column lengths are given by

$$\begin{cases} \mathscr{C}_i(\imath_{\check{\mathcal{O}}}) = \dfrac{\mathscr{R}_{2i}(\check{\mathcal{O}}) + 1}{2}, & 1 \leqslant i \leqslant \dfrac{l-1}{2}, \\ \mathscr{C}_i(\jmath_{\check{\mathcal{O}}}) = \dfrac{\mathscr{R}_{2i-1}(\check{\mathcal{O}}) - 1}{2}, & 1 \leqslant i \leqslant \dfrac{l+1}{2}, \end{cases}$$

if l is odd, and

$$\begin{cases} \mathscr{C}_i(\imath_{\check{\mathcal{O}}}) = \dfrac{\mathscr{R}_{2i-1}(\check{\mathcal{O}})+1}{2}, & 1 \leqslant i \leqslant \dfrac{l}{2}, \\ \mathscr{C}_i(\jmath_{\check{\mathcal{O}}}) = \dfrac{\mathscr{R}_{2i}(\check{\mathcal{O}})-1}{2}, & 1 \leqslant i \leqslant \dfrac{l}{2}, \end{cases}$$

if l is even.

For any Young diagram $\imath$, we introduce the set $\mathrm{BOX}(\imath)$ of boxes of $\imath$ as the following subset of $\mathbb{N}^+ \times \mathbb{N}^+$:

$$\mathrm{BOX}(\imath) := \{(i,j) \in \mathbb{N}^+ \times \mathbb{N}^+ \mid j \leqslant \mathscr{R}_i(\imath)\}.$$

We introduce five symbols $\bullet$, s, r, c, and d, and make the following definition.

Definition 4.1. A painting on a Young diagram $\imath$ is a map

$$\mathcal{P} : \mathrm{Box}(\imath) \to \{\bullet, s, r, c, d\}$$

with the following properties:

- $\mathcal{P}^{-1}(S)$ is the set of boxes of a Young diagram when $S = \{\bullet\}, \{\bullet, s\}, \{\bullet, s, r\}$, or $\{\bullet, s, r, c\}$;
- when $S = \{s\}$ or $\{r\}$, every row of $\imath$ has at most one box in $\mathcal{P}^{-1}(S)$;
- when $S = \{c\}$ or $\{d\}$, every column of $\imath$ has at most one box in $\mathcal{P}^{-1}(S)$.

Definition 4.2. Define $\mathrm{PBP}(\check{\mathcal{O}})$ to be the set of all pairs $(\mathcal{P}, \mathcal{Q})$, where $\mathcal{P}$ and $\mathcal{Q}$ are paintings on $\imath_{\check{\mathcal{O}}}$ and $\jmath_{\check{\mathcal{O}}}$, respectively, subject to the following conditions:

- $\mathcal{P}^{-1}(\bullet) = \mathcal{Q}^{-1}(\bullet)$;
- the image of $\mathcal{P}$ is contained in

$$\begin{cases} \{\bullet, r, c, d\}, & \text{if } l \text{ is odd}, \\ \{\bullet, s, r, c, d\}, & \text{if } l \text{ is even}. \end{cases}$$

- the image of $\mathcal{Q}$ is contained in

$$\begin{cases} \{\bullet, s\}, & \text{if } l \text{ is odd}, \\ \{\bullet\}, & \text{if } l \text{ is even}. \end{cases}$$

Let $\tau = (\mathcal{P}, \mathcal{Q}) \in \mathrm{PBP}(\check{\mathcal{O}})$. We associate a classical group G_τ as follows.

If l is odd, define $G_\tau := \mathrm{Sp}_{m-1}(\mathbb{R})$.

If l is even, define the signature (p_τ, q_τ) by counting the various symbols appearing in $(\imath_{\check{\mathcal{O}}}, \mathcal{P})$, $(\jmath_{\check{\mathcal{O}}}, \mathcal{Q})$:

$$\begin{cases} p_\tau := (\#\bullet) + 2(\#r) + (\#c) + (\#d), \\ q_\tau := (\#\bullet) + 2(\#s) + (\#c) + (\#d). \end{cases}$$

Here

$$\#\bullet := \#\big(\mathcal{P}^{-1}(\bullet)\big) + \#\big(\mathcal{Q}^{-1}(\bullet)\big) \quad (\#\text{ indicates the cardinality of a finite set}),$$

and the other terms are similarly defined. Define $G_\tau := \mathrm{O}(p_\tau, q_\tau)$. In addition, define $\varepsilon_\tau \in \mathbb{Z}/2\mathbb{Z}$ such that $\varepsilon_\tau = 0$ if and only if the symbol d occurs in the first column of $\mathcal{P}$ or $\mathcal{Q}$.

If $l > 1$, we define $\check{\mathcal{O}}'$ to be the Young diagram obtained from $\check{\mathcal{O}}$ by removing the first row. The descent map

$$\nabla : \mathrm{PBP}(\check{\mathcal{O}}) \to \mathrm{PBP}(\check{\mathcal{O}}')$$

is defined in [6, SECTION 2] and plays a crucial role in our construction of special unipotent representations.

Example. Let

$$\check{\mathcal{O}} = \begin{array}{|c|c|c|c|c|c|c|}\hline \ & \ & \ & \ & \ & \ & \ \\ \hline \ & \ & \ & \ & \ & \ & \ \\ \hline \ & \ & \ & \ & \ & \ & \ \\ \hline \ & \ & \ & \multicolumn{4}{c}{} \\ \cline{1-3} \end{array}.$$

Then

$$(\imath_{\check{\mathcal{O}}}, \jmath_{\check{\mathcal{O}}}) = \left(\begin{array}{|c|c|}\hline \ & \ \\ \hline \ & \ \\ \hline \ & \ \\ \hline \ & \ \\ \hline \end{array}\, , \begin{array}{|c|c|}\hline \ & \ \\ \hline \ & \multicolumn{1}{c}{} \\ \cline{1-1} \ & \multicolumn{1}{c}{} \\ \cline{1-1} \end{array} \right)$$

and

$$\check{\mathcal{O}}' = \begin{array}{|c|c|c|c|c|c|c|}\hline \ & \ & \ & \ & \ & \ & \ \\ \hline \ & \ & \ & \ & \ & \ & \ \\ \hline \ & \ & \ & \multicolumn{4}{c}{} \\ \cline{1-3} \end{array}.$$

Also let

$$\tau = (\mathcal{P}, \mathcal{Q}) = \left(\begin{array}{|c|c|}\hline \bullet & \bullet \\ \hline \bullet & s \\ \hline \bullet & s \\ \hline r & d \\ \hline \end{array}\, , \begin{array}{|c|c|}\hline \bullet & \bullet \\ \hline \bullet & \multicolumn{1}{c}{} \\ \cline{1-1} \bullet & \multicolumn{1}{c}{} \\ \cline{1-1} \end{array} \right) \in \mathrm{PBP}(\check{\mathcal{O}}).$$

Then $G_\tau = \mathrm{O}(11, 13)$, $\varepsilon_\tau = 1$, and

$$\nabla(\tau) = (\mathcal{P}', \mathcal{Q}') = \left(\begin{array}{|c|}\hline \bullet \\ \hline \bullet \\ \hline \bullet \\ \hline d \\ \hline \end{array}\, , \begin{array}{|c|c|}\hline \bullet & s \\ \hline \bullet & \multicolumn{1}{c}{} \\ \cline{1-1} \bullet & \multicolumn{1}{c}{} \\ \cline{1-1} \end{array} \right) \in \mathrm{PBP}(\check{\mathcal{O}}').$$

Define $\mathrm{PP}(\check{\mathcal{O}})$ to be the set of all $i \in \mathbb{N}^+$ such that

$$\mathbf{R}_i(\check{\mathcal{O}}) > \mathbf{R}_{i+1}(\check{\mathcal{O}}) > 0 \quad \text{and} \quad i \equiv l \pmod 2.$$

Put

$$\mathrm{PBP}^{\mathrm{ext}}(\check{\mathcal{O}}) := \mathrm{PBP}(\check{\mathcal{O}}) \times \big\{\wp \subset \mathrm{PP}(\check{\mathcal{O}})\big\}.$$

For each $(\tau, \wp) \in \mathrm{PBP}^{\mathrm{ext}}(\check{\mathcal{O}})$, we will construct a representation $\pi_{\tau,\wp}$ of G_τ.

5. SPECIAL UNIPOTENT REPRESENTATIONS OF CLASSICAL LIE GROUPS

As in Section 4, let $\check{\mathcal{O}}$ be a nonempty Young diagram which satisfies the good parity condition (4.1) and $(\tau, \wp) \in \mathrm{PBP}^{\mathrm{ext}}(\check{\mathcal{O}})$. Let $G := G_\tau$, whose complexification $G_{\mathbb{C}}$ equals $\mathrm{Sp}_{m-1}(\mathbb{C})$ or $\mathrm{O}_m(\mathbb{C})$, respectively, when l is odd or even. The Langlands dual of $G_{\mathbb{C}}$ is defined to be $\mathrm{O}_m(\mathbb{C})$. Identify $\check{\mathcal{O}}$ with the corresponding nilpotent $\mathrm{O}_m(\mathbb{C})$-orbit in $\mathfrak{o}_m(\mathbb{C})$. Take an $\mathfrak{sl}_2$-triple $(\check{e}, \check{h}, \check{f})$ in $\mathfrak{o}_m(\mathbb{C})$ such that $\check{e} \in \check{\mathcal{O}}$. Then $\frac{1}{2}\check{h}$ is a semisimple element of $\mathfrak{o}_m(\mathbb{C})$, which determines a character $\chi(\check{\mathcal{O}}) : \mathcal{U}(\mathfrak{g})^{G_{\mathbb{C}}} \to \mathbb{C}$ in the usual way [4,8]. By a well-known result of Dixmier [9, SECTION 3], we know that there is a unique maximal G-stable ideal of $\mathcal{U}(\mathfrak{g})$ that contains the kernel of $\chi(\check{\mathcal{O}})$. Write $I_{\check{\mathcal{O}}}$ for this ideal. The associated variety of $I_{\check{\mathcal{O}}}$ is the closure of a nilpotent orbit $\mathcal{O} \in \mathrm{Nil}_{G_{\mathbb{C}}}(\mathfrak{g})$ which is called the Barbasch–Vogan dual of $\check{\mathcal{O}}$. Following Barbasch and Vogan [2,8], an irreducible Casselman–Wallach representation π of G is said to be special unipotent attached to $\check{\mathcal{O}}$ if $I_{\check{\mathcal{O}}}$ annihilates π. Write $\mathrm{Unip}_{\check{\mathcal{O}}}(G)$ for the set of isomorphism classes of irreducible Casselman–Wallach representations of G that are special unipotent attached to $\check{\mathcal{O}}$.

Put

$$\mathrm{Unip}(\check{\mathcal{O}}) := \begin{cases} \mathrm{Unip}_{\check{\mathcal{O}}}(\mathrm{Sp}_{m-1}(\mathbb{R})), & \text{if } l \text{ is odd,} \\ \bigsqcup_{p,q\in\mathbb{N},p+q=m} \mathrm{Unip}_{\check{\mathcal{O}}}(\mathrm{O}(p,q)), & \text{if } l \text{ is even.} \end{cases}$$

We have the following result on the counting of special unipotent representations.

Theorem 5.1 ([6,7]). *Let $\check{\mathcal{O}}$ be a nonempty Young diagram which satisfies the good parity condition* (4.1). *Then*

$$\#\big(\mathrm{Unip}(\check{\mathcal{O}})\big) = \begin{cases} \#(\mathrm{PBP}^{\mathrm{ext}}(\check{\mathcal{O}})), & \text{if } l \text{ is odd,} \\ 2\#(\mathrm{PBP}^{\mathrm{ext}}(\check{\mathcal{O}})), & \text{if } l \text{ is even.} \end{cases}$$

For each $(\tau, \wp) \in \mathrm{PBP}^{\mathrm{ext}}(\check{\mathcal{O}})$, we shall now construct an irreducible Casselman–Wallach representation $\pi_{\tau,\wp}$ of G by induction on l. First assume that $l = 1$, namely the Young diagram $\check{\mathcal{O}}$ has only one row. Then $G = \mathrm{Sp}_{m-1}(\mathbb{R})$, and the set $\mathrm{PBP}^{\mathrm{ext}}(\check{\mathcal{O}})$ has a unique element. In this case, we define $\pi_{\tau,\wp}$ to be the trivial representation of G.

Now assume that the Young diagram of $\check{\mathcal{O}}$ has at least two rows. Write $\tau' := \nabla(\tau) \in \mathrm{PBP}(\check{\mathcal{O}}')$, and define

$$\wp' := \{i \in \mathbb{N}^+ \mid i + 1 \in \wp\} \subset \mathrm{PP}(\check{\mathcal{O}}').$$

Write $m' := |\check{\mathcal{O}}'|$ and $G' := G_{\tau'}$. Note that G and G' form a reductive dual pair in $\mathrm{Sp}(W)$, where W is a real symplectic space of dimension $(m-1)m'$ or $m(m'-1)$, respectively, when l is odd or even. Let $J = (G \times G') \ltimes \mathrm{H}(W)$ and let ω be as in Section 1. If G is an even orthogonal group, we assume G acts trivially on the one-dimensional space ω_X (the coinvariant space of X), for every G-stable Lagrangian subspace X of W. Similar assumption is made when G' is an even orthogonal group.

By induction hypothesis, we have an irreducible Casselman–Wallach representation $\pi_{\tau',\wp'}$ of G'. We define

$$\pi_{\tau,\wp} := \begin{cases} \Theta_{G'}^{G}(\pi_{\tau',\wp'}^{\vee} \otimes \det^{\varepsilon_\wp}), & \text{if } l \text{ is odd,} \\ \Theta_{G'}^{G}(\pi_{\tau',\wp'}^{\vee}) \otimes (1_{p_\tau,q_\tau}^{+,-})^{\varepsilon_\tau}, & \text{if } l \text{ is even.} \end{cases} \tag{5.1}$$

Here $1_{p_\tau,q_\tau}^{+,-}$ denotes the character of $\mathrm{O}(p_\tau,q_\tau)$ whose restriction to $\mathrm{O}(p_\tau) \times \mathrm{O}(q_\tau)$ equals $1 \otimes \det$ (1 stands for the trivial character), and $\varepsilon_\wp$ denote the element in $\mathbb{Z}/2\mathbb{Z}$ such that

$$\varepsilon_\wp = 1 \Leftrightarrow 1 \in \wp.$$

It turns out that the representation $\pi_{\tau,\wp}$ remains unchanged if we replace $\Theta_{G'}^{G}$ by $\theta_{G'}^{G}$ or $\bar{\theta}_{G'}^{G}$ in (5.1).

Theorem 5.2 ([6]). *Let $\check{\mathcal{O}}$ be a nonempty Young diagram which satisfies the good parity condition* (4.1).

(a) *For every $(\tau,\wp) \in \mathrm{PBP}^{\mathrm{ext}}(\check{\mathcal{O}})$, the representation $\pi_{\tau,\wp}$ of G_τ is irreducible, unitarizable, and special unipotent attached to $\check{\mathcal{O}}$.*

(b) *Suppose that l is odd so that $G = \mathrm{Sp}_{m-1}(\mathbb{R})$. Then the map*

$$\mathrm{PBP}^{\mathrm{ext}}(\check{\mathcal{O}}) \to \mathrm{Unip}_{\check{\mathcal{O}}}(G), \quad (\tau,\wp) \mapsto \pi_{\tau,\wp}$$

is bijective.

(c) *Suppose that l is even, and p, q are nonnegative integers with $p+q=m$. Then the map*

$$\big\{(\tau,\wp) \in \mathrm{PBP}^{\mathrm{ext}}(\check{\mathcal{O}}) \mid (p_\tau,q_\tau) = (p,q)\big\} \times \mathbb{Z}/2\mathbb{Z} \to \mathrm{Unip}_{\check{\mathcal{O}}}\big(\mathrm{O}(p,q)\big), \quad (\tau,\wp,\epsilon) \mapsto \pi_{\tau,\wp} \otimes \det^{\epsilon}$$

is bijective.

We remark that the unitarizability of $\pi_{\tau,\wp}$ in part (a) of Theorem 5.2 follows from the preservation of unitarity (Theorem 2.4). Furthermore the computation of the associated cycles of $\pi_{\tau,\wp}$, in particular Theorem 3.2, plays a critical role in the proof of Theorem 5.2. By Theorem 5.2, we have explicitly constructed all special unipotent representations in $\mathrm{Unip}_{\check{\mathcal{O}}}(G)$, when all row lengths of $\check{\mathcal{O}}$ are odd. If some row lengths of $\check{\mathcal{O}}$ are even, then these even row lengths must come in pairs. In this case, the set $\mathrm{Unip}_{\check{\mathcal{O}}}(G)$ of the special unipotent representations attached to $\check{\mathcal{O}}$ is similarly defined, and via irreducible unitary parabolic inductions, the construction of representations in $\mathrm{Unip}_{\check{\mathcal{O}}}(G)$ is reduced to the case when all row lengths of $\check{\mathcal{O}}$ are odd (see [7]). In the same approach, we may parameterize and construct all special unipotent representations of the real classical groups $\mathrm{GL}_n(\mathbb{R})$, $\mathrm{GL}_n(\mathbb{C})$, $\mathrm{GL}_n(\mathbb{H})$, $\mathrm{U}(p,q)$, $\mathrm{O}(p,q)$, $\mathrm{Sp}_{2n}(\mathbb{R})$, $\mathrm{O}^*(2n)$, $\mathrm{Sp}(p,q)$, $\mathrm{O}_n(\mathbb{C})$, $\mathrm{Sp}_{2n}(\mathbb{C})$, as well as all metaplectic special unipotent representations of $\widetilde{\mathrm{Sp}}_{2n}(\mathbb{R})$ and $\mathrm{Sp}_{2n}(\mathbb{C})$. See [5] for the notion of metaplectic special unipotent representations.

We thus have the following result which confirms the Arthur–Barbasch–Vogan conjecture [2, INTRODUCTION] for real classical groups.

Theorem 5.3 ([6]). *All special unipotent representations of the real classical groups are unitarizable; all metaplectic special unipotent representations of* $\widetilde{\mathrm{Sp}}_{2n}(\mathbb{R})$ *and* $\mathrm{Sp}_{2n}(\mathbb{C})$ *are also unitarizable.*

Remark. The unitarizability of special unipotent representations for quasisplit classical groups is independently due to Adams, Arancibia Robert, and Mezo [1].

The authors would like to conclude by noting the prescient remark of A. A. Kirillov in a survey article on the orbit method in 1999 [20]: Howe duality – a new branch of representation theory where the orbit method has not yet been used to the fullest.

ACKNOWLEDGMENTS

We thank Roger Howe for sharing his mathematical insight and for his encouragement.

FUNDING

B. Sun was supported in part by National Key R&D Program of China (2020YFA0712600), National Natural Science Foundation of China (11688101, 11621061), and Kunpeng program of Zhejiang Province.

C.-B. Zhu was supported in part by MOE AcRF Tier 1 grant R-146-000-314-114, and Provost's Chair grant C-146-000-052-001 in NUS.

REFERENCES

[1] J. Adams, N. Arancibia Robert, and P. Mezo, Equivalent definitions of Arthur packets for real classical groups. 2021, arXiv:2108.05788.

[2] J. Adams, D. Barbasch, and D. A. Vogan, *The Langlands classification and irreducible characters for real reductive groups*. Progr. Math. 104, Birkhäuser, 1991.

[3] J. Arthur, On some problems suggested by the trace formula. In *Lie group representations, II* (College Park, Md.), pp. 1–49, Lecture Notes in Math. 1041, Springer, 1984.

[4] J. Arthur, Unipotent automorphic representations: conjectures. In *Orbites unipotentes et représentations, II, Astérisque* **171–172** (1989), 13–71.

[5] D. Barbasch, J.-J. Ma, B. Sun, and C.-B. Zhu, On the notion of metaplectic Barbasch–Vogan duality. 2020, arXiv:2010.16089.

[6] D. Barbasch, J.-J. Ma, B. Sun, and C.-B. Zhu, Special unipotent representations of real classical groups: construction and unitarity. 2021, arXiv:1712.05552.

[7] D. Barbasch, J.-J. Ma, B. Sun, and C.-B. Zhu, Special unipotent representations of real classical groups: counting. 2022, https://blog.nus.edu.sg/matzhucb/research/.

[8] D. Barbasch and D. A. Vogan, Unipotent representations of complex semisimple groups. *Ann. of Math.* **121** (1985), 41–110.

[9] W. Borho, Recent advances in enveloping algebras of semisimple Lie-algebras. *Séminaire Bourbaki*, Exp. No. **489**, (1976/77), 1–18.

[10] A. Daszkiewicz, W. Kraskiewicz, and T. Przebinda, Nilpotent orbits and complex dual pairs. *J. Algebra* **190** (1997), no. 2, 518–539.

[11] R. Goodman and N. R. Wallach, *Symmetry, Representations, and Invariants.* Grad. Texts in Math. 255, Springer, 2009.

[12] H. He, Unitary representations and theta correspondence for type I classical groups. *J. Funct. Anal.* **199** (2003), no. 1, 92–121.

[13] H. He, Unipotent representations and quantum induction. 2007, arXiv:math/0210372.

[14] R. Howe, θ-series and invariant theory. In *Automorphic forms, representations and L-functions*, pp. 275–285, Proc. Sympos. Pure Math. 33, Amer. Math. Soc., 1979.

[15] R. Howe, Small unitary representations of classical groups. In *Group representations, ergodic theory, operator algebras, and mathematical physics*, pp. 121–150, Math. Sci. Res. Inst. Publ. 6, Springer, New York, 1987.

[16] R. Howe, Remarks on classical invariant theory. *Trans. Amer. Math. Soc.* **313** (1989), 539–570.

[17] R. Howe, Transcending classical invariant theory. *J. Amer. Math. Soc.* **2** (1989), 535–552.

[18] M. Kashiwara and M. Vergne, On the Segal–Shale–Weil Representations and Harmonic Polynomials. *Invent. Math.* **44** (1978), 1–48.

[19] A. A. Kirillov, Unitary representations of nilpotent Lie groups. *Uspekhi Mat. Nauk* **17** (1962), no. 4, 57–110.

[20] A. A. Kirillov, Merits and demerits of the orbit method. *Bull. Amer. Math. Soc.* **36** (1999), no. 4, 433–488.

[21] B. Kostant, Quantization and unitary representations. In *Lectures in modern nalysis and applications III*, pp. 87–208, Lecture Notes in Math. 170, Springer, 1970.

[22] H. Kraft and C. Procesi, On the geometry of conjugate classes in classical groups. *Comment. Math. Helv.* **57** (1982), 539–602.

[23] S. S. Kudla, Splitting metaplectic covers of dual reductive pairs. *Israel J. Math.* **87** (1994), 361–401.

[24] E. Lapid and S. Rallis, On the local factors of representations of classical groups. In *Automorphic representations, L-functions and applications: progress and prospects*, pp. 309–359, Ohio State Univ. Math. Res. Inst. Publ. 11, de Gruyter, Berlin, 2005.

[25] J.-S. Li, Singular unitary representations of classical groups. *Invent. Math.* **97** (1989), 237–255.

[26] J.-S. Li, Theta lifting for unitary representations with nonzero cohomology. *Duke Math. J.* **61** (1990), no. 3, 913–937.

[27] H. Y. Loke and J.-J. Ma, Invariants and K-spectrums of local theta lifts. *Compos. Math.* **151** (2015), no. 1, 179–206.

[28] C. Moeglin, M.-F. Vigneras, and J.-L. Waldspurger, *Correspondances de Howe sur un corps p-adique*. Lecture Notes in Math. 1291, Springer, 1987.

[29] K. Nishiyama, H. Ochiai, K. Taniguchi, H. Yamashita, and S. Kato, Nilpotent orbits, associated cycles and Whittaker models for highest weight representations. *Astérisque* **273** (2001), 1–163.

[30] K. Nishiyama, H. Ochiai, and C.-B. Zhu, Theta lifting of nilpotent orbits for symmetric pairs. *Trans. Amer. Math. Soc.* **358** (2006), 2713–2734.

[31] K. Nishiyama and C.-B. Zhu, Theta lifting of unitary lowest weight modules and their associated cycles. *Duke Math. J.* **125** (2004), no. 3, 415–465.

[32] I. Piatetski-Shapiro and S. Rallis, ϵ factor of representations of classical groups. *Proc. Natl. Acad. Sci. USA* **83** (1986), no. 13, 4589–4593.

[33] D. A. Vogan, Representations of reductive Lie groups. In *Proceedings of the International Congress of Mathematicians* (Berkeley, California), pp. 245–266, Amer. Math. Soc., 1986.

[34] D. A. Vogan, *Unitary representations of reductive Lie groups*. Ann. of Math. Stud. 118, Princeton University Press, 1987.

[35] D. A. Vogan, Associated varieties and unipotent representations. In *Harmonic analysis on reductive groups* (Bowdoin College, 1989), edited by W. Barker and P. Sally, pp. 315–388, Progr. Math. 101, Birkhäuser (Boston–Basel–Berlin), 1991.

[36] D. A. Vogan, The method of coadjoint orbits for real reductive groups. In *Representation theory of Lie groups* (Park City, UT, 1998), pp. 179–238, IAS/Park City Math. Ser. 8, Amer. Math. Soc., 2000.

[37] N. R. Wallach, *Real Reductive Groups I*. Academic Press, Inc., 1988.

[38] N. R. Wallach, *Real Reductive Groups II*. Academic Press, Inc., 1992.

[39] A. Weil, Sur certain group d'operateurs unitaires. *Acta Math.* **111** (1964), 143–211.

[40] H. Weyl, *The Classical Groups*. Princeton University Press, 1939.

BINYONG SUN

Institute for Advanced Study in Mathematics, Zhejiang University, Hangzhou, China, and Academy of Mathematics and Systems Science, Chinese Academy of Sciences, Beijing, China, sunbinyong@zju.edu.cn

CHEN-BO ZHU

Department of Mathematics, National University of Singapore, 10 Lower Kent Ridge Road, Singapore 119076, matzhucb@nus.edu.sg

This Page Intentionally Left Blank

QUANTUM SYMMETRIC PAIRS

WEIQIANG WANG

ABSTRACT

This is a survey of some recent progress on quantum symmetric pairs and applications. The topics include quasi-K-matrices, $\imath$Schur duality, canonical bases, super Kazhdan–Lusztig theory, $\imath$Hall algebras, current presentations for affine $\imath$quantum groups, and braid group actions.

MATHEMATICS SUBJECT CLASSIFICATION 2020

Primary 17B37; Secondary 20G42, 17B10

KEYWORDS

Quantum symmetric pairs, canonical bases, super Kazhdan–Lusztig theory, q-Schur dualities, Hall algebras, braid group actions

INTERNATIONAL CONGRESS OF MATHEMATICIANS 2022 JULY 6–14

Proc. Int. Cong. Math. 2022, Vol. 4, pp. 3080–3102
DOI 10.4171/ICM2022/76

1. INTRODUCTION

1.1. Quantum groups

According to Drinfeld and Jimbo, a quantum group $\mathbf{U} = \mathbf{U}_v(\mathfrak{g})$ is the quantum deformation (as Hopf algebras) of the universal enveloping algebra $U(\mathfrak{g})$, for a semisimple or Kac–Moody Lie algebra $\mathfrak{g}$. Since their inception in 1985, quantum groups have found numerous applications to diverse areas, including mathematical physics, representation theory, algebraic combinatorics, and low dimensional topology.

We offer a (personal) Top Ten List of highlights in quantum groups, viewed as a part of Lie theory, as follows:

(1) Definition [27,31]

(2) (Quasi-)R-matrix [27,55]

(3) Canonical basis [53–55] [32]

(4) Quantum Schur duality [31]

(5) Super type A Kazhdan–Lusztig theory [18] [22] [19]

(6) Hall algebra [60] [17]

(7) Current presentation of affine quantum groups [27] [14,23]

(8) Braid group action [53,55]

(9) Categorification [34,61]

(+) ○ ○ ○ ○ ○ ○

We apologize beforehand for omitting many important constructions for quantum groups in the above list, and your favorite construction may likely fall into the black holes in Item (+). Also, the references listed above are mostly samples of the original contributions, and there are often dozens or hundreds of additional works which are not cited.

The list (1)–(9) is so arranged that the topics are to be matched with the $\imath$-generalizations which we shall describe later in the same ordering.

1.2. Quantum symmetric pairs

Recall that a symmetric pair $(\mathfrak{g}, \mathfrak{g}^\theta)$ consists of a semisimple Lie algebra $\mathfrak{g}$ and an involution θ on $\mathfrak{g}$. The classification of irreducible symmetric pairs is equivalent to the classification of real forms of complex simple Lie algebras (which goes back to É. Cartan); for example, they can now be classified in terms of Satake diagrams.

Let $\mathbf{U} = \mathbf{U}_v(\mathfrak{g})$ be a quantum group (of finite type) with comultiplication Δ in (2.2). According to Gail Letzter [39–41], a quantum symmetric pair $(\mathbf{U}, \mathbf{U}^\imath)$ consists of a Drinfeld–Jimbo quantum group $\mathbf{U}$ and its (right) coideal subalgebra $\mathbf{U}^\imath$ (i.e., $\Delta : \mathbf{U}^\imath \to \mathbf{U}^\imath \otimes \mathbf{U}$) which specializes at $v \mapsto 1$ to $U(\mathfrak{g}^\theta)$. Starting with the Satake diagrams, Letzter constructed the corresponding quantum symmetric pairs. The $\mathbf{U}^\imath$ comes with parameters, which reflects the fact that there is a family of (explicit) embeddings of $\mathbf{U}^\imath$ into $\mathbf{U}$; see (2.4). A generalization of quantum symmetric pairs of Kac–Moody type was carried out by Kolb in [35], now a standard reference in the subject; Kolb's conventions are compatible with those in Lusztig's book [55]. As we may deal with $\mathbf{U}^\imath$ alone, we shall call $\mathbf{U}^\imath$ an *ıquantum group*.

Letzter's foundational work on quantum symmetric pairs was motivated by harmonic analysis on quantum symmetric spaces, generalizing earlier examples given by Koornwinder, Gavrilik–Klimyk, Noumi, and others. Letzter's work was ahead of her time.

1.3. Goal

We take the view that ıquantum groups are a vast generalization of quantum groups (as real reductive groups are a generalization of compact or complex reductive groups).

Example 1.1 (Quantum groups as ıquantum groups). Consider the diagonal symmetric pair $(\mathfrak{g} \times \mathfrak{g}, \mathfrak{g}^\triangle)$, where $\mathfrak{g}^\triangle$ is a diagonal copy of $\mathfrak{g}$. Similarly, we have a quantum symmetric pair of diagonal type $(\mathbf{U} \otimes \mathbf{U}, \mathbf{U})$; the embedding is given by $(\omega \otimes 1)\Delta : \mathbf{U} \to \mathbf{U} \otimes \mathbf{U}$ (see [8]), where one checks that the image of $\mathbf{U}$ is a coideal subalgebra of $\mathbf{U} \otimes \mathbf{U}$.

The goal of this report is to survey some recent advances on quantum symmetric pairs and ıquantum groups, generalizing Items (1)–(9) above. The $\imath$-analogs of the constructions in Items (2)–(5) were initiated by Huanchen Bao and the author in [9], where it was proposed that all fundamental (algebraic, geometric, categorical) constructions in quantum groups should be generalized to ıquantum groups.

The good news is that all Items (1)–(9) admit genuine $\imath$-generalizations, while the bad news, or the exciting news for an optimist, is that many $\imath$-generalizations are not yet in full generality. A reader might well be tempted to try one's hands in developing these $\imath$-generalizations in greater generality. He or she is encouraged to pick his or her own favorite construction in quantum groups in Item (+); even better, to supply its missing $\imath$-generalization.

All the known constructions indicate that the $\imath$-generalizations (when done right!) look natural and inevitable, though proofs are often much lengthier and challenging. The complications in ıquantum groups are often caused by

(i) absence of triangular decompositions;

(ii) presence of many rank 1 types *and* parameters;

(iii) presence of nonuniform Serre-type relations;

(iv) hidden (nonobvious) integral forms.

1.4. A quick overview

A Serre-type presentation for an ıquantum group $\mathbf{U}^\imath$ is due to Letzter [40] in the finite type setting, and has been generalized since then to Kac–Moody type in various forms (cf. [2, 20, 24, 35, 38]). This can be viewed as an $\imath$-analog of the construction in Item (1).

Let us provide some details on the constructions in Items (2)–(5).

As a generalization of Jimbo–Schur duality [31], an $\imath$Schur duality between a *quasi-split* type AIII $\imath$quantum group $\mathbf{U}^\imath$ and type B Hecke algebra was formulated in [9] (and [4,11]). Recently, the Jimbo–Schur duality and quasisplit $\imath$Schur duality have been uniformly generalized using a general $\imath$quantum group of type AIII by Yaolong Shen and the author in [62]. This has led unexpectedly to quasiparabolic Kazhdan–Luszig bases associated to (possibly nonparabolic) reflection subgroups of the type B Weyl group, extending the classic constructions of Kazhdan–Lusztig [33] and Deodhar [25].

The $\imath$canonical basis and $\imath$Schur duality (in quasisplit type AIII setting) were motivated by and played a key role in formulating a super type B Kazhdan–Lusztig theory (which was a decades old open problem) [9]; see also Bao [4] for a super type D formulation. Our approach was inspired by Brundan–Kazhdan–Lusztig conjecture in the super type A setting [18] and its proof by Cheng, Lam, and the author [22].

The $\imath$-analog of quasi-R-matrix, known as quasi-K-matrix nowadays, was formulated by Bao and the author in [9] as a key step in the construction of $\imath$canonical bases, and a proof for its existence in great generality has been given by Balagovic–Kolb in [3]; see Appel-Vlaar [1] for a more recent reformulation and generalization. The quasi-K-matrix further leads to a construction of the K-matrix [3,9], which was shown in [3] to provide solutions to the reflection equation.

As an extension of Lusztig's approach [53,54], a theory of $\imath$canonical bases arising from quantum symmetric pairs has been systematically developed by Bao and the author in [8,10]. The $\imath$canonical bases on based $\mathbf{U}$-modules (viewed as $\mathbf{U}^\imath$-modules) and on the modified $\imath$quantum groups are established; also see [5,43] and [29] for a geometric approach in type AIII setting. The $\imath$canonical basis in split rank 1, also known as $\imath$divided powers [9,16], has found applications in the works with Xinhong Chen and Ming Lu [20,21,52]. H. Watanabe [65,66] has developed a crystal approach (à la Kashiwara [32]) to $\imath$canonical bases of $\mathbf{U}^\imath$-modules, for some quasisplit finite types.

Motivated by earlier constructions of Bridgeland [17] and then M. Gorsky [30] (extending the foundational work of Ringel [60]), Lu–Peng and Lu developed semiderived Hall algebras associated to 1-Gorenstein algebras (see [44] and [50, APPENDIX A]). Lu and the author introduced in [50,52] the $\imath$Hall algebras, i.e., semiderived Hall algebras associated to the $\imath$quiver algebras, to realize the (*universal*) $\imath$quantum groups $\tilde{\mathbf{U}}^\imath$. This is a conceptual $\imath$-analog of (6), generalizing [17].

In [49], Lu and the author have formulated a Drinfeld-type presentation for affine $\imath$quantum groups of split ADE type; this has been generalized to split BCFG type by Weinan Zhang [67]. This provides an $\imath$-analog of (7). Lu and Ruan are developing $\imath$Hall algebras of the weighted projective lines [45] to realize the affine $\imath$quantum groups in the new presentation, generalizing the rank-1 construction in [46].

The braid group actions associated to relative Weyl groups on (mostly quasisplit) finite type $\mathbf{U}^\imath$ were obtained by Kolb–Pellegrini in [37] (via computer computation), where the existence of such an action on an arbitrary $\imath$quantum group of finite type was conjectured. Relative braid group actions on $\tilde{\mathbf{U}}^\imath$ (of quasisplit Kac–Moody type) were obtained in [47] via reflection functors in $\imath$Hall algebras, and it becomes clear that the universal $\imath$quantum groups provide a conceptual framework for braid group actions. We announce an intrinsic approach developed by W. Zhang and the author in [64] to relative braid group action on $\tilde{\mathbf{U}}^\imath$ of *arbitrary* finite type (and on $\mathbf{U}$-modules) with explicit formulas, providing a conceptual $\imath$-analog of (8). The quasi-K-matrix plays a basic role in our formulation.

This survey is organized as follows. The $\imath$-counterparts of the quantum group highlights (1)–(8) will be formulated in Sections 2–9 below. In Section 10, we discuss additional topics in $\imath$quantum groups including the $\imath$-analog of (9), and list some open problems.

2. QUANTUM SYMMETRIC PAIRS: DEFINITION

2.1. Quantum groups

We mostly follow notations in the book [55]. Denote by $\mathbb{N}$ the set of nonnegative integers. Let $(Y, X, \langle\cdot,\cdot\rangle, \dots)$ be a root datum of type $(\mathbb{I}, \cdot)$; cf. [55]. The quantum group $\mathbf{U}$ associated with this root datum $(Y, X, \langle\cdot,\cdot\rangle, \dots)$ is the associative $\mathbb{Q}(v)$-algebra generated by E_i, F_i for $i \in \mathbb{I}$ and K_μ for $\mu \in Y$, subject to standard relations which can be found in [55]. Let W denote the Weyl group generated by simple reflections s_i for $i \in \mathbb{I}$. The $\mathbb{Q}(v)$-algebra $\mathbf{U}$ admits a Chevalley involution ω such that

$$\omega(E_i) = F_i, \quad \omega(F_i) = E_i, \quad \omega(K_\mu) = K_{-\mu}, \quad \text{for } i \in \mathbb{I},\ \mu \in Y. \tag{2.1}$$

For any $i \in \mathbb{I}$, we set $v_i = v^{\frac{i \cdot i}{2}}$. For $i \in \mathbb{I}$, $n, s \in \mathbb{N}$ with $0 \le s \le n$, we define $[n]_i = \frac{v_i^n - v_i^{-n}}{v_i - v_i^{-1}}$ and $[s]_i^! = \prod_{k=1}^s [k]_i$, and $\begin{bmatrix} n \\ s \end{bmatrix}_i = \frac{[n]_i^!}{[s]_i^! [n-s]_i^!}$.

Let $\mathbf{U}^+$, $\mathbf{U}^0$, and $\mathbf{U}^-$ be the $\mathbb{Q}(v)$-subalgebra of $\mathbf{U}$ generated by $E_i (i \in \mathbb{I})$, $K_\mu (\mu \in Y)$, and $F_i (i \in \mathbb{I})$, respectively. Let $\mathcal{A} = \mathbb{Z}[v, v^{-1}]$. We let ${}_\mathcal{A}\mathbf{U}^-$ (respectively, ${}_\mathcal{A}\mathbf{U}^+$) denote the $\mathcal{A}$-subalgebra of $\mathbf{U}^-$ (respectively, $\mathbf{U}^+$) generated by all divided powers $F_i^{(a)} = F_i^s / [s]_i^!$ (respectively, $E_i^{(a)}$). With $\tilde{K}_{\pm i} := K_{\pm \frac{i \cdot i}{2} i}$, the coproduct $\Delta : \mathbf{U} \to \mathbf{U} \otimes \mathbf{U}$ is given by

$$\Delta(E_i) = E_i \otimes 1 + \tilde{K}_i \otimes E_i, \quad \Delta(F_i) = 1 \otimes F_i + F_i \otimes \tilde{K}_{-i}, \quad \Delta(K_\mu) = K_\mu \otimes K_\mu. \tag{2.2}$$

Let $X^+ = \{\lambda \in X \mid \langle i, \lambda \rangle \in \mathbb{N}, \forall i \in \mathbb{I}\}$ be the set of dominant integral weights. By $\lambda \gg 0$ we shall mean that the integers $\langle i, \lambda \rangle$ for all i are sufficiently large. The Verma module $M(\lambda)$ of highest weight $\lambda \in X$ has a unique simple quotient $\mathbf{U}$-module $L(\lambda)$ with a highest weight vector η_λ. We define a $\mathbf{U}$-module ${}^\omega L(\lambda)$, which has the same underlying vector space as $L(\lambda)$ but with the action twisted by the Chevalley involution ω in (2.1); then ${}^\omega L(\lambda)$ is simple of lowest weight $-\lambda$ with lowest weight vector denoted by $\xi_{-\lambda}$. For $\lambda \in X^+$, we let ${}_\mathcal{A}L(\lambda) = {}_\mathcal{A}\mathbf{U}^- \eta_\lambda$ and ${}^\omega_\mathcal{A}L(\lambda) = {}_\mathcal{A}\mathbf{U}^+ \xi_{-\lambda}$ be the $\mathcal{A}$-submodules of $L(\lambda)$ and ${}^\omega L(\lambda)$.

There is a canonical basis $\mathbf{B}$ on $\mathbf{U}^-$, inducing a canonical basis on $\mathbf{U}^+$ via the standard isomorphism $\mathbf{U}^- \cong \mathbf{U}^+$. For each $\lambda \in X^+$, there is a subset $\mathbf{B}(\lambda)$ of $\mathbf{B}$ so that $\{b^- \eta_\lambda \mid b \in \mathbf{B}(\lambda)\}$ forms a canonical basis of $L(\lambda)$. For $w \in W$, let $\eta_{w\lambda}$ denote the unique canonical basis element of weight $w\lambda$.

Let $\dot{\mathbf{U}} = \bigoplus_{\zeta \in X} \dot{\mathbf{U}} \mathbf{1}_\zeta$ be the modified quantum group and ${}_\mathcal{A}\dot{\mathbf{U}}$ be its $\mathcal{A}$-form. Then $\dot{\mathbf{U}}$ admits a canonical basis $\dot{\mathbf{B}} = \{b_1 \diamondsuit_\zeta b_2 \mid (b_1, b_2) \in \mathbf{B} \times \mathbf{B}, \zeta \in X\}$, compatible with canonical bases on ${}^{\omega}L(\lambda) \otimes L(\mu)$, for $\lambda, \mu \in X^+$; cf. [55, PART IV]. For any $\mathbb{I}_\bullet \subset \mathbb{I}$, let $\mathbf{U}_{\mathbb{I}_\bullet}$ be the $\mathbb{Q}(v)$-subalgebra of $\mathbf{U}$ generated by F_i, E_i, and $K_i (i \in \mathbb{I}_\bullet)$. Let $\mathbf{B}_{\mathbb{I}_\bullet}$ be the canonical basis of $\mathbf{U}^-_{\mathbb{I}_\bullet}$.

2.2. Satake diagrams and admissible pairs

Let τ be an involution of the Cartan datum $(\mathbb{I}, \cdot)$; we allow $\tau = \mathrm{Id}$. We further assume that τ extends to an involution on X and an involution on Y, respectively, such that the perfect bilinear pairing is invariant under the involution τ. Given a finite type $\mathbb{I}_\bullet \subset \mathbb{I}$, let $W_{\mathbb{I}_\bullet} = \langle s_i \mid i \in \mathbb{I}_\bullet \rangle$ be the parabolic subgroup of W with longest element $w_\bullet$, and let $\rho_\bullet$ (and $\rho^\vee_\bullet$) be the half-sum of all positive roots (and coroots) in the root system $R_\bullet$ (and $R^\vee_\bullet$). Set $\mathbb{I}_\circ = \mathbb{I} \backslash \mathbb{I}_\bullet$. A pair $(\mathbb{I} = \mathbb{I}_\bullet \cup \mathbb{I}_\circ, \tau)$ is called *admissible* (cf. [35, DEFINITION 2.3]) if $\tau(\mathbb{I}_\bullet) = \mathbb{I}_\bullet$, the actions of τ and $-w_\bullet$ on $\mathbb{I}_\bullet$ coincide, and $\langle \rho^\vee_\bullet, j' \rangle \in \mathbb{Z}$ whenever $\tau j = j \in \mathbb{I}_\circ$. We regard admissible pairs as a synonyms for Satake diagrams.

Note that $\theta = -w_\bullet \circ \tau$ is an involution of X and Y. For any $\lambda \in X$ (or Y), we shall write $\lambda^\tau = \tau(\lambda)$, $\lambda^\theta = \theta(\lambda)$. Following [8], we introduce the $\imath$-weight and $\imath$-root lattices

$$X_\imath = X/\breve{X}, \quad \text{where } \breve{X} = \{\lambda - \lambda^\theta \mid \lambda \in X\},$$
$$Y^\imath = \{\mu \in Y \mid \mu^\theta = \mu\}.$$

For $\lambda \in X$, we denote its image in $X_\imath$ by $\overline{\lambda}$. The involution τ of $\mathbb{I}$ induces an isomorphism of the $\mathbb{Q}(v)$-algebra $\mathbf{U}$, denoted also by τ, which sends $E_i \mapsto E_{\tau i}$, $F_i \mapsto F_{\tau i}$, and $K_\mu \mapsto K_{\tau\mu}$.

2.3. Quantum symmetric pairs

We review the definition of quantum symmetric pair $(\mathbf{U}, \mathbf{U}^\imath)$, where $\mathbf{U}^\imath$ is a coideal subalgebra of $\mathbf{U}$, following [35]; also see [2, 3, 8]. Letzter's convention is a little different.

An *ıquantum group* $\mathbf{U}^\imath$ is the $\mathbb{Q}(v)$-subalgebra of $\mathbf{U}$ generated by

$$B_i := F_i + \varsigma_i T_{w_\bullet}(E_{\tau i}) \tilde{K}_i^{-1} + \kappa_i \tilde{K}_i^{-1} \ (i \in \mathbb{I}_\circ), \quad K_\mu (\mu \in Y^\imath), \quad F_i, E_i \ (i \in \mathbb{I}_\bullet), \tag{2.3}$$

where $\varsigma_i \in \mathbb{Q}(v)^\times$, $\kappa_i \in \mathbb{Q}(v)$, for $i \in \mathbb{I}_\circ$, are parameters, and $T_w = T''_{w,+1}$ denotes a braid group operator as in [55]. Denote by $\mathbf{U}^{\imath 0}$ the $\mathbb{Q}(v)$-subalgebra of $\mathbf{U}^\imath$ generated by K_μ ($\mu \in Y^\imath$), and denote the embedding via (2.3) by

$$\imath : \mathbf{U}^\imath \to \mathbf{U}, \quad x \mapsto x^\imath. \tag{2.4}$$

The parameters ς_i, κ_i are required to satisfy Conditions (2.5)–(2.6):

$$\kappa_i = 0 \quad \text{unless “}\tau i = i, \langle i, j' \rangle = 0 \ \forall j \in \mathbb{I}_\bullet, \& \langle k, i' \rangle \in 2\mathbb{Z} \ \forall \tau k = k = w_\bullet k \in \mathbb{I}_\circ\text{”}; \tag{2.5}$$

$$\varsigma_i = \varsigma_{\tau i} \quad \text{if } i \cdot \theta(i) = 0. \tag{2.6}$$

Conditions (2.5)–(2.6) ensure the quantum Iwasawa decomposition and hence $\mathbf{U}^\imath$ have the expected size [35, 40]. By definition, the algebra $\mathbf{U}^\imath$ contains $\mathbf{U}_{\mathbb{I}_\bullet}$ as a subalgebra.

In the remainder of this paper, we shall impose the following additional conditions on parameters besides (2.5)–(2.6), as required for the construction of quasi-K-matrix and

$\imath$canonical basis:

$$\varsigma_i, \kappa_i \in \mathbb{Z}[v, v^{-1}], \quad \text{for } i \in \mathbb{I}_\circ, \tag{2.7}$$

$$\overline{\kappa_i} = \kappa_i, \quad \varsigma_{\tau i} = (-1)^{\langle 2\rho_\bullet^\vee, i' \rangle} v_i^{-\langle i, 2\rho_\bullet + w_\bullet \tau i' \rangle} \overline{\varsigma_i}. \tag{2.8}$$

Conditions in (2.8) appeared in [3].

The nontrivial Serre presentation of $\mathbf{U}^\imath$ was given by Letzter [40] in finite type setting, and further generalized in different forms [2, 20, 35, 38]. Certain Serre–Lusztig (or higher-order Serre) relations for $\mathbf{U}^\imath$ have been obtained in [21] via $\imath$divided powers (see Example 4.3).

Example 2.1. We call $\mathbf{U}^\imath$ *quasisplit* if $\mathbb{I}_\bullet = \emptyset$ and *split* if, in addition, $\tau = \mathrm{Id}$. A split $\imath$quantum group $\mathbf{U}^\imath$ is the subalgebra of $\mathbf{U}$ generated by $B_i = F_i + \varsigma_i E_i \tilde{K}_i^{-1} + \kappa_i \tilde{K}_i^{-1}$ $(i \in \mathbb{I})$; often $\kappa_i = 0$ thanks to (2.5). Relations in split $\mathbf{U}^\imath$ are given in (9.1)–(9.2) (with $\mathbb{K}_i = -v_i^2 \varsigma_i$) for ADE type and in (9.9)–(9.10) (with $k = l = 0$) for BCFG type.

3. (QUASI) K-MATRICES

As predicted in [9] and established in [2,36] (also cf. [20]), there is a unique antilinear bar involution of the $\mathbb{Q}$-algebra $\mathbf{U}^\imath$, denoted by $\psi_\imath$, such that $\psi_\imath(v) = v^{-1}$ and

$$\psi_\imath(B_i) = B_i \ (i \in \mathbb{I}_\circ), \quad \psi_\imath(F_j) = F_j, \quad \psi_\imath(E_j) = E_j \ (i \in \mathbb{I}_\bullet),$$
$$\psi_\imath(K_\mu) = K_{-\mu} \ (\mu \in Y^\imath).$$

The formulation of the *quasi-K-matrix* Υ below (called an *intertwiner* earlier) was due to Bao and the author [9]; its existence in great generality has been established in [3] (also cf. [8, REMARK 4.9]) with additional technicality removed in [10,36]. The quasi-K-matrix for a quantum symmetric pair of diagonal type reduces to the quasi-R-matrix (cf. [55]).

Theorem 3.1 ([3,8,9]). *There exists a unique family of elements $\Upsilon_\mu \in \mathbf{U}^+_\mu$, for $\mu \in \mathbb{N}\mathbb{I}$, such that $\Upsilon_0 = 1$ and $\Upsilon = \sum_{\mu \in \mathbb{N}\mathbb{I}} \Upsilon_\mu$ satisfies the following identity:*

$$\psi_\imath(u)\Upsilon = \Upsilon \psi(u), \quad \textit{for all } u \in \mathbf{U}^\imath. \tag{3.1}$$

Moreover, $\Upsilon_\mu = 0$, unless $\mu^\theta = -\mu \in X$. (Recall that $\theta = -w_\bullet \circ \tau$.)

A quasi-K-matrix was originally introduced in order to construct a new bar involution and then $\imath$canonical bases on based $\mathbf{U}$-modules (see Section 4 below). On the other hand, a suitable twisting of Υ by elements in $\mathbf{U}^{\imath 0}$ provides certain $\mathbf{U}^\imath$-module isomorphisms [9, §2.5], [3] (also see [8, §4.5] for a different formulation), nowadays known as the K-matrix. It is shown in [3] that the K-matrix provides solutions to the reflection equation, an $\imath$-analog of the Yang–Baxter equation. There has been a reformulation of quasi-K-matrix in [1] without referring explicitly to the bar involution of $\mathbf{U}^\imath$; this has the advantage of making sense of quasi-K-matrices with general parameters satisfying (2.5)–(2.6).

4. CANONICAL BASES ARISING FROM QUANTUM SYMMETRIC PAIRS

4.1. Based modules

The quasi-K-matrix Υ in (a completion of) $\mathbf{U}^+$ induces a $\mathbb{Q}(v)$-linear map $\Upsilon : M \otimes L(\lambda) \to M \otimes L(\lambda)$, for any $\lambda \in X^+$ and any weight $\mathbf{U}$-module M with weights bounded above. A $\mathbf{U}^\imath$-module M equipped with an antilinear involution $\psi_\imath$ is called *$\imath$-involutive* if $\psi_\imath(um) = \psi_\imath(u)\psi_\imath(m)$, for all $u \in \mathbf{U}^\imath$, $m \in M$. Let (M, B) be a based $\mathbf{U}$-module [55, IV] with weights bounded above. We denote the bar involution on M by ψ. Then M is an $\imath$-involutive $\mathbf{U}^\imath$-module with involution (see [9])

$$\psi_\imath := \Upsilon \circ \psi.$$

Assume that (M, B) is a based $\mathbf{U}$-module with weights bounded above such that Υ preserves the $\mathcal{A}$-submodule ${}_\mathcal{A}M$. Then the $\mathbb{Q}(v)$-linear map $\psi_\imath = \Upsilon \circ \psi$ (and subsequently, Υ) preserves the $\mathcal{A}$-submodule ${}_\mathcal{A}M \otimes_\mathcal{A} {}_\mathcal{A}L(\lambda)$, for any $\lambda \in X^+$; see [10]. In particular, the involution $\psi_\imath$ on the $\imath$-involutive $\mathbf{U}^\imath$-module $L(\lambda_1) \otimes \cdots \otimes L(\lambda_\ell)$ preserves the $\mathcal{A}$-submodule ${}_\mathcal{A}L(\lambda_1) \otimes_\mathcal{A} \cdots \otimes_\mathcal{A} {}_\mathcal{A}L(\lambda_\ell)$, where $\lambda_i \in X^+$ for $1 \le i \le \ell$. For $(\mathbf{U}, \mathbf{U}^\imath)$ of finite type, a stronger statement holds [10], namely $\Upsilon_\mu \in {}_\mathcal{A}\mathbf{U}^+$, for each μ. This generalizes the integrality of quasi-R-matrix of finite type due to Lusztig [55, 24.1.6].

Define a partial order $\le$ on the weight lattice X such that $\lambda \le \lambda'$ if and only if $\lambda' - \lambda \in \mathbb{N}\mathbb{I}$. For an element x in $\mathbf{U}$ or in a $\mathbf{U}$-module M of weight $\mu \in X$, we write $|x| = \mu$.

Theorem 4.1 ([8,10]). *Let (M, B) be a based $\mathbf{U}$-module with weights bounded above. Assume that the involution $\psi_\imath = \Upsilon\psi$ of M preserves the $\mathcal{A}$-submodule ${}_\mathcal{A}M$. Then,*

(1) *the $\mathbf{U}^\imath$-module M admits a unique basis (called $\imath$canonical basis) $B^\imath := \{b^\imath \mid b \in B\}$ which is $\psi_\imath$-invariant and of the form*

$$b^\imath \in b + \sum_{b' \in B, |b| < |b'|} v^{-1}\mathbb{Z}[v^{-1}]b';$$

(2) *$B^\imath$ forms an $\mathcal{A}$-basis for the $\mathcal{A}$-lattice ${}_\mathcal{A}M$ (generated by B), and forms a $\mathbb{Z}[v^{-1}]$-basis for the $\mathbb{Z}[v^{-1}]$-lattice $\mathcal{M}$ (generated by B).*

In particular, $L(\lambda_1) \otimes \cdots \otimes L(\lambda_\ell)$, where $\lambda_i \in X^+$ for all i, admits an $\imath$canonical basis. Theorem 4.1 was further generalized in [10, 12] to provide an $\imath$canonical basis on $N \otimes L(\lambda)$, for a based $\mathbf{U}^\imath$-module N and $\lambda \in X^+$.

4.2. Canonical bases on modified $\imath$quantum groups

Following [55, IV], we can define a modified $\imath$quantum group $\dot{\mathbf{U}}^\imath$ (an associative $\mathbb{Q}(q)$-algebra structure without unit) such that $\dot{\mathbf{U}}^\imath = \bigoplus_{\lambda \in X_\imath} \mathbf{U}^\imath \mathbf{1}_\lambda$; see [10]. In contrast to quantum groups, the $\mathcal{A}$-form of $\dot{\mathbf{U}}^\imath$ is far from being obvious (even for rank 1).

For $\lambda, \mu \in X^+$ and $w \in W$, denoting $\eta_\lambda^\bullet := \eta_{w_\bullet \lambda}$, we introduce the following $\mathbf{U}$-submodule:

$$L^\imath(\lambda, \mu) := \mathbf{U}(\eta_{w_\bullet \lambda} \otimes \eta_\mu) \subset L(\lambda) \otimes L(\mu),$$

which can be shown to be a based **U**-module and hence admits an $\imath$canonical basis by Theorem 4.1. Let $\zeta = w_\bullet\lambda + \mu$ and $\zeta_\imath = \overline{\zeta}$. The following hold [8, 10]:

- ▷ The $\imath$canonical basis of $L^\imath(\lambda,\mu)$ is of the form $\mathbf{B}^\imath(\lambda,\mu) = \{(b_1\diamondsuit_{\zeta_\imath} b_2)^\imath_{w_\bullet\lambda,\mu} \mid (b_1,b_2) \in \mathbf{B}_{\mathbb{I}_\bullet} \times \mathbf{B}\}\backslash\{0\}$, where $(b_1\diamondsuit_{\zeta_\imath} b_2)^\imath_{w_\bullet\lambda,\mu}$ is $\psi_\imath$-invariant and lies in
$$(b_1\diamondsuit_\zeta b_2)(\eta^\bullet_\lambda \otimes \eta_\mu) + \sum_{|b'_1|+|b'_2|\le |b_1|+|b_2|} v^{-1}\mathbb{Z}[v^{-1}](b'_1\diamondsuit_\zeta b'_2)(\eta^\bullet_\lambda \otimes \eta_\mu).$$
- ▷ We have a projective system $\{L^\imath(\lambda+\nu^\tau,\mu+\nu)\}_{\nu\in X^+}$ of $\mathbf{U}^\imath$-modules, where
$$\pi_{\nu+\nu_1,\nu_1} : L^\imath(\lambda+\nu^\tau+\nu_1^\tau,\mu+\nu+\nu_1) \to L^\imath(\lambda+\nu^\tau,\mu+\nu), \quad \nu,\nu_1 \in X^+,$$
is the unique homomorphism of $\mathbf{U}^\imath$-modules mapping $\eta^\bullet_{\lambda+\nu^\tau+\nu_1^\tau} \otimes \eta_{\mu+\nu+\nu_1}$ to $\eta^\bullet_{\lambda+\nu^\tau} \otimes \eta_{\mu+\nu}$. The K-matrix [3, 8, 9] plays a role here.

The following theorem generalizes [55, CHAP. 25]. For quantum symmetric pairs of diagonal type in Example 1.1 or a trivial pair (**U**, **U**), it reduces to Lusztig's setting.

Theorem 4.2 ([8] [10, THEOREM 7.2]). *Let $\zeta_\imath \in X_\imath$ and $(b_1,b_2) \in B_{\mathbb{I}_\bullet} \times B$.*

(1) *There is a unique element $u = b_1\diamondsuit^\imath_{\zeta_\imath} b_2 \in \dot{\mathbf{U}}^\imath$ such that*
$$u(\eta^\bullet_\lambda \otimes \eta_\mu) = (b_1\diamondsuit_{\zeta_\imath} b_2)^\imath_{w_\bullet\lambda,\mu} \in L^\imath(\lambda,\mu),$$
$$\text{for all } \lambda,\mu \gg 0 \text{ with } \overline{w_\bullet\lambda+\mu} = \zeta_\imath.$$

(2) *The element $b_1\diamondsuit^\imath_{\zeta_\imath} b_2$ is $\psi_\imath$-invariant.*

(3) *The set $\dot{\mathbf{B}}^\imath = \{b_1\diamondsuit^\imath_{\zeta_\imath} b_2 \mid \zeta_\imath \in X_\imath, (b_1,b_2) \in B_{\mathbb{I}_\bullet} \times B\}$ forms a $\mathbb{Q}(v)$-basis of $\dot{\mathbf{U}}^\imath$.*

Example 4.3 ($\imath$Divided powers). Consider the quantum symmetric pair of split rank 1 $(\mathbf{U},\mathbf{U}^\imath) = (\mathbf{U}_v(\mathfrak{sl}_2), \mathbb{Q}(v)[B_i])$ associated to $\mathbb{I} = \{i\}$, via the embedding $\mathbf{U}^\imath \to \mathbf{U}$, $B_i \mapsto F_i + \varsigma_i E_i K_i^{-1}$. It is a new phenomenon of $\imath$quantum groups [9] that there are two different modified forms of $\mathbf{U}^\imath$, denoted by $\dot{\mathbf{U}}^\imath\mathbf{1}_{\bar{0}}$ and $\dot{\mathbf{U}}^\imath\mathbf{1}_{\bar{1}}$, depending on a parity $X_\imath = \{\bar{0},\bar{1}\}$, which are compatible with the parity of highest weights of finite-dimensional simple **U**-modules.

We define the *$\imath$divided powers* of B to be
$$B^{(m)}_{i,\bar{1}} = \frac{1}{[m]^!_i}\begin{cases} B_i \prod_{j=1}^{\ell}(B_i^2 - v_i\varsigma_i[2j-1]_i^2) & \text{if } m = 2\ell+1,\\ \prod_{j=1}^{\ell}(B_i^2 - v_i\varsigma_i[2j-1]_i^2) & \text{if } m = 2\ell,\end{cases}$$
$$B^{(m)}_{i,\bar{0}} = \frac{1}{[m]^!_i}\begin{cases} B_i \prod_{j=1}^{\ell}(B_i^2 - v_i\varsigma_i[2j]_i^2) & \text{if } m = 2\ell+1,\\ B_i^2 \prod_{j=1}^{\ell-1}(B_i^2 - v_i\varsigma_i[2j]_i^2) & \text{if } m = 2\ell.\end{cases}$$
These formulas (with parity swapped) appeared first in [9, CONJECTURE 4.13] (in terms of B_i in (2.3) with $\kappa_i = 1$, $\varsigma_i = v_i^{-1}$); see [20] for application to Serre relations for $\imath$quantum groups.

Set the parameter $\varsigma_i = v_i^{-1}$. Then $\mathbf{B}_{\bar{0}} := \{B^{(m)}_{i,\bar{0}} \mid m \ge 0\}$ (and respectively, $\mathbf{B}_{\bar{1}} := \{B^{(m)}_{i,\bar{1}} \mid m \ge 0\}$) forms the $\imath$canoical basis for the modified $\imath$quantum group $\dot{\mathbf{U}}^\imath\mathbf{1}_{\bar{0}}$

(and respectively, $\dot{\mathbf{U}}^\imath \mathbf{1}_{\bar{1}}$). Let $L(n)$ be the simple $\mathbf{U}$-module of highest weight $n \in \mathbb{N}$ with highest weight vector η. Then, for n even, $\mathbf{B}_{\bar{0}}\eta = \{B_{i,\bar{0}}^{(m)}\eta \mid n \geq m \geq 0\}$ forms the $\imath$canonical basis for $L(n)$ and $B_{i,\bar{0}}^{(m)}\eta = 0$, for $m > n$; similar claims hold for $L(n)$ with n odd and $\mathbf{B}_{\bar{1}}$; see **[16]**.

5. QUANTUM SCHUR DUALITIES

5.1. Quasi-parabolic Kazhdan–Lusztig bases

Let W_d be the Weyl group of type B_d generated by simple reflections s_i, for $0 \leq i \leq d-1$. It contains the symmetric group $\mathfrak{S}_d$ as a subgroup generated by s_i, for $1 \leq i \leq d-1$. For $p \in v^{\mathbb{Z}}$, the Hecke algebra $\mathscr{H}_{B_d}$ of type B_d is a $\mathbb{Q}(v)$-algebra generated by H_i $(0 \leq i \leq d-1)$ such that $(H_0 + p^{-1})(H_0 - p) = 0$, $(H_i + v^{-1})(H_i - v) = 0$ for $i \geq 1$, and braid relations hold; $\mathscr{H}_{B_d}$ admits a bar involution $\bar{\ }$ such that $\overline{v} = v^{-1}$ and $\overline{H_i} = H_i^{-1}$, for all i.

For $x \in \mathbb{R}$ and $m \in \mathbb{N}$, we denote $[x..x+m] = \{x, x+1, \ldots, x+m\}$. For $a \in \mathbb{Z}_{\geq 1}$, we denote by $\mathbb{I}_a = [\frac{1-a}{2}..\frac{a-1}{2}]$. For $r, m \in \mathbb{N}$ (not both zero), we introduce a new notation $\mathbb{I}_{r|m|r} := \mathbb{I}_{2r+m}$ to indicate a fixed set partition, $\mathbb{I}_{r|m|r} = \mathbb{I}_\circ^- \cup \mathbb{I}_\bullet \cup \mathbb{I}_\circ^+$, where

$$\mathbb{I}_\circ^+ = \left[\frac{m+1}{2}..r + \frac{m-1}{2}\right], \quad \mathbb{I}_\bullet = \left[\frac{1-m}{2}..\frac{m-1}{2}\right], \quad \mathbb{I}_\circ^- = -\mathbb{I}_\circ^+. \tag{5.1}$$

We view $f \in \mathbb{I}_{r|m|r}^d$ as a map $f : \{1, \ldots, d\} \to \mathbb{I}_{r|m|r}$, and identify $f = (f(1), \ldots, f(d))$. We define a right action of W_d on $\mathbb{I}_{r|m|r}^d$ such that, for $f \in \mathbb{I}_{r|m|r}^d$ and $0 \leq j \leq d-1$,

$$f \cdot s_j = \begin{cases} (\ldots, f(j+1), f(j), \ldots), & \text{if } j > 0, \\ (-f(1), f(2), \ldots, f(d)), & \text{if } j = 0,\ f(1) \in \mathbb{I}_\circ^- \cup \mathbb{I}_\circ^+, \\ (f(1), f(2), \ldots, f(d)), & \text{if } j = 0,\ f(1) \in \mathbb{I}_\bullet. \end{cases}$$

Let $p \in v^{\mathbb{Z}}$. Consider the $\mathbb{Q}(v)$-vector space $\mathbb{V} = \bigoplus_{a \in \mathbb{I}_{r|m|r}} \mathbb{Q}(v) u_a$. Given $f = (f(1), \ldots, f(d)) \in \mathbb{I}_{r|m|r}^d$, we denote $M_f = u_{f(1)} \otimes u_{f(2)} \otimes \cdots \otimes u_{f(d)}$. We shall call f a weight and $\{M_f \mid f \in \mathbb{I}_{r|m|r}^d\}$ the standard basis for $\mathbb{V}^{\otimes d}$. There is a right action of the Hecke algebra $\mathscr{H}_{B_d}$ on $\mathbb{V}^{\otimes d}$ as follows (see **[62]**):

$$M_f \cdot H_i = \begin{cases} M_{f\cdot s_i} + (v - v^{-1}) M_f, & \text{if } f(i) < f(i+1),\ i > 0, \\ M_{f\cdot s_i}, & \text{if } f(i) > f(i+1),\ i > 0, \\ v M_f, & \text{if } f(i) = f(i+1),\ i > 0, \\ M_{f\cdot s_i} + (p - p^{-1}) M_f, & \text{if } f(1) \in \mathbb{I}_\circ^+,\ i = 0, \\ M_{f\cdot s_i}, & \text{if } f(1) \in \mathbb{I}_\circ^-,\ i = 0, \\ p M_f, & \text{if } f(1) \in \mathbb{I}_\bullet,\ i = 0. \end{cases}$$

A weight $f \in \mathbb{I}_{r|m|r}^d$ is called *antidominant* if $\frac{m-1}{2} \geq f(1) \geq f(2) \geq \cdots \geq f(d)$; in this case we have $f(j) \in \mathbb{I}_\circ^- \cup \mathbb{I}_\bullet$, for all j. Denote $\mathbb{I}_{r|m|r}^{d,-} = \{f \in \mathbb{I}_{r|m|r}^d \mid f \text{ is antidominant}\}$. Decompose $\mathbb{V}^{\otimes d}$ into a direct sum of cyclic submodules generated by M_f, for antidominant weights f: $\mathbb{V}^{\otimes d} = \bigoplus_{f \in \mathbb{I}_{r|m|r}^{d,-}} \mathbb{M}_f$, where $\mathbb{M}_f = M_f \mathscr{H}_{B_d}$. Denote by $\mathcal{O}_f$ the orbit of f under the action of W_d on $\mathcal{I}^d_{r|m|r}$. The $\mathscr{H}_{B_d}$-module $\mathbb{M}_f$ admits a standard basis $\{M_g \mid g \in \mathcal{O}_f\}$.

The stabilizer subgroup of $f \in \mathcal{I}^{d,-}_{r|m|r}$ in W_d is of the form

$$W_f = W_{m_1} \times \cdots \times W_{m_k} \times S_{m_{k+1}} \times \cdots \times S_{m_l},$$

with all $m_i > 0$ and $W_{m_1} \times \cdots \times W_{m_k}$ corresponding to the components of f in $\mathcal{I}_\bullet$. Note that the stabilizer subgroup W_f is not a parabolic subgroup of W_d when $k \geq 2$. This phenomenon does not occur in the setting of [9,12]. We call the summands $\mathbb{M}_f$ of $\mathbb{V}^{\otimes d}$ *quasipermutation modules*. Clearly, for $f, f' \in \mathcal{I}^{d,-}_{r|m|r}$, we have $\mathbb{M}_f \cong \mathbb{M}_{f'}$, if $W_f = W_{f'}$. If W_f is not parabolic, $\mathbb{M}_f$ is in general not an induced module as those considered in parabolic KL setting [25].

Let $f \in \mathcal{I}^{d,-}_{r|m|r}$. Denote by ${}^fW$ the set of minimal length right coset representatives in $W_f \backslash W_d$. We define a $\mathbb{Q}$-linear map $\psi_\imath$ on the module $\mathbb{M}_f$ (which has a basis $M_{f\cdot\sigma}$, for $\sigma \in {}^fW$) by $\psi_\imath(v) = v^{-1}$, $\psi_\imath(M_{f\cdot\sigma}) = M_f \bar{H}_\sigma$, $\forall \sigma \in {}^fW$. It can be shown [62] (more difficult than the parabolic case in [25]) that the map $\psi_\imath$ on $\mathbb{M}_f$ is compatible with the bar operator on the Hecke algebra, i.e., $\psi_\imath(xh) = \psi_\imath(x)\bar{h}$, for all $x \in \mathbb{M}_f, h \in \mathscr{H}_{B_d}$. In particular, $\psi_\imath^2 = \mathrm{Id}$. We shall call $\psi_\imath$ the bar involution on $\mathbb{M}_f$.

Theorem 5.1 ([62]). *Suppose $p \in v^{\mathbb{Z}}$ and let $f \in \mathcal{I}^{d,-}_{r|m|r}$. Then for each $\sigma \in {}^fW$, there exists a unique element $C_\sigma \in \mathbb{M}_f$ such that*

$$\psi_\imath(C_\sigma) = C_\sigma \quad \text{and} \quad C_\sigma \in M_{f\cdot\sigma} + \sum_{w \in {}^fW, w<\sigma} v^{-1}\mathbb{Z}[v^{-1}]M_{f\cdot w}.$$

Similarly, there exist elements $C^*_\sigma \in \mathbb{M}_f$, for $\sigma \in {}^fW$, characterized by $\psi_\imath(C^*_\sigma) = C^*_\sigma$ and $C^*_\sigma \in M_{f\cdot\sigma} + \sum_{w\in {}^fW, w<\sigma} v\mathbb{Z}[v]M_{f\cdot w}$. The basis $\{C_\sigma \mid \sigma \in {}^fW\}$ is called a *quasiparabolic KL basis* for $\mathbb{M}_f$; the basis $\{C^*_\sigma \mid \sigma \in {}^fW\}$ is called a *dual quasiparabolic KL basis* for $\mathbb{M}_f$. Depending on the choice of f, the canonical basis can be type B or type A parabolic KL basis [25,33], or neither.

5.2. A type B ıSchur duality

Set $n = \frac{m}{2} \in \frac{1}{2}\mathbb{N}$. We consider the quantum group $\mathbf{U} = \mathbf{U}_v(\mathfrak{sl}_{2r+m})$ of type A_{2r+m-1}, where $\mathbb{I} := [1-n-r..n+r-1]$. We view $\mathbb{V}$ as a natural representation of $\mathbf{U}$, and so $\mathbb{V}^{\otimes d}$ is a $\mathbf{U}$-module via the comultiplication Δ. We consider the Satake diagram of type AIII with $m-1 = 2n-1$ black nodes and r pairs of white nodes, and a diagram involution τ:

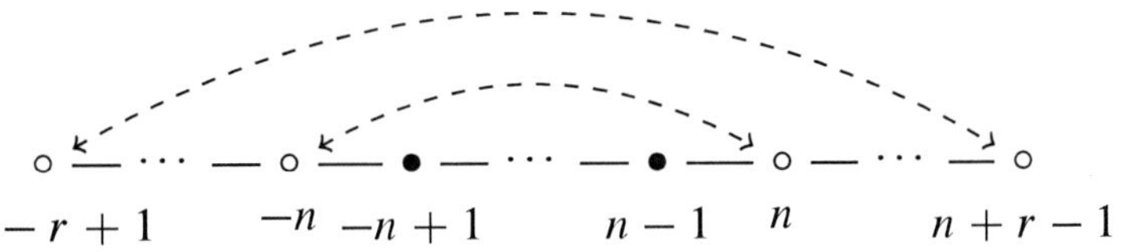

(In case $n = 0$, the black nodes are dropped; the nodes n and $-n$ are identified and fixed by τ.) The involution τ on $\mathbb{I}$ sends $i \mapsto \tau(i) = -i$, for all i, and it induces an involution of $\mathbf{U}$, denoted again by τ, by permuting the indices of its generators $E_i, F_i, K_i^{\pm 1}$.

Let $\mathbb{I}_\bullet = [1-n..n-1]$ be the set of all black nodes in $\mathbb{I}$ and $\mathbb{I}_\circ = \mathbb{I}\backslash\mathbb{I}_\bullet$. Associated to the Satake diagram $(\mathbb{I} = \mathbb{I}_\bullet \cup \mathbb{I}_\circ, \tau)$, we have a quantum symmetric pair $(\mathbf{U}, \mathbf{U}^\imath)$ of type AIII. Recall that $p \in v^{\mathbb{Z}}$. We shall fix the parameters to be

$$\begin{cases} \varsigma_i = 1 \quad (\text{for } i \neq \pm n), \\ \varsigma_n = (-1)^{m-1} v^m p^{-1}, \quad \varsigma_{-n} = p, \quad \text{if } m = 2n \in \mathbb{Z}_{\geq 1}, \end{cases} \tag{5.2}$$

$$\begin{cases} \varsigma_0 = v^{-1}, \quad \varsigma_i = 1 \quad (\text{for } i \neq 0), \\ \kappa_0 = (p - p^{-1})/(v - v^{-1}), \quad \text{if } m = 0. \end{cases} \tag{5.3}$$

It can be shown [62] that the actions of $\mathbf{U}^\imath$ and $\mathscr{H}_{B_d}$ on $\mathbb{V}^{\otimes d}$ commute with each other:

$$\mathbf{U}^\imath \overset{\Psi}{\curvearrowright} \mathbb{V}^{\otimes d} \overset{\Phi}{\curvearrowleft} \mathscr{H}_{B_d}.$$

Moreover, $\Psi(\mathbf{U}^\imath)$ and $\Phi(\mathscr{H}_{B_d})$ form double centralizers in $\mathrm{End}(\mathbb{V}^{\otimes d})$. This duality has been called *ıSchur duality* (which goes back to [9] in a quasisplit case).

There exists a unique antilinear bar involution $\psi_\imath : \mathbb{V}^{\otimes d} \to \mathbb{V}^{\otimes d}$ such that $\psi_\imath(M_f) = M_f$, for $f \in \mathcal{I}^{d,-}_{r|m|r}$, and it is compatible with the bar involutions on $\mathscr{H}_{B_d}$ and $\mathbf{U}^\imath$, that is, for $u \in \mathbf{U}^\imath$, $v \in \mathbb{V}^{\otimes d}$, and $h \in \mathscr{H}_{B_d}$, $\psi_\imath(uvh) = \psi_\imath(u)\psi_\imath(v)\bar{h}$. Recall that $\mathbb{V}^{\otimes d}$ is a direct sum of the quasipermutation modules $\mathbb{M}_f$ of $\mathscr{H}_{B_d}$. The union of the (dual) quasiparabolic KL bases on the summands $\mathbb{M}_f$ provides a (dual) quasiparabolic KL basis on $\mathbb{V}^{\otimes d}$.

Theorem 5.2 ([62]). *The (dual) ıcanonical basis on $\mathbb{V}^{\otimes d}$ (viewed as a $\mathbf{U}^\imath$-module) coincides with the (dual) quasiparabolic KL basis on $\mathbb{V}^{\otimes d} = \bigoplus_f \mathbb{M}_f$ (viewed as an $\mathscr{H}_{B_d}$-module).*

The quasiparabolic KL polynomials are by definition the transition matrix entries from the quasiparabolic KL to the standard basis. An inversion formula for parabolic KL polynomials (theorems of Kazhdan–Lusztig and Douglass) can be generalized to the quasiparabolic cases.

Remark 5.3. In case when $r = 0$, ıSchur duality reduces to Jimbo–Schur duality [31] between $\mathbf{U}$ and Hecke algebra of type A, and Theorem 5.2 goes back to a result of Frenkel–Khovanov–Kirillov. In case when $m = 0, 1$, it reduces to the quasisplit ıSchur duality [4,9,11], which has applications to Kazhdan–Lusztig theory of classical type.

6. APPLICATION TO SUPER KAZHDAN–LUSZTIG THEORY

6.1. The BGG category

Consider the BGG category $\mathcal{O}$ of $\mathfrak{g}$-modules, where $\mathfrak{g} = \mathfrak{n}^- \oplus \mathfrak{h} \oplus \mathfrak{n}$ is a simple or reductive Lie (super)algebra over $\mathbb{C}$. There is a duality functor $\vee : \mathcal{O} \to \mathcal{O}$ sending $M = \bigoplus_{\mu\in\mathfrak{h}^*} M_\mu$ to $M^\vee := \bigoplus_{\mu\in\mathfrak{h}^*} M^*_\mu$. Let $M(\lambda)$ be the Verma module with highest weight λ and $L(\lambda)$ be its unique irreducible quotient. It is known that a simple module $L(\lambda)$ and a tilting module $T(\lambda)$ (of highest weight λ) are self-dual with respect to $\vee$.

For semisimple Lie algebras, the linkage is controlled by the dot action of the Weyl group, and the BGG category $\mathcal{O}$ admits a block decomposition according to the central characters. For a (or any) regular block $\mathcal{O}^0$, its Grothendieck group is identified with the Weyl group algebra. If we further identify the Verma module basis in $[\mathcal{O}^0]$ with the standard basis in the Hecke algebra (specialized at $v = 1$), then the Kazhdan–Lusztig conjecture (a theorem of Beilinson–Bernstein and Brylinski–Kashiwara) states that the simple module basis corresponds to the dual canonical basis (and the tilting module basis corresponds to the canonical basis).

For general linear or orthosymplectic Lie superalgebras, the linkage in the BGG category is no longer controlled by the Weyl group, and so the formulation of Kazhdan–Lusztig theory via Hecke algebras breaks down.

6.2. Super type BCD character formulas

Let us treat the super type B case, $\mathfrak{g} = \mathfrak{osp}(2m+1|2n)$, in detail. With respect to a standard Dynkin diagram

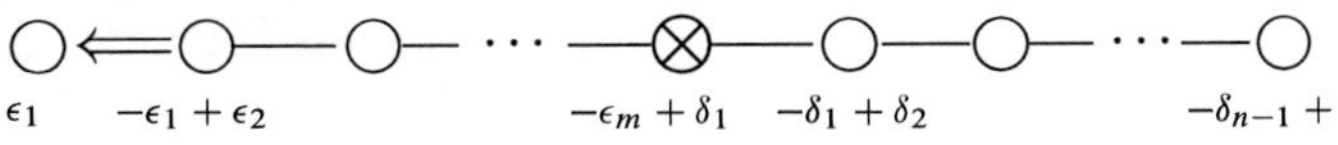

we have the Weyl vector

$$\rho = \frac{1}{2}\epsilon_1 + \frac{3}{2}\epsilon_2 + \cdots + (m - \frac{1}{2})\epsilon_m - (m - \frac{1}{2})\delta_1 - (m - \frac{3}{2})\delta_2 - \cdots - (m - n + \frac{1}{2})\delta_n.$$

There exists a ρ-shift bijection for the set of integer weights

$$X^{m|n} := \bigoplus_{i=1}^{m} \mathbb{Z}\epsilon_i \oplus \bigoplus_{j=1}^{n} \mathbb{Z}\delta_j \xrightarrow[\rho\text{-shift}]{\cong} \left(\frac{1}{2} + \mathbb{Z}\right)^{m+n}, \quad \lambda \mapsto f_\lambda,$$

where f_λ is defined via $\lambda + \rho = \sum_{i=1}^{m} f_\lambda(i)\epsilon_i + \sum_{j=1}^{n} f_\lambda(m+j)\delta_j$. Similarly, there exists a bijection for the set of half-integer weights

$$X^{m|n}_{\frac{1}{2}} := \bigoplus_{i=1}^{m} \left(\frac{1}{2} + \mathbb{Z}\right)\epsilon_i \oplus \bigoplus_{j=1}^{n} \left(\frac{1}{2} + \mathbb{Z}\right)\delta_j \xrightarrow[\rho\text{--shift}]{\cong} \mathbb{Z}^{m+n}, \quad \lambda \mapsto f_\lambda.$$

Denote by $\mathcal{O}^{m|n}_{\mathfrak{b}}$ (respectively, $\mathcal{O}^{m|n}_{\mathfrak{b},\frac{1}{2}}$) the BGG category which contains the Verma modules $M(\lambda)$, tilting modules $T(\lambda)$ and simple modules $L(\lambda)$, parametrized by the weights $\lambda \in X^{m|n}$ (respectively, $\lambda \in X^{m|n}_{\frac{1}{2}}$).

Recall from Section 5.2 the quasisplit quantum symmetric pair $(\mathbf{U}_v(\mathfrak{sl}_N), \mathbf{U}^\imath)$ of type AIII, where we fix $p = v$ in (5.2)–(5.3). Recall the natural representation $\mathbb{V}$ with basis $\{u_i \mid i \in [\frac{1-N}{2} .. \frac{N-1}{2}]\}$, for N even and odd, allowing $N = \infty$ with parity! Then we can identify the indexing set $[\frac{1-N}{2} .. \frac{N-1}{2}]$ with $\frac{1}{2} + \mathbb{Z}$ for $N = \infty$ (even), and with $\mathbb{Z}$ for $N = \infty$ (odd).

By Theorem 4.1, the $\mathbf{U}_v(\mathfrak{sl}_\infty)$-module $\mathbb{V}^{\otimes m} \otimes \mathbb{V}^{*\otimes n}$ (regarded as $\mathbf{U}^\imath$-module with $p = v$) admits an $\imath$canonical basis, denoted by $\{C^\imath_f\}$, and a dual $\imath$canonical basis, denoted by $\{L^\imath_f\}$, where $f \in (\frac{1}{2} + \mathbb{Z})^{m+n}$ or $\mathbb{Z}^{m+n}$, respectively.

Define the following $\mathbb{Z}$-module isomorphisms:

$$\Psi_{\mathfrak{b}} : [\mathcal{O}_{\mathfrak{b}}^{m|n}] \to \mathbb{V}_{\mathbb{Z}}^{\otimes m} \otimes \mathbb{V}_{\mathbb{Z}}^{*\otimes n}, \quad [M(\lambda)] \mapsto M_{f_\lambda} \quad (\lambda \in X^{m|n}),$$
$$\Psi_{\mathfrak{b},\frac{1}{2}} : [\mathcal{O}_{\mathfrak{b},\frac{1}{2}}^{m|n}] \to \mathbb{V}_{\mathbb{Z}}^{\otimes m} \otimes \mathbb{V}_{\mathbb{Z}}^{*\otimes n}, \quad [M(\lambda)] \mapsto M_{f_\lambda} \quad (\lambda \in X_{\frac{1}{2}}^{m|n}). \tag{6.1}$$

A basic fact here [9] is that the generators B_i in $\mathbf{U}^\imath$ act on $[\mathcal{O}_{\mathfrak{b}}^{m|n}]$ and $[\mathcal{O}_{\mathfrak{b},\frac{1}{2}}^{m|n}]$ by translation functors, and the above $\mathbb{Z}$-module isomorphisms become $\mathbf{U}_{\mathbb{Z}}^\imath$-module isomorphisms at $v = 1$. (A similar observation on translation functors and B_i is valid for $p = 1$ [4], and it was made independently in [28] in the nonsuper setting.)

Theorem 6.1 ([9]). *The $\mathbb{Z}$-module isomorphism $\Psi_{\mathfrak{b}}$ (respectively, $\Psi_{\mathfrak{b},\frac{1}{2}}$) in (6.1) sends*

$$[L(\lambda)] \mapsto L_{f_\lambda}^\imath, \quad [T(\lambda)] \mapsto C_{f_\lambda}^\imath, \quad \textit{for } \lambda \in X^{m|n} \textit{ (and respectively, } \lambda \in X_{\frac{1}{2}}^{m|n}\textit{)}.$$

Remark 6.2 (Super type A Kazhdan–Lusztig theory). Consider the BGG category $\mathcal{O}^{m|n}$ of modules over the general linear Lie superalgebra $\mathfrak{g} = \mathfrak{gl}(m|n)$ of integer weights. We have an almost identical $\mathbb{Z}$-module isomorphism $\Psi : [\mathcal{O}^{m|n}] \to \mathbb{V}_{\mathbb{Z}}^{\otimes m} \otimes \mathbb{V}_{\mathbb{Z}}^{*\otimes n}$ as in (6.1), which match the Verma basis with the standard basis. Then Brundan–Kazhdan–Lusztig conjecture [18] (proved by Cheng, Lam, and the author in [22]) states that the simple module basis (respectively, tilting module basis) is mapped by Ψ to Lusztig dual canonical basis (respectively, canonical basis). There has been a second proof in [19] using ideas of categorification.

Example 6.3. Take $n = 0$ and $m = 1$, so that $\mathfrak{g} = \mathfrak{so}_3 \cong \mathfrak{sl}_2$. If the standard basis (= canonical basis) $\{u_i \mid i \in \mathbb{Z}\}$ for $\mathbb{V}$ is indexed by $\mathbb{Z}$, then $\mathbb{V}$ admits an $\imath$canonical basis $\{u_0, u_{-i}, u_i + v^{-1}u_{-i} \mid i \in \mathbb{Z}_{>0}\}$ and a dual $\imath$canonical basis $\{u_0, u_{-i}, u_i - vu_{-i} \mid i \in \mathbb{Z}_{>0}\}$.

Theorem 6.1 can be adapted to the type D Lie superalgebra $\mathfrak{g} = \mathfrak{osp}(2m|2n)$, by setting the parameter $p = 1$ (instead of $p = v$) in (5.2)–(5.3) (see Bao [4]). Thanks to Theorem 5.2, Theorem 6.1 for $m = 0$ amounts to a reformulation for the type B Kazhdan–Luszitg conjecture [33]. For further extension to Kazhdan–Lusztig theory for super parabolic BGG categories, see [12].

7. HALL ALGEBRAS

The Drinfeld double quantum group $\tilde{\mathbf{U}} = \tilde{\mathbf{U}}(\mathfrak{g})$ is the $\mathbb{Q}(v)$-algebra generated by E_i, F_i, K_i, K_i' $(i \in \mathbb{I})$ subject to relations in [50, (6.1)–(6.5)] similar to those in $\mathbf{U}$. Denote the Cartan matrix for $\mathfrak{g}$ by $(c_{ij})_{i,j\in\mathbb{I}}$. Following [50], we define the *universal $\imath$quantum group* $\tilde{\mathbf{U}}^\imath$ associated to a Satake diagram $(\mathbb{I} = \mathbb{I}_\bullet \cup \mathbb{I}_\circ, \tau)$ as the $\mathbb{Q}(v)$-subalgebra of $\tilde{\mathbf{U}}$ generated by $\tilde{\mathbf{U}}_{\mathbb{I}_\bullet}$ and $\{B_i, \tilde{k}_i \mid i \in \mathbb{I}_\circ\}$, with identifications

$$B_i \mapsto F_i + \tilde{T}_{w_\bullet}(E_{\tau i})K_i', \quad \tilde{k}_i \mapsto K_i K_{\tau i}', \quad i \in \mathbb{I}_\circ. \tag{7.1}$$

Denote the embedding by $\imath : \tilde{\mathbf{U}}^\imath \to \tilde{\mathbf{U}}$, $x \mapsto x^\imath$. One checks that $\tilde{\mathbf{U}}^\imath$ is a right coideal subalgebra of $\tilde{\mathbf{U}}$. The $\imath$quantum group $\mathbf{U}^\imath$ (with $Y = \mathbb{N}\mathbb{I}$) can be obtained from $\tilde{\mathbf{U}}^\imath$ via a central reduction.

In the remainder of this section, we shall only consider $\tilde{\mathbf{U}}^\imath$ of quasisplit types, i.e., $\mathbb{I}_\bullet = \emptyset$. Let $Q = (Q_0, Q_1)$ be a virtually acyclic quiver; see [52, DEF. 4.4]. This is a mild generalization of acyclic quivers, allowing a generalized Kronecker subquiver. Throughout the paper, we shall identify $Q_0 = \mathbb{I}$. An *ıquiver* (Q, τ) consists of a (virtually acyclic) quiver Q and an involution τ of Q; we allow the *trivial* involution Id. We work over a finite field $\mathbb{F}_q$. An involution τ of Q induces an involution of the path algebra $\mathbb{F}_q Q$, also denoted by τ.

Let $\overline{Q}$ be a new quiver obtained from Q by adding a loop ε_i at the vertex $i \in Q_0$ if $\tau i = i$, and adding an arrow $\varepsilon_i : i \to \tau i$ for each $i \in Q_0$ if $\tau i \neq i$; the ε_i are in purple color below. The *ıquiver algebra* $\Lambda^\imath$ associated to (Q, τ) can be defined in terms of the quiver $\overline{Q}$ with relations, cf. [50, PROP. 2.6], that is, $\Lambda^\imath \cong \mathbb{F}_q \overline{Q} / \overline{I}$, where $\overline{I}$ is generated by $\varepsilon_i \varepsilon_{\tau i}$ for each $i \in Q_0$ and $\varepsilon_i \alpha - \tau(\alpha)\varepsilon_j$ for each arrow $\alpha : j \to i$ in Q_1.

Rank 1 or 2 subquivers of the quiver $\overline{Q}$ associated to a general virtually acyclic quiver Q look like as follows (where $\overline{Q}$ is obtained from Q by adding arrows ε's):

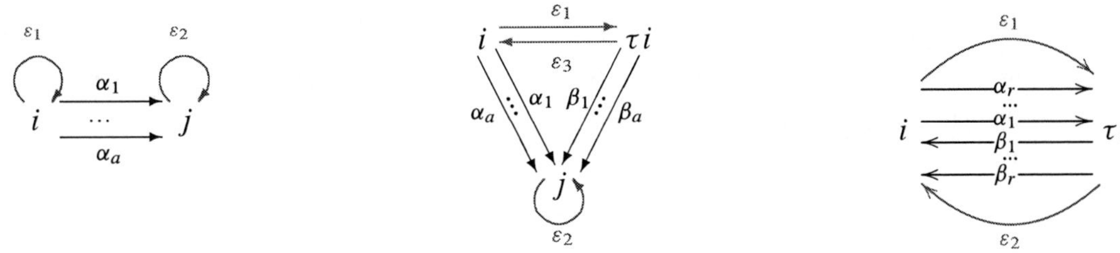

Denote by $\mathrm{mod}^{\mathrm{nil}}(\Lambda^\imath)$ the category of finite-dimensional nilpotent $\Lambda^\imath$-modules. Denote by S_i the 1-dimensional $\Lambda^\imath$-module supported at $i \in \mathbb{I}$, and $\mathbb{K}_i$ the 2-dimensional module "supported at ε_i." The algebra $\Lambda^\imath$ is a 1-Gorenstein algebra and hence admits favorable homological properties. In particular, the subcategory $\mathcal{P}^{\leq 1}(\Lambda^\imath)$ of $\mathrm{mod}^{\mathrm{nil}}(\Lambda^\imath)$ consisting of modules of projective dimension at most 1 admits clean characterization.

Let $\mathcal{H}(\Lambda^\imath)$ be the Ringel–Hall algebra of $\mathrm{mod}^{\mathrm{nil}}(\Lambda^\imath)$ over $\mathbb{Q}(\sqrt{q})$, that is, the $\mathbb{Q}(\sqrt{q})$-vector space whose basis is formed by the isoclasses $[M]$ of objects $M \in \mathrm{mod}^{\mathrm{nil}}(\Lambda^\imath)$, with multiplication defined by $[M] \diamond [N] = \sum_{[L] \in \mathrm{Iso}(\mathrm{mod}(\Lambda^\imath))} \frac{|\mathrm{Ext}^1(M,N)_L|}{|\mathrm{Hom}(M,N)|}[L]$. Then, the semiderived Hall algebra $\mathcal{S}\mathcal{D}\mathcal{H}(\Lambda^\imath)$ of $\Lambda^\imath$ is defined in terms of localization of a quotient algebra of the Ringel–Hall algebra $\mathcal{H}(\Lambda^\imath)$ with respect to $\mathcal{P}^{\leq 1}(\Lambda^\imath)$, and the *ıHall algebra* $\tilde{\mathcal{H}}(\mathbb{F}_q Q, \tau)$ is defined to be $\mathcal{S}\mathcal{D}\mathcal{H}(\Lambda^\imath)$ with a new multiplication via twisting by an Euler form; see [50, APPENDIX A] [52] for precise definitions.

Let $\mathbb{I}_\tau$ be a set of representatives of the τ-orbits on $\mathbb{I}_\circ$.

Theorem 7.1 ([50,52]). *Let (Q, τ) be a virtually acyclic ıquiver. Then there exists a $\mathbb{Q}(\sqrt{q})$-algebra monomorphism $\widetilde{\psi} : \tilde{\mathbf{U}}^\imath_{|v=\sqrt{q}} \to \tilde{\mathcal{H}}(\mathbb{F}_q Q, \tau)$, which sends*

$$B_j \mapsto \frac{-1}{q-1}[S_j], \quad \text{if } j \in \mathbb{I}_\tau, \quad \tilde{k}_i \mapsto -q^{-1}[\mathbb{K}_i], \quad \text{if } \tau i = i \in \mathbb{I},$$
$$B_j \mapsto \frac{\sqrt{q}}{q-1}[S_j], \quad \text{if } j \notin \mathbb{I}_\tau, \quad \tilde{k}_i \mapsto \sqrt{q}^{\frac{-c_{i,\tau i}}{2}}[\mathbb{K}_i], \quad \text{if } \tau i \neq i \in \mathbb{I}.$$

This theorem for diagonal ıquivers $(Q \sqcup Q, \text{swap})$ specializes to Bridgeland's Hall algebra realization of Drinfeld double quantum groups in [17]. The above monomorphism becomes an isomorphism for Dynkin $\imath$quivers. Reflection functors on $\imath$Hall algebras provide a conceptual approach to braid group actions on $\tilde{\mathbf{U}}^\imath$ (of quasisplit type); see [47,48].

8. RELATIVE BRAID GROUP ACTIONS

Let $(\tilde{\mathbf{U}}, \tilde{\mathbf{U}}^\imath)$ be the universal quantum symmetric pair associated to a Satake diagram $(\mathbb{I} = \mathbb{I}_\bullet \cup \mathbb{I}_\circ, \tau)$. It is shown in [64] that there exists a quasi-K-matrix $\widetilde{\Upsilon}$ associated to $(\tilde{\mathbf{U}}, \tilde{\mathbf{U}}^\imath)$ satisfying an intertwining relation like (3.1). For $i \in \mathbb{I}_\circ$, denote by $\widetilde{\Upsilon}_i$ the quasi-K-matrix associated to the rank-1 Satake subdiagram $\mathbb{I}_{\bullet,i} := \mathbb{I}_\bullet \cup \{i, \tau i\}$ in the setting of $(\tilde{\mathbf{U}}, \tilde{\mathbf{U}}^\imath)$.

Let $W_{\bullet,i}$ be the parabolic subgroup of the Weyl group W of $\mathfrak{g}$ generated by s_i, for $i \in \mathbb{I}_{\bullet,i}$, with longest element $w_{\bullet,i}$. Define $\mathbf{r}_i \in W_{\bullet,i}$ by $\mathbf{r}_i = w_{\bullet,i} w_\bullet$; it is clear that $\mathbf{r}_i = \mathbf{r}_{\tau i}$. Recall that $\mathbb{I}_\tau$ denotes a set of representatives of the τ-orbits on $\mathbb{I}_\circ$. The relative Weyl group associated to the symmetric pair is identified with the subgroup $\boldsymbol{W} = \langle \mathbf{r}_i \mid i \in \mathbb{I}_\tau \rangle$ of W. There is a notion of the relative braid group associated to $\boldsymbol{W}$. The existence of such a relative braid group action on an $\imath$quantum group $\mathbf{U}^\imath$ was conjectured in [37]. The conjecture was verified therein via a computation for (mostly quasisplit) finite type; for an alternative approach via $\imath$Hall algebras, see [47,48].

There is a braid group action associated to W on the Drinfeld double $\tilde{\mathbf{U}}$ (see [48]), a variant of the braid group action on $\mathbf{U}$ in [55]. We shall need a suitably rescaled variant, denoted by $\tilde{T}_i^{-1}$, for $i \in \mathbb{I}$, which again satisfies the braid group relations. In particular, an automorphism $\tilde{T}_{\mathbf{r}_i}^{-1}$ of $\tilde{\mathbf{U}}$, for $i \in \mathbb{I}_\tau$, is defined. We announce a new conceptual approach developed by W. Zhang and the author to relative braid group actions.

Theorem 8.1 ([64]). *Let $(\tilde{\mathbf{U}}, \tilde{\mathbf{U}}^\imath)$ be a universal quantum symmetric pair of arbitrary finite type. Then there exists an automorphism $\tilde{\mathbf{T}}_i^{-1}$ of $\tilde{\mathbf{U}}^\imath$, for $i \in \mathbb{I}_\tau$, which satisfies the intertwining relation*

$$\tilde{\mathbf{T}}_i^{-1}(x)\, \widetilde{\Upsilon}_i = \widetilde{\Upsilon}_i\, \tilde{T}_{\mathbf{r}_i}^{-1}(x^\imath), \quad \textit{for all } x \in \tilde{\mathbf{U}}^\imath.$$

Moreover, the automorphisms $\tilde{\mathbf{T}}_i^{-1}$, for $i \in \mathbb{I}_\tau$, satisfy the relative braid group relations.

The approach in [64] has additional consequences. Explicit compact formulas for the action of $\tilde{\mathbf{T}}_i^{-1}$ on the generators of $\tilde{\mathbf{U}}^\imath$ are obtained. The relative braid group action on $\tilde{\mathbf{U}}^\imath$ gives rise to compatible relative braid group actions on $\mathbf{U}^\imath$ and $\mathbf{U}$-modules (viewed as $\mathbf{U}^\imath$-modules). Along the way, we prove the conjecture of Dobson–Kolb [26] on factorization of quasi-K-matrices of arbitrary finite type.

9. A CURRENT PRESENTATION OF AFFINE TYPE

In this section, we consider universal $\imath$quantum groups $\tilde{\mathbf{U}}^\imath$ of split affine type, that is, $\mathbb{I}_\bullet = \emptyset$, $\tau = \mathrm{Id}$, and the Cartan matrix $(c_{ij})_{i,j\in\mathbb{I}}$ is of untwisted affine type. By definition, $\tilde{\mathbf{U}}^\imath$ is a subalgebra of $\tilde{\mathbf{U}}$; alternatively, $\tilde{\mathbf{U}}^\imath$ is the $\mathbb{Q}(v)$-algebra generated by $B_i, \mathbb{K}_i^{\pm 1}$ $(i \in \mathbb{I})$, subject to the following relations: $\mathbb{K}_i$ are central, and

$$B_i B_j - B_j B_i = 0, \quad \text{if } c_{ij} = 0, \tag{9.1}$$

$$B_i^2 B_j - [2]_i B_i B_j B_i + B_j B_i^2 = -v_i^{-1} B_j \mathbb{K}_i, \quad \text{if } c_{ij} = -1. \tag{9.2}$$

We omit here the more complicated Serre-type relations between B_i, B_j for $c_{ij} = -2, -3$; they can be read off from setting $k = l = 0$ in (9.9)–(9.10) below. The $\mathbb{K}_i$ (which is natural from $\imath$Hall algebra viewpoint) is related to $\tilde{k}_i$ used earlier by $\mathbb{K}_i = -v_i^2 \tilde{k}_i$.

Associated to the affine Lie algebra $\mathfrak{g}$, we denote by $\mathfrak{g}_0, \mathbb{I}_0, X_0, W_0$ the underlying semisimple Lie algebra, its simple roots, weight lattice, and finite Weyl group. Recall the (extended) affine Weyl group $W^e = W_0 \ltimes X_0$. There are automorphisms $\tilde{\mathbf{T}}_i$ of $\tilde{\mathbf{U}}^\imath$, for $i \in \mathbb{I}$, which arise naturally from $\imath$Hall algebras [47, 48]; also see Section 8. They give rise to the action of an affine braid group associated to W^e for the affine Lie algebra $\mathfrak{g}$. In particular, we have automorphisms $\tilde{\mathbf{T}}_w$ of $\tilde{\mathbf{U}}^\imath$, for $w \in W^e$.

Define a sign function $o(\cdot) : \mathbb{I}_0 \to \{\pm 1\}$ such that $o(i)o(j) = -1$ whenever $c_{ij} < 0$. Define v-root vectors $B_{i,k}, \acute{\Theta}_{i,m}, \Theta_{i,m}$ in $\tilde{\mathbf{U}}^\imath$ for $i \in \mathbb{I}_0$, $k \in \mathbb{Z}$ and $m \geq 1$ by [49, 67]

$$B_{i,k} = o(i)^k \tilde{\mathbf{T}}_{\omega_i}^{-k}(B_i),$$

$$\acute{\Theta}_{i,m} = o(i)^m \Big(-B_{i,m-1} \tilde{\mathbf{T}}_{\omega_i'}(B_i) + v_i^2 \tilde{\mathbf{T}}_{\omega_i'}(B_i) B_{i,m-1} + (v_i^2 - 1) \sum_{p=0}^{m-2} B_{i,p} B_{i,m-p-2} \mathbb{K}_i^{-1} \mathbb{K}_\delta \Big),$$

$$\Theta_{i,m} = \acute{\Theta}_{i,m} - \sum_{a=1}^{\lfloor \frac{m-1}{2} \rfloor} (v_i^2 - 1) v_i^{-2a} \acute{\Theta}_{i,m-2a} \mathbb{K}_{a\delta} - \delta_{m,ev} v_i^{1-m} \mathbb{K}_{\frac{m}{2}\delta}.$$

A version of the v-root vectors for $\mathbf{U}^\imath$ in affine rank-1 case (known as q-Onsager algebra) was constructed earlier in [13].

Let ${}^{\mathrm{Dr}}\tilde{\mathbf{U}}^\imath$ be the $\mathbb{Q}(v)$-algebra generated by $\mathbb{K}_i^{\pm 1}$, $C^{\pm 1}$, $H_{i,m}$, and $B_{i,l}$, where $i \in \mathbb{I}_0$, $m \in \mathbb{Z}_{\geq 1}$, $l \in \mathbb{Z}$, subject to the following relations, for $m, n \in \mathbb{Z}_{\geq 1}$ and $k, l \in \mathbb{Z}$:

$$\mathbb{K}_i, C \text{ are central}, \quad [H_{i,m}, H_{j,n}] = 0, \quad \mathbb{K}_i \mathbb{K}_i^{-1} = 1, \quad C C^{-1} = 1, \tag{9.3}$$

$$[H_{i,m}, B_{j,l}] = \frac{[mc_{ij}]_i}{m} B_{j,l+m} - \frac{[mc_{ij}]_i}{m} B_{j,l-m} C^m, \tag{9.4}$$

$$[B_{i,k}, B_{j,l+1}]_{v_i^{-c_{ij}}} - v_i^{-c_{ij}} [B_{i,k+1}, B_{j,l}]_{v_i^{c_{ij}}} = 0, \text{ if } i \neq j, \tag{9.5}$$

$$\begin{aligned} &[B_{i,k}, B_{i,l+1}]_{v_i^{-2}} - v_i^{-2} [B_{i,k+1}, B_{i,l}]_{v_i^2} \\ &= v_i^{-2} \Theta_{i,l-k+1} C^k \mathbb{K}_i - v_i^{-4} \Theta_{i,l-k-1} C^{k+1} \mathbb{K}_i + v_i^{-2} \Theta_{i,k-l+1} C^l \mathbb{K}_i \\ &\quad - v_i^{-4} \Theta_{i,k-l-1} C^{l+1} \mathbb{K}_i, \end{aligned} \tag{9.6}$$

$$[B_{i,k}, B_{j,l}] = 0, \quad \text{if } c_{ij} = 0, \tag{9.7}$$

$$\sum_{s=0}^{2} (-1)^s \begin{bmatrix} 2 \\ s \end{bmatrix}_i B_{i,k}^{2-s} B_{j,l} B_{i,k}^s = -v_i^{-1} B_{j,l} \mathbb{K}_i C^k, \quad \text{if } c_{ij} = -1, \tag{9.8}$$

$$\sum_{s=0}^{3}(-1)^s \begin{bmatrix} 3 \\ s \end{bmatrix}_i B_{i,k}^{3-s} B_{j,l} B_{i,k}^s = -v_i^{-1}[2]_i^2 (B_{i,k}B_{j,l} - B_{j,l}B_{i,k})\mathbb{K}_i C^k, \quad \text{if } c_{ij} = -2, \tag{9.9}$$

$$\begin{aligned} &\sum_{s=0}^{4}(-1)^s \begin{bmatrix} 4 \\ s \end{bmatrix}_i B_{i,k}^{4-s} B_{j,l} B_{i,k}^s \\ &\quad = -v_i^{-1}\big(1 + [3]_i^2\big)(B_{j,l}B_{i,k}^2 + B_{i,k}^2 B_{j,l})\mathbb{K}_i C^k \\ &\qquad + v_i^{-1}[4]_i\big(1 + [2]_i^2\big) B_{i,k}B_{j,l}B_{i,k}\mathbb{K}_i C^k - v_i^{-2}[3]_i^2 B_{j,l}\mathbb{K}_i^2 C^{2k}, \quad \text{if } c_{ij} = -3, \end{aligned} \tag{9.10}$$

where $\Theta_{i,m}$ are related to $H_{i,m}$ by

$$1 + \sum_{m\geq 1}(v_i - v_i^{-1})\Theta_{i,m}u^m = \exp\Big((v_i - v_i^{-1})\sum_{m\geq 1} H_{i,m}u^m\Big).$$

(The ${}^{\mathrm{Dr}}\tilde{\mathbf{U}}^\imath$ is denoted by ${}^{\mathrm{Dr}}\tilde{\mathbf{U}}^\imath_{\mathrm{red}}$ in [67].)

Below is an $\imath$-analog of the Drinfeld presentation of affine quantum groups [27] (proved in [14,23]).

Theorem 9.1 ([49,67]). *There is a $\mathbb{Q}(v)$-algebra isomorphism $\Phi : {}^{\mathrm{Dr}}\tilde{\mathbf{U}}^\imath \to \tilde{\mathbf{U}}^\imath$, which sends*

$$B_{i,k} \mapsto B_{i,k}, \quad H_{i,m} \mapsto H_{i,m}, \quad \Theta_{i,m} \mapsto \Theta_{i,m}, \quad \mathbb{K}_i \mapsto \mathbb{K}_i, \quad C \mapsto \mathbb{K}_\delta,$$

for $i \in \mathbb{I}_0$, $k \in \mathbb{Z}$, $m \geq 1$.

Remark 9.2. More involved Serre relations among $B_{i,k}$, $B_{j,l}$, $B_{i,k'}$ generalizing the relations (9.8)–(9.9) are available; see [49,67]. They can be shown to be equivalent to (9.8)–(9.9), when combined with other relations (9.3)–(9.7) above.

It is straightforward to pass the v-root vectors and Drineld-type presentation of $\tilde{\mathbf{U}}^\imath$ to $\mathbf{U}^\imath$ with arbitrary parameters by central reduction. This current (or Drinfeld-type) presentation will be extended beyond split types in a future work.

10. OPEN PROBLEMS

"There's no use trying, one can't believe i... things."
"Why, sometimes I've believed as many as six i... things before breakfast."

—*Alice in Wonderland*

The open problems in the following six (interconnected) directions on $\imath$quantum groups look most appealing to us:

(1) *Positivity of $\imath$canonical basis*
Positivity of $\imath$canonical basis holds in the (affine) type AIII setting [29, 43]. We conjecture that the $\imath$canonical bases arising from the $\imath$quantum groups of

(quasi)split ADE type (with parameters suitably specified; see Example 4.3) exhibit various positivity properties. Recently Lusztig extended his earlier construction of total positivity to symmetric spaces in [56]. It would be interesting to strengthen this construction by connecting to $\imath$canonical basis (with positivity).

(2) *$\imath$Quiver varieties and geometric realizations of $\imath$quantum groups*
Geometric realizations of quantum groups are obtained in [15, 57, 58, 63]. The works [5, 29] can be regarded as the $\imath$-generalizations of [15]. Li [42] provides an $\imath$-analog of some Nakajima quiver varieties. We observe, however, that the diagram involutions used in that work are in line with Vogan diagrams instead of Satake diagrams. A fresh start is needed to construct general $\imath$quiver varieties (allowing Satake diagrams with black nodes and non-Dynkin types). The geometric realization of $\imath$quantum groups à la [57, 58, 63] remains to be carried out. Lu and the author provided in [51] a realization of $\tilde{\mathbf{U}}^\imath$ via Nakajima–Keller–Scherotzke quiver varieties, generalizing F. Qin's approach for quantum groups [59].

(3) *$\imath$Categorification*
There has been a KLR-type categorification of one family of modified $\imath$quantum groups of type AIII by Bao, Shan, Webster, and the author [7]. The categorification of the split rank-1 $\imath$quantum group (see Example 4.3) will be a fundamental new step, allowing the $\imath$categorification to move forward. The $\imath$categorification shall have applications to modular representation theory.

(4) *$\imath$Hall algebra*
So far, the $\imath$Hall algebras can only realize the *quasisplit* $\imath$quantum groups; see [52]. It is desirable to extend the $\imath$Hall algebras to a greater generality allowing Satake diagrams with black nodes, and also to understand categorically the embedding $\tilde{\mathbf{U}}^\imath \to \tilde{\mathbf{U}}$, as well as the coideal structure ($\Delta : \tilde{\mathbf{U}}^\imath \to \tilde{\mathbf{U}}^\imath \otimes \tilde{\mathbf{U}}$).

(5) *Representations of affine $\imath$quantum groups*
There have been numerous results in finite-dimensional representations of affine quantum groups and connections to other areas, presented by V. Chari and many others. One hopes that the Drinfeld-type presentation of affine $\imath$quantum groups (see [49,67]) can stimulate the development of their finite-dimensional representations.

(6) *$\imath$Quantum groups at roots of 1*
Building on Lusztig's constructions for quantum groups at roots of 1, Bao and Sale [6] have taken a first step in formulating small quantum symmetric pairs. More can be expected in this direction in light of Lusztig's program.

It is hoped that $\imath$quantum groups may find more applications in mathematical physics, geometric and modular representation theory, quantum topology, and algebraic combinatorics.

Just as generalizing the study of compact or complex Lie groups to real Lie groups and symmetric spaces, we hope to have convinced the reader that it is good to generalize various fundamental constructions from quantum groups to $\imath$quantum groups.

It is time for the reader to come up with his or her own favorite item (+) in the list of highlights for quantum groups in the Introduction, and supply its missing $\imath$-generalization!

ACKNOWLEDGMENTS

This survey represents part of what the author has learned about $\imath$quantum groups from his friends, many collaborators, and students, to whom he is very grateful; special thanks go to Huanchen Bao, Stefan Kolb, Ming Lu, and Weinan Zhang. The dramatic development and rapid expansion in this research area in recent years have been made possible largely due to their ideas, insights, generosity, and tireless efforts. We thank Huanchen Bao, Ming Lu, and Hideya Watanabe for careful reading of the paper and helpful comments.

FUNDING

This work was partially supported by the NSF grant DMS-2001351.

REFERENCES

[1] A. Appel and B. Vlaar, Universal K-matrix for quantum Kac–Moody algebras. 2020, arXiv:2007.09218.

[2] M. Balagovic and S. Kolb, The bar involution for quantum symmetric pairs. *Represent. Theory* **19** (2015), 186–210.

[3] M. Balagovic and S. Kolb, Universal K-matrix for quantum symmetric pairs. *J. Reine Angew. Math.* **747** (2019), 299–353.

[4] H. Bao, Kazhdan–Lusztig theory of super type D and quantum symmetric pairs. *Represent. Theory* **21** (2017), 247–276.

[5] H. Bao, J. Kujawa, Y. Li, and W. Wang, Geometric Schur duality of classical type. *Transform. Groups* **23** (2018), 329–389.

[6] H. Bao and T. Sale, Quantum symmetric pairs at roots of 1. *Adv. Math.* **380** (2021), 107576.

[7] H. Bao, P. Shan, W. Wang, and B. Webster, Categorification of quantum symmetric pairs I. *Quantum Topol.* **9** (2018), 643–714.

[8] H. Bao and W. Wang, Canonical bases arising from quantum symmetric pairs. *Invent. Math.* **213** (2018), 1099–1177.

[9] H. Bao and W. Wang, A new approach to Kazhdan–Lusztig theory of type B via quantum symmetric pairs. *Astérisque* **402** (2018), vii+134 pp. arXiv:1310.0103.

[10] H. Bao and W. Wang, Canonical bases arising from quantum symmetric pairs of Kac–Moody type. *Compos. Math.* **157** (2021), 1507–1537.

[11] H. Bao, W. Wang, and H. Watanabe, Multiparameter quantum Schur duality of type B. *Proc. Amer. Math. Soc.* **146** (2018), 3203–3216.

[12] H. Bao, W. Wang, and H. Watanabe, Canonical bases for tensor products and super Kazhdan–Lusztig theory. *J. Pure Appl. Algebra* **224** (2020), no. 8, 106347, 9 pp.

[13] P. Baseilhac and S. Kolb, Braid group action and root vectors for the q-Onsager algebra. *Transform. Groups* **25** (2020), 363–389.

[14] J. Beck, Braid group actions and quantum affine algebras. *Comm. Math. Phys.* **165** (1994), 555–568.

[15] A. Beilinson, G. Lusztig, and R. MacPherson, A geometric setting for the quantum deformation of GL_n. *Duke Math. J.* **61** (1990), 655–677.

[16] C. Berman and W. Wang, Formulae of $\imath$-divided powers in $\mathbf{U}_q(\mathfrak{sl}_2)$. *J. Pure Appl. Algebra* **222** (2018), 2667–2702.

[17] T. Bridgeland, Quantum groups via Hall algebras of complexes. *Ann. of Math.* **177** (2013), 739–759.

[18] J. Brundan, Kazhdan–Lusztig polynomials and character formulae for the Lie superalgebra $\mathfrak{gl}(m|n)$. *J. Amer. Math. Soc.* **16** (2003), 185–231.

[19] J. Brundan, I. Losev, and B. Webster, Tensor product categorifications and the super Kazhdan–Lusztig conjecture. *Int. Math. Res. Not. IMRN* **2017** (2017), no. 20, 6329–6410.

[20] X. Chen, M. Lu, and W. Wang, A Serre presentation for the $\imath$quantum groups. *Transform. Groups* **26** (2021), 827–857. arXiv:1810.12475.

[21] X. Chen, M. Lu, and W. Wang, Serre–Lusztig relations for $\imath$quantum groups. *Comm. Math. Phys.* **382** (2021), 1015–1059.

[22] S.-J. Cheng, N. Lam, and W. Wang, Brundan–Kazhdan–Lusztig conjecture for general linear Lie superalgebras. *Duke J. Math.* **164** (2015), 617–695.

[23] I. Damiani, Drinfeld realization of affine quantum algebras: the relations. *Publ. Res. Inst. Math. Sci.* **48** (2012), 661–733.

[24] H. de Clercq, Defining relations for quantum symmetric pair coideals of Kac–Moody type. *J. Combin. Alg.* **5** (2021), 297–367.

[25] V. Deodhar, On some geometric aspects of Bruhat orderings II. The parabolic analogue of Kazhdan–Lusztig polynomials. *J. Algebra* **111** (1987), 483–506.

[26] L. Dobson and S. Kolb, Factorisation of quasi K-matrices for quantum symmetric pairs. *Selecta Math. (N.S.)* **25** (2019), 63.

[27] V. Drinfeld, Quantum groups. In *Proceedings of the International Congress of Mathematicians (Berkeley, CA, 1986), Vols. 1, 2*, pp. 798–820, AMS, 1987.

[28] M. Ehrig and C. Stroppel, Nazarov–Wenzl algebras, coideal subalgebras and categorified skew Howe duality. *Adv. Math.* **331** (2018), 58–142.

[29] Z. Fan, C. Lai, Y. Li, L. Luo, and W. Wang, Affine flag varieties and quantum symmetric pairs. *Mem. Amer. Math. Soc.* **265** (2020), no. 1285, v+123 pp.

[30] M. Gorsky, Semi-derived and derived Hall algebras for stable categories. *Int. Math. Res. Not. IMRN* **2018**, 138–159.

[31] M. Jimbo, A q-analogue of $U(\mathfrak{gl}(N+1))$, Hecke algebra, and the Yang–Baxter equation. *Lett. Math. Phys.* **11** (1986), 247–252.

[32] M. Kashiwara, On crystal bases of the Q-analogue of universal enveloping algebras. *Duke Math. J.* **63** (1991), 456–516.

[33] D. Kazhdan and G. Lusztig, Representations of Coxeter groups and Hecke algebras. *Invent. Math.* **53** (1979), 165–184.

[34] M. Khovanov and A. Lauda, A diagrammatic approach to categorification of quantum groups. I. *Represent. Theory* **13** (2009), 309–347.

[35] S. Kolb, Quantum symmetric Kac–Moody pairs. *Adv. Math.* **267** (2014), 395–469.

[36] S. Kolb, The bar involution for quantum symmetric pairs – hidden in plain sight. 2021, arXiv:2104.06120.

[37] S. Kolb and J. Pellegrini, Braid group actions on coideal subalgebras of quantized enveloping algebras. *J. Algebra* **336** (2011), 395–416.

[38] S. Kolb and M. Yakimov, Defining relations of quantum symmetric pair coideal subalgebras. *Forum Math. Sigma* **9** (2021), Paper No. e67, 38 pp.

[39] G. Letzter, Symmetric pairs for quantized enveloping algebras. *J. Algebra* **220** (1999), 729–767.

[40] G. Letzter, Quantum symmetric pairs and their zonal spherical functions. *Transform. Groups* **8** (2003), 261–292.

[41] G. Letzter, Cartan subalgebras for quantum symmetric pair coideals. *Represent. Theory* **23** (2019), 99–153.

[42] Y. Li, Quiver varieties and symmetric pairs. *Represent. Theory* **23** (2019), 1–56.

[43] Y. Li and W. Wang, Positivity vs negativity of canonical bases. *Bull. Inst. Math. Acad. Sin. (N.S.)* **13** (2018), 143–198.

[44] M. Lu and L. Peng, Semi-derived Ringel–Hall algebras and Drinfeld doubles. *Adv. Math.* **383** (2021), 107668.

[45] M. Lu and S. Ruan, $\imath$Hall algebras of weighted projective lines and quantum symmetric pairs. 2021, arXiv:2110.02575.

[46] M. Lu, S. Ruan, and W. Wang, $\imath$Hall algebra of the projective line and q-Onsager algebra. 2020, arXiv:2010.00646.

[47] M. Lu and W. Wang, Hall algebras and quantum symmetric pairs II: reflection functors. *Comm. Math. Phys.* **381** (2021), 799–855.

[48] M. Lu and W. Wang, Braid group symmetries on quasi-split $\imath$quantum groups via $\imath$Hall algebras. 2021, arXiv:2107.06023.

[49] M. Lu and W. Wang, A Drinfeld type presentation of affine $\imath$quantum groups I: split ADE type. *Adv. Math.* **393** (2021), 108111.

[50] M. Lu and W. Wang, Hall algebras and quantum symmetric pairs I: foundations. *Proc. Lond. Math. Soc.* (to appear), arXiv:1901.11446.

[51] M. Lu and W. Wang, Hall algebras and quantum symmetric pairs III: quiver varieties. *Adv. Math.* **393** (2021), 108071.

[52] M. Lu and W. Wang, Hall algebras and quantum symmetric pairs of Kac–Moody type. 2020, arXiv:2006.06904.

[53] G. Lusztig, Canonical bases arising from quantized enveloping algebras. *J. Amer. Math. Soc.* **3** (1990), 447–498.

[54] G. Lusztig, Canonical bases in tensor products. *Proc. Natl. Acad. Sci.* **89** (1992), 8177–8179.

[55] G. Lusztig, *Introduction to quantum groups, Modern Birkhäuser Classics*, Reprint of the 1993 Edition. Birkhäuser, Boston, 2010.

[56] G. Lusztig, Total positivity in symmetric spaces. 2021, arXiv:2107.13447.

[57] D. Maulik and A. Okounkov, Quantum Groups and Quantum Cohomology. *Astérisque* **408** (2018), ix+209 pp. arXiv:1211.1287.

[58] H. Nakajima, Quiver varieties and finite dimensional representations of quantum affine algebras. *J. Amer. Math. Soc.* **14** (2000), 145–238.

[59] F. Qin, Quantum groups via cyclic quiver varieties I. *Compos. Math.* **152** (2016), 299–326.

[60] C. M. Ringel, Hall algebras and quantum groups. *Invent. Math.* **101** (1990), 583–591.

[61] R. Rouquier, 2-Kac–Moody algebras. 2008, arXiv:0812.5023.

[62] Y. Shen and W. Wang, $\imath$Schur duality and Kazhdan–Lusztig basis expanded. 2021, arXiv:2108.00630.

[63] E. Vasserot, Affine quantum groups and equivariant K-theory. *Transform. Groups* **3** (1998), 269–299.

[64] W. Wang and W. Zhang, An intrinsic approach to relative braid group symmetries on $\imath$quantum groups. 2022, arXiv:2201.01803.

[65] H. Watanabe, Global crystal bases for integrable modules over a quantum symmetric pair of type AIII. *Represent. Theory* **25** (2021), 27–66.

[66] H. Watanabe, Based modules over the $\imath$quantum group of type AI. 2021, arXiv:2103.12932.

[67] W. Zhang, A Drinfeld type presentation of affine $\imath$quantum groups II: split BCFG type. 2021, arXiv:2102.03203.

WEIQIANG WANG

Department of Mathematics, University of Virginia, Charlottesville, VA 22904, USA, ww9c@virginia.edu

8. ANALYSIS

SPECIAL LECTURE

CONVEX GEOMETRY AND ITS CONNECTIONS TO HARMONIC ANALYSIS, FUNCTIONAL ANALYSIS AND PROBABILITY THEORY

KEITH BALL

ABSTRACT

Convex geometry and analysis have connections to many areas of the mathematical sciences: PDEs, discrete geometry, optimization, theoretical computer science, and mathematical economics. No article could even scratch the surface of all of these. Instead, we shall begin by describing how the development of the subject was influenced over the last 50 years by two other fields, harmonic and functional analysis, and then discuss the subtle and still somewhat mysterious way in which convex domains exhibit properties that we normally expect to see within probability theory.

MATHEMATICS SUBJECT CLASSIFICATION 2020

Primary 52A20; Secondary 52A21, 42B99, 60F05, 46B07

KEYWORDS

Convex geometry, isoperimetric inequality, harmonic analysis, probability, central limit theorem, optimal transport

ICM INTERNATIONAL CONGRESS OF MATHEMATICIANS 2022 JULY 6–14

Proc. Int. Cong. Math. 2022, Vol. 4, pp. 3104–3139
DOI 10.4171/ICM2022/65

INTRODUCTION

The task I was set in this article was to discuss convex geometry and analysis and their connections to other fields. As pointed out in the abstract, it would be impossible to write a readable article that even began to exhaust such a broad remit. Naturally, I have opted to explain the connections between convex geometry and the areas that I am most familiar with and, in order to make the article accessible to as large an audience as possible, I have included a few pages of introduction describing the classical theory. The three main sections cover the subject's connections with harmonic analysis, functional analysis, and probability theory, respectively. Some of the material goes back several decades but helps to provide a context for the more recent material: again the aim is to make the article widely accessible to nonspecialists.

It is natural to begin a discussion of convex geometry with the isoperimetric inequality: the statement that if you wish to enclose the largest volume with a given surface area, the optimal shape is a Euclidean ball. Equivalently, if a set $K \subset \mathbf{R}^d$ is measurable then

$$dv_d^{1/d}|K|^{(d-1)/d} \leq |\partial K|^1$$

where v_d is the volume of the Euclidean ball of radius 1. Formally, this is not a statement involving convexity since it applies to all sufficiently nice sets, but it is clear that in spirit we are talking about convex sets. It may not be any easier to give a *formal* proof that the optimizers are convex than to prove the full inequality; but it is *intuitively* clear that if your set has gaps, then you can push bits of it together so as to decrease its surface without changing its volume.

Isoperimetric principles of one sort or another appear all over mathematics: in particular, their generalizations in the form of large deviation inequalities play a crucial role in probability theory. Section 3 of this article discusses the influence of functional analysis on convex geometry, and in this section we shall describe how deviation principles are used to prove one of the most celebrated results in convex geometry, Dvoretzky's Theorem, which guarantees that all convex bodies have almost ellipsoidal sections of quite high dimension. The section will also discuss the reverse Santaló inequality of Bourgain and Milman and how this grew out of the interaction between functional analysis and geometry.

Quite early in the 20th century it was realized that the isoperimetric inequality can be extended from sets to functions to give the Gagliardo–Nirenberg–Sobolev inequality. If a measurable $f : \mathbf{R}^d \to \mathbf{R}$ has a gradient almost everywhere then

$$dv_d^{1/d}\left(\int_{\mathbf{R}^d} |f|^{d/(d-1)}\right)^{(d-1)/d} \leq \int_{\mathbf{R}^d} \|\nabla f\|_2.$$

It is in this spirit that we shall look at links between convex geometry and harmonic analysis in Section 2 of the article. We shall discuss a convolution inequality of Brascamp and Lieb that belongs firmly in harmonic analysis but which dovetails perfectly with a geometric

1 Throughout this article we shall use the modulus sign $|\cdot|$ to denote the volume measure of the appropriate dimension.

principle of Fritz John to prove the reverse form of the isoperimetric inequality found by the present author. We shall also discuss the beautiful monotone transportation map of Brenier and how Barthe used it to prove the Brascamp–Lieb inequality. Finally, we shall briefly discuss the quite extensive recent work on the stability of the isoperimetric inequality, in particular in the work of Fusco, Figalli, Jerison, Maggi, and Pratelli.

The final section, Section 4 of the article, focusses on a remarkable correspondence between convex geometry and probability. In linking geometry and harmonic analysis, we shall frequently switch between a convex domain and its indicator function. If the domain has volume 1 then its indicator is automatically the density of a random vector in $\mathbf{R}^d$. So at a trivial level there is obviously a reinterpretation of geometry[2] in terms of probability. But probability theory is much more than just analysis with total measure 1. Central to it is the concept of independence and a wealth of related ideas: filtrations, conditional expectations, and so on. Over the last three decades, it has become increasingly clear that the uniform measure on a convex domain exhibits properties that we would expect from the joint law of *independent* random variables: for example, the central limit theorem for convex domains that was conjectured by the present author and proved by Klartag. The background to these developments was a collection of conjectures made in the late 1980s and early 1990s, and on which quite a lot of progress has been made in the last 20 years. One of the motivations for these conjectures is their relationship to a lovely problem in theoretical computer science: the difficulty of computing volumes of convex sets. So we shall mention the algorithms of Dyer, Frieze, Kannan, Lovász, Simonovits, Applegate, Vempala, and Lee, which depend upon the rate at which Markov chains diffuse inside a convex body. This section also includes Paouris' decay estimate for the Euclidean norm on a convex set, the stochastic localization technique of Eldan and a very recent development by Chen. Being more recent, this material has not yet been highly digested, and so this section is much less polished than the earlier ones. Section 1 of the article will recall some standard facts from convex geometry that we shall refer to throughout the article.

Since this article cannot touch on all of the many areas in which convex analysis appears, we shall say nothing about the combinatorial theory of polytopes and its relation to the topology of complex varieties and very little about the huge field of optimization. An excellent starting point on polytopes is the article by Henk, Richter-Gebert, and Ziegler **[67]**. We shall also not mention the relationship between polyhedra and lattice points described in loving detail in the book by Barvinok **[16]**. If my selection of topics has a unifying theme, it is (as by now the reader will have guessed) the isoperimetric inequality.

1. THE FUNDAMENTALS OF CONVEX GEOMETRY

The aim of this section is to describe some of the most basic ideas in convex geometry. The list is far from exhaustive: the topics are selected mainly so that I can refer to them in the subsequent sections of the article.

2 Or at least the kind of geometry we are talking about.

1.1. The Brunn–Minkowski inequality

The basic Sobolev inequality mentioned in the introduction is one way to generalize the isoperimetric inequality, but there is another rather different generalization, which constitutes the most fundamental relation between volume and the linear structure of space. For a set A with a nice enough boundary, we can compute its surface area by considering the volume of its neighborhoods

$$A_\varepsilon = \{x : \|x - y\| \le \varepsilon, \text{ for some } y \in A\}.$$

We then compute the surface area as the "derivative" of volume

$$|\partial A| = \lim_{\varepsilon \to 0} \frac{|A_\varepsilon| - |A|}{\varepsilon}.$$

The ε-neighborhood can be described in a different way as

$$A + B(\varepsilon) = \{x + u : x \in A, u \in B(\varepsilon)\}$$

where $B(\varepsilon)$ is the Euclidean ball of radius ε. Therefore the isoperimetric inequality will follow from a sufficiently strong estimate from below for the volume of the sumset $A + B(\varepsilon)$. The Brunn–Minkowski inequality provides such an estimate for the sum of any two (let us say compact) sets.

Theorem 1 (Brunn–Minkowski). *Suppose A and B are compact sets in $\mathbf{R}^d$. Then*

$$|A + B|^{1/d} \ge |A|^{1/d} + |B|^{1/d}.$$

The inequality can be reformulated in terms of convex combinations of sets (rather than sums). For compact sets A and B in $\mathbf{R}^d$ and $\lambda \in (0, 1)$,

$$\big|(1-\lambda)A + \lambda B\big|^{1/d} \ge (1-\lambda)|A|^{1/d} + \lambda|B|^{1/d}$$

and, by using the arithmetic/geometric mean inequality, we can deduce a multiplicative version, which has a number of advantages,

$$\big|(1-\lambda)A + \lambda B\big| \ge |A|^{1-\lambda}|B|^{\lambda}. \tag{1.1}$$

Among other things, this formulation has a natural generalization to functions that was found by Prékopa and Leindler (see, for example, [101]) and which can be very easily proved by induction on the dimension d. (This induction argument seems to have appeared first in [30].) The original inequality does not lend itself to such a proof because in order to deduce the d-dimensional result for (indicator functions of) sets, you need to apply the $(d-1)$-dimensional result for more general functions.

Theorem 2 (Prékopa–Leindler). *If $f, g, m : \mathbf{R}^d \to [0, \infty)$ are measurable, $\lambda \in (0, 1)$, and, for each x and y,*

$$m\big((1-\lambda)x + \lambda y\big) \ge f(x)^{1-\lambda} g(y)^{\lambda},$$

then

$$\int m \ge \left(\int f\right)^{1-\lambda} \left(\int g\right)^{\lambda}.$$

A function $f : \mathbf{R}^d \to [0, \infty)$ is called *logarithmically concave* if its logarithm is concave (with the usual convention regarding $-\infty$). Equivalently, f is logarithmically concave if it satisfies

$$f\big((1-\lambda)x + \lambda y\big) \geq f(x)^{1-\lambda} f(y)^{\lambda}$$

for all $x, y \in \mathbf{R}^d$ and $\lambda \in [0, 1]$. The Prékopa–Leindler inequality ensures that if f is such a function then its marginals are too, and from this it follows that convolutions of logarithmically concave functions are also logarithmically concave. The class of such functions thus constitutes a natural extension of the class of indicator functions of convex sets, but which is closed under the most common operations applied to densities in probability theory.

There is a rather odd consequence (or variant) of the Brunn–Minkowski inequality found by Busemann in [33]. Suppose K is a symmetric convex body, or equivalently the unit ball of a norm on $\mathbf{R}^d$. For each unit vector θ, look at the intersection of K with the $(d-1)$-dimensional subspace $\theta^{\perp}$ orthogonal to θ. Then the function

$$\theta \mapsto \left|K \cap \theta^{\perp}\right|^{-1}$$

that measures the reciprocal of the $(d-1)$-dimensional volume of the intersection, extends to a norm on $\mathbf{R}^d$. So this is a precise way to say that if you pick two nearby sections of K then a section between them cannot have volume much smaller than they do. Busemann's Theorem has a simple, but surprisingly useful extension [9]:

Theorem 3 (Busemann–Ball). *Let $f : \mathbf{R}^d \to [0, \infty)$ be an even logarithmically concave function whose integral is finite and strictly positive. Then for each $p \geq 1$, the function*

$$x \mapsto \left(\int_0^{\infty} f(rx) r^{p-1}\, dr\right)^{-1/p}$$

defines a norm on $\mathbf{R}^d$.

By generating a norm (and hence a convex set) from a logarithmically concave function, the theorem automatically transfers information about convex sets to logarithmically concave functions. Working with functions provides the flexibility to take marginals and convolutions without much affecting what is true. Many of the well-known inequalities for convex sets have analogues for logarithmically concave functions that can be proved in similar ways.

1.2. Fritz John's Theorem

In a famous paper [69] from 1948 on optimization problems, Fritz John gave an example that turned out to be extremely prescient and which has become one of the standard tools in understanding convex domains. The theorem characterizes the ellipsoid of largest volume inside a convex domain in terms of the geometric structure of the contact points between the ellipsoid and the surface of the body. There are two versions, one for symmetric bodies and one for general ones. To get the feel of the theorem, we will just quote the simpler symmetric version. (Throughout the article we will often quote results just for symmetric sets or even functions. In all cases they hold without the symmetry assumption, but the statements are often more complicated and the proofs need no additional ideas.)

Theorem 4 (John). *Let K be a symmetric convex body in $\mathbf{R}^d$. Then K contains a unique ellipsoid of largest volume. This ellipsoid is the standard Euclidean ball B_2^d if and only if the ball is indeed included in K and there are unit vectors $u_1, u_2, \ldots, u_m$ on the surface of K and positive weights $c_1, c_2, \ldots, c_m$ for which*

$$\sum_1^m c_i \langle u_i, x\rangle^2 = \|x\|^2 \tag{1.2}$$

for every $x \in \mathbf{R}^d$.

The condition shows that the contact points behave somewhat like an orthogonal basis. That forces their directions to be distributed in a well-spread-out way: they cannot all lie too close to a subspace of dimension less than d. By applying the identity to an orthonormal basis and summing, we get that

$$\sum c_i = d. \tag{1.3}$$

Plainly, K is included in the set

$$\{x : |\langle u_i, x\rangle| \leq 1, \text{ for all } i\}$$

and hence for any $x \in K$,

$$\|x\|^2 \leq \sum_1^m c_i = d.$$

Consequently, K not only includes the ball of radius 1 but is included in the ball of radius $\sqrt{d}$. Thus John's Theorem provides a way to use a linear map to make a convex body "as round as possible": choose the largest ellipsoid inside K and map that to the standard Euclidean ball.

1.3. The Blaschke–Santaló inequality and symmetrization

A crucial role is played in functional analysis by duality. The theory extends from symmetric convex bodies, the unit balls of norms on $\mathbf{R}^d$, to more general convex sets. The polar of a body K is

$$K^\circ = \{y : \langle y, x\rangle \leq 1, \text{ for all } x \in K\}.$$

The fundamental fact here is the Blaschke–Santaló inequality [21] and [184].

Theorem 5 (Blaschke–Santaló). *If K is a symmetric convex body then the product of the volumes of K and its polar K° is no more than that for the Euclidean balls v_d^2.*

As it stands the statement cannot be true for arbitrary (nonsymmetric) sets because if the origin is not inside the set, the polar will be unbounded. However, there is an extension to general sets, in which one first shifts the set K to the optimal position in space before taking the polar K°.

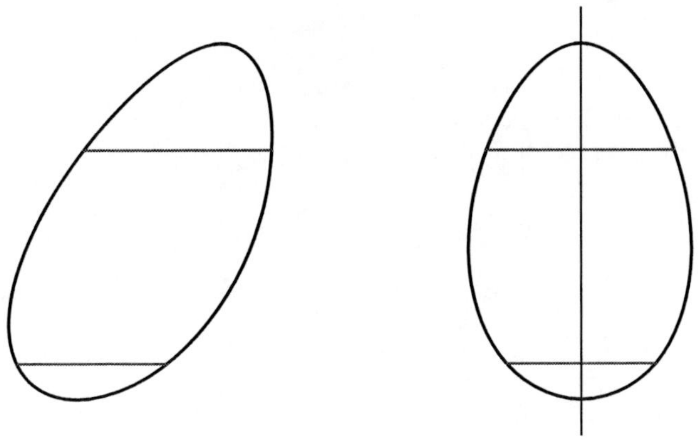

FIGURE 1
Symmetrizing a convex set

A classical and very lovely way to establish inequalities such as the isoperimetric inequality is by means of Steiner symmetrization. If U is a 1-codimensional subspace of $\mathbf{R}^d$ then we can *symmetrize* K with respect to U in the following way. For each line perpendicular to U, consider its intersection with K. Now shift that line segment so that it sits symmetrically either side of the subspace U; see Figure 1. Plainly, the new set has the same volume as K and Steiner showed that it has a smaller surface area. To establish the isoperimetric inequality, you need to show that, by repeatedly symmetrizing a set in different subspaces, you can (in the limit) turn it into a ball. This was done in a famous article by De Giorgi **[41]**.

One can generalize this idea of symmetrization to subspaces of dimension other than $d-1$. If U is a subspace then we replace K by the set of all points of the form

$$u + (v - w)/2,$$

where u is a point in U, v and w are in the orthogonal complement $U^{\perp}$, and $u+v$ and $u+w$ are in K. It was found by Saint-Raymond **[103]** that, by using this type of symmetrization, one can give a proof of the Blaschke–Santaló inequality. (See also **[9]**.) If you symmetrize K and its polar in orthogonal subspaces then the polar of the symmetrization contains the symmetrization of the polar. Therefore the product of the volumes goes up when you symmetrize. A lovely generalization of this argument to nonsymmetric sets was found by Meyer and Pajor **[87]**.

1.4. Lévy's inequality

Isoperimetric inequalities hold in many manifolds although there are not too many examples where the exact optimizers are known. One case in which the optimum *is* known is that of the sphere S^{d-1} in Euclidean d-dimensional space. We use the rotation-invariant probability measure σ_{d-1} on the sphere to measure "volume." Lévy proved that among compact subsets of the sphere with a given measure, those with the smallest boundary are the spherical caps; see Figure 2.

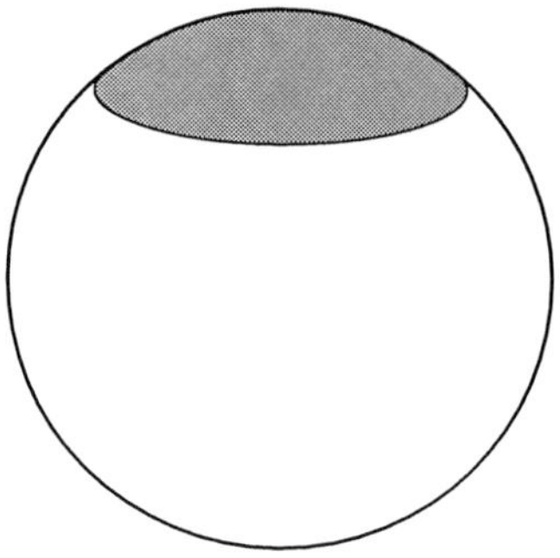

FIGURE 2
A spherical cap

The inequality can be extended. For each $\varepsilon > 0$, the subset of S^{d-1} of a given measure whose ε-neighborhood has smallest measure is a spherical cap. It was shown by Benyamini that this can be proved by a kind of symmetrization argument: the so-called 2-point symmetrization first introduced by Wolontis **[110]**. Benyamini's argument is included in the article **[51]**. The process is this. You start with a subset of the sphere. Choose a direction (let us say downwards) and, for each line in that direction that meets the sphere, ask whether the two points where it meets the sphere lie in the set. If both do, neither do, or just the bottom one does, then leave them alone. But if just the top one belongs to the set, then move it to the bottom. You thus compress the set as much as possible into the southern hemisphere. So this is actually a "compression" argument rather than a "symmetrization" argument, but they are clearly very similar in spirit.

Lévy's inequality implies a deviation estimate on the sphere something like the following:

Theorem 6. *Suppose* $A \subset S^{d-1}$ *and*

$$\sigma_{d-1}(A) \geq 1/2.$$

Then its ε*-neighborhood has probability at least*

$$\sigma_{d-1}(A_\varepsilon) \geq (1 - 2e^{-d\varepsilon^2/2}).$$

1.5. Differentiability

A classic and much loved text on convex analysis is that by Rockafellar **[102]**. His book was written with optimization in mind and so proceeds in a very different direction from this article, but we shall want one famous fact from that source. It contrasts appealingly with the warning we impress upon our students in their second or third analysis courses, namely that convergence of functions does not imply convergence of their derivatives.

Theorem 7. *If* $\phi : \mathbf{R}^d \to \mathbf{R}$ *is convex then* f *has a gradient almost everywhere. Indeed, the gradient exists outside a set of Hausdorff dimension at most* $d - 1$*. If* ϕ_k *are convex functions* *converging locally uniformly to ϕ then wherever the derivatives exist we also have*

$$\nabla \phi_k \to \nabla \phi.$$

In fact, much more is true. There are a number of ways to make sense of the idea that convex functions are *twice* differentiable almost everywhere. It cannot be true in the classical sense because the function might fail to be differentiable on a dense set. However, if we are content to ask just for a second-order Taylor expansion instead of the existence of a classical second derivative then the Busemann–Feller–Alexandrov Theorem [1, 34] guarantees its existence almost everywhere. As you would expect, this Hessian of a convex function will be a positive semidefinite map almost everywhere. An excellent account of the various forms of twice differentiability, and several new arguments are contained in the article by Bianchi, Colesanti, and Pucci [20].

2. CONNECTIONS WITH HARMONIC ANALYSIS

The aim of this section is to describe a number of geometric principles that have been established by using very precise inequalities from harmonic analysis and how these methods then fed back into the study of the original inequalities. An important role is played here by the monotone transport of Brenier which became a powerful tool in PDEs and a subject of considerable attention in the late 1990s and early part of this century. A good reference is the monograph of Villani [109].

2.1. The reverse isoperimetric inequality

In 1937 Behrend [18] asked a rather natural question about reversing the isoperimetric inequality. If a set looks like a scattering of dust it can have huge surface area but very small volume. Even if the set is convex there is no upper bound for the surface in terms of the volume, because the set could be a pancake. Behrend's question was this: suppose you are allowed to apply a linear map to your convex set which preserves the volume but makes the surface area as small as possible. For which convex set is the minimal surface area largest? The natural conjecture is that in each dimension, the simplex is the solution to this max–min problem.

In 1961 Petty [96] found a characterization of the optimal affine image for each convex body.

Theorem 8 (Petty). *A convex body $K \subset \mathbf{R}^d$ has the least surface area among all its affine images of the same volume, if and only if for every $x \in \mathbf{R}^d$,*

$$\frac{d}{|\partial K|} \int_{\partial K} \langle n, x \rangle^2 = \|x\|^2, \tag{2.1}$$

where the integral is taken with respect to the area measure on the boundary of K and n is the unit normal at each point of the boundary.

The condition is clearly very similar to the Fritz John condition in Theorem 4 and, in fact, there are a number of results with the same general "shape"; see [61]. In spite of this attractive characterization, Behrend's question was not answered until 1990 [11], and it turned out that the affine image which was best adapted to solving the problem was actually the one characterized by John rather than the optimal one for surface area. It is an elementary exercise to check that if a convex $K \subset \mathbf{R}^d$ includes the Euclidean ball of radius 1, then $|\partial K| \leq d|K|$ with equality if K is a polytope whose facets all touch the ball (and in many other cases). Since there is equality for a regular solid simplex, we can prove the reverse isoperimetric inequality by showing that the regular simplex has the largest volume among all bodies whose ellipsoid of maximal volume is the Euclidean unit ball. (Among symmetric convex bodies, the cube has the largest volume ratio; the proof is similar but a bit simpler.)

Theorem 9 (Ball). *Suppose K is a convex set in $\mathbf{R}^d$, the ellipsoid of largest volume in K is the Euclidean ball $B(1)$, and T is a regular simplex with the same maximal ellipsoid. Then*

$$|K| \leq |T|,$$

and consequently,

$$\frac{|\partial K|}{|K|^{(d-1)/d}} \leq \frac{|\partial T|}{|T|^{(d-1)/d}}.$$

As mentioned in the introduction, the proof of this theorem depended upon a convolution inequality of Brascamp and Lieb. The famous inequality of Young for convolutions states that if $f, g, h : \mathbf{R} \to [0, \infty)$ are measurable and $1/p + 1/q + 1/r = 2$ then

$$\int_{\mathbf{R}} f * g * h \leq \|f\|_p \|g\|_q \|h\|_r.$$

The inequality holds on any locally compact group using integrals with respect to Haar measure. On compact groups where constant functions belong to all L_p-spaces, the inequality is sharp, but for the real line it is not. The sharp version was found for certain exponents by Beckner [17] and in full generality by Brascamp and Lieb [29]. The extremal functions for the inequality are Gaussian densities rather than constant functions.

The key to proving the reverse isoperimetric inequality, Theorem 9, was to recognize that the Brascamp–Lieb inequality dovetails perfectly with Fritz John's Theorem. The appropriate formulation is this.

Theorem 10 (Brascamp–Lieb). *Suppose that unit vectors (u_i) in $\mathbf{R}^d$ and weights (c_i) satisfy the John condition*

$$\sum_1^m c_i \langle u_i, x\rangle^2 = \|x\|_2^2$$

for all $x \in \mathbf{R}^d$. Then if (f_i) are nonnegative measurable functions on $\mathbf{R}$,

$$\int_{\mathbf{R}^d} \prod_1^m f_i(\langle u_i, x\rangle)^{c_i} \leq \prod_1^m \left(\int_{\mathbf{R}} f_i\right)^{c_i}.$$

Some feel for the inequality can be gained by observing that there is obviously equality if the (f_i) are identical Gaussian densities. If $f_i(t) = e^{-t^2}$ for each i then

$$\prod f_i(\langle u_i, x\rangle)^{c_i} = \exp\Big(-\sum c_i \langle u_i, x\rangle^2\Big) = e^{-\|x\|_2^2}$$

by the Fritz John condition (1.2). The integral of this function is the same as the product

$$\prod\left(\int_{\mathbf{R}} f_i\right)^{c_i} = \left(\int e^{-t^2}\right)^d$$

by equation (1.3). There is something pleasingly counterintuitive about the fact that we prove an inequality for which the simplex is extremal by using a result from harmonic analysis in which Gaussian densities are extremal. The resolution of the paradox lies in the fact that the Brascamp–Lieb inequality is sharp (whatever the f_i) if the vectors u_i form an orthonormal basis.

It is natural to conjecture an extension of the Brascamp–Lieb inequality in which one replaces the rank-one projections $x \mapsto \langle u_i, x\rangle u_i$ by orthogonal projections of higher rank. This generalized inequality was proved by Lieb in a later article **[81]**.

2.2. Monotone transport

A few years after the proof of the reverse isoperimetric inequality, Barthe **[15]** gave an elegant new proof of the generalized Brascamp–Lieb inequality using the optimal transportation map discovered by Brenier. A map $T : \mathbf{R}^d \to \mathbf{R}^d$ transports a probability measure μ on $\mathbf{R}^d$ to a probability measure ν if, for each measurable set $A \subset \mathbf{R}^d$, we have

$$\mu\big(T^{-1}(A)\big) = \nu(A). \tag{2.2}$$

Among all maps that do so, one can ask for the one that minimizes the total cost

$$\int_{\mathbf{R}^d} c(x, Tx)\, d\mu(x)$$

for some cost function c. (So $c(x, y)$ is the cost of moving unit measure from x to y.) Brenier **[32]** realized that for one very specific cost function, $c(x, y) = \|x - y\|^2$, the square of the Euclidean distance, the optimal map exists under only very weak hypotheses about the measures and has a very special form: it is the gradient of a convex function. (This map is called a monotone transport map by analogy with the 1-dimensional case in which the derivative of a convex function is monotone.) A version for still more general measures was found by McCann **[86]**. For our purposes, the following theorem, which has a very simple proof **[12]**, gives a good enough picture:

Theorem 11 (Brenier–McCann). *If μ and ν are probability measures on $\mathbf{R}^d$, ν has compact support and μ assigns no mass to any set of Hausdorff dimension $d - 1$, then there is a convex function $\phi : \mathbf{R}^d \to \mathbf{R}$, so that $T = \nabla\phi$ transports μ to ν.*

The hypothesis on μ corresponds precisely to the conclusion of the differentiability Theorem 7 since we need ϕ to be differentiable almost everywhere with respect to μ.

Barthe's proof of the Brascamp–Lieb inequality involves transporting the given densities to Gaussian densities and checking that the integral of the product increases as a result. The latter depends crucially upon the fact that the transportation map is the gradient of a convex function and hence that its Hessian is positive semidefinite symmetric. So the argument constitutes a kind of symmetrization technique in which we do not know exactly what map we are using but we do have an inequality for its Hessian. We introduced monotone transport because of Barthe's proof of the Brascamp–Lieb inequality but it has an obvious alternative point of contact with convex analysis: it involves the gradient of a convex function. The contact is actually much closer. If $K \subset \mathbf{R}^d$ is a convex body then, by Theorem 7, it has a well-defined outward unit normal almost everywhere on its surface. By the divergence theorem, the integral of this normal over the surface is zero. The Gauss map which takes a point on the surface to the normal at that point, transports the surface area measure to a measure ν on the sphere satisfying

$$\int_{S^{d-1}} \theta \, d\nu(\theta) = 0. \tag{2.3}$$

A beautiful classical theorem of Minkowski goes in the other direction, just like Theorem 11.

Theorem 12 (Minkowski existence theorem). *Suppose that ν is a finite measure on S^{d-1} for which equation* (2.3) *holds and whose support spans $\mathbf{R}^d$. Then there is a convex body K whose Gauss map transports its surface measure to ν.*

This theorem may look a bit different from Theorem 11 because in this case we appear to build the surface measure at the same time as the convex set instead of being given both measures to begin with. But in reality we are effectively given the surface measure, namely it is $(d-1)$-dimensional Hausdorff measure.

Another elegant approach to the Brascamp–Lieb inequality, this one using heat flow methods, was found by Carlen, Lieb, and Loss **[37]** and Bennett, Carbery, Christ, and Tao **[19]**. In the latter article the formulation of the inequality that matches John's Theorem is called the "geometric form" of the inequality. Following Barthe's argument, monotone transportation was also used to give very elegant proofs of a number of other geometric inequalities (with best possible constants). For the purposes of this article, the obvious paper to mention is that of Cordero-Erausquin, Nazaret, and Villani **[40]** where they study the Sobolev and Gagliardo–Nirenberg inequalities. The original proofs of the best constants were found by Aubin **[6]** and Talenti **[108]** and by Del Pino and Dolbeault **[42]**.

2.3. Projections and surface area

If K is a symmetric convex body in $\mathbf{R}^d$ and for each unit vector θ we consider the $(d-1)$-dimensional volume $|P_\theta(K)|$ of the projection of K onto the orthogonal complement of the span of θ, then it is easy to see that the map

$$\theta \mapsto |P_\theta(K)|$$

extends to a norm on $\mathbf{R}^d$. The unit ball of this norm has volume

$$V_K = d v_d \int_{S^{d-1}} \frac{1}{|P_\theta K|^d} \, d\sigma_{d-1}$$

and, unlike the surface area, this quantity is unchanged if we apply a linear map of determinant 1 to K. On the other hand, the surface area of K is (apart from the obvious constant) the average of the volumes of the projections

$$|\partial K| = \frac{d v_d}{v_{d-1}} \int_{S^{d-1}} |P_\theta K| \, d\sigma_{d-1}.$$

So it is natural to ask whether there is a strong form of the isoperimetric inequality guaranteeing that V_K is minimized over bodies of a given volume by the Euclidean ball. This is, indeed, the case and was proved by Petty [97]. Petty's projection inequality has been considerably generalized in the work of Lutwak, Yang, and Zhang [83] and Haberl and Schuster [65].

The corresponding reverse question was solved by Zhang [111] who proved that V_K is *maximized* over bodies of a given volume by the simplex. Whereas Petty's Theorem strengthens the isoperimetric inequality, Zhang's Theorem does not follow from Theorem 9 with the correct constant because for the simplex the volume $|P_\theta K|$ is not constant as a function of θ, and so there is strict inequality in the Hölder inequality that one would wish to invoke.

2.4. Stability

Whenever one has an important inequality like the isoperimetric inequality, it is natural to ask about its stability. If a set has small surface area, must it resemble a Euclidean ball? Again, strictly speaking, this is not necessarily a problem about convex sets: the question might well make sense for other sets, as long as we specify carefully what we mean by "resemble." The most famous classical result in this direction is that of Bonnesan [24] from 1924 which is, indeed, specific to convex sets. He proved that for a convex region C in the plane with area A and perimeter P, there are concentric discs D_1 and D_2 with radii r_1 and r_2 for which $D_1 \subset C \subset D_2$ and with

$$P^2 \geq 4\pi\big(A + (r_2 - r_1)^2\big).$$

Thus if C almost satisfies the isoperimetric inequality $P^2 \geq 4\pi A$ with equality, its boundary can be sandwiched between two very similar circles. This result was not extended to higher dimensions until 1989 when Fuglede [55] showed that for a convex body K in $\mathbf{R}^d$, the Hausdorff distance of K from the closest Euclidean ball can be estimated by a certain power of the gap between $|\partial K|$ and the surface area of a ball of the same volume. This sort of conclusion is clearly impossible without the convexity assumption since a general set could have a tiny piece far from the rest of it, which contributes very little to either the volume or the surface area.

However, if we choose to measure the distance of a set from a ball by the volume of their symmetric difference then we can drop the convexity. In a couple of articles, in particular [66], Hall proved the following:

Theorem 13 (Hall). *For each d, there is a constant $C(d)$ so that if A is a measurable set in $\mathbf{R}^d$ then there is a Euclidean ball B of the same volume as K for which*

$$|A \,\triangle\, B|^4 \leq C(d)\big(|\partial A| - |\partial B|\big).$$

Hall conjectured that the exponent 4 in this theorem was not optimal and that 2 was the correct exponent. This was proved by Fusco, Maggi, and Pratelli in [57]. Normally, the only reasonable way to prove a stability estimate for a sharp inequality is to take a proof of the inequality and to "watch carefully" what it does, so as to track how much the quantities on each side change as you run through the proof. (Since the stability result implies the original inequality, your argument had better give a proof of the original, so you might as well start with such a proof.) Each of the stability results mentioned so far, tracks the Steiner symmetrization proof of the isoperimetric inequality which was mentioned in Section 1.3.

Knöthe [76] found a different approach to the Brunn–Minkowski inequality and (a fortiori) to the isoperimetric inequality. He used a measure transport map satisfying an equation like (2.2), but whose derivative at each point is upper triangular rather than positive semidefinite symmetric. His argument works just as well with the Brenier map except that to make it rigorous you need some regularity for the map, and this is more difficult to establish in the case of monotone transport. (The main reference here is the subtle regularity theory of Cafarelli [36] for the Monge–Ampére equation.) In their article [50], Figalli, Maggi, and Pratelli obtain stability results by tracking the transportation proof of the isoperimetric inequality instead of the symmetrization proof. The main thrust of their article is that the latter method works for "anisotropic" isoperimetric inequalities. It was pointed out by Gromov [92] that Knöthe's argument works even when you measure surface area in a direction-dependent (anisotropic) way, whereas the symmetrization argument cannot possibly work because the extremal cases are no longer Euclidean balls.

As a consequence of the Brunn–Minkowski inequality, we know that the map

$$\varepsilon \mapsto |A + B(\varepsilon)|^{1/d}$$

is concave in ε. If we control the surface area of A then we control the derivative of the map at 0 and hence its value at each ε. Therefore one can strengthen the stability estimates for the isoperimetric inequality by showing that A must look like a ball under the weaker assumption that the volume $|A + B(\varepsilon)|^{1/d}$ is not too large. This was done by Figalli and Jerison [49]. There are two very comprehensive surveys of all of these stability results, namely by Fusco [56] and Maggi [84]. To some extent, the existence of these surveys, by authors heavily involved in the developments, has prompted me to give relatively brief descriptions.

3. APPLICATIONS OF FUNCTIONAL ANALYSIS

In the 1970s and 1980s researchers in geometric functional analysis began to focus on quantitative problems in finite-dimensional normed spaces rather than problems in infinite dimensions that were, at least in spirit, qualitative. The work led to many new results in convex geometry and, perhaps most strikingly, the reverse Santaló inequality of Bourgain and Milman. These developments really began with Dvoretzky's Theorem about a decade earlier.

3.1. Dvoretzky's Theorem

We shall say that a symmetric convex body K is t-equivalent to an ellipsoid if for some ellipsoid $\mathcal{E}$,

$$\mathcal{E} \subset K \subset t\mathcal{E}.$$

This is the same as saying that the normed space with unit ball K is isomorphic to Euclidean space with a constant of isomorphism t. Dvoretzky [44] answered a question of Grothendieck by proving that high-dimensional convex bodies have fairly high-dimensional slices that are almost indistinguishable from ellipsoids.

Theorem 14 (Dvoretzky). *For each positive integer k and each $\varepsilon > 0$, there is an integer d so that any d-dimensional symmetric convex body has a k-dimensional slice that is $(1+\varepsilon)$-equivalent to an ellipsoid.*

This theorem was one of the earliest triumphs of the probabilistic method: the use of probability theory to construct (or demonstrate the existence of) mathematical objects with special properties. About 12 years later, Milman [89] found a different proof which gives the optimal dependence of k on d. (The fact that it *is* optimal is shown, for example, by the cube.)

Theorem 15 (Milman). *For each $\varepsilon > 0$, there is a constant $c(\varepsilon) > 0$ so that any d-dimensional symmetric convex body has a k-dimensional slice that is $(1+\varepsilon)$-equivalent to an ellipsoid with*

$$k \geq c(\varepsilon) \log d.$$

Milman's proof is now the most familiar. It proceeds in several steps. Assume (by applying a linear transformation) that the Euclidean unit ball is the ellipsoid of largest volume inside the convex body K. Then if $\|\cdot\|_K$ is the norm whose unit ball is K, we have that $\|x\|_K \leq \|x\|_2$ for every x in $\mathbf{R}^d$ giving a bound on the Lipschitz constant of $\|\cdot\|_K$ as a function on the Euclidean sphere. Using the first step, you check that if $\|x\|_K$ is roughly constant on a reasonably dense finite subset of the sphere in some k-dimensional subspace, then it will be roughly constant on the whole sphere in that subspace. A simple argument shows that the sphere in $\mathbf{R}^k$ has a fairly dense subset with only about 4^k points. Now using the Lipschitz property again and Levy's isoperimetric inequality, Theorem 6, you show that $\|\cdot\|_K$ is roughly constant on a huge part of the Euclidean sphere in $\mathbf{R}^d$. Now choose a space at random from among all k-dimensional subspaces and conclude that with high probability the norm, restricted to 4^k points in this subspace, will be almost constant.

The proof seems complete, but a moment's thought shows that there is a point glossed over. We transformed K to make it as round as possible in the hope that there would then be large sets on the sphere where $\|\cdot\|_K$ is almost constant. However, so far we have only used the fact that K includes the ball of radius 1. That would still be true if K were a huge set with no similarity to a ball whatsoever. To make the details of the argument work, we need to know that the average of the norm over the Euclidean sphere is not too small (using the fact that the Euclidean ball is the ellipsoid of *maximal* volume). Thus we have a final step using a result of Dvoretzky and Rogers which shows that if the ellipsoid of maximal volume inside K is the Euclidean unit ball then for some c independent of d,

$$\int_{S^{d-1}} \|\theta\|_K \, d\sigma_{d-1}(\theta) \geq c\sqrt{\log d}.$$

Dvoretzky's original argument is more complicated. Dvoretzky introduced the first and last steps but did not apply the discretization. Instead, he showed "directly" that the norm is almost constant on a k-dimensional subspace. As a result, instead of considering neighborhoods of substantial subsets of the sphere, he was forced to consider neighborhoods of sets that meet a substantial proportion of the k-dimensional subspaces. Milman's method threw into sharper relief the idea that on a space like the sphere, which satisfies a deviation principle, any Lipschitz function will be almost equal to its average on a huge part of the space. This viewpoint led to a series of important results in the 1980s which will be discussed in Section 3.3, but in the next short subsection we shall say a bit more about Euclidean slices.

3.2. Sections of ℓ_p balls

In 1974 Kašin [71] showed that the finite-dimensional L_1 spaces, ℓ_1^d, have Euclidean subspaces of much higher dimension than is guaranteed by Dvoretzky's Theorem.

Theorem 16 (Kašin). *For each d, there is a subspace of ℓ_1^d of dimension at least $d/2$ which is 32-isomorphic to a Euclidean space.*

He used the fact that the unit ball of $B_1^d = \{(x_i)_1^d : \sum |x_i| \leq 1\}$ contains a Euclidean ball of radius $1/\sqrt{d}$ whose volume is as large as $1/2^d$ times the volume of B_1^d. This remarkable fact, that the unit ball of ℓ_1^d has almost spherical slices of dimension proportional to d, was reproved in [51] using Milman's approach to Dvoretzky's Theorem.

A familiar phenomenon in functional analysis is that the L_p spaces for $p < 2$ behave very differently from those for $p > 2$: several important identities in Hilbert space become inequalities in L_p, in one direction for $p < 2$ but in the other direction for $p > 2$. Kašin's argument can be used to show that for $p < 2$ the space ℓ_p^d contains subspaces of proportional dimension that are almost Euclidean. However, for $p > 2$ the correct dependence is $d^{2/p}$ as shown in [51]. This fact was the starting point for Bourgain's remarkable solution to the Λ_p problem on subspaces of L_p spanned by trigonometric characters [26].

3.3. The reverse Santaló inequality

In 1939 Mahler asked a very natural question, prompted by applications in the geometry of numbers. We already saw that the product of the volumes of a symmetric convex body and its polar cannot be more than for the Euclidean ball v_d^2. Mahler asked whether the minimum occurs for the pair consisting of the cube and the so-called cross-polytope, the unit balls of ℓ_∞^d and ℓ_1^d, respectively, for which the product is $4^d/d!$. He also asked whether the minimum over all (not necessarily symmetric) bodies occurs for the simplex. The precise questions are still open but, for example, the products v_d^2 and $4^d/d!$ have a ratio of about $(\pi/2)^d$, so for most purposes it is enough to have an estimate

$$|K|.|K^\circ| \geq c^d v_d^2$$

for some positive constant c. Such an estimate was proved in a well-known article of Bourgain and Milman [28].

Theorem 17 (Bourgain–Milman). *There is a constant $c > 0$ so that if K is a symmetric convex body and K° is its polar then*

$$|K|.|K^\circ| \geq \left(\frac{c}{d}\right)^d.$$

The assumption of symmetry was removed in subsequent works, but the ideas involved in proving the more general statement do not really add anything to the original.

The original proof of the theorem used a subtle, but rather technical, estimate of Milman's [90] (which he called the lower M^*-estimate), together with the theory of type and cotype developed principally by Kwapień, Maurey, and Pisier; see in particular [85, 98]. A crucial result of the latter is Pisier's estimate for the norm of the Rademacher projection on a finite-dimensional space [99]. If K is a convex body that is t-equivalent to an ellipsoid in the sense of Section 3.1 then there is a linear image $\tilde{K}$ so that the norms whose unit balls are $\tilde{K}$ and its polar satisfy

$$\int_{S^{d-1}} \|\theta\|_{\tilde{K}}\, d\sigma_{d-1}(\theta) \int_{S^{d-1}} \|\phi\|_{\tilde{K}^\circ}\, d\sigma_{d-1}(\phi) \leq C(1 + \log t) \tag{3.1}$$

for some constant C. As alluded to in Section 2.3, an upper estimate for

$$\int_{S^{d-1}} \|\theta\|_{\tilde{K}}\, d\sigma_{d-1}(\theta)$$

yields a lower estimate for the volume of K because of Hölder's inequality. So Pisier's result gives an estimate

$$|K|.|K^\circ| \geq \left(\frac{c}{d \log d}\right)^d$$

which is much stronger than the estimate that follows from John's Theorem, but contains an extra $\log d$ that is not present in the Bourgain–Milman Theorem. Milman's lower M^*-estimate demonstrates the existence of a high-dimensional subspace on which $\|\cdot\|_{\tilde{K}}$ is controlled in terms of the quantity

$$\int_{S^{d-1}} \|\phi\|_{\tilde{K}^\circ}\, d\sigma_{d-1}(\phi).$$

This and the very weak (logarithmic) dependence of the integral on the distance of a normed space from Euclidean made possible an iterative argument: apply a linear map that makes the integral small, find a subspace much closer to Euclidean and repeat. Shortly after Theorem 17, Milman proved another result in the same spirit: his reverse Brunn–Minkowski inequality [91]. The reverse Santaló and reverse Brunn–Minkowski inequalities are explained at length in the books [5, 100].

In the years following the Bourgain–Milman Theorem, there have been a number of other proofs using very different methods. Kuperberg [77] found one that uses topology, and Nazarov [93] presented a method using complex analysis. Quite recently an "elementary" proof was found by Giannopoulos, Paouris, and Vritsiou [60] which is very much in the spirit of convex geometry. This will be discussed in Section 4.6 below.

4. THE PROBABILISTIC PICTURE

The aim of this last section is to explain the somewhat cryptic assertion made in the the introduction that convex domains exhibit many of the properties we expect of the joint densities of independent random variables. In the previous section it was explained how the probabilistic method appears in the proofs of subtle geometric results. But here we are after a more intimate connection between geometry and probability. Instead of probability being a tool for proving geometric facts, we want to see classical probability theory actually mimicked by the geometry.

4.1. The cube and the Gaussian isoperimetric inequality

To begin gently, observe that the indicator function of the cube $[-1/2, 1/2]^n$ is exactly the joint density of n independent random variables, each one uniformly distributed on the interval $[-1/2, 1/2]$. If our convex set happened to be a rectangle then it would be the joint density of independent random variables, but we have to choose the coordinate system carefully. If a long thin rectangle is not aligned with the coordinates then its coordinates are highly dependent; see Figure 3.

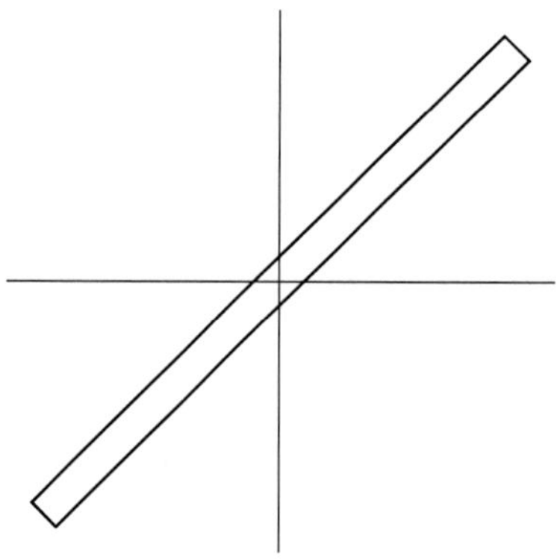

FIGURE 3
A rectangle misaligned

For a general convex domain, there is no natural choice of a coordinate system, so, in order to witness its similarity to a joint density, we must first transform the domain in such a way that the choice of coordinate system does not really matter. If (X_i) are not just independent but also *identically distributed* then all marginals of the joint distribution have the same variance: the vector $(X_1, \ldots, X_d)$ has the property that for any unit vector (θ_i),

$$\mathrm{E}\Big(\sum \theta_i X_i\Big)^2 = \mathrm{E}X_1^2.$$

So, given a symmetric convex domain, we start by applying the linear map which makes its inertia tensor a multiple of the identity. We call a symmetric domain K *isotropic* if

$$\int_K \langle u, x\rangle^2 \, du = L^2 \|x\|_2^2 \tag{4.1}$$

for some L and every vector x. Note that this condition again resembles the conditions of John (1.2) and Petty (2.1). (If the domain is not symmetric, we also shift it so that $\int_K x = 0$.) For an isotropic body, it makes sense to ask whether its indicator looks like the joint density of IID random variables.

The cube is the only example of a convex domain which *exactly* corresponds to IID random variables, but the model we want to keep in mind is that of the standard Gaussian density on $\mathbf{R}^d$, namely

$$x \mapsto g(x) = \frac{1}{(\sqrt{2\pi})^d} \exp(-\|x\|_2^2/2).$$

Although this is not the density of a convex set, it is logarithmically concave and is the joint density of independent 1-dimensional Gaussians. There is an isoperimetric principle for the Gaussian density established independently by Borell [25] and by Sudakov and Tsirel'son [106]. We write

$$\gamma(A) = \int_A g$$

for the Gaussian measure of a measurable set A in $\mathbf{R}^d$.

Theorem 18 (Borell–Sudakov–Tsirel'son). *Suppose A is a measurable subset of $\mathbf{R}^d$ and H is a half-space with the same Gaussian measure, $\gamma(A) = \gamma(H)$. Then for each $\varepsilon > 0$ the ε-neighborhoods of these sets satisfy*

$$\gamma(A_\varepsilon) \geq \gamma(H_\varepsilon).$$

Both articles establish the theorem by using the isoperimetric inequality on the sphere and a limiting process. There are other, direct, proofs; a particularly elegant one was found by Bobkov [22]. From this isoperimetric principle, we can immediately obtain a deviation estimate: if $\gamma(A) = 1/2$ then the t-neighborhood of A has large measure:

$$\gamma(A_t) \geq 1 - 1/2e^{-t^2/2}.$$

It follows from this, or can easily be checked by simple calculation, that most of the mass of the Gaussian density in $\mathbf{R}^d$ lies in a spherical shell of constant thickness (the constant being independent of dimension) and radius about $\sqrt{d}$. In other words, most of the mass lies in a shell much thinner than its radius.

As explained, the aim of this final section of the article is to discuss the extent to which convex domains exhibit features like those of the Gaussian density. To set the scene, let us remark that we already have a deviation inequality for the Euclidean ball: Theorem 6 works just as well for the solid ball as for the sphere. Moreover, Pisier noticed that we can get a similar inequality for the cube by transporting the Gaussian measure to Lebesgue measure on the cube. However, one cannot hope to obtain such a sub-Gaussian deviation principle for a general convex domain. The unit ball of ℓ_1^d has volume $2^d/d!$ so the scaled copy that has volume 1 is about d times as large. Its marginal in a coordinate direction decays like $(1 - x/d)^d$ and this is only subexponential, rather than sub-Gaussian. For a general convex domain, a subexponential deviation estimate can be provided by a Poincaré inequality which estimates the smallest nontrivial eigenvalue $\lambda_1(K)$ of the Neumann Laplacian on the domain. In an influential paper [38] Cheeger showed that (on any compact Riemannian manifold) this eigenvalue cannot be too small if there is an isoperimetric inequality for subsets in the manifold. On the other hand, Gromov and Milman [62] showed that a lower bound for λ_1 does imply a subexponential deviation estimate. The statement that $\lambda_1(K)$ is the first nontrivial eigenvalue can be written as an inequality: if $s : K \to \mathbf{R}$ is differentiable and perpendicular to the trivial constant eigenfunction

$$\int_K s = 0,$$

then

$$\lambda_1(K) \int_K s^2 \leq \int_K \|\nabla s\|_2^2. \tag{4.2}$$

Observe that if K is isotropic with constant L as in equation (4.1), then for any unit vector e we have

$$\int_K \langle u, e\rangle^2 \, du = L^2.$$

So by taking s to be the function $s : u \mapsto \langle u, e\rangle$ whose gradient everywhere is the vector e of length 1, we conclude that $\lambda_1(K)$ cannot be larger than $1/L^2$.

The Laplacian with respect to Gaussian measure, namely the operator

$$s \mapsto -\frac{\nabla.(g\nabla s)}{g} = -\Delta s + \langle x, \nabla s\rangle,$$

satisfies a Poincaré inequality with constant 1: the linear functions $x \mapsto \langle u, x\rangle$ are the eigenfunctions of this operator that have the smallest nonzero eigenvalue. The conjectures discussed below are intended to capture the extent to which convex domains share with Gaussian densities the properties of having Gaussian marginals, satisfying a Poincaré inequality or concentrating mass in a thin shell. Before addressing the conjectures that frame our probabilistic picture of convex domains, it is helpful to discuss sections of convex bodies.

4.2. Sections of convex bodies

From now on we shall assume that K is a symmetric convex domain in $\mathbf{R}^d$ with volume 1 which is isotropic, that is,

$$\int_K \langle u, x\rangle^2 \, du = L^2 \|x\|_2^2 \tag{4.3}$$

for all $x \in \mathbf{R}^d$. As a consequence of the Brunn–Minkowski inequality, each marginal density of K is logarithmically concave, and that means we can relate its maximum value to its "variance" L^2. Hensley [68] pointed out that this implies that if H is a 1-codimensional subspace of $\mathbf{R}^d$ then the slice $H \cap K$ has volume between, say, $1/(4L)$ and $1/L$. The obvious question is "How big is L?". We can apply (4.3) to an orthonormal basis to get

$$\int_K \|u\|_2^2 = dL^2,$$

and so it is clear that L is *minimized* if K is a Euclidean ball. In this case L is approximately $1/\sqrt{2\pi e}$. So the question is how *large* L can be? It is tempting to think that a convex domain of volume 1 must obviously have some 1-codimensional slices as large as those of the Euclidean ball of the same volume and use Hensley's result to deduce that L must be at most a constant independent of dimension. In 1956 Busemann and Petty [35] asked a general version of this question: Is it true that if K and B are symmetric convex bodies and every 1-codimensional slice through the center of K has $(d-1)$-dimensional volume smaller than the corresponding slice of B, then K itself must have smaller d-dimensional volume than B? The answer is no, and the simplest counterexamples take B to be the Euclidean ball, so there is no hope of estimating L in this way. A negative answer in 12 dimensions was provided in 1975 by Larman and Rogers [78] with K a random perturbation of the Euclidean ball. Some years later I proved that each 1-codimensional slice of the cube has volume at most $\sqrt{2}$ and, as a result, if the dimension is at least 10, the unit cube has all its slices smaller than those of the Euclidean ball of volume 1; see [8] and [10].

The Busemann–Petty problem is now solved in all dimensions. Lutwak [82] showed that the problem can be reformulated in terms of intersection bodies (the unit balls of the norms generated by Busemann's Theorem 3) and using this Gardner [58] proved that the problem has a positive solution in 3 dimensions, while Zhang [112] proved it for 4 dimensions. For dimension 5 and above, the solution is negative, and a unified treatment of the problem can be given using ideas of Koldobsky; see the article by Gardner, Koldobsky, and Schlumprecht [59]. (In the special case in which K is the Euclidean ball, the answer to the Busemann–Petty question is yes in all dimensions: every convex body has a slice as *small* as those of the Euclidean ball of the same volume.) So there remains the question: Is there an upper bound, independent of dimension, for the variance of isotropic convex domains in Euclidean space? This will be the subject of the first conjecture in the next subsection.

4.3. The conjectures

The aim of this subsection is to describe three conjectures that have motivated much of the work in high-dimensional geometry over the last two decades, each of which describes a sense in which the indicators of convex domains look like the densities of independent random variables or, more specifically, like Gaussian densities.

Conjecture 19 (Bourgain's slicing conjecture). *There is a constant M independent of dimension so that if K is an isotropic symmetric convex domain of volume 1 in $\mathbf{R}^d$ then*

$$\int_K \|u\|_2^2 \le M^2 d.$$

While this conjecture is usually attributed to Bourgain, I personally do not think he actually believed it. The question of just how large the integral can be remains open but there has been dramatic recent progress that will be discussed in Section 4.6. Note that the conjecture is equivalent to the following tantalisingly simple statement: there is a constant $\delta > 0$ so that every convex body of volume 1 has a 1-codimensional slice of volume at least δ; hence the name of the conjecture. As explained in the previous section, you cannot hope to prove the conjecture by showing that every convex body of volume 1 has a slice as large as the Euclidean ball of the same volume. What is trivial is that every convex domain of volume 1 must have width at most that of the Euclidean ball of the same volume, in at least one direction, and this is about $\sqrt{d}$. By the Cavalieri principle, it must have a slice of volume at least $1/\sqrt{d}$. On the face of it, this conjecture appears to be saying much less than that a convex domain looks like a Gaussian density; this point will be taken up in Section 4.4 below.

The second conjecture was made by the present author in the mid-1990s (and later published in a joint article with Anttila and Perissinaki [3]) and also by Brehm and Voigt [31]. Roughly speaking, it says the following:

Conjecture 20 (The central limit problem). *Let K be an isotropic, convex domain of volume 1 in $\mathbf{R}^d$. Then in all but a small proportion of directions, the 1-dimensional marginals of K are approximately Gaussian.*

(The precise formulation stipulates that as $d \to \infty$ the proportion decreases to 0 and the distance of the marginals from Gaussian also decreases to 0.)

The third and final conjecture concerns the Poincaré constant for an isotropic convex domain. It has been known for a century that for a bounded connected domain $\Omega \subset \mathbf{R}^d$ there is a spectral gap for the Neumann Laplacian on Ω. The gap can be very small if Ω is a dumbbell-shaped domain because then s can be equal to -1 on one of the weights, 1 on the other, and only have a nonzero gradient on the narrow bar that joins the two weights. Even if Ω is convex, the constant can be large if the set is long and thin since then s can take large values at the two ends by changing only very slowly along the length of Ω. However, for an *isotropic* convex set K, there is a bound depending only upon dimension. As was remarked earlier the spectral gap cannot be more than $1/L^2$. In their article [70], Kannan, Lovász, and Simonovits conjectured that this is the correct order.

Conjecture 21 (Kannan–Lovász–Simonovits). *There is a constant C independent of dimension so that if K is an isotropic, convex domain of volume 1 then for any differentiable s on K with $\int_K s = 0$,*

$$\int_K s^2 \le CL^2 \int_K \|\nabla s\|_2^2 \tag{4.4}$$

where L is the "slicing constant" of K, that is,

$$\int_K \langle u, x\rangle^2 \, du = L^2 \|x\|_2^2.$$

Thus the conjecture is that for convex domains, linear functions are approximately the worst for the spectral gap problem, just as they are for the Gaussian density.

In the same article Kannan, Lovász, and Simonovits gave a better bound for the spectral gap than the trivial one that can be deduced just from a bound on the diameter of K. To do so, they used a localization method for proving inequalities originally employed by Payne and Weinberger [95], which is roughly as follows. Among convex sets, the cones are "extremal" in the sense that they are only just convex. As you scan along a cone in $\mathbf{R}^d$ the $(d-1)$-dimensional volume of the slices is given by $x \mapsto l(x)^{d-1}$ where l is a real-valued linear function. You have a pair of functions f and g and you want to contradict the claim that they both have positive integral (thereby proving the inequality you want). Assume that they do. Choose a hyperplane that cuts space into two pieces on which the integrals of f are the same and pick the piece on which the integral of g is the larger. Thus you have found a smaller region on which both functions have positive integral. Keep doing this and take a "limit." It is possible to show that the limiting region is an infinitesimal truncated cone: formally, a line segment and a weight function of the form $l(x)^{d-1}$ for some linear function l. You now just have to prove the inequality you want in this 1-dimensional setting. The estimate given by Kannan, Lovász, and Simonovits is

$$\int_K s^2 \le CdL^2 \int_K \|\nabla s\|_2^2$$

with an additional factor of the dimension d, instead of inequality (4.4). The same estimate was later established by Bobkov [23] using entropy arguments.

The next subsection explains the probabilistic picture of geometry that grew out of these conjectures and how this led to a study of the entropy of logarithmically concave random variables. The following subsections will then describe the state of play on each of the three conjectures. There are a number of books and survey articles on these topics, for example, [2,75].

4.4. The probabilistic picture clarified

It was remarked in Section 1.1 that the class of logarithmically concave densities is an extension of the class of convex domains which has the virtue of being closed under the most common operations of probability theory. Facts about convex domains usually transfer to this larger class: for example, using Theorem 3 it is not hard to show that if M is a bound in the slicing conjecture for a given dimension then eM works for logarithmically concave densities in the same dimension (see [9]). When studying a convex domain, it makes sense to consider its indicator function which takes the value 1 on the domain. But when looking at more general densities, it is not natural to normalize by fixing the value at a point. In the context of probability theory, it is clearly much more natural to consider probability densities $f : \mathbf{R}^d \to \mathbf{R}$ for which the covariance matrix is the identity; so for all x,

$$\int_{\mathbf{R}^d} f(u)\langle u, x\rangle^2 = \|x\|_2^2.$$

Once you rescale in this way, two of the conjectures read differently. The central limit problem, of course, stays as it is: we still want marginals to look Gaussian, just with a different variance.

The KLS conjecture now becomes as simple as it could be, namely the slicing constant L simply disappears from the statement.

Conjecture 22 (KLS probabilistic version). *There is a constant C independent of dimension so that if $f : \mathbf{R}^d \to [0, \infty)$ is an even logarithmically concave probability density whose covariance matrix is the identity, then for any differentiable $s : \mathbf{R}^d \to \mathbf{R}$ with compact support and $\int sf = 0$,*

$$\int s^2 f \le C \int \|\nabla s\|_2^2 f.$$

The slicing problem states that in passing from the convex body normalization to the probabilistic normalization, we haven't had to rescale too much. So it now specifies that the value $f(0)$ cannot be more than M^d for some constant M. However, the quantity $f(0)$ looks rather unnatural in the probabilistic setting: among other things, it is very unstable under probabilistic operations such as convolution. Fortunately, it can be replaced by a proxy that is much more appropriate. It is not difficult to check that for an even logarithmically concave density the entropy, $\mathrm{Ent}(f) = -\int_{\mathbf{R}^d} f \log f$, satisfies

$$-\log f(0) \leq \mathrm{Ent}(f) \leq -\log f(0) + cd$$

for some constant c. (Within a few days of my pointing this out, Fradelizi showed me a neat argument that gives the optimal constant, $c = 1$. He included it into an article some years later [54].) Therefore the slicing problem now reads

Conjecture 23 (Slicing, probabilistic version). *There is a constant C independent of dimension so that if $f : \mathbf{R}^d \to [0, \infty)$ is an even logarithmically concave probability density whose covariance matrix is the identity, then*

$$\mathrm{Ent} f \geq -Cd.$$

Among random vectors with a given covariance matrix, the Gaussian has the largest entropy. The gap between the entropy of a random vector on $\mathbf{R}^d$ with density f and the entropy of the Gaussian is a well-known and very natural measure of how far the random vector is from being Gaussian. So this version of the slicing problem shows clearly that it does *indeed* constitute a statement about logarithmically concave densities being similar to Gaussians.

This entropic formulation of the problem was the motivation behind a series of articles [4, 13] of Artstein, Barthe, Naor, and the author, which used a local version of the Brunn–Minkowski inequality to find a new formula for the entropy (or more precisely, the Fisher information) of a marginal distribution. This did not solve the slicing problem, but led us to (among other things) the solution of an old problem in information theory: Is the central limit theorem driven by an analogue of the second law of thermodynamics? Since the Gaussian has the largest entropy among random variables with a given variance, it makes sense to ask whether the central limit theorem can be "explained" by the fact that normalized sums of IID random variables have increasing entropy that drives them to the Gaussian.

Theorem 24 (Artstein–Ball–Barthe–Naor). *If (X_i) are IID square-integrable random variables then the entropies*

$$\mathrm{Ent}\left(\frac{1}{\sqrt{n}} \sum_1^n X_i\right)$$

increase with n.

This theorem constitutes an application of convex geometry to information theory rather than the other way around, but the machinery developed in [13] did have one surprising consequence for geometry. Following the work of Bakry and Emery [7], most studies of entropy consider the evolution of a random vector along the Ornstein–Uhlenbeck semigroup. This is a semigroup $\{P_t\}_{t\geq 0}$ of convolution operators on L_1 which can be defined as follows. If f is the density of a random vector X then $P_t f$ is the density of the random vector $X_t = \sqrt{e^{-2t}}X + \sqrt{1-e^{-2t}}G$ where G is a standard Gaussian independent of X. Thus the semigroup evolves the original random vector towards the Gaussian. The logarithmic Sobolev inequality of Gross [63] ensures that the rate of decrease of the entropy gap between the random vector X_t at time t and the Gaussian limit is at least a certain multiple of the gap:

$$2(\mathrm{Ent}G - \mathrm{Ent}X_t) \leq -\frac{\partial}{\partial t}(\mathrm{Ent}G - \mathrm{Ent}X_t).$$

As a result the entropy gap decays to zero at least as fast as the exponential e^{-2t}. One consequence of the methods found in [13] is that if we start with a random variable X with a logarithmically concave density $f = e^{-\phi}$ for which the Laplacian

$$s \mapsto -\frac{\nabla.(f\nabla s)}{f} = -\Delta s + \langle \nabla\phi, \nabla s\rangle$$

itself has a spectral gap, then the entropic convergence is enhanced. So if the KLS conjecture holds for a particular density f, we get more rapid convergence of the entropy gap to zero. With some care, this can be used to show that the initial entropy gap was not too large to start with and hence yield an estimate for the slicing constant for f. The argument appears in [14].

Theorem 25 (Ball–Nguyen). *Let $f : \mathbf{R}^d \to \mathbf{R}$ be an isotropic even logarithmically concave probability density satisfying a Poincaré inequality*

$$\int s^2 f \leq C \int \|\nabla s\|_2^2 f$$

for differentiable s with compact support satisfying $\int sf = 0$. Then the slicing constant of f is at most e^{16C}.

Some years later Eldan and Klartag [47] gave a much tighter estimate for the relationship between the spectral gap and the slicing bound, not for each convex domain but "globally." They showed that an estimate C in the KLS conjecture for *all* convex domains transfers to an estimate of some constant multiple of C in the slicing problem for all domains. The most interesting thing about the argument is that they apply the spectral gap property to a more natural function, the Euclidean norm, than in the case of the theorem for individual domains stated above. Their result has acquired new significance following a recent article of Chen [39] and will be taken up in Section 4.6. The last three sections deal with what is known on the three conjectures stated above.

4.5. The central limit problem

It was shown by Sudakov [105] and by Diaconis and Freedman [43] that an isotropic probability measure μ on high-dimensional space will have Gaussian marginals, in the sense of the central limit problem stated above, as long as the measure satisfies a thin shell estimate of the kind that Gaussian measure satisfies. If

$$\int_{\mathbf{R}^d} \langle u, x\rangle^2 \, d\mu(u) = L^2 \|x\|_2^2$$

for all x then the typical radius of the random vector with law μ is $L\sqrt{d}$. The thin shell condition states that most of the mass of μ lies in a shell of roughly this radius, whose thickness is significantly smaller, namely

$$\mu\big(\big|\|x\|_2 - L\sqrt{d}\big| > \varepsilon L\sqrt{d}\big) < \varepsilon.$$

A measure satisfying this will have Gaussian marginals in most directions up to an error of ε plus a term depending only on d (that is, $o(1)$ as $d \to \infty$). In the case of the uniform measure on a convex set, one can obtain essentially optimal estimates (in terms of ε and d) as in **[3]**. This thin shell condition is clearly implied by a Poincaré inequality for the measure, so the KLS conjecture is stronger than both the central limit conjecture and the slicing problem. The KLS conjecture would give the thin shell property with ε of the order of $1/\sqrt{d}$ which is the best one could possibly hope for. It was explained in Section 4.3 that Kannan, Lovász, and Simonovits had used localization to obtain a bound on the spectral gap for isotropic convex sets or logarithmically concave functions. That bound does not provide an estimate for ε that tends to 0 as the dimension grows.

The central limit problem was solved in 2006 by Klartag **[73]** and shortly afterwards a completely different proof was given by Fleury, Guédon, and Paouris **[53]**. The key idea in Klartag's article is to show that typical marginals of the body of fairly high dimension, say $\log d$, have almost exactly rotation-invariant densities, a kind of Dvoretzky Theorem for marginal densities instead of sections. (The possibility of such an approximate rotation invariance was suggested by Gromov during the 1980s.) For a rotation invariant density, the thin shell property is a 1-dimensional question that is easily solved. Then, since the 1-dimensional marginals of the original body are the 1-dimensional marginals of the $\log d$-dimensional ones, they must be almost Gaussian.

A year or so before the proof of the central limit theorem for convex domains, Paouris **[94]** proved an optimal decay estimate for the Euclidean norm, which is clearly related to the thin shell estimate but ultimately has rather different consequences.

Theorem 26 (Paouris). *Suppose K is an isotropic, symmetric convex body of volume* 1 *in* $\mathbf{R}^d$ *and*

$$\int_K \|x\|_2^2 = L^2 d.$$

Then the volume of the part of K where $\|x\|_2$ is significantly larger than $L\sqrt{d}$ decays as

$$\big|\{x \in K : \|x\|_2 \geq cL\sqrt{d}t\}\big| < e^{-\sqrt{d}t}$$

for some constant c independent of dimension and K, and for all $t > 1$.

The restriction that t should be larger than 1 means that the theorem does not yield a concentration of volume in a shell but it gives an excellent decay rate for large radii. The proof of this theorem depends upon a delicate analysis of the integrals

$$\int_{S^{d-1}} \|\theta\|_K^q \, d\sigma_{d-1}(\theta)$$

of powers of the norm corresponding to K. This in turn depends upon a study of the norm restricted to subspaces in the same spirit as the Bourgain–Milman Theorem, the new ingredient here being that a crucial role is played by subspaces of dimension $\sqrt{d}$. The optimality (apart from the value of c) is shown by sets like the unit ball of ℓ_1^d as remarked earlier.

The proof of the central limit problem given by Fleury, Guédon, and Paouris uses a variant of the methods of Theorem 26. Each of these two articles gives an estimate for the thin shell problem with ε only logarithmic in the dimension. Klartag [74] quickly gave a power-type estimate and this was improved by Fleury [52], and again by Guédon and E. Milman [64] by combining the techniques from [53,73]. Then in [79] Lee and Vempala showed how to use the stochastic localization method of Eldan to estimate the thin shell bound. At the time this gave a bound of the form $\varepsilon = 1/d^{1/4}$, but the recent work of Chen reduces this almost to $1/\sqrt{d}$.

4.6. The slicing conjecture

Bourgain not only posed the slicing problem but gave the first significant estimate. As remarked above, there is a trivial bound of $\sqrt{d}$ for slicing constants in dimension d. In [27] Bourgain improved this dramatically to $d^{1/4} \log d$.

Theorem 27 (Bourgain). *For some M independent of dimension, if K is an isotropic symmetric convex body of volume 1 then*

$$\frac{1}{d} \int_K \|x\|_2^2 \leq M(d^{1/4} \log d)^2.$$

Bourgain's argument used much of the functional-analytic machinery that had just become available: Pisier's estimate (3.1), Talagrand's majorizing measure theorem [107], and an interpolation argument using the Brunn–Minkowski inequality.

In the late 1980s when the problem was first discussed, there was considerable interest in the question of whether the slicing constant is an isomorphic invariant for normed spaces. Suppose we have two t-equivalent norms on $\mathbf{R}^d$ with unit balls J and K which are isotropic and have volume 1. If the slicing constant of K is L, must it be that the constant for J is at most some fixed function of t times as large as L? Klartag [72] showed some 15 years later that, in this form at least, the question is a bit of a red herring.

Theorem 28 (Klartag). *If K is a convex body and $\varepsilon > 0$, there is another body T which is $(1+\varepsilon)$-equivalent to K and whose slicing constant is at most $C/\sqrt{\varepsilon}$.*

So every convex body is quite similar to one with a bounded slicing constant. Thus the only way that the slicing constant *can* be an isomorphic invariant is that it is essentially the same for all convex bodies. Klartag pointed out that by combining this with Paouris' estimate, Theorem 26, one can eliminate the $\log d$ factor in Bourgain's Theorem to give an estimate for slicing constants of $d^{1/4}$.

Klartag's argument is very surprising. He considers the following logarithmically concave function:

$$y \mapsto e^{\langle x,y \rangle}$$

restricted to K, for different choices of x. Using the theory of monotone transport discussed in Section 2.2, he shows that for an appropriate x, the slicing constant for this function can be bounded in terms of the volume product $|K|.|K^\circ|$ and then invokes a nonsymmetric version of Theorem 3 to create a convex set from the logarithmically concave function. To get Theorem 28, he used the reverse Santaló inequality (Theorem 17). Some years later Giannopoulos, Paouris, and Vritsiou realized that the final step could be avoided in a rather dramatic way. It is possible to estimate the volume product of a body in terms of its slicing constant. In itself that is not very surprising: the slicing conjecture is a strong statement, and already in [9] there was a very simple proof that the slicing conjecture implies Milman's reverse Brunn–Minkowski inequality. But the key point here is that the powers of the volume product in Klartag's Theorem and in the reverse direction are different, and this means that they can be combined to yield a proof of the reverse Santaló inequality. That in turn can be fed back into Klartag's Theorem. So there is now an "elementary" approach to Theorem 17.

It was remarked in Section 4.4 that the KLS conjecture implies the slicing conjecture and that Eldan and Klartag [47] had sharpened the dependence. Their argument applies the spectral gap property to the function $\|x\|_2$ and so what they actually prove is that estimates in the slicing conjecture can be deduced from estimates for the thin shell bound discussed in the previous subsection. Their original argument involved the construction of a Riemannian metric related to a convex body which seems to have little or nothing to do with the other ideas discussed in this article, although one can see a link to the proof of Theorem 28. Subsequently, the machinery of stochastic localization developed by Eldan, which will be discussed in the next section, was used by Eldan and Lehec [48] to give an alternative proof. The best estimates currently known in the slicing problem have just been improved dramatically as will be explained at the end of the next subsection.

4.7. The KLS conjecture

It was remarked in Section 4.1 that for a convex set, a bound on the spectral gap is equivalent to an isoperimetric inequality for subsets, that is, a bound on the Cheeger constant. The Cheeger constant for a domain K is the minimum over all subsets A of K of the ratio

$$\frac{|\partial A|}{|A|.|K-A|}$$

where the "surface area" $|\partial A|$ of A includes only the part of the surface inside K. So the constant tells you how small can be the area of a (curved) cut that divides the set into roughly equal pieces. In the paper [88] of E. Milman, there is a detailed explanation of the relationship between the Cheeger constant, the spectral gap, and (on the face of it) much weaker notions for convex domains.

The interest of computer scientists in the spectral gap problem for convex domains arose because of the problem of effective computation of volume for convex sets. To compute volume deterministically, even to within a factor that is exponential in the dimension, is computationally hard. However, in [45] Dyer, Frieze, and Kannan found a randomized algorithm which involved running a random walk inside the set: sampling from a set is more or less equivalent to computing its volume. There has been a succession of improvements in the run time of the algorithm by Lovász, Simonovits, Applegate, Vempala, Lee, and the original authors over a period of 30 years. A very helpful survey of the history is provided by Lee and Vempala [80]. Some of these improvements involve choosing enhanced random walks, but others depend upon getting better estimates for the geometry of the domains being sampled. In order to sample effectively, you want the random walk to mix throughout the domain quickly and the barrier to that happening will be a bottleneck: a way of cutting the domain into substantial pieces with a cut whose area is small. If your domain has this dumbbell shape then a random walk can get trapped in one of the weights. This is exactly the Cheeger constant problem. So better estimates in the KLS conjecture immediately imply better run times in the algorithm.

A very new approach to the problem was found by Eldan, [46]. Instead of *convolving* with a Gaussian as in the Ornstein–Uhlenbeck process above, Eldan's method can be thought of as the apparently simpler one of multiplying by a Gaussian density. But it is a random density and the aim is to show that the *typical* product behaves as you would like. If you multiply by a Gaussian with large variance you do not really change the logarithmically concave density: if you multiply by a Gaussian with small variance you get essentially a Gaussian, for which you know everything. The problem is to keep track of how quantities change in going from one to the other. The key is to effect the multiplications by a stochastic process, what Eldan called stochastic localization. The process is governed by a stochastic differential equation, but Eldan explains that this can be thought of in the following way. At each infinitesimal step, you multiply the density by a linear function whose gradient has a random direction. A linear function such as $x \mapsto 1 - x_1$ puts greater weight on one half of the density, thus mimicking the localization technique described in Section 4.3. A familiar fact in analysis is that the product of two "complementary" linear functions

$$(1 - x_1)(1 + x_1) = 1 - x_1^2$$

gives you a hump which is the first step towards a Gaussian.

When he introduced the method, Eldan used it to show that the thin shell property for a logarithmically concave density implies the KLS conjecture up to a factor that is only a power of $\log d$. As explained above, Lee and Vempala modified the technique to prove an estimate $\varepsilon = 1/d^{1/4}$ for the thin shell problem and hence by [47] a bound of $d^{1/4}$ for the slicing conjecture, the same as Klartag's. At that point it was tempting to wonder whether this might be the correct order of the worst slicing constant on the grounds that two (or even three) completely different methods gave (essentially) the same bound. However, in a recent remarkable breakthrough, Chen [39] found a way to use stochastic localization to get a bound for the KLS conjecture, the thin shell problem and the slicing problem, which is $O(d^{\alpha})$ for every $\alpha > 0$.

Theorem 29 (Eldan–Chen). *For every $\alpha > 0$, there is a constant $C(\alpha)$ so that for every symmetric, isotropic convex domain $K \subset \mathbf{R}^d$ of volume 1 and every differentiable $s : K \to \mathbf{R}$ with $\int_K s = 0$, we have*

$$\int_K s^2 \leq C(\alpha) d^\alpha \int_K \|\nabla s\|_2^2.$$

4.8. Conclusion

I introduced the final section by suggesting that convex bodies mimic classical probability theory. But with hindsight one should perhaps see the situation differently. Independence and convexity each, in their different ways, force a measure to be "genuinely" high-dimensional, as opposed to being a low-dimensional measure that accidentally lies in a high-dimensional space. What makes a measure roughly Gaussian is the high-dimensionality. How is it that the extra freedom in high dimensions creates what appears to be more order and predictability instead of less? I admit to being biased but surely a clue is given by Theorem 24: the disorder that comes from high-dimensionality is the sort of disorder that is found in physical systems, namely the disorder of high entropy. And increased entropy presents to low-dimensional human eyes as uniformity and regularity.

ACKNOWLEDGMENTS

I am grateful to Franck Barthe, Sergei Bobkov, Andrea Colesanti, Ronen Eldan, Matthieu Fradelizi, David Jerison, Emanuel Milman, Assaf Naor, Grigouris Paouris, and Gaoyong Zhang for helpful discussions during the writing of this article.

REFERENCES

[1] A. D. Aleksandrov, Almost everywhere existence of the second differential of a convex function and related properties of convex surfaces. *Uch. Zap. Leningrad Gos. Univ. Math. Ser.* **37** (1939), 3–35 (in Russian).

[2] D. Alonso-Gutierrez and J. Bastero, *Approaching the Kannan–Lovász–Simonovits and variance conjectures*. Lecture Notes in Math. 2131, Springer, Berlin, 2015.

[3] M. Anttila, K. M. Ball, and I. Perissinaki, The central limit problem for convex bodies. *Trans. Amer. Math. Soc.* **355** (2003), 4723–4735.

[4] S. Artstein, K. M. Ball, F. Barthe, and A. Naor, A solution of Shannon's problem on the monotonicity of entropy. *J. Amer. Math. Soc.* **17** (2004), 975–982.

[5] S. Artstein-Avidan, A. A. Giannopoulos, and V. D. Milman, *Asymptotic geometric analysis: Part 1*. Math. Surveys Monogr. 202, Amer. Math. Soc., Providence, RI, 2015.

[6] T. Aubin, Problèmes isopérimétriques et espaces de Sobolev. *J. Differential Geom.* **11** (1976), 573–598.

[7] D. Bakry and M. Émery, Diffusions hypercontractives. In *Séminaire de probabilités, XIX, 1984/84*, edited by J. Azéma and M. Yor, pp. 177–206, Lecture Notes in Math. 1123, Springer, Berlin, 1985.

[8] K. M. Ball, Cube slicing in $\mathbf{R}^n$. *Proc. Amer. Math. Soc.* **97** (1986), 465–473.

[9] K. M. Ball, *Isometric problems in ℓ_p and sections of convex sets*. PhD thesis, University of Cambridge, 1987.

[10] K. M. Ball, Some remarks on the geometry of convex sets. In *Geometric aspects of functional analysis*, edited by V. D. Milman and G. Schechtman, pp. 224–231, Lecture Notes in Math. 1317, Springer, Berlin, 1988.

[11] K. M. Ball, Volume ratios and a reverse isoperimetric inequality. *J. Lond. Math. Soc.* **44** (1991), 351–359.

[12] K. M. Ball, An elementary introduction to monotone transportation. In *Geometric aspects of functional analysis*, edited by V. D. Milman and G. Schechtman, pp. 41–52, Lecture Notes in Math. 1850, Springer, Berlin, 2004.

[13] K. M. Ball, F. Barthe, and A. Naor, Entropy jumps in the presence of a spectral gap. *Duke Math. J.* **119** (2003), 41–64.

[14] K. M. Ball and V. H. Nguyen, Entropy jumps for isotropic log-concave random vectors and spectral gap. *Studia Math.* **213** (2012), 81–96.

[15] F. Barthe, Inégalités de Brascamp–Lieb et convexité. *C. R. Acad. Sci. Paris* **324** (1997), 885–888.

[16] A. Barvinok, *A course in convexity*. Grad. Stud. Math. 54, AMS, Providence, RI, 2002.

[17] W. Beckner, Inequalities in Fourier analysis. *Ann. of Math.* **102** (1975), 159–182.

[18] F. Behrend, Über einige Affinvarianten konvexer Bereiche. *Math. Ann.* **113** (1937), 713–747.

[19] J. Bennett, A. Carbery, M. Christ, and T. Tao, The Brascamp–Lieb inequalities: finiteness, structure and extremals. *Geom. Funct. Anal.* **17** (2005), 1343–1415.

[20] G. Bianchi, A. Colesanti, and C. Pucci, On the second differentiability of convex surfaces. *Geom. Dedicata* **60** (1996), 39–48.

[21] W. Blaschke, Über affine Geometrie VII: Neue Extremeingenschaften von Ellipse und Ellipsoid. *Ber. Verh. Sächs. Akad. Wiss. Math. Phys. Kl.* **69** (1917), 412–420.

[22] S. Bobkov, A functional form of the isoperimetric inequality for the Gaussian measure. *J. Funct. Anal.* **135** (1996), 39–49.

[23] S. G. Bobkov, Isoperimetric and analytic inequalities for log-concave probability measures. *Ann. Probab.* **27** (1999), 1903–1921.

[24] T. Bonnesen, Über die isoperimetrische Eigenschaft des Kreises auf der Kugeloberfläche und in der Ebene. *Math. Ann.* **60** (1905), 117–136.

[25] C. Borell, The Brunn–Minkowski inequality in Gauss space. *Invent. Math.* **30** (1975), 205–216.

[26] J. Bourgain, Bounded orthogonal systems and the Λ_p-set problem. *Acta Math.* **162** (1989), 227–245.

[27] J. Bourgain, On the distribution of polynomials on high dimensional convex sets. In *Geometric aspects of functional analysis*, edited by J. Lindenstrauss and V. D. Milman, pp. 127–137, Lecture Notes in Math. 1469, Springer, Berlin, 1991.

[28] J. Bourgain and V. Milman, New volume ratio properties for convex bodies in $\mathbf{R}^n$. *Invent. Math.* **88** (1987), 319–340.

[29] H. J. Brascamp and E. H. Lieb, Best constants in Young's inequality, its converse and its generalization to more than three functions. *Adv. Math.* **20** (1976), 151–173.

[30] H. J. Brascamp and E. H. Lieb, On extensions of the Brunn–Minkowski and Prékopa–Leindler theorems, including inequalities for log-concave functions, and with an application to the diffusion equation. *J. Funct. Anal.* **22** (1976), 366–389.

[31] U. Brehm and J. Voigt, Asymptotics of cross sections for convex bodies. *Beitr. Algebra Geom.* **41** (2000), 437–454.

[32] J. Brenier, Polar factorization and monotone rearrangement of vector-valued functions. *Comm. Pure Appl. Math.* **44** (1991), 375–417.

[33] H. Busemann, A theorem on convex bodies of the Brunn–Minkowski type. *Proc. Natl. Acad. Sci. USA* **35** (1949), 27–31.

[34] H. Busemann and W. Feller, Krümmungseigenschaften konvexer Flächen. *Acta Math.* **66** (1935), 1–47.

[35] H. Busemann and C. M. Petty, Problems on convex bodies. *Math. Scand.* **4** (1956), 88–94.

[36] L. Caffarelli, Regularity of mappings with a convex potential. *J. Amer. Math. Soc.* **5** (1992), 99–104.

[37] E. Carlen, E. H. Lieb, and M. Loss, A sharp analog of Young's inequality on S^n and related entropy inequalities. *J. Geom. Anal.* **14** (2004), 487–520.

[38] J. Cheeger, A lower bound for the smallest eigenvalue of the Laplacian. In *Problems in analysis (Papers dedicated to Salomon Bochner, 1969)*, edited by R. C. Gunning, pp. 195–199, PUP, Princeton, NJ, 1970.

[39] Y. Chen, An almost constant lower bound of the isoperimetric coefficient in the KLS conjecture. 2021, arXiv:2011.13661v2.

[40] D. Cordero-Erausquin, B. Nazaret, and C. Villani, A mass transportation approach to sharp Sobolev and Gagliardo–Nirenberg inequalities. *Adv. Math.* **182** (2004), 307–332.

[41] E. De Giorgi, Sulla proprietà isoperimetrica dell'ipersfera, nella classe degli insiemi aventi frontiera orientata di misura finita. *Atti Accad. Naz. Lincei. Mem. Cl. Sci. Fis. Mat. Nat. Sez.* **5** (1958), 33–44.

[42] M. Del Pino and J. Dolbeault, Best constants for Galgliardo–Nirenberg inequalities and applications to nonlinear diffusions. *J. Math. Pures Appl.* **81** (2002), 847–875.

[43] P. Diaconis and D. Freedman, Asymptotics of graphical projection pursuit. *Ann. Statist.* **12** (1984), 793–815.

[44] A. Dvoretzky, A theorem on convex bodies and applications to Banach spaces. *Proc. Natl. Acad. Sci.* **45** (1959), 223–226.

[45] M. E. Dyer, A. M. Frieze, and R. Kannan, A random polynomial time algorithm for approximating the volume of convex bodies. *J. ACM* **38** (1991), 1–17.

[46] R. Eldan, Thin shell implies spectral gap up to a polylog via a stochastic localization scheme. *Geom. Funct. Anal.* **23** (2013), 532–569.

[47] R. Eldan and B. Klartag, Approximately Gaussian marginals and the hyperplane conjecture. In *Proceedings of a workshop on concentration, functional inequalities and isoperimetry*, pp. 55–68, Contemp. Math. 545, Amer. Math. Soc., Providence, RI, 2011.

[48] R. Eldan and J. Lehec, Bounding the norm of a log-concave vector via thin-shell estimates. In *Geometric aspects of functional analysis*, edited by B. Klartag and E. Milman, pp. 107–122, Lecture Notes in Math. 2116, Springer, Berlin, 2014.

[49] A. Figalli and D. Jerison, Quantitative stability for sumsets in $\mathbf{R}^n$. *J. Eur. Math. Soc. (JEMS)* **17** (2015), 1079–1106.

[50] A. Figalli, F. Maggi, and A. Pratelli, A mass transportation approach to quantitative isoperimetric inequalities. *Invent. Math.* **182** (2010), 167–211.

[51] T. Figiel, J. Lindenstrauss, and V. Milman, The dimension of almost spherical sections of convex bodies. *Acta Math.* **139** (1977), 53–94.

[52] B. Fleury, Concentration in a thin Euclidean shell for log-concave measures. *J. Funct. Anal.* **259** (2010), 832–841.

[53] B. Fleury, O. Guédon, and G. Paouris, A stability result for mean width of ℓ_p-centroid bodies. *Adv. Math.* **214** (2007), 865–877.

[54] M. Fradelizi and M. Meyer, Increasing functions and inverse Santaló inequality for unconditional functions. *Positivity* **12** (2008), 407–420.

[55] B. Fuglede, Stability in the isoperimetric problem for convex or nearly spherical domains in $\mathbf{R}^n$. *Trans. Amer. Math. Soc.* **314** (1989), 619–638.

[56] N. Fusco, The quantitative isoperimetric inequality and related topics. *Bull. Math. Sci.* **5** (2015), 517–607.

[57] N. Fusco, F. Maggi, and A. Pratelli, The sharp quantitative isoperimetric inequality. *Ann. of Math.* **168** (2008), 941–980.

[58] R. Gardner, A positive answer to the Busemann–Petty problem in three dimensions. *Ann. of Math.* **140** (1994), 435–447.

[59] R. Gardner, A. Koldobsky, and T. Schlumprecht, An analytic solution to the Busemann–Petty problem on sections of convex bodies. *Ann. of Math.* **149** (1999), 691–703.

[60] A. Giannopoulos, G. Paouris, and B. Vritsiou, The isotropic position and the reverse Santaló inequality. *Israel J. Math.* **203** (2014), 1–22.

[61] A. Giannopoulos and M. Papadimitrakis, Isotropic surface area measures. *Mathematika* **46** (1999), 1–13.

[62] M. Gromov and V. D. Milman, A topological application of the isoperimetric inequality. *Amer. J. Math.* **105** (1983), 843–854.

[63] L. Gross, Logarithmic Sobolev inequalities. *Amer. J. Math.* **97** (1975), 1061–1083.

[64] O. Guédon and E. Milman, Interpolating thins-shell and sharp large deviation estimates for isotropic log-concave measures. *Geom. Funct. Anal.* **21** (2011), 1043–1068.

[65] C. Haberl and F. E. Schuster, General L_p affine isoperimetric inequalities. *J. Differential Geom.* **83** (2009), 1–26.

[66] R. R. Hall, A quantitative isoperimetric inequality in n-dimensional space. *J. Reine Angew. Math.* **428** (1992), 161–176.

[67] M. Henk, J. Richter-Gebert, and G. Ziegler, Basic properties of convex polytopes. In *Handbook of discrete and computational geometry*, edited by J. E. Goodman, J. O'Rourke, and C. D. Toth, pp. 383–413, Chapman and Hall/CRC, 2017.

[68] D. Hensley, Slicing convex bodies – bounds for slice area in terms of the bodies' covariance. *Proc. Amer. Math. Soc.* **79** (1980), 619–625.

[69] F. John, Extremum problems with inequalities as subsidiary conditions. In *Studies and essays presented to R. Courant on his 60th birthday*, pp. 187–204, Interscience, NY, 1948.

[70] R. Kannan, L. Lovász, and M. Simonovits, Isoperimetric problems for convex bodies and a localization lemma. *Discrete Comput. Geom.* **13** (1995), 541–559.

[71] B. Kašin, The widths of certain finite-dimensional sets and classes of smooth functions. *Izv. Ross. Akad. Nauk Ser. Mat.* **41** (1977), 334–351.

[72] B. Klartag, On convex perturbations with a bounded isotropic constant. *Geom. Funct. Anal.* **16** (2006), 1274–1290.

[73] B. Klartag, A central limit theorem for convex sets. *Invent. Math.* **168** (2007), 91–131.

[74] B. Klartag, Power-law estimates for the central limit theorem for convex sets. *J. Funct. Anal.* **245** (2007), 284–310.

[75] B. Klartag and V. Milman, The slicing problem by Bourgain (to appear).

[76] H. Knöthe, Contributions to the theory of convex bodies. *Michigan Math. J.* **4** (1957), 39–52.

[77] G. Kuperberg, From the Mahler conjecture to Gauss linking integrals. *Geom. Funct. Anal.* **18** (2008), 870–892.

[78] D. G. Larman and C. A. Rogers, The existence of a centrally symmetric convex body with central sections that are unexpectedly small. *Mathematika* **22** (1975), 164–175.

[79] Y. T. Lee and S. S. Vempala, Eldan's stochastic localization and the KLS hyperplane conjecture: An improved lower bound for expansion. In *2017 IEEE 58th Annual Symposium on Foundations of Computer Science (FOCS)*, pp. 998–1007, IEEE, 2017.

[80] Y. T. Lee and S. S. Vempala, The Kannan–Lovász–Simonovits conjecture. In *Current developments in math. 2017*, edited by D. Jerison, M. Kisin, P. Seidel, R. Stanley, H.-T. Yau, and S.-T. Yau, pp. 1–36, International Press, Boston, 2019.

[81] E. H. Lieb, Gaussian kernels have only Gaussian maximizers. *Invent. Math.* **102** (1990), 179–208.

[82] E. Lutwak, Intersection bodies and dual mixed volumes. *Adv. Math.* **71** (1988), 232–261.

[83] E. Lutwak, D. Yang, and G. Zhang, L_p affine isoperimetric inequalities. *J. Differential Geom.* **56** (2000), 111–132.

[84] F. Maggi, Some methods for studying stability in isoperimetric type problems. *Bull. Amer. Math. Soc. (N.S.)* **45** (2008), 367–408.

[85] B. Maurey, Type, cotype and K-convexity. In *Handbook of the geometry of Banach spaces*, edited by W. B. Johnson and J. Lindenstrauss, pp. 1299–1332, North-Holland, Amsterdam, 2003.

[86] R. McCann, Existence and uniqueness of measure preserving maps. *Duke Math. J.* **80** (1995), 309–323.

[87] M. Meyer and A. Pajor, On the Blaschke–Santaló inequality. *Arch. Math.* **55** (1990), 82–93.

[88] E. Milman, On the role of convexity in isoperimetry, spectral gap and concentration. *Invent. Math.* **177** (2009), 1–43.

[89] V. Milman, A new proof of A. Dvoretzky's theorem on cross-sections of convex bodies. *Funktsional. Anal. i Prilozhen.* **5** (1971), 28–37 (in Russian).

[90] V. Milman, Random subspaces of proportional dimension of finite dimensional normed spaces: approach through the isoperimetric inequality. In *Banach spaces*, edited by N. Kalton and E. Saab, pp. 106–115, Lecture Notes in Math. 1166, Springer, Berlin, 1985.

[91] V. Milman, Inégalité de Brunn–Minkowski inverse et applications à la théorie locale des espaces normés. *C. R. Acad. Sci. Paris Sér. I Math.* **302** (1986), 25–28.

[92] V. Milman and G. Schechtman, *Asymptotic theory of finite-dimensional normed spaces. With an apendix by M. Gromov*. Lecture Notes in Math. 1200, Springer, Berlin, 1986.

[93] F. Nazarov, The Hörmander proof of the Bourgain–Milman theorem. In *Geometric aspects of functional analysis*, edited by B. Klartag, S. Mendelson, and V. D. Milman, pp. 335–343, Lecture Notes in Math. 2050, Springer, Berlin, 2012.

[94] G. Paouris, Concentration of mass on convex bodies. *Geom. Funct. Anal.* **16** (2006), 1021–1049.

[95] L. E. Payne and H. F. Weinberger, An optimal Poincaré inequality for convex domains. *Arch. Ration. Mech. Anal.* **5** (1960), 286–292.

[96] C. M. Petty, Surface area of a convex body under affine transformations. *Proc. Amer. Math. Soc.* **12** (1961), 824–828.

[97] C. Petty, Projection bodies. In *Proc. of the colloquium on convexity, Copenhagen, 1965*, pp. 234–241, Kobenhavns Univ. Mat. Inst, Copenhagen, 1967.

[98] G. Pisier, Sur les espaces de Banach qui ne contiennent pas uniformément de ℓ_1^n. *C. R. Acad. Sci. Paris* **277** (1973), 991–994.

[99] G. Pisier, Sur les espaces de Banach K-convexes. In *Séminaire d'Analyse Fonctionelle, 1979–1980*, pp. 1–15, 11, Ecole Polytech, Palaiseau, 1980.

[100] G. Pisier, *The volume of convex bodies and Banach space geometry*. Cambridge Tracts in Math. 94, CUP, Cambridge, 1997.

[101] A. Prékopa, Logarithmic concave measures with application to stochastic programming. *Acta Sci. Math.* **32** (1971), 301–316.

[102] R. T. Rockafellar, *Convex analysis*. PMS 28, PUP, Princeton, 1997.

[103] J. Saint-Raymond, Sur le volume des corps convexes symétriques. In *Séminaire d'Initiation à l'Analyse: 20th Year: 1980/1981*, edited by G. Choquet, M. Rogalski, and J. Saint-Raymond, Publ. Math. Univ. Pierre et Marie Curie 46, Univ. Paris VI, Paris, 1981.

[104] L. A. Santaló, Un invariante afin para los cuerpos convexsos del espacio des n dimensiones. *Port. Math.* **8** (1949), 155–161.

[105] V. N. Sudakov, Typical distributions of linear functionals in finite-dimensional spaces of high dimension. *Sov. Math., Dokl.* **19** (1978), 1578–1582.

[106] V. N. Sudakov and B. S. Tsirel'son, Extremal properties of half-spaces for spherically invariant measures. *J. Sov. Math.* **9** (1978), 9–18. Translated from *Zap. Nauchn. Sem. S.-Peterburg. Otdel. Mat. Inst. Steklov. (POMI)* **41** (1974), 14–24.

[107] M. Talagrand, Regularity of Gaussian processes. *Acta Math.* **159** (1987), 99–149.

[108] G. Talenti, Best constants in Sobolev inequality. *Ann. Mat. Pura Appl.* **110** (1976), 353–372.

[109] C. Villani, *Topics in optimal transportation*. Grad. Stud. Math. 58, AMS, Providence, RI, 2003.

[110] V. Wolontis, Properties of conformal invariants. *Amer. J. Math.* **74** (1952), 587–606.

[111] G. Zhang, Restricted chord projection and affine inequalities. *Geom. Dedicata* **39** (1991), 213–222.

[112] G. Zhang, A positive solution to the Busemann–Petty problem in $\mathbf{R}^4$. *Ann. of Math.* **149** (1999), 535–543.

KEITH BALL

Mathematics Institute, University of Warwick, Coventry CV4 7AL, UK,
k.m.ball@warwick.ac.uk

This Page Intentionally Left Blank

8. ANALYSIS

MOMENT METHODS ON COMPACT GROUPS: WEINGARTEN CALCULUS AND ITS APPLICATIONS

BENOÎT COLLINS

ABSTRACT

A fundamental property of compact groups and compact quantum groups is the existence and uniqueness of a left and right invariant probability—the Haar measure. This is a very natural playground for classical and quantum probability, provided that it is possible to compute its moments. Weingarten calculus addresses this question in a systematic way. The purpose of this manuscript is to survey recent developments, describe some salient theoretical properties of Weingarten functions, as well as applications of this calculus to random matrix theory, quantum probability and algebra, mathematical physics, and operator algebras.

MATHEMATICS SUBJECT CLASSIFICATION 2020

Primary 15B52; Secondary 60B20, 28C10, 43A05, 46L54, 22D20

KEYWORDS

Haar measure on compact groups, Weingarten calculus, asymptotic freeness, quantum information theory, quantum groups, Schur–Weyl duality, Norm estimates, random tensors

INTERNATIONAL CONGRESS OF MATHEMATICIANS
2022 JULY 6–14

Proc. Int. Cong. Math. 2022, Vol. 4, pp. 3142–3164
DOI 10.4171/ICM2022/114

Published by EMS Press

1. INTRODUCTION

One of the key properties of a compact group G is that it admits a unique left and right invariant probability measure μ_G. It is called the *Haar measure*, and we refer to **[15]** for reference. In other words, $\mu_G(G) = 1$, and for any Borel subset A of G and $g \in G$, $\mu_G(Ag) = \mu_G(gA) = \mu_G(A)$, where $Ag = \{hg, h \in A\}$ and $gA = \{gh, h \in A\}$. The left and right invariance together with the uniqueness of μ_G readily imply that $\mu_G(A^{-1}) = \mu_G(A)$. The standard proofs of existence of the Haar measure are not constructive. In the more general context of locally compact groups, a left (resp. right) invariant measure exists, too. It is finite if and only if the group is compact and uniqueness is up to a nonnegative scalar multiple. In addition, the left and right Haar measures need not be the same. For locally compact groups, a classical proof of existence imitates the construction of the Lebesgue measure on $\mathbb{R}$ and resorts to outer measures. In the specific case of compact groups, a fixed-point argument can be applied. Either way, in both cases, the proof of existence is not constructive, in the sense that it does not tell us *how to integrate functions*. Weingarten calculus is about addressing this problem systematically. *Which functions* one wants to integrate needs, of course, to be clarified. We focus on the case of matrix groups, for which there are very natural candidates, namely polynomials in coordinate functions.

We recast this problem as the question of *computing the moments* of the Haar measure. Recall that, for a real random variable X, its moments are by definition the sequence $\mathbb{E}(X^k), k \geq 0$—whenever they are defined. If the variable is vector-valued in $\mathbb{R}^n$, i.e., $X = (X_1, \dots, X_n)$, then the moments are the numbers $\mathbb{E}(X_1^{k_1} \cdots X_n^{k_n}), k_1, \dots k_n \geq 0$. Naturally, the existence of moments is not granted and is subject to the integrability of the functions. In the case of matrix compact groups, we have $G \subset \mathbb{M}_n(\mathbb{C}) = \mathbb{R}^{2n^2}$ therefore we may consider that the random variable we are studying is a random vector in $\mathbb{R}^{2n^2}$ whose distribution is the Haar measure with respect to the above inclusion. In this sense, we are really considering a moment problem. For this reason, we do not consider only coordinate functions, but also their complex conjugates in our moment problem.

The goal of this note is to provide an account of Weingarten calculus and in particular its multiple applications, with emphasis on the moment aspects and applications. From the point of view of the theory, there have been many approaches to computing integral of functions with respect to the Haar measure. We enumerate here a few important ones:

(1) Historically, the first nontrivial functions computed are arguably Fourier transforms, e.g., the Harish-Chandra integral **[51]**. The literature is huge and started from the initial papers of Harish-Chandra and Itzykson Zuber until now, however, we do not elaborate too much on this field as we focus on polynomial integrals. These techniques involve representation theory, symplectic geometry, and complex analysis. We refer to **[62]** for a recent approach and to the bibliography therein for references.

(2) Geometric techniques are natural because, when compact groups are manifolds, the measure can be described locally with differential geometry. They are effi-

cient for small groups. We refer, for example, to [3] for such techniques and gaussianization methods, with application to quantum groups. Geometry is also useful to compute specific functions, such as polynomials in one row or column with respect to orthogonal or unitary groups.

(3) Probability, changes of variables, and stochastic calculus are natural tools to try to compute the moments of Haar measures. For example, Rains in [68] used Brownian motion on compact groups and the fact that the Haar measure is the unique invariant measure to compute a complete set of relations. Subsequently, Lévy, Dahlqvist, Kemp, and the author have made progress on understanding the unitary multiplicative Brownian version of Weingarten calculus in [21,56].

(4) Representation theory has always been ubiquitous in the quest for calculating the Haar measure. A first important set of applications can be found by [45], but results were already available by [17,44,73].

(5) Combinatorial interpretations of the Haar measure in some specific cases were initiated in [18]. In another direction, there was the notable work of [18]. Subsequently, new combinatorial techniques were developed in [20,24], and we refer to [26] for substantial generalizations. We also refer [59] for modern interpretations and applications to geometric group theory.

As for the applications, they can be found in a considerable number of areas, including theoretical physics (2D quantum gravity, matrix integrals, random tensors), mathematical physics (quantum information theory, quantum spin chains), operator algebras (free probability), probability (limit theorems), representation theory, statistics, finance, machine learning, and group theory. The foundations of Weingarten calculus, as well as its applications, keep expanding rapidly, and this manuscript is a subjective snapshot of the-state-of-the-art. This introduction is followed by Section 2 that contains the foundations and theoretical results about the Weingarten functions. Section 3 investigates "simple" asymptotics of Weingarten functions and applications to random matrix theory. Section 4 deals with "higher order" asymptotics and applications to mathematical physics. Section 5 considers "uniform" asymptotics and applications to functional analysis, whereas the last section contains concluding remarks and perspectives.

2. WEINGARTEN CALCULUS

2.1. Notation

On the complex matrix algebra $\mathbb{M}_n(\mathbb{C})$, we denote by $\overline{A}$ the entrywise conjugate of a matrix A and $A^* = \overline{A}^t$ the adjoint. In the sequel we work with a compact matrix group G, i.e., a subgroup of $\mathrm{GL}_n(\mathbb{C})$ of invertible complex matrices, that is compact for the induced topology. It is known that such a group is conjugate inside $\mathrm{GL}_n(\mathbb{C})$ to the unitary group $\mathcal{U}_n = \{U, UU^* = U^*U = 1_n\}$. Writing an element U of $\mathcal{U}_n$ as a matrix

$U = (u_{ij})_{i,j \in \{1,\dots,n\}}$, we view the entries u_{ij} as polynomial functions $\mathcal{U}_n \to \mathbb{C}$. As functions, they form a $*$-algebra—the $*$-operation being the complex conjugation. By construction, they are separating for $\mathcal{U}_n$, therefore, by Weierstrass' theorem, the $*$-algebra generated by $u_{ij}, i, j \in \{1, \dots, n\}$, which is the algebra of polynomial functions on $\mathcal{U}_n$, is a dense subalgebra for the sup norm in the algebra of continuous functions on G.

By Riesz' theorem, understanding the Haar measure boils down to understanding $\int_{U \in G} f(U) d\mu_G(U)$ for any continuous function. By density and linearity, it is actually enough to be able to calculate systematically

$$\int_{U \in G} u_{i_1 j_1} \dots u_{i_k j_k} \overline{u_{i'_1 j'_1} \dots u_{i'_{k'} j'_{k'}}} d\mu_G(U).$$

No answer was known in full generality until a systematic development was initiated in [20, 40]. However, in the particular case of $\mathcal{U}_n, \mathcal{O}_n$, an algorithm to calculate a development in large n was devised in [44, 73], with further improvements by [66], and character expansions were obtained in [17], however, these approaches are largely independent. Likewise, Woronowicz obtained a formula for the moments of characters in the case of quantum groups in [74]. Interestingly, motivated by probability questions, the same formula was rediscovered independently by Diaconis–Shashahani [45] in the particular case of compact matrix groups.

2.2. Fundamental formula

Although the partial answers to the question of computing moments were rather involved, the general answer turns out, in hindsight, to be surprisingly simple, so we describe it here. We also refer to [32] for an invitation to the theory. We first start with the following notation: for an element $U = (u_{ij}) \in G \subset \mathbb{M}_n(\mathbb{C})$, $\overline{U}$ is the entrywise conjugate, i.e., $\overline{U} = (\overline{U_{ij}})$. Since U is unitary, $\overline{U}$ is unitary, too. We denote by $V = \mathbb{C}^n$ the fundamental representation of G, and $\overline{V}$ the contragredient representation. For a general representation W of G, $\mathrm{Fix}(G, W)$ is the vector subspace of W of fixed points under the action of G, i.e., $\mathrm{Fix}(G, W) = \{x \in W, \forall U \in G, Ux = x\}$. Finally, we fix two integers k, k', and set

$$Z_G = \int_{U \in G} U^{\otimes k} \otimes \overline{U}^{\otimes k'} d\mu_G(U),$$

and abbreviate $\mathrm{Fix}(G, V^{\otimes k} \otimes \overline{V}^{\otimes k'})$ into $\mathrm{Fix}(G, k, k')$.

Proposition 2.1. *The matrix Z_G is the orthogonal projection onto* $\mathrm{Fix}(G, k, k')$.

Proof. Since the distribution of U and UU' is the same for any fixed $U' \in G$, it implies that for any $U \in G$, $Z_G = Z_G \cdot U^{\otimes k} \otimes \overline{U^{\otimes k'}}$. Integrating once more over U gives the fact that Z_G is a projection. The fact that the map $U \to U^{-1} = U^*$ preserves the Haar measure implies that $Z_G = Z_G^*$. From the definition of invariance, for $x \in \mathrm{Fix}(G, k, k')$ and for any $U \in G$ one has $U^{\otimes k} \otimes \overline{U^{\otimes k'}} \cdot x = x$. Integrating with respect to the Haar measure of G gives $Z_G \cdot x = x$. Finally, take x outside $\mathrm{Fix}(G, k, k')$. It means that there exists U such that

$$U^{\otimes k} \otimes \overline{U^{\otimes k'}} x \neq x.$$

However, $||U^{\otimes k} \otimes \overline{U^{\otimes k'}} x||_2 = ||x||_2$. Thanks to the strict convexity of the Euclidean ball, after averaging over the Haar measure, we necessarily get $||Z_G x||_2 < ||x||_2$, which implies that x is not in $\mathrm{Im}(Z_G)$. Therefore we proved that $\mathrm{Im}(Z_G) = \mathrm{Fix}(G, k, k')$. ■

From this, we can deduce an integration formula as soon as we have a generating family $y_1, \dots, y_l$ for $\mathrm{Fix}(G, k, k')$ (for any k, k'). Let

$$\mathrm{Gr} = (g_{ij})_{i,j \in \{1,\dots,l\}}$$

be its Gram matrix, i.e., $g_{ij} = \langle y_i, y_j \rangle$ and $W = (w_{ij})$ the pseudoinverse of Gr. Let $E_1, \dots, E_n$ be the canonical orthonormal basis of $V = \mathbb{C}^n$. Let k be a number and we consider the tensor space $V^{\otimes k}$ with its canonical orthogonal basis $E_I = e_{i_1} \otimes \cdots \otimes e_{i_k}$, where $I = (i_1, \dots, i_k)$ is a multiindex in $\{1, \dots, n\}^k$. Let $I = (i_1, \dots, i_k, i'_1, \dots, i'_{k'})$, $J = (j_1, \dots, j_k, j'_1, \dots, j'_{k'})$ be $k + k'$-indices, i.e.. elements of $\{1, \dots, n\}^{k+k'}$. Then

Theorem 2.2.

$$\begin{aligned}\int_{U \in G} u_{i_1 j_1} \dots u_{i_k j_k} \overline{u_{i'_1 j'_1} \dots u_{i'_{k'} j'_{k'}}} d\mu_G(U) &= \langle Z_G, E_I \otimes E_J \rangle \\ &= \sum_{i,j \in \{1,\dots,l\}} \langle E_I, y_i \rangle \langle y_j, E_J \rangle w_{ij}.\end{aligned}$$

2.3. Examples with classical groups

For interesting applications to be derived, the following conditions must be met:

(1) $y_1, \dots, y_l$ must be easy to describe.

(2) Gr should be easy to compute—and if possible, its inverse, the Weingarten matrix, too.

(3) $\langle E_I, y_i \rangle$ should be easy to compute.

Let us describe some fundamental examples. Let $P_2(k)$ be the collection of *pair partitions* on $\{1, \dots, k\}$ ($P_2(k)$ is empty if k is odd, and its cardinal is $1 \cdots (k-1) = k!!$ if k is even). Typically, a partition $\pi \in P_2(k)$ consists of $k/2$ blocks of cardinal 2, $\pi = \{V_1, \dots, V_{k/2}\}$, and we call $\delta_{\pi,I}$ the multiindex Kronecker function whose value is 1 if, for any block $V = \{k < k'\}$ of π, $i_k = i_{k'}$, and zero in all other cases. Likewise, we call $E_\pi = \sum_I E_I \delta_{\pi,I}$.

In [40], we obtained a complete solution to computing moments of Haar integrals for $\mathcal{O}_n, \mathcal{U}_n, \mathcal{S}p_n$. The following theorem describes this method. For convenience, we stick to the case of $\mathcal{O}_n, \mathcal{U}_n$.

Theorem 2.3. *The entries of* Gr *are* $\langle E_\pi, E_{\pi'} \rangle = n^{\mathrm{loops}(\pi,\pi')}$, *and we have* $\langle E_I, E_\pi \rangle = \delta_{\pi,I}$.

- *(The orthogonal case) For* $\mathcal{O}_n$, $E_\pi, \pi \in P_2(k)$ *is a generating family of the image of* $Z_{\mathcal{O}_n}$.
- *(The unitary case) Thanks to commutativity, and setting* $2k' = k$, *we consider the subset of* $P_2(k)$ *of pair partitions such that each block pairs one of the first* k'

elements with one of the last k' elements. This set is in natural bijection with the permutations $S_{k'}$, and it is the generating family of the image of $Z_{\mathcal{U}_n}$.

Proof. The first two points are direct calculations. The last two points are a reformulation of Schur–Weyl duality, respectively in the case of the unitary group and of the orthogonal group (see. e.g., [46]). ∎

2.4. Example with Quantum groups

We finish the general theory of Weingarten calculus with a quick excursion through compact matrix quantum groups. For the theory of compact quantum groups, we refer to [74,75]. The subtlety for quantum groups is that in general we can not capture all representations with just $U^{\otimes k} \otimes \overline{U^{\otimes k'}}$ because U and $\overline{U}$ fail to commute in general. The theory of Tannaka–Krein duality for compact quantum groups is completely developed, and in order to get a completely general formula, we must instead consider

$$U^{\otimes k_1} \otimes \overline{U^{\otimes k_1'}} \otimes \cdots \otimes U^{\otimes k_p} \otimes \overline{U^{\otimes k_p'}}.$$

Let us just illustrate the theory with the *free quantum orthogonal group* O_n^+. It was introduced by Wang in [72], and its Tannaka–Krein dual was computed by Banica in [1]. Its algebra of polynomial functions $\mathbb{C}(O_n^+)$ is the noncommutative unital $*$-algebra generated by n^2 self-adjoint elements u_{ij} that satisfy the relation $\sum_k u_{ik}u_{jk} = \delta_{ij}1$ and $\sum_k u_{ki}u_{kj} = \delta_{ij}1$. Note that the abelianized version of this unital $*$-algebra is the $*$-algebra of polynomial functions on $\mathcal{O}_n$, which explains why it is called the free orthogonal quantum group. There exists a unital $*$-algebra homomorphism, called the coproduct $\Delta : \mathbb{C}(O_n^+) \to \mathbb{C}(O_n^+) \otimes \mathbb{C}(O_n^+)$ defined on generators by $\Delta u_{ij} = \sum_k u_{ik} \otimes u_{kj}$, and a unique linear functional $\mu : \mathbb{C}(O_n^+) \to \mathbb{C}$ such that $\mu(1) = 1$ and

$$(\mu \otimes Id)\Delta = 1\mu, \qquad (Id \otimes \mu)\Delta = 1\mu.$$

This functional is known as the Haar state, and it extends the notion of Haar measure on compact groups. Although the whole definition is completely algebraic, the proofs rely on functional analysis and operator algebras.

However, the calculation of the Haar state is purely algebraic and just relies on the notion of noncrossing pair partitions, denoted by $NC_2(k)$, which are a subset of $P_2(k)$ defined as follows. A partition π of $P_2(k)$ is noncrossing—and therefore in $NC_2(k)$ if any two of its blocks $\{i, j\}$ and $\{i', j'\}$ fail to satisfy the crossing relations $i < j, i' < j'$, $i < i' < j < j'$. This notion was found to be of crucial use for free probability by Speicher, see, e.g., [64]. The following theorem is a particular case of a series of results that can be found in [4]:

Theorem 2.4. *In the case of O_n^+, for $U^{\otimes k}$, the complete solution follows from the following result: $E_\pi, \pi \in NC_2(k)$ is a generating family of the image of $Z_{O_n^+}$.*

Note that, since $U = \overline{U}$, it is enough to consider $U^{\otimes k}$ to compute fully the Haar measure. We refer to [2,5,6] for applications of classical Weingarten functions to quantum groups, and to [3,4,8] for further developments of quantum Weingarten theory.

2.5. Representation theoretic formulas

A representation theoretic approach to Weingarten calculus is available for many families of groups, including unitary, orthogonal, and symplectic groups. Here we only describe the unitary group, and for the others, we refer to [30,60].

Call S_k the symmetric group and consider its *group algebra* $\mathbb{C}[S_k]$—the unital $*$-algebra whose basis as a vector space is $\lambda_\sigma, \sigma \in S_k$, and endowed with the multiplication $\lambda_\sigma \lambda_\tau = \lambda_{\sigma\tau}$ and the $*$-structure $\lambda_\sigma^* = \lambda_{\sigma^{-1}}$. We follow standard representation-theoretic notation, see, e.g., [19] and $\lambda \vdash k$ denotes a Young diagram λ has k boxes; $\lambda \vdash k$ enumerates both the conjugacy classes of S_k and its irreducible representations. The symmetric group S_k acts on the set $\{1, \dots, k\}$, and in turn, by leg permutation on $(\mathbb{C}^n)^{\otimes k}$, which induces an algebra morphism $\mathbb{C}[S_k] \to \mathbb{M}_n(\mathbb{C})^{\otimes k}$. By Schur–Weyl duality, λ describes also irreducible polynomial representations of the unitary group $\mathcal{U}_n$ if its length is less than n and in this context, V_λ stands for the associated representation of the unitary group. For a permutation $\sigma \in S_k$, we call $\#\sigma$ the number of cycles (or loops) in its cycle product decomposition (counting fixed points). Consider the function

$$G = \sum_{\sigma \in S_k} n^{\#\sigma} \lambda_\sigma,$$

and its pseudoinverse $W = G^{-1} = \sum_{\sigma \in S_k} w(\sigma)\lambda_\sigma$. The following result was observed by the author and Śniady in [40] and it provides the link between representation theory and Weingarten calculus:

Theorem 2.5. *G is positive in $\mathbb{C}[S_k]$. In addition, we have $w(\sigma, \tau) = w(\tau\sigma^{-1})$, which we rename as $\mathrm{Wg}(n, \tau\sigma^{-1})$, and the following character expansion:*

$$\mathrm{Wg}(n, \sigma) = \frac{1}{k!^2} \sum_{\lambda \vdash k} \frac{\chi_\lambda(e)^2 \chi_\lambda(\sigma)}{\dim V_\lambda}.$$

Proof. Consider the action of S_k on $(\mathbb{C}^n)^{\otimes k}$ by leg permutation. It extends to a unital $*$-algebra morphism $\phi : \mathbb{C}[S_k] \to \mathbb{M}_n(\mathbb{C})^{\otimes k}$. By inspection, for $A \in \mathbb{C}[S_k]$, $\mathrm{Tr}[\phi(A)] = \tau(GA)$, where τ is the regular trace $\tau(\lambda_g) = \delta_{g,e}$. The positivity of τ implies that of G which proves positivity. The remaining points follow from the fact that G is central and by a character formula. ∎

2.6. Combinatorial formulations

Let us write formally $n^{-k} G = \lambda_e + \sum_{\sigma \in S_k - \{e\}} n^{\#\sigma - k} \lambda_\sigma$. It follows that as a power series in n^{-1},

$$n^k W = \lambda_e + \sum_{p \geq 1} (-1)^p \Big(\sum_{\sigma \in S_k - \{e\}} n^{\#\sigma - k} \lambda_\sigma \Big)^p.$$

Reading through the coefficients of this series gives a combinatorial formula for Wg in the unitary case. Such formulas were first found in [20], and we refer to [26] for substantial generalizations. See also [59] for other interpretations, as well as [18].

However, this formula is signed, and therefore impractical for the quest of uniform asymptotics. In a series of works, Novak and coworkers in [47–49, 61] came with a very interesting solution to this problem which we describe below. It relies on Jucys Murphy elements, which are the following elements of $\mathbb{C}[S_k]$: $J_i = \sum_{j>i} \lambda_{(ij)}$. The following important result was observed:

$$G = (n + J_1) \cdots (n + J_{k-1}).$$

This follows from the fact that every permutation σ has a unique factorization as

$$\sigma = (i_1 j_1) \cdots (i_l j_l)$$

with the property $i_p < j_p$ and $j_p < j_{p+1}$.

This prompts us to define $P(\sigma, l)$ to be the set of solutions to the equation $\sigma = (i_1 j_1) \cdots (i_l j_l)$ with $i_p < j_p$, $j_p \leq j_{p+1}$. The number of solutions to this problem is related to Hurwitz numbers, for details we refer, for example, to [26] and to the above references. From this we have the following theorem:

Theorem 2.6. *For $\sigma \in S_k$, we have the expansion*

$$\mathrm{Wg}(n, \sigma) = n^{-k} \sum_{l \geq 0} \# P(\sigma, l)(-n^{-1})^l. \tag{2.1}$$

The first strategy to compute the Weingarten formula was initiated in [73]. Let us outline it. We can write $\mathrm{Wg}(n, \sigma) = \int u_{11} \cdots u_{kk} \overline{u_{1\sigma 1} \cdots u_{k\sigma k}}$. Indeed, when considering the integral on the right-hand side in Theorems 2.2 and 2.3, the only pairing appearing corresponds to $\mathrm{Wg}(n, \sigma)$. Replacing the first row index of u and $\overline{u}$ by i and summing over i, we are to evaluate

$$\begin{aligned}\sum_{i=1}^{n} \int u_{i1} \cdots u_{kk} \overline{u_{i\sigma(1)} \cdots u_{k\sigma(k)}} &= \delta_{1\sigma(1)} \int u_{22} \cdots u_{kk} \overline{u_{2\sigma(2)} \cdots u_{k\sigma(k)}} \\ &= n\,\mathrm{Wg}(n, \sigma) + \sum_{i=2}^{l} \mathrm{Wg}\big(n, (1i)\sigma\big),\end{aligned} \tag{2.2}$$

where the first equality follows from orthogonality and the second from repeated use of the Weingarten formula. The second line provides an iterative technique to compute $\mathrm{Wg}(n, \sigma)$ both numerically and combinatorially. Historically, this is the idea of Weingarten, and in [73] he proved that the collection of all relations obtained above determine uniquely Wg for k fixed, n large enough.

In [31], we revisited his argument and figured out that these equations can be interpreted as a fixed point problem and a path counting formula, both formally and numerically. We got theoretical mileage from this approach and obtained new theoretical results, such as

Theorem 2.7. *All unitary Weingarten functions and their derivatives are monotone on (k, ∞).*

The unavoidability of Weingarten's historical argument becomes blatant when one studies quantum Weingarten function. Partial results about their asymptotics were obtained in [8], however, the asymptotics were not optimal for all entries. On the other hand, motivated by the study of planar algebras, Vaughan Jones asked us the following question: considering the canonical basis of the Temperley Lieb algebra $\mathrm{TL}_k(n)$, are the coefficients of the dual basis all nonzero when expressed in the canonical basis? For notations, we refer to our paper [16]. One motivation for this question is that the dual element of the identity is a multiple of the Jones–Wenzl projection.

Observing that this question is equivalent, up to a global factor, to the problem of computing the Weingarten function for O_n^+, and realizing that representation theory did not give tractable formulas in this case, we revisited the original idea of Weingarten and proved the following result, answering a series of open questions of Jones:

Theorem 2.8. *The quantum O_n^+ Weingarten function is never zero on the noncritical interval $[2, \infty)$, and it is monotone.*

Our proof actually provides explicit formulas for a Laurent expansion of the free Wg in the neighborhood of $n = \infty$, as a generating series of paths on graphs.

3. ASYMPTOTICS AND PROPERTIES OF WEINGARTEN FUNCTIONS

In this section, we are interested in the following problem. For a given permutation $\sigma \in S_k$, what is the behavior as $n \to \infty$ of $\mathrm{Wg}(n, \sigma)$? This function is rational as soon as $n \geq k$, and even elementary observations about its asymptotics have nontrivial applications in analysis. In the forthcoming subsections, we refine iteratively our study of the asymptotics, and derive each time new applications. Similar results have been obtained for most sequences of classical compact groups, but we focus here mostly on $\mathcal{U}_n$ and $\mathcal{O}_n$, and refer to the literature for other compact groups.

3.1. First order for identity Weingarten coefficients and Borel theorems

Let us first setup notations related to *noncommutative probability spaces* and of *convergence in distribution* in a noncommutative sense. A noncommutative probability space (NCPS) is a unital $*$-algebra $\mathcal{A}$ together with a state τ ($\tau : \mathcal{A} \to \mathbb{C}$ is linear, $\tau(1) = 1$, and $\tau(xx^*) \geq 0$ for any x). In general, we will assume *traciality*, $\tau(ab) = \tau(ba)$ for all a, b.

Assume we have a family of NCPS $(\mathcal{A}_n, \tau_n)$, a limiting object $(\mathcal{A}, \tau)$ and a d-tuple $(x_n^1, \dots, x_n^d) \in \mathcal{A}_n^d$. We say that this d-tuple of *noncommutative random variables converges in distribution* to $(x^1, \dots, x^d) \in \mathcal{A}^d$ iff for any sequence $i_1, \dots, i_k$ of indices in $\{1, \dots, d\}$,

$$\tau_n(x_n^{i_1} \cdots x_n^{i_k}) \to \tau(x^{i_1} \cdots x^{i_k}).$$

In the abelian case this corresponds to a convergence in moments (which is not in general the convergence in distribution); however, in the noncommutative framework, it is usually called convergence in noncommutative distribution, cf. [71]. The following result was proved in [42] in the classical case, and [4] in the quantum case:

Theorem 3.1. *Consider a sequence of vectors* $(A_1^n, \dots, A_r^n)$ *in* $\mathbb{M}_n(\mathbb{R})$ *such that the matrix* $(\mathrm{tr}(A_i A_j^t))$ *converges to* A*, and a* $\mathcal{O}_n$*-Haar distributed random variable* U_n*. Then, as* $n \to \infty$*, the sequence random vectors*

$$\big(\mathrm{Tr}(A_1^n U_n), \dots, \mathrm{Tr}(A_r^n U_n)\big)$$

converges in moments (and in distribution) to a Gaussian real vector of covariance A*. If we assume instead* U_n *to be in* O_n^+*, then* $(\mathrm{Tr}(A_1^n U_n), \dots, \mathrm{Tr}(A_r^n U_n))$ *converges in noncommutative distribution to a free semicircular family of covariance* A*.*

The proof relies on two ingredients. Firstly, for all examples considered so far, $\mathrm{Gr} = n^k \cdot 1_l(1 + O(n^{-1}))$, which implies that $W = \mathrm{Gr}^{-1} = n^{-k} 1_l(1 + O(n^{-1}))$. By inspection, it turns out that in the above theorem, the only entries of W that contribute asymptotically are the diagonal ones, and one can conclude with the classical (resp. the free) Wick theorem.

3.2. Other leading orders for Weingarten coefficients

The asymptotics obtained in the previous section are sharp only for the diagonal coefficients, however, they already yield nontrivial limit theorems. For more refined theorems, it is, however, necessary to obtain sharp asymptotics for all Weingarten coefficients. In the case of $\mathcal{U}_n$, sharp asymptotics can be deduced from the following

Theorem 3.2. *In the case of the full cycle in* S_k*, we have the following explicit formula:*

$$\mathrm{Wg}\big(n, (1 \cdots k)\big) = \frac{(-1)^{k+1} c_k}{(n-k+1) \cdots (n+k-1)},$$

where $c_k = (k+1)^{-1}\binom{2k}{k}$ *is the Catalan number. In addition,* Wg *is almost multiplicative in the following sense: if* σ *is a disjoint product of two permutations* $\sigma = \sigma_1 \sqcup \sigma_2$ *then*

$$\mathrm{Wg}(n, \sigma) = \mathrm{Wg}(n, \sigma_1)\, \mathrm{Wg}(n, \sigma_2)\big(1 + O(n^{-2})\big).$$

This result defines recursively a function $\mathrm{Moeb} : \bigsqcup_{k \geq 1} S_k \mapsto \mathbb{Z} - \{0\}$ satisfying

$$\mathrm{Wg}(n, \sigma) = n^{-k-|\sigma|}\mathrm{Moeb}(\sigma)\big(1 + O(n^{-2})\big).$$

This function was actually already introduced by Biane in [12], and it is closely related to Speicher's noncrossing Möbius function on the incidence algebra of the lattice of noncrossing partitions—see, e.g., [64]. Similar results are available for the orthogonal and symplectic group, we refer to [41]. Finally, let us mention that the asymptotics Weingarten function for the unitary group are the object of intense study; see, for example, [59, 69].

3.3. Classical asymptotic freeness

Weingarten calculus allows answering the following

Question 1. Given two families $(A_i^{(n)})_{i \in I}$ and $(B_j^{(n)})_{j \in J}$ of matrices in $\mathbb{M}_n(\mathbb{C})$, what is the joint behavior of $(A_i^{(n)})_{i \in I} \sqcup (U_n B_j^{(n)} U_n^*)_{j \in J}$, where U_n is invariant according to the Haar measure on $\mathcal{U}_n$?

The notion of behavior has to be clarified, and it will be refined in the same time as we refine our estimates of the Weingarten function. For now, we assume that $(A_i^{(n)})_{i\in I}$ and $(B_j^{(n)})_{j\in J}$ have asymptotic moments, namely, for any sequence $i_1, \dots, i_l$,

$$\operatorname{tr} A_{i_1}^{(n)} \cdots A_{i_l}^{(n)}$$

admits a finite limit, and likewise for $(B_j^{(n)})_{j\in J}$ (note that our standing notation is $\operatorname{tr} = n^{-1} \operatorname{Tr}$). In this specific context, the question becomes:

Question 2. Does the enlarged family $(A_i^{(n)})_{i\in I} \sqcup (U_n B_j^{(n)} U_n^*)_{j\in J}$ have asymptotic moments?

Let us note that since the moments are random, the question admits variants, namely, does the enlarged family have asymptotic moments *in expectation, almost surely*? The answer turns out to be *yes*—irrespective of the variant chosen—and the above asymptotics allow us to deduce the joint behavior of random matrices in large dimension. We recall that a family of unital $*$-subalgebras $\mathcal{A}_i, i \in I$ of an NCPS $(\mathcal{A}, \tau)$ is *free* iff for any $l \in \mathbb{N}_*, i_1, \dots, i_l \in I$, $i_1 \neq i_2, \dots, i_{l-1} \neq i_l, \tau(x_1 \cdots x_l) = 0$ as soon as (i) $\tau(x_j) = 0$ and (ii) $x_j \in A_{i_j}$. Asymptotic freeness holds when a family has a limit distribution and the limiting distribution generates free $*$-subalgebras.

Theorem 3.3. *The answer to Question 2 is yes. The limit of the union is determined by the relation of asymptotic freeness, and the convergence is almost sure.*

The proof relies on calculating moments, together with our knowledge of the asymptotics of the Weingarten function. In the next theorem, we observe that different types of "asymptotic behavior," such as the existence of a limiting point spectrum, are also preserved under enlargement of the family. The theorem below is a particular case of a results to be found in [27]:

Theorem 3.4. *Let $\lambda_{i,n}$ be sequences of complex numbers such that $\lim_n \lambda_{i,n} = 0$. Let $\Lambda_{i,n} = \operatorname{diag}(\lambda_{i,1}, \dots, \lambda_{i,n})$ and $A_{j,n}$ be random matrices with the property that (i) $(A_{j,n})_j$ converges in NC distribution as $n \to \infty$ and (ii) $(UA_{j,n}U^*)_j$ has the same distribution as $(A_{j,n})_j$ as a d-tuple of random matrices. Let P be a noncommutative polynomial. Then the eigenvalues of $P(\Lambda_{i,n}, A_{j,n})$ converge almost surely.*

The proof is also based on Weingarten calculus and moment formula. The limiting distribution is of a new type—involving pure point spectrum—and we call it *cyclic monotone convergence*.

3.4. Quantum asymptotic freeness

Finally, let us discuss another seemingly completely unrelated application, to asymptotic representation theory. The idea is to replace classical randomness by quantum randomness. To keep the exposition simple, we stick to the case of the unitary group, although more general results are true for more general Lie groups, see [41]. Call E_{ij} the canonical matrix entries of $\mathbb{M}_n(\mathbb{C})$, and e_{ij} the generators of the enveloping Lie algebra $\mathfrak{U}(\mathrm{GL}_n(\mathbb{C}))$

of $\mathrm{GL}_n(\mathbb{C})$, namely, the unital $*$-algebra generated by e_{ij} and the relations $e_{ij}^* = e_{ji}$ and $[e_{ij}, e_{kl}] = \delta_{jk} e_{il} - \delta_{il} e_{kj}$. The map $E_{ij} \to e_{ij}$ can be factored through all Lie algebra representations of $\mathcal{U}_n$, and we are interested in the following variants of its Choi matrix

$$A_n^{(1)} = \sum_{ij} E_{ij} \otimes e_{ij} \otimes 1, \quad A_n^{(2)} = \sum_{ij} E_{ij} \otimes 1 \otimes e_{ij} \in \mathbb{M}_n(\mathbb{C}) \otimes \mathfrak{U}\big(\mathrm{GL}_n(\mathbb{C})\big)^{\otimes 2}.$$

In [39], thanks—among others—to asymptotics of Weingarten functions, we proved the following, extending considerably the results of [11].

Theorem 3.5. *For each n, take λ_n, μ_n two Young diagrams corresponding to a polynomial representations V_{λ_n}, V_{μ_n} of $\mathrm{GL}_n(\mathbb{C})$. Assume that both dimensions tend to infinity as $n \to \infty$ and consider the traces on $\chi_{\lambda_n}, \chi_{\mu_n}$ on $\mathfrak{U}(\mathrm{GL}_n(\mathbb{C}))$. Assume that A_n converges in noncommutative distribution in Voiculescu's sense both for $\mathrm{tr} \otimes \chi_{\lambda_n}$ and $\mathrm{tr} \otimes \chi_{\mu_n}$. Then $A_m^{(1)}, A_n^{(2)}$ are asymptotically free with respect to $\mathrm{tr} \otimes \chi_{\lambda_n} \otimes \chi_{\mu_n}$.*

4. MULTIPLICATIVITY AND APPLICATIONS TO MATHEMATICAL PHYSICS

4.1. Higher-order freeness

The asymptotic multiplicativity of the Weingarten function states that

$$\mathrm{Wg}(\sigma_1 \sqcup \sigma_1) = \mathrm{Wg}(\sigma_1)\,\mathrm{Wg}(\sigma_2)\big(1 + O(n^{-2})\big)$$

and it is very far reaching. The fact that the error term $O(n^{-2})$ is summable in n allows in [20] to use a Borel–Cantelli lemma and prove almost sure convergence of moments for random matrices, cf. [70] for the original proof.

A more systematic understanding of the error term is possible and has deep applications in random matrix theory. It requires the notion of classical cumulants that we recall now. Let X be a random variable, the cumulant $C_p(X)$ is defined formally by

$$C(t) = \log \mathbb{E}(\exp tX) = \sum_{p \geq 1} t^p \frac{C_p(X)}{p!}.$$

For instance, the second cumulant $C_2(X) = \mathbb{E}(X^2) - \mathbb{E}(X)^2$ is the variance of the probability distribution of X. Cumulant $C_p(X)$ is well defined as soon as X has moments up to order p, and it is an n-homogeneous function in X, therefore we can polarize it and define a p-linear symmetric function $(X_1, \dots, X_p) \to C_p(X_1, \dots, X_p)$. For any partition π of p elements with blocks $B \in \pi$, we define $C_\pi(X_1, \dots, X_p) = \prod_{B \in \pi} C(\prod_{i \in B} X_i)$. We are now in the position to write the expectations in term of the cumulants:

$$\mathbb{E}\left(\prod_{i=1}^p X_i\right) = \sum_{\pi \in P(p)} C_\pi.$$

The equation can be inverted through the Möbius inversion formula. Asymptotic freeness considers the case where moments have a limit, whereas higher-order asymptotic freeness considers the case where things are known about the fluctuations of the moments: in addition to the existence of $\lim_n \operatorname{tr} A_{i_1}^{(n)} \cdots A_{i_l}^{(n)}$, we assume the existence of

$$\lim_n n^{2k-2} C_k(A_{i_{11}}^{(n)} \cdots A_{i_{l_1 1}}^{(n)}, \ldots, A_{i_{1k}}^{(n)} \cdots A_{i_{l_k k}}^{(n)})$$

for any sequence of indices. We call this set of limits the *higher order limit*. In [33], we proved

Theorem 4.1. *The extended family $(A_i^{(n)})_{i \in I} \sqcup (UB_j^{(n)}U^*)_{j \in J}$ admits a higher order limit. In addition, there exists a combinatorial rule to construct the joint asymptotic correlations from the asymptotic correlations of each family.*

This rule extends freeness, is called *higher order freeness*. Subsequent work was done in the case of orthogonal invariance by Mingo and Redelmeier.

4.2. Matrix integrals

Historically, matrix integrals have been studied before higher order freeness. However, from the point of view of formal expansion, higher order freeness supersedes matrix integrals. In [20], we proved the following

Theorem 4.2. *Let A be a noncommutative polynomial in formal variables $(Q_i)_{i \in I}$, formal unitaries U_j, $j \in J$ and their adjoint. Consider in $\mathbb{M}_n(\mathbb{C})$ matrices $(Q_i^{(n)})_{i \in I}$ admitting a joint limiting distribution as $n \to \infty$, and in i.i.d. Haar distributed $(U_j^{(n)})_{j \in J}$ and their adjoint. Evaluating A in these matrices in the obvious sense, we obtain a random matrix A_n and consider the Taylor expansion around zero of the function*

$$z \to n^{-2} \log E\big(\exp(zn^2 A_n)\big) = \sum_{q \geq 1} a_q^{(n)} z^q.$$

Then, for all q, $\lim_n a_q^{(n)}$ exists and depends only on the polynomial and the limiting distribution of $Q_i^{(n)}$.

In [24], we upgraded this result in the case where A_n is selfadjoint, and proved that there exists a *real* neighborhood of zero on which the convergence holds uniformly. The complex convergence remains a difficult problem, as a uniform understanding of the higher genus expansion must be obtained. Novak made a recent breakthrough in this direction, in the case of the HCIZ integral, see [65].

4.3. Random tensors

Let us revisit Question 1, under the assumption that U has *more structure*, i.e., less randomness. Our model is a tensor structure, namely, $U = U_1 \otimes \cdots \otimes U_D$ where $U_i \in \mathbb{M}_n(\mathbb{C})$ are i.i.d. In other words, we are interested in the symmetries under conjugation by elements of the group $\mathcal{U}_n^{\otimes D}$. The joint moments of a matrix are a complete invariant of global symmetry under $\mathcal{U}_n$-conjugation in $\mathbb{M}_n(\mathbb{C})$, however, for $U_1 \otimes \cdots \otimes U_D$-invariance in $\mathbb{M}_n(\mathbb{C})^{\otimes D}$, one needs more invariants, generated, for $\sigma_1, \ldots, \sigma_D \in S_k$, by

$$\operatorname{Tr}_{\sigma_1,\ldots,\sigma_D}(A) = \sum_{i_{11},\ldots,i_{Dk},j_{11},\ldots,j_{Dk}} A_{i_{11}\ldots i_{D1},j_{11}\ldots j_{D1}} \cdots A_{i_{1k}\ldots i_{Dk},j_{1k}\ldots j_{Dk}} \times \delta_{i_{11},j_{1\sigma_1(1)}} \cdots \delta_{i_{1k},j_{1\sigma_1(k)}} \cdots \delta_{i_{D1},j_{D\sigma_D(1)}} \cdots \delta_{i_{Dk},j_{D\sigma_D(k)}}. \tag{4.1}$$

In the case of higher tensors, thanks to the Weingarten calculus, we unveil many new inequivalent asymptotic regimes for higher order tensors. These questions are addressed in a series of projects with Gurau and Lionni, starting with [26]. We study the asymptotic expansion of the Fourier transform of the tensor valued Haar measure—a tensor extension of the Harish-Chandra integral to tensors, and considerably extend the single tensor case. Just as the HCIZ integral can be seen as a generating function for monotone Hurwitz numbers, which count certain weighted branched coverings of the 2-sphere, the integral studied in [26] leads to a generalization of monotone Hurwitz numbers, which count weighted branched coverings of a collection of 2-spheres that "touch" at one common nonbranch node.

4.4. Quantum information theory

Quantum information theory has been a powerful source of problems in random matrix theory in the last two decades, and their tensor structure has made it necessary to resort to moment techniques. The goal of this section is to elaborate on a few salient cases. One starting point is the paper [53] where the authors compute moments of the output of random quantum channels. We just recall here strictly necessary definitions, and refer to [36] for details. A *quantum channel* Φ is a linear map $\mathbb{M}_n(\mathbb{C}) \to \mathbb{M}_k(\mathbb{C})$ that preserves the nonnormalized trace, and that is completely positive, i.e., $\Phi \otimes \mathrm{Id}_l : \mathbb{M}_n \otimes \mathbb{M}_l(\mathbb{C}) \to \mathbb{M}_k(\mathbb{C}) \otimes \mathbb{M}_l(\mathbb{C})$ is positive for any integer l. It follows from Stinespring theorem that for any quantum channel, there exists an integer p and an isometry $U : \mathbb{C}^n \to \mathbb{C}^k \otimes \mathbb{C}^p$ such that $\Phi(X) = (\mathrm{Id}_k \otimes \mathrm{Tr}_p)UXU^*$.

The set of *density matrices* D_n consists in the selfadjoint matrices whose eigenvalues are nonnegative and whose trace is 1. For $A \in D_n$, we define its *von Neumann entropy* $H(A)$ as $\sum_{i=1}^n -\lambda_i(A)\log\lambda_i(A)$ with the convention that $0\log 0 = 0$ and the eigenvalues of A are $\lambda_1(A) \geq \cdots \geq \lambda_n(A)$. The *minimum output entropy* of a quantum channel Φ is defined as $H_{\min}(\Phi) = \min_{A\in D_n} H(\Phi(A))$, and a crucial question in QIT was whether one can find Φ_1, Φ_2 such that

$$H_{\min}(\Phi_1 \otimes \Phi_2) < H_{\min}(\Phi_1) + H_{\min}(\Phi_2).$$

For the statement, implications and background, we refer to [36]. An answer to this question was given in [52] and it relies on random methods, which motivates us to consider quantum channels obtained from Haar unitaries. A description of $\Phi(D_n)$ in some appropriate large n limit has been found in [9], and the minimum in the limit of the entropy was found in [10]. In the meantime, the image under tensor product of random channels $\Phi_1 \otimes \Phi_2$ of appropriate matrices (known as Bell states) had to be computed. To achieve this, we had to develop a graphical version of Weingarten calculus in [34].

We consider the case where k is a fixed integer, and $t \in (0,1)$ is a fixed number. For each n, we consider a random unitary matrix $U \in \mathbb{M}_{nk}(\mathbb{C})$, and a projection q_n of $\mathbb{M}_{nk}(\mathbb{C})$

of rank p_n such that $p_n/(nk) \sim t$ as $n \to \infty$. Our model of a random quantum channel is $\Phi : \mathbb{M}_{p_n}(\mathbb{C}) \to \mathbb{M}_n(\mathbb{C})$ given by $\Phi(X) = \mathrm{tr}_k(UXU^*)$, where $\mathbb{M}_{p_n}(\mathbb{C}) \simeq q_n \mathbb{M}_{nk}(\mathbb{C}) q_n$. By Bell we denote the Bell state on $\mathbb{M}_{p_n}(\mathbb{C})^{\otimes 2}$. In [34], we proved

Theorem 4.3. *Almost surely, as $n \to \infty$, the random matrix $\Phi \otimes \overline{\Phi}(\mathrm{Bell}) \in \mathbb{M}_{n^2}(\mathbb{C})$ has nonzero eigenvalues converging towards*

$$\gamma^{(t)} = \Big(t + \frac{1-t}{k^2}, \underbrace{\frac{1-t}{k^2}, \ldots, \frac{1-t}{k^2}}_{k^2-1 \text{ times}}\Big).$$

This result plays an important result in the understanding of phenomena underlying the subadditivity of the minimum output entropy, and relies heavily on Weingarten calculus, and in particular a graphical interpretation thereof. Much more general results in related areas of quantum information theory have been obtained in [22, 28, 35, 37, 38, 43].

5. UNIFORM ESTIMATES AND APPLICATIONS TO ANALYSIS

5.1. A motivating question

The previous sections show that when the degree of a polynomial is fixed, very precise asymptotics can be obtained in the limit of large dimension. For the purpose of analysis, an important question is whether such estimates hold uniformly. About 20 years ago, Gilles Pisier asked me the following question: given k i.i.d. Haar unitaries $U_1^{(n)}, \ldots, U_k^{(n)} \in \mathcal{U}_n$, what is the large dimension behavior of the real random variable

$$t_n = ||U_1^{(n)} + \cdots + U_k^{(n)}||_\infty,$$

where $||\cdot||_\infty$ stands for the operator norm? It follows from asymptotic freeness results that almost surely $\liminf t_n \geq 2\sqrt{k-1}$ as soon as $k \geq 2$. Setting $X_n = U_1^{(n)} + \cdots + U_k^{(n)}$, it would be in principle enough to estimate

$$\mathbb{E}\big(\mathrm{Tr}\big((X_n X_n^*)^{l(n)}\big)\big)$$

for $l(n) \gg \log n$. However, there are two major hurdles: (i) uniform estimates of Weingarten calculus would be needed; (ii) unlike in the multimatrix model case, the combinatorics grow exponentially and a direct moment approach is not possible. Both hurdles require developing specific tools, which we describe in the sequel.

One notion on which we rely heavily is that of *strong convergence*. Given a multiatrix model that admits a joint limiting distribution in Voiculescu's sense, we say that it *converges strongly* iff the operator norm of any polynomial P, evaluated in the matrices of the model—thus yielding the random matrix P_n—satisfies

$$\lim_n ||P_n|| = \lim_\ell \big(\lim_n n^{-1}\,\mathrm{Tr}\big((P_n P_n^*)^\ell\big)\big)^{(2\ell)^{-1}}.$$

In other words, the operator norm of any matrix model obtained from a noncommutative polynomial converges to the operator norm of the limiting object. Strong convergence was established in [50] in the case of Gaussian random matrices. Subsequently, the author and

Male solved the counterpart for Haar unitary matrices in [29], with no explicit speed of convergence. This result was refined further by Parraud [67] with explicit speeds of convergence, relying on ideas of [25]. The strongest result concerning strong convergence of random unitaries can be found in [14]:

Theorem 5.1. $(\overline{U}_i^{\otimes q_-} \otimes U_i^{\otimes q_+})_{i=1,\dots,d}$ *are strongly asymptotically free as* $n \to \infty$ *on the orthogonal of fixed point spaces.*

This means that strong asymptotic freeness does not hold at the sole level of the fundamental representation of $\mathcal{U}_n$, but with respect to any sequence of representation associated to a nontrivial (λ, μ). In other words, the only obstructions to strong freeness are the dimension one irreducible representations of $\mathcal{U}_n$. We need a *linearization step*, popularized by [50] to evaluate the norm of $\sum_{i=-d}^{d} a_i \otimes X_i^{(n)}$, where $X_{-i}^{(n)} = X_i^{(n)*}$ and $X_0^{(n)} = Id$. Although this first simplification step was sufficient to obtain strong convergence for i.i.d. GUE—i.e., matrices with high symmetries—thanks to analytic techniques, this turns out to be insufficient when one has to resort to moment methods. In [13], we initiated techniques based on a *operator version* of *nonbacktracking theory*, which we generalized in [14]. We outline one key feature here.

We consider $(b_1, \dots, b_l)$ elements in $\mathcal{B}(\mathcal{H})$ where $\mathcal{H}$ is a Hilbert space. We assume that the index set is endowed with an involution $i \mapsto i^*$ (and $i^{**} = i$ for all i). The *nonbacktracking operator* associated to the ℓ-tuple of matrices $(b_1, \dots, b_l)$ is the operator on $\mathcal{B}(\mathcal{H} \otimes \mathbb{C}^l)$ defined by

$$B = \sum_{j \neq i^*} b_j \otimes E_{ij}. \tag{5.1}$$

The following theorem allows leveraging moments techniques on linearization of noncommutative polynomials through the study of B:

Theorem 5.2. *Let* $\lambda \in \mathbb{C}$ *satisfy* $\lambda^2 \notin \cup_{i \in \{1,\dots,l\}} \operatorname{spec}(b_i b_{i^*})$. *Define the operator* A_λ *on* $\mathcal{H}$ *through*

$$A_\lambda = b_0(\lambda) + \sum_{i=1}^{\ell} b_i(\lambda), \quad b_i(\lambda) = \lambda b_i (\lambda^2 - b_{i^*} b_i)^{-1},$$

and

$$b_0(\lambda) = -1 - \sum_{i=1}^{\ell} b_i (\lambda^2 - b_{i^*} b_i)^{-1} b_{i^*}.$$

Then $\lambda \in \sigma(B)$ *if and only if* $0 \in \sigma(A_\lambda)$.

5.2. Centering and uniform Weingarten estimates

In order to use Theorem 5.2, one has to understand the spectral radius of the operator B and therefore, evaluate $\tau(B^T B^{*T})$ with T growing with the matrix dimension, and this can be done through moment methods as soon as we have uniform estimates on Weingarten functions. The first uniform estimate was obtained in [23] and had powerful applications to the study of area laws in mathematical physics, however, it was not sufficient for norm estimates, and it was superseded by [31]:

Theorem 5.3. *For any $\sigma \in S_k$ and $n > \sqrt{6}k^{7/4}$,*

$$\frac{1}{1-\frac{k-1}{n^2}} \leq \frac{n^{k+|\sigma|}\operatorname{Wg}(n,\sigma)}{\operatorname{Moeb}(\sigma)} \leq \frac{1}{1-\frac{6k^{7/2}}{n^2}}.$$

In addition, the left-hand side inequality is valid for any $n \geq k$.

This result already enables us to prove Theorem 5.1 in the case where $q_- \neq q_+$ because there are no fixed points there. Let us now outline how to tackle the case $q_- = q_+$, which is interesting because it has fixed points. To handle fixed points, we need to introduce the *centering* of a random variable X, namely $[X] = X - E(X)$. For a symbol $\varepsilon \in \{\cdot, -\}$ and $z \in \mathbb{C}$, we take the notation that $z^\varepsilon = z$ if $\varepsilon = \cdot$ and $z^\varepsilon = \overline{z}$ if $\varepsilon = -$. We want to compute, for $U = (U_{ij})$ Haar distributed on $\mathcal{U}_n$, expressions of the form $\mathbb{E}\prod_{t=1}^{T}[\prod_{l=1}^{k_t} U_{x_{tl}y_{tl}}^{\varepsilon_{tl}}]$ in a meaningful way. We can write a Weingarten formula:

$$\mathbb{E}\prod_{t=1}^{T}\left[\prod_{l=1}^{k_t} U_{x_{tl}y_{tl}}^{\varepsilon_{tl}}\right] = \sum_{\sigma,\tau\in P_2(k_1+\cdots+k_T)} \delta_{\sigma,x}\delta_{\tau,y}\operatorname{Wg}(\sigma,\tau;k_1,\ldots,k_T),$$

where the function Wg depends on the pairings and the partition. We say that a block of the partition $\{\{1,\ldots,k_1\},\ldots,\{k_1+\cdots+k_{T-1}+1,\ldots,k_1+\cdots+k_T\}\}$ is *lonesome* with respect to the pairing (σ,τ) iff the group generated by σ,τ stabilizes it. In **[14]**, we prove

Theorem 5.4. Wg *decays as n^{-k} where $k = (k_1+\cdots+k_T)/2 + d(\sigma,\tau) + 2\#$lonesome blocks, and this estimate is uniform on $k \sim \operatorname{Poly}(n)$.*

This theorem, together a comparison with Gaussian vectors, allows proving Theorem 5.1.

6. PERSPECTIVES

Understanding better how to integrate over compact groups is a fascinating problem which is connected to many questions in various branches of mathematics and other scientific fields. We conclude this manuscript by a brief and completely subjective list of perspectives:

(1) *Uniform measures on (quantum) symmetric spaces*
Viewing a group as a compact manifold, can one extend the Weingarten calculus to other surfaces? Some substantial work has been done algebraically in this direction by Matsumoto **[60]** in the case of symmetric spaces, see as well **[42]** for the asymptotic version. It would be interesting to study extensions of Matsumoto's results for compact quantum symmetric spaces.

(2) *Surfaces and geometric group theory*
An important observation by Magee and Puder is that if G is a compact subgroup of $\mathcal{U}_n$, the Haar measure on G^k yields a random representation of the free group F_k on $\mathcal{U}_n$ whose law is invariant under *outer automorphisms* of F_k.

This motivated them to compute, in [59], the expectation of the trace of nontrivial words in $(U_1, \dots, U_k) \in \mathcal{U}_n{}^k$. In addition to refining known asymptotics, they used the properties of the Weingarten function to solve nontrivial problems about the orbits of F_k under the action by its outer conjugacy group. In a different vein, Magee has very recent achieved a breakthrough by obtaining the first steps of Weingarten calculus for representations of some one relator groups [57,58].

(3) *Other applications to representation theory*
The problem of calculating Weingarten functions on $\mathrm{SU}(n)$ efficiently is hard, and even more so when the degree is high in comparison to n. A striking example is $\int_{U \in \mathrm{SU}(n)} \prod_{i,j=1}^{n} u_{ij}$. It was established in [55] that proving that this integral is nonzero is equivalent to the Alon–Tarski conjecture.
More generally, this raises the question of computing efficiently integrals of high degree on classical groups (typically, of degree $\geq n$ or $\geq n^2$). Weingarten calculus as developed in this manuscript is not well adapted to this task. Some results in this direction have been obtained by Novak [65] the author, and Cioppa.

(4) *More tensors and norm estimates*
In [13,14], we obtained strong convergence for an arbitrary finite number of tensors of random unitaries or random permutations. It turns out that the result can be relaxed a bit to allow the number of legs to vary slowly to infinity as the dimension of the group goes to infinity. This points to a double limit problem, and we wonder to which extent the number of legs of tensors and the size of the matrix can be independent. In the extreme case, could strong freeness hold for a given finite group but a number of tensors tending to infinity? Many variants of this problem exist, e.g., taking i.i.d. copies of unitaries instead of the same.
Likewise, an important question is the behavior of $U_i \otimes U_j$ for arbitrary indices—not only $i = j$ as in [13,14]. As observed by Hayes in [54], this is a possible approach towards the Peterson–Thom conjecture in operator algebras, and it seems plausible that Weingarten calculus could help to solve this problem.

(5) *Maximizing functionals over groups*
Given a polynomial function $f : G \to \mathbb{C}$, finding $m = \max_{U \in G} |f(U)|$ could provide approaches to various conjectures in analysis or algebra. In general, finding the argmax is a problem intricately linked to the conjectures, and Haar integration could yield nonconstructive approaches. Indeed,

$$l^{-1} \log \int \left(f(U)\overline{f(U)}\right)^l d\mu_G(U) \sim 2 \log m,$$

and the left-hand side could in principle be approached with Weingarten calculus. Let us mention the example of the Hadamard conjecture. It states that for any $4/n$, there exists an orthogonal matrix in $\mathbb{M}_n(\mathbb{R})$ whose entries are ± 1.

An approach to this problem would be to show that the minimum of the polynomial function $f(U) = \sum_{ij} u_{ij}^4$ on $\mathcal{O}_n$ is 1. We refer to [7] for attempts with Weingarten calculus. We also believe that some important problems in operator algebra could be approached that way (e.g., the problem of the nonexistence of hyperlinear group).

ACKNOWLEDGMENTS

It has always been extremely stimulating to work with people from very diverse backgrounds: coauthors, graduate students, as well as postdoctoral fellows. I want to thank them all for our collaborations. I would like to thank Charles Bordenave, Mike Brannan, Luca Lionni, Sho Matsumoto, Akihiro Miyagawa, Ion Nechita, and Jonathan Novak for reading carefully my manuscript and for many suggestions of improvements.

FUNDING

This work was partially supported JSPS Kakenhi 17H04823, 20K20882, 21H00987. Most of the original results presented in this note have been supported by JSPS Kakenhi, NSERC grants and ANR grants, while the author was working at either of the following places: CNRS (Lyon 1), the University of Ottawa, or Kyoto University.

REFERENCES

[1] T. Banica, Le groupe quantique compact libre $U(n)$. *Comm. Math. Phys.* **190** (1997), no. 1, 143–172.

[2] T. Banica, S. T. Belinschi, M. Capitaine, and B. Collins, Free Bessel laws. *Canad. J. Math.* **63** (2011), no. 1, 3–37.

[3] T. Banica, J. Bichon, and B. Collins, Quantum permutation groups: a survey. In *Noncommutative harmonic analysis with applications to probability*, pp. 13–34, Banach Center Publ. 78, Polish Acad. Sci. Inst. Math, Warsaw, 2007.

[4] T. Banica and B. Collins, Integration over compact quantum groups. *Publ. Res. Inst. Math. Sci.* **43** (2007), no. 2, 277–302.

[5] T. Banica and B. Collins, Integration over quantum permutation groups. *J. Funct. Anal.* **242** (2007), no. 2, 641–657.

[6] T. Banica and B. Collins, Integration over the Pauli quantum group. *J. Geom. Phys.* **58** (2008), no. 8, 942–961.

[7] T. Banica, B. Collins, and J.-M. Schlenker, On polynomial integrals over the orthogonal group. *J. Combin. Theory Ser. A* **118** (2011), no. 3, 778–795.

[8] T. Banica, S. Curran, and R. Speicher, Stochastic aspects of easy quantum groups. *Probab. Theory Related Fields* **149** (2011), no. 3–4, 435–462.

[9] S. Belinschi, B. Collins, and I. Nechita, Eigenvectors and eigenvalues in a random subspace of a tensor product. *Invent. Math.* **190** (2012), no. 3, 647–697.

[10] S. T. Belinschi, B. Collins, and I. Nechita, Almost one bit violation for the additivity of the minimum output entropy. *Comm. Math. Phys.* **341** (2016), no. 3, 885–909.

[11] P. Biane, Representations of unitary groups and free convolution. *Publ. Res. Inst. Math. Sci.* **31** (1995), no. 1, 63–79.

[12] P. Biane, Representations of symmetric groups and free probability. *Adv. Math.* **138** (1998), no. 1, 126–181.

[13] C. Bordenave and B. Collins, Eigenvalues of random lifts and polynomials of random permutation matrices. *Ann. of Math. (2)* **190** (2019), no. 3, 811–875.

[14] C. Bordenave and B. Collins, Strong asymptotic freeness for independent uniform variables on compact groups associated to non-trivial representations. 2020, arXiv:2012.08759.

[15] N. Bourbaki, *Integration. II. Chapters 7–9.* Elem. Math. (Berlin), Springer, Berlin, 2004. Translated from the 1963 and 1969 French originals by S. K. Berberian.

[16] M. Brannan and B. Collins, Dual bases in Temperley–Lieb algebras, quantum groups, and a question of Jones. *Quantum Topol.* **9** (2018), no. 4, 715–748.

[17] E. Brézin and D. J. Gross, The external field problem in the large N limit of QCD. *Phys. Lett. B* **97** (1980), no. 1, 120–124.

[18] P. W. Brouwer and C. W. J. Beenakker, Diagrammatic method of integration over the unitary group, with applications to quantum transport in mesoscopic systems. *J. Math. Phys.* **37** (1996), no. 10, 4904–4934.

[19] T. Ceccherini-Silberstein, F. Scarabotti, and F. Tolli, *Representation theory of the symmetric groups.* Cambridge Stud. Adv. Math. 121, Cambridge University Press, Cambridge, 2010. The Okounkov–Vershik approach, character formulas, and partition algebras.

[20] B. Collins, Moments and cumulants of polynomial random variables on unitary groups, the Itzykson–Zuber integral, and free probability. *Int. Math. Res. Not.* **17** (2003), 953–982.

[21] B. Collins, A. Dahlqvist, and T. Kemp, The spectral edge of unitary Brownian motion. *Probab. Theory Related Fields* **170** (2018), no. 1–2, 49–93.

[22] B. Collins, M. Fukuda, and I. Nechita, On the convergence of output sets of quantum channels. *J. Operator Theory* **73** (2015), no. 2, 333–360.

[23] B. Collins, C. E. González-Guillén, and D. Pérez-García, Matrix product states, random matrix theory and the principle of maximum entropy. *Comm. Math. Phys.* **320** (2013), no. 3, 663–677.

[24] B. Collins, A. Guionnet, and E. Maurel-Segala, Asymptotics of unitary and orthogonal matrix integrals. *Adv. Math.* **222** (2009), no. 1, 172–215.

[25] B. Collins, A. Guionnet, and F. Parraud, On the operator norm of non-commutative polynomials in deterministic matrices and iid GUE matrices. 2019, arXiv:1912.04588. Accepted in *Camb. J. Math.*

[26] B. Collins, R. Gurau, and L. Lionni, The tensor Harish-Chandra–Itzykson–Zuber integral I: Weingarten calculus and a generalization of monotone Hurwitz numbers. 2020, arXiv:2010.13661.

[27] B. Collins, T. Hasebe, and N. Sakuma, Free probability for purely discrete eigenvalues of random matrices. *J. Math. Soc. Japan* **70** (2018), no. 3, 1111–1150.

[28] B. Collins, P. Hayden, and I. Nechita, Random and free positive maps with applications to entanglement detection. *Int. Math. Res. Not. IMRN* **3** (2017), 869–894.

[29] B. Collins and C. Male, The strong asymptotic freeness of Haar and deterministic matrices. *Ann. Sci. Éc. Norm. Supér. (4)* **47** (2014), no. 1, 147–163.

[30] B. Collins and S. Matsumoto, On some properties of orthogonal Weingarten functions. *J. Math. Phys.* **50** (2009), no. 11, 113516, 14 pp.

[31] B. Collins and S. Matsumoto, Weingarten calculus via orthogonality relations: new applications. *ALEA Lat. Am. J. Probab. Math. Stat.* **14** (2017), no. 1, 631–656.

[32] B. Collins, S. Matsumoto, and J. Novak, The Weingarten calculus. 2021, arXiv:2109.14890.

[33] B. Collins, J. A. Mingo, P. Śniady, and R. Speicher, Second order freeness and fluctuations of random matrices. III. Higher order freeness and free cumulants. *Doc. Math.* **12** (2007), 1–70.

[34] B. Collins and I. Nechita, Random quantum channels I: Graphical calculus and the Bell state phenomenon. *Comm. Math. Phys.* **297** (2010), no. 2, 345–370.

[35] B. Collins and I. Nechita, Gaussianization and eigenvalue statistics for random quantum channels (III). *Ann. Appl. Probab.* **21** (2011), no. 3, 1136–1179.

[36] B. Collins and I. Nechita, Random matrix techniques in quantum information theory. *J. Math. Phys.* **57** (2016), no. 1, 015215, 34 pp.

[37] B. Collins, I. Nechita, and D. Ye, The absolute positive partial transpose property for random induced states. *Random Matrices Theory Appl.* **1** (2012), no. 3, 1250002, 22 pp.

[38] B. Collins, I. Nechita, and K. Życzkowski, Area law for random graph states. *J. Phys. A* **46** (2013), no. 30, 305302, 18 pp.

[39] B. Collins, J. Novak, and P. Śniady, Semiclassical asymptotics of $\mathrm{GL}_N(C)$ tensor products and quantum random matrices. *Selecta Math. (N.S.)* **24** (2018), no. 3, 2571–2623.

[40] B. Collins and P. Śniady, Integration with respect to the Haar measure on unitary, orthogonal and symplectic group. *Comm. Math. Phys.* **264** (2006), no. 3, 773–795.

[41] B. Collins and P. Śniady, Representations of Lie groups and random matrices. *Trans. Amer. Math. Soc.* **361** (2009), no. 6, 3269–3287.

[42] B. Collins and M. Stolz, Borel theorems for random matrices from the classical compact symmetric spaces. *Ann. Probab.* **36** (2008), no. 3, 876–895.

[43] B. Collins, Z. Yin, and P. Zhong, The PPT square conjecture holds generically for some classes of independent states. *J. Phys. A* **51** (2018), no. 42, 425301, 19 pp.

[44] B. de Wit and G. 't Hooft, Nonconvergence of the $1/N$ expansion for $\mathrm{SU}(N)$ gauge fields on a lattice. *Phys. Lett. B* **69** (1977), no. 1, 61–64.

[45] P. Diaconis and M. Shahshahani, On the eigenvalues of random matrices. J. Appl. Probab. **31A** (1994), 49–62.

[46] R. Goodman and N. R. Wallach, *Symmetry, representations, and invariants*. Grad. Texts in Math. 255, Springer, Dordrecht, 2009.

[47] I. P. Goulden, M. Guay-Paquet, and J. Novak, Monotone Hurwitz numbers in genus zero. *Canad. J. Math.* **65** (2013), no. 5, 1020–1042.

[48] I. P. Goulden, M. Guay-Paquet, and J. Novak, Polynomiality of monotone Hurwitz numbers in higher genera. *Adv. Math.* **238** (2013), 1–23.

[49] I. P. Goulden, M. Guay-Paquet, and J. Novak, Monotone Hurwitz numbers and the HCIZ integral. *Ann. Math. Blaise Pascal* **21** (2014), no. 1, 71–89.

[50] U. Haagerup and S. Thorbjørnsen, A new application of random matrices: $\mathrm{Ext}(C^*_{\mathrm{red}}(F_2))$ is not a group. *Ann. of Math. (2)* **162** (2005), no. 2, 711–775.

[51] Harish-Chandra, Differential operators on a semisimple Lie algebra. *Amer. J. Math.* **79** (1957), 87–120.

[52] M. B. Hastings, Superadditivity of communication capacity using entangled inputs. *Nat. Phys.* **5** (2009), no. 4, 255–257.

[53] P. Hayden and A. Winter, Counterexamples to the maximal p-norm multiplicity conjecture for all $p > 1$. *Comm. Math. Phys.* **284** (2008), no. 1, 263–280.

[54] B. Hayes, A random matrix approach to the Peterson–Thom conjecture. 2020, arXiv:2008.12287.

[55] S. Kumar and J. M. Landsberg, Connections between conjectures of Alon–Tarsi, Hadamard–Howe, and integrals over the special unitary group. *Discrete Math.* **338** (2015), no. 7, 1232–1238.

[56] T. Lévy, Schur–Weyl duality and the heat kernel measure on the unitary group. *Adv. Math.* **218** (2008), no. 2, 537–575.

[57] M. Magee, Random unitary representations of surface groups I: asymptotic expansions. 2021, arXiv:2101.00252.

[58] M. Magee, Random unitary representations of surface groups II: the large n limit. 2021, arXiv:2101.03224.

[59] M. Magee and D. Puder, Matrix group integrals, surfaces, and mapping class groups I: $U(n)$. *Invent. Math.* **218** (2019), no. 2, 341–411.

[60] S. Matsumoto, Weingarten calculus for matrix ensembles associated with compact symmetric spaces. *Random Matrices Theory Appl.* **2** (2013), no. 2, 1350001, 26 pp.

[61] S. Matsumoto and J. Novak, Jucys–Murphy elements and unitary matrix integrals. *Int. Math. Res. Not. IMRN* **2** (2013), 362–397.

[62] C. McSwiggen, A new proof of Harish-Chandra's integral formula. *Comm. Math. Phys.* **365** (2019), no. 1, 239–253.

[63] J. A. Mingo and R. Speicher, *Free probability and random matrices*. Fields Inst. Monogr. 35, Springer, New York; Fields Institute for Research in Mathematical Sciences, Toronto, ON, 2017.

[64] A. Nica and R. Speicher, *Lectures on the combinatorics of free probability*. London Math. Soc. Lecture Note Ser. 335, Cambridge University Press, Cambridge, 2006.

[65] J. Novak, On the complex asymptotics of the HCIZ and BGW integrals. 2020, arXiv:2006.04304.

[66] K. H. O'Brien and J. B. Zuber, A note on $U(N)$ integrals in the large N limit. *Phys. Lett. B* **144** (1984), no. 5–6, 407–408.

[67] F. Parraud, On the operator norm of non-commutative polynomials in deterministic matrices and iid Haar unitary matrices. 2020, arXiv:2005.13834. Accepted in PTRF.

[68] E. M. Rains, Combinatorial properties of Brownian motion on the compact classical groups. *J. Theoret. Probab.* **10** (1997), no. 3, 659–679.

[69] D. Stanford, Z. Yang, and S. Yao, Subleading Weingartens. 2021, arXiv:2107.10252.

[70] D. Voiculescu, A strengthened asymptotic freeness result for random matrices with applications to free entropy. *Int. Math. Res. Not.* **1** (1998), 41–63.

[71] D. V. Voiculescu, K. J. Dykema, and A. Nica, *A noncommutative probability approach to free products with applications to random matrices, operator algebras and harmonic analysis on free groups*. Free random variables, CRM Monogr. Ser. 1, American Mathematical Society, Providence, RI, 1992.

[72] S. Wang, Free products of compact quantum groups. *Comm. Math. Phys.* **167** (1995), no. 3, 671–692.

[73] D. Weingarten, Asymptotic behavior of group integrals in the limit of infinite rank. *J. Math. Phys.* **19** (1978), no. 5, 999–1001.

[74] S. L. Woronowicz, Compact matrix pseudogroups. *Comm. Math. Phys.* **111** (1987), no. 4, 613–665.

[75] S. L. Woronowicz, Tannaka–Kreĭn duality for compact matrix pseudogroups. Twisted SU(N) groups. *Invent. Math.* **93** (1988), no. 1, 35–76.

BENOÎT COLLINS

Mathematics Department, Kyoto University, Kyoto, Japan, collins@math.kyoto-u.ac.jp

This Page Intentionally Left Blank

ANALYSIS ON SIMPLE LIE GROUPS AND LATTICES

MIKAEL DE LA SALLE

ABSTRACT

We present a simple tool to perform analysis with groups such as $\mathrm{SL}_n(\mathbf{R})$ and $\mathrm{SL}_n(\mathbf{Z})$, that has been introduced by Vincent Lafforgue in his study of nonunitary representations and strong property (T), in connection with the Baum–Connes conjecture. It has been later applied in various contexts: operator algebras, Fourier analysis, geometry of Banach spaces, and dynamics. The idea is to first restrict to compact subgroups and then exploit how they sit inside the whole group.

MATHEMATICS SUBJECT CLASSIFICATION 2020

Primary 22E46; Secondary 46B28, 46B20, 46L07, 43A75, 43A50

KEYWORDS

Rank zero reduction, higher-rank lattices, Banach space representations, group von Neumann algebras, Fourier synthesis, expander graphs

ICM INTERNATIONAL CONGRESS OF MATHEMATICIANS 2022 JULY 6–14

Proc. Int. Cong. Math. 2022, Vol. 4, pp. 3166–3188
DOI 10.4171/ICM2022/160

Published by EMS Press

This text is devoted to the presentation of a single idea that has been useful to answer several analysis questions on higher rank simple Lie groups and lattices. This idea originates from Vincent Lafforgue's work [43], and can be summarized as *rank* 0 *reduction*.

One of the basic tools to study Lie algebras is that of $\mathfrak{sl}_2$ triples. In the context a semisimple Lie groups, it is often used in the following form: to understand a possibly complicated Lie group, one restricts to its subgroups locally isomorphic to $SL_2(\mathbf{R})$ (and there are plenty by the Jacobson-Morozov theorem), the simplest noncompact semisimple Lie group. This idea, that we could call *rank* 1 *reduction* because $SL_2(\mathbf{R})$ has rank 1, can be powerful. For example, it allows obtaining precise information on the unitary representations of higher rank simple Lie groups [34, 57]. But, as we shall see later in this text, there are situations where rank 1 reduction is not efficient. Rank 2 reduction has also become a standard tool in the study of higher rank simple Lie groups, as every real simple Lie group of rank ≥ 2 contains a subgroup locally isomorphic to one of the rank 2 groups $SL_3(\mathbf{R})$ of $SP_2(\mathbf{R})$. See, for example, [2, I.1.6].

Rank 0 reduction, the subject of this survey, consists in studying a Lie group through its compact subgroups. The idea is to first restrict to the compact subgroups of the Lie group G and do analysis there. It is perhaps surprising that there are nontrivial general things to say (see Proposition 1.1). And then, in a second step, by analyzing the relative positions of the various cosets of compact groups in G, it is possible to promote the local phenomena that have been discovered in the first step to global phenomena in G.

In the whole text, except for the brief and last Section 5, we only consider for simplicity the Lie group $SL_3(\mathbf{R})$ and its subgroup $SL_3(\mathbf{Z})$. In the first section, we illustrate rank 0 reduction in the simplest meaningful setting of unitary representations of $SL_3(\mathbf{R})$: we shall see that this idea provides a new proof of Kazhdan's celebrated theorem that $SL_3(\mathbf{R})$ has property (T). In the next sections we show several other applications of this idea for $SL_3(\mathbf{R})$ or $SL_3(\mathbf{Z})$, each time with a brief history of the problems. In Section 2, devoted to Fourier analysis for noncommutative groups, we explain how this same idea can show that Fourier synthesis (the reconstruction of a "function" from its Fourier series) is in a way impossible for $SL_3(\mathbf{R})$ and $SL_3(\mathbf{Z})$. We then interpret these Fourier analysis statements in terms of approximation properties of the von Neumann algebra of $SL_3(\mathbf{Z})$ and its noncommutative L_p spaces. Section 3 is devoted to strong property (T): we study nonunitary representations of $SL_3(\mathbf{R})$ and $SL_3(\mathbf{Z})$ on Hilbert spaces and see that the same idea allows proving some form of property (T) for them. Applications of strong property (T) are also described, in particular we explain how strong property (T) appears as a key tool in the resolution by Brown, Fisher, and Hurtado of Zimmer's conjecture for group actions of high rank lattices on manifolds of small dimension. Section 4 is devoted to group actions on Banach spaces: we investigate how much of strong property (T) can be proved for representations on more general Banach spaces. Finally, in Section 5 we survey how these ideas have been used for other semisimple Lie groups or algebraic groups over other local fields.

1. A PROOF OF PROPERTY (T) FOR $\mathrm{SL}_3(\mathbf{R})$

Throughout this text, by *representation* of a locally compact group G on a Banach space X, we will always mean a group homomorphism π from G to the group of invertible continuous linear maps on X that is continuous for the strong operator topology: for every $\xi \in X$, $g \mapsto \pi(g)\xi$ is continuous. A unitary representation is when π takes values in the unitary group of a Hilbert space.

A topological group G has Kazhdan's property (T) whenever the trivial representation is isolated in its unitary dual for the Fell topology. This means that every unitary representation π of G on a Hilbert space $\mathcal{H}$ which almost has invariant vectors (i.e., there is a net ξ_i of unit vectors in $\mathcal{H}$ such that $\lim_i \|\pi(g)\xi_i - \xi_i\| = 0$ uniformly on compact subsets of G), has a nonzero invariant vector. Because of its numerous applications, property (T) has become a central concept in many areas of mathematics such as geometric and analytic group theory, operator algebras, ergodic theory, etc; see **[2]**.

The purpose of this introductory section is to give a detailed proof of Kazhdan's celebrated theorem **[35]** that the group $\mathrm{SL}_3(\mathbf{R})$ has property (T). We do not give one of the classical proofs (which rely in a way or another on the pair $\mathrm{SL}_2 \subset \mathrm{SL}_3$) but a proof due to Lafforgue **[43]** (which relies on the pair $\mathrm{SO}(3) \subset \mathrm{SL}_3(\mathbf{R})$). We denote $G = \mathrm{SL}_3(\mathbf{R})$, $K = \mathrm{SO}(3) \subset G$ being the maximal compact subgroup. Then the polar decomposition asserts that every element of G can be written as a product $g = kak'$ for $k, k' \in K$ and a diagonal matrix a with positive entries in nonincreasing order. In other words, it identifies the double classes $K\backslash G/K$ with the *Weyl chamber* $\Lambda = \{(r,s,t) \in \mathbf{R}^3, r \geq s \geq t, r+s+t = 0\}$ via the identification of (r,s,t) with the class $KD(r,s,t)K$ of

$$D(r,s,t) = \begin{pmatrix} e^r & 0 & 0 \\ 0 & e^s & 0 \\ 0 & 0 & e^t \end{pmatrix}.$$

We also introduce the subgroup $U \simeq \mathrm{SO}(2) \subset K$ of block-diagonal matrices

$$U = \left\{ \begin{pmatrix} 1 & 0 & 0 \\ 0 & * & * \\ 0 & * & * \end{pmatrix} \right\} \cap K.$$

The double classes $U\backslash K/U$ are then parametrized by $[-1,1]$, the parametrization being given by $UkU \mapsto k_{1,1}$. For every $\delta \in [-1,1]$, let k_δ denote a representative of the corresponding double class, for example,

$$k_\delta = \begin{pmatrix} \delta & -\sqrt{1-\delta^2} & 0 \\ \sqrt{1-\delta^2} & \delta & 0 \\ 0 & 0 & 1 \end{pmatrix}.$$

Step 1. The first step is the observation that U-biinvariant matrix coefficients of unitary representations of K are Hölder $\frac{1}{2}$-continuous on $(-1,1)$. In particular, we have

Proposition 1.1. *For every unitary representation π of K on a Hilbert space $\mathcal{H}$ and every $\pi(U)$-invariant unit vectors $\xi, \eta \in \mathcal{H}$, we have*

$$\left|\langle \pi(k_\delta)\xi, \eta\rangle - \langle \pi(k_0)\xi, \eta\rangle\right| \leq 2\sqrt{|\delta|}.$$

Proof. By the Peter–Weyl theorem, it is enough to prove the inequality for the irreducible representations of SO(3). For the nth irreducible representation of SO(3) (the degree n spherical harmonics), the quantity $\langle \pi(k_\delta)\xi, \eta\rangle$ is equal to $\pm P_n(\delta)$, the value at δ of the nth Legendre polynomial normalized by $P_n(1) = 1$, see, for example, [22]. So we have to prove that $\sup_n |P_n(\delta) - P_n(0)| \le 2\sqrt{|\delta|}$. By bounding

$$\left|P_n(\delta) - P_n(0)\right| \le \min\Bigl(\left|P_n(0)\right| + \left|P_n(\delta)\right|, |\delta| \max_{t\in[0,1]} \left|P_n'(t\delta)\right|\Bigr)$$

and using the Bernstein inequality $|P_n(x)| \le \min(1, \sqrt{\frac{2}{\pi n}}(1-x^2)^{-\frac{1}{4}})$ [70, THEOREM 7.3.3] and the formula

$$(1-x^2)P_n'(x) = -nxP_n(x) + nP_{n-1}(x)$$

expressing P_n' in terms of P_n and P_{n-1}, one deduces the proposition. ■

Step 2. The next step is to deduce regularity properties of K-biinvariant matrix coefficients of unitary representations of G. The proof is short, but there are important things happening.

Proposition 1.2. *Let π be a unitary representation of G on a Hilbert space $\mathcal{H}$, and ξ, η be $\pi(K)$-invariant unit vectors. Then for every $g_1, g_2 \in G$,*

$$\left|\langle \pi(g_1)\xi, \eta\rangle - \langle \pi(g_2)\xi, \eta\rangle\right| \le 100 \min\bigl(\|g_1\|, \|g_1^{-1}\|, \|g_2\|, \|g_2^{-1}\|\bigr)^{-\frac{1}{2}}.$$

Proof. We may regard the matrix coefficient $g \mapsto \langle \pi(g)\xi, \eta\rangle$ as the map $c\colon \Lambda \to \mathbf{C}$ given by $c(r,s,t) = \langle \pi(D(r,s,t))\xi, \eta\rangle$. For every $(r,s,t) \in \Lambda$, the matrix $D(-t, \frac{t}{2}, \frac{t}{2})$ commutes with U, so the unit vectors $\pi(D(-t, \frac{t}{2}, \frac{t}{2}))\xi$ and $\pi(D(t, -\frac{t}{2}, -\frac{t}{2}))\eta$ are U-invariant. We can therefore apply Proposition 1.1. With $\delta = \frac{\sinh(r+\frac{t}{2})}{\sinh(-\frac{3t}{2})}$, we obtain

$$\left|c(r,s,t) - c\left(-\frac{t}{2}, -\frac{t}{2}, t\right)\right| \le 2\sqrt{\delta} \le 2e^{\frac{r}{2}+t} = 2e^{-\frac{r}{2}-s}. \tag{1.1}$$

Applying this to the representation $g \mapsto \pi((g^T)^{-1})$, we also obtain

$$\left|c(r,s,t) - c\left(r, -\frac{r}{2}, -\frac{r}{2}\right)\right| \le 2e^{\frac{t}{2}+s}. \tag{1.2}$$

These two inequalities are best understood on a picture (see Figure 1): (1.1) expresses that the amplitude of c is very small (exponentially small in the distance to the origin) on lines of slope $-\frac{1}{2}$ in the region $s \ge -1$, whereas (1.2) expresses that the amplitude of c is very small on vertical lines in the region $s \le 0$. We can join any two points of the Weyl chamber by a zigzag path as in Figure 1, and combining these estimates we deduce

$$\left|c(r,s,t) - c(r',s',t')\right| \le 100 \max\bigl(e^{-\frac{\min(r,-t)}{2}}, e^{-\frac{\min(r',-t')}{2}}\bigr),$$

which is exactly the proposition. ■

If π is a representation of G on a Hilbert space $\mathcal{H}$ and $g \in G$, let us denote by $\pi(KgK)$ the bounded operator on $\mathcal{H}$ mapping ξ to $\iint_{K\times K} \pi(kgk')\xi \, dk\, dk'$, where the integrals are with respect to the Haar probability measure on K. We also say that a vector $\xi \in \mathcal{H}$ is *harmonic* if $\pi(KgK)\xi = \xi$ for every $g \in G$ (the terminology is justified by [24]).

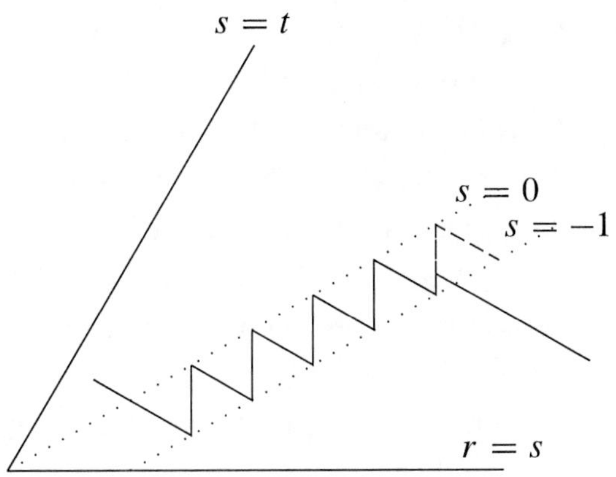

FIGURE 1
The zigzag path in the Weyl chamber Λ.

Corollary 1.3. *If π is a unitary representation of G, then*

$$\left\|\pi(KgK) - P\right\|_{B(\mathcal{H})} \leq 100 \min\left(\|g\|, \|g^{-1}\|\right)^{-\frac{1}{2}}, \tag{1.3}$$

where P is a projection on the space of harmonic vectors.

Proof. Taking the supremum over all K-invariant unit vectors in Proposition 1.2, we obtain that $(\pi(KgK))_{g \in G}$ is Cauchy in $B(\mathcal{H})$, and that its limit P satisfies (1.3). Then P is clearly the identity on the space of harmonic vectors, and the following computation shows that the image of P is made of harmonic vectors:

$$\pi(KgK)P\xi dk = \lim_{g' \to \infty} \pi(KgK)\pi(Kg'K)\xi = \lim_{g' \to \infty} \int_K \pi(Kgkg'K)\xi = P\xi.$$

So P is indeed a projection (and even *the* orthogonal projection) on the space of harmonic vectors. ∎

Step 3. The last step, which can be summarized as *harmonic implies invariant*, is the conclusion of the proof of property (T). Let π be a unitary representation of G with almost invariant vectors. Evaluating $\pi(KgK)$ at almost invariant vectors, we see that $\pi(KgK)$ has norm 1 for every $g \in G$. The limit P in Corollary 1.3 therefore also has norm 1, which means that there is a nonzero harmonic vector ξ. For every $g \in G$, the equality $\xi = \iint \pi(kgk')\xi dk dk'$ expresses ξ as an average of vectors of the same norm $\pi(kg'k)\xi$. By strict convexity of Hilbert spaces, we have that $\pi(kgk')\xi = \xi$ for every k, k', in particular ξ is $\pi(g)$-invariant. So ξ is a nonzero invariant vector. This proves that $G = \mathrm{SL}_3(\mathbf{R})$ has property (T).

For further reference, we can rephrase Proposition 1.1 in terms of the operators $T_\delta \colon L_2(S^2) \to L_2(S^2)$ defined by $T_\delta f(x)$ is the average of f on the circle $\{y \in S^2 | \langle x, y\rangle = \delta\}$. Identifying S^2 with $\mathrm{SO}(3)/\mathrm{SO}(2)$, we see that Proposition 1.2 is equivalent to

$$\|T_\delta - T_0\| \leq 2\sqrt{|\delta|}. \tag{1.4}$$

1.1. Comments on the proofs

The distinct steps in the proof of property (T) for $SL_3(\mathbf{R})$ have different nature. The first step is analytic and deals with harmonic analysis on a compact group, and more precisely on the pair $(U \subset K)$ of compact groups. The second step is geometric/combinatorial. What happens is that one studies the various K-equivariant embeddings of the sphere S^2 (identified with $K/U = SO(3)/SO(2)$) into the symmetric space $G/K = SL_3(\mathbf{R})/SO(3)$. The relative position between a pair of such embeddings gives rise to an embedding of $U \backslash K/U = [-1, 1]$ inside the Weyl chamber $K\backslash G/K$. There are three types of such embeddings: the segments in Figure 1; others, that are of no use for us, are segments parallel to the line $s = 0$. Combining these embeddings allows us to explore the whole Weyl chamber of $SL(3, \mathbf{R})$, and to prove that K-biinvariant matrix coefficients of unitary representations of G are Hölder $\frac{1}{2}$-continuous away from the boundary of Λ. The crucial fact that leads to (1.1) and (1.2) expresses that the interesting embeddings are exponentially distorted in the distance to the origin in the Weyl chamber, and hence that this exploration allows us to escape to infinity *in finite time*. The last step is rather obvious, but will become much more involved later.

It is informative to make similar computations for rank 1 simple Lie groups G which contain a subgroup isomorphic to SO(3) (for example, for SO(3, 1)). In that case, one gets also lots of embeddings of $[-1, 1]$ inside the Weyl chamber $[0, \infty)$ of G, and also enough to explore the whole Weyl chamber and prove Hölder $\frac{1}{2}$-regularity in the interior of the Weyl chamber. The difference (and the reason why this does not create a contradiction by proving that SO(3, 1) has property (T)!) is that these embeddings are almost isometric, and so it takes an infinite time to explore the whole Weyl chamber. In a sense, only the segments going straight but slowly to infinity exist in rank one. Those that are used in the zigzag argument, which take less direct routes but are faster, only appear in higher rank.

The fact that all the analysis is done at the level of the compact groups U, K is very important, because harmonic analysis for compact groups is much better understood than for arbitrary groups (see, for example, the very easy results in Lemmas 2.2, 3.1, 3.6). This is what permits using a similar approach for other objects than coefficients of unitary representations, and proving rigidity results in various other linear settings. This is the content of the remaining of this survey.

1.2. Induction and property (T) for $SL_3(\mathbf{Z})$

For later reference, we recall the classical argument why property (T) for $SL_3(\mathbf{R})$ implies property (T) for $SL_3(\mathbf{Z})$. We uses Minkowski's theorem that $SL_3(\mathbf{Z})$ is a lattice in $SL_3(\mathbf{R})$. A lattice in a locally compact group G is a discrete subgroup such that the quotient G/Γ carries a finite G-invariant Borel probability measure. Equivalently, there is a Borel probability measure μ on G whose image in G/Γ is G-invariant. The proof that property (T) passes to lattices uses induction of representations. If π is a unitary representation of Γ on a Hilbert space $\mathcal{H}$, the space of the induced representation is the space $L_2(G, \mu; \mathcal{H})^\Gamma$ of measurable functions $G \to \mathcal{H}$ that satisfy $f(g\gamma) = \pi(\gamma^{-1})f(g)$ for every $g \in G$ and γ

in Γ, and such that $\int_G \|f(g)\|^2 d\mu(g) < \infty$. It is equipped with a unitary representation of G by left-translation $g \cdot f = f(g^{-1}\cdot)$. If π almost has invariant vectors, then so does the induced representation. By property (T) for G, it has a nonzero invariant vector. This vector is a constant function with values in $\mathcal{H}^\pi$. So π has invariant nonzero vectors.

2. FOURIER SERIES, APPROXIMATION PROPERTIES, OPERATOR ALGEBRAS

2.1. Fourier series, absence of Fourier synthesis

If $1 < p < \infty$, it is very well known (this follows immediately from Marcel Riesz's theorem that the Hilbert transform $f \mapsto \sum_{n \geq 0} \hat{f}(n) e^{2i\pi \cdot}$ is bounded on L_p) that the Fourier series of every function $f \in L_p(\mathbf{R}/\mathbf{Z})$ converges in L_p:

$$\lim_N \left\| f - \sum_{n=-N}^{N} \hat{f}(n) e^{2i\pi \cdot} \right\|_p = 0.$$

This is not true for $p = 1$ or $p = \infty$, but there are more clever *summation methods*, for example, Fejér's method: if we denote $W_N(n) = \max(1 - \frac{|n|}{N}, 0)$, then

$$\lim_N \left\| f - \sum_n W_N(n) \hat{f}(n) e^{2i\pi \cdot} \right\|_p = 0,$$

and this time the convergence holds for L_1, and even L_∞, if f is continuous. All this remains true on the torus $(\mathbf{R}/\mathbf{Z})^d$ of arbitrary dimension, and more generally on the Pontryagin dual $\widehat{\Gamma}$ of every discrete abelian group.

When Γ is a nonabelian discrete group, $\widehat{\Gamma}$ does not make sense as a group, but the spaces $C(\widehat{\Gamma})$, $L_\infty(\widehat{\Gamma})$, and $L_p(\widehat{\Gamma})$ have a very natural meaning: they are respectively the reduced C^*-algebra $C^*_\lambda(\Gamma)$, the von Neumann algebra $\mathcal{L}\Gamma$, and the noncommutative L_p space **[64]** of the von Neumann algebra of $\mathcal{L}\Gamma$. Recall that if λ is the left regular representation (by left-translation) of Γ on $\ell_2(\Gamma)$, $C^*_\lambda(\lambda)$ is the norm closure in $B(\ell_2(\Gamma))$ of the linear span of $\lambda(\Gamma)$, $\mathcal{L}\Gamma$ is its closure for the weak-operator topology, and $L_p(\mathcal{L}\Gamma)$ its completion for the norm $x \mapsto \langle |x|^p \delta_e, \delta_e \rangle^{\frac{1}{p}}$. The elements of each of these spaces admit a Fourier series $f = \sum_\gamma \hat{f}(\gamma)\lambda(\gamma)$. This is a formal series, whose convergence is not clear in general (except in L_2). One can wonder in that case whether, as in the case when Γ is abelian, there exist Fourier summation methods in this context, i.e., a sequence of functions $W_N \colon \Gamma \to \mathbf{C}$ with finite support such that

$$f = \lim_N \sum_\gamma W_N(\gamma) \hat{f}(\gamma) \lambda(\gamma), \tag{2.1}$$

for every f in $C^*_\lambda(\Gamma)$ or $L_p(\mathcal{L}\Gamma)$ (convergence in norm). It is well known that this is the case when Γ is amenable. According to a celebrated (and surprising at that time) result by Haagerup **[28]**, this is also true when Γ is a nonabelian free group. This result has inspired a massive research program on approximation properties of group operator algebras. Indeed, by the Banach–Steinhaus theorem, if (2.1) holds, then the maps $f \mapsto \sum_\Gamma W_N(\gamma)\hat{f}(\gamma)\lambda(\gamma)$, called Fourier multipliers, are uniformly bounded, have finite rank, and converge pointwise to the identity. In [10] de Cannière and Haagerup realized that in the case of free groups, the convergence even holds for every f in the C^*-algebra $B(\ell^2) \otimes C^*_\lambda(\Gamma)$, that is, when the Fourier coefficients $\hat{f}(\gamma)$ are bounded operators on ℓ^2. When this holds, Γ is said to be weakly amenable. Moreover, a function $W\colon \Gamma \to \mathbf{C}$ is said to be a completely bounded Fourier multiplier if the map $f \mapsto \sum_\gamma W(\gamma)\hat{f}(\gamma) \otimes \lambda(\gamma)$ is bounded on $B(\ell^2) \otimes C^*_\lambda(\Gamma)$. So weak amenability comes with a constant, which is the smallest common upper bound on these completely bounded norms, among all sequences of finitely supported multipliers achieving (2.1). Mutatis mutandis, completely bounded Fourier multipliers and weak amenability also make sense for locally compact groups, and importantly restriction to closed subgroups and induction from lattices work well for completely bounded Fourier multipliers. In particular, the weak amenability constant coincide for a group and a lattice [29]. The major achievement in this direction was obtained in a series a work by Haagerup with de Cannière and Cowling [10, 16, 29], that led to the exact computation of the weak amenability constant of all simple Lie groups. Excluding exceptional groups, it is equal to 1 for $\mathrm{SO}(n,1)$, $\mathrm{SU}(n,1)$, $2n-1$ for $\mathrm{SP}(n,1)$, and is infinite for higher rank groups.

In [32], Haagerup and Kraus discovered a strange phenomenon: it might happen that there is no sequence W_N satisfying (2.1) for every $f \in B(\ell^2) \otimes C^*_\lambda(\Gamma)$, but there exists such a generalized sequence (or net).[1] A group for which such a net exists is said to have *the approximation property* of Haagerup and Kraus, or simply AP.

So groups without the AP are groups in which no L_∞-summation method exists whasotever for operator coefficients. It has been difficult to produce such groups. It was known [32] that nonexact groups [59] would be such examples, but they are difficult to construct. Haagerup and Kraus had conjectured that $\mathrm{SL}_3(\mathbf{Z})$ was another example. This conjecture turned out to be delicate because classical approaches to rigidity in higher-rank lattices, which rely on the subgroup $\mathrm{SL}_2(\mathbf{Z}) \ltimes \mathbf{Z}^2$ or its relative $\mathrm{SL}_2(\mathbf{R}) \ltimes \mathbf{R}^2$, are inefficient by the strange phenomenon described above. It is the ideas described in Section 1 that allowed to settle it.

Theorem 2.1 ([47]). $\mathrm{SL}_3(\mathbf{Z})$ *does not have the approximation property of Haagerup and Kraus.*

The original proof of this theorem was not direct and went through L_p Fourier theory (Theorem 2.3), but, thanks to several simplifications [30, 66, 71], a very easy proof is now known. We have already justified that, by induction, we can as well prove the theorem for $\mathrm{SL}_3(\mathbf{R})$. The starting point is the following straightforward lemma (which is known to hold more generally if [36] and only if [4] the locally compact group K is amenable), which provides a characterization of completely bounded Fourier multipliers of compact groups.

1 Remembering that the Banach–Steinhaus theorem is false for nets might help imagine how such a statement could be true. An example is given by $\mathrm{SL}_2(\mathbf{Z}) \ltimes \mathbf{Z}^2$: it has the AP as the semidirect product of two weakly amenable groups and AP is stable by group extensions [32]. That it is not weakly amenable was proved in [29] and also reproved in [60].

Lemma 2.2. *Let K be a compact group. A function $\varphi: K \to \mathbf{C}$ defines a completely bounded Fourier multiplier of $C^*_\lambda(K)$ if and only if φ is a matrix coefficient of a unitary representation of K.*

Proposition 1.1 therefore says that SO(2)-biinvariant completely bounded Fourier multipliers of $C^*_\lambda(\mathrm{SO}(3))$ are Hölder-$\frac{1}{2}$ continuous in the interior of $\mathrm{SO}(2)\backslash\mathrm{SO}(3)/\mathrm{SO}(2)$. The same proof as Proposition 1.2 then produces a map from $\mathrm{SO}(3)\backslash\mathrm{SL}_3(\mathbf{R})/\mathrm{SO}(3)$ into the dual of the space of completely bounded Fourier multipliers of $\mathrm{SL}_3(\mathbf{R})$, which satisfies the Cauchy criterion. Its limit vanishes on compactly supported functions and takes the value 1 on the identity multiplier. This is exactly a Hahn–Banach-type separation, which says that $\mathrm{SL}_3(\mathbf{R})$ does not have the AP.

The same argument can also be used to say something about L^p Fourier summability for some finite p and to obtain the following, which is an equivalent form of the main result in [47].

Theorem 2.3 ([47]). *For every $4 < p < \infty$ or $1 \leq p < \frac{4}{3}$, there is $f \in L_p(\mathcal{L}\mathrm{SL}_3(\mathbf{Z}) \otimes B(\ell_2))$ such that, for every finitely supported $W : \mathrm{SL}_3(\mathbf{Z}) \to \mathbf{C}$,*

$$\Big\| f - \sum_\gamma W(\gamma)\hat{f}(\gamma)\lambda(\gamma) \Big\|_p \geq 1.$$

Again, the proof has the same structure as in Section 1. To initiate the first step, we need to investigate in more details the spectral decomposition of the operators T_δ in (1.4). For a Hilbert space $\mathcal{H}$, the Schatten p-class $S_p(\mathcal{H})$ is the space of operators T on $\mathcal{H}$ such that $\|T\|_{S_p} := (\mathrm{Tr}(|T|^p))^{\frac{1}{p}} < \infty$. It can be shown [47] that the operators T_δ, $\delta \in (-1, 1)$ belong to $S_p(L_2(S^2))$ if $p > 4$, and there is a constant C such that for every $\delta, \delta' \in [-1/2, 1/2]$,

$$\|T_\delta - T_0\|_{S_p} \leq \frac{C}{(p-4)^{1/p}} |\delta|^{\frac{1}{2}-\frac{2}{p}}. \tag{2.2}$$

With this inequality, we can run the argument of Section 1 and obtain a form of Theorem 2.3 for $\mathrm{SL}_3(\mathbf{R})$. However, the induction procedure for completely bounded L_p Fourier multipliers, which works well when $p = \infty$ [29, 32], is problematic for $1 < p \neq 2 < \infty$. It is known to work well mainly for amenable groups [12, 13, 56]. The solution is to work with Herz–Schur multipliers on $S_p(L_2(G))$, that is, operators of the form $A = (A_{g,h}) \in S_p(L_2(G)) \mapsto (W(gh^{-1})A_{g,h})$ for functions $W: G \to \mathbf{C}$, for which induction works well. Fortunately, for compact groups (and even amenable groups [12, 56]) Schur S_p and Fourier L_p multipliers coincide.

The preceding sketch is the only proof I know of Theorem 2.3. It can be shown that (2.2) is optimal, and that $T_\delta - T_0$ does not belong to S_4 for any $\delta \neq 0$. So this idea cannot work for $2 < p \leq 4$, and it is an intriguing question whether this condition is really needed for Theorem 2.3. As we will explain in Section 5, a positive answer to this question would allow distinguishing the von Neumann algebra of $\mathrm{SL}_3(\mathbf{Z})$ and $\mathrm{PSL}_n(\mathbf{Z})$ for $n > 4$, and would confirm a conjecture of Connes. The challenge is to construct nontrivial L_p Fourier multipliers for $\mathrm{SL}_3(\mathbf{Z})$. A first step is to do so for $\mathrm{SL}_3(\mathbf{R})$. Together with Parcet and Ricard

[62], we made some progress on that recently by proving a satisfactory *local* form of the Hörmander–Mikhlin multiplier theorem for $SL_n(\mathbf{R})$.

2.2. Approximation properties for Banach spaces and operator algebras

In his thesis [26], Grothendieck initiated the study of tensor product of topological vector spaces, and realized the tight connection with the Banach's approximation property AP: a Banach space E has the approximation property (AP) if the identity operator belongs to the closure of the space of finite rank operators, for the topology of uniform convergence on compact sets. He was even led to conjecture that all Banach space have the AP (this would make the theory of tensor product simpler!). Later in his *Résumé* [25] he changed his mind and actually expected that Banach spaces exist, which fail the AP. Récoltes et Semailles [27] contains a moving part, where Grothendieck explains how much he suffered from the year he spent on working on this problem without progress. It was only much later than the first example of Banach space was *constructed* by Enflo [21], followed by many other examples. See the survey [11] for a list. Let me emphasize: the examples of Enflo, as most other examples, are obtained from delicate combinatorial constructions, and in particular they are not Banach spaces, the existence of which was known to Grothendieck. There is, to my knowledge, only one example of a Banach space that is both natural (not obtained from an *ad hoc* construction) and known to fail the AP, namely the space $B(\ell_2)$ of all bounded operators on ℓ_2 [69]. This space is not reflexive and we are lacking separable and natural examples of space with the AP. For some time in the 1970s, a candidate of such a satisfactory example was $C^*_\lambda(\mathbb{F}_2)$, the reduced C^*-algebra on the nonabelian free group with two generators, and the hope was that the lack of AP could be explained by the non amenability of $\mathbb{F}_2$. Haagerup broke this hope in [28] by proving, better, that $C^*_\lambda(\mathbb{F}_2)$ has the metric AP (the identity belongs to the closure of the norm 1 finite rank operators). More precisely, as explained in the previous section, $\mathbb{F}_2$ is weakly amenable with constant 1. However, two serious candidates of natural separable Banach spaces without the AP remain in the same vein: $C^*(\mathbb{F}_2)$, the full C^*-algebra of $\mathbb{F}_2$ and $C^*_\lambda(SL_3(\mathbf{Z}))$. It is hoped that the first lacks the AP for the simple reason that $\mathbb{F}_2$ is nonamenable, and the latter for reasons related to the ideas in Section 1.

This last conjecture has not been settled, but its weakening in the sense of operator spaces is known. Indeed, there are natural variants of Grothendieck's approximation property in the category of operator spaces rather than Banach spaces (replacing bounded operators by completely bounded operators [63]), that we denote by OAP and CBAP. The CBAP, meaning that the identity on E belongs to the closure of the finite rank operators with completely bounded norm bounded above by some constant C, comes with a constant (the best such C). Given a discrete group Γ, the fundamental observation of Haagerup [29] (and Kraus for OAP [32]) is that the CBAP/OAP for $C^*_\lambda(\Gamma)$ (and respectively its variants for dual spaces for the von Neumann algebra $\mathcal{L}(\Gamma)$) can be achieved with finitely supported Fourier multipliers. As a consequence, the weak amenability constant of a discrete group Γ is an invariant of its von Neumann algebra, and so is the AP of Haagerup and Kraus. For example, the computation of the weak amenability constant of simple Lie groups [10, 16, 29] discussed above allows distinguishing the von Neumann algebra of a lattice in $SP(n, 1)$ (for $n \neq 3$, $n \neq 11$) from the von Neumann algebra of a lattice in a simple Lie group that is not locally $\mathrm{SP}(n,1)$. Also, Theorem 2.1 can be rephrased as $C^*_\lambda(\mathrm{SL}_3(\mathbf{Z}))$ *does not have the approximation property in the category of operator spaces*. This was the first example of an exact C^*-algebra without the OAP. Similarly, Theorem 2.3 can be rephrased as $L_p(\mathcal{L}\mathrm{SL}_3(\mathbf{Z}))$ *does not have the OAP for* $p > 4$. For $4 < p < 80$, this was the first example of a noncommutative L_p space without the CBAP.

3. NONUNITARY REPRESENTATIONS: LAFFORGUE'S STRONG PROPERTY (T)

This section is devoted to non unitary representations on Hilbert spaces.

3.1. Strong property (T) for $\mathrm{SL}_3(\mathbf{R})$

It is useful at this point to have a look back at the proof of property (T) for $\mathrm{SL}_3(\mathbf{R})$ in Section 1 and see precisely where the assumption that the representations are unitary was used. Step 1 is about the compact group SO(3), and therefore only uses that the restriction to SO(3) is unitary. This is not a very strong assumption, as compact groups are unitarizable:

Lemma 3.1. *Every representation of a compact group on a Hilbert space is similar to a unitary representation.*

In particular, every representation of $\mathrm{SL}_3(\mathbf{R})$ is similar to a representation whose restriction to SO(3) is unitary.

Step 2 did not use that π is unitary in a strong way, and remains true (with a different constant and the exponent $-\frac{1}{2}$ replaced by $2\alpha - \frac{1}{2}$) if

$$\alpha = \alpha(\pi) := \limsup_{|t|\to\infty} \frac{1}{|t|} \log\left\|\pi\left(D\left(-t, \frac{t}{2}, \frac{t}{2}\right)\right)\right\| < \frac{1}{4}.$$

So if π is a representation of $\mathrm{SL}_3(\mathbf{R})$ on a Hilbert space such that $\alpha(\pi) < \frac{1}{4}$, we obtain that $\pi(KgK)$ converges in norm to a projection on the space of harmonic vectors. Step 3 apparently relies more fundamentally on the fact that π is unitary. However, it is still true when $\alpha(\pi) < \frac{1}{4}$ that harmonic vectors are invariant. This requires new ideas, which make step 3 the most involved part, see **[43]** or the presentation in **[65]**. To summarize, the condition $\alpha(\pi) < \frac{1}{4}$ is enough to guarantee that $\pi(KgK)$ converges in norm to a projection on the invariant vectors. In the terminology of Vincent Lafforgue **[43]**, $\mathrm{SL}_3(\mathbf{R})$ has strong property (T), which is a form of property (T) for representations on Hilbert spaces with small exponential growth rate. Let us spell out the definition.

If G is a locally compact function, a length function $\ell : G \to \mathbf{R}_+$ is a function that is bounded on compact subsets, that is, symmetric $\ell(g) = \ell(g^{-1})$ and subadditive $\ell(gh) \leq \ell(g) + \ell(h)$. Let us denote $\mathcal{C}_\ell(G)$ the completion of $C_c(G)$ for the norm given by $\|f\|_{\mathcal{C}_\ell(G)} = \sup \|\pi(f)\|$ where the supremum is over all representations of G on a Hilbert space such that $\|\pi(g)\| \leq e^{\ell(g)}$ for every $g \in G$. It is a Banach algebra for the convolution. An element $p \in \mathcal{C}_\ell(G)$ is called a *Kazhdan projection* if $\pi(p)$ is a projection on the invariant vectors $\mathcal{H}^\pi$ of π for every such representation. Kazhdan projections have been investigated in [20, 67]. If a Kazhdan projection exists, $\pi(p)$ is the unique G-equivariant projection on $\mathcal{H}^\pi$, so p is unique [67].

Definition 3.2 ([43]). G has strong property (T) if for every length function ℓ, there is $s > 0$ such that for every $C > 0$, $\mathcal{C}_{s\ell+C}(G)$ has a Kazhdan projection.

A posteriori, a group with strong property (T) is necessarily compactly generated (this is even a consequence of property (T)), and it is enough to check the definition for ℓ the word-length with respect to a compact generating set.

We have just explained the proof of the following theorem.

Theorem 3.3 ([43]). $\mathrm{SL}_3(\mathbf{R})$ *has strong property (T).*

Moreover, the Kazhdan projection belongs to the closure of $C_c(G)_+ := \{f \in C_c(G) \mid f(g) \geq 0 \forall g \in G\}$. Some authors [6–9] add this precision to the definition because, as we will see below, this is crucial for some applications.

Vincent Lafforgue's original motivation for this definition was his work on the Baum–Connes conjecture. Indeed, strong property (T) (and variants of it) are the natural obstructions for applying some of his ideas (and in particular the ideas in [46]) to groups such as $\mathrm{SL}_3(\mathbf{Z})$. We refer to [45] for more on the link with the Baum–Connes conjecture.

3.2. Strong property (T) for $\mathrm{SL}_3(\mathbf{Z})$

Theorem 3.3 also holds for $\mathrm{SL}_3(\mathbf{Z})$, but the proof turned out to be delicate. In particular, we do not see how to prove in general that strong property (T) passes to lattices.

Theorem 3.4 ([68]). $\mathrm{SL}_3(\mathbf{Z})$ *has strong property (T).*

The way Theorem 3.4 is proved is by introducing and working with representation-like objects, where one is only allowed to compose once and that I call *two-step representations*.

Definition 3.5. A two-step representation of a topological group G is a tuple $(X_0, X_1, X_2, \pi_0, \pi_1)$ where X_0, X_1, X_2 are Banach spaces and $\pi_i \colon G \to B(X_i, X_{i+1})$ are strongly continuous[2] maps such that

$$\pi_1(gg')\pi_0(g'') = \pi_1(g)\pi_0(g'g'') \quad \text{for every } g, g', g'' \in G.$$

In this case we will denote by $\pi \colon G \to B(X_0, X_2)$ the continuous map satisfying $\pi(gg') = \pi_1(g)\pi_0(g')$ for every $g, g' \in G$.

The reason for this introduction is that two-step representations appear naturally when we induce non unitary representations. Indeed, let $\Gamma \subset G$ be a lattice. For every probability measure μ on G as in Section 1.2, we can consider the space $L_2(G, \mu; \mathcal{H})^\Gamma$. When Γ is a cocompact lattice (that is G/Γ is compact), Lafforgue [43] observed that it

2 In other words, for every $x \in X_i$, the map $g \in G \mapsto \pi(g)x \in X_{i+1}$ is continuous.

is possible to choose μ in such a way that the induced representation remains by bounded operators on $L_2(G,\mu;\mathcal{H})^\Gamma$, with small exponential growth if the original representation of Γ had small exponential growth. Therefore, strong property (T) passes to cocompact lattices [43]. When Γ is not cocompact and π is not uniformly bounded, there does not seem to be any choice of μ for which the representation is by well-defined (bounded) operators. However, in the particular case of $SL_3(\mathbf{Z}) \subset SL_3(\mathbf{R})$, if μ is well chosen and if the original representation π of has small enough exponential growth, it possible to show that the representation $g \cdot f = f(g^{-1}\cdot)$ is bounded from $L_p(G,\mu;\mathcal{H})^\Gamma \to L_q(G,\mu;\mathcal{H})^\Gamma$ whenever $\frac{1}{q} - \frac{1}{p} \geq \frac{1}{2}$. This uses some strong exponential integrability properties of $SL_3(\mathbf{Z})$ in $SL_3(\mathbf{R})$, which rely on the celebrated Lubotzky–Mozes–Raghunathan theorem [52]. In particular, we obtain a two-step representation with $X_0 = L_\infty(G,\mu;\mathcal{H})^\Gamma$, $X_1 = L_2(G,\mu;\mathcal{H})^\Gamma$, and $X_2 = L_1(G,\mu;\mathcal{H})^\Gamma$. So Theorem 3.4 is a consequence of a form of strong property (T) for two-step representations of $SL_3(\mathbf{R})$ where X_1 is a Hilbert space. This is done following the strategy in Section 1. The starting point is again a straightforward statement, which asserts that two-step representations of compact groups are governed by usual representations:

Lemma 3.6. *If $(X_0, X_1, X_2, \pi_0, \pi_1)$ is a two-step representation of a compact group K where X_1 is a Hilbert space, then there is a constant C such that, for every $f \in C_c(K)$,*

$$\|\pi(f)\|_{B(X_0,X_2)} \leq C\|\lambda(f)\|_{C^*_\lambda(K)}.$$

3.3. Applications of strong property (T)

Let us end this section with two applications of strong property (T). We will see in the next section other applications of variants of strong property (T) for representations on Banach spaces. All applications have in common that strong (T) is used as a way to systematically find and locate fixed point. The first is a result about vanishing of first cohomology spaces for representations with small exponential growth. If π is a representation of G on a space $\mathcal{H}$, we denote by $H^1(G,\pi)$ the quotient of the space of cocycles

$$Z^1(G,\pi) := \{b \in C(G,\mathcal{H}) \mid \forall g_1, g_2 \in G, b(g_1g_2) = b(g_1) + \pi(g_1)b(g_2)\}$$

by the subspace of coboundaries

$$B^1(G,\pi) = \{g \mapsto \pi(g)\xi - \xi \mid \xi \in \mathcal{H}\}.$$

Hence $H^1(G,\pi)$ parametrizes the continuous affine actions of G on $\mathcal{H}$ with linear part π, up to a change of origin. Delorme and Guichardet have proved that a second countable locally compact group G has property (T) if and only if $H^1(G,\pi) = 0$ for every unitary representation π. The following result has the same flavor, but the proof is different. The idea is that any $b \in Z^1(G,\pi)$ gives rise to a representation $\begin{pmatrix} \pi(g) & b(g) \\ 0 & 1 \end{pmatrix}$ with the same exponential growth rate on $\mathcal{H} \oplus \mathbf{C}$.

Proposition 3.7 ([44]). *If G has strong property (T) and ℓ is a length function on G, there is $s > 0$ such that $H^1(G,\pi) = 0$ for every representation with exponential growth rate $\leq s$.*

This proposition can be used to show that strong property (T) is incompatible with hyperbolic geometry, and in particular that infinite Gromov-hyperbolic groups do not have strong property (T). Indeed, in [43], for every group G acting with infinite orbits on a Gromov-hyperbolic graph with bounded degree, a representation with quadratic growth rate on a Hilbert space is constructed with $H^1(G,\pi) \neq 0$. By Proposition 3.7, such a group cannot have strong property (T).

Another notable application of strong property (T) was found in the resolution of most cases of Zimmer's conjecture by Brown, Fisher, and Hurtado (see [5,23]). The following is a particular case of their result.

Theorem 3.8 ([7]). *Let G be a locally compact group with length function ℓ and $\alpha : G \to \operatorname{Diff}(M)$ an action by C^∞ diffeomorphisms on a compact Riemannian manifold with subexponential growth of derivatives:*

$$\forall \varepsilon > 0, \quad \sup_{g\in G} e^{-\varepsilon\ell(g)} \sup_{x\in M} \left\| D_x\alpha(g)\right\| < \infty.$$

If G has strong property (T) with Kazhdan projections in the closure of $C_c(G)_+$, then for every k, α preserves C^k Riemannian metrics on M. In particular, for the original metric, α has bounded derivatives.

The idea is to use strong property (T) for the representation on the Hilbert space of signed metrics on M with Sobolev norms $W^{n,2}$, which take into account the L^2-norms of all derivatives of order $\leq n$. Here n is an arbitrary positive integer. Strong property (T) allows constructing such invariant signed metrics. The fact that the Kazhdan projection belongs to the closure of nonnegative functions is used to ensure that these signed metrics can be taken positive. The Sobolev embedding theorems say that, for n large, these metrics become smoother.

4. BANACH SPACE REPRESENTATIONS

4.1. Banach spaces versions of property (T)

The last two decades have seen important developments in the study of group actions on Banach spaces, initiated by a number of more or less simultaneous investigations [1, 14, 15, 42, 43, 72].

The work [1] (and also [14, 43]) has proposed to study different possible generalizations of property (T) with Hilbert spaces replaced by Banach spaces. If one adopts the definition in terms of almost invariant vectors, one gets property (T_X). Let $\mathcal{E}$ be a class of Banach spaces.

Definition 4.1 (Bader, Furman, Gelander, Monod [1]). A locally compact group G has property ($T_{\mathcal{E}}$) if for every isometric representation $\pi : G \to O(X)$ on a space X in $\mathcal{E}$, the quotient representation $G \to O(X/X^{\pi(G)})$ does not almost have invariant vectors. Here, $X^{\pi(G)}$ denotes the closed subspace of X consisting of vectors that are fixed by π.

When the quotient representation $G \to O(X/X^{\pi(G)})$ does not almost have invariant vectors, we say that π has spectral gap.

Compact groups have T_X with respect to all Banach spaces, but for other locally compact groups we have to impose conditions on a Banach space to hope to have T_X. For example, the action by left-translation of $C_0(G)$ has spectral gap if and only G is compact.

Adapting the equivalent characterization of property (T) in cohomological terms, we obtain:

Definition 4.2 ([1]). A locally compact group G has property $F_{\mathcal{E}}$ if every action of G by affine isometries on a space in $\mathcal{E}$ has a fixed point.

It still holds that for σ-compact groups, $F_{\mathcal{E}}$ implies $(T_{\mathcal{E}})$, but the converse is not true. For example, Pansu's computation of L_p-cohomology of rank 1 symmetric spaces [61] says that $\mathrm{Sp}(n,1)$ does not have F_{L_p} for $p > 4n+2$, whereas as every group with property (T) [1], it has T_{L_p} for every $1 \leq p < \infty$. Pansu's result has been generalized by Yu [72] who showed that every Gromov-hyperbolic group has a proper isometric action on an L_p space for every large p. We refer to [17,18,48,53] and [19,40,58] for recent progresses on fixed points properties for actions on L_p spaces and other Banach spaces. So studying for which spaces a group has F_X is a way to quantify the strength of property (T). The following conjecture therefore is another indication that $\mathrm{SL}_3(\mathbf{Z})$ has a very strong form of property (T).

Conjecture 4.3 ([1]). *Any action by isometries of* $\mathrm{SL}_3(\mathbf{Z})$ *(or, more generally, a lattice in a connected simple Lie group of real rank* ≥ 2*) on a uniformly convex Banach space has a fixed point.*

4.2. Expander graphs

The study of group actions on Banach spaces is also related to questions about the possible interactions between geometry of finite graphs and of Banach spaces. Given a finite graph $\mathcal{G}$ and a Banach space X, the X-valued Poincaré constant $\rho(\mathcal{G}, X)$ is the smallest constant ρ such that, for every function f from the vertex set of $\mathcal{G}$ to X, the Poincaré inequality

$$\inf_{\xi \in X} \|f - \xi\|_2 \leq \rho \|\nabla f\|_2$$

holds, where ∇f is the function on the edge set of $\mathcal{G}$ taking the value $f(x) - f(y)$ at the edge (y, x). A sequence of bounded degree graphs with size going to infinity is said to be expanding with respect to X if $\inf_n \rho(\mathcal{G}_n, X) > 0$. When X is the line, or a Hilbert space, or an L_p space for $1 \leq p < \infty$, we recover the usual notion of expander graph. On the opposite, there are no expanders with respect to ℓ_∞, and more generally with respect to a space containing arbitrarily large copies of ℓ_∞^n (such space are called *spaces with trivial cotype*). A sequence $\mathcal{G}_n$ that is expanding with respect to every uniformly convex Banach space is called a sequence of *superexpanders* [54]. There are two sources of examples known: one that we will discuss in the last section, coming from quotients of arithmetic groups over nonarchimedean local fields [44] and relying on the ideas from Section 1, and one coming from zigzag products [54].

When a sequence of graphs $\mathcal{G}_n$ are Cayley graphs of (Γ_n, S_n) where Γ_n is a sequence finite quotients of a group Γ with size going to infinity and S_n the image of a fixed generating set of Γ in Γ_n, the fact that $\mathcal{G}_n$ are expanders with respect to X is equivalent to the fact that the representation π on $\ell_2(\bigcup_n \Gamma_n; X)$ has spectral gap. This if, for example, the case if Γ has $(\mathrm{T}_{\ell_2(\mathbf{N};X)})$. It follows from this discussion is that Conjecture 4.3 is stronger than the following conjecture:

Conjecture 4.4. *The sequence of Cayley graphs of* $\mathrm{SL}_3(\mathbf{Z}/n\mathbf{Z})$ *with respect to the elementary matrices is a sequence of superexpanders.*

It is conceivable that they are even expanders with respect to every Banach space of nontrivial cotype. The existence of such expanders is still unknown. On the opposite, it is also unknown whether there exist expanders that are not expanders with respect to all Banach spaces of nontrivial cotype. This is revealing of how little such questions are understood.

4.3. Strong Banach property (T)

Let $\mathcal{E}$ be a class of Banach spaces. If, in Definition 3.2, we allow representations on a Banach space in $\mathcal{E}$ rather than only on a Hilbert space, we say that G as strong property (T) with respect to $\mathcal{E}$. So there are as many Banach space versions of strong property (T) that classes of Banach spaces. In [44], Lafforgue uses the terminology strong Banach property (T) to mean that G has strong property (T) with respect to every class $\mathcal{E}$ in which ℓ_1 is not finitely representable. This is essentially the largest possible class, because no noncompact group can have strong property (T) with respect to $L_1(G)$.

Strong property (T) with respect to Banach spaces has the same kind of applications than for Hilbert spaces. First as in Proposition 3.7, strong property (T) with respect to $X \oplus \mathbf{C}$ implies that $H^1(G, \pi) = 0$ for every representation on X with small enough exponential growth. We also have a variant of Theorem 3.8. If G is assumed to have strong property (T) with respect to all subspaces of L_p spaces for all $2 \leq p < \infty$, then it is enough to assume that the action is by C^{1+s}-diffeomorphisms to ensure that for every $\varepsilon > 0$ α preserves of metric of regularity $C^{s-\varepsilon}$ [7], and enough to assume that the action is by C^1-diffeomorphisms to ensure that for every $p < \infty$ α preserves a measurable metric that is L^p-integrable [6].

The sketch of proof of strong property (T) for $\mathrm{SL}_3(\mathbf{R})$ and $\mathrm{SL}_3(\mathbf{Z})$ is Section 3 applies in the same way, provided that there are constants $C, \theta > 0$ such that

$$\forall \delta \in [-1, 1], \quad \|T_\delta - T_0\|_{B(L_2(S^2;X))} \leq C|\delta|^{\theta/2}. \tag{4.1}$$

Theorem 4.5 ([43, 68]). $\mathrm{SL}_3(\mathbf{R})$ *and* $\mathrm{SL}_3(\mathbf{Z})$ *have strong property (T) with respect to X for every Banach space satisfying* (4.1) *for some* $C, \theta > 0$.

It is expected that all uniformly convex Banach spaces spaces satisfy (4.1) for some C, θ. More precisely, it should be true that spaces satisfying (4.1) are exactly the spaces of nontrivial Rademacher type. This would settle Conjecture 4.3 and 4.4.

If follows from the Riesz–Thorin theorem that (4.1) holds for L_p spaces for $1 < p < \infty$ and therefore for every subspace of an L_p space, and more generally for every θ-Hilbertian space. In [65], exploiting the stronger summability property from (2.2), I showed that (4.1) holds whenever $d_n(X) = O(n^{\frac{1}{4}-\varepsilon})$ as $n \to \infty$. Here $d_n(X)$ denotes the supremum over all subspaces E of X of dimension n of the Banach–Mazur distance to the Euclidean space of the same dimension. For example, this holds if X has type p and cotype q with $\frac{1}{p} - \frac{1}{q} < \frac{1}{4}$.

5. OTHER GROUPS

There is no general theory yet, in which the strategy from Section 1 for $SL_3(\mathbf{R})$ fits, but there are other examples of groups for which such tools have been developed: $SL_3(\mathbf{F})$ for a nonarchimedean local field [43], $SP_2(\mathbf{R})$ and its universal cover [30, 31, 37, 38], $SP_2(\mathbf{F})$ [50, 51], $SL_n(\mathbf{F})$ and $SL_n(\mathbf{R})$ for $n \geq 4$ [47] and [39, 41], $SO(n, 1)$ [62], and finally lattices in locally finite affine buildings of type $\tilde{A}_2$ [49].

Let me expand a bit, starting with the real Lie groups. The group $SP_2(\mathbf{R})$ is the group of 4×4 matrices which preserve the standard symplectic form $\omega(x, y) = y_1 x_3 + y_2 x_4 - x_1 y_3 - x_2 y_4$. The Bernstein inequalities used in Proposition 1.1 were generalized in [33] to Jacobi polynomials, which appear as spherical functions for other Gelfand pairs than $SO(2) \subset SO(3)$. In [30, 31], Haagerup and de Laat then generalized Theorem 2.1 to $SP_2(\mathbf{R})$ and its universal cover, respectively. The analogue of Theorem 2.3 was obtained in [37], but for $p > 12$ (improved to $p > 10$ in [38] by refining the Bernstein inequalities from [33]), and strong property (T) was proved in [38] for the Lie groups or their cocompact lattices, and in [68] for their non cocompact lattices, for a class of Banach spaces that is more restrictive than for $SL_3(\mathbf{R})$. By rank 2 reduction all these results extend to all real connected semisimple Lie groups all of whose simple factors have rank ≥ 2, and all their lattices.

When $\mathbf{F}$ is a nonarchimedean local field, in the same way, to obtain all almost $\mathbf{F}$-simple algebraic group of split rank at least two, it was enough to consider SL_3 and SP_2. Lafforgue's original article [43] already contained a proof of strong property (T) for $SL_3(\mathbf{F})$. Steps 2 and 3 are almost identical to the real case, but the first step is very different because maximal compact subgroups of $SL_3(\mathbf{F})$ are very different from those in $SL_3(\mathbf{R})$. For example, when $\mathbf{F} = \mathbf{Q}_p$, a maximal compact subgroup is $SL_3(\mathbf{Z}_p)$, which contains large nilpotent groups (the groups of upper-triangular matrices). This difference turns out to play a rôle for Banach space versions of strong property (T). Indeed, exploiting these large nilpotent groups and the good understanding of abelian Fourier analysis on vector-valued L_p spaces [3], Lafforgue [44] was able to make the first step work for representations of the maximal compact subgroups of $SL_3(\mathbf{F})$ on arbitrary Banach space of nontrivial type. For SP_2, the same was proved by Liao in [50]. As a consequence, all almost $\mathbf{F}$-simple algebraic group of split rank at least two and their lattices have strong property (T) with respect to Banach spaces of nontrivial type, and Conjectures 4.3 and 4.4 hold for $SL_3(\mathbf{Z})$ replaced by such lattices.

The Fourier analysis and approximation results from Section 2 have also been obtained for nonarchimedean local fields, in [47] for SL_3 and [51] for SP_2. The results are identical for $SL_3(\mathbf{F})$ and $SL_3(\mathbf{R})$. Interestingly, for SP_2 a difference appears: the condition $p > 10$ becomes better, namely $p > 4$, in the nonarchimedean case [51].

The fact that both in the real and nonarchimedean case the proofs do not allow taking p down to 2 in Theorem 2.3 has been a motivation for finding other groups for which the restriction on p is smaller. And indeed, this is the case for $\mathrm{SL}_n(\mathbf{F})$ [47] and $\mathrm{SL}_n(\mathbf{R})$ [39] for large n. Let us focus on $\mathrm{SL}_n(\mathbf{R})$ because the situation is closer to Section 1. In that case, a satisfactory replacement for Step 1 is to work with the pair $\mathrm{SO}(d) \subset \mathrm{SO}(d+1)$. In the sphere picture, we are studying the operators T_δ defined as for S^2 but in higher dimension $S^d = \mathrm{SO}(d+1)/\mathrm{SO}(d)$. In [39], we proved that the map $\delta \in (-1,1) \mapsto T_\delta \in S_p$ is Hölder continuous for every $p > 2 + \frac{2}{d-1}$. Step 2 is more delicate. Taking $n \geq d+1$ and seeing $\mathrm{SO}(d+1) \subset \mathrm{SL}_n(\mathbf{R})$, by considering all possible matrices $D, D' \in \mathrm{SL}_n(\mathbf{R})$, the maps $k \in \mathrm{SO}(d+1) \mapsto DkD' \in \mathrm{SL}_n(\mathbf{R})$ pass to maps from the segment $[-1,1] = \mathrm{SO}(d)\backslash\mathrm{SO}(d+1)/\mathrm{SO}(d)$ into the Weyl chamber $\Lambda_n := \mathrm{SO}(n)\backslash\mathrm{SL}_n(\mathbf{R})/\mathrm{SO}(n)$. The problem is that all these maps take values in a fixed $(d-2)$-codimensional subset of Λ_n, so it is hopeless to efficiently connect any two points in the Weyl chamber by such moves as in Figure 1. Even worse, when $n < 2d-1$, the unions of these segments have only bounded connected components. Fortunately, when $n \geq 2d-1$, these connected components merge to form an unbounded component, and a weaker form of efficient exploration of this unbounded component is possible. Putting everything together, we obtain that $\mathrm{SL}_{2d-1}(\mathbf{Z})$ satisfies Theorem 2.3 for every $p > 2 + \frac{2}{d-1}$. Equivalently, the noncommutative L_p space of the von Neumann algebra of $\mathrm{SL}_{2d-1}(\mathbf{Z})$ does not have the CBAP for every $p > 2 + \frac{2}{d-1}$. If $d \geq 5$, by rank $2d-1$ reduction, we obtain that the same is true for every Γ is a simple Lie group of rank at least $2d-1$. In particular, if $L_p(\mathcal{L}\mathrm{SL}_3(\mathbf{Z}))$ had the CBAP for some $p > 2$, this would distinguish the von Neumann algebras of $\mathrm{SL}_3(\mathbf{Z})$ and $\mathrm{SL}_{2d-1}(\mathbf{Z})$ for large d. Even more optimistically, if conversely $L_p(\mathcal{L}\mathrm{SL}_{2d-1}(\mathbf{Z}))$ had the CBAP for $2 \leq p \leq \frac{2}{d-1}$, this would distinguish the von Neumann algebras of $\mathrm{SL}_n(\mathbf{Z})$ for odd n.

In [41] the exploration procedure of the Weyl chamber of $\mathrm{SL}_n(\mathbf{R})$ was refined, and this allowed proving strong property (T) for $\mathrm{SL}_n(\mathbf{R})$ with respect to classes of Banach spaces that become larger with n: if a Banach space X satisfies that, for some $\beta < \frac{1}{2}$, $d_k(X) = O(k^\beta)$ as $k \to \infty$, then X has strong property (T) with respect to X for every $n \geq \frac{C}{\frac{1}{2}-\beta}$. The property that $d_k(X) = o(k^{\frac{1}{2}})$ characterizes the Banach spaces with nontrivial type [55], and it is an old problem whether they automatically satisfy the stronger condition $d_k(X) = o(k^{\frac{1}{2}})$. This is, for example, the case if X has type 2.

So far, we have only talked about higher rank groups, and rank 0 reduction was used to prove strong rigidity results. But rank 0 reduction can also say something about rank 1 groups, which are not rigid in the same way as higher rank group. The following is an example.

Proposition 5.1 ([62]). *Every $\mathrm{SO}(n)$-biinvariant matrix coefficient of every representation of $\mathrm{SO}(n,1)$ on a Hilbert space is of class $C^{\frac{n}{2}-1-\varepsilon}$ outside of $\mathrm{SO}(n)$, for every $\varepsilon > 0$.*

We do not know if the regularity $\frac{n}{2} - 1$ is optimal, but the linear order of n is, already for unitary representations. For odd n, $\varepsilon = 0$ is allowed.

ACKNOWLEDGMENTS

I would like to thank Vincent Lafforgue for his beautiful discoveries, and all my other collaborators as well.

REFERENCES

[1] U. Bader, A. Furman, T. Gelander, and N. Monod, Property (T) and rigidity for actions on Banach spaces. *Acta Math.* **198** (2007), no. 1, 57–105.

[2] B. Bekka, P. de la Harpe, and A. Valette, *Kazhdan's property (T)*. New Math. Monogr. 11, Cambridge University Press, Cambridge, 2008.

[3] J. Bourgain, A Hausdorff–Young inequality for B-convex Banach spaces. *Pacific J. Math.* **101** (1982), no. 2, 255–262.

[4] M. Bożejko, Positive definite bounded matrices and a characterization of amenable groups. *Proc. Amer. Math. Soc.* **95** (1985), no. 3, 357–360.

[5] A. Brown, Lattice subgroups acting on manifolds. In *Proc. Int. Cong. Math. 2022, Vol. 5*, pp. 3388–3411, EMS Press, Berlin, 2022.

[6] A. Brown, D. Damjanovic, and Z. Zhang, C^1 actions on manifolds by lattices in lie groups. 2018, arXiv:1801.04009.

[7] A. Brown, D. Fisher, and S. Hurtado, Zimmer's conjecture: Subexponential growth, measure rigidity, and strong property (T). 2016, arXiv:1608.04995.

[8] A. Brown, D. Fisher, and S. Hurtado, Zimmer's conjecture for actions of $SL(m,\mathbf{Z})$. *Invent. Math.* **221** (2020), no. 3, 1001–1060.

[9] A. Brown, D. Fisher, and S. Hurtado, Zimmer's conjecture for non-uniform lattices and escape of mass. 2021, arXiv:2105.14541.

[10] J. de Cannière and U. Haagerup, Multipliers of the Fourier algebras of some simple Lie groups and their discrete subgroups. *Amer. J. Math.* **107** (1985), no. 2, 455–500.

[11] P. G. Casazza, Approximation properties. In *Handbook of the geometry of Banach spaces, Vol. I*, pp. 271–316, North-Holland, Amsterdam, 2001.

[12] M. Caspers and M. de la Salle, Schur and Fourier multipliers of an amenable group acting on non-commutative L^p-spaces. *Trans. Amer. Math. Soc.* **367** (2015), no. 10, 6997–7013.

[13] M. Caspers, J. Parcet, M. Perrin, and E. Ricard, Noncommutative de Leeuw theorems. *Forum Math. Sigma* **3** (2015), e21, 59.

[14] I. Chatterji, C. Druţu, and F. Haglund, Kazhdan and Haagerup properties from the median viewpoint. *Adv. Math.* **225** (2010), no. 2, 882–921.

[15] Y. de Cornulier, R. Tessera, and A. Valette, Isometric group actions on Banach spaces and representations vanishing at infinity. *Transform. Groups* **13** (2008), no. 1, 125–147.

[16] M. Cowling and U. Haagerup, Completely bounded multipliers of the Fourier algebra of a simple Lie group of real rank one. *Invent. Math.* **96** (1989), no. 3, 507–549.

[17] A. Czuroń, Property $F\ell_q$ implies property $F\ell_p$ for $1 < p < q < \infty$. *Adv. Math.* **307** (2017), 715–726.

[18] A. Czuroń and M. Kalantar, On fixed point property for l_p-representations of Kazhdan groups. 2020, arXiv:2007.15168.

[19] C. Druţu and J. M. Mackay, Random groups, random graphs and eigenvalues of p-Laplacians. *Adv. Math.* **341** (2019), 188–254.

[20] C. Druţu and P. W. Nowak, Kazhdan projections, random walks and ergodic theorems. *J. Reine Angew. Math.* **754** (2019), 49–86.

[21] P. Enflo, A counterexample to the approximation problem in Banach spaces. *Acta Math.* **130** (1973), 309–317.

[22] J. Faraut, *Analysis on Lie groups*. Cambridge Stud. Adv. Math. 110, Cambridge University Press, Cambridge, 2008.

[23] D. Fisher, Rigidity, lattices, and invariant measures beyond homogeneous dynamics. In *Proc. Int. Cong. Math. 2022, Vol. 5*, pp. 3484–3507, EMS Press, Berlin, 2022.

[24] R. Godement. Une généralisation du théorème de la moyenne pour les fonctions harmoniques. *C. R. Acad. Sci. Paris* **234** (1952), 2137–2139

[25] A. Grothendieck, Résumé de la théorie métrique des produits tensoriels topologiques. *Bol. Soc. Mat. São Paulo* **8** (1953), 1–79.

[26] A. Grothendieck, Produits tensoriels topologiques et espaces nucléaires. *Mem. Amer. Math. Soc.* **16** (1955), Chapter 1: 196 pp.; Chapter 2: 140.

[27] A. Grothendieck, *Récoltes et Semailles I, II. Réflexions et témoignage sur un passé de mathématicien*, Collection Tel, Gallimard, 2022.

[28] U. Haagerup, An example of a nonnuclear C^*-algebra, which has the metric approximation property. *Invent. Math.* **50** (1978/79), no. 3, 279–293.

[29] U. Haagerup, Group C^*-algebras without the completely bounded approximation property. *J. Lie Theory* **26** (2016), no. 3, 861–887.

[30] U. Haagerup and T. de Laat, Simple Lie groups without the approximation property. *Duke Math. J.* **162** (2013), no. 5, 925–964.

[31] U. Haagerup and T. de Laat, Simple Lie groups without the approximation property II. *Trans. Amer. Math. Soc.* **368** (2016), no. 6, 3777–3809.

[32] U. Haagerup and J. Kraus, Approximation properties for group C^*-algebras and group von Neumann algebras. *Trans. Amer. Math. Soc.* **344** (1994), no. 2, 667–699.

[33] U. Haagerup and H. Schlichtkrull, Inequalities for Jacobi polynomials. *Ramanujan J.* **33** (2014), no. 2, 227–246.

[34] R. Howe and E.-C. Tan, *Nonabelian harmonic analysis*. Universitext, Springer, New York, 1992.

[35] D. A. Kazhdan, On the connection of the dual space of a group with the structure of its closed subgroups. *Funktsional. Anal. i Prilozhen.* **1** (1967), 71–74.

[36] M. G. Kreĭn, Hermitian-positive kernels in homogeneous spaces. II. *Ukr. Mat. Zh.* **2** (1950), 10–59.

[37] T. de Laat, Approximation properties for noncommutative L^p-spaces associated with lattices in Lie groups. *J. Funct. Anal.* **264** (2013), no. 10, 2300–2322.

[38] T. de Laat and M. de la Salle, Strong property (T) for higher-rank simple Lie groups. *Proc. Lond. Math. Soc. (3)* **111** (2015), no. 4, 936–966.

[39] T. de Laat and M. de la Salle, Approximation properties for noncommutative L^p-spaces of high rank lattices and nonembeddability of expanders. *J. Reine Angew. Math.* **737** (2018), 49–69.

[40] T. de Laat and M. de la Salle, Banach space actions and L^2-spectral gap. *Anal. PDE* **14** (2021), no. 1, 45–76.

[41] T. de Laat, M. Mimura, and M. de la Salle, On strong property (T) and fixed point properties for Lie groups. *Ann. Inst. Fourier (Grenoble)* **66** (2016), no. 5, 1859–1893.

[42] V. Lafforgue, K-théorie bivariante pour les algèbres de Banach et conjecture de Baum-Connes. *Invent. Math.* **149** (2002), no. 1, 1–95.

[43] V. Lafforgue, Un renforcement de la propriété (T). *Duke Math. J.* **143** (2008), no. 3, 559–602.

[44] V. Lafforgue, Propriété (T) renforcée banachique et transformation de Fourier rapide. *J. Topol. Anal.* **1** (2009), no. 3, 191–206.

[45] V. Lafforgue, Propriété (T) renforcée et conjecture de Baum–Connes. In *Quanta of maths*, pp. 323–345, Clay Math. Proc. 11, Amer. Math. Soc., Providence, RI, 2010.

[46] V. Lafforgue, La conjecture de Baum–Connes à coefficients pour les groupes hyperboliques. *J. Noncommut. Geom.* **6** (2012), no. 1, 1–197.

[47] V. Lafforgue and M. de la Salle, Noncommutative L^p-spaces without the completely bounded approximation property. *Duke Math. J.* **160** (2011), no. 1, 71–116.

[48] O. Lavy and B. Olivier, Fixed-point spectrum for group actions by affine isometries on L_p-spaces. *Ann. Inst. Fourier (Grenoble)* **71** (2021), no. 1, 1–26.

[49] J. Lécureux, S. Witzel, and M. de la Salle, Strong property (T), weak amenability and ℓ^p-cohomology for $\tilde{A}_2$-groups. 2020, arXiv:2010.07043.

[50] B. Liao, Strong Banach property (T) for simple algebraic groups of higher rank. *J. Topol. Anal.* **6** (2014), no. 1, 75–105.

[51] B. Liao, Approximation properties for p-adic symplectic groups and lattices. 2015, arXiv:1509.04814.

[52] A. Lubotzky, S. Mozes, and M. S. Raghunathan, Cyclic subgroups of exponential growth and metrics on discrete groups. *C. R. Acad. Sci. Paris Sér. I Math.* **317** (1993), no. 8, 735–740.

[53] A. Marrakchi and M. de la Salle, Isometric actions on L_p-spaces: dependence on the value of p. 2020, arXiv:2001.02490.

M. DE LA SALLE

[54] M. Mendel and A. Naor, Nonlinear spectral calculus and super-expanders. *Publ. Math. Inst. Hautes Études Sci.* **119** (2014), 1–95.

[55] V. D. Milman and H. Wolfson, Minkowski spaces with extremal distance from the Euclidean space. *Israel J. Math.* **29** (1978), no. 2–3, 113–131.

[56] S. Neuwirth and E. Ricard, Transfer of Fourier multipliers into Schur multipliers and sumsets in a discrete group. *Canad. J. Math.* **63** (2011), no. 5, 1161–1187.

[57] H. Oh, Uniform pointwise bounds for matrix coefficients of unitary representations and applications to Kazhdan constants. *Duke Math. J.* **113** (2002), no. 1, 133–192.

[58] I. Oppenheim, Garland's method with Banach coefficients. 2020, arXiv:2009.01234.

[59] N. Ozawa, Amenable actions and applications. In *Proceedings of the International Congress of Mathematicians (ICM), Madrid, Spain, August 22–30, 2006. volume II: Invited lectures*, pp. 1563–1580, European Mathematical Society (EMS), Zürich, 2006.

[60] N. Ozawa, Examples of groups which are not weakly amenable. *Kyoto J. Math.* **52** (2012), no. 2, 333–344.

[61] P. Pansu, Cohomologie L^p des variétés à courbure négative, cas du degré 1. Conference on Partial Differential Equations and Geometry (Torino, 1988). *Rend. Sem. Mat. Univ. Politec.* (1990), 95–120, Special Issue, 1989.

[62] J. Parcet, E. Ricard, and M. de la Salle, Fourier multipliers in $\mathrm{SL}_n(\mathbf{R})$. *Duke Math. J.* **171** (2022), no. 6, 1235–1297.

[63] G. Pisier, *Introduction to operator space theory*. London Math. Soc. Lecture Note Ser. 294, Cambridge University Press, Cambridge, 2003.

[64] G. Pisier and Q. Xu, Non-commutative L^p-spaces. In *Handbook of the geometry of Banach spaces, Vol. 2*, pp. 1459–1517, North-Holland, Amsterdam, 2003.

[65] M. de la Salle, Towards strong Banach property (T) for $\mathrm{SL}(3, \mathbf{R})$. *Israel J. Math.* **211** (2015), no. 1, 105–145.

[66] M. de la Salle, *Rigidity and malleability aspects of groups and their representations*. Habilitation thesis, 2016.

[67] M. de la Salle, A local characterization of Kazhdan projections and applications. *Comment. Math. Helv.* **94** (2019), no. 3, 623–660.

[68] M. de la Salle, Strong property (T) for higher-rank lattices. *Acta Math.* **223** (2019), no. 1, 151–193.

[69] A. Szankowski, $B(H)$ does not have the approximation property. *Acta Math.* **147** (1981), no. 1–2, 89–108.

[70] G. Szegő, *Orthogonal polynomials*. Fourth edn., American Mathematical Society, Providence, R.I., 1975.

[71] I. Vergara, The p-approximation property for simple Lie groups with finite center. *J. Funct. Anal.* **273** (2017), no. 11, 3463–3503.

[72] G. Yu, Hyperbolic groups admit proper affine isometric actions on l^p-spaces. *Geom. Funct. Anal.* **15** (2005), no. 5, 1144–1151.

MIKAEL DE LA SALLE
CNRS, Université de Lyon, Institut Camille Jordan, Villeurbanne, France,
delasalle@math.univ-lyon1.fr

This Page Intentionally Left Blank

WEIGHTED FOURIER EXTENSION ESTIMATES AND APPLICATIONS

XIUMIN DU

ABSTRACT

We describe some recent results on weighted Fourier extension estimates and their applications in PDEs and geometric measure theory.

MATHEMATICS SUBJECT CLASSIFICATION 2020

Primary 42B37; Secondary 42B20

KEYWORDS

Shrodinger maximal function, Fourier restriction, weighted Fourier extension, Falconer distance set problem

INTERNATIONAL CONGRESS OF MATHEMATICIANS 2022 JULY 6–14

Proc. Int. Cong. Math. 2022, Vol. 4, pp. 3190–3201
DOI 10.4171/ICM2022/28
Published by EMS Press

The goal of this paper is to go over some recent work by the author and her collaborators on Schrödinger maximal estimates, weighted Fourier extension estimates, and Falconer distance set problem. The study of Schrödinger maximal estimates arises from the pointwise convergence problem of the solution to the Schrödinger equation raised by Carleson in the late 1970s. Weighted Fourier extension estimates are closely related to the classical Fourier restriction problem raised by Stein and have various applications in PDEs and geometric measure theory. Falconer distance set problem was introduced by Falconer in the early 1980s and remains to be a difficult and wide open question in geometric measure theory. Two major special cases of the general weighted Fourier extension estimates apply to Schrödinger maximal estimates and Falconer distance set problem.

1. SCHRÖDINGER MAXIMAL ESTIMATES

The solution to the free Schrödinger equation

$$\begin{cases} iu_t - \Delta u = 0, & (x,t) \in \mathbb{R}^n \times \mathbb{R}, \\ u(x,0) = f(x), & x \in \mathbb{R}^n, \end{cases}$$

is

$$e^{it\Delta} f(x) = (2\pi)^{-n} \int e^{i(x\cdot\xi + t|\xi|^2)} \hat{f}(\xi)\, d\xi.$$

It is not hard to show that the solution $e^{it\Delta} f(x)$ converges to the initial data f in L^2 as the time t tends to 0. However, the problem of pointwise convergence is much harder. About 40 years ago, Carleson [10] proposed the question of identifying the optimal exponent s for which $\lim_{t\to 0} e^{it\Delta} f(x) = f(x)$ almost everywhere whenever f lies in the Sobolev space $H^s(\mathbb{R}^n)$. He proved himself the convergence for $s \geq 1/4$ when $n = 1$. Dahlberg and Kenig [12] then showed that this result is sharp. The higher-dimensional case has since been studied intensely [4–6, 9, 11, 13, 15, 16, 20, 30–32, 36, 38, 39, 41, 42]. Recently, Bourgain [6] gave counterexamples showing that $s \geq \frac{n}{2(n+1)}$ is necessary for the pointwise convergence to hold. In collaboration with Guth and Li, and then with Zhang, we proved that Bourgain's bound is also sufficient (up to endpoint).

Theorem 1.1 ($n = 2$, Du, Guth, and Li [15]; $n \geq 3$, Du and Zhang [20]). *Let $n \geq 2$. For every $f \in H^s(\mathbb{R}^n)$ with $s > \frac{n}{2(n+1)}$, $\lim_{t\to 0} e^{it\Delta} f(x) = f(x)$ almost everywhere.*

The pointwise convergence problem can be approached via a standard smooth approximation argument. Indeed, the convergence holds uniformly for Schwartz functions because of their rapid decay nature. Since the space of Schwartz functions is dense in the Sobolev space, to prove that $\lim_{t\to 0} e^{it\Delta} f(x) = f(x)$ holds a.e. for any f in $H^s(\mathbb{R}^n)$, it is enough to show that the associated maximal function $\sup_{0<t\leq 1} |e^{it\Delta} f(x)|$ is bounded from $H^s(\mathbb{R}^n)$ to $L^p(B^n(c,1))$ for some $p \geq 1$ and any unit ball $B^n(c,1)$ in $\mathbb{R}^n$. Such estimates are called the *Schrödinger maximal estimates*. More precisely, results in Theorem 1.1 are consequences of the following two theorems on Schrödinger maximal estimates.

Theorem 1.2 (Du, Guth, and Li [15]). *For any $s > \frac{1}{3}$, the following bound holds: for any function $f \in H^s(\mathbb{R}^2)$,*

$$\Big\| \sup_{0<t\le 1} |e^{it\Delta} f| \Big\|_{L^3(B^2(0,1))} \le C_s \|f\|_{H^s(\mathbb{R}^2)}.$$

Theorem 1.3 (Du and Zhang [20]). *Let $n \ge 3$. For any $s > \frac{n}{2(n+1)}$, the following bound holds: for any function $f \in H^s(\mathbb{R}^n)$,*

$$\Big\| \sup_{0<t\le 1} |e^{it\Delta} f| \Big\|_{L^2(B^n(0,1))} \le C_s \|f\|_{H^s(\mathbb{R}^n)}.$$

Comparing the two Schrödinger maximal estimates above, we note that one can derive Theorem 1.3 in the case $n = 2$ from Theorem 1.2 using Hölder's inequality. Despite the fact that Theorem 1.3 in the cases $n = 1, 2$ can recover the almost sharp results of pointwise convergence problem, the sharp estimates for the L^2-norm of the Schrödinger maximal function are not as strong as the previous sharp estimates for the L^p-norm ($p = 4$ when $n = 1$ is due to Kenig, Ponce, and Vega [29], and $p = 3$ when $n = 2$ is due to Du, Guth, and Li [15]). Based on these results, it is natural to ask the following:

Question 1.4. Consider the Schrödinger maximal estimates of the form

$$\Big\| \sup_{0<t\le 1} |e^{it\Delta} f| \Big\|_{L^p(B^n(0,1))} \le C_s \|f\|_{H^s(\mathbb{R}^n)}. \tag{1.1}$$

(1) Let $n \ge 3$. Determine the optimal $p = p(n)$ for which (1.1) holds for any $s > \frac{n}{2(n+1)}$.

(2) Let $n \ge 3$ and fix $p > 2$. Identify the optimal range of $s = s(n, p)$ for which (1.1) holds.

Via a localization argument, Littlewood–Paley decomposition, and parabolic rescaling, the above question can be reduced to the problem of identifying the sharp exponent $\boldsymbol{\gamma(n, p)}$, which is the optimal γ such that

$$\Big\| \sup_{0<t\le R} |e^{it\Delta} f| \Big\|_{L^p(B^n(0,R))} \lessapprox R^\gamma \|f\|_{L^2}, \quad \forall f : \operatorname{supp} \hat{f} \subset B^n(0,1). \tag{1.2}$$

Here $A \lessapprox B$ means $A \le C_\varepsilon R^\varepsilon B$ for any $\varepsilon > 0$, $R \ge 1$. The known results [6,12,15,20,29] can be summarized as

$$\gamma(n, p) = \max\left\{ n\left(\frac{1}{p} - \frac{n}{2(n+1)}\right), 0 \right\} \tag{1.3}$$

for any $p \ge 1$ when $n = 1, 2$, and $1 \le p \le 2$ when $n \ge 3$.

It remains an interesting problem to determine $\gamma(n, p)$ for $p > 2$ when $n \ge 3$. It seemed possible that (1.3) should hold for any $p \ge 1$ and $n \ge 1$. However, we disproved this for a certain range of p when $n \ge 3$ by examining Bourgain's example [6] in all intermediate dimensions:

Theorem 1.5 (Du et al. [19]). *Let $n \ge 3$ and $p > 2$. Then*

$$\gamma(n, p) \ge \max_{m \in \mathbb{Z}, 1 \le m \le n} \left[\frac{n+m}{2} \left(\frac{1}{p} - \frac{1}{2} \right) + \frac{m}{2(m+1)} \right].$$

Let us look at two special cases of Theorem 1.5: by a direct calculation,

- if $\gamma_{n,p} = n(\frac{1}{p} - \frac{n}{2(n+1)})$, then $p \le p_0(n) := 2 + \frac{4}{(n-1)(n+2)}$;
- if $\gamma_{n,p} = 0$, then $p \ge p_1(n) := \max_{m\in\mathbb{Z}, 1\le m\le n}[2 + \frac{4}{n-1+m+n/m}]$.

Note that $p_0(n) < \frac{2(n+1)}{n} < p_1(n)$ when $n \ge 3$. Therefore, (1.3) fails for $p_0(n) < p < p_1(n)$ when $n \ge 3$.

To establish (1.2), it is helpful to consider a more general setting, in which it is convenient to use induction on scales. By formalizing the being locally constant property, one can treat $|e^{it\Delta} f(x)|$ essentially as a constant on each unit cube, and therefore the left-hand side of (1.2) is equivalent to $\|e^{it\Delta} f\|_{L^p(X)}$, where X is a union of lattice unit cubes in $B^{n+1}(0, R)$ such that each lattice vertical thin tube of dimensions $1 \times \cdots \times 1 \times R$ contains exactly one unit cube from X. In particular, the set X satisfies the condition that $|X \cap B_r| \lesssim r^n$ for any ball B_r of radius $r \ge 1$. Based on this observation, we are led to consider a slight generalization of (1.2):

Question 1.6. Let $\mathcal{X}_{n+1,R}$ denote the collection of subsets X such that each X in $\mathcal{X}_{n+1,R}$ is a union of lattice unit cubes in $B^{n+1}(0, R)$ satisfying $|X \cap B_r| \le r^n$ for any ball B_r of radius $r \ge 1$.

Let $n \ge 3$ and fix $p > 2$. Determine the sharp exponent $\tilde{\gamma}(n, p)$, which is the optimal γ for which the following holds:

$$\|e^{it\Delta} f\|_{L^p(X)} \lessapprox R^{\gamma} \|f\|_2, \quad \forall X \in \mathcal{X}_{n+1,R}, \ \forall f : \operatorname{supp} \hat{f} \subset B^n(0, 1). \tag{1.4}$$

The argument from Du, Guth, and Li [15] can be adapted to establish $\tilde{\gamma}(n = 2, p = 3) = 0$, which in turn determines $\tilde{\gamma}(n = 2, p)$ for all $p \ge 1$:

$$\tilde{\gamma}(n = 2, p) = \begin{cases} 2(\frac{1}{p} - \frac{1}{3}), & 1 \le p \le 3, \\ 0, & p \ge 3. \end{cases}$$

For general $n \ge 3$, the fractal L^2 Fourier extension estimate from Du and Zhang [20] gives the sharp exponent $\tilde{\gamma}(n, p = 2) = \frac{n}{2(n+1)}$, which also determines $\tilde{\gamma}(n, p)$ for all $1 \le p \le 2$:

$$\tilde{\gamma}(n, p) = n\left(\frac{1}{p} - \frac{n}{2(n+1)}\right), \quad \forall n \ge 3, \forall 1 \le p \le 2.$$

The main ingredients in the work of [15] include the polynomial partitioning method adapted by Guth to Fourier restriction problem [25] and Bourgain–Demeter's l^2 decoupling theorem [7]. The method of polynomial partitioning identifies algebraic structures where the Schrödinger solutions are most concentrated, and reduces the original 3-dimensional problem to an essentially 2-dimensional one. The reduced problem can then be solved by a bilinear refined Strichartz estimate, which is derived via decoupling and induction on scales.

The proof of the fractal L^2 Fourier extension estimate in [20] uses a broad–narrow analysis [5,7,8,26]. In the broad case, there are $n + 1$ transverse frequency caps in $B^n(0, 1)$ making significant contributions, and we can apply either Bennett–Carbery–Tao's multilinear restriction estimates [2] or multilinear refined Strichartz estimates of Du et al. [16]. In the narrow case, we invoke l^2 decoupling [7] in dimension n and use an induction on scales argument which is rooted in the proof of the refined Strichartz estimates [15].

It remains a wide open and challenging problem to determine $\tilde{\gamma}(n, p)$ for $n \geq 3$ and $p > 2$.

2. WEIGHTED FOURIER EXTENSION ESTIMATES

In this section, we discuss the general weighted Fourier extension estimates, which include the Schrödinger maximal estimate as a major special case. Besides their own independent interest, such general estimates have various applications in PDEs and geometric measure theory.

Definition 2.1. Let $0 < \alpha \leq d$.

(1) We say that μ is an *α-dimensional measure* in $B^d(0,1)$ if it is a positive Borel measure, supported in the unit ball $B^d(0,1)$, that satisfies

$$c_\alpha(\mu) := \sup_{x \in \mathbb{R}^d, r>0} \frac{\mu(B(x,r))}{r^\alpha} < \infty.$$

(2) We say that H is an *α-dimensional weight* in $\mathbb{R}^d$ if it is a nonnegative measurable function on $\mathbb{R}^d$ that satisfies

$$\int_{B(x,r)} H(x)\, dx \leq r^\alpha, \quad \forall x \in \mathbb{R}^d, \ \forall r \geq 1.$$

(3) Let $\mathcal{X}_{d,\alpha,R}$ denote the collection of subsets X such that each X in $\mathcal{X}_{d,\alpha,R}$ is a union of lattice unit cubes in $B^d(0,R)$ satisfying

$$\left|X \cap B(x,r)\right| \leq r^\alpha, \quad \forall x \in \mathbb{R}^d, \ \forall r \geq 1.$$

Let S denote either the unit sphere $\mathbb{S}^{d-1}$ or the truncated paraboloid $\mathbb{P}^{d-1}$. Let $d\sigma$ be the induced Lebesgue measure on S. Consider the Fourier extension operator

$$Ef(x) = E_S f(x) = \int_S e^{i\omega \cdot x} f(\omega)\, d\sigma(\omega).$$

Note that $E_{\mathbb{P}^{d-1}} f$ corresponds to free Schrödinger solutions. We are interested in the following *weighted Fourier extension estimates*:

Question 2.2. Determine the sharp exponent $\boldsymbol{\gamma(d, \alpha, p)}$, which is the optimal γ for which the following two equivalent estimates hold:

(1) For any α-dimensional weight H in $\mathbb{R}^d$ and any function $f \in L^2(S, d\sigma)$,

$$\|Ef\|_{L^p(B^d(0,R); H dx)} \lessapprox R^\gamma \|f\|_2; \tag{2.1}$$

(2) For any subset $X \in \mathcal{X}_{d,\alpha,R}$ and any function $f \in L^2(S, d\sigma)$,

$$\|Ef\|_{L^p(X)} \lessapprox R^\gamma \|f\|_2. \tag{2.2}$$

To see the equivalence of the two estimates (2.1) and (2.2), one direction is easy: given $X \in \mathcal{X}_{d,\alpha,R}$, the characteristic function of X is an α-dimensional weight in $\mathbb{R}^d$; the other direction was proved in Du and Zhang [20] where the being locally constant property and dyadic pigeonhole argument play key roles. The advantage of the expression (2.2) is that it allows us to take into account geometric structures more directly.

The case $\alpha = n = d - 1$ of (2.2) is a generalization of Schrödinger maximal estimates as described in Section 1. Estimates (2.1) for general α are related to the study of spherical average Fourier decay rates of fractal measures [3, 17, 20–23, 33–35, 40, 43]. For α around $\frac{d}{2}$, estimates (2.1) have drawn particular interest because of their application to Falconer's distance set problem [17, 20–24, 43]. The techniques in the proof of Theorems 1.2 and 1.3 are used to establish the following results:

Theorem 2.3. *Let H be an α-dimensional weight in $\mathbb{R}^d$.*

(1) (Du et al. [17]). *For $\frac{3}{2} < \alpha \leq 2$,*

$$\|Ef\|_{L^3(B^3(0,R);Hdx)} \lessapprox \|f\|_2, \quad \forall f \in L^2(S, d\sigma). \tag{2.3}$$

(2) (Du and Zhang [20]). *For $d \geq 3$ and $\frac{d}{2} < \alpha < d$,*

$$\|Ef\|_{L^2(B^d(0,R);Hdx)} \lessapprox R^{\frac{\alpha}{2d}} \|f\|_2, \quad \forall f \in L^2(S, d\sigma). \tag{2.4}$$

In particular, (2.3) gives that $\gamma(d = 3, \alpha = 2, p = 3) = 0$, which in turn determines the exact $\gamma(d = 3, \alpha = 2, p)$ for all $p \geq 1$. For $\frac{3}{2} < \alpha < 2$, it is unknown, but expected, that $\gamma(d = 3, \alpha, p) = 0$ for some p smaller than 3.

Due to a recent example in Du [14], (2.4) is sharp when $d - 1 \leq \alpha < d$, and it determines the exact value of $\gamma(d, \alpha, p)$ for all $d - 1 \leq \alpha < d$ and $1 \leq p \leq 2$. For $\alpha < d - 1$, it is expected that there is still room to improve the estimate (2.4). The key feature of the examples from [14] is that for $\alpha \in [m, m + 1]$ the corresponding examples are concentrated around hyperplanes of dimension m or $m + 1$. This explains in some way why (2.4) only gives sharp results for large α. When working towards answering Question 2.2 for small α, we need to explore new methods which can reduce the original question that in a much lower dimension.

Theorem 2.3 gives certain state-of-the-art results for several problems in PDEs and geometric measure theory, including size of the divergence set of Schrödinger solutions, Fourier decay rates of fractal measures, and Falconer distance set problem.

2.1. Divergence set of Schrödinger solutions

A natural refinement of Carleson's problem was initiated by Sjögren and Sjölin [38]: determine the size of the divergence set, in particular, consider

$$\alpha_n(s) := \sup_{f \in H^s(\mathbb{R}^n)} \dim\Big\{x \in \mathbb{R}^n : \lim_{t \to 0} e^{it\Delta} f(x) \neq f(x)\Big\},$$

where dim denotes the Hausdorff dimension. It is known that

$$\alpha_n(s) = \begin{cases} n, & \text{for } s \leq \frac{n}{2(n+1)} \quad \text{(Bourgain [6], Lucà and Rogers [32])}, \\ n - 2s, & \text{for } \frac{n}{4} \leq s \leq \frac{n}{2} \quad \text{(Žubrinić [44], Barceló et al. [1])}. \end{cases}$$

An open problem is to determine $\alpha_n(s)$ for $n \geq 2$ and $\frac{n}{2(n+1)} < s < \frac{n}{4}$. For α in this range, the best known lower bounds (examples) are due to Lucà and Rogers [31, 32], and the best known upper bounds follow from Theorem 2.3:

$$\alpha_n(s) \leq n + 1 - \frac{2(n+1)s}{n} \quad \text{(Du and Zhang [20])}.$$

An improvement of estimate (2.4) will give a better upper bound of $\alpha_n(s)$. More precisely, if

$$\|Ef\|_{L^2(B^{n+1}(0,R);Hdx)} \lessapprox R^{\gamma}\|f\|_2, \quad \forall\alpha\text{-dimensional weight } H,\ \forall f \in L^2(\mathbb{P}^n, d\sigma),$$

then we have $\alpha_n(s) \leq \alpha_0$, where α_0 is the root for α to the equation $\gamma + \frac{n-\alpha}{2} = s$ [33].

2.2. Fourier decay rates of fractal measures

Let $\beta_d(\alpha, S)$ denote the average Fourier decay rate of fractal measures, which is defined as the supremum of the numbers β for which

$$\|\widehat{\mu}(R\cdot)\|^2_{L^2(S,d\sigma)} \lesssim c_\alpha(\mu)\|\mu\|R^{-\beta}, \tag{2.5}$$

whenever $R > 1$ and μ is an α-dimensional measure in $B^d(0,1)$. The problem of identifying the value of $\beta_d(\alpha, \mathbb{S}^{d-1})$ was proposed by Mattila [35].

In dimension two, the exact decay rates are known:

$$\beta_2(\alpha, S) = \begin{cases} \alpha, & \alpha \in (0, 1/2] \quad \text{(Mattila [34])}, \\ 1/2, & \alpha \in [1/2, 1] \quad \text{(Mattila [34])}, \\ \alpha/2, & \alpha \in [1, 2] \quad \text{(Wolff [43])}. \end{cases}$$

In higher dimensions, it is known that $\beta_d(\alpha, S) = \alpha$ in the range $\alpha \in (0, \frac{d-1}{2}]$, and

$$\beta_d(\alpha, \mathbb{P}^{d-1}) = \frac{(d-1)\alpha}{d} \quad \text{for } d \geq 3 \text{ and } d-1 \leq \alpha < d \quad \text{(Du and Zhang [20], Du [14])}.$$

In other cases, $\beta_d(\alpha, S)$ is still a mystery. The current best lower bounds are

$$\beta_d(\alpha, S) \geq \begin{cases} \alpha, & \alpha \in (0, \frac{d-1}{2}] \quad \text{(Mattila [34])}, \\ \frac{d-1}{2}, & \alpha \in [\frac{d-1}{2}, \frac{d}{2}] \quad \text{(Mattila [34])}, \\ \frac{(d-1)\alpha}{d}, & \alpha \in [\frac{d}{2}, d] \quad \text{(Du et al. [17, } d=3\text{], Du and Zhang [20, } d \geq 4\text{])}. \end{cases}$$

For upper bounds, the author's recent work [14] includes a summary. New upper bounds are obtained in [14] for $\beta_d(\alpha, \mathbb{S}^{d-1})$ with $d \geq 4$, $\alpha > \frac{d}{2}$, and for $\beta_d(\alpha, \mathbb{P}^{d-1})$ with $d \geq 3$, $\alpha > \frac{d-1}{2}$.

By a duality argument and Hölder's inequality, the weighted Fourier extension estimates (2.1) and $\beta_d(\alpha, S)$ are related as follows: if

$$\|Ef\|_{L^p(B^d(0,R);Hdx)} \lessapprox R^{\gamma}\|f\|_2, \quad \forall\alpha\text{-dimensional weight } H,\ \forall f \in L^2(S, d\sigma),$$

then $\beta_d(\alpha, S) \geq 2(\frac{\alpha}{p} - \gamma)$. Therefore, in order to determine the exact $\beta_d(\alpha, S)$, it is of particular interest to study Question 2.2 for fractional dimension $\alpha \in (\frac{d-1}{2}, d-1)$.

2.3. Falconer's distance set problem

The Falconer distance set conjecture, which is a famously difficult problem in geometric measure theory, is a continuous version of the celebrated Erdős distinct distance conjecture whose two-dimensional case was resolved by Guth and Katz [28]. The study of the Falconer problem is naturally related to Fourier restriction theory, projection theory of fractal measures, and incidence geometry. It has attracted a great amount of attention over the decades and has seen some very recent breakthroughs. See [17,18,20,27] and the references therein for more details.

Let $E \subset \mathbb{R}^d$ be a compact set. Its *distance set* $\Delta(E)$ is defined by $\Delta(E) := \{|x-y| : x, y \in E\}$.

Conjecture 2.4 (Falconer [24]). *Let $d \geq 2$ and $E \subset \mathbb{R}^d$ be a compact set. Then*

$$\dim(E) > \frac{d}{2} \Rightarrow \big|\Delta(E)\big| > 0.$$

Here $|\cdot|$ denotes the Lebesgue measure and $\dim(\cdot)$ *is the Hausdorff dimension.*

Different methods have been invented to lower the dimensional threshold. To name a few landmarks: in 1985, Falconer [24] showed that $|\Delta(E)| > 0$ if $\dim(E) > \frac{d}{2} + \frac{1}{2}$. This dimensional threshold has since been lowered gradually. It was further lowered by Wolff [43] to $\frac{4}{3}$ in the case $d = 2$, and by Erdoğan [22] to $\frac{d}{2} + \frac{1}{3}$ when $d \geq 3$. These records were recently broken with the following state-of-the-art thresholds:

$$\begin{cases} \frac{5}{4}, & d = 2 \quad \text{(Guth et al. [27])}, \\ \frac{9}{5}, & d = 3 \quad \text{(Du et al. [17])}, \\ \frac{d^2}{2d-1} = \frac{d}{2} + \frac{1}{4} + \frac{1}{8d-4}, & d \geq 3 \text{ and } d \text{ is odd} \quad \text{(Du and Zhang [20])}, \\ \frac{d}{2} + \frac{1}{4}, & d \geq 4 \text{ and } d \text{ is even} \quad \text{(Du et al. [18])}. \end{cases}$$

By a classical analytic approach of Mattila [34], we can approach Falconer's problem via Fourier decay rates of fractal measures and thus via weighted Fourier extension estimates. This is the route taken in many prior works, including [17,20,22,43]. More precisely, if

$$\|Ef\|_{L^p(B^d(0,R);Hdx)} \lessapprox R^\gamma \|f\|_2, \quad \forall \alpha\text{-dimensional weight } H,\ \forall f \in L^2(S, d\sigma),$$

then $|\Delta(E)| > 0$ if $\dim(E) > \alpha_0$, where α_0 is the root for α to the equation $\alpha = d - 2(\frac{\alpha}{p} - \gamma)$.

In a recent breakthrough by Guth et al. [27], they studied the two-dimensional Falconer problem, and developed a new method that modifies the original Mattila's approach. Their argument consists primarily of two steps. First, prune the natural Frostman measure μ on E by removing "bad" wave packets at different scales, and show that the error introduced in the pruning process can be controlled. Second, apply a refined decoupling inequality to estimate some L^2 quantity involving the pruned good measure.

The above arguments do not readily extend to higher dimensions. In [27], to verify that the pruned measure is close enough to the original Frostman measure, one applies a radial projection theorem of Orponen [37] that assumes the measure has dimension $\alpha > d - 1$. However, when $d \geq 3$, this condition fails to hold if α is close enough to $\frac{d}{2}$.

In a recent work Du et al. [18], we overcame this difficulty by introducing another ingredient into the process: orthogonal projections of the original measure. Combining orthogonal projections and Orponen's radial projection theorem, we were able to remove certain *bad* part from the original measure and approach Falconer's distance set problem via the following:

Question 2.5. Prove weighted L^2 Fourier extension estimates for *good* functions:

$$\|Ef\|_{L^2(B^d(0,R);Hdx)} \lessapprox R^{\gamma}\|f\|_2, \quad \forall\alpha\text{-dimensional weight } H,\ \forall \text{ good } f \in L^2(S, d\sigma).$$

By the techniques from [18], we can define good functions as follows: we say $f \in L^2(S, d\sigma)$ is *good* if in its wave packet decomposition $f = \sum_{T\in\mathbb{T}} f_T$ (here for each wave packet f_T, Ef_T is essentially supported on a tube T of dimensions $R^{1/2} \times \cdots \times R^{1/2} \times R$, and f_T is supported on a cap $\theta = \theta(T) \subset S$ of radius $R^{-1/2}$), for each tube $T \in \mathbb{T}$ with $f_T \neq 0$,

$$\int_T H(x)\,dx \lessapprox \begin{cases} R^{\alpha}R^{-\frac{d}{4}}, & d \text{ is even,} \\ R^{\alpha}R^{-\frac{d-1}{4}}, & d \text{ is odd.} \end{cases} \tag{2.6}$$

Note that since H is α-dimensional, we have that the total weight H on $B^d(0,R)$ is $\leq R^{\alpha}$. Condition (2.6) says that a function f is good if the weight H on each relative tube from the wave packet decomposition of f is just a small proportion of the total weight. Roughly speaking, all the relative tubes are *light*. To further improve the current results for Falconer's distance set problem, one may explore other tools from geometric measure theory which could help removing more *bad* parts from the original measure and so the functions under consideration are *good* at various scales in contrast with (2.6).

FUNDING

The author was partially supported by NSF grant DMS-2107729.

REFERENCES

[1] J. A. Barceló, J. Bennett, A. Carbery, and K. M. Rogers, On the dimension of divergence sets of dispersive equations. *Math. Ann.* **349** (2011), 599–622.

[2] J. Bennett, A. Carbery, and T. Tao, On the multilinear restriction and Kakeya conjectures. *Acta Math.* **196** (2006), 261–302.

[3] J. Bourgain, Hausdorff dimension and distance sets. *Israel J. Math.* **87** (1994), no. 1–3, 193–201.

[4] J. Bourgain, In *Some new estimates on oscillatory integrals*, pp. 83–112, Princeton Math. Ser. 42, Princeton Univ. Press, Princeton, NJ, 1995.

[5] J. Bourgain, On the Schrödinger maximal function in higher dimension. *Proc. Steklov Inst. Math.* **280** (2013), no. 1, 46–60.

[6] J. Bourgain, A note on the Schrödinger maximal function. *J. Anal. Math.* **130** (2016), 393–396.

[7] J. Bourgain and C. Demeter, The proof of the l^2 decoupling conjecture. *Ann. of Math. (2)* **182** (2015), no. 1, 351–389.

[8] J. Bourgain and L. Guth, Bounds on oscillatory integral operators based on multilinear estimates. *Geom. Funct. Anal.* **21** (2011), no. 6, 1239–1295.

[9] A. Carbery, Radial Fourier multipliers and associated maximal functions. In *Recent progress in Fourier analysis (El Escorial, 1983)*, pp. 49–56, North-Holl. Math. Stud. 111, Notas Mat., 101, North-Holland, Amsterdam, 1985.

[10] L. Carleson, Some analytic problems related to statistical mechanics. In *Euclidean Harmonic Analysis (Proc. Sem., Univ. Maryland, College Park, MD, 1979)*, pp. 5–45, Lecture Notes in Math. 779, Springer, Berlin, Heidelberg, 1979.

[11] M. Cowling, Pointwise behavior of solutions to Schrödinger equations. In *Harmonic Analysis (Cortona, 1982)*, pp. 83–90, Lecture Notes in Math. 992, Springer, Berlin, 1983.

[12] B. E. J. Dahlberg and C. E. Kenig, A note on the almost everywhere behavior of solutions to the Schrödinger equation. In *Harmonic Analysis (Minneapolis, MN, 1981)* pp. 205–209, Lecture Notes in Math. 908, Springer, Berlin, Heidelberg, 1981.

[13] C. Demeter and S. Guo, Schrödinger maximal function estimates via the pseudoconformal transformation. 2016, arXiv:1608.07640.

[14] X. Du, Upper bounds for Fourier decay rates of fractal measures. *J. Lond. Math. Soc. (2)* **102** (2020), no. 3, 1318–1336.

[15] X. Du, L. Guth, and X. Li, A sharp Schrödinger maximal estimate in $\mathbb{R}^2$. *Ann. of Math. (2)* **186** (2017), no. 2, 607–640.

[16] X. Du, L. Guth, X. Li, and R. Zhang, Pointwise convergence of Schrödinger solutions and multilinear refined Strichartz estimate. *Forum Math. Sigma* **6** (2018), e14.

[17] X. Du, L. Guth, Y. Ou, H. Wang, B. Wilson, and R. Zhang, Weighted restriction estimates and application to Falconer distance set problem. *Amer. J. Math.* **143** (2021), no. 1, 175–211.

[18] X. Du, A. Iosevich, Y. Ou, H. Wang, and R. Zhang, An improved result for Falconer's distance set problem in even dimensions. *Math. Ann.* **380** (2021), no. 3–4, 1215–1231.

[19] X. Du, J. Kim, H. Wang, and R. Zhang, Lower bounds for estimates of the Schrödinger maximal function. *Math. Res. Lett.* **27** (2020), no. 3, 687–692.

[20] X. Du and R. Zhang, Sharp L^2 estimate of the Schrödinger maximal function in higher dimensions. *Ann. of Math. (2)* **189** (2019), no. 3, 837–861.

[21] M. B. Erdoğan, A note on the Fourier transform of fractal measures. *Math. Res. Lett.* **11** (2004), no. 2–3, 299–313.

[22] M. B. Erdoğan, A bilinear Fourier extension theorem and applications to the distance set problem. *Int. Math. Res. Not.* **23** (2005), 1411–1425.

[23] M. B. Erdoğan, On Falconer's distance set conjecture. *Rev. Mat. Iberoam.* **22** (2006), no. 2, 649–662.

[24] K. J. Falconer, On the Hausdorff dimensions of distance sets. *Mathematika* **32** (1985), no. 2, 206–212.

[25] L. Guth, A restriction estimate using polynomial partitioning. *J. Amer. Math. Soc.* **29** (2016), no. 2, 371–413.

[26] L. Guth, Restriction estimates using polynomial partitioning II. *Acta Math.* **221** (2018), no. 1, 81–142.

[27] L. Guth, A. Iosevich, Y. Ou, and H. Wang, On Falconer's distance set problem in the plane. *Invent. Math.* **219** (2020), no. 3, 779–830.

[28] L. Guth and N. H. Katz, On the Erdős distinct distance problem in the plane. *Ann. of Math. (2)* **181** (2015), no. 1, 155–190.

[29] C. E. Kenig, G. Ponce, and L. Vega, Oscillatory integrals and regularity of dispersive equations. *Indiana Univ. Math. J.* **40** (1991), no. 1, 33–69.

[30] S. Lee, On pointwise convergence of the solutions to Schrödinger equations in $\mathbb{R}^2$. *Int. Math. Res. Not.* **2006** (2006), 1–21, Art. ID 32597.

[31] R. Lucà and K. Rogers, Coherence on fractals versus pointwise convergence for the Schrödinger equation. *Comm. Math. Phys.* **351** (2017), no. 1, 341–359.

[32] R. Lucà and K. Rogers, A note on pointwise convergence for the Schrödinger equation. *Math. Proc. Cambridge Philos. Soc.* (2017).

[33] R. Lucà and K. Rogers, Average decay for the Fourier transform of measures with applications. *J. Eur. Math. Soc. (JEMS)* **21** (2019), no. 2, 465–506.

[34] P. Mattila, Spherical averages of Fourier transforms of measures with finite energy; dimensions of intersections and distance sets. *Mathematika* **34** (1987), no. 2, 207–228.

[35] P. Mattila, Hausdorff dimension, projections, and the Fourier transform. *Publ. Mat.* **48** (2004), no. 1, 3–48.

[36] A. Moyua, A. Vargas, and L. Vega, Schrödinger maximal function and restriction properties of the Fourier transform. *Int. Math. Res. Not.* **16** (1996), 793–815.

[37] T. Orponen, On the dimension and smoothness of radial projections. *Anal. PDE* **12** (2019), no. 5, 1273–1294.

[38] P. Sjögren and P. Sjölin, Convergence properties for the time-dependent Schrödinger equation. *Ann. Acad. Sci. Fenn.* **14** (1989), no. 1, 13–25.

[39] P. Sjölin, Regularity of solutions to the Schrödinger equation. *Duke Math. J.* **55** (1987), no. 3, 699–715.

[40] P. Sjölin, Estimates of spherical averages of Fourier transforms and dimensions of sets. *Mathematika* **40** (1993), no. 2, 322–330.

[41] T. Tao and A. Vargas, A bilinear approach to cone multipliers. II. Applications. *Geom. Funct. Anal.* **10** (2000), no. 1, 216–258.

[42] L. Vega, Schrödinger equations: pointwise convergence to the initial data. *Proc. Amer. Math. Soc.* **102** (1988), no. 4, 874–878.

[43] T. Wolff, Decay of circular means of Fourier transforms of measures. *Int. Math. Res. Not.* **10** (1999), 547–567.

[44] D. Žubrinić, Singular sets of Sobolev functions. *C. R. Math. Acad. Sci. Paris* **334** (2002), 539–544.

XIUMIN DU

Mathematics Department, Northwestern University, Evanston, IL 60208, USA, xdu@northwestern.edu

NONCOMMUTATIVE ERGODIC THEORY OF HIGHER RANK LATTICES

CYRIL HOUDAYER

ABSTRACT

We survey recent results regarding the study of dynamical properties of the space of positive definite functions and characters of higher rank lattices. These results have several applications to ergodic theory, topological dynamics, unitary representation theory, and operator algebras. The key novelty in our work is a dynamical dichotomy theorem for equivariant faithful normal unital completely positive maps between noncommutative von Neumann algebras and the space of bounded measurable functions defined on the Poisson boundary of semisimple Lie groups.

MATHEMATICS SUBJECT CLASSIFICATION 2020

Primary 22D25; Secondary 22E40, 37B05, 46L10, 46L55

KEYWORDS

Characters, higher rank lattices, Poisson boundaries, semisimple Lie groups, unitary representations, von Neumann algebras

INTERNATIONAL CONGRESS OF MATHEMATICIANS 2022 JULY 6–14

Proc. Int. Cong. Math. 2022, Vol. 4, pp. 3202–3223
DOI 10.4171/ICM2022/39

Published by EMS Press

1. INTRODUCTION AND MAIN RESULTS

In order to explain the motivation for our work and to state our main results, we set up the following terminology regarding *higher rank lattices*.

Terminology. Let G be any connected semisimple real Lie group with finite center, no nontrivial compact factor, and real rank $\mathrm{rk}_{\mathbf{R}}(G) \geq 2$. Let $\Gamma < G$ be any *irreducible lattice*, meaning that Γ is a discrete subgroup of G with finite covolume such that $N \cdot \Gamma$ is a dense subgroup of G for every noncentral closed normal subgroup $N \lhd G$. In what follows, if all the above conditions are satisfied, then we simply say that $\Gamma < G$ is a *higher rank lattice*.

The following examples of higher rank lattices are particular cases of general results due to Borel–Harish-Chandra [11].

Examples. For every $d \geq 2$, the special linear group $\mathrm{SL}_d(\mathbf{R})$ is a connected simple real Lie group with finite center $\mathscr{Z}(\mathrm{SL}_d(\mathbf{R})) = \{\pm 1_d\}$ and real rank $\mathrm{rk}_{\mathbf{R}}(\mathrm{SL}_d(\mathbf{R})) = d - 1$.

(1) For every $d \geq 3$, $\mathrm{SL}_d(\mathbf{Z}) < \mathrm{SL}_d(\mathbf{R})$ is a higher rank lattice.

(2) For every $d \geq 2$ and every square free integer $q \in \mathbf{N}$,

$$\Gamma := \{(g, g^\sigma) \mid g \in \mathrm{SL}_d(\mathbf{Z}[\sqrt{q}])\} < \mathrm{SL}_d(\mathbf{R}) \times \mathrm{SL}_d(\mathbf{R}) := G$$

is a higher rank lattice, where σ is the order-2 automorphism of $\mathbf{Q}(\sqrt{q})$.

The main inspiration for our work is Margulis' celebrated *normal subgroup theorem* which states that for any higher rank lattice $\Gamma < G$, any normal subgroup $N \lhd \Gamma$ is either finite and contained in $\mathscr{Z}(\Gamma)$ or N has finite index in Γ (see [41, THEOREM IV.4.9]). Margulis' remarkable strategy to prove the normal subgroup theorem consists of two "halves": the *amenability half* and the *property* (T) *half*. Indeed, assuming that $N \lhd \Gamma$ is a noncentral normal subgroup, to prove that the quotient group Γ/N is finite, Margulis showed that Γ/N is both amenable and has property (T). The proof of the amenability half relies on Margulis' *factor theorem* which states that any measurable Γ-factor of the homogeneous space G/P, where $P < G$ is a minimal parabolic subgroup, is measurably isomorphic to a G-factor whence of the form G/Q, where $P < Q < G$ is an intermediate parabolic subgroup (see [41, THEOREM IV.2.11]). Margulis' strategy has been used to prove a normal subgroup theorem for various classes of irreducible lattices in product groups (see [4, 16, 24, 49]) and to understand the structure of point stabilizers of ergodic probability measure preserving actions of higher rank lattices (see [23, 50]). More recently, Margulis' strategy has been adapted to the noncommutative setting to study *characters* of higher rank lattices (see [22, 46]).

In that respect, for any countable discrete group Λ, we denote by $\mathscr{P}(\Lambda)$ the space of positive definite functions $\varphi : \Lambda \to \mathbf{C}$ normalized so that $\varphi(e) = 1$. Then $\mathscr{P}(\Lambda) \subset \ell^\infty(\Lambda)$ is a weak-$*$ compact convex subset. Thanks to the Gelfand–Naimark–Segal (GNS) construction, to any positive definite function $\varphi \in \mathscr{P}(\Lambda)$ corresponds a triple $(\pi_\varphi, \mathscr{H}_\varphi, \xi_\varphi)$, where $\pi_\varphi : \Lambda \to \mathscr{U}(\mathscr{H}_\varphi)$ is a unitary representation and $\xi_\varphi \in \mathscr{H}_\varphi$ is a unit vector such that the linear span of $\pi_\varphi(\Lambda)\xi_\varphi$ is dense in $\mathscr{H}_\varphi$ and

$$\forall \gamma \in \Lambda, \quad \varphi(\gamma) = \langle \pi_\varphi(\gamma)\xi_\varphi, \xi_\varphi \rangle.$$

We consider the conjugation action $\Lambda \curvearrowright \mathscr{P}(\Lambda)$ defined by

$$\forall \gamma, g \in \Lambda, \forall \varphi \in \mathscr{P}(\Lambda), \quad (\gamma\varphi)(g) := \varphi(\gamma^{-1} g \gamma).$$

A fixed point $\varphi \in \mathscr{P}(\Lambda)$ for the conjugation action is called a *character*. We denote by $\mathrm{Char}(\Lambda) \subset \mathscr{P}(\Lambda)$ the weak-$*$ compact convex subset of all characters. Any countable discrete group Λ always admits at least two characters: the *trivial* character 1_Λ and the *regular* character δ_e. The GNS representation of the regular character δ_e coincides with the left regular representation $\lambda : \Lambda \to \mathscr{U}(\ell^2(\Lambda))$. An important source of characters comes from ergodic theory. Indeed, for any probability measure preserving action $\Lambda \curvearrowright (X, \nu)$ on a standard probability space, the function $\varphi_\nu : \Lambda \to \mathbf{C} : \gamma \mapsto \nu(\mathrm{Fix}(\gamma))$ defines a character. The action $\Lambda \curvearrowright (X, \nu)$ is (essentially) free if and only if the above character φ_ν is equal to δ_e.

For any unitary representation $\pi : \Lambda \to \mathscr{U}(\mathscr{H}_\pi)$, we consider the unital C^*-algebra

$$\mathrm{C}^*_\pi(\Lambda) := \mathrm{C}^*\big(\{\pi(\gamma) \mid \gamma \in \Lambda\}\big) \subset \mathrm{B}(\mathscr{H}_\pi)$$

endowed with the conjugation action $\mathrm{Ad}(\pi) : \Lambda \curvearrowright \mathrm{C}^*_\pi(\Lambda)$. We then regard the state space $\mathfrak{S}(\mathrm{C}^*_\pi(\Lambda))$ as a Λ-invariant weak-$*$ compact convex subset of $\mathscr{P}(\Lambda)$ via the mapping $\mathfrak{S}(\mathrm{C}^*_\pi(\Lambda)) \hookrightarrow \mathscr{P}(\Lambda) : \psi \mapsto \psi \circ \pi$. When $\pi = \lambda$ is the left regular representation, $\mathrm{C}^*_\lambda(\Lambda)$ is the *reduced* group C^*-algebra which is endowed with the canonical faithful trace τ_Λ defined by $\tau_\Lambda : \mathrm{C}^*_\lambda(\Lambda) \to \mathbf{C} : a \mapsto \langle a\delta_e, \delta_e \rangle$.

Given unitary representations $\pi_i : \Lambda \to \mathscr{U}(\mathscr{H}_i)$, $i = 1, 2$, we say that π_2 is *weakly contained* in π_1 if the map $\pi_1(\Lambda) \to \pi_2(\Lambda) : \pi_1(\gamma) \mapsto \pi_2(\gamma)$ is well defined and extends to a $*$-homomorphism $\mathrm{C}^*_{\pi_1}(\Lambda) \to \mathrm{C}^*_{\pi_2}(\Lambda)$. Following **[5]**, we say that a unitary representation $\pi : \Lambda \to \mathscr{U}(\mathscr{H}_\pi)$ is *amenable* if the trivial representation 1_Λ is weakly contained in $\pi \otimes \overline{\pi}$. If π contains a finite dimensional subrepresentation, then π is amenable. If Λ has property (T), then conversely any amenable representation $\pi : \Lambda \to \mathscr{U}(\mathscr{H}_\pi)$ contains a finite-dimensional subrepresentation.

We now present in a unified way the main results we obtained in **[13]** (joint work with R. Boutonnet) and **[3]** (joint work with U. Bader, R. Boutonnet, and J. Peterson). Our first main result deals with the *existence of characters*. It is a fixed point theorem for the affine action of higher rank lattices on their space of positive definite functions.

Theorem A (**[3, 13]**). *Let $\Gamma < G$ be any higher rank lattice. Then any nonempty Γ-invariant weak-$*$ compact convex subset $\mathscr{C} \subset \mathscr{P}(\Gamma)$ contains a character.*

Our second main result deals with the *classification of characters* of higher rank lattices. Bekka **[6]** obtained the first character rigidity results in the case $\Gamma = \mathrm{SL}_d(\mathbf{Z})$ for $d \geq 3$. More recently, using a different approach based on Margulis' strategy discussed above, Peterson **[46]** obtained character rigidity results for arbitrary higher rank lattices (see also **[22]** for the case of irreducible lattices in certain product groups). The operator-algebraic framework we developed in **[3, 13]** enables us to obtain a new and more conceptual proof of Peterson's character rigidity results **[46]**.

Theorem B (Peterson, **[46]**). *Let $\Gamma < G$ be any higher rank lattice. Then any character $\varphi \in \mathrm{Char}(\Gamma)$ is either supported on $\mathscr{Z}(\Gamma)$ or its GNS representation π_φ is amenable.*

In case G has a simple factor with property (T), *any character* $\varphi \in \mathrm{Char}(\Gamma)$ *is either supported on* $\mathscr{Z}(\Gamma)$ *or its GNS representation* π_φ *contains a finite dimensional subrepresentation.*

Theorem B generalizes Margulis' normal subgroup theorem [41] and Stuck–Zimmer's stabilizer rigidity theorem [50]. Also, Theorem B solved a conjecture formulated by Connes (see [38]). For other recent results regarding classification of characters, we refer the reader to [7, 9, 22, 40, 47].

Combining Theorems A and B, we obtain new results regarding the simplicity and the unique trace property for the C^*-algebra $\mathrm{C}^*_\pi(\Gamma)$ associated with an arbitrary nonamenable (resp. weakly mixing) unitary representation $\pi : \Gamma \to \mathscr{U}(\mathscr{H}_\pi)$. In particular, Corollary C provides a far reaching generalization of the results obtained by Bekka–Cowling–de la Harpe [8] for the reduced C^*-algebra $\mathrm{C}^*_\lambda(\Gamma)$.

Corollary C ([3, 13]). *Let* $\Gamma < G$ *be any higher rank lattice. Let* $\pi : \Gamma \to \mathscr{U}(\mathscr{H}_\pi)$ *be any unitary representation. Then* $\mathrm{C}^*_\pi(\Gamma)$ *admits a trace.*

Assume, moreover, that G has trivial center. If π *is not amenable, then* λ *is weakly contained in* π *and the unique* $*$*-homomorphism* $\Theta : \mathrm{C}^*_\pi(\Gamma) \to \mathrm{C}^*_\lambda(\Gamma) : \pi(\gamma) \mapsto \lambda(\gamma)$ *satisfies the following properties:*

(1) $\tau_\Gamma \circ \Theta$ *is the unique trace on* $\mathrm{C}^*_\pi(\Gamma)$.

(2) $\ker(\Theta)$ *is the unique proper maximal ideal of* $\mathrm{C}^*_\pi(\Gamma)$.

In case G has property (T), *the above properties hold as soon as* π *does not contain any nonzero finite-dimensional subrepresentation.*

In case Λ is a countable discrete group and $\pi = \lambda$, Hartman–Kalantar [34] initiated the study of the noncommutative dynamical system $\Lambda \curvearrowright \mathrm{C}^*_\lambda(\Lambda)$ and obtained a new characterization of the simplicity and the unique trace property for $\mathrm{C}^*_\lambda(\Lambda)$ (see also [15, 39]).

As a byproduct of our operator algebraic methods, we also obtain a topological analogue of Stuck–Zimmer's stabilizer rigidity theorem [50]. In particular, our next result gives a positive answer to a recent problem raised by Glasner–Weiss [32].

Theorem D ([3, 13]). *Let* $\Gamma < G$ *be any higher rank lattice and assume that G has trivial center. Let* $\Gamma \curvearrowright X$ *be any minimal action on a compact space. Then at least one of the following assertions holds:*

(1) *There exists a* Γ*-invariant Borel probability measure on* X.

(2) *The action* $\Gamma \curvearrowright X$ *is topologically free.*

In case G has a simple factor with property (T), *either* X *is finite or the action* $\Gamma \curvearrowright X$ *is topologically free.*

As we will explain, all the main results stated above are consequences of a dynamical dichotomy theorem for Γ-equivariant faithful normal unital completely positive (ucp) maps $\Phi : M \to \mathrm{L}^\infty(G/P)$, where M is an arbitrary von Neumann algebra endowed with an ergodic action $\Gamma \curvearrowright M$ (see Theorem E and Theorem 3.8 below). This dynamical dichotomy theorem is one of the key novelties of our operator algebraic framework. Both its statement and its proof rely on von Neumann algebra theory and depend heavily on whether the connected semisimple real Lie group G is simple or not.

In our first joint work [13], we dealt with the case where G is simple and $\mathrm{rk}_{\mathbf{R}}(G) \geq 2$ (and, more generally, where all its simple factors G_i satisfy $\mathrm{rk}_{\mathbf{R}}(G_i) \geq 2$). In that case, we obtained the following noncommutative analogue of Nevo–Zimmer's structure theorem [43, 44]. We denote by $P < G$ a minimal parabolic subgroup and whenever $P < Q < G$ is an intermediate parabolic subgroup, we denote by $p_Q : G/P \to G/Q : gP \mapsto gQ$ the canonical factor map and by $p_Q^* : \mathrm{L}^\infty(G/Q) \to \mathrm{L}^\infty(G/P) : f \mapsto f \circ p_Q$ the corresponding unital normal embedding.

Theorem E ([13]). *Let $\Gamma < G$ be any higher rank lattice and assume that G is simple. Let M be any von Neumann algebra, $\Gamma \curvearrowright M$ any ergodic action, and $\Phi : M \to \mathrm{L}^\infty(G/P)$ any Γ-equivariant faithful normal ucp map. Then the following dichotomy holds:*

- *Either $\Phi(M) = \mathbf{C}1$,*
- *Or there exist a proper parabolic subgroup $P < Q < G$ and a Γ-equivariant unital normal embedding $\iota : \mathrm{L}^\infty(G/Q) \hookrightarrow M$ such that $\Phi \circ \iota = p_Q^*$.*

Theorem E extends the work of Nevo–Zimmer in two ways. Firstly, we deal with arbitrary von Neumann algebras M (instead of measure spaces (X, ν)) and secondly, we deal with Γ-actions $\Gamma \curvearrowright M$ (instead of G-actions $G \curvearrowright (X, \nu)$). We refer to Section 3 for the correspondence between equivariant ucp maps and stationary states. The remarkable feature of Theorem E is that when $\Phi : M \to \mathrm{L}^\infty(G/P)$ is not invariant, there is a nontrivial Γ-invariant *commutative* von Neumann subalgebra $M_0 \subset M$ such that $M_0 \cong \mathrm{L}^\infty(G/Q)$. This allows us to exploit the dynamical properties of the ergodic action $\Gamma \curvearrowright G/Q$ and the fact that every noncentral element $\gamma \in \Gamma \setminus \mathscr{Z}(\Gamma)$ acts (essentially) freely on G/Q. In particular, in [13], we used Theorem E to derive all the main results stated above.

In our second joint work [3], we dealt with the case where G is not simple. We point out that when G has a rank one simple factor (e.g., $G = \mathrm{SL}_2(\mathbf{R}) \times \mathrm{SL}_2(\mathbf{R})$), a Nevo–Zimmer structure theorem does not hold, and the method used in [13] to prove the main results does not apply. Instead, we proceeded as follows [3]. Firstly, for *any* higher rank lattice $\Gamma < G$, we formulated a general dynamical dichotomy theorem *invariant vs. singular* for Γ-equivariant faithful normal ucp maps $\Phi : M \to \mathrm{L}^\infty(G/P)$ (see Theorem 3.8 below) and we showed that all the main results stated above can be derived from this general dichotomy theorem. In case G is simple, the dynamical dichotomy Theorem 3.8 is a straightforward consequence of Theorem E. Secondly, to prove the dynamical dichotomy Theorem 3.8 in the case where G is not simple, we developed a new method based on the product structure in G. In that respect, the tools developed in [13] and [3] are complementary.

For any nonsingular action $\Lambda \curvearrowright (X, \nu)$ on a standard probability space, we denote by $\mathrm{L}(\Lambda \curvearrowright X)$ the corresponding *group measure space* von Neumann algebra (see Section 2).

For higher rank latttices $\Gamma < G$ where G is simple with trivial center, we present yet another application of Theorem E that appeared in [2] (joint work with U. Bader and R. Boutonnet). The next result can be regarded as a noncommutative analogue of Margulis' factor theorem.

Corollary F ([2]). *Let $\Gamma < G$ be any higher rank lattice and assume that G is simple with trivial center. Let $\mathrm{L}(\Gamma) \subset M \subset \mathrm{L}(\Gamma \curvearrowright G/P)$ be any intermediate von Neumann subalgebra. Then there is a unique intermediate parabolic subgroup $P < Q < G$ such that*

$$M = \mathrm{L}(\Gamma \curvearrowright G/Q).$$

In Section 2, we give some preliminary background on Poisson boundaries, semisimple Lie groups and operator algebras. In Section 3, we introduce the notion of boundary structures for von Neumann algebras and state the dynamical dichotomy theorem. We also outline the main steps of the proof of Theorem E. In Section 4, we sketch the proofs of Theorems A, B and Corollary C based on the dynamical dichotomy theorem. We also discuss open problems related to our main results. In Section 5, we discuss Corollary F and its relevance for Connes' rigidity conjecture.

Remark. In this survey article, we only consider higher rank lattices in connected semisimple *real* Lie groups to simplify the exposition and to focus on the main ideas. However, we point out that all our main results do hold for higher rank lattices in semisimple algebraic groups defined over arbitrary local fields. We refer the reader to [2,3] for more general statements and further details.

2. PRELIMINARIES

2.1. Poisson boundaries

Let H be any locally compact second countable (lcsc) group. We say that a Borel probability measure $\mu \in \mathrm{Prob}(H)$ is *admissible* if the following conditions are satisfied:

(1) μ is absolutely continuous with respect to the Haar measure;

(2) $\mathrm{supp}(\mu)$ generates H as a semigroup;

(3) $\mathrm{supp}(\mu)$ contains a neighborhood of the identity element $e \in H$.

We say that a bounded measurable function $F : H \to \mathbf{C}$ is (right) *μ-harmonic* if

$$\forall g \in H, \quad F(g) = \int_H F(gh)\, \mathrm{d}\mu(h).$$

Any μ-harmonic function is continuous. We denote by $\mathrm{Har}^\infty(H, \mu) \subset \mathrm{C}_b(H)$ the space of all (right) μ-harmonic functions. The left translation action $\lambda : H \curvearrowright \mathrm{C}_b(H)$ leaves the subspace $\mathrm{Har}^\infty(H, \mu)$ globally invariant.

Let (X, ν) be any standard probability space endowed with a measurable action $H \curvearrowright X$. We say that (X, ν) is a *(H, μ)-space* if ν is *μ-stationary*, that is, $\mu * \nu = \nu$. For any (H, μ)-space (X, ν), define the *Poisson transform* $\Psi_\mu : \mathrm{L}^\infty(X, \nu) \to \mathrm{Har}^\infty(H, \mu)$ by the formula

$$\forall f \in \mathrm{L}^\infty(X,\nu), \forall g \in H, \quad \Psi_\mu(f)(g) = \int_X f(gx)\,\mathrm{d}\nu(x).$$

The mapping $\Psi_\mu : \mathrm{L}^\infty(X,\nu) \to \mathrm{Har}^\infty(H,\mu)$ is H-equivariant, unital, positive, and contractive.

Theorem 2.1 (Furstenberg, [26]). *There exists a unique (H,μ)-space (B,ν_B) for which the Poisson transform $\Psi_\mu : \mathrm{L}^\infty(B,\nu_B) \to \mathrm{Har}^\infty(H,\mu)$ is bijective.*

The (H,μ)-space (B,ν_B) is called the (H,μ)-*Poisson boundary*. For a construction of the (H,μ)-space (B,ν_B), we also refer to [4, 25]. The (H,μ)-space (B,ν_B) enjoys remarkable ergodic theoretic properties. In that respect, let $(E, \|\cdot\|)$ be any separable continuous isometric Banach H-module and $\mathscr{C} \subset E^*$ any nonempty H-invariant weak-$*$ compact convex subset. Denote by $\mathrm{Bar} : \mathrm{Prob}(\mathscr{C}) \to \mathscr{C}$ the H-equivariant continuous barycenter map. A point $c \in \mathscr{C}$ is μ-*stationary* if $\mathrm{Bar}(\iota_{c*}\mu) = c$ where $\iota_c : H \to \mathscr{C} : g \mapsto gc$ is the orbit map associated with $c \in \mathscr{C}$. By Markov–Kakutani's fixed point theorem, the subset $\mathscr{C}_\mu \subset \mathscr{C}$ of all μ-stationary points in $\mathscr{C}$ is not empty.

The following theorem due to Furstenberg provides the existence (and uniqueness) of boundary maps (see also [4, SECTION 2]).

Theorem 2.2 (Furstenberg, [26]). *Let $c \in \mathscr{C}_\mu$ be any μ-stationary point. Then there exists an (essentially) unique H-equivariant measurable map $\beta : B \to \mathscr{C}$ such that*

$$\mathrm{Bar}(\beta_*\nu_B) = c.$$

We say that $\beta : B \to \mathscr{C}$ is the H-equivariant boundary map *associated with $c \in \mathscr{C}_\mu$.*

2.2. Semisimple Lie groups

Let G be any connected semisimple real Lie group with finite center and no nontrivial compact factors. Fix an Iwasawa decomposition $G = KAV$, where $K < G$ is a maximal compact subgroup, $A < G$ is a Cartan subgroup, and $V < G$ is a unipotent subgroup. Denote by $L := \mathscr{Z}_G(A)$ the centralizer of A in G and set $P := LV$. Then $P < G$ is a minimal parabolic subgroup. Since $K \curvearrowright G/P$ is transitive, G/P is a compact homogeneous space, and there exists a unique K-invariant Borel probability measure $\nu_P \in \mathrm{Prob}(G/P)$. The measure class of ν_P coincides with the unique G-invariant measure class on G/P.

Example 2.3. Assume that $G = \mathrm{SL}_d(\mathbf{R})$ for $d \geq 2$. Then we may take $K = \mathrm{SO}_d(\mathbf{R})$, $A < G$ the subgroup of diagonal matrices, and $V < G$ the subgroup of strict upper triangular matrices. In that case, $P = AV < G$ is the subgroup of upper triangular matrices. The homogeneous space G/P is the *full flag variety* which consists of all flags $\{0\} \subset W_1 \subset \cdots \subset W_d = \mathbf{R}^d$, where $W_i \subset \mathbf{R}^d$ is a vector subspace such that $\dim_{\mathbf{R}}(W_i) = i$ for every $1 \leq i \leq d$.

Observe that, for any left K-invariant Borel probability measure $\mu_G \in \mathrm{Prob}(G)$, the probability measure $\mu_G * \nu_P$ is K-invariant on G/P and so $\mu_G * \nu_P = \nu_P$, that is,

$(G/P, \nu_P)$ is a (G, μ_G)-space. Furstenberg [27] proved the following fundamental result describing the Poisson boundary of semisimple Lie groups.

Theorem 2.4 (Furstenberg, [27]). *Let $\mu_G \in \mathrm{Prob}(G)$ be any K-invariant admissible Borel probability measure. Then $(G/P, \nu_P)$ is the (G, μ_G)-Poisson boundary.*

For lattices $\Gamma < G$ in connected semisimple *real* Lie groups as above, Furstenberg [28] also showed that $(G/P, \nu_P)$ can be regarded as the (Γ, μ_Γ)-Poisson boundary with respect to a well chosen probability measure $\mu_\Gamma \in \mathrm{Prob}(\Gamma)$ (see also [25] and the references therein).

Theorem 2.5 (Furstenberg, [28]). *Let $\Gamma < G$ be any lattice. Then there exists a probability measure $\mu_\Gamma \in \mathrm{Prob}(\Gamma)$ with full support such that $(G/P, \nu_P)$ is the (Γ, μ_Γ)-Poisson boundary.*

We call a probability measure $\mu_\Gamma \in \mathrm{Prob}(\Gamma)$ as in Theorem 2.5 a *Furstenberg measure*. Combining Theorems 2.4 and 2.5, we have

$$\mathrm{Har}^\infty(G, \mu_G) \underset{G\text{-equiv.}}{\cong} \mathrm{L}^\infty(G/P, \nu_P) \underset{\Gamma\text{-equiv.}}{\cong} \mathrm{Har}^\infty(\Gamma, \mu_\Gamma).$$

A combination of Theorem 2.5 and [33] implies that, for any intermediate parabolic subgroup $P < Q < G$, the map

$$\delta \circ p_Q : G/P \to \mathrm{Prob}(G/Q) : gP \mapsto \delta_{gQ} \tag{2.1}$$

is the (essentially) unique Γ-equivariant measurable mapping $\zeta : G/P \to \mathrm{Prob}(G/Q)$.

2.3. Operator algebras

A C^*-*algebra* A is a Banach $*$-algebra endowed with a complete norm $\|\cdot\|$ that satisfies the C^*-identity $\|a^*a\| = \|a\|^2$, for every $a \in A$. Any C^*-algebra A admits a faithful isometric $*$-representation on a Hilbert space $\pi : A \to \mathrm{B}(\mathscr{H})$. After identifying A with $\pi(A)$, we may regard $A \subset \mathrm{B}(\mathscr{H})$ as a concrete C^*-algebra. Unless stated otherwise, all C^*-algebras and all linear mappings between C^*-algebras are always assumed to be unital.

We denote by $\mathfrak{S}(A)$ the *state space* of A. Then $\mathfrak{S}(A) \subset \mathrm{Ball}(A^*)$ is a weak-$*$ compact convex subset. We say that an action $\sigma : H \curvearrowright A$ is *continuous* if the action map $H \times A \to A : (g, a) \mapsto \sigma_g(a)$ is continuous. We then simply say that A is a H-C^*-algebra. The continuous action $H \curvearrowright A$ induces a weak-$*$ continuous affine action $H \curvearrowright \mathfrak{S}(A)$. We may apply the results from Section 2.1 to the H-invariant weak-$*$ compact convex set $\mathscr{C} = \mathfrak{S}(A)$. When $\mu \in \mathrm{Prob}(H)$ is an admissible Borel probability measure, we denote by $\mathfrak{S}_\mu(A) \subset \mathfrak{S}(A)$ the nonempty weak-$*$ compact convex subset of all μ-stationary states.

Examples 2.6. We will consider the following examples of C^*-algebras:

(1) For any compact metrizable space X, the space $\mathrm{C}(X)$ of all continuous functions on X endowed with the uniform norm $\|\cdot\|_\infty$ is a commutative C^*-algebra. Any commutative C^*-algebra arises this way. We identify the set $\mathrm{Prob}(X)$ of

Borel probability measures on X with the state space $\mathfrak{S}(\mathrm{C}(X))$ via the continuous mapping $\mathrm{Prob}(X) \to \mathfrak{S}(\mathrm{C}(X)) : \nu \mapsto \int_X \cdot \, d\nu$. Any continuous action by homeomorphisms $H \curvearrowright X$ naturally gives rise to a continuous action $H \curvearrowright \mathrm{C}(X)$ in the above sense.

(2) For any countable discrete group Λ and any unitary representation $\pi : \Lambda \to \mathcal{U}(\mathcal{H}_\pi)$, define the C^*-algebra

$$\mathrm{C}^*_\pi(\Lambda) := \mathrm{C}^*\big(\{\pi(\gamma) \mid \gamma \in \Lambda\}\big) \subset \mathrm{B}(\mathcal{H}_\pi)$$

and consider the conjugation action $\mathrm{Ad}(\pi) : \Lambda \curvearrowright \mathrm{C}^*_\pi(\Lambda)$. The state space $\mathfrak{S}(\mathrm{C}^*_\pi(\Lambda))$ is a Λ-invariant weak-$*$ compact convex subset of $\mathscr{P}(\Lambda)$ via the mapping $\mathfrak{S}(\mathrm{C}^*_\pi(\Lambda)) \hookrightarrow \mathscr{P}(\Lambda) : \psi \mapsto \psi \circ \pi$. If $\pi = \lambda$ is the left regular representation, then $\mathrm{C}^*_\lambda(\Lambda)$ is the *reduced* group C^*-algebra. Moreover, the state $\tau_\Lambda : \mathrm{C}^*_\lambda(\Lambda) \to \mathbf{C} : a \mapsto \langle a\delta_e, \delta_e\rangle$ is a faithful trace.

A *von Neumann algebra* (or W^**-algebra*) M is a unital C^*-algebra which admits a faithful unital $*$-representation $\pi : M \to \mathrm{B}(\mathcal{H})$ such that $\pi(M) \subset \mathrm{B}(\mathcal{H})$ is closed with respect to the weak (equivalently strong) operator topology. After identifying M with $\pi(M)$, we may regard $M \subset \mathrm{B}(\mathcal{H})$ as a concrete von Neumann algebra. By von Neumann's bicommutant theorem, a unital $*$-subalgebra $M \subset \mathrm{B}(\mathcal{H})$ is a von Neumann algebra if and only if M is equal to its own bicommutant M'', that is, $M = M''$. There is a unique Banach space predual M_* such that $M = (M_*)^*$. The ultraweak topology on M coincides with the weak-$*$ topology arising from the identification $M = (M_*)^*$. A linear mapping between von Neumann algebras is *normal* if it is continuous with respect to the ultraweak topology. We say that an action $\sigma : H \curvearrowright M$ is *continuous* if the corresponding action map $H \times M_* \to M_* : (g, \varphi) \mapsto \varphi \circ \sigma_g^{-1}$ is continuous (see, e.g., [52, PROPOSITION X.1.2]). We then simply say that M is a H-von Neumann algebra. The action $H \curvearrowright M$ is *ergodic* if the fixed point von Neumann subalgebra $M^H = \{x \in M \mid \forall g \in H, \sigma_h(x) = x\}$ is trivial.

Examples 2.7. We will consider the following examples of von Neumann algebras:

(1) For any standard probability space (X, ν), the space $\mathrm{L}^\infty(X, \nu)$ of all ν-equivalence classes of (essentially) bounded measurable functions endowed with the (essential) uniform norm $\|\cdot\|_\infty$ is a commutative von Neumann algebra. Any commutative von Neumann algebra arises this way. Any nonsingular action $H \curvearrowright (X, \nu)$ naturally gives rise to a continuous action $H \curvearrowright \mathrm{L}^\infty(X, \nu)$ in the above sense. When no confusion is possible, we simply write $\mathrm{L}^\infty(X) = \mathrm{L}^\infty(X, \nu)$.

(2) For any countable discrete group Λ and any nonsingular action $\Lambda \curvearrowright (X, \nu)$ on a standard probability space, define the *group measure space von Neumann algebra*

$$\mathrm{L}(\Lambda \curvearrowright X) := \big\{f \otimes 1, \pi(\gamma) \mid f \in \mathrm{L}^\infty(X), \gamma \in \Lambda\big\}'' \subset \mathrm{B}\big(\mathrm{L}^2(X, \nu) \otimes \ell^2(\Lambda)\big)$$

where $\pi : \Lambda \to \mathscr{U}(\mathrm{L}^2(X,\nu) \otimes \ell^2(\Lambda))$ is the unitary representation defined by

$$\forall \xi \in \mathrm{L}^2(X,\nu), \forall \gamma, h \in \Lambda, \quad \pi(\gamma)(\xi \otimes \delta_h) = \sqrt{\frac{\mathrm{d}(\nu \circ \gamma^{-1})}{\mathrm{d}\nu}}\, \xi \circ \gamma^{-1} \otimes \delta_{\gamma h}.$$

When (X,ν) is a singleton, the von Neumann algebra $\mathrm{L}(\Lambda \curvearrowright X)$ coincides with the *group von Neumann algebra* $\mathrm{L}(\Lambda)$. When the action $\Lambda \curvearrowright (X,\nu)$ is (essentially) free and ergodic, the von Neumann algebra $\mathrm{L}(\Lambda \curvearrowright X)$ is a factor whose type coincides with the type of the action (see, e.g., [53, THEOREM XIII.1.7]).

A von Neumann algebra $M \subset \mathrm{B}(\mathscr{H})$ is *amenable* if there exists a norm-one projection $\mathrm{E} : \mathrm{B}(\mathscr{H}) \to M$. By Connes' fundamental result [17], M is amenable if and only if M is *approximately finite dimensional*, that is, there exists an increasing net of finite-dimensional subalgebras $M_i \subset M$ such that $\bigvee_{i\in I} M_i = M$.

3. DYNAMICAL DICHOTOMY FOR BOUNDARY STRUCTURES

3.1. Boundary structures

For any C*-algebra $A \subset \mathrm{B}(\mathscr{H})$ and any $n \geq 1$, $\mathrm{M}_n(A) := \mathrm{M}_n(\mathbf{C}) \otimes A \subset \mathrm{B}(\mathscr{H}^{\oplus n})$ is naturally a C*-algebra. Let A, B be any C*-algebras. A linear map $\Phi : A \to B$ is said to be *unital completely positive* (ucp) if Φ is unital and if for every $n \geq 1$, the linear map $\Phi^{(n)} : \mathrm{M}_n(A) \to \mathrm{M}_n(B) : [a_{ij}]_{ij} \mapsto [\Phi(a_{ij})]_{ij}$ is positive. Any unital $*$-homomorphism $\pi : A \to B$ is a ucp map. When A or B is commutative, any unital positive linear map $\Phi : A \to B$ is automatically ucp (see, e.g., [45, THEOREMS 3.9 AND 3.11]).

Definition 3.1 ([3]). Let $\Gamma < G$ be any higher rank lattice and M any Γ-von Neumann algebra with separable predual. A Γ-*boundary structure* $\Phi : M \to \mathrm{L}^\infty(G/P)$ is a Γ-equivariant faithful normal ucp map. We say that Φ is *invariant* if $\Phi(M) = \mathbf{C}1$.

We will simply say that $\Phi : M \to \mathrm{L}^\infty(G/P)$ is a *boundary structure* instead of a Γ-boundary structure when it is understood that M is a Γ-von Neumann algebra. In this survey, we only deal with higher rank lattices in connected semisimple *real* Lie groups. In that setting, the notion of boundary structure is equivalent to the notion of stationary state. Indeed, fix a Furstenberg measure $\mu_\Gamma \in \mathrm{Prob}(\Gamma)$ so that $(G/P, \nu_P)$ is the (Γ, μ_Γ)-Poisson boundary (see Theorem 2.5).

- If $\Phi : M \to \mathrm{L}^\infty(G/P)$ is a boundary structure, then $\varphi := \nu_P \circ \Phi \in M_*$ is a faithful normal μ_Γ-stationary state on M. Moreover, if Φ is invariant, then φ is Γ-invariant.
- Conversely, let $\varphi \in M_*$ be any faithful normal μ_Γ-stationary state on M. Define the Γ-equivariant faithful normal ucp map

$$\Phi : M \to \mathrm{Har}^\infty(\Gamma, \mu_\Gamma) : x \mapsto \big(\gamma \mapsto \varphi(\gamma^{-1}x)\big).$$

Since $\mathrm{Har}^\infty(\Gamma, \mu_\Gamma) \cong \mathrm{L}^\infty(G/P, \nu_P)$ as Γ-operator systems, we may further regard $\Phi : M \to \mathrm{L}^\infty(G/P)$ as a boundary structure such that $\varphi = \nu_P \circ \Phi$. If φ is Γ-invariant, then Φ is invariant.

Remark 3.2. The notion of boundary structure was developed in [3] to replace the notion of stationary state used in [13] in order to deal with higher rank lattices in semisimple algebraic groups defined over *arbitrary* local fields.

It is useful to restrict boundary structures to separable C^*-subalgebras. Let M be any Γ-von Neumann algebra with separable predual. A globally Γ-invariant separable ultraweakly dense C^*-subalgebra $A \subset M$ is called a *separable model* for the action $\Gamma \curvearrowright M$. If $\Phi : M \to \mathrm{L}^\infty(G/P)$ is a boundary structure, then $(\nu_P \circ \Phi)|_A \in \mathfrak{S}_{\mu_\Gamma}(A)$ and the restriction $\Phi|_A : A \to \mathrm{L}^\infty(G/P)$ gives rise to the Γ-equivariant boundary map $\beta : G/P \to \mathfrak{S}(A) : b \mapsto \beta_b$ such that $\mathrm{Bar}(\beta_* \nu_P) = (\nu_P \circ \Phi)|_A$, where

$$\forall a \in A, \quad \Phi(a)(b) = \beta_b(a).$$

We present several examples of boundary structures.

Example 3.3 (Boundary structure arising from unitary representations). Let $\pi : \Gamma \to \mathscr{U}(\mathscr{H}_\pi)$ be any unitary representation and set $A := \mathrm{C}^*_\pi(\Gamma)$. Choose an extremal μ_Γ-stationary state $\varphi \in \mathfrak{S}_{\mu_\Gamma}(A)$ and consider the GNS triple $(\pi_\varphi, \mathscr{H}_\varphi, \xi_\varphi)$. Denote by $\beta : G/P \to \mathfrak{S}(A) : b \mapsto \beta_b$ the Γ-equivariant boundary map associated with $\varphi \in \mathfrak{S}_{\mu_\Gamma}(A)$. By duality, we may consider the Γ-equivariant ucp map $\Phi : A \to \mathrm{L}^\infty(G/P) : a \mapsto (b \mapsto \beta_b(a))$ which satisfies $\nu_P \circ \Phi = \varphi$. Set $M := \pi_\varphi(A)'' = (\pi_\varphi \circ \pi)(\Gamma)''$. By extremality, the conjugation action $\mathrm{Ad}(\pi_\varphi \circ \pi) : \Gamma \curvearrowright M$ is ergodic. Moreover, the Γ-equivariant ucp map

$$\pi_\varphi(A) \to \mathrm{L}^\infty(G/P) : \pi_\varphi(a) \mapsto \Phi(a)$$

is well defined and extends to a boundary structure $\Phi : M \to \mathrm{L}^\infty(G/P)$. We refer to [13, PROOF OF THEOREM A] for further details.

Example 3.4 (Boundary structure arising from characters). Let $\varphi \in \mathrm{Char}(\Gamma)$ be any extremal character. Simply denote by $(\pi, \mathscr{H}, \xi)$ the GNS triple associated with $\varphi \in \mathrm{Char}(\Gamma)$. Denote by $J : \mathscr{H} \to \mathscr{H} : \pi(\gamma)\xi \mapsto \pi(\gamma)^*\xi$ the canonical conjugation. Following [46], define the *noncommutative Poisson boundary* $\mathscr{B}$ as the von Neumann algebra of all ν_P-equivalence classes of (essentially) bounded measurable functions $f : G/P \to \mathrm{B}(\mathscr{H})$ satisfying $f(\gamma b) = \mathrm{Ad}(J\pi(\gamma)J)(f(b))$ for every $\gamma \in \Gamma$ and almost every $b \in G/P$. Observe that $\mathbf{C}1 \otimes \pi(\Gamma)'' \subset \mathscr{B}$. Since P is amenable, $\mathscr{B}$ is an amenable von Neumann algebra. By extremality, the conjugation action $\mathrm{Ad}(\pi) : \Gamma \curvearrowright \mathscr{B}$ is ergodic. Moreover,

$$\Phi : \mathscr{B} \to \mathrm{L}^\infty(G/P) : f \mapsto \big(b \mapsto \langle f(b)\xi, \xi\rangle\big)$$

is a boundary structure. When $\varphi = \delta_e$ is the regular character, the noncommutative Poisson boundary $\mathscr{B}$ coincides with the group measure space von Neumann algebra $\mathrm{L}(\Gamma \curvearrowright G/P)$ and the boundary structure $\Phi : \mathrm{L}(\Gamma \curvearrowright G/P) \to \mathrm{L}^\infty(G/P)$ is the canonical Γ-equivariant conditional expectation. We refer to [13, PROOF OF THEOREM C] for further details.

Example 3.5 (Boundary structure arising from topological dynamics). Let $\Gamma \curvearrowright X$ be any minimal action on a compact metrizable space. Choose an extremal μ_Γ-stationary Borel probability measure $\nu \in \mathrm{Prob}_{\mu_\Gamma}(X)$. By minimality, we have $\mathrm{supp}(\nu) = X$. Denote by $\beta : G/P \to \mathrm{Prob}(X) : b \mapsto \beta_b$ the Γ-equivariant boundary map associated with $\nu \in \mathrm{Prob}_{\mu_\Gamma}(X)$. By duality, we may consider the Γ-equivariant ucp map $\Phi : \mathrm{C}(X) \to \mathrm{L}^\infty(G/P) : f \mapsto (b \mapsto \beta_b(f))$ which satisfies $\nu_P \circ \Phi = \nu$. By extremality, the nonsingular action $\Gamma \curvearrowright (X, \nu)$ is ergodic. Moreover, $\Phi : \mathrm{C}(X) \to \mathrm{L}^\infty(G/P)$ extends to a boundary structure $\Phi : \mathrm{L}^\infty(X, \nu) \to \mathrm{L}^\infty(G/P)$.

The notion of boundary structure is well adapted to induction. Indeed, let $\Phi : M \to \mathrm{L}^\infty(G/P)$ be any Γ-boundary structure. Denote by $\hat{M} := \mathrm{Ind}_\Gamma^G(M) \cong \mathrm{L}^\infty(G/\Gamma) \mathbin{\overline{\otimes}} M$ the induced G-von Neumann algebra. Since G/P is a G-space, we have $\mathrm{Ind}_\Gamma^G(\mathrm{L}^\infty(G/P)) \cong \mathrm{L}^\infty(G/\Gamma) \mathbin{\overline{\otimes}} \mathrm{L}^\infty(G/P)$, where $G \curvearrowright G/\Gamma \times G/P$ acts diagonally. Denote by $\nu_\Gamma \in \mathrm{Prob}(G/\Gamma)$ the unique G-invariant Borel probability measure. Then the map $\widehat{\Phi} := \nu_\Gamma \otimes \Phi : \hat{M} \to \mathrm{L}^\infty(G/P)$ is a G-equivariant faithful normal ucp map. We then refer to $\widehat{\Phi}$ as the *induced G-boundary structure*. Note that Φ is invariant if and only if $\widehat{\Phi}$ is invariant.

This framework provides a more conceptual approach to the *stationary induction* considered in [13, SECTION 4]. Let $\mu_G \in \mathrm{Prob}(G)$ be any K-invariant admissible Borel probability measure. Let φ be any faithful normal μ_Γ-stationary state on M and define the corresponding Γ-boundary structure $\Phi : M \to \mathrm{L}^\infty(G/P)$ such that $\nu_P \circ \Phi = \varphi$. Consider the induced G-boundary structure $\widehat{\Phi} : \hat{M} \to \mathrm{L}^\infty(G/P)$. Then $\widehat{\varphi} := \nu_P \circ \widehat{\Phi}$ is a faithful normal μ_G-stationary state on $\hat{M}$. Moreover, φ is Γ-invariant if and only if $\widehat{\varphi}$ is G-invariant.

3.2. The dynamical dichotomy theorem for boundary structures

Let A be any separable C*-algebra. We say that $\phi, \psi \in \mathfrak{S}(A)$ are *pairwise singular* and write $\phi \perp \psi$ if there exists a sequence $(a_k)_k$ in A such that $0 \leq a_k \leq 1$ for every $k \in \mathbf{N}$ and for which $\lim_k \phi(a_k) = 0 = \lim_k \psi(1 - a_k)$. This notion naturally extends the notion of pairwise singularity of Borel probability measures on metrizable compact spaces. Observe that for any unital C*-subalgebra $B \subset A$ and any states $\phi, \psi \in \mathfrak{S}(A)$, if $\phi|_B \perp \psi|_B$, then $\phi \perp \psi$. We introduce the following terminology.

Definition 3.6 ([3]). Let $\Gamma < G$ be any higher rank lattice. Let M be any Γ-von Neumann algebra with separable predual and $\Phi : M \to \mathrm{L}^\infty(G/P)$ any boundary structure. We say that Φ is *singular* if there exists a separable model $A \subset M$ for the action $\Gamma \curvearrowright M$ such that the corresponding Γ-equivariant boundary map $\beta : G/P \to \mathfrak{S}(A) : b \mapsto \beta_b$ satisfies the following property:

$$\text{For every } \gamma \in \Gamma \setminus \mathscr{Z}(\Gamma), \text{ for almost every } b \in G/P, \quad \beta_{\gamma b} \perp \beta_b. \tag{3.1}$$

The notion of singularity for boundary structures is quite robust. If $\Phi : M \to \mathrm{L}^\infty(G/P)$ is singular, then for *every* separable model $A \subset M$, the corresponding Γ-equivariant boundary map $\beta : G/P \to \mathfrak{S}(A) : b \mapsto \beta_b$ satisfies (3.1) (see [3, PROPOSITION 4.10]). This implies the following useful fact. If $M_0 \subset M$ is a Γ-invariant von Neumann subalgebra and if the restriction $\Phi|_{M_0} : M_0 \to \mathrm{L}^\infty(G/P)$ is singular, then Φ is singular as well.

In case the action $\Gamma \curvearrowright M$ is given by conjugation, singular boundary structures enjoy the following useful vanishing property.

Proposition 3.7 ([3]). *Let M be any von Neumann algebra with separable predual and $\pi : \Gamma \to \mathscr{U}(M)$ any unitary representation. Consider the conjugation action $\mathrm{Ad}(\pi) : \Gamma \curvearrowright M$. Let $\Phi : M \to \mathrm{L}^\infty(G/P)$ be any singular boundary structure. Then for every $\gamma \in \Gamma \setminus \mathscr{Z}(\Gamma)$, we have $\Phi(\pi(\gamma)) = 0$.*

Proof. The proof is similar to [34, LEMMA 2.2]. We may choose a separable model $A \subset M$ for the conjugation action $\mathrm{Ad}(\pi) : \Gamma \curvearrowright M$ such that $\pi(\Gamma) \subset A$. Denote by $\beta : G/P \to \mathfrak{S}(A)$ the Γ-equivariant boundary map arising from $\Phi|_A$. Let $\gamma \in \Gamma \setminus \mathscr{Z}(\Gamma)$ be any element. Choose a conull measurable subset $Y \subset G/P$ such that for every $b \in Y$, we have $\beta_{\gamma b} = \gamma \beta_b = \beta_b \circ \mathrm{Ad}(\pi(\gamma)^*)$ and $\beta_{\gamma b} \perp \beta_b$. Let $b \in Y$ be any point and choose a sequence $(a_k)_k$ in A such that $0 \leq a_k \leq 1$ for every $k \in \mathbf{N}$ and for which $\lim_k \beta_{\gamma b}(a_k) = 0 = \lim_k \beta_b(1 - a_k)$. Then Cauchy–Schwarz inequality implies that

$$\begin{aligned} \left|\beta_b\big((1-a_k)\pi(\gamma)\big)\right| &= \left|\beta_b\big((1-a_k)^{1/2} \cdot (1-a_k)^{1/2}\pi(\gamma)\big)\right| \\ &\leq \beta_b(1-a_k)^{1/2} \to 0. \end{aligned}$$

Likewise, Cauchy–Schwarz inequality implies that

$$\begin{aligned} \left|\beta_b\big(a_k\pi(\gamma)\big)\right| &= \left|\beta_{\gamma b}\big(\pi(\gamma)a_k^{1/2} \cdot a_k^{1/2}\big)\right| \\ &\leq \beta_{\gamma b}(a_k)^{1/2} \to 0. \end{aligned}$$

Then $\beta_b(\pi(\gamma)) = \beta_b((1-a_k)\pi(\gamma)) + \beta_b(a_k\pi(\gamma)) \to 0$ and so $\beta_b(\pi(\gamma)) = 0$. Since this holds true for every $b \in Y$, it follows that $\Phi(\pi(\gamma)) = 0$. ∎

As we mentioned in the Introduction, the following dynamical dichotomy theorem *invariant vs. singular* for boundary structures is the key novelty in our operator-algebraic framework.

Theorem 3.8 ([3,13]). *Let M be any ergodic Γ-von Neumann algebra with separable predual and $\Phi : M \to \mathrm{L}^\infty(G/P)$ any boundary structure. Then Φ is either invariant or singular.*

The proof of Theorem 3.8 depends heavily upon whether the ambient connected semisimple real Lie group G is simple or not.

In case G is simple, let us explain why Theorem E implies Theorem 3.8. Let $\Phi : M \to \mathrm{L}^\infty(G/P)$ be any noninvariant boundary structure. By Theorem E, there exist a proper parabolic subgroup $P < Q < G$ and a Γ-equivariant unital normal embedding $\iota : \mathrm{L}^\infty(G/Q) \hookrightarrow M$ such that $\Phi \circ \iota = p_Q^*$. Set $M_0 := \iota(\mathrm{L}^\infty(G/Q)) \subset M$. Then $A := \iota(\mathrm{C}(G/Q)) \subset M_0$ is a separable model for the action $\Gamma \curvearrowright M_0$ and the Γ-equivariant boundary map corresponding to $\Phi|_A$ is exactly $\delta \circ p_Q : G/P \to \mathrm{Prob}(G/Q) : gP \mapsto \delta_{gQ}$. Since any element $\gamma \in \Gamma \setminus \mathscr{Z}(\Gamma)$ acts (essentially) freely on G/Q (see, e.g., [13, LEMMA 6.2]), it follows that the restriction $\Phi|_{M_0} : M_0 \to \mathrm{L}^\infty(G/P)$ is singular. Thus, Φ is singular.

In case G is not simple, we have to use a different approach. Following [3], we outline the main steps of the proof of Theorem 3.8 in the particular case where $G = G_1 \times G_2$ with

G_1 and G_2 noncompact connected simple Lie groups with finite center. This particular case already contains all the main conceptual difficulties. Let $i = 1, 2$. Denote by $p_i : G \to G_i$ the canonical factor map. Denote by $P_i < G_i$ a minimal parabolic subgroup and set $P = P_1 \times P_2 < G$. By Theorem 2.4, (G_i, ν_{P_i}) is the (G_i, μ_i)-Poisson boundary with respect to appropriate Borel probability measures $\mu_i \in \mathrm{Prob}(G_i)$ and $\nu_{P_i} \in \mathrm{Prob}(G_i/P_i)$. Let $\Phi : M \to \mathrm{L}^\infty(G/P)$ be any noninvariant boundary structure. Our goal is to show that Φ is singular.

Step 1: Induction. Denote by $\widehat{\Phi} : \hat{M} \to \mathrm{L}^\infty(G/P)$ the induced G-boundary structure. Since Φ is not invariant, $\widehat{\Phi}$ is not invariant either, that is, $\widehat{\Phi}(\hat{M}) \neq \mathbf{C}1$.

Step 2: Reduction to the von Neumann algebra of G_1-continuous elements. Exploiting the product structure $G = G_1 \times G_2$ and up to permuting the indices, we show that the restriction of $\widehat{\Phi}$ to the G_2-fixed point von Neumann subalgebra $\hat{M}^{G_2} \subset \hat{M}$ still satisfies $\Phi(\hat{M}^{G_2}) \neq \mathbf{C}1$ and, moreover, $\widehat{\Phi}(\hat{M}^{G_2}) \subset \mathrm{L}^\infty(G_1/P_1)$. Exploiting that $\Gamma < G_1 \times G_2$ is irreducible and that (G_i, ν_{P_i}) is the (G_i, μ_i)-Poisson boundary, $i = 1, 2$, we show that the G_1-von Neumann algebra $\hat{M}^{G_2}$ is Γ-isomorphic to the Γ-von Neumann subalgebra $M_1 \subset M$ of all elements $x \in M$ for which the action map $\Gamma \to M : \gamma \mapsto \sigma_\gamma(x)$ extends continuously to G_1. We say that $M_1 \subset M$ is the von Neumann subalgebra of G_1-*continuous elements*. Moreover, under the identification $\hat{M}^{G_2} = M_1$, we naturally have the identification $\widehat{\Phi}|_{\hat{M}^{G_2}} = \Phi|_{M_1}$. Then M_1 is a G_1-ergodic von Neumann algebra and $\Phi|_{M_1} : M_1 \to \mathrm{L}^\infty(G_1/P_1)$ is a G_1-boundary structure such that $\Phi(M_1) \neq \mathbf{C}1$. Since $\Phi|_{M_1}$ is the restriction to M_1 of the Γ-boundary structure Φ, it suffices to show that $\Phi|_{M_1}$ is singular.

Step 3: Singularity of $\Phi|_{M_1}$. We may choose a separable model $A_1 \subset M_1$ for the continuous action $G_1 \curvearrowright M_1$. Since $G_1 \curvearrowright G_1/P_1$ is transitive, $\Phi|_{A_1} : A_1 \to \mathrm{L}^\infty(G_1/P_1)$ gives rise to a G_1-equivariant continuous boundary map $\beta : G_1/P_1 \to \mathfrak{S}(A_1) : b \mapsto \beta_b$. Since the action $G_1 \curvearrowright M_1$ is ergodic, the P_1-invariant state $\psi := \beta_{P_1} \in \mathfrak{S}(A_1)$ is extremal among P_1-invariant states. We then show that for every $g \in G_1$, either $g\psi \perp \psi$ or $g\psi = \psi$. Since ψ is not G_1-invariant on A_1, the stabilizer $Q_1 = \mathrm{Stab}_{G_1}(\psi)$ is a proper parabolic subgroup such that $P_1 < Q_1$. Since any element $g \in G_1 \setminus \mathscr{Z}(G_1)$ acts (essentially) freely on G_1/Q_1 and since $p_1(\Gamma \setminus \mathscr{Z}(\Gamma)) \subset G_1 \setminus \mathscr{Z}(G_1)$, it follows that the restriction $\Phi|_{M_1}$ is singular. Thus, Φ is singular.

3.3. Outline of the proof of Theorem E

Following [13], we outline the main steps of the proof of Theorem E. We may assume that G is a connected simple real Lie group with trivial center and real rank $\mathrm{rk}_{\mathbf{R}}(G) \geq 2$. Recall that G admits an Iwasawa decomposition $G = KAV$ where $K < G$ is a maximal compact subgroup, $A < G$ is a Cartan subgroup, and $V < G$ is a unipotent subgroup. Set $P = LV$ where $L = \mathscr{Z}_G(A)$ and note that $P < G$ is a minimal parabolic subgroup. Likewise, write $\overline{P} = L\overline{V}$ for the opposite minimal parabolic subgroup. Note that $G = \langle P, \overline{V} \rangle$ and the map $\overline{V} \to G/P : \bar{v} \mapsto \bar{v}P$ defines a measurable isomorphism. Let $\Phi : M \to \mathrm{L}^\infty(G/P)$ be any noninvariant boundary structure. Our goal is to show that there exist a proper parabolic subgroup $P < Q < G$ and a Γ-equivariant unital normal embedding $\iota : \mathrm{L}^\infty(G/Q) \hookrightarrow M$.

Step 1: Induction. Exactly as in the proof of Theorem 3.8, denote by $\widehat{\Phi} : \hat{M} \to \mathrm{L}^\infty(G/P)$ the induced G-boundary structure which satisfies $\widehat{\Phi}(\hat{M}) \neq \mathbf{C}1$. We may choose a separable model $A \subset \hat{M}$ for the continuous action $G \curvearrowright \hat{M}$. Since $G \curvearrowright G/P$ is transitive, $\widehat{\Phi}|_A : A \to \mathrm{L}^\infty(G/P)$ gives rise to a G-equivariant continuous boundary map $\beta : G/P \to \mathfrak{S}(A) : b \mapsto \beta_b$. Denote by $\psi := \beta_P \in \mathfrak{S}(A)$ the corresponding P-invariant state, consider the GNS representation $\pi_\psi : A \to \mathrm{B}(\mathscr{H}_\psi)$ and set $N := \pi_\psi(A)''$. Then N is a P-von Neumann algebra and we may consider the induced G-von Neumann algebra $\mathrm{Ind}_P^G(N)$. Moreover, we may regard $\hat{M} \subset \mathrm{Ind}_P^G(N)$ as a G-von Neumann subalgebra. We point out that the normal state $\psi \in N_*$ need not be faithful on N. Since we only give a sketch of the proof, we will assume that ψ is faithful on N. We refer to [13, THEOREM 5.1] for the general proof.

Step 2: Construction of a well behaved von Neumann subalgebra. In this step, we build upon Nevo–Zimmer's proof of [44, THEOREM 1]. Since $\widehat{\Phi}(\hat{M}) \neq \mathbf{C}1$, the P-invariant state $\psi \in \mathfrak{S}(A)$ is not G-invariant whence not $\overline{V}$-invariant. Using the real rank assumption $\mathrm{rk}_{\mathbf{R}}(G) \geq 2$, there is a strict intermediate parabolic subgroup $P < P_0 < G$ with Levi decomposition $P_0 = L_0 V_0$, where $L_0 = \mathscr{Z}_G(A_0)$, $A_0 < A$, and $V_0 < V$, such that $\psi \in \mathfrak{S}(A)$ is not $\overline{V}_0$-invariant. Choose a nontrivial element $s \in A_0$ so that s (resp. s^{-1}) acts by conjugation as a contracting automorphism of V_0 (resp. $\overline{V}_0$). By Mautner phenomenon, any s-fixed element in N is necessarily V_0-fixed. Since the subgroup $\langle s, V_0 \rangle$ is normal in P, it follows that $N^s \subset N$ is a P-invariant von Neumann subalgebra. Using these assumptions, we show that $\mathscr{M}_0 = \hat{M} \cap \mathrm{Ind}_P^G(N^s)$ is a G-von Neumann subalgebra such that $\widehat{\Phi}(\mathscr{M}_0) \neq \mathbf{C}1$.

Step 3: From noncommutative to commutative. We reached the point where we can no longer rely on Nevo–Zimmer's argument [44]. Indeed, $\mathscr{M}_0$ is not commutative and so we cannot use the Gauss map trick from [44, SECTION 3] to conclude. We adopt the following new strategy. Using the induction in two steps, we may write $\mathrm{Ind}_P^G(N^s) = \mathrm{Ind}_{P_0}^G(\mathrm{Ind}_P^{P_0}(N^s)) \cong \mathrm{L}^\infty(\overline{V}_0, \mathrm{Ind}_P^{P_0}(N^s))$. Regarding $\mathscr{M}_0 \subset \mathrm{L}^\infty(\overline{V}_0, \mathrm{Ind}_P^{P_0}(N^s))$ as a G-von Neumann subalgebra, denote by $\mathscr{N}_0 \subset \mathrm{Ind}_P^{P_0}(N^s)$ the von Neumann subalgebra generated by the essential values of all the elements $f \in \mathscr{A}_0$, where $\mathscr{A}_0 \subset \mathscr{M}_0$ is an ultraweakly dense separable C*-subalgebra. On the one hand, by construction, we have $\mathscr{M}_0 \subset \mathrm{L}^\infty(\overline{V}_0, \mathscr{N}_0) \cong \mathrm{L}^\infty(\overline{V}_0) \mathbin{\overline{\otimes}} \mathscr{N}_0$. On the other hand, exploiting that s^{-1} acts by conjugation as a contracting automorphism of $\overline{V}_0$, we show that $\mathbf{C}1 \mathbin{\overline{\otimes}} \mathscr{N}_0 \subset \mathscr{M}_0$. Therefore, we have the inclusions $\mathbf{C}1 \mathbin{\overline{\otimes}} \mathscr{N}_0 \subset \mathscr{M}_0 \subset \mathrm{L}^\infty(\overline{V}_0) \mathbin{\overline{\otimes}} \mathscr{N}_0$. By considering the centers, we also have $\mathbf{C}1 \mathbin{\overline{\otimes}} \mathscr{Z}(\mathscr{N}_0) \subset \mathscr{Z}(\mathscr{M}_0) \subset \mathrm{L}^\infty(\overline{V}_0) \mathbin{\overline{\otimes}} \mathscr{Z}(\mathscr{N}_0)$. Since $G \curvearrowright \mathscr{M}_0$ is faithful and since s acts identically on $\mathbf{C}1 \mathbin{\overline{\otimes}} \mathscr{N}_0$, we have $\mathbf{C}1 \mathbin{\overline{\otimes}} \mathscr{N}_0 \subsetneq \mathscr{M}_0$. Exploiting Ge–Kadison's splitting theorem [30], we show that $\mathbf{C}1 \mathbin{\overline{\otimes}} \mathscr{Z}(\mathscr{N}_0) \neq \mathscr{Z}(\mathscr{M}_0)$. This further implies that $\Phi(\mathscr{Z}(\mathscr{M}_0)) \neq \mathbf{C}1$. We may now apply Nevo–Zimmer's result [44, THEOREM 1] to the commutative G-von Neumann algebra $\mathscr{Z}(\mathscr{M}_0)$ to obtain a proper parabolic subgroup $P < Q < G$ and a G-equivariant unital normal embedding $\iota : \mathrm{L}^\infty(G/Q) \hookrightarrow \mathscr{Z}(\mathscr{M}_0) \subset \hat{M}$.

Step 4: Back to the lattice. We use the simple argument given in [2]. Since the action $G \curvearrowright \mathrm{C}(G/Q)$ is $\|\cdot\|$-continuous, by G-equivariance, it follows that $\iota(\mathrm{C}(G/Q))$ is contained in the C*-subalgebra $\mathrm{C}_b^*(G, M)^\Gamma \subset \hat{M}$ of all Γ-equivariant bounded continuous functions f :

$G \to M$. Consider the evaluation $*$-homomorphism $\mathrm{ev} : \mathrm{C}^*_b(G, M)^\Gamma \to M : f \mapsto f(e)$. Then $\mathrm{ev} \circ \iota : \mathrm{C}(G/Q) \to M$ is a Γ-equivariant $*$-homomorphism which extends uniquely to a Γ-equivariant unital normal embedding $\mathrm{L}^\infty(G/Q) \hookrightarrow M$ thanks to (2.1).

4. PROOFS OF THE MAIN RESULTS

In this section, following [3, 13], we explain how to use the dynamical dichotomy Theorem 3.8 to prove the main results stated in the Introduction. We fix a higher rank lattice $\Gamma < G$ and a Furstenberg measure $\mu_\Gamma \in \mathrm{Prob}(\Gamma)$ so that $(G/P, \nu_P)$ is the (Γ, μ_Γ)-Poisson boundary (see Theorem 2.5). Since the proofs of Theorems A and D are similar, we only give the proofs of Theorems A, B and Corollary C.

Proof of Theorem A. Denote by $\pi : \Gamma \to \mathscr{U}(\mathscr{H}_\pi)$ the *universal* unitary representation, meaning that π is equal to the orthogonal direct sum of all cyclic unitary representations of Γ. Then $A := \mathrm{C}^*_\pi(\Gamma)$ coincides with the full C^*-algebra $\mathrm{C}^*(\Gamma)$ and we may use the identification $\mathfrak{S}(A) = \mathscr{P}(\Gamma)$. Let $\mathscr{C} \subset \mathfrak{S}(A)$ be any nonempty Γ-invariant weak-$*$ compact convex subset. We claim that any μ_Γ-stationary state $\varphi \in \mathscr{C}$ is Γ-invariant. More generally, we claim that any μ_Γ-stationary state on A is Γ-invariant. By Krein–Milman theorem, it suffices to show that any extremal μ_Γ-stationary $\varphi \in \mathfrak{S}(A)$ is Γ-invariant. Letting $M = \pi_\varphi(A)''$, consider the boundary structure $\Phi : M \to \mathrm{L}^\infty(G/P)$ such that $\nu_P \circ \Phi = \varphi$ as in Example 3.3. By Theorem 3.8, Φ is either invariant or singular. If Φ is invariant, then $\varphi \in \mathrm{Char}(\Gamma)$. If Φ is singular, then Proposition 3.7 implies that for every $\gamma \in \Gamma \setminus \mathscr{Z}(\Gamma)$, we have $\Phi(\pi(\gamma)) = 0$ and so $\varphi(\gamma) = 0$. Then φ is supported on $\mathscr{Z}(\Gamma)$ and so $\varphi \in \mathrm{Char}(\Gamma)$. ■

Proof of Theorem B. Let $\varphi \in \mathrm{Char}(\Gamma)$ be any character. We may assume that φ is an extremal character. Denote by $\mathscr{B}$ the noncommutative Poisson boundary and consider the boundary structure $\Phi : \mathscr{B} \to \mathrm{L}^\infty(G/P)$ as in Example 3.4. By Theorem 3.8, Φ is either invariant or singular. If Φ is invariant, then for every $f \in \mathscr{B}$, the function $\Phi(f) : G/P \to \mathbf{C} : b \mapsto \langle f(b)\xi, \xi\rangle$ is (essentially) constant. Since $\mathbf{C}1 \otimes \pi_\varphi(\Gamma)'' \subset \mathscr{B}$ and since the linear span of $\pi_\varphi(\Gamma)\xi_\varphi$ is dense in $\mathscr{H}_\varphi$, it easily follows that every $f \in \mathscr{B}$ is (essentially) constant as a function $G/P \to \mathrm{B}(\mathscr{H}_\varphi)$. This further implies that $\mathscr{B} = \mathbf{C}1 \otimes \pi_\varphi(\Gamma)''$ and so $\pi_\varphi(\Gamma)''$ is amenable. This further implies that π_φ is amenable. If Φ is singular, then for every $\gamma \in \Gamma \setminus \mathscr{Z}(\Gamma)$ we have $\Phi(\pi(\gamma)) = 0$ and so $\varphi(\gamma) = 0$. Thus, φ is supported on $\mathscr{Z}(\Gamma)$.

If G has property (T), then Γ has property (T). If φ is not supported on $\mathscr{Z}(\Gamma)$, then π_φ is amenable and so π_φ necessarily contains a finite-dimensional subrepresentation. In the more general case where G has a simple factor with property (T), we refer to the proof of [3, PROPOSITION 7.5]. ■

Proof of Corollary C. Let $\pi : \Gamma \to \mathscr{U}(\mathscr{H}_\pi)$ be any unitary representation and set $A := \mathrm{C}^*_\pi(\Gamma)$. Regarding $\mathfrak{S}(A) \subset \mathscr{P}(\Gamma)$ as a Γ-invariant weak-$*$ compact convex subset, Theorem A implies that the C^*-algebra A admits a trace.

Assume, moreover, that G has trivial center and that π is not amenable. Let $\varphi \in \mathfrak{S}(A)$ be any trace. Regarding $\varphi \in \mathrm{Char}(\Gamma)$ as a character, the GNS representation π_φ is weakly contained in π and so π_φ is not amenable. Theorem A implies that $\varphi = \delta_e$. Then $\pi_\varphi = \lambda$ is the left regular representation and λ is weakly contained in π. Denote by $\Theta : C^*_\pi(\Gamma) \to C^*_\lambda(\Gamma)$ the unique $*$-homomorphism such that $\Theta(\pi(\gamma)) = \lambda(\gamma)$ for every $\gamma \in \Gamma$. Then $\tau_\Gamma \circ \Theta$ is the unique trace on $A = C^*_\pi(\Gamma)$. Let $J \lhd A$ be any proper ideal and define $\rho : A \to A/J$. Then the unitary representation $\rho \circ \pi$ is weakly contained in π and so $\rho \circ \pi$ is not amenable. The previous reasoning implies that λ is weakly contained in $\rho \circ \pi$ and so there is a $*$-homomorphism $\overline{\Theta} : A/J \to C^*_\lambda(\Gamma)$ such that $\Theta = \overline{\Theta} \circ \rho$. Then $J = \ker(\rho) \subset \ker(\Theta)$. Therefore, $\ker(\Theta) \lhd C^*_\pi(\Gamma)$ is the unique maximal proper ideal. ■

We now discuss open problems in relation with our main results. Let $(E, \|\cdot\|)$ be any separable Banach Γ-module and $\mathscr{C} \subset E^*$ any nonempty Γ-invariant weak-$*$ compact convex subset. Let $\mu \in \operatorname{Prob}(\Gamma)$ be any probability measure. By analogy with [29], we say that the affine action $\Gamma \curvearrowright \mathscr{C}$ is *μ-stiff* if any μ-stationary point is invariant. The proof of Theorem A shows that for any Furstenberg measure $\mu_\Gamma \in \operatorname{Prob}(\Gamma)$, the affine action $\Gamma \curvearrowright \mathscr{P}(\Gamma)$ is μ_Γ-stiff. It is natural to ask whether the stiffness property holds for more general probability measures $\mu \in \operatorname{Prob}(\Gamma)$.

Problem 4.1. Let $\mu \in \operatorname{Prob}(\Gamma)$ be *any* probability measure such that $\langle \operatorname{supp}(\mu) \rangle = \Gamma$. Is the action $\Gamma \curvearrowright \mathscr{P}(\Gamma)$ μ-stiff?

Problem 4.1 requires a new strategy as we can no longer identify the (Γ, μ)-Poisson boundary with $(G/P, \nu_P)$. We note that in case Γ has trivial center, it is showed in [34] that for any probability measure $\mu \in \operatorname{Prob}(\Gamma)$ such that $\langle \operatorname{supp}(\mu) \rangle = \Gamma$, the affine action $\Gamma \curvearrowright \mathfrak{S}(C^*_\lambda(\Gamma))$ is μ-stiff and the canonical trace τ_Γ is the only μ-stationary state on $C^*_\lambda(\Gamma)$. Problem 4.1 is particularly relevant in case $\Gamma := \mathrm{SL}_d(\mathbf{Z}) < \mathrm{SL}_d(\mathbf{R}) := G$ for $d \geq 3$. To draw a parallel with homogeneous dynamics, we point out that the affine action $\mathrm{SL}_d(\mathbf{Z}) \curvearrowright \operatorname{Prob}(\mathbf{T}^d)$ is μ-stiff for every $d \geq 2$ and every probability measure $\mu \in \operatorname{Prob}(\mathrm{SL}_d(\mathbf{Z}))$ such that $\langle \operatorname{supp}(\mu) \rangle = \mathrm{SL}_d(\mathbf{Z})$ [12]. We refer the reader to [12] and [10] for more general results regarding measure rigidity.

We say that a countable discrete group Λ is *character rigid* if for any extremal character $\varphi \in \operatorname{Char}(\Lambda)$, either φ is supported on $\mathscr{Z}(\Lambda)$ or the GNS representation π_φ is finite dimensional. Assuming that G has a simple factor with property (T), Theorem B says that Γ is character rigid. It is showed in [47] that $\mathrm{SL}_d(\mathbf{Z}[S^{-1}])$ is character rigid for every $d \geq 2$ and every nonempty set of primes S. In view of Margulis' normal subgroup theorem which holds for arbitrary higher rank lattices, the next problem is of fundamental importance.

Problem 4.2. Let $\Gamma < G$ be *any* higher rank lattice. Is Γ character rigid?

Problem 4.2 is also discussed in [31, QUESTION 2.1] for characters arising from ergodic probability measure preserving actions $\Gamma \curvearrowright (X, \nu)$, in connection with Stuck–Zimmer's results [50]. To answer positively Problem 4.2, it would suffice to prove for every extremal character $\varphi \in \operatorname{Char}(\Gamma)$ that is not supported on $\mathscr{Z}(\Gamma)$, the tracial GNS factor $\pi_\varphi(\Gamma)''$ has property (T) in the sense of Connes–Jones [20]. This would correspond to the *property* (T) *half* in Margulis' normal subgroup theorem.

5. NONCOMMUTATIVE FACTOR THEOREM AND CONNES' RIGIDITY CONJECTURE

Connes [18] obtained the first rigidity result in von Neumann algebras by showing that for any icc (infinite conjugacy classes) countable discrete group Λ with property (T), the type II_1 factor $M = \mathrm{L}(\Lambda)$ has countable outer automorphism group $\mathrm{Out}(M)$ and countable fundamental group $\mathscr{F}(M)$[1]. This result prompted Connes to state the following bold conjecture (see [19, PROBLEM V.B.1]).

Connes' rigidity conjecture. *Let Λ_1 and Λ_2 be any icc countable discrete groups with property* (T) *such that* $\mathrm{L}(\Lambda_1) \cong \mathrm{L}(\Lambda_2)$. *Show that* $\Lambda_1 \cong \Lambda_2$.

We say that a discrete group Λ is W^*-*superrigid* if whenever Υ is another discrete group such that $\mathrm{L}(\Lambda) \cong \mathrm{L}(\Upsilon)$, we have $\Lambda \cong \Upsilon$. Using [20], Connes' rigidity conjecture asks whether every icc countable discrete group Λ with property (T) is W^*-superrigid.

A first deep result towards Connes' rigidity conjecture was obtained by Cowling–Haagerup [21] where they showed that for every $n \geq 2$ and every lattice Λ in the rank one connected simple Lie group $\mathrm{Sp}(n, 1)/\{\pm 1\}$, the type II_1 factor $\mathrm{L}(\Lambda)$ retains the integer $n \geq 2$. For the last two decades, Popa's *deformation/rigidity* theory [48] has led to tremendous progress regarding classification and rigidity results for group (resp. group measure space) von Neumann algebras. In particular, the first examples of W^*-superrigid groups were obtained by Ioana–Popa–Vaes [37]. The examples constructed in [37] are generalized wreath products groups and so they do not have property (T). It is still an open problem to find an example of a W^*-superrigid countable discrete group Λ with property (T). For other recent results regarding classification and rigidity results for von Neumann algebras, we refer the reader to the surveys [35, 36, 54].

Connes' rigidity conjecture is particularly relevant for the class of higher rank lattices. In this context, celebrated strong rigidity results by Mostow and Margulis (see [42] and [41, CHAPTER VI]) show that whenever $\Gamma_i < G_i$ is a higher rank lattice, where G_i has trivial center, $i = 1, 2$, if $\Gamma_1 \cong \Gamma_2$, then $G_1 \cong G_2$. In view of the strong rigidity results by Mostow and Margulis, we state the following version of Connes' rigidity conjecture for higher rank lattices.

Conjecture. *For every $i = 1, 2$, let G_i be any connected simple real Lie group with trivial center and real rank* $\mathrm{rk}_{\mathbf{R}}(G_i) \geq 2$ *and $\Gamma_i < G_i$ any lattice. If* $\mathrm{L}(\Gamma_1) \cong \mathrm{L}(\Gamma_2)$, *then $G_1 \cong G_2$.*

Popa's *deformation/rigidity* theory cannot be used to tackle the above conjecture because higher rank lattices are somehow "too rigid." As suggested by Connes himself (see the discussion in [38, SECTION 4]), it is natural to try and develop a strategy building upon the works of Furstenberg, Margulis, and Zimmer. In what follows, we assume that G is a connected simple real Lie group with trivial center and real rank $\mathrm{rk}_{\mathbf{R}}(G) \geq 2$. We fix a minimal parabolic subgroup $P < G$. Let $\Gamma < G$ be any lattice (e.g., $\Gamma := \mathrm{PSL}_d(\mathbf{Z}) < \mathrm{PSL}_d(\mathbf{R}) := G$

1 The *fundamental group* $\mathscr{F}(M)$ of a type II_1 factor M is defined as the subgroup of $\mathbf{R}^*_+$ that consists of all $\frac{\tau(p)}{\tau(q)}$, where $p, q \in M$ are projections for which $pMp \cong qMq$.

for $d \geq 3$). Then Γ is icc and the group von Neumann algebra $L(\Gamma)$ is a type II_1 factor. Moreover, the nonsingular action $\Gamma \curvearrowright (G/P, \nu_P)$ is (essentially) free and ergodic and the corresponding von Neumann factor $L(\Gamma \curvearrowright G/P)$ is amenable and of type III_1 (see, e.g., [14, PROPOSITION 4.7]). We now give the proof of Corollary F by combining Theorem E with Suzuki's results [51].

Proof of Corollary F. Denote by $E : L(\Gamma \curvearrowright G/P) \to L^\infty(G/P)$ the canonical Γ-equivariant conditional expectation. Let $L(\Gamma) \subset M \subset L(\Gamma \curvearrowright G/P)$ be any intermediate von Neumann subalgebra and consider the boundary structure $\Phi = E|_M : M \to L^\infty(G/P)$. Note that the conjugation action $\Gamma \curvearrowright L(\Gamma \curvearrowright G/P)$ is ergodic. By Theorem E, there are two cases to consider.

Firstly, assume that Φ is invariant. Following Examples 2.7(2), we simply denote by $u_\gamma \in L(\Gamma) \subset L(\Gamma \curvearrowright G/P)$ the canonical unitaries implementing the action $\Gamma \curvearrowright G/P$. For every $x \in M$, write $x = \sum_{\gamma \in \Gamma} x_\gamma u_\gamma$ for its Fourier expansion, where $x_\gamma = E(x u_\gamma^*)$ for every $\gamma \in \Gamma$. Since $E|_M = \Phi$ is invariant and since $L(\Gamma) \subset M$, it follows that $x_\gamma \in \mathbf{C}1$ for every $\gamma \in \Gamma$ and every $x \in M$. This implies that $M = L(\Gamma)$ (see, e.g., [1, LEMMA 6.8]).

Secondly, assume that Φ is not invariant. There there exist a proper parabolic subgroup $P < Q < G$ and a Γ-equivariant unital normal embedding $\iota : L^\infty(G/Q) \hookrightarrow M$ such that $E \circ \iota = p_Q^*$. This further implies that $L(\Gamma \curvearrowright G/Q) = L(\Gamma) \vee L^\infty(G/Q) \subset M$. Since the nonsingular action $\Gamma \curvearrowright (G/Q, \nu_Q)$ is (essentially) free (see, e.g., [13, LEMMA 6.2]), a combination of [51, THEOREM 3.6] and [41, THEOREM IV.2.11] implies that there exists a parabolic subgroup $P < R < Q$ such that $M = L(\Gamma \curvearrowright G/R)$. ∎

It is well known that there are exactly $2^{\mathrm{rk}_{\mathbf{R}}(G)}$ intermediate parabolic subgroups $P < Q < G$. Thus, Corollary F implies that there are exactly $2^{\mathrm{rk}_{\mathbf{R}}(G)}$ intermediate von Neumann subalgebras $L(\Gamma) \subset M \subset L(\Gamma \curvearrowright G/P)$. In particular, the inclusion $L(\Gamma) \subset L(\Gamma \curvearrowright G/P)$ retains the real rank $\mathrm{rk}_{\mathbf{R}}(G)$. We believe this result could be useful to tackle Connes' rigidity conjecture and to show that the group von Neumann algebra $L(\Gamma)$ retains the real rank $\mathrm{rk}_{\mathbf{R}}(G)$.

ACKNOWLEDGMENTS

I would like to express my deepest gratitude to my coauthors for our exciting joint works. I also thank Adrian Ioana for his useful remarks.

FUNDING

This work was partially supported by Institut Universitaire de France.

REFERENCES

[1] H. Ando, U. Haagerup, C. Houdayer, and A. Marrakchi, Structure of bicentralizer algebras and inclusions of type III factors. *Math. Ann.* **376** (2020), 1145–1194.

[2] U. Bader, R. Boutonnet, and C. Houdayer, Charmenability of higher rank arithmetic groups. 2021, arXiv:2112.01337.

[3] U. Bader, R. Boutonnet, C. Houdayer, and J. Peterson, Charmenability of arithmetic groups of product type. 2020, arXiv:2009.09952.

[4] U. Bader and Y. Shalom, Factor and normal subgroup theorems for lattices in products of groups. *Invent. Math.* **163** (2006), 415–454.

[5] B. Bekka, Amenable unitary representations of locally compact groups. *Invent. Math.* **100** (1990), 383–401.

[6] B. Bekka, Operator-algebraic superridigity for $SL_n(\mathbf{Z})$, $n \geq 3$. *Invent. Math.* **169** (2007), 401–425.

[7] B. Bekka, Character rigidity of simple algebraic groups. *Math. Ann.* **378** (2020), 1223–1243.

[8] B. Bekka, M. Cowling, and P. de la Harpe, Some groups whose reduced C^*-algebra is simple. *Publ. Math. Inst. Hautes Études Sci.* **80** (1994), 117–134.

[9] B. Bekka and C. Francini, Characters of algebraic groups over number fields. 2020, arXiv:2002.07497.

[10] Y. Benoist and J.-F. Quint, Mesures stationnaires et fermés invariants des espaces homogènes. *Ann. of Math.* **174** (2011), 1111–1162.

[11] A. Borel and Harish-Chandra, Arithmetic subgroups of algebraic groups. *Ann. of Math.* **75** (1962), 485–535.

[12] J. Bourgain, A. Furman, E. Lindenstrauss, and S. Mozes, Stationary measures and equidistribution for orbits of nonabelian semigroups on the torus. *J. Amer. Math. Soc.* **24** (2011), 231–280.

[13] R. Boutonnet and C. Houdayer, Stationary characters on lattices of semisimple Lie groups. *Publ. Math. Inst. Hautes Études Sci.* **133** (2021), 1–46.

[14] L. Bowen and A. Nevo, Pointwise ergodic theorems beyond amenable groups. *Ergodic Theory Dynam. Systems* **33** (2013), 777–820.

[15] É. Breuillard, M. Kalantar, M. Kennedy, and N. Ozawa, C^*-simplicity and the unique trace property for discrete groups. *Publ. Math. Inst. Hautes Études Sci.* **126** (2017), 35–71.

[16] M. Burger and S. Mozes, Lattices in product of trees. *Publ. Math. Inst. Hautes Études Sci.* **92** (2001), 151–194.

[17] A. Connes, Classification of injective factors. Cases II_1, II_∞, III_λ, $\lambda \neq 1$. *Ann. of Math.* **74** (1976), 73–115.

[18] A. Connes, A factor of type II_1 with countable fundamental group. *J. Operator Theory* **4** (1980), 151–153.

[19] A. Connes, *Noncommutative geometry*. Academic Press, Inc., San Diego, CA, 1994, xiv+661 pp.

[20] A. Connes and V. F. R. Jones, Property T for von Neumann algebras. *Bull. Lond. Math. Soc.* **17** (1985), 57–62.

[21] M. Cowling and U. Haagerup, Completely bounded multipliers of the Fourier algebra of a simple Lie group of real rank one. *Invent. Math.* **96** (1989), 507–549.

[22] D. Creutz and J. Peterson, Character rigidity for lattices and commensurators. 2013, arXiv:1311.4513.

[23] D. Creutz and J. Peterson, Stabilizers of ergodic actions of lattices and commensurators. *Trans. Amer. Math. Soc.* **369** (2017), 4119–4166.

[24] D. Creutz and Y. Shalom, A normal subgroup theorem for commensurators of lattices. *Groups Geom. Dyn.* **8** (2014), 789–810.

[25] A. Furman, Random walks on groups and random transformations. In *Handbook of dynamical systems*, Vol. 1A, pp. 931–1014, North-Holland, Amsterdam, 2002.

[26] H. Furstenberg, Non commuting random products. *Trans. Amer. Math. Soc.* **108** (1963), 377–428.

[27] H. Furstenberg, A Poisson formula for semi-simple Lie groups. *Ann. of Math.* **77** (1963), 335–386.

[28] H. Furstenberg, Poisson boundaries and envelopes of discrete groups. *Bull. Amer. Math. Soc.* **73** (1967), 350–356.

[29] H. Furstenberg, Stiffness of group actions. In *Lie groups and ergodic theory* (Mumbai, 1996), pp. 105–117, Tata Inst. Fund. Res. Stud. Math. 14, Tata Inst. Fund. Res., Bombay, 1998.

[30] L. Ge and R. Kadison, On tensor products for von Neumann algebras. *Invent. Math.* **123** (1996), 453–466.

[31] T. Gelander, A view on invariant random subgroups and lattices. In *Proceedings of the International Congress of Mathematicians–Rio de Janeiro 2018. Vol. II. Invited lectures*, pp. 1321–1344, World Sci. Publ., Hackensack, NJ, 2018.

[32] E. Glasner and B. Weiss, Uniformly recurrent subgroups. In *Recent trends in ergodic theory and dynamical systems*, pp. 63–75, Contemp. Math. 631, Amer. Math. Soc., Providence, RI, 2015.

[33] I. Ya. Goldsheid and G. A. Margulis, Lyapunov indices of a product of random matrices. *Russian Math. Surveys* **44** (1989), 11–71.

[34] Y. Hartman and M. Kalantar, Stationary C^*-dynamical systems. *J. Eur. Math. Soc. (JEMS)* (to appear), arXiv:1712.10133.

[35] A. Ioana, Classification and rigidity for von Neumann algebras. In *European Congress of Mathematics*, pp. 601–625, Eur. Math. Soc., Zürich, 2013.

[36] A. Ioana, Rigidity for von Neumann algebras. In *Proceedings of the International Congress of Mathematicians–Rio de Janeiro 2018. Vol. III. Invited lectures*, pp. 1639–1672, World Sci. Publ., Hackensack, NJ, 2018.

[37] A. Ioana, S. Popa, and S. Vaes, A class of superrigid group von Neumann algebras. *Ann. of Math.* **178** (2013), 231–286.

[38] V. F. R. Jones, Ten problems. In *Mathematics: frontiers and perspectives*, pp. 79–91, Amer. Math. Soc., Providence, RI, 2000.

[39] M. Kalantar and M. Kennedy, Boundaries of reduced C^*-algebras of discrete groups. *J. Reine Angew. Math.* **727** (2017), 247–267.

[40] O. Lavi and A. Levit, Characters of the group $\mathrm{EL}_d(R)$ for a commutative Noetherian ring R. 2020, arXiv:2007.15547.

[41] G. A. Margulis, *Discrete subgroups of semisimple Lie groups*. Ergeb. Math. Grenzgeb. (3) 17, Springer, Berlin, 1991, x+388 pp.

[42] G. D. Mostow, *Strong rigidity of locally symmetric spaces*. Ann. of Math. Stud. 78, Princeton University Press, Princeton, NJ; University of Tokyo Press, Tokyo, 1973, v+195 pp.

[43] A. Nevo and R. J. Zimmer, Homogenous projective factors for actions of semi-simple Lie groups. *Invent. Math.* **138** (1999), 229–252.

[44] A. Nevo and R. J. Zimmer, A structure theorem for actions of semisimple Lie groups. *Ann. of Math.* **156** (2002), 565–594.

[45] V. Paulsen, *Completely bounded maps and operator algebras*. Cambridge Stud. Adv. Math. 78, Cambridge University Press, Cambridge, 2002, xii+300 pp.

[46] J. Peterson, Character rigidity for lattices in higher-rank groups. 2014, preprint, https://math.vanderbilt.edu/peters10/rigidity.pdf.

[47] J. Peterson and A. Thom, Character rigidity for special linear groups. *J. Reine Angew. Math.* **716** (2016), 207–228.

[48] S. Popa, Deformation and rigidity for group actions and von Neumann algebras. In *International Congress of Mathematicians. Vol. I*, pp. 445–477, Eur. Math. Soc., Zürich, 2007.

[49] Y. Shalom, Rigidity of commensurators and irreducible lattices. *Invent. Math.* **141** (2000), 1–54.

[50] G. Stuck and R. J. Zimmer, Stabilizers for ergodic actions of higher rank semi-simple groups. *Ann. of Math.* **139** (1994), 723–747.

[51] Y. Suzuki, Complete descriptions of intermediate operator algebras by intermediate extensions of dynamical systems. *Comm. Math. Phys.* **375** (2020), 1273–1297.

[52] M. Takesaki, *Theory of operator algebras*. II. Encyclopaedia Math. Sci. 125, Oper. Algebr. Noncommut. Geom. 6, Springer, Berlin, 2003, xxii+518 pp.

[53] M. Takesaki, *Theory of operator algebras*. III. Encyclopaedia Math. Sci. 127, Oper. Algebr. Noncommut. Geom. 8, Springer, Berlin, 2003, xxii+548 pp.

[54] S. Vaes, Rigidity for von Neumann algebras and their invariants. In *Proceedings of the International Congress of Mathematicians. Volume III*, pp. 1624–1650, Hindustan Book Agency, New Delhi, 2010.

CYRIL HOUDAYER

Université Paris-Saclay, Institut Universitaire de France, CNRS, Laboratoire de mathématiques d'Orsay, 91405 Orsay, France, cyril.houdayer@universite-paris-saclay.fr

ON SOME PROPERTIES OF SPARSE SETS: A SURVEY

MALABIKA PRAMANIK

ABSTRACT

Sparse sets are, by definition, sets that are small, either in cardinality, measure, dimension, or density. Curves, surfaces, and other submanifolds are standard examples of sparse sets in Euclidean space. However, many sparse sets naturally occurring in ergodic and geometric measure theory, such as Cantor-like sets or self-similar fractals, lack the regularity of the aforementioned objects. Despite this deficiency, many sparse sets are rich in arithmetic, geometric, and analytic properties that can be viewed as working substitutes for smoothness. This has led to a vibrant line of inquiry into the governing principles behind certain phenomena that are typically associated with submanifolds and that have the potential for ubiquity in far more general contexts. Structural and analytical properties of sparse sets, whether discrete or continuous, lie at the center of many problems in harmonic analysis, fractal geometry, combinatorics, and number theory. This is a survey of a few such problems that the author has worked on.

MATHEMATICS SUBJECT CLASSIFICATION 2020

Primary 42-02; Secondary 28-02, 28A80, 58C40, 05D10, 26A24

KEYWORDS

Hausdorff and Fourier dimension, differentiation theorems, growth of Laplace-Beltrami eigenfunctions on manifolds, Euclidean Ramsey theory

INTERNATIONAL CONGRESS OF MATHEMATICIANS 2022 JULY 6–14

Proc. Int. Cong. Math. 2022, Vol. 4, pp. 3224–3248
DOI 10.4171/ICM2022/153

1. INTRODUCTION

Many problems in harmonic analysis and geometric measure theory, including restriction, Bochner–Riesz, Kakeya, Falconer conjectures, maximal operator bounds, and oscillatory integral estimates, are at heart questions involving size and structural properties of sparse sets. In classical formulations of the problems, these sparse sets are often lower-dimensional surfaces in Euclidean space, such as lines, curves, other submanifolds, zero sets of polynomials or real-analytic functions. For example, the famous Kakeya conjecture aims to quantify the size of a possibly small (i.e., Lebesgue-null) set that contain lines in every direction; intersection properties of lines or thin tubes (which are thickened versions of lines) are essential to its analysis. The restriction problem is a statement about the Lebesgue integrability of the Fourier transform of certain classes of functions in $\mathbb{R}^d$ after being restricted to a sphere; the latter hypersurface provides the geometric basis for this problem. Decay rates of multidimensional oscillatory integrals and integral operators are tied to critical points of their phases; if the phase function is polynomial or analytic in nature, the set of critical points is a semialgebraic set whose structure determines the asymptotic behavior of the integral. These avenues of research naturally use well-developed analytic and geometric notions, such as smoothness or curvature, of the underlying surfaces. A more recent research direction in harmonic analysis has been devoted to investigating extensions of classical results that are normally associated to surfaces and manifolds to the context of more general sets. More precisely, do similar results continue to hold for arbitrary Euclidean sets, possibly of fractional dimension, where tools like smoothness or curvature are unavailable, or have to be replaced by appropriate generalizations? To what extent are such results transferable between the discrete and continuum settings, and which features are unique? This in turn has led to a deeper investigation of the structure and content of sparse sets.

This article, arranged in three distinct and notationally self-contained parts, gives an overview of part of the author's work to date in this area, undertaken with many collaborators over the last decade and a half. The common theme is the study of sparse sets. Each section contains a brief motivation for the problems considered, and a statement of the main results. Proofs are delegated to the publications where the results appear. An in-depth discussion of the surrounding literature had to be scaled back due to space constraints, but the bibliography contains some important landmarks in the subject, as well as more exhaustive surveys.

2. MAXIMAL AVERAGES AND DIFFERENTIATION

2.1. Motivation of the problem

There is a vast literature on maximal and averaging operators over families of lower-dimensional submanifolds of $\mathbb{R}^d$. The aim is to quantify the behavior of such operators; for instance, what are the Lebesgue mapping properties of the maximal operator? What is the Lebesgue or Sobolev regularity of the averaging operator? For concreteness, we will focus here on the case of maximal operators over rescaled copies of a single submanifold. Assuming that the submanifold in question is sufficiently smooth, an important issue turns out to be its curvature. Roughly speaking, curved submanifolds admit nontrivial maximal estimates, whereas flat submanifolds do not. A fundamental and representative positive result is the *spherical maximal theorem*, due to Stein [90] for $d \geq 3$ and Bourgain [9] for $d = 2$. Recall that the spherical maximal operator in $\mathbb{R}^d$ is defined to be

$$\mathfrak{M}_{\mathbb{S}^{d-1}} f(x) := \sup_{r>0} \int_{\mathbb{S}^{d-1}} |f(x+ry)| d\sigma(y), \tag{2.1}$$

where σ is the normalized Lebesgue measure on the unit sphere $\mathbb{S}^{d-1} \subseteq \mathbb{R}^d$.

Theorem 2.1 (Stein [90], Bourgain [9]). *For any $d \geq 2$, the maximal operator $\mathfrak{M}_{\mathbb{S}^{d-1}}$ is bounded on $L^p(\mathbb{R}^d)$ for $p > \frac{d}{d-1}$, and this range of p is optimal.*

It is well known that Theorem 2.1 fails in all dimensions $d \geq 2$ if the sphere $\mathbb{S}^{d-1}$ is replaced by a polygonal line or the surface of a polytope. These geometric objects, while still piecewise smooth, do not have any curvature. In intermediate cases such as conical surfaces, which are flat along their generating rays but curved in other directions, maximal estimates may still be available but with weaker exponents. Many results of this type are known under varying smoothness and curvature conditions. We refer the reader to [19, 45, 46, 66–68, 88, 89, 91, 92] for an introduction to this prolific area of research and further references.

Stein's proof of the spherical maximal theorem for $d \geq 3$ exploits curvature only through the decay of the Fourier transform of the surface measure on the manifold, as do many of the other results just mentioned. Let us recall that for any finite measure μ on $\mathbb{R}^d$, its Fourier transform is defined as

$$\widehat{\mu}(\xi) := \int_{\mathbb{R}^d} e^{-ix\cdot\xi} d\mu(x). \tag{2.2}$$

In the case of the sphere, the Fourier transform decays like $|\xi|^{-\frac{d-1}{2}}$ at infinity; similar estimates hold for other convex hypersurfaces of codimension 1 with nonvanishing Gaussian curvature. The decay estimates are weaker for manifolds with flat directions, which in turn result in the restricted range of exponents in maximal and averaging estimates mentioned above.

The connection between the Fourier decay of a measure and the maximal average associated with it is exemplified in the following classical result of Rubio de Francia [73]. Given a measure μ, let us define its corresponding maximal operator

$$\mathfrak{M}_{\mu} f(x) := \sup_{r>0} \int \left|f(x+ry)\right| d\mu(y), \tag{2.3}$$

which is the supremal average of $|f(x+\cdot)|$ over all possible dilates of μ. If μ is the surface measure of $\mathbb{S}^{d-1}$, then $\mathfrak{M}_{\mu}$ coincides with the spherical maximal operator defined in (2.1).

Theorem 2.2 (Rubio de Francia [73]). *Suppose that μ is a compactly supported Borel measure on $\mathbb{R}^d$, $d \geq 1$, such that*

$$\left|\widehat{\mu}(\xi)\right| \leq C\left(1+|\xi|\right)^{-a} \tag{2.4}$$

for some $a > \frac{1}{2}$. Then the maximal operator $\mathfrak{M}_{\mu}$, defined as in (2.3) *is bounded on $L^p(\mathbb{R}^d)$ for $p > (2a+1)/(2a)$.*

Theorem 2.2 implies Theorem 2.1 for $d \geq 3$, since then the surface measure σ on the sphere obeys the above assumption with $a = \frac{d-1}{2} > \frac{1}{2}$. However, this Fourier decay is insufficient for $d = 2$. In other words, Theorem 2.2 fails to capture the circular maximal estimate in $\mathbb{R}^2$, for which $a = \frac{1}{2}$ just misses the stated range. Instead, Bourgain's proof of the circular maximal theorem for $d = 2$ relies more directly on the geometry involved. The relevant geometric information concerns intersections of pairs of δ-thickened circles, in other words, annuli of width δ. In an arrangement of many such annuli, most pairwise intersections are of the order δ^2, which is smaller by an order of magnitude than each annulus itself. While larger intersections are possible, Bourgain's proof shows that they do not occur frequently. An alternative proof of the circular maximal theorem that exploits finer properties of Fourier integral operators rather than Fourier decay can be found in [65].

2.2. Main results

The above class of results does not offer an easy extension to dimension one. Indeed, it is not clear a priori what a one-dimensional theory might look like, given that the real line has no nontrivial lower-dimensional submanifolds. However, given any $\varepsilon > 0$, there are many singular measures on $\mathbb{R}$ supported on sets of Hausdorff dimension $1 - \varepsilon$. Viewing ε as an analogue of "codimension," it is natural to ask whether by imposing additional structure on these sets that would assume the role of curvature, one might obtain L^p estimates similar to those in Theorem 2.1 for the associated maximal operators and for a range $p > p_\varepsilon$, where $p_\varepsilon \searrow 1$ as $\varepsilon \to 0$. Theorem 2.3 below, joint with Izabella Łaba, provides an affirmative answer to this question. Theorem 2.3 may be interpreted as the limiting situation as $\varepsilon \to 0$ (compare with Theorem 2.1 as $d \to \infty$) where the maximal range $(1, \infty]$ of p is achieved for a single set S of zero Lebesgue measure.

Theorem 2.3 ([59]). *For every $0 \leq \varepsilon < \frac{1}{3}$, there exists a probability measure $\mu = \mu(\varepsilon)$ supported on a Lebesgue-null set S of Hausdorff dimension $1 - \varepsilon$ such that $\mathfrak{M}_\mu$ is bounded on $L^p(\mathbb{R})$ for all $p > \frac{1+\varepsilon}{1-\varepsilon}$.*

The result above is one of many similar ones involving the operator on restricted sets and restricted scales. The interested reader is referred to [59] for other analogous statements concerning $\mathfrak{M}_\mu$ and its variants. As a consequence of Theorem 2.3, we obtain a differentiation theorem for μ that answers a question of Aversa and Preiss [2, 3, 69].

Theorem 2.4. *For $0 \leq \varepsilon < \frac{1}{3}$, let $\mu = \mu(\varepsilon)$ be the measure specified in Theorem 2.3. Then for every $f \in L^p(\mathbb{R})$ with $p \in ((1+\varepsilon)/(1-\varepsilon), \infty)$, we have*

$$\lim_{r \to 0} \left| \int f(x + ry) d\mu(y) - f(x) \right| = 0 \quad \text{for a.e. } x \in \mathbb{R}. \tag{2.5}$$

Thus for $\varepsilon = 0$, the measure $\mu = \mu(0)$ is supported on a full-dimensional set, is singular with respect to Lebesgue, and yet differentiates L^p in the sense of (2.5), as does the Lebesgue measure. Further, the maximal operator $\mathfrak{M}_\mu$ is bounded on the same Lebesgue spaces L^p (namely, $p \in (1, \infty)$) as the one-dimensional Hardy–Littlewood maximal function. However, unlike the Lebesgue measure, the measure $\mu = \mu(0)$ fails to differentiate L^1, as shown by Preiss [59, SECTION 8].

The measures $\mu = \mu(\varepsilon)$ in our results are constructed by randomizing a Cantor-type iteration. More precisely, we describe a random mechanism for building the nested Cantor iterates S_k as a union of finitely many intervals. The measure μ is then shown to be the weak-$*$ limit of the natural probability measures $1_{S_k}/|S_k|$, which are supported on $S = \bigcap_k S_k$.

It turns out that the proof of Theorem 2.3 does not use any Fourier decay conditions. Instead, the proof relies on geometric arguments akin to those in Bourgain's proof of the circular maximal theorem. The right substitute for Fourier decay turns out to be a correlation condition between affine copies of the sets S_k, providing the needed bound on the size of multiple intersections analogous to those arising in Bourgain's argument. Readers familiar with the proof of Theorem 2.1 for $d = 2$ or other similar results will recognize the correlation condition as a bound on the integrand (interpreted as the correlation function) in the expression for the L^n-norm of the dual linearized and discretized maximal operator, for large integer values of n. The proof attempts to minimize this integrand whenever possible through randomization arguments.

The threshold exponent $p_0 = (1+\varepsilon)/(1-\varepsilon)$ is suboptimal in general. Shmerkin and Suomala [82] have improved the range of p for random measures μ associated to an Ahlfors-regular variant of fractal percolation. Another improvement in a different direction is due to Łaba [57], who has obtained slightly weaker estimates for $\mathfrak{M}_\mu$, but for a much larger class of measures μ; in particular, her results apply to measures μ with self-similar supports of arbitrarily small Hausdorff dimension and no Fourier decay. Determining the Lebesgue boundedness of $\mathfrak{M}_\mu$ where μ is the Cantor–Lebesgue measure on the standard middle-third Cantor set remains an open problem.

3. SPARSE RESTRICTION OF LAPLACE–BELTRAMI EIGENFUNCTIONS

The study of eigenfunctions of Laplacians lies at the interface of several areas of mathematics, including analysis, geometry, mathematical physics, and number theory. These special functions arise in physics and in partial differential equations as modes of periodic vibration of drums and membranes. In quantum mechanics, they represent the stationary energy states of a free quantum particle on a Riemannian manifold.

Let (M, g) denote a compact, connected, n-dimensional Riemannian manifold without boundary, and $-\Delta_g$ the positive Laplace–Beltrami operator on M. It is well known [87, CHAPTER 3] that the spectrum of this operator is nonnegative and discrete. Let us denote its eigenvalues by $\{\lambda_j^2 : j \geq 0\}$, and the corresponding eigenspaces by $\mathbb{E}_j$. Without loss of generality, the positive square roots of the distinct eigenvalues can be arranged in increasing order, with

$$0 = \lambda_0 < \lambda_1 < \lambda_2 < \cdots \lambda_j < \cdots \to \infty.$$

It is a standard fact [87, CHAPTER 3] that each $\mathbb{E}_j$ is finite-dimensional. Further, the space $L^2(M, dV_g)$ (consisting of functions on M that are square-integrable with respect to the canonical volume measure dV_g) admits an orthogonal decomposition in terms of $\mathbb{E}_j$:

$$L^2(M, dV_g) = \bigoplus_{j=0}^{\infty} \mathbb{E}_j.$$

One of the fundamental questions surrounding Laplace–Beltrami eigenfunctions targets their concentration phenomena, via high-energy asymptotics or high-frequency behavior. There are many avenues for this study, as exemplified in [1, 11, 13, 20, 32, 41, 61, 74, 79, 83, 99, 100]. One such approach involves studying the growth of the L^p-norms of these eigenfunctions as the eigenvalue goes to infinity. My joint work with Suresh Eswarathasan [27], the main focus of this section, lies in this category. Specifically, we describe the $L^2(M) \to L^p(\Gamma)$ mapping property of a certain spectral projector (according to the spectral decomposition above), where Γ is a Lebesgue-null subset of M. In particular, Γ does not enjoy any smooth structure, a point of departure from prior work where this feature was heavily exploited. We begin by reviewing the current research landscape that will help place the main results in context.

3.1. Motivation of the problem

The Weyl law in spectral theory provides an L^∞-bound on eigenfunctions on M [43]. The first results that establish L^p eigenfunction bounds for $p < \infty$ are due to Sogge [86].

Theorem 3.1 ([86]). *Given any manifold M as above and $p \in [2, \infty]$, there exists a constant $C = C(M, p) > 0$ such that the following inequality holds for all $\lambda \geq 1$:*

$$\|\varphi_\lambda\|_{L^p(M)} \leq C(1+\lambda)^{\delta(n,p)} \|\varphi_\lambda\|_{L^2(M)}, \quad \textit{with} \tag{3.1}$$

$$\delta(n, p) = \begin{cases} \dfrac{n-1}{4} - \dfrac{n-1}{2p}, & \textit{if } 2 \leq p \leq \dfrac{2(n+1)}{n-1}, \\ \dfrac{n-1}{2} - \dfrac{n}{p}, & \textit{if } \dfrac{2(n+1)}{n-1} \leq p \leq \infty \end{cases}. \tag{3.2}$$

Here φ_λ is any eigenfunction of $-\Delta_g$ corresponding to the eigenvalue λ^2. The bound is sharp for the n-dimensional unit sphere $M = \mathbb{S}^n$, equipped with the surface measure.

Historically, an important motivation and source of inspiration for this line of investigation has been the Fourier restriction problem, which explores the behavior of the Fourier transform when restricted to curved surfaces in Euclidean spaces. In fact, the Stein–Tomas L^2-restriction theorem [96], originating in Euclidean harmonic analysis, was a key ingredient in an early proof of Theorem 3.1 for the sphere. Indeed, Theorem 3.1 may be viewed as a form of discrete restriction on M where the frequencies are given by the spectrum of the manifold, see, for example, [85]. Conversely, it is possible to recover the L^2-restriction theorem for the sphere from a spectral projection theorem such as Theorem 3.1 applied to the n-dimensional flat torus. The lecture notes of Yung [98, SECTION 2] contain a discussion of these implications.

Theorem 3.1 permits a number of independent proofs. For an argument that involves well-known oscillatory integral estimates of Hörmander applied to the smooth spectral projector (denoted $\rho(\sqrt{-\Delta_g} - \lambda)$), we refer the reader to the treatise **[87]**. The semiclassical approach of Koch, Tataru, and Zworski **[54]** has also yielded many powerful applications.

Finer information on eigenfunction growth may be obtained through L^p-bounds on φ_λ when restricted to smooth submanifolds of M. One expects φ_λ to assume large values on small sets. Thus its L^p-norm on a Lebesgue-null set such as a submanifold, if meaningful, is typically expected to be larger in comparison with the L^p-norm taken over the entire manifold M, as given by Theorem 3.1. The first step in this direction is due to Reznikov **[78]**, who studied eigenfunction restriction phenomena on hyperbolic surfaces via representation-theoretic tools. The most general results to date on restricted norms of Laplace eigenfunctions are by Burq, Gérard, and Tzvetkov **[14]**, and independently by Hu **[44]**. The work of Tacy **[95]** has extended these results to the setting of a semiclassical pseudodifferential operator (not merely the Laplacian) on a Riemannian manifold, while removing logarithmic losses at a critical threshold. Another particular endpoint result is due to Chen and Sogge **[18]**. We have summarized below the currently known best eigenfunction restriction estimates for a general manifold, combined from this body of work and for easy referencing later.

Theorem 3.2 (**[14, 44, 95]**). *Let $\Sigma \subset M$ be a smooth d-dimensional submanifold of M, equipped with the canonical measure $d\sigma$ that is naturally obtained from the metric g. Then for each $p \in [2, \infty]$, there exists a constant $C = C(M, \Sigma, p) > 0$ such that for any $\lambda \geq 1$ and any Laplace eigenfunction φ_λ associated with the eigenvalue λ^2, the following estimate holds:*

$$\|\varphi_\lambda\|_{L^p(\Sigma, d\sigma)} \leq C(1+\lambda)^{\delta(n,d,p)} \|\varphi_\lambda\|_{L^2(M, dV_g)}. \tag{3.3}$$

The exponent $\delta(n, d, p)$ admits a multipart description. Specifically,

$$\delta(n, n-1, p) = \begin{cases} \frac{n-1}{4} - \frac{n-2}{2p}, & \text{for } 2 \leq p \leq \frac{2n}{n-1}, \\ \frac{n-1}{2} - \frac{n-1}{p}, & \text{for } \frac{2n}{n-1} \leq p \leq \infty. \end{cases} \tag{3.4}$$

For $d \neq n-1$,

$$\delta(n, d, p) = \frac{n-1}{2} - \frac{d}{p}, \quad \text{for } 2 \leq p \leq \infty \text{ and } (d, p) \neq (n-2, 2). \tag{3.5}$$

For $(d, p) = (n-2, 2)$, the exponent $\delta(n, d, p)$ is still given by (3.5); however, there is an additional logarithmic factor $\log^{1/2}(\lambda)$ appearing in the right-hand side of inequality (3.3).

The proofs in **[14]** and **[18]** use a delicate analysis of oscillatory representations of the smoothed spectral projector $\chi(\sqrt{-\Delta_g} - \lambda)$ restricted to submanifolds Σ, combined with refined estimates influenced by the considered geometry. Alternatively, **[44]** uses general mapping properties for Fourier integral operators with prescribed degenerate canonical relations to obtain bounds for the oscillatory integral operators in question. There are several recurrent features in these proofs; namely, stationary phase methods, arguments involving integration by parts and operator-theoretic convolution inequalities. This methodology heavily relies on the fact that the underlying measures are induced by Lebesgue, which in turn is a consequence of M and Σ being smooth manifolds. We wanted to explore the accessibility of this machinery in the absence of smoothness, and to find working substitutes when such methods are unavailable.

There is a common theme in Theorems 3.1 and 3.2 above; namely, the left-hand side of both inequalities (3.1) and (3.3) involves the L^p-norm of an eigenfunction φ_λ but over different submanifolds of M (including M itself), and with respect to natural measures on these submanifolds. An interesting feature of the exponents $\delta(n, p)$ and $\delta(n, d, p)$ is that for large p, they are both of the form $(n-1)/2 - \alpha/p$, where

$$\begin{aligned} \alpha &= \text{dimension of the space on which the } L^p\text{-norm of } \varphi_\lambda \text{ is measured} \\ &= \begin{cases} \dim(M) = n & \text{in Theorem 3.1,} \\ \dim(\Sigma) = d & \text{in Theorem 3.2.} \end{cases} \end{aligned} \tag{3.6}$$

In view of this commonality in (3.2), (3.4), and (3.5), we pose the following question:

- *Given an arbitrary Borel set $\Gamma \subseteq \Sigma$, does there exist a measure μ supported on Σ with respect to which we can estimate the growth of the eigenfunctions φ_λ?*

The nontrivial situation arises when Γ is Lebesgue-null, i.e., μ is singular with respect to the canonical measure on Σ. The optimal scenario would be to obtain bounds that reflect the dimensionality of the set Γ in the same way that Theorems 3.1 and 3.2 do. We answer this by presenting the main results of our article [27].

3.2. Main results

Given a compact n-dimensional Riemannian manifold (M, g), let $\Sigma \subseteq M$ be a smooth embedded submanifold of dimension $1 \leq d \leq n$, equipped with the restricted Riemannian metric naturally endowed by g. Let $(U, \mathfrak{u})$ be a local coordinate chart on Σ, where $U \subseteq \mathbb{R}^d$ is an open set containing $[0,1]^d$ and $\mathfrak{u} : U \to \mathfrak{u}(U) \hookrightarrow \Sigma$ is a smooth embedding. Given any $\varepsilon \in [0, 1)$, let $E \subseteq [0,1]^d$ be an arbitrary Borel set of Hausdorff dimension $\dim_{\mathbb{H}}(E) = d(1-\varepsilon)$. We refer the reader to the classical textbook of Mattila [64, CHAPTER 4] for the definitions and properties of Hausdorff dimension of sets in Euclidean spaces. The Borel set $E \subseteq [0,1]^d$ generates a corresponding Borel set $\Gamma = \Gamma[E]$ in Σ by setting $\Gamma := \mathfrak{u}(E)$. Conversely, every Borel subset Γ in $\mathfrak{u}([0,1]^d) \subseteq \Sigma$ can be identified with a set $E = \mathfrak{u}^{-1}(\Gamma) \subseteq [0,1]^d$. Similarly, any measure ν supported on Γ corresponds with a measure $\mu = \mathfrak{u}^*\nu$ on E via the pull-back $\mathfrak{u}^*$, i.e.,

$$\nu(A) := \mu\big(\mathfrak{u}^{-1}(A)\big) \quad \text{for all Borel sets } A \subseteq \Sigma. \tag{3.7}$$

The converse is also true; any Borel measure μ generates another measure ν on Γ through its push-forward, given by the same relation (3.7). Since $\mathfrak{u}$ is a diffeomorphism, and thus bi-Lipschitz, it preserves Hausdorff dimension [28, COROLLARY 2.4]; hence $\dim_{\mathbb{H}}(\Gamma) = \dim_{\mathbb{H}}(E) = d(1-\varepsilon)$. Let us define our critical exponent

$$p_0 = p_0(n, d, \varepsilon) := \frac{4d(1-\varepsilon)}{n-1}. \tag{3.8}$$

Our result below, representative of the class of results presented in [27], specifies a family of restricted eigenfunction estimates for every such set Γ.

Theorem 3.3 ([27]). *Let M, Σ, Γ and p_0 be as above. Then for every $\kappa > 0$ sufficiently small, there exists a probability measure $\nu^{(\kappa)}$ supported on Γ such that for all $\lambda \geq 1$ and all $p \in [2, \infty]$, we have the eigenfunction estimate*

$$\|\varphi_\lambda\|_{L^p(\Gamma,\nu^{(\kappa)})} \leq C_{\kappa,p}(1+\lambda)^{\theta_p}\Theta(\lambda;\kappa,p)\|\varphi_\lambda\|_{L^2(M,dV_g)}. \tag{3.9}$$

Here φ_λ denotes any L^2-eigenfunction associated with the eigenvalue λ^2 for the Laplace–Beltrami operator $-\Delta_g$ on M. For $p_0 > 2$, the exponent θ_p is given by

$$\theta_p = \theta_p(n,d,\varepsilon) := \begin{cases} \frac{n-1}{4} & \text{if } 2 \leq p \leq p_0 - \frac{4\kappa}{n-1}, \\ \frac{n-1}{2} - \frac{d(1-\varepsilon)}{p} & \text{if } p \geq p_0 - \frac{4\kappa}{n-1}. \end{cases} \tag{3.10}$$

The function Θ represents a power loss of κ/p beyond the critical threshold:

$$\Theta(\lambda;\kappa,p) := \begin{cases} 1 & \text{if } 2 \leq p \leq p_0 - \frac{4\kappa}{n-1}, \\ (1+\lambda)^{\frac{\kappa}{p}}(\log\lambda)^{\frac{1}{p}} & \text{if } p = p_0 - \frac{4\kappa}{n-1}, \\ (1+\lambda)^{\frac{\kappa}{p}} & \text{if } p > p_0 - \frac{4\kappa}{n-1}. \end{cases}$$

For $p_0 \leq 2$ and $2 \leq p \leq \infty$, we set

$$\theta_p = \theta_p(n,d,\varepsilon) := \frac{n-1}{2} - \frac{d(1-\varepsilon)}{p} \quad \text{and} \quad \Theta(\lambda;\kappa,p) = (1+\lambda)^{\frac{\kappa}{p}}.$$

The positive constant $C_{\kappa,p}$ in (3.9) may depend on $n, d, \varepsilon, \kappa$, and p, but is independent of λ. The probability measure $\nu^{(\kappa)}$ in (3.9), given by Frostman's lemma, obeys the following volume growth condition: there exists $C_\kappa > 0$ such that for all $x \in \Sigma$ and $r > 0$,

$$\nu^{(\kappa)}\big(B_g(x;r)\big) \leq C_\kappa r^{d(1-\varepsilon)-\kappa}, \tag{3.11}$$

where $B_g(x;r) \subseteq M$ denotes the Riemannian ball centered at x of radius r.

The estimate (3.9) is sharp for $p \geq \max(2, p_0)$, except possibly for the infinitesimal blow-up factor of $(1+\lambda)^{\kappa/p}$. More precisely, for every $\varepsilon \in [0,1)$, $d \leq n$, and $p \geq \max(2, p_0)$, the bound in (3.9) is realized, ignoring subpolynomial losses, for certain sets of dimension $d(1-\varepsilon)$ in $M = \mathbb{S}^n$. The estimate is not sharp for $2 \leq p < p_0$, when $p_0 > 2$. This is an artifact of our proof strategy.

For an arbitrary Borel set Γ, the information available about measures supported on it is limited. As a result, the measure $\nu^{(\kappa)}$ that realizes (3.9) varies with κ in general. Thus we are able to prove (3.9) only for all $\kappa > 0$ and not for $\kappa = 0$. On the other hand, if $\Gamma = M$ or if Γ is a submanifold of M, it follows from [14, 44, 87, 95] that there is a natural Lebesgue-induced measure on Γ for which (3.9) does hold with $\kappa = 0$. We show in [27] is that such an improvement holds in a generic sense. In Theorem 1.7 and Corollary 1.8 of [27], we provide a large class of sparse subsets Γ that are not submanifolds, each supporting *a single probability measure ν that obeys* (3.11) *for all $\kappa > 0$*, even though $C_\kappa \nearrow \infty$ as $\kappa \searrow 0$. For this measure ν, we show that a stronger version of (3.9) holds, with $\kappa = 0$. However,

Θ is then replaced by a function of slow growth in the range $p \geq \max(2, p_0)$. A precise functional form for Θ that quantifies the infinitesimal blowup is provided.

For $p \geq \max(2, p_0)$, the exponent θ_p in Theorem 3.3 is of the same form alluded to in (3.6), namely $\theta_p = (n-1)/2 - \alpha/p$ with $\alpha = d(1-\varepsilon) = \dim_{\mathbb{H}}(\Gamma)$. Thus our result may be viewed as a natural interpolation between the global estimates in **[86]** and the smooth restriction estimates in **[14]**, bridging the estimates across a family of arbitrary Borel sets with continuously varying Hausdorff dimensions.

To the best of our knowledge, Theorem 3.3 is the first result of its kind in several distinct categories. First, it offers eigenfunction bounds restricted to *any Borel subset of positive Hausdorff dimension*, for every manifold M and every smooth submanifold Σ therein. Second, even for integers m, our result produces new sets of dimension m, for example, with $(n, d, \varepsilon) = (2, 2, 1/2)$, that are not necessarily contained in any m-dimensional submanifold, and yet capture the same eigenfunction growth bounds as smooth submanifolds of the same dimension, up to subpolynomial losses. Third, when $\varepsilon = 0$, our result provides examples of singular measures supported on submanifolds with respect to which the eigenfunctions obey the same L^p growth bounds, up to any prescribed κ-loss, as with the induced Lebesgue measure on the same submanifold.

The work of Burq, Gérard, and Tzvetkov **[14, THEOREM 2]** shows that when $n = 2$, $d = 1$ and Γ is a curve of nonvanishing geodesic curvature, Theorem 3.2 admits a significant improvement; namely the growth exponent $\delta(2, 1, p)$ can be replaced by the smaller exponent $\tilde{\delta}(2, 1, p) = 1/3 - 1/(3p)$ in the range $2 \leq p \leq 4$. The correct analogue of the nonvanishing curvature condition for an arbitrary sparse set Γ that would lead to similar improvements for Theorem 3.3 is as yet unknown.

4. CONFIGURATIONS IN SPARSE SETS

Another related field of research, at the interface of harmonic analysis, geometric measure theory, and fractal geometry, is the study of patterns or configurations in sparse sets. Questions here are typically of the following type: *Under what conditions must a small set contain a given pattern? Can it contain many? Can a large set avoid specified patterns? How can one quantify the patterns contained in a set?* Stated in this level of generality, these questions lack precision, both in the quantification of size and in the specification of patterns. "Large" could be interpreted in the context of cardinality, Lebesgue measure, asymptotic or Banach density, Hausdorff, Minkowski or Fourier dimension. "Patterns" could be geometric in nature, for example, arithmetic or geometric progressions, equilateral triangles, parallelograms; alternatively, they could be algebraic, such as solutions of certain equations. This line of investigation has a particularly rich history in number theory and additive combinatorics where the ambient space is often the space of integers, or subsets thereof. It has expanded into an active research area in the continuum setting within the last two decades. While the questions often look similar in the discrete and continuous regimes, the answers are sometimes very different.

4.1. Existence and avoidance of linear patterns

Can large sets avoid many patterns? Regardless of the many possible variants of such a question, it would seem that a natural answer would be "no," with any reasonable definition. Indeed, there is a large body of work that supports this intuition; see [5,15,35–37, 42,58].

However, there are also many results in the literature that challenge this intuition, especially when slight variations in the notions of size lead to very different conclusions regarding the existence of patterns. For example, in the discrete setting, a classical result of Behrend [4] and Salem and Spencer [78] says that for any $\varepsilon > 0$ and all sufficiently large positive integers M, there exists a set $X_M \subseteq [M] := \{0, 1, 2, \ldots, M-1\}$ such that $\#(X_M) > M^{1-\varepsilon}$ and X_M contains no nontrivial three-term arithmetic progression. This is in sharp contrast with the celebrated results of Roth [71,72] and Szemerédi [93,94], which state that for any $k \geq 3$ and any $c > 0$, there exists $M_0 \geq 1$ such that for $M \geq M_0$, any set $X_M \subseteq [M]$ obeying $\#(X_M) \geq cM$ contains a nontrivial k-term arithmetic progression. The work of Ruzsa [75,76] on subsets of integers avoiding nontrivial solutions of linear equations has been particularly influential in subsequent research in additive combinatorics.

Similar results exist in the continuum as well. For instance, one can deduce from the Lebesgue density theorem that any set in $\mathbb{R}$ with a Lebesgue density point contains a nontrivial affine copy of any finite configuration. This conclusion applies therefore to any set of positive Lebesgue measure. On the other hand, Keleti [52] constructs a compact subset $E \subseteq [0, 1]$ with Hausdorff dimension 1 but Lebesgue measure zero such that there does not exist any nontrivial solution of $x - y = z - w$, with $x < y \leq z < w$ and $(x, y, z, w) \in E^4$. In particular, E avoids all three-term arithmetic progressions. Subsequent results [23,30,40, 52,53,62,63] have explored the issue of avoidance further, providing examples of sets of large Hausdorff dimension that omit increasingly general families of algebraic and geometric patterns. Let us recall from [64, THEOREM 8.8] or [28, SECTION 4.1] that the *Hausdorff dimension* $\dim_{\mathbb{H}}(A)$ of a Borel set $A \subseteq \mathbb{R}^n$ is the supremum of exponents $\alpha > 0$ with the following property: there exists a probability measure μ supported on A such that for some positive, finite constant C_1,

$$\mu\big(B(x,r)\big) \leq C_1 r^{\alpha} \quad \text{for all } x \in \mathbb{R}^n,\ r > 0. \tag{4.1}$$

These results suggest a general rule of thumb: large Hausdorff dimension is usually not enough to ensure that a set contains a specified family of patterns.

On the other hand, the situation is expected to be different for sets A of large *Fourier dimension*. The Fourier dimension $\dim_{\mathbb{F}}(A)$ of a Borel set $A \subseteq \mathbb{R}^n$ is defined as the supremum of exponents $\beta \leq n$ obeying the following condition: there exist a probability measure μ supported on A and a positive finite constant C_2 such that

$$\big|\widehat{\mu}(\xi)\big| \leq C_2\big(1 + |\xi|\big)^{-\beta/2} \quad \text{for all } \xi \in \mathbb{R}^n, \quad \text{where } \widehat{\mu}(\xi) := \int e^{-ix\xi}\, d\mu(x). \tag{4.2}$$

Frostman's lemma [64, P. 168] states that $\dim_{\mathbb{F}}(A) \leq \dim_{\mathbb{H}}(A)$ for any Borel set A. This inequality implies that sets of large Fourier dimension form a smaller subclass within the class of sets of large Hausdorff dimension. It gives rise to the intuition that such sets are more likely to enjoy additional properties rooted in the Fourier decay of the supporting measures; in particular, they could possibly contain a richer class of patterns. A Borel set whose Fourier dimension equals its Hausdorff dimension is called a *Salem set*. One therefore hopes that a Salem set of large dimension may contain patterns that a non-Salem set of the same Hausdorff dimension does not. While this naive expectation turns out to be false in general (more on this in Section 4.2), there is a core of truth in this heuristic principle. In joint work with Yiyu Liang [60], we have made this precise. Our results in this direction form the main content of this subsection and the next.

The intuition that large Salem sets are richer in structure than their non-Salem counterparts of the same dimension is perhaps also due to the known examples of such sets. Salem sets are ubiquitous among random sets. Many random constructions yield sets that are, on the one hand, often (almost surely) Salem and, on the other hand, embody verifiable algebraic or geometric structure. The first such random construction is due to Salem himself [77]; many subsequent random constructions have appeared in [7,8,16,17,25,49,50,58,81]. Deterministic examples of Salem sets are comparatively fewer [38,39,47,48,51], but they arise naturally in number theory [6,12,24] and are rich in arithmetic patterns as well. The work of Körner [55,56], which explicitly addresses the relation between the rate of decay of the Fourier transform of a measure and possible algebraic relations within its support, is perhaps closest to the main theme of this section.

In the results to be stated shortly, we will provide a quantitative formulation of the heuristic principle that Salem sets possess richer structure, in the specific context of translation-invariant linear patterns. More precisely, we will be concerned with algebraic patterns that occur as a nontrivial zero of some function in the class

$$\mathscr{F} = \mathscr{F}(\mathbb{N}) := \bigcup_{v=2}^{\infty} \mathscr{F}_v(\mathbb{N}), \quad \text{where} \tag{4.3}$$

$$\mathscr{F}_v(\mathbb{N}) := \left\{ f(x_0, \dots, x_v) := m_0 x_0 - \sum_{i=1}^{v} m_i x_i \;\middle|\; \begin{array}{l} m_0, \dots, m_v \in \mathbb{N}, \quad m_0 = \sum_{i=1}^{v} m_i, \\ \gcd(m_0, m_1, \dots, m_v) = 1 \end{array} \right\}. \tag{4.4}$$

Here $v \in \mathbb{N} \setminus \{1\}$ and $\mathbb{N} := \{1, 2, \dots\}$.

Definition 4.1. Let us briefly review the patterns whose existence or avoidance we will explore in this section.

- Given $f \in \mathscr{F}_v(\mathbb{N})$, a vector $x = (x_0, x_1, \dots, x_v) \in \mathbb{R}^{v+1}$ is said to be a *zero of* f if it obeys the equation $f(x_0, \dots, x_v) = 0$. Such a vector x will also be referred to as a *solution* of the equation $f(x_0, \dots, x_v) = 0$.
- A zero $x = (x_0, \dots, x_v) \in \mathbb{R}^{v+1}$ of a function $f \in \mathscr{F}_v(\mathbb{N})$ is said to be *nontrivial* if the entries of x are all distinct. All other zeros of f are called *trivial*. The terms "trivial" and "nontrivial" apply with the same definition to solutions of equations of the form $f = 0$ as well.

- Given a set $E \subseteq \mathbb{R}$, we say that E *contains a nontrivial zero of* $f \in \mathscr{F}_v(\mathbb{N})$ if there exists $x = (x_0, x_1, \ldots, x_v) \in E^{v+1}$ with all distinct entries such that $f(x) = 0$. If no such $x \in E^{v+1}$ exists, we say that E *avoids all nontrivial zeros of* f.

- A set $E \subseteq \mathbb{R}$ is said to contain *a nontrivial translation-invariant rational linear pattern* if it contains a nontrivial zero of some $f \in \mathscr{F}$.

A three-term arithmetic progression (x_0, x_1, x_2) with nonzero common difference is a simple example of a nontrivial translation-invariant rational linear pattern, since it is a nontrivial zero of the function $f(x_0, x_1, x_2) = 2x_0 - (x_1 + x_2)$. If a vector

$$x = (x_0, \ldots, x_v) \in \mathbb{R}^{v+1}$$

is a trivial zero of some $f \in \mathcal{F}_v(\mathbb{N})$ with $v \geq 3$ but has at least two distinct entries, then the vector $y = (y_0, \ldots, y_{v'})$ consisting of the distinct entries of x provides a nontrivial zero of some $g \in \mathcal{F}_{v'}(\mathbb{N})$, $v' < v$.

In the following subsection, we provide answers to variants of the following question: *given $\mathscr{F}^* \subseteq \mathscr{F}(\mathbb{N})$, how large a set $E \subset \mathbb{R}$, in the sense of Fourier dimension, can one construct that avoids all the nontrivial zeros of all $f \in \mathscr{F}^*$? Alternatively, are sets of large enough Fourier dimension guaranteed to contain a nontrivial zero of some $f \in \mathscr{F}^*$?* The requirement $m_0 = \sum_{i=1}^{v} m_i$ in $\mathcal{F}_v(\mathbb{N})$ is designed to avoid trivial answers; without this assumption, one can always find an avoiding interval (of positive Lebesgue measure) centered around 1.

4.2. Main results

In [58, THEOREM 1.2], we showed, in joint work with Izabella Łaba, that if a compact set $A \subseteq [0, 1]$ supports a probability measure μ obeying a ball condition of the type (4.1) and a Fourier decay condition of the type (4.2), then A contains a nontrivial three-term arithmetic progression, provided (a) $\beta > 2/3$, (b) the constants C_1 and C_2 are appropriately controlled, and (c) the exponent α is sufficiently close to 1, depending on C_1, C_2, and β. The article [58, SECTION 7] also contains a large class of examples of Salem sets that verify the hypotheses of [58, THEOREM 1.2]. This leads to a natural question whether the technical growth conditions (b) on C_1, C_2 are truly necessary, and whether progressions exist in any set of large enough Fourier dimension. This naive expectation is, however, false. Shmerkin [80, THEOREMS A AND B] has recently proved the existence of a compact full-dimensional Salem set contained in [0, 1] that avoids all nontrivial arithmetic progressions. The existence of such a Salem set seems, at first glance, to contradict the conventional belief that such sets should enjoy richer structure.

4.2.1. Rational linear patterns

Our next three results show that even though a Salem set of large dimension can avoid a specific linear pattern (or even finitely many) given by $\mathcal{F}$, it cannot avoid all of them.

Theorem 4.2 ([60]). *Given $v \in \mathbb{N}$, $v \geq 2$, let $E \subseteq [0, 1]$ be a closed set satisfying $\dim_{\mathbb{F}}(E) > \frac{2}{v+1}$, i.e., there exist some $\beta > \frac{1}{v+1}$, a probability measure μ supported on E,* *and some positive constant C such that*

$$\left|\hat{\mu}(\xi)\right| \leq C\left(1+|\xi|\right)^{-\beta}. \tag{4.5}$$

Then E contains a nontrivial zero of some $f \in \mathscr{F}_v(\mathbb{N})$ defined in (4.4). *In other words, there exists $\{m_0, \ldots, m_v\} \subseteq \mathbb{N}$ satisfying $m_0 = \sum_{i=1}^{v} m_i$, such that E contains a nontrivial solution of the equation*

$$\sum_{i=1}^{v} m_i x_i = m_0 x_0. \tag{4.6}$$

Corollary 4.3 ([60]). *Let $E \subseteq [0,1]$ be a closed set of positive Fourier dimension. Then E contains a nontrivial translation-invariant rational linear pattern, in the sense of Definition* 4.1.

We compare Theorem 4.2 with earlier results of Körner [55, 56]. For instance, in [56, LEMMA 2.3] he shows that if E is a subset of the unit circle $\mathbb{T} = \mathbb{R}/\mathbb{Z}$ with $\dim_{\mathbb{F}}(E) > \frac{2}{(v+1)}$, then there exist integers $m_0, m_1, \ldots, m_v \in \mathbb{Z}$, not all zero, and distinct points $x_0, x_1, \ldots, x_v \in E$ such that

$$m_0 x_0 = m_1 x_1 + \cdots + m_v x_v \pmod 1. \tag{4.7}$$

A priori, one does not know the number of integers m_j that are zero in the above equation, the signs of the nonzero integers m_j and whether the equation is translation-invariant. On the other hand, the linear equations stemming from $\mathscr{F}_v(\mathbb{N})$ and underlying Theorem 4.2 are exact (not modulo integers), and the coefficients $m_0, \ldots, m_v$ are all positive with the further constraint $m_0 = m_1 + \cdots + m_v$. Körner [56, THEOREM 2.4] also constructs a set $E \subseteq \mathbb{T}$ of Fourier dimension $1/v$ with the following property: there does not exist any nonzero vector $(m_0, \ldots, m_v) \in \mathbb{Z}^{v+1}$ for which the equation (4.7) admits a nontrivial solution consisting of distinct points $x_0, x_1, \ldots, x_v \in E$. Körner's construction, based on a Baire category argument, is nonexplicit. We ask the interested reader to compare Körner's construction of an avoiding set with the avoidance results in this paper (Theorems 4.5, 4.6, and 4.10). The sets that we construct avoid more restricted classes of equations but are of larger Fourier dimension.

As another point of contrast, we mention a construction of Keleti [53] that provides, for any countable set $T \subseteq (0,1)$, a subset $E \subseteq [0,1]$ of Hausdorff dimension 1 that does not contain any triple of distinct points $\{x, y, z\}$ such that $tx + (1-t)y = z$ for any $t \in T$. Choosing $v = 2$ and $T = \mathbb{Q} \cap (0,1)$, the set of rationals in $(0,1)$, we observe that $\dim_{\mathbb{F}}$ in Theorem 4.2 cannot be replaced by $\dim_{\mathbb{H}}$. Generalizing Keleti's result, Mathé [63] proves the existence of a rationally independent set in $\mathbb{R}$ of full Hausdorff dimension. We recall that a set $E \subseteq \mathbb{R}$ is *rationally independent* if for any integer $v \geq 2$ and any choice of distinct points $x_1, x_2, \ldots, x_v \in E$,

$$\sum_{j=1}^{v} a_j x_j = 0 \quad \text{with } \{a_1, \ldots, a_v\} \subseteq \mathbb{Z} \quad \text{implies} \quad a_1 = a_2 = \cdots = a_v = 0.$$

Theorem 4.2 implies that such sets cannot be Salem. Indeed, any set $E \subseteq \mathbb{R}$ of positive Fourier dimension will support a probability measure μ that satisfies (4.5) for some $v \in \mathbb{N}$

and some $\beta > 1/(v+1)$. By Theorem 4.2, it will contain a rationally dependent $(v+1)$-tuple of distinct points that obeys a relation of the form (4.6).

Corollary 4.4 ([60]). *There can be no rationally independent set in $\mathbb{R}$ of positive Fourier dimension.*

However, it is possible for a large Salem set to avoid nontrivial zeros of any *finite* sub-collection of $\mathscr{F}$, as our next result illustrates.

Theorem 4.5 ([60]). *Let $\mathscr{F}$ be as in* (4.3). *Given any finite collection $\mathscr{G} \subseteq \mathscr{F}$, there exists a set $E \subseteq [0,1]$ with $\dim_{\mathbb{F}} E = 1$ such that E contains no nontrivial zero of any $f \in \mathscr{G}$.*

Corollary 4.4 and Theorem 4.5 lead to a natural question: *does there exist a full-dimensional Salem set that avoids the nontrivial zeros of some countably infinite sub-collection of $\mathscr{F}$?* We answer this question in the affirmative; see Theorem 4.10. At the moment, we do not know how to characterize such subcollections.

Our next result attempts to strike a balance of a different sort between Theorems 4.2 and 4.5. While Theorem 4.2 dictates that a Salem set of large Fourier dimension must necessarily contain a nontrivial zero $(x_0, x_1, \ldots, x_v)$ of some function $f \in \mathcal{F}_v$ and some $v \geq 2$, it a priori does not specify the diameter or spread of such a solution,

$$\operatorname{diam}(x_0, \ldots, x_v) = \max\{|x_i - x_j|;\ i, j \in \{0, 1, \ldots, v\}, i \neq j\},$$

which could in principle be very small; in other words, the nontrivial solution could be "almost trivial." We now show that it is possible to construct a full-dimensional Salem set that prohibits, in a quantifiable way, nontrivial zeros from being almost trivial.

Theorem 4.6 ([60]). *There exists a set $E \subseteq [0,1]$, $\dim_{\mathbb{F}} E = 1$ with the following property. For every $v \geq 2$ and every $f \in \mathcal{F}_v(\mathbb{N})$ defined as in* (4.4), *there exists $\kappa > 0$ such that whenever there exists a $(v+1)$-tuple $(x_0, x_1, \ldots, x_v) \in E^{v+1}$ with*

$$\operatorname{diam}(x_0, x_1, \ldots, x_v) < \kappa \quad \text{and} \quad f(x_0, x_1, \ldots, x_v) = 0, \tag{4.8}$$

we have that $x_0 = x_1 = \cdots = x_v$. In particular, a nontrivial zero of f in E, if it exists, would obey $\operatorname{diam}(x_0, \ldots, x_v) \geq \kappa$.

In addition, the constant $\kappa = \kappa_N$ can be chosen uniformly for all $f \in \mathscr{F}$ whose coefficients are bounded by N.

4.2.2. General linear patterns

The statements of Theorems 4.2 and 4.5 lead to an interesting possibility. Let $\mathscr{F}_v(\mathbb{R}_+)$ denote the class of translation-invariant linear functions in $(v+1)$ variables with real positive coefficients. Then $\mathscr{F}_v(\mathbb{R}_+)$ can be identified with the $(v-1)$-dimensional set

$$\mathscr{T}_v = \{\mathbf{t} \in (0,1)^{v-1} : t_1 + t_2 + \cdots + t_{v-1} < 1\}, \tag{4.9}$$

which is a half-space of $\mathbb{R}^{v-1}$ restricted to the open unit cube, via the map

$$\mathbf{t} = (t_1, \dots, t_{v-1}) \in \mathscr{T} \mapsto f_{\mathbf{t}} \in \mathscr{F}_v(\mathbb{R}_+), \quad \text{where}$$

$$f_{\mathbf{t}}(x) = x_0 - (t_1 x_1 + t_2 x_2 + \cdots + t_v x_v), \quad t_v = 1 - \sum_{i=1}^{v-1} t_i.$$

Under this map, the class $\mathscr{F}_v(\mathbb{N})$ is identified with the positive rationals in $\mathscr{T}_v$, and hence is of Hausdorff dimension zero. On the other hand, $\mathscr{F}_v(\mathbb{R}_+)$ is of positive $(v-1)$-dimensional Lebesgue measure. One is then led to ask: *Given a collection $\overline{\mathscr{F}} \subseteq \mathscr{F}_v(\mathbb{R}_+)$ that is of positive Lebesgue measure or large Hausdorff dimension under this identification, does there exist a set $E \subseteq \mathbb{R}$ of large Fourier dimension that avoids all nontrivial zeros of $\overline{\mathscr{F}}$?* In our next two theorems, we answer this question in the affirmative, in the special case of trivariate equations, where $v = 2$ and $\overline{\mathscr{F}}$ can be viewed as a subset of $(0, 1)$. In Theorem 4.7, the class $\overline{\mathscr{F}}$ is identified with a union of intervals, in Theorem 4.8 with collections of badly approximable numbers.

Theorem 4.7 ([60]). *Let us fix any $p \in \mathbb{N}$ with $p \geq 2$ and any $\alpha \in (0, 1)$. Then there exist some $\kappa = \kappa(p, \alpha) > 0$ and $E \subseteq [0, 1]$ with $\dim_{\mathbb{F}}(E) \geq \alpha$ such that E contains no nontrivial solution of*

$$tx + (1-t)y = z \quad \text{for all } t \in \bigcup_{q=1}^{p-1} \left(\frac{q}{p} - \kappa, \frac{q}{p} + \kappa \right).$$

For fixed constants $0 < \tau, c \leq 1$, let us define the collection $\mathscr{E}_{c,\tau}$ of badly approximable numbers as follows:

$$\mathscr{E}_{c,\tau} := \left\{ t \in (0,1) : \left| t - \frac{q}{p} \right| > \frac{c}{p^{1+\tau}}, \text{ for all } \frac{q}{p} \in \mathbb{Q}, p \in \mathbb{N}, q \in \mathbb{Z}, \gcd(p,q) = 1 \right\}. \tag{4.10}$$

Sets of this type have applications in number theory, and their sizes have been widely studied. For example, if $\tau = 1$, then the Hausdorff dimension of $\mathscr{E}_{c,\tau}$ is of the order of $1 - O_c(1)$ as $c \to 0$. We refer the reader to [84, THEOREM 1.3] and the bibliography in this article for a survey of such results.

Theorem 4.8 ([60]). *For every $\varepsilon_0 \in (0, \frac{1}{2})$, there exists a set $E \subseteq [0, 1]$ with $\dim_{\mathbb{F}}(E) = \frac{1}{1+\tau}$ such that E contains no nontrivial solution of*

$$tx + (1-t)y = z, \quad \text{for any } t \in \mathscr{E}_{c,\tau} \cap (\varepsilon_0, 1 - \varepsilon_0).$$

The combined strategies of Theorems 4.7 and 4.8 imply the following corollary.

Corollary 4.9 ([60]). *Let us fix $0 < \tau$, $c \leq 1$, $\varepsilon_0 \in (0, \frac{1}{2})$, and $p \in \mathbb{N} \setminus \{1\}$. Then for all sufficiently large $M \in \mathbb{N}$ and $\kappa = \frac{1}{2pM}$, there exists $E \subseteq [0, 1]$ with $\dim_{\mathbb{F}} E = \frac{1}{1+\tau}$ such that E contains no nontrivial solution of*

$$tx + (1-t)y = z, \quad \text{for any } t \in \left[C \cap (\varepsilon_0, 1 - \varepsilon_0) \right] \cup \left[\bigcup_{q=1}^{p-1} \left(\frac{q}{p} - \kappa, \frac{q}{p} + \kappa \right) \right].$$

The sets of forbidden coefficients t in Theorems 4.7 and 4.8 are large, as a consequence of which the avoiding sets we obtain are not of full dimension. *Is it possible to construct a full-dimensional Salem set for which the set of forbidden coefficients is still quantifiably large?* Our next result provides an affirmative answer to this question, while also addressing the question posed after Theorem 4.5.

Theorem 4.10 ([60]). *There exists an infinite set $\mathfrak{C} \subseteq (0,1)$ and $E \subseteq [0,1]$ with $\dim_{\mathbb{F}} E = 1$ such that E contains no nontrivial solution of*

$$tx + (1-t)y = z \quad \text{for any } t \in \mathfrak{C}. \tag{4.11}$$

The set $\mathfrak{C}$ contains infinitely many rationals and uncountably many irrationals.

It is natural to ask whether there exists a version of Shmerkin's theorem [88] or Theorem 4.5 for a finite but arbitrary collection of equations in $\mathscr{F}_v(\mathbb{R}_+)$; for instance, does there exist a full-dimensional Salem set E that contains no nontrivial solution of $tx + (1-t)y = z$, for any prespecified irrational $t \in (0,1)$? We are currently unable to provide an answer to this question. Also, the proof techniques of this paper are not immediately generalizable to other types of translation invariant equations, for example, when

$$\sum_{i=1}^{m} s_i x_i = \sum_{j=1}^{n} t_j y_j \quad \text{with} \sum_{i=1}^{m} s_i = \sum_{j=1}^{n} t_j = 1, \quad 0 < s_i, t_j < 1 \text{ and } m, n \geq 2,$$

or when the equation is nonlinear, say $x_3 - x_1 = (x_2 - x_1)^2$. We are pursuing these directions in ongoing work.

4.3. A Roth-type result for dense Euclidean sets

A main objective of Ramsey theory is the study of geometric configurations in large, but otherwise arbitrary sets. A typical problem in this area reads as follows: given a set S, a family $\mathcal{F}$ of subsets of S and a positive integer r, is it true that any r-coloring of S yields some monochromatic configuration from $\mathcal{F}$? More precisely, for any partition of $S = S_1 \cup \cdots \cup S_r$ into r disjoint subsets, does there exist $i \in \{1, 2, \ldots, r\}$ and $F \in \mathcal{F}$ such that $F \subseteq S_i$? In discrete (respectively Euclidean) Ramsey theory, S is generally $\mathbb{Z}^d$ (respectively $\mathbb{R}^d$), and sets in $\mathcal{F}$ are geometric in nature. For example, if X is a fixed finite subset of $\mathbb{R}^d$, such as a collection of equally spaced collinear points or vertices of an isosceles right triangle, then $\mathcal{F} = \mathcal{F}(X)$ could be the collection of all isometric copies or all homothetic copies of X in S. A coloring theorem refers to a choice of S and $\mathcal{F}$ for which the answer to the above-mentioned question is yes. Such theorems are often consequences of sharper, more quantitative statements known as density theorems. A fundamental result with $S = \mathbb{N} = \{1, 2, \ldots\}$ is Szemerédi's theorem [93] (already mentioned in Section 4.1), which states that if $E \subseteq \mathbb{N}$ has positive upper density, i.e.,

$$\limsup_{N \to \infty} \frac{|E \cap \{1, \ldots, N\}|}{N} > 0,$$

then E contains a k-term arithmetic progression for every k. This in particular implies van der Waerden's theorem [34, 97], which asserts that given $r \geq 1$, any r-coloring of $\mathbb{N}$ must produce a k-term monochromatic progression, i.e., a homothetic copy of $\{1, 2, \ldots, k\}$. This subsection is devoted to joint work with Brian Cook and Akos Magyar [22], where we are concerned with certain density theorems in Ramsey theory over $\mathbb{R}^d$.

4.3.1. Motivation of the problem

A basic and representative result of the type we are interested in states that, with $d \geq 2$, a set $A \subseteq \mathbb{R}^d$ of positive upper Banach density contains all large distances, i.e., for every sufficiently large $\lambda \geq \lambda_0(A)$ there are points $x, x+y \in A$ such that $\|y\|_2 = \lambda$. Recall that the positive upper Banach density of A is defined as

$$\bar{\delta}(A) := \lim_{N\to\infty} \sup_{x\in\mathbb{R}^d} \frac{|A \cap (x+[0,N]^d)|}{N^d}.$$

The quoted result was obtained independently, along with various generalizations, by a number of authors, for example, Furstenberg, Katznelson, and Weiss [31], Falconer and Marstrand [29], and Bourgain [10].

To paraphrase, the above result shows that for any two-point configuration $X \in \mathbb{R}^d$ we are guaranteed the existence, up to congruence, of *all sufficiently large* dilates of X inside of A. The term configuration simply refers to a finite point set. From this point of view, it is a natural question to ask is whether similar statements exist that involve configurations with a greater number of points. If one looks for *some* (rather than every) sufficiently large dilate of a given configuration, such results are well known in the discrete regime of the integer lattice, under suitable assumptions of largeness on the underlying set. These results can often be translated to existence of configurations in the Euclidean setting as well. For instance, Roth's theorem [71] in the integers states that a subset of $\mathbb{Z}$ of positive upper density contains a three-term arithmetic progression $\{x, x+y, x+2y\}$, and it easily implies that a measurable set $A \subseteq \mathbb{R}$ of positive upper density contains a three-term progression whose gap size can be arbitrarily large. Results ensuring *all sufficiently large* dilates of a configuration in a set of positive Banach density are stronger, and their proofs typically more difficult. Bourgain [10] shows that if $X = \{x_1, \dots, x_k\}$ is any nondegenerate k-point simplex in $\mathbb{R}^d$, $d \geq k \geq 2$ (i.e., if $\{x_2 - x_1, \dots, x_k - x_1\}$ spans a $(k-1)$-dimensional space), then any subset of $\mathbb{R}^d$ of positive upper Banach density contains a congruent copy of λX for *all sufficiently large* λ.

On the other hand, a simple example given in [10] shows that there is a set $A \subseteq \mathbb{R}^d$ in any dimension $d \geq 1$, such that the gap lengths of all 3-progressions in A do not contain all sufficiently large numbers. In other words, the result of [10] is false for the degenerate configuration $X = \{0, e_1, 2e_1\}$, where e_1 is the canonical unit vector in the x_1-direction. More precisely, the counterexample provided in [10] is the set A of points $x \in \mathbb{R}^d$ such that $\big|\|x\|_2^2 - m\big| \leq \frac{1}{10}$ for some $m \in \mathbb{N}$. The parallelogram identity

$$2\|y\|_2^2 = \|x\|_2^2 + \|x+2y\|_2^2 - 2\|x+y\|_2^2$$

then dictates that $\big|\|y\|_2^2 - \frac{\ell}{2}\big| \leq \frac{4}{10}$ (for some $\ell \in \mathbb{N}$) for any progression $\{x, x+y, x+2y\} \subseteq A$. Thus the squares of the gap lengths are restricted to lie close to the half-integers, and therefore cannot realize all sufficiently large numbers.

The counterexample above has an interesting connection with a result in Euclidean Ramsey theory due to Erdős et al. [26]. Let us recall [33] that a finite point set X is said to be *Ramsey* if for every $r \geq 1$, there exists $d = d(X, r)$ such that any r-coloring of $\mathbb{R}^d$ contains a congruent copy of X. A result in [26] states that every Ramsey configuration X is spherical, i.e., the points in X lie on an Euclidean sphere. (The converse statement is currently an open conjecture due to Graham [33]). Since a set of three collinear points is nonspherical, it is natural to ask whether Bourgain-type counterexamples exist for any nonspherical X. This question was posed by Furstenberg and answered in the affirmative by Graham [33].

Theorem 4.11 (Graham [33]). *Let X be a finite nonspherical set. Then for any $d \geq 2$, there exist a set $A \subseteq \mathbb{R}^d$ with $\bar{\delta}(A) > 0$ and a set $\Lambda \subset \mathbb{R}$ with $\underline{\delta}(\Lambda) > 0$ so that A contains no congruent copy of λX for any $\lambda \in \Lambda$.*

4.3.2. Main result

It is interesting to observe that while Bourgain's counterexample prevents an existence theorem for three term arithmetic progressions of all sufficiently large *Euclidean* gaps, it does not exclude the validity of such a result when the gaps are measured using some other metric on $\mathbb{R}^d$ that does not obey the parallelogram law. In [22] we prove that such results do indeed exist for the l^p metrics $\|y\|_p := (\sum_{i=1}^d |y_i|^p)^{1/p}$ for all $1 < p < \infty$, $p \neq 2$. In this sense, a counterexample as described above is more the exception rather than the rule.

Variations of our arguments also work for other metrics given by specific classes of positive homogeneous polynomials of degree at least 4 and those generated by symmetric convex bodies with special structure. Results of the first type were obtained in the finite field setting by Cook and Magyar [21]. Also, the arguments here can be applied to obtain similar results for certain other degenerate point configurations.

Theorem 4.12 ([22]). *Let $1 < p < \infty$, $p \neq 2$. Then there exists a constant $d_p \geq 2$ such that for $d \geq d_p$ the following holds. Any measurable set $A \subseteq \mathbb{R}^d$ of positive upper Banach density contains a three-term arithmetic progression $\{x, x+y, x+2y\} \subseteq A$ with gap $\|y\|_p = \lambda$ for all sufficiently large $\lambda \geq \lambda(A)$.*

The result is sharp in the range of p. Easy variants of the example in [10] show that Theorem 4.12 and in fact even the two-point results of [10,29,31] cannot be true for $p = 1$ and $p = \infty$. Indeed, if $A = \mathbb{Z}^d + \varepsilon_0[-1,1]^d$ for some small $\varepsilon_0 > 0$, then, on the one hand, A is of positive upper Banach density. On the other hand, if $x, x+y \in A$ for some $y \neq 0$, then both $\|y\|_\infty$ and $\|y\|_1$ are restricted to lie within distance $O(\varepsilon_0)$ from some positive integer.

Indeed, counterexamples similar to [10] and the above can be constructed for norms given by a symmetric, convex body, a nontrivial part of whose boundary is either flat or coincides with an l^2-sphere. An appropriate formulation of a positive result for a general norm, and indeed the measurement of failure of the parallelogram law for such norms, remains an interesting open question.

We do not know whether the p-dependence of the dimensional threshold d_p stated in the theorem is an artifact of our proof. In our analysis, d_p grows without bound as $p \nearrow \infty$, while other implicit constants involved in the proof blow up near $p = 1$ and $p = 2$. It would be of interest to determine whether Theorem 4.12 holds for all $d \geq 2$ for the specified values of p.

Since three distinct collinear points cannot lie on an l^p-sphere for any $p \in (1, \infty)$, Theorem 4.12 shows that a result of the type considered by Graham in [33] is in general false for an l^p-sphere if $p \neq 1, 2, \infty$. Thus any connection between Ramsey-like properties and the notion of sphericality appears to be a purely l^2 phenomenon.

ACKNOWLEDGMENTS

I am deeply grateful to all my collaborators, whose generous insights have shaped my work and mathematical tastes. Conversations with the undergraduate and graduate students and postdoctoral fellows at UBC have been a constant source of joy and inspiration that have propelled my research—my warm thanks to all of them. Parts of the work took place in workshops and conferences hosted by the Fields Institute (2008), Banff International Research Station (2010, 2019), Mathematical Sciences Research Institute (2017), and Park City Mathematics Institute (2018). The creative space and facilities enabled by these institutes are gratefully acknowledged.

FUNDING

The author's work described in this article was partially supported by three consecutive Discovery grants (2007–2022) from the Natural Sciences and Engineering Research Council of Canada (NSERC), a Ruth E. Michler Fellowship from the Association for Women in Mathematics and Cornell University (2015), a scholarship from the Peter Wall Institute of Advanced Study (2018–2019) and a Simons Fellowship (2019–2020).

REFERENCES

[1] N. Anantharaman, Entropy and the localization of eigenfunctions. *Ann. of Math. (2)* **168** (2008), no. 2, 435–475.

[2] V. Aversa and D. Preiss, Hearts density theorems. *Real Anal. Exchange* **13** (1987), no. 1, 28–32.

[3] V. Aversa and D. Preiss, *Sistemi di derivazione invarianti per affinita*. Preprint (unpublished), Complesso Universitario di Monte S. Angelo, Napoli, 1995.

[4] F. A. Behrend, On sets of integers which contain no three terms in arithmetical progression. *Proc. Nat. Acad. Sci.* **32** (1946), 331–332.

[5] M. Bennett, A. Iosevich, and K. Taylor, Finite chains inside thin subsets of $\mathbb{R}^d$. *Anal. PDE* **9** (2016), no. 3, 597–614.

[6] A. S. Besicovitch, Sets of fractional dimensions (IV): on rational approximation to real numbers. *J. Lond. Math. Soc.* **9** (1934), no. 2, 126–131.

[7] C. Bluhm, Random recursive construction of Salem sets. *Ark. Mat.* **34** (1996), 51–63.

[8] C. Bluhm, On a theorem of Kaufman: Cantor-type construction of linear fractal Salem sets. *Ark. Mat.* **36** (1998), no. 2, 307–316.

[9] J. Bourgain, Averages in the plane over convex curves and maximal operators. *J. Anal. Math.* **47** (1986), 69–85.

[10] J. Bourgain, A Szemerédi type theorem for sets of positive density in $\mathbb{R}^k$. *Israel J. Math.* **54** (1986), no. 3, 307–316.

[11] J. Bourgain and Z. Rudnick, Restriction of toral eigenfunctions to hypersurfaces and nodal sets. *Geom. Funct. Anal.* **22** (2012), 878–937.

[12] J. D. Bovey and M. M. Dodson, The Hausdorff dimension of systems of linear forms. *Acta Arith.* **45** (1986), no. 4, 337–358.

[13] N. Burq, P. Gérard, and N. Tzetkov, Multilinear eigenfunction estimates and global existence for the three dimensional nonlinear Schrödinger equations. *Ann. Sci. Éc. Norm. Supér. (4)* **38** (2005), 255–301.

[14] N. Burq, P. Gérard, and N. Tzetkov, Restrictions of the Laplace–Beltrami eigenfunctions to submanifolds. *Duke Math. J.* **138** (2007), no. 3, 445–487.

[15] V. Chan, I. Łaba, and M. Pramanik, Point configurations in sparse sets. *J. Anal. Math.* **128** (2016), no. 1, 289–335.

[16] X. Chen, Sets of Salem type and sharpness of the L^2-Fourier restriction theorem. *Trans. Amer. Math. Soc.* **368** (2016), no. 3, 1959–1977.

[17] X. Chen and A. Seeger, Convolution powers of Salem measures with applications. *Canad. J. Math.* **69** (2017), no. 2, 284–320.

[18] X. Chen and C. Sogge, A few endpoint geodesic restriction estimates for eigenfunctions. *Comm. Math. Phys.* **329** (2014), 435–459.

[19] M. Christ, A. Nagel, E. M. Stein, and S. Wainger, Singular and maximal Radon transforms: analysis and geometry. *Ann. of Math.* **150** (1999), 489–577.

[20] Y. Colin de Verdière, Ergodicité et fonctions propres du Laplacien. *Comm. Math. Phys.* **102** (1985), 497–502.

[21] B. Cook and Á. Magyar, On restricted arithmetic progressions over finite fields. *Online J. Anal. Comb.* **7** (2012), 1–10.

[22] B. Cook, Á. Magyar, and M. Pramanik, A Roth-type theorem for dense subsets of $\mathbb{R}^d$. *Bull. Lond. Math. Soc.* **49** (2017), no. 4, 676–689.

[23] J. Denson, M. Pramanik, and J. Zahl, Large sets avoiding rough patterns. 2019, arXiv:1904.02337. Springer volume "Harmonic Analysis and Applications", edited by Michail Rassias.

[24] H. G. Eggleston, Sets of fractional dimensions which occur in some problems of number theory. *Proc. Lond. Math. Soc.* **54** (1952), no. 2, 42–93.

[25] F. Ekström, Fourier dimension of random images. *Ark. Mat.* **54** (2016), no. 2, 455–471.

[26] P. Erdős, R. L. Graham, P. Montgomery, B. L. Rothschild, J. H. Spencer, and E. G. Straus, Euclidean Ramsey theorems. *J. Combin. Theory Ser. A* **14** (1973), 341–363.

[27] S. Eswarathasan and M. Pramanik, Restriction of Laplace–Beltrami eigenfunctions to arbitrary sets on manifolds. *Int. Math. Res. Not.* **rnaa167** (2020), DOI 10.1093/imrn/rnaa167.

[28] K. Falconer, *Fractal geometry: mathematical foundations and applications*. 1st edn., Wiley Brothers, 1990.

[29] K. Falconer and J. Marstrand, Plane sets with positive density at infinity contain all large distances. *Bull. Lond. Math. Soc.* **18** (1986), no. 5, 471–474.

[30] R. Fraser and M. Pramanik, Large sets avoiding patterns. *Anal. PDE* **11** (2018), no. 5, 1083–1111.

[31] H. Furstenberg, Y. Katznelson, and B. Weiss, Ergodic theory and configurations in sets of positive density. In *Mathematics of Ramsey theory*, pp. 184–198, Springer, Berlin–Heidelberg, 1990.

[32] P. Gérard and É. Leichtnam, Ergodic properties of eigenfunctions for the Dirichlet problem. *Duke Math. J.* **71** (1993), 559–607.

[33] R. L. Graham, Recent trends in Euclidean Ramsey theory. *Discrete Math.* **136** (1994), 119–127.

[34] R. L. Graham, B. L. Rothschild, and J. H. Spencer, *Ramsey theory*. 2nd edn., John Wiley, New York, 1990.

[35] A. Greenleaf and A. Iosevich, On triangles determined by subsets of the Euclidean plane, the associated bilinear operators and applications to discrete geometry. *Anal. PDE* **5** (2012), no. 2, 397–409.

[36] A. Greenleaf, A. Iosevich, B. Liu, and E. Palsson, A group-theoretic viewpoint on Erdős–Falconer problems and the Mattila integral. *Rev. Mat. Iberoam.* **31** (2015), no. 3, 799–810.

[37] A. Greenleaf, A. Iosevich, and M. Pramanik, On necklaces inside thin subsets of $\mathbb{R}^d$. *Math. Res. Lett.* **24** (2017), no. 2, 347–362.

[38] K. Hambrook, Explicit Salem sets in $\mathbb{R}^2$. *Adv. Math.* **311** (2017), 634–648.

[39] K. Hambrook, Explicit Salem sets and applications to metrical Diophantine approximation. *Trans. Amer. Math. Soc.* **371** (2019), no. 6, 4353–4376.

[40] V. Harangi, T. Keleti, G. Kiss, P. Maga, A. Mathé, P. Mattila, and B. Strenner, How large dimension guarantees a given angle? *Monatsh. Math.* **171** (2013), no. 2, 169–187.

[41] B. Helffer, A. Martinez, and D. Robert, Ergodicité et limite semi-classique. *Comm. Math. Phys.* **109** (1987), 313–326.

[42] K. Henriot, I. Łaba, and M. Pramanik, On polynomial configurations in fractal sets. *Anal. PDE* **9** (2016), no. 5, 1153–1184.

[43] L. Hörmander, Spectral function of an elliptic operator. *Acta Math.* **121** (1968), 193–218.

[44] R. Hu, L^p norm estimates of eigenfunctions restricted to submanifolds. *Forum Math.* **21** (2009), 1021–1052.

[45] A. Iosevich and E. Sawyer, Maximal averages over surfaces. *Adv. Math.* **132** (1997), 46–119.

[46] A. Iosevich and E. Sawyer, Three problems motivated by the average decay of the Fourier transform. In *Harmonic analysis at Mount Holyoke (South Hadley, MA, 2001)*, pp. 205–215, Contemp. Math. 320, Amer. Math. Soc., Providence, RI, 2003.

[47] V. Jarník, Diophantischen Approximationen und Hausdorffsches Mass. *Mat. Sb.* **36** (1929), 371–382.

[48] V. Jarník, Über die simultanen diophantischen Approximationen (German). *Math. Z.* **33** (1931), no. 1, 505–543.

[49] J. P. Kahane, Images browniennes des ensembles parfaits. *C. R. Acad. Sci. Paris, Sér. A–B* **263** (1966), A613–A615.

[50] J. P. Kahane, Images d'ensembles parfaits par des séries de Fourier gaussiennes. *C. R. Acad. Sci. Paris, Sér. A–B* **263** (1966), A678–A681.

[51] R. Kaufman, On the theorem of Jarník and Besicovitch. *Acta Arith.* **39** (1981), 265–267.

[52] T. Keleti, A 1-dimensional subset of the reals that intersects each of its translates in at most a single point. *Real Anal. Exchange* **24** (1998/1999), no. 2, 843–844.

[53] T. Keleti, Construction of one-dimensional subsets of the reals not containing similar copies of given patterns. *Anal. PDE* **1** (2008), no. 1, 29–33.

[54] H. Koch, D. Tataru, and M. Zworski, Semiclassical L^p estimates. *Ann. Henri Poincaré* **5** (2007), 885–916.

[55] T. Körner, Measure on independent sets, a quantitative version of a theorem of Rudin. *Proc. Amer. Math. Soc.* **135** (2007), no. 12, 3823–3832.

[56] T. Körner, Fourier transforms of measures and algebraic relations on their support. *Ann. Inst. Fourier (Grenoble)* **59** (2009), no. 4, 1291–1319.

[57] I. Łaba, Maximal operators and decoupling for $\Lambda(p)$ Cantor measures. 2018, arXiv:1808.05657.

[58] I. Łaba and M. Pramanik, Arithmetic progressions in sets of fractional dimension. *Geom. Funct. Anal.* **19** (2009), no. 2, 429–456.

[59] I. Łaba and M. Pramanik, Maximal operators and differentiation theorems for sparse sets. *Duke Math. J.* **158** (2011), no. 3, 347–410.

[60] Y. Liang and M. Pramanik, Fourier analysis and avoidance of linear patterns. 2020, arXiv:2006.10941.

[61] E. Lindenstrauss, Invariant measures and arithmetic quantum unique ergodicity. *Ann. of Math. (2)* **163** (2006), 165–219.

[62] P. Maga, Full dimensional sets without given patterns. *Real Anal. Exchange* **36** (2010/2011), 79–90.

[63] A. Máthé, Sets of large dimension not containing polynomial configurations. *Adv. Math.* **316** (2017), 691–709.

[64] P. Mattila, Geometry of sets and measures in Euclidean space, Cambridge Stud. Adv. Math., Cambridge Univ. Press, Cambridge, 1995.

[65] G. Mockenhaupt, A. Seeger, and C. Sogge, Wave front sets, local smoothing and Bourgain's circular maximal theorem. *Ann. of Math.* **136** (1992), 207–218.

[66] A. Nagel, A. Seeger, and S. Wainger, Averages over convex hypersurfaces. *Amer. J. Math.* **115** (1993), 903–927.

[67] D. H. Phong and E. M. Stein, Hilbert integrals, singular integrals, and Radon transforms I. *Acta Math.* **157** (1986), 99–157.

[68] D. H. Phong and E. M. Stein, Hilbert integrals, singular integrals, and Radon transforms II. *Invent. Math.* **86** (1986), 75–113.

[69] D. Preiss, Lectures given in Ravello, 1985. Unpublished.

[70] A. Reznikov, Norms of geodesic restrictions for eigenfunctions on hyperbolic surfaces and representation theory. Unpublished update to a work from 2005. 2010, arXiv:math/0403437v3.

[71] K. Roth, On certain sets of integers. *J. Lond. Math. Soc.* **28** (1953), 104–109.

[72] K. Roth, Irregularities of sequences relative to arithmetic progressions, IV. *Period. Math. Hungar.* **2** (1972), 301–306.

[73] J. L. Rubio de Francia, Maximal functions and Fourier transforms. *Duke Math. J.* **53** (1986), 395–404.

[74] Z. Rudnick and P. Sarnak, The behaviour of eigenstates of arithmetic hyperbolic manifolds. *Comm. Math. Phys.* **161** (1994), no. 1, 195–213.

[75] I. Ruzsa, Solving a linear equation in a set of integers I. *Acta Arith.* **65** (1993), no. 3, 259–282.

[76] I. Ruzsa, Solving a linear equation in a set of integers II. *Acta Arith.* **72** (1995), no. 4, 385–397.

[77] R. Salem, On singular monotonic functions whose spectrum has a given Hausdorff dimension. *Ark. Mat.* **1** (1951), no. 4, 353–365.

[78] R. Salem and D. C. Spencer, On sets of integers which contain no three terms in arithmetical progression. *Proc. Natl. Acad. Sci. USA* **28** (1942), 561–563.

[79] P. Sarnak, Arithmetic quantum chaos. In *The Schur lectures (Tel Aviv, 1992)*, pp. 183–236, Israel Math. Conf. Proc. 8, Bar-Ilan Univ., Ramat Gan, Israel, 1995.

[80] P. Shmerkin, Salem sets with no arithmetic progressions. *Int. Math. Res. Not.* **7** (2017), 1929–1941.

[81] P. Shmerkin and V. Suomala, Spatially independent martingales, intersections, and applications. *Mem. Amer. Math. Soc.* **251** (2018), no. 1195, v+102 pp.

[82] P. Shmerkin and V. Suomala, New bounds on Cantor maximal operators. 2021, arXiv:2106.14818.

[83] A. I. Shnirelman, Ergodic properties of eigenfunctions. *Uspekhi Mat. Nauk* **29** (1974), no. 6, 181–182 (in Russian).

[84] D. Simmons, A Hausdorff measure version of the Jarník–Schmidt theorem in Diophantine approximation. *Math. Proc. Cambridge Philos. Soc.* **164** (2018), no. 3, 413–459.

[85] C. Sogge, Oscillatory integrals and spherical harmonics. *Duke Math. J.* **53** (1986), 43–65.

[86] C. Sogge, Concerning the L^p norms of spectral clusters of second-order elliptic operators on compact manifolds. *J. Funct. Anal.* **77** (1988), 123–138.

[87] C. Sogge, *Fourier integrals in classical analysis*. Cambridge Tracts in Math., Cambridge Univ. Press, Cambridge, 1993.

[88] C. Sogge and E. M. Stein, Averages of functions over hypersurfaces in $\mathbb{R}^n$. *Invent. Math.* **82** (1985), 543–556.

[89] C. Sogge and E. M. Stein, Averages over hypersurfaces II. *Invent. Math.* **86** (1986), 233–242.

[90] E. M. Stein, Maximal functions: spherical means. *Proc. Natl. Acad. Sci. USA* **73** (1976), 2174–2175.

[91] E. M. Stein, *Harmonic analysis*. Princeton Univ. Press, Princeton, 1993.

[92] E. M. Stein and S. Wainger, Problems in harmonic analysis related to curvature. *Bull. Amer. Math. Soc.* **84** (1978), 1239–1295.

[93] E. Szemerédi, On sets of integers containing no 4 elements in arithmetic progression. *Acta Math. Sci. Hung.* **20** (1969), 89–104.

[94] E. Szemerédi, On sets of integers containing no k elements in arithmetic progression. *Acta Arith. Collection of articles in memory of Juriĭ Vladimirovič Linnik* **27** (2015), 199–245.

[95] M. Tacy, Semiclassical L^p estimates of quasimodes on submanifolds. *Comm. Partial Differential Equations* **35** (2010), no. 8, 1538–1562.

[96] P. Tomas, A restriction theorem for the Fourier transform. *Bull. Amer. Math. Soc.* **81** (1975), 477–478.

[97] B. L. van der Waerden, Beweis einer Baudetschen Vermutung. *Nieuw Arch. Wiskd.* **15** (1927), 212–216.

[98] P. Yung, *Spectral projection theorems on compact manifolds*. Lecture notes, available under "Notes/Slides" at https://maths-people.anu.edu.au/~plyung/Sogge.pdf.

[99] S. Zelditch, Uniform distribution of eigenfunctions on compact hyperbolic surfaces. *Duke Math. J.* **55** (1987), 919–941.

[100] S. Zelditch and M. Zworski, Ergodicity of eigenfunctions for ergodic billiards. *Comm. Math. Phys.* **175**, 673–682 (199.

MALABIKA PRAMANIK

Department of Mathematics, University of British Columbia, 1984 Mathematics Road, Vancouver BC V6T 1Z2, Canada, malabika@math.ubc.ca

This Page Intentionally Left Blank

THE NUMBER OF CLOSED IDEALS IN THE ALGEBRA OF BOUNDED OPERATORS ON LEBESGUE SPACES

GIDEON SCHECHTMAN

ABSTRACT

We survey some of the recent progress in determining the number of two-sided closed ideals in the Banach algebras of bounded linear operators on Lebesgue spaces, $L_p[0, 1]$. In particular, we discuss two recent results: the first of Johnson, Pisier, and the author, showing that there are a continuum of such ideal in the case of $p = 1$; the second a result of Johnson and the author, showing that in the case $1 < p < \infty$, $p \neq 2$, there are exactly 2 to the continuum such ideals.

MATHEMATICS SUBJECT CLASSIFICATION 2020

Primary 47L20; Secondary 46E30

KEYWORDS

Operator ideals, L_p spaces

ICM INTERNATIONAL CONGRESS OF MATHEMATICIANS 2022 JULY 6–14

Proc. Int. Cong. Math. 2022, Vol. 4, pp. 3250–3265
DOI 10.4171/ICM2022/7
Published by EMS Press

1. INTRODUCTION

For a Banach space X over the real or complex field, we denote by $L(X)$ the Banach algebra of bounded linear operators on X. The wider subject of study here is the structure of the class of closed two-sided ideals in this algebra. We recall that a closed ideal here is a closed linear subspace M of $L(X)$ such that if $T \in M$ and $A, B \in L(X)$ then $ATB \in M$. The research we shall concentrate on describing here is concerned with the modest aim of deciding what the number of such different closed ideals is when X is one of the Lebesgue spaces, $L_p[0,1]$, $1 \le p \le \infty$.

Recall that for a measure space $(\Omega, \mathcal{F}, \mu)$ and $1 \le p < \infty$, $L_p(\Omega, \mathcal{F}, \mu)$ denotes the Banach spaces of all (equivalence classes of) $\mathcal{F}$ measurable functions, f, such that $\|f\|_p = (\int_\Omega |f|^p d\mu)^{1/p} < \infty$. By $L_\infty(\Omega, \mathcal{F}, \mu)$ we denote the space of essentially bounded functions with the sup norm. Of particular interest will be the case of $\Omega = \mathbb{N}$ with the counting measure, in which case we will denote the space by ℓ_p, and the case when $(\Omega, \mathcal{F}, \mu)$ is the interval $[0,1]$ with Lebesgue measure, which we will denote by $L_p[0,1]$. We recall that for $1 \le p < \infty$, any infinite-dimensional separable $L_p(\Omega, \mathcal{F}, \mu)$ space is isomorphic to either $L_p[0,1]$ or ℓ_p. Of course, $L_2[0,1]$ and ℓ_2 are isometric. Also, $L_\infty[0,1]$ and ℓ_∞ are known to be isomorphic. We also denote by c_0 the subspace of ℓ_∞ of sequences tending to zero.

Most probably, the first result concerning the structure of ideals of $L(X)$ is the influential work of Calkin [7] who showed that if X is a separable Hilbert space then the only nontrivial (that is, different from the whole space and $\{0\}$) closed ideal here is the ideal of compact operators. This result was generalized in [11] to the other separable classical sequence spaces ℓ_p, $1 \le p < \infty$, and c_0. There were more results for special cases, including some natural nonseparable ones. Pietsch's book [17, CHAPTER 5] contains a survey of the results obtained by 1980. Pietsch also points out that, following a known result, in $L(L_p[0,1])$, $1 < p < \infty$, $p \ne 2$, there are countably many different closed ideals and raises the question of how many ideals are there in $L(L_p[0,1])$, $1 \le p < \infty$, and some other classical spaces.

The subject of the structure of the set of ideals in $L(X)$ laid dormant for a while but gained new drive since the beginning of this century. We shall not survey most of these developments. We refer to the very good introduction in [3] for a survey of the known results up to a couple of years ago. Let us just say that there are very few spaces X for which we have a complete knowledge of all the ideals in $L(X)$. From now on we shall concentrate only on the question of the number of closed ideals in $L(X)$ for X being a separable L_p or some related space.

Note that if P is a (always bounded, linear) projection onto a subspace Y of X and Y is isomorphic to its square, $Y \oplus Y$, then the set

$$\{APB;\ A, B \in L(X)\}$$

is a closed ideal in $L(X)$. This is easy to verify. (The requirement that Y is isomorphic to its square comes to ensure that this set is a closed subspace of $L(X)$.) If P_1, P_2 are two such projections onto Y_1, Y_2, respectively, and if there is no isomorphism of X which carries Y_1 onto Y_2 then the two ideals generated are different. In particular, this is the case if

Y_1 and Y_2 are not isomorphic. By the time [17] was written, it was known [19] that there are countably many mutually nonisomorphic complemented subspaces (i.e., ranges of projections) of $L_p[0,1]$, $1 < p < \infty$, $p \neq 2$, each isomorphic to its square (in particular, they are infinite-dimensional). So the reasoning above, appearing in [17], yields infinitely many different closed ideals in $L_p[0,1]$, $1 < p < \infty$, $p \neq 2$. Pietsch asked in his book whether there are uncountably many such ideals and also what the number of closed ideals in $L_1[0,1]$ is (at that time only finitely many closed ideals were known). Shortly afterward, [6] produced $\aleph_1$ mutually nonisomorphic complemented subspaces of $L_p[0,1]$, each isomorphic to its square, raising the number of different ideals in $L(L_p[0,1])$ to $\aleph_1$. In his book Pietsch also noticed (building again on known complemented subspaces) that in the algebra of bounded operators on $C(0,1)$, the space of continuous functions on the unit interval, there are $\aleph_1$ different closed ideals. We remark in passing that in the situation above (where the different ideals in each case are related to complemented subspaces) the ideals constructed are not even mutually isomorphic as Banach algebras. This follows immediately from a result of Eidelheit [8]: If Y and Z are Banach spaces such that the algebras $L(Y)$ and $L(Z)$ are isomorphic as Banach algebras then Y and Z are isomorphic as Banach spaces (and trivially vice versa).

The main purpose of this note is to focus on two recent advancements in this direction in which the author was involved. In [13] we built a continuum of different closed ideals in $L(L_1[0,1])$, in $L(C(0,1))$, and also in $L(L_\infty[0,1])$. (In these results and also in the other we survey here, we are not using the continuum hypothesis, so the cardinality of the continuum may be larger than $\aleph_1$.) The result is stated as Theorem 5.1 below.

The result of [15] may be more surprising: The number of different closed ideals in each of $L(L_p[0,1])$, $1 < p < \infty$, $p \neq 2$, is $2^{\mathfrak{c}}$ ($\mathfrak{c} = 2^{\aleph_0}$ is the cardinality of the continuum). The upper bound is simple. The problem is to produce $2^{\mathfrak{c}}$ ideals. The result is stated as Theorem 4.1 below.

The proofs of the two results are quite different, but a common feature is that the constructions and proofs (that the constructed ideals are really different) boils down to inequalities in the quantitative finite-dimensional world and involve probabilistic and/or harmonic-analytical methods.

Except for these two papers, there are more results, by others, involving related questions of Pietsch, that we shall only report on here but will not go into the detailed constructions. Pietsch asked whether for $1 \leq p < q < \infty$, $L(\ell_p \oplus \ell_q)$ contains infinitely many different closed ideals. This was solved by Schlumprecht and Zsák [20] producing a continuum of such ideals in these spaces as well as in $L(\ell_p \oplus c_0)$, $1 \leq p < \infty$. Later, building in part on the method in [15], Freeman, Schlumprecht, and Zsák in [9, 10] showed that there are $2^{\mathfrak{c}}$ different closed ideals in the spaces $L(\ell_p \oplus \ell_q)$, $1 < p < q < \infty$, as well as in $L(\ell_p \oplus c_0)$, $L(\ell_p \oplus \ell_\infty)$, $L(\ell_p \oplus \ell_1)$, $1 < p < \infty$.

In all the questions and answers described above, two ideals are considered different if they are different as sets. There are, of course, weaker distinctions one can consider. A natural one is to consider two ideals to be different if they are not isomorphic as Banach algebras, i.e., are not homomorphic by a homomorphism which is continuous in both directions. In Corollary 6.7 we report on a recent observation of Bill Johnson, Chris Phillips, and the author, based of Eidelheit's [8], showing that this seemingly weaker distinction still gives the same results.

Another question from [17] was whether ℓ_p, $1 \le p < \infty$ and c_0 are the only spaces X in which the only nontrivial closed ideal in $L(X)$ is the ideal of compact operators. This turned out not to be the case. Solving an old problem of Lindenstrauss, Argyros, and Haydon, [1] built a Banach space in which every operator is a multiple of the identity plus a compact operator from which it easily follows that the only nontrivial closed ideal in the space of operators on this space is the ideal of compact operators.

In Section 2 we survey what was previously known about closed ideals in the space of operators on separable L_p spaces, $1 \le p < \infty$. We also define the notions of small and large ideals. Section 3 deals with a criterion for Banach spaces X ensuring the existence of $2^{\mathfrak{c}}$ different closed ideals, which turns out to be relevant for a construction of that many ideals in $L(L_p[0,1])$, $1 < p < \infty$. The criterion is in terms of the existence of a certain operator on the space X. Section 4 is devoted to the construction of such an operator. Here the presentation is different than in the original paper and, since we think it may be useful in the future, is given in more detail. Section 5 deals with the case of $L(L_1[0,1])$ and is independent of the previous sections. In the final section we gather some remarks and open problems.

2. OLD IDEALS

Here we survey what was known before [13, 15, 20]. It is not needed for reading the next three sections which contain newer results. We begin with a few simple observations about (always two-sided) closed ideals in $L(X)$ for a general (infinite-dimensional) Banach space X. Since any Banach space admits a rank-one operator and since for any two rank-one operators $R_1, R_2 \in L(X)$ there are $S, T \in L(X)$ with $R_2 = SR_1T$, every ideal in $L(X)$ contains any rank-one operator and, since it is a subspace, all finite rank operators. So any closed ideal in $L(X)$ contains the closure of the finite rank operators, $\mathcal{F}(X)$. If X has the approximation property, as any L_p space and all the other classical Banach spaces have, then $\mathcal{F}(X)$ is equal to the ideal of compact operators, $\mathcal{K}(X)$. Since X is infinite-dimensional $\mathcal{K}(X)$ is a proper ideal. As we already mentioned, for some X including ℓ_p, $1 \le p < \infty$, and c_0, $\mathcal{K}(X)$ is the only proper closed ideal in $L(X)$. Another closed "small" proper ideal presented in every space (although sometimes coincides with the compact operators) is that of the strictly singular operators, $\mathcal{S}(X)$, i.e., the set of all operators on $L(X)$ which are not an isomorphism when restricted to any infinite-dimensional subspace. We call a closed ideal in $L(X)$ *small* if it is contained in $\mathcal{S}(X)$, otherwise we call it *large*. This distinction is not always useful but in $L(L_p[0,1])$ and, in particular, in $L(L_1[0,1])$ it contributes to the understanding of the structure of the class of closed ideals as we shall see. It also gives rise to some open problems.

In $L(L_1[0,1])$ consider the set I_{ℓ_1} of operators that factor through ℓ_1. It turns out that this is a closed ideal. It is, of course, large (as ℓ_1 is isometric to a complemented subspace of $L_1[0,1]$). It follows from known results (see the introduction in [13] for this and other unexplained reasoning in this section concerning $L(L_1[0,1])$) that I_{ℓ_1} contains $\mathcal{S}(L_1[0,1])$ and is contained in any large ideal. In particular, any large ideal in $L(L_1[0,1])$ contains any small ideal.

Except for $\mathcal{K}(L_1[0,1])$ and $\mathcal{S}(L_1[0,1])$, there is another classical small ideal: This is the set of Danford–Pettis operators (operators which send weakly compact sets onto norm compact sets). As for classical large ideals, except for I_{ℓ_1}, there is only one other known large ideal in $L(L_1[0,1])$. This is the maximal proper ideal which turns out to be the set of all operators which are not isomorphisms when restricted to a subspace of $L_1[0,1]$ isomorphic to $L_1[0,1]$ (the fact that this is an ideal is not trivial at all). The continuum of ideals produced in Section 5 are all small. We do not know if there are infinitely many large ideals in $L(L_1[0,1])$.

For $L(L_p[0,1])$, $1 < p < \infty$, $p \neq 2$, the break point between small and large ideals is not as sharp as for $p = 1$. Except for the ideal I_{ℓ_p} of all operators which factor through ℓ_p, there is another incomparable minimal large ideal. This is the closure of the operators which factor through ℓ_2, denoted $\overline{\Gamma_2}$. It turns out that every large ideal contains one of these two ideals. However, $\overline{\Gamma_2}$ does not contain the strictly singular operators in $L(L_p[0,1])$, $1 < p < \infty$, $p \neq 2$. So not every large ideal contains all the small ideals. We refer to the introduction in [15] for this and other unexplained reasoning here. As was already remarked above, there were $\aleph_1$ large closed ideals known in $L(L_p[0,1])$ for quite a while (an ideal generated by a projection onto an infinite-dimensional complemented subspace is clearly large). There is also the maximal ideal of all operators not preserving an isomorphic copy of $L_p[0,1]$ which is clearly large. (Again the fact that this is an ideal is not simple.) Schlumprecht and Zsák [20] produced a continuum of small ideals in these $L(L_p[0,1])$ algebras. Prior to [20], only a finite number of such small ideals were known. As is exposed in Sections 3 and 4 below, in [15] we produced $2^{\mathfrak{c}}$ large ideals, as well as $2^{\mathfrak{c}}$ small ideals in $L(L_p[0,1])$, $1 < p < \infty$, $p \neq 2$.

3. A CRITERION FOR HAVING MANY CLOSED IDEALS

This section is taken almost verbatim from [15, SECTION 2].

Recall first the notion of unconditional basis for a Banach space X. A sequence $\{e_i\}_{i=1}^{\infty}$ is said to be a *(Schauder) basis* for X if any $x \in X$ has a unique representation as $x = \sum_{i=1}^{\infty} a_i e_i$ for some coefficients $\{a_i\}$. The basis $\{e_i\}_{i=1}^{\infty}$ is said to be *K-unconditional* if for all signs $\{\varepsilon_i\}_{i=1}^{\infty} \in \{-1,1\}^{\mathbb{N}}$ and all $\sum_{i=1}^{\infty} a_i e_i \in X$,

$$\left\| \sum_{i=1}^{\infty} \varepsilon_i a_i e_i \right\| \leq K \left\| \sum_{i=1}^{\infty} a_i e_i \right\|.$$

Note that given a subset $\mathbb{M}$ of $\mathbb{N}$, the natural projection $P_{\mathbb{M}}$, given by $P_{\mathbb{M}}(\sum_{i=1}^{\infty} a_i e_i) = \sum_{i \in \mathbb{M}} a_i e_i$, is of norm at most K. We also denote its range, the closed linear span of $\{e_i\}_{i \in \mathbb{M}}$ by $[e_i]_{i \in \mathbb{M}}$. It is also true and easy to show that an unconditional basis is a Schauder basis in any order.

The theorem below is stated for a 1-unconditional basis, enough for our purposes, but can easily be generalized for any K-unconditional basis.

There is a continuum of infinite subsets of the natural numbers $\mathbb{N}$, each two of which have only a finite intersection. Denote some fixed such continuum by $\mathcal{C}$. For a finite-dimensional normed space E, we denote by $d(E)$ the Banach–Mazur distance of E to a Euclidean space, i.e., if the dimension of E is k then

$$d(E) = \inf\{\|A\|\|B\| \; ; \; A : \ell_2^k \to E, \; B : E \to \ell_2^k, AB = I_E\}.$$

Also, recall that for an operator $T : X \to Y$ between two normed spaces, $\gamma_2(T)$ denotes its factorization constant through a Hilbert space:

$$\gamma_2(T) = \inf\{\|A\|\|B\| \; ; \; A : H \to Y, B : X \to H, \; T = AB, \; H \text{ a Hilbert space}\}.$$

If T is of rank k, then $\gamma_2(T) \le k^{1/2}\|T\|$ because every k dimensional normed space is $k^{1/2}$-isomorphic to ℓ_2^k. Note that $d(E)$ is just $\gamma_2(I_E)$, where I_E is the identity operator on E.

Theorem 3.1. *Let X be a Banach space with a 1-unconditional basis $\{e_i\}$, let Y be a Banach space, and let $T : X \to Y$ be an operator of norm at most 1 satisfying:*

(a) *For some $\eta > 0$ and for every M, there is a finite-dimensional subspace E of X such that $d(E) > M$ and $\|Tx\| > \eta\|x\|$ for all $x \in E$.*

(b) *For some constant Γ and every m, there is an n such that every m-dimensional subspace E of $[e_i]_{i\ge n}$ satisfies $\gamma_2(T_{|E}) \le \Gamma$.*

Then there exist natural numbers $1 = p_1 < q_1 < p_2 < q_2 < \cdots$ such that, denoting for each k, $G_k := [e_i]_{i=p_k}^{q_k}$, and defining for each $\alpha \in \mathcal{C}$, the operator $P_\alpha : X \to [G_k]_{k\in\alpha}$ to be the natural basis projection, and setting $T_\alpha := TP_\alpha$, we have the following:

If $\alpha_1, \dots, \alpha_s \in \mathcal{C}$ (possibly with repetitions) and $\alpha \in \mathcal{C} \setminus \{\alpha_1, \dots, \alpha_s\}$, then for all $A_1, \dots, A_s \in L(Y)$ and all $B_1, \dots, B_s \in L(X)$,

$$\left\| T_\alpha - \sum_{i=1}^{s} A_i T_{\alpha_i} B_i \right\| \ge \eta/2. \tag{3.1}$$

Since we do not have anything to add to the original proof of this theorem, we refer the interested reader to **[15]** for the not-so-hard proof.

Theorem 3.1 provides a criterion for having $2^{\mathfrak{c}}$ different closed ideals in a space satisfying the assumptions of the theorem.

Corollary 3.2. *Let X be a Banach space with a 1-unconditional basis $\{e_i\}$ and assume there is an operator $T : X \to X$ of norm at most 1 satisfying (a) and (b) of Theorem 3.1. Then $L(X)$ has exactly $2^{\mathfrak{c}}$ different closed ideals.*

Proof. Indeed, for any nonempty proper subset $\mathcal{A}$ of $\mathcal{C}$, let $I_{\mathcal{A}}$ be the ideal generated by $\{T_\alpha\}_{\alpha\in\mathcal{A}}$, i.e., all operators of the form $\sum_{i=1}^{s} A_i T_{\alpha_i} B_i$ with $s \in \mathbb{N}$, $A_i, B_i \in L(X)$, $\alpha_i \in \mathcal{A}$, $i = 1, \dots, s$. Since we allow repetition of the T_{α_i}, it is easy to see that this really defines a (nonclosed) ideal. We will show that, when $\mathcal{A}$ ranges over the nonempty proper subsets of $\mathcal{C}$, $\overline{I_{\mathcal{A}}}$ define different closed ideals.

Let $\mathcal{B}$ be a subset of $\mathcal{C}$ different from $\mathcal{A}$ and assume, without loss of generality, that $\mathcal{B} \not\subset \mathcal{A}$. Let $\alpha \in \mathcal{B} \setminus \mathcal{A}$. Then, by Theorem 3.1, $T_\alpha \notin \overline{I_{\mathcal{A}}}$. Consequently, $\overline{I_{\mathcal{A}}}$ and $\overline{I_{\mathcal{B}}}$ are different.

Since the density character of $L(X)$, for any separable X, is at most the continuum, it is easy to see that, for any separable space X, $L(X)$ has at most $2^{\mathfrak{c}}$ different closed ideals. ■

Remark 3.3. If Y is a Banach space that contains a complemented subspace X with the properties of Corollary 3.2 then, clearly, $L(Y)$ also has $2^{\mathfrak{c}}$ different closed ideals. The same is true also for any space isomorphic to such a Y. Also, the assumption that T has norm at most 1 can be weakened to just requiring that T is bounded.

Remark 3.4. By the discussion just before Corollary 6.7 below, if Y is as in the previous remark then $L(Y)$ actually has $2^{\mathfrak{c}}$ closed ideals, each two of which are not isomorphic as Banach algebras. That is, there is no homomorphism between them which is continuous in both directions.

Maybe the simplest examples of spaces X that satisfy the hypotheses of Corollary 3.2 (and thus $L(X)$ has $2^{\mathfrak{c}}$ different closed ideals) are $(\sum \ell_{r_i}^{n_i})_2$ for $r_i \uparrow 2$ and n_i satisfying $n_i^{\frac{1}{r_i}-\frac{1}{2}} \to \infty$. These spaces satisfy the assumptions with T being the identity. Verifying (a) is simple with E being one of the spaces $\ell_{r_i}^{n_i}$ for i large enough. Verifying (b) is a bit more involved and, since as we shall shortly remark that this space is not good for our purposes, we shall not enter into the reasoning here. (The main point is that the distance of the worst m-dimensional subspace of L_r from a Euclidean space tends to 1 when r tends to 2.) Unfortunately, $(\sum \ell_{r_i}^{n_i})_2$ for $r_i \uparrow 2$ and $n_i^{\frac{1}{r_i}-\frac{1}{2}} \to \infty$ does not embed isomorphically as a complemented subspace into any L_p, $p < \infty$, so this example is not good for our purposes. Actually, at least for some sequences $\{(r_i, n_i)\}$ with the above properties, $(\sum \ell_{r_i}^{n_i})_2$ does not even embed isomorphically into any L_p space, $p < \infty$. That this is true, for example, if each $(r, n) \in \{(r_i, n_i)\}$ repeats n times, follows from Corollary 3.4 in [16].

In the next section we show how to get complemented subspaces of the reflexive L_p spaces that satisfy the hypotheses of Corollary 3.2.

4. A SPECIAL OPERATOR AND THE CASE OF REFLEXIVE LEBESGUE SPACES

In order to apply the criterion in Theorem 3.1 and deduce by Corollary 3.2, the existence of $2^{\mathfrak{c}}$ different closed ideals in $L_p[0,1]$, $1 < p < \infty$, $p \neq 2$, it is enough, by Remark 3.3, to find a complemented subspace of a space isomorphic to $L_p[0,1]$ having a 1-unconditional basis and an operator on it satisfying (a) and (b) of Theorem 3.1. In [15] this is done by using a certain complemented subspace of $L_p[0,1]$, $1 < p < \infty$, $p \neq 2$, and a certain operator on it (which is a variant of an operator the authors used in a previous paper [14] for a different purpose). The complemented subspace, X_p, is a span of independent,

3-valued, symmetric random variables. The space X_p which was investigated by Rosenthal starting with [18] was very influential in studying the geometry of L_p spaces. The operator is a certain diagonal operator between two such X_p spaces (followed by an injection of the second space into the first). This is where probabilistic inequalities, alluded to in the introduction, enter into the reasoning.

Here we shall describe the construction in a different way (although if one digs into the roots of the two constructions, they amount to basically the same operator). We think the presentation here may be cleaner and thus more accessible for further applications.

We begin with a nontraditional representation of (a space isomorphic to) $L_p[0,1]$. For $2 < p \le \infty$, define M_p to be $L_p(0,\infty) \cap L_2(0,\infty)$ with norm

$$\|f\|_{M_p} = \max\{\|f\|_{L_p(0,\infty)}, \|f\|_{L_2(0,\infty)}\}.$$

For $1 \le q < 2$, we define M_q to be $L_q(0,\infty) + L_2(0,\infty)$ with norm

$$\|f\|_{M_q} = \inf\{\|g\|_{L_q(0,\infty)} + \|h\|_{L_2(0,\infty)}\ ;\ f = g + h\}.$$

Here, $L_r(0,\infty)$ denotes the space of functions, f, on $(0,\infty)$ with $\|f\|_r = (\int_0^\infty |f(t)|^r\, dt)^{1/r} < \infty$. (Also, $M_2 := L_2(0,\infty)$.)

Note that M_p, $1 \le p \le \infty$, are rearrangement-invariant spaces, i.e., the norm of f depends only on the distribution of $|f|$. Also it is easy to prove that for $1 \le q < \infty$, the dual of M_q is M_p where, $\frac{1}{q} + \frac{1}{p} = 1$. In [12, CHAPTER 1] it is proved that, for $1 < p < \infty$, $p \ne 2$, M_p is isomorphic to $L_p[0,1]$. We remark in passing that this is done based on Rosenthat's [18] and is where probabilistic inequalities are used. In the presentation below, probability will not appear anymore. So, $L_p[0,1]$, $1 < p < \infty$, $p \ne 2$, has two different isomorphic representations as rearrangement-invariant function spaces on $(0,\infty)$. It is also proved in [12] that these are the only two such representations, a fact we will not use here. For $p = 1$ and $p = \infty$, M_p is not isomorphic to $L_p[0,1]$.

If $q < r < 2$ then the function $f_r(t) = t^{-1/r}$ is in M_q. Indeed,

$$\|f_r\|_{M_q} \le \|f_r \mathbf{1}_{(0,1)}\|_q + \|f_r \mathbf{1}_{[1,\infty)}\|_2 < \infty.$$

If $f^1, f^2, \dots$ are disjoint functions on $(0,\infty)$, each (when restricted to its support) with the same distribution as f_r, then $\{f^i\}_{i=1}^\infty$ is isometrically equivalent in M_q to the unit vector basis of ℓ_r. This actually holds in any rearrangement-invariant function space on $(0,\infty)$ containing the function f_r and follows from the simple fact that if $\sum_{i=1}^\infty |a_i|^r = 1$ then $|\sum_{i=1}^\infty a_i f^i|$ has the same distribution as f_r.

For $1 \le q < r < 2$ and $s > 1$, define $D_s : M_q \to M_q$ by $D_s f(t) = s^{1/r} f(st)$. Note that

$$D_s f_r = f_r, \tag{4.1}$$

for all $f \in L_2(0,\infty)$,

$$\|D_s f\|_2 = s^{\frac{1}{r}-\frac{1}{2}} \|f\|_2, \tag{4.2}$$

and for all $f \in L_q(0,\infty)$,

$$\|D_s f\|_q = s^{\frac{1}{r}-\frac{1}{q}} \|f\|_q. \tag{4.3}$$

Also, $D^* : M_p \to M_p$, $p = q/(q-1)$, is given by

$$D_s^* g(t) = s^{\frac{1}{r}-1} g(t/s). \tag{4.4}$$

Given $0 < \delta < 1$, put $s = s(\delta) = \delta^{\frac{rq}{q-r}}$ and define $r = r(\delta)$ by $s^{\frac{1}{r}-\frac{1}{2}} = 2$. Note that $\delta \searrow 0$ implies that $s(\delta) \nearrow \infty$ and $r(\delta) \nearrow 2$. Also for all $f \in L_2(0,\infty)$,

$$\|D_{s(\delta)} f\|_2 = 2\|f\|_2, \tag{4.5}$$

and for all $f \in L_q(0,\infty)$,

$$\|D_{s(\delta)} f\|_q = \delta\|f\|_q. \tag{4.6}$$

Let $\{\Omega_{i,j}\}_{i,j=1}^{\infty}$ be a partition of $(0,\infty)$ into disjoint measurable sets of infinite measure. For each i, j, let $\varphi_{i,j} : (0,\infty) \to \Omega_{i,j}$ be a one-to-one and onto measure-preserving transformation. Let $\delta_i \searrow 0$ and put $s_i = s_i(\delta_i)$, $r_i = r_i(\delta_i)$. Define $f_{i,j} : \Omega_{i,j} \to \mathbb{R}^+$ by

$$f_{i,j}\big(\varphi_{i,j}^{-1}(t)\big) = t^{-1/r_i}, \quad t \in (0,\infty),$$

and $D_{i,j} : M_q(\Omega_{i,j}) \to M_q(\Omega_{i,j})$ by

$$D_{i,j} f\big(\varphi_{i,j}(t)\big) = s_i^{1/r_i} f\big(\varphi_{i,j}^{-1}(s_i t)\big).$$

Define also $D : M_q \to M_q$ by $D_{|M_q(\Omega_{i,j})} = D_{i,j}$. Then (denoting by $f_{i,j}$ also the function which is equal to $f_{i,j}$ on $\Omega_{i,j}$ and zero elsewhere),

$$D(f_{i,j}) = f_{i,j}. \tag{4.7}$$

In particular, for each i, D is the identity on the span of $\{f_{i,j}\}_{j=1}^{\infty}$ which is isometric to ℓ_{r_i}. For all $f \in L_2(0,\infty)$,

$$\|Df\|_2 = 2\|f\|_2, \tag{4.8}$$

and for all $f \in L_q(\bigcup_{i=i_0}^{\infty} \bigcup_{j=1}^{\infty} \Omega_{i,j})$,

$$\|Df\|_q \le \delta_{i_0}\|f\|_q. \tag{4.9}$$

Note that (4.8) and (4.9) imply that D is bounded (by 2) on M_q.

Let $\varepsilon_{i,j}$, $i, j = 1, 2, \ldots$, be an arbitrary sequence of positive numbers and, for each i, j, let $A_1^{i,j}, \ldots, A_{n_{i,j}}^{i,j}$ be a sequence of disjoint sets in $\Omega_{i,j}$ such that the distance of $f_{i,j}$ from the span of $\{\mathbf{1}_{A_k^{i,j}}\}_{k=1}^{n_{i,j}}$ is at most $\varepsilon_{i,j}$, $i, j = 1, 2, \ldots$

Let $m_i \in \mathbb{N}$ be such that $m_i^{\frac{1}{r_i}-\frac{1}{2}} \nearrow \infty$ (recall that $m_i^{\frac{1}{r_i}-\frac{1}{2}}$ is the Banach–Mazur distance of $\ell_{r_i}^{m_1}$ to a Euclidean space) and pick the $\varepsilon_{i,j}$'s to be such that for each i the span of $\{\mathbf{1}_{A_k^{i,j}}\}_{k=1,j=1}^{n_{i,j}\ m_i}$ contains a sequence $\{g_{i,j}\}_{j=1}^{m_i}$ which is a, say, $1/4$-perturbation of $\{f_{i,j}\}_{j=1}^{m_i}$:

$$\left\| \sum_{i=1}^{m_i} a_j f_{i,j} - \sum_{i=1}^{m_i} a_j g_{i,j} \right\|_{M_q} < \frac{1}{4}\left(\sum_{j=1}^{m_i} |a_j|^{r_i}\right)^{1/r_i} \tag{4.10}$$

for all $\{a_j\}_{j=1}^{m_i}$. The properties of D then assure that it preserves a 2-isomorph of $\ell_{r_i}^{m_i}$ up to constant 3.

The space $X = X_q$ that we will use the criterion of Theorem 3.1 on is the span, in $M_q, 1 \le q < 2$, of $\{\mathbf{1}_{A_k^{i,j}}\}_{k=1,j=1,i=1}^{n_{i,j}\ m_i\ \ \infty}$ with

$$x_{i,j,k} = \mathbf{1}_{A_k^{i,j}}/\|\mathbf{1}_{A_k^{i,j}}\|_{M_q}, \quad i = 1, 2, \ldots, \ j = 1, \ldots, m_i, \ k = 1, \ldots, n_{i,j},$$

as its 1-unconditional basis. (We used the notation X_q in the beginning of this section for seemingly different spaces. The two spaces are actually isomorphic, a fact we will not use here.) It is easy to see that X_q is complemented in M_q. Actually, the conditional expectation – replacing the values of a function f by their averaged values on each of the sets $A_k^{i,j}$ – is a norm-one projection.

The operator T we would like to use is basically D defined above, restricted to X_q. There is a slight problem here as D does not map X_q back to X_q. This is easy to rectify. Note first that for each $\varepsilon > 0$ the sum of the measures of the sets in $\{A_k^{i,j}\}_{k=1,j=1,i=1}^{n_{i,j}\ m_i\ \ \infty}$ which are of measure smaller than ε is infinite. Otherwise, as is easily verified, $\{x_{i,j,k}\}$ is equivalent in some order to the natural basis of ℓ_q, ℓ_2, or $\ell_q \oplus \ell_2$. But none of these three bases contains block bases 3-equivalent to the natural basis of $\ell_{r_i}^{m_i}$ with $m_i^{\frac{1}{r_i}-\frac{1}{2}} \nearrow \infty$. Now, $D\mathbf{1}_{A_k^{i,j}}$, $i = 1, 2, \ldots, \ j = 1, \ldots, m_i, \ k = 1, \ldots, n_{i,j}$ are disjoint characteristic functions, $\mathbf{1}_{B_k^{i,j}}$, $i = 1, 2, \ldots, \ j = 1, \ldots, m_i, \ k = 1, \ldots, n_{i,j}$. By the property of the $A_k^{i,j}$'s each $B_{k'}^{i',j'}$ is equal in distribution to a disjoint union of sets from $\{A_k^{i,j}\}_{k=1,j=1,i=1}^{n_{i,j}\ m_i\ \ \infty}$, and one can choose the sets in such a manner that each set in $\{A_k^{i,j}\}_{k=1,j=1,i=1}^{n_{i,j}\ m_i\ \ \infty}$ appears at most once in these representations. It follows that DX is isometric to a subspace of X. So the operator T we will use is D restricted to X, followed by this isometry. By the sentence following (4.10), T satisfies (a) of Theorem 3.1.

The fact that T satisfies (b) follows from (4.8) and (4.9). Indeed, let E be an m-dimensional subspace of $M_q(\bigcup_{i=i_0}^{\infty} \bigcup_{j=1}^{\infty} \Omega_{i,j})$. (The subspace of M_q containing all functions supported on $\bigcup_{i=i_0}^{\infty} \bigcup_{j=1}^{\infty} \Omega_{i,j}$.) It is enough to show that if δ_0 is small enough (depending only on m) then D restricted to E has γ_2 norm at most 6. This will clearly imply that T, which is basically the restriction of D to X_q, satisfies (b).

The dual of $M_q(\bigcup_{i=i_0}^{\infty} \bigcup_{j=1}^{\infty} \Omega_{i,j})$ is $M_p(\bigcup_{i=i_0}^{\infty} \bigcup_{j=1}^{\infty} \Omega_{i,j})$, $\frac{1}{q} + \frac{1}{p} = 1$. So there is a subspace F of $M_p(\bigcup_{i=i_0}^{\infty} \bigcup_{j=1}^{\infty} \Omega_{i,j})$ of dimension $k(m)$ depending only on m which 2-norms E. Simple duality properties of the γ_2 norm imply that it is enough to prove that if δ_0 is small enough (depending only on m) then D^* restricted to F has γ_2 norm at most 3.

Now M_p is naturally a subspace of $L_p(0,\infty) \oplus_\infty L_2(0,\infty)$ and $D^* : M_p \to M_p$ is the restriction of the operator $K : L_p(0,\infty) \oplus_\infty L_2(0,\infty) \to L_p(0,\infty) \oplus_\infty L_2(0,\infty)$ given by

$$K(f, g) = (D^* f, D^* g).$$

We will denote by P_1 and P_2 the natural projections onto the first and second components of $L_p(0,\infty) \oplus_\infty L_2(0,\infty)$, respectively. Equations (4.8) and (4.9) imply that for all $f \in L_2(0,\infty)$,

$$\|D^* f\|_2 = 2\|f\|_2, \tag{4.11}$$

and for all $f \in L_p(\bigcup_{i=i_0}^{\infty} \bigcup_{j=1}^{\infty} \Omega_{i,j})$,

$$\|D^* f\|_p \le \delta_{i_0} \|f\|_p. \tag{4.12}$$

The standard inequality, $\gamma_2(S) \leq \|S\|k^{1/2}$ for any operator of rank k, implies that if $\delta_0 < k(m)^{-1/2}$ then the γ_2 norm of KP_2 restricted to F is smaller than 1. Since KP_2 has γ_2 norm 2, we get that $\gamma_2(K_{|F}) < 3$. This implies (using another simple property of the γ_2 norm) that $\gamma_2(D^*_{|F}) < 3$.

The discussion above, Corollary 3.2 and Remark 3.3 imply the main theorem below for $1 < p < 2$. The case $2 < p < \infty$ follows by duality.

Theorem 4.1 (JS). *For $1 < p < \infty$, $p \neq 2$, $L(L_p[0,1])$ has exactly $2^{\mathfrak{c}}$ different closed ideals.*

Remark 4.2. The proof also gives that M_1 (which is not isomorphic to an L_1 space) has exactly $2^{\mathfrak{c}}$ different closed ideals. By duality, also M_∞ has at least $2^{\mathfrak{c}}$ different closed ideals.

Remark 4.3. By Corollary 6.7 below, Theorem 4.1 and Remark 4.2 can be strengthened by interpreting the word "different" to mean mutually nonisomorphic as Banach algebras. That is, no two ideals admit an homomorphism between them which is continuous in both directions.

5. THE NONREFLEXIVE CLASSICAL SPACES

Here we deal mostly with the number of closed ideals in $L(L_1[0,1])$. The result is less impressive than that in the previous section as we only prove the existence of a continuum of such ideals. On the other hand, the leap from previous results may seem larger, compared with the case of $L_p[0,1]$, $p > 1$, as prior to [13] only a finite number of such ideals were known. We also deal here with the spaces $L(C(0,1))$ and $L(L_\infty[0,1])$.

The result is:

Theorem 5.1 (JPS). *There exists a family $\{\mathcal{J}_p;\ 2 < p < \infty\}$ of (nonclosed) ideals in $L(L_1[0,1])$ such that their closures $\overline{\mathcal{J}_p}$ are distinct ideals in $L(L_1[0,1])$. The spaces $L(C(0,1))$ and $L(L_\infty[0,1])$ also have a continuum of closed ideals.*

Remark 5.2. As with the case of $L(L_p[0,1])$, by Corollary 6.7 below, Theorem 5.1 can be strengthened by interpreting the word "distinct" to mean mutually nonisomorphic as Banach algebras. That is, no two of those ideals admit an homomorphism between them which is continuous in both directions.

We do not have much to add to the actual proof in [13]. We will only sketch the construction and comment on the idea of the proof. The gist of the construction is the simple Lemma 5.3, which we bring in full and try to explain its relevance.

In the discussion below, we replace $L_1[0,1]$ with its isometric copy, $L_1(\mathbb{T})$. Recall that a set of characters on the circle group, $\mathbb{T}$, equipped with the normalized Lebesgue measure, is called a Λ_p set, $2 < p < \infty$, if the L_p norm on the closed linear span of this set of characters is equivalent to the L_2 norm. For each $2 < p < \infty$, we will build a sequence of characters of the circle group $\{\gamma_j^p\}_{j=1}^\infty$ which form a Λ_p set and is "as dense as possible" in a certain precise way. We then let J_p be the formal identity from ℓ_1 to this set viewed in

$L_1(\mathbb{T})$, i.e., $J_p : \ell_1 \to L_1(\mathbb{T})$, $J_p e_i = \gamma_i^p$. Each ideal $\mathcal{J}_p$, $2 < p < \infty$, in the statement of Theorem 5.1 will be the set of all operators which factor through J_p, i.e.,

$$\mathcal{J}_p = \{AJ_pB;\ B : L_1(\mathbb{T}) \to \ell_1,\ A : L_1(\mathbb{T}) \to L_1(\mathbb{T})\}.$$

To show that the closures of the $\mathcal{J}_p$'s are different, we show that for $q > p > 2$, $J_q P$ is not in $\mathcal{J}_p$, where P is a norm-one projection from $L_1(\mathbb{T})$ onto (an isometric copy of) ℓ_1.

For $1 \le r < \infty$ and $M \in \mathbb{N}$, we denote L_r over a finite set of cardinality M equipped with the normalized counting measure by L_r^M. We recall that for each $p > 2$ there exists a positive C depending only on p, and for each $N \in \mathbb{N}$ there are vectors $\{v_i\}_{i=1}^N$ in $L_p^{N^{p/2}}$ such that

(1) $\|\sum_{i=1}^N a_i v_i\|_p \le C(\sum_{i=1}^N |a_i|^2)^{1/2}$, and

(2) $\min_{1\le i\le N} \|v_i\|_1 \ge 1$.

One can take the v_i's to be characters in the span of the first $N^{p/2}$ characters (a space isomorphic with constant depending only on p to $L_p^{N^{p/2}}$). This follows from the solution of Bourgain to the Λ_p problem **[5]**. The existence of the v_i's also follows from easier and earlier probabilistic construction of **[4]** which does not yield characters but is good enough for our purposes. The dimension $N^{p/2}$ is best possible, up to constants depending only on p. The next lemma shows this in greater generality.

Lemma 5.3. *Let $1 \le p < q < \infty$, $\{v_1, \dots, v_N\} \subset L_q(\mathbb{T})$, and let $T : L_1(\mathbb{T}) \to L_1^{N^{\frac{p}{2}}}$ be an operator. Suppose that C and ϵ satisfy*

(1) $\max_{|\epsilon_i|=1} \|\sum_{i=1}^N \epsilon_i v_i\|_q \le CN^{1/2}$, *and*

(2) $\min_{1\le i\le N} \|Tv_i\|_1 \ge \epsilon$.

Then $\|T\| \ge (\epsilon/C)N^{\frac{q-p}{2q}}$.

Lemma 5.3 and the discussion preceding it should be interpreted in the following way: For each $2 < p < \infty$ and N, there is a nicely bounded operator $J_p^N : \ell_1^N \to L_1^{N^{p/2}}$. But for $q > p$, J_q^N does not factor well through J_p^N.

The actual operator J_p is built by gluing together infinitely many J_p^N's for an increasing sequence of N's. Also, we repeat each block infinitely often to ensure that $\mathcal{J}_p$ is a subspace, a requirement in the definition of an ideal. The discussion in the previous paragraph hints at the proof that, for $q > p$, J_q does not factor through J_p. We will not repeat the actual construction and proof here and refer the interested reader to the original paper. We do reproduce the proof of Lemma 5.3 here, as we believe it should be useful elsewhere and we would like to emphasize its relative simplicity.

Proof of Lemma 5.3. Take u_i^* in $L_\infty^{N^{\frac{p}{2}}} = (L_1^{N^{\frac{p}{2}}})^*$ with $|u_i^*| \equiv 1$ so that $\langle u_i^*, Tv_i\rangle = \|Tv_i\|_1 \geq \epsilon$. Then

$$\begin{aligned}
\epsilon N &= \sum_{i=1}^{N} \langle T^* u_i^*, v_i\rangle := \frac{1}{2\pi}\int_0^{2\pi} \sum_{i=1}^{N} (T^* u_i^*)(a) v_i(a)\, da \\
&\leq \frac{1}{2\pi}\int_0^{2\pi} \sup_{a\in[0,1]} \left|\sum_{i=1}^{N} (T^* u_i^*)(a) v_i(b)\right| db \\
&=: \frac{1}{2\pi}\int_0^{2\pi} \left\|\sum_{i=1}^{N} v_i(b) T^* u_i^*\right\|_{L_\infty[0,1]} db \\
&\leq \|T\| \frac{1}{2\pi}\int_0^{2\pi} \left\|\sum_{i=1}^{N} v_i(b) u_i^*\right\|_{L_\infty^{N^{\frac{p}{2}}}} db \\
&\leq \|T\| N^{\frac{p}{2q}} \frac{1}{2\pi}\int_0^{2\pi} \left(\int_{[N^{\frac{p}{2}}]} \left|\sum_{i=1}^{N} u_i^*(c) v_i(b)\right|^q dc\right)^{\frac{1}{q}} db \\
&\leq \|T\| N^{\frac{p}{2q}} \left(\int_{[N^{\frac{p}{2}}]} \frac{1}{2\pi}\int_0^{2\pi} \left|\sum_{i=1}^{N} u_i^*(c) v_i(b)\right|^q db\, dc\right)^{\frac{1}{q}} \\
&\leq C\|T\| N^{\frac{p+q}{2q}}. \qquad \blacksquare
\end{aligned}$$

The results stated in Theorem 5.1 for $L(C(0,1))$ and $L(L_\infty[0,1])$ are proved by not completely trivial reasoning from the L_1 case. We will not repeat the arguments here.

6. REMARKS AND OPEN PROBLEMS

The main problem left open here is

Problem 6.1. How many different closed ideals are there in $L(L_1[0,1])$, $L(C(0,1))$, and $L(L_\infty[0,1])$?

Another problem concerning ideals in $L(L_1[0,1])$ comes from the fact that the continuum of ideals built in [13] and discussed in Section 5 are all small.

Problem 6.2. Are there infinitely many large ideals in $L(L_1[0,1])$?

This, of course, is very much connected with the question of what the complemented subspaces of $L_1[0,1]$ are. We repeat the well-known simplest question in this direction here.

Problem 6.3. Are there infinite-dimensional complemented subspaces of $L_1[0,1]$ which are not isomorphic to either ℓ_1 or $L_1[0,1]$?

Remark 6.4. The ideals constructed in Section 4, based on Corollary 3.2, turn out to be all large. In [13] we also build $2^{\mathfrak{c}}$ small ideals in $L(L_p[0,1])$, $1 < p < \infty$, $p \neq 2$.

Remark 6.5. 1. One can strengthen the conclusion of Corollary 3.2 by getting an antichain of $2^{\mathfrak{c}}$ closed ideals in $L(X)$, i.e., a collection of $2^{\mathfrak{c}}$ closed ideals, no two of which are included one in the other. For that, one just uses a collection of $2^{\mathfrak{c}}$ subsets of $\mathcal{C}$, no two of which are included one in the other.

2. Similarly, one gets a collection of $\mathfrak{c}$ different closed ideals in $L(X)$ that form a chain (by taking a chain of subsets of $\mathcal{C}$ of that cardinality). It is also easy to show by a density argument that, for any separable X, this is the maximal cardinality of any chain of closed ideals in $L(X)$.

3. Consequently, $L(L_p[0,1])$, $1 < p < \infty$, $p \neq 2$ contains an antichain of cardinality $2^{\mathfrak{c}}$ of closed ideals. It also contains a chain of length $\mathfrak{c}$ of different closed ideals.

4. The construction surveyed in Section 5 also produces a chain of length $\mathfrak{c}$ of closed ideals in $L(L_1[0,1])$.

Next we would like to discuss a stronger notion of distinction between closed ideals (and Banach algebras, in general). We say that two Banach algebras A and B are *isomorphically homomorphic* if there is an injective and surjective homomorphism from A onto B which is continuous in both directions. In the literature on Banach algebras, the isomorphism of Banach algebras is sometimes understood to be an isometry, i.e., preserving the norm. We use the ad hoc term *isomorphic homomorphism* to emphasize that we only require the homomorphism to be bounded (equivalently, continuous) in both directions. One could ask

Question 6.6. Let $1 \leq p < \infty$, $p \neq 2$. How many closed ideals are there in $L(L_p[0,1])$, each two of which are not isomorphically homomorphic?

Eidelheit [8] proved that if X and Y are Banach spaces such that $L(X)$ and $L(Y)$ are isomorphically homomorphic then X and Y are isomorphic Banach spaces. It follows that the $\aleph_1$ ideals in $L(L_p[0,1])$, $1 < p < \infty$, $p \neq 2$ coming from nonmutually isomorphic complemented subspaces of $L_p[0,1]$ are mutually nonisomorphically homomorphic. Going a bit deeper into the proof of [8], Johnson, Phillips, and the author showed that if $\mathcal{I}$ and $\mathcal{J}$ are two closed ideals in $L(X)$ which are isomorphically homomorphic then $\mathcal{I} = \mathcal{J}$. The proof will appear elsewhere. This, together with Theorems 5.1, 4.1 and the results of [9, 10], gives

Corollary 6.7. 1. *$L(L_1[01,])$ contains a continuum of mutually nonisomorphically homomorphic closed ideals.*

2. *For $1 < p < \infty$, $p \neq 2$, $L(L_p[0,1])$ contains exactly $2^{\mathfrak{c}}$ mutually nonisomorphically homomorphic closed ideals.*

3. *Each of the spaces $L(\ell_p \oplus \ell_q)$, $1 < p < q < \infty$, $L(\ell_p \oplus c_0)$, $L(\ell_p \oplus \ell_\infty)$, $L(\ell_p \oplus \ell_1)$, $1 < p < \infty$, contains $2^{\mathfrak{c}}$ mutually nonisomorphically homomorphic closed ideals.*

We do not know the answer to the relevant question in the Banach space category:

Problem 6.8. Let $1 \leq p < \infty$, $p \neq 2$. How many closed ideals are there in $L(L_p[0,1])$, each two of which are not isomorphic as Banach spaces?

A result of Arias and Farmer [2] states that for every infinite-dimensional complemented subspace X of $L_p[0,1]$, $1 < p < \infty$, which is not isomorphic to a Hilbert space, $L(X)$ is isomorphic (as a Banach space) to $L(L_p[0,1])$. So all the ideals coming from complemented subspaces of $L_p[0,1]$ are isomorphic.

Next we repeat the main problem concerning complemented subspaces of $L_p[0,1]$, $1 < p < \infty$.

Problem 6.9. Is there a continuum of complemented subspaces of $L_p[0,1]$, $1 < p < \infty$, $p \neq 2$, which are mutually nonisomorphic?

There was very little progress on new constructions of complemented subspaces of $L_p[0,1]$, $1 < p < \infty$, $p \neq 2$, since [6], which contains a list of still open problems. We would like to repeat one of them as it may appeal to the Harmonic Analysis community. The mutually nonisomorphic $\aleph_1$ complemented subspaces of $L_p[0,1]$ constructed in [6] are all translation-invariant subspaces of L_p over the Cantor group $\{-1,1\}^{\mathbb{N}}$ endowed with the natural product measure (which is isometric to $L_p[0,1]$). The projections onto them are translation-invariant operators (it is easy to prove that if there is a bounded projection onto a translation-invariant subspace then the translation-invariant one is also bounded), i.e., idempotent multipliers in $L_p(\{-1,1\}^{\mathbb{N}})$. This produces $\aleph_1$ quite nontrivial multipliers. Pełczynski asked whether a similar phenomenon happens on other groups, in particular on $\mathbb{T}$.

Problem 6.10. Are there uncountably many mutually nonisomorphic complemented translation invariant subspaces of $L_p(\mathbb{T})$?

FUNDING

This work was partially supported by the Israel Science Foundation.

REFERENCES

[1] S. Argyros and R. G. Haydon, A hereditarily indecomposable L_∞-space that solves the scalar-plus-compact problem. *Acta Math.* **206** (2011), no. 1, 1–54.

[2] A. Arias and J. Farmer, On the structure of tensor products of lp-spaces. *Pacific J. Math.* **175** (1996), no. 1, 13–37.

[3] K. Beanland, T. Kania, and N. J. Laustsen, Closed ideals of operators on the Tsirelson and Schreier spaces. *J. Funct. Anal.* **279** (2020), no. 8, 108668, 28 pp.

[4] G. Bennett, L. E. Dor, V. Goodman, W. B. Johnson, and C. M. Newman, On uncomplemented subspaces of L_p, $1 < p < 2$. *Israel J. Math.* **26** (1977), no. 2, 178–187.

[5] J. Bourgain, Bounded orthogonal systems and the $\Lambda(p)$-set problem. *Acta Math.* **162** (1989), no. 3–4, 227–245.

[6] J. Bourgain, H. P. Rosenthal, and G. Schechtman, An ordinal L^p-index for Banach spaces, with application to complemented subspaces of L^p. *Ann. of Math. (2)* **114** (1981), no. 2, 193–228.

[7] J. W. Calkin, Two-sided ideals and congruences in the ring of bounded operators in Hilbert space. *Ann. of Math. (2)* **42** (1941), 839–873.

[8] M. Eidelheit, On isomorphisms of rings of linear operators. *Studia Math.* **9** (1940), 97–105.

[9] D. Freeman, T. Schlumprecht, and A. Zsák, Closed ideals of operators between the classical sequence spaces. Bull. Lond. Math. Soc. 49(5) (2017), 859–-876.

[10] D. Freeman, T. Schlumprecht, and A, Zsák, Banach spaces for which the space of operators has $2^{\mathfrak{c}}$ closed ideals. Forum Math. Sigma 9 (2021), Paper No. e27, 20 pp.

[11] I. C. Gohberg, A. S. Markus, and I. A. Feldman, Normally solvable operators and ideals associated with them. (Russian) *Bul. Acad. Ştiinţe Repub. Mold.* no. **10** (1960), no. 76, 51–70.

[12] W. B. Johnson, B. Maurey, G. Schechtman, and L. Tzafriri, Symmetric structures in Banach spaces. *Mem. Amer. Math. Soc.* **19** (1979), no. 217.

[13] W. B. Johnson, G. Pisier, and G. Schechtman, Ideals in $L(L_1)$. *Math. Ann.* **376** (2020), no. 1–2, 693–705.

[14] W. B. Johnson and G. Schechtman, Multiplication operators on $L(L_p)$ and ℓ_p-strictly singular operators. *J. Eur. Math. Soc. (JEMS)* **10** (2008), no. 4, 1105–1119.

[15] W. B. Johnson and G. Schechtman, The number of closed ideals in $L(L_p)$. *Acta Math.* **227** (2021), 103–113.

[16] S. Kwapień and C. Schütt, Some combinatorial and probabilistic inequalities and their application to Banach space theory. II. *Studia Math.* **95** (1989), no. 2, 141–154.

[17] A. Pietsch, *Operator ideals.* Translated from German by the author. N.-Holl. Math. Libr. 20. North-Holland Publishing Co., Amsterdam-New York, 1980. 451 pp. ISBN: 0-444-85293-X.

[18] H. P. Rosenthal, On the subspaces of L_p $(p > 2)$ spanned by sequences of independent random variables. *Israel J. Math.* **8** (1970), 273–303.

[19] G. Schechtman, Examples of $\mathcal{L}_p$ spaces $(1 < p \neq 2 < \infty)$. *Israel J. Math.* **22** (1975), no. 2, 138–147.

[20] T. Schlumprecht and A. Zsák, The algebra of bounded linear operators on $\ell_p \oplus \ell_q$ has infinitely many closed ideals. *J. Reine Angew. Math.* **735** (2018), 225–247.

GIDEON SCHECHTMAN

Department of Mathematics, Weizmann Institute of Science, Rehovot, Israel 76100, gideon@weizmann.ac.il

SLICES AND DISTANCES: ON TWO PROBLEMS OF FURSTENBERG AND FALCONER

PABLO SHMERKIN

ABSTRACT

We survey the history and recent developments around two decades-old problems that continue to attract a great deal of interest: the slicing $\times 2$, $\times 3$ conjecture of H. Furstenberg in ergodic theory, and the distance set problem in geometric measure theory introduced by K. Falconer. We discuss some of the ideas behind our solution of Furstenberg's slicing conjecture, and recent progress in Falconer's problem. While these two problems are on the surface rather different, we emphasize some common themes in our approach: analyzing fractals through a combinatorial description in terms of "branching numbers," and viewing the problems through a "multiscale projection" lens.

MATHEMATICS SUBJECT CLASSIFICATION 2020

Primary 28A80; Secondary 05D99, 11K55, 28A75, 28A78, 37C45

KEYWORDS

Hausdorff dimension, Furstenberg conjectures, self-similarity, projections, distance sets

INTERNATIONAL CONGRESS OF MATHEMATICIANS 2022 JULY 6–14

Proc. Int. Cong. Math. 2022, Vol. 4, pp. 3266–3290
DOI 10.4171/ICM2022/97

Published by EMS Press

1. INTRODUCTION

In this article we survey recent progress on the following two old conjectures. Hausdorff dimension is denoted $\dim_H$.

Conjecture 1.1 (Furstenberg's slicing conjecture, [18]). *Let* $X, Y \subset [0,1)$ *be closed and invariant under* T_a, T_b *respectively, where* $T_m(x) = mx \bmod 1$ *is multiplication by* m *on the circle. Assume that* $\log a/\log b$ *is irrational. Then*

$$\dim_H\big((X \times Y) \cap \ell\big) \le \max\big(\dim_H(X) + \dim_H(Y) - 1, 0\big)$$

for all lines ℓ *that are neither vertical nor horizontal.*

Conjecture 1.2 (Falconer's distance set problem, originating in [16]). *Let* $X \subset \mathbb{R}^d$, $d \ge 2$ *be a Borel set with* $\dim_H(X) \ge d/2$. *Let* $\Delta(X) = \{|x - y| : x, y \in X\}$. *Then*

$$\dim_H\big(\Delta(X)\big) = 1.$$

We discuss the history and motivation behind these conjectures in Sections 3 and 4, respectively. Conjecture 1.1 was resolved by the first author [48] and, simultaneously, independently, and with a strikingly different proof, by M. Wu [57]. Many related problems remain open. Conjecture 1.2 is open in all dimensions.

At first sight, Conjectures 1.1 and 1.2 appear to be rather different (other than both involving Hausdorff dimension). A key difference is that Furstenberg's conjecture deals with sets with a rigid arithmetic structure, while Falconer's conjecture involves arbitrary Borel sets. A more subtle but also crucial distinction is that Furstenberg's conjecture is *linear* in nature (it concerns linear slices of $X \times Y$), while Falconer's conjecture deals with Euclidean distances and curvature plays a key rôle in all partial progress towards it.

Nevertheless, we will see that there are some similar ideas in our own approach to these two problems. We will recast both in terms of *projections*. To handle these projections, we use in both cases a combinatorial approach to the study of fractals through their branching structure. Bourgain's celebrated discretized projection theorem [6,7] (or its proof) makes an appearance in our work on both conjectures.

In Section 2, we discuss a key uniformization lemma, and Bourgain's discretized sumset, sum-product, and projection theorems. In Section 3, we put Furstenberg's slicing conjecture into context and give an impressionistic account of our solution. In Section 4, we discuss Falconer's problem and some of our recent progress towards it (obtained partly in collaboration with T. Keleti and with H. Wang). Along the way, we will touch upon the closely related and vast field of *projection theory* in geometric measure theory.

A word on notation. Given two positive quantities A, B, the notation $A \lesssim B$ means that $A \le CB$ for some constant $C > 0$, while $A \lesssim_x B$ means that $A \le C(x)B$, where again $C(x) > 0$. We write $A \gtrsim B$ for $B \lesssim A$ and $A \sim B$ for $A \lesssim B \lesssim A$, and likewise with subindices. We denote positive constants whose value is not too important by c, C and as before indicate their dependencies with subindices.

2. A GLIMPSE OF BOURGAIN'S DISCRETIZED GEOMETRY

2.1. Uniform sets and uniformization

Even though the statements of Conjectures 1.1 and 1.2 involve Hausdorff dimension, most approaches discretize the problem at a small scale δ. Given a bounded set $X \subset \mathbb{R}^d$, let $|X|_\delta$ be the number of δ-mesh cubes $\prod_{i=1}^d [k_i\delta, (k_i+1)\delta)$ intersecting X. If X has the property that $|X|_\delta \le \delta^{-s}$ for arbitrarily small values of δ, then $\dim_H(X) \le s$. If, on the other hand, $|X|_\delta \ge \delta^{-s}$ for all small δ, it does not quite follow that $\dim_H(X) \ge s$—what is technically true is that the lower Minkowski dimension of X is at least s. For the sake of simplicity, we will ignore this distinction, and consider the growth rate of $|X|_\delta$ as a good proxy for the (Hausdorff) dimension of X. In this discussion, there is no loss of generality in restricting δ to dyadic numbers 2^{-m} or even (2^T)-adic numbers $2^{-T\ell}$ once the integer T has been fixed.

Let $\mathcal{D}_\delta$ denote the family of δ-mesh cubes in $\mathbb{R}^d$. If $X \subset \mathbb{R}^d$ is a union of cubes in $\mathcal{D}_{2^{-m}}$, we say that X is a 2^{-m}-set. For $X \subset \mathbb{R}^d$, we denote the set of cubes in $\mathcal{D}_\delta$ intersecting X by $\mathcal{D}_\delta(X)$.

Let $X \subset [0,1)^d$. Given a $T \in \mathbb{N}$ (which we consider fixed) and $\ell \ge 1$, we can view $(\mathcal{D}_{2^{-Tj}}(X))_{j=0}^{\ell-1}$ as a tree, with $[0,1)^d$ as the root and descendance given by inclusion. This tree provides a combinatorial description of X at resolution $2^{-T\ell}$. In general, the tree may be very irregular, with different vertices having different numbers of offspring. In many situations, the set X is easier to study if one knows that the tree is *spherically symmetric*, meaning that the number of offspring is constant at each level of the tree (but can still change from level to level).

Definition 2.1. A set $X \subset [0,1)^d$ is $(T, (N_j)_{j=0}^{\ell-1})$-uniform if

$$Q \in \mathcal{D}_{2^{-jT}}(X) \Longrightarrow \left|\mathcal{D}_{2^{-(j+1)T}}(X \cap Q)\right| = N_j, \quad j = 0, 1, \dots, \ell-1.$$

If X is $(T, (N_j)_{j=0}^{\ell-1})$-uniform for some $(N_j)_{j=0}^{\ell-1}$, then we also say that X is $(T;\ell)$-uniform.

We emphasize that what is fixed at each scale is the *number* of offspring; the particular set of N_j subcubes is still allowed to depend on the parent cube of level j. The following *uniformization lemma* says that by taking T large, and at the price of replacing X by a large subset, we may always assume that X is $(T;\ell)$-uniform.

Lemma 2.2. *Fix $T, \ell \in \mathbb{N}$ and write $m = T\ell$. Let $X \subset [0,1)^d$ be a 2^{-m}-set. Then X contains a $(T;\ell)$-uniform subset X' with*

$$\left|X'\right| \ge (2T)^{-\ell}|X| = 2^{(-\log(2T)/T)m}|X|.$$

Proof. We begin from the bottom of the tree, setting $X^{(\ell)} := X$. Once $X^{(j+1)}$ is constructed, we let

$$X^{(j,k)} = \bigcup\left\{X^{(j+1)} \cap Q : \left|Q \cap X^{(j+1)}\right|_{2^{-(j+1)T}} \in \left[2^k+1, 2^{k+1}\right]\right\}, \quad k = 0, \dots, T-1.$$

Since k takes T values, we can pick $k = k_j$ such that $|X^{(j,k)}| \ge |X^{(j+1)}|/T$. By removing at most half of the cubes in $\mathcal{D}_{2^{-(j+1)T}}(X^{(j+1)})$ from each of the sets $Q \cap X^{(j+1)}$

making up $X^{(j,k)}$, we obtain a set $X^{(j)} \subset X^{(j+1)}$ such that $|X^{(j)}| \geq |X^{(j+1)}|/(2T)$ and $|Q \cap X^{(j)}|_{2^{-(j+1)T}} = 2^k$ for all $Q \in \mathcal{D}_{jT}(X^{(j)})$. We see inductively that $|Q \cap X^{(j)}|_{2^{-(j'+1)T}}$ is constant over all $Q \in \mathcal{D}_{j'T}(X^{(j)})$, for all $j' = j, j+1, \dots, \ell - 1$. The lemma follows by taking $X' = X^{(0)}$. ■

We make some remarks on this statement and its proof. Firstly, this is just the simplest example of a flexible and powerful multiscale pigeonholing argument. For example, instead of (or additionally to) uniformizing the branching numbers N_j, we can pigeonhole any property of $Q \cap X$, $Q \in \mathcal{D}_{Tj}(X)$, that depends only on the behavior at scale $2^{-T(j+1)}$ and can be partitioned into a number C_T of classes, with $\log(C_T)/T \to 0$ as $T \to \infty$. Secondly, these ideas can also be used to "uniformize" a measure μ—an additional first step in this case is to pigeonhole a "μ-large" $2^{-T\ell}$-set X such that the density of $\mu|_X$ is roughly constant; we can then invoke the argument for sets. Here "μ-large" could simply mean that $\mu(X)$ is large, but sometimes it is convenient to look at other quantities like $\|\mu|_X\|_{L^q}$. Lastly, we can iterate such a uniformization lemma to decompose X (or μ) into a union of finitely many "large" uniform subsets X_i, plus a "small" remaining set X_{bad}; see, e.g., [33, COROLLARY 3.5].

2.2. Bourgain's sumset theorem

Let $X \subset [0,1)$ be a 2^{-m}-set for some large m. We are interested in understanding how the size of the arithmetic sum $X + X = \{x + y : x, y \in X\}$ relates to the structure of X. If X is an interval, then $|X + X|_\delta \sim |X|_\delta$. There are many "fractal" sets which satisfy $|X + X|_\delta \leq 2^{\varepsilon m}|X|_\delta$ with $\varepsilon > 0$ arbitrarily small: fix a large $T \in \mathbb{N}$, an even larger $\ell \gg T$, and $J \subset \{0, 1, \dots, \ell - 1\}$. Let X_J be the set of points in $[0,1)$ whose base 2^T-expansion has a digit zero at position $j + 1$ for $j \in J$, but is otherwise arbitrary. Then $X_J + X_J$ has the same structure, except that there could be carries; however, because T is large, these carries will not substantially increase the size of $X_J + X_J$. More precisely,

$$|X_J + X_J| \leq 2^{\ell - |J|}|X_J| \leq 2^{\varepsilon m}|X_J|, \quad \text{where } \varepsilon = 1/T,$$

where as usual we write $m = T\ell$. Note that even though X_J may not look macroscopically like an interval, there is a sequence of scales at which it looks like a union of intervals of the same length, and the left endpoints of these intervals form an arithmetic progression.

The set X_J is $(T; (N_j)_{j=0}^{\ell-1})$-uniform, with $N_j = 1$ if $j \in J$, and $N_j = 2^T$ otherwise. Bourgain's sumset theorem, which is implicit in [7], and stated in this form in [48, COROLLARY 3.10], asserts that having a small sumset forces this kind of branching structure:

Theorem 2.3. *Given $\delta > 0$ there are $\varepsilon > 0$, $T \in \mathbb{N}$, such that the following holds for all sufficiently large $\ell \in \mathbb{N}$.*

Let $m = \ell T$. Suppose X is a 2^{-m}-set with $|X + X|_{2^{-m}} \leq 2^{\varepsilon m}|X|_{2^{-m}}$. Then X contains a $(T, (N_j)_{j=0}^{\ell-1})$-uniform subset X' such that:

(i) $|X'|_{2^{-m}} \geq 2^{-\delta m}|X|_{2^{-m}}$,

(ii) *for each j, either $N_j = 1$, or $N_j \geq 2^{(1-\delta)T}$.*

In other words, up to passing to a large subset, 2^{-m}-sets with sub-exponential doubling locally look, depending on the scale, like an interval or a point. This is an example of an "inverse theorem" in (discretized) additive combinatorics, in which from a purely combinatorial fact (small doubling) one deduces strong structural information. We will encounter another (related) inverse theorem in Section 3.6. We emphasize that Theorem 2.3 does *not* characterize sets with small doubling—even if X is uniform with either full or no branching at each scale, if the locations of the (single) offspring cubes at the scales wit no branching do not have any arithmetic structure, it may well happen that $|X+X|_{2^{-m}}$ is far larger than $|X|_{2^{-m}}$.

2.3. Bourgain's discretized sum-product and projection theorems

A heuristic principle of great reach asserts that if X is a subset of some ring, then either the sumset $X+X$ or the product set $X \cdot X$ must be substantially larger than X, unless X itself looks like a subring. For example, it is a longstanding conjecture of Erdős and Szemerédi that if $X \subset \mathbb{Z}$, then $\max\{|X+X|, |X \cdot X|\} \gtrsim_\varepsilon |X|^{2-\varepsilon}$—in other words, either the sumset or the product set must be as large as possible. See [47] for the best bound at the time of writing, and further discussion.

When dealing with products, it is more convenient to work with subsets of $[1,2)$ rather than $[0,1)$. Again, if $X=[a,b) \subset [1,2)$, then both $|X+X|_\delta$ and $|X \cdot X|_\delta$ are comparable to $|X|_\delta$. Heuristically, one would expect that if $X \subset [1,2)$ does not look roughly like an interval at scales in $[\delta, 1]$, then either $|X+X|_\delta$ or $|X \cdot X|_\delta$ is substantially larger than $|X|_\delta$. This is the content of Bourgain's discretized sum-product theorem, which confirmed a conjecture of Katz and Tao [30]:

Theorem 2.4 ([6,7,9]). *Given $0<\alpha<1$ and $\beta>0$ there are $\kappa(\alpha,\beta)>0$, $\eta=\eta(\alpha,\beta)>0$ such that the following holds for $\delta \le \delta_0(\alpha,\beta)$. Let $X \subset [1,2]$ satisfy $|X|_\delta \ge \delta^{-\alpha}$ and*

$$\big|X \cap [t,t+r]\big|_\delta \le \delta^{-\kappa} r^\beta |X|_\delta, \quad t \in [1,2], r \in [\delta, 1]. \tag{2.1}$$

Then

$$\max\big\{|X+X|_\delta, |X \cdot X|_\delta\big\} \ge \delta^{-\alpha-\eta}.$$

Hypothesis (2.1) is known as a *nonconcentration* assumption, and it quantifies the fact that X "does not look like an interval." Note that because of the factor $\delta^{-\kappa}$, it is vacuous at scales close to 1 or δ. Bourgain [6] first proved this theorem under the stronger assumption that (2.1) holds with α in place of β (so that the nonconcentration exponent matches the size of the set). Bourgain and Gamburd [9] then proved it as stated, and used it to establish a spectral gap for subgroups of SU(2) satisfying a diophantine condition. Under the assumption $\beta=\alpha$, Guth, Katz, and Zahl [24] recently found a simpler proof with an explicit value: any $\eta<\frac{\alpha(1-\alpha)}{4(7+3\alpha)}$ works (with κ depending also on η).

In [7], Bourgain proved a discretized projection theorem that can be seen as a far more flexible form of Theorem 2.4. Let $\Pi_x(a,b)=a+bx$.

Theorem 2.5 ([7, THEOREM 2]). *Given $0<\alpha<2$ and $\beta>0$, there are $\kappa(\alpha,\beta)>0$ and $\eta=\eta(\alpha,\beta)>0$ such that the following holds for $\delta \le \delta_0(\alpha,\beta)$. Let $E \subset [0,1]^2$ satisfy*

$|E|_\delta \geq \delta^{-\alpha}$ *and*

$$\left|E \cap B(x,r)\right|_\delta \leq \delta^{-\kappa} r^\beta |E|_\delta, \quad x \in [0,1]^2, r \in [\delta,1].$$

Let $X \subset [1,2]$ *be a set satisfying* (2.1).

Then there is a set $X_0 \subset X$ *with* $|X \setminus X_0|_\delta \leq \delta^\kappa |X|_\delta$*, such that if* $x \in X_0$ *then*

$$\left|\Pi_x(E')\right|_\delta \geq \delta^{-\alpha/2-\eta} \quad \text{for all } E' \subset E, \ \left|E'\right|_\delta \geq \delta^\kappa |E|_\delta.$$

This is not quite the form the theorem was stated in [7] but is formally equivalent; see W. He's article [25] for this formulation and an extension of Theorem 2.5 to projections from $\mathbb{R}^d \to \mathbb{R}^k$. Taking $E = X \times X$ with $|X|_\delta = \delta^{-\gamma}$, we obtain in particular $|X + X \cdot X|_\delta \gtrsim \delta^{-\gamma-\eta}$, which is close to Theorem 2.4. One can in fact recover Theorem 2.4 from Theorem 2.5, see [7, PROOF OF THEOREM 1].

The proof of Theorem 2.5 relies on Theorem 2.3. An intermediate step in the proof is showing that if $Y \subset [1,2)$ satisfies the nonconcentration assumption (2.1), then $|Y + xY|_\delta$ is large for some $x \in X$. If this does not hold, then it is easy to see that $|Y + Y|_\delta$ is also small. The structural information on Y provided by Theorem 2.3 can then be used (very nontrivially!) to show that in fact $|Y + xY|_\delta$ must be large for some $x \in X$.

Theorem 2.5 has striking applications, for example, to equidistribution of linear random walks in the torus [8] and bounds for the dimensions of Kakeya sets in $\mathbb{R}^3$ [31]. We discuss a nonlinear version of the theorem and applications to the Falconer distance set problem in Section 4.2. For later reference, we conclude this discussion with a Hausdorff dimension version of Theorem 2.5. We note however that it is the discretized version that gets used in the applications.

Theorem 2.6 ([7, THEOREM 4]). *Given* $0 < \alpha < 2$ *and* $\beta > 0$*, there is* $\eta = \eta(\alpha, \beta) > 0$ *such that for any Borel set* $E \subset \mathbb{R}^2$ *with* $\dim_H(E) \geq \alpha$,

$$\dim_H\left\{x \in \mathbb{R} : \dim_H(\Pi_x E) < \frac{\alpha}{2} + \eta\right\} \leq \beta.$$

3. FURSTENBERG'S SLICING PROBLEM

3.1. Furstenberg's principle and rigidity result

Recall that integers $a, b \in \mathbb{N}$ are called *multiplicatively dependent* (denoted $a \sim b$) if $\log a / \log b \in \mathbb{Q}$ or, equivalently, a and b are powers of a common integer. Otherwise, we say that a and b are *multiplicatively independent*, and denote it by $a \nsim b$. If $a \sim b$, say $a = m^{a'}$, $b = m^{b'}$, then there is a straightforward relationship between the expansion of a real number x to bases a and b: they are both essentially the expansion to base m, looking at it in blocks of a' and b' digits at a time. In the 1960s, H. Furstenberg proposed a series of conjectures which, in different ways, aim to capture the heuristic principle that, on the other hand, *expansions in multiplicatively independent bases have no common structure*.

Recall that if $a \in \mathbb{N}_{\geq 2}$, we let $T_a : [0,1) \to [0,1)$, $x \mapsto ax \bmod 1$ denote multiplication by a on the circle. A set $X \subset [0,1)$ is T_a-invariant if $T_a X \subset X$. Since the map T_a shifts the a-ary expansion of a real number, a proper, closed, infinite T_a-invariant subset of

$[0, 1)$ can be thought of as being structured to base a. (The full circle $[0, 1)$ and finite rational orbits $\{j/m\}_{j=1}^{m-1}$ are trivially invariant under all T_a.) In 1967, Furstenberg [17] proved that no proper infinite closed subset of the circle can be invariant under T_a and T_b if $a \not\sim b$. This was the first concrete verification of the above heuristic principle, and gave birth to the vast and ongoing area of *rigidity* in ergodic theory. Furstenberg's $\times 2$, $\times 3$ problem asks whether the natural analog of this result also holds for measures, and is one of the most fundamental open questions in ergodic theory and beyond. He also proposed a number of other conjectures involving T_a-invariant sets, that we discuss next.

3.2. Furstenberg's sumset, slice, and orbit conjectures

In this section $a, b \geq 2$ are multiplicatively independent, and $X, Y \subset [0, 1)$ are closed and invariant under T_a, T_b. According to Furstenberg's principle, such sets X, Y should have no common structure. Furstenberg's rigidity result established a rough form of this: X and Y cannot be identical, unless trivial. Furstenberg conjectured that X and Y should be not just distinct but "geometrically independent," obeying dimensional relationships analogous to those of linear planes in general position. Since these are fractal sets (T_a-invariance can be seen as a kind of self-similarity, and it is well known that $\dim_H(X) < 1$ unless $X = [0, 1)$), it is natural to use Hausdorff dimension.

Furstenberg's *sumset conjecture* (which originated in the 1960s but was never stated in print) asserts that

$$\dim_H(X + Y) = \min\big(\dim_H(X) + \dim_H(Y), 1\big),$$

while Furstenberg's *slice* or *intersection conjecture*, stated as Conjecture 1 in [18], states that

$$\dim_H(X \cap Y) \leq \max\big(\dim_H(X) + \dim_H(Y) - 1, 0\big).$$

As pointed out in [18], this latter conjecture easily implies that $X \neq Y$ (unless trivial), recovering the rigidity result. While stopping short of proving the conjecture, Furstenberg in [18] introduced some ideas that are at the heart of modern progress in the area, including what are now known as *CP-chains*, a class of Markov chains where the transitions consist in "zooming in" dyadically towards typical points for the measures (see [19] for an elegant formulation of the theory). Using CP-chains, he showed that if $\dim_H(X \cap Y) > \gamma$, then for almost all reals u there is a line ℓ_u with slope u such that $\dim_H((X \times Y) \cap \ell_u) > \gamma$; moreover, there is an ergodic dynamical system on (measures supported on) linear fibers of $X \times Y$ of dimension $> \gamma$.

After partial progress in [44], the sumset conjecture was fully resolved by M. Hochman and the author in [29], using CP-chains as a key tool. In this work we also introduced the method of *local entropy averages* to bound from below the entropy and dimension of projected images; we will come back to this in Section 4.4. A simple, purely combinatorial proof was recently obtained by D. Glasscock, J. Moreira, and F. Richter [21].

The slice conjecture was resolved around 10 years later, independently by the author [48] and M. Wu [57]. Wu's proof is also based on CP-chains and the ideas from [18], but introduces a key new ergodic-theoretic insight. A simple conceptual proof, also based on the

CP-chains from [18], was recently obtained by T. Austin [4]. By adapting Wu's method, H. Yu [58] gave a more elementary and quantitative proof in the case $\dim_H(X) + \dim_H(Y) < 1$. Our proof follows a different approach, based on additive combinatorics and multifractal analysis—we will describe some of the main ideas in the rest of this section. All the proofs yield also Conjecture 1.1, which was also implicitly stated in [18]. They all also imply the sumset conjecture. Applications of the slice conjecture to number-theoretic problems involving integers with restricted digit expansions were given in [10, 20].

A further conjecture of Furstenberg [18, CONJECTURE 2], in the authors' view among the hardest and most beautiful in mathematics, asserts that for every irrational $x \in [0, 1)$, if $\mathcal{O}_{m,x} = \overline{\{T_m^n x\}_{n\in\mathbb{N}}}$, is the closure of the orbit of x under T_m, then

$$\dim_H(\mathcal{O}_{a,x}) + \dim_H(\mathcal{O}_{b,x}) \geq 1.$$

This fits into the theme of lack of common structure for expansions to bases a, b: it says that such expansions of an irrational number cannot simultaneously have "low complexity," as measured by the dimension of the orbit closure. In particular, if the orbit closure under T_a has "minimal complexity" (dimension 0), then the T_b-orbit must be dense, meaning that every possible b-ary block appears in the base b expansion of x. This conjecture is wide open; even proving that either $\dim_H(\mathcal{O}_{a,x})$ or $\dim_H(\mathcal{O}_{b,x})$ has positive dimension seems to require completely new ideas. However, it is a formal consequence of the slicing conjecture that the set of x for which the orbit conjecture fails has Hausdorff dimension zero. Unfortunately, this says nothing about points x for which $\dim_H(\mathcal{O}_{a,x}) = 0$, since all such points form a zero-dimensional set. Recently, B. Adamczewski and C. Faverjon [1] showed that an irrational number cannot be automatic in bases a and b; being automatic is a computational notion of "simplicity," and so this can be seen as a first verification that an irrational number cannot be "too simple" in two multiplicatively independent bases.

3.3. L^q dimensions, self-similarity, and the dimension of slices

Let $\mathcal{P}(X)$ denote the family of Borel probability measures on a metric space X. Given $\mu \in \mathcal{P}(\mathbb{R}^d)$, the L^q dimensions $\{D_\mu(q)\}_{q>1}$ are a family of indices measuring the degree of singularity of μ through its q-moments:

$$D_\mu(q) = D(\mu, q) = \liminf_{\delta\to 0} \frac{\log \sum_{Q\in\mathcal{D}_\delta} \mu(Q)^q}{(q-1)\log\delta}.$$

(It is also possible to define $D_\mu(q)$ for $q < 1$, but we do not need this here.) The normalizing factor $1/(q-1)$ ensures that $D_\mu(q) \in [0, d]$. If μ has an L^q density, then $D_\mu(q) = 1$ but $D_\mu(q) < 1$ is possible even for other absolutely continuous measures. For any fixed μ, the function D_μ is nonincreasing, so it makes sense to define

$$D_\mu(\infty) = D(\mu, \infty) = \lim_{q\to\infty} D_\mu(q).$$

It is not hard to show that $D_\mu(\infty)$ is the supremum of the s such that $\mu(B(x,r)) \leq Cr^s$ for some constant $C = C(\mu, s)$ and all closed balls $B(x, r)$. Such s are also called *Frostman exponents* of μ. The function $\tau_\mu(q) = (q-1)D_\mu(q)$ is known as the *L^q-spectrum* of μ. It is always concave. In particular, both τ_μ and D_μ are differentiable outside of a countable set of q. See [34, SECTION 3] for proofs of these facts and further background on the L^q spectrum and dimension.

We are interested in upper bounds for the dimension of slices. The next very simple but key lemma relates this problem to lower bounds on $D_\mu(\infty)$ for suitable measures μ. Given a map $\pi : X \to Y$ and $\mu \in \mathcal{P}(X)$, we denote the push-forward measure by $\pi\mu = \mu \circ \pi^{-1}$.

Lemma 3.1. *Suppose $\pi : \mathbb{R}^d \to \mathbb{R}$ is a Lipschitz map. Let $\mu \in \mathcal{P}([0,1]^d)$ be such that $\mu(B(x,r)) \ge cr^\alpha$ for all $x \in X := \mathrm{supp}(\mu)$, $r \in (0,1]$. If $D(\pi\mu, \infty) \ge \beta$, then*

$$\dim_{\mathrm{H}}\big(X \cap \pi^{-1}(y)\big) \le \alpha - \beta \quad \text{for all } y \in \mathbb{R}.$$

Proof. Fix a small $\varepsilon > 0$ and $y \in \mathbb{R}$. Let $(x_j)_{j=1}^M$ be a maximal ε-separated subset of $X \cap \pi^{-1}(y)$, and let $A = \bigcup_{j=1}^M B(x_j, \varepsilon/2)$. Since the balls are disjoint, $\mu(A) \ge cM(\varepsilon/2)^\alpha$. On the other hand, πA is contained in an interval of size $\lesssim \varepsilon$ and hence, for any $\eta > 0$,

$$\mu(A) \le (\pi\mu)(\pi A) \lesssim_\eta \varepsilon^{\beta-\eta}.$$

Comparing the bounds, $M \lesssim_{c,\eta} \varepsilon^{\beta-\alpha-\eta}$. Now $X \cap \pi^{-1}(y) \subset \bigcup_{j=1}^M B(x_j, \varepsilon)$ by the maximality of (x_j), and hence $X \cap \pi^{-1}(y)$ can be covered by $\lesssim_{c,\eta} \varepsilon^{\beta-\alpha-\eta}$ balls of radius ε. Letting $\eta \to 0$, we get the claim. ■

In order to connect this lemma to the slice conjecture, our next step is to look at measures defined on invariant sets. A set $X \subset \mathbb{R}$ is *self-similar* if there are finitely many contracting similarity transformations $f_i(x) = r_i x + t_i$, $i \in I$ with $0 < r_i < 1$, such that $X = \bigcup_{i \in I} f_i(X)$. The family $(f_i)_{i \in I}$ is called an *iterated function system (IFS)* and X is its *attractor*. For simplicity, from now we assume that we are in the homogeneous case, meaning that all the contractions r_i are equal.

A closed T_a-invariant set X needs not be self-similar in the sense above. However, it is easy to see [48, P. 378] that for every $\varepsilon > 0$ there is a set $X' \supset X$ with $\dim_{\mathrm{H}}(X') < \dim_{\mathrm{H}}(X) + \varepsilon$, which is the attractor of an IFS of the form $\{a^{-m}(x+j)\}_{j \in J}$, where m and the "digit set" $J \subset \{0, \dots, a^m - 1\}$ depend on ε. Hence, in order to establish Conjecture 1.1, we may assume that X, Y are self-similar of this special form. Since the assumption $a \not\sim b$ is not affected by taking powers, we assume that $m = 1$ for simplicity.

Given a homogeneous IFS $\mathcal{I} = \{rx + t_i\}_{i \in I}$, let $\Delta = \Delta_{\mathcal{I}} = \frac{1}{|I|}\sum_{i \in I} \delta_{t_i}$, where δ_t denotes a unit mass at t, and define the (natural) *self-similar measure*

$$\mu = \mu_{\mathcal{I}} = *_{n=0}^{\infty} S_{r^n} \Delta,$$

where $S_u = ux$ scales by u. In other words, μ is the push-forward of $\prod_{n=0}^{\infty} \Delta$ under $(x_n)_{n=0}^{\infty} \mapsto \sum_{n=0}^{\infty} x_n r^n$. Then μ is supported on the attractor X, and it easy to check that for $\alpha = \log|I|/\log(1/r)$,

$$\mu\big(B(x,r)\big) \ge cr^\alpha, \quad x \in X, r \in (0,1].$$

The parameter α is the *similarity dimension* of the IFS $\mathcal{I}$; if the pieces $(f_i(X))_{i \in I}$ are disjoint, then it equals $\dim_{\mathrm{H}}(X)$, but it is a well known open problem to understand when equality holds in the overlapping situation; see [26] and P. Varjú's survey in this volume for progress on this problem.

Fix closed T_a, T_b-invariant self-similar sets X, Y as above, and let μ_X, μ_Y be the corresponding self-similar measures, defined in terms of atomic measures Δ_X, Δ_Y. Let $\alpha = \dim_H(X)$, $\beta = \dim_H(Y)$. As we have seen,

$$(\mu_X \times \mu_Y)\big(B(p,r)\big) \geq c r^{\alpha+\beta}, \quad p \in X \times Y, r \in (0,1].$$

Recall that $\Pi_u(x,y) = x + uy$. Then $\Pi_u(\mu_X \times \mu_Y) = \mu_X * S_u \mu_Y$. By the above discussion and Lemma 3.1, in order to prove Conjecture 1.1, it is enough to show:

Theorem 3.2.

$$D(\mu_X * S_u \mu_Y, \infty) = \min(\alpha+\beta, 1) \quad \textit{for all } u \neq 0. \tag{3.1}$$

This recasts the slice conjecture into a problem concerning projections and self-similarity. This is convenient, since a lot was previously known about this topic. For example, (3.1) was known to hold for Hausdorff dimension in place of L^∞ dimension [29] and even for L^q dimension for $q \in (1,2]$ [38]. However, these results used in an essential way the known fact that for arbitrary measures μ, ν, equation (3.1) with $q \in (1,2]$ in place of ∞ holds for almost every u. This is not true for $q > 2$ and hence new ideas were needed. While the setting is different, the inspiration came from M. Hochman's work on self-similarity, see the survey [28] for an overview.

3.4. Dynamical self-similarity and exponential separation

While μ_X and μ_Y are self-similar measures in the sense described in Section 3.3, the convolution $\mu_X * S_u \mu_Y$ is not strictly self-similar since $a \not\sim b$. However, it satisfies a more flexible notion that we term *dynamical self-similarity*. Suppose $a < b$, and let us define $\mathbb{G} = [0, \log b)$, $\mathbf{T} : \mathbb{G} \to \mathbb{G}$, $x \mapsto x + \log a \bmod (\log b)$. For each $x \in \mathbb{G}$, let

$$\Delta(x) = \begin{cases} \Delta_X * S_{e^x} \Delta_Y & \text{if } x \in [0, \log a), \\ \Delta_X & \text{if } x \in [\log a, \log b). \end{cases} \tag{3.2}$$

These are finitely supported measures. It is easy to check (see [48, §1.4]) that

$$\nu_x := \mu_X * S_{e^x} \mu_Y = *_{n=0}^{\infty} S_{a^{-n}}\big(\Delta(\mathbf{T}^n x)\big).$$

This is what we mean by dynamical self-similarity: ν_x has a structure analogous to that of μ_X, μ_Y, but the discrete measure Δ now depends on the scale and is driven by the dynamics of $\mathbf{T}$. Note that

$$\nu_x = \nu_{x,n} * S_{a^{-n}} \nu_{\mathbf{T}^n x}, \quad \text{where } \nu_{x,n} = *_{j=0}^{n-1} \Delta(\mathbf{T}^j x). \tag{3.3}$$

This says that ν_x is a convex combination of scaled down copies, not quite of itself (as in the strictly self-similar case), but of the related measures $\nu_{\mathbf{T}^n x}$.

In the proof of Theorem 3.2, dynamical self-similarity plays a central rôle. Another key feature is *exponential separation*. The measures $\nu_{x,n}$ defined in (3.3) are purely atomic;

let $\mathcal{A}_x(n)$ denote the set of its atoms. Then

$$\left|\mathcal{A}_x(n)\right| \le \prod_{j=0}^{n-1} \left|\operatorname{supp}\left(\Delta(\mathbf{T}^j x)\right)\right|.$$

Let $M_x(n)$ denote the minimal separation between two elements of $\mathcal{A}_x(n)$, defined to be 0 if the inequality above is strict. We claim that there is a number $c > 0$ such that

$$M_x(n) \ge c^n \quad \text{for } n \ge n_0(x), \text{ for Lebesgue almost all } x \in \mathbb{G}. \tag{3.4}$$

Indeed, the distance between two elements of $\mathcal{A}_x(n)$ has the form $ia^{-n} + e^x jb^{-n}$ for some $|i| < a^n$, $|j| < b^n$, and i, j are not both 0. If $j = 0$, then $i \ne 0$ and the distance is $\ge a^{-n}$. Otherwise, $x \mapsto ia^{-n} + e^x jb^{-n}$ has derivative $\ge b^{-n}$ in absolute value, and so is $\ge c^n$ in absolute value outside of a set of x of measure $2(cb)^n$. Since there are $\lesssim (ab)^n$ pairs i, j, we see that $M_x(n) \ge c^n$ outside of a set of measure $\lesssim (c \cdot ab^2)^n$. Hence if $c < (ab^2)^{-1}$, then the Borel–Cantelli lemma yields (3.4).

Exponential separation was introduced in the self-similar setting by Hochman **[26]**. The way we apply it will be conceptually similar. However, in the strictly self-similar setting, this condition is often hard to check (or fails) for concrete examples, while as we have seen, in the dynamical setting the one-dimensional group $\mathbb{G}$ makes the verification straightforward.

A final ingredient of the proof of Theorem 3.2 is *unique ergodicity*: the dynamical system $(\mathbb{G}, \mathbf{T})$ is isomorphic to a $(\log a/\log b)$-rotation on the circle. Because $a \nsim b$, Lebesgue measure on the circle is the only $\mathbf{T}$-invariant measure on $\mathbb{G}$: this is the point in the proof where the hypothesis $a \nsim b$ gets used. As we will see, this will be crucial in obtaining information for *every* $x \in \mathbb{G}$ out of seemingly weaker information for *almost every* $x \in \mathbb{G}$.

In the rest of this section, we indicate how dynamical self-similarity, exponential separation and unique ergodicity enter into the proof of Theorem 3.2. The theorem extends to a more general setting in which appropriate versions of these three properties hold (plus some additional technical assumptions): see **[48, §1.5]**.

3.5. A subadditive cocycle and the rôle of unique ergodicity

Fix $q \in (1, \infty)$, recall that $\nu_x = \mu_X * S_{e^x}\mu_Y = \Pi_{e^x}(\mu_X \times \mu_Y)$, and let

$$\phi_{q,n}(x) = \log\Bigg(\sum_{I \in \mathcal{D}_{2^{-n}}} \nu_x(I)^q\Bigg), \quad x \in \mathbb{G},$$

where here and below logarithms are to base 2. In order to establish Theorem 3.2, it is enough to show that

$$\liminf_{n\to\infty} \frac{\phi_{q,n}(x)}{-(q-1)n} \ge \min(\alpha+\beta, 1), \quad \text{for all } x \in \mathbb{G}. \tag{3.5}$$

Indeed, it is rather easy to check that for any $x \in \mathbb{G}$

$$\limsup_{n\to\infty} \frac{\phi_{q,n}(x)}{-(q-1)n} \le \min(\alpha+\beta, 1),$$

and so (3.5) yields $D(\mu_X * S_u\mu_Y, q) = \max(\alpha+\beta, 1)$ (and the limit in the definition of L^q dimension exists), from where the claim follows by taking $q \to \infty$. A priori this is only true for $u = e^x \in [1, b)$, but using self-similarity it is not hard to extend it to every $u \ne 0$.

Dynamical self-similarity and the convexity of t^q imply (see [48, PROP. 4.6])

$$\phi_{q,n+m}(x) \leq C_q + \phi_{q,n}(x) + \phi_{q,m}(\mathbf{T}^n x).$$

Hence $(\phi_{q,n} + C_q)_n$ is a *subadditive cocycle* over the dynamical system $(\mathbb{G}, \mathbf{T})$. The functions $\phi_{q,n}$ are continuous on $\mathbb{G}$ except at $x = \log a$. The unique ergodicity of $(\mathbb{G}, \mathbf{T})$ can then be seen to imply ([48, §4.2]) that there is a number $D(q)$ such that

$$\liminf_{n\to\infty} \frac{\phi_{q,n}(x)}{-(q-1)n} = D(q) \quad \text{for } \textit{all } x \in \mathbb{G},$$
$$\lim_{n\to\infty} \frac{\phi_{q,n}(x)}{-(q-1)n} = D(q) \quad \text{for almost all } x \in \mathbb{G}.$$

Hence the task is now to show that $D(q) = \max(\alpha + \beta, 1)$. This is a really crucial point, because one only needs to compute the almost sure limit $D(q)$ in order to reach a conclusion valid for *every* x. This is also the strategy from [38] in the case $q \leq 2$; the almost sure statement follows in that case by classical projection results, while the more involved argument discussed below is required when $q > 2$. Because the L^q dimension is continuously decreasing in q, it is easy to check that $D = D_{\nu_x}$ (as a function) for almost all x; in particular, D is differentiable outside of a countable set.

3.6. An inverse theorem for the L^q norms of convolutions

So far, discretized additive combinatorics has not entered the picture. As indicated earlier, the proof of Theorem 3.2 is inspired by Hochman's work on self-similar sets and measures [26]. Hochman [26, THEOREM 2.7] proved an inverse theorem for the entropy of convolutions of general measures on $\mathbb{R}$, then applied it to self-similar measures, and concluded that under exponential separation they have the "expected" dimension; again we refer to [28] for a survey of these ideas. We follow a parallel strategy; in particular, we rely on a new inverse theorem for the L^q norms of convolutions.

If ν is finitely supported, we denote $\|\nu\|_q^q = \sum_x \nu(x)^q$ for $q \in (1, \infty)$. If μ, ν are supported on $2^{-m}\,\mathbb{Z} \cap [0, 1)$ then, by Young's inequality (which in this setting is just the convexity of t^q),

$$\|\mu * \nu\|_q \leq \|\mu\|_q \|\nu\|_1. \tag{3.6}$$

We are interested in understanding what happens when we are close to equality, in an exponential sense (up to $2^{-\varepsilon m}$ factors). This is the case if μ is the uniform measure on $2^{-m}\,\mathbb{Z} \cap [0, 1)$, or if ν is supported on a single atom, but also in some "fractal" situations. For example, if $\mu = \nu$ is the uniform measure on the (left endpoints of the intervals making up the) sets X_J from Section 2.2; it is also possible to construct similar examples with μ different from ν. Our inverse theorem asserts, roughly speaking, that if we are close to equality in (3.6), then *locally* either μ looks very uniform or ν looks like an atom.

Theorem 3.3 ([48, THEOREM 2.1]). *For each $q > 1$, $\delta > 0$, there are $T \in \mathbb{N}$, $\varepsilon > 0$ such that the following holds for $\ell \geq \ell_0(q, \delta)$. Let $m = \ell T$ and let $\mu, \nu \in \mathcal{P}(2^{-m}\,\mathbb{Z} \cap [0, 1))$. Suppose*

$$\|\mu * \nu\|_q \geq 2^{-\varepsilon m} \|\mu\|_q.$$

Then there exist sets $X \subset \operatorname{supp}\mu$ *and* $Y \subset \operatorname{supp}\nu$, *so that*

(i) $\|\mu|_X\|_q \geq 2^{-\delta m}\|\mu\|_q$ *and* $\|\nu|_Y\|_1 = \nu(Y) \geq 2^{-\delta m}$;

(ii) $\mu(x_1) \leq 2\mu(x_2)$ *for all* $x_1, x_2 \in X$; *and* $\nu(y_1) \leq 2\nu(y_2)$ *for all* $y_1, y_2 \in Y$;

(iii) X *and* Y *are* $(T;\ell)$*-uniform; let* $(N_j)_{j=0}^{\ell-1}$, $(N_j')_{j=0}^{\ell-1}$ *be the associated sequences;*

(iv) *For each* $0 \leq i < \ell$, *either* $N_j \geq 2^{(1-\delta)T}$ *or* $N_j' = 1$ *(or both).*

The reader will note the analogy with Theorem 2.3, especially in the case $\mu = \nu$. In fact, Theorem 2.3 is a central component of the proof of Theorem 3.3. In order to pass from the size of sumsets to the L^q norm of convolutions, we use the celebrated Balog–Szemerédi–Gowers (BSG) Theorem, see [54, §2.5]. Simplifying slightly, the BSG Theorem asserts that if $\|\mu * \mu\|_2 \geq K^{-1}\|\mu\|_2$ for $\mu \in \mathcal{P}(\mathbb{Z})$, then there is a set X such that $\mu(X) \geq K^{-C}$ and $|X + X| \leq K^C|X|$, where $C > 0$ is universal. To be more precise, this holds if μ is the uniform measure on some set X_0. In the case $\mu = \nu$ and $q = 2$, the claim is little more than the BSG Theorem combined with Theorem 2.3 and some dyadic pigeonholing. To deal with the general case, we appeal to an asymmetric version of BSG, [54, THEOREM 2.35], while the general case $q \in (1, \infty)$ can be reduced to the case $q = 2$ by an application of Hölder's inequality [48, LEMMA 3.4]. We remark that the theorem fails at $q = 1$ and $q = \infty$ due to lack of strict convexity; this is the reason why, even though we are ultimately interested in L^∞ dimensions, we work with L^q dimensions throughout the proof.

While motivated by the slice conjecture, Theorem 3.3 is a result in geometric measure theory. In [46], E. Rossi and the author applied it to the growth of L^q dimension under convolution. It also features in two recent results of T. Orponen [41,42] concerning projections of planar sets outside of a zero-dimensional set of directions.

3.7. Conclusion of the proof: sketch

We indicate very briefly how the proof of Theorem 3.2 (and hence of Conjecture 1.1) is concluded. Given a measure μ on $\mathbb{R}$, we let $\mu^{(m)}$ be the purely atomic measure with

$$\mu^{(m)}(j2^{-m}) = \mu([j2^{-m}, (j+1)2^{-m})).$$

Thus, $\mu^{(m)}$ is a discrete approximation to μ at scale 2^{-m}. Note that $\phi_{q,m}(x) = \log \|\nu_x^{(m)}\|_q^q$. The inverse theorem is used to show:

Theorem 3.4 ([48, THEOREM 5.1]). *Fix* $q \in (1, \infty)$ *such that* D *is differentiable at* q *and* $D(q) < 1$. *For every* $\sigma > 0$ *there is* $\varepsilon = \varepsilon(\sigma, q) > 0$ *such that if* $m \geq m_0(\sigma, q)$, *and* $\rho \in \mathcal{P}(2^{-m}\mathbb{Z} \cap [0,1))$ *satisfies* $\|\rho\|_q \leq 2^{-\sigma m}$, *then*

$$\left\|\nu_x^{(m)} * \rho\right\|_q \leq 2^{-(D(q)+\varepsilon)m}, \quad x \in \mathbb{G}.$$

The assumption $\|\rho\|_q \leq 2^{-\sigma m}$ says that ρ is not too close to being atomic in the L^q sense. Since $D(q) = D_{\nu_x}(q)$ for almost all x, the theorem says that convolving with any quantitatively nonatomic measure results in a smoothening of the L^q norm of ν_x at small scales (unless $D(q) = 1$, in which case ν_x was already "maximally smooth"). This is, again, a dynamical, L^q version of a result of Hochman, [26, COROLLARY 5.5]. Heuristically, this is deduced from Theorem 3.3 as follows: assuming the conclusion fails, let X, Y be the sets provided by the inverse theorem. Using that $\|\rho\|_q \le 2^{-\sigma m}$, one can see that Y has positive branching ($N'_j > 1$) for a positive proportion of scales j. Then by (iv), X must have "almost full branching" ($N_j \ge 2^{(1-\delta)T}$) at those scales. But the dynamical self-similarity of ν_x can be used to rule this out, since it implies that ν_x should have "roughly constant branching," which is less than full since $D(q) < 1$. Making this precise is one of the biggest hurdles in the proof of Theorem 3.2; it relies on ideas from multifractal analysis, in particular, the fact that if $D'(q)$ exists then $\|\mu_x^{(m)}\|_q$ is heavily concentrated on points of mass $\approx 2^{T'(q)m}$, where $T = (q-1)D$.

Once Theorem 3.4 is in hand, the rest of the proof of Theorem 3.2 is a fairly straightforward adaptation of Hochman's arguments. Theorem 3.4 is used to show that (always assuming $D'(q)$ exists and $D(q) < 1$)

$$\lim_{n\to\infty} \frac{\log \|\nu_{x,n}^{(Rn)}\|_q^q}{(q-1)n\log(1/a)} = D(q) \quad \text{for any } R > \log a \text{ and almost all } x \in \mathbb{G},$$

where $\nu_{x,n}$ is the discrete approximation to ν_x defined in (3.3). See [48, PROPOSITION 5.2]. Now the exponential separation (3.4) comes into play: if R is taken large enough in terms of c, then the atoms of $\nu_{x,n}$ are 2^{-Rn}-separated for $n \ge n_0(x)$, and this easily yields

$$\log\big\|\nu_{x,n}^{(Rn)}\big\|_q^q = \log \|\nu_{x,n}\|_q^q = (1-q)\sum_{j=0}^{n-1} \log\big|\operatorname{supp}\big(\Delta(\mathbf{T}^j x)\big)\big|.$$

Recalling (3.2), the ergodic theorem can then be used to conclude that if $D(q) < 1$, then $D(q) = \alpha + \beta$, completing the proof.

3.8. Extensions and open problems

3.8.1. Slices of McMullen carpets

The set $X \times Y$ in Conjecture 1.1 is invariant under the toral endomorphism $T_a \times T_b$, but there are many closed invariant sets under $T_a \times T_b$ which are not cartesian products. The simplest class are McMullen carpets: given $J \subset \{0,\dots,a-1\} \times \{0,\dots,b-1\}$, let

$$E_J = \left\{ \left(\sum_{n=1}^{\infty} x_n a^{-n}, \sum_{n=1}^{\infty} y_n b^{-n} \right) : (x_n, y_n) \in J \text{ for all } n \right\}.$$

If $J = J_1 \times J_2$ then we are in the setting of Conjecture 1.1, but otherwise the methods of [48, 57] do not directly apply. One new difficulty is that these carpets often have different Hausdorff, Minkowski, and Assouad dimension, while these all coincide in the product case. Nevertheless, by modifying the method of Wu, A. Algom [2] proved an upper bound for the dimension of linear slices of McMullen carpets, that reduces to Conjecture 1.1 in the product case. The bound was recently improved further by A. Algom and M. Wu [3], but the optimal result remains elusive.

3.8.2. Bernoulli convolutions

Given $\lambda \in (1/2, 1)$, we define the *Bernoulli convolution* (BC) $\nu_\lambda = *_{n=0}^{\infty} S_{\lambda^n}\Delta$, where $\Delta = \frac{\delta_{-1}+\delta_1}{2}$. This is the simplest family of overlapping self-similar measures, yet it remains a major open problem with deep connections to number theory to elucidate their properties. BCs are extensively discussed in [28] and in P. Varjú's article in this volume, so here we only point out that the method of proof discussed in this section also yields that $D(\nu_\lambda, \infty) = 1$ for all λ with exponential separation (a set of Hausdorff co-dimension zero) and ν_λ is absolutely continuous with a density in L^q for all $q \in (1, \infty)$, for all λ outside of a (nonexplicit) set of exceptions of zero Hausdorff dimension. See [48, SECTION 9]. In a major breakthrough, P. Varjú [55] proved that ν_λ has Hausdorff dimension 1 (which is weaker than $D(\nu_\lambda, q) = 1$ if $q > 1$) for *all* transcendental λ. It remains a challenge to extend Varjú's result to L^∞ and even to L^q dimensions.

3.8.3. Higher dimensions

A natural higher-dimensional version of Conjecture 1.1 involves slicing the product of closed sets $(X_i)_{i=1}^d$ invariant under $(T_{a_i})_{i=1}^d$, with affine subspaces. As another application of the dynamical self-similarity framework, we have:

Theorem 3.5. *Let X_i be closed, T_{a_i}-invariant sets, $i = 1, \dots, d$, with $a_i \nsim a_j$ for $i \neq j$. Then*

$$\dim_H\big((X_1 \times \cdots \times X_d) \cap H\big) \leq \max\big(\dim_H(X_1) + \cdots + \dim_H(X_d) - 1, 0\big)$$

for all affine hyperplanes $H \subset \mathbb{R}^d$ not containing a line in a coordinate direction.

The case $d = 2$ is Conjecture 1.1. The higher-dimensional case follows in a similar way, using [48, THEOREM 1.11] and Lemma 3.1 for projections from $\mathbb{R}^d$ to $\mathbb{R}$, although verifying the exponential separation assumption takes a little bit of work, see [52]. We underline that it seems hard to prove such a result using the approaches of [4, 57]. To be more precise, it is possible but under the more restrictive assumption that $(1/\log a_i)_{i=1}^d$ are linearly independent over $\mathbb{Q}$. This is unknown in most cases, for example, for 2, 3, 5.

What about slicing with lower dimension subspaces? For this, we need to consider projections $\Pi : \mathbb{R}^d \to \mathbb{R}^k$ and in turn this requires an inverse theorem for convolutions in $\mathbb{R}^k$. This is necessarily more challenging because there is a new obstruction to smoothening of convolutions: having the measures (locally) concentrated on lower-dimensional subspaces. Nevertheless, Hochman [27] proved an inverse theorem for the entropy of convolutions in arbitrary dimension. In [52], using Hochman's result, we derive an L^q version, and use it to deduce a generalization of Theorem 3.5 to slices with planes of arbitrary dimension.

4. FALCONER'S DISTANCE SET PROBLEM

4.1. Introduction

We now discuss Conjecture 1.2. It is a natural continuous analog of the P. Erdős distinct distances conjecture, stating that N points in $\mathbb{R}^d$ determine $\gtrsim_{d,\varepsilon} N^{2/d-\varepsilon}$ distinct distances. Erdős' conjecture was famously resolved in the plane by L. Guth and N. Katz [23], but the techniques they used seem hard to extend to the continuous setting. As shown already by Falconer [16], the measurability condition in Conjecture 1.2 is crucial.

From now on fix a Borel set $X \subset \mathbb{R}^d$. Falconer [16] proved that $|\Delta(X)| > 0$ provided $\dim_{\mathrm{H}}(X) > (d+1)/2$ (here and below, $|\cdot|$ denotes Lebesgue measure, as well as cardinality). In the plane, the threshold $3/2$ was lowered successively to $13/9$ by J. Bourgain [5], to $4/3$ by T. Wolff [56], and recently to $5/4$ by L. Guth, A. Iosevich, Y. Ou, and H. Wang [22]. There have been parallel developments in higher dimensions [11–13, 15]. These results use deep methods from restriction theory in harmonic analysis; the connection to restriction was made by P. Mattila [36], through what has become known as the *Mattila integral.* B. Liu [35] found a pinned version of the Mattila integral, that is, with $\Delta(X)$ replaced by $\Delta_y(X) = \{|x-y| : x \in X\}$. As a result, all the previous results are also valid for pinned distance sets. Summarizing, the current world records are [11–13, 22]: let

$$\alpha_d = \begin{cases} \frac{d}{2} + \frac{1}{4} & \text{if } d \text{ is even,} \\ \frac{d}{2} + \frac{1}{4} + \frac{1}{8d-4} & \text{if } d \text{ is odd.} \end{cases}$$

Then for a Borel set $X \subset \mathbb{R}^d$ with $\dim_{\mathrm{H}}(X) > \alpha_d$ there is $y \in X$ such that $|\Delta_y(X)| > 0$.

What if we assume $\dim_{\mathrm{H}}(X) = d/2$ instead? Falconer [16] proved that in this case $\dim_{\mathrm{H}}(\Delta(X)) \geq 1/2$. There are at least three reasons why this is a natural barrier to overcome: (i) If R was a $1/2$-dimensional Borel subring of the reals, then the distance set of $X = R \times \cdots \times R \subset \mathbb{R}^d$ would be contained in a locally Lipschitz image of R. By the product formula for dimension, $\dim_{\mathrm{H}}(X) \geq d/2$, so if R existed then Falconer's bound would be sharp. As it turns out, no such Borel subring exists [14], but this was an open problem for nearly 40 years. (ii) For a natural single-scale version of the problem, the exponent $1/2$ is actually sharp. This is the "train track" example introduced by N. Katz and T. Tao [30]: given a small scale $\delta > 0$, let $X \subset [0,1]^2$ be the union of $\sim \delta^{-1/2}$ equally spaced vertical rectangles of size $\delta \times \delta^{1/2}$, with a $\delta^{1/2}$ space between consecutive rectangles. See [30, FIGURE 1]. Then $|X|_\delta \sim \delta^{-1}$ and

$$\left|X \cap B(x,r)\right|_\delta \sim r|X|_\delta, \quad x \in X, r \in [\delta, 1].$$

Hence X looks very much like a set of dimension 1 (even Ahlfors regular) down to resolution δ. Yet, the set of distances between two separated rectangles is contained in an interval of length $\lesssim \delta$, and this can be used to show that $|\Delta(X)|_\delta \sim \delta^{-1/2}$. (iii) Finally, if the Euclidean norm is replaced by the ℓ_∞ norm, then again it is not hard to see that the threshold $1/2$ is sharp, so any improvement must exploit the curvature of the Euclidean norm. We also emphasize that even though the harmonic analytic methods described above also yield dimension estimates when $\dim_{\mathrm{H}}(X) \leq \alpha_d$, they do not say anything for $\dim_{\mathrm{H}}(X) = d/2$.

Despite these challenges, we have:

Theorem 4.1 (Katz–Tao [30], Bourgain [6]). *There is a universal $\eta > 0$ such that if $X \subset \mathbb{R}^2$ is a Borel set with $\dim_{\mathrm{H}}(X) \geq 1$, then $\dim_{\mathrm{H}}(\Delta(X)) \geq 1/2 + \eta$.*

Katz and Tao [30] proved that the discretized sum-product conjecture (Theorem 2.4) implies the above theorem. As we saw, Bourgain [6] then proved Theorem 2.4. In order to avoid "train track" examples, Katz and Tao had as an intermediate step a "discretized bilinear" version of Falconer's problem. This approach does not seem to extend to pinned distance sets. The value of η, although effective in principle, is hard to track down and would in any event be tiny (recall that the conjecture is $\eta = 1/2$).

4.2. A nonlinear version of Bourgain's projection theorem

There is a formal analogy between Theorems 2.6 and 4.1: both provide an "η-impovement" over a natural barrier, and as we saw they are both connected to discretized sum-product. We take this analogy further. We can view $\{\Delta_y(x) = |x-y|\}_{y\in X}$ as a family of (nonlinear) projections. One can then ask if it satisfies an estimate similar to that of Theorem 2.5. It turns out that it does:

Theorem 4.2 ([50, THEOREM 1.1]). *Given $\alpha \in (0,2)$, $\beta > 0$, there is $\eta = \eta(\alpha,\beta) > 0$ such that the following holds: let $X \subset \mathbb{R}^2$ be a Borel set with $\dim_{\mathrm{H}}(X) \geq \alpha$. Then*

$$\dim_{\mathrm{H}}(\mathcal{E}(X,\eta)) \leq \beta, \quad \text{where } \mathcal{E}(X,\eta) = \left\{ y \in \mathbb{R}^2 : \dim_{\mathrm{H}}(\Delta_y(X)) < \frac{\alpha}{2} + \eta \right\}. \tag{4.1}$$

In particular, taking $\beta < 1 = \alpha$, this provides a pinned version of Theorem 4.1.

Theorem 4.2 follows from a general scheme that can be seen as a nonlinear extension and refinement of Bourgain's projection theorem and its higher rank generalization by W. He. See [50] for further discussion and precise statements. This scheme yields Theorem 4.2 also for smooth norms of everywhere positive Gaussian curvature and ℓ^p norms for $p \in (1,\infty)$, as well as some partial extensions to higher dimensions; see [50, THEOREM 1.1].

Using the nonlinear adaptation of Bourgain's projection theorem (along with many other ideas), O. Raz and J. Zahl [45] have recently obtained a further refinement of Theorem 4.2. They show that for every $\alpha \in (0,2)$ there is $\eta = \eta(\alpha) > 0$ such that the set $\mathcal{E}(X,\eta)$ from (4.1) is *flat*, which roughly means that it is contained in a union of a small set of lines, see [45, DEFINITION 1.4]. This is optimal since they also observe that Theorem 4.2 is sharp in the sense that $\eta \to 0$ as $\beta \to 0$, but the sets that witness this are contained in a line (or a union of a small family of lines). Raz and Zahl also obtain a related single-scale distance set estimate involving only three noncollinear vantage points:

Theorem 4.3 ([45, THEOREM 1.9]). *Given $\alpha \in (0,2)$, there is $\eta = \eta(\alpha) > 0$ such that if $E \subset [0,1]^2$ satisfies the nonconcentration estimate*

$$\left|E \cap B(p,r)\right|_\delta \leq \delta^{-\eta} r^{\alpha} |E|_\delta, \quad p \in [0,1]^2, r \in [\delta,1],$$

and $y_1, y_2, y_3 \in [0,1]^2$ span a triangle of area $\geq \delta^\eta$, then $\max_{i=1}^3 |\Delta_{y_i} X|_\delta \geq \delta^{-\alpha/2-\eta}$.

Note that the quantitative noncollinearity hypothesis prevents the train-track almost counterexamples discussed above. These are just special cases of general theorems involving nonlinear projections and Blaschke curvature, see [45] for further details.

4.3. Explicit estimates and sets of equal Hausdorff and packing dimension

The improvements upon the natural threshold $1/2$ that we have seen so far all involve a tiny and unknown parameter η. The following were the first explicit bounds in the near critical regime:

Theorem 4.4 (T. Keleti and P. Shmerkin, [33]). *Let $E \subset \mathbb{R}^2$ be a Borel set with $\dim_{\mathrm{H}}(E) > 1$. Then $\dim_{\mathrm{H}}(\Delta(E)) > 37/54 \approx 0.685$, and there is $y \in E$ such that $\dim_{\mathrm{H}}(\Delta_y(E)) > 2/3$ and $\dim_{\mathrm{P}}(\Delta_y(E)) = (2+\sqrt{3})/4 \approx 0.933$.*

Here $\dim_{\mathrm{P}}$ is packing dimension; we refer to [37, §5.9–5.10] for its definition and basic properties, and recall only that it lies between Hausdorff and Minkowski (box) dimensions. See [35, 51] for some further improvements, always assuming $\dim_{\mathrm{H}}(E) > 1$.

The first explicit estimates in the critical case $\dim_{\mathrm{H}}(E) = d/2$ were obtained only very recently by the author and H. Wang [53, THEOREMS 1.1 AND 1.2]:

Theorem 4.5. *Let $E \subset \mathbb{R}^d$ be a Borel set with $\dim_{\mathrm{H}}(E) = d/2$ where $d = 2$ or 3. Then $\sup_{y\in E} \dim_{\mathrm{H}}(\Delta_y E) \geq \alpha_d$, where $\alpha_2 = (\sqrt{5}-1)/2 \approx 0.618$ and $\alpha_3 > 0.57$.*

While these are the best currently known estimates for general Borel sets, for sets of equal Hausdorff and packing dimension we are able to prove the full strength of Falconer's conjecture [53, THEOREM 1.4]:

Theorem 4.6. *Let $E \subset \mathbb{R}^d$, $d \geq 2$, be a Borel set with $\dim_{\mathrm{H}}(E) = \dim_{\mathrm{P}}(E) = d/2$. Then $\sup_{y\in E} \dim_{\mathrm{H}}(\Delta_y E) = 1$, and if E has positive $d/2$-dimensional Hausdorff measure then the supremum is attained.*

If $\dim_{\mathrm{H}}(E) = \dim_{\mathrm{P}}(E) = \alpha$, then for each $\varepsilon > 0$ there is $\mu \in \mathcal{P}(E)$ such that

$$r^{\alpha+\varepsilon} \lesssim_\varepsilon \mu\big(B(x,r)\big) \lesssim_\varepsilon r^{\alpha-\varepsilon}, \quad r \in (0,1], x \in \operatorname{supp}(\mu).$$

Thus we can interpret this condition as a rough or approximate version of Ahlfors regularity (which corresponds to the case $\varepsilon = 0$).

In the plane, Theorem 4.6 has several predecessors. In an influential article, Orponen [39] proved that if E is Ahlfors regular of dimension 1, then the *packing* dimension of $\Delta(E)$ is 1. In [49], assuming that $\dim_{\mathrm{H}}(E) > 1$, we showed that there is $y \in E$ with $\dim_{\mathrm{H}}(\Delta_y E) = 1$; this result was recovered and made more quantitative in [33]. Extending the proof to the critical case $\dim_{\mathrm{H}}(E) = 1$ and to higher dimensions required new ideas; we sketch some of them in Section 4.6.

4.4. A multiscale formula for the entropy of projections

A common theme through the proofs of Theorems 4.2, 4.4, 4.5, and 4.6 is the use of a lower bound for the entropy of projections in terms of multiscale decompositions. Recall that the Shannon entropy of $\mu \in \mathcal{P}(\mathbb{R}^d)$ with respect to a partition $\mathcal{A}$ of $\mathbb{R}^d$ is

$$H(\mu; \mathcal{A}) = \sum_{A\in\mathcal{A}} \mu(A) \log\big(1/\mu(A)\big).$$

This quantity measures how uniform the measure μ is among the atoms $A \in \mathcal{A}$. A basic property is that $H(\mu; \mathcal{A}) \leq \log |\mathcal{A}|$, and in particular

$$H_\delta(\mu) := H(\mu; \mathcal{D}_\delta) \leq \log |\operatorname{supp} \mu|_\delta. \tag{4.2}$$

Given a measure μ and a set X with $\mu(X) > 0$, we denote $\mu_X = \frac{1}{\mu(X)} \mu|_X \in \mathcal{P}(X)$. Finally, fix a C^2 map $F : U \supset [0,1]^d \to \mathbb{R}$ with no singular points, and let $\theta(x) = \nabla F(x)/|\nabla F(x)|$.

Proposition 4.7 ([50, PROPOSITION A.1]). *Let $\mu \in \mathcal{P}([0,1)^d)$, let $1 > \delta_0 > \delta_1 > \cdots > \delta_J = \delta$ be a sequence with $\delta_j^2 \leq \delta_{j+1}$, $0 \leq j < J$. Let F be as above. Then, denoting orthogonal projection in direction θ by $P_\theta(x) = \langle x, \theta \rangle$,*

$$H_\delta(F\mu) \geq -C_{F,d} J + \int \sum_{j=1}^{J} H_{\delta_{j+1}}(P_{\theta(x)} \mu_{\mathcal{D}_{\delta_j}(x)}) \, d\mu(x). \tag{4.3}$$

A *local* variant of this formula is a key element in the proof of Furstenberg's sumset conjecture in [29]. Orponen [39] introduced this approach to the distance set problem. The method was further refined in [33, 49]—these papers highlighted the importance of choosing the scales δ_j depending on the combinatorics of the measure μ, a point to which we will come back shortly. Thanks to (4.2), the formula (4.3) provides a lower bound on box-counting numbers $|F(\operatorname{supp} \mu)|_\delta$. In order to obtain Hausdorff dimension estimates, one needs a more robust (and technical) variant; we refer to [50, APPENDIX A] for details and here we stick with (4.3) for simplicity.

In all our applications the number of scales J is bounded while $\delta \to 0$, and so the error term is negligible. Note that if $Q \in \mathcal{D}_{\delta_j}$, then $P_\theta Q$ is an interval of length $\lesssim_d \delta_j$, and hence $H_{\delta_{j+1}}(P_\theta \mu_Q) \leq \log(\delta_j/\delta_{j+1}) + C_d$.

Why is Proposition 4.7 useful? A key feature is that it linearizes the nonlinear projection F; the hypothesis $\delta_j^2 \leq \delta_{j+1}$ comes from linearization, and can be dropped if F is linear. Another advantage is that it replaces the single projection $F\mu$ by an average of projections, taken over x and, crucially, over the scales (δ_j).

4.5. Theorems for radial and linear projections, and choice of scales

We sketch how Proposition 4.7 is used to prove the bound $\dim_H(\Delta^y E) > 2/3$ from Theorem 4.4. By Frostman's Lemma [37, THEOREM 8.8], there are $\mu, \nu \in \mathcal{P}(E)$ with

$$\mu(B(x,r)), \nu(B(y,r)) \lesssim r^\alpha, \quad x, y \in \mathbb{R}^2, r > 0,$$

where $\alpha > 1$, and $X = \operatorname{supp}(\mu)$, $Y = \operatorname{supp}(\nu)$ are disjoint. We apply Proposition 4.7 to the family $(\Delta_y)_{y \in Y}$ and μ. Since $\nabla \Delta_y(x) = \pi_x(y) := |y - x|/(y - x)$, Equation (4.3) becomes

$$H_\delta(\Delta_y \mu) \geq -CJ + \int \sum_{j=1}^{J} H_{\delta_{j+1}}(P_{\pi_x(y)} \mu_{\mathcal{D}_{\delta_j}(x)}) \, d\mu(x). \tag{4.4}$$

The scales δ_j will eventually be chosen in such a way that $\delta_{j+1} \leq \delta^c \delta_j$ for some small constant c; in particular, this ensures that $J \leq \lceil c^{-1} \rceil$ is bounded as $\delta \to 0$.

A radial projection theorem of Orponen **[40]** yields $\int \|\pi_x \nu\|_p^p \, d\mu(x) < \infty$ for some $p = p(\alpha) > 1$; it is here that the hypothesis $\dim_{\mathrm{H}}(E) > 1$ gets used. Restricting μ, we may thus assume that $\|\pi_x \nu\|_p \lesssim 1$ for all $x \in X$. Hölder's inequality and a quantitative form of Marstrand's projection theorem **[37, THEOREM 9.7]** yield that, for any $x \in X$ and $\rho \in \mathcal{P}(\mathbb{R}^2)$,

$$\pi_x \nu\{\theta \in S^1 : \|P_\theta \rho\|_2^2 \geq \delta^{-\varepsilon} I_1(\rho)\} \leq 2^{c_p \varepsilon}, \tag{4.5}$$

where $I_1(\rho) = \int |x-y|^{-1} d\rho(x) d\rho(y)$ is the 1-energy of ρ. Once the scales δ_j are fixed, we write $\mu_{x,j}$ for the convolution of $\mu_{\mathcal{D}_{\delta_j}(x)}$ with a bump function at scale δ_{j+1}, scaled up by a factor δ_j^{-1}. Applying (4.5) to $\rho = \mu_{x,j}$ for fixed x and $1 \leq j \leq J$, using that J is bounded, and then Fubini, we eventually obtain a point $y \in Y$ and a set $X' \subset X$ with $\mu(X') \geq 1/2$, such that

$$\|P_{\pi_x(y)} \mu_{x,j}\|_2^2 \leq \delta^{-\varepsilon} I_1(\mu_{x,j}), \quad x \in X', 1 \leq j \leq J.$$

Jensen's inequality can be used to bound

$$H_{\delta_{j+1}}(P_{\pi_x(y)} \mu_{x,j}) \geq \log(\delta_j/\delta_{j+1}) - \log \|P_{\pi_x(y)} \mu_{x,j}\|_2^2 - C.$$

Putting everything together, (4.4) becomes (denoting a negligible error term by err)

$$H_\delta(\Delta_y \mu) \geq \log(1/\delta) - \sum_{j=1}^{J} \int_{X'} \log\big(I_1(\mu_{x,j})\big) \, d\mu(x) - \mathrm{err}. \tag{4.6}$$

Now the task has become to choose the scales δ_j (depending on μ!) subject to the constrains $\delta_j^{1/2} \leq \delta_{j+1} \leq \delta^c \delta_j$, in such a way that $\log(I_1(\mu_{x,j}))$ is minimized on average. This is a combinatorial problem that becomes more tractable by first uniformizing μ by applying (the measure version of) Lemma 2.2 and using the branching numbers N_j as the combinatorial input. The issue to deal with is that, even though μ is α-dimensional, many of the measures $\mu_{x,j}$ can be nearly atomic (if $N_j \approx 1$), which causes the 1-energy to explode, so one seeks to merge the scales at which this happens with coarser scales at which μ looks like a large dimensional set. The value $2/3$ is the outcome of this combinatorial problem. Note that if μ is (roughly) Ahlfors regular, then so are the measures $\mu_{x,j}$, and then $\log(I_1(\mu_{x,j}))$ is uniformly small—so (4.6) also yields $\dim_{\mathrm{H}}(\Delta^y E) = 1$ in this case.

We underline that even though linearization is at the core of this approach, it is still crucial that the distance map is *nonlinear*, as this is what generates a rich set of directions $\pi_x(y) = \frac{d}{dx} \Delta_y(x)$ to work with—curvature is still key!

The proofs of Theorems 4.1, 4.5, and 4.6 follow a similar approach, but they each involve different radial and linear projection theorems. For example, Theorem 4.1 relies (unsurprisingly) on Theorem 2.5 and a *different* radial projection bound of Orponen **[40]**. One feature of Theorems 4.5 and 4.6 is that they depend on *new* radial and linear projection theorems; we briefly describe them in the next section, in the planar case.

4.6. Improving Kaufman's projection theorem, and radial projections

Let $E \subset \mathbb{R}^2$ be as in Theorems 4.5 or 4.6. Fix $\varepsilon > 0$ and, as before, let μ, ν be Frostman measures on E with exponents $1 - \varepsilon$, and disjoint supports X, Y. If either μ or ν

gives positive mass to a line, then E intersects that line in dimension $\geq 1-\varepsilon$, which makes the distance set estimate immediate. So we may assume that μ, ν give zero mass to all lines.

As our discussion in Section 4.5 suggests, it is key to understand radial projections $(\pi_y)_{y\in Y}$ first, and for this we use (again) Proposition 4.7. Note that $\frac{d}{dx}\pi_y(x) = \pi_y(x)^{\perp}$: this means that in order to estimate radial projections in this way, we need to rely on *a priori* radial projection bounds. This opens the door to bootstrapping arguments, and this is exactly what is done to prove Theorems 4.5 and 4.6.

To start the bootstrapping, we need an a priori radial projection estimate for measures of dimension ≤ 1 that give zero mass to lines (note that if both measures are supported on the same line, the radial projections $\pi_x\nu$ are atomic for all $x \in X$; this is why we excluded this case at the beginning of the argument). This is provided by a result of Orponen from **[40]**, that we alluded to earlier in connection with Theorem 4.1. In our setting it asserts that for a set of x of μ-measure $\geq 1/2$, the radial projection $\pi_x\nu$ satisfies a Frostman condition of exponent $1/2-\varepsilon$ (more precisely this holds after restricting ν further, depending on x).

The goal is to apply Proposition 4.7 to $(\pi_y)_{y\in Y}$ to bootstrap the parameter $1/2-\varepsilon$ to 1 and to $(\sqrt{5}-1)/2$ in the Ahlfors regular and general case, respectively. Following the scheme of Section 4.5, we end up needing a certain linear projection theorem, that we discuss next. A classical projection theorem of R. Kaufman **[32]** from 1968 asserts that if $X \subset \mathbb{R}^2$ is a Borel set and $s < \min(1, \dim_{\mathrm{H}}(X))$, then

$$\dim_{\mathrm{H}}\{\theta \in S^1 : \dim_{\mathrm{H}}(P_\theta X) \leq s\} \leq s.$$

It is natural to conjecture that Kaufman's theorem is not optimal, in that the bound s on the right-hand side can be lowered, depending on s and $\dim_{\mathrm{H}}(X)$. When $s \leq \dim_{\mathrm{H}}(X)/2 + \eta(\dim_{\mathrm{H}}(X))$, such improvement follows from Theorem 2.6, but the general case was established only very recently by T. Orponen and the author:

Theorem 4.8 (**[43, THEOREM 1.2]**). *Given $s \in (0,1)$, $t \in (s,2]$, there is $\varepsilon = \varepsilon(s,t) > 0$ such that if $X \subset \mathbb{R}^2$ is a Borel set with $\dim_{\mathrm{H}}(X) \geq t$, then*

$$\dim_{\mathrm{H}}\{\theta \in S^1 : \dim_{\mathrm{H}}(P_\theta X) \leq s\} \leq s-\varepsilon.$$

The proof uses many of the ingredients we have discussed in this survey: Bourgain's projection theorem, the uniformization lemma, and choosing the scales depending on the given measure. There are also new ideas, including an "incidence version" of Proposition 4.7 and a dichotomy between the "roughly Ahlfors regular" and "far from Ahlfors regular" situations, each requiring different arguments.

A quantitative version of Theorem 4.8 (see **[43, THEOREM 1.3]**) provides the input necessary to complete the bootstrapping step in the proofs of the planar cases of Theorems 4.5 and 4.6. To be more precise, so far we have been considering radial projections, but because $\frac{d}{dx}\pi_y$ and $\frac{d}{dx}\Delta_y$ are rotations of each other, the argument for distance sets can be completed in parallel. The golden mean $(\sqrt{5}-1)/2$ arises as the outcome of the combinatorial problem of optimizing the choice of scales (after uniformization).

ACKNOWLEDGMENTS

Thanks to Péter Varjú, Hong Wang, and Josh Zahl for comments and corrections on earlier versions of the manuscript.

FUNDING

This work was partially supported by an NSERC discovery grant and by Project PICT 2015-3675 (ANPCyT).

REFERENCES

[1] B. Adamczewski and C. Faverjon, Mahler's method in several variables and finite automata. 2020, arXiv:2012.08283.

[2] A. Algom, Slicing theorems and rigidity phenomena for self-affine carpets. *Proc. Lond. Math. Soc. (3)* **121** (2020), no. 2, 312–353.

[3] A. Algom and M. Wu, Improved versions of some Furstenberg type slicing theorems for self-affine carpets. 2021, arXiv:2107.02068.

[4] T. Austin, A new dynamical proof of the Shmerkin–Wu theorem. *J. Mod. Dyn.* **18** (2022), 1–11.

[5] J. Bourgain, Hausdorff dimension and distance sets. *Israel J. Math.* **87** (1994), no. 1–3, 193–201.

[6] J. Bourgain, On the Erdős-Volkmann and Katz-Tao ring conjectures. *Geom. Funct. Anal.* **13** (2003), no. 2, 334–365.

[7] J. Bourgain, The discretized sum-product and projection theorems. *J. Anal. Math.* **112** (2010), 193–236.

[8] J. Bourgain, A. Furman, E. Lindenstrauss, and S. Mozes, Stationary measures and equidistribution for orbits of nonabelian semigroups on the torus. *J. Amer. Math. Soc.* **24** (2011), no. 1, 231–280.

[9] J. Bourgain and A. Gamburd, On the spectral gap for finitely-generated subgroups of SU(2). *Invent. Math.* **171** (2008), no. 1, 83–121.

[10] S. A. Burrell and H. Yu, Digit expansions of numbers in different bases. *J. Number Theory* **226** (2021), 284–306.

[11] X. Du, L. Guth, Y. Ou, H. Wang, B. Wilson, and R. Zhang, Weighted restriction estimates and application to Falconer distance set problem. *Amer. J. Math.* **143** (2021), no. 1, 175–211.

[12] X. Du, A. Iosevich, Y. Ou, H. Wang, and R. Zhang, An improved result for Falconer's distance set problem in even dimensions. *Math. Ann.* **380** (2021), no. 3–4, 1215–1231.

[13] X. Du and R. Zhang, Sharp L^2 estimates of the Schrödinger maximal function in higher dimensions. *Ann. of Math. (2)* **189** (2019), no. 3, 837–861.

[14] G. A. Edgar and C. Miller, Borel subrings of the reals. *Proc. Amer. Math. Soc.* **131** (2003), no. 4, 1121–1129.

[15] M. B. Erdoğan, A bilinear Fourier extension theorem and applications to the distance set problem. *Int. Math. Res. Not.* **23** (2005), 1411–1425.

[16] K. J. Falconer, On the Hausdorff dimensions of distance sets. *Mathematika* **32** (1985), no. 2, 206–212.

[17] H. Furstenberg, Disjointness in ergodic theory, minimal sets, and a problem in Diophantine approximation. *Math. Syst. Theory* **1** (1967), 1–49.

[18] H. Furstenberg, Intersections of Cantor sets and transversality of semigroups. In *Problems in analysis (Sympos. Salomon Bochner, Princeton Univ., Princeton, N.J., 1969)*, pp. 41–59, Princeton University Press, 1970.

[19] H. Furstenberg, Ergodic fractal measures and dimension conservation. *Ergodic Theory Dynam. Systems* **28** (2008), no. 2, 405–422.

[20] D. Glasscock, J. Moreira, and F. Richter, Additive and geometric transversality of fractal sets in the integers. 2021, arXiv:2007.05480v2.

[21] D. Glasscock, J. Moreira, and F. Richter, A combinatorial proof of a sumset conjecture of Furstenberg. 2021, arXiv:2107.10605.

[22] L. Guth, A. Iosevich, Y. Ou, and H. Wang, On Falconer's distance set problem in the plane. *Invent. Math.* **219** (2020), no. 3, 779–830.

[23] L. Guth and N. H. Katz, On the Erdős distinct distances problem in the plane. *Ann. of Math. (2)* **181** (2015), no. 1, 155–190.

[24] L. Guth, N. H. Katz, and J. Zahl, On the discretized sum-product problem. *Int. Math. Res. Not. IMRN* **13** (2021), 9769–9785.

[25] W. He, Orthogonal projections of discretized sets. *J. Fractal Geom.* **7** (2020), no. 3, 271–317.

[26] M. Hochman, On self-similar sets with overlaps and inverse theorems for entropy. *Ann. of Math. (2)* **180** (2014), no. 2, 773–822.

[27] M. Hochman, On self-similar sets with overlaps and inverse theorems for entropy in $\mathbb{R}^d$. *Mem. Amer. Math. Soc.* (in press).

[28] M. Hochman, Dimension theory of self-similar sets and measures. In *Proceedings of the International Congress of Mathematicians—Rio de Janeiro 2018. Vol. III. Invited lectures*, pp. 1949–1972, World Sci. Publ., Hackensack, NJ, 2018.

[29] M. Hochman and P. Shmerkin, Local entropy averages and projections of fractal measures. *Ann. of Math. (2)* **175** (2012), no. 3, 1001–1059.

[30] N. H. Katz and T. Tao, Some connections between Falconer's distance set conjecture and sets of Furstenburg type. *New York J. Math.* **7** (2001), 149–187.

[31] N. H. Katz and J. Zahl, An improved bound on the Hausdorff dimension of Besicovitch sets in $\mathbb{R}^3$. *J. Amer. Math. Soc.* **32** (2019), no. 1, 195–259.

[32] R. Kaufman, On Hausdorff dimension of projections. *Mathematika* **15** (1968), 153–155.

[33] T. Keleti and P. Shmerkin, New bounds on the dimensions of planar distance sets. *Geom. Funct. Anal.* **29** (2019), no. 6, 1886–1948.

[34] K.-S. Lau and S.-M. Ngai, Multifractal measures and a weak separation condition. *Adv. Math.* **141** (1999), no. 1, 45–96.

[35] B. Liu, An L^2-identity and pinned distance problem. *Geom. Funct. Anal.* **29** (2019), no. 1, 283–294.

[36] P. Mattila, Spherical averages of Fourier transforms of measures with finite energy; dimension of intersections and distance sets. *Mathematika* **34** (1987), no. 2, 207–228.

[37] P. Mattila, *Geometry of sets and measures in Euclidean spaces*. Cambridge Stud. Adv. Math. 44, Cambridge University Press, Cambridge, 1995.

[38] F. Nazarov, Y. Peres, and P. Shmerkin, Convolutions of Cantor measures without resonance. *Israel J. Math.* **187** (2012), 93–116.

[39] T. Orponen, On the distance sets of Ahlfors–David regular sets. *Adv. Math.* **307** (2017), 1029–1045.

[40] T. Orponen, On the dimension and smoothness of radial projections. *Anal. PDE* **12** (2019), no. 5, 1273–1294.

[41] T. Orponen, On arithmetic sums of Ahlfors-regular sets. *Geom. Funct. Anal.* (in press).

[42] T. Orponen, On the Assouad dimension of projections. *Proc. Lond. Math. Soc. (3)* **122** (2021), no. 2, 317–351.

[43] T. Orponen and P. Shmerkin, On the Hausdorff dimension of Furstenberg sets and orthogonal projections in the plane. 2021, arXiv:2106.03338.

[44] Y. Peres and P. Shmerkin, Resonance between Cantor sets. *Ergodic Theory Dynam. Systems* **29** (2009), no. 1, 201–221.

[45] O. E. Raz and J. Zahl, On the dimension of exceptional parameters for nonlinear projections, and the discretized Elekes-Rónyai theorem. 2021, arXiv:2108.07311.

[46] E. Rossi and P. Shmerkin, On measures that improve L^q dimension under convolution. *Rev. Mat. Iberoam.* **36** (2020), no. 7, 2217–2236.

[47] M. Rudnev and S. Stevens, An update on the sum-product problem. 2020, arXiv:2005.11145.

[48] P. Shmerkin, On Furstenberg's intersection conjecture, self-similar measures, and the L^q norms of convolutions. *Ann. of Math. (2)* **189** (2019), no. 2, 319–391.

[49] P. Shmerkin, On the Hausdorff dimension of pinned distance sets. *Israel J. Math.* **230** (2019), no. 2, 949–972.

[50] P. Shmerkin, A nonlinear version of Bourgain's projection theorem. *J. Eur. Math. Soc. (JEMS)* (in press).

[51] P. Shmerkin, Improved bounds for the dimensions of planar distance sets. *J. Fractal Geom.* **8** (2021), no. 1, 27–51.

[52] P. Shmerkin, An inverse theorem and dynamical self-similarity in $\mathbb{R}^d$, 2021, work in progress.

[53] P. Shmerkin and H. Wang, On the distance sets spanned by sets of dimension $d/2$ in $\mathbb{R}^d$, 2021, arXiv:2112.09044.

[54] T. Tao and V. Vu, *Additive combinatorics*. Cambridge Stud. Adv. Math. 105, Cambridge University Press, Cambridge, 2006.

[55] P. P. Varjú, On the dimension of Bernoulli convolutions for all transcendental parameters. *Ann. of Math. (2)* **189** (2019), no. 3, 1001–1011.

[56] T. Wolff, Decay of circular means of Fourier transforms of measures. *Int. Math. Res. Not.* **10** (1999), 547–567.

[57] M. Wu, A proof of Furstenberg's conjecture on the intersections of $\times p$- and $\times q$-invariant sets. *Ann. of Math. (2)* **189** (2019), no. 3, 707–751.

[58] H. Yu, An improvement on Furstenberg's intersection problem. *Trans. Amer. Math. Soc.* **374** (2021), no. 9, 6583–6610.

PABLO SHMERKIN

Department of Mathematics, The University of British Columbia, 1984 Mathematics Road, Vancouver, BC, V6T 1Z2, Canada, pshmerkin@math.ubc.ca

This Page Intentionally Left Blank

QUANTITATIVE INVERTIBILITY OF NON-HERMITIAN RANDOM MATRICES

KONSTANTIN TIKHOMIROV

Dedicated to Prof. Nicole Tomczak-Jaegermann

ABSTRACT

The problem of estimating the smallest singular value of random square matrices is important in connection with matrix computations and analysis of the spectral distribution. In this survey, we consider recent developments in the study of quantitative invertibility in the non-Hermitian setting, and review some applications of this line of research.

MATHEMATICS SUBJECT CLASSIFICATION 2020

Primary 60B20; Secondary 65F05, 15A18

KEYWORDS

Random matrices, condition number, spectrum

INTERNATIONAL CONGRESS OF MATHEMATICIANS 2022 JULY 6–14

Proc. Int. Cong. Math. 2022, Vol. 4, pp. 3292–3313
DOI 10.4171/ICM2022/10

Published by EMS Press

1. INTRODUCTION

Given an $N \times n$ $(N \geq n)$ matrix A, its singular values are defined as square roots of the eigenvalues of the positive semidefinite $n \times n$ matrix A^*A:

$$s_i(A) := \sqrt{\lambda_i(A^*A)}, \quad i = 1, 2, \dots, n,$$

where we assume the nonincreasing ordering $\lambda_1(A^*A) \geq \lambda_2(A^*A) \geq \dots \geq \lambda_n(A^*A)$. The classical Courant–Fischer–Weyl theorem provides a variational formula

$$s_i(A) = \min_{E:\dim(E)=n-i+1} \max_{x \in E,\, \|x\|_2=1} \|Ax\|_2, \quad 1 \leq i \leq n,$$

where the minimum taken over all linear subspaces E of the specified dimension. In particular, *the smallest* and *the largest* singular values of A can be computed as

$$s_{\min}(A) = s_n(A) = \min_{x:\|x\|_2=1} \|Ax\|_2, \quad s_{\max}(A) = s_1(A) = \max_{x:\|x\|_2=1} \|Ax\|_2.$$

Additionally, if the matrix A is square $(N = n)$ and invertible then $s_{\min}(A) = \frac{1}{s_{\max}(A^{-1})}$.

The magnitude of the smallest singular value of square random matrices has attracted much attention due to the special role it plays in several questions of theoretical significance and in applications. In particular, the ratio of the largest and smallest singular values of a square matrix – *the condition number* – is systematically used in numerical analysis as a measure of sensitivity to round-off errors. Further, for certain random matrix models, bounds on the spectral norm of the matrix' resolvent (or, equivalently, the smallest singular value of diagonal shifts of the matrix) is a crucial point in the study of the spectral distribution. We refer to Sections 2 and 3 of the survey for a discussion of those directions.

In this survey, we consider *quantitative invertibility* of random *non-Hermitian* square matrices, including matrices with independent entries and adjacency matrices of random regular digraphs. The main objective in this line of research is to obtain bounds on the probability $\mathbb{P}\{s_{\min}(A) \leq t\}$ as a function of t, of the dimension, and, possibly, of some parameters of the model under consideration, such as the variance profile of the matrix or its mean.

One approach to the problem, which can be named analytical, is based on comparing the distribution of $s_{\min}(A)$ with the distribution of the smallest singular value of a corresponding Gaussian random matrix. The latter is very well understood **[25]** since explicit formulas for the joint distribution of the singular values of Gaussian matrices are available **[46]**. We refer to **[14, 85]** for results of that type.

Another approach, which is the focus of this survey, falls in the category of *non-asymptotic* methods **[75]** and is based on a combination of techniques originated within asymptotic geometric analysis. It often produces very strong probability estimates, although typically lacks the precision of the analytical methods. The major features of this approach are (a) reducing the estimation of $s_{\min}$ to estimating distances between random vectors and random linear subspaces associated with the matrix, and (b) using concentration (Bernstein-type) and anti-concentration (Littlewood–Offord-type) inequalities. Often, this approach also involves constructing discretizations of certain subsets of $\mathbb{R}^n$ or $\mathbb{C}^n$ (*ε-nets*) and estimating their cardinalities. We will give a description of the features by considering multiple examples from the literature.

Because of some differences in methodology, and because we wish to emphasize the importance of the matrix invertibility for numerical analysis and in the study of the spectral distribution, this survey does not cover nonquantitative results on the singularity of random matrices. We note that estimating the singularity probability for several models of *discrete* random matrices is a major topic within the combinatorial random matrix theory [10,21,47,50,63,83]. In the last few years there has been a significant progress in this research direction (also, as corollaries of quantitative results), in particular, the problem of estimating the singularity probability of adjacency matrices of random regular (di)graphs [38,62,65], of Bernoulli random matrices [43,57,92] and, more generally, discrete matrices with i.i.d. entries [44], as well as of random symmetric matrices [11–13,29]. We refer to a recent survey [97] for a discussion and further references.

The rest of the survey is organized as follows. Sections 2 and 3 provide motivation for studying quantitative invertibility of non-Hermitian random matrices, and a brief account of known results. In Section 4, we give an overview of the methodology, starting with the result of Rudelson and Vershynin [72] as a main illustration. We then discuss novel additions to the methodology made in the past ten years, which allowed making progress on several important problems in the random matrix theory. Finally, in Section 5, we discuss some open problems.

Let us recall some notions which will be used further.

A random variable X on $\mathbb{R}$ or $\mathbb{C}$ is called *subgaussian* if $\mathbb{E}\exp(|X|^2/K^2) < \infty$ for some number $K > 0$. The smallest value of K such that $\mathbb{E}\exp(|X|^2/K^2) - 1 \le 1$, is called *the subgaussian moment* of X. Any gaussian random variable is also subgaussian; further, all bounded random variables are subgaussian.

Given a sequence of *random* Borel probability measures $(\mu_m)_{m=1}^{\infty}$ and a random probability measure μ on $\mathbb{C}$, we say that μ_m converge *weakly in probability* to μ if for every bounded continuous function f on $\mathbb{C}$,

$$\lim_{m\to\infty} \mathbb{P}\left\{\left|\int f\, d\mu_m - \int f\, d\mu\right| > \varepsilon\right\} = 0, \quad \forall \varepsilon > 0.$$

We will denote by $\|\cdot\|$ the spectral norm of a matrix. The standard Euclidean norm in $\mathbb{R}^n$ or $\mathbb{C}^n$ will be denoted by $\|\cdot\|_2$. We will write $\mathrm{dist}(S, T)$ for the Euclidean distance between two subsets S and T of $\mathbb{R}^n$ or $\mathbb{C}^n$. By $S^{n-1}(\mathbb{R})$ or $S^{n-1}(\mathbb{C})$ we denote the unit Euclidean sphere in $\mathbb{R}^n$ or $\mathbb{C}^n$, respectively. The constants will be denoted by C, c', etc.

2. QUANTITATIVE INVERTIBILITY IN MATRIX COMPUTATIONS

In this section, we discuss the importance of estimating the smallest singular value in numerical analysis, and provide a brief overview of related results on random matrices.

2.1. The condition number in numerical analysis

For an $n \times n$ invertible matrix $A \in \mathbb{C}^{n\times n}$, *the condition number of* A is defined as

$$\kappa(A) := \|A\| \|A^{-1}\| = \frac{s_{\max}(A)}{s_{\min}(A)}.$$

Consider a system of n linear equations in n variables, represented in the matrix–vector form as $Ax = b$. If the system is *well conditioned*, i.e., the condition number of the coefficient matrix A is small, a perturbation of the matrix or the coefficient vector does not strongly affect the solution. In particular, the round-off errors in matrix computations such as the Gaussian elimination, do not significantly distort the solution vector.

As an example of well-known theoretical guarantees, we mention an estimate on *the relative distance* between the solution of $Ax = b$ and the solution of a perturbed system

$$(A + F)y = (b + f).$$

The terms $F \in \mathbb{C}^{n\times n}$ and $f \in \mathbb{C}^n$ can be thought of as consequences of measurement or round-off errors. It is not difficult to check that under the assumption that $\delta := \max(\frac{\|F\|}{\|A\|}, \frac{\|f\|_2}{\|b\|_2})$ is small, the relative distance $\frac{\|y-x\|_2}{\|x\|_2}$ satisfies

$$\frac{\|y - x\|_2}{\|x\|_2} = O\big(\delta\kappa(A)\big)$$

(see, in particular, [33, SECTION 2.6.2], [80, SECTION 4]). In the specific setting when the system $Ax = b$ is solved using the Gaussian elimination with partial pivoting and the perturbation of the system is due to round-off errors, Wilkinson [98] showed that the relative distance between the computed and actual solutions can be bounded above by $n^{O(1)}\varepsilon\kappa(A)\rho$. Here *the growth factor* ρ is defined as $\rho := \frac{\max_{k=0,1,\dots;i,j\le n} |a_{ij}^{(k)}|}{\max_{i,j\le n} |a_{ij}|}$, with $a_{ij}^{(k)}$ being the (i, j)th element of the matrix $A^{(k)}$ obtained from A after k iterations of the Gaussian elimination process, and ε is the precision of the machine (see also [77, 94]).

Whereas the condition number of A characterizes sensitivity of the corresponding system of linear equations to small perturbations, *the eigenvector condition number* quantifies the stability of the spectrum and eigenvectors of A. The eigenvector condition number of a diagonalizable matrix $A \in \mathbb{C}^{n\times n}$ is defined as

$$\kappa_V(A) := \min_{W\in\mathbb{C}^{n\times n}: W^{-1}AW \text{ is diagonal}} \kappa(W) = \min_{W\in\mathbb{C}^{n\times n}: W^{-1}AW \text{ is diagonal}} \frac{s_{\max}(W)}{s_{\min}(W)}.$$

Clearly, $\kappa_V(A) = 1$ if and only if A is *unitarily diagonalizable* (normal). A classical stability result for a matrix spectrum using the eigenvector condition number is the Bauer–Fike theorem [8]. According to the theorem, given a diagonalizable matrix A and its perturbation $A + F$, the distance between any eigenvalue μ of $A + F$ and the spectrum of A can be estimated as

$$\min_{\lambda\in\mathrm{Spec}(A)} |\mu - \lambda| \le \kappa_V(A)\|F\|.$$

Moreover, stability of matrix functions under perturbations of the argument can be quantified using the eigenvector condition number (see [36, SECTION 3.3]). Here, we refer to a related line of research dealing with *the approximate diagonalization* of matrices, namely approximating a matrix with one having a small eigenvector condition number (see [2,3,22,41] and references therein). A connection between $\kappa_V(A)$ and quantitative invertibility of diagonal shifts of A is established through the notion of a pseudospectrum. *An ε-pseudospectrum of A*, denoted $\mathrm{Spec}_\varepsilon(A)$, is defined as the set of all points $z \in \mathbb{C}$ with $s_{\min}(A - z\,\mathbf{Id}) < \varepsilon$. It can be shown (see [23, LEMMA 9.2.11]) that for a diagonalizable matrix A with D being a corresponding diagonal matrix, $\mathrm{Spec}(D) + \kappa_V(A)^{-1}\varepsilon U \subset \mathrm{Spec}_\varepsilon(A) \subset \mathrm{Spec}(D) + \kappa_V(A)\varepsilon U$, where U is the unit disk of the complex plane.

2.2. Related results on random matrices

Randomness is a natural approach to simulate typical matrices observed in applications. For example, the LINPACK benchmark for measuring the computing power involves systems of linear equations with a randomly generated coefficient matrix [24]. Condition numbers of random square matrices with the computational perspective were first considered by von Neumann and Goldstine [96]. Rigorous results were obtained much later, notably by Edelman [25] for Gaussian random matrices (see also Szarek [82]). We note here that for *sufficiently dense* random matrices with i.i.d. entries satisfying certain moment conditions, estimating the largest singular values up to a constant multiple can be accomplished by a simple combination of Bernstein-type inequalities and an ε-net argument (see, for example, [75]), and with precision up to $(1 \pm o(1))$ multiple via the trace method [30,79,100]. Further, we only discuss estimates for the smallest singular value.

The average-case quantitative analysis of the matrix invertibility, when a typical matrix is modeled as a random matrix with independent entries and with matching first two moments, has been developed in multiple works. We refer, in particular, to papers [14, 85] employing the analytical approach, as well as works [5, 7, 37, 43, 44, 57–60, 67, 69, 71, 72, 89–92] based on the reduction to distance estimates and use of concentration/anticoncentration inequalities. Some of those results are mentioned below.

In [72], Rudelson and Vershynin showed that given a random $n \times n$ matrix A with i.i.d. real entries of zero mean, unit variance, and a bounded *subgaussian moment*, the smallest singular value of A satisfies

$$\mathbb{P}\{s_{\min}(A) \le n^{-1/2}t\} \le C(t + c^n), \quad t > 0,$$

where the constants $C > 0$ and $c \in (0, 1)$ may only depend on the subgaussian moment (in fact, the statement is preserved if A is shifted by a nonrandom matrix with the spectral norm of order $O(\sqrt{n})$). The moment assumptions and the requirement that the entries are equidistributed were relaxed in later works [58,59,67]. On the other hand, in the special case of a matrix A with i.i.d. entries taking values $+1$ and -1 with probability $1/2$, it was proved in [92] that for any $\varepsilon > 0$,

$$\mathbb{P}\{s_{\min}(A) \le n^{-1/2}t\} \le Ct + C(1/2 + \varepsilon)^n, \quad t > 0,$$

where $C > 0$ is only allowed to depend on ε (see the introduction to [92], as well as [97], for a discussion of this result in the context of the combinatorial random matrix theory). An even stronger result is available when A has i.i.d. *discrete* entries which are not uniformly distributed on their support [44]: for every $\varepsilon > 0$ and assuming n is sufficiently large,

$$\mathbb{P}\{s_{\min}(A) \le n^{-1/2}t\} \le Ct + (1+\varepsilon)\mathbb{P}\{\text{two rows or columns of } A \text{ are colinear}\}, \quad t > 0,$$

with $C > 0$ depending only on the individual entry's distribution (see [44] for the statement in its full strength). In the setting when A has i.i.d. Bernoulli(p) entries and p is allowed to depend on n, its was shown in [7, 37, 57] that, as long as $p \le c$ for a small universal constant $c > 0$, for every $\varepsilon > 0$ and assuming n is sufficiently large,

$$\mathbb{P}\{s_{\min}(A) \le n^{-C}t\} \le t + (1+\varepsilon)\mathbb{P}\{\text{a row or a column of } A \text{ is zero}\}, \quad t > 0,$$

where $C > 0$ is a universal constant. We refer to [37] for a generalization to matrix rank estimates, as well as work [43] for sharp bounds in the setting of constant $p \in (0, 1/2)$, and [5] for stronger quantitative estimates in a certain range for the parameter p.

Put forward by Spielman and Teng [81], *the smoothed analysis* of the condition number is concerned with quantitative invertibility of a typical matrix in a small neighborhood of a fixed matrix (with possibly a very large spectral norm). A basic model of that type is of the form $A + M$, where M is a nonrandom matrix, and A has i.i.d. entries. The result of Sankar–Spielman–Teng [78] provided a bound for the smallest singular value of a shifted *Gaussian* real random matrix with i.i.d. standard normal entries, *independent* from the shift:

$$\mathbb{P}\{s_{\min}(A+M) \le tn^{-1/2}\} \le Ct, \quad t > 0,\ M \in \mathbb{R}^{n\times n},$$

for a certain universal constant $C > 0$ (see also [3, SECTION 2.3]). Analogous estimates for a broader class of random matrices with continuous distribution were later obtained in [91]. On the other hand, it was observed that for certain discrete random matrices, such as random sign (Bernoulli) matrices, no shift-independent small ball probability bounds for $s_{\min}(A+M)$ are possible [42, 87, 91]. In particular, it is shown in [42] that, assuming A has i.i.d. entries taking values ± 1 with probability $1/4$ and zero with probability $1/2$, for every $L \ge 1$ and every positive integer K,

$$\sup_{M:\|M\|\le n^L} \mathbb{P}\{s_{\min}(A+M) \le Cn^{-KL}\} \ge cn^{-K(K-1)/4},$$

where $C, c > 0$ may only depend on L and K. The smoothed analysis of the matrix condition number for discrete distributions was carried out in works [39, 42, 87, 88] (see also references therein). The following result was proved in [87]. Let $K, B, \varepsilon > 0$ and $L \ge 1/2$ be arbitrary parameters. Then, for all sufficiently large n, given an $n \times n$ random matrix A with i.i.d. centered entries of unit variance and the subgaussian moment bounded above by B, and given a nonrandom matrix M with $\|M\| \le n^L$, one has

$$\mathbb{P}\{s_{\min}(A+M) \le n^{-(2K+1)L}\} \le n^{-K+\varepsilon}.$$

In [42], it is shown that the above small ball probability bound can be significantly improved to match the average-case result of Rudelson and Vershynin [72], under the assumption that a positive fraction of the singular values of M are of order $O(\sqrt{n})$. More specifically, for every $\tilde{c} \in (0,1)$ and $\tilde{C} > 0$, and any fixed matrix M with $s_{n-\lfloor \tilde{c}n \rfloor}(M) \le \tilde{C}\sqrt{n}$, one has

$$\mathbb{P}\{s_{\min}(A+M) \le tn^{-1/2}\} \le C(t+c^n),$$

where $C > 0$ and $c \in (0,1)$ may only depend on $\tilde{c}$, $\tilde{C}$, and the subgaussian moment B. Under much weaker assumptions on the shift M, though at a price of precision, quantitative bounds for $s_{\min}(A+M)$ were obtained in [39].

3. INVERTIBILITY AND SPECTRUM

Given a square $n \times n$ matrix A_n, denote by μ_{A_n} its normalized spectral measure (spectral distribution):

$$\mu_{A_n} := \frac{1}{n}\sum_{i=1}^{n} \delta_{\lambda_i(A_n)}.$$

For real and complex Gaussian matrices with i.i.d. standard entries (*the Ginibre ensemble*), explicit formulas for the joint distribution of the eigenvalues are known [26, 31, 51]. Those, in turn, were used by Mehta [61], Silverstein (unpublished; see [9, SECTION 3]) and Edelman [26] to derive convergence results for the spectral distribution in the Gaussian case.

In the non-Gaussian setting, where no similar formulas are available, Girko [32] proposed a *Hermitization* argument based on the identity

$$\begin{aligned}\frac{1}{n}\sum_{i=1}^{n} \log\left|z-\lambda_i(A_n)\right| &= \frac{1}{n}\log\sqrt{\det\bigl((A_n - z\,\mathbf{Id})(A_n - z\,\mathbf{Id})^*\bigr)} \\ &= \frac{1}{n}\sum_{i=1}^{n}\log s_i(A_n - z\,\mathbf{Id}),\end{aligned}$$

which relates the spectrum to the singular values of the matrix resolvent. A modern form of the argument can be summarized as follows (see [9, LEMMA 4.3], as well as *the replacement principle* in [86]). Assume that a sequence of random matrices $(A_n)_{n=1}^{\infty}$ is such that for almost every $z \in \mathbb{C}$, the sequence of measures $\mu_{\sqrt{(A_n - z\,\mathbf{Id})(A_n - z\,\mathbf{Id})^*}}$ converges weakly in probability to a nonrandom probability measure μ_z. Assume further that the logarithm is *uniformly integrable in probability* with respect to $(\mu_{\sqrt{(A_n - z\,\mathbf{Id})(A_n - z\,\mathbf{Id})^*}})_{n=1}^{\infty}$ for almost every $z \in \mathbb{C}$, that is,

$$\lim_{t\to\infty}\sup_n \mathbb{P}\left\{\frac{1}{n}\sum_{i\le n:|\log s_i(A_n - z\,\mathbf{Id})|>t}\left|\log s_i(A_n - z\,\mathbf{Id})\right| > \varepsilon\right\} = 0, \quad \forall \varepsilon > 0. \tag{3.1}$$

Then there is a measure μ on $\mathbb{C}$ such that the sequence $(\mu_{A_n})_{n=1}^{\infty}$ converges to μ weakly in probability; moreover, the measure μ can be characterized in terms of $(\mu_z)_{z\in\mathbb{C}}$. We refer to [9] for proofs, as well as a detailed historical account of the study of the spectral distribution of non-Hermitian random matrices, up to 2000s.

In view of the uniform integrability requirement (3.1), strong quantitative estimates for small singular values of matrices $A_n - z\,\mathbf{Id}$ are an essential part of the Hermitization argument. In the setting of matrices with i.i.d. non-Gaussian entries, first rigorous estimates on the small singular values of $A_n - z\,\mathbf{Id}$ sufficient for the argument to go through were obtained for a class of continuous distributions by Bai [1], who applied the estimates to study the limiting spectral distribution in that setting. As the techniques to quantify invertibility of more general classes of matrices became available through the works of Tao–Vu [89], Rudelson [69], and Rudelson–Vershynin [72], the result of Bai was consequently generalized in works [34,66,84,86]. The *strong circular law* under minimal moment assumptions proved in [86] can be formulated as follows. Let ξ be a complex-valued random variable of zero mean and unit absolute second moment, and let $(A_n)_{n=1}^{\infty}$ be a sequence of random matrices, where each A_n is $n \times n$ with i.i.d. entries equidistributed with ξ. Then the sequence of spectral distributions $(\mu_{\frac{1}{\sqrt{n}}A_n})_{n=1}^{\infty}$ converges weakly almost surely to the uniform probability measure on the unit disk of the complex plane.

In the context of the circular law, the most studied model of *sparse* random matrices is of the form $A_n = B_n \odot M_n$, where B_n is the random matrix with i.i.d. Bernoulli(p_n) entries, M_n is independent from B_n and has i.i.d. entries equidistributed with a random variable ξ of unit variance, and "$\odot$" denotes the Hadamard (entrywise) product of matrices. In the regime $p_n \geq n^{-1+\varepsilon}$ for a fixed $\varepsilon > 0$, the (weak) circular law has been established in [99] following earlier works [34,84] dealing with additional moment assumptions.

In an even sparser regime, estimating the smallest singular value of $A_n - z\,\mathbf{Id}$ presents significant challenges, and further progress has only been made recently in [6,70]. In [70], it is proved that, assuming ξ is a real-valued random variable with unit variance, $np_n \leq n^{1/8}$, and np_n tends to infinity with n, and assuming the matrices $A_n = B_n \odot M_n$ are defined as in the previous paragraph, the sequence of spectral distributions $(\mu_{\frac{1}{\sqrt{np_n}}A_n})_{n=1}^{\infty}$ converges weakly in probability to the uniform measure on the unit disk of $\mathbb{C}$. A central technical result of [70] is the following quantitative bound for $s_{\min}(A_n - z\,\mathbf{Id})$: under the assumption that $|z| \leq np_n$ and $|\Im(z)| \geq 1$,

$$\mathbb{P}\{s_{\min}(A_n - z\,\mathbf{Id}) \leq \exp(-C \log^3 n)\} \leq C(np_n)^{-c},$$

where $C, c > 0$ may only depend on the c.d.f. of ξ.

Quantitative invertibility and spectrum of adjacency matrices of random regular directed graphs have been considered in multiple works in the past [4,15,17,18,53–55]. Given integers n and d, a *d-regular digraph* on vertices $\{1, 2, \ldots, n\}$ is a directed graph in which every vertex has d incoming edges and d outgoing edges. Here, we focus on the model when no multiedges are allowed, but the graph may have loops (the latter condition is not conventional). For each n, denote by $A_{n,d}$ the adjacency matrix of a random graph uniformly distributed on the set of all d-regular digraphs on $\{1, 2, \ldots, n\}$ (we allow d to depend on n). The first results on invertibility for this model were obtained by Cook [18]. The circular law for the sequence of spectral measures $(\mu_{\frac{1}{\sqrt{d(1-d/n)}}A_{n,d}})_{n=1}^{\infty}$ has been established in [17] under the assumption $\min(d, n-d) \geq \log^{96} n$. Later, in [53–55], the range $\omega(1) = d \leq \log^{96} n$ was treated. Either of the two results relies heavily on the estimates of the smallest singular values of $A_{n,d} - z\,\mathbf{Id}$. In particular, the main theorem of [53] is the following statement:

assuming $C \le d \le n/\log^2 n$ and $|z| \le d/6$,

$$\mathbb{P}\{s_{\min}(A_{n,d} - z\,\mathbf{Id}) < n^{-6}\} \le \frac{C \log^2 d}{\sqrt{d}},$$

where $C > 0$ is a universal constant.

The invertibility of *structured* random matrices and applications to the study of limiting spectral distribution have been considered, in particular, in [16,19,20,40,76]. A basic model of interest here is of the form $A_n = U_n \odot M_n - z\,\mathbf{Id}$, where M_n is a matrix with i.i.d. entries having zero mean and unit variance, $z \in \mathbb{C}$ is some complex number, U_n is a nonrandom matrix with nonnegative real entries encoding *the standard deviation profile*, and "$\odot$" denotes the Hadamard (entrywise) product of matrices. Note that A_n has mutually independent entries, with $\sqrt{\operatorname{Var} a_{ij}} = u_{ij}$, $1 \le i, j \le n$. In [76], the invertibility (and, more generally, the singular spectrum) of $U_n \odot M_n$ was studied in connection with the problem of estimation of matrix permanents. In particular, strong quantitative bounds on $s_{\min}(U_n \odot M_n)$ were obtained in the setting when M_n is the standard real Gaussian matrix, and U_n is a *broadly connected* profile (see [76, SECTION 2]). A significant progress in the study of structured random matrices was made by Cook in [16], who extended the result of [76] to non-Gaussian matrices, and obtained a polynomial lower bound on $s_{\min}(U_n \odot M_n - z\,\mathbf{Id})$ under very general assumptions on U_n. Namely, assuming that all entries of U_n are in the interval $[0, C]$, that $z \in [c\sqrt{n}, C\sqrt{n}]$ for some constants $c, C > 0$, and that the entries of M_n have a bounded $(4+\varepsilon)$-moment, the main result of [16] asserts that

$$\mathbb{P}\{s_{\min}(U_n \odot M_n - z\,\mathbf{Id}) \le n^{-\beta}\} \le n^{-\alpha}$$

for some $\alpha, \beta > 0$ depending only on c, C, ε, and the value of the $(4+\varepsilon)$-moment. In [19], this estimate was applied to derive limiting laws for the spectral distributions, under some additional assumptions on U_n. One of the results of [19] is *the circular law for doubly stochastic variance profiles*: provided that $\sum_{i=1}^n (U_n)_{ij}^2 = \sum_{i=1}^n (U_n)_{ji}^2 = n$, $1 \le j \le n$, and $\sup_n \max_{i,j} (U_n)_{ij} < \infty$, the sequence of spectral distributions $(\mu_{\frac{1}{\sqrt{n}} U_n \odot M_n})_{n=1}^\infty$ converges weakly in probability to the uniform measure on the unit disc of $\mathbb{C}$.

The setting of *sparse* structured matrices is not well understood. For results in that direction, we refer to a recent paper [40] dealing with the invertibility and spectrum of *block band matrices*.

4. METHODOLOGY

We start this section with a brief outline of [72] which will serve as an illustration of nonasymptotic methods (at the same time, we note that the argument of [72] is strongly influenced by earlier works, in particular, by Tao–Vu [89] and Rudelson [69]). The proof of the main theorem in [72] relies on four major components: sphere partitioning, invertibility via distance, ε-net arguments, and Littlewood–Offord-type inequalities.

Let A be an $n \times n$ matrix with i.i.d. real entries having zero mean and unit variance, and assume for simplicity that the entries are K-subgaussian for some constant $K > 0$.

A vector $x \in \mathbb{R}^n$ is called *m-sparse* if the size of its support is at most m. We will denote the set of all m-sparse vectors by $\mathrm{Sparse}_n(m)$. The proof of [72, THEOREM 3.1] starts with splitting $S^{n-1}(\mathbb{R})$ into sets of *compressible* and *incompressible* vectors,

$$\mathrm{Comp}_n(\delta, \rho) := \{x \in S^{n-1}(\mathbb{R}) : \mathrm{dist}(x, \mathrm{Sparse}_n(\delta n)) < \rho\};$$
$$\mathrm{Incomp}_n(\delta, \rho) := \{x \in S^{n-1}(\mathbb{R}) : \mathrm{dist}(x, \mathrm{Sparse}_n(\delta n)) \geq \rho\}.$$

Here, $\delta, \rho \in (0, 1)$ are small constants. The variational formula for $s_{\min}(A)$ allows writing

$$\mathbb{P}\{s_{\min}(A) \leq s\} \leq \mathbb{P}\{\|Ax\|_2 \leq s \text{ for some } x \in \mathrm{Comp}_n(\delta, \rho)\} \\ + \mathbb{P}\{\|Ax\|_2 \leq s \text{ for some } x \in \mathrm{Incomp}_n(\delta, \rho)\}, \quad s > 0.$$

If both δ and ρ are sufficiently small, the set of compressible vectors has small *covering numbers*, which allows applying an ε-net argument. More specifically, it can be checked that for every $\varepsilon \in (3\rho, 1/2]$, there is a discrete subset $\mathcal{N} \subset \mathrm{Comp}_n(\delta, \rho)$ of size at most $(\frac{C}{\varepsilon\delta})^{\delta n}$ such that for every $x \in \mathrm{Comp}_n(\delta, \rho)$, we have $\mathrm{dist}(x, \mathcal{N}) \leq \varepsilon$ (i.e., $\mathcal{N}$ is an ε-net in $\mathrm{Comp}_n(\delta, \rho)$ with respect to the Euclidean metric). Consequently, for every $L > 0$,

$$\mathbb{P}\{\|Ax\|_2 \leq s \text{ for some } x \in \mathrm{Comp}_n(\delta, \rho)\} \\ \leq \mathbb{P}\{\|Ay\|_2 \leq s + \varepsilon L\sqrt{n} \text{ for some } y \in \mathcal{N}\} + \mathbb{P}\{\|A\| > L\sqrt{n}\} \\ \leq |\mathcal{N}| \sup_{z \in S^{n-1}(\mathbb{R})} \mathbb{P}\{\|Az\|_2 \leq s + \varepsilon L\sqrt{n}\} + \mathbb{P}\{\|A\| > L\sqrt{n}\}.$$

For any $z \in S^{n-1}(\mathbb{R})$, the vector Az has i.i.d. subgaussian components with unit variances, and a standard Laplace transform argument implies that, as long as $s + \varepsilon L\sqrt{n}$ is much less than $\sqrt{n}$, the probability $\mathbb{P}\{\|Az\|_2 \leq s + \varepsilon L\sqrt{n}\}$ is exponentially small in n. Moreover, for a sufficiently large constant L, the probability $\mathbb{P}\{\|A\| > L\sqrt{n}\}$ is exponentially small in n. Therefore, an appropriate choice of parameters $\delta, \rho, \varepsilon, L$ yields

$$\mathbb{P}\{\|Ax\|_2 \leq s \text{ for some } x \in \mathrm{Comp}_n(\delta, \rho)\} \leq 2\exp(-cn), \quad s = o(\sqrt{n}).$$

We refer to [72] and [75] for details regarding the above computations. Let us note also that the idea of sphere partitioning was applied a few years earlier in [56] dealing with rectangular random matrices.

The incompressible vectors are treated using the *invertibility via distance* argument, which is based on the observation that for any incompressible vector x, a constant proportion of its components are of order $\Omega(n^{-1/2})$ by the absolute value. For every $1 \leq i \leq n$, denote by $H_i(A)$ the linear span of columns of A except the ith,

$$H_i(A) := \mathrm{Span}\{\mathrm{Col}_j(A),\ j \neq i\}.$$

Then for arbitrary vector x and arbitrary "threshold" $\tau > 0$ with $\{i : |x_i| \geq \tau\} \neq \emptyset$, we have

$$\|Ax\|_2 \geq \max_{1 \leq i \leq n} \left(|x_i| \mathrm{dist}(\mathrm{Col}_i(A), H_i(A))\right) \geq \tau \max_{i:|x_i| \geq \tau} \mathrm{dist}(\mathrm{Col}_i(A), H_i(A)),$$

and hence for any $s > 0$, $\mathbf{1}_{\{\|Ax\|_2 \leq s\}} \leq \frac{1}{|\{i:|x_i| \geq \tau\}|} \sum_{i=1}^n \mathbf{1}_{\{\mathrm{dist}(\mathrm{Col}_i(A), H_i(A)) \leq s/\tau\}}$. This, combined with Markov's inequality and the fact that every (δ, ρ)–ncompressible vector is *spread*, i.e., has at least δn components of magnitude at least $\rho n^{-1/2}$, gives for $t > 0$,

$$\mathbb{P}\big\{\exists\, x \in \mathrm{Incomp}_n(\delta,\rho) : \|Ax\|_2 \le t n^{-1/2}\big\} \le \frac{1}{\delta n}\sum_{i=1}^{n} \mathbb{P}\big\{\mathrm{dist}\big(\mathrm{Col}_i(A), H_i(A)\big) \le t/\rho\big\} \tag{4.1}$$

(see **[72, LEMMA 3.5]**). Since the distribution of A is invariant under column permutations, the last relation can be rewritten as

$$\begin{aligned}\mathbb{P}\big\{\|Ax\|_2 \le t n^{-1/2} \text{ for some } x \in \mathrm{Incomp}_n(\delta,\rho)\big\} &\le \frac{1}{\delta}\mathbb{P}\big\{\mathrm{dist}\big(\mathrm{Col}_n(A), H_n(A)\big) \le t/\rho\big\} \\ &\le \frac{1}{\delta}\mathbb{P}\big\{\big|\big\langle \mathrm{Col}_n(A), Y_n(A)\big\rangle\big| \le t/\rho\big\},\end{aligned}$$

where $Y_n(A)$ denotes a unit normal to $H_n(A)$ measurable with respect to $\sigma(H_n(A))$.

The most involved part of **[72]** is the analysis of anticoncentration of $\langle \mathrm{Col}_n(A), Y_n(A)\rangle$. Recall that *the Lévy concentration function* $\mathcal{L}(Z,t)$ of a real variable Z is defined as

$$\mathcal{L}(Z,t) := \sup_{r\in\mathbb{R}} \mathbb{P}\big\{|Z - r| \le t\big\}, \quad t \ge 0.$$

The relationship between the magnitude of $\mathcal{L}(\sum_{i=1}^n a_i Z_i, t)$ for a linear combination of random variables $\sum_{i=1}^n a_i Z_i$ and the structure of the coefficient vector $(a_1,\dots,a_n)$ has been studied in numerous works, starting from an inequality of Erdős–Littlewood–Offord **[27, 52]**; we refer, in particular, to works **[28, 48, 49, 68]**, as well as **[89]** and a survey **[64]** for a more recent account of *the Littlewood–Offord theory* and its applications to the matrix invertibility.

To characterize the structure of a coefficient vector in regard to anticoncentration, the notion of the *essential least common denominator* (LCD) has been introduced in **[72]**. We quote a slightly modified definition from **[73]**:

$$\mathrm{LCD}(a) := \inf\big\{\theta > 0 : \mathrm{dist}\big(\theta a, \mathbb{Z}^n\big) < \min\big(\gamma\|\theta a\|_2, \alpha\sqrt{n}\big)\big\}, \quad a \in \mathbb{R}^n.$$

Here, α, γ are small positive constants. The Littlewood–Offord-type inequality used in **[72, 73]** can be stated as follows. If $Z_1, Z_2, \dots, Z_n$ are i.i.d. real-valued random variables with $\mathbb{P}\{|Z_i - \mathbb{E}Z_i| < \beta\} \le 1 - \beta$ for some $\beta > 0$ then for any unit vector $a \in \mathbb{R}^n$,

$$\mathcal{L}\bigg(\sum_{i=1}^{n} a_i Z_i, t\bigg) \le Ct + \frac{C}{\mathrm{LCD}(a)} + 2\exp(-cn), \quad t > 0, \tag{4.2}$$

where $C > 0$ may only depend on β, γ and $c > 0$ only on α, β (see **[73]** for a proof). Using an ε-net argument, the authors of **[72]** show that, with probability exponentially close to one, the random unit normal vector $Y_n(A)$ has an exponentially large LCD. This implies

$$\begin{aligned}\mathbb{P}\big\{\big|\big\langle \mathrm{Col}_n(A), Y_n(A)\big\rangle\big| \le s\big\} &\le \mathbb{P}\big\{\mathrm{LCD}\big(Y_n(A)\big) < \exp(c'n)\big\} + Cs + 2\exp(-c'n) \\ &\le Cs + 3\exp(-c''n), \quad s > 0.\end{aligned}$$

The combination of all the ingredients now gives the final estimate

$$\mathbb{P}\big\{s_{\min}(A) \le t n^{-1/2}\big\} \le \tilde{C}t + \tilde{C}\exp(-\hat{c}n), \quad t > 0,$$

matching, by the order of magnitude and up to the exponentially small additive term, the known asymptotics of $s_{\min}$ of Gaussian random matrices [25, 82].

In the remaining part of this section, we will consider some of the novel additions to the methodology made over the past years. To avoid technical details as much as possible, we will refer to compressible vectors, as well as all related notions from the literature, as *almost sparse* vectors, and to incompressible vectors and their relatives as *spread* vectors.

Invertibility over almost sparse vectors. In the setting of *dense* random matrices as described above, the set of almost sparse vectors AlSp_n can be treated by a simple ε-net argument since anticoncentration estimates for $\|Az\|_2$ for an *arbitrary* vector $z \in S^{n-1}$ are able to overpower the cardinality of the ε-net $\mathcal{N}$ in AlSp_n. In the case of sparse and certain models of structured random matrices, such an argument may not be sufficient since the product $|\mathcal{N}| \sup_{z \in S^{n-1}(\mathbb{R})} \mathbb{P}\{\|Az\|_2 \le s\}$ may become infinitely large even for small $s > 0$. We consider two (related) approaches to this problem from the literature.

The first is based on further subdividing AlSp_n into a few subsets $T_1, T_2, \ldots$ according to the size of set of vector's components of nonnegligible magnitude, and applying an ε-net argument within each subset. Anticoncentration estimates for Az for vectors $z \in T_i$ then compete with the cardinality of an ε-net on the set T_i rather than on the entire collection AlSp_n, which, for certain models, allows the proof to go through. We refer, in particular, to [76, SECTION 4] and [16, SECTION 3] for an application of this strategy to structured random matrices; as well as [17, PROPOSITION 3.1] dealing with adjacency matrices of random d-regular digraphs.

The second approach consists in identifying a class of nonrandom matrices $\mathcal{C}$ such that, for every $M \in \mathcal{C}$ and every almost sparse vector $z \in S^{n-1}$, Mz has a nonnegligible Euclidean norm, and then showing that, with probability close to one, $A \in \mathcal{C}$. As an example, consider a collection of matrices M such that for every nonempty subset $I \subset [n]$ with $|I| \le m$, there is a row $\mathrm{Row}_i(M)$ with $|\mathrm{supp}\,\mathrm{Row}_i(M) \cap I| = 1$. Then, it is not difficult to check that, for every nonzero m-sparse vector z, one has $Mz \neq 0$. It can further be verified that a random matrix A with i.i.d. Bernoulli(p) elements and $n^{-1}\mathrm{polylog}(n) \le p \le cm^{-1}$ belongs to this class with probability tending to one as $n \to \infty$ [5]. The construction can be made robust to treat almost sparse vectors, and can be further elaborated to deal with diagonal shifts of very sparse matrices [5, 53, 70].

Invertibility via distance. Relation (4.1) discovered in [72] can be applied to any model of randomness. However, this relation is not completely satisfactory when either (a) there are strong probabilistic dependencies between $\mathrm{Col}_i(A)$ and $H_i(A)$ which make estimating $\mathbb{P}\{\mathrm{dist}(\mathrm{Col}_i(A), H_i(A)) \le t\}$ challenging, or (b) invertibility over the almost sparse vectors cannot be treated with a desired precision using approaches based on ε-net arguments or on conditioning on a particular structure of the matrix. Here, we consider some developments of the invertibility via distance argument made in the contexts of d-regular random digraphs and smoothed analysis of the condition number.

Let $A_{n,d}$ be the adjacency matrix of a uniform random d-regular directed graph on n vertices. The regularity condition implies that for every $1 \le i \le n$, $\mathrm{Col}_i(A_{n,d})$ is a function of $\{\mathrm{Col}_j(A_{n,d})\}_{j\neq i}$, creating issues with applying the original version of the argument from [72]. In [18], Cook proposed a modification of the argument based on considering distances between the matrix columns and random subspaces of the form $H_{i_1,i_2,+}(A_{n,d}) := \mathrm{Span}\{\mathrm{Col}_j(A_{n,d}),\ j\neq i_1,i_2;\ \mathrm{Col}_{i_1}(A_{n,d})+\mathrm{Col}_{i_2}(A_{n,d})\}$, for $i_1\neq i_2$. This was later applied in [17,53]. Here, we quote [53, LEMMA 4.2]: denoting by $S(\rho,\delta)$ the collection of all unit vectors x in $\mathbb{C}^n$ with $\inf_{\lambda\in\mathbb{C}} |\{i\le n : |x_i-\lambda| > \rho n^{-1/2}\}| > \delta n$, one has

$$\mathbb{P}\Big\{\inf_{x\in S(\rho,\delta)} \big\|(A_{n,d}-z\,\mathbf{Id})x\big\|_2 \le tn^{-1/2}\Big\}$$
$$\le \frac{1}{\delta n^2}\sum_{\substack{i_1,i_2\in[n],\\ i_1\neq i_2}} \mathbb{P}\big\{\mathrm{dist}\big(\mathrm{Col}_{i_1}(A_{n,d}-z\,\mathbf{Id}),\, H_{i_1,i_2,+}(A_{n,d}-z\,\mathbf{Id})\big)\le t/\rho\big\}.$$

Conditioned on a realization of $\mathrm{Col}_j(A_{n,d})$, $j\neq i_1,i_2$ (hence, also $Y := \mathrm{Col}_{i_1}(A_{n,d}) + \mathrm{Col}_{i_2}(A_{n,d})$), the support of the i_1th column of $A_{n,d}$ is uniformly distributed on the collection of d-subsets Q satisfying $\{j\le n : Y_j = 2\}\subset Q\subset \mathrm{supp}\,Y$. In the regime $d\to\infty$ as n tends to infinity, this is "sufficient randomness" for a satisfactory bound on $s_{\min}(A_{n,d}-z\,\mathbf{Id})$ required by the Hermitization argument [17,53].

We remark here that another version of the argument for matrices with dependencies based on evaluation of certain quadratic forms, introduced in [95], has been used in a non-Hermitian setting in [74] to estimate the smallest singular value of unitary and orthogonal perturbations of fixed matrices, which in turn is an important ingredient of *the single ring theorem* [35,74]. We refer to [74] for details.

In [91], a variant of the invertibility via the distance argument was developed to deal with nonrandom shifts of matrices with continuous distributions. The main observation of [91] is that the distances $\mathrm{dist}(\mathrm{Col}_i(A), H_i(A))$, $1\le i\le n$, are highly correlated, which allows for a more efficient analysis than the first moment method estimate (4.1). The invertibility via distance is applied in [91] to the entire sphere rather than the set of spread vectors. As an illustration of the principle, we consider a simpler setting of centered random matrices when the argument is still able to produce new results. Assuming A is an $n\times n$ real random matrix with i.i.d. entries having zero mean, unit variance, and the distribution density bounded above by ρ, for every $t>0$ and $1\le k\le n$, one has $\mathbb{P}\{\exists\, I\subset[n] : |I|\ge k,\ \mathrm{dist}(\mathrm{Col}_i(A), H_i(A))\le t\ \ \forall i\in I\}\le C_\rho t(n/k)^{5/11}$, where $C_\rho>0$ may only depend on ρ (see [91, PROP. 3.8]). This, combined with the simple consequence of the *negative second moment identity*

$$s_{\min}(A)\ge\bigg(\sum_{i=1}^n \mathrm{dist}\big(\mathrm{Col}_i(A), H_i(A)\big)^{-2}\bigg)^{-1/2},$$

implies an estimate $\mathbb{P}\{s_{\min}(A)\le tn^{-1/2}\}\le C'_\rho t$, $t>0$, which does not carry the c^n additive term inevitable when an ε-net-based approach is used. We refer to [91] for the more involved setting of noncentered random matrices.

Alternatives to the LCD. Functions of coefficient vectors different from the essential least common denominator have been introduced in the literature to deal with anticoncentration in the context of sparse and inhomogeneous random matrices, and matrices with dependencies. Here, we review some of them (for non-Hermitian models only).

The original notion of LCD is not applicable to the study of linear combinations of nonidentically distributed variables: in fact, given any vector $a \in S^{n-1}(\mathbb{R})$ with an exponentially large LCD, one can easily construct mutually independent variables $Z_1, \ldots, Z_n$ with $\mathcal{L}(Z_i, 1) \le 1/2$, $1 \le i \le n$, and such that $\mathcal{L}(\sum_{i=1}^n a_i Z_i, 0) = \Omega(n^{-1/2})$. Given a random vector X in $\mathbb{R}^n$ and denoting by $\bar{X}$ the difference $X - X'$ (where X' is an independent copy of X), the *randomized least common denominator* with respect to X is defined by

$$\mathrm{RLCD}^X(a) := \inf\bigl\{\theta > 0 : \mathbb{E}\mathrm{dist}^2\bigl((\theta a_1 \bar{X}_1, \ldots, \theta a_n \bar{X}_n), \mathbb{Z}^n\bigr) < \min\bigl(\gamma \|\theta a\|_2^2, \alpha n\bigr)\bigr\}, \quad a \in \mathbb{R}^n.$$

The notion was introduced in [59] to deal with inhomogeneous random matrices with different entry distributions. The small ball probability inequality (4.2) from [72,73] extends to the non-i.i.d. setting with the RLCD taking place of the original notion. We refer to [59] for details.

Strong quantitative invertibility results for matrices with fixed rowsums and adjacency matrices of d-regular digraphs obtained recently in [93] and [45], respectively, rely on a modification of the LCD which allows treating linear combinations of Bernoulli variables conditioned on their sum. Specifically, in [93] the notion of the *combinatorial least common denominator* CLCD is defined by

$$\mathrm{CLCD}(a) := \inf\bigl\{\theta > 0 : \mathrm{dist}\bigl(\theta(a_i - a_j)_{i<j}, \mathbb{Z}^{\binom{n}{2}}\bigr) < \min\bigl(\gamma \bigl\|\theta(a_i - a_j)_{i<j}\bigr\|_2, \alpha n\bigr)\bigr\}, \quad a \in \mathbb{R}^n,$$

where $(a_i - a_j)_{i<j}$ denotes a vector in $\mathbb{R}^{\binom{n}{2}}$ with the (i, j)th coordinate equal to $a_i - a_j$, $1 \le i < j \le n$. It is further shown that for the random vector $(Z_1, Z_2, \ldots, Z_n)$ uniformly distributed on the collection of 0/1 vectors with exactly $n/2$ ones, an analog of the anticoncentration inequality (4.2) holds, with LCD replaced by CLCD. A modification of the notion, called QCLCD, was further considered in [45]. We refer to that paper for details.

Another functional – *the degree of unstructuredness* UD – was introduced in [57] to study the invertibility of sparse Bernoulli random matrices. The main observation exploited in [57] is that, for $p = o(1)$, linear combinations of i.i.d. Bernoulli(p) random variables $\sum_{i=1}^n a_i Z_i$ are often more concentrated than corresponding linear combinations of *dependent* 0/1 variables conditioned to sum to a fixed number of order $\Theta(pn)$. In [57], the argument proceeds by conditioning on the size of the support of a column of the matrix and estimating the anticoncentration of $\mathrm{dist}(\mathrm{Col}_i(A), H_i(A)) = |\langle \mathrm{Col}_i(A), Y_i(A)\rangle|$ in terms of the degree of unstructuredness of the unit random normal $Y_i(A)$. The definition of UD is technically involved, and we do not provide it here; see [57] for details.

Average-case analysis of anticoncentration. The average-case study of Littlewood–Offord-type inequalities for linear combinations $\sum_{i=1}^n a_i Z_i$, introduced in the random matrix context in [92], was a crucial element in some of recent advances on quantitative invertibility of random discrete matrices [43, 44, 92], which helped resolve some long-standing problems in the combinatorial random matrix theory. The main idea of [92] is, rather than attempting to obtain an explicit description of vectors a such that $\sum_{i=1}^{n} a_i Z_i$ is strongly anticoncentrated, to consider the linear combination for a *randomly chosen* coefficient vector (with an appropriately defined notion of randomness). This approach allowed strengthening the invertibility results available through the use of the LCD. As an example, we consider a simplified version of the main technical result of [92]. Let $\varepsilon \in (0, 1/2)$, $M \geq 1$. Then there exist $n_0 = n_0(\varepsilon, M)$ depending on ε, M and $L_0 = L_0(\varepsilon) > 0$ depending *only* on ε (and not on M) with the following property. Take $n \geq n_0$, $1 \leq N \leq (1/2+\varepsilon)^{-n}$, and let $\mathcal{A} := (\{-2N, \dots, -N-1\} \cup \{N+1, \dots, 2N\})^n$. Assume that a random vector $a = (a_1, \dots, a_n)$ is uniformly distributed on $\mathcal{A}$. Then

$$\mathbb{P}_a\{\mathcal{L}_Z(a_1 Z_1 + \cdots + a_n Z_n, \sqrt{n}) > L_0 N^{-1}\} \leq e^{-Mn}.$$

Here, $\mathcal{L}_Z(\cdot, \cdot)$ denotes the Lévy concentration function with respect to the randomness of $(Z_1, \dots, Z_n)$, a vector with independent ± 1 components. The main point of the statement is that the parameter L_0 controlling the anticoncentration of the linear combination does not depend on M, i.e., the proportion of the coefficient vectors in $\mathcal{A}$ such that the anticoncentration of $a_1 Z_1 + \cdots + a_n Z_n$ is weak, becomes *superexponentially small* in n as $n \to \infty$.

Matrices with heavy entries. For the invertibility of (dense) random matrices with independent entries assuming only finite second moments, we refer to [58, 59, 67].

5. OPEN PROBLEMS

We conclude this survey with a selection of open research problems.

Refined smoothed analysis of invertibility. Recall that a standard model in the setting of the smoothed analysis of the condition number is of the form $A + M$, where A is an $n \times n$ random matrix with i.i.d. entries, and M is a nonrandom shift.

Problem 1 (Shift-independent estimates for matrices with continuous distributions). Let ξ be a real random variable with zero mean, unit variance, and bounded distribution density. Let A be an $n \times n$ matrix with i.i.d. entries equidistributed with ξ. It is true that for every nonrandom matrix M,

$$\mathbb{P}\{s_{\min}(A+M) \leq t n^{-1/2}\} \leq Ct, \quad t > 0,$$

where $C > 0$ may only depend on the c.d.f. of ξ (and not on n)?

For partial results on the above problem, see [78, 91].

Problem 2 (Optimal dependence of $s_{\min}(A+M)$ on the norm of the shift in the discrete setting). Let A be an $n \times n$ matrix with i.i.d. ± 1 entries, and let $T, t > 0$ be parameters. For any $\varepsilon, L > 0$, estimate $\sup_{M:\|M\| \leq T} \mathbb{P}\{s_{\min}(A+M) \leq t\}$ up to a multiplicative error $O(n^{\varepsilon})$ and an additive error $O(n^{-L})$, that is, find an explicit function $f(n, T, t)$ such that

$$n^{-\varepsilon} f(n, T, t) - C n^{-L} \leq \sup_{M:\|M\| \leq T} \mathbb{P}\{s_{\min}(A+M) \leq t\} \leq n^{\varepsilon} f(n, T, t) + C n^{-L},$$

where $C > 0$ may only depend on ε and L.

For the best known partial results on the above problem, see [42, 87].

Problem 3 (Dependence of $s_{\min}(A+M)$ on M in the Gaussian setting). Let A be an $n \times n$ matrix with i.i.d. standard real Gaussian entries. Find an estimate on $\mathbb{E}s_{\min}(A+M)$ in terms of the singular spectrum of M.

One can assume in the above problem that M is a diagonal matrix with the ith diagonal element $s_i(M)$, $1 \le i \le n$. Note that A may either improve or degrade the invertibility of M.

Invertibility and spectrum of very sparse matrices. Here, we consider the problem of identifying the limiting spectral distribution for non-Hermitian matrices with *constant* average number of nonzero elements in a row/column.

Problem 4 (The oriented Kesten–McKay law; see [9, SECTION 7]). Let $d \ge 3$. For each n, let $A_{n,d}$ be the adjacency matrix of a uniform random d-regular directed graph on n vertices. Prove that the sequence of spectral distributions $(\mu_{A_{n,d}})_{n=1}^{\infty}$ converges weakly to the probability measure on $\mathbb{C}$ with the density function

$$\rho_d(z) := \frac{1}{\pi}\frac{d^2(d-1)}{(d^2-|z|^2)^2}\mathbf{1}_{\{|z|<\sqrt{d}\}}.$$

Assuming the standard Hermitization approach to the above problem, the following is the crucial (perhaps the main) step of the argument:

Problem 5. Let $d \ge 3$ and let $(A_{n,d})_{n=1}^{\infty}$ be as above. Prove that for almost every $z \in \mathbb{C}$ and every $\varepsilon > 0$,

$$\lim_{n\to\infty} \mathbb{P}\big\{s_{\min}(A_{n,d} - z\,\mathbf{Id}) \le \exp(-\varepsilon n)\big\} = 0.$$

Problem 6 (Spectrum of directed Erdős–Renyi graphs of constant average degree). Let $\alpha > 0$. For each $n \ge \alpha$, let A_n be an $n \times n$ random matrix with i.i.d. Bernoulli(α/n) entries. Does a sequence of spectral distributions (μ_{A_n}) converge weakly to a nonrandom probability measure?

As in the case of regular digraphs, assuming the Hermitization argument, the following problem constitutes an important step in understanding the asymptotics of the spectrum:

Problem 7. For each $n \ge \alpha$, let A_n be an $n \times n$ random matrix with i.i.d. Bernoulli(α/n) entries. Is it true that for almost every $z \in \mathbb{C}$ and every $\varepsilon > 0$,

$$\lim_{n\to\infty} \mathbb{P}\big\{s_{\min}(A_n - z\,\mathbf{Id}) \le \exp(-\varepsilon n)\big\} = 0?$$

Invertibility and spectrum of structured random matrices. The spectrum of structured random matrices in the absence of expansion-like properties (such as *broad connectivity* [76] or *robust irreducibility* [19, 20]) is not well understood as of now. In particular, a full description of the class of inhomogeneous matrices with independent entries with spectral convergence to the circular law seems to be out of reach of modern methods.

Problem 8. Give a complete description of sequences of standard deviation profiles $(U_n)_{n=1}^{\infty}$ satisfying the following condition: assuming that ξ is any random variable with zero mean and unit variance, and that for each n, M_n is an $n \times n$ matrix with i.i.d. entries equidistributed with ξ, the sequence of spectral distributions $(\mu_{U_n \odot M_n})$ converges weakly in probability to the uniform measure on the unit disc of $\mathbb{C}$.

A natural class of profiles considered, in particular, in [19, 20], are *doubly stochastic* profiles. One may expect that those profile sequences, under some weak assumption on the magnitude of the maximal entry, should be sufficient for the circular law to hold:

Problem 9. Assume that for each n, the standard deviation profile U_n satisfies

$$\sum_{i=1}^{n} (U_n)_{ij}^2 = \sum_{i=1}^{n} (U_n)_{ji}^2 = 1, \quad 1 \le j \le n,$$

and that for some $\varepsilon > 0$, $\limsup_n \max_{ij}((U_n)_{ij} n^{\varepsilon}) = 0$. Is it true that, with M_n as in the above problem, the sequence $(\mu_{U_n \odot M_n})$ converges weakly in probability to the uniform measure on the unit disc of $\mathbb{C}$?

Note that the above setting allows sparse matrices (cf. [19, THEOREM 2.4]). Solving the above problem, if approached with Girko's Hermitization procedure, requires satisfactory bounds on the smallest singular values of $U_n \odot M_n - z\,\mathbf{Id}$.

FUNDING

This work was partially supported by the Sloan Research Fellowship and by NSF grant DMS 2054666.

REFERENCES

[1] Z. D. Bai, Circular law. *Ann. Probab.* **25** (1997), no. 1, 494–529.

[2] J. Banks, J. Garza Vargas, A. Kulkarni, and N. Srivastava, Overlaps, eigenvalue gaps, and pseudospectrum under real Ginibre and absolutely continuous perturbations. 2020, arXiv:2005.08930.

[3] J. Banks, A. Kulkarni, S. Mukherjee, and N. Srivastava, *Gaussian regularization of the seudospectrum and Davies' conjecture*. CPAM, to appear.

[4] A. Basak, N. Cook, and O. Zeitouni, Circular law for the sum of random permutation matrices. *Electron. J. Probab.* **23** (2018), Paper No. 33, 51 pp.

[5] A. Basak and M. Rudelson, Invertibility of sparse non-Hermitian matrices. *Adv. Math.* **310** (2017), 426–483.

[6] A. Basak and M. Rudelson, The circular law for sparse non-Hermitian matrices. *Ann. Probab.* **47** (2019), no. 4, 2359–2416.

[7] A. Basak and M. Rudelson, Sharp transition of the invertibility of the adjacency matrices of sparse random graphs. *Probab. Theory Related Fields* **180** (2021), no. 1–2, 233–308.

[8] F. L. Bauer and C. T. Fike, Norms and exclusion theorems. *Numer. Math.* **2** (1960), 137–141.

[9] C. Bordenave and D. Chafaï, Around the circular law. *Probab. Surv.* **9** (2012), 1–89.

[10] J. Bourgain, V. H. Vu, and P. M. Wood, On the singularity probability of discrete random matrices. *J. Funct. Anal.* **258** (2010), no. 2, 559–603.

[11] M. Campos, M. Jenssen, M. Michelen, and J. Sahasrabudhe, Singularity of random symmetric matrices revisited. 2020, arXiv:2011.03013.

[12] M. Campos, M. Jenssen, M. Michelen, and J. Sahasrabudhe, The singularity probability of a random symmetric matrix is exponentially small. 2021, arXiv:2105.11384.

[13] M. Campos, L. Mattos, R. Morris, and N. Morrison, On the singularity of random symmetric matrices. *Duke Math. J.* **170** (2021), no. 5, 881–907.

[14] Z. Che and P. Lopatto, Universality of the least singular value for sparse random matrices. *Electron. J. Probab.* **24** (2019), Paper No. 9, 53 pp.

[15] N. Cook, The circular law for random regular digraphs with random edge weights. *Random Matrices Theory Appl.* **6** (2017), no. 3, 1750012, 23 pp.

[16] N. Cook, Lower bounds for the smallest singular value of structured random matrices. *Ann. Probab.* **46** (2018), no. 6, 3442–3500.

[17] N. Cook, The circular law for random regular digraphs. *Ann. Inst. Henri Poincaré Probab. Stat.* **55** (2019), no. 4, 2111–2167.

[18] N. A. Cook, On the singularity of adjacency matrices for random regular digraphs. *Probab. Theory Related Fields* **167** (2017), no. 1–2, 143–200.

[19] N. Cook, W. Hachem, J. Najim, and D. Renfrew, Non-Hermitian random matrices with a variance profile (I): deterministic equivalents and limiting ESDs. *Electron. J. Probab.* **23** (2018), Paper No. 110, 61 pp.

[20] N. Cook, W. Hachem, J. Najim, and D. Renfrew, Non-Hermitian random matrices with a variance profile (II): properties and examples. 2020, arXiv:2007.15438.

[21] K. P. Costello, T. Tao, and V. Vu, Random symmetric matrices are almost surely nonsingular. *Duke Math. J.* **135** (2006), no. 2, 395–413.

[22] E. B. Davies, Approximate diagonalization. *SIAM J. Matrix Anal. Appl.* **29** (2007), no. 4, 1051–1064.

[23] E. B. Davies, *Linear operators and their spectra*. Cambridge Stud. Adv. Math. 106, Cambridge University Press, Cambridge, 2007.

[24] J. J. Dongarra, P. Luszczek, and A. Petitet, The LINPACK Benchmark: past, present and future. *Concurrency Comput.: Pract. Exp.* **15** (2003), 803–820.

[25] A. Edelman, Eigenvalues and condition numbers of random matrices. *SIAM J. Matrix Anal. Appl.* **9** (1988), no. 4, 543–560.

[26] A. Edelman, The probability that a random real Gaussian matrix has k real eigenvalues, related distributions, and the circular law. *J. Multivariate Anal.* **60** (1997), no. 2, 203–232.

[27] P. Erdős, On a lemma of Littlewood and Offord. *Bull. Amer. Math. Soc.* **51** (1945), 898–902.

[28] C. G. Esseen, On the Kolmogorov–Rogozin inequality for the concentration function. *Z. Wahrsch. Verw. Gebiete* **5** (1966), 210–216.

[29] A. Ferber and V. Jain, Singularity of random symmetric matrices—a combinatorial approach to improved bounds. *Forum Math. Sigma* **7** (2019), Paper No. e22, 29 pp.

[30] S. Geman, A limit theorem for the norm of random matrices. *Ann. Probab.* **8** (1980), no. 2, 252–261.

[31] J. Ginibre, Statistical ensembles of complex, quaternion, and real matrices. *J. Math. Phys.* **6** (1965), 440–449.

[32] V. L. Girko, The circular law. *Teor. Veroyatn. Primen.* **29** (1984), no. 4, 669–679.

[33] G. H. Golub and C. F. Van Loan, *Matrix computations*. 4th edn., Johns Hopkins Stud. Math. Sci., Johns Hopkins University Press, Baltimore, 2013.

[34] F. Götze and A. Tikhomirov, The circular law for random matrices. *Ann. Probab.* **38** (2010), no. 4, 1444–1491.

[35] A. Guionnet, M. Krishnapur, and O. Zeitouni, The single ring theorem. *Ann. of Math. (2)* **174** (2011), no. 2, 1189–1217.

[36] N. J. Higham, *Functions of matrices*. SIAM, Philadelphia, 2008.

[37] H. Huang, Rank of sparse Bernoulli matrices. 2020, arXiv:2009.13726.

[38] J. Huang, Invertibility of adjacency matrices for random d-regular graphs. 2018, arXiv:1807.06465.

[39] V. Jain, Quantitative invertibility of random matrices: a combinatorial perspective. *Discrete Anal.*, to appear.

[40] V. Jain, I. Jana, K. Luh, and S. O'Rourke, Circular law for random block band matrices with genuinely sublinear bandwidth. 2020, arXiv:2008.03850.

[41] V. Jain, A. Sah, and M. Sawhney, On the real Davies' conjecture. 2020, arXiv:2005.08908.

[42] V. Jain, A. Sah, and M. Sawhney, On the smoothed analysis of the smallest singular value with discrete noise. *Bull. Lond. Math. Soc.*, to appear.

[43] V. Jain, A. Sah, and M. Sawhney, Sharp invertibility of random Bernoulli matrices. 2020, arXiv:2010.06553.

[44] V. Jain, A. Sah, and M. Sawhney, Singularity of discrete random matrices. 2020, arXiv:2010.06554.

[45] V. Jain, A. Sah, and M. Sawhney, The smallest singular value of dense random regular digraphs. 2020, arXiv:2008.04755.

[46] A. T. James, The distribution of the latent roots of the covariance matrix. *Ann. Math. Stat.* **31** (1960), 151–158.

[47] J. Kahn, J. Komlós, and E. Szemerédi, On the probability that a random ± 1-matrix is singular. *J. Amer. Math. Soc.* **8** (1995), no. 1, 223–240.

[48] H. Kesten, A sharper form of the Doeblin–Lévy–Kolmogorov–Rogozin inequality for concentration functions. *Math. Scand.* **25** (1969), 133–144.

[49] A. Kolmogorov, Sur les propriétés des fonctions de concentrations de M. P. Lévy. *Ann. Inst. Henri Poincaré* **16** (1958), 27–34.

[50] J. Komlós, On the determinant of (0, 1) matrices. *Studia Sci. Math. Hungar.* **2** (1967), 7–21.

[51] N. Lehmann and H.-J. Sommers, Eigenvalue statistics of random real matrices. *Phys. Rev. Lett.* **67** (1991), no. 8, 941–944.

[52] J. E. Littlewood and A. C. Offord, On the number of real roots of a random algebraic equation. III. *Rec. Math. [Mat. Sbornik] N.S.* **12** (1943), no. 54, 277–286.

[53] A. E. Litvak, A. Lytova, K. Tikhomirov, N. Tomczak-Jaegermann, and P. Youssef, The smallest singular value of a shifted d-regular random square matrix. *Probab. Theory Related Fields* **173** (2019), no. 3–4, 1301–1347.

[54] A. E. Litvak, A. Lytova, K. Tikhomirov, N. Tomczak-Jaegermann, and P. Youssef, Structure of eigenvectors of random regular digraphs. *Trans. Amer. Math. Soc.* **371** (2019), no. 11, 8097–8172.

[55] A. E. Litvak, A. Lytova, K. Tikhomirov, N. Tomczak-Jaegermann, and P. Youssef, Circular law for sparse random regular digraphs. *J. Eur. Math. Soc. (JEMS)* **23** (2021), no. 2, 467–501.

[56] A. E. Litvak, A. Pajor, M. Rudelson, and N. Tomczak-Jaegermann, Smallest singular value of random matrices and geometry of random polytopes. *Adv. Math.* **195** (2005), no. 2, 491–523.

[57] A. E. Litvak and K. Tikhomirov, Singularity of sparse Bernoulli matrices. *Duke Math. J.*, to appear.

[58] G. V. Livshyts, The smallest singular value of heavy-tailed not necessarily i.i.d. random matrices via random rounding. *J. Anal. Math.* **145** (2021), 257–306.

[59] G. V. Livshyts, K. Tikhomirov, and R. Vershynin, The smallest singular value of inhomogeneous square random matrices. *Ann. Probab.* **49** (2021), no. 3, 1286–1309.

[60] K. Luh, Complex random matrices have no real eigenvalues. *Random Matrices Theory Appl.* **7** (2018), no. 1, 1750014, 17 pp.

[61] M. L. Mehta, *Random matrices and the statistical theory of energy levels*. Academic Press, New York, 1967.

[62] A. Mészáros, The distribution of sandpile groups of random regular graphs. *Trans. Amer. Math. Soc.* **373** (2020), no. 9, 6529–6594.

[63] H. H. Nguyen, Inverse Littlewood–Offord problems and the singularity of random symmetric matrices. *Duke Math. J.* **161** (2012), no. 4, 545–586.

[64] H. H. Nguyen and V. H. Vu, Small ball probability, inverse theorems, and applications. In *Erdős centennial*, pp. 409–463, Bolyai Soc. Math. Stud. 25, János Bolyai Math. Soc., Budapest, 2013.

[65] H. H. Nguyen and M. M. Wood, Cokernels of adjacency matrices of random r-regular graphs. 2018, arXiv:1806.10068.

[66] G. Pan and W. Zhou, Circular law, extreme singular values and potential theory. *J. Multivariate Anal.* **101** (2010), no. 3, 645–656.

[67] E. Rebrova and K. Tikhomirov, Coverings of random ellipsoids, and invertibility of matrices with i.i.d. heavy-tailed entries. *Israel J. Math.* **227** (2018), no. 2, 507–544.

[68] B. A. Rogozin, On the increase of dispersion of sums of independent random variables. *Teor. Veroyatn. Primen.* **6** (1961), 106–108.

[69] M. Rudelson, Invertibility of random matrices: norm of the inverse. *Ann. of Math. (2)* **168** (2008), no. 2, 575–600.

[70] M. Rudelson and K. Tikhomirov, The sparse circular law under minimal assumptions. *Geom. Funct. Anal.* **29** (2019), no. 2, 561–637.

[71] M. Rudelson and R. Vershynin, The least singular value of a random square matrix is $O(n^{-1/2})$. *C. R. Math. Acad. Sci. Paris* **346** (2008), no. 15–16, 893–896.

[72] M. Rudelson and R. Vershynin, The Littlewood–Offord problem and invertibility of random matrices. *Adv. Math.* **218** (2008), no. 2, 600–633.

[73] M. Rudelson and R. Vershynin, Smallest singular value of a random rectangular matrix. *Comm. Pure Appl. Math.* **62** (2009), no. 12, 1707–1739.

[74] M. Rudelson and R. Vershynin, Invertibility of random matrices: unitary and orthogonal perturbations. *J. Amer. Math. Soc.* **27** (2014), no. 2, 293–338.

[75] M. Rudelson and R. Vershynin, Non-asymptotic theory of random matrices: extreme singular values. In *Proceedings of the International Congress of Mathematicians. Volume III*, pp. 1576–1602, Hindustan Book Agency, New Delhi, 2010.

[76] M. Rudelson and O. Zeitouni, Singular values of Gaussian matrices and permanent estimators. *Random Structures Algorithms* **48** (2016), no. 1, 183–212.

[77] A. Sankar, *Smoothed analysis of Gaussian elimination*. PhD Thesis, 2004.

[78] A. Sankar, D. A. Spielman, and S.-H. Teng, Smoothed analysis of the condition numbers and growth factors of matrices. *SIAM J. Matrix Anal. Appl.* **28** (2006), no. 2, 446–476.

[79] Y. Seginer, The expected norm of random matrices. *Combin. Probab. Comput.* **9** (2000), no. 2, 149–166.

[80] S. Smale, On the efficiency of algorithms of analysis. *Bull. Amer. Math. Soc. (N.S.)* **13** (1985), no. 2, 87–121.

[81] D. A. Spielman and S.-H. Teng, Smoothed analysis of algorithms. In *Proceedings of the International Congress of Mathematicians, Vol. I (Beijing, 2002)*, pp. 597–606, Higher Ed. Press, Beijing.

[82] S. J. Szarek, Condition numbers of random matrices. *J. Complexity* **7** (1991), no. 2, 131–149.

[83] T. Tao and V. Vu, On the singularity probability of random Bernoulli matrices. *J. Amer. Math. Soc.* **20** (2007), no. 3, 603–628.

[84] T. Tao and V. Vu, Random matrices: the circular law. *Commun. Contemp. Math.* **10** (2008), no. 2, 261–307.

[85] T. Tao and V. Vu, Random matrices: the distribution of the smallest singular values. *Geom. Funct. Anal.* **20** (2010), no. 1, 260–297.

[86] T. Tao and V. Vu, Random matrices: universality of ESDs and the circular law. *Ann. Probab.* **38** (2010), no. 5, 2023–2065.

[87] T. Tao and V. Vu, Smooth analysis of the condition number and the least singular value. *Math. Comp.* **79** (2010), no. 272, 2333–2352.

[88] T. Tao and V. Vu, The condition number of a randomly perturbed matrix. In *STOC'07—Proceedings of the 39th Annual ACM Symposium on Theory of Computing*, pp. 248–255, ACM, New York, 2007.

[89] T. Tao and V. H. Vu, Inverse Littlewood–Offord theorems and the condition number of random discrete matrices. *Ann. of Math. (2)* **169** (2009), no. 2, 595–632.

[90] K. Tatarko, An upper bound on the smallest singular value of a square random matrix. *J. Complexity* **48** (2018), 119–128.

[91] K. Tikhomirov, Invertibility via distance for noncentered random matrices with continuous distributions. *Random Structures Algorithms* **57** (2020), no. 2, 526–562.

[92] K. Tikhomirov, Singularity of random Bernoulli matrices. *Ann. of Math. (2)* **191** (2020), no. 2, 593–634.

[93] T. Tran, The smallest singular value of random combinatorial matrices. 2019, arXiv:1909.04219.

[94] L. N. Trefethen and R. S. Schreiber, Average-case stability of Gaussian elimination. *SIAM J. Matrix Anal. Appl.* **11** (1990), no. 3, 335–360.

[95] R. Vershynin, Invertibility of symmetric random matrices. *Random Structures Algorithms* **44** (2014), no. 2, 135–182.

[96] J. von Neumann and H. H. Goldstine, Numerical inverting of matrices of high order. *Bull. Amer. Math. Soc.* **53** (1947), 1021–1099.

[97] V. H. Vu, Recent progress in combinatorial random matrix theory. *Probab. Surv.* **18** (2021), 179–200.

[98] J. H. Wilkinson, Error analysis of direct methods of matrix inversion. *J. Assoc. Comput. Mach.* **8** (1961), 281–330.

[99] P. M. Wood, Universality and the circular law for sparse random matrices. *Ann. Appl. Probab.* **22** (2012), no. 3, 1266–1300.

[100] Y. Q. Yin, Z. D. Bai, and P. R. Krishnaiah, On the limit of the largest eigenvalue of the large-dimensional sample covariance matrix. *Probab. Theory Related Fields* **78** (1988), no. 4, 509–521.

KONSTANTIN TIKHOMIROV

School of Mathematics, 686 Cherry street, Atlanta GA 30332, ktikhomirov6@gatech.edu

ABSTRACT CLASSIFICATION THEOREMS FOR AMENABLE C^*-ALGEBRAS

STUART WHITE

ABSTRACT

Operator algebras are subalgebras of the bounded operators on a Hilbert space. They divide into two classes: C^*-algebras and von Neumann algebras according to whether they are required to be closed in the norm or weak-operator topology, respectively.

In the 1970s Alain Connes identified the appropriate notion of amenabilty for von Neumann algebras, and used it to obtain a deep internal finite-dimensional approximation structure for these algebras. This structure is exactly what is needed for classification, and one of many consequences of Connes' theorem is the uniqueness of amenable II_1 factors, and later a complete classification of all simple amenable von Neumann algebras acting on separable Hilbert spaces.

The Elliott classification programme aims for comparable structure and classification results for C^*-algebras using operator K-theory and traces. The definitive unital classification theorem was obtained in 2015. This is a combination of the Kirchberg–Phillips theorem and the large scale activity in the stably finite case by numerous researchers over the previous 15–20 years. It classifies unital simple separable amenable C^*-algebras satisfying two extra hypotheses: a universal coefficient theorem which computes KK-theory in terms of K-theory and a regularity hypothesis excluding exotic high-dimensional behaviour. Today the regularity hypothesis can be described in terms of tensor products ($\mathcal{Z}$-stability). These hypotheses are abstract, and there are deep tools for verifying the universal coefficient theorem and $\mathcal{Z}$-stability in examples.

This article describes the unital classification theorem, its history and context, together with the new abstract approach to this result developed in collaboration with Carrión, Gabe, Schafhauser, and Tikuisis. This method makes a direct connection to the von Neumann algebraic results, and does not need to obtain any kind of approximation structure inside C^*-algebras en route to classification. The companion survey [32] focuses on the role of the $\mathcal{Z}$-stability hypothesis, and the associated work on "regularity."

MATHEMATICS SUBJECT CLASSIFICATION 2020

Primary 46L35; Secondary 46L05, 46L55, 37A55

KEYWORDS

Operator algebras, C^*-algebras, classification

INTERNATIONAL CONGRESS OF MATHEMATICIANS 2022 JULY 6–14

Proc. Int. Cong. Math. 2022, Vol. 4, pp. 3314–3338
DOI 10.4171/ICM2022/183

1. OPERATOR ALGEBRAS

This survey aims to describe how the classification theorems for simple amenable C^*-algebras parallel Connes' celebrated von Neumann algebra classification results from the 1970s. The first four sections set the context and are written for a nonexpert. Section 5 focuses on the two key hypotheses required for the classification theorem. The last four sections give a flavour of some of the ideas involved in these results, and require increasingly more background.

Operator algebras originate in the mathematical foundations of quantum mechanics. There are two main classes: C^*-algebras and von Neumann algebras. A C^*-algebra is a complex Banach algebra A equipped with an involution * satisfying the fundamental C^*-identity, $\|x^*x\| = \|x\|^2$ for all $x \in A$. This seemingly innocuous formula binds the algebraic and analytic aspects of the definition together. It is the source of a surprising amount of rigidity: the norm on a C^*-algebra is unique and determined algebraically through spectral data. C^*-algebras are based on abstraction of the bounded operators on a Hilbert space $\mathcal{H}$, and so examples arise from Hilbert space representations. For example, the left regular representation $\lambda : G \to \mathcal{B}(\ell^2(G))$ of a discrete group G is given by $\lambda_g(f)(h) := f(g^{-1}h)$ for $g, h \in G$ and $f : G \to \mathbb{C}$ in $\ell^2(G)$. To this we associate the *reduced group C^*-algebra* of G, $C^*_\lambda(G)$ as the C^*-algebra generated by the unitaries $(\lambda_g)_{g\in G}$. Group C^*-algebras provide a framework linking operator algebras with group representation theory and harmonic analysis.

Norm convergence of bounded operators is a restrictive condition, and there are a number of finer topologies on $\mathcal{B}(\mathcal{H})$: the strong operator topology of pointwise convergence, the weak operator topology of pointwise weak convergence, and the weak*-topology from the duality with the trace class operators. Von Neumann algebras are the *-subalgebras of $\mathcal{B}(\mathcal{H})$ which are closed in these topologies (they all give *-subalgebras the same closure). The group von Neumann algebra $L(G)$ is the von Neumann algebra generated by the image of the left regular representation, so $L(G)$ is the weak operator closure of $C^*_\lambda(G)$.

The distinction between the norm and strong or weak operator topologies creates a profound difference between the flavour of C^*-algebras and von Neumann algebras. The former are topological in nature, while the latter are measure-theoretic. This is evident in the abelian case: via the Gelfand transform, every commutative C^*-algebra arises as the algebra $C_0(X)$ of continuous functions vanishing at infinity on a locally compact Hausdorff space X. While uniform limits of continuous functions are continuous, pointwise limits are only guaranteed to be measurable, and a general abelian von Neumann algebra is $L^\infty(X,\mu)$ for some measure space (X,μ). In the setting of discrete abelian groups G, these spaces come from the Fourier transform: $C^*(G) \cong C_0(\hat{G})$ and $L(G) \cong L^\infty(\hat{G})$, where $\hat{G}$ is the Fourier dual group. Considering $G = \mathbb{Z}$ and $\mathbb{Z}^2$, we have $C^*_\lambda(\mathbb{Z}) \ncong C^*_\lambda(\mathbb{Z}^2)$ as the Fourier duals $\mathbb{T}$ and $\mathbb{T}^2$ are not homeomorphic. The circle and torus are measurably indistinguishable, so $L(\mathbb{Z}) \cong L(\mathbb{Z}^2)$.

Another important family of examples arises from dynamics. Given a group action $G \curvearrowright X$, we obtain an induced action α on the relevant abelian operator algebra, $C_0(X)$ or

$L^\infty(X)$, according to whether X is topological or measurable. Reminiscent of the semidirect product construction from group theory, the reduced crossed product is a C^*-algebra $C_0(X) \rtimes_r G$ or von Neumann algebra $L^\infty(X) \rtimes G$ generated by the relevant abelian algebra $C_0(X)$ or $L^\infty(X)$ and unitaries $(u_g)_{g\in G}$ that implement the induced action: $u_g f u_g^* = \alpha_g(f)$ for $g \in G$ and f in the commutative subalgebra. This construction is set up so that the unitaries u_g generate $C^*_\lambda(G)$ or $L(G)$, respectively. These operator algebras provide tools for examining actions whose quotient spaces are non-Hausdorff, such as the action $\mathbb{Z} \curvearrowright \mathbb{T}$ of rotation by an irrational multiple θ of 2π leading to the famous irrational rotation C^*-algebras $A_\theta = C(\mathbb{T}) \rtimes_{r,\theta} \mathbb{Z}$.

2. PROJECTIONS AND APPROXIMATE FINITE DIMENSIONALITY

Classification results for operator algebras go back to the foundational work of Murray and von Neumann on projections. This relativises the classification of closed subspaces of an infinite-dimensional Hilbert space by their dimension to an operator algebra A. Projections $p, q \in A$ are equivalent (written $p \sim q$) if there exists $v \in A$ with $v^*v = p$ and $vv^* = q$; p is sub-equivalent to q ($p \precsim q$) if there exists $q_0 \leq q$ with $p \sim q_0$. Akin to Dedekind's definitions for sets, a projection is called infinite if it is equivalent to a proper sub-projection of itself, and finite otherwise. Likewise a unital operator algebra is infinite or finite according to the behaviour of its unit. This theory is particularly clean for von Neumann algebras where $p \precsim q$ and $q \precsim p$ imply $p \sim q$, and is at its crispest in the case of *factors*. A factor is a von Neumann algebra with trivial centre; these are precisely the simple von Neumann algebras. Factors are the irreducible building blocks of von Neumann algebras and the set of equivalence classes of projections in a factor is always totally ordered.

Murray and von Neumann used this total ordering to loosely divide factors into types. The type I factors have a discrete order $\{0, 1, 2, \ldots, n\}$ or $\{0, 1, 2, \ldots\} \cup \{\infty\}$. The only examples are bounded operators on a Hilbert space. The interest begins with the type II factors, where the projections form a continuum. These subdivide further into II_1 factors, where the unit is finite, and II_∞ factors where it is infinite. After rescaling, the Murray–von Neumann equivalence classes of projections in a II_1 factor $\mathcal{M}$ identify with $[0, 1]$ and are determined by the *trace*. That is, there is a unique positive linear functional τ with $\tau(1_{\mathcal{M}}) = 1$ satisfying the trace identity $\tau(xy) = \tau(yx)$. Then projections $p, q \in \mathcal{M}$ satisfy $p \precsim q$ if and only if $\tau(p) \leq \tau(q)$. Every II_∞ factor is a von Neumann tensor product of a II_1 factor and $\mathcal{B}(\mathcal{H})$, so for many purposes the study of II_∞ factors reduces to that of II_1 factors. Finally, there are the type III factors, which are *purely infinite*: all non-zero projections are infinite (and in the separably acting case equivalent).

Murray and von Neumann constructed an example of a II_1 factor $\mathcal{R}$ as a suitable completion of the algebraic tensor product of infinitely many copies of the algebra M_2 of 2×2 matrices, where the trace comes from extending the product of the normalised traces on the matrix algebras. This led them to isolate a key internal approximation property: *hyperfiniteness*. A von Neumann algebra is *hyperfinite* if finite families of operators can be approximated in strong operator topology by finite dimensional subalgebras (in $\mathcal{R}$ one uses finite tensor products of copies of M_2 to perform these approximations).

Murray and von Neumann's celebrated uniqueness theorem from [23] shows that all separably acting hyperfinite II_1 factors are isomorphic. In particular, there is nothing special about the 2×2 matrices above. Any choice of matrix size would lead to the same hyperfinite II_1 factor, denoted $\mathcal{R}$.

Examples of II_1 factors appear naturally from groups and dynamics. The von Neumann algebra of an infinite discrete group G is a II_1 factor if and only if G is an ICC group: one where all non-trivial elements have infinite conjugacy classes. When G is an inductive limit of finite groups, $L(G)$ will be hyperfinite. If G is additionally ICC (such as the group S_∞ of all finitely supported permutations of the natural numbers) then $L(G)$ will be isomorphic to $\mathcal{R}$ by Murray and von Neumann's uniqueness theorem. Approximating an irrational rotation by a rational rotation one gets hyperfiniteness for the II_1 factors $L^\infty(\mathbb{T}) \rtimes \mathbb{Z}$ associated to irrational rotations.

For C^*-algebras, *approximate finite dimensionality* (AF) is defined analogously to hyperfiniteness working with the operator norm in place of the strong topology. This changes things significantly: the infinite C^*-tensor products $M_{2^\infty} := \bigotimes_{n=1}^\infty M_2$ and $M_{3^\infty} := \bigotimes_{n=1}^\infty M_3$ are not isomorphic. Indeed, there is no unital embedding of M_2 into M_{3^∞}, as this would give rise (by a perturbation argument) to a unital embedding of M_2 into some M_{3^n}, which is impossible. The difference is that norm-close projections are equivalent, whereas strong-operator-close projections need not be. This line of reasoning led Glimm to classify all such C^*-algebra infinite tensor products of matrices (known as uniformly hyperfinite (UHF) algebras) by the supernatural number consisting of the infinite product of primes appearing in the matrix sizes. Later, Bratelli examined isomorphisms between separable AF C^*-algebras in terms of diagrammatic data for inclusions of finite-dimensional C^*-algebras, and Elliott gave an algebraic classification which turned out to be in terms of K-theory.

Operator K-theory is a non-commutative generalisation of Atiyah and Hirzebruch's topological K-theory. For a unital C^*-algebra A, $K_0(A)$ is defined as the Grothendieck group of the abelian semigroup of projections in matrices over A up to equivalence (addition is given by a diagonal direct sum). Elliott's invariant for unital AF-algebras is then $(K_0(A), K_0(A)_+, [1_A]_0)$, where $K_0(A)_+$ is the positive cone arising from projections in matrices over A, and $[1_A]_0$ is the class of the unit. The unital case of Elliott's theorem shows that for separable unital AF C^*-algebras A and B, any isomorphism $\Phi : K_0(A) \to K_0(B)$ with $\Phi(K_0(A)_+) = K_0(B)_+$ and $\Phi([1_A]_0) = [1_B]_0$ comes from an isomorphism $\phi : A \to B$.

The K_0-functor preserves inductive limits, so one can easily compute $K_0(M_{2^\infty})$ as the dyadic rationals $\{m/2^n : m \in \mathbb{Z}, n = 0, 1, \dots\}$ with positivity inherited from the positive reals, and $[1_{M_{2^\infty}}]_0 = 1$, whereas $K_0(M_{3^\infty})$ is the triadic rationals.

3. CONNES' THEOREM

While Murray and von Neumann's uniqueness theorem is a beautiful result, it is not easy to apply directly; obtaining hyperfiniteness explicitly is out of reach in many examples. Connes resolved this issue in a landmark abstract characterisation from the 1970s.

Theorem 3.1 (Connes, [5]). *Let $\mathcal{M} \subset \mathcal{B}(\mathcal{H})$ be a von Neumann algebra. The following are equivalent:*

(1) *$\mathcal{M}$ is injective, i.e., there is a linear contraction $\Phi : \mathcal{B}(\mathcal{H}) \to \mathcal{M}$ with $\Phi(x) = x$ for $x \in \mathcal{M}$.*

(2) *$\mathcal{M}$ is semidiscrete.*

(3) *$\mathcal{M}$ is hyperfinite.*

Injectivity is so named because it is equivalent to injectivity of $\mathcal{M}$ in the category of von Neumann algebras and completely positive and contractive (cpc) maps. The definition is independent of the representation of $\mathcal{M}$. Semidiscreteness is a finite-dimensional approximation property, though a priori of a much weaker nature than hyperfiniteness. It asks for point weak* approximations of the identity map by cpc maps factoring through finite-dimensional C^*-algebras—such maps preserve the adjoint and order structure (at all levels of matrix amplification), but are not required to preserve the product. Both injectivity and semidiscreteness are much easier to access than hyperfiniteness in examples; indeed, the equivalence of amenablility of a discrete group G with both injectivity and semidiscreteness of $L(G)$ significantly predates Connes' theorem. Likewise all actions of amenable groups on injective von Neumann algebras produce injective crossed products.

The marriage of Connes' structural theorem with Murray and von Neumann's uniqueness theorem gives a definitive classification result—there is a unique injective separable II_1 factor—with readily verifiable hypotheses. All the von Neumann algebras associated to countably infinite discrete amenable ICC groups are isomorphic. Moreover, Connes was able to apply Theorem 3.1 to his earlier work on type III factors, giving an almost complete classification of separable injective factors. The puzzle was completed 10 years later when Haagerup established the uniqueness of the hyperfinite III_1 factor [13].

Theorem 3.2 (Connes, Haagerup, Murray–von Neumann). *There is a complete classification of separable injective factors.*

The impact of Theorem 3.1 goes well beyond classification results. It shows that hyperfiniteness passes to subalgebras of the hyperfinite II_1 factor, which is vital to Jones' theory of subfactors, and directly inspired the classification of amenable equivalence relations by Connes, Feldman, and Weiss.

Connes proved Theorem 3.1 in the II_1 factor case, and deduced the other cases from this. A central component of his argument is a deep generalisation of characterisations of amenability in terms of invariant means and Følner sets to traces on operator algebras. As such it is highly non-constructive, and it remains completely out of reach to describe hyperfiniteness of $L(G)$ explicitly in terms of Følner sets for an amenable group G. Both Connes' theorem and the techniques involved remain completely instrumental today; for example they play a major role in Popa's deformation-rigidity theory.

Various aspects of Connes' arguments, and the later proofs of Theorem 3.1 by Haagerup and Popa have heavily influenced developments for C^*-algebras. Here I highlight two ingredients of Connes' proof for later comparison.

(1) Connes approximates the trace on a separable injective II_1 factor $\mathcal{M}$ externally by a sequence of approximately trace preserving, approximately multiplicative cpc maps (in normalised Hilbert–Schmidt norm) to matrices. Such a sequence is conveniently encoded by an embedding of $\mathcal{M}$ into the ultrapower $\mathcal{R}^\omega$ of the hyperfinite II_1 factor.

(2) It is clear that the infinite algebraic tensor product of matrices satisfies $\bigotimes_{n=1}^\infty M_2 \cong \bigotimes_{n=1}^\infty M_2 \otimes \bigotimes_{n=1}^\infty M_2$. This persists in the weak operator closure used to obtain $\mathcal{R}$, which becomes idempotent for the von Neumann tensor product, $\mathcal{R} \cong \mathcal{R} \otimes \mathcal{R}$. This is known as *self-absorption*. A major aspect of Connes' proof is to show that $\mathcal{R}$ is a tensorial unit for all separable injective II_1 factors $\mathcal{M}$, i.e., $\mathcal{M} \cong \mathcal{M} \otimes \mathcal{R}$. This condition had been previously developed by McDuff, who characterised these *McDuff factors* by the existence of approximately central matrix subalgebras **[22]**.

4. SIMPLE NUCLEAR C^*-ALGEBRAS AND THE ELLIOTT CLASSIFICATION PROGRAMME

In a nutshell, the Elliott classification programme aims for a C^*-analogue of the Connes, Haagerup, and Murray and von Neumann classification theorem. Ideally, we seek complete classification results, with abstract hypotheses that are widely verifiable in a range of examples. A fundamental question is: *which C^*-algebras should be classified, and by what data*? This is illustrated over the next two sections by the crossed products $C(X) \rtimes_r G$ associated to the action of a discrete group G on a compact Hausdorff space X.

Semidiscreteness has a direct analogue for C^*-algebras—the *completely positive approximation property* (CPAP). The only change to the definition is to use the point-norm topology in the C^*-setting. The CPAP is in turn equivalent to *nuclearity*, which is characterised by the uniqueness of a C^*-norm on the algebraic tensor product $A \odot B$ for all C^*-algebras B. These conditions give the appropriate notion of amenability for C^*-algebras. For instance, a discrete group G is amenable if and only if $C^*_\lambda(G)$ is nuclear and nuclearity is preserved by actions of amenable groups. Moreover, as a surprising application of Connes' theorem, a C^*-algebra A has the CPAP if and only if its bidual A^{**} (which is naturally a von Neumann algebra) is semidiscrete.

The most naive attempt to generalise Connes' theorem fails spectacularly: nuclear C^*-algebras will rarely be AF (for example, $C[0,1]$ is nuclear, but certainly not AF). Nonetheless, there are many situations where nuclear C^*-algebras have properties akin to those of injective von Neumann algebras, after making some subtle adjustments to allow for topological phenomena. For example, separable nuclear C^*-algebras acting on the same separable Hilbert spaces whose unit balls are close in the Hausdorff metric are spatially isomorphic **[4]**. This is analogous to existing results for injective von Neumann algebras **[3]**—in both cases amenability is the key hypothesis—but one has to accept somewhat less strong control on the spatial isomorphism in the C^*-setting.

For group actions, $C(X) \rtimes_r G$ is nuclear whenever G is amenable. One can also define amenable actions of non-amenable groups, such as hyperbolic groups acting on their boundary. Then $C(X) \rtimes_r G$ is nuclear precisely when $G \curvearrowright X$ is amenable.

Within the class of nuclear C^*-algebras, the classification programme has mainly focused on simple C^*-algebras. From one point of view, these are analogous to factors—the simple von Neumann algebras—but, on the other hand (and unlike von Neumann algebras) we cannot decompose a general C^*-algebra in terms of simple algebras. Nevertheless, simple C^*-algebras have proved a very fertile ground. For example, the construction of the Cuntz algebras $\mathcal{O}_n$ led Cuntz to identify the C^*-analogue of type III factors—*purely infinite simple C^*-algebras*. These have an abundance of projections (simple purely infinite C^*-algebras have real rank zero, and hence are the closed linear span of their projections) and any two nonzero projections p and q satisfy $p \precsim q$ and $q \precsim p$. In a von Neumann factor, this would force $p \sim q$, but this need not hold in a C^*-algebra. In fact, the equivalence classes of projections can be very complex: any pair (G_0, G_1) of countable abelian groups can appear as the K-theory of a separable simple nuclear purely infinite C^*-algebra. Two particularly important examples are $\mathcal{O}_2$ and $\mathcal{O}_\infty$ which have K-theories $(0, 0)$ and $(\mathbb{Z}, 0)$, respectively. In the setting of amenable group actions $G \curvearrowright X$, paradoxicality can be used to obtain simple purely infinite nuclear C^*-algebras.

The Elliott programme to classify separable simple nuclear C^*-algebras began in earnest after twin breakthroughs in the 1990s. In the infinite setting, Kirchberg's revolutionary work led to the Kirchberg–Phillips theorem: a complete K-theoretic classification of simple separable nuclear and purely infinite C^*-algebras—now called *Kirchberg algebras*—satisfying Rosenberg and Schochet's universal coefficient theorem (see Section 5).

Theorem 4.1 (Unital Kirchberg–Phillips theorem **[16,24]**). *Let A and B be unital Kirchberg algebras satisfying Rosenberg and Schochet's universal coefficient theorem. Then $A \cong B$ if and only if there is an isomorphism $\alpha_* : K_*(A) \xrightarrow{\cong} K_*(B)$ with $\alpha_0([1_A]_0) = [1_B]_0$.*

In the finite setting, Elliott's classification of A$\mathbb{T}$-algebras (inductive limits of C^*-algebras $C(\mathbb{T}, F)$, where F is finite dimensional) of real rank zero by ordered K-theory, combined with the Elliott–Evans theorem that irrational rotation C^*-algebras are A$\mathbb{T}$, to spark substantial work on inductive limit C^*-algebras. Thomsen soon realised that ordered K-theory alone was not enough to classify all simple A$\mathbb{T}$ algebras. Traces are also needed.

As with von Neumann algebras, a trace τ on a simple C^*-algebra A ensures finiteness: if $v^*v \leq vv^*$, then $\tau(vv^* - v^*v) = 0$, forcing $vv^* = v^*v$. Moreover, $M_n(A)$ will be finite for all $n \in \mathbb{N}$, i.e., A is *stably finite*. A deep theorem of Haagerup shows that stably finite simple nuclear (and, more generally, exact) C^*-algebras always admit traces. So one can detect stable finiteness by traces. However, there is not a direct dichotomy between the finite and infinite: Rørdam produced a finite simple nuclear C^*-algebra A which is not stably finite [25]. Such an algebra does not have a von Neumann counterpart.

We write $T(A)$ for the set of traces on A. When A is unital, $T(A)$ is compact in the weak*-topology, and convex Moreover $T(A)$ is a Choquet simplex (this ensures that $T(A)$ can be written as an inverse limit of finite simplices, a fact which is very useful in producing C^*-algebras with given trace spaces). We view $T(A)$ as a family of non-commutative measures, and in many examples it can be determined explicitly. Indeed, traces on simple crossed products $C(X) \rtimes_r G$ correspond to G invariant measures on X. In this way, the irrational rotation C^*-algebras have a unique trace coming from the Lebesgue measure on the circle.

It is often convenient to work with the space $\operatorname{Aff} T(A)$ of continuous affine functions $T(A) \to \mathbb{R}$. By Kadison duality, this is canonically dual to $T(A)$. (One reason for working with $A \mapsto \operatorname{Aff} T(A)$ is that makes the entire classification invariant covariant). Every trace induces a well-defined order-preserving map $K_0(A) \to \mathbb{R}$ mapping $[1_A]_0$ to 1, and so there is a *pairing* between K-theory and traces. At the level of affine functions this is given by

$$\rho_A : K_0(A) \to \operatorname{Aff} T(A), \quad \rho_A(x)(\tau) = \tau(x), \quad x \in K_0(A),\ \tau \in T(A). \tag{4.1}$$

Putting these ingredients together, the *Elliott invariant* of a unital C^*-algebra A is

$$\operatorname{Ell}(A) = \big(K_0(A), K_0(A)_+, [1_A]_0, K_1(A), \operatorname{Aff} T(A), \rho_A\big), \tag{4.2}$$

and Elliott conjectured that this should classify all non-elementary simple separable unital nuclear C^*-algebras, analogously to the classification of injective factors [7].

For the irrational rotation algebras $A_\theta = C(\mathbb{T}) \rtimes_{r,\theta} \mathbb{Z}$, one has $K_*(A_\theta) \cong (\mathbb{Z}^2, \mathbb{Z}^2)$. The unique trace τ embeds $K_0(A) \subset \mathbb{R}$ by $\tau(m,n) = m + \theta n$ for $m, n \in \mathbb{Z}$ and here the trace determines the positive cone on $K_0(A)$: $x \geq 0$ in $K_0(A)$ if and only if $x = 0$ or $\tau(x) > 0$. A computation then shows $A_\theta \cong A_\phi$ if and only if $\phi \in \pm\theta + \mathbb{Z}$.

For irrational rotation algebras both $K_0(A)_+$ and the trace pairing carry the same information. In generality, neither of $K_0(A)_+$ and $(T(A), \rho_A)$ can be recovered from the other. However, in all cases where classification has been obtained $K_0(A)_+$ is determined by $T(A)$ as above. Thus, although the historical evolution of the Elliott invariant includes $K_0(A)_+$ (this dates back to the classification of AF algebras, whereas traces were only added to the invariant later), with hindsight it is perhaps natural to work with the smaller invariant $KT_u(A)$ consisting of $(K_0(A), [1_A]_0, K_1(A), T(A), \rho_A)$. I prefer this approach for a number of reasons: it makes it clear that the order structure on K_0 does not play an explicit role; the range of KT_u on simple separable unital and nuclear C^*-algebras is completely understood (it remains a very challenging problem to determine all possible orders on $K_0(A)_+$ when they are not given by the trace pairing); and KT_u interacts very cleanly with the crucial tensorial absorption condition of $\mathcal{Z}$-stability (see Section 5.2 below). I choose to use the Elliott invariant in the main classification theorems, reflecting both the important role Ell has played historically and the amazing vision Elliott showed in his conjecture, and subsequent programme.

5. THE UNITAL CLASSIFICATION THEOREM

Rørdam's examples (and precursors by Villadsen) showed that there are highly exotic simple nuclear C^*-algebras with phenomena that have no von Neumann algebraic counterpart. Toms later refined these ideas, constructing simple nuclear C^*-algebras which can never be classified by reasonably tractable data **[31]**. Thus we need additional, more subtle, hypotheses to divide the classifiable stably finite simple nuclear C^*-algebras from the exotic. By the late 1990s two necessary conditions were known:

(1) A satisfies the universal coefficient theorem;

(2) A is stable under tensoring by the Jiang–Su algebra $\mathcal{Z}$, i.e., $A \cong A \otimes \mathcal{Z}$.

5.1. The universal coefficient theorem

Kasparov's bivariant KK-theory is one of the fundamental tools in the classification of C^*-algebras, providing a tool unifying K-theory and extension theory ($K_0(A) \cong KK(\mathbb{C}, A)$ and $\operatorname{Ext}(A) \cong KK(A, C_0(\mathbb{R}))$). In fact, Kirchberg and Phillips both showed that equivalence in KK-theory (viewed as a very weak kind of homotopy equivalence) gives rise to an isomorphism for Kirchberg algebras. Thus the Kirchberg–Phillips theorem has the flavour of a homotopy rigidity result.

Theorem 5.1 (Classification of unital Kirchberg algebras by KK-theory). *Let A and B be unital Kirchberg algebras. Then $A \cong B$ if and only if there is a KK-equivalence $\alpha \in KK(A, B)$ with $\alpha \cdot [1_A]_0 = [1_B]_0$.*

The key tool for computing KK-theory is Rosenberg and Schochet's universal coefficient theorem (UCT) from **[26]**. The Kasparov product gives a map $KK(A, B) \to \operatorname{Hom}(K_*(A), K_*(B))$, and a C^*-algebra A satisfies the UCT when this map fits into a short exact sequence

$$0 \to \operatorname{Ext}^1_{\mathbb{Z}}\big(K_*(A), K_{*+1}(B)\big) \to KK(A, B) \to \operatorname{Hom}\big(K_*(A), K_*(B)\big) \to 0 \qquad (5.1)$$

for all separable C^*-algebras B. In particular, when both A and B satisfy the UCT, an isomorphism $K_*(A) \cong K_*(B)$ lifts to a KK-equivalence; this is how one returns from Theorem 5.1 to the K-theoretic classification of Theorem 4.1.

Rosenberg and Schochet established their universal coefficient theorem for abelian C^*-algebras, and then showed that the class of nuclear C^*-algebras satisfying the UCT is closed under various natural operations (in particular, all C^*-inductive limits covered by various classification results lie in the UCT class). It is a major and rather pressing open problem whether all separable nuclear C^*-algebras satisfy the UCT, but for the purposes of applying the classification theorem to concrete examples, this is rarely a difficulty. Pretty much all separable nuclear C^*-algebras that can be explicitly described are known to satisfy the UCT, often through Tu's result (building on Higson and Kasparov's work on the Baum–Connes conjecture) that all C^*-algebras associated to amenable groupoids satisfy the UCT. In particular, all crossed products $C(X) \rtimes_r G$ coming from amenable actions satisfy the UCT.

Moreover as the examples realising the invariant for Kirchberg algebras all satisfy the UCT, the UCT is necessary for a K-theoretic classification.

5.2. $\mathcal{Z}$-stability

The Cuntz algebra $\mathcal{O}_\infty$ satisfies $K_*(\mathcal{O}_\infty) \cong K_*(\mathbb{C}) = (\mathbb{Z}, 0)$. Since $\mathcal{O}_\infty$ satisfies the UCT, one can apply the Künneth formula to obtain $K_*(A \otimes \mathcal{O}_\infty) \cong K_*(A)$ for all separable A. So classification predicts that a Kirchberg algebra A should be isomorphic to $A \otimes \mathcal{O}_\infty$, i.e., A is *$\mathcal{O}_\infty$-stable*. This was confirmed in one of Kirchberg's famous absorption theorems:

Theorem 5.2 (Kirchberg's absorption theorems [16,17]). *Let A be a separable simple nuclear C^*-algebra. Then:*

(1) *A is purely infinite if and only if $A \otimes \mathcal{O}_\infty \cong A$;*

(2) $A \otimes \mathcal{O}_2 \cong \mathcal{O}_2$.

Just as Connes goes via $\mathcal{R}$-stability of injective II_1 factors en-route to hyperfiniteness, the published approaches to the classification of Kirchberg algebras all use the $\mathcal{O}_\infty$-absorption theorem in a crucial way. With hindsight $\mathcal{O}_\infty$-absorption is the key hypothesis which enables the classification of Kirchberg algebras. The absorption theorem then enables classification to be accessed via the more elementary condition of pure infiniteness. This view point is strengthened by Kirchberg's subsequent classification (by ideal related KK-theory) of all separable nuclear $\mathcal{O}_\infty$-stable C^*-algebras.

Returning to stably finite C^*-algebras, it is natural to ask *what is the right analogue of the hyperfinite* II_1 *factor*? A naive first answer might be the CAR-algebra, M_{2^∞}, but this is not canonical. Or one could try the universal UHF-algebra $\mathcal{Q} = \bigotimes_{n=2}^\infty M_n$, but this is too big: M_{2^∞} is a tensorial unit for $\mathcal{Q}$, not the other way round. No UHF-algebra is a tensorial unit for all its fellows. We want a stably finite unital simple C^*-algebra with K-theory $(\mathbb{Z}, 0)$ generated by the class of the unit and a unique trace, so that (assuming the Künneth formula) it will be a tensorial unit at the level of K-theory and traces, just as $\mathcal{O}_\infty$ is for Kirchberg algebras. One such C^*-algebra is $\mathbb{C}$. Are there others?

In the mid-1990s, Elliott constructed infinite dimensional stably finite simple separable nuclear C^*-algebras with arbitrary K-theory/trace pairings and so implicitly obtained a C^*-algebra with the properties above. A few years later Jiang and Su tackled this question more explicitly from the view point of tensorial absorption, giving another construction of what we now call the Jiang–Su algebra $\mathcal{Z}$ (though at the time it would not have been obvious that Elliott and Jiang and Su produced the same algebra) proving $\mathcal{Z} \cong \mathcal{Z} \otimes \mathcal{Z}$. Accordingly, it makes sense to consider *$\mathcal{Z}$-stable* C^*-algebras—those A for which $A \cong A \otimes \mathcal{Z}$—and, moreover, through the later abstract framework of strongly self-absorbing algebras, Winter showed that $\mathcal{Z}$-stability is, in a precise sense, the minimal tensorial absorption hypothesis akin to the McDuff property of a II_1 factor.

While $\mathcal{Z}$ is a little tricky to construct, $\mathcal{Z}$-stability of a separable C^*-algebra can be described without direct reference to $\mathcal{Z}$ and in a comparable fashion to McDuff's characterisation of $\mathcal{R}$-stable II_1 factors in terms of approximately central matrix subalgebras.

Via Matui and Sato's breakthrough [20], this is particularly clean for a stably finite simple separable nuclear C^*-algebra which is $\mathcal{Z}$-stable when it contains tracially large approximately central cones over matrices. This is vital in Kerr's approach to detecting $\mathcal{Z}$-stability of crossed products [14] (leading to the recent touchstone result that all free minimal actions of elementary amenable groups on finite dimensional spaces have $\mathcal{Z}$-stable crossed product [15]). I will describe $\mathcal{Z}$ and $\mathcal{Z}$-stability further in the companion survey article [32].

Just as the UCT is necessary for classification, so too is $\mathcal{Z}$-stability. Not only are the models realising all K-theory/trace pairings $\mathcal{Z}$-stable, but tensoring by $\mathcal{Z}$ acts as the identity on KT_u, and, when A is exact, the order on $K_0(A \otimes \mathcal{Z})$ is always given by traces. Moreover, combining work of Kirchberg and Rørdam, for a unital separable nuclear C^*-algebra without traces, $\mathcal{Z}$-stability and $\mathcal{O}_\infty$-stability are equivalent. In particular, $\mathcal{Z}$-stability is a generalisation of the classifiability hypothesis for Kirchberg algebras.

5.3. The unital classification theorem, dichotomy, and Toms–Winter regularity

The combined efforts of large numbers of researchers over close to 30 years have culminated in the definitive classification theorems for simple nuclear C^*-algebras, providing the topological counterpart to Connes' theorem. The two subtle hypotheses in the previous subsections are sufficient as well as necessary. We state this in the unital case (and call C^*-algebras satisfying the hypothesis of the following theorem *classifiable*). In particular, all crossed products $C(X) \rtimes_r G$ arising from free minimal actions of countable elementary amenable groups on compact metrisable spaces of finite covering dimension are classifiable.

Theorem 5.3 (The unital classification theorem). *Let A and B be unital separable simple nuclear C^*-algebras which are $\mathcal{Z}$-stable and satisfy the UCT. Then A and B are isomorphic if and only if* $\mathrm{Ell}(A) \cong \mathrm{Ell}(B)$.

As noted above the unital classification theorem is accompanied by a "range of the invariant theorem": any K-theory/trace pairing can arise. This can be used to establish properties of all classifiable C^*-algebras through models. For example, all stably finite classifiable C^*-algebras are approximately subhomogeneous (with at most 2-dimensional building blocks); inductive limit structure is a consequence of classification. Recently Li used this to show that all classifiable C^*-algebras arise from twisted étale groupoids [18].

Although the unital classification theorem covers both stably finite and purely infinite C^*-algebras, the two cases are handled separately. A beautiful dichotomy theorem of Kirchberg shows that any unital simple exact C^*-algebra which is a tensor product of two infinite-dimensional C^*-algebras (such as a $\mathcal{Z}$-stable nuclear C^*-algebra) is either purely infinite or stably finite. Moreover, the presence or absence of traces decides in which camp a classifiable C^*-algebra is found. The unital classification theorem is then the combination of the Kirchberg–Phillips theorem, and the stably finite case of the unital classification theorem. The rest of the article focuses on the stably finite situation.

The stably finite unital classification theorem was originally obtained in 2015 by combining [8, 11, 12, 30] (and with $\mathcal{Z}$-stability being replaced with the at the time stronger hypothesis of finite nuclear dimension, of which a tiny bit more below). These in turn build on decades of work—the unital classification theorem is the collective result of the entire C^*-research community—but in this millennium two names stand out: Lin and Winter. They drove two major strands of activity in parallel: classification through tracial approximations and regularity through dimension and $\mathcal{Z}$-stability. The cross fertilisations between these directions have been the source of many of the breakthroughs which have peppered the route to Theorem 5.3.

On the classification side, inspired by Popa's local quantisation technique (a C^*-version of the ideas in his proof of injectivity implies hyperfiniteness) Lin introduced the notion of tracial approximations in the early 2000s. These are a kind of internal approximation by subalgebras whose unit is a projection which is uniformly large in trace. Weakening the approximation in this way allows more algebras to be reached; the class of tracially AF-algebras is larger than the class of AF-algebras. Lin and collaborators then massively developed these ideas in a huge body of work culminating in [11, 12].

Meanwhile, Winter and his collaborators developed non-commutative versions of covering dimension (decomposition rank, and then nuclear dimension) for C^*-algebras through refined versions of the completely positive approximation property. This is one of the central concepts in the Toms–Winter conjecture, and combining Winter's $\mathcal{Z}$-stability theorem from [33] with the recent [2] (which completes a line of work going back to Matui and Sato's breakthrough [21]) a simple separable unital nuclear non-elementary C^*-algebra is $\mathcal{Z}$-stable if and only if it has finite nuclear dimension. In the setting of the action of a group G on a finite-dimensional compact metrizable space X, one can directly estimate the nuclear dimension of $C(X) \rtimes G$ when G is nilpotent, but one cannot expect direct estimates to work much more generally. In contrast, $\mathcal{Z}$-stability can be obtained much more generally: now when G is elementary amenable. A detailed discussion of regularity is out of scope here; I will discuss this further in [32].

The two strands come together in a number of landmarks, such as Winter's strategy [34] for converting a strong form of classification for UHF-stable C^*-algebras to the classification of $\mathcal{Z}$-stable C^*-algebras. The later result is used in the monumental work of Gong, Lin, and Niu [11, 12] to classify all $\mathcal{Z}$-stable C^*-algebras A with the property that for a UHF-algebra U, $A \otimes U$ has a certain 1-dimensional tracial approximation. As they show, such algebras exhaust the invariant, so the challenge is to access these very general approximations. This is achieved in [8] using [30] by combining finite nuclear dimension and the UCT.

The ideas in the 2015 proof of the unital classification theorem and the regularity programme are described in more detail in Winter's survey [35]. In the rest of this article, I will outline some ingredients in a new abstract and short(er) approach to the stably finite unital classification theorem which is joint work with Carrión, Gabe, Schafhauser, and Tikuisis [1], which makes more direct contact with von Neumann classification results.

6. ELLIOTT INTERTWINING: CLASSIFYING C^*-ALGEBRAS BY CLASSIFYING MAPS

The route towards the unital classification theorem, like many classification results before it, is through a classification of maps together with an Elliott intertwining argument. This overarching technique goes back to Elliott's classification of AF-algebras, and aspects can even be seen in Murray and von Neumann's uniqueness theorem. We start out by revisiting the classification of AF-algebras via a framework which applies much more generally.

6.1. Classifying AF algebras

We consider the classification of countable inductive limits of finite-dimensional algebras by ordered K_0, dividing this into three steps.

Step 1. Classify maps from finite-dimensional C^*-algebras. A C^*-algebra B has *cancelation* if K_0 determines Murray–von Neumann equivalence of projections, i.e., $[p]_0 = [q]_0 \implies p \sim q$. This ensures that unital *-homomorphisms from finite dimensional algebras into B are classified up to unitary equivalence by ordered K_0 and the class of the unit. As ever, classification of maps consists of two components: *existence* (here that any unital ordered K_0-morphism is realised by a unital *-homomorphism) and *uniqueness* (here two *-homomorphisms are unitarily equivalent if and only if they induce the same map on K_0).

Step 2. Intertwine to classify maps from inductive limits. Step 1 can be boosted by taking inductive limits to classify maps from a countable inductive limit $A = \overline{\bigcup_{n=1}^{\infty} A_n}$ of finite dimensional C^*-algebras into a C^*-algebra B with cancelation. The invariant remains ordered K_0 and the class of the unit, but one can only expect uniqueness up to *approximate unitary equivalence* ($\phi, \psi : A \to B$ are approximately unitary equivalent, written $\phi \approx_{\mathrm{au}} \psi$, when for all finite subsets $\mathcal{F} \subset A$ and $\epsilon > 0$ there exists a unitary $u \in B$ with $\|u\phi(x)u^* - \psi(x)\| < \epsilon$ for $x \in \mathcal{F}$).

That the invariant gives uniqueness of maps $A \to B$ up to approximate unitary equivalence is immediate from the uniqueness up to unitary equivalence of maps $A_n \to B$ in step 1. Existence needs a (one-sided) Elliott intertwining argument. Given a homomorphism $\Phi : (K_0(A), K_0(A)_+, [1_A]_0) \to (K_0(B), K_0(B)_+, [1_B]_0)$, construct compatible *-homomorphisms $\phi_n : A_n \to B$ inductively. Existence in step 1 gives a map $\tilde{\phi}_n$ implementing Φ on $K_0(A_n)$. Then uniqueness gives a unitary conjugate ϕ_n of $\tilde{\phi}_n$ agreeing with the previously defined ϕ_{n-1} on A_{n-1}. The resulting map defined on $\bigcup_{n=1}^{\infty} A_n$ extends by continuity to a *-homomorphism ϕ inducing Φ.

Step 3. Symmetrise assumptions to classify separable AF-algebras. The following abstract form of Elliott's intertwining argument converts a classification of maps up to approximate unitary equivalence to a classification of algebras.

Proposition 6.1 (Elliott's two-sided intertwining argument). *Suppose that A and B are separable unital C^*-algebras and there are *-homomorphisms $\phi : A \to B$ and $\psi : B \to A$ such* *that $\psi \circ \phi \approx_{\mathrm{au}} \mathrm{id}_A$ and $\phi \circ \psi \approx_{\mathrm{au}} \mathrm{id}_B$. Then A and B are isomorphic. Moreover, ϕ and ψ are approximately unitarily equivalent to mutually inverse isomorphisms.*

In particular, if a functor F classifies unital maps on a class $\mathcal{A}$ of separable unital C^*-algebras up to approximate unitary equivalence, then F also classifies $\mathcal{A}$. Indeed, given $A, B \in \mathcal{A}$ and an isomorphism $\Phi : F(A) \to F(B)$, the existence component of classification gives *-homomorphisms $\phi : A \to B$ and $\psi : B \to A$ with $F(\phi) = \Phi$ and $F(\psi) = \Phi^{-1}$, and the uniqueness component shows that ϕ and ψ satisfy the conditions of Proposition 6.1.

This process classifies those C^*-algebras which are both inductive limits of finite dimensional C^*-algebras and have cancelation by ordered K_0 together with the unit. But the latter hypothesis is readily seen to be automatic for an AF-algebra, so in fact it classifies countable inductive limits of finite dimensional C^*-algebras.

6.2. Reducing the unital classification theorem to the classification of approximately multiplicative maps

Let us now return to the general setting, and follow the same 3 step strategy.

Step 1. Classify approximately multiplicative maps $A \to B$. When $A = \overline{\bigcup_{n=1}^{\infty} A_n}$ is an AF-algebra as in Section 6.1, a sequence (ϕ_n) of *-homomorphisms $A_n \to B$ can be viewed as an approximately multiplicative map on A. Indeed, each ϕ_n has a cpc extension to A which is approximately multiplicative in that $\|\phi_n(x)\phi_n(y) - \phi_n(xy)\| \to 0$ for $x, y \in A$. Such approximately multiplicative cpc maps $A \to B$ provide a starting point for classification results in the general setting.

A uniqueness theorem is of the form: for all finite subsets $\mathcal{F} \subset A$ and $\varepsilon > 0$, there exists a finite subset $\mathcal{G} \subset A$ and $\delta > 0$ such that any two cpc maps $\phi, \psi : A \to B$ which are $(\mathcal{G}, \delta)$-approximately multiplicative, and approximately agree on the invariant are approximately unitary equivalent on $\mathcal{F}$ up to ε. Such statements (and their counterparts for existence) can very quickly become a morass of quantifiers, so it is convenient to use sequence algebras or ultraproducts (just as Connes did in his proof that injectivity implies hyperfiniteness).

Definition 6.2. The *sequence algebra B_∞* of a C^*-algebra B is the quotient $\ell^\infty(B)/c_0(B)$. It is typical to use representative bounded sequences in B to denote elements of B_∞.

Reindexing—the art of turning approximate statements into exact ones—is a key feature of sequence algebras and ultraproducts. For example, when A is a separable C^*-algebra, and B a unital C^*-algebra, *-homomorphisms $\phi, \psi : A \to B_\infty$ are approximately unitary equivalent if and only if they are unitarily equivalent. So we aim to classify maps $A \to B_\infty$ up to unitary equivalence. Such a result cleanly encodes a classification of approximately multiplicative maps up to approximate unitary equivalence.

Step 2. Intertwine to classify maps $A \to B$. Using separability of A and an intertwining argument we boost the classification of approximately multiplicative maps to a classification of *-homomorphisms $A \to B$. There is a very clean way to do this through *intertwining through reparameterisations*. Under very mild conditions (namely that inclusions $B \to B_\infty$

induce an injective map at the level of invariants), this uses classification into B_∞ to show that if $\theta : A \to B_\infty$ looks like it factors through B (i.e., it has an invariant factoring through B), then it is approximately unitary equivalent to a map that really does factor through B.

One can also use this approach for finite von Neumann algebras (suitably adjusted to $\|\cdot\|_2$-approximations and ultrapowers). Here the classification of maps from finite-dimensional C^*-algebras into finite von Neumann algebras $\mathcal{N}$ by traces (essentially Murray and von Neumann's analysis of projections) gives rise to a classification of maps $\mathcal{M} \to \mathcal{N}^\omega$ when $\mathcal{M}$ is separable and hyperfinite, and $\mathcal{N}^\omega$ is a tracial ultrapower of $\mathcal{N}$. Then a one-sided intertwining classifies maps $\mathcal{M} \to \mathcal{N}$, and a two-sided intertwining gives Murray and von Neumann's uniqueness of the hyperfinite II_1 factor (avoiding a number of explicit perturbation results).

Moreover, via Connes's theorem we have:

Theorem 6.3. *Maps from a separable nuclear C^*-algebra A to a finite von Neumann algebra $\mathcal{N}$ are classified up to strong operator approximate unitary equivalence by traces.*

The point is that any map $A \to \mathcal{N}$ will factor through a tracial von Neumann completion of A which is injective, so hyperfinite. Then the previous paragraph applies to give Theorem 6.3. We use this well-known result explicitly in our proof of the unital classification theorem, and it is the only point in the argument where we use some form of internal approximation by subalgebras (namely that the finite part of A^{**} is hyperfinite).

Step 3. Symmetrise assumptions to classify C^*-algebras. In both steps 1 and 2, the assumptions on the domain A and codomain B are likely to be quite different, and there may also be assumptions on the map (such as nuclearity). Now we symmetrise all the assumptions (requiring that the identity maps on all algebras under consideration satisfy any morphism assumptions) and obtain a classification of algebras using the two-sided Elliott intertwining argument (Proposition 6.1).

The upshot of this section is that the unital classification theorem can be expected to follow from a classification of maps $A \to B_\infty$. The rest of the article examines this.

7. THE TOTAL INVARIANT FOR CLASSIFYING APPROXIMATE MULTIPLICATIVE MAPS

Examples from the 1990s show that K-theory and traces are not enough to classify *-homomorphisms. For example, the tensor flip

$$\sigma : \mathcal{O}_3 \otimes \mathcal{O}_3 \to \mathcal{O}_3 \otimes \mathcal{O}_3; \quad x \otimes y \mapsto y \otimes x \tag{7.1}$$

on the Cuntz algebra $\mathcal{O}_3$ is trivial on K-theory. However, $\sigma \otimes \mathrm{id}_{\mathcal{O}_3}$ does not act trivially on K_0, and so is not approximately inner. This section discusses the additional ingredients which must be added to the invariant to obtain uniqueness theorems.

The underlying obstruction behind the example in (7.1) is found in K-theory with coefficients. Introduced by Schochet, the groups $K_*(A; \mathbb{Z}/n\mathbb{Z})$ fit into a natural six-term exact sequence

$$\begin{array}{ccccc} K_0(A) & \longrightarrow & K_0(A;\mathbb{Z}/n\mathbb{Z}) & \longrightarrow & K_1(A) \\ {\scriptstyle \times n}\uparrow & & & & \downarrow{\scriptstyle \times n} \\ K_0(A) & \longleftarrow & K_1(A;\mathbb{Z}/n\mathbb{Z}) & \longleftarrow & K_1(A) \end{array} \tag{7.2}$$

and so provide a framework for studying torsion in K-theory (namely the kernel of the maps of multiplication by n) at the C^*-algebraic level. An efficient way to define $K(A;\mathbb{Z}/n\mathbb{Z})$ is as $K_*(A \otimes C_n)$ for any separable nuclear C^*-algebra C_n in the UCT class with $K_*(C_n) \cong (\mathbb{Z}/n\mathbb{Z}\ , 0)$, such as $C_n = \mathcal{O}_{n+1}$. The *total K-theory* of A, $\underline{K}(A)$ is the combination of $K_*(A)$ and $\bigoplus_{n\geq 2} K_*(A;\mathbb{Z}/n\mathbb{Z})$ together with the natural maps in (7.2) (and other natural maps connecting the groups with different coefficients). Each of the groups $K_i(A;\mathbb{Z}/n\mathbb{Z})$ is determined by $K_*(A)$ but in an unnatural fashion. It is on morphisms $\phi : A \to B$ where $\underline{K}(\phi)$ carries more information than $K_*(\phi)$, e.g., $K_*(\sigma_n) = K_*(\mathrm{id}_{\mathcal{O}_n\otimes\mathcal{O}_n})$ while $\underline{K}(\sigma_n) \neq \underline{K}(\mathrm{id}_{\mathcal{O}_n\otimes\mathcal{O}_n})$.

While KK-theory determines whether UCT-Kirchberg algebras are isomorphic, it is a little too refined for detecting approximate unitary equivalence: maps $\phi, \psi : A \to B$ with $\phi \approx_{\mathrm{au}} \phi$ can differ in $KK(A, B)$. Rørdam (in the UCT case) and later Dadarlat (in general) identified a quotient $KL(A, B)$ of $KK(A, B)$ which is constant on approximate unitary equivalence classes of morphisms; for Kirchberg algebras, the converse holds and $KL(A, B)$ determines approximate unitary equivalence. One computes $KL(A, B)$ through Dadarlat and Loring's universal multicoefficient theorem [6]: $KL(A, B) \cong \mathrm{Hom}(\underline{K}(A), \underline{K}(B))$ whenever A has the UCT. In this way, total K-theory classifies morphisms between UCT-Kirchberg algebras. This works in much more generality (see, for example, Gabe's retreatment of Kirchberg's $\mathcal{O}_\infty$-stable classification [10]).

Theorem 7.1. *Let A be a separable exact unital C^*-algebra satisfying the UCT, and let B be a unital simple $\mathcal{O}_\infty$-stable C^*-algebra. Then unital full nuclear *-homorphisms $A \to B_\infty$ are classified up to approximate unitary equivalence by total K-theory (with maps preserving the class of the unit in K_0).*

In the stably finite setting, one needs yet further information. Examples due to Nielsen and Thomsen in the setting of $A\mathbb{T}$-algebras show the importance of a certain algebraic K_1-group. Given a unital C^*-algebra A, equip $U_\infty(A) = \bigcup_{n=1}^\infty U(M_n(A))$ with the inductive limit topology. The map $U_\infty(A) \to K_1(A)$ factors through the abelianisation $U_\infty(A)/DU_\infty(A)$, where $DU_\infty(A)$ is the group generated by commutators in $U_\infty(A)$, but this functor is not invariant under approximate unitary equivalence of morphisms. The solution is to form the *Hausdorffised unitary algebraic K_1-group*, $\overline{K}_1^{\mathrm{alg}}(A)$ of A as $U_\infty(A)/\overline{DU_\infty(A)}$. We write $\not{a}_A : \overline{K}_1^{\mathrm{alg}}(A) \to K_1(A)$ for the canonical quotient map.

The group $\overline{K}_1^{\mathrm{alg}}(A)$ was systematically studied by Thomsen [29] who used the de la Harpe–Skandalis determinant to relate it to K-theory and traces through a natural map $\mathrm{Th}_A : \mathrm{Aff}\, T(A) \to \overline{K}_1^{\mathrm{alg}}(A)$ which fits into a sequence

$$\mathrm{Aff}\, T(A) \stackrel{\mathrm{Th}_A}{\to} \overline{K}_1^{\mathrm{alg}}(A) \stackrel{\not{a}_A}{\to} K_1(A). \tag{7.3}$$

The kernel of Th_A is precisely the closure of $\rho_A(K_0(A))$ in Aff $T(A)$, so that $\ker \not{d}_A \cong \mathrm{Aff}\, T(A)/\overline{\rho_A(K_0(A))}$. This is a divisible group, so there is a non-canonical splitting

$$\overline{K}_1^{\mathrm{alg}}(A) \cong K_1(A) \oplus \ker \not{d}_A. \tag{7.4}$$

Given a *-homomorphism $\phi : A \to B$, there is no reason why $\overline{K}_1^{\mathrm{alg}}(\phi)$ should respect the splittings. In general, there is a *rotation map* $r_\phi : K_1(A) \to \ker \not{d}_B$ so that, with respect to decompositions (7.4), $\overline{K}_1^{\mathrm{alg}}(\phi) : K_1(A) \oplus \ker \not{d}_A \to K_1(B) \oplus \ker \not{d}_B$ is given by

$$\overline{K}_1^{\mathrm{alg}}(\phi) = \begin{pmatrix} K_1(\phi) & 0 \\ r_\phi & \overline{K}_1^{\mathrm{alg}}(\phi)|_{\ker \not{d}_A} \end{pmatrix}. \tag{7.5}$$

Example 7.2. Consider the crossed product $A = (\bigotimes_{-\infty}^{\infty} \mathcal{Z}) \rtimes \mathbb{Z}$, where the action is given by a Bernoulli shift on the tensor product. The Pimsner–Voiculescu 6-term exact sequence can be used to calculate $K_*(A) \cong (\mathbb{Z}, \mathbb{Z})$, with $K_1(A)$ being generated by the canonical unitary u implementing the action. Moreover, A has a unique trace (from the unique trace on $\mathcal{Z}$). So $\ker \not{d}_A \cong \mathrm{Aff}\, T(A)/\overline{\rho_A(K_0(A))} \cong \mathbb{R}/\mathbb{Z} \cong \mathbb{T}$, and $\overline{K}_1^{\mathrm{alg}}(A) \cong \mathbb{Z} \oplus \mathbb{T}$. The elements $\{\lambda 1_A : \lambda \in \mathbb{T}\}$ give representatives of $\ker \not{d}_A$ (by an easy de la Harpe–Skandalis determinant calculation).

There are just two automorphisms of $K_*(A)$ fixing $[1_A]_0$: the identity and $(m, n) \mapsto (m, -n)$. These can be paired with any rotation map $\mathbb{Z} \to \mathbb{T}$. The combination of $\mathrm{id}_{K_*(A)}$ with rotation map λ is implemented by the automorphism of A fixing $\bigotimes_{-\infty}^{\infty} \mathcal{Z}$ and sending u to λu, while the automorphism of A which reverses the order of the infinite tensor product and sends $u \mapsto \lambda u^*$ acts as the flip on $K_1(A)$ with rotation map λ.

It turns out that the extra data described above is now enough for uniqueness. It is formalised in the total invariant.

Definition 7.3. The *total invariant* $\underline{K}T_u(A)$ of a unital C^*-algebra A consists of $\underline{K}(A)$, Aff $T(A)$, and $\overline{K}_1^{\mathrm{alg}}(A)$, together with all the natural maps between these objects.

While it is necessary to adjoin $\underline{K}(\cdot)$ and $\overline{K}_1^{\mathrm{alg}}(\cdot)$ to Elliott's invariant to obtain uniqueness of morphisms, doing so increases the difficulty of proving the corresponding existence result. We must now determine exactly which maps between these invariants arise from *-homomorphisms. In addition to the pairing maps $\rho_\bullet$, the maps $\mathrm{Th}_\bullet$ and $\not{d}_\bullet$, it turns out that there are natural maps

$$\zeta_A^{(n)} : K_0(A; \mathbb{Z}/n\mathbb{Z}) \to \overline{K}_1^{\mathrm{alg}}(A), \quad n \geq 2, \tag{7.6}$$

relating total K-theory and $\overline{K}_1^{\mathrm{alg}}$. Compatibility with the $\zeta_\bullet^{(n)}$ is an extra obstruction for maps $(\underline{K}(A), \mathrm{Aff}\, T(A), \overline{K}_1^{\mathrm{alg}}(A)) \to (\underline{K}(B), \mathrm{Aff}\, T(B), \overline{K}_1^{\mathrm{alg}}(B))$ to come from a *-homomorphism. We use the last clause of Definition 7.3 to regard the maps $\zeta_\bullet^{(n)}$ as part of $\underline{K}T_u$ so that by definition $\underline{K}T_u$-morphisms are compatible with $\zeta_\bullet^{(n)}$. This completes the total invariant—no more compatibility requirements are needed for an existence theorem.

The maps $\zeta_A^{(n)}$ are a little fiddly to set up in general (see [1, SECTION 3], which also sets out how they interact with the other natural maps making up $\underline{K}T_u$), but they are readily identified in straightforward examples. For example, under the identifications $K_0(\mathcal{Z}; \mathbb{Z}/n\mathbb{Z}) \cong$

$\mathbb{Z}/n\mathbb{Z}$, and $\overline{K}_1^{\mathrm{alg}}(\mathcal{Z}) \cong \mathbb{T}$, the maps $\zeta_{\mathcal{Z}}^{(n)}$ are just the inclusions of the nth roots of unity into the circle. Moreover, the maps $\zeta_{\bullet}^{(n)}$ do not play a role when $K_1(A)$ is torsion free; in this case compatibility with the $\zeta_{\bullet}^{(n)}$ is automatic from the other compatibility requirements.

Everything is now in place to state a general version of the classification of unital approximate morphisms. Note how the hypotheses found in the unital classification theorem split up amongst the domain, codomain, and morphism in Theorem 7.4.

Theorem 7.4 (Stably finite classification of approximately multiplicative maps [1]). *Let A be a separable unital nuclear C^*-algebra satisfying the UCT, and let B be a unital simple $\mathcal{Z}$-stable nuclear C^*-algebra with $T(B) \neq \emptyset$. Then the total invariant $\underline{K}T_u$ classifies full unital nuclear maps $A \to B_\infty$ up to unitary equivalence.*

Once Theorem 7.4 has been established, the Elliott intertwining techniques discussed in Section 6 can be used to obtain classification results for algebras. Applying Step 2 of Section 6 to Theorem 7.4 classifies unital nuclear maps $A \to B$, and then symmetrising assumptions following Step 3 of Section 6 classifies the algebras in the unital classification theorem. But the invariant is $\underline{K}T_u$ not KT_u or Ell. So the final ingredient in the unital classification theorem is to extend an isomorphism $KT_u(A) \cong KT_u(B)$ to $\underline{K}T_u(A) \cong \underline{K}T_u(B)$. The extension to $\underline{K}$-theory, and $\overline{K}_1^{\mathrm{alg}}(\cdot)$ are purely algebraic results appearing in earlier classification work. A last little detail is required to correct these extensions and ensure compatibility with the $\zeta_{\bullet}^{(n)}$ when $K_1(A)$ has torsion. Such an extension is highly non-canonical (and typically far from unique).

8. QUASIDIAGONALITY

It is easier to construct approximately multiplicative maps (existence in Step 6.1) as compared with a *-homomorphism (existence in Step 6.2). This is exemplified by contrasting *quasidiagonality* with embeddings into the universal UHF-algebra $\mathcal{Q}$. Voiculescu showed that quasidiagonality of a C^*-algebra A can be viewed as an external approximation property: the existence of approximately multiplicative, approximately isometric cpc maps from A into matrix algebras. When A is separable, these can be packaged into to an embedding of A into $\mathcal{Q}_\infty$ (this characterises quasidiagonality when A is nuclear).

Many C^*-algebras are quasidiagonal; a deep theorem of Voiculescu shows that quasidiagonality is invariant under homotopy, so that all cones $C_0(0,1] \otimes A$ are quasidiagonal—a result Kirchberg uses in his $\mathcal{O}_2$-embedding theorem. On the other hand, as $\mathcal{Q}$ has a faithful trace, no cone over a simple purely infinite C^*-algebra embeds in $\mathcal{Q}$. As one cannot model an infinite projection in a matrix algebra, quasidiagonal C^*-algebras are stably finite. The Blackadar–Kirchberg problem asks whether this is the only obstruction for nuclear C^*-algebras: are all stably finite nuclear C^*-algebras quasidiagonal? This question parallels Connes' important observation that injective II_1 factors always embed into $\mathcal{R}^\omega$. Moreover, the constructions of stably finite simple separable nuclear C^*-algebras which exhaust the

Elliott invariant are all quasidiagonal. Finding an abstract source of quasidiagonality is necessary for stably finite classification theorems.

This was achieved for simple stably finite nuclear C^*-algebras with the UCT in the quasidiagonality theorem [30]. The idea is to use traces as a kind of measuring device, by showing that all traces on A are quasidiagonal. One definition of quasidiagonality of $\tau \in T(A)$ is the existence of a sequence $(\theta_n)_n$ of approximately multiplicative cpc maps $A \to \mathcal{Q}$ with $\tau(x) = \lim_n \tau_{\mathcal{Q}}(\theta_n(x))$ for all $x \in A$.

Theorem 8.1 (The quasidiagonality theorem [30]). *Let A be a separable nuclear C^*-algebra satisfying the UCT. Then all faithful amenable traces on A are quasidiagonal. Accordingly, stably finite simple separable nuclear C^*-algebras satisfying the UCT are quasidiagonal.*

With hindsight the quasidiagonality theorem has turned out to be just the right level of difficulty to isolate and simplify fundamental tools in classification. The original proof was inspired by a stable uniqueness across the interval technique from the tracial approximation approach to classification, and the quasidiagonality theorem is then used to construct tracial approximations from abstract conditions in [8]. A major breakthrough was subsequently made by Schafhauser [27]. Reframing the problem in terms of liftings, he gave a conceptual new proof using Ext-groups. This idea provides the main framework for our approach to the classification of approximate morphisms (as outlined in the next section).

To sketch Schafhauser's plan, we begin with the trace-kernel extension. At this point is preferable to work with ultrapowers rather than sequence algebras, so let $\omega \in \beta\mathbb{N} \setminus \mathbb{N}$ be a free ultrafilter. We form $\mathcal{Q}_\omega$ as the quotient of $\ell^\infty(\mathcal{Q})$ by those sequences (x_n) with $\lim_{n\to\omega} \|x_n\| = 0$; it behaves analogously to $\mathcal{Q}_\infty$. The ultrapower $\mathcal{R}^\omega$ is the quotient of $\ell^\infty(\mathcal{R})$ by those sequences with $\lim_{n\to\omega} \tau(x_n^* x_n) = 0$; the point is that this is a von Neumann algebra, whereas a sequence algebra version is not. Since $\mathcal{Q}$ is weakly dense in $\mathcal{R}$, Kaplansky's density theorem gives rise to a surjection of $\mathcal{Q}_\omega$ onto $\mathcal{R}^\omega$ with kernel J.

Given a trace τ on a separable nuclear C^*-algebra A, one has an embedding $\theta : A \to \mathcal{R}^\omega$ realising τ (by Theorem 6.3). Via the Choi–Effros lifting theorem in one direction, and Theorem 6.3 in the other, τ is quasidiagonal if and only if θ lifts to $\tilde{\theta} : A \to \mathcal{Q}_\omega$,

$$\begin{array}{ccccccccc} & & & & & & A & & \\ & & & & & \overset{\tilde{\theta}}{\swarrow} & \downarrow \theta & & \\ 0 & \longrightarrow & J & \longrightarrow & \mathcal{Q}_\omega & \longrightarrow & \mathcal{R}^\omega & \longrightarrow & 0. \end{array} \tag{8.1}$$

Forming the pullback extension

$$\begin{array}{ccccccccc} \eta : 0 & \longrightarrow & J & \longrightarrow & E & \longrightarrow & A & \longrightarrow & 0 \\ & & \| & & \downarrow & & \downarrow & & \\ 0 & \longrightarrow & J & \longrightarrow & \mathcal{Q}_\omega & \longrightarrow & \mathcal{R}^\omega & \longrightarrow & 0, \end{array} \tag{8.2}$$

liftability of θ is equivalent to the existence of a *-homomorphism splitting $A \to E$ of η.

Extension theory provides the ideal tool for tackling problems of this nature, as η induces a class in $\mathrm{Ext}(A, J)$. However, there is a problem, the trace-kernel ideal J appears somewhat unwieldy. In particular, it is neither stable nor σ-unital, which is a deterrent to using Ext. Schafhauser's key observation is that comparison properties of $\mathcal{Q}$ ensure that J is *separably stable*: for every separable C^*-subalgebra $J_0 \subset J$, there is a stable separable J_1 with $J_0 \subset J_1 \subset J$. With a fair bit of care, this is enough stability to use Ext.

A computation using the UCT and the K-theory of $\mathcal{R}^\omega$ readily shows that $[\eta] = 0$ in $\mathrm{Ext}(A, J)$. This does not yet mean that η splits, but rather that after adjoining a further trivial extension η_1, say, the sum $\eta \oplus \eta_1$ splits. So the final step is to ensure that η is *absorbing* so that $\eta \cong \eta \oplus \eta_1$. This is achieved using an abstract Weyl–von Neumann/Voiculescu-type theorem of Elliott and Kucerovsky [9] which heavily exploits Kirchberg's work for infinite C^*-algebras. When A is non-unital, absorption is a consequence of the faithfulness of τ via injectivity of θ. There is an important detail when A (and hence θ) are unital. In this case η can never be absorbing and we can only ask for absorption of unital extensions. The trick is to pass to a non-unital 2×2 matrix amplification to replace θ by a non-unital map. This de-unitization idea recurs extensively in the classification of approximate morphisms.

9. CLASSIFICATION OF APPROXIMATELY MULTIPLICATIVE MAPS

We end with a brief discussion of some ingredients in the classification of approximately multiplicative maps. For the rest of the article, let A and B be as in Theorem 7.4.

Crudely the plan is to solve the classification problem at the von Neumann level, and lift this back to the C^*-setting. Slightly more precisely, we look for a quotient $\mathcal{R}_B$ of B_∞ into which we can classify maps $A \to B_\infty$ by traces. This will fit into a short exact sequence

$$0 \to J_B \to B_\infty \to \mathcal{R}_B \to 0. \tag{9.1}$$

We then try and classify unital lifts of a given unital *-homomorphism $\theta : A \to \mathcal{R}_B$, i.e., characterise when a lift $\tilde{\theta} : A \to B_\infty$ of θ exists, and classify these up to unitary equivalence. Successfully combining these steps will classify maps $A \to B_\infty$.

When B has unique trace τ, it is natural to take $\mathcal{R}_B$ to be the II_1 factor ultrapower $(\pi_\tau(B)'')^\omega$, which is a quotient of B_∞ by Kaplansky's density theorem. Via Connes' theorem, unital *-homomorphisms $\theta : A \to \mathcal{R}_B$ are classified up to unitary equivalence by the trace they induce on A (Theorem 6.3). Assuming additionally that A has the UCT and B is $\mathcal{Q}$-stable with $K_1(B) = 0$, and working with B_ω in place of B_∞ Schafhauser classified lifts of a given $\theta : A \to \mathcal{R}_B$ by K_0. Combining these two statements and then intertwining gives a classification of maps $A \to B$ by K_0 and traces. Symmetrising hypotheses in the spirit of Section 6 gives the first truly abstract proof of a stably finite classification theorem. While the hypotheses are quite stringent, they are powerful enough to show that a separable exact C^*-algebra satisfying the UCT and with a faithful trace embeds into a monotracial AF-algebra [28]. It is vital that the ideal J_B is separably stable, which one gets from $\mathcal{Q}$-stability of B (the need for separable stability also forces the use of B_ω when we work with an ultrapower quotient).

Outside the unique trace setting, it is tempting to take $\mathcal{R}_B$ to be a suitable von Neumann ultrapower of B^{**}_{fin}, as Connes' theorem would classify maps $A \to \mathcal{R}_B$ by traces. However, using positive elements $(x_n)_{n=1}^\infty$ in B_∞ for which $\lim_n \tau(x_n) = 0$ pointwise but not uniformly in τ, one can easily obstruct separable stability of the resulting J_B. A more refined choice of quotient is needed to handle traces in a uniform fashion. Such constructions came to the fore through Matui and Sato's work [20]. This is the point where Schafhauser's abstract classification machinery merges with the Toms–Winter regularity programme. Write $\|x\|_{2,T(B)} = \sup_{\tau\in T(B)} \tau(x^*x)^{1/2}$. Then we define the *uniform tracial sequence algebra* by

$$B^\infty = \ell^\infty(B)/\Big\{(x_n)_{n=1}^\infty \in \ell^\infty(B) : \lim_{n\to\infty} \|x_n\|_{2,T(B)} = 0\Big\}. \qquad (9.2)$$

In this way, B_∞ quotients onto B^∞ leading to the *uniform trace-kernel extension*

$$0 \to J_B \xrightarrow{j_b} B_\infty \xrightarrow{q_b} B^\infty \to 0. \qquad (9.3)$$

This is the right framework to classify unital maps $A \to B^\infty$ by traces and their lifts back to B_∞ by the other aspects of $\underline{KT}_u$. The former uses regularity techniques, while the latter uses abstract classification. The essential point is that $\mathcal{Z}$-stability of B gives separable stability of J_B allowing $KK(A, J_B)$ to be used. In what follows, I pretend that J_B is stable.

9.1. Classifying unital maps $A \to B^\infty$

When B has a unique trace, B^∞ is not quite the von Neumann algebra ultrapower $(\pi_\tau(B)'')^\omega$ used in Schafhauser's unique trace UHF-stable argument. But for the purposes of classifying maps from separable nuclear C^*-algebras A, there is no real difference. The real challenge comes when B has infinitely many extremal traces, particularly if the extreme boundary $\partial_e T(B)$ is not compact. In this case, for each $\tau \in T(B^\infty)$, Connes' theorem classifies maps $\theta : A \to \pi_\tau(B^\infty)''$. We must glue these together to a classification of maps $A \to B^\infty$ by traces.

Problems of this nature have been at the heart of work on Toms–Winter regularity conjecture, and a general strategy for gluing properties from each $\pi_\tau(B^\infty)''$ together to obtain global statements which hold uniformly in all traces was developed in [2]. These techniques give B^∞ a "von Neumann-like" flavour when B is nuclear and $\mathcal{Z}$-stable, and in particular they can be used to obtain the required classification of maps $A \to B^\infty$. Consequently, given two maps $\phi_1, \phi_2 : A \to B_\infty$ which agree on traces, the compositions $q_B \circ \phi_1$ and $q_B \circ \phi_2$ are unitarily equivalent via a unitary $u \in B^\infty$ say. For each trace τ on B^∞, we can write $\pi_\tau(u)$ as an exponential e^{ih_τ} for a self-adjoint $h_\tau \in \pi_\tau(B^\infty)''$. Another application of the gluing procedure can be used to find a single self-adjoint $h \in B^\infty$ with $u = e^{ih}$. Therefore, u lifts to a unitary in B_∞, and by conjugating by a lift of u, we can assume that $q_B \circ \phi_1 = q_B \circ \phi_2$. In this way, the remainder of the uniqueness problem for a pair of maps $\phi_1, \phi_2 : A \to B_\infty$, becomes a question about the uniqueness of lifts of the common *-homomorphism $q_B \circ \phi_1 = q_B \circ \phi_2 : A \to B^\infty$ back to B_∞.

In fact, the full force of $\mathcal{Z}$-stability is not needed, and one can get away with a weaker central sequence condition in the spirit of Murray and von Neumann's property Γ. There is a lot going on behind the scenes here, and I will describe the ideas behind these techniques a bit further in the more regularity focused companion survey [32].

9.2. Classifying unital lifts

In the second part we are given a unital map $\theta : A \to B^\infty$ and aim to classify lifts back to B_∞. A necessary condition for a lift is the existence of $\kappa \in KK(A, B_\infty)$ with $[q_B]\kappa = [\theta]$ in KK. We can produce these κ using the universal (multi)coefficient theorem from the total K-theory component of a map $\underline{K}T_u(A) \to \underline{K}T_u(B^\infty)$.

Given such a κ, small modifications of Schafhauser's proof of the quasidiagonality theorem produces some lift $\psi_- : A \to B_\infty$ of θ. But this might not have $[\psi_-] = \kappa$ in $KK(A, B_\infty)$, and it must be corrected so that it does. By construction, $\kappa - [\psi_-]$ will map to 0 under the map $KK(A, B_\infty) \to KK(A, B^\infty)$ induced by (9.3) so half-exactness of $KK(A, \cdot)$ gives that $\kappa - [\psi_-]$ is in the image of $KK(A, j_B) : KK(A, J_B) \to KK(A, B_\infty)$. Write $\kappa - [\psi_-] = KK(A, j_B)(\lambda)$ for some $\lambda \in KK(A, J_B)$.

Cuntz's quasihomomorphism picture of KK-theory is particularly well suited to C^*-classification problems. This defines $KK(A, J_B)$ as homotopy classes of *Cuntz-pairs*: maps $(\phi_+, \phi_-) : A \to \mathcal{M}(J_B)$ such that $\phi_+(x) - \phi_-(x) \in J_B$. In order to translate between KK-theory and *-homomorphisms into B_∞, we need KK-existence and uniqueness theorems, both of which rely on absorption. The existence theorem says that if $\phi_- : A \to \mathcal{M}(J_B)$ is absorbing, then given any $\lambda \in KK(A, J_B)$ we can find $\phi_+ : A \to \mathcal{M}(J_B)$ such that the Cuntz-pair (ϕ_+, ϕ_-) realises λ. This works in vast generality and has been regularly used in classification (in our situation all one needs is the separable-stability to work with $KK(A, J_B)$). Following the map ψ_- above by the natural map $B_\infty \to \mathcal{M}(J_B)$ gives rise to $\phi_- : A \to \mathcal{M}(J_B)$. Using the Elliott–Kucerovsky theorem (and modulo the de-unitisation trick alluded to at the end of Section 8, which is suppressed here), ϕ_- is absorbing. Thus we can find $\phi_+ : A \to \mathcal{M}(J_B)$ such that (ϕ_+, ϕ_-) forms a Cuntz-pair representing λ. A fairly standard pullback calculation then produces a map $\psi_+ : A \to B_\infty$ also lifting θ (as a consequence of (ϕ_+, ϕ_-) being a Cuntz-pair) so that $KK(A, j_B)(\lambda) = [\psi_+] - [\psi_-]$. Therefore ψ_+ realizes the element $\kappa \in KK(A, B_\infty)$.

How unique is ψ_+? Given two lifts $\psi_1, \psi_2 : A \to B_\infty$ of θ, we obtain a Cuntz-pair $(\phi_1, \phi_2) : A \to \mathcal{M}(J_B)$ representing a class in $KK(A, J_B)$. A KK- or KL-uniqueness theorem is designed to give asymptotic or approximate unitary equivalence of absorbing ϕ_1 and ϕ_2 when $[\phi_1, \phi_2]$ vanishes in $KK(A, J_B)$ and $KL(A, J_B)$, respectively. While KK-existence holds very generally, KK- and KL-uniqueness are more subtle, going back to Dadarlat and Eilers in the setting of $KK(A, \mathcal{K})$, and it is currently unclear how generally such results can hold. For us, $\mathcal{Z}$-stability of B is the key ingredient through a *$\mathcal{Z}$-stable KL-uniqueness theorem* developed in [1] (extending a $\mathcal{Q}$-stable KK-uniqueness theorem from [28]). This gives approximate unitary equivalence (with unitaries in the unitisation of $J_B \otimes \mathcal{Z}$) of the $\mathcal{Z}$-stabilisations $\phi_1 \otimes 1_{\mathcal{Z}}, \phi_2 \otimes 1_{\mathcal{Z}} : A \to \mathcal{M}(J_B) \otimes \mathcal{Z}$ from $[\phi_1, \phi_2] = 0$ in $KL(A, J_B)$. Using separable $\mathcal{Z}$-stability of B_∞ and the fact we work in a sequence algebra, this gives unitary equivalence of ψ_1 and ψ_2. So lifts are classified by $KL(A, J_B)$, which fits into an exact sequence

$$\ker KL(A, j_B) \to KL(A, J_B) \to KL(A, B_\infty). \tag{9.4}$$

Dadarlat and Loring's universal (multi)coefficient theorem (obtained from the UCT) computes $KL(A, B_\infty) \cong \mathrm{Hom}(\underline{K}(A), \underline{K}(B_\infty))$. We need to interpret $\ker KL(A, j_B)$ in terms of $\overline{K}_1^{\mathrm{alg}}$ and in particular the rotation maps r_ϕ from (7.5).

This is achieved through an isomorphism

$$R_{A,B} : \ker KL(A, j_B) \to \mathrm{Hom}\big(K_1(A)/\mathrm{Tor}\big(K_1(A)\big), \ker \not{a}_{B_\infty}\big) \tag{9.5}$$

with the property that $R_{A,B}([\psi_1, \psi_2]) \circ t_A = r_{\psi_1} - r_{\psi_2}$ when $(\psi_1, \psi_2) : A \to B_\infty$ realise a class in $\ker KL(A, j_B)$. Here $t_A : K_1(A) \to K_1(A)/\mathrm{Tor}(K_1(A))$ is the quotient map, which removes torsion from $K_1(A)$. While the individual rotation maps r_{ψ_1} and r_{ψ_2} depend on a choice of decomposition in (7.4), when ψ_1 and ψ_2 agree on KT_u the difference $r_{\psi_1} - r_{\psi_2}$ does not. In this case $(r_{\psi_1} - r_{\psi_2}) \circ \not{a}_A = \overline{K}_1^{\mathrm{alg}}(\psi_1) - \overline{K}_1^{\mathrm{alg}}(\psi_2) : \overline{K}_1^{\mathrm{alg}}(A) \to \overline{K}_1^{\mathrm{alg}}(B_\infty)$. Then, given $\phi_1, \phi_2 : A \to B_\infty$ agreeing on $\underline{KT}_u$, by successively using traces (to reduce to the case that $q_B \circ \phi_1 = q_B \circ \phi_2$), and then total K-theory, and $\overline{K}_1^{\mathrm{alg}}$ (to see that ϕ_1 and ϕ_2 induce the same class in $KL(A, J_B)$), these tools combine to give unitary equivalence of ϕ_1 and ϕ_2.

The pairing maps $\zeta_\bullet^{(n)}$ from (7.6) are required for existence. When we attempt to realise maps $\alpha : \underline{K}(A) \to \underline{K}(B_\infty)$, $\beta : \overline{K}_1^{\mathrm{alg}}(A) \to \overline{K}_1^{\mathrm{alg}}(B_\infty)$, and $\gamma : \mathrm{Aff}\, T(A) \to \mathrm{Aff}\, T(B_\infty)$, one first constructs $\theta : A \to B^\infty$ using γ. Then one lifts to $\phi : A \to B_\infty$ realising a lift κ of α. As both $\overline{K}_1^{\mathrm{alg}}(\phi)$ and (α, β, γ) are $\underline{KT}_u$-morphisms, one can use compatibility with $\zeta_\bullet^{(n)}$ to show that the rotation map induced by $\beta - \overline{K}_1^{\mathrm{alg}}(\phi)$ vanishes on $\mathrm{Tor}(K_1(A))$. This enables $R_{A,B}$ to be used to modify the behaviour of ϕ on $\overline{K}_1^{\mathrm{alg}}(A)$.

The isomorphism $R_{A,B}$ is an abstract sequence algebra version of the rotation map computations developed by Lin (see [19], for an example of the use of a rotation map in an asymptotic classification result). It is built in two steps. First, the UCT gives an isomorphism $\ker(KL(A, j_B))$ to $\ker \mathrm{Hom}(\underline{K}(A), \underline{K}(j_B))$. The second part of the isomorphism is then a direct computation, which relies heavily on the "von Neumann like" structure of B^∞ (in particular, $K_*(B^\infty) \cong (\mathrm{Aff}\, T(B^\infty), 0)$) via the techniques in [2] hinted at in Section 9.1.

ACKNOWLEDGMENTS

I would like to thank my collaborators on [1, 2]: José Carrión, Jorge Castillejos, Sam Evington, Jamie Gabe, Chris Schafhauser, Aaron Tikuisis, and Wilhelm Winter. I have learnt a lot from you all. I would like to thank Bruce Blackadar, José Carrión, Jamie Gabe, Shanshan Hua, Robert Neagu, and Chris Schafhauser for their very helpful comments on an earlier draft of this article. The paper [1] was initiated at a BIRS workshop in 2017, and has benefitted massively from discussions at the American Institute of Mathematics as part of their SQuaRE programme. I thank both institutes, and their funders for their support.

FUNDING

This work was partially supported by EPSRC EP/R025061/1-2.

REFERENCES

[1] J. Carrión, J. Gabe, C. Schafhauser, A. Tikuisis, and S. White, Classifying $*$-homomorphisms I: simple nuclear C^*-algebras. Manuscript in preparation.

[2] J. Castillejos, S. Evington, A. Tikuisis, S. White, and W. Winter, Nuclear dimension of simple C^*-algebras. *Invent. Math.* **224** (2021), no. 1, 245–290.

[3] E. Christensen, Perturbations of operator algebras. *Invent. Math.* **43** (1977), no. 1, 1–13.

[4] E. Christensen, A. Sinclair, R. Smith, S. White, and W. Winter, Perturbations of nuclear C^*-algebras. *Acta Math.* **208** (2012), no. 1, 93–150.

[5] A. Connes, Classification of injective factors. Cases II_1, II_∞, III_λ, $\lambda \neq 1$. *Ann. of Math. (2)* **104** (1976), no. 1, 73–115.

[6] M. Dadarlat and T. Loring, A universal multicoefficient theorem for the Kasparov groups. *Duke Math. J.* **84** (1996), no. 2, 355–377.

[7] G. Elliott, The classification problem for amenable C^*-algebras. In *Proceedings of the International Congress of Mathematicians, Vol. 1, 2 (Zürich, 1994)*, pp. 922–932, Birkhäuser, Basel, 1995.

[8] G. Elliott, G. Gong, H. Lin, and Z. Niu, On the classification of simple amenable C^*-algebras with finite decomposition rank, II. 2016, arXiv:1507.03437.

[9] G. Elliott and D. Kucerovsky, An abstract Voiculescu–Brown–Douglas–Fillmore absorption theorem. *Pacific J. Math.* **198** (2001), no. 2, 385–409.

[10] J. Gabe, Classification of $\mathcal{O}_\infty$-stable C^*-algebras. *Mem. Amer. Math. Soc.*, to appear, 2021, arXiv:1910.06504.

[11] G. Gong, H. Lin, and Z. Niu, A classification of finite simple amenable $\mathcal{Z}$-stable C^*-algebras, I: C^*-algebras with generalized tracial rank one. *C. R. Math. Acad. Sci. Soc. R. Can.* **42** (2020), no. 3, 63–450.

[12] G. Gong, H. Lin, and Z. Niu, A classification of finite simple amenable $\mathcal{Z}$-stable C^*-algebras, II: C^*-algebras with rational generalized tracial rank one. *C. R. Math. Acad. Sci. Soc. R. Can.* **42** (2020), no. 4, 451–539.

[13] U. Haagerup, Connes' bicentralizer problem and uniqueness of the injective factor of type III_1. *Acta Math.* **158** (1987), no. 1–2, 95–148.

[14] D. Kerr, Dimension, comparison, and almost finiteness. *J. Eur. Math. Soc. (JEMS)* **22** (2020), no. 11, 3697–3745.

[15] D. Kerr and P. Naryshkin, Elementary amenability and almost finiteness. 2021, arXiv:2107.05273.

[16] E. Kirchberg, Exact C^*-algebras, tensor products, and the classification of purely infinite algebras. In *Proceedings of the International Congress of Mathematicians, Vol. 1, 2 (Zürich, 1994)*, pp. 943–954, Birkhäuser, Basel, 1995.

[17] E. Kirchberg and N. C. Phillips, Embedding of exact C^*-algebras in the Cuntz algebra $\mathcal{O}_2$. *J. Reine Angew. Math.* **525** (2000), 17–53.

[18] X. Li, Every classifiable simple C^*-algebra has a Cartan subalgebra. *Invent. Math.* **219** (2020), no. 2, 653–699.

[19] H. Lin, Asymptotic unitary equivalence and classification of simple amenable C^*-algebras. *Invent. Math.* **183** (2011), no. 2, 385–450.

[20] H. Matui and Y. Sato, Strict comparison and $\mathcal{Z}$-absorption of nuclear C^*-algebras. *Acta Math.* **209** (2012), no. 1, 179–196.

[21] H. Matui and Y. Sato, Decomposition rank of UHF-absorbing C^*-algebras. *Duke Math. J.* **163** (2014), no. 14, 2687–2708.

[22] D. McDuff, Central sequences and the hyperfinite factor. *Proc. Lond. Math. Soc. (3)* **21** (1970), 443–461.

[23] F. Murray and J. von Neumann, On rings of operators. IV. *Ann. of Math. (2)* **44** (1943), 716–808.

[24] N. C. Phillips, A classification theorem for nuclear purely infinite simple C^*-algebras. *Doc. Math.* **5** (2000), 49–114.

[25] M. Rørdam, A simple C^*-algebra with a finite and an infinite projection. *Acta Math.* **191** (2003), no. 1, 109–142.

[26] J. Rosenberg and C. Schochet, The Künneth theorem and the universal coefficient theorem for Kasparov's generalized K-functor. *Duke Math. J.* **55** (1987), no. 2, 431–474.

[27] C. Schafhauser, A new proof of the Tikuisis–White–Winter theorem. *J. Reine Angew. Math.* **759** (2020), 291–304.

[28] C. Schafhauser, Subalgebras of simple AF-algebras. *Ann. of Math. (2)* **192** (2020), no. 2, 309–352.

[29] K. Thomsen, Traces, unitary characters and crossed products by $\mathbb{Z}$. *Publ. Res. Inst. Math. Sci.* **31** (1995), no. 6, 1011–1029.

[30] A. Tikuisis, S. White, and W. Winter, Quasidiagonality of nuclear C^*-algebras. *Ann. of Math. (2)* **185** (2017), no. 1, 229–284.

[31] A. Toms, On the classification problem for nuclear C^*-algebras. *Ann. of Math. (2)* **167** (2008), no. 3, 1029–1044.

[32] S. White, $\mathcal{Z}$-stability, tracial completions and regularity for C^*-algebras. Manuscript in preparation.

[33] W. Winter, Nuclear dimension and $\mathcal{Z}$-stability of pure C^*-algebras. *Invent. Math.* **187** (2012), no. 2, 259–342.

[34] W. Winter, Localizing the Elliott conjecture at strongly self-absorbing C^*-algebras. *J. Reine Angew. Math.* **692** (2014), 193–231.

[35] W. Winter, Structure of nuclear $C*$-algebras: from quasidiagonality to classification and back again. In *Proceedings of the International Congress of Mathematicians—Rio de Janeiro 2018. Vol. III. Invited lectures*, pp. 1801–1823, World Sci. Publ., Hackensack, NJ, 2018.

STUART WHITE

Mathematical Institute, University of Oxford, Andrew Wiles Building, Radcliff Observatory Quarter, Woodstock Road, Oxford, OX2 6GG, UK, stuart.white@maths.ox.ac.uk

This Page Intentionally Left Blank

ASYMPTOTIC BEHAVIORS OF RANDOM WALKS ON COUNTABLE GROUPS

TIANYI ZHENG (郑天一)

ABSTRACT

In this note we survey some topics in random walks on countable groups. The main focus is on quantitative estimates for random walk characteristics on amenable groups, in connections to geometric and algebraic properties of the underlying groups.

MATHEMATICS SUBJECT CLASSIFICATION 2020

Primary 20F69; Secondary 60B15, 46B85

KEYWORDS

Random walks on groups, volume growth, harmonic functions

INTERNATIONAL CONGRESS OF MATHEMATICIANS 2022 JULY 6–14

Proc. Int. Cong. Math. 2022, Vol. 4, pp. 3340–3365
DOI 10.4171/ICM2022/138

Published by EMS Press

1. INTRODUCTION: A BRIEF REVIEW OF SOME HISTORY

Random walks on general countable groups were introduced by H. Kesten in his thesis titled *Symmetric Random Walks on Groups* **[63]**. Let Γ be a countable group and let μ be a probability measure on Γ. We say that μ is symmetric if $\mu(g) = \mu(g^{-1})$ for all $g \in \Gamma$, and μ is nondegenerate if the support of μ generates Γ as a semigroup. Consider a random walk on Γ in which every step consists of right multiplication by a random group element distributed according to μ. In other words, take a sequence of independent random variables $(Y_n)_{n=1}^{\infty}$ on Γ distributed as μ. Let $W_0 = \mathrm{id}_\Gamma$, $W_n = Y_1 \cdots Y_n$. We refer to the process $(W_n)_{n=0}^{\infty}$ as a μ-random walk on Γ. The distribution of W_n is the nth convolution power $\mu^{(n)}$.

When Γ is generated by a finite subset $S \subset \Gamma$, consider the Cayley graph of (Γ, S), which is a graph with vertex set Γ and edge set $\{(g, gs) : g \in \Gamma, s \in S\}$. The word length $|g|_S$ of a group element is the smallest integer $n \geq 0$ for which there exist $s_1, \ldots, s_n \in S \cup S^{-1}$ such that $g = s_1 \cdots s_n$. A random walk with step distribution μ supported on $S \cup S^{-1}$ can be visualized as a random nearest neighbor exploration process on the Cayley graph.

The questions considered in **[62,63]** regard the relation between the spectrum of the associated linear operator P_μ on $\ell^2(\Gamma)$ and the structure of the group Γ, where $P_\mu f(x) = \sum_{y \in \Gamma} f(xy)\mu(y)$. The original definition of amenability, introduced by von Neumann to explain the Hausdorff–Banach–Tarski paradox, says that Γ is amenable if there is a Γ-invariant mean on $\ell^\infty(\Gamma)$. Kesten's characterization of amenability **[62]** states that when μ is a nondegenerate symmetric probability measure on Γ, the spectral radius $\lambda(\Gamma, \mu) = \lim_{n \to \infty} \mu^{(2n)}(\mathrm{id})^{1/2n}$ is 1 if and only if Γ is amenable. When Γ is amenable, one might further ask what is the behavior of the spectral distribution of P_μ near 1, or of the decay of return probability $\mu^{(2n)}(\mathrm{id})$ when n goes to infinity. Since the pioneering work of Varopoulos, such questions are studied using both analytic and geometric tools. In particular, there are close relations between the behavior of the heat semigroup $(P_\mu^n)_{n=0}^{\infty}$ on $\ell^2(G)$ and geometric properties captured through Sobolev-type inequalities, see the survey **[98]** and the monograph **[99]**.

A real-valued function f on Γ is called μ-harmonic if it satisfies the mean value property that $f(x) = \sum_{y \in \Gamma} f(xy)\mu(y)$ for all $x \in \Gamma$. When all bounded μ-harmonic functions are constant, we say (Γ, μ) has the *Liouville property*. A theory for the non-Liouville case, in the more general context of locally compact second countable (abbreviated as *lcsc*) groups is initiated by Furstenberg **[37,39,40]**, where a measure-theoretical object called *Poisson boundary* (also called Poisson–Furstenberg boundary) was introduced to represent the space of bounded μ-harmonic functions. In particular, the Poisson boundary of (G, μ) is trivial if and only if (G, μ) has the Liouville property. Note that the measure-theoretical Poisson–Furstenberg boundary is different from the topological Furstenberg boundary which is also introduced in **[37]**.

Entropy is a crucial quantity in the study of Poisson boundaries. Let $(W_n)_{n=0}^{\infty}$ be a μ-random walk on a countable group Γ. Denote by $H_\mu(n)$ the (Shannon) entropy of W_n,

$$H_\mu(n) = H(W_n) = -\sum_{g\in\Gamma} \mu^{(n)}(g) \log \mu^{(n)}(g). \tag{1.1}$$

The limit $\mathbf{h}_\mu = \lim_{n\to\infty} H_\mu(n)/n$ is called the (Avez) asymptotic entropy of the μ-random walk. The celebrated *entropy criterion for the Liouville property*, due to Avez [7], Derriennic [28] and Kaimanovich–Vershik [60], states that if μ has finite entropy, then the asymptotic entropy is 0 if and only if (Γ, μ) is Liouville.

Suppose now Γ is generated by a finite subset $S \subset \Gamma$. When the measure μ has finite first moment with respect to the word norm $|\cdot|_S$, that is, $\sum |g|_S \mu(g) < \infty$, we may consider the speed function (also called rate of escape or drift), defined as

$$L_\mu(n) = \mathbb{E}\big[|W_n|_S\big] = \sum_{g\in\Gamma} |g|_S \mu^{(n)}(g). \tag{1.2}$$

The limit $\mathbf{l}_\mu = \lim_{n\to\infty} L_\mu(n)/n$ is called the asymptotic speed/drift. By Kingman's subadditive ergodic theorem, we have $|W_n|_S/n \to \mathbf{l}_\mu$ when $n \to \infty$ almost surely.

By the so called "fundamental inequality" that $\mathbf{h}_\mu \le v\mathbf{l}_\mu$ (see, e.g., [19]), where v is the asymptotic volume growth rate of (Γ, S), we have that $\mathbf{l}_\mu = 0$ implies $\mathbf{h}_\mu = 0$. A theorem of Karlsson and Ledrappier [61] combined with the entropy criterion imply the following speed criterion: for a nondegenerate *centered* step distribution μ on Γ with finite first moment, the asymptotic speed $\mathbf{l}_\mu = 0$ if and only if (Γ, μ) is Liouville. In the special case where μ is a nondegenerate symmetric probability measure with finite support, the speed criterion is proved earlier in [96] by showing a general off-diagonal estimate for transition probabilities $P^n(x, y)$, where P is a reversible Markov operator on a countable state space. A generalization and improvement of this estimate is given in [22] (with a simpler proof) and is now called the Varopoulos–Carne inequality. Applied to a μ-random walk on Γ, the inequality reads: let $S = \operatorname{supp}\mu$ and take the word distance on the Cayley graph (Γ, S), then

$$P_\mu^n(x, y) \le 2e^{-d_S(x,y)^2/2n}. \tag{1.3}$$

When (Γ, μ) is not Liouville, the Poisson boundary identification problem asks if one can find an "explicit" Γ-space X, a σ-algebra $\mathcal{B}$ on X and a probability measure ν on $\mathcal{B}$, such that $(X, \mathcal{B}, \nu)$ is isomorphic to the Poisson boundary of (Γ, μ) via a Γ-equivariant measurable isomorphism. The entropy criterions of Kaimanovich [57] provide powerful tools for the identification problem. As remarked in [59], a majority of known examples of nontrivial boundary behaviors of random walks on countable groups fall into one of the following two classes:

(i) Convergence of random walk sample paths to some suitable geometric boundary, in the presence of hyperbolicity or nonpositive curvature.

(ii) Pointwise stabilization of some notion of "configurations" along random walk sample paths.

Based on the limit behavior of the random walks observed, one can define a space $(X, \mathcal{B}, \nu)$ which is a quotient of the Poisson boundary (such a space is called a μ-boundary). Roughly speaking, the entropy criterion says that the candidate space $(X, \mathcal{B}, \nu)$ is isomorphic to the Poisson boundary if the random walk $(W_n)_{n=0}^{\infty}$ conditioned on its "limit" in X, has asymptotic entropy 0 almost surely (see **[57, SECTION 4]** for a precise statement). The ray and strip criterions in **[57]** provide more checkable sufficient conditions to ensure the conditional asymptotic entropy given $(X, \mathcal{B}, \nu)$ is 0 almost surely.

Prototype examples of (i) are random walks on nonelementary Gromov word hyperbolic groups, where the geometric boundary is the visual boundary. In this case for a step distribution μ of finite entropy and finite log-moment, the visual boundary equipped with the hitting distribution is identified as the Poisson boundary of the μ-random walk in **[57]**. This type of geometric boundary identification holds for a wide class of groups acting on hyperbolic spaces (not necessarily proper or locally compact), see the work of Maher and Tiozzo **[76]** and references therein.

Prototype examples of (ii) are random walks on the so called lamplighter groups $(\mathbb{Z}/2\mathbb{Z}) \wr \mathbb{Z}^d = (\oplus_{\mathbb{Z}^d} \mathbb{Z}/2\mathbb{Z}) \rtimes \mathbb{Z}^d$. Symmetric random walks on $(\mathbb{Z}/2\mathbb{Z}) \wr \mathbb{Z}^d$, $d \geq 3$, are considered in **[60]** as first examples of random walks on amenable groups with nontrivial Poisson boundary. A crucial observation there is that for some suitably chosen step distribution μ, when the projected random walk on $\mathbb{Z}^d$ is transient, the configuration in $\oplus_{\mathbb{Z}^d} \mathbb{Z}/2\mathbb{Z}$ stabilizes pointwise along the random walk trajectory almost surely. The question whether the μ-boundary from pointwise stabilization is the full Poisson boundary remained open, until resolved positively in the work of Erschler **[33]** for $d \geq 5$ and in Lyons and Peres **[73]** for all $d \geq 3$.

In the rest of this note we survey in more details some aspects around asymptotic behaviors of random walks. Because of limitations of space and the author's knowledge, this survey is not intended to be comprehensive; rather only a small selection of topics are discussed.

2. A FEW THEMES

Here are some loosely phrased questions that have emerged from the study of random walks on groups:

(1) Can some random walk behaviors (for classes of random walks on a group, say finite range nondegenerate symmetric random walks, random walks satisfying some moment condition, etc.) be deemed *group invariant*?

(2) What properties of the group can be characterized in terms of random walk behaviors?

(3) Can random walk behavior be used to understand groups and their actions?

In each of these directions of research, many natural questions remain open.

2.1. Stability problems

Question 1 is often casted as *stability* problems. It is interesting both from the point of view of understanding random walk behaviors, and searching for group invariants arising from stochastic processes on them. Regarding the behavior of return probabilities, the following stability theorem is established by Pittet and Saloff-Coste [87] using comparison of Dirichlet form techniques (for the definition of Dirichlet form in this context see Section 3.1). Given two nonincreasing functions $f, g : \mathbb{N} \to \mathbb{R}$, we say they are equivalent if there is a constant $C \geq 1$ such that $g(Cx)/C \leq f(x) \leq Cg(x/C)$. We say a probability measure μ on Γ has finite α-moment (with respect to the word distance) if $\sum_{g\in\Gamma} |g|_S^\alpha \mu(g) < \infty$.

Theorem 2.1 ([87]). *The equivalence class of the decay function $n \mapsto \mu^{(2n)}(\mathrm{id})$, where μ is the step distribution of a nondegenerate symmetric random walk of finite second moment on Γ, is a quasiisometry invariant.*

The finite second moment condition in the theorem is necessary. For example, on $\mathbb{Z}$ consider an α-stable-like measure $\mu_\alpha(x) = \frac{c_\alpha}{(1+|x|)^{\alpha+1}}$, where $\alpha \in (0, 2)$ and c_α is the normalizing constant so that μ_α has total mass 1. Then the decay function behaves like $\mu_\alpha^{(2n)}(0) \simeq n^{-\frac{1}{\alpha}}$, which is not equivalent to the decay function of symmetric simple random walk on $\mathbb{Z}$.

In [16], Bendikov and Saloff-Coste considered the question of fastest decay of return probability under a given moment condition. Given $\alpha \in (0, 2)$ and a constant $C > 0$, let $\mathcal{S}_{\Gamma,\alpha}$ be the set of all symmetric probability measures μ on Γ such that $\sum_{g\in\Gamma} |g|_S^\alpha \mu(g) \leq C$. Consider the following function of fastest decay under α-moment condition:

$$\Phi_{\Gamma,\alpha} : n \mapsto \inf\{\mu^{(2n)}(\mathrm{id}),\ \mu \in \mathcal{S}_{\Gamma,\alpha}\}.$$

It turns out to be convenient to consider the version defined using weak α-moment as well. For some specific classes of groups the behavior of $\Phi_{\Gamma,\alpha}$ can be understood rather well, see [16, 90]. However, the question whether the equivalence class of the function $\Phi_{\Gamma,\alpha}$ is a quasiisometry invariant remains open.

In contrast to the decay of return probabilities, where tools such as comparison of Dirichlet forms are available, for entropy and speed functions stability is a well-known open problem for general amenable groups. For instance, fix the group Γ, it is an important open question whether the behavior of the entropy function $n \mapsto H_\mu(n)$ or the speed function $n \mapsto \mathbb{E}[|X_n|_S]$, is stable among nondegenerate, symmetric, finitely supported step distributions on Γ. Note that it is known that the Liouville property is not stable under quasiisometry for general graphs, see [17, 75].

2.2. Characterizations in terms of random walks

Kesten's characterization of amenability cited earlier can be viewed as a first result in the direction of Question 2. Kesten asked in [64] for a characterization of recurrent groups. What are the finitely generated groups that can carry a nondegenerate recurrent symmetric random walk? This problem was settled by Varopoulos in the 1980s, invoking Gromov's polynomial growth theorem.

Theorem 2.2 ([97]). *Suppose Γ is a finitely generated group and there exists a nondegenerate symmetric probability measure μ on Γ such that the μ-random walk is recurrent. Then Γ is a finite extension of the trivial group $\{\mathrm{id}\}$, $\mathbb{Z}$, or $\mathbb{Z}^2$.*

The growth function is an important geometric invariant of a group. Let Γ be a finitely generated group and S be a finite generating set of Γ. The growth function $v_{\Gamma,S}(n)$ counts the number of elements with word length $|\gamma|_S \leq n$, that is,

$$v_{\Gamma,S}(n) = \left|\{\gamma \in \Gamma : |\gamma|_S \leq n\}\right|.$$

A finitely generated group Γ is of *polynomial growth* if there exists $0 < D < \infty$ and a constant $C > 0$ such that $v_{\Gamma,S}(n) \leq Cn^D$. It is of *exponential growth* if there exist $\lambda > 1$ and $c > 0$ such that $v_{\Gamma,S}(n) \geq c\lambda^n$. If $v_{\Gamma,S}(n)$ is subexponential, but Γ is not of polynomial growth, we say Γ is a group of *intermediate growth*. By Gromov's theorem [46], any group of polynomial growth is virtually nilpotent. For some classes of groups, the growth is either polynomial or exponential (for example, for solvable groups by results of Milnor and Wolf [79, 101], and for linear groups as it follows from the Tits alternative [93]).

The first examples of groups of intermediate growth are constructed by Grigorchuk in [43], answering a question of Milnor. These groups are indexed by infinite strings ω in $\{0, 1, 2\}^\infty$: for any such ω, four automorphisms $a, b_\omega, c_\omega, d_\omega$ of the rooted binary tree are associated. The group G_ω is generated by $S = \{a, b_\omega, c_\omega, d_\omega\}$. By [43], if ω is eventually constant then G_ω is virtually abelian (hence of polynomial growth); otherwise G_ω is of intermediate growth. The group G_ω is periodic (also called torsion) if and only if ω contains all three letters 0, 1, 2 infinitely often. A special example of such a string is $(012)^\infty$; the corresponding group is called the *first Grigorchuk group*, which was introduced and shown to be an infinite torsion group Grigorchuk in [45]. First examples of simple groups of intermediate growth are constructed in the recent work of Nekrashevych [83, 84].

A key point in Varopoulos' proof is that a volume growth lower bound (geometric property of the underlying group) implies an upper bound for the decay of return probabilities (analytic property of the heat semigroup). More precisely, suppose that there are constants $c, d > 0$ such that the volume growth of the group Γ satisfies $v_{\Gamma,S}(n) \geq cn^d$ for all n, then there is a constant $c_1 > 0$ such that $\mu^{(2n)}(\mathrm{id}) \leq c_1 n^{-\frac{d}{2}}$. Since, by Gromov's theorem [46] and its version in van den Dries and Wilkie [94], a group of weak polynomial growth is virtually nilpotent, this estimate leads to a proof of Theorem 2.2.

We now discuss some results that characterize properties of the underlying group Γ in terms of boundary behaviors of random walks. In [37] Furstenberg proved that if Γ is nonamenable, then for any nondegenerate step distribution μ on Γ, the Poisson boundary of (Γ, μ) is nontrivial; and the converse to this statement was conjectured to be true as well. This conjecture was proved independently by Kaimanovich and Vershik [60], Rosenblatt [88]:

Theorem 2.3 ([60, 88]). *A countable group Γ is amenable if and only if there is a nondegenerate symmetric random walk on Γ with trivial Poisson boundary.*

It is classical that for any step distribution on a virtually nilpotent group, the associated Poisson boundary is trivial, see [30]. In [60] it is conjectured that on any group of exponential growth, there exists a symmetric step distribution (of infinite support in general) with nontrivial Poisson boundary. The first result on nontrivial boundary behavior of random walks on intermediate growth groups is due to Erschler [32]. Note that by the entropy criterion for the Liouville property, any finite range random walk on a group of intermediate growth has trivial Poisson boundary. Thus to observe nontrivial boundary behavior, it is necessary to take random walk step distributions with infinite support.

Theorem 2.4 ([32]). *Let $\Gamma = G_\omega$ be a Grigorchuk group with $\omega \in \{0,1\}^\infty$, where ω contains infinite numbers of* 0 *and* 1. *Then Γ admits a symmetric measure μ of finite entropy such that the Poisson boundary of (Γ, μ) is nontrivial.*

We mention that the nontrivial limit behavior observed in Theorem 2.4 is of type (ii) as described in the Introduction.

The longstanding problem which groups admit random walks with nontrivial Poisson boundary is completely settled in the recent work of Frisch, Hartman, Tamuz, and Vahidi Ferdowsi [36]. Recall that Γ has the infinite conjugacy class property (ICC) if each of its nontrivial elements has an infinite conjugacy class. Note that in some works the definition of ICC also requires the group to be nontrivial. For a finitely generated group Γ, having no ICC quotient except the trivial one $\{\mathrm{id}\}$ is equivalent to being virtually nilpotent.

Theorem 2.5 ([36]). *Let Γ be a countable group. The following are equivalent:*

(i) *Γ has a quotient group $\bar{\Gamma}$ such that $\bar{\Gamma}$ is a nontrivial ICC group.*

(ii) *There is a probability measure μ on Γ with nontrivial Poisson boundary.*

The direction (ii) implies (i) in the statement is known from the earlier work of Jaworski in [52]. The direction (i) implies (ii) is proved in [36] by a novel construction of step distributions with nontrivial Poisson boundary, directly using the ICC property. Moreover, the measure μ can be taken to be a symmetric measure with finite Shannon entropy whose support generates Γ. Theorem 2.5 solves the aforementioned conjecture of Kaimanovich and Vershik positively; and moreover brings in the key insight that the algebraic condition of having a nontrivial ICC quotient plays a crucial role. We mention that the similar problem for nondiscrete locally compact groups remains open: for instance, how to characterize a locally compact group G which admits a step distribution μ with nontrivial Poisson boundary, where μ is absolutely continuously with respect to the Haar measure of G. For this formulation, an answer is known for *connected* compactly generated lcsc groups by [50]: the characterization is that G is not of polynomial growth. The question is open for totally disconnected locally compact groups. One may also drop the constraint on μ and formulate the problem for any step distribution on the group G. Dropping the assumption on μ may change possible boundary behaviors even on polynomial growth groups, see [51, 53].

2.3. Random walks as tools

Now we turn to Question 3. Random walks provide a natural tool to study stationary measures. Consider an action of Γ on a compact space X by homeomorphims and let μ be a probability measure on Γ. Denote by $P(X)$ the space of probability measures on the Borel σ-field of X. Then by a standard compact argument, there always exists a μ-stationary measure $\nu \in P(X)$, that is, ν satisfies $\sum_{g\in\Gamma} g.\nu\mu(g) = \nu$ where $g.\nu(A) = \nu(g^{-1}.A)$. One fundamental observation in [37] is that the martingale convergence theorem implies that almost surely, along the μ-random walk $(W_n)_{n=0}^{\infty}$, the sequence of measures $(W_n.\nu)$ converges in the weak topology of $P(X)$. In particular, the limit measures give rise to a Γ-equivariant map from the Poisson boundary of (Γ, μ) to $P(X)$. This map is sometimes referred to as the affine boundary map associated to $\Gamma \curvearrowright (X, \nu)$. Ideas of using Poisson boundaries to answer algebraic questions appear in the work of Furstenberg on lattice envelopes [38]. Furstenberg's ideas inspire the use of boundary theory in later works on rigidity phenomena, which we do not touch on here.

Next we discuss some works in which random walks are used as tools to prove amenability of groups. It is shown in von Neumann's work that finite groups and abelian groups are amenable and that the class of amenable groups (AG) is closed under four standard operations: taking (i) subgroups, (ii) quotients, (iii) group extensions, and (iv) direct unions. The term *elementary amenable* is coined by Day: denote by EG be the smallest class of groups which contains all finite groups and all abelian groups and is closed under operations (i)–(iv).

A finitely generated group Γ of subexponential growth is amenable: subexponential growth implies that there exists a subsequence of balls $B(\mathrm{id}, r_i)$ forming a Følner sequence. Chou shows in [23] that a finitely generated group in EG is either virtually nilpotent or of exponential growth; a torsion group in EG is locally finite; and a finitely generated simple group in EG is finite. Thus Grigorchuk groups of intermediate growth are in AG but not EG. As in [27], denote by SG the closure of all groups of subexponential growth under (i)–(iv). It is clear that $EG \subseteq SG \subseteq AG$.

A first example to separate SG and AG is shown in Bartholdi and Virág [14]. The example is called the Basilica group, which was first studied in [44]. The Basilica group B is a two-generated group acting on the rooted binary tree. One key idea in [14] is certain random walk on B enjoys self-similar properties compatible with the wreath recursions down the tree. The self-similarity allows one to efficiently use the recursion on the tree to study behaviors of the random walk. In [14] it is shown that the return probability of such a random walk on B decays subexponentially, thus B is amenable by Kesten's criterion. The idea of self-similar random walks is later extended to larger classes of groups acting on trees in [2, 13, 20, 58]. It is later understood that for proving amenability, one crucial property is that the induced random walk on certain orbital Schreier graph is recurrent. For instance, amenability of B can be shown in the unified framework of "extensive amenability," which emerges from the seminal work [54] and is developed in [55, 56].

For the rest of this subsection we focus on the relation of random walks with nontrivial Poisson boundary to volume growth of the group. The most direct way to obtain a growth lower bound is to exhibit distinct elements within a given radius. For instance, if Γ contains two elements a, b such that they generate a free semigroup, then thIS semigroup provides 2^n distinct elements within radius $n \max\{|a|_S, |b|_S\}$. In general, it can be rather challenging to find explicit elements within a given distance, see [11,71] on the first Grigorchuk group. As a consequence of the entropy criterion and Shannon's theorem, random walks with nontrivial Poisson boundary on Γ can provide lower bounds on volume growth of Γ. Heuristically, instead of exhibiting many distinct elements in a ball, one constructs random walks with positive asymptotic entropy, which indirectly imply there must be sufficiently many points in balls. The quantitative relation between the tail decay of the step distribution μ and growth of Γ can be made precise:

Lemma 2.6. *Suppose Γ admits a μ-random walk with nontrivial Poisson boundary, where the probability measure μ has finite entropy $H(\mu) < \infty$ and finite α-moment, for some $\alpha \in (0, 1]$. Then there is a constant $c > 0$ such that the volume function satisfies*

$$v_{\Gamma,S}(r) \geq \exp(cn^{\alpha}).$$

See [35, LEMMA 2.1] for a more general statement. To avoid possible periodicity issues, we may always assume that $\mu(\{\mathrm{id}_\Gamma\}) > 0$: changing μ to a convex combination of μ and the δ-mass at id does not change the space of harmonic functions. To make use of Lemma 2.6, one first constructs a random walk μ with finite entropy, which is designed to guarantee that there exists a tail event A of the μ-random walk whose probability is not in $\{0, 1\}$. Then we have:

Observation of one nontrivial tail event for μ-random walk

$\Updownarrow$

Poisson boundary of (G, μ) is nontrivial

$\Downarrow$

Volume lower bound from the moment condition satisfied by μ.

The random walks with nontrivial Poisson boundary on the Grigorchuk group $G_{(01)^\infty}$ constructed by Erschler in Theorem 2.4 yield lower bounds which match rather tightly with upper bounds. More precisely, by [32, THEOREMS 2 AND 3], the growth function of $G = G_{(01)^\infty}$ satisfies

$$\exp\left(\frac{n}{\log^{2+\epsilon}(n)}\right) \lesssim v_{G,S}(n) \lesssim \exp\left(\frac{n}{\log^{1-\epsilon}(n)}\right), \quad \text{for any } \epsilon > 0.$$

The construction in [32] uses the fact that $G_{(01)^\infty}$ contains an infinite dihedral group. For the Grigorchuk group $G_{(012)^\infty}$, which is a torsion group by [45], random walks with nontrivial Poisson boundary and tight control over the tail decay are constructed in [35].

Theorem 2.7 ([35]). *Let $\alpha_0 = \frac{\log 2}{\log \lambda_0} \approx 0.7674$, where λ_0 is the positive root of the polynomial $X^3 - X^2 - 2X - 4$. For any $\epsilon > 0$, there exists a constant $C_\epsilon > 0$ and a nondegenerate symmetric probability measure μ on $G = G_{(012)^\infty}$ of finite entropy and nontrivial Poisson boundary, where the tail decay of μ satisfies that for all $r \geq 1$,*

$$\mu(\{g : l_S(g) \geq r\}) \leq C_\epsilon r^{-\alpha_0+\epsilon}.$$

As a consequence, for any $\epsilon > 0$, there exists a constant $c_\epsilon > 0$ such that for all $n \geq 1$,

$$v_{G,S}(n) \geq \exp(c_\epsilon n^{\alpha_0 - \epsilon}).$$

The volume lower bound in Theorem 2.7 matches up in exponent with the growth upper bound in [10]. In particular, combined with the upper bound we conclude that the volume exponent of the first Grigorchuk group $G = G_{(012)^\infty}$ exists and is equal to α_0, that is,

$$\lim_{n\to\infty} \frac{\log \log v_{G,S}(n)}{\log n} = \alpha_0.$$

A detailed sketch of the construction of measure μ stated in Theorem 2.7 can be found in [35, INTRODUCTION].

The method also applies to other period strings ω in $\{0, 1, 2\}^\infty$, which contains all three symbols infinitely often, to show the corresponding Grigorchuk group G_ω has a volume exponent $\alpha_\omega \in (0, 1)$, see [35, THEOREM B]. When ω is not periodic, it is shown in [43] that for some choices of ω, the growth of G_ω exhibits oscillating behavior: limit of $\log \log v_{G_\omega,S}(n)/ \log n$ does not exist. Indeed, the family $\{G_\omega\}$ provides a continuum of mutually nonequivalent growth functions. This statement is shown in [43] with the introduction of the space of marked groups, see more discussion in Section 3.2. In [35, THEOREMS C AND 8.5] it is shown that for a large collection of nonperiodic ω, the oscillating growth function of G_ω can be estimated with good precision. In particular, combined with the upper bounds from [12], the estimates show that given any $\alpha \leq \beta$ in the interval $[\alpha_0, 1]$, where α_0 is the growth exponent of $G_{(012)^\infty}$, there is an $\omega \in \{0, 1, 2\}^\infty$ with

$$\liminf_{n\to\infty} \frac{\log \log v_{G_\omega,S}(n)}{\log n} = \alpha \quad \text{and} \quad \limsup_{n\to\infty} \frac{\log \log v_{G_\omega,S}(n)}{\log n} = \beta.$$

We mention that it is an open problem whether one can find an example of intermediate growth group whose lower growth exponent is strictly less than α_0. Such questions are related to Grigorchuk's gap conjecture, see [42].

3. QUANTITATIVE BEHAVIOR OF RANDOM WALK CHARACTERISTICS

In this section we focus on quantitative estimates for random walks on groups. Here are some of the key interrelated aspects:

(i) What are the *spectral properties* of the convolution operator $f \mapsto f * \mu$ when μ is a symmetric probability measure on Γ? What is the behavior of the probability of return of a symmetric random walk driven by step distribution μ?

(ii) What is the asymptotic *entropic behavior*, that is, the behavior of $n \mapsto H_\mu(n)$ as n tends to infinity? Here $H_\mu(n)$ is the Shannon entropy of W_n as in (1.1).

(iii) What is the *escape behavior* of transient random walks captured in terms of some given distance function on the group, say, in the form of average displacement as in (1.2) or more refined descriptions?

(iv) What is the structure of sets μ-harmonic functions (bounded, positive, of polynomial growth, of a given growth type, slow or fast, etc.)?

In this section we mainly focus on topics around (i)–(iii); unbounded harmonic functions are discussed in the next section.

3.1. Isoperimetric profiles

We first introduce some notations for (i). Let R_μ be the right convolution operator $\ell^2(\Gamma) \to \ell^2(\Gamma)$ defined as $R_\mu(f)(x) = (f * \mu)(x) = \sum_{g\in\Gamma} f(xg^{-1})\mu(g)$. The return probability to identity at time n is given by $\mu^{(n)}(\mathrm{id}) = \langle R_\mu^n \delta_{\mathrm{id}}, \delta_{\mathrm{id}}\rangle_{\ell^2(\Gamma)}$. When μ is symmetric, R_μ is a self-adjoint operator. Denote by $E_\lambda^{R_\mu} = \mathbf{1}_{(-\infty,\lambda]}(R_\mu)$ the spectral projections of R_μ. The spectral measure of R_μ is given by

$$N_{R_\mu}((-\infty,\lambda]) = \langle E_\lambda^{R_\mu}(\delta_{\mathrm{id}}), \delta_{\mathrm{id}}\rangle_{\ell^2(\Gamma)}.$$

The relation between return probabilities and the spectral measure is expressed through the transform

$$\mu^{(2n)}(\mathrm{id}) = \int_{\mathbb{R}} \lambda^{2n} dN_{R_\mu}(\lambda) = \int_{-1}^{1} \lambda^{2n} dN_{R_\mu}(\lambda).$$

The last equality is because R_μ is a Markov operator. When Γ is an infinite amenable group, one may draw information on the behavior of the spectral measure near 1 from the decay of return probabilities, using Tauber–Karamata theorems for Laplace transforms.

The decay of return probability $\mu^{(2n)}(\mathrm{id})$ is closely related to isoperimetric profiles of R_μ, which in the discrete setting can be introduced for more general reversible Markov operators. Let V be a countable set, typically the vertex set of a graph, and let $P: V \times V \to [0,1]$ be the transition probabilities of a reversible Markov chain on V. Denote by π a reversing measure for P, that is, $\pi(x)P(x,y) = \pi(y)P(y,x)$ for all $x,y \in V$. Consider the associated Dirichlet form

$$\mathcal{E}_P(f_1,f_2) = \frac{1}{2}\sum_{x,y}\big(f_1(y)-f_1(x)\big)\big(f_2(y)-f_2(x)\big)\pi(x)P(x,y),$$

which is a bilinear form on $\mathrm{Dom}(\mathcal{E}_P) = \{f \in L^2(V,\pi) : \mathcal{E}_P(f,f) < \infty\}$. The L^2-*isoperimetric profile* of P, also called the *spectral profile*, is defined as

$$\begin{aligned}\Lambda_{2,P} &: \mathbb{R}_+ \to [0,1],\\ &v \mapsto \inf\{\lambda_P(\Omega) : \Omega \subseteq V, \pi(\Omega) \le v\},\end{aligned}$$

where $\lambda_P(\Omega)$ is the lowest eigenvalue of the Laplacian operator $I - P$ with Dirichlet boundary condition in Ω,

$$\lambda_P(\Omega) = \inf\{\mathcal{E}_P(f,f) : \mathrm{supp}(f) \subseteq \Omega, \|f\|_{L^2(V,\pi)} = 1\}.$$

The L^1-isoperimetric profile is defined analogously. Using an appropriate coarea formula, $\Lambda_{1,P}$ can equivalently be defined more geometrically as

$$\Lambda_{1,P}(v) = \inf\left\{\frac{\sum_{x,y\in V}\mathbf{1}_\Omega(x)\mathbf{1}_{V\setminus\Omega}(y)\pi(x)P(x,y)}{\pi(\Omega)} : \pi(\Omega) \le v\right\},$$

where the quantity $\sum_{x,y\in V} \mathbf{1}_\Omega(x)\mathbf{1}_{V\setminus\Omega}(y)\pi(x)P(x,y)$ measures the size of the boundary of Ω with respect to P.

Between the L^1- and L^2-isoperimetric profiles, we have the following inequality, often referred to as Cheeger's inequality (see, e.g., [68]):

$$\frac{1}{2}\Lambda_{1,P}^2 \leq \Lambda_{2,P} \leq \Lambda_{1,P}. \tag{3.1}$$

The nonobvious direction $\Lambda_{2,P} \geq \frac{1}{2}\Lambda_{1,P}^2$ is useful for transferring L^1-expansion inequalities to spectral profile lower bounds. The Coulhon–Saloff-Coste inequality [26], which implies, for example,

$$\Lambda_{1,R_\mathbf{u}}\big(2v_{\Gamma,S}(r)\big) \geq \frac{1}{2r},$$

where $\mathbf{u}$ is the uniform measure on $S \cup S^{-1}$ and $v_{\Gamma,S}$ is the volume function function of (Γ, S), can be proved by an elementary mass displacement type argument. Sharp L^1-expansion inequalities on $\mathbb{Z}^d$ can be derived from Loomis–Whitney inequalities, see, e.g., [72, SECTION 6.6]; further connections between isoperimetric inequalities and entropy inequalities (and consequences such as Loomis–Whitney, Harper inequalities) are investigated in [48].

The use of Nash-type inequalities to estimate return probabilities in the discrete setting was introduced in [95]. It turns out Nash inequalities are equivalent to Faber–Krahn-type inequalities, where the latter is of the form $\Lambda_{2,P}(v) \geq f(v)$ for some positive function f. In fact, in very general settings, it is known that various forms of functional inequalities are equivalent, see [8] and references therein. Comparison of forms, considered in [29] for random walks on finite groups, is a useful tool to deduce isoperimetric inequalities for a Markov operator P of interest from known results on other Markov operators.

Through a series of works by Coulhon and Grigor'yan [24, 25], it is shown that under some mild conditions, the asymptotic decay of return probability $\sup_{x\in V} P^{2n}(x,x)$ and the L^2-isoperimetric profile of P determine each other. More precisely, suppose that $\pi^* = \inf_{x\in V} \pi(x) > 0$. Let $\gamma(t)$ be the function defined by the equation

$$t = \int_{\pi*}^{\psi(t)} \frac{dv}{\Lambda_{2,P}(v)v}, \tag{3.2}$$

then under a mild regularity assumption, the return probabilities satisfy

$$\sup_{x\in V} \frac{P^{2n}(x,x)}{\pi(x)} \simeq \frac{1}{\psi(2n)}. \tag{3.3}$$

For the Markov operator R_μ, where μ is a probability measure on a countable amenable group, a precise formula relating the behavior of the spectral measure N_{R_μ} near 1 and Λ_{R_μ} near infinity is obtained in [15], under the assumption that $\Lambda_{R_\mu} \circ \exp$ is doubling near infinity.

For classes of groups where explicit estimates of return probabilities and isoperimetric profiles are known, the read may consult [15, TABLE 1] and pointers to references there. In addition to the Table, for free solvable groups see [89], and for discrete subgroups of upper triangular matrices over a local field see [92].

A consequence of (3.2) and (3.3) is the following isoperimetry test for transience. Suppose we have a Faber–Krahn inequality for an irreducible reversible Markov operator P of the form $\Lambda_{2,P}(s) \geq f(s)$, $s \in [1,\infty)$, where f is a continuous positive decreasing function on $[1,\infty)$. If

$$\int_1^\infty \frac{ds}{s^2 f(s)} < \infty, \tag{3.4}$$

then the Markov chain with transition operator P is transient, see [41, THEOREM 6.12]. Recall the type (ii) boundary behavior described in the Introduction, which often relies on the transience of induced random walks on certain orbits. In particular, isoperimetry of induced random walks on Schreier graphs play an important role in the construction of random walks with nontrivial Poisson boundary in Theorem 2.7.

3.2. The space of marked groups and realization problems

The speed function realization problem, which is often attributed to Vershik, asks what kind of functions can be realized as the speed function of a simple random walk on some finitely generated group. By realizing a given function f as a speed function, we mean finding (Γ,μ) such that the speed of the μ-random walk satisfies $f(n)/C \leq L_\mu(n) \leq Cf(n)$ for some constant $C \geq 1$. Similar realization problems can be posed for other random walk characteristics such as the entropy function $H_\mu(n)$; and for geometric invariants such as the growth function $v_{\Gamma,S}$.

A complete solution to a realization problem consists of two parts. The first part identifies constraints on the functions; the second part shows that all functions under the necessary constraints can be realized. Consider the speed function $L_\mu(n)$, where μ is a nondegenerate symmetric probability measure of finite support on Γ. Then the triangle inequality for the norm $|\cdot|_S$ and translation invariance imply that $L_\mu(n)$ is a subadditive function, $L_\mu(n+m) \leq L_\mu(n) + L_\mu(m)$. Another constraint is known as the "universal diffusive lower bound", which is proved in Lee and Peres [69] building on an earlier idea of Erschler: there is a universal constant $c > 0$ such that for any infinite amenable group Γ equipped with a finite generating set S, for any symmetric probability measure μ on Γ whose support contains S, we have

$$L_\mu(n) \geq c\sqrt{p_* n}, \quad \text{where } p_* = \min_{\gamma \in S} \mu(\gamma).$$

For more discussion on the connection of such a general bound to harmonic embeddings into a Hilbert space, see Section 4. The diffusive lower bound is achieved, for example, by simple random walk on $\mathbb{Z}$: take $\Gamma = \mathbb{Z}$ and $\mu(\pm 1) = \frac{1}{2}$. Given these constraints, the speed function realization problem asks what functions between $\sqrt{n}$ and n can be realized as speed function of finite range symmetric random walks on groups.

Theorem 3.1 ([21]). *There exists a universal constant $C > 1$ such that the following holds. For any function $f : [1,\infty) \to [1,\infty)$ such that $f(1) = 1$ and $x/f(x)$ is nondecreasing, there exist a group Δ equipped with a finite generating set T and a nondegenerate symmetric probability measure μ on Δ of finite support such that*

- *the speed and entropy functions satisfy* $L_\mu(n) \simeq_C H_\mu(n) \simeq_C \sqrt{n} f(\sqrt{n})$;
- *the* L^p*-isoperimetric profile satisfies* $\Lambda_{p,R_\mu}(v) \simeq_C (\frac{f(\log(e+v))}{\log(e+v)})^p$ *for any* $p \in [1,2]$;
- *the return probability satisfies* $-\log(\mu^{(2n)}(\mathrm{id})) \simeq_C w(n)$, *where* $w(n)$ *is determined by* $n = \int_1^{w(n)} (\frac{s}{f(s)})^2 ds$.

When the function f *is sublinear, that is,* $\lim_{x\to\infty} f(x)/x = 0$, *the group* Δ *can be chosen to be elementary amenable with asymptotic dimension 1.*

Here the notation $f \simeq_C g$ means $g(x)/C \le f(x) \le Cg(x)$. In particular, the first item gives a satisfying answer to the speed function realization problem. The statement for speed function realization between $n^{3/4}$ and n^γ, $\gamma < 1$, is obtained earlier by Amir and Virag in [4]. The group Δ constructed in Theorem 3.1 is of exponential growth. It is an open problem for the entropy function, whether a nondegenerate symmetric random walk μ on a group Γ of exponential growth always satisfies $H_\mu(n) \gtrsim \sqrt{n}$.

As cited in Section 2.3, the space of marked groups is introduced by Grigorchuk in [43] to show that there are $2^{\aleph_0}$ groups with pairwise inequivalent growth functions. A k-marked group is a pair (Γ, T), where $T = (t_1, \dots, t_k)$ is an ordered k-tuple in Γ^k which generates Γ. Equivalently, let $\mathbf{F}_k$ be the rank k free group; (Γ, T) corresponds to the kernel of the homomorphism $\mathbf{F}_k \to \Gamma$ which sends the jth free generator of $\mathbf{F}_k$ to t_j, $1 \le j \le k$. Denote by $\mathcal{M}_k$ the space of k-marked groups. The product topology on $2^{\mathbf{F}_k}$ induces a topology on $\mathcal{M}_k$, via the identification described above. This topology on $\mathcal{M}_k$ is sometimes called the Cayley–Grigorchuk topology, as two marked groups are close if their labeled Cayley graphs agree on a large ball around the identity. Under this topology, $\mathcal{M}_k$ is a metrizable compact Hausdorff space.

The product operation in $\mathcal{M}_k$ is called a diagonal product: consider a collection of marked groups $((\Gamma_i, T_i))_{i\in I}$ their diagonal product, denoted by $\bigotimes_{i\in I}(\Gamma_i, T_i)$, is the quotient of $\mathbf{F}_k$ with kernel $\bigcap_{i\in I} \ker(\mathbf{F}_k \to \Gamma_i)$. In some situations it is possible to understand well the structure of a diagonal product. Consider a converging sequence of marked groups $((\Gamma_i, T_i))_{i=1}^\infty$ in $\mathcal{M}_k$ and denote by $(\Gamma_0, T_0) = \lim_{i\to\infty}(\Gamma_i, T_i)$. Then the limit (Γ_0, T_0) is a marked quotient of the diagonal product $\Delta = \bigotimes_{i=1}^\infty(\Gamma_i, T_i)$. When the sequence Γ_i consists of finite groups, Δ is a FC-central extension of Γ_0. The construction in [21] takes diagonal product of a sequence of marked groups which converges to a wreath product of the form $\Gamma_0 = (A \times B) \wr \mathbb{Z}$ where A, B are finite groups. The sequence is chosen so that one can understand what elements in $\ker(\Delta \to \Gamma_0)$ are, and, moreover, explicitly estimate the word length of such elements with respect to the marking on Δ. The flexibility in the construction allows proving Theorem 3.1.

3.3. Relations between random walk characteristics

The three random walk characteristic functions, namely the decay of return probabilities, the entropy function, and the speed function post constraints on each other.

3.3.1. Between speed and entropy

As a consequence of the Varopoulos–Carne inequality and the fundamental inequality mentioned earlier, for a symmetric probability measure μ on G with finite support, entropy and speed satisfy

$$\frac{1}{n}\left(\frac{1}{4}L_\mu(n)\right)^2 - 1 \leq H_\mu(n) \leq (v+\varepsilon)L_\mu(n) + \log n + C, \tag{3.5}$$

where v is the exponential volume growth rate of $(G, \operatorname{supp}\mu)$, $C > 0$ is an absolute constant, see [4, 31]. For example, if we know $H_\mu(n) \simeq n^\theta$ (that is, the entropy exponent is θ), then the speed function is constrained by $n^\theta \lesssim L_\mu(n) \lesssim n^{(1+\theta)/2}$.

In [1], the joint realization problem of speed and entropy is considered. The construction in [21] with a sequence of expanders as input and the one with finite dihedral groups allows showing that for any entropy exponent $\theta \in [1/2, 1]$, all speed exponents allowed by the constraint (3.5) can be realized. That is, for any $\theta \in [\frac{1}{2}, 1]$ and $\gamma \in [\frac{1}{2}, 1]$ satisfying $\theta \leq \gamma \leq \frac{1}{2}(\theta+1)$, there exists a finitely generated group G and a symmetric probability measure μ of finite support on G, such that the random walk on G with step distribution μ has entropy exponent θ and speed exponent γ, see [21, COROLLARY 1.3, PROPOSITION 3.17]. The case where both exponents θ, γ belong to $[\frac{3}{4}, 1]$ was treated by Amir [1].

3.3.2. Between return probabilities and entropy

Let μ be a symmetric probability measure of finite entropy on a group G. In [86, 90], the following connection between return probability and entropy is shown. Let μ be a symmetric probability measure of finite entropy on Γ. Then:

- if $-\log \mu^{(2n)}(\mathrm{id})/n^{1/2} \to 0$ as $n \to \infty$, then the pair (G, μ) has the Liouville property;
- furthermore, if $-\log \mu^{(2n)}(\mathrm{id}) \lesssim n^\beta$ where $\beta \in (0, 1/2)$, then the entropy function satisfies

$$H_\mu(n) \lesssim n^{\frac{\beta}{1-\beta}},$$

see [86, THEOREMS 1.1 AND 3.2]. The sharpness of this bound, which turns a return probability lower estimate into an entropy upper estimate, is demonstrated on a family of groups called bubble groups, which are considered in [67].

If instead of slow decay of return probabilities, one has estimates on the spectral profiles of balls, $\lambda_{R_\mu}(B(\mathrm{id}, r)) \lesssim r^{-\theta}$, then, by [86, THEOREM 1.6], the α-moment of displacement of the μ-random walk $(W_n)_{n=1}^\infty$ satisfies

$$\mathbb{E}\Big[\max_{1\leq k\leq n} |W_k|_S^\alpha\Big] \leq C n^{\alpha/\theta}, \quad \text{for any } \alpha \in (0, \theta).$$

3.4. Connection to metric embeddings

The study of embeddings of finitely generated groups (viewed as a metric space with word distance on its Cayley graph) into Hilbert space was initiated by Gromov [47]. In the seminal work [102], G. Yu proved that groups that admit coarse embeddings into Hilbert space satisfy the coarse Baum–Connes conjecture.

Distortion of embeddings of finite metric spaces has been extensively studied in the theory of Banach spaces. Similar to the notion of distortion, Guentner and Kaminker [49] introduce a natural quasiisometry invariant that characterizes how close to bi-Lipschitz can an embedding of an infinite group into a Banach space be. Let Γ be a group generated by a finite set S and equipped with the associated left-invariant word metric d_S. For a Banach space X let $\alpha_X^*(\Gamma)$ be the supremum over all $\alpha \geq 0$ such that there exists a Lipschitz mapping $f : \Gamma \to X$ and $c > 0$ such that for all $x, y \in \Gamma$ we have $\|f(x) - f(y)\|_X \geq c d_S(x, y)^\alpha$. Similarly, one can define the equivariant compression exponent $\alpha_X^\#(\Gamma)$ by restricting to equivariant Lipschitz maps, namely $\|f(gx) - f(gy)\|_X = \|f(x) - f(y)\|_X$ for all $g, x, y \in G$. When the target space is the classical Lebesgue space $L_p([0,1])$, we write $\alpha_p^*(\Gamma)$ and $\alpha_p^\#(\Gamma)$ for the compression exponents.

The idea of connecting the notion of Markov type, which is an important metric invariant introduced by K. Ball [9], to Banach compression exponent of infinite groups first appears in Austin, Naor, and Peres [6]. For wreath products, in [81] an explicit formula for the Hilbert compression exponent of $H \wr \mathbb{Z}$ is shown, assuming that the lamp group H satisfies $\alpha_2^\#(H) = \frac{1}{2\beta^*(H)}$, where $\beta^*(H)$ is the supremum of upper speed exponent of symmetric random walk of bounded step distribution on H. Further, in [82] which significantly extends the method in [6,81], the L^p-compression exponent of $\mathbb{Z} \wr \mathbb{Z}$ was determined for $p \geq 1$. In [21] it is shown that for any $p \in [1,2]$ and a finitely generated infinite group H, the equivariant L_p-compression exponent of the wreath product $H \wr \mathbb{Z}$ is

$$\alpha_p^\#(H \wr \mathbb{Z}) = \min\left\{\frac{\alpha_p^\#(H)}{\alpha_p^\#(H) + (1 - \frac{1}{p})}, \alpha_p^\#(H)\right\}.$$

When applying the Markov-type method, one has the flexibility of choosing which Markov chains to consider: for instance, α-stable like random walks in [82] and jumping processes confined on finite subsets of $H \wr \mathbb{Z}$ in [21].

It is known that distortion of metric embeddings can be captured by Poincaré inequalities of general forms. In particular, the Markov-type inequalities mentioned above can be viewed as a special form of Poincaré inequalities. Other types of obstructions to low distortion embeddings can be observed in the metric geometry of finitely generated groups. The construction of diagonal product Δ with infinite dihedral groups as input in [21] contains scaled ℓ^∞-cubes of growing sizes in Δ. Sharp estimates of distortion of embeddings of ℓ^∞-cubes into L^p-spaces are provided by the deep work of Mendel and Naor on metric cotype in [77]. Explicit evaluation of compression exponents of such diagonal products yields the following. With certain choice of parameters, such groups also provide the first examples where L_p-compression exponent, $p > 2$, is strictly larger than the Hilbert compression exponent. It might be interesting to investigate this collection of groups in the program on quasiisometric rigidity of solvable groups.

Theorem 3.2 ([21]). *For any $\frac{2}{3} \leq \alpha \leq 1$, there exists a 3-step solvable group Δ such that for any $p \in [1,2]$,*

$$\alpha_p^*(\Delta) = \alpha_p^\#(\Delta) = \alpha.$$

Further, there exists a 3-step solvable group Δ_1 such that for all $p \in (2,\infty)$,

$$\alpha_p^\#(\Delta_1) \geq \frac{3p-4}{4p-5} > \alpha_2^\#(\Delta_1) = \frac{2}{3}.$$

4. UNBOUNDED HARMONIC FUNCTIONS AND EQUIVARIANT EMBEDDINGS INTO HILBERT SPACES

Besides bounded harmonic functions one may consider other classes of harmonic functions and their relation to random walks. In this section we focus on harmonic functions of at most linear growth on amenable groups. Unlike the boundary theory associated with bounded harmonic functions, there is no systematic theory developed for this class of harmonic functions. Throughout this section, let Γ be a finitely generated group and take a symmetric probability measure μ of finite generating support on Γ.

Let $\pi : \Gamma \to \mathcal{U}(\mathcal{H})$ be a unitary representation of Γ on a separable Hilbert space $\mathcal{H}$. A map $b : \Gamma \to \mathcal{H}$ is a 1-*cocycle* if $b(gh) = b(g) + \pi_g b(h)$ for all $g, h \in \Gamma$. Because of the cocycle equality, given a probability measure μ on G, b is μ-harmonic if $\sum_{s\in\Gamma} b(s)\mu(s) = 0$. A μ-harmonic 1-cocycle $b : G \to \mathcal{H}$ is also referred to as an *equivariant harmonic embedding* of G into $\mathcal{H}$.

As a special case of results in [80] (for the finitely presented case) and [66], a finitely generated group G does not have Kazhdan's Property (T) if and only if it admits a nonconstant equivariant μ-harmonic embedding into a Hilbert space. For an exposition of the proof in the setting of finitely generated groups, see [65, APPENDIX]. In the amenable case, nontrivial μ-harmonic embeddings can be constructed more explicitly by using μ-random walks, see, for example, [69, SECTION 3] and [34].

One may ask about properties of unitary representations associated with nonconstant μ-harmonic 1-cocycles. In [91] Shalom introduced the following notions in connection to the large-scale geometry of the groups. We say Γ has Property H_{FD} (H_{F}, or H_{T}, respectively) if for every nonconstant μ-harmonic 1-cocycle $b : G \to \mathcal{H}$, the associated representation π has a finite-dimensional (finite, or trivial, respectively) subrepresentation. We say π is *weakly mixing* if it does not admit any finite-dimensional subrepresentations. It is clear that Properties H_{FD}, H_{F}, and H_{T} are in increasing strength, while the sharpest one of them, H_{T}, implies that all μ-harmonic 1-cocyles are homomorphisms to $\mathcal{H}$.

4.1. Martingale small-ball probabilities

The existence of a nontrivial equivariant μ-harmonic embedding $b : \Gamma \to \mathcal{H}$ implies the a diffusive lower bound for speed of a μ-random walk on G, see [69]. In this subsection, we review bounds on small-ball probabilities of the martingale $b(W_t)$, which provide additional information about the behavior of the random walk. Note that from the cocycle equality, $\|b(gs) - b(g)\|_{\mathcal{H}} = \|b(s)\|_{\mathcal{H}}$, in particular the map b is Lipschitz.

Consider a martingale $(X_t)_{t=0}^{\infty}$ with respect to filtration $(\mathcal{F}_t)_{t=0}^{\infty}$ taking values in a Hilbert space $\mathcal{H}$. Under the assumption of bounded increments and that the conditional variances $\mathbb{E}[\|X_{t+1} - X_t\|^2 | \mathcal{F}_t]$ are constant, general small-ball probabilities estimates are proved independently in [70] and [5]. Applied to the martingale $(b(W_t))_{t=0}^{\infty}$, where $(W_t)_{t=0}^{\infty}$ is a μ-random walk on Γ, we have:

Theorem 4.1 ([5,70]). *For any nonconstant equivariant μ-harmonic embedding $b : G \to \mathcal{H}$, there is a constant $C > 0$, such that for all $t, r \geq 1$,*

$$\mathbb{P}\left(\|b(W_t)\| \leq r\right) \leq \frac{Cr}{\sqrt{t}}.$$

Note that this general bound cannot be improved, since, for instance, it is sharp for a simple random walk (W_t) on $\mathbb{Z}$ where $\mu(\pm 1) = 1/2$ and $b : \mathbb{Z} \to \mathbb{R}$ given by $b(z) = z$. Note that in this example the representation associated with b is trivial.

When $b : G \to \mathcal{H}$ is a nonconstant harmonic 1-cocycle with weakly mixing representation π, the martingale $X_t = b(W_t)$ satisfies an asymptotic orthogonality condition: for large k, the direction of the increment $X_{t+k} - X_t$ is almost orthogonal to X_t. More precisely, when the representation π is weakly mixing, there exists a sequence of nonincreasing constants $(\epsilon_k)_{k \in \mathbb{N}}$ such that $\lim_{k \to \infty} \epsilon_k = 0$ and for any $t, k \in \mathbb{N}$, the martingale $X_t = b(W_t)$ satisfies

$$\frac{1}{k}\mathbb{E}\left[\left\langle \frac{X_t}{\|X_t\|}, X_{t+k} - X_t \right\rangle^2 | \mathcal{F}_t\right] \leq \epsilon_k \quad \text{almost surely.} \tag{4.1}$$

This claim can be deduced directly from [85, LEMMA], which is a step in Ozawa's functional analytic proof of the Gromov polynomial growth theorem.

It turns out for $\mathcal{H}$-valued martingales with bounded increments and asymptotic orthogonality property (4.1), one can obtain superpolynomial decay bounds for small-ball probabilities, following a classical Foster–Lyapunov drift-type supermartingale argument. Applied to the martingale $(b(W_t))_{t=0}^{\infty}$, we have the following superpolynomial decay estimate for small-ball probabilities. If $b : G \to \mathcal{H}$ is a nontrivial harmonic 1-cocycle with weakly mixing representation π, then for any $\beta > 0$, there exists a constant $C > 0$ depending on β such that

$$\mathbb{P}\left(\|b(W_t)\| \leq r\right) \leq C\left(\frac{r}{\sqrt{t}}\right)^{\beta} \quad \text{for all } t, r \geq 1. \tag{4.2}$$

4.2. Some open problems

Since its introduction in [91], it is believed that for amenable groups, Property H_{FD} is a rather strong property only satisfied by certain "small" groups. This is reflected in the state that the only known examples of groups with Property H_{FD} are nilpotent groups, polycyclic groups, wreath product $F \wr \mathbb{Z}$ with finite F, and certain extensions of such groups. There are very limited known tools to establish that a given group has Property H_{FD}. If Γ embeds as a lattice in a nondiscrete locally compact group G, then one might use the representation theory of G: this approach is carried out in [91] to establish that polycyclic groups have Property H_{FD}. Probabilistic approaches using random walks, see [85], require strong coupling properties. For instance, it is still not known whether the wreath product $F \wr \mathbb{Z}^2$, where F is finite, has Property H_{FD}.

In this subsection, we discuss two problems on general amenable groups, in both situations knowing that the group Γ does not have Property H_{F} implies positive answers. These problems provide motivations to understand better the class of groups with Property H_{F}.

4.2.1. Dimension of the space of linear growth harmonic functions

Problem 4.2 ([78]). Let $\mathrm{HF}_1(\Gamma, \mu)$ denote the space of μ-harmonic functions on Γ whose growth is bounded by a linear function. Is it true that $\mathrm{HF}_1(\Gamma, \mu)$ is finite dimensional if and only if Γ is of polynomial growth?

For more background and discussions around this question see [78, INTRODUCTION]. The "if" direction is known: it is a step in Kleiner's proof [65] of Gromov's polynomial growth theorem. Towards the "only if" direction, when Γ does not have Property H_{FD}, there exists an irreducible unitary representation $\pi : \Gamma \to \mathcal{H}$ which is weakly mixing and associated with a nonconstant μ-harmonic cocycle $b : \Gamma \to \mathcal{H}$. Consider the subspace of Lipschitz μ-harmonic functions $\{f_v\}_{v \in \mathcal{H}}$ given by $f_v(g) = \langle b(g), v\rangle$. One can check that since π is weakly mixing, the space $\{f_v\}_{v\in\mathcal{H}}$ is infinite dimensional. When Γ has Property H_{FD} but not H_{F}, then it virtually admits a solvable group of exponential growth as a quotient group, see [91, PROPOSITION 4.2.3]. In this case applying the results in [78] to a solvable quotient group then lifting back to Γ show that $\mathrm{HF}_1(\Gamma, \mu)$ is infinite dimensional. Therefore the problem remains open only for groups with Property H_{F}.

4.2.2. Occupation time of balls

For a transient μ-random walk $W = (W_n)_{n=1}^{\infty}$ on Γ, one can consider the occupation time of a finite set, that is, the total amount of time the random walk spent in the given set. Of particular interest is the occupation time of balls:

$$\mathcal{N}_W(r) := \left|\{n \in \mathbb{N} : W_n \in B_S(\mathrm{id}, r)\}\right| = \sum_{\gamma \in B_S(\mathrm{id},r)} G_\mu(\mathrm{id}, \gamma),$$

where $B_S(\mathrm{id}, r)$ is the set of vertices within graph distance r to the identity on the Cayley graph (Γ, S), and $G_\mu(x, y) = \sum_{n=0}^{\infty} \mu^{(n)}(x^{-1}y)$ is the Green function of the μ-random walk.

Problem 4.3 ([74]). If $(W_n)_{n=0}^{\infty}$ is a transient symmetric random walk on a finitely generated group G, then

$$\mathbb{E}\big(\mathcal{N}_W(r)\big) \lesssim r^2,$$

where $\mathcal{N}_W(r)$ is the occupation time of the ball $B_S(\mathrm{id}, r)$ as defined above.

The motivation for the conjectured quadratic bound is as follows. Let τ_r be the first exit time of the ball $B(\mathrm{id}, r)$ of a random walk starting at the identity. Take a equivariant μ-harmonic embedding $b : G \to \mathcal{H}$ normalized such that $\mathbb{E}\|b(W_1)\|^2 = 1$. Applying the optional stopping theorem to the martingale $\|b(W_t)\|^2 - t$, we deduce that $\mathbb{E}(\tau_r) \lesssim r^2$.

Heuristically, since the random walk is assumed to be transient, once it has left a ball $B_S(\mathrm{id}, Cr)$, where C is a large constant, the chances that it comes back to $B(\mathrm{id}, r)$ is small. Hence conjecturally, the expected occupation time of balls should admit a quadratic upper bound as well.

In [74], it is shown that in general, for any nondegenerate symmetric transient random walk $(W_n)_{n=0}^{\infty}$ on a finitely generated infinite group Γ, we have

$$\mathbb{E}\big(\mathcal{N}_W(r)\big) \lesssim r^2 \sqrt{\log v_{\Gamma,S}(r)}, \tag{4.3}$$

where $v_{\Gamma,S}(r)$ is the volume growth function of (Γ, S). In particular, on groups of exponential volume growth it yields the upper bound $r^{5/2}$. The bound $\mathbb{E}(\mathcal{N}_W(r)) \lesssim r^3$ on exponential growth groups is shown in [18] relying only on the Varopoulos bound that on such groups $\mu^{(2n)}(\mathrm{id}) \lesssim e^{-n^{1/3}}$.

We mention a connection of polynomial upper bounds on occupation times of balls to positive μ-harmonic functions. See [100, CHAPTER IV] for a treatment of the Martin boundary, which is a topological boundary representing positive harmonic functions. The bound $\mathbb{E}(\mathcal{N}_W(r)) \le Cr^D$ implies that the minimum of the Green function of the μ-random walk satisfies

$$\min_{\gamma \in B(\mathrm{id},r)} G_\mu(\mathrm{id}, \gamma) \le \frac{Cr^D}{|B(\mathrm{id}, r)|}.$$

In particular, when Γ is of exponential growth, by the classical bound $\mathbb{E}(\mathcal{N}_W(r)) \lesssim r^3$, the minimum of the Green function in $B(\mathrm{id}, r)$ decays exponentially in r. By translation invariance, we can write the Green function as a telescoping product

$$\frac{G_\mu(\mathrm{id}, y_1 y_2 \cdots y_n)}{G_\mu(\mathrm{id}, \mathrm{id})} = \prod_{i=0}^{n-1} \frac{G_\mu(y_i, y_{i+1} \cdots y_n)}{G_\mu(y_{i+1}, y_{i+1} \cdots y_n)}, \qquad \text{where } y_0 = \mathrm{id},\ y_i \in \Gamma.$$

Then an argument by contradiction shows that exponential decay of $\min_{\gamma \in B(\mathrm{id},r)} G_\mu(\mathrm{id}, \gamma)$ in r implies that there exists $s \in S^2$ and a sequence $(\gamma_n)_{n=0}^{\infty}$ in Γ going to infinity such that $\lim_{n\to\infty} G_\mu(s, \gamma_n)/G_\mu(\mathrm{id}, \gamma_n) < 1$. This shows that the Martin kernel $K(\cdot, \xi)$ is not constant in the first coordinate for some point ξ in the Martin boundary; equivalently, there are nonconstant positive μ-harmonic functions on Γ. Thus it gives another proof (though similar in spirit) of the result in [3], for any nondegenerate symmetric probability measure μ on a group Γ of exponential growth.

For Problem 4.3, when when Γ does not have Property H_{FD}, we can apply the estimate (4.2) to a nonconstant μ-harmonic cocycle $b : \Gamma \to \mathcal{H}$ with weakly mixing π. Indeed, choose any $\beta > 2$, since b is C-Lipschitz, we have

$$\mathbb{E}\big(\mathcal{N}_W(r)\big) \le r^2 + \sum_{t=r^2}^{\infty} \mathbb{P}\big(\|b(W_t)\| \le Cr\big) \le r^2 + C' \sum_{t=r^2}^{\infty} \Big(\frac{r}{\sqrt{t}}\Big)^{\beta} \le C'' r^2.$$

When Γ has Property H_{FD} but not H_{F}, then one can verify the quadratic bound by directly examining the random walk on a virtual quotient which is solvable of exponential growth. Therefore the problem is open only for groups with Property H_{F} that are not virtually nilpotent.

ACKNOWLEDGMENTS

The author thanks Laurent Saloff-Coste for helpful discussions.

FUNDING

This work was partially supported by an Alfred P. Sloan research fellowship.

REFERENCES

[1] G. Amir, On the joint behaviour of speed and entropy of random walks on groups. *Groups Geom. Dyn.* **11** (2017), no. 2, 455–467.

[2] G. Amir, O. Angel, and B. Virág, Amenability of linear-activity automaton groups. *J. Eur. Math. Soc. (JEMS)* **15** (2013), no. 3, 705–730.

[3] G. Amir and G. Kozma, Every exponential group supports a positive harmonic function. Arxiv preprint, 2018.

[4] G. Amir and B. Virág, Speed exponents of random walks on groups. *Int. Math. Res. Not. IMRN* **9** (2017), 2567–2598.

[5] S. N. Armstrong and O. Zeitouni, Local asymptotics for controlled martingales. *Ann. Appl. Probab.* **26** (2016), no. 3, 1467–1494.

[6] T. Austin, A. Naor, and Y. Peres, The wreath product of $\mathbb{Z}$ with $\mathbb{Z}$ has Hilbert compression exponent $\frac{2}{3}$. *Proc. Amer. Math. Soc.* **137** (2009), no. 1, 85–90.

[7] A. Avez, Harmonic functions on groups. In *Differential geometry and relativity*, pp. 27–32, Math. Phys. Appl. Math. 3, Reidel, Dordrecht, 1976.

[8] D. Bakry, T. Coulhon, M. Ledoux, and L. Saloff-Coste, Sobolev inequalities in disguise. *Indiana Univ. Math. J.* **44** (1995), no. 4, 1033–1074.

[9] K. Ball, Markov chains, Riesz transforms and Lipschitz maps. *Geom. Funct. Anal.* **2** (1992), no. 2, 137–172.

[10] L. Bartholdi, The growth of Grigorchuk's torsion group. *Int. Math. Res. Not.* **20** (1998), 1049–1054.

[11] L. Bartholdi, Lower bounds on the growth of a group acting on the binary rooted tree. *Internat. J. Algebra Comput.* **11** (2001), no. 1, 73–88.

[12] L. Bartholdi and A. Erschler, Groups of given intermediate word growth. *Ann. Inst. Fourier (Grenoble)* **64** (2014), no. 5, 2003–2036.

[13] L. Bartholdi, V. A. Kaimanovich, and V. V. Nekrashevych, On amenability of automata groups. *Duke Math. J.* **154** (2010), no. 3, 575–598.

[14] L. Bartholdi and B. Virág, Amenability via random walks. *Duke Math. J.* **130** (2005), no. 1, 39–56.

[15] A. Bendikov, C. Pittet, and R. Sauer, Spectral distribution and L^2-isoperimetric profile of Laplace operators on groups. *Math. Ann.* **354** (2012), no. 1, 43–72.

[16] A. Bendikov and L. Saloff-Coste, Random walks driven by low moment measures. *Ann. Probab.* **40** (2012), no. 6, 2539–2588.

[17] I. Benjamini, Instability of the Liouville property for quasi-isometric graphs and manifolds of polynomial volume growth. *J. Theoret. Probab.* **4** (1991), no. 3, 631–637.

[18] S. Blachère and S. Brofferio, Internal diffusion limited aggregation on discrete groups having exponential growth. *Probab. Theory Related Fields* **137** (2007), no. 3–4, 323–343.

[19] S. Blachère, P. Haïssinsky, and P. Mathieu, Asymptotic entropy and Green speed for random walks on countable groups. *Ann. Probab.* **36** (2008), no. 3, 1134–1152.

[20] J. Brieussel, Behaviors of entropy on finitely generated groups. *Ann. Probab.* **41** (2013), no. 6, 4116–4161.

[21] J. Brieussel and T. Zheng, Speed of random walks, isoperimetry and compression of finitely generated groups. *Ann. of Math. (2)* **193** (2021), no. 1, 1–105.

[22] T. K. Carne, A transmutation formula for Markov chains. *Bull. Sci. Math. (2)* **109** (1985), no. 4, 399–405.

[23] C. Chou, Elementary amenable groups. *Illinois J. Math.* **24** (1980), no. 3, 396–407.

[24] T. Coulhon, Ultracontractivity and Nash type inequalities. *J. Funct. Anal.* **141** (1996), no. 2, 510–539.

[25] T. Coulhon and A. Grigor'yan, On-diagonal lower bounds for heat kernels and Markov chains. *Duke Math. J.* **89** (1997), no. 1, 133–199.

[26] T. Coulhon and L. Saloff-Coste, Isopérimétrie pour les groupes et les variétés. *Rev. Mat. Iberoam.* **9** (1993), no. 2, 293–314.

[27] P. de la Harp, R. I. Grigorchuk, and T. Chekerini-Sil'berstaĭn, Amenability and paradoxical decompositions for pseudogroups and discrete metric spaces. *Tr. Mat. Inst. Steklova* **224** (1999), 68–111.

[28] Y. Derriennic, Quelques applications du théorème ergodique sous-additif. *Astérisque* **74** *Soc. Math. France, Paris*, (1980), 183–201.

[29] P. Diaconis and L. Saloff-Coste, Comparison techniques for random walk on finite groups. *Ann. Probab.* **21** (1993), no. 4, 2131–2156.

[30] E. B. Dynkin and M. B. Maljutov, Random walk on groups with a finite number of generators. *Dokl. Akad. Nauk SSSR* **137** (1961), 1042–1045.

[31] A. Erschler, On drift and entropy growth for random walks on groups. *Ann. Probab.* **31** (2003), no. 3, 1193–1204.

[32] A. Erschler, Boundary behavior for groups of subexponential growth. *Ann. of Math.* (2004), 1183–1210.

[33] A. Erschler, Poisson–Furstenberg boundary of random walks on wreath products and free metabelian groups. *Comment. Math. Helv.* **86** (2011), no. 1, 113–143.

[34] A. Erschler and N. Ozawa, Finite-dimensional representations constructed from random walks. *Comment. Math. Helv.* **93** (2018), no. 3, 555–586.

[35] A. Erschler and T. Zheng, Growth of periodic Grigorchuk groups. *Invent. Math.* **219** (2020), no. 3, 1069–1155.

[36] J. Frisch, Y. Hartman, O. Tamuz, and P. Vahidi Ferdowsi, Choquet–Deny groups and the infinite conjugacy class property. *Ann. of Math. (2)* **190** (2019), no. 1, 307–320.

[37] H. Furstenberg, A Poisson formula for semi-simple Lie groups. *Ann. of Math. (2)* **77** (1963), 335–386.

[38] H. Furstenberg, Poisson boundaries and envelopes of discrete groups. *Bull. Amer. Math. Soc.* **73** (1967), 350–356.

[39] H. Furstenberg, Boundaries of Lie groups and discrete subgroups. In *Actes du Congrès International des Mathématiciens (Nice, 1970), Tome 2*, pp. 301–306, Gauthier-Villars, Nice, 1970.

[40] H. Furstenberg, Random walks and discrete subgroups of Lie groups. In *Advances in probability and related topics, Vol. 1*, pp. 1–63, Dekker, New York, 1971.

[41] A. Grigor'yan, *Introduction to analysis on graphs*. Univ. Lecture Ser. 71, American Mathematical Society, Providence, RI, 2018.

[42] R. Grigorchuk, On the gap conjecture concerning group growth. *Bull. Math. Sci.* **4** (2014), no. 1, 113–128.

[43] R. I. Grigorchuk, Degrees of growth of finitely generated groups, and the theory of invariant means. *Izv. Math.* **25** (1985), no. 2, 259–300.

[44] R. I. Grigorchuk and A. Żuk, On a torsion-free weakly branch group defined by a three state automaton. *Internat. J. Algebra Comput.* **12** (2002), 223–246.

[45] R. I. Grigorčuk, On Burnside's problem on periodic groups. *Funktsional. Anal. i Prilozhen.* **14** (1980), no. 1, 53–54.

[46] M. Gromov, Groups of polynomial growth and expanding maps. *Publ. Math. Inst. Hautes Études Sci.* **53** (1981), 53–73.

[47] M. Gromov, Asymptotic invariants of infinite groups. In *Geometric group theory, Vol. 2 (Sussex, 1991)*, pp. 1–295, London Math. Soc. Lecture Note Ser. 182, Cambridge Univ. Press, Cambridge, 1993.

[48] M. Gromov, Entropy and isoperimetry for linear and non-linear group actions. *Groups Geom. Dyn.* **2** (2008), no. 4, 499–593.

[49] E. Guentner and J. Kaminker, Exactness and uniform embeddability of discrete groups. *J. Lond. Math. Soc. (2)* **70** (2004), no. 3, 703–718.

[50] W. Jaworski, Strong approximate transitivity, polynomial growth, and spread out random walks on locally compact groups. *Pacific J. Math.* **170** (1995), no. 2, 517–533.

[51] W. Jaworski, On shifted convolution powers and concentration functions in locally compact groups. In *Probability on algebraic structures (Gainesville, FL, 1999)*, pp. 23–41, Contemp. Math. 261, Amer. Math. Soc., Providence, RI, 2000.

[52] W. Jaworski, Countable amenable identity excluding groups. *Canad. Math. Bull.* **47** (2004), no. 2, 215–228.

[53] W. Jaworski and C. R. E. Raja, The Choquet–Deny theorem and distal properties of totally disconnected locally compact groups of polynomial growth. *New York J. Math.* **13** (2007), 159–174.

[54] K. Juschenko and N. Monod, Cantor systems, piecewise translations and simple amenable groups. *Ann. of Math.* **178** (2013), no. EPFL–ARTICLE-179598, 775–787.

[55] K. Juschenko, V. Nekrashevych, and M. de la Salle, Extensions of amenable groups by recurrent groupoids. *Invent. Math.* **206** (2016), no. 3, 837–867.

[56] K. Juschenko, N. Matte Bon, N. Monod, and M. de la Salle, Extensive amenability and an application to interval exchanges. *Ergodic Theory Dynam. Systems* **38** (2018), no. 1, 195–219.

[57] V. A. Kaimanovich, The Poisson formula for groups with hyperbolic properties. *Ann. of Math. (2)* **152** (2000), no. 3, 659–692.

[58] V. A. Kaimanovich, "Münchhausen trick" and amenability of self-similar groups. *Internat. J. Algebra Comput.* **15** (2005), no. 5–6, 907–937.

[59] V. A. Kaimanovich, Thompson's group F is not Liouville. In *Groups, Graphs and Random Walks*, pp. 300–342, London Math. Soc. Lecture Note Ser. 436, Cambridge University Press, Cambridge, 2017.

[60] V. A. Kaĭmanovich and A. M. Vershik, Random walks on discrete groups: boundary and entropy. *Ann. Probab.* **11** (1983), no. 3, 457–490.

[61] A. Karlsson and F. Ledrappier, Linear drift and Poisson boundary for random walks. *Pure Appl. Math. Q.* **3** (2007), no. 4, 1027–1036. Special Issue: In honor of Grigory Margulis. Part 1.

[62] H. Kesten, Full Banach mean values on countable groups. *Math. Scand.* **7** (1959), 146–156.

[63] H. Kesten, Symmetric random walks on groups. *Trans. Amer. Math. Soc.* **92** (1959), 336–354.

[64] H. Kesten, The Martin boundary of recurrent random walks on countable groups. In *Proc. fifth Berkeley sympos. math. statist. and probability (Berkeley, CA, 1965/1966). Vol. II: Contributions to probability theory, part 2*, pp. 51–74, Univ. California Press, Berkeley, CA, 1967.

[65] B. Kleiner, A new proof of Gromov's theorem on groups of polynomial growth. *J. Amer. Math. Soc.* **23** (2010), no. 3, 815–829.

[66] N. J. Korevaar and R. M. Schoen, Global existence theorems for harmonic maps to non-locally compact spaces. *Comm. Anal. Geom.* **5** (1997), no. 2, 333–387.

[67] M. Kotowski and B. Virág, Non-Liouville groups with return probability exponent at most $1/2$. *Electron. Commun. Probab.* **20** (2015), no. 12.

[68] G. F. Lawler and A. D. Sokal, Bounds on the L^2 spectrum for Markov chains and Markov processes: a generalization of Cheeger's inequality. *Trans. Amer. Math. Soc.* **309** (1988), no. 2, 557–580.

[69] J. R. Lee and Y. Peres, Harmonic maps on amenable groups and a diffusive lower bound for random walks. *Ann. Probab.* **41** (2013), no. 5, 3392–3419.

[70] J. R. Lee, Y. Peres, and C. K. Smart, A Gaussian upper bound for martingale small-ball probabilities. *Ann. Probab.* **44** (2016), no. 6, 4184–4197.

[71] Y. G. Leonov, A lower bound for the growth function of periods in Grigorchuk groups. *Mat. Stud.* **8** (1997), no. 2, 192–197, 237.

[72] R. Lyons and Y. Peres, *Probability on trees and networks*. Camb. Ser. Stat. Probab. Math. 42, Cambridge University Press, New York, 2016.

[73] R. Lyons and Y. Peres, Poisson boundaries of lamplighter groups: proof of the Kaimanovich–Vershik conjecture. *J. Eur. Math. Soc. (JEMS)* **23** (2021), no. 4, 1133–1160.

[74] R. Lyons, Y. Peres, X. Sun, and T. Zheng, Occupation measure of random walks and wired spanning forests in balls of Cayley graphs. *Ann. Fac. Sci. Toulouse Math. (6)* **29** (2020), no. 1, 97–109.

[75] T. Lyons, Instability of the Liouville property for quasi-isometric Riemannian manifolds and reversible Markov chains. *J. Differential Geom.* **26** (1987), no. 1, 33–66.

[76] J. Maher and G. Tiozzo, Random walks on weakly hyperbolic groups. *J. Reine Angew. Math.* **742** (2018), 187–239.

[77] M. Mendel and A. Naor, Metric cotype. *Ann. of Math. (2)* **168** (2008), no. 1, 247–298.

[78] T. Meyerovitch and A. Yadin, Harmonic functions of linear growth on solvable groups. *Israel J. Math.* **216** (2016), no. 1, 149–180.

[79] J. Milnor, Growth of finitely generated solvable groups. *J. Differential Geom.* **2** (1968), 447–449.

[80] N. Mok, Harmonic forms with values in locally constant Hilbert bundles. In *Proceedings of the Conference in Honor of Jean-Pierre Kahane (Orsay, 1993)*, pp. 433–453, CRC Press, Boca Raton, FL, 1995. Special Issue.

[81] A. Naor and Y. Peres, Embeddings of discrete groups and the speed of random walks. *Int. Math. Res. Not. IMRN* (2008).

[82] A. Naor and Y. Peres, L_p compression, traveling salesmen, and stable walks. *Duke Math. J.* **157** (2011), no. 1, 53–108.

[83] V. Nekrashevych, Simple groups of dynamical origin. *Ergodic Theory Dynam. Systems* (2017), 1–26.

[84] V. Nekrashevych, Palindromic subshifts and simple periodic groups of intermediate growth. *Ann. of Math. (2)* **187** (2018), no. 3, 667–719.

[85] N. Ozawa, A functional analysis proof of Gromov's polynomial growth theorem. *Ann. Sci. Éc. Norm. Supér. (4)* **51** (2018), no. 3, 549–556.

[86] Y. Peres and T. Zheng, On groups, slow heat kernel decay yields liouville property and sharp entropy bounds. *Int. Math. Res. Not.* **rny034** (2018).

[87] C. Pittet and L. Saloff-Coste, On the stability of the behavior of random walks on groups. *J. Geom. Anal.* **10** (2000), no. 4, 713–737.

[88] J. Rosenblatt, Ergodic and mixing random walks on locally compact groups. *Math. Ann.* **257** (1981), no. 1, 31–42.

[89] L. Saloff-Coste and T. Zheng, Random walks on free solvable groups. *Math. Z.* **279** (2015), no. 3–4, 811–848.

[90] L. Saloff-Coste and T. Zheng, Random walks and isoperimetric profiles under moment conditions. *Ann. Probab.* **44** (2016), no. 6, 4133–4183.

[91] Y. Shalom, Harmonic analysis, cohomology, and the large-scale geometry of amenable groups. *Acta Math.* **192** (2004), no. 2, 119–185.

[92] R. Tessera, Isoperimetric profile and random walks on locally compact solvable groups. *Rev. Mat. Iberoam.* **29** (2013), no. 2, 715–737.

[93] J. Tits, Free subgroups in linear groups. *J. Algebra* **20** (1972), 250–270.

[94] L. van den Dries and A. J. Wilkie, Gromov's theorem on groups of polynomial growth and elementary logic. *J. Algebra* **89** (1984), no. 2, 349–374.

[95] N. T. Varopoulos, Isoperimetric inequalities and Markov chains. *J. Funct. Anal.* **63** (1985), no. 2, 215–239.

[96] N. T. Varopoulos, Long range estimates for Markov chains. *Bull. Sci. Math. (2)* **109** (1985), no. 3, 225–252.

[97] N. T. Varopoulos, Théorie du potentiel sur des groupes et des variétés. *C. R. Acad. Sci. Paris Sér. I Math.* **302** (1986), no. 6, 203–205.

[98] N. T. Varopoulos, Analysis and geometry on groups. In *Proceedings of the International Congress of Mathematicians, Vol. I, II (Kyoto, 1990)*, pp. 951–957, Math. Soc. Japan, Tokyo, 1991.

[99] N. T. Varopoulos, L. Saloff-Coste, and T. Coulhon, *Analysis and geometry on groups*. Cambridge Tracts in Math. 100, Cambridge University Press, Cambridge, 1992.

[100] W. Woess, *Random walks on infinite graphs and groups*. Cambridge Tracts in Math. 138, Cambridge University Press, Cambridge, 2000.

[101] J. A. Wolf, Growth of finitely generated solvable groups and curvature of Riemaniann manifolds. *J. Differential Geom.* **2** (1968), 421–446.

[102] G. Yu, The coarse Baum–Connes conjecture for spaces which admit a uniform embedding into Hilbert space. *Invent. Math.* **139** (2000), no. 1, 201–240.

TIANYI ZHENG (郑天一)

Department of Mathematics, University of California San Diego, 9500 Gilman Dr., La Jolla, CA 92093, USA, tzheng2@math.ucsd.edu

This Page Intentionally Left Blank

LIST OF CONTRIBUTORS